W9-BUD-524

FOURTEENTH EDITION

OPERATIONS MANAGEMENT

Sustainability and Supply Chain Management

JAY
HEIZER
Jesse H. Jones Professor of Business Administration
Texas Lutheran University

BARRY
RENDER
Charles Harwood Professor of Operations Management
Graduate School of Business
Rollins College

CHUCK
MUNSON
Professor of Operations Management
Carson College of Business
Washington State University

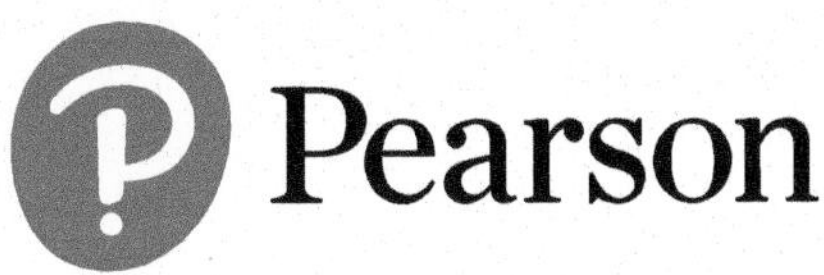

Please contact https://support.pearson.com/getsupport/s/ with any queries on this content
Cover Image Courtesy of Correct Craft Holding Company, LLC

Library of Congress Cataloging-in-Publication Data

Names: Heizer, Jay, author. | Render, Barry, author. | Munson, Chuck, author.
Title: Operations management: sustainability and supply chain management / Jay Heizer, Jesse H. Jones Professor of Business Administration, Texas Lutheran University, Barry Render, Charles Harwood Professor of Operations Management, Graduate School of Business, Rollins College, Chuck Munson, Professor of Operations Management, Carson College of Business, Washington State University.
Description: Fourteenth Edition. | Hoboken, NJ: Pearson, 2022. | Revised edition of the authors' Operations management, [2020] | Summary: "The goal of this text and its accompanying online resources is to present students a broad introduction to the field of operations in a realistic, practical, and applied manner. We want students to understand how operations work within an organization by seeing first-hand what goes on behind the scenes at a concert or major sports event, placing an order through Amazon.com, boarding a flight on Alaska Airlines, or taking a cruise with Celebrity Cruises. This text along with its online learning tools and resources offers behind-the-scenes views that no other product on the market provides and one that students tell us they value because they gain a true understanding of operations"– Provided by publisher.
Identifiers: LCCN 2021041402 (print) | LCCN 2021041403 (ebook) | ISBN 9780137476442 (hardcover) | ISBN 9780137476411 (ebook) | ISBN 9780137476343 (ebook) | ISBN 9780137476312 (ebook)
Subjects: LCSH: Production management.
Classification: LCC TS155 .H3725 2022 (print) | LCC TS155 (ebook) | DDC 658.5–dc23
LC record available at https://lccn.loc.gov/2021041402
LC ebook record available at https://lccn.loc.gov/2021041403

3 2022

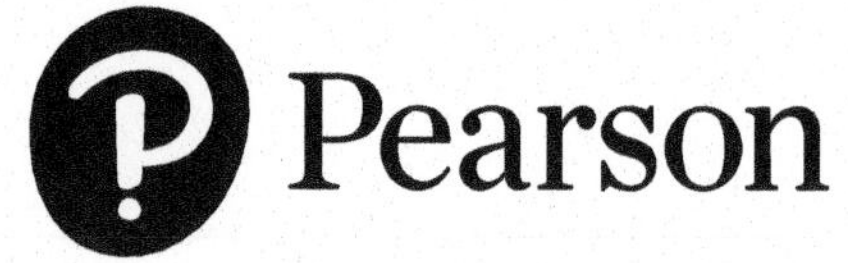

ISBN-10: 0-13-747644-2
ISBN-13: 978-0-13-747644-2

To Kay Heizer, always at my side

J.H.

To Donna, Charlie, and Jesse

B.R.

To Kim, Christopher, and Mark Munson for their unwavering support

C.M.

Pearson's Commitment to Diversity, Equity, and Inclusion

Pearson is dedicated to creating bias-free content that reflects the diversity of all learners. We embrace the many dimensions of diversity, including but not limited to race, ethnicity, gender, socioeconomic status, ability, age, sexual orientation, and religious or political beliefs.

Education is a powerful force for equity and change in our world. It has the potential to deliver opportunities that improve lives and enable economic mobility. As we work with authors to create content for every product and service, we acknowledge our responsibility to demonstrate inclusivity and incorporate diverse scholarship so that everyone can achieve their potential through learning. As the world's leading learning company, we have a duty to help drive change and live up to our purpose to help more people create a better life for themselves and to create a better world.

Our ambition is to purposefully contribute to a world where:

- Everyone has an equitable and lifelong opportunity to succeed through learning.
- Our educational products and services are inclusive and represent the rich diversity of learners.
- Our educational content accurately reflects the histories and experiences of the learners we serve.
- Our educational content prompts deeper discussions with students and motivates them to expand their own learning (and worldview).

We are also committed to providing products that are fully accessible to all learners. As per Pearson's guidelines for accessible educational Web media, we test and retest the capabilities of our products against the highest standards for every release, following the WCAG guidelines in developing new products for copyright year 2022 and beyond. You can learn more about Pearson's commitment to accessibility at https://www.pearson.com/us/accessibility.html.

While we work hard to present unbiased, fully accessible content, we want to hear from you about any concerns or needs with this Pearson product so that we can investigate and address them.

- Please contact us with concerns about any potential bias at https://www.pearson.com/report-bias.html.
- For accessibility-related issues, such as using assistive technology with Pearson products, alternative text requests, or accessibility documentation, email the Pearson Disability Support team at disability.support@pearson.com.

Brief Table of Contents

ONLINE TUTORIALS

Table of Contents

PART TWO Designing Operations 157

ONLINE TUTORIALS

Preface

Operations is an exciting area of management that has a profound effect on productivity. The goal of this text and its accompanying online resources is to present students a broad introduction to the field of operations in a realistic, practical, and applied manner. We want students to understand how operations work within an organization by seeing first-hand what goes on behind the scenes at a concert or major sports event, placing an order through Amazon.com, boarding a flight on Alaska Airlines, or taking a cruise with Celebrity Cruises. This text along with its online learning tools and resources offers behind-the-scenes views that no other product on the market provides and one that students tell us they value because they gain a true understanding of operations.

New to This Edition

With each edition, we work to gather feedback from instructors and students to enhance our text and online instructional resources. Based on that feedback, we have added the following new features and improvements.

New Coverage and Examples

We provide a detailed list of chapter-by-chapter changes below, but here are a few important topics we are discussing in the 14th edition.

- Industry 4.0, sometimes called the Fourth Industrial Revolution, with extensive digitalization and pervasive impact on OM, is now introduced throughout the text via discussion, photos, and cases. For instance, an introduction is provided in Chapter 1 and illustrated in Chapter 7's discussion of sensors and vision technology. It is further discussed in Chapter 9's digitalization of layout and Chapter 10's digital monitoring of workers. In Chapter 11, we also added a new section called "Digitalization and the Internet of Things," noting how sensors are impacting the logistics process.
- COVID-19 had, of course, a major impact of lives, education, work habits, and global supply chains. Not every aspect of business will return to normal immediately, and we address the fallouts throughout the text. For example, Figure 2.1 shows the impact of COVID on world trade. We note COVID's impact on forecasting in Chapter 4. We added an online case study, "Global Chemical's COVID-19 Capacity Decision," in Supplement 7. We revamped coverage of office layout changes due to COVID in Chapter 9. Finally, we added a new case study, "Premiere Bicycles' COVID Problem," and a discussion of supermarket supply chain shortages in Chapter 11.
- Every chapter supplement and module now opens with a photo vignette that provides a motivational image of OM as applied to that topic.
- There are 42 "OM in Action" boxes, which are real-world examples and tell interesting tales of OM in business throughout the text. Thirteen are new to this edition and deal with such topics as 3-D printed steaks (Chapter 5), an automated sushi restaurant (Chapter 7), bitcoin (Chapter 8), and scheduling in the NBA (Chapter 15).

Product Design at Nautique Boat Company	Video Case

For nearly 100 years, Florida-based Nautique Boat Company has built innovative boats in a very competitive market. Nautique, as the premier boat for waterskiing, wakeboarding, and wake surfing, is on the cutting edge of style, customer satisfaction, and performance. There is continuing and rapid change in this industry, which sees substantial input from imaginative, experimenting customers. Success means integrating customer feedback, technological change, and creative engineering talent into a dynamic, but ongoing, product line. As number one in its market, Nautique is a vivid example of what it takes to be a creative leader.

From the introduction of waterskiing in the 1920s to barefoot skiing to wake boarding in the 1990s and wake surfing in the 2000s, Nautique has led. While these new sport expectations were placed on boat performance, changes in marine engineering and technology were also taking place. Ski boats were initially made of wood, changing to fiberglass in the late 1950s, with tracking fins added in the 1960s. The 1990s brought longer and wider boats with hull changes, slopping transoms, spray relief, and flight control towers. This was followed by *Total Wake Control*, which allows users to instantly switch from wakeboarding to wake surfing. Indeed, the wake can now be customized, providing ramp-style and vertical-style wakes behind the same boat. Simultaneously, carpet, trim, and color options and a variety of powerful engines, as well as major electronic innovations such as sonar, GPS, and sophisticated sound systems with strategically placed speakers, have been introduced. Nautique's design team is now developing the boat of the future—the electric boat—with the prototype in testing.

To maintain Nautique's innovative prowess, Chief Designer Steve Carlton (shown in photo) has organized his 40-person department into three teams: model integration, design, and engineering. In addition to creating, developing, and designing new innovations, Nautique also plans on two to four new models each year, with remodeling or updating every 2 years.

To facilitate efficient development and design integrity, Carlton's design and development team uses sophisticated CAD software from Rhino and SolidWorks. NX software is also available for tooling, and StrataSys 3D printers facilitate prototyping small parts. Design typically begins with sketches and clay models, progressing to CAD drawings and then to a full-scale wooden model, followed by a full-size boat for testing in Nautique's private lake.

Each design innovation requires ongoing discussion with the in-house engineering staff to provide design integrity. Similarly, the team sends staff to the factory floor to work with production personnel to ensure production capability and with external suppliers to ensure raw material and component quality and delivery. This coordination with external suppliers and internal production by the design group has the goal of facilitating a smooth transition to production.

Video Cases—Nautique Boat Company

With each edition, we offer integrated Video Cases as a valuable teaching tool for students. These short videos help readers see and understand operations in action within a variety of industries. With this edition, we are pleased to take you behind the scenes of Nautique Boat Company, maker of the iconic Ski Nautique and other premium pleasure boats. This fascinating organization opened its doors for us to examine and share with you leading-edge OM in the boating industry.

These new videos provide an inside look at:

- Operations strategy at Nautique (Chapter 2);
- How Nautique designs a new product (Chapter 5);
- Supply chain issues facing Nautique (Supplement 11); and
- Inventory management at Nautique (Chapter 12).

In addition, we continue to offer our previous Video Cases that cover: Celebrity Cruises, Alaska Airlines, the Orlando Magic basketball team, Frito-Lay, Darden/Red Lobster Restaurants, Hard Rock Cafe, Arnold Palmer Hospital, and Wheeled Coach Ambulances.

We take the integration of our video case studies seriously, and for this reason, all 46 videos are created by the authors to explicitly match text content and terminology.

VIDEO CASES	Ch 1: Operations & Productivity	Ch 2: Operations Strategy	Ch 3: Project Management	Ch 4: Forecasting	Ch 5: Design of Goods & Services	Sup 5: Sustainability	Ch 6: Managing Quality	Sup 6: Statistical Process Control	Ch 7: Process Strategies	Sup 7: Capacity & Constraint Mgt	Ch 8: Location Strategies	Ch 9: Layout Strategies	Ch 10: HR & Work Measurement	Ch 11: Supply Chain Management	Sup 11: Supply Chain Mgt Analysis	Ch 12: Inventory Management	Ch 13: Aggregate Planning & S&OP	Ch 14: MRP and ERP	Ch 15: Short-Term Scheduling	Ch 16: Lean Operations	Ch 17: Maintenance & Reliability	Module B: Linear Programming
Alaska Airlines							✓		✓				✓							✓		✓
Arnold Palmer Hospital			✓				✓		✓	✓		✓		✓						✓		
Celebrity Cruises	✓				✓	✓	✓									✓						
Darden Restaurants		✓						✓			✓			✓								
Frito-Lay	✓					✓		✓								✓					✓	
Hard Rock Cafe	✓	✓	✓	✓							✓		✓						✓			
Nautique Boat Company		✓			✓										✓	✓						
Orlando Magic (Amway Center)				✓		✓											✓	✓	✓			
Wheeled Coach Ambulance									✓			✓				✓		✓				

New Videos to Help Students Build Their Own Excel Spreadsheets

Excel use in the Operations Management course is more and more important. Instructors often ask their students to develop their own Excel spreadsheet models. We include "Creating Your Own Excel Spreadsheets" examples toward the end of numerous chapters to illustrate how students can build their own spreadsheets to solve OM problems, and in this edition, we've created 12 new videos to accompany these examples.

More Homework Problems—Quantity, Algorithmic, and Conceptual

We know that a vast selection of quality homework problems, ranging from easy to challenging (denoted by one to four dots), is critical for both instructors and students. Instructors need a broad selection of problems to choose from for homework, quizzes, and exams—without reusing the same set from semester to semester. We take pride in having more problems—by far, with 850—than any other OM text.

For this edition, we have added scores of new algorithmic problems and concept questions in **MyLab Operations Management**!

Algorithmic Test Bank Questions

About 200 numerical multiple choice test bank questions have been converted to algorithmic so that every student sees different numbers and a different set of answers for these questions.

Detailed Chapter-by-Chapter Changes

Chapter 1: Operations and Productivity

- New chapter opener features OM career opportunities with real job listings for five jobs.
- New section covering Industry 4.0 (the 4th industrial revolution).
- New author-created video to accompany "Creating Your Own Excel Spreadsheets."
- Uber Technologies case study rewritten to bring it up-to-date with recent company changes.

Chapter 2: Operations Strategy in a Global Environment

- New chapter opener/photo features Super Air Nautique product differentiation.
- Global Company Profile on Boeing completely rewritten to focus on the whole product line, not just the 787. This includes new photos.
- Figure 2.1 updated to show USMCA and the impact of containerization and COVID on world trade.
- NAFTA material updated to include USMCA.
- Figure 2.5, Product Life Cycle, has mostly new product examples.
- New author-created video to accompany "Creating Your Own Excel Spreadsheets."
- Competitive ranking homework problem uses a new index.
- New video case: "Strategy at Nautique Boat Company."

Chapter 3: Project Management

- New chapter opener photo features Celebrity Cruises ship construction.
- Overlays for Figure 3.10 have been removed and replaced by a revamped figure.
- Section on Microsoft Project is more generic and now called "Using Software to Manage Projects." Program 3.3 is deleted.
- Four homework problems (3.4, 3.16, 3.26, 3.27) have been changed to make them more assignable in quizzes and tests.

Chapter 4: Forecasting

- New chapter opener/photo features Yamaha products, illustrating seasonal forecasts.
- Material on COVID's impact on forecasting is added.
- Forecasting technique called "stagger charts" is added.
- New author-created video to accompany "Creating Your Own Excel Spreadsheets."
- Three very large homework problems have been shortened to make them more assignable in quizzes and tests.

Chapter 5: Design of Goods and Services

- New chapter opening photo/caption shows McDonald's revamped Happy Meal.
- New Global Company Profile features Nautique Boats' product design (with new photos).
- Overlays for Example 1 have been removed and the figure redrawn.
- New OM in Action box features 3D printed steaks.
- New homework problems 5.26–5.27 have been added.
- New video case study: "Product Design at Nautique Boat Company."

Chapter 6: Managing Quality

- New chapter opening/photo caption features Alaska Airlines.
- Update from ISO 9001:2015 to ISO 9004:2018.
- Six Sigma material shortened and simplified.
- Figure 6.4 (Taguchi) redrawn for clarity.
- New discussion question 18 relates to the OM in Action Box on Boeing 787.

Supplement 6: Statistical Process Control

- New explanation of *p*-chart sample size restrictions.
- Example S4 is revised.
- New homework problems: S6.14, S6.15, S6.20, S6.25, S6.26, S6.27, S6.31, S6.35, S6.38, and S6.47
- Extensive revision to the treatment of process ratio and capability index, including Example S7.
- New author-created video to accompany "Creating Your Own Excel Spreadsheets."
- New case study: "PEI Potato Purveyors."
- New case study: "Alabama Airlines' On-Time Schedule."

Chapter 7: Process Strategies

- New chapter opening photo of process control at an Australian steel mill.
- Major revision of the treatment of mass customization.
- Chapter revisions focus on technology, including: the section on Production Technology now includes a discussion of sensors, a new OM in Action box on Tyson Chicken's use of vision technology, and a new OM in Action box on an automated sushi restaurant.
- Revised homework problems 7.13–7.17 are now specific enough to be assigned in MyLab.

Supplement 7: Capacity and Constraint Management

- Rewritten OM in Action box on airline capacity to note impacts of the pandemic.
- Updated Arnold Palmer Hospital capacity case with 3 years of recent data, which creates a new solution.
- Added the online case "Global Chemical's COVID-19 Capacity Decision."
- New author-created video to accompany "Creating Your Own Excel Spreadsheets."

Chapter 8: Location Strategies

- New chapter opener on Geographic Information Systems used in site location.
- Updated the Global Company Profile on FedEx.
- Replaced Table 8.1's "Competitive Index" with a new index called "Ease of Doing Business."
- Added OM in Action box: "Bitcoin Goes to Where the Power Is Cheap."
- Revised Table 8.2—corruption rankings—with 2020 data.
- New author-created video to accompany "Creating Your Own Excel Spreadsheets."
- Added 3 new homework problems: 8.25, 8.33, 8.34.
- New case study: "National Assembly Services."

Chapter 9: Layout Strategies

- New chapter opener/photo features Hard Rock's retail layout.
- New material on digitalization of layout.
- Material on Muther's Grid moved from office layout section to process layout, with two new assignable questions/problems.
- Office layout revised to reflect the realities of COVID-19.
- New homework problem 9.24.

Chapter 10: Human Resources, Job Design, and Work Measurement

- New chapter-opening photo features Alaska Airlines' CPR training for crew.
- The Global Company Profile on NASCAR has been updated.
- New OM in Action box called "The Rise of the Exoskeleton."
- New material on use of sensors in helping workers ergonomically.
- New section introducing Digital Monitoring Techniques.
- Four new homework problems: 10.38, 10.43, 10.45, 10.46.

Chapter 11: Supply Chain Management

- New opening photo relating supermarket supply chain shortages during COVID.
- New material on Korean coalitions called "chaebols."
- New section on the Omnichannel strategy.
- New material on blockchain, with an example from Carrefour's chicken supply chain, including graphic flowchart in Figure 11.3.
- New section on Digitalization and the Internet of Things (IoT).
- New photo/material on sensors in the logistics process.
- New OM in Action box: "New York City Chokes on Deliveries from Online Orders."
- New case study: "Premiere Bicycles' COVID Problem."

Supplement 11: Supply Chain Management Analytics

- New OM in Action Box: "The Recurring Bullwhip Effect."
- New Figure S11.3 on RFID at Walmart.
- New homework problem S11.15.
- New video case study: "Supply Chain Issues at Nautique Boat Company."
- New additional case study: "JIT after a Catastrophe" appears in MyLab.

Chapter 12: Inventory Management

- New chapter opening featuring inventory control at LEGO.
- Updated Global Company Profile on Amazon.com.
- New OM in Action box on Mattel.
- New OM in Action box on apparel companies' unsold clothes.
- New author-created video to accompany "Creating Your Own Excel Spreadsheets."
- ABC homework problems 12.5–12.6 rewritten with new data.
- Inventory Control at Wheeled Coach case study moved to MyLab.
- New video case study: "Inventory Control at Nautique Boat Co."

Chapter 13: Aggregate Planning and S&OP

- New chapter-opening photo features John Deere.

Chapter 14: Material Requirements Planning (MRP) and ERP

- New opening photo featuring Nautique Boat Company's MRP system.
- Figure 14.1 on the planning process revised.
- Major rewrite and reorganization of material on Lot sizing, MRP management and MRPII, with a new section on finite scheduling in MRP.
- Problems 14.11 and 14.12 modified.
- New additional case study: "OSI's Attempt at ERP."

Chapter 15: Short-Term Scheduling

- New chapter-opening photo featuring scheduling on Celebrity Cruises.
- New OM in Action box: "NBA's Scheduling Secret."
- Two new homework problems (15.31 and 15.32) on cyclical scheduling.

Chapter 16: Lean Operations

- New chapter-opening photo featuring a lean restaurant.
- New section covering the topic of Activity-Based Costing.

Chapter 17: Maintenance and Reliability

- New chapter-opening photo spread featuring the importance of maintenance at four companies.
- New homework problem added.

Module A: Decision Making Tools

- New author-created video to accompany "Creating Your Own Excel Spreadsheets."

Module B: Linear Programming

- New OM in Action box: "Art and Science of Scheduling in the NFL."
- Revision of Example B2 to address alternate optimal solutions.
- New author-created video to accompany "Creating Your Own Excel Spreadsheets."

Module D: Waiting-Line Models

- New OM in Action box: "The High Cost of Long ER Waits."
- Revision of Example D2 and the New England Foundry case study.
- New section covering the topic of The Psychology of Waiting.

Module E: Learning Curves

- New OM in Action box: "The Navy's Learning Curve Challenge."
- Revised Table E.3 (Learning Curve Coefficients) to add a fourth decimal place of accuracy. This change creates an extra level of accuracy in grading MyLab problems.
- All examples and Solved Problems have slightly revised answers.

Module F: Simulation

- New author-created video to accompany "Creating Your Own Excel Spreadsheets."

Module G: Applying Analytics to Big Data in Operations Management

- Additional discussion of the five Vs of big data.
- Expanded the graphing tips.
- New case study: "Labor Concerns at Zapco Industries."

In Every Chapter

- There is a new section at the end of the Rapid Reviews titled, "ADDITIONAL MYLAB OPERATIONS MANAGEMENT RESOURCES." Each of these can contain up to six topics: (1) Videos for Creating Your Own Excel Spreadsheets, (2) Additional Case Studies, (3) Southwestern University Integrated Case Studies, (4) Multiple Choice Case Questions, (5) Recent Graduate Videos, and (6) Simulations.

Solving Teaching and Learning Challenges

Our text is now in its 14th edition, meaning it has been educating and challenging students for over 40 years. We have served close to 2 million readers with the comprehensive learning package that has made the book the best-selling Operations Management text in the U.S. and global marketplaces. We created the learning system with the goal of teaching and preparing your students with employable skills. Here is how:

OM in Action Bitcoin Goes to Where the Power Is Cheap[1]

Home to hydroelectric dams that harness the flow of the Columbia River, north central Washington has some of the cheapest power in the U.S. That has made the largely rural area best known for its apple orchards a magnet for bitcoin miners, who use powerful specialized computers to generate new units of cryptocurrencies—a process that requires vast amounts of electricity to run and cool thousands of machines. "If you ask the guys at UPS or FedEx what they're delivering to Wenatchee, I think they'd tell you it's a whole bunch of bitcoin mining machines," says that town's mayor.

Mining operations can squeeze into small spaces. Shoebox-size computer servers that suck up as much power as 1,000 homes can be packed into a 25-by-25-foot room. Miners have popped up in unexpected places in the area: an old laundromat, a former warehouse, and apartments. There are already at least 30 *known* cryptocurrency-mining operations in north central Washington.

These aren't the first businesses to come to the region for its cheap power. Aluminum smelters once flocked here. In more recent years, companies including Microsoft and Dell have built data-storage centers. Electricity in the region costs 2 to 4 cents per kwh compared with more than 10 cents nationwide. Some residents and officials hope that mining will be the first step toward transforming the area into a business hub for *blockchain* technology, bringing new jobs.

Others worry these miners will drain the area of the surplus power that helps keep rates low. Here is why: Comparative power usage rates (per sq. ft. per year): school—10, home—12, hotel—18, hospital—32, grocery store—40, **computer data center**—2,100!

Real-world examples on page after page. Each chapter opens with a two-page "Global Company Profile" that describes how the featured firm, be it Boeing, Alaska Air, or NASCAR, achieves competitive advantage using the OM techniques and tools of that chapter. We continue with "OM in Action" boxes, each telling a short story about current OM issues and how an organization tackled them. Finally, every chapter concludes with one or more case studies, including 46 case studies of real companies with accompanying videos. Readers will deal with the cruise industry at Celebrity Cruises, the healthcare business at Arnold Palmer Hospital, boat manufacturing at Nautique Boats, and food/entertainment at Hard Rock Cafes, among many others. These cases help students connect the OM concepts to real-world scenarios.

A wealth of examples, solved problems, and homework problems. The only way to understand the analytics of OM are to practice them. Each of these examples contains an "added insight" feature to further understanding. In each chapter we provide numerous step-by-step examples of how to tackle a real problem. We also include a section called "Solved Problems" at the end of each chapter, which provides another look at the topic with similar, detailed solutions. In addition, these 91 Solved Problems are supplemented with "virtual office hour" videos (5 to 15 minutes each) with the authors carefully walking students through each problem. Then come the 850 homework problems, with answers to even-numbered problems in the appendix.

"Rapid Reviews" at the end of each chapter. Students often need an outline to help study what they have learned in the chapter. Our two-page Rapid Reviews contain a summary of all key points, formulas, definitions, and concepts in that chapter. The Rapid Reviews conclude with five or six multiple-choice questions (answers provided) to test students' understanding and help prepare them for a quiz.

Ethical dilemmas facing OM decision makers. Near the end of each chapter, we pose a tough question: What should a manager do when faced with an ethical dilemma (for example, poorly paid workers in a developing country, animal slaughterhouses, pollution, corruption in a company that is a customer of *your* firm, and so on)? This is the world we live in, and we want to prepare students to face questions that may be out of their comfort zone.

The 10 OM strategy decisions. We have structured this text around the 10 Operations Management decisions that organizations make. For each of the decisions, we introduce the decision, its objective, and its relevant tools and techniques for successful decision making. Starting with Chapter 5 and continuing through Chapter 17, these decisions are:

(1) What products or services our firm makes (Chapter 5) [clearly the first decision]
(2) Then how we address quality (Chapter 6)
(3) How we create our goods or services (Chapter 7)
(4) Where we locate our facilities (Chapter 8)
(5) How we lay out our factory, warehouse, office, or retail store (Chapter 9)
(6) How we design jobs, then train and evaluate our employees (Chapter 10)
(7) How we establish and manage supply chains (Chapter 11)
(8) How we control inventory (Chapters 12, 14 and 16)
(9) How we schedule employees, machines, and operations (Chapters 13 and 15)
(10) How we keep our operations up and running (Chapter 17)

MyLab Operations Management

MyLab Operations Management is the teaching and learning platform that empowers *every* student. When combined with educational content written by the authors, **MyLab Operations Management** helps deliver the learning outcomes to which students and instructors aspire.

Operations Management Simulations

Five operations management simulations give students hands-on experience in real-world roles, helping them make decisions, think critically, and link course concepts to on-the-job application.

By receiving real-time, dynamic feedback from stakeholders, students see the impact of their choices and can gauge their performance against individual, peer, and system metrics. Results of these simulations are recorded in the MyLab Gradebook.

The five simulations are:

- Project Management (Chapter 3)
- Forecasting (Chapter 4)
- Quality Management (Chapter 6)
- Supply Chain Management (Chapter 11)
- Inventory Management (Chapter 12)

Students tell us that they enjoy learning OM through these simulations!

A Powerful Homework and Test Manager

Problems from the textbook can be assigned to students via a robust platform, which allows instructors to manage, create, and import online homework assignments, quizzes, and tests that are automatically graded. Instructors can choose from a wide range of assignment options, including time limits, proctoring, and maximum number of attempts allowed. The bottom line: **MyLab Operations Management** means more learning and less time grading.

Learning Aids

Right at the time of learning, students can access Learning Aids such as Help Me Solve This, Videos from the authors of similar problems being solved, Ask My Instructor, and eText Pages. These all provide students feedback and assistance when they need it most.

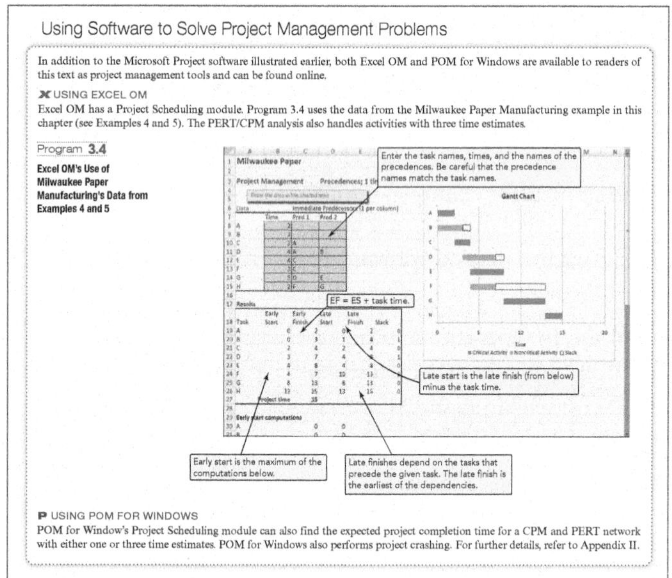

Using Software to Solve Project Management Problems

In addition to the Microsoft Project software illustrated earlier, both Excel OM and POM for Windows are available to readers of this text as project management tools and can be found online.

USING EXCEL OM

Excel OM has a Project Scheduling module. Program 3.4 uses the data from the Milwaukee Paper Manufacturing example in this chapter (see Examples 4 and 5). The PERT/CPM analysis also handles activities with three time estimates.

Program 3.4

Excel OM's Use of Milwaukee Paper Manufacturing's Data from Examples 4 and 5

USING POM FOR WINDOWS

POM for Window's Project Scheduling module can also find the expected project completion time for a CPM and PERT network with either one or three time estimates. POM for Windows also performs project crashing. For further details, refer to Appendix II.

Decision Support Software

We provide two decision support software programs, Excel OM for Windows and Mac, and POM for Windows, to help solve homework problems and case studies. More information on these packages can be found on the Student Download Page at **www.pearsonhighered.com/heizer**.

Jay, Barry, & Chuck's OM Blog

As a complement to this text, we have created a companion blog with coordinated features to help teach the OM course. There are teaching tips, highlights of OM items in the news (along with class discussion questions and links), video tips, guest posts by instructors using our text, and much more—all arranged by chapter. To learn more about any chapter topics, visit *www.heizerrenderOM.wordpress.com*. As instructors prepare their lectures and syllabus, they can scan our blog for discussion ideas, teaching tips, and classroom exercises. Over 1,000,000 visitors indicate that instructors have found the blog useful in their course.

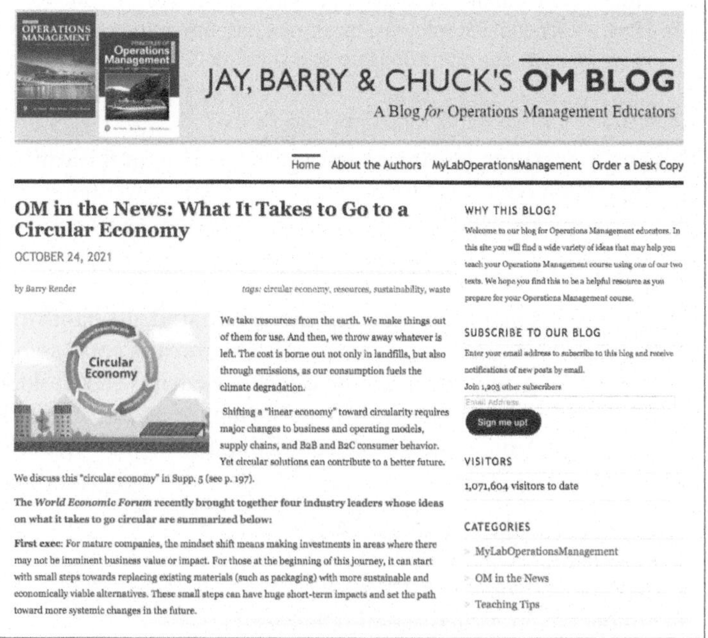

JAY, BARRY & CHUCK'S OM BLOG

A Blog *for* Operations Management Educators

Home About the Authors MyLabOperationsManagement Order a Desk Copy

OM in the News: What It Takes to Go to a Circular Economy

OCTOBER 24, 2021

by Barry Render — *tags:* circular economy, resources, sustainability, waste

We take resources from the earth. We make things out of them for use. And then, we throw away whatever is left. The cost is borne out not only in landfills, but also through emissions, as our consumption fuels the climate degradation.

Shifting a "linear economy" toward circularity requires major changes to business and operating models, supply chains, and B2B and B2C consumer behavior. Yet circular solutions can contribute to a better future. We discuss this "circular economy" in Supp. 5 (see p. 197).

The *World Economic Forum* recently brought together four industry leaders whose ideas on what it takes to go circular are summarized below:

First exec: For mature companies, the mindset shift means making investments in areas where there may not be imminent business value or impact. For those at the beginning of this journey, it can start with small steps towards replacing existing materials (such as packaging) with more sustainable and economically viable alternatives. These small steps can have huge short-term impacts and set the path toward more systemic changes in the future.

WHY THIS BLOG?

Welcome to our blog for Operations Management educators. In this site you will find a wide variety of ideas that may help you teach your Operations Management course using one of our two texts. We hope you find this to be a helpful resource as you prepare for your Operations Management course.

SUBSCRIBE TO OUR BLOG

Enter your email address to subscribe to this blog and receive notifications of new posts by email.

Join 1,203 other subscribers

Email Address

Sign me up!

VISITORS

1,071,604 visitors to date

CATEGORIES

- MyLabOperationsManagement
- OM in the News
- Teaching Tips

Additional Resources

Resources available to instructors and students at www.pearsonhighered.com/heizer	Features of the Resource
Online Tutorials	These additional supplemental chapters cover the following topics: • Statistical Tools for Managers • Acceptance Sampling • The Simplex Method of Linear Programming • The MODI and VAM Methods of Solving Transportation Problems • Vehicle Routing and Scheduling
Excel Data Files	The data files are prepared for specific examples and allow users to solve all the marked text examples without reentering any data.
Excel OM	Excel OM is our exclusive user-friendly Excel add-in. Excel OM automatically creates worksheets to model and solve problems. This software is great for student homework, what-if analysis, and classroom demonstrations.
POM for Windows	POM for Windows is a powerful tool for easily solving OM problems.
Active Models	Active Models are Excel-based OM simulations, designed to help students understand the quantitative methods shown in the textbook examples. Students may change the data to see how the changes affect the answers.

Additional instructor resources are available at **www.pearson.com**. These include the following:

Instructor Resource	Features of the Supplement
Instructor's Resource Manual authored by Chuck Munson	• Chapter summary • Class Discussion Ideas • Active Classroom Learning Exercises • Company Videos discussion • Cinematic Ticklers • Jay, Barry, and Chuck's OM Blog • PowerPoint Slides discussion • Additional Assignment Ideas • Online Resources and Other Supplementary Materials
Instructor's Solutions Manual	The Instructor's Solutions Manual, written by the authors, contains the answers to all of the discussion questions, Ethical Dilemmas, Active Models, and cases in the text, as well as worked-out solutions to all the end-of-chapter problems, additional homework problems, and additional case studies.
Test Bank	• More than 1,500 multiple-choice, true-or-false, and essay questions • Keyed by learning objective • Classified according to difficulty level • AACSB learning standard identified (Ethical Understanding and Reasoning; Analytical Thinking Skills; Integration of Real-World Business Experiences; Diverse and Multicultural Work; Reflective Thinking; Application of Knowledge)
Computerized TestGen	TestGen allows instructors to • customize, save, and generate classroom tests. • edit, add, or delete questions from the Test Item Files. • analyze test results. • organize a database of tests and student results. New to this edition: Hundreds of multiple choice questions have been converted to *algorithmic* questions, with different answers for each student.
PowerPoints	An extensive set of ADA-compliant PowerPoint presentations is available for each chapter, with well over 2,000 slides in all.

Acknowledgments

We thank the many individuals who were kind enough to assist us in this endeavor. The following professors provided insights that guided us in this edition (their names are in bold) and in prior editions:

ALABAMA

John Mittenthal
University of Alabama
Philip F. Musa
University of Alabama at Birmingham
William Petty
University of Alabama
Doug Turner
Auburn University

ALASKA

Paul Jordan
University of Alaska

ARIZONA

Susan K. Norman
Northern Arizona University
Scott Roberts
Northern Arizona University
Vicki L. Smith-Daniels
Arizona State University
Susan K. Williams
Northern Arizona University

CALIFORNIA

Jean-Pierre Amor
University of San Diego
Moshen Attaran
California State University–Bakersfield
Ali Behnezhad
California State University–Northridge
Joe Biggs
California Polytechnic State University
Lesley Buehler
Ohlone College
Manny Fernandez
Bakersfield College
Rick Hesse
Pepperdine
Jianli Hu
Cerritos College
Ravi Kathuria
Chapman University
Richard Martin
California State University–Long Beach
Ozgur Ozluk
San Francisco State University
Zinovy Radovilsky
California State University–Hayward
Robert Saltzman
San Francisco State University
Robert J. Schlesinger
San Diego State University
V. Udayabhanu
San Francisco State University
Rick Wing
San Francisco State University

COLORADO

Peter Billington
Colorado State University–Pueblo
Gregory Stock
University of Colorado at Colorado Springs

CONNECTICUT

David Cadden
Quinnipiac University
Larry A. Flick
Norwalk Community Technical College

FLORIDA

Joseph P. Geunes
University of Florida
Rita Gibson
Embry-Riddle Aeronautical University
Donald Hammond
University of South Florida
Wende Huehn–Brown
St. Petersburg College
Andrew Johnson
University of Central Florida
Adam Munson
University of Florida
Ronald K. Satterfield
University of South Florida
Theresa A. Shotwell
Florida A&M University
Jeff Smith
Florida State University

GEORGIA

John H. Blackstone
University of Georgia
Johnny Ho
Columbus State University
John Hoft
Columbus State University
John Miller
Mercer University
Nikolay Osadchiy
Emory University
Spyros Reveliotis
Georgia Institute of Technology

ILLINOIS

Suad Alwan
Chicago State University
Lori Cook
DePaul University
Matt Liontine
University of Illinois–Chicago
Zafar Malik
Governors State University

INDIANA

Barbara Flynn
Indiana University
B.P. Lingeraj
Indiana University
Frank Pianki
Anderson University
Stan Stockton
Indiana University
Jerry Wei
University of Notre Dame
Jianghua Wu
Purdue University
Xin Zhai
Purdue University

IOWA

Debra Bishop
Drake University
Kevin Watson
Iowa State University
Lifang Wu
University of Iowa

KANSAS

William Barnes
Emporia State University
George Heinrich
Wichita State University
Sue Helms
Wichita State University

Hugh Leach
Washburn University
M.J. Riley
Kansas State University
Teresita S. Salinas
Washburn University
Avanti P. Sethi
Wichita State University

KENTUCKY

Wade Ferguson
Western Kentucky University
Aman Gupta
Embry-Riddle Aeronautical University
Kambiz Tabibzadeh
Eastern Kentucky University

LOUISIANA

Carolyn Borne
Louisiana State University
Roy Clinton
University of Louisiana at Monroe
L. Wayne Shell (retired)
Nicholls State University

MARYLAND

Eugene Hahn
Salisbury University
Samuel Y. Smith, Jr.
University of Baltimore

MASSACHUSETTS

Peter Ittig
University of Massachusetts
Jean Pierre Kuilboer
University of Massachusetts–Boston
Dave Lewis
University of Massachusetts–Lowell
Mike Maggard (retired)
Northeastern University
Peter Rourke
Wentworth Institute of Technology
Keivan Sadeghzadeh
University of Massachusetts–Dartmouth
Daniel Shimshak
University of Massachusetts–Boston
Ernest Silver
Curry College

MICHIGAN

Darlene Burk
Western Michigan University
Sima Fortsch
University of Michigan–Flint
Damodar Golhar
Western Michigan University
Dana Johnson
Michigan Technological University
Doug Moodie
Michigan Technological University

MINNESOTA

Rick Carlson
Metropolitan State University
John Nicolay
University of Minnesota
Michael Pesch
St. Cloud State University
Manus Rungtusanatham
University of Minnesota
Kingshuk Sinha
University of Minnesota
Peter Southard
University of St. Thomas
Wenqing Zhang
University of Minnesota, Duluth

MISSOURI

Shahid Ali
Rockhurst University
Stephen Allen
Truman State University
Sema Alptekin
University of Missouri–Rolla
Gregory L. Bier
University of Missouri–Columbia
James Campbell
University of Missouri–St. Louis
Wooseung Jang
University of Missouri–Columbia
Mary Marrs
University of Missouri–Columbia
A. Lawrence Summers
University of Missouri

NEBRASKA

Zialu Hug
University of Nebraska–Omaha

NEVADA

Joel D. Wisner
University of Nevada, Las Vegas

NEW HAMPSHIRE

Dan Bouchard
Granite State College

NEW JERSEY

Daniel Ball
Monmouth University
Leon Bazil
Stevens Institute of Technology
Mark Berenson
Montclair State University
Linguo Gong
Rider University
Grace Greenberg
Rider University
Joao Neves
The College of New Jersey
Leonard Presby
William Paterson University
Faye Zhu
Rowan University

NEW MEXICO

William Kime
University of New Mexico

NEW YORK

Michael Adams
SUNY Old Westbury
Theodore Boreki
Hofstra University
John Drabouski
DeVry University
Richard E. Dulski
Daemen College
Jonatan Jelen
Baruch College
Beate Klingenberg
Marist College
Purushottam Meena
New York Institute of Technology
Donna Mosier
SUNY Potsdam
Elizabeth Perry
SUNY Binghamton
William Reisel
St. John's University
Abraham Seidmann
University of Rochester
Kaushik Sengupta
Hofstra University
Girish Shambu
Canisius College
Rajendra Tibrewala
New York Institute of Technology

NORTH CAROLINA

Coleman R. Rich
Elon University
Ray Walters
Fayetteville Technical Community College

OHIO

Victor Berardi
Kent State University
Lance Chen
University of Dayton

Mark Jacobs
University of Dayton
Ruth Seiple
University of Cincinnati
Andrew R. Thomas
University of Akron

OKLAHOMA

Wen-Chyuan Chiang
University of Tulsa

OREGON

Anne Deidrich
Warner Pacific College
Gordon Miller
Portland State University
John Sloan
Oregon State University

PENNSYLVANIA

Jeffrey D. Heim
Pennsylvania State University
James F. Kimpel
University of Pittsburgh
Ian M. Langella
Shippensburg University
Prafulla Oglekar
LaSalle University
Stanford Rosenberg
LaRoche College
Edward Rosenthal
Temple University
Susan Sherer
Lehigh University
Howard Weiss
Temple University

RHODE ISLAND

Laurie E. Macdonald
Bryant College
Susan Sweeney
Providence College

SOUTH CAROLINA

Jerry K. Bilbrey
Anderson University
Luis Borges
College of Charleston
Larry LaForge
Clemson University
Emma Jane Riddle
Winthrop University

TENNESSEE

Hugh Daniel
Lipscomb University
Cliff Welborn
Middle Tennessee State University

TEXAS

Phillip Flamm
Texas Tech University
Gregg Lattier
Lee College
Arunachalam Narayanan
Texas A&M University
Ranga V. Ramasesh
Texas Christian University
Mohan Rao
Texas A&M University–Corpus Christi
Cecelia Temponi
Texas State University
John Visich-Disc
University of Houston
Dwayne Whitten
Texas A&M University
David Widdifield
University of Texas at Dallas
Bruce M. Woodworth
University of Texas–El Paso

UTAH

William Christensen
Dixie State College of Utah
Shane J. Schvaneveldt
Weber State University
Madeline Thimmes (retired)
Utah State University

VIRGINIA

Sidhartha Das
George Mason University
Cheryl Druehl
George Mason University
Andy Litteral
University of Richmond
Arthur C. Meiners, Jr.
Marymount University
Michael Plumb
Tidewater Community College
Yu (Amy) Xia
College of William and Mary

WASHINGTON

Mark McKay
University of Washington
Chris Sandvig
Western Washington University
John Stec
Oregon Institute of Technology
Scott Swenson
Washington State University

WASHINGTON, DC

Prabir K Bagchi
The George Washington University
Narendrea K. Rustagi
Howard University

WEST VIRGINIA

Charles Englehardt
Salem International University
Daesung Ha
Marshall University
James S. Hawkes
University of Charleston

WISCONSIN

James R. Gross
University of Wisconsin–Oshkosh
Marilyn K. Hart (retired)
University of Wisconsin–Oshkosh
Niranjan Pati
University of Wisconsin–La Crosse
X. M. Safford
Milwaukee Area Technical College
Rao J. Taikonda
University of Wisconsin–Oshkosh

WYOMING

Cliff Asay
University of Wyoming

INTERNATIONAL

Steven Harrod
Technical University of Denmark
Wolfgang Kersten
Hamburg University of Technology (TUHH)
Robert D. Klassen
University of Western Ontario
Ronald Lau
Hong Kong University of Science and Technology
Brent Snider
University of Calgary

In addition, we appreciate the wonderful people at Pearson Education who provided both help and advice: Lynn Huddon, our content strategy manager; Krista Mastroianni, our product manager; Ashley DePace, our product marketer; Courtney Kamauf and Nancy Chen for their fantastic and dedicated work on MyLab Operations Management; Yasmita Hota, our content producer; and Kristin Jobe and Allison Campbell, our project managers at Integra. We are truly blessed to have such a fantastic team of experts directing, guiding, and assisting us.

In this edition, we were thrilled to be able to include one of the country's premier boat manufacturers, Nautique Boats, in our ongoing Video Case Study series. This was possible because of the wonderful efforts of President Greg Meloon and his superb management team. This included Kris Honigosky (VP, Operations), Steve Carlton (Chief Designer and Director of Product Design and Development), Paula Sleiman (Industrial Engineer Supervisor), Drew Pope (Materials and Supply Chain Manager), Tim Sochar (Customer Service Representative), and Erica Marrero (Project Manager). We are grateful to all of these fine people, as well as the many others that participated in the development of the videos and cases during our trips to the Orlando headquarters.

We also appreciate the efforts of colleagues who have helped to shape the entire learning package that accompanies this text. Professor Howard Weiss (Temple University) developed the Active Models, Excel OM, and POM for Windows software; Professor Mahesh Srinivasan (University of Akron) updated the PowerPoint presentations; and Professor M. Khurrum S. Bhutta (Ohio University) updated the test bank. Beverly Amer produced and directed the video series; Professors Keith Willoughby (Bucknell University) and Ken Klassen (Brock University) contributed the two Excel-based simulation games; and Professor Gary LaPoint (Syracuse University) developed the Microsoft Project crashing exercise and the dice game for SPC. We have been fortunate to have been able to work with all these people.

We wish you a pleasant and productive introduction to operations management.

JAY HEIZER

Texas Lutheran University
1000 W. Court Street
Seguin, TX 78155
Email: jheizer@tlu.edu

BARRY RENDER

Graduate School of Business
Rollins College
Winter Park, FL 32789
Email: profrender@gmail.com

CHUCK MUNSON

Carson College of Business
Washington State University
Pullman, WA 99164-4746
Email: munson@wsu.edu

ABOUT THE AUTHORS

JAY HEIZER

The Jesse H. Jones Professor Emeritus of Business Administration, Texas Lutheran University, Seguin, Texas. He received his B.B.A. and M.B.A. from the University of North Texas and his Ph.D. in Management and Statistics from Arizona State University. He was previously a member of the faculty at the University of Memphis, the University of Oklahoma, Virginia Commonwealth University, where he was department chair, and the University of Richmond. He has also held visiting positions at Boston University, George Mason University, the Czech Management Center, and the Otto-Von-Guericke University, Magdeburg.

Dr. Heizer's industrial experience is extensive. He learned the practical side of operations management as a machinist apprentice at Foringer and Company, as a production planner for Westinghouse Airbrake, and at General Dynamics, where he worked in engineering administration. In addition, he has been actively involved in consulting in the OM and MIS areas for a variety of organizations, including Philip Morris, Firestone, Dixie Container Corporation, Columbia Industries, and Tenneco. He holds the CPIM certification from APICS/ASCM—the Association for Supply Chain Management.

Professor Heizer has co-authored five books and has published more than 30 articles on a variety of management topics. His papers have appeared in the *Academy of Management Journal, Journal of Purchasing, Personnel Psychology, Production & Inventory Control Management, APICS—The Performance Advantage, Journal of Management History, IIE Solutions,* and *Engineering Management,* among others. He has taught operations management courses in undergraduate, graduate, and executive programs.

BARRY RENDER

The Charles Harwood Professor Emeritus of Operations Management, Crummer Graduate School of Business, Rollins College, Winter Park, Florida. He received his B.S. in Mathematics and Physics at Roosevelt University, and his M.S. in Operations Research and Ph.D. in Quantitative Analysis at the University of Cincinnati. He previously taught at George Washington University, University of New Orleans, Boston University, and George Mason University, where he held the Mason Foundation Professorship in Decision Sciences and was Chair of the Decision Sciences Department. Dr. Render has also worked in the aerospace industry for General Electric, McDonnell Douglas, and NASA.

Professor Render has co-authored 10 textbooks for Pearson, including *Managerial Decision Modeling with Spreadsheets, Quantitative Analysis for Management, Service Management, Introduction to Management Science*, and *Cases and Readings in Management Science. Quantitative Analysis for Management*, now in its 13th edition, is a leading text in that discipline in the United States and globally. Dr. Render's more than 100 articles on a variety of management topics have appeared in *Decision Sciences, Production and Operations Management, Interfaces, Information and Management, Journal of Management Information Systems, Socio-Economic Planning Sciences, IIE Solutions*, and *Operations Management Review*, among others. Dr. Render has been honored as an AACSB Fellow and was twice named a Senior Fulbright Scholar. He served as Software Review Editor for Decision Line for six years and as Editor of the *New York Times* Operations Management special issues for five years. For nine years, Dr. Render was President of Management Science Associates of Virginia, Inc., whose technology clients included the FBI, NASA, the U.S. Navy, Fairfax County, Virginia, and C&P Telephone. Dr. Render was selected by Roosevelt University as the recipient of the St. Claire Drake Award for Outstanding Scholarship. He also received the Rollins College MBA Student Award for Best Overall Course and was named Professor of the Year by full-time MBA students.

CHUCK MUNSON

Professor of Operations Management and Carson College of Business Ph.D. Program Director, Washington State University, Pullman, Washington. He received his BSBA *summa cum laude* in finance, along with his MSBA and Ph.D. in operations management, from Washington University in St. Louis. For three years, he worked as a financial analyst for Contel Telephone Corporation.

Professor Munson has served as a senior editor for *Production and Operations Management,* and he serves on the editorial review board of five other journals. He has published more than 30 articles in such journals as *Production and Operations Management, IIE Transactions, Decision Sciences, Naval Research Logistics, European Journal of Operational Research, International Journal of Production Economics, Journal of the Operational Research Society,* and *Annals of Operations Research.* He is editor of the book *The Supply Chain Management Casebook: Comprehensive Coverage and Best Practices in SCM,* and he has co-authored the research monograph *Quantity Discounts: An Overview and Practical Guide for Buyers and Sellers*. He is also coauthor of *Managerial Decision Modeling: Business Analytics with Spreadsheets* (4th edition), published by deGruyter.

Dr. Munson has taught operations management core and elective courses at the undergraduate, MBA, and Ph.D. levels at Washington State University. He has also conducted several teaching workshops at international conferences and for Ph.D. students at Washington State University. His major awards include winning the Sahlin Faculty Excellence Award for Instruction (Washington State University's top teaching award); being a Founding Board Member of the Washington State University President's Teaching Academy; winning the WSU College of Business Outstanding Teaching Award (twice), Research Award, and Service Award (twice); and twice being named the WSU MBA Professor of the Year.

Operations and Productivity

CHAPTER 1

CHAPTER OUTLINE

GLOBAL COMPANY PROFILE: *Hard Rock Café*

About 40% of *all* jobs are in operations management.

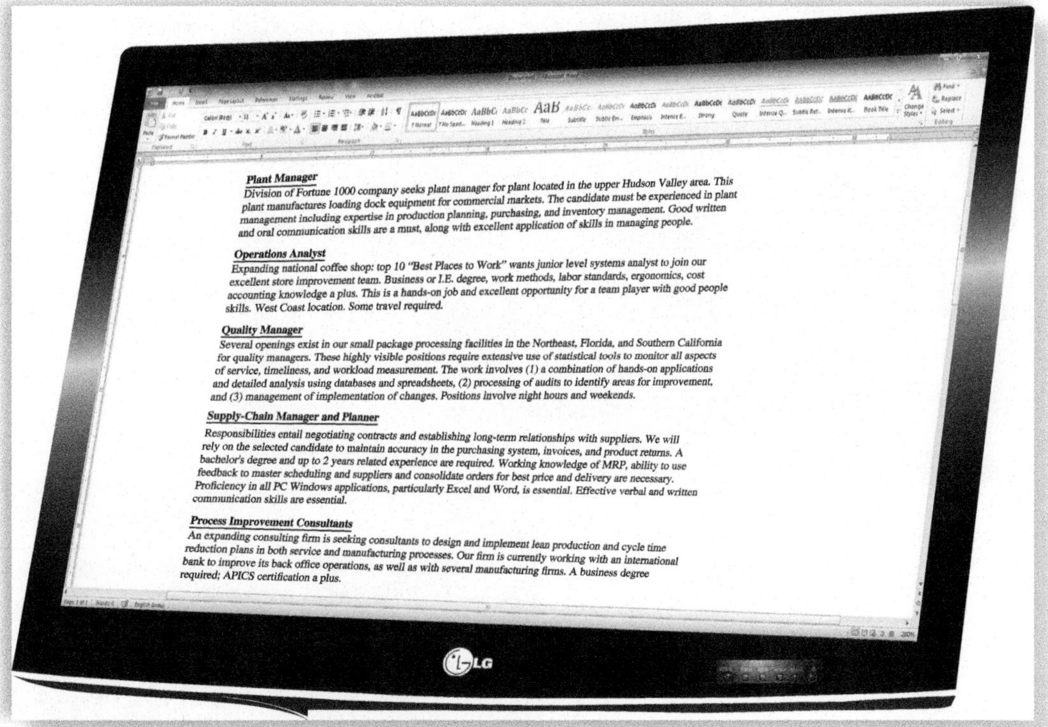

Plant Manager
Division of Fortune 1000 company seeks plant manager for plant located in the upper Hudson Valley area. This plant manufactures loading dock equipment for commercial markets. The candidate must be experienced in plant management including expertise in production planning, purchasing, and inventory management. Good written and oral communication skills are a must, along with excellent application of skills in managing people.

Operations Analyst
Expanding national coffee shop: top 10 "Best Places to Work" wants junior level systems analyst to join our excellent store improvement team. Business or I.E. degree, work methods, labor standards, ergonomics, cost accounting knowledge a plus. This is a hands-on job and excellent opportunity for a team player with good people skills. West Coast location. Some travel required.

Quality Manager
Several openings exist in our small package processing facilities in the Northeast, Florida, and Southern California for quality managers. These highly visible positions require extensive use of statistical tools to monitor all aspects of service, timeliness, and workload measurement. The work involves (1) a combination of hands-on applications and detailed analysis using databases and spreadsheets, (2) processing of audits to identify areas for improvement, and (3) management of implementation of changes. Positions involve night hours and weekends.

Supply-Chain Manager and Planner
Responsibilities entail negotiating contracts and establishing long-term relationships with suppliers. We will rely on the selected candidate to maintain accuracy in the purchasing system, invoices, and product returns. A bachelor's degree and up to 2 years related experience are required. Working knowledge of MRP, ability to use feedback to master scheduling and suppliers and consolidate orders for best price and delivery are necessary. Proficiency in all PC Windows applications, particularly Excel and Word, is essential. Effective verbal and written communication skills are essential.

Process Improvement Consultants
An expanding consulting firm is seeking consultants to design and implement lean production and cycle time reduction plans in both service and manufacturing processes. Our firm is currently working with an international bank to improve its back office operations, as well as with several manufacturing firms. A business degree required; APICS certification a plus.

10 OM STRATEGY DECISIONS

- Design of Goods and Services
- Managing Quality
- Process Strategies
- Location Strategies
- Layout Strategies
- Human Resources
- Supply Chain Management
- Inventory Management
- Scheduling
- Maintenance

GLOBAL COMPANY PROFILE
Hard Rock Cafe

Operations Management at Hard Rock Cafe

Operations managers throughout the world are producing products every day to provide for the well-being of society. These products take on a multitude of forms. They may be washing machines at Whirlpool, motion pictures at DreamWorks, rides at Disney World, or food at Hard Rock Cafe. These firms produce thousands of complex products every day—to be delivered as the customer ordered them, when the customer wants them, and where the customer wants them. Hard Rock does this for over 35 million guests worldwide every year. This is a challenging task, and the operations manager's job, whether at Whirlpool, DreamWorks, Disney, or Hard Rock, is demanding.

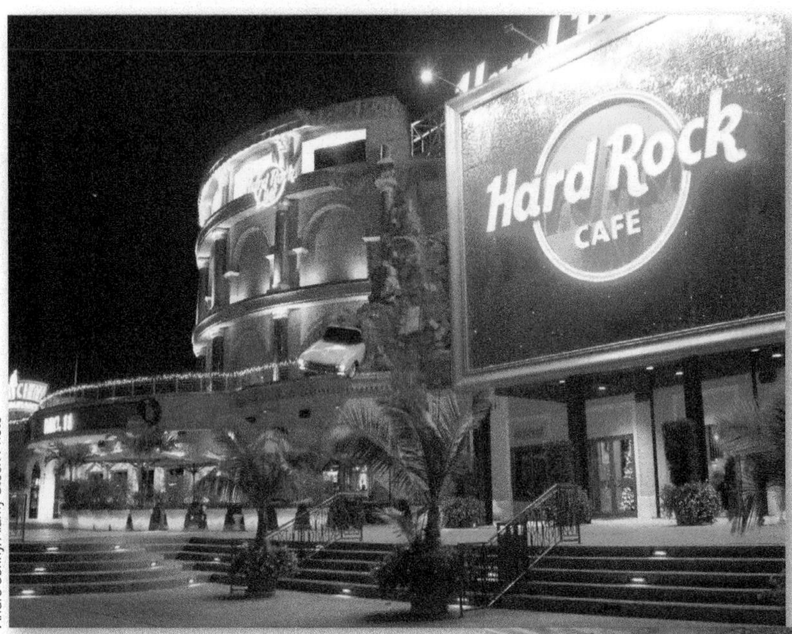

Andre Jenny/Alamy Stock Photo

Hard Rock Cafe in Orlando, Florida, prepares over 3,500 meals each day. Seating more than 1,500 people, it is one of the largest restaurants in the world. But Hard Rock's operations managers serve the hot food hot and the cold food cold.

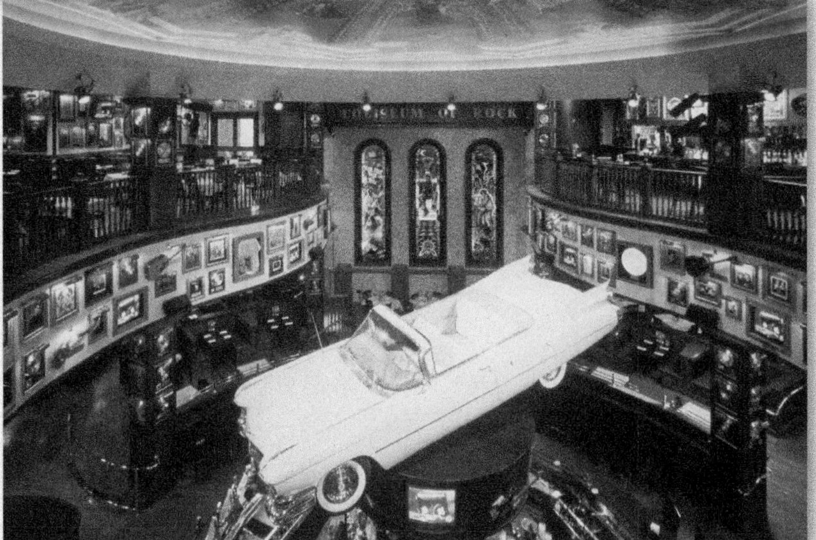

Hulton Archive/Handout/Getty Images

Operations managers are interested in the attractiveness of the layout, but they must be sure that the facility contributes to the efficient movement of people and material with the necessary controls to ensure that proper portions are served.

Eamonn McCormack/WireImage/Getty Images

Lots of work goes into designing, testing, and costing meals. Then suppliers deliver quality products on time, every time, for well-trained cooks to prepare quality meals. But none of that matters unless an enthusiastic waitstaff, such as the one shown here, holding guitars previously owned by members of U2, is doing its job.

Efficient kitchen layouts, motivated personnel, tight schedules, and the right ingredients at the right place at the right time are required to delight the customer.

Jack Picone/Alamy Stock Photo

Orlando-based Hard Rock Cafe opened its first restaurant in London in 1971, making it over 50 years old and the granddaddy of theme restaurants. Although other theme restaurants have come and gone, Hard Rock is still going strong, with 25 hotels, 185 restaurants, and 12 casinos in more than 74 countries—and new restaurants opening each year. Hard Rock made its name with rock music memorabilia, having started when Eric Clapton, a regular customer, marked his favorite bar stool by hanging his guitar on the wall in the London cafe. Now Hard Rock has 70,000 items and millions of dollars invested in memorabilia. To keep customers coming back time and again, Hard Rock creates value in the form of good food and entertainment.

The operations managers at Hard Rock Cafe at Universal Studios in Orlando provide more than 3,500 custom products—in this case meals—every day. These products are designed, tested, and then analyzed for cost of ingredients, labor requirements, and customer satisfaction. On approval, menu items are put into production—and then only if the ingredients are available from qualified suppliers. The production process, from receiving, to cold storage, to grilling or baking or frying, and a dozen other steps, is designed and maintained to yield a quality meal. Operations managers, using the best people they can recruit and train, also prepare effective employee schedules and design efficient layouts.

Managers who successfully design and deliver goods and services throughout the world understand operations. In this text, we look not only at how Hard Rock's managers create value but also how operations managers in other services, as well as in manufacturing, do so. Operations management is demanding, challenging, and exciting. It affects our lives every day. Ultimately, operations managers determine how well we live.

LEARNING OBJECTIVES

STUDENT TIP Let's begin by defining what this course is about.

What Is Operations Management?

LO 1.1 *Define* operations management

VIDEO 1.1 Operations Management at Hard Rock

VIDEO 1.2 Operations Management at Frito-Lay

VIDEO 1.3 Celebrity Cruises: Operations Management at Sea

Production
The creation of goods and services.

Operations management (OM)
Activities that relate to the creation of goods and services through the transformation of inputs to outputs.

Operations management (OM) is a discipline that applies to restaurants like Hard Rock Cafe as well as to factories like Ford and Whirlpool. The techniques of OM apply throughout the world to virtually all productive enterprises. It doesn't matter if the application is in an office, a hospital, a restaurant, a department store, or a factory—the production of goods and services requires operations management. And the *efficient* production of goods and services requires effective applications of the concepts, tools, and techniques of OM that we introduce in this book.

As we progress through this text, we will discover how to manage operations in an economy in which both customers and suppliers are located throughout the world. An array of informative examples, charts, text discussions, and pictures illustrates concepts and provides information. We will see how operations managers create the goods and services that enrich our lives.

In this chapter, we first define *operations management*, explaining its heritage and exploring the exciting role operations managers play in a huge variety of organizations. Then we discuss production and productivity in both goods- and service-producing firms. This is followed by a discussion of operations in the service sector and the challenge of managing an effective and efficient production system.

Production is the creation of goods and services. **Operations management (OM)** is the set of activities that creates value in the form of goods and services by transforming inputs into outputs. Activities creating goods and services take place in all organizations. In manufacturing firms, the production activities that create goods are usually quite obvious. In them, we can see the creation of a tangible product such as a Sony TV or a Harley-Davidson motorcycle.

In an organization that does not create a tangible good or product, the production function may be less obvious. We often call these activities *services*. The services may be "hidden" from the public and even from the customer. The product may take such forms as the transfer of funds from a savings account to a checking account, the transplant of a liver, the filling of an empty seat on an airplane, or the education of a student. Regardless of whether the end product is a good or service, the production activities that go on in the organization are often referred to as operations, or *operations management*.

STUDENT TIP Operations is one of the three functions that every organization performs.

Organizing to Produce Goods and Services

To create goods and services, all organizations perform three functions (see Figure 1.1). These functions are the necessary ingredients not only for production but also for an organization's survival. They are:

1. *Marketing,* which generates the demand, or at least takes the order for a product or service (nothing happens until there is a sale).
2. *Production/operations,* which creates, produces, and delivers the product.
3. *Finance/accounting,* which tracks how well the organization is doing, pays the bills, and collects the money.

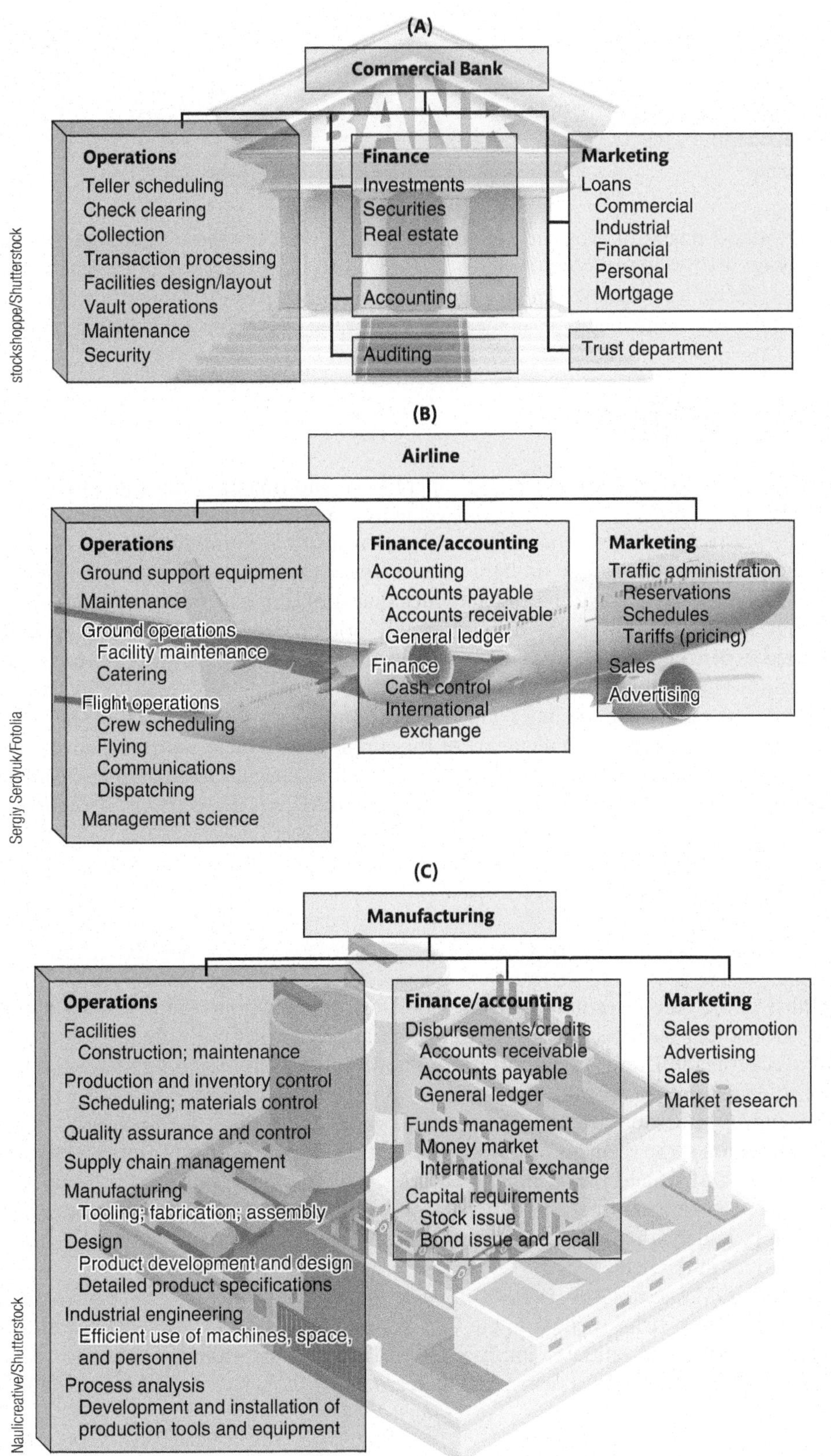

Figure 1.1

Organization Charts for Two Service Organizations and One Manufacturing Organization

(A) a bank, (B) an airline, and (C) a manufacturing organization. The blue areas are OM activities.

STUDENT TIP

The areas in blue indicate the significant role that OM plays in both manufacturing and service firms.

Figure 1.2

Soft Drink Supply Chain

A supply chain for a bottle of Coke requires a beet or sugar cane farmer, a syrup producer, a bottler, a distributor, and a retailer, each adding value to satisfy a customer. Only with collaborations between all members of the supply chain can efficiency and customer satisfaction be maximized. The supply chain, in general, starts with the provider of basic raw materials and continues all the way to the final customer at the retail store.

Universities, places of worship, and businesses all perform these functions. Even a volunteer group such as Meals on Wheels is organized to perform these three basic functions. Figure 1.1 shows how a bank, an airline, and a manufacturing firm organize themselves to perform these functions. The blue-shaded areas show the operations functions in these firms.

The Supply Chain

Supply chain
A global network of organizations and activities that supplies a firm with goods and services.

Through the three functions—marketing, operations, and finance—value for the customer is created. However, firms seldom create this value by themselves. Instead, they rely on a variety of suppliers who provide everything from raw materials to accounting services. These suppliers, when taken together, can be thought of as a *supply chain*. A **supply chain** (see Figure 1.2) is a global network of organizations and activities that supply a firm with goods and services.

As our society becomes more technologically oriented, we see increasing specialization. Specialized expert knowledge, instant communication, and cheaper transportation also foster specialization and worldwide supply chains. It just does not pay for a firm to try to do everything itself. The expertise that comes with specialization exists up and down the supply chain, adding value at each step. When members of the supply chain collaborate to achieve high levels of customer satisfaction, we have a tremendous force for efficiency and competitive advantage. Competition in the 21st century is not between companies; it is between *supply chains.*

STUDENT TIP Good OM managers are scarce, and as a result, career opportunities and pay are excellent.

Why Study OM?

We study OM for four reasons:

1. OM is one of the three major functions of any organization, and it is integrally related to all the other business functions. All organizations market (sell), finance (account), and produce (operate), and it is important to know how the OM activity functions. Therefore, we study *how people organize themselves for productive enterprise*.
2. We study OM because we want to know *how goods and services are produced*. The production function is the segment of our society that creates the products and services we use.
3. We study OM to *understand what operations managers do*. Regardless of your job in an organization, you can perform better if you understand what operations managers do. In addition, understanding OM will help you explore the numerous and lucrative career opportunities in the field.
4. We study OM *because it is such a costly part of an organization*. A large percentage of the revenue of most firms is spent in the OM function. Indeed, OM provides a major opportunity for an organization to improve its profitability and enhance its service to society. Example 1 considers how a firm might increase its profitability via the production function.

Example 1

EXAMINING THE OPTIONS FOR INCREASING CONTRIBUTION

Fisher Technologies is a small firm that must double its dollar contribution to fixed cost and profit in order to be profitable enough to purchase the next generation of production equipment. Management has determined that if the firm fails to increase contribution, its bank will not make the loan and the equipment cannot be purchased. If the firm cannot purchase the equipment, the limitations of the old equipment will force Fisher to go out of business and, in doing so, put its employees out of work and discontinue producing goods and services for its customers.

APPROACH ▶ Table 1.1 shows a simple profit-and-loss statement and three strategic options (marketing, finance/accounting, and operations) for the firm. The first option is a *marketing option*, where excellent marketing management may increase sales by 50%. By increasing sales by 50%, contribution will in turn increase 71%. But increasing sales 50% may be difficult; it may even be impossible.

TABLE 1.1 Options for Increasing Contribution

	CURRENT	MARKETING OPTION[a] INCREASE SALES REVENUE 50%	FINANCE/ ACCOUNTING OPTION[b] REDUCE FINANCE COSTS 50%	OM OPTION[c] REDUCE PRODUCTION COSTS 20%
Sales	$100,000	$150,000	$100,000	$100,000
Costs of goods	–80,000	–120,000	–80,000	–64,000
Gross margin	20,000	30,000	20,000	36,000
Finance costs	–6,000	–6,000	–3,000	–6,000
Subtotal	14,000	24,000	17,000	30,000
Taxes at 25%	–3,500	–6,000	–4,250	–7,500
Contribution[d]	$ 10,500	$ 18,000	$ 12,750	$ 22,500

[a]Increasing sales 50% increases contribution by $7,500, or 71% (7,500/10,500).
[b]Reducing finance costs 50% increases contribution by $2,250, or 21% (2,250/10,500).
[c]Reducing production costs 20% increases contribution by $12,000, or 114% (12,000/10,500).
[d]Contribution to fixed cost (excluding finance costs) and profit.

The second option is a *finance/accounting option*, where finance costs are cut in half through good financial management. But even a reduction of 50% is still inadequate for generating the necessary increase in contribution. Contribution is increased by only 21%.

The third option is an *OM option*, where management reduces production costs by 20% and increases contribution by 114%.

SOLUTION ▶ Given the conditions of our brief example, Fisher Technologies has increased contribution from $10,500 to $22,500. It may now have a bank willing to lend it additional funds.

INSIGHT ▶ The OM option not only yields the greatest improvement in contribution but also may be the only feasible option. Increasing sales by 50% and decreasing finance cost by 50% may both be virtually impossible. Reducing operations cost by 20% may be difficult but feasible.

LEARNING EXERCISE ▶ What is the impact of only a 15% decrease in costs in the OM option? [Answer: A $19,500 contribution; an 86% increase.]

Example 1 underscores the importance of the effective operations activity of a firm. Development of increasingly effective operations is the approach taken by many companies as they face growing global competition.

What Operations Managers Do

All good managers perform the basic functions of the management process. The management process consists of *planning, organizing, staffing, leading*, and *controlling*. Operations managers apply this management process to the decisions they make in the OM function. The **Ten strategic OM decisions** are introduced in Table 1.2. Successfully addressing each of these decisions requires planning, organizing, staffing, leading, and controlling.

Ten Strategic OM Decisions
Design of goods and services
Managing quality
Process strategies
Location strategies
Layout strategies
Human resources
Supply-chain management
Inventory management
Scheduling
Maintenance

Where Are the OM Jobs? How does one get started on a career in operations? The ten strategic OM decisions identified in Table 1.2 are made by individuals who work in the disciplines shown in the blue areas of Figure 1.1. Business students who know their accounting,

STUDENT TIP An operations manager must successfully address the 10 decisions around which this text is organized.

LO 1.2 *Identify* the 10 strategic decisions of operations management

TABLE 1.2 Ten Strategic Operations Management Decisions

DECISION	CHAPTER(S)
1. *Design of goods and services:* Defines much of what is required of operations in each of the other OM decisions. For instance, product design usually determines the lower limits of cost and the upper limits of quality, as well as major implications for sustainability and the human resources required.	5, Supplement 5
2. *Managing quality and statistical process control:* Determines the customer's quality expectations and establishes policies and procedures to identify and achieve that quality.	6, Supplement 6
3. *Process and capacity strategies:* Determines how a good or service is produced (i.e., the process for production) and commits management to specific technology, quality, human resources, and capital investments that determine much of the firm's basic cost structure.	7, Supplement 7
4. *Location strategies:* Requires judgments regarding nearness to customers, suppliers, and talent, while considering costs, infrastructure, logistics, and government.	8
5. *Layout strategies:* Requires integrating capacity needs, personnel levels, technology, and inventory requirements to determine the efficient flow of materials, people, and information.	9
6. *Human resources, job design and work measurement:* Determines how to recruit, motivate, and retain personnel with the required talent and skills. People are an integral and expensive part of the total system design.	10
7. *Supply chain management:* Decides how to integrate the supply chain into the firm's strategy, including decisions that determine what is to be purchased, from whom, and under what conditions.	11, Supplement 11
8. *Inventory management:* Considers inventory ordering and holding decisions and how to optimize them as customer satisfaction, supplier capability, and production schedules are considered.	12, 14, 16
9. *Scheduling:* Determines and implements intermediate- and short-term schedules that effectively and efficiently use both personnel and facilities while meeting customer demands.	13, 15
10. *Maintenance:* Requires decisions that consider facility capacity, production demands, and personnel necessary to maintain a reliable and stable process.	17

statistics, finance, and OM have an opportunity to assume entry-level positions in all of these areas. As you read this text, identify disciplines that can assist you in making these decisions. Then take courses in those areas. The more background an OM student has in accounting, statistics, information systems, and mathematics, the more job opportunities will be available. About 40% of *all* jobs are in OM.

LO 1.3 *Identify* career opportunities in operations management

The following professional organizations provide various certifications that may enhance your education and be of help in your career:

- Association for Supply Chain Management (ASCM/APICS) **(www.ascm.org)**
- American Society for Quality (ASQ) **(www.asq.org)**
- Institute for Supply Management (ISM) **(www.ismworld.org)**
- Project Management Institute (PMI) **(www.pmi.org)**
- Council of Supply Chain Management Professionals **(www.cscmp.org)**

The Heritage of Operations Management

The field of OM is relatively young, but its history is rich and interesting. Our lives and the OM discipline have been enhanced by the innovations and contributions of numerous individuals. We now introduce a few of these people, and we provide a summary of significant events in operations management in Figure 1.3.

Eli Whitney (1800) is credited for the early popularization of interchangeable parts, which was achieved through standardization and quality control. Through a contract he signed with

Everett Collection/Newscom

Cost Focus		Quality Focus	Customization Focus	Globalization Focus
Early Concepts 1776–1880 Labor Specialization (Smith, Babbage) Standardized Parts (Whitney) **Scientific Management Era 1880–1910** Gantt Charts (Gantt) Motion & Time Studies (Gilbreth) Process Analysis (Taylor) Queuing Theory (Erlang)	**Mass Production Era 1910–1980** Moving Assembly Line (Ford/Sorensen) Statistical Sampling (Shewhart) Economic Order Quantity (Harris) Linear Programming (Dantzig) Material Requirements Planning (MRP)	**Lean Production Era 1980–1995** Just-in-Time (JIT) Computer-Aided Design (CAD) Electronic Data Interchange (EDI) Total Quality Management (TQM) Baldrige Award Empowerment Kanbans	**Mass Customization Era 1995–2005** Internet/E-Commerce Enterprise Resource Planning International Quality Standards (ISO) Finite Scheduling Supply Chain Management Mass Customization Build-to-Order Radio Frequency Identification (RFID)	**Globalization Era 2005–2025** Global Supply Chains and Logistics Containerization of Shipping Growth of Transnational Organizations Sustainability Ethics in the Global Workplace Internet of Things (IoT) Digital Operations Industry 4.0

Figure **1.3**

Significant Events in Operations Management

the U.S. government for 10,000 muskets, he was able to command a premium price because of their interchangeable parts.

Frederick W. Taylor (1881), known as the father of scientific management, contributed to personnel selection, planning and scheduling, motion study, and the now popular field of ergonomics. One of his major contributions was his belief that management should be much more resourceful and aggressive in the improvement of work methods. Taylor and his colleagues, Henry L. Gantt and Frank and Lillian Gilbreth, were among the first to systematically seek the best way to produce.

Another of Taylor's contributions was the belief that management should assume more responsibility for:

1. Matching employees to the right job.
2. Providing the proper training.
3. Providing proper work methods and tools.
4. Establishing legitimate incentives for work to be accomplished.

By 1913, Henry Ford and Charles Sorensen combined what they knew about standardized parts with the quasi-assembly lines of the meatpacking and mail-order industries and added the revolutionary concept of the assembly line, where workers stood still and material moved.

Quality control is another historically significant contribution to the field of OM. Walter Shewhart (1924) combined his knowledge of statistics with the need for quality control and provided the foundations for statistical sampling in quality control. W. Edwards Deming

(1950) believed, as did Frederick Taylor, that management must do more to improve the work environment and processes so that quality can be improved.

Operations management will continue to progress as contributions from other disciplines, including *industrial engineering, statistics, management, analytics,* and *economics*, improve decision making.

Innovations from the *physical sciences* (biology, anatomy, chemistry, physics) have also contributed to advances in OM. These innovations include new adhesives, faster integrated circuits, gamma rays to sanitize food products, and specialized glass for iPhones and plasma TVs. Innovation in products and processes often depends on advances in the physical sciences.

Especially important contributions to OM have come from *information technology*, which we define as the systematic processing of data to yield information. Information technology—with digitalization, wireless links, Internet, and e-commerce—is reducing costs and accelerating communication.

Decisions in operations management require individuals who are well versed in analytical tools, in information technology, and often in the biological or physical sciences. In this textbook, we look at the diverse ways a student can prepare for a career in operations management.

STUDENT TIP Services are especially important because almost 80% of all jobs are in service firms.

Operations for Goods and Services

Services Economic activities that typically produce an intangible product (such as education, entertainment, lodging, government, financial, and health services).

Manufacturers produce a tangible product, while service products are often intangible. But many products are a combination of a good and a service, which complicates the definition of a service. Even the U.S. government has trouble generating a consistent definition. Because definitions vary, much of the data and statistics generated about the service sector are inconsistent. However, we define **services** as including repair and maintenance, government, food and lodging, transportation, insurance, trade, financial, real estate, education, legal, medical, entertainment, and other professional occupations.

The operation activities for both goods and services are often very similar. For instance, both have quality standards, are designed and produced on a schedule that meets customer demand, and are made in a facility where people are employed. However, some major differences *do* exist between goods and services. These are presented in Table 1.3.

TABLE 1.3 Differences Between Goods and Services

CHARACTERISTICS OF SERVICES	CHARACTERISTICS OF GOODS
Intangible: Ride in an airline seat	Tangible: The seat itself
Produced and consumed simultaneously: Beauty salon produces a haircut that is consumed as it is produced	Product can usually be kept in inventory (beauty care products)
Unique: Your investments and medical care are unique	Similar products produced (iPads, earbuds)
High customer interaction: Often what the customer is paying for (consulting, education)	Limited customer involvement in production
Inconsistent product definition: Auto insurance changes with age and type of car	Product standardized (iPhone)
Often knowledge based: Legal, education, and medical services are hard to automate	Standard tangible product tends to make automation feasible
Services dispersed: Service may occur at retail store, local office, house call, or via Internet	Product typically produced at a fixed facility
Quality may be hard to evaluate: Consulting, education, and medical services	Many aspects of quality for tangible products are easy to evaluate (strength of a bolt)
Reselling is unusual: Musical concert or medical care	Product often has some residual value

LO 1.4 *Explain* the distinction between goods and services

We should point out that in many cases, the distinction between goods and services is not clear-cut. In reality, almost all services and almost all goods are a mixture of a service and a tangible product. Even services such as consulting may require a tangible report. Similarly, the sale of most goods includes a service. For instance, many products have the service components of financing and delivery (e.g., automobile sales). Many also require after-sale training and maintenance (e.g., office copiers and machinery). "Service" activities may also be an integral part of production. Human resource activities, logistics, accounting, training, field service, and repair are all service activities, but they take place within a manufacturing organization. Very few services are "pure," meaning they have no tangible component. Counseling may be one of the exceptions.

Growth of Services

Services constitute the largest economic sector in postindustrial societies. Until about 1900, most Americans were employed in agriculture. Increased agricultural productivity allowed people to leave the farm and seek employment in the city. Similarly, manufacturing employment has decreased for the past 60 years. The changes in agriculture, manufacturing, and service employment as a percentage of the workforce are shown in Figure 1.4. Although the *number* of people employed in manufacturing has decreased since 1950, each person is now producing almost 20 times more than in 1950. Moreover, manufacturing's influence on employment and the economic system extends well beyond direct manufacturing to include the related advances in technology, as well as the warehousing and logistics necessary to move products through the supply chain to the ultimate consumer. The total manufacturing impact accounts for about 30% of gross domestic product (GDP).

Services became the dominant employer in the early 1920s, with manufacturing employment peaking at about 32% in 1950. The huge productivity increases in agriculture and manufacturing have allowed more of our economic resources to be devoted to services. Consequently, much of the world can now enjoy the pleasures of education, health services, entertainment, and myriad other things that we call services. Examples of firms and percentage of employment in the U.S. service sector are shown in Table 1.4. Table 1.4 also provides employment percentages for the nonservice sectors of manufacturing, construction, agriculture, and mining on the bottom four lines.

Service sector
The segment of the economy that includes trade, financial, lodging, education, legal, medical, and other professional occupations.

Service Pay

Although there is a common perception that service industries are low paying, in fact, many service jobs pay very well. Operations managers in the maintenance facility of an airline are very well paid, as are the operations managers who supervise computer services to the financial community. About 42% of all service workers receive wages above the national average. However, the service-sector average is driven down because 14 of the U.S. Department of Commerce categories of the 33 service industries do indeed pay below the all-private industry average. Of these, retail trade, which pays only 61% of the national private industry average, is large. But even considering the retail sector, the average wage of all service workers is about 96% of the average of all private industries.

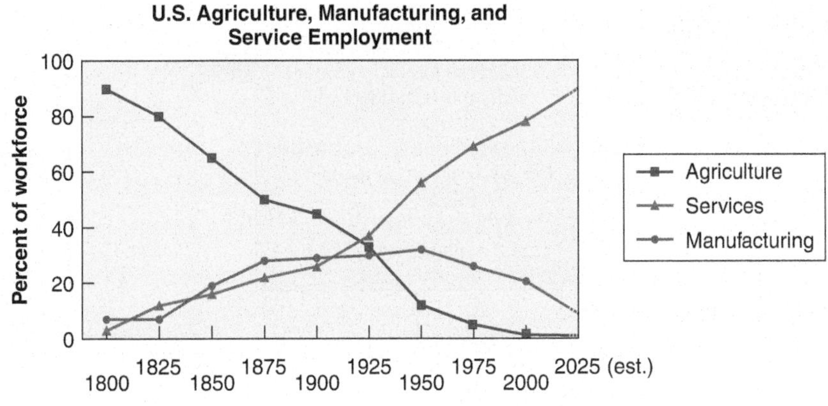

Figure **1.4**

U.S. Agriculture, Manufacturing, and Service Employment

Source: U.S. Bureau of Labor Statistics.

TABLE 1.4 Examples of Organizations in Each Sector

SECTOR	EXAMPLE	PERCENTAGE OF ALL JOBS	
Service Sector			
Education, Medical, Other	San Diego State University, Arnold Palmer Hospital	14.9	85.7
Trade (retail, wholesale), Transportation, Utilities	Walgreen's, Walmart, Nordstrom, Alaska Airlines	17.0	
Information, Publishers, Broadcast	IBM, Bloomberg, Pearson, ESPN	1.8	
Professional, Legal, Business Services, Associations	Snelling and Snelling, Waste Management, American Medical Association, Ernst & Young	17.2	
Finance, Insurance, Real Estate	Citicorp, American Express, Prudential, Aetna	10.8	
Leisure, Lodging, Entertainment	Red Lobster, Motel 6, Celebrity Cruises	10.2	
Government (Fed, State, Local)	U.S., State of Alabama, Cook County	13.8	
Manufacturing Sector	General Electric, Ford, U.S. Steel, Intel		7.9
Construction Sector	Bechtel, McDermott		4.6
Agriculture	King Ranch		1.4
Mining Sector	Homestake Mining		0.4
Grand Total			100.0

Source: Bureau of Labor Statistics, 2020.

STUDENT TIP Why is productivity important? Because it determines our standard of living.

The Productivity Challenge

Productivity
The ratio of outputs (goods and services) divided by one or more inputs (such as labor, capital, or management).

The creation of goods and services requires changing resources into goods and services. The more efficiently we make this change, the more productive we are and the more value is added to the good or service provided. **Productivity** is the ratio of outputs (goods and services) divided by the inputs (resources, such as labor and capital) (see Figure 1.5). The operations manager's job is to enhance (improve) this ratio of outputs to inputs. Improving productivity means improving efficiency.[1]

This improvement can be achieved in two ways: reducing inputs while keeping output constant or increasing output while keeping inputs constant. Both represent an improvement in productivity. In an economic sense, inputs are labor, capital, and management, which are integrated into a production system. Management creates this production system, which provides the conversion of inputs to outputs. Outputs are goods and services, including such diverse

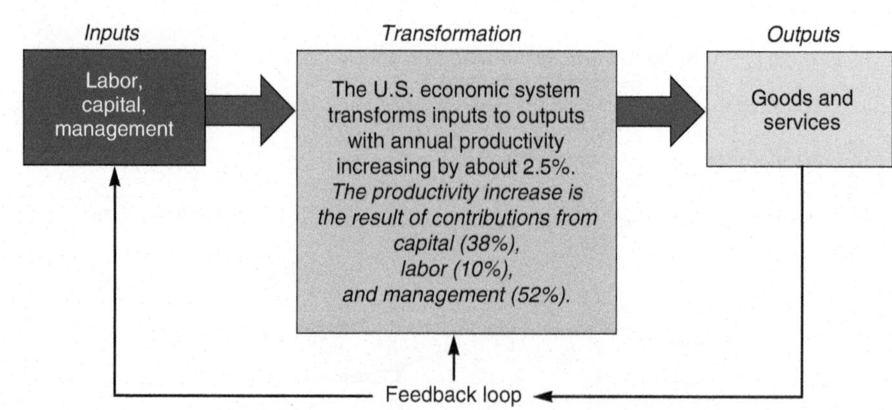

Figure 1.5
The Economic System Adds Value by Transforming Inputs to Outputs
An effective feedback loop evaluates performance against a strategy or standard. It also evaluates customer satisfaction and sends signals to managers controlling the inputs and transformation process.

items as guns, butter, education, improved judicial systems, and ski resorts. *Production* is the making of goods and services. High production may imply only that more people are working and that employment levels are high (low unemployment), but it does not imply high *productivity*.

LO 1.5 *Explain* the difference between production and productivity

Measurement of productivity is an excellent way to evaluate a country's ability to provide an improving standard of living for its people. *Only through increases in productivity can the standard of living improve.* Moreover, only through increases in productivity can labor, capital, and management receive additional payments. If returns to labor, capital, or management are increased without increased productivity, prices rise. On the other hand, downward pressure is placed on prices when productivity increases because more is being produced with the same resources.

The benefits of increased productivity are illustrated in the *OM in Action* box, "Improving Productivity at Starbucks."

For well over a century (from about 1869), the U.S. has been able to increase productivity at an average rate of almost 2.5% per year. Such growth has doubled U.S. wealth every 30 years. However, U.S. annual productivity growth in the early part of the 21st century is slightly below the 2.5% range for the economy as a whole and in recent years has been trending down.

In this text, we examine how to improve productivity through operations management. Productivity is a significant issue for the world and one that the operations manager is uniquely qualified to address.

Productivity Measurement

LO 1.6 *Compute* single-factor productivity

The measurement of productivity can be quite direct. Such is the case when productivity is measured by labor-hours per ton of a specific type of steel. Although labor-hours is a common measure of input, other measures such as capital (dollars invested), materials (tons of ore), or energy (kilowatts of electricity) can be used.[3] An example of this can be summarized in the following equation:

$$\text{Productivity} = \frac{\text{Units produced}}{\text{Input used}} \tag{1-1}$$

For example, if units produced = 1,000 and labor-hours used is 250, then:

$$\text{Single-factor productivity} = \frac{\text{Units produced}}{\text{Labor-hours used}} = \frac{1,000}{250} = 4 \text{ units per labor-hour}$$

OM in Action — Improving Productivity at Starbucks[2]

"This is a game of seconds..." says Silva Peterson, whom Starbucks has put in charge of saving seconds. Her team of 10 analysts is constantly asking themselves: "How can we shave time off this?"

Peterson's analysis suggested that there were some obvious opportunities. First, stop requiring signatures on credit-card purchases under $25. This sliced 8 seconds off the transaction time at the cash register.

Then analysts noticed that Starbucks' largest cold beverage, the Venti size, required two bending and digging motions to scoop up enough ice. The scoop was too small. Redesign of the scoop provided the proper amount in one motion and cut 14 seconds off the average time of 1 minute.

Third were new espresso machines; with the push of a button, the machines grind coffee beans and brew. This allowed the server, called a "barista" in Starbucks's vocabulary, to do other things. The savings: about 12 seconds per espresso shot.

As a result, operations improvements at Starbucks outlets have increased the average transactions per hour to 11.7—a 46% increase—and yearly volume by $250,000, to about $1 million. The result: a 27% improvement in overall productivity—about 4.5% per year. In the service industry, a 4.5% per year increase is very tasty.

Kondor83/Shutterstock

Single-factor productivity
Indicates the ratio of goods and services produced (outputs) to one resource (input).

Multifactor productivity
Indicates the ratio of goods and services produced (outputs) to many or all resources (inputs).

The use of just one resource input to measure productivity, as shown in Equation (1-1), is known as **single-factor productivity**. However, a broader view of productivity is **multifactor productivity**, which includes all inputs (e.g., capital, labor, material, energy). Multifactor productivity is also known as *total factor productivity*. Multifactor productivity is calculated by combining the input units as shown here:

$$\text{Multifactor productivity} = \frac{\text{Output}}{\text{Labor} + \text{Material} + \text{Energy} + \text{Capital} + \text{Miscellaneous}} \tag{1-2}$$

To aid in the computation of multifactor productivity, the individual inputs (the denominator) can be expressed in dollars and summed as shown in Example 2.

Example 2

COMPUTING SINGLE-FACTOR AND MULTIFACTOR GAINS IN PRODUCTIVITY

Collins Title Insurance Ltd. wants to evaluate its labor and multifactor productivity with a new computerized title-search system. The company has a staff of four, each working 8 hours per day (for a payroll cost of $640/day) and overhead expenses of $400 per day. Collins processes and closes on 8 titles each day. The new computerized title-search system will allow the processing of 14 titles per day. Although the staff, their work hours, and pay are the same, the overhead expenses are now $800 per day.

APPROACH ▶ Collins uses Equation (1-1) to compute labor productivity and Equation (1-2) to compute multifactor productivity.

SOLUTION ▶

LO 1.7 *Compute* multifactor productivity

$$\text{Labor productivity with the old system: } \frac{8 \text{ titles per day}}{32 \text{ labor-hours}} = .25 \text{ title per labor-hour}$$

$$\text{Labor productivity with the new system: } \frac{14 \text{ titles per day}}{32 \text{ labor-hours}} = .4375 \text{ title per labor-hour}$$

$$\text{Multifactor productivity with the old system: } \frac{8 \text{ titles per day}}{\$640 + \$400} = .0077 \text{ title per dollar}$$

$$\text{Multifactor productivity with the new system: } \frac{14 \text{ titles per day}}{\$640 + \$800} = .0097 \text{ title per dollar}$$

Labor productivity has increased from .25 to .4375. The change is $(.4375 - .25)/.25 = 0.75$, or a 75% increase in labor productivity. Multifactor productivity has increased from .0077 to .0097. This change is $(.0097 - .0077)/.0077 = 0.26$, or a 26% increase in multifactor productivity.

INSIGHT ▶ Both the labor (single-factor) and multifactor productivity measures show an increase in productivity. However, the multifactor measure provides a better picture of the increase because it includes all the costs connected with the increase in output.

LEARNING EXERCISE ▶ If the overhead goes to $960 (rather than $800), what is the multifactor productivity? [Answer: .00875.]

RELATED PROBLEMS ▶ 1.1, 1.2, 1.4, 1.5, 1.6, 1.7, 1.8, 1.9, 1.10, 1.11, 1.13, 1.14, 1.17

Use of productivity measures aids managers in determining how well they are doing. But results from the two measures can be expected to vary. If labor productivity growth is entirely the result of capital spending, measuring just labor distorts the results. Multifactor productivity is usually better, but more complicated. Labor productivity is the more popular measure. The multifactor-productivity measures provide better information about the trade-offs among factors, but substantial measurement problems remain. Some of these measurement problems are:

1. *Quality* may change while the quantity of inputs and outputs remains constant. Compare a smart LED TV of this decade with a black-and-white TV of the 1950s. Both are TVs, but few people would deny that the quality has improved. The unit of measure—a TV—is the same, but the quality has changed.
2. *External elements* may cause an increase or a decrease in productivity for which the system under study may not be directly responsible. A more reliable electric power service may greatly improve production, thereby improving the firm's productivity because of this support system rather than because of managerial decisions made within the firm.
3. *Precise units of measure* may be lacking. Not all automobiles require the same inputs: Some cars are subcompacts, others are 911 Turbo Porsches.

Productivity measurement is particularly difficult in the service sector, where the end product can be hard to define. For example, economic statistics ignore the quality of your haircut, the outcome of a court case, or the service at a retail store. In some cases, adjustments are made for the quality of the product sold but *not* the quality of the sales presentation or the advantage of a broader product selection. Productivity measurements require specific inputs and outputs, but a free economy is producing worth—what people want—which includes convenience, speed, and safety. Traditional measures of outputs may be a very poor measure of these other measures of worth. Note the quality-measurement problems in a law office, where each case is different, altering the accuracy of the measure "cases per labor-hour" or "cases per employee."

Productivity Variables

As we saw in Figure 1.5, productivity increases are dependent on three **productivity variables**:

Productivity variables
The three factors critical to productivity improvement—labor, capital, and the art and science of management.

1. *Labor*, which contributes about 10% of the annual increase.
2. *Capital*, which contributes about 38% of the annual increase.
3. *Management*, which contributes about 52% of the annual increase.

These three factors are critical to improved productivity. They represent the broad areas in which managers can take action to improve productivity.

LO 1.8 *Identify* the critical variables in enhancing productivity

Labor Improvement in the contribution of labor to productivity is the result of a healthier, better-educated, and better-nourished labor force. Some increase may also be attributed to a shorter workweek. Historically, about 10% of the annual improvement in productivity is attributed to improvement in the quality of labor. Three key variables for improved labor productivity are:

1. Basic education appropriate for an effective labor force.
2. Diet of the labor force.
3. Social overhead that makes labor available, such as transportation and sanitation.

Illiteracy and poor diets are a major impediment to productivity, costing countries up to 20% of their productivity. Infrastructure that yields clean drinking water and sanitation is also an opportunity for improved productivity, as well as an opportunity for better health, in much of the world.

In developed nations, the challenge becomes *maintaining and enhancing the skills of labor* in the midst of rapidly expanding technology and knowledge. Recent data suggest that the average American 17-year-old knows significantly less mathematics than the average Japanese at the same age, and about half cannot answer the questions in Figure 1.6. Moreover, about one-third of American job applicants tested for basic skills were deficient in reading, writing, or math.

Overcoming shortcomings in the quality of labor while other countries have a better labor force is a major challenge. Perhaps improvements can be found not only through increasing competence of labor but also via *better utilized labor with a stronger commitment*. Training, motivation, team building, and the human resource strategies discussed in Chapter 10, as well as improved education, may be among the many techniques that will contribute to increased labor productivity. Improvements in labor productivity are possible; however, they can be expected to be increasingly difficult and expensive.

Figure 1.6

About Half of the 17-Year-Olds in the U.S. Cannot Correctly Answer Questions of This Type

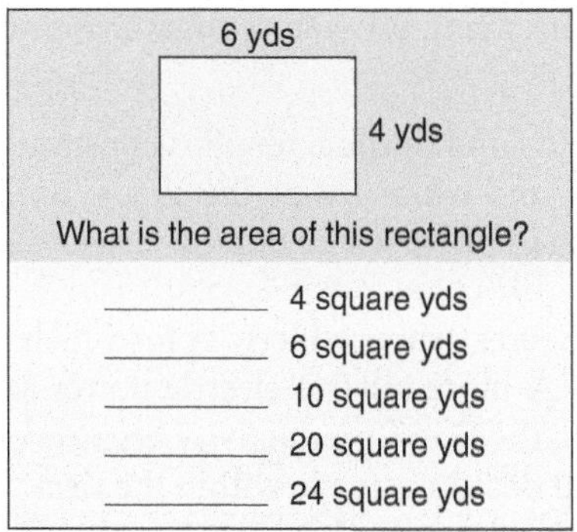

6 yds
4 yds
What is the area of this rectangle?
_____ 4 square yds
_____ 6 square yds
_____ 10 square yds
_____ 20 square yds
_____ 24 square yds

If $9y + 3 = 6y + 15$ then $y =$

_____ 1 _____ 4

_____ 2 _____ 6

Which of the following is true about 84% of 100?

_____ It is greater than 100

_____ It is less than 100

_____ It is equal to 100

Capital Human beings are tool-using animals. Capital investment provides those tools. Capital investment has increased in the U.S. every year except during a few very severe recession periods. Annual capital investment in the U.S. has increased at an annual rate of 1.5% after allowances for depreciation.

Inflation and taxes increase the cost of capital, making capital investment increasingly expensive. When the capital invested per employee drops, we can expect a drop in productivity. Using labor rather than capital may reduce unemployment in the short run, but it also makes economies less productive and therefore lowers wages in the long run. Capital investment is often a necessary, but seldom a sufficient, ingredient in the battle for increased productivity.

The trade-off between capital and labor is continually in flux. The higher the cost of capital or perceived risk, the more projects requiring capital are "squeezed out": they are not pursued because the potential return on investment for a given risk has been reduced. Managers adjust their investment plans to changes in capital cost and risk.

Management Management is a factor of production and an economic resource. Management is responsible for ensuring that labor and capital are effectively used to increase productivity. Management accounts for over half of the annual increase in productivity. This

Andrzej Thiel/Fotolia

Guy Shapira/Shutterstock

The effective use of capital often means finding the proper trade-off between investment in capital assets (automation, left) and human assets (a manual process, right). While there are risks connected with any investment, the cost of capital and physical investments is fairly clear-cut, but the cost of employees has many hidden costs including fringe benefits, social insurance, and legal constraints on hiring, employment, and termination.

increase includes improvements made through the use of knowledge and the application of technology.

Using knowledge and technology is critical in postindustrial societies. Consequently, post-industrial societies are also known as **knowledge societies**. Knowledge societies are those in which much of the labor force has migrated from manual work to technical and information-processing tasks requiring ongoing education. The required education and training are important high-cost items that are the responsibility of operations managers as they build organizations and workforces. The expanding knowledge base of contemporary society requires that managers use *technology and knowledge effectively.*

Knowledge society
A society in which much of the labor force has migrated from manual work to work based on knowledge.

More effective use of capital also contributes to productivity. It falls to the operations manager, as a productivity catalyst, to select the best new capital investments as well as to improve the productivity of existing investments.

The productivity challenge is difficult. A country cannot be a world-class competitor with second-class inputs. Poorly educated labor, inadequate capital, and dated technology are second-class inputs. High productivity and high-quality outputs require high-quality inputs, including good operations managers.

Productivity and the Service Sector

The service sector provides a special challenge to the accurate measurement of productivity and productivity improvement. The traditional analytical framework of economic theory is based primarily on goods-producing activities. Consequently, most published economic data relate to goods production. But the data do indicate that, as our contemporary service economy has increased in size, we have had slower growth in productivity.

Productivity of the service sector has proven difficult to improve because service-sector work is:

1. Typically labor intensive (e.g., counseling, teaching).
2. Frequently focused on unique individual attributes or desires (e.g., investment advice).
3. Often an intellectual task performed by professionals (e.g., medical diagnosis).
4. Often difficult to mechanize and automate (e.g., a haircut).
5. Often difficult to evaluate for quality (e.g., performance of a law firm).

The more intellectual and personal the task, the more difficult it is to achieve increases in productivity. Low-productivity improvement in the service sector is also attributable to the growth of low-productivity activities in the service sector. These include activities not previously a part of the measured economy, such as child care, food preparation, house cleaning, and laundry service. These activities have moved out of the home and into the measured economy as more and more women have joined the workforce. Inclusion of these activities

Olaf Jandke/Agencja Fotograficzna Caro/Alamy Stock Photo

Siemens, a multi-billion-dollar German conglomerate, has long been known for its apprentice programs in its home country. Because education is often the key to efficient operations in a technological society, Siemens has spread its apprentice-training programs to its U.S. plants. These programs are laying the foundation for the highly skilled workforce that is essential for global competitiveness.

OM in Action — Taco Bell Improves Productivity and Goes Green to Lower Costs[4]

Founded in 1962 by Glenn Bell, Taco Bell seeks competitive advantage via low cost. Like many other services, Taco Bell relies on its operations management to improve productivity and reduce cost.

Its menu and meals are designed to be easy to prepare. Taco Bell has shifted a substantial portion of food preparation to suppliers who could perform food processing more efficiently than a stand-alone restaurant. Ground beef is precooked prior to arrival and then reheated, as are many dishes that arrive in plastic boil bags for easy sanitary reheating. Similarly, tortillas arrive already fried and onions prediced. Efficient layout and automation has cut to 8 seconds the time needed to prepare tacos and burritos and has cut time in the drive-through lines by 1 minute. These advances have been combined with training and empowerment to increase the span of management from one supervisor for 5 restaurants to one supervisor for 30 or more.

Operations managers at Taco Bell have cut in-store labor by 15 hours per day and reduced floor space by more than 50%. The result is a store that can average 164 seconds for each customer, from drive-up to pull-out.

More recently, Taco Bell completed the rollout of its new Grill-to-Order kitchens by installing water- and energy-saving grills that conserve 300 million gallons of water and 200 million kilowatt hours of electricity each year. This "green"-inspired cooking method also saves the company's 5,800 restaurants $17 million per year.

Effective operations management has resulted in productivity increases that support Taco Bell's low-cost strategy. Taco Bell is now the fast-food low-cost leader with a 58% share of the Mexican fast-food market.

Bob Pardue-Signs/Alamy Stock Photo

has probably resulted in lower measured productivity for the service sector, although, in fact, actual productivity has probably increased because these activities are now more efficiently produced than previously.

However, despite the difficulty of improving productivity in the service sector, improvements are being made. And this text presents a multitude of ways to make these improvements. Indeed, what can be done when management pays attention to how work actually gets done is astonishing!

Although the evidence indicates that all industrialized countries have the same problem with service productivity, the U.S. remains the world leader in overall productivity *and* service productivity. Retailing is twice as productive in the U.S. as in Japan, where laws protect shopkeepers from discount chains. The U.S. telephone industry is at least twice as productive as Germany's. The U.S. banking system is also 33% more efficient than Germany's banking oligopolies. However, because productivity is central to the operations manager's job and because the service sector is so large, we take special note in this text of how to improve productivity in the service sector. (See, for instance, the *OM in Action* box, "Taco Bell Improves Productivity and Goes Green to Lower Costs.")

Current Challenges in Operations Management

Operations managers work in an exciting and dynamic environment. This environment is the result of a variety of challenging forces, from globalization of world trade to the transfer of ideas, products, and money at electronic speeds. Let's look at some of these challenges:

STUDENT TIP One of the reasons OM is such an exciting discipline is that an operations manager is confronted with ever-changing issues, from technology, to global supply chains, to sustainability.

- *Globalization:* The rapid decline in the cost of communication and transportation has made markets global. Similarly, resources in the form of capital, materials, talent, and labor are also now global. As a result, countries throughout the world are contributing to globalization as they vie for economic growth. Operations managers are rapidly seeking creative designs, efficient production, and high-quality goods via international collaboration.
- *Supply-chain partnering:* Shorter product life cycles, demanding customers, and fast changes in technology, materials, and processes require supply-chain partners to be in tune with the needs of end users. And because suppliers may be able to contribute unique expertise, operations managers are outsourcing and building long-term partnerships with critical players in the supply chain.

- *Sustainability:* Operations managers' continuing battle to improve productivity is concerned with designing products and processes that are ecologically sustainable. This means designing green products and packaging that minimize resource use, can be recycled or reused, and are generally environmentally friendly.
- *Technological change:* **Industry 4.0** is the name given to the new digital world. Some consider Industry 4.0 the fourth industrial revolution, hence the name. Why is this considered the fourth industrial revolution? The first industrial revolution included the harnessing of water and steam power in the late 1700s, leading to rapid mechanization and division of labor. This was followed quickly by electricity and the second industrial revolution's assembly lines and mass production. The third, in the 20th century, yielded communication between man and machine with computers, automation, and robots. Finally, the fourth, *Industry 4.0*, is the widespread use of precision sensors and digital signals—in a word, *digitalization*. From raw materials to design to manufacturing to logistics, services, and ultimately the end consumer, digital signals surround us. Moreover, digitalization suggests connecting this massive amount of data *in real time*. Harnessing the vast and growing array of digital signals is a huge opportunity for operations management, but it is also a significant challenge.

Industry 4.0
The fourth industrial revolution with widespread real-time digitalization.

- *Mass customization:* Once managers recognize the *world* as the marketplace, the cultural and individual differences become quite obvious. In a world where consumers are increasingly aware of innovation and options, substantial pressure is placed on firms to respond in a creative way. And OM must rapidly respond with product designs and flexible production processes that cater to the individual whims of consumers. The goal is to produce customized products, whenever and wherever needed.
- *Lean operations:* Lean is the management model sweeping the world and providing the standard against which operations managers must compete. Lean can be thought of as the driving force in a well-run operation, where the customer is satisfied, employees are respected, and waste does not exist. The theme of this text is to build organizations that are more efficient, where management creates enriched jobs that help employees engage in continuous improvement, and where goods and services are produced and delivered when and where the customer desires them. These ideas are also captured in the phrase *Lean*.

These challenges must be successfully addressed by today's operations managers. This text will provide you with the foundations necessary to meet those challenges.

Ethics, Social Responsibility, and Sustainability

The systems that operations managers build to convert resources into goods and services are complex. And they function in a world where the physical and social environment is evolving, as are laws and values. These dynamics present a variety of challenges that come from the conflicting perspectives of **stakeholders**, such as customers, distributors, suppliers, owners, lenders, employees, and community. Stakeholders, as well as government agencies at various levels, require constant monitoring and thoughtful responses.

Stakeholders
Those with a vested interest in an organization, including customers, distributors, suppliers, owners, lenders, employees, and community members.

Identifying ethical and socially responsible responses while developing sustainable processes that are also effective and efficient productive systems is not easy. Managers are also challenged to:

- Develop and produce safe, high-quality green products
- Train, retain, and motivate employees in a safe workplace
- Honor stakeholder commitments

Managers must do all this while meeting the demands of a very competitive and dynamic world marketplace. If operations managers have a *moral awareness and focus on increasing productivity in this system*, then many of the ethical challenges will be successfully addressed. The organization will use fewer resources, the employees will be committed, the market will be satisfied, and the ethical climate will be enhanced. Throughout this text, we note ways in which operations managers can take ethical and socially responsible actions while successfully addressing these challenges of the market. We also conclude each chapter with an *Ethical Dilemma* exercise.

Summary

Operations, marketing, and finance/accounting are the three functions basic to all organizations. The operations function creates goods and services. Much of the progress of operations management has been made in the twentieth century, but since the beginning of time, humankind has been attempting to improve its material well-being. Operations managers are key players in the battle to improve productivity.

As societies become increasingly affluent, more of their resources are devoted to services. In the U.S., more than 85% of the workforce is employed in the service sector. Productivity improvements and a sustainable environment are difficult to achieve, but operations managers are the primary vehicle for making improvements.

Key Terms

Production (p. 4)
Operations management (OM) (p. 4)
Supply chain (p. 6)
Ten strategic OM decisions (p. 7)
Services (p. 10)
Service sector (p. 11)
Productivity (p. 12)
Single-factor productivity (p. 14)
Multifactor productivity (p. 14)
Productivity variables (p. 15)
Knowledge society (p. 17)
Industry 4.0 (p. 19)
Stakeholders (p. 19)

Ethical Dilemma

The American car battery industry boasts that its recycling rate now exceeds 95%, the highest rate for any commodity. However, with changes brought about by specialization and globalization, parts of the recycling system are moving offshore. This is particularly true of automobile batteries, which contain lead. The Environmental Protection Agency (EPA) is contributing to the offshore flow with newly implemented standards that make domestic battery recycling increasingly difficult and expensive. The result is a major increase in used batteries going to Mexico, where environmental standards and control are less demanding than they are in the U.S. One in five batteries is now exported to Mexico. There is seldom difficulty finding buyers because lead is expensive and in worldwide demand. While U.S. recyclers operate in sealed, mechanized plants, with smokestacks equipped with scrubbers and plant surroundings monitored for traces of lead, this is not the case in most Mexican plants. The harm from lead is legendary, with long-run residual effects. Health issues include high blood pressure, kidney damage, detrimental effects on fetuses during pregnancy, neurological problems, and arrested development in children.

Given the two scenarios below, what action do you take?

a) You own an independent auto repair shop and are trying to safely dispose of a few old batteries each week. (Your battery supplier is an auto parts supplier who refuses to take your old batteries.)
b) You are manager of a large retailer responsible for disposal of thousands of used batteries each day.

Discussion Questions

1. Why should one study operations management?
2. What are some career opportunities in the operations management discipline?
3. Identify four people who have contributed to the theory and techniques of operations management.
4. Briefly describe the contributions of the four individuals identified in the preceding question.
5. Figure 1.1 outlines the operations, finance/accounting, and marketing functions of three organizations. Prepare a chart similar to Figure 1.1 outlining the same functions for one of the following:
   a) a newspaper
   b) a drugstore
   c) a college library
   d) a summer camp
   e) a small costume-jewelry factory
6. What are the three basic functions of a firm?
7. Identify the 10 strategic operations management decisions.
8. Apply the 10 OM decisions to Amazon. (*Hint:* As a starting point, read the Global Profile that begins Chapter 12.)
9. Name four areas that are significant to improving labor productivity.
10. The U.S., and indeed much of the rest of the world, has been described as a "knowledge society." How does this affect productivity measurement and the comparison of productivity between the U.S. and other countries?
11. What are the measurement problems that occur when one attempts to measure productivity?
12. Mass customization and rapid product development were identified as challenges to modern manufacturing operations. What is the relationship, if any, between these challenges? Can you cite any examples?
13. What are the five reasons productivity is difficult to improve in the service sector?
14. Describe some of the actions taken by Taco Bell to increase productivity that have resulted in Taco Bell's ability to serve "twice the volume with half the labor."
15. As a library or Internet assignment, find the U.S. productivity rate (increase) last year for the (a) national economy, (b) manufacturing sector, and (c) service sector.

Using Software for Productivity Analysis

This section presents three ways to solve productivity problems with computer software. First, you can create your own Excel spreadsheets to conduct productivity analysis. Second, you can use the Excel OM software that comes with this text and can be found online. Third, POM for Windows is another program that is available with this text and can be found online.

CREATING YOUR OWN EXCEL SPREADSHEETS

Program 1.1 illustrates how to build an Excel spreadsheet for the data in Example 2.

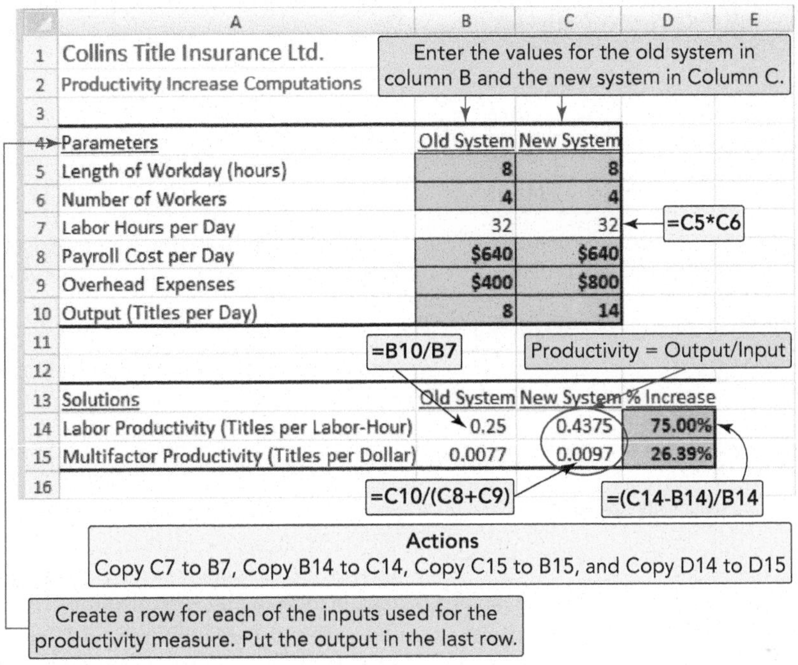

Program **1.1**

USING EXCEL OM

Excel OM is an Excel "add-in" with 26 Operations Management decision support "Templates." To access the templates, double-click on the *Excel OM* tab at the top of the page; then in the menu bar choose the appropriate chapter (in this case Chapter 1) from either the "Chapter" or "Alphabetic" tab on the left. Each of Excel OM's 26 modules includes instructions for that particular module. The instructions can be turned on or off via the "instruction" tab in the menu bar.

USING POM FOR WINDOWS

POM for Windows is decision support software that includes 25 Operations Management modules. The modules are accessed by double-clicking on *Module* in the menu bar, and then double-clicking on the appropriate (in this case *Productivity*) item. Instructions are provided for each module just below the menu bar. Please refer to Appendix II for further details.

Solved Problems

SOLVED PROBLEM 1.1

Productivity can be measured in a variety of ways, such as by labor, capital, energy, material usage, and so on. At Modern Lumber, Inc., Art Binley, president and producer of apple crates sold to growers, has been able, with his current equipment, to produce 240 crates per 100 logs. He currently purchases 100 logs per day, and each log requires 3 labor-hours to process. He believes that he can hire a professional buyer who can buy a better-quality log at the same cost. If this is the case, he can increase his production to 260 crates per 100 logs. His labor-hours will increase by 8 hours per day.

What will be the impact on productivity (measured in crates per labor-hour) if the buyer is hired?

SOLUTION

a) Current labor productivity

$$= \frac{240 \text{ crates}}{100 \text{ logs} \times 3 \text{ hours/log}} = \frac{240}{300} = .8 \text{ crates per labor-hour}$$

b) Labor productivity with buyer

$$= \frac{260 \text{ crates}}{(100 \text{ logs} \times 3 \text{ hours/log}) + 8 \text{ hours}} = \frac{260}{308} = .844 \text{ crates per labor-hour}$$

Using current productivity (.80 from [a]) as a base, the increase will be 5.5% (.844/.8=1.055, or a 5.5% increase).

SOLVED PROBLEM 1.2

Art Binley has decided to look at his productivity from a multifactor (total factor productivity) perspective (refer to Solved Problem 1.1). To do so, he has determined his labor, capital, energy, and material usage and decided to use dollars as the common denominator. His total labor-hours are now 300 per day and will increase to 308 per day. His capital and energy costs will remain constant at $350 and $150 per day, respectively. Material costs for the 100 logs per day are $1,000 and will remain the same. Because he pays an average of $10 per hour (with fringes), Binley determines his productivity increase as follows:

SOLUTION

CURRENT SYSTEM	
Labor:	300 hrs. @10 = 3,000
Material:	100 logs/day 1,000
Capital:	350
Energy:	150
Total Cost:	$4,500

Multifactor productivity of current system:
= 240 crates/$4,500 = .0533 crates/dollar

SYSTEM WITH PROFESSIONAL BUYER	
308 hrs. @10 =	$3,080
	1,000
	350
	150
	$4,580

Multifactor productivity of proposed system:
= 260 crates/$4,580 = .0568 crates/dollar

Using current productivity (.0533) as a base, the increase will be .066. That is, .0568/.0533 = 1.066, or a 6.6% increase.

Problems

Note: **PX** means the problem may be solved with POM for Windows and/or Excel OM.

Problems 1.1 to 1.17 relate to The Productivity Challenge

• **1.1** Chuck Sox makes wooden boxes in which to ship motorcycles. Chuck and his three employees invest a total of 40 hours per day making the 120 boxes.

a) What is their productivity?
b) Chuck and his employees have discussed redesigning the process to improve efficiency. If they can increase the rate to 125 per day, what will be their new productivity?
c) What will be their unit *increase* in productivity per hour?
d) What will be their percentage change in productivity? **PX**

• **1.2** Carbondale Casting produces cast bronze valves on a 10-person assembly line. On a recent day, 160 valves were produced during an 8-hour shift.

a) Calculate the labor productivity of the line.
b) John Goodale, the manager at Carbondale, changed the layout and was able to increase production to 180 units per 8-hour shift. What is the new labor productivity per labor-hour?
c) What is the percentage of productivity increase? **PX**

• **1.3** This year, Druehl, Inc., will produce 57,600 hot water heaters at its plant in Delaware in order to meet expected global demand. To accomplish this, each laborer at the plant will work 160 hours per month. If the labor productivity at the plant is 0.15 hot water heaters per labor-hour, how many laborers are employed at the plant?

• **1.4** Lori Cook produces "Final Exam Care Packages" for resale by her sorority. She is currently working a total of 5 hours per day to produce 100 care packages.

a) What is Lori's productivity?
b) Lori thinks that by redesigning the package, she can increase her total productivity to 133 care packages per day. What will be her new productivity?
c) What will be the percentage increase in productivity if Lori makes the change? **PX**

•• **1.5** George Kyparisis makes bowling balls in his Miami plant. With recent increases in his costs, he has a newfound interest in efficiency. George is interested in determining the productivity of his organization. He would like to know if his organization is maintaining the manufacturing average of 3% increase in productivity per year. He has the following data representing a month from last year and an equivalent month this year:

	LAST YEAR	NOW
Units produced	1,000	1,000
Labor (hours)	300	275
Resin (pounds)	50	45
Capital invested ($)	10,000	11,000
Energy (BTU)	3,000	2,850

Show the productivity percentage change for each category and then determine the improvement for labor-hours, the typical standard for comparison. **PX**

•• **1.6** George Kyparisis (using data from Problem 1.5) determines his costs to be as follows:

- *Labor:* $10 per hour
- *Resin:* $5 per pound
- *Capital expense:* 1% per month of investment
- *Energy:* $0.50 per BTU

Show the percent change in productivity for one month last year versus one month this year, on a multifactor basis with dollars as the common denominator. **PX**

• **1.7** Hokey Min's Kleen Karpet cleaned 65 rugs in October, consuming the following resources:

Labor:	520 hours at $13 per hour
Solvent:	100 gallons at $5 per gallon
Machine rental:	20 days at $50 per day

a) What is the labor productivity per dollar?
b) What is the multifactor productivity? **PX**

•• **1.8** Lillian Fok is president of Lakefront Manufacturing, a producer of bicycle tires. Fok makes 1,000 tires per day with the following resources:

Labor:	400 hours per day @ $12.50 per hour
Raw material:	20,000 pounds per day @ $1 per pound
Energy:	$5,000 per day
Capital costs:	$10,000 per day

a) What is the labor productivity per labor-hour for these tires at Lakefront Manufacturing?
b) What is the multifactor productivity for these tires at Lakefront Manufacturing?
c) What is the percent change in multifactor productivity if Fok can reduce the energy bill by $1,000 per day without cutting production or changing any other inputs? **PX**

••• **1.9** Brown's, a local bakery, is worried about increased costs—particularly energy. Last year's records can provide a fairly good estimate of the parameters for this year. Wende Brown, the owner, does not believe things have changed much, but she did invest an additional $3,000 for modifications to the bakery's ovens to make them more energy efficient. The modifications were supposed to make the ovens at least 15% more efficient. Brown has asked you to check the energy savings of the new ovens and also to look over other measures of the bakery's productivity to see if the modifications were beneficial. You have the following data to work with:

	LAST YEAR	NOW
Production (dozen)	1,500	1,500
Labor (hours)	350	325
Capital investment ($)	15,000	18,000
Energy (BTU)	3,000	2,750

PX

•• **1.10** Munson Performance Auto, Inc., modifies 375 autos per year. The manager, Adam Munson, is interested in obtaining a measure of overall performance. He has asked you to provide him with a multifactor measure of last year's performance as a benchmark for future comparison. You have assembled the following data. Resource inputs were labor, 10,000 hours; 500 suspension and engine modification kits; and energy, 100,000 kilowatt-hours. Average labor cost last year was $20 per hour, kits cost $1,000 each, and energy costs were $3 per kilowatt-hour. What do you tell Mr. Munson? **PX**

•• **1.11** Lake Charles Seafood makes 500 wooden packing boxes for fresh seafood per day, working in two 10-hour shifts. Due to increased demand, plant manager Jasmine Hines-Allen has decided to operate three 8-hour shifts instead. The plant is now able to produce 650 boxes per day.
a) Calculate the company's productivity before the change in work rules and after the change.
b) What is the percentage increase in productivity?
c) If production is increased to 700 boxes per day, what is the new productivity? **PX**

••• **1.12** Charles Lackey operates a bakery in Idaho Falls, Idaho. Because of its excellent product and excellent location, demand has increased by 25% in the last year. On far too many occasions, customers have not been able to purchase the bread of their choice. Because of the size of the store, no new ovens can be added. At a staff meeting, one employee suggested ways to load the ovens differently so that more loaves of bread can be baked at one time. This new process will require that the ovens be loaded by hand, requiring additional manpower. This is the only thing to be changed. If the bakery makes 1,500 loaves per month with a labor productivity of 2.344 loaves per labor-hour, how many workers will Lackey need to *add*? (*Hint:* Each worker works 160 hours per month.)

•• **1.13** Refer to Problem 1.12. The pay will be $8 per hour for employees. Charles Lackey can also improve the yield by purchasing a new blender. The new blender will mean an increase in his investment. This added investment has a cost of $100 per month, but he will achieve the same output (an increase to 1,875) as the change in labor-hours. Which is the better decision?
a) Show the productivity change, in loaves per dollar, with an increase in labor cost (from 640 to 800 hours).
b) Show the new productivity, in loaves per dollar, with only an increase in investment ($100 per month more).
c) Show the percentage productivity change for labor and investment.

••• **1.14** Refer to Problems 1.12 and 1.13. If Charles Lackey's utility costs remain constant at $500 per month, labor at $8 per hour, and cost of ingredients at $0.35 per loaf, but Charles does not purchase the blender suggested in Problem 1.13, what will the productivity of the bakery be? What will be the percentage increase or decrease?

•• **1.15** In December, General Motors produced 6,600 customized vans at its plant in Detroit. The labor productivity at this plant is known to have been 0.10 vans per labor-hour during that month. 300 laborers were employed at the plant that month.
a) How many hours did the average laborer work that month?
b) If productivity can be increased to 0.11 vans per labor-hour, how many hours would the average laborer work that month?

•• **1.16** Susan Williams runs a small Flagstaff job shop where garments are made. The job shop employs eight workers. Each worker is paid $10 per hour. During the first week of March, each worker worked 45 hours. Together, they produced a batch of 132 garments. Of these garments, 52 were "seconds" (meaning that they were flawed). The seconds were sold for $90 each at a factory outlet store. The remaining 80 garments were sold to retail outlets at a price of $198 per garment. What was productivity, in dollars per labor-hour, at this job shop during the first week of March?

••• **1.17** As part of a study for the Department of Labor Statistics, Shakira Dominguez has been assigned the task of evaluating the improvement in productivity of small businesses. Data for one of the small businesses she is to evaluate are shown in the table. The data are the monthly average of last year and the monthly average this year. Determine the multifactor productivity with dollars as the common denominator for:

a) Last year.
b) This year.
c) Then determine the percentage change in productivity for the monthly average last year versus the monthly average this year on a multifactor basis.

- *Labor:* $8 per hour
- *Capital:* 0.83% per month of investment
- *Energy:* $0.60 per BTU

	LAST YEAR	THIS YEAR
Production (dozen)	1,500	1,500
Labor (hours)	350	325
Capital investment ($)	15,000	18,000
Energy (BTU)	3,000	2,700

CASE STUDIES

Uber Technologies, Inc.

The $100 billion firm Uber Technology, Inc., has unsettled the traditional taxi business. In over 84 countries and 900 cities around the world, Uber and similar companies are challenging the existing taxi business model. Uber and its growing list of competitors, Lyft and Curb in America, along with fledging rivals in Europe, Asia, and India, think their smartphone apps can provide a new and improved way to call a taxi. This disruptive business model uses an app to arrange rides between riders and cars, theoretically a nearby car, which is tracked by the app. The Uber system also provides a history of rides, routes, and fees as well as automatic billing for over 78 million monthly active users worldwide. In addition, driver and rider are also allowed to evaluate each other. The services are increasingly popular, worrying established taxi services in cities from New York to Berlin, and from Rio de Janeiro to Bangkok. In many markets, Uber has proven to be the best, fastest, and most reliable way to find a ride. Consumers worldwide are endorsing the system as a replacement for the usual taxi ride. As the most established competitor in the field, Uber is putting more cars on the road, meaning faster pickup times, which should attract even more riders, which in turn attracts even more drivers, and so on. This growth cycle may speed the demise of the existing taxi businesses as well as provide substantial competition for firms with a technology-oriented model similar to Uber's.

Uber is a software company. It does not own any cars, but it is the largest taxi company in the world. The Uber business model bypasses taxi ownership and a number of regulations, while at the same time offering better service and lower fees than traditional taxis. However, the traditional taxi industry is fighting back, and regulations are mounting. The regulations vary by country and city, but increasingly special licensing, testing, and inspections have been imposed. Part of the fee charged to riders does not go to the driver, but to Uber, as there are real overhead costs. Uber's costs, depending on the locale, may include insurance, background checks for drivers, vetting of vehicles, software development and maintenance, and centralized billing. Additionally, with 5 billion transactions per minute, computer costs are significant. How these overhead costs compare to traditional taxi costs is yet to be determined. Therefore, improved efficiency may not be immediately obvious, and contract provisions are significant (see www.uber.com/legal/terms/us).

In addition to growing regulations, a complicating factor in the model is finding volunteer drivers at inopportune times. A sober driver and a clean car at 1:00 a.m. New Year's Eve does cost more. Consequently, Uber has introduced a dynamic pricing model. Dynamic pricing means a higher price, sometimes much higher, than normal, but dynamic pricing has proven necessary to ensure that cars and drivers are available at unusual times. Customers are quoted the fares in advance.

Discussion Questions

1. The market has decided that Uber and its immediate competitors are adding efficiency to our society. How is Uber providing that added efficiency?
2. Do you think the Uber model will work in the trucking industry?
3. In what other areas/industries might the Uber model be used?
4. What are some disadvantages of the Uber model?

Frito-Lay: Operations Management in Manufacturing

Video Case 

Frito-Lay, the massive Dallas-based subsidiary of PepsiCo, has 55 plants and 55,000 employees in North America. Seven of Frito-Lay's 41 brands exceed $1 billion in sales: Fritos, Lay's, Cheetos, Ruffles, Tostitos, Doritos, and Walker's Potato Chips. Operations is the focus of the firm—from designing products for new markets, to meeting changing consumer preferences, to adjusting to rising commodity costs, to subtle issues involving flavors and preservatives—OM is under constant cost, time, quality, and market pressure. Here is a look at how the 10 decisions of OM are applied at this food processor.

In the food industry, product development kitchens experiment with new products, submit them to focus groups, and perform test marketing. Once the product specifications have been set, processes capable of meeting those specifications and the necessary quality standards are created. At Frito-Lay, quality begins at the farm, with onsite inspection of the potatoes used in Ruffles and the corn used in Fritos. Quality continues throughout the manufacturing process, with visual inspections and with statistical process control of product variables such as oil, moisture, seasoning, salt, thickness, and weight. Additional quality evaluations are conducted throughout shipment, receipt, production, packaging, and delivery.

The production process at Frito-Lay is designed for large volumes and small variety, using expensive special-purpose equipment, and with swift movement of material through the facility. Product-focused facilities, such as Frito-Lay's, typically have high capital costs, tight schedules, and rapid processing. Frito-Lay's facilities are located regionally to aid in the rapid delivery of products because freshness is a critical issue. Sanitary issues and necessarily fast processing of products put a premium on an efficient layout. Production lines are designed for balanced throughput and high utilization. Cross-trained workers, who handle a variety of production lines, have promotion paths identified for their particular skill set. The company rewards employees with medical, retirement, and education plans. Its turnover is very low.

The supply chain is integral to success in the food industry; vendors must be chosen with great care. Moreover, the finished food product is highly dependent on perishable raw materials. Consequently, the supply chain brings raw material (potatoes, corn, etc.) to the plant securely and rapidly to meet tight production schedules. For instance, from the time that potatoes are picked in St. Augustine, Florida, until they are unloaded at the Orlando plant, processed, packaged, and shipped from the plant is under 12 hours. The requirement for fresh product requires on-time, just-in-time deliveries combined with both low raw material and finished goods inventories. The continuous-flow nature of the specialized equipment in the production process permits little work-in-process inventory. The plants usually run 24/7. This means that there are four shifts of employees each week.

Tight scheduling to ensure the proper mix of fresh finished goods on automated equipment requires reliable systems and effective maintenance. Frito-Lay's workforce is trained to recognize problems early, and professional maintenance personnel are available on every shift. Downtime is very costly and can lead to late deliveries, making maintenance a high priority.

Discussion Questions*

1. From your knowledge of production processes and from the case and the video, identify how each of the 10 decisions of OM is applied at Frito-Lay.
2. How would you determine the productivity of the production process at Frito-Lay?
3. How are the 10 decisions of OM different when applied by the operations manager of a production process such as Frito-Lay versus a service organization such as Hard Rock Cafe? See the Hard Rock Cafe video case below.)

*You may wish to view the video that accompanies this case before addressing these questions.

Hard Rock Cafe: Operations Management in Services

In its 50-plus years of existence, Hard Rock has grown from a modest London pub to a global power managing 185 restaurants, 25 hotels, and 12 casinos. This puts Hard Rock firmly in the service industry—a sector that employs over 75% of the people in the U.S. Hard Rock moved its world headquarters to Orlando, Florida, in 1988 and has expanded to more than 50 locations throughout the U.S., serving over 100,000 meals each day. Hard Rock chefs are modifying the menu from classic American—burgers and chicken wings—to include higher-end items such as stuffed veal chops and lobster tails. Just as taste in music changes over time, so does Hard Rock Cafe, with new menus, layouts, memorabilia, services, and strategies.

At Orlando's Universal Studios, a traditional tourist destination, Hard Rock Cafe serves more than 3,500 meals each day. The cafe employs about 400 people. Most are employed in the restaurant, but some work in the retail shop. Retail is now a standard and increasingly prominent feature in Hard Rock Cafes (since close to 48% of revenue comes from this source). Cafe employees include kitchen and waitstaff, hosts, and bartenders. Hard Rock employees are not only competent in their job skills but are also passionate about music and have engaging personalities. Cafe staff is scheduled down to 15-minute intervals to meet seasonal and daily demand changes in the tourist environment of Orlando. Surveys are done on a regular basis to evaluate quality of food and service at the cafe. Scores are rated on a 1-to-7 scale, and if the score is not a 7, the food or service is a failure.

Hard Rock is adding a new emphasis on live music and is redesigning its restaurants to accommodate the changing tastes. Since Eric Clapton hung his guitar on the wall to mark his favorite bar stool, Hard Rock has become the world's leading collector and exhibitor of rock 'n' roll memorabilia, with changing exhibits at its cafes throughout the world. The collection includes 70,000 pieces, valued at $40 million. In keeping with the times, Hard Rock also maintains a Web site, www.hardrock.com, which receives more than 100,000 hits per week, and a weekly cable television program on VH1. Hard Rock's brand recognition, at 92%, is one of the highest in the world.

Discussion Questions*

1. From your knowledge of restaurants, from the video, from the *Global Company Profile* that opens this chapter, and from the case itself, identify how each of the 10 OM strategy decisions is applied at Hard Rock Cafe.
2. How would you determine the productivity of the kitchen staff and waitstaff at Hard Rock?
3. How are the 10 OM strategy decisions different when applied to the operations manager of a service operation such as Hard Rock versus an automobile company such as Ford Motor Company?

*You may wish to view the video that accompanies this case before addressing these questions.

Celebrity Cruises: Operations Management at Sea

Video Case 

On any given day, Celebrity Cruises, Inc. has tens of thousands of passengers at sea on more than a dozen spectacular ships, spanning 7 continents and 75 countries. With this level of capital investment along with the responsibility for the happiness and safety of so many passengers, excellence in operations is required. To make it all work, the 10 operations management decisions must be executed flawlessly. From product design (which encompasses the ship's layout, the food, and 300 destinations), to scheduling, supply chain, inventory, personnel, maintenance, and the processes that hold them together, OM is critical.

Cruise lines require precise scheduling of ships, with down-to-the-minute docking and departure times. In addition to ship and port scheduling, some 2,000 plus crew members must be scheduled. And there are many schedule variations. Entertainers may arrive and leave at each port, while officers may have a schedule of 10 weeks on and 10 weeks off. Other crew members have onboard commitments varying from 4 to 9 months.

With $400 million invested in a ship and more than 5,000 lives involved in a cruise, detailed processes to ensure maintenance and reliability are vital. The modern ship is a technological marvel with hundreds of electronic monitors operating 24/7 to track everything from ship speed and location, to sea depth, to shipboard power demand and cabin temperature.

Celebrity's ship layout, destinations, and routing are adjusted to meet seasonal demands and the expectations of its premium market segment. With destinations from Alaska to Europe to Asia, crews are recruited worldwide, with as many as 70 nationalities represented. Instilling a quality culture requires an aggressive quality service orientation and, of course, meticulous cleanliness and attention to detail. Processes for food preparation, laundry, quality, and maintenance are complete and detailed.

A cruise ship, as a moving city, requires a comprehensive and precise supply chain that replenishes everything from food to fuel to soap and water. Land-based buyers support Celebrity's annual food and beverage purchases that exceed $110 million. Included in these expenditures are weekly shipments of 6 to 10 containers from the Miami headquarters destined for ships in European ports. An onboard staff organizes inventories to support this massive operation. The logistics effort includes hedging the weekly use of 24,000 gallons of fuel per ship with purchases 6 years into the future. Reliable global supply chains have been developed that deliver the required inventory on a tight time frame.

These crucial shipboard systems typically represent the best of operations management. Such is the case at Celebrity Cruises.

Discussion Questions*

1. Describe how the 10 OM decisions are implemented at Celebrity Cruises, Inc.
2. Identify how the 10 OM decisions at Celebrity Cruises differ from those decisions at a manufacturing firm.
3. Identify how the 10 OM decisions at Celebrity Cruises differ from those decisions at a retail store.
4. How are hotel operations on a ship different from those at a land-based hotel?

*You may wish to view the video that accompanies this case before addressing these questions.

Endnotes

1. *Efficiency* means doing the job well—with a minimum of resources and waste. Note the distinction between being *efficient*, which implies doing the job well, and *effective*, which means doing the right thing. A job well done—say, by applying the 10 strategic decisions of operations management—helps us be *efficient*; developing and using the correct strategy helps us be *effective*.
2. Sources: *Businessweek* (August 23–30, 2012); *Fortune* (October 30, 2014); and **QZ.com/Starbucks**.
3. The quality and time period are assumed to remain constant.
4. Sources: *Businessweek* (May 5, 2011) and J. Hueter and W. Swart, *Interfaces* (Vol. 28, no. 1).

Bibliography

Goldstone, L. *Drive! Henry Ford, George Selden, and the Race to Invent the Auto Age*. New York: Ballantine, 2016.

Hounshell, D. A. *From the American System to Mass Production 1800–1932: The Development of Manufacturing*. Baltimore: Johns Hopkins University Press, 1985.

Lewis, W. W. *The Power of Productivity*. Chicago: University of Chicago Press, 2005.

Malone, T. W., R. J. Laubacher, and T. Johns, "The Age of Hyperspecialization." *Harvard Business Review* 89, no. 7 (July–August 2011): 56–65.

Maroto, A., and L. Rubalcaba. "Services Productivity Revisited." *The Service Industries Journal* 28, no. 3 (April 2008): 337.

McLaughlin, P. "Measuring Productivity." *Management Services* 58, no. 4 (Winter 2014): 31–37.

Rao, M. P. "Optimal Base-Period Data for Productivity Measurement." *International Journal of Operations & Production Management* 13, no. 8 (1993): 37–44.

Sahay, B. S. "Multi-factor Productivity Measurement Model for Service Organization." *International Journal of Productivity and Performance Management* 54, no. 1–2 (2005): 7–23.

San, G., T. Huang, and L. Huang. "Does Labor Quality Matter on Productivity Growth?" *Total Quality Management and Business Excellence* 19, no. 10 (October 2008): 1043.

Sprague, L. G. "Evolution of the Field of Operations Management." *Journal of Operations Management* 25, no. 2 (March 2007): 219–238.

Tangen, S. "Demystifying Productivity and Performance." *International Journal of Productivity and Performance Measurement* 54, no. 1–2 (2005): 34–47.

Taylor, F. W. *The Principles of Scientific Management*. New York: Harper & Brothers, 1911.

Wren, D. A., and A. G. Bedelan. *The Evolution of Management Thought*, 7th ed. New York: Wiley, 2017.

Chapter 1 *Rapid* Review

Main Heading	Review Material	MyLab Operations Management
WHAT IS OPERATIONS MANAGEMENT?	■ ***Production***—The creation of goods and services ■ ***Operations management (OM)***—Activities that relate to the creation of goods and services through the transformation of inputs to outputs	Concept Questions: 1.1–1.5 **VIDEOS 1.1, 1.2 and 1.3** OM at Hard Rock OM at Frito-Lay Celebrity Cruises: Operations Management at Sea
ORGANIZING TO PRODUCE GOODS AND SERVICES	All organizations perform three functions to create goods and services: **1.** *Marketing*, which generates demand **2.** *Production/operations,* which creates the product **3.** *Finance/accounting,* which tracks how well the organization is doing, pays the bills, and collects the money	Concept Questions: 2.1–2.6
THE SUPPLY CHAIN	■ **Supply chain**—A global network of organizations and activities that supply a firm with goods and services	Concept Questions: 3.1–3.4
WHY STUDY OM?	We study OM for four reasons: **1.** To learn how people organize themselves for productive enterprise **2.** To learn how goods and services are produced **3.** To understand what operations managers do **4.** Because OM is a costly part of an organization	Concept Questions: 4.1–4.2
WHAT OPERATIONS MANAGERS DO	**Ten strategic OM decisions** are required of operations managers: **1.** Design of goods and services **2.** Managing quality **3.** Process strategies **4.** Location strategies **5.** Layout strategies **6.** Human resources **7.** Supply chain management **8.** Inventory management **9.** Scheduling **10.** Maintenance About 40% of *all* jobs are in OM. Operations managers possess job titles such as plant manager, quality manager, process improvement consultant, and operations analyst.	Concept Questions: 5.1–5.6
THE HERITAGE OF OPERATIONS MANAGEMENT	Significant events in modern OM can be classified into six eras: **1.** Early concepts (1776–1880)—Labor specialization (Smith, Babbage), standardized parts (Whitney) **2.** Scientific management (1880–1910)—Gantt charts (Gantt), motion and time studies (Gilbreth), process analysis (Taylor), queuing theory (Erlang) **3.** Mass production (1910–1980)—Assembly line (Ford/Sorensen), statistical sampling (Shewhart), economic order quantity (Harris), linear programming (Dantzig), PERT/CPM (DuPont), material requirements planning **4.** Lean production (1980–1995)—Just-in-time, computer-aided design, electronic data interchange, total quality management, Baldrige Award, empowerment, kanbans **5.** Mass customization (1995–2005)—Internet/e-commerce, enterprise resource planning, international quality standards, finite scheduling, supply-chain management, mass customization, build-to-order, radio frequency identification (RFID) **6.** Globalization era (2005–2025)—Global supply chains and logistics, growth of transnational organizations, sustainability, ethics in the global workplace, Internet of Things (IoT), digital operations, Industry 4.0	Concept Questions: 6.1–6.6
OPERATIONS FOR GOODS AND SERVICES	■ **Services**—Economic activities that typically produce an intangible product (such as education, entertainment, lodging, government, financial, and health services). Almost all services and almost all goods are a mixture of a service and a tangible product. ■ **Service sector**—The segment of the economy that includes trade, financial, lodging, education, legal, medical, and other professional occupations. Services now constitute the largest economic sector in postindustrial societies. The huge productivity increases in agriculture and manufacturing have allowed more of our economic resources to be devoted to services. Many service jobs pay very well.	Concept Questions: 7.1–7.5

Main Heading	Review Material	MyLab Operations Management
THE PRODUCTIVITY CHALLENGE	■ **Productivity**—The ratio of outputs (goods and services) divided by one or more inputs (such as labor, capital, or management) High production means producing many units, while high productivity means producing units efficiently. Only through increases in productivity can the standard of living of a country improve. U.S. productivity has averaged a 2.5% increase per year for over a century. $$\text{Single-factor productivity} = \frac{\text{Units produced}}{\text{Input used}} \quad (1\text{-}1)$$ ■ **Single-factor productivity**—Indicates the ratio of goods and services produced (outputs) to one resource (input). ■ **Multifactor productivity**—Indicates the ratio of goods and services produced (outputs) to many or all resources (inputs). Multifactor productivity $$= \frac{\text{Output}}{\text{Labor} + \text{Material} + \text{Energy} + \text{Capital} + \text{Miscellaneous}} \quad (1\text{-}2)$$ Measurement problems with productivity include: (1) the quality may change, (2) external elements may interfere, and (3) precise units of measure may be lacking. ■ **Productivity variables**—The three factors critical to productivity improvement are labor (10%), capital (38%), and management (52%). ■ **Knowledge society**—A society in which much of the labor force has migrated from manual work to work based on knowledge	Concept Questions: 8.1–8.6 Problems: 1.1–1.17 Virtual Office Hours for Solved Problems: 1.1, 1.2
CURRENT CHALLENGES IN OPERATIONS MANAGEMENT	Some of the current challenges for operations managers include: ■ Globalization; international collaboration ■ Supply-chain partnering; joint ventures; alliances ■ Sustainability; green products; recycle, reuse ■ Technological change; digitalization ■ Mass customization; customized products ■ Lean operations; continuous improvement and elimination of waste	Concept Questions: 9.1–9.5
ETHICS, SOCIAL RESPONSIBILITY, AND SUSTAINABILITY	Among the many ethical challenges facing operations managers are (1) efficiently developing and producing safe, quality products; (2) maintaining a clean environment; (3) providing a safe workplace; and (4) honoring stakeholder commitments. ■ **Stakeholders**—Those with a vested interest in an organization	Concept Question: 10.1
ADDITIONAL MYLAB OPERATIONS MANAGEMENT RESOURCES	✔ Video for Creating Your Own Excel Spreadsheets (Example 2) ✔ Additional Case Studies (National Air Express and Zychol Chemicals Corp.) ✔ Multiple Choice Case Questions (Zychol Chemicals Corp.) ✔ Recent Graduate Video: Jeremy Knowles, Risk Analyst, Genesis Financial Solutions	

Self Test

LO 1.1 Productivity increases when:
- **a)** inputs increase while outputs remain the same.
- **b)** inputs decrease while outputs remain the same.
- **c)** outputs decrease while inputs remain the same.
- **d)** inputs and outputs increase proportionately.
- **e)** inputs increase at the same rate as outputs.

LO 1.2 A strategy that is *not* one of the 10 strategic operations management decisions is:
- **a)** maintenance.
- **b)** human resources, job design and work measurement.
- **c)** location strategies.
- **d)** design of goods and services.
- **e)** advertising strategies.

LO 1.3 Operations management jobs comprise approximately _____% of all jobs.

LO 1.4 Services often:
- **a)** are tangible.
- **b)** are standardized.
- **c)** are knowledge based.
- **d)** are low in customer interaction.
- **e)** have consistent product definition.

LO 1.5 Productivity:
- **a)** can use many factors as the numerator.
- **b)** is the same thing as production.
- **c)** increases at about 0.5% per year.
- **d)** is dependent upon labor, management, and capital.
- **e)** is the same thing as effectiveness.

LO 1.6 Single-factor productivity:
- **a)** remains constant.
- **b)** is never constant.
- **c)** usually uses labor as a factor.
- **d)** seldom uses labor as a factor.
- **e)** uses management as a factor.

LO 1.7 Multifactor productivity:
- **a)** remains constant.
- **b)** is never constant.
- **c)** usually uses substitutes as common variables for the factors of production.
- **d)** seldom uses labor as a factor.
- **e)** always uses management as a factor.

LO 1.8 Productivity increases each year in the U.S. are a result of three factors:
- **a)** labor, capital, management
- **b)** engineering, labor, capital
- **c)** engineering, capital, quality control
- **d)** engineering, labor, data processing
- **e)** engineering, capital, data processing

Answers: LO 1.1. b; LO 1.2. e; LO 1.3 40; LO 1.4. c; LO 1.5. d; LO 1.6. c; LO 1.7. c; LO 1.8. a.

Operations Strategy in a Global Environment

CHAPTER 2

CHAPTER OUTLINE

Courtesy of Correct Craft Holding Company, LLC

Ski Nautique practices a differentiation strategy based on innovation, performance, and looking forward. The company wants to create the marketplace and give consumers what they want before they even know they want it.

GLOBAL COMPANY PROFILE
Boeing

Boeing's Global Strategy Yields Competitive Advantage

Boeing is the world's largest aerospace company and leading manufacturer of commercial jetliners, defense systems, and space systems. As America's biggest manufacturing exporter, the company supplies airlines and government customers in more than 150 countries. Boeing products include commercial and military aircraft, satellites, weapons, electronic and defense systems, launch and space exploration systems, and advanced communication systems.

View Pictures/Alastair Philip Wiper/VIEW/Newscom

Components from Boeing's worldwide supply chain come together on assembly lines in Charleston, South Carolina. Although components come from throughout the world, about 35% of the 787 structure comes from Japanese companies.

Some of the International Suppliers of Boeing 787 Components

SUPPLIER	HQ COUNTRY	COMPONENT
Latecoere	France	Passenger doors
Labinel	France	Wiring
Dassault	France	Design and product life cycle management software
Messier-Bugatti	France	Electric brakes
Thales	France	Electrical power conversion system
Messier-Dowty	France	Landing gear structure
Diehl	Germany	Interior lighting
Cobham	UK	Fuel pumps and valves
Rolls-Royce	UK	Engines
Smiths Aerospace	UK	Central computer system
BAE Systems	UK	Electronics
Alenia Aeronautica	Italy	Upper center fuselage
Toray Industries	Japan	Carbon fiber for wing and tail units
Fuji Heavy Industries	Japan	Center wing box
Kawasaki Heavy Ind.	Japan	Forward fuselage, fixed sections of wing
Teijin Seiki	Japan	Hydraulic actuators
Mitsubishi Heavy Ind.	Japan	Wing box
Chengdu Aircraft	China	Rudder
Hafei Aviation	China	Parts
Korean Airlines	South Korea	Wingtips
Saab	Sweden	Cargo and access doors

Boeing has a long tradition of aerospace leadership and innovation. It has been the premier manufacturer of commercial jetliners for decades. Today, the company manufactures the 737, 777, and 787 airplane families and the Boeing Business Jet. More than 10,000 Boeing-built commercial jetliners are in service worldwide, which is almost half the world fleet. The company also offers the most complete family of planes designed for freight. About 90 percent of the world's cargo is carried on board Boeing planes.

Its broad portfolio includes KC-46 aerial refueling aircraft, AH-64 Apache helicopters, the 702 family of satellites, CST-100 Starliner spacecraft, and the Echo Voyager.

With corporate offices in Chicago, Boeing employs more than 153,000 people. Its market

Stocktrek Images/Getty Images

An Apache helicopter landing on a U.S. aircraft carrier.

Boeing's Starliner spacecraft being launched by a NASA rocket.

NASA Space flight

Elaine Thompson/AP/Shutterstock

Boeing's collaborative technology enables a "virtual workspace" that allows Washington-based engineers, as well as partners in Australia, Japan, Italy, and Canada and across the United States, to make concurrent design changes to aerospace vehicles in real time.

success plays a key role in supporting high-value aerospace jobs across its supply chain and across the U.S., working with more than 12,000 businesses creating 1,000,000-plus supplier-related jobs.

Boeing's supply chain is also global, with over 300 suppliers in dozens of countries. Its newest jet, the 787 Dreamliner, was designed in collaboration with 20 foreign suppliers, some of whom are shown in the table. The expectation is that countries that have a stake in the Dreamliner are more likely to buy from Boeing than from its European competitor, Airbus.

This enormous global supply chain delivers more than a billion parts and subassemblies to Boeing plants every year.

LEARNING OBJECTIVES

LO 2.1 ***Define*** mission and strategy 35

LO 2.2 ***Identify*** and explain three strategic approaches to competitive advantage 36

LO 2.3 ***Understand*** the significance of key success factors and core competencies 42

LO 2.4 ***Use*** factor rating to evaluate both country and outsource providers 47

LO 2.5 ***Identify*** and explain four global operations strategy options 48

A Global View of Operations and Supply Chains

Today's successful operations manager has a global view of operations strategy. Since the early 1990s, nearly 3 billion people in developing countries have overcome the cultural, religious, ethnic, and political barriers that constrain productivity. And now they are all players on the global economic stage. As these barriers disappeared, simultaneous advances were being made in technology, tariff reductions, reliable shipping, and inexpensive communication. These changes mean that, increasingly, firms find their customers and suppliers located around the world. The unsurprising result is the growth of world trade (see Figure 2.1), global capital markets, and the international movement of people. This means increasing economic integration and interdependence of countries—in a word, globalization. In response, organizations are hastily extending their distribution channels and supply chains globally. The result is innovative strategies where firms compete not just with their own expertise but with the talent in their entire global supply chain. For instance:

- Boeing is competitive because both its sales and supply chain are worldwide.
- Italy's Benetton moves inventory to stores around the world faster than its competition with rapid communication and by building exceptional flexibility into design, production, and distribution.
- Sony purchases components from a supply chain that extends to Thailand, Malaysia, and elsewhere around the world for assembly of its electronic products, which in turn are distributed around the world.
- Volvo, considered a Swedish company, was purchased by a Chinese company, Geely, but Volvo assembles cars in Sweden, Belgium, Malaysia, and China.
- China's Haier (pronounced "higher"), from its South Carolina plant, produces compact refrigerators (it has one-third of the U.S. market) and refrigerated wine cabinets (it has half of the U.S. market). Haier also controls 10% of the large appliance market worldwide and owns the GE appliance division, which employs 6,000 in Appliance Park, Kentucky.

Figure **2.1**

Growth of World Trade as a Percentage of World GDP

Sources: World Bank; World Trade Organization; and IMF.

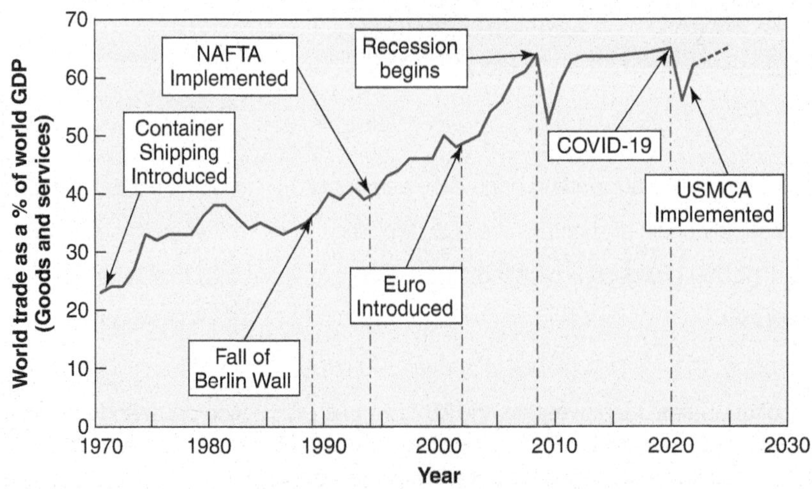

Globalization means customers, talent, and suppliers are worldwide. The new standards of global competitiveness impact quality, variety, customization, convenience, timeliness, and cost. Globalization strategies contribute efficiency, adding value to products and services, but they also complicate the operations manager's job. Complexity, risk, and competition are intensified, forcing companies to adjust for a shrinking world.

We have identified six reasons domestic business operations decide to change to some form of international operation. They are:

1. Improve the supply chain.
2. Reduce costs and exchange rate risk.
3. Improve operations.
4. Understand markets.
5. Improve products.
6. Attract and retain global talent.

Let us examine, in turn, each of the six reasons.

Improve the Supply Chain The supply chain can often be improved by locating facilities in countries where unique resources are available. These resources may be human resource expertise, low-cost labor, or raw material. For example, auto-styling studios from throughout the world have migrated to the auto mecca of southern California to ensure the necessary talent in contemporary auto design. Similarly, a perfume manufacturer wants a presence in Grasse, France, where much of the world's perfume essences are prepared from the flowers of the Mediterranean.

Reduce Costs and Exchange Rate Risk Many international operations seek to reduce risks associated with changing currency values (exchange rates) as well as take advantage of the tangible opportunities to reduce their direct costs. Less stringent government regulations on a wide variety of operations practices (e.g., environmental control, health and safety) can also reduce indirect costs.

Shifting low-skilled jobs to another country has several potential advantages. First, and most obviously, the firm may reduce costs. Second, moving the lower-skilled jobs to a lower-cost location frees higher-cost workers for more valuable tasks. Third, reducing wage costs allows the savings to be invested in improved products and facilities (and the retraining of existing workers, if necessary) at the home location.

Many firms use *financial hedging* by purchasing currency options to protect against negative exchange rate changes. Firms can also pursue **operational hedging** by intentionally building extra capacity in different countries and then shifting production from country to country as costs or exchange rates vary. This allows them to finesse currency risks and costs as economic conditions dictate.

Operational hedging
Maintaining excess capacity in different countries and shifting production levels among those countries as costs and exchange rates change.

The United States and Mexico have created **maquiladoras** (free trade zones) that allow manufacturers to cut their costs by paying only for the value added by Mexican workers. If a U.S. manufacturer, such as Caterpillar, brings a $1,000 engine to a maquiladora operation for assembly work costing $200, tariff duties will be charged only on the $200 of work performed in Mexico.

Maquiladoras
Mexican factories located along the U.S.–Mexico border that receive preferential tariff treatment.

Trade agreements also help reduce tariffs and thereby reduce the cost of operating facilities in foreign countries. The **World Trade Organization (WTO)** has helped reduce tariffs from 40% in 1940 to less than 3% today. Two major trade agreements for North America are **NAFTA and USMCA**. These agreements seek to phase out all trade and tariff barriers among Canada, Mexico, and the U.S. Other trade agreements that are accelerating global trade include APEC (the Pacific Rim countries), SEATO (Australia, New Zealand, Japan, Hong Kong, South Korea, New Guinea, and Chile), MERCOSUR (Argentina, Brazil, Paraguay, and Uruguay), and CAFTA (Central America, Dominican Republic, and United States).

World Trade Organization (WTO)
An international organization that promotes world trade by lowering barriers to the free flow of goods across borders.

NAFTA and USMCA
Free trade agreements between Canada, Mexico, and the United States.

Another trading group is the **European Union (EU)**.[1] The European Union has reduced trade barriers among the participating European nations through standardization and a common currency, the euro. However, this major U.S. trading partner, with over 450 million consumers, is also placing some of the world's most restrictive conditions on products sold in the EU. Everything from recycling standards to automobile bumpers to hormone-free farm products must meet EU standards, complicating international trade.

European Union (EU)
A European trade group that has 27 member states.

Improve Operations Operations learn from better understanding of management innovations in different countries. For instance, the Japanese have improved inventory management, the Germans are aggressively using robots, and the Scandinavians have contributed to improved ergonomics throughout the world.

Another reason to have international operations is to reduce response time to meet customers' changing product and service requirements. Providing quick and adequate service is often improved by locating facilities in the customer's home country.

Understand Markets Because international operations require interaction with foreign customers, suppliers, and other competitive businesses, international firms inevitably learn about opportunities for new products and services. Europe led the way with cell phone innovations, and then the Japanese and Indians led with cell phone fads. Knowledge of markets not only helps firms understand where the market is going but also helps firms diversify their customer base, add production flexibility, and smooth the business cycle.

Another reason to go into foreign markets is the opportunity to extend the *life cycle* (i.e., stages a product goes through; see Chapter 5) of an existing product. While some products in the U.S. are in a "mature" stage of their product life cycle, they may represent state-of-the-art products in developing countries.

Improve Products Learning does not take place in isolation. Firms serve themselves and their customers well when they remain open to the free flow of ideas. For example, Toyota and BMW will manage joint research and share development costs on battery research for the next generation of green cars. Similarly, international learning in operations is taking place as South Korea's Samsung and Germany's Robert Bosch join to produce lithium-ion batteries to the benefit of both.

Attract and Retain Global Talent Global organizations can attract and retain better employees by offering more employment opportunities. They need people in all functional areas and areas of expertise worldwide. Global firms can recruit and retain good employees because they provide both greater growth opportunities and insulation against unemployment during times of economic downturn. During economic downturns in one country or continent, a global firm has the means to relocate unneeded personnel to more prosperous locations.

A worldwide strategy places added burdens on operations management. Because of economic and lifestyle differences, designers must target products to each market. For instance, clothes washers sold in northern countries must spin-dry clothes much better than those in warmer climates, where consumers are likely to line-dry them. Similarly, as shown here, Whirlpool refrigerators sold in Bangkok are manufactured in bright colors because they are often put in living rooms.

Kraipit Phanvut/Sipa Press

So, to recap, successfully achieving a competitive advantage in our shrinking world means maximizing all the possible opportunities, from tangible to intangible, that international operations can offer.

Cultural and Ethical Issues

While there are great forces driving firms toward globalization, many challenges remain. One of these challenges is reconciling differences in social and cultural behavior. With issues ranging from bribery, to child labor, to the environment, managers sometimes do not know how to respond when operating in a different culture. What one country's culture deems acceptable may be considered unacceptable or illegal in another. It is not by chance that there are fewer female managers in the Middle East than in India.

In the last decade, changes in international laws, agreements, and codes of conduct have been applied to define ethical behavior among managers around the world. The WTO, for example, helps to make uniform the protection of both governments and industries from foreign firms that engage in unethical conduct. Even on issues where significant differences between cultures exist, as in the area of bribery or the protection of intellectual property, global uniformity is slowly being accepted by many nations.

Despite cultural and ethical differences, we live in a period of extraordinary mobility of capital, information, goods, and even people. We can expect this to continue. The financial sector, the telecommunications sector, and the logistics infrastructure of the world are healthy institutions that foster efficient and effective use of capital, information, and goods. Globalization, with all its opportunities and risks, is here. It must be embraced as managers develop missions and strategies.

Determining Missions and Strategies

STUDENT TIP
Getting an education and managing an organization both require a mission and strategy.

An effective operations management effort must have a *mission* so it knows where it is going and a *strategy* so it knows how to get there. This is the case for a small domestic organization as well as a large international organization.

Mission

Economic success, indeed survival, is the result of identifying missions to satisfy a customer's needs and wants. We define the organization's **mission** as its purpose—what it will contribute to society. Mission statements provide boundaries and focus for organizations and the concept around which the firm can rally. The mission states the rationale for the organization's existence. Developing a good strategy is difficult, but it is much easier if the mission has been well defined. Figure 2.2 provides examples of mission statements.

Mission
The purpose or rationale for an organization's existence.

LO 2.1 *Define* mission and strategy

Once an organization's mission has been decided, each functional area within the firm determines its supporting mission. By *functional area* we mean the major disciplines required by the firm, such as marketing, finance/accounting, and production/operations. Missions for each function are developed to support the firm's overall mission. Then within that function lower-level supporting missions are established for the OM functions. Figure 2.3 provides such a hierarchy of sample missions.

Strategy

With the mission established, strategy and its implementation can begin. **Strategy** is an organization's action plan to achieve the mission. Each functional area has a strategy for achieving its mission and for helping the organization reach the overall mission. These strategies exploit opportunities and strengths, neutralize threats, and avoid weaknesses. In the following sections, we will describe how strategies are developed and implemented.

Strategy
How an organization expects to achieve its missions and goals.

LO 2.2 *Identify* and explain three strategic approaches to competitive advantage

VIDEO 2.1
Strategy at Nautique Boat Company

Firms achieve missions in three conceptual ways: (1) differentiation, (2) cost leadership, and (3) response. This means operations managers are called on to deliver goods and services that are (1) *better*, or at least different, (2) *cheaper*, and (3) more *responsive*. Operations managers translate these *strategic concepts* into tangible tasks to be accomplished. Any one or combination of these three strategic concepts can generate a system that has a unique advantage over competitors. Much of the remainder of this text is devoted to the challenging task of translating strategy into execution.

Achieving Competitive Advantage Through Operations

Competitive advantage
The creation of a unique advantage over competitors.

Each of the three strategies provides an opportunity for operations managers to achieve competitive advantage. **Competitive advantage** implies the creation of a system that has a unique advantage over competitors. The idea is to create customer value in an efficient and sustainable way. Pure forms of these strategies may exist, but operations managers will more likely be called on to implement some combination of them. Let us briefly look at how managers achieve competitive advantage via *differentiation, low cost*, and *response*.

STUDENT TIP
For many organizations, the operations function provides *the* competitive advantage.

Competing on Differentiation

Safeskin Corporation is number one in latex exam gloves because it has differentiated itself and its products. It did so by producing gloves that were designed to prevent allergic reactions about which doctors were complaining. When other glove makers caught up, Safeskin developed hypoallergenic gloves. Then it added texture to its gloves. Then it developed a synthetic disposable glove for those allergic to latex—always staying ahead of the competition. Safeskin's strategy is to develop a reputation for designing and producing reliable state-of-the-art gloves, thereby differentiating itself.

Differentiation
Distinguishing the offerings of an organization in a way that the customer perceives as adding value.

Differentiation is concerned with providing *uniqueness*. A firm's opportunities for creating uniqueness are not located within a particular function or activity but can arise in virtually everything the firm does. Moreover, because most products include some service, and most services include some product, the opportunities for creating this uniqueness are limited only by imagination. Indeed, **differentiation** should be thought of as going beyond both physical characteristics and service attributes to encompass everything about the product or service that influences the value that the customers derive from it. Therefore, effective operations managers

Figure **2.2**

Mission Statements for Three Organizations

Source: Mission statement from Merck. Copyright © by Merck & Co., Inc. Reprinted with permission.

Merck
The mission of Merck is to provide society with superior products and services—innovations and solutions that improve the quality of life and satisfy customer needs—to provide employees with meaningful work and advancement opportunities and investors with a superior rate of return.
PepsiCo
Our mission is to be the world's premier consumer products company focused on convenient foods and beverages. We seek to produce financial rewards to investors as we provide opportunities for growth and enrichment to our employees, our business partners, and the communities in which we operate. And in everything we do, we strive for honesty, fairness, and integrity.
Arnold Palmer Hospital
Arnold Palmer Hospital for Children provides state of the art, family-centered healthcare focused on restoring the joy of childhood in an environment of compassion, healing, and hope.

Sample Company Mission	
To manufacture and service an innovative, growing, and profitable worldwide microwave communications business that exceeds our customers' expectations.	
Sample Operations Management Mission	
To produce products consistent with the company's mission as the worldwide low-cost manufacturer.	
Sample OM Department Missions	
Product design	To design and produce products and services with outstanding quality and inherent customer value.
Quality management	To attain the exceptional value that is consistent with our company mission and marketing objectives by close attention to design, supply chain, production, and field service opportunities.
Process design	To determine, design, and develop the production process and equipment that will be compatible with low-cost product, high quality, and a good quality of work life.
Location	To locate, design, and build efficient and economical facilities that will yield high value to the company, its employees, and the community.
Layout design	To achieve, through skill, imagination, and resourcefulness in layout and work methods, production effectiveness and efficiency while supporting a high quality of work life.
Human resources	To provide a good quality of work life, with well-designed, safe, rewarding jobs, stable employment, and equitable pay, in exchange for outstanding individual contribution from employees at all levels.
Supply chain management	To collaborate with suppliers to develop innovative products from stable, effective, and efficient sources of supply.
Inventory	To achieve low investment in inventory consistent with high customer service levels and high facility utilization.
Scheduling	To achieve high levels of throughput and timely customer delivery through effective scheduling.
Maintenance	To achieve high utilization of facilities and equipment by effective preventive maintenance and prompt repair of facilities and equipment.

Figure 2.3

Sample Missions for a Company, the Operations Function, and Major OM Departments

assist in defining everything about a product or service that will influence the potential value to the customer. This may be the convenience of a broad product line, product features, or a service related to the product. Such services can manifest themselves through convenience (store location, curbside pickup, home delivery, etc.), training, product delivery and installation, or repair and maintenance services.

In the service sector, one option for extending product differentiation is through an *experience*. Differentiation by experience in services is a manifestation of the growing "experience economy." The idea of **experience differentiation** is to engage the customer—to use people's five senses so they become immersed, or even an active participant, in the product. Disney does this with the Magic Kingdom. People no longer just go on a ride; they are immersed in the Magic Kingdom—surrounded by dynamic visual and sound experiences that complement the physical ride. Some rides further engage the customer with virtual reality or changing air flow and smells, as well as having them steer the ride or shoot at targets or villains. Even movie theaters are moving in this direction with surround sound, moving seats, changing "smells," and mists of "rain," as well as multimedia inputs to story development.

Experience differentiation
Engaging a customer with a product through imaginative use of the five senses, so the customer "experiences" the product.

VIDEO 2.2
Hard Rock's Global Strategy

Theme restaurants, such as Hard Rock Cafe, likewise differentiate themselves by providing an "experience." Hard Rock engages the customer with classic rock music, big-screen rock videos, memorabilia, and staff who can tell stories. In many instances, a full-time guide is available to explain the displays, and there is always a convenient retail store so the guest can take home a tangible part of the experience. The result is a "dining experience" rather than just a meal. In a less dramatic way, both Starbucks and your local supermarket deliver an experience when they provide music and the aroma of fresh coffee or freshly baked bread.

Competing on Cost

Southwest Airlines has been a consistent moneymaker while other U.S. airlines have lost billions. Southwest has done this by fulfilling a need for low-cost and short-hop flights. Its operations strategy has included use of secondary airports and terminals, first-come, first-served seating, few fare options, smaller crews flying more hours, and snacks-only or no-meal flights.

In addition, and less obviously, Southwest has very effectively matched capacity to demand and effectively utilized this capacity. It has done this by designing a route structure that matches the capacity of its Boeing 737, the only plane in its fleet. Second, it achieves more air miles than other airlines through faster turnarounds—its planes are on the ground less.

One driver of a low-cost strategy is a facility that is effectively utilized. Southwest and others with low-cost strategies understand this and use financial resources effectively. Identifying the optimum size (and investment) allows firms to spread overhead costs, providing a cost advantage. For instance, Walmart continues to pursue its low-cost strategy with superstores, open 24 hours a day. For 20 years, it has successfully grabbed market share. Walmart has driven down store overhead costs, shrinkage, and distribution costs. Its rapid transportation of goods, reduced warehousing costs, and direct shipment from manufacturers have resulted in high inventory turnover and made it a low-cost leader.

Likewise, Franz Colruyt, a Belgian discount food retailer, is also an aggressive cost cutter. Colruyt cuts overhead by using converted factory warehouses, movie theaters, and garages as outlets. Customers find no background music, shopping bags, or bright lights: all have been eliminated to cut costs. Walmart and Colruyt are winning with a low-cost strategy.

Low-cost leadership
Achieving maximum value, as perceived by the customer.

Low-cost leadership entails achieving maximum *value* as defined by your customer. It requires examining each of the 10 OM decisions in a relentless effort to drive down costs while meeting customer expectations of value. A low-cost strategy does *not* imply low value or low quality.

Competing on Response

Response
A set of values related to rapid, flexible, and reliable performance.

The third strategy option is response. Response is often thought of as *flexible* response, but it also refers to *reliable* and *quick* response. Indeed, we define **response** as including the entire range of values related to timely product development and delivery, as well as reliable scheduling and flexible performance.

Flexible response may be thought of as the ability to match changes in a marketplace where design innovations and volumes fluctuate substantially.

Hewlett-Packard is an exceptional example of a firm that has demonstrated flexibility in both design and volume changes in the volatile world of personal computers. HP's products often have a life cycle of months, and volume and cost changes during that brief life cycle are dramatic. However, HP has been successful at institutionalizing the ability to change products and volume to respond to dramatic changes in product design and costs—thus building a *sustainable competitive advantage*.

The second aspect of response is the *reliability* of scheduling. One way the German machine industry has maintained its competitiveness despite having the world's highest labor costs is through reliable response. This response manifests itself in reliable scheduling. German machine firms have meaningful schedules—and they perform to these schedules. Moreover, the results of these schedules are communicated to the customer, and the customer can, in turn,

Darren Hauck/AP Images

Response strategy wins orders at Super Fast Pizza. Using a wireless connection, orders are transmitted to $20,000 kitchens in vans. The driver, who works solo, receives a printed order, goes to the kitchen area, pulls premade pizzas from the cooler, and places them in the oven—it takes about 1 minute. The driver then delivers the pizza—sometimes even arriving before the pizza is ready.

rely on them. Consequently, the competitive advantage generated through reliable response has value to the end customer.

The third aspect of response is *quickness*. Johnson Electric Holdings, Ltd., with headquarters in Hong Kong, makes 83 million tiny motors each month. The motors go in cordless tools, household appliances, and personal care items such as hair dryers; dozens are found in each automobile. Johnson's major competitive advantage is speed: speed in product development, speed in production, and speed in delivery.

Whether it is a production system at Johnson Electric or a pizza delivered in 5 minutes by Pizza Hut, the operations manager who develops systems that respond quickly and reliably can have a competitive advantage.

In practice, differentiation, low cost, and response can increase productivity and generate a sustainable competitive advantage. Proper implementation of the ten decisions by operations managers (see Figure 2.4) will allow these advantages to be achieved.

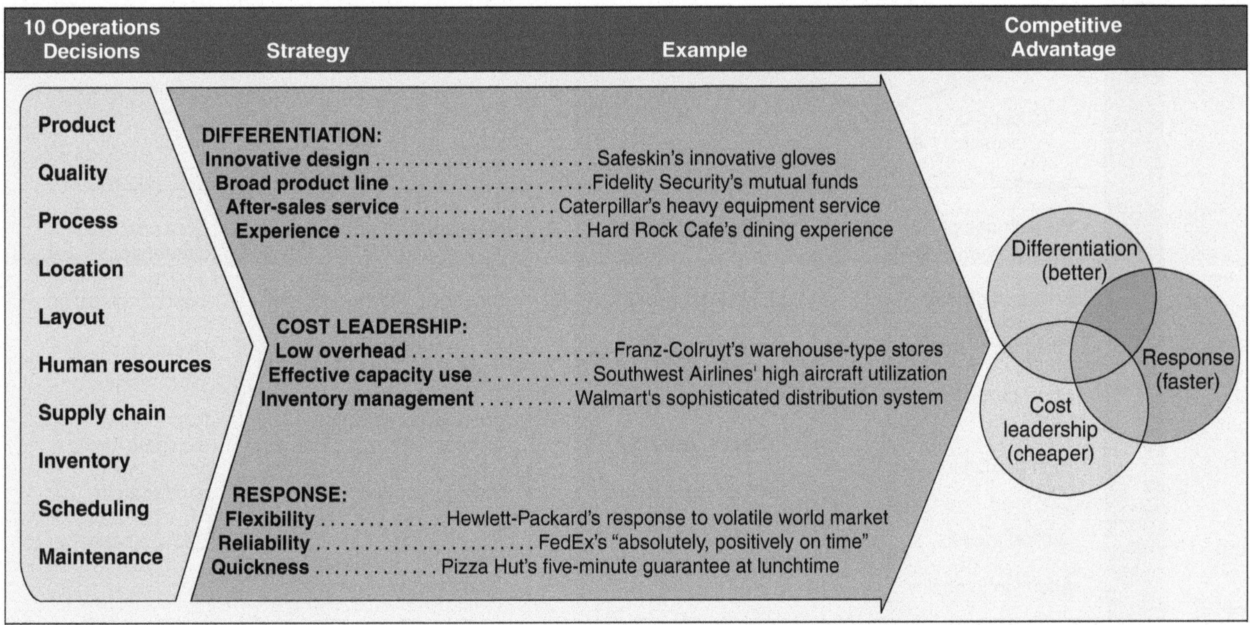

Figure 2.4

Achieving Competitive Advantage Through Operations

Issues in Operations Strategy

Resources view
A method managers use to evaluate the resources at their disposal and manage or alter them to achieve competitive advantage.

Value-chain analysis
A way to identify those elements in the product/service chain that uniquely add value.

Five forces model
A method of analyzing the five forces in the competitive environment.

Whether the OM strategy is differentiation, cost, or response (as shown in Figure 2.4), OM is a critical player. Therefore, prior to establishing and attempting to implement a strategy, some alternate perspectives may be helpful. One perspective is to take a **resources view**. This means thinking in terms of the financial, physical, human, and technological resources available and ensuring that the potential strategy is compatible with those resources. Another perspective is Porter's value-chain analysis.[2] **Value-chain analysis** is used to identify activities that represent strengths, or potential strengths, and may be opportunities for developing competitive advantage. These are areas where the firm adds its unique *value* through product research, design, human resources, supply-chain management, process innovation, or quality management. Porter also suggests analysis of competitors via what he calls his **five forces model**.[3] These potential competing forces are immediate rivals, potential entrants, customers, suppliers, and substitute products.

In addition to the competitive environment, the operations manager needs to understand that the firm is operating in a system with many other external factors. These factors range from economic, to legal, to cultural. They influence strategy development and execution and require constant scanning of the environment.

The firm itself is also undergoing constant change. Everything from resources, to technology, to product life cycles is in flux. Consider the significant changes required within the firm as its products move from introduction, to growth, to maturity, and to decline (see Figure 2.5). These internal changes, combined with external changes, require strategies that are dynamic.

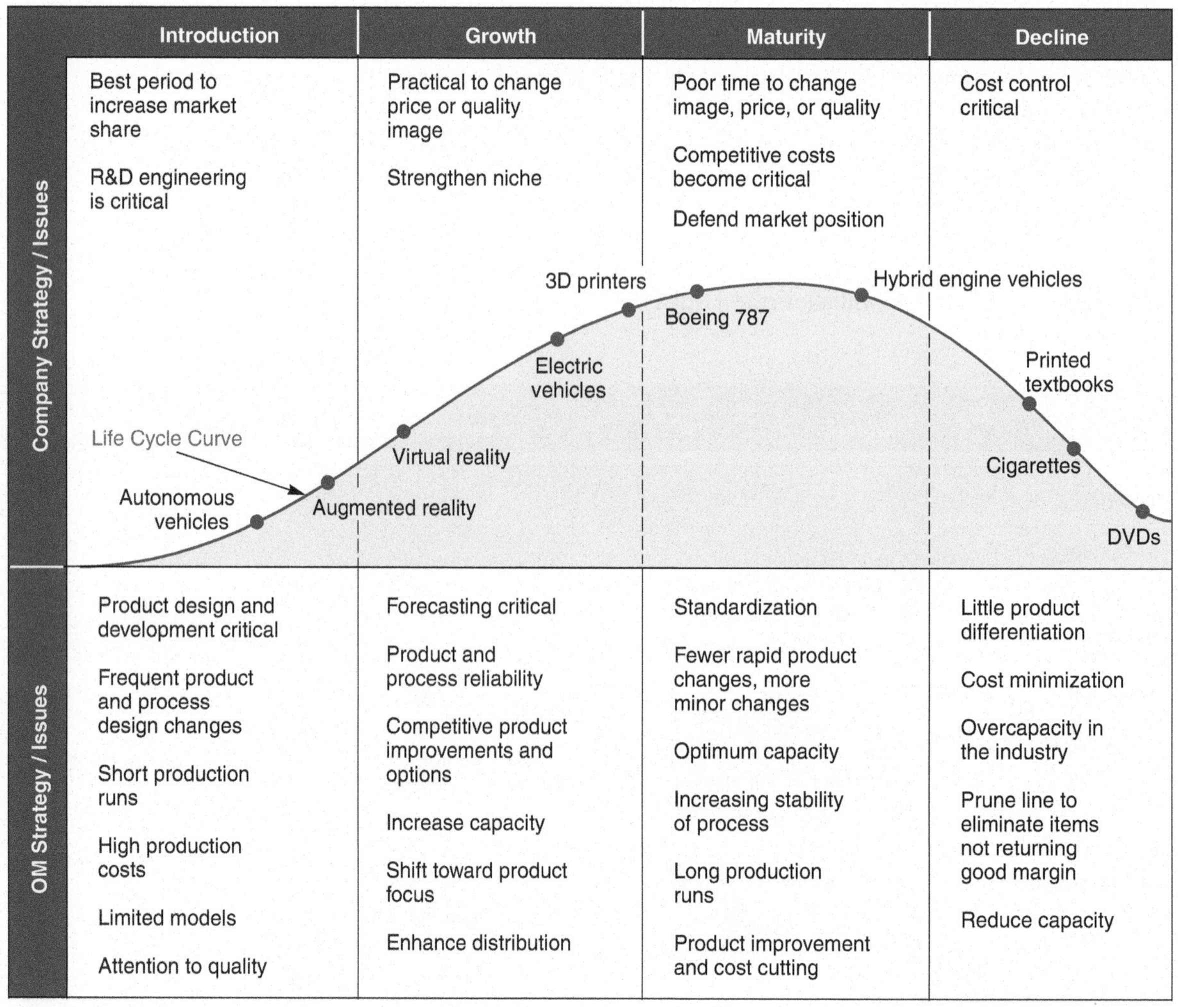

Introduction
Growth
Maturity
Decline
Company Strategy / Issues
Best period to increase market share
R&D engineering is critical
Practical to change price or quality image
Strengthen niche
Poor time to change image, price, or quality
Competitive costs become critical
Defend market position
Cost control critical
Life Cycle Curve
Autonomous vehicles
Augmented reality
Virtual reality
Electric vehicles
3D printers
Boeing 787
Hybrid engine vehicles
Printed textbooks
Cigarettes
DVDs
OM Strategy / Issues
Product design and development critical
Frequent product and process design changes
Short production runs
High production costs
Limited models
Attention to quality
Forecasting critical
Product and process reliability
Competitive product improvements and options
Increase capacity
Shift toward product focus
Enhance distribution
Standardization
Fewer rapid product changes, more minor changes
Optimum capacity
Increasing stability of process
Long production runs
Product improvement and cost cutting
Little product differentiation
Cost minimization
Overcapacity in the industry
Prune line to eliminate items not returning good margin
Reduce capacity

Figure 2.5

Strategy and Issues During a Product's Life Cycle

In this chapter's *Global Company Profile*, Boeing provides an example of how strategy must change as technology and the environment change. Boeing builds planes and rockets from carbon fiber, using a global supply chain. Like many other OM strategies, Boeing's strategy has changed with technology and globalization. Microsoft has also had to adapt quickly to a changing environment. Faster processors, new computer languages, changing customer preferences, increased security issues, the Internet, the cloud, and Google have all driven changes at Microsoft. These forces have moved Microsoft's product strategy from operating systems to office products, to Internet service provider, and now to integrator of computers, cell phones, games, and television via the cloud. Also notice, as discussed in the *OM In Action box*, "Amazon Updates Sears' Strategy," how Sears has languished while Amazon has embraced the new digital world to build a new worldwide multi-billion dollar business.

The more thorough the analysis and understanding of both the external and internal factors, the more likely that a firm can find the optimum use of its resources. Once a firm understands itself and the environment, a SWOT analysis, which we discuss next, is in order.

Strategy Development and Implementation

STUDENT TIP
A SWOT analysis provides an excellent model for evaluating a strategy.

SWOT analysis
A method of determining internal strengths and weaknesses and external opportunities and threats.

A **SWOT analysis** is a formal review of internal strengths and weaknesses and external opportunities and threats. Beginning with SWOT analyses, organizations position themselves, through their strategy, to have a competitive advantage. A firm may have excellent design skills or great talent at identifying outstanding locations. However, it may recognize limitations of its manufacturing process or in finding good suppliers. The idea is to maximize opportunities and minimize threats in the environment while maximizing the advantages of the organization's strengths and minimizing the weaknesses. Any preconceived ideas about mission are then reevaluated to ensure they are consistent with the SWOT analysis. Subsequently, a strategy for achieving the mission is developed. This strategy is continually evaluated against the value provided customers and competitive realities. The process is shown in Figure 2.6. From this process, key success factors are identified.

OM in Action Amazon Updates Sears' Strategy[4]

A century ago, a retail giant that shipped millions of products by mail moved swiftly into the brick-and-mortar business, changing retail forever. Is that happening again? A look at Sears' strategy predicts nearly everything Amazon is doing.

In the last few years, Amazon has opened numerous physical bookstores. It bought Whole Foods and its 400 grocery locations, and it announced a partnership with Kohl's to allow returns at the physical retailer's stores. Amazon's corporate strategy is following a familiar playbook—that of Sears. Sears might seem like a zombie today, but it's easy to forget how transformative the company was 100 years ago. To understand Amazon's evolution, strategy, and future, can we look to Sears?

Mail was an internet before the Internet. After the Civil War, the telegraph, rail, and parcel delivery made it possible to shop via catalog at home and have items delivered to your door. Americans browsed catalogs for everything from food, to books, to houses. Merchants sent the parcels by rail. Then Sears made the successful transition to a brick-and-mortar giant. Like Amazon among its online rivals, Sears was not the country's first mail-order retailer, but it became the largest. Like Amazon, it started with a single product category—watches, rather than books. Like Amazon, the company grew to include a range of products, including guns, gramophones, cars, and groceries.

By building a large base of fiercely loyal consumers, Sears was able to buy more cheaply from manufacturers. It managed its deluge of orders with massive warehouses. But the company's brick-and-mortar transformation was astonishing. At the start of 1925, there were no Sears stores. By 1929, there were 300. Like Amazon today, the company used its position to enter adjacent businesses. To supplement its huge auto-parts business, Sears started selling car insurance under the Allstate brand. Perhaps Sear's shift from selling products to services is analogous to the creation of Amazon Web Services—or Amazon's TV shows. The growth of both companies was the result of a strategic vision and a focus on operations efficiency, with an eye on the changes in demographics, technology, and logistics. But Sears failed to adjust to a rapidly changing environment. Will Amazon fair better?

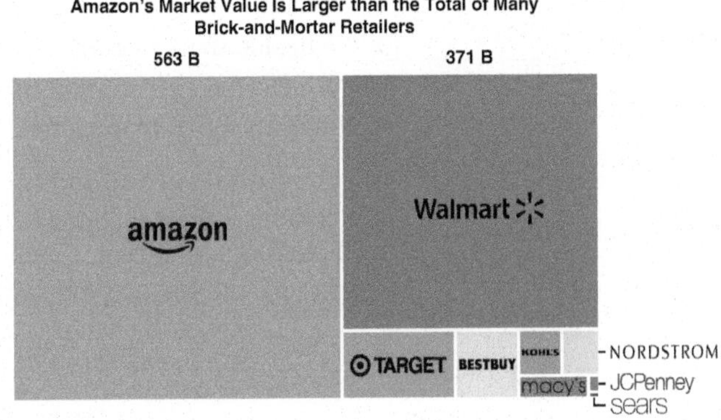

Figure 2.6
Strategy Development Process

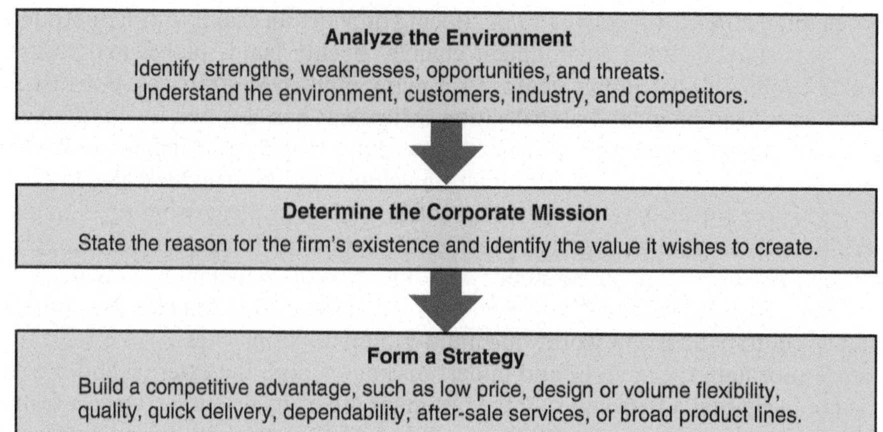

Analyze the Environment
Identify strengths, weaknesses, opportunities, and threats.
Understand the environment, customers, industry, and competitors.
Determine the Corporate Mission
State the reason for the firm's existence and identify the value it wishes to create.
Form a Strategy
Build a competitive advantage, such as low price, design or volume flexibility,
quality, quick delivery, dependability, after-sale services, or broad product lines.

Key Success Factors and Core Competencies

Key success factors (KSFs)
Activities or factors that are *key* to achieving competitive advantage.

Core competencies
A set of outstanding skills, talents, and capabilities that differentiates an organization from its competition.

Because no firm does everything exceptionally well, a successful strategy requires determining the firm's key success factors and core competencies. **Key success factors (KSFs)** are those activities that are necessary for a firm to achieve its goals. Key success factors can be so significant that a firm must get them right to survive. A KSF for McDonald's, for example, is layout. Without an effective drive-through and an efficient kitchen, McDonald's cannot be successful. KSFs are often necessary, but not sufficient for competitive advantage. On the other hand, **core competencies** are the set of unique skills, talents, and capabilities that a firm does at a world-class standard. They allow a firm to set itself apart and develop a competitive advantage. Organizations that prosper identify their core competencies and nurture them. While McDonald's KSFs may include layout, its core competency may be consistent quality. Honda Motors' core competence is gas-powered engines—engines for automobiles, motorcycles, lawn mowers, generators, snow blowers, and more. The idea is to build KSFs and core competencies that provide a competitive advantage and support a successful strategy and mission. A core competency may be the ability to perform the KSFs or a combination of KSFs. The operations manager begins this inquiry by asking:

LO 2.3 *Understand* the significance of key success factors and core competencies

- "What tasks must be done particularly well for a given strategy to succeed?"
- "Which activities provide a competitive advantage?"
- "Which elements contain the highest likelihood of failure, and which require additional commitment of managerial, monetary, technological, and human resources?"

Only by identifying and strengthening key success factors and core competencies can an organization achieve sustainable competitive advantage. In this text we focus on the 10 strategic OM decisions that typically include the KSFs. These decisions, plus major decision areas for marketing and finance, are shown in Figure 2.7.

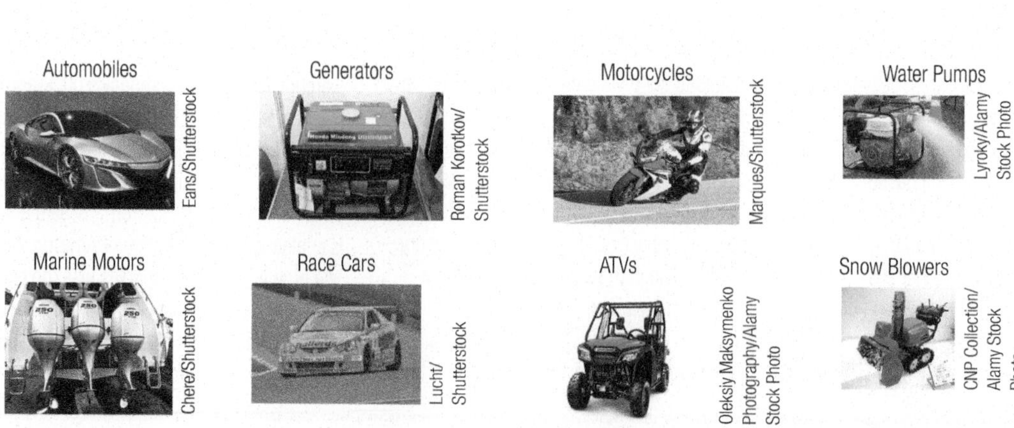

Honda's core competence is the design and manufacture of gas-powered engines. This competence has allowed Honda to become a leader in the design and manufacture of a wide range of gas-powered products. Tens of millions of these products are produced and shipped around the world.

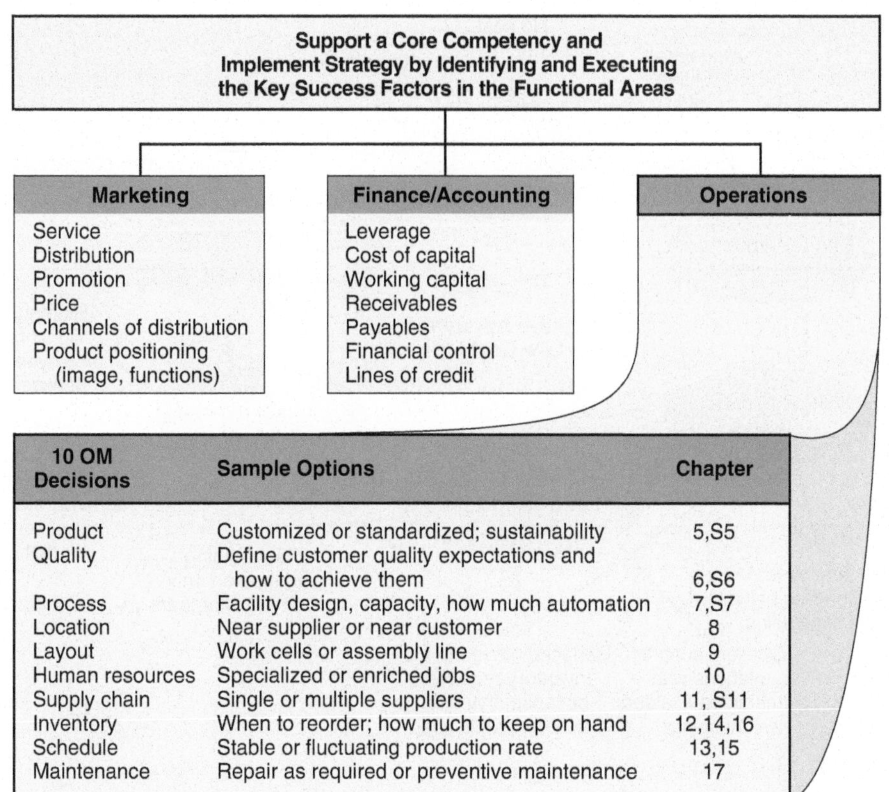

10 OM Decisions	Sample Options	Chapter
Product	Customized or standardized; sustainability	5,S5
Quality	Define customer quality expectations and how to achieve them	6,S6
Process	Facility design, capacity, how much automation	7,S7
Location	Near supplier or near customer	8
Layout	Work cells or assembly line	9
Human resources	Specialized or enriched jobs	10
Supply chain	Single or multiple suppliers	11, S11
Inventory	When to reorder; how much to keep on hand	12,14,16
Schedule	Stable or fluctuating production rate	13,15
Maintenance	Repair as required or preventive maintenance	17

Figure **2.7**

Implement Strategy by Identifying and Executing Key Success Factors That Support Core Competencies

STUDENT TIP

These 10 decisions are used to implement a specific strategy and yield a competitive advantage.

Integrating OM with Other Activities

Whatever the KSFs and core competencies, they must be supported by the related activities. One approach to identifying the activities is an **activity map**, which links competitive advantage, KSFs, and supporting activities. For example, Figure 2.8 shows how Southwest Airlines, whose core competency is operations, built a set of integrated activities to support its low-cost competitive advantage. Notice how the KSFs support operations and in turn are supported by other activities. The activities fit together and reinforce each other. In this way, all of the areas support the company's objectives. For example, short-term scheduling in the airline industry is dominated by volatile customer travel patterns. Day-of-week preference, holidays, seasonality, college schedules, and so on all play roles in changing flight schedules. Consequently, airline scheduling, although an OM activity, is tied to marketing. Effective scheduling in the trucking industry is reflected in the amount of time trucks travel loaded. But maximizing the time trucks travel loaded requires the integration of information from deliveries completed, pickups pending, driver availability, truck maintenance, and customer priority. Success requires integration of all of these activities.

Activity map
A graphical link of competitive advantage, KSFs, and supporting activities.

The better the activities are integrated and reinforce each other, the more sustainable the competitive advantage. By focusing on enhancing its core competence and KSFs with a supporting set of activities, firms such as Southwest Airlines have built successful strategies.

Building and Staffing the Organization

Once a strategy, KSFs, and the necessary integration have been identified, the second step is to group the necessary activities into an organizational structure. Then, managers must staff the organization with personnel who will get the job done. The manager works with subordinate managers to build plans, budgets, and programs that will successfully implement strategies that achieve missions. Firms tackle this organization of the operations function in a variety of ways. The organization charts shown in Chapter 1 (Figure 1.1) indicate the way some firms have organized to perform the required activities. *The operations manager's job is to implement an OM strategy, provide competitive advantage, and increase productivity.*

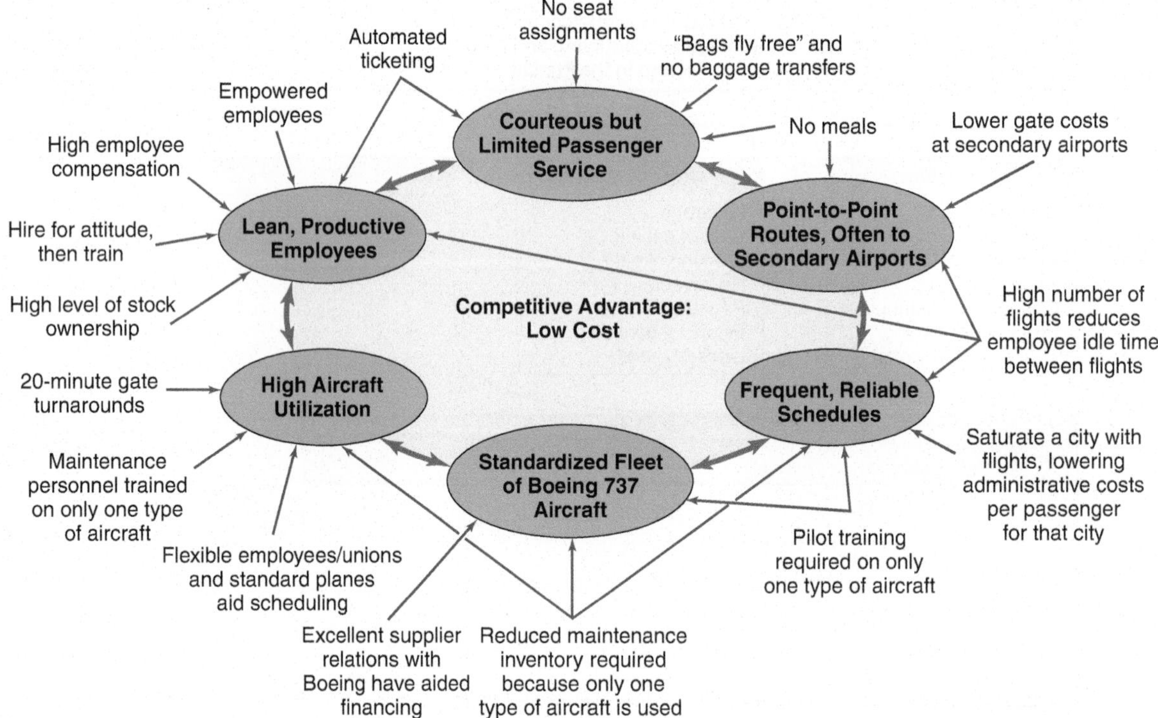

Figure 2.8

Activity Mapping of Southwest Airlines' Low-Cost Competitive Advantage

To achieve a low-cost competitive advantage, Southwest has identified a number of key success factors (connected by red arrows) and support activities (shown by blue arrows). As this figure indicates, Southwest's low-cost strategy is highly dependent on a very well-run operations function.

Implementing the 10 Strategic OM Decisions

As mentioned earlier, the challenging task of implementing the 10 strategic OM decisions is influenced by a variety of issues—from missions and strategy to key success factors and core competencies—while addressing such issues as product mix and product life cycle in a competitive environment. Because each product brings its own mix of attributes, the importance and method of implementation of the 10 strategic OM decisions will vary. Throughout this text, we discuss how these decisions are implemented in ways that provide competitive advantage. How this might be done for two drug companies, one seeking competitive advantage via differentiation and the other via low cost, is shown in Table 2.1.

Strategic Planning, Core Competencies, and Outsourcing

As organizations develop missions, goals, and strategies, they identify their strengths—what they do as well as or better than their competitors—as their *core competencies*. By contrast, *non-core activities*, which can be a sizable portion of an organization's total business, are good candidates for outsourcing. **Outsourcing** is transferring activities that have traditionally been internal to external suppliers.

Outsourcing
Transferring a firm's activities that have traditionally been internal to external suppliers.

Outsourcing is not a new concept, but it does add complexity and risk to the supply chain. Because of its potential, outsourcing continues to expand. The expansion is accelerating due to three global trends: (1) increased technological expertise, (2) more reliable and cheaper transportation, and (3) the rapid development and deployment of advancements in telecommunications and computers. This rich combination of economic advances is contributing to both lower cost and more specialization. As a result more firms are candidates for outsourcing of non-core activities.

VIDEO 2.3
Outsourcing Offshore at Darden

Outsourcing implies an agreement (typically a legally binding contract) with an external organization. The classic make-or-buy decision, concerning which products to make and which

TABLE 2.1 Operations Strategies of Two Drug Companies*

COMPETITIVE ADVANTAGE	BRAND NAME DRUGS, INC. PRODUCT DIFFERENTIATION STRATEGY	GENERIC DRUG CORP. LOW-COST STRATEGY
Product selection and design	Heavy R&D investment; extensive labs; focus on development in a broad range of drug categories	Low R&D investment; focus on development of generic drugs
Quality	Quality is major priority, standards exceed regulatory requirements	Meets regulatory requirements on a country-by-country basis, as necessary
Process	Product and modular production process; tries to have long product runs in specialized facilities; builds capacity ahead of demand	Process focused; general production processes; "job shop" approach, short-run production; focus on high utilization
Location	Still located in city where it was founded	Operates in low-tax, low-labor-cost environment
Layout	Layout supports automated product-focused production	Layout supports process-focused "job shop" practices
Human resources	Hire the best; nationwide searches	Very experienced top executives hired to provide direction; other personnel paid below industry average
Supply chain	Long-term supplier relationships	Tends to purchase competitively to find bargains
Inventory	Maintains high finished goods inventory primarily to ensure all demands are met	Process focus drives up work-in-process inventory; finished goods inventory tends to be low
Scheduling	Centralized production planning	Many short-run products complicate scheduling
Maintenance	Highly trained staff; extensive parts inventory	Highly trained staff to meet changing process and equipment demands

*Notice how the 10 decisions are altered to build two distinct strategies in the same industry.

to buy, is the basis of outsourcing. When firms such as Apple find that their core competency is in creativity, innovation, and product design, they may want to outsource manufacturing.

Outsourcing manufacturing is an extension of the long-standing practice of *subcontracting* production activities, which when done on a continuing basis is known as *contract manufacturing*. Contract manufacturing is becoming standard practice in many industries, from computers to automobiles. For instance, Johnson & Johnson, like many other big drug companies whose core competency is research and development, often farms out manufacturing to contractors. On the other hand, Sony's core competency is electromechanical design of chips. This is its core competency, but Sony is also one of the best in the world when it comes to rapid response and specialized production of these chips. Therefore, Sony finds that it wants to be its own *manufacturer*, while specialized providers come up with major innovations in such areas as software, human resources, and distribution. These areas are the providers' business, not Sony's, and the provider may very well be better at it than Sony.

Keith Dannemiller/Alamy Stock Photo

Contract manufacturers such as Flextronics provide outsourcing service to IBM, Cisco Systems, HP, Microsoft, Sony, Nortel, Ericsson, and Sun, among many others. Flextronics is a high-quality producer that has won over 450 awards, including the Malcolm Baldrige Award. One of the side benefits of outsourcing is that client firms such as IBM can actually improve their performance by using the competencies of an outstanding firm like Flextronics. But there are risks involved in outsourcing.

Other examples of outsourcing non-core activities include:

- DuPont routing legal services to the Philippines
- IBM handling travel services and payroll, and Hewlett-Packard providing IT services to P&G
- ADP managing payroll services for thousands of organizations
- Accenture providing consulting and back office services for 95 of the Global Fortune 100
- Blue Cross sending hip resurfacing surgery patients to India

Managers evaluate their strategies and core competencies and ask themselves how to use the assets entrusted to them. Do they want to be the company that does low-margin work at 3%–4% or the innovative firm that makes a 30%–40% margin? PC and iPad contract manufacturers in China and Taiwan earn 3%–4%, but Apple, which innovates, designs, and sells, has a margin 10 times as large. (See the *OM in Action box*, "China Outsources Too—to Ethiopia.")

The Theory of Comparative Advantage

Theory of comparative advantage
A theory which states that countries benefit from specializing in (and exporting) goods and services in which they have relative advantage, and they benefit from importing goods and services in which they have a relative disadvantage.

The motivation for international outsourcing comes from the **theory of comparative advantage**. This theory focuses on the economic concept of relative advantage. Accordingly, if an external provider, regardless of its geographic location, can perform activities more productively than the purchasing firm, then the external provider should do the work. This allows the purchasing firm to focus on what it does best—its core competencies. Consistent with the theory of comparative advantage, outsourcing continues to grow. But outsourcing the wrong activities can be a disaster. And even outsourcing non-core activities has risks.

STUDENT TIP
The substantial risk of outsourcing requires managers to invest in the effort to make sure they do it right.

Risks of Outsourcing

Risk management starts with a realistic analysis of uncertainty and results in a strategy that minimizes the impact of these uncertainties. Indeed, outsourcing *is* risky, with roughly half of all outsourcing agreements failing because of inadequate planning and analysis. Timely delivery and quality standards can be major problems, as can underestimating increases in inventory and logistics costs. Additionally, companies that outsource customer service tend to see a drop in customer satisfaction.

OM in Action — China Outsources Too—to Ethiopia[5]

With many workers in the Haijian International shoe factory in China complaining about excessive hours and seeking higher pay, that company is sending thousands of their jobs to Ethiopia. Like many Chinese firms, Haijian faces scrutiny from labor activists for how it treats workers. The focus of the activists points to changing labor conditions in China as manufacturers try to get more work out of an increasingly expensive labor pool. But deep economic and demographic shifts imply that a lot of low-end work—like making shoes—doesn't offer huge profit in China.

Today, Chinese workers are less cheap and less willing. More young people are going to college and want office jobs. The blue-collar workforce is aging. Long workdays in a factory no longer appeal to those older workers, even with the promise of overtime pay. Such tensions are fueling the drive of Haijian to outsource .... in this case to Ethiopia.

In many respects, China's economy is maturing. The number of people who turn 18 each year and do not enroll in college—the group that might consider factory work—had plummeted to 10.5 million by 2015 from 18.5 million in 2000. Wages have increased ninefold since the late 1990s. Haijian peaked at 26,000 employees in China in 2006. Staffing is now down to 7,000–8,000, thanks to automation and the shift to Ethiopia. Citing labor costs and the country's foreign investment push, Haijian has built a sprawling complex of factories on the southern outskirts of Ethiopia's capital, Addis Ababa, with 5,000 employees. When finished in 2021, the Addis Ababa complex will be ringed by a replica of the Great Wall of China.

Kay Nietfeld/dpa picture alliance/Alamy Stock Photo

TABLE 2.2 Potential Advantages and Disadvantages of Outsourcing

ADVANTAGES	DISADVANTAGES
Cost savings	Increased logistics and inventory costs
Gaining outside expertise that comes with specialization	Loss of control (quality, delivery, etc.)
Improving operations and service	Potential creation of future competition
Maintaining a focus on core competencies	Negative impact on employees
Accessing outside technology	Risks may not manifest themselves for years

When outsourcing is overseas, additional issues must be considered. These issues include financing, labor skills and availability, and the general business environment. Managers can find substantial efficiencies in outsourcing non-core activities, but they must be cautious in outsourcing those elements of the product or service that provide a competitive advantage. Table 2.2 identifies some potential advantages and disadvantages of outsourcing. The next section provides a methodology that helps analyze the outsourcing decision process.

Rating Outsource Providers

LO 2.4 *Use* factor rating to evaluate both country and outsource providers

Research indicates that the most common reason for the failure of outsourcing agreements is that the decisions are made without sufficient analysis. The *factor-rating method* provides an objective way to evaluate outsource providers. We assign points for each factor to each provider and then importance weights to each of the factors. We now apply the technique in Example 1 to compare outsourcing providers being considered by a firm.

Example 1

RATING PROVIDER SELECTION CRITERIA

National Architects, Inc., a San Francisco–based designer of high-rise office buildings, has decided to outsource its information technology (IT) function. Three outsourcing providers are being actively considered: one in the U.S., one in India, and one in Israel.

APPROACH ▶ National's VP–Operations, Susan Cholette, has made a list of seven criteria she considers critical. After putting together a committee of four other VPs, she has rated each firm (boldface type, on a 1–5 scale, with 5 being highest) and has also placed an importance weight on each of the factors, as shown in Table 2.3.

TABLE 2.3 Factor Ratings Applied to National Architects' Potential IT Outsourcing Providers

		OUTSOURCE PROVIDERS		
FACTOR (CRITERION)[6]	IMPORTANCE WEIGHT	BIM (U.S.)	S.P.C. (INDIA)	TELCO (ISRAEL)
1. Can reduce operating costs	.2	.2 × **3** = .6	.2 × **3** = .6	.2 × **5** = 1.0
2. Can reduce capital investment	.2	.2 × **4** = .8	.2 × **3** = .6	.2 × **3** = .6
3. Skilled personnel	.2	.2 × **5** = 1.0	.2 × **4** = .8	.2 × **3** = .6
4. Can improve quality	.1	.1 × **4** = .4	.1 × **5** = .5	.1 × **2** = .2
5. Can gain access to technology not in company	.1	.1 × **5** = .5	.1 × **3** = .3	.1 × **5** = .5
6. Can create additional capacity	.1	.1 × **4** = .4	.1 × **2** = .2	.1 × **4** = .4
7. Aligns with policy/philosophy/culture	.1	.1 × **2** = .2	.1 × **3** = .3	.1 × **5** = .5
Total Weighted Score		3.9	3.3	3.8

SOLUTION ▶ Susan multiplies each rating by the weight and sums the products in each column to generate a total score for each outsourcing provider. She selects BIM, which has the highest overall rating.

INSIGHT ▶ When the total scores are as close (3.9 vs. 3.8) as they are in this case, it is important to examine the sensitivity of the results to inputs. For example, if one of the importance weights or factor scores changes even marginally, the final selection may change. Management preference may also play a role here.

LEARNING EXERCISE ▶ Susan decides that "Skilled personnel" should instead get a weight of 0.1 and "Aligns with policy/philosophy/culture" should increase to 0.2. How do the total scores change? [Answer: BIM = 3.6, S.P.C. = 3.2, and Telco = 4.0, so Telco would be selected.]

RELATED PROBLEMS ▶ 2.8–2.12

EXCEL OM Data File **Ch02Ex1.xls** can be found online.

Global Operations Strategy Options

International business
A firm that engages in cross-border transactions.

Multinational corporation (MNC)
A firm that has extensive involvement in international business, owning or controlling facilities in more than one country.

As we suggested early in this chapter, many operations strategies now require an international dimension. An international business is any firm that engages in international trade or investment. A multinational corporation (MNC) is a firm with *extensive* international business involvement. MNCs buy resources, create goods or services, and sell goods or services in a variety of countries. The term *multinational corporation* applies to most of the world's large, well-known businesses. Certainly IBM is a good example of an MNC. It imports electronics components to the U.S. from over 50 countries, exports to over 130 countries, has facilities in 45 countries, and earns more than half its sales and profits abroad.

Operations managers of international and multinational firms approach global opportunities with one of four strategies: *international, multidomestic, global*, or *transnational* (see Figure 2.9). The matrix of Figure 2.9 has a vertical axis of cost reduction and a horizontal axis of local responsiveness. Local responsiveness implies quick response and/or the differentiation necessary for the local market. The operations manager must know how to position the firm in this matrix. Let us briefly examine each of the four strategies.

LO 2.5 *Identify* and explain four global operations strategy options

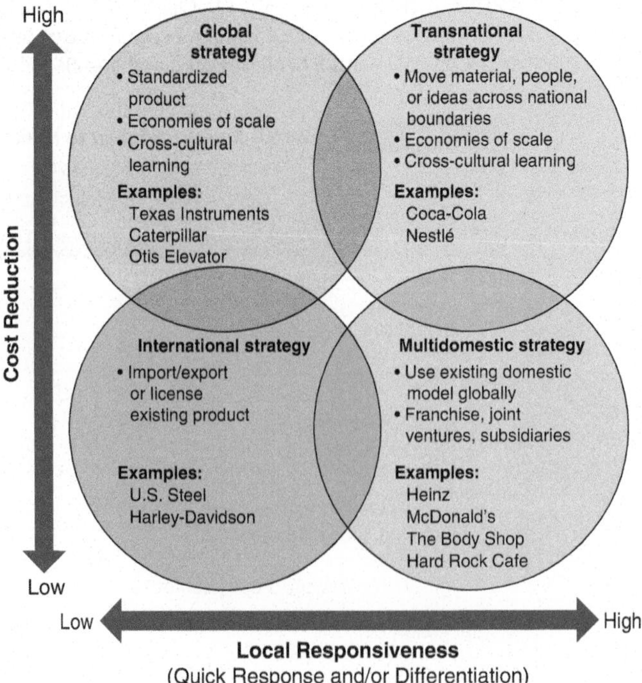

Figure **2.9**

Four International Operations Strategies

Source: Based on M. Hitt, R. D. Ireland, and R. E. Hoskisson, *Strategic Management: Concepts, Competitiveness, and Globalization*, 8th ed. (Cincinnati: Southwestern College Publishing).

An **international strategy** uses exports and licenses to penetrate the global arena. This strategy is the least advantageous, with little local responsiveness and little cost advantage. But an international strategy is often the easiest because exports can require little change in existing operations, and licensing agreements often leave much of the risk to the licensee.

International strategy
A strategy in which global markets are penetrated using exports and licenses.

The **multidomestic strategy** has decentralized authority with substantial autonomy at each business. These are typically subsidiaries, franchises, or joint ventures with substantial independence. The advantage of this strategy is maximizing a competitive response for the local market; however, the strategy has little or no cost advantage. Many food producers, such as Heinz, use a multidomestic strategy to accommodate local tastes because global integration of the production process is not critical. The concept is one of "we were successful in the home market; let's export the management talent and processes, not necessarily the product, to accommodate another market."

Multidomestic strategy
A strategy in which operating decisions are decentralized to each country to enhance local responsiveness.

A **global strategy** has a high degree of centralization, with headquarters coordinating the organization to seek out standardization and learning between plants, thus generating economies of scale. This strategy is appropriate when the strategic focus is cost reduction but has little to recommend it when the demand for local responsiveness is high. Caterpillar, the world leader in earth-moving equipment, and Texas Instruments, a world leader in semiconductors, pursue global strategies. Caterpillar and Texas Instruments find this strategy advantageous because the end products are similar throughout the world. Earth-moving equipment is the same in Nigeria as in Iowa.

Global strategy
A strategy in which operating decisions are centralized and headquarters coordinates the standardization and learning between facilities.

A **transnational strategy** exploits economies of scale and learning, as well as pressure for responsiveness, by recognizing that core competence does not reside in just the "home" country but can exist anywhere in the organization. *Transnational* describes a condition in which material, people, and ideas cross—or *transgress*—national boundaries. These firms have the potential to pursue all three operations strategies (i.e., differentiation, low cost, and response). Such firms can be thought of as "world companies" whose country identity is not as important as their interdependent network of worldwide operations. Nestlé is a good example of such a company. Although it is legally Swiss, 95% of its assets are held and 98% of its sales are made outside Switzerland. Fewer than 10% of its workers are Swiss.

Transnational strategy
A strategy that combines the benefits of global-scale efficiencies with the benefits of local responsiveness.

Washington Imaging/Alamy Stock Photo

Bernd Wüstneck/dpa/picture-alliance/Newscom

In a continuing fierce worldwide battle, both Komatsu and Caterpillar seek global advantage in the heavy equipment market. As Komatsu (left) moved west to the UK, Caterpillar (right) moved east, with 13 facilities and joint ventures in China. Both firms are building equipment throughout the world as cost and logistics dictate. Their global strategies allow production to move as markets, risk, and exchange rates suggest.

Summary

Global operations provide an increase in both the challenges and opportunities for operations managers. Although the task is difficult, operations managers can and do improve productivity. They build and manage global OM functions and supply chains that contribute in a significant way to competitiveness. Organizations identify their strengths and weaknesses. They then develop effective missions and strategies that account for these strengths and weaknesses and complement the opportunities and threats in the environment. If this procedure is performed well, the organization can have competitive advantage through some combination of product differentiation, low cost, and response.

Increasing specialization provides economic pressure to build organizations that focus on core competencies and to outsource the rest. But there is also a need for planning outsourcing to make it beneficial to all participants. In this increasingly global world, competitive advantage is often achieved via a move to international, multidomestic, global, or transnational strategies.

Effective use of resources, whether domestic or international, is the responsibility of the professional manager, and professional managers are among the few in our society who *can* achieve this performance. The challenge is great, and the rewards to the manager and to society are substantial.

Key Terms

Operational hedging (p. 33)
Maquiladoras (p. 33)
World Trade Organization (WTO) (p. 33)
NAFTA and USMCA (p. 33)
European Union (EU) (p. 33)
Mission (p. 35)
Strategy (p. 35)
Competitive advantage (p. 36)
Differentiation (p. 36)
Experience differentiation (p. 37)
Low-cost leadership (p. 38)
Response (p. 38)
Resources view (p. 40)
Value-chain analysis (p. 40)
Five forces model (p. 40)
SWOT analysis (p. 41)
Key success factors (KSFs) (p. 42)
Core competencies (p. 42)
Activity map (p. 43)
Outsourcing (p. 44)
Theory of comparative advantage (p. 46)
International business (p. 48)
Multinational corporation (MNC) (p. 48)
International strategy (p. 49)
Multidomestic strategy (p. 49)
Global strategy (p. 49)
Transnational strategy (p. 49)

Ethical Dilemma

As a manufacturer of athletic shoes whose image—indeed performance—is widely regarded as socially responsible, you find your costs increasing. Traditionally, your athletic shoes have been made in Indonesia and China. Although the ease of doing business in those countries has been improving, wage rates have also been increasing. The labor-cost differential between your current suppliers and a contractor who will get the shoes made in Vietnam now exceeds $1 per pair. Your sales next year are projected to be 10 million pairs, and your analysis suggests that this cost differential is not offset by any other tangible costs; you face only the political risk and potential damage to your commitment to social responsibility. Thus, this $1 per pair savings should flow directly to your bottom line. There is no doubt that the Vietnamese government remains repressive and is a long way from a democracy. Perhaps more significantly, you will have little or no control over working conditions, sexual harassment, and pollution. What do you do, and on what basis do you make your decision?

Lionel derimais/Alamy Stock Photo

Discussion Questions

1. Based on the descriptions and analyses in this chapter, would Boeing be better described as a global firm or a transnational firm? Discuss.
2. List six reasons to internationalize operations.
3. Coca-Cola is called a global product. Does this mean that Coca-Cola is formulated in the same way throughout the world? Discuss.
4. Define *mission*.
5. Define *strategy*.
6. Describe how an organization's *mission* and *strategy* have different purposes.
7. Identify the mission and strategy of your automobile repair garage. What are the manifestations of the 10 strategic OM decisions at the garage? That is, how is each of the 10 decisions accomplished?
8. As a library or Internet assignment, identify the mission of a firm and the strategy that supports that mission.
9. How does an OM strategy change during a product's life cycle?

10. There are three primary ways to achieve competitive advantage. Provide an example, not included in the text, of each. Support your choices.
11. Given the discussion of Southwest Airlines in the text, define an *operations* strategy for that firm now that it has purchased AirTran.
12. How must an operations strategy integrate with marketing and accounting?
13. How would you summarize outsourcing trends?
14. What potential cost-saving advantages might firms experience by using outsourcing?
15. What internal issues must managers address when outsourcing?
16. How should a company select an outsourcing provider?
17. What are some of the possible consequences of poor outsourcing?
18. What global operations strategy is most descriptive of McDonald's?

Using Software to Solve Outsourcing Problems

Excel, Excel OM, and POM for Windows may be used to solve many of the problems in this chapter.

CREATING YOUR OWN EXCEL SPREADSHEETS

Program 2.1 illustrates how to build an Excel spreadsheet for the data in Example 1. In this example the factor rating method is used to compare National Architects' three potential outsourcing providers.

This program provides the data inputs for seven important factors, including their weights (0.0–1.0) and ratings (1–5 scale where 5 is the highest rating) for each country. As we see, BIM is most highly rated, with a 3.9 score, versus 3.3 for S.P.C. and 3.8 for Telco.

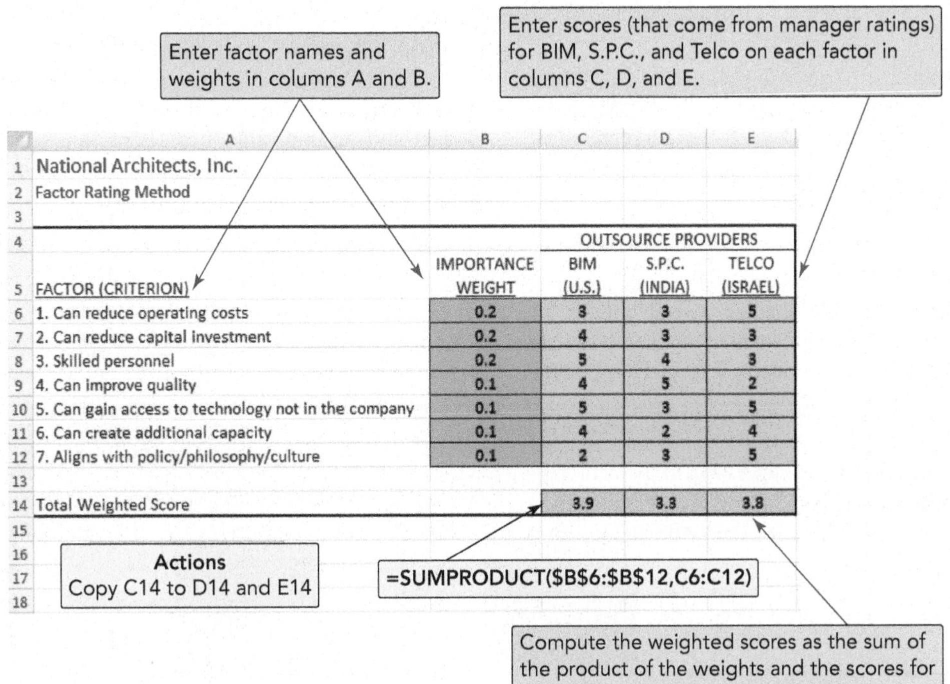

	A	B	C	D	E
1	National Architects, Inc.				
2	Factor Rating Method				
3					
4			OUTSOURCE PROVIDERS		
5	FACTOR (CRITERION)	IMPORTANCE WEIGHT	BIM (U.S.)	S.P.C. (INDIA)	TELCO (ISRAEL)
6	1. Can reduce operating costs	0.2	3	3	5
7	2. Can reduce capital investment	0.2	4	3	3
8	3. Skilled personnel	0.2	5	4	3
9	4. Can improve quality	0.1	4	5	2
10	5. Can gain access to technology not in the company	0.1	5	3	5
11	6. Can create additional capacity	0.1	4	2	4
12	7. Aligns with policy/philosophy/culture	0.1	2	3	5
13					
14	Total Weighted Score		3.9	3.3	3.8

Program **2.1**

Using Excel to Develop a Factor Rating Analysis, With Data from Example 1

USING EXCEL OM

Excel OM (found online) may be used to solve Example 1 (with the Factor Rating module).

USING POM FOR WINDOWS

POM for Windows also includes a factor rating module. For details, refer to Appendix II. POM for Windows is found online and can solve all problems labeled with a **P**.

Solved Problems

SOLVED PROBLEM 2.1

The global tire industry continues to consolidate. Michelin buys Goodrich and Uniroyal and builds plants throughout the world. Bridgestone buys Firestone, expands its research budget, and focuses on world markets. Goodyear spends almost 4% of its sales revenue on research. These three aggressive firms have come to dominate the world tire market, with total market share approaching 60%. And the German tire maker Continental

AG has strengthened its position as fourth in the world, with a dominant presence in Germany and a research budget of 6%. Against this formidable array, the old-line Italian tire company Pirelli SpA is challenged to respond effectively. Although Pirelli still has almost 5% of the market, it is a relatively small player in a tough, competitive business.

And although the business is reliable even in recessions, as motorists still need replacement tires, the competition is getting stronger. The business rewards companies that have large market shares and long production runs. Pirelli, with its small market share and 1,200 specialty tires, has neither. However, Pirelli has some strengths: an outstanding reputation for tire research and excellent high-performance tires, including supplying specially engineered tires for performance automobiles, Ducati motorcycles, and Formula 1 racing teams. In addition, Pirelli's operations managers complement the creative engineering with world-class innovative manufacturing processes that allow rapid changeover to different models and sizes of tires.

Use a SWOT analysis to establish a feasible strategy for Pirelli.

SOLUTION

First, find an opportunity in the world tire market that avoids the threat of the mass-market onslaught by the big-three tire makers. Second, use the internal marketing strength represented by Pirelli's strong brand name supplying Formula 1 racing and a history of winning World Rally Championships. Third, maximize the innovative capabilities of an outstanding operations function. This is a classic differentiation strategy, supported by activity mapping that ties Pirelli's marketing strength to research and its innovative operations function.

To implement this strategy, Pirelli is differentiating itself with a focus on higher-margin performance tires and away from the low-margin standard tire business. Pirelli has established deals with luxury brands Jaguar, BMW, Maserati, Ferrari, Bentley, and Lotus Elise and established itself as a provider of a large share of the tires on new Porsches and S-class Mercedes. Pirelli also made a strategic decision to divest itself of other businesses. As a result, the vast majority of the company's tire production is now high-performance tires. People are willing to pay a premium for Pirellis.

The operations function continued to focus its design efforts on performance tires and developing a system of modular tire manufacture that allows much faster switching between models. This modular system, combined with billions of dollars in new manufacturing investment, has driven batch sizes down to as small as 150 to 200, making small-lot performance tires economically feasible. Manufacturing innovations at Pirelli have streamlined the production process, moving it from a 14-step process to a 3-step process.

Pirelli still faces a threat from the big three going after the performance market, but the company has bypassed its weakness of having a small market share with a substantial research budget and an innovative operations function. The firm now has 19 plants in 13 countries and a presence in more than 160 countries, with sales exceeding $6 billion.[7]

SOLVED PROBLEM 2.2

DeHoratius Electronics, Inc., is evaluating several options for sourcing a critical processor for its new modem. Three sources are being considered: Hi-Tech in Canada, Zia in Hong Kong, and Zaragoza in Spain. The owner, Nicole DeHoratius, has determined that only three criteria are critical. She has rated each firm on a 1–5 scale (with 5 being highest) and has also placed an importance weight on each of the factors, as shown below:

FACTOR (CRITERION)	IMPORTANCE WEIGHT	OUTSOURCE PROVIDERS					
		HI-TECH (CANADA)		ZIA (HONG KONG)		ZARAGOZA (SPAIN)	
		Rating	Wtd. Score	Rating	Wtd. Score	Rating	Wtd. Score
1. Cost	.5	3	1.5	3	1.5	5	2.5
2. Reliability	.2	4	.8	3	.6	3	.6
3. Competence	.3	5	1.5	4	1.2	3	.9
Totals	1.0		3.8		3.3		4.0

SOLUTION

Nicole multiplies each rating by the weight and sums the products in each column to generate a total score for each outsourcing provider. For example the weighted score for Hi-Tech equals $(.5 \times 3) + (.2 \times 4) + (.3 \times 5) = 1.5 + .8 + 1.5 = 3.8$. She selects Zaragoza, which has the highest overall rating.

Problems

Note: **PX** means the problem may be solved with POM for Windows and/or Excel OM.

Problems 2.1–2.3 relate to A Global View of Operations and Supply Chains

•• **2.1** Match the product with the proper parent company and country in the table below:

PRODUCT	PARENT COMPANY	COUNTRY
Arrow Shirts	a. Volkswagen	1. France
Braun Household Appliances	b. Bidermann International	2. Great Britain
Volvo Autos	c. Bridgestone	3. Germany
Firestone Tires	d. Campbell Soup	4. Japan
Godiva Chocolate	e. Haier	5. U.S.
Häagen-Dazs Ice Cream (USA)	f. Tata	6. Switzerland
Jaguar Autos	g. Procter & Gamble	7. China
GE Appliances	h. Michelin	8. India
Lamborghini Autos	i. Nestlé	
Goodrich Tires	j. Geely	
Alpo Pet Foods		

•• **2.2** Based on the corruption perception index developed by Transparency International (**www.transparency.org**), rank the following countries from most corrupt to least: Venezuela, Denmark, the U.S., Switzerland, and China.

•• **2.3** Based on the Ease of Doing Business rankings developed by the World Bank (**www.doingbusiness.org/en/rankings**), rank the following countries from the easiest to do business in to the most difficult: Mexico, Switzerland, the U.S., and China.

Problems 2.4 and 2.5 relate to Achieving Competitive Advantage Through Operations

• **2.4** The text provides three primary strategic approaches (differentiation, cost, and response) for achieving competitive advantage. Provide an example of each not given in the text. Support your choices. (*Hint:* Note the examples provided in the text.)

•• **2.5** Within the food service industry (restaurants that serve meals to customers, but not just fast food), find examples of firms that have sustained competitive advantage by competing on the basis of (1) cost leadership, (2) response, and (3) differentiation. Cite one example in each category; provide a sentence or two in support of each choice. Do not use fast-food chains for all categories. (*Hint:* A "99¢ menu" is very easily copied and is not a good source of sustained advantage.)

Problem 2.6 relates to Issues in Operations Strategy

••• **2.6** Identify how changes within an organization affect the OM strategy for a company. For instance, discuss what impact the following internal factors might have on OM strategy:

a) Maturing of a product.
b) Technology innovation in the manufacturing process.
c) Changes in laptop computer design that builds in wireless technology.

Problem 2.7 relates to Strategy Development and Implementation

••• **2.7** Identify how changes in the external environment affect the OM strategy for a company. For instance, discuss what impact the following external factors might have on OM strategy:

a) Major increases in oil prices.
b) Water- and air-quality legislation.
c) Fewer young prospective employees entering the labor market.
d) Inflation versus stable prices.
e) Legislation moving health insurance from a pretax benefit to taxable income.

Problems 2.8–2.12 relate to Strategic Planning, Core Competencies, and Outsourcing

•• **2.8** Claudia Pragram Technologies, Inc., has narrowed its choice of outsourcing provider to two firms located in different countries. Pragram wants to decide which one of the two countries is the better choice, based on risk-avoidance criteria. She has polled her executives and established four criteria. The resulting ratings for the two countries are presented in the table below, where 1 is a lower risk and 3 is a higher risk.

SELECTION CRITERION	ENGLAND	CANADA
Price of service from outsourcer	2	3
Nearness of facilities to client	3	1
Level of technology	1	3
History of successful outsourcing	1	2

The executives have determined four criteria weightings: Price, with a weight of 0.1; Nearness, with 0.6; Technology, with 0.2; and History, with 0.1.

a) Using the factor-rating method, which country would you select?
b) Double each of the weights used in part (a) (to 0.2, 1.2, 0.4, and 0.2, respectively). What effect does this have on your answer? Why? **PX**

•• **2.9** Ranga Ramasesh is the operations manager for a firm that is trying to decide which one of four countries it should research for possible outsourcing providers. The first step is to select a country based on cultural risk factors, which are critical to eventual business success with the provider. Ranga has reviewed outsourcing provider directories and found that the four countries in the table that follows have an ample number of providers from which they can choose. To aid in the country selection step, he has enlisted the aid of a cultural expert, John Wang, who has provided ratings of the various criteria in the table. The resulting ratings are on a 1 to 10 scale, where 1 is a low risk and 10 is a high risk.

John has also determined six criteria weightings: Trust, with a weight of 0.4; Quality, with 0.2; Religious, with 0.1; Individualism, with 0.1; Time, with 0.1; and Uncertainty, with 0.1. Using the factor-rating method, which country should Ranga select? **PX**

CULTURE SELECTION CRITERION	MEXICO	PANAMA	COSTA RICA	PERU
Trust	1	2	2	1
Society value of quality work	7	10	9	10
Religious attitudes	3	3	3	5
Individualism attitudes	5	2	4	8
Time orientation attitudes	4	6	7	3
Uncertainty avoidance attitudes	3	2	4	2

•• **2.10** Fernando Garza's firm wishes to use factor rating to help select an outsourcing provider of logistics services.

a) With weights from 1–5 (5 highest) and ratings 1–100 (100 highest), use the following table to help Garza make his decision:

		RATING OF LOGISTICS PROVIDERS		
CRITERION	WEIGHT	OVERNIGHT SHIPPING	WORLDWIDE DELIVERY	UNITED FREIGHT
Quality	5	90	80	75
Delivery	3	70	85	70
Cost	2	70	80	95

b) Garza decides to increase the weights for quality, delivery, and cost to 10, 6, and 4, respectively. How does this change your conclusions? Why?
c) If Overnight Shipping's ratings for each of the factors increase by 10%, what are the new results? **PX**

••• **2.11** Walker Accounting Software is marketed to small accounting firms throughout the U.S. and Canada. Owner George Walker has decided to outsource the company's help desk and is considering three providers: Manila Call Center (Philippines), Delhi Services (India), and Moscow Bell (Russia). The following table summarizes the data Walker has assembled. Which outsourcing firm has the best rating? (Higher weights imply higher importance and higher ratings imply more desirable providers.) **PX**

CRITERION	IMPORTANCE WEIGHT	PROVIDER RATINGS		
		MANILA	DELHI	MOSCOW
Flexibility	0.5	5	1	9
Trustworthiness	0.1	5	5	2
Price	0.2	4	3	6
Delivery	0.2	5	6	6

•••• **2.12** Rao Technologies, a California-based high-tech manufacturer, is considering outsourcing some of its electronics production. Four firms have responded to its request for bids, and CEO Mohan Rao has started to perform an analysis on the scores his OM team has entered in the table below.

FACTOR	WEIGHT	RATINGS OF OUTSOURCE PROVIDERS			
		A	B	C	D
Labor	*w*	5	4	3	5
Quality procedures	30	2	3	5	1
Logistics system	5	3	4	3	5
Price	25	5	3	4	4
Trustworthiness	5	3	2	3	5
Technology in place	15	2	5	4	4
Management team	15	5	4	2	1

Weights are on a scale from 1 through 30, and the outsourcing provider scores are on a scale of 1 through 5. The weight for the labor factor is shown as a *w* because Rao's OM team cannot agree on a value for this weight. For what range of values of *w*, if any, is company C a recommended outsourcing provider, according to the factor-rating method?

Problem 2.13 relates to Global Operations Strategy Options

•• **2.13** Does Boeing practice a multinational operations strategy, a global operations strategy, or a transnational operations strategy? Support your choice with specific references to Boeing's operations and the characteristics of each type of organization.

CASE STUDIES

Rapid-Lube

A huge market exists for automobile tune-ups, oil changes, and lubrication service for more than 250 million vehicles on U.S. roads. Some of this demand is filled by full-service auto dealerships, some by Walmart and Firestone, and some by other tire/service dealers. However, Rapid-Lube, Mobil-Lube, Jiffy Lube, and others have also developed strategies to accommodate this opportunity.

Rapid-Lube stations perform oil changes, lubrication, and interior cleaning in a spotless environment. The buildings are clean, usually painted white, and often surrounded by neatly trimmed landscaping. To facilitate fast service, cars can be driven through three abreast. At Rapid-Lube, the customer is greeted by service representatives who are graduates of Rapid-Lube U. The Rapid-Lube school is not unlike McDonald's Hamburger University near Chicago or Holiday Inn's training school in Memphis. The greeter takes the order, which typically includes fluid checks (oil, water, brake fluid, transmission fluid, differential grease) and the necessary lubrication, as well as filter changes for air and oil. Service personnel in neat uniforms then move into action. The standard three-person team has one person checking fluid levels under the hood, another assigned interior vacuuming and window cleaning, and the third in the garage pit, removing the oil filter, draining the oil, checking the differential and transmission, and lubricating as necessary. Precise task assignments and good training are designed to move the car into and out of the bay in 10 minutes. The business model is to charge no more, and hopefully less, than gas stations, automotive repair chains, and auto dealers, while providing better and faster service.

Discussion Questions

1. What constitutes the mission of Rapid-Lube?
2. How does the Rapid-Lube operations strategy provide competitive advantage? (*Hint:* Evaluate how Rapid-Lube's traditional competitors perform the 10 decisions of operations management vs. how Rapid-Lube performs them.)
3. Is it likely that Rapid-Lube has increased productivity over its more traditional competitors? Why? How would we measure productivity in this industry?

Strategy at Nautique Boat Company

Video Case 

Nautique Boat Company, headquartered near the Orlando International Airport in Florida, is the world's premier manufacturer of luxury ski boats, wakeboarding boats, and wake-surfing boats. With a 95-year history, its name became iconic in 1961 with the introduction of the Ski Nautique line of boats. The firm has seen 400% growth in the past 15 years and now produces over 3,000 boats per year. They range in price from the $80,000 Ski Nautique to the nearly $300,000 Paragon G23 Coastal Edition. They're all part of the company's strategy of *differentiation.*

"We are the team creating the standards that others are judged by—the flagship," says president Greg Meloon. "Our customers often have the desire and passion to own a Nautique their whole lives. We maintain a legacy of excellence as a privately-held company and take a long-term approach. So we're going to build a boat that lasts. And it all comes down to details."

Nautique's mission is to: "Cultivate a world-class results-driven team that delivers exceptional products and creates a lifetime of memorable experiences."

To achieve its competitive advantage, Nautique has identified these six operations management key success factors:

1. *Human resources:* With a workforce of 600 employees, Nautique constantly trains and promotes employees for well-defined career paths. Additionally, to maintain low turnover,

the firm uses a 10-hour/4-day workweek, competitive wages, and a supportive family atmosphere. Nautique strives to be "an employer of choice."

2. *Product development:* As a quality boat manufacturer, Nautique starts with a 40-plus person design team, the largest in the industry. It uses state-of-the-art computer-aided design, 3D printing, and a huge engineering staff to create new cutting-edge models every year.
3. *Supply chain:* Since a product is only as good as the parts that go into it, Nautique has close ties with its major suppliers. It even sends its engineers to suppliers to advise them on how to increase *their* quality and efficiencies. Nautique has also joined with other independent boat makers, in the American Boatbuilders Association, allowing it to attain economies of scale in procurement against billion-dollar competitors. This means Nautique sources from the best partners, for the best prices, for on-schedule delivery.
4. *Production:* Nautique's production of premium boats means establishing the best manufacturing practices, through a highly engaged team, exceeding customer expectations. It operates 5 workstations, moving each boat forward to the next station every 38 to 44 minutes depending upon the production schedule.
5. *Quality:* Beginning with visits to vendor facilities, Nautique specifies only the highest-quality raw materials and components with quality inspections at every workstation. As part of the comprehensive quality assurance system, every Nautique boat is taken to Nautique's private lake for rigorous inspection and testing.
6. *Information technology:* Nautique delivers IT support to meet the plant's needs through software such as Rhino 3D modeling, Solidworks CAD, and an EPICOR ERP system. These systems are supplemented with dynamic, timely, and creative information displays throughout the plant.

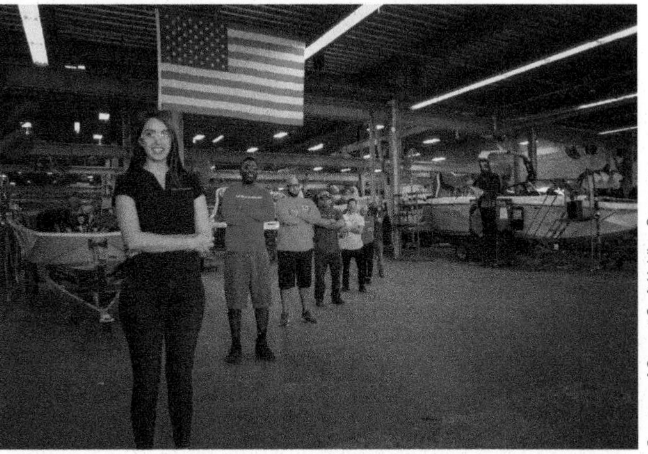
Courtesy of Correct Craft Holding Company

Discussion Questions*

1. Using an activity map similar to that for Southwest Airlines in Figure 2.8, create such a map for Nautique.
2. Identify the strengths, weaknesses, opportunities, and threats that are unique to Nautique's strategy.
3. How has Nautique evolved over the decades?
4. What impact might an industry-wide evolution to electric boats have on Nautique's operations and strategy?

*The Global Company Profile featuring Nautique Boat Company (which opens Chapter 5) provides further background on Nautique's strategy, as does the video that accompanies this case. You may wish to review both prior to answering these questions.

Hard Rock Cafe's Global Strategy

Video Case

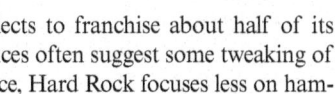

Hard Rock brings the concept of the "experience economy" to its cafe operation. The strategy incorporates a unique "experience" into its operations. This innovation is somewhat akin to mass customization in manufacturing. At Hard Rock, the experience concept is to provide not only a custom meal from the menu but a dining event that includes a unique visual and sound experience not duplicated anywhere else in the world. This strategy is succeeding. Other theme restaurants have come and gone while Hard Rock continues to grow. As Professor C. Markides of the London Business School says, "The trick is not to play the game better than the competition, but to develop and play an altogether different game."* At Hard Rock, the different game is the experience game.

From the opening of its first cafe in London in 1971, during the British rock music explosion, Hard Rock has been serving food and rock music with equal enthusiasm. Hard Rock Cafe has 108 U.S. locations, over a dozen in Europe and the remainder scattered throughout the world, from Bangkok and Beijing to Beirut. New construction, leases, and investment in remodeling are long term; so a global strategy means special consideration of political risk, currency risk, and social norms in a context of a brand fit. Although Hard Rock is one of the most recognized brands in the world, this does not mean its cafe is a natural everywhere. Special consideration must be given to the supply chain for the restaurant and its accompanying retail store. About 48% of a typical cafe's sales are from merchandise.

The Hard Rock Cafe business model is well defined, but because of various risk factors and differences in business practices and employment law, Hard Rock elects to franchise about half of its cafes. Social norms and preferences often suggest some tweaking of menus for local taste. For instance, Hard Rock focuses less on hamburgers and beef and more on fish and lobster in its British cafes.

Because 70% of Hard Rock's guests are tourists, recent years have found it expanding to "destination" cities. While this has been a winning strategy for decades, allowing the firm to grow from one London cafe to over 200 facilities in 74 countries, it has made Hard Rock susceptible to economic fluctuations that hit the tourist business hardest. So Hard Rock is signing long-term leases for locations in cities that are not standard tourist destinations. At the same time, menus are being upgraded. Hopefully, repeat business from locals in these cities will smooth demand and make Hard Rock less dependent on tourists.

Discussion Questions†

1. Identify the strategy changes that have taken place at Hard Rock Cafe since its founding in 1971.
2. As Hard Rock Cafe has changed its strategy, how has its responses to some of the 10 decisions of OM changed?
3. Where does Hard Rock fit in the four international operations strategies outlined in Figure 2.9? Explain your answer.

*Constantinos Markides, "Strategic Innovation," *MIT Sloan Management Review* 38, no. 3: 9.

†You may wish to view the video that accompanies the case before addressing these questions.

Outsourcing Offshore at Darden

Video Case

Darden Restaurants, owner of popular brands such as Olive Garden, Bahama Breeze, and Longhorn Grill, serves more than 320 million meals annually in more than 1,500 restaurants across the U.S. and Canada. To achieve competitive advantage via its supply chain, Darden must achieve excellence at each step. With purchases from 35 countries, and seafood products with a shelf life as short as 4 days, this is a complex and challenging task.

Those 320 million meals annually mean 40 million pounds of shrimp and huge quantities of tilapia, swordfish, and other fresh purchases. Fresh seafood is typically flown to the U.S. and monitored each step of the way to ensure that 34°F is maintained.

Darden's purchasing agents travel the world to find competitive advantage in the supply chain. Darden personnel from supply chain and development, quality assurance, and environmental relations contribute to developing, evaluating, and checking suppliers. Darden also has seven native-speaking representatives living on other continents to provide continuing support and evaluation of suppliers. All suppliers must abide by Darden's food standards, which typically exceed FDA and other industry standards. Darden expects continuous improvement in durable relationships that increase quality and reduce cost.

Darden's aggressiveness and development of a sophisticated supply chain provide an opportunity for outsourcing. Much food preparation is labor intensive and is often more efficient when handled in bulk. This is particularly true where large volumes may justify capital investment. For instance, Tyson and Iowa Beef prepare meats to Darden's specifications much more economically than can individual restaurants. Similarly, Darden has found that it can outsource both the cutting of salmon to the proper portion size and the cracking/peeling of shrimp more cost-effectively offshore than in U.S. distribution centers or individual restaurants.

Discussion Questions*

1. What are some outsourcing opportunities in a restaurant?
2. What supply chain issues are unique to a firm sourcing from 35 countries?
3. Examine how other firms or industries develop international supply chains as compared to Darden.
4. Why does Darden outsource harvesting and preparation of much of its seafood?

*You may wish to view the video that accompanies this case study before answering these questions.

Endnotes

1. The 27 members of the European Union (EU) as of 2021 were Austria, Belgium, Bulgaria, Cyprus, Croatia, Czech Republic, Denmark, Estonia, Finland, France, Germany, Greece, Hungary, Ireland, Italy, Latvia, Lithuania, Luxembourg, Malta, the Netherlands, Poland, Portugal, Romania, Slovakia, Slovenia, Spain, and Sweden. Not all have adopted the euro. In addition, Iceland, Macedonia, Montenegro, and Turkey are candidates for entry into the European Union.
2. M. E. Porter, *Competitive Advantage: Creating and Sustaining Superior Performance*.New York: The Free Press, 1985.
3. M. E. Porter, *Competitive Strategy: Techniques for Analyzing Industries and Competitors*. New York: The Free Press, 1980, 1998.
4. Sources: *The Atlantic* (September 25, 2017); *Entrepreneur* (February 18, 2016); and *USA Today* (November 23, 2017).
5. Sources: *The Wall Street Journal* (June 1, 2017) and (July 13, 2015); and *Businessweek* (March 5, 2018).
6. These seven major criteria are based on a survey of 165 procurement executives, as reported in J. Schildhouse, *Inside Supply Management* (December 2005): 22–29.
7. Sources: Based on *The Economist* (January 8, 2011): 65; **www.pirelli.com**; and **RubberNews.com**.

Bibliography

Beckman, S. L., and D. B. Rosenfield. *Operations Strategy: Competing in the 21st Century*. New York: McGraw-Hill, 2008.

Bravard, J., and R. Morgan. *Smarter Outsourcing*. Upper Saddle River, NJ: Pearson, 2006.

Flynn, B. B., R. G. Schroeder, and E. J. Flynn. "World Class Manufacturing." *Journal of Operations Management* 17, no. 3 (March 1999): 249–269.

Gooderham, P. N., B. Grøgaard, and K. Foss. *Global Strategy and Management: Theory and Practice*. Northampton, MA: Edward Elgar Publishing, Inc., 2019.

Gordon, J. S. *An Empire of Wealth: The Epic History of American Economic Power*. New York: HarperCollins, 2005.

Greenwald, B. C., and J. Kahn. *Globalization: The Irrational Fear That Someone in China Will Take Your Job*. New York: Wiley, 2009.

Kaplan, R. S., and D. P. Norton. *Strategy Maps*. Boston: Harvard Business School Publishing, 2003.

Kathuria, R., M. P. Joshi, and S. Dellande. "International Growth Strategies of Service and Manufacturing Firms." *International Journal of Operations and Production Management* 28, no. 10 (2008): 968.

Lee, H. L., and C.-Y. Lee. *Building Supply Chain Excellence in Emerging Economies*. Secaucus, NJ: Springer, 2007.

Levinson, M. *The Box: How the Shipping Container Made the World Smaller and the World Economy Bigger*, 2nd ed. Princeton, NJ: Princeton University Press, 2016.

Porter, M., and N. Siggelkow. "Contextuality Within Activity Systems and Sustainability of Competitive Advantage." *Academy of Management Perspectives* 22, no. 2 (May 2008): 34–36.

Rudberg, M., and B. M. West. "Global Operations Strategy." *Omega* 36, no. 1 (February 2008): 91.

Skinner, W. "Manufacturing Strategy: The Story of Its Evolution." *Journal of Operations Management* 25, no. 2 (March 2007): 328–334.

Slack, N., and M. Lewis. *Operations Strategy*, 5th ed. Boston: Pearson, 2017.

Wolf, M. *Why Globalization Works*. London: Yale University Press, 2004.

Chapter 2 *Rapid* Review

Main Heading	Review Material	MyLab Operations Management
A GLOBAL VIEW OF OPERATIONS AND SUPPLY CHAINS	Domestic business operations decide to change to some form of international operations for six main reasons: 1. Improve supply chain 2. Reduce costs and exchange rate risks 3. Improve operations 4. Understand markets 5. Improve products 6. Attract and retain global talent ■ **Operational hedging**—Maintaining excess capacity in different countries and shifting production levels among those countries as costs and exchange rates change. ■ **Maquiladoras**—Mexican factories located along the U.S.–Mexico border that receive preferential tariff treatment. ■ **World Trade Organization (WTO)**—An international organization that promotes world trade by lowering barriers to the free flow of goods across borders. ■ **NAFTA and USMCA**—Free trade agreements between Canada, Mexico, and the United States. ■ **European Union (EU)**—A European trade group that has 27 member states.	Concept Questions: 1.1–1.6 Problems: 2.1–2.3
DETERMINING MISSIONS AND STRATEGIES	An effective operations management effort must have a *mission* so it knows where it is going and a *strategy* so it knows how to get there. ■ **Mission**—The purpose or rationale for an organization's existence. ■ **Strategy**—How an organization expects to achieve its missions and goals. The three strategic approaches to competitive advantage are: **1.** Differentiation **2.** Cost leadership **3.** Response	Concept Questions: 2.1–2.4 **VIDEO 2.1** Strategy at Nautique Boat Company
ACHIEVING COMPETITIVE ADVANTAGE THROUGH OPERATIONS	■ **Competitive advantage**—The creation of a unique advantage over competitors. ■ **Differentiation**—Distinguishing the offerings of an organization in a way that the customer perceives as adding value. ■ **Experience differentiation**—Engaging the customer with a product through imaginative use of the five senses, so the customer "experiences" the product. ■ **Low-cost leadership**—Achieving maximum value, as perceived by the customer. ■ **Response**—A set of values related to rapid, flexible, and reliable performance.	Concept Questions: 3.1–3.6 Problems: 2.4–2.5 **VIDEO 2.2** Hard Rock's Global Strategy
ISSUES IN OPERATIONS STRATEGY	■ **Resources view**—A view in which managers evaluate the resources at their disposal and manage or alter them to achieve competitive advantage. ■ **Value-chain analysis**—A way to identify the elements in the product/service chain that uniquely add value. ■ **Five forces model**—A way to analyze the five forces in the competitive environment. Forces in Porter's five forces model are (1) immediate rivals, (2) potential entrants, (3) customers, (4) suppliers, and (5) substitute products. Different issues are emphasized during different stages of the product life cycle: ■ ***Introduction***—Company strategy: Best period to increase market share, R&D engineering is critical. OM strategy: Product design and development critical, frequent product and process design changes, short production runs, high production costs, limited models, attention to quality. ■ ***Growth***—Company strategy: Practical to change price or quality image, strengthen niche. OM strategy: Forecasting critical, product and process reliability, competitive product improvements and options, increase capacity, shift toward product focus, enhance distribution. ■ ***Maturity***—Company strategy: Poor time to change image or price or quality, competitive costs become critical, defend market position. OM strategy: Standardization, less rapid product changes (more minor changes), optimum capacity, increasing stability of process, long production runs, product improvement and cost cutting. ■ ***Decline***—Company strategy: Cost control critical. OM strategy: Little product differentiation, cost minimization, overcapacity in the industry, prune line to eliminate items not returning good margin, reduce capacity.	Concept Questions: 4.1–4.6 Problem: 2.6

Main Heading	Review Material	MyLab Operations Management
STRATEGY DEVELOPMENT AND IMPLEMENTATION	■ **SWOT analysis**—A method of determining internal strengths and weaknesses and external opportunities and threats. ■ **Key success factors (KSFs)**—Activities or factors that are key to achieving competitive advantage. ■ **Core competencies**—A set of unique skills, talents, and activities that a firm does particularly well. A core competence may be a combination of KSFs. ■ **Activity map**—A graphical link of competitive advantage, KSFs, and supporting activities.	Concept Questions: 5.1–5.6 Problem: 2.7 Virtual Office Hours for Solved Problem: 2.1
STRATEGIC PLANNING, CORE COMPETENCIES, AND OUTSOURCING	■ **Outsourcing**—Procuring from external sources services or products that are normally part of an organization. ■ **Theory of comparative advantage**—The theory which states that countries benefit from specializing in (and exporting) products and services in which they have relative advantage and importing goods in which they have a relative disadvantage. Perhaps half of all outsourcing agreements fail because of inappropriate planning and analysis. Potential risks of outsourcing include: ■ A drop in quality or customer service ■ Political backlash that results from outsourcing to foreign countries ■ Negative impact on employees ■ Potential future competition ■ Increased logistics and inventory costs The most common reason given for outsourcing failure is that the decision was made without sufficient understanding and analysis. The factor-rating method is an excellent tool for dealing with both country risk assessment and provider selection problems.	Concept Questions: 6.1–6.6 Problems: 2.8–2.12 Virtual Office Hours for Solved Problem: 2.2 **VIDEO 2.3** Outsourcing Offshore at Darden
GLOBAL OPERATIONS STRATEGY OPTIONS	■ **International business**—A firm that engages in cross-border transactions. ■ **Multinational corporation (MNC)**—A firm that has extensive involvement in international business, owning or controlling facilities in more than one country. The four operations strategies for approaching global opportunities can be classified according to local responsiveness and cost reduction: ■ **International strategy**—A strategy in which global markets are penetrated using exports and licenses with little local responsiveness. ■ **Multidomestic strategy**—A strategy in which operating decisions are decentralized to each country to enhance local responsiveness. ■ **Global strategy**—A strategy in which operating decisions are centralized and headquarters coordinates the standardization and learning between facilities. ■ **Transnational strategy**—A strategy that combines the benefits of global-scale efficiencies with the benefits of local responsiveness. These firms transgress national boundaries.	Concept Questions: 7.1–7.6 Problem 2.13
ADDITIONAL MYLAB OPERATIONS MANAGEMENT RESOURCES	✔ Video for Creating Your Own Excel Spreadsheets (Example 1) ✔ Multiple Choice Case Questions (Rapid-Lube)	

Self Test

LO 2.1 A mission statement is beneficial to an organization because it:
- **a)** is a statement of the organization's purpose.
- **b)** provides a basis for the organization's culture.
- **c)** identifies important constituencies.
- **d)** details specific income goals.
- **e)** ensures profitability.

LO 2.2 The three strategic approaches to competitive advantage are _____, _____, and _____.

LO 2.3 Core competencies are those strengths in a firm that include:
- **a)** specialized skills.
- **b)** unique production methods.
- **c)** proprietary information/knowledge.
- **d)** things a company does better than others.
- **e)** all of the above.

LO 2.4 Evaluating outsourcing providers by comparing their weighted average scores involves:
- **a)** factor-rating analysis.
- **b)** cost-volume analysis.
- **c)** transportation model analysis.
- **d)** linear regression analysis.
- **e)** crossover analysis.

LO 2.5 A company that is organized across international boundaries, with decentralized authority and substantial autonomy at each business via subsidiaries, franchises, or joint ventures, has:
- **a)** a global strategy.
- **b)** a transnational strategy.
- **c)** an international strategy.
- **d)** a multidomestic strategy.

Answers: LO 2.1. a; LO 2.2. differentiation, cost leadership, response; LO 2.3. e; LO 2.4. a; LO 2.5. d.

Project Management

CHAPTER 3

CHAPTER OUTLINE

GLOBAL COMPANY PROFILE: *Bechtel Group*

OASIS OF THE SEAS/SPLASH NEWS/ALAMY STOCK PHOTO

Celebrity Cruises' *Oasis of the Seas*, one of the world's largest and most expensive cruise liners, is shown here under construction in Turku, Finland. The 6,000-passenger ship includes a beach pool, shopping mall, and spa. Managing this project uses the same project management techniques as managing the remodeling of a store, installing a new production line, or implementing a new computer system.

CHAPTER 3

GLOBAL COMPANY PROFILE
Bechtel Group

Project Management Provides a Competitive Advantage for Bechtel

Over a century old, the San Francisco–based Bechtel Group (**www.bechtel.com**) is the world's premier manager of massive construction and engineering projects. Known for billion-dollar projects, Bechtel is famous for its construction feats on the Hoover Dam and over 25,000 other projects in 160 countries. With 55,000 employees and revenues over $22 billion, Bechtel is the world's largest project manager.

Courtesy of Bay Area Rapid Transit District (BART)

San Francisco's Bay Area Rapid Transit (BART) system, with 400,000 riders a day, was another engineering challenge for Bechtel. BART included a 3.6-mile underwater passage, constructed in 57 sections that lie on the bay floor as deep as 135 feet beneath the surface.

Steve Hebert/Bechtel National, Inc./PRN/Newscom

In addition to major construction projects, Bechtel used its project management skills to provide emergency response to major catastrophes as it did here in the wake of Hurricane Katrina.

- Bechtel's first job outside the U.S. was building the Mene Grande pipeline in Venezuela in 1940. In 1947, Bechtel began construction on what was then the world's longest oil pipeline, the Trans-Arabian Pipeline, which began in Saudi Arabia, ran across Jordan and Syria, and ended in Lebanon.
- Major projects in the 1950s included Canada's Trans Mountain Oil Pipeline and the Bay Area Rapid Transit (BART) system.
- During the 1960s and 1970s, Bechtel was involved in constructing 40% of the nuclear plants in the United States. But by the end of the 1970s, the company had moved from nuclear power construction to nuclear cleanup projects, including Three Mile Island in 1979.
- In 1976, the company began work on the industrial city of Jubail in Saudi Arabia, transforming the area from a small village to a city with more than a quarter of a million people.
- In the 1980s, Bechtel handled the project management of the 1984 Los Angeles Summer Olympics. In 1987, Bechtel was awarded a contract for managing construction of the undersea tunnel linking the UK and France called the "Chunnel."

Courtesy of Bechtel Corporation

Managing massive construction projects such as this is the strength of Bechtel. With large penalties for late completion and incentives for early completion, good project managers are worth their weight in gold.

- In 1991, Bechtel broke ground on Boston's Central Artery/Tunnel Project or "Big Dig," the largest and most complex urban transportation project ever undertaken in the U.S.
- As a result of the Gulf war, Bechtel took on the task of extinguishing oil well fires in Kuwait in 1991—part of the overall effort to rebuild that country's infrastructure.
- Following Hurricane Katrina in 2005, Bechtel was hired by FEMA to build temporary housing. It delivered 35,000 trailers in less than a year for displaced residents.
- More recently, the firm completed one of the largest and most complex rail construction projects in the U.S., Washington D.C.'s Silver Line metro. This included 6 miles of elevated track and a tunnel beneath one of the busiest retail areas in the country.

When companies or countries seek out firms to manage massive projects, they go to Bechtel, which, again and again, through outstanding project management, has demonstrated its competitive advantage.

Courtesy of Get Link

The Channel Tunnel, or commonly called the Chunnel, runs some 32 miles (51 kilometers) from a terminal near Folkestone on the English side to Calais in France. Bechtel helped manage the project to its successful completion, providing management, technical, and construction expertise.

Joe Cavaretta/AP Images

Bechtel was the construction contractor for the Hoover Dam. This dam, on the Colorado River, is the highest in the Western Hemisphere.

LEARNING OBJECTIVES

LO 3.1 ***Use*** a Gantt chart for scheduling 65
LO 3.2 ***Draw*** AOA and AON networks 69
LO 3.3 ***Complete*** forward and backward passes for a project 72
LO 3.4 ***Determine*** a critical path 76
LO 3.5 ***Calculate*** the variance of activity times 79
LO 3.6 ***Crash*** a project 83

STUDENT TIP
Wherever your career takes you, one of the most useful tools you can have, as a manager, is the ability to manage a project.

The Importance of Project Management

When Bechtel, the subject of the opening *Global Company Profile*, begins a project, it quickly has to mobilize substantial resources, often consisting of manual workers, construction professionals, cooks, medical personnel, and even security forces. Its project management team develops a supply chain to access materials to build everything from ports to bridges, dams, and monorails. Bechtel is just one example of a firm that faces modern phenomena: growing project complexity and collapsing product/service life cycles. This change stems from awareness of the strategic value of time-based competition and a quality mandate for continuous improvement. Each new product/service introduction is a unique event—a project. In addition, projects are a common part of our everyday life. We may be planning a wedding or a surprise birthday party, remodeling a house, or preparing a semester-long class project.

VIDEO 3.1
Project Management at Hard Rock's Rockfest

Scheduling projects can be a difficult challenge for operations managers. The stakes in project management are high. Cost overruns and unnecessary delays occur due to poor scheduling and poor controls.

Projects that take months or years to complete are usually developed outside the normal production system. Project organizations within the firm may be set up to handle such jobs and are often disbanded when the project is complete. On other occasions, managers find projects just a part of their job. The management of projects involves three phases (see Figure 3.1):

1. *Planning:* This phase includes goal setting, defining the project, and team organization.
2. *Scheduling:* This phase relates people, money, and supplies to specific activities and relates activities to each other.
3. *Controlling:* Here the firm monitors resources, costs, quality, and budgets. It also revises or changes plans and shifts resources to meet time and cost demands.

We begin this chapter with a brief overview of these functions. Three popular techniques to allow managers to plan, schedule, and control—Gantt charts, PERT, and CPM—are also described.

Project Planning

Projects can be defined as a series of related tasks directed toward a major output. In some firms a **project organization** is developed to make sure existing programs continue to run smoothly on a day-to-day basis while new projects are successfully completed.

Project organization
An organization formed to ensure that programs (projects) receive the proper management and attention.

For companies with multiple large projects, such as a construction firm, a project organization is an effective way of assigning the people and physical resources needed. It is a temporary organization structure designed to achieve results by using specialists from throughout the firm.

The project organization may be most helpful when:

1. Work tasks can be defined with a specific goal and deadline.
2. The job is unique or somewhat unfamiliar to the existing organization.
3. The work contains complex interrelated tasks requiring specialized skills.

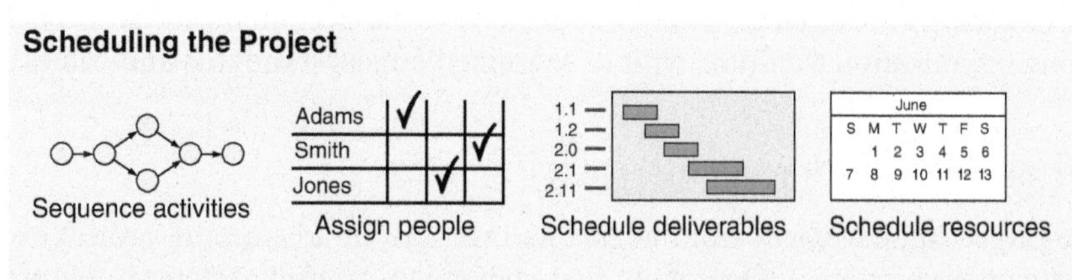

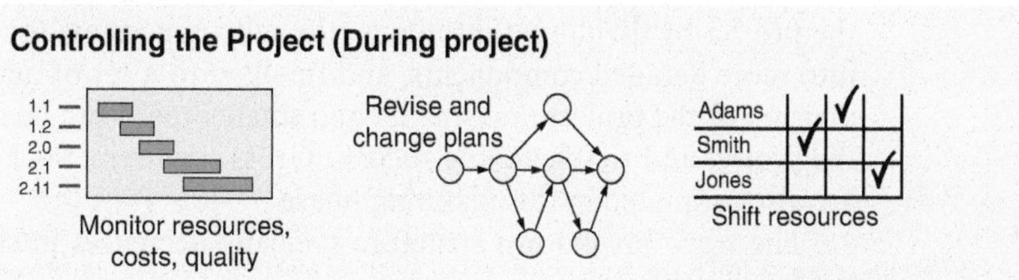

Controlling the Project (During project)

1.1 — 1.2 — 2.0 — 2.1 — 2.11 —

Monitor resources, costs, quality

Revise and change plans

Adams
Smith
Jones

Shift resources

Figure 3.1

Project Planning, Scheduling, and Controlling

STUDENT TIP

Managers must "make the plan and then work the plan."

4. The project is temporary but critical to the organization.
5. The project cuts across organizational lines.

The Project Manager

An example of a project organization is shown in Figure 3.2. Project team members are temporarily assigned to a project and report to the project manager. The manager heading the project coordinates activities with other departments and reports directly to top management. Project managers receive high visibility in a firm and are responsible for making sure that (1) all necessary activities are finished in proper sequence and on time; (2) the project comes in within budget; (3) the project meets its quality goals; and (4) the people assigned to the project receive the motivation, direction, and information needed to do their jobs. This means that project managers should be good coaches and communicators, and be able to organize activities from a variety of disciplines.

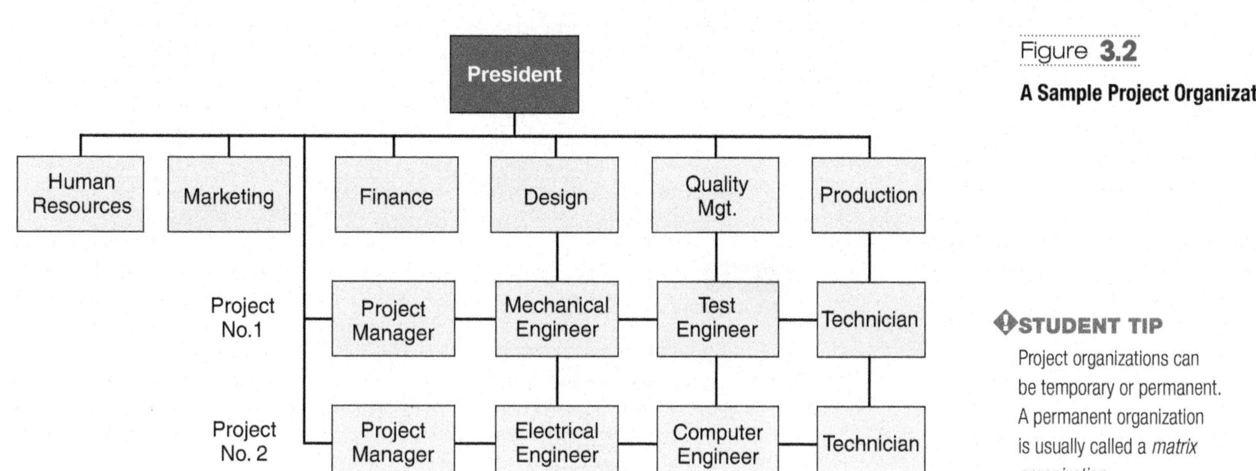

Figure 3.2

A Sample Project Organization

STUDENT TIP

Project organizations can be temporary or permanent. A permanent organization is usually called a *matrix organization*.

Ethical Issues Faced in Project Management Project managers not only have high visibility but they also face ethical decisions on a daily basis. How they act establishes the code of conduct for the project. Project managers often deal with (1) offers of gifts from contractors, (2) pressure to alter status reports to mask the reality of delays, (3) false reports for charges of time and expenses, and (4) pressures to compromise quality to meet bonuses or avoid penalties related to schedules.

Using the Project Management Institute's **(www.pmi.org)** ethical codes is one means of trying to establish standards. These codes need to be accompanied by good leadership and a strong organizational culture, with its ingrained ethical standards and values.

Work Breakdown Structure

Work breakdown structure (WBS)
A hierarchical description of a project into more and more detailed components.

The project management team begins its task well in advance of project execution so that a plan can be developed. One of its first steps is to carefully establish the project's objectives, then break the project down into manageable parts. This **work breakdown structure (WBS)** defines the project by dividing it into its major subcomponents (or tasks), which are then subdivided into more detailed components, and finally into a set of activities and their related costs. The division of the project into smaller and smaller tasks can be difficult, but is critical to managing the project and to scheduling success. Gross requirements for people, supplies, and equipment are also estimated in this planning phase.

The work breakdown structure typically decreases in size from top to bottom and is indented like this:

Level
1. Project
2. Major tasks in the project
3. Subtasks in major tasks
4. Activities (or "work packages") to be completed

This hierarchical framework can be illustrated with the development of Microsoft's operating system Windows 11. As we see in Figure 3.3, the project, creating a new operating system, is labeled 1.0. The first step is to identify the major tasks in the project (level 2). Three examples would be software design (1.1), cost management plan (1.2), and system testing (1.3). Two major subtasks for 1.1 are development of graphical user interfaces (GUIs) (1.1.1) and creating compatibility with previous versions of Windows (1.1.2). The major subtasks for 1.1.2 are level-4 activities, such as creating a team to handle compatibility with Windows 10 (1.1.2.1),

Figure 3.3
Work Breakdown Structure

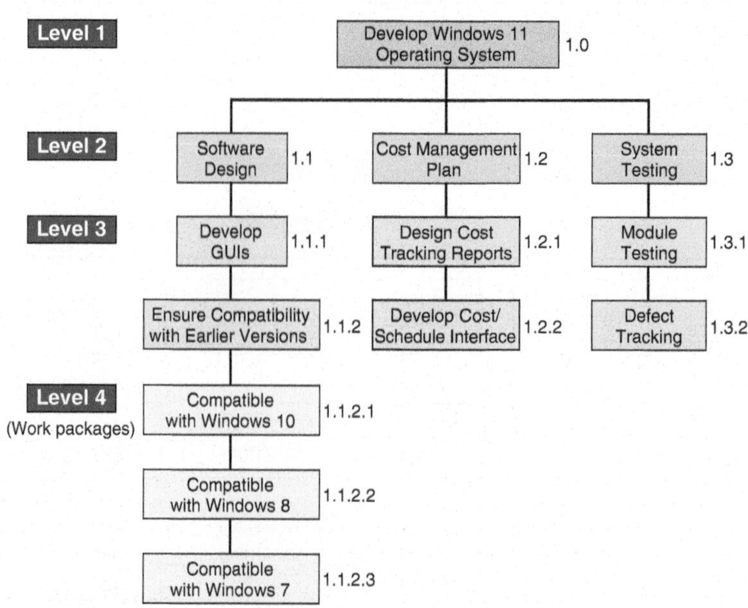

creating a team for Windows 8 (1.1.2.2), and creating a team for Windows 7 (1.1.2.3). There are usually many level-4 activities.

Project Scheduling

Project scheduling involves sequencing and allotting time to all project activities. At this stage, managers decide how long each activity will take and compute the resources needed at each stage of production. Managers may also chart separate schedules for personnel needs by type of skill (management, engineering, or pouring concrete, for example) and material needs.

One popular project scheduling approach is the Gantt chart. **Gantt charts** are low-cost means of helping managers make sure that (1) activities are planned, (2) order of performance is documented, (3) activity time estimates are recorded, and (4) overall project time is developed. As Figure 3.4 shows, Gantt charts are easy to understand. Horizontal bars are drawn for each project activity along a time line. This illustration of a routine servicing of a Delta jetliner during a 40-minute layover shows that Gantt charts also can be used for scheduling repetitive operations. In this case, the chart helps point out potential delays. The *OM in Action* box on Delta provides additional insights.

Gantt charts
Planning charts used to schedule resources and allocate time.

On simple projects, scheduling charts such as these permit managers to observe the progress of each activity and to spot and tackle problem areas. Gantt charts, though, do not adequately illustrate the interrelationships between the activities and the resources.

STUDENT TIP
Gantt charts are simple and visual, making them widely used.

PERT and CPM, the two widely used network techniques that we shall discuss shortly, *do* have the ability to consider precedence relationships and interdependency of activities. On complex projects, the scheduling of which is almost always computerized, PERT and CPM thus have an edge over the simpler Gantt charts. Even on huge projects, though, Gantt charts can be used as summaries of project status and may complement the other network approaches.

LO 3.1 *Use* a Gantt chart for scheduling

To summarize, whatever the approach taken by a project manager, project scheduling serves several purposes:

1. It shows the relationship of each activity to others and to the whole project.
2. It identifies the precedence relationships among activities.
3. It encourages the setting of realistic time and cost estimates for each activity.
4. It helps make better use of people, money, and material resources by identifying critical bottlenecks in the project.

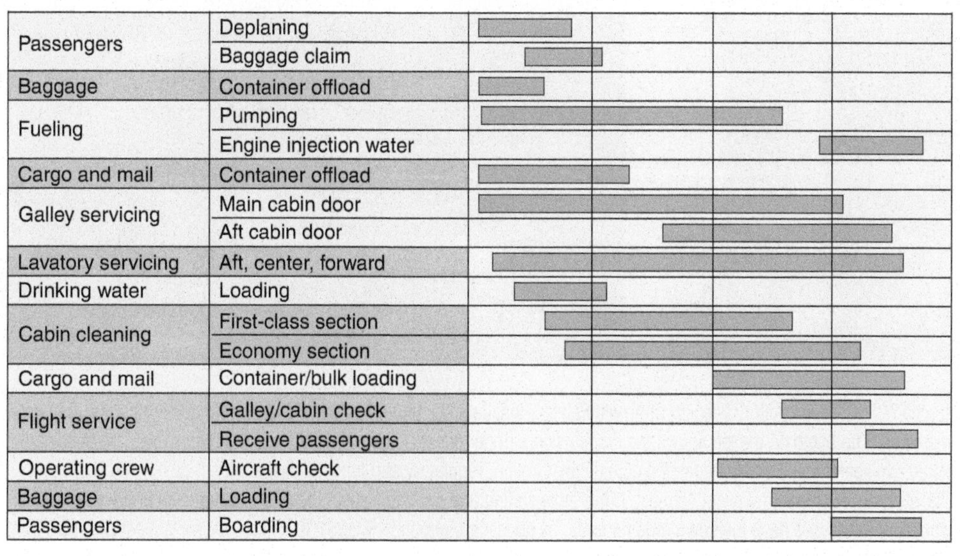

Figure 3.4

Gantt Chart of Service Activities for a Delta Jet During a 40-Minute Layover

Delta saves $50 million a year with this turnaround time, which is a reduction from its traditional 60-minute routine.

OM in Action Delta's Ground Crew Orchestrates a Smooth Takeoff[1]

Flight 574's engines screech its arrival as the jet lumbers down Richmond's taxiway with 140 passengers arriving from Atlanta. In 40 minutes, the plane is to be airborne again.

However, before this jet can depart, there is business to attend to: passengers, luggage, and cargo to unload and load; thousands of gallons of jet fuel and countless drinks to restock; cabin and restrooms to clean; toilet holding tanks to drain; and engines, wings, and landing gear to inspect.

The 10-person ground crew knows that a miscue anywhere—a broken cargo loader, lost baggage, misdirected passengers—can mean a late departure and trigger a chain reaction of headaches from Richmond to Atlanta to every destination of a connecting flight.

Carla Sutera, the operations manager for Delta's Richmond International Airport, views the turnaround operation like a pit boss awaiting a race car. Trained crews are in place for Flight 574 with baggage carts and tractors, hydraulic cargo loaders, a truck to load food and drinks, another to lift the cleanup crew, another to put fuel on, and a fourth to take water off. The "pit crew" usually performs so smoothly that most passengers never suspect the proportions of the effort. Gantt charts, such as the one in Figure 3.4, aid Delta and other airlines with the staffing and scheduling that are needed for this task.

Jeff Topping/Getty Images News/Getty Images

Project Controlling

VIDEO 3.2
Project Management at Arnold Palmer Hospital

The control of projects, like the control of any management system, involves close monitoring of resources, costs, quality, and budgets. Control also means using a feedback loop to revise the project plan and having the ability to shift resources to where they are needed most. Computerized PERT/CPM reports and charts are widely available today from scores of competing software firms. Some of the more popular of these programs are Oracle Primavera (by Oracle), MindView (by Match Ware), HP Project (by Hewlett-Packard), Fast Track (by AEC Software), Project Portfolio Management (by SAP), and Microsoft Project (by Microsoft), which we illustrate in this chapter.

STUDENT TIP
To use project management software, you first need to understand the next two sections in this chapter.

These programs produce a broad variety of reports, including (1) detailed cost breakdowns, (2) labor requirements, (3) cost and hour summaries, (4) raw material and expenditure forecasts, (5) variance reports, (6) time analysis reports, and (7) work status reports.

Courtesy Arnold Palmer Medical Center

Courtesy Arnold Palmer Medical Center

Construction of the new 11-story building at Arnold Palmer Hospital in Orlando, Florida, was an enormous project for the hospital administration. The photo on the left shows the first six floors under construction. The photo on the right shows the building as completed two years later. Prior to beginning actual construction, regulatory and funding issues added, as they do with most projects, substantial time to the overall project. Cities have zoning and parking issues; the EPA has drainage and waste issues; and regulatory authorities have their own requirements, as do issuers of bonds. The $100 million, 4-year project at Arnold Palmer Hospital is discussed in the Video Case Study at the end of this chapter.

Controlling projects can be difficult. The stakes are high; cost overruns and unnecessary delays can occur due to poor planning, scheduling, and controls. Some projects are "well-defined," whereas others may be "ill-defined." Projects typically only become well-defined after detailed extensive initial planning and careful definition of required inputs, resources, processes, and outputs. Well-established projects where constraints are known (e.g., buildings and roads) and engineered products (e.g., airplanes and cars) with well-defined specifications and drawings may fall into this category. Well-defined projects are assumed to have changes small enough to be managed without substantially revising plans. This is called the waterfall approach, where the project progresses smoothly, in a step-by-step manner, through each phase to completion.

Waterfall projects
Projects that progress smoothly in a step-by-step manner until completed.

But many projects, such as software development (e.g., 3-D games) and new technology (e.g., landing the Mars land rover) are ill-defined. These projects require what is known as an agile style of management with collaboration and constant feedback to adjust to the many unknowns of the evolving technology and project specifications. In agile projects, the project manager creates a general plan, based on overall requirements and a broad perspective of the solution. From that moment, the end project is tackled iteratively and incrementally, with each increment building on the steps preceding it. This approach emphasizes small activities (or, in the case of software projects, small chunks of code) to meet limited objectives. These activities are then delivered to end users quickly so problems can be identified and corrected. There are numerous checkpoints and feedback loops to track progress.

Agile projects
Ill-defined projects requiring collaboration and constant feedback to adjust to project unknowns.

Most projects fall somewhere between waterfall and agile.

Project Management Techniques: PERT and CPM

Program evaluation and review technique (PERT) and the critical path method (CPM) were both developed in the 1950s to help managers schedule, monitor, and control large and complex projects. CPM arrived first, as a tool developed to assist in the building and maintenance of chemical plants at DuPont. Independently, PERT was developed in 1958 for the U.S. Navy.

Program evaluation and review technique (PERT)
A project management technique that employs three time estimates for each activity.

Critical path method (CPM)
A project management technique that uses only one time factor per activity.

The Framework of PERT and CPM

PERT and CPM both follow six basic steps:

1. Define the project and prepare the work breakdown structure.
2. Develop the relationships among the activities. Decide which activities must precede and which must follow others.
3. Draw the network connecting all the activities.
4. Assign time and/or cost estimates to each activity.
5. Compute the *longest* time path through the network. This is called the critical path.
6. Use the network to help plan, schedule, monitor, and control the project.

Critical path
The computed *longest* time path(s) through a network.

Step 5, finding the critical path, is a major part of controlling a project. The activities on the critical path represent tasks that will delay the entire project if they are not completed on time. Managers can gain the flexibility needed to complete critical tasks by identifying noncritical activities and replanning, rescheduling, and reallocating labor and financial resources.

Although PERT and CPM differ to some extent in terminology and in the construction of the network, their objectives are the same. Furthermore, the analysis used in both techniques is very similar. The major difference is that PERT employs three time estimates for each activity. These time estimates are used to compute expected values and standard deviations for the activity. CPM makes the assumption that activity times are known with certainty and hence requires only one time factor for each activity.

For purposes of illustration, the rest of this section concentrates on a discussion of PERT. Most of the comments and procedures described, however, apply just as well to CPM.

PERT and CPM are important because they can help answer questions such as the following about projects with thousands of activities:

1. When will the entire project be completed?
2. What are the critical activities or tasks in the project—that is, which activities will delay the entire project if they are late?
3. Which are the noncritical activities—the ones that can run late without delaying the whole project's completion?
4. What is the probability that the project will be completed by a specific date?
5. At any particular date, is the project on schedule, behind schedule, or ahead of schedule?
6. On any given date, is the money spent equal to, less than, or greater than the budgeted amount?
7. Are there enough resources available to finish the project on time?
8. If the project is to be finished in a shorter amount of time, what is the best way to accomplish this goal at the least cost?

Network Diagrams and Approaches

Activity-on-node (AON)
A network diagram in which nodes designate activities.

Activity-on-arrow (AOA)
A network diagram in which arrows designate activities.

The first step in a PERT or CPM network is to divide the entire project into significant activities in accordance with the work breakdown structure. There are two approaches for drawing a project network: **activity-on-node (AON)** and **activity-on-arrow (AOA)**. Under the AON convention, *nodes* designate activities. Under AOA, *arrows* represent activities. Activities consume time and resources. The basic difference between AON and AOA is that the nodes in an AON diagram represent activities. In an AOA network, the nodes represent the starting and finishing times of an activity and are also called *events*. So nodes in AOA consume neither time nor resources.

Although both AON and AOA are popular in practice, many of the project management software packages, including Microsoft Project, use AON networks. For this reason, although we illustrate both types of networks in the next examples, we focus on AON networks in subsequent discussions in this chapter.

Example 1

PREDECESSOR RELATIONSHIPS FOR POLLUTION CONTROL AT MILWAUKEE PAPER

Milwaukee Paper Manufacturing had long delayed the expense of installing advanced computerized air pollution control equipment in its facility. But when the board of directors adopted a new proactive policy on sustainability, it did not just *authorize* the budget for the state-of-the-art equipment. It directed the plant manager, Julie Ann Williams, to complete the installation in time for a major announcement of the policy, on Earth Day, exactly 16 weeks away! Under strict deadline from her bosses, Williams needs to be sure that installation of the filtering system progresses smoothly and on time.

Given the following information, develop a table showing activity precedence relationships.

APPROACH ▶ Milwaukee Paper has identified the eight activities that need to be performed for the project to be completed. When the project begins, two activities can be simultaneously started: building the internal components for the device (activity A) and the modifications necessary for the floor and roof (activity B). The construction of the collection stack (activity C) can begin when the internal components are completed. Pouring the concrete floor and installation of the frame (activity D) can be started as soon as the internal components are completed and the roof and floor have been modified.

After the collection stack has been constructed, two activities can begin: building the high-temperature burner (activity E) and installing the pollution control system (activity F). The air pollution device can be installed (activity G) after the concrete floor has been poured, the frame has been installed, and the high-temperature burner has been built. Finally, after the control system and pollution device have been installed, the system can be inspected and tested (activity H).

SOLUTION ▶ Activities and precedence relationships may seem rather confusing when they are presented in this descriptive form. It is therefore convenient to list all the activity information in a table, as shown in Table 3.1. We see in the table that activity A is listed as an *immediate predecessor* of activity C. Likewise, both activities D and E must be performed prior to starting activity G.

TABLE 3.1 Milwaukee Paper Manufacturing's Activities and Predecessors

ACTIVITY	DESCRIPTION	IMMEDIATE PREDECESSORS
A	Build internal components	—
B	Modify roof and floor	—
C	Construct collection stack	A
D	Pour concrete and install frame	A, B
E	Build high-temperature burner	C
F	Install pollution control system	C
G	Install air pollution device	D, E
H	Inspect and test	F, G

INSIGHT ▶ To complete a network, all predecessors must be clearly defined.

LEARNING EXERCISE ▶ What is the impact on this sequence of activities if Environmental Protection Agency (EPA) approval is required after *Inspect and Test?* [Answer: The immediate predecessor for the new activity would be H, *Inspect and Test*, with *EPA approval* as the last activity.]

Activity-on-Node Example

Note that in Example 1, we only list the *immediate* predecessors for each activity. For instance, in Table 3.1, because activity A precedes activity C, and activity C precedes activity E, the fact that activity A precedes activity E is *implicit*. This relationship need not be explicitly shown in the activity precedence relationships.

LO 3.2 *Draw* AOA and AON networks

When there are many activities in a project with fairly complicated precedence relationships, it is difficult for an individual to comprehend the complexity of the project from just the tabular information. In such cases, a visual representation of the project, using a *project network*, is convenient and useful. A project network is a diagram of all the activities and the precedence relationships that exist between these activities in a project. Example 2 illustrates how to construct an AON project network for Milwaukee Paper Manufacturing.

It is convenient to have the project network start and finish with a unique node. In the Milwaukee Paper example, it turns out that a unique activity, H, is the last activity in the project. We therefore automatically have a unique ending node.

In situations in which a project has multiple ending activities, we include a "dummy" ending activity. We illustrate this type of situation in Solved Problem 3.1 at the end of this chapter.

Example 2

AON GRAPH FOR MILWAUKEE PAPER

Draw the AON network for Milwaukee Paper, using the data in Example 1.

APPROACH ▶ In the AON approach, we denote each activity by a node. The lines, or arrows, represent the precedence relationships between the activities.

Dummy activity
An activity having no time that is inserted into a network to maintain the logic of the network.

SOLUTION ▶ In this example, there are two activities (A and B) that do not have any predecessors. We draw separate nodes for each of these activities, as shown in Figure 3.5. Although not required, it is usually convenient to have a unique starting activity for a project. We have therefore included a *dummy activity* called *Start* in Figure 3.5. This **dummy activity** does not really exist and takes up zero time and resources. Activity *Start* is an immediate predecessor for both activities A and B, and it serves as the unique starting activity for the entire project.

Figure 3.5
Beginning AON Network for Milwaukee Paper

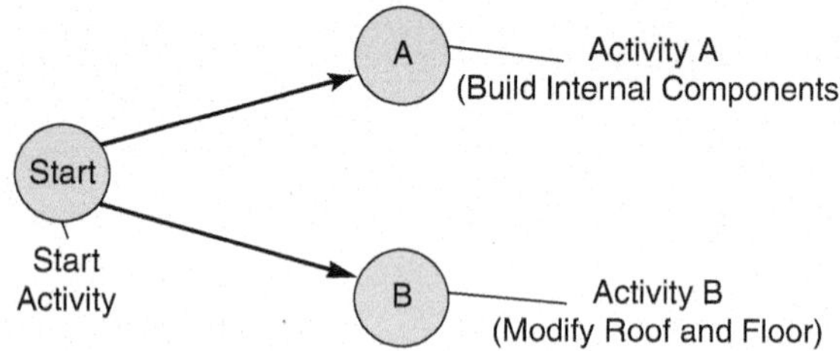

We now show the precedence relationships using lines with arrow symbols. For example, an arrow from activity Start to activity A indicates that Start is a predecessor for activity A. In a similar fashion, we draw an arrow from Start to B.

Next, we add a new node for activity C. Because activity A precedes activity C, we draw an arrow from node A to node C (see Figure 3.6). Likewise, we first draw a node to represent activity D. Then, because activities A and B both precede activity D, we draw arrows from A to D and from B to D (see Figure 3.6).

Figure 3.6
Intermediate AON Network for Milwaukee Paper

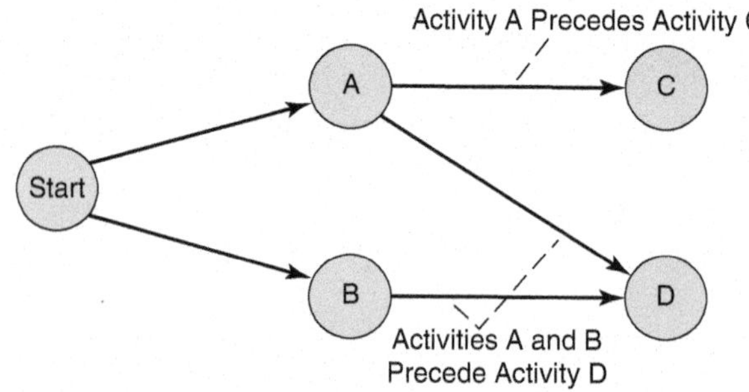

We proceed in this fashion, adding a separate node for each activity and a separate line for each precedence relationship that exists. The complete AON project network for the Milwaukee Paper Manufacturing project is shown in Figure 3.7.

INSIGHT ▶ Drawing a project network properly takes some time and experience. We would like the lines to be straight and arrows to move to the right when possible.

Figure 3.7
Complete AON Network for Milwaukee Paper

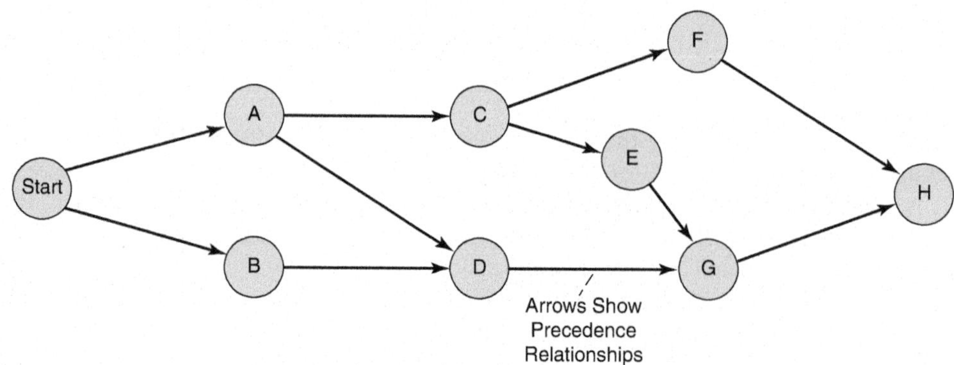

LEARNING EXERCISE ▶ If *EPA Approval* occurs after *Inspect and Test*, what is the impact on the graph? [Answer: A straight line is extended to the right beyond H (with a node I added) to reflect the additional activity.]

RELATED PROBLEMS ▶ 3.4a, 3.5, 3.8, 3.9, 3.10, 3.12, 3.13, 3.14

Activity-on-Arrow Example

In an AOA project network we can represent activities by arrows. A node represents an *event*, which marks the start or completion time of an activity. We usually identify an event (node) by a number.

Example 3

ACTIVITY-ON-ARROW FOR MILWAUKEE PAPER

Draw the complete AOA project network for Milwaukee Paper's problem.

APPROACH ▶ Using the data from Table 3.1 in Example 1, draw one activity at a time, starting with A.

SOLUTION ▶ We see that activity A starts at event 1 and ends at event 2. Likewise, activity B starts at event 1 and ends at event 3. Activity C, whose only immediate predecessor is activity A, starts at node 2 and ends at node 4. Activity D, however, has two predecessors (i.e., A and B). Hence, we need both activities A and B to end at event 3, so that activity D can start at that event. However, we cannot have multiple activities with common starting and ending nodes in an AOA network. To overcome this difficulty, in such cases, we may need to add a dummy line (activity) to enforce the precedence relationship. The dummy activity, shown in Figure 3.8 as a dashed line, is inserted between events 2 and 3 to make the diagram reflect the precedence between A and D. The remainder of the AOA project network for Milwaukee Paper's example is also shown.

Figure **3.8**

Complete AOA Network (with Dummy Activity) for Milwaukee Paper

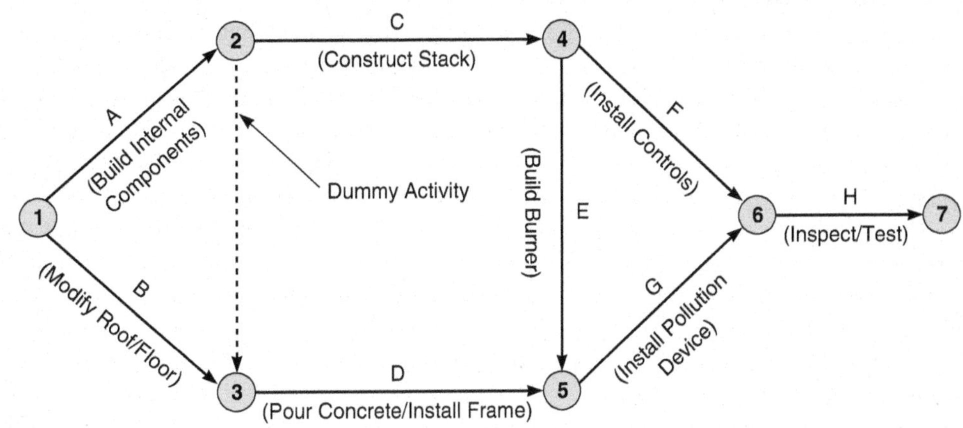

STUDENT TIP

The dummy activity consumes no time, but note how it changes precedence. Now activity D cannot begin until *both* B and the dummy are complete.

INSIGHT ▶ Dummy activities are common in AOA networks. They do not really exist in the project and take zero time.

LEARNING EXERCISE ▶ A new activity, *EPA Approval*, follows activity H. Add it to Figure 3.8. [Answer: Insert an arrowed line from node 7, which ends at a new node 8, and is labeled I (EPA Approval).]

RELATED PROBLEMS ▶ 3.4b, 3.6, 3.7, 3.11a

Determining the Project Schedule

Look back at Figure 3.7 (in Example 2) for a moment to see Milwaukee Paper's completed AON project network. Once this project network has been drawn to show all the activities and their precedence relationships, the next step is to determine the project schedule. That is, we need to identify the planned starting and ending time for each activity.

Let us assume Milwaukee Paper estimates the time required for each activity, in weeks, as shown in Table 3.2. The table indicates that the total time for all eight of the company's activities is 25 weeks. However, because several activities can take place simultaneously, it is clear that the total project completion time may be less than 25 weeks. To find out just how long the project will take, we perform the **critical path analysis** for the network.

Critical path analysis
A process that helps determine a project schedule.

TABLE 3.2 Time Estimates for Milwaukee Paper Manufacturing

ACTIVITY	DESCRIPTION	TIME (WEEKS)
A	Build internal components	2
B	Modify roof and floor	3
C	Construct collection stack	2
D	Pour concrete and install frame	4
E	Build high-temperature burner	4
F	Install pollution control system	3
G	Install air pollution device	5
H	Inspect and test	2
	Total time (weeks)	25

STUDENT TIP Does this mean the project will take 25 weeks to complete? No. Don't forget that several of the activities are being performed at the same time. It would take 25 weeks if they were done sequentially.

As mentioned earlier, the critical path is the *longest* time path through the network. To find the critical path, we calculate two distinct starting and ending times for each activity. These are defined as follows:

Earliest start (ES) = earliest time at which an activity can start, assuming all predecessors have been completed

Earliest finish (EF) = earliest time at which an activity can be finished

Latest start (LS) = latest time at which an activity can start so as to not delay the completion time of the entire project

Latest finish (LF) = latest time by which an activity has to finish so as to not delay the completion time of the entire project

We use a two-pass process, consisting of a forward pass and a backward pass, to determine these time schedules for each activity. The early start and finish times (ES and EF) are determined during the forward pass. The late start and finish times (LS and LF) are determined during the backward pass.

Forward pass
A process that identifies all the early times.

LO 3.3 *Complete* forward and backward passes for a project

Forward Pass

To clearly show the activity schedules on the project network, we use the notation shown in Figure 3.9. The ES of an activity is shown in the top left corner of the node denoting that activity. The EF is shown in the top right corner. The latest times, LS and LF, are shown in the bottom-left and bottom-right corners, respectively.

Figure **3.9**

Notation Used in Nodes for Forward and Backward Pass

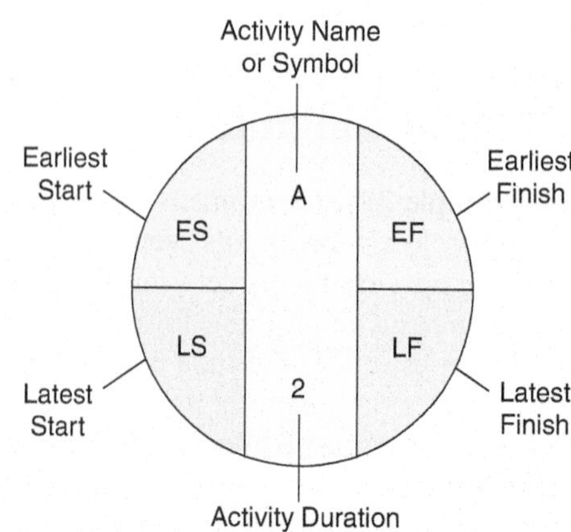

Earliest Start Time Rule Before an activity can start, *all* its immediate predecessors must be finished:

- If an activity has only a single immediate predecessor, its ES equals the EF of the predecessor.
- If an activity has multiple immediate predecessors, its ES is the maximum of all EF values of its predecessors. That is:

$$ES = \text{Max}\ \{\text{EF of all immediate predecessors}\} \qquad (3\text{-}1)$$

STUDENT TIP
All predecessor activities must be completed before an acitivity can begin.

Earliest Finish Time Rule The earliest finish time (EF) of an activity is the sum of its earliest start time (ES) and its activity time. That is:

$$EF = ES + \text{Activity time} \qquad (3\text{-}2)$$

Example 4

COMPUTING EARLIEST START AND FINISH TIMES FOR MILWAUKEE PAPER

Calculate the earliest start and finish times for the activities in the Milwaukee Paper Manufacturing project.

APPROACH ▶ Use Table 3.2, which contains the activity times. Complete the project network for the company's project, along with the ES and EF values for all activities.

SOLUTION ▶ With the help of Figure 3.10, we describe how these values are calculated.

Because activity Start has no predecessors, we begin by setting its ES to 0. That is, activity Start can begin at time 0, which is the same as the beginning of week 1. If activity Start has an ES of 0, its EF is also 0, since its activity time is 0.

Next, we consider activities A and B, both of which have only Start as an immediate predecessor. Using the earliest start time rule, the ES for both activities A and B equals zero, which is the EF of activity Start. Now, using the earliest finish time rule, the EF for A is 2 $(= 0 + 2)$, and the EF for B is 3 $(= 0 + 3)$.

Since activity A precedes activity C, the ES of C equals the EF of A $(= 2)$. The EF of C is therefore 4 $(= 2 + 2)$.

We now come to activity D. Both activities A and B are immediate predecessors for D. Whereas A has an EF of 2, activity B has an EF of 3. Using the earliest start time rule, we compute the ES of activity D as follows:

$$\text{ES of D} = \text{Max}\{\text{EF of } A,\ \text{EF of } B\} = \text{Max}\,(2, 3) = 3$$

The EF of D equals 7 $(= 3 + 4)$. Next, both activities E and F have activity C as their only immediate predecessor. Therefore, the ES for both E and F equals 4 $(= \text{EF of C})$. The EF of E is 8 $(= 4 + 4)$, and the EF of F is 7 $(= 4 + 3)$.

Activity G has both activities D and E as predecessors. Using the earliest start time rule, its ES is therefore the maximum of the EF of D and the EF of E. Hence, the ES of activity G equals 8 $(= \text{maximum of } 7 \text{ and } 8)$, and its EF equals 13 $(= 8 + 5)$.

Finally, we come to activity H. Because it also has two predecessors, F and G, the ES of H is the maximum EF of these two activities. That is, the ES of H equals 13 (= maximum of 13 and 7). This implies that the EF of H is 15 $(= 13 + 2)$. Because H is the last activity in the project, this also implies that the earliest time in which the entire project can be completed is 15 weeks.

INSIGHT ▶ The ES of an activity that has only one predecessor is simply the EF of that predecessor. For an activity with more than one predecessor, we must carefully examine the EFs of all immediate predecessors and choose the largest one.

LEARNING EXERCISE ▶ A new activity I, *EPA Approval*, takes 1 week. Its predecessor is activity H. What are I's ES and EF? [Answer: 15, 16]

RELATED PROBLEMS ▶ 3.15, 3.16, 3.19c, 3.26c

EXCEL OM Data File **Ch03Ex4.xls** can be found online.

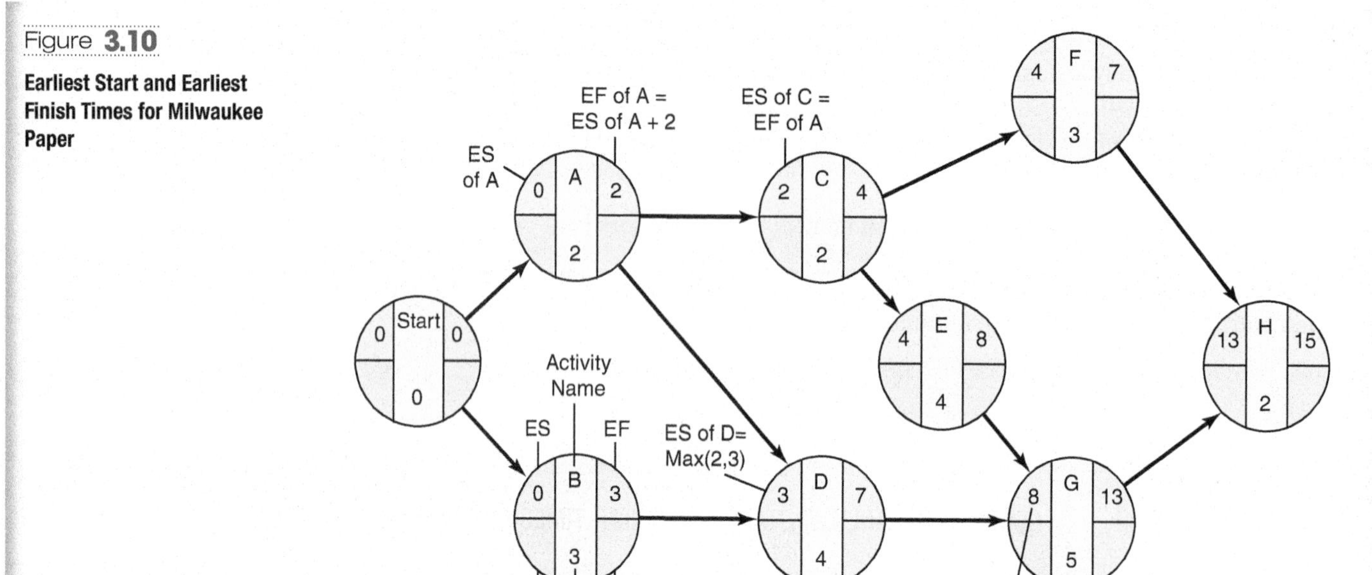

Figure 3.10
Earliest Start and Earliest Finish Times for Milwaukee Paper

Although the forward pass allows us to determine the earliest project completion time, it does not identify the critical path. To identify this path, we need to now conduct the backward pass to determine the LS and LF values for all activities.

Backward Pass

Backward pass
An activity that finds all the late start and late finish times.

Just as the forward pass began with the first activity in the project, the **backward pass** begins with the last activity in the project. For each activity, we first determine its LF value, followed by its LS value. The following two rules are used in this process.

Latest Finish Time Rule This rule is again based on the fact that before an activity can start, all its immediate predecessors must be finished:

- If an activity is an immediate predecessor for just a single activity, its LF equals the LS of the activity that immediately follows it.
- If an activity is an immediate predecessor to more than one activity, its LF is the minimum of all LS values of all activities that immediately follow it. That is:

$$\text{LF} = \text{Min}\{\text{LS of all immediate following activities}\} \quad (3\text{-}3)$$

Latest Start Time Rule The latest start time (LS) of an activity is the difference of its latest finish time (LF) and its activity time. That is:

$$\text{LS} = \text{LF} - \text{Activity time} \quad (3\text{-}4)$$

Example 5

COMPUTING LATEST START AND FINISH TIMES FOR MILWAUKEE PAPER

Calculate the latest start and finish times for each activity in Milwaukee Paper's pollution project.

APPROACH ▶ Use Figure 3.10 as a beginning point. Figure 3.11 shows the complete project network for Milwaukee Paper, along with added LS and LF values for all activities. In what follows, we see how these values were calculated.

Figure 3.11

Latest Start and Latest Finish Times for Milwaukee General Hospital

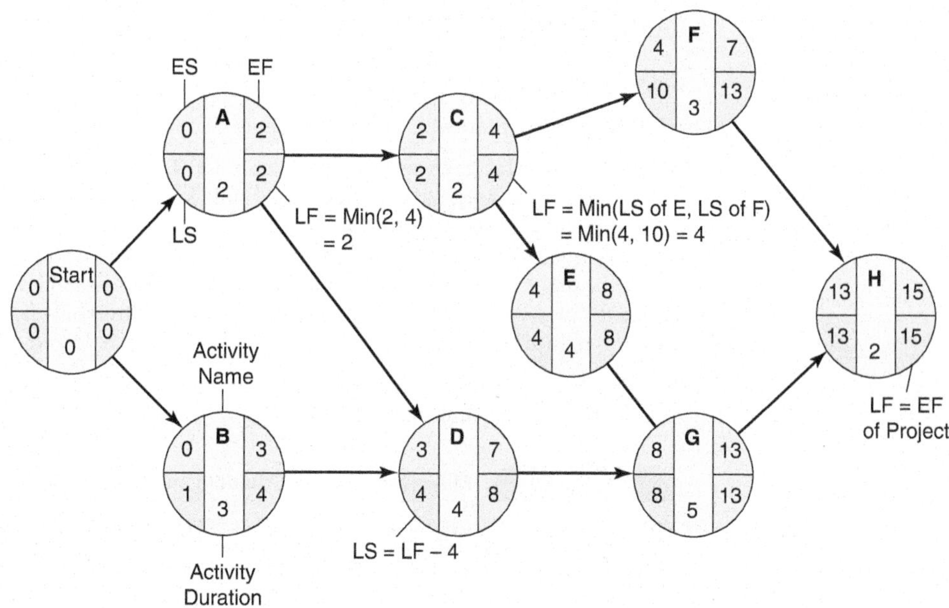

SOLUTION ▶ We begin by assigning an LF value of 15 weeks for activity H. That is, we specify that the latest finish time for the entire project is the same as its earliest finish time. Using the latest start time rule, the LS of activity H is equal to 13 (= 15 – 2).

Because activity H is the lone succeeding activity for both activities F and G, the LF for both F and G equals 13. This implies that the LS of G is 8 (= 13 – 5), and the LS of F is 10 (= 13 – 3).

Proceeding in this fashion, we see that the LF of E is 8 (= LS of G), and its LS is 4 (= 8 – 4). Likewise, the LF of D is 8 (= LS of G), and its LS is 4 (= 8 – 4).

We now consider activity C, which is an immediate predecessor to two activities: E and F. Using the latest finish time rule, we compute the LF of activity C as follows:

$$\text{LF of C} = \text{Min}\{\text{LS of E, LS of F}\} = \text{Min}(4, 10) = 4$$

The LS of C is computed as 2 (= 4 – 2). Next, we compute the LF of B as 4 (= LS of D) and its LS as 1 (= 4 – 3).

We now consider activity A. We compute its LF as 2 (= minimum of LS of C and LS of D). Hence, the LS of activity A is 0 (= 2 – 2). Finally, both the LF and LS of activity Start are equal to 0.

INSIGHT ▶ The LF of an activity that is the predecessor of only one activity is just the LS of that following activity. If the activity is the predecessor to more than one activity, its LF is the smallest LS value of all activities that follow immediately.

LEARNING EXERCISE ▶ A new activity I, *EPA Approval*, takes 1 week. Its predecessor is activity H. What are I's LS and LF? [Answer: 15, 16]

RELATED PROBLEMS ▶ 3.15, 3.19c

Calculating Slack Time and Identifying the Critical Path(s)

After we have computed the earliest and latest times for all activities, it is a simple matter to find the amount of **slack time** that each activity has. Slack is the length of time an activity can be delayed without delaying the entire project. Mathematically:

Slack time
Free time for an activity. Also referred to as free float or free slack.

$$\text{Slack} = \text{LS} - \text{ES} \quad \text{or} \quad \text{Slack} = \text{LF} - \text{EF} \tag{3-5}$$

Example 6

CALCULATING SLACK TIMES FOR MILWAUKEE PAPER

Calculate the slack for the activities in the Milwaukee Paper project.

APPROACH ▶ Start with the data in Figure 3.11 in Example 5 and develop Table 3.3 one line at a time.

SOLUTION ▶ Table 3.3 summarizes the ES, EF, LS, LF, and slack time for all of the firm's activities. Activity B, for example, has 1 week of slack time because its LS is 1 and its ES is 0 (alternatively, its LF is 4 and its EF is 3). This means that activity B can be delayed by up to 1 week, and the whole project can still be finished in 15 weeks.

On the other hand, activities A, C, E, G, and H have *no* slack time. This means that none of them can be delayed without delaying the entire project. Conversely, if plant manager Julie Ann Williams wants to reduce the total project times, she will have to reduce the length of one of these activities.

INSIGHT ▶ Slack may be computed from either early/late starts or early/late finishes. The key is to find which activities have zero slack.

TABLE 3.3 Milwaukee Paper's Schedule and Slack Times

ACTIVITY	ACTIVITY TIME	EARLIEST START ES	EARLIEST FINISH EF	LATEST START LS	LATEST FINISH LF	SLACK LS – ES	ON CRITICAL PATH
A	2	0	2	0	2	0	Yes
B	3	0	3	1	4	1	No
C	2	2	4	2	4	0	Yes
D	4	3	7	4	8	1	No
E	4	4	8	4	8	0	Yes
F	3	4	7	10	13	6	No
G	5	8	13	8	13	0	Yes
H	2	13	15	13	15	0	Yes

LEARNING EXERCISE ▶ A new activity I, *EPA Approval*, follows activity H and takes 1 week. Is it on the critical path? [Answer: Yes, it's LS – ES = 0]

RELATED PROBLEMS ▶ 3.8d, 3.11, 3.15d, 3.19c

ACTIVE MODEL 3.1 This example is further illustrated in Active Model 3.1 found online.

LO 3.4 *Determine* a critical path

The activities with zero slack are called *critical activities* and are said to be on the critical path. The critical path is a continuous path through the project network that:

- Starts at the first activity in the project (Start in our example).
- Terminates at the last activity in the project (H in our example).
- Includes only critical activities (i.e., activities with no slack time).

Example 7

SHOWING CRITICAL PATH WITH BLUE ARROWS

Show Milwaukee Paper's critical path and find the project completion time.

APPROACH ▶ We use Table 3.3 and Figure 3.12. Figure 3.12 indicates that the total project completion time of 15 weeks corresponds to the longest path in the network. That path is Start-A-C-E-G-H in network form. It is shown with thick blue arrows.

INSIGHT ▶ The critical path follows the activities with slack = 0. This is considered the longest path through the network.

LEARNING EXERCISE ▶ Why are activities B, D, and F not on the path with the thick blue line? [Answer: They are not critical and have slack values of 1, 1, and 6 weeks, respectively.]

RELATED PROBLEMS ▶ 3.5–3.11, 3.16, 3.19b, 3.21a

Figure 3.12

Critical Path and Slack Times for Milwaukee General Hospital

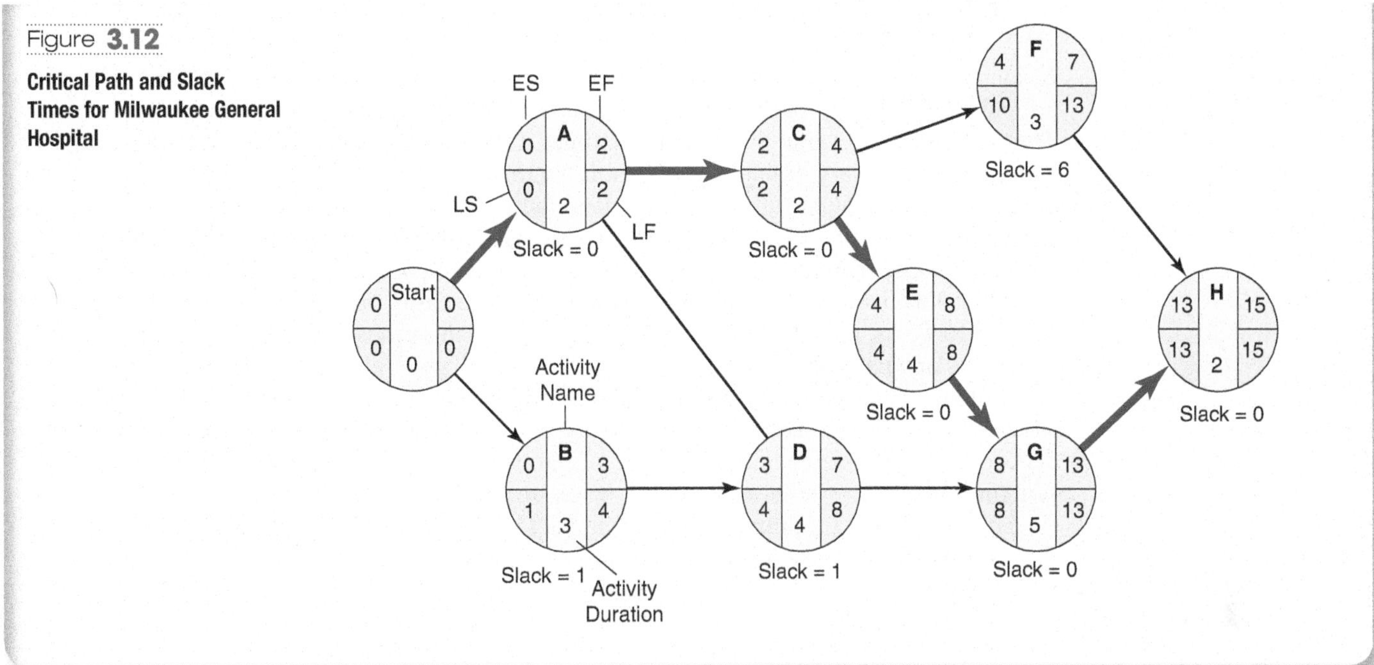

Total Slack Time Look again at the project network in Figure 3.12. Consider activities B and D, which have slack of 1 week each. Does it mean that we can delay *each* activity by 1 week, and still complete the project in 15 weeks? The answer is no.

Let's assume that activity B is delayed by 1 week. It has used up its slack of 1 week and now has an EF of 4. This implies that activity D now has an ES of 4 and an EF of 8. Note that these are also its LS and LF values, respectively. That is, activity D also has no slack time now. Essentially, the slack of 1 week that activities B and D had is, for that path, *shared* between them. Delaying either activity by 1 week causes not only that activity, but also the other activity, to lose its slack. This type of a slack time is referred to as *total slack*. Typically, when two or more noncritical activities appear successively in a path, they share total slack.

Christina Kennedy/Alamy Stock Photo

To plan, monitor, and control the huge number of details involved in sponsoring a rock festival attended by more than 100,000 fans, managers use Microsoft Project and the tools discussed in this chapter. The *Video Case Study* "Managing Hard Rock's Rockfest," at the end of the chapter, provides more details of the management task.

Variability in Activity Times

In identifying all earliest and latest times so far, and the associated critical path(s), we have adopted the CPM approach of assuming that all activity times are known and fixed constants. That is, there is no variability in activity times. However, in practice, it is likely that activity completion times vary depending on various factors.

For example, building internal components (activity A) for Milwaukee Paper Manufacturing is estimated to finish in 2 weeks. Clearly, supply-chain issues such as late arrival of materials, absence of key personnel, and so on could delay this activity. Suppose activity A actually ends up taking 3 weeks. Because A is on the critical path, the entire project will now be delayed by 1 week to 16 weeks. If we had anticipated completion of this project in 15 weeks, we would obviously miss our Earth Day deadline.

Although some activities may be relatively less prone to delays, others could be extremely susceptible to delays. For example, activity B (modify roof and floor) could be heavily dependent on weather conditions. A spell of bad weather could significantly affect its completion time.

STUDENT TIP PERT's ability to handle three time estimates for each activity enables us to compute the probability that we can complete the project by a target date.

This means that we cannot ignore the impact of variability in activity times when deciding the schedule for a project. PERT addresses this issue.

Three Time Estimates in PERT

In PERT, we employ a probability distribution based on three time estimates for each activity, as follows:

Optimistic time
The "best" activity completion time that could be obtained in a PERT network.

Pessimistic time
The "worst" activity time that could be expected in a PERT network.

Most likely time
The most probable time to complete an activity in a PERT network.

Optimistic time (a) = time an activity will take if everything goes as planned. In estimating this value, there should be only a small probability (say, 1/100) that the activity time will be $< a$.

Pessimistic time (b) = time an activity will take assuming very unfavorable conditions. In estimating this value, there should also be only a small probability (also 1/100) that the activity time will be $> b$.

Most likely time (m) = most realistic estimate of the time required to complete an activity.

When using PERT, we often assume that activity time estimates follow the beta probability distribution (see Figure 3.13). This continuous distribution is often appropriate for determining the expected value and variance for activity completion times.

To find the *expected activity time*, t, the beta distribution weights the three time estimates as follows:

$$t = (a + 4m + b)/6 \tag{3-6}$$

That is, the most likely time (m) is given four times the weight as the optimistic time (a) and pessimistic time (b). The time estimate t computed using Equation (3-6) for each activity is used in the project network to compute all earliest and latest times.

To compute the *dispersion* or *variance of activity completion time*, we use the formula:[2]

$$\text{Variance} = [(b - a)/6]^2 \tag{3-7}$$

Figure **3.13**

Beta Probability Distribution with Three Time Estimates

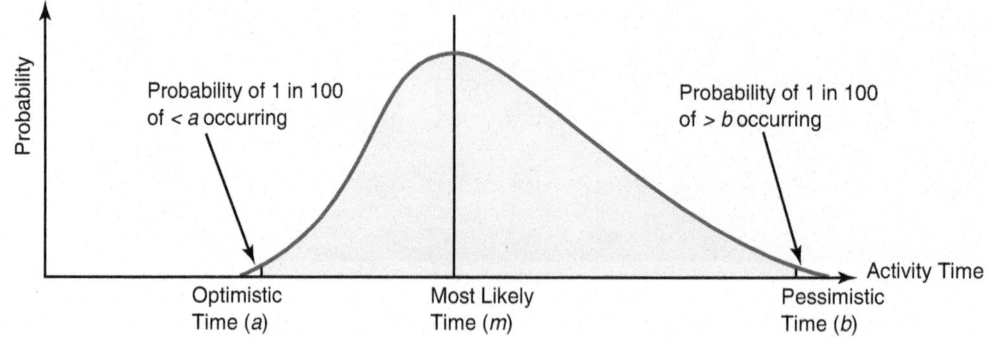

Example 8

EXPECTED TIMES AND VARIANCES FOR MILWAUKEE PAPER

Julie Ann Williams and the project management team at Milwaukee Paper want an expected time and variance for Activity F (Installing the Pollution Control System) where:

$$a = 1 \text{ week}, \; m = 2 \text{ weeks}, \; b = 9 \text{ weeks}$$

APPROACH ▶ Use Equations (3–6) and (3–7) to compute the expected time and variance for F.

SOLUTION ▶ The expected time for Activity F is:

$$t = \frac{a + 4m + b}{6} = \frac{1 + 4(2) + 9}{6} = \frac{18}{6} = 3 \text{ weeks}$$

LO 3.5 *Calculate* the variance of activity times

The variance for Activity F is:

$$\text{Variance} = \left[\frac{(b-a)}{6}\right]^2 = \left[\frac{(9-1)}{6}\right]^2 = \left(\frac{8}{6}\right)^2 = \frac{64}{36} = 1.78$$

INSIGHT ▶ Williams now has information that allows her to understand and manage Activity F. The expected time is, in fact, the activity time used in our earlier computation and identification of the critical path.

LEARNING EXERCISE ▶ Review the expected times and variances for all of the other activities in the project. These are shown in Table 3.4.

STUDENT TIP Can you see why the variance is higher in some activities than in others? Note the spread between the optimistic and pessimistic times.

TABLE 3.4 Time Estimates (in weeks) for Milwaukee Paper's Project

ACTIVITY	OPTIMISTIC a	MOST LIKELY m	PESSIMISTIC b	EXPECTED TIME $t = (a + 4m + b)/6$	VARIANCE $[(b - a)/6]^2$
A	1	2	3	2	$[(3 - 1)/6]^2 = 4/36 = .11$
B	2	3	4	3	$[(4 - 2)/6]^2 = 4/36 = .11$
C	1	2	3	2	$[(3 - 1)/6]^2 = 4/36 = .11$
D	2	4	6	4	$[(6 - 2)/6]^2 = 16/36 = .44$
E	1	4	7	4	$[(7 - 1)/6]^2 = 36/36 = 1.00$
F	1	2	9	3	$[(9 - 1)/6]^2 = 64/36 = 1.78$
G	3	4	11	5	$[(11 - 3)/6]^2 = 64/36 = 1.78$
H	1	2	3	2	$[(3 - 1)/6]^2 = 4/36 = .11$

RELATED PROBLEMS ▶ 3.17a, b, 3.18, 3.19a, 3.20a, 3.26a, 3.27a

EXCEL OM Data File **Ch03Ex8.xls** can be found online.

Probability of Project Completion

The critical path analysis helped us determine that Milwaukee Paper's expected project completion time is 15 weeks. Julie Ann Williams knows, however, that there is significant variation in the time estimates for several activities. Variation in activities that are on the critical path can affect the overall project completion time—possibly delaying it. This is one occurrence that worries the plant manager considerably.

PERT uses the variance of critical path activities to help determine the variance of the overall project. Project variance is computed by summing variances of *critical* activities:

$$\sigma_p^2 = \text{Project variance} = \Sigma(\text{variances of activities on critical path}) \qquad (3\text{-}8)$$

Example 9

COMPUTING PROJECT VARIANCE AND STANDARD DEVIATION FOR MILWAUKEE PAPER

Milwaukee Paper's managers now wish to know the project's variance and standard deviation.

APPROACH ▶ Because the activities are independent, we can add the variances of the activities on the critical path and then take the square root to determine the project's standard deviation.

SOLUTION ▶ From Example 8 (Table 3.4), we have the variances of all of the activities on the critical path. Specifically, we know that the variance of activity A is 0.11, variance of activity C is 0.11, variance of activity E is 1.00, variance of activity G is 1.78, and variance of activity H is 0.11.

Compute the total project variance and project standard deviation:

$$\text{Project variance } \left(\sigma_p^2\right) = 0.11 + 0.11 + 1.00 + 1.78 + 0.11 = 3.11$$

which implies:

$$\text{Project standard deviation } (\sigma_p) = \sqrt{\text{Project variance}} = \sqrt{3.11} = 1.76 \text{ weeks}$$

INSIGHT ▶ Management now has an estimate not only of expected completion time for the project but also of the standard deviation of that estimate.

LEARNING EXERCISE ▶ If the variance for activity A is actually 0.30 (instead of 0.11), what is the new project standard deviation? [Answer: 1.817.]

RELATED PROBLEMS ▶ 3.17e, 3.24

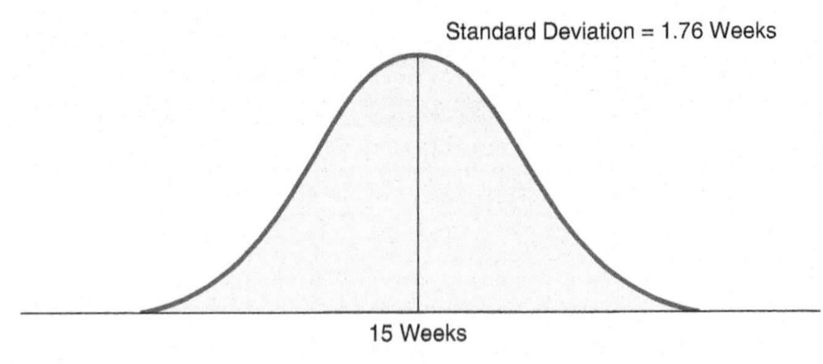

Figure **3.14**

Probability Distribution for Project Completion Times at Milwaukee Paper

How can this information be used to help answer questions regarding the probability of finishing the project on time? PERT makes two more assumptions: (1) total project completion times follow a normal probability distribution, and (2) activity times are statistically independent. With these assumptions, the bell-shaped normal curve shown in Figure 3.14 can be used to represent project completion dates. This normal curve implies that there is a 50% chance that the manufacturer's project completion time will be less than 15 weeks and a 50% chance that it will exceed 15 weeks.

Example 10

PROBABILITY OF COMPLETING A PROJECT ON TIME

Julie Ann Williams would like to find the probability that her project will be finished on or before the 16-week Earth Day deadline.

APPROACH ▶ To do so, she needs to determine the appropriate area under the normal curve. This is the area to the left of the 16th week.

SOLUTION ▶ The standard normal equation can be applied as follows:

STUDENT TIP Here is a chance to review your statistical skills and use of a normal distribution table (Appendix I).

$$\begin{aligned} Z &= \left(\text{Due date} - \text{Expected date of completion}\right)/\sigma_p \\ &= (16 \text{ weeks} - 15 \text{ weeks})/1.76 \text{ weeks} = 0.57 \end{aligned} \quad (3\text{-}9)$$

where Z is the number of standard deviations the due date or target date lies from the mean or expected date.

Referring to the Normal Table in Appendix I (alternatively using the Excel formula = NORMSDIST(0.57)), we find a Z-value of 0.57 to the right of the mean indicates a probability of 0.7157. Thus, there is a 71.57% chance that the pollution control equipment can be put in place in 16 weeks or less. This is shown in Figure 3.15.

Figure 3.15

Probability That Milwaukee Paper Will Meet the 16-Week Deadline

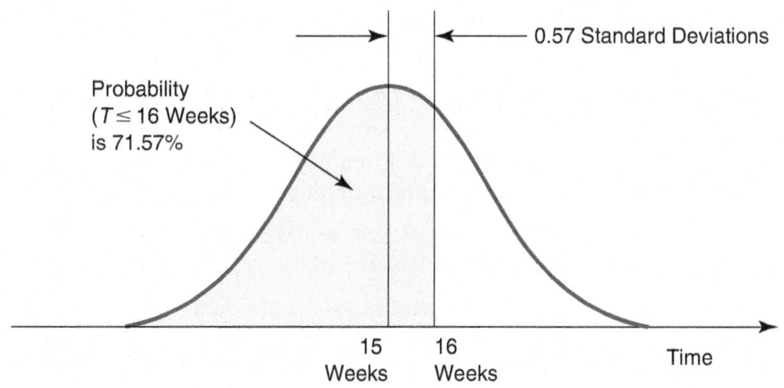

INSIGHT ▶ The shaded area to the left of the 16th week (71.57%) represents the probability that the project will be completed in less than 16 weeks.

LEARNING EXERCISE ▶ What is the probability that the project will be completed on or before the 17th week? [Answer: About 87.2%.]

RELATED PROBLEMS ▶ 3.17f, 3.19d, 3.20d, 3.21b, 3.23, 3.25, 3.26b,c,d, 3.27b

Determining Project Completion Time for a Given Confidence Level Let's say Julie Ann Williams is worried that there is only a 71.57% chance that the pollution control equipment can be put in place in 16 weeks or less. She thinks that it may be possible to plead with the board of directors for more time. However, before she approaches the board, she wants to arm herself with sufficient information about the project. Specifically, she wants to find the deadline by which she has a 99% chance of completing the project. She hopes to use her analysis to convince the board to agree to this extended deadline, even though she is aware of the public relations damage the delay will cause.

Clearly, this due date would be greater than 16 weeks. However, what is the exact value of this new due date? To answer this question, we again use the assumption that Milwaukee Paper's project completion time follows a normal probability distribution with a mean of 15 weeks and a standard deviation of 1.76 weeks.

Example 11

COMPUTING PROBABILITY FOR ANY COMPLETION DATE

Julie Ann Williams wants to find the due date that gives her company's project a 99% chance of *on-time* completion.

APPROACH ▶ She first needs to compute the Z-value corresponding to 99%, as shown in Figure 3.16. Mathematically, this is similar to Example 10, except the unknown is now the due date rather than Z.

Figure 3.16

Z-Value for 99% Probability of Project Completion at Milwaukee Paper

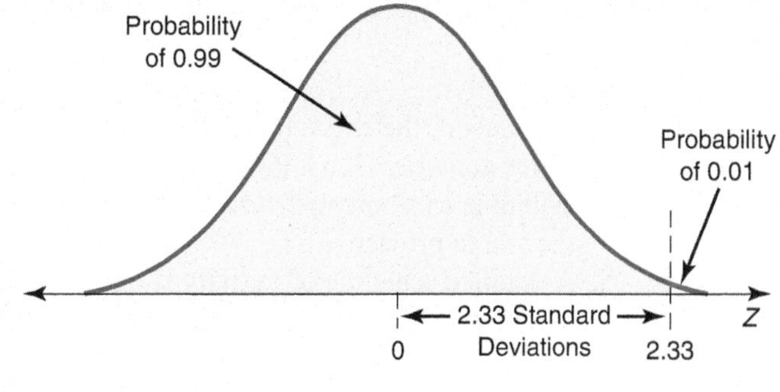

SOLUTION ▶ Referring again to the Normal Table in Appendix I (alternatively using the Excel formula =NORMSINV(0.99)), we identify a Z-value of 2.33 as being closest to the probability of 0.99. That is, Julie Ann Williams's due date should be 2.33 standard deviations above the mean project completion time. Starting with the standard normal equation [see Equation (3-9)], we can solve for the due date and rewrite the equation as:

$$\begin{aligned} \text{Due date} &= \text{Expected completion time} + (Z \times \sigma_p) \\ &= 15 + (2.33 \times 1.76) = 19.1 \text{ weeks} \end{aligned} \qquad (3\text{-}10)$$

INSIGHT ▶ If Williams can get the board to agree to give her a new deadline of 19.1 weeks (or more), she can be 99% sure of finishing the project by that new target date.

LEARNING EXERCISE ▶ What due date gives the project a 95% chance of on-time completion? [Answer: About 17.9 weeks.]

RELATED PROBLEMS ▶ 3.21c, 3.23e

Variability in Completion Time of Noncritical Paths In our discussion so far, we have focused exclusively on the variability in the completion times of activities on the critical path. This seems logical because these activities are, by definition, the more important activities in a project network. However, when there is variability in activity times, it is important that we also investigate the variability in the completion times of activities on *noncritical* paths.

Consider, for example, activity D in Milwaukee Paper's project. Recall from Figure 3.12 (in Example 7) that this is a noncritical activity, with a slack time of 1 week. We have therefore not considered the variability in D's time in computing the probabilities of project completion times. We observe, however, that D has a variance of 0.44 (see Table 3.4 in Example 8). In fact, the pessimistic completion time for D is 6 weeks. This means that if D ends up taking its pessimistic time to finish, the project will not finish in 15 weeks, even though D is not a critical activity.

For this reason, when we find probabilities of project completion times, it may be necessary for us to not focus only on the critical path(s). Indeed, some research has suggested that expending project resources to reduce the variability of activities not on the critical path can be an effective element in project management. We may need also to compute these probabilities for noncritical paths, especially those that have relatively large variances. It is possible for a noncritical path to have a smaller probability of completion within a due date, when compared with the critical path. Determining the variance and probability of completion for a noncritical path is done in the same manner as Examples 9 and 10.

What Project Management Has Provided So Far Project management techniques have thus far been able to provide Julie Ann Williams with several valuable pieces of management information:

1. The project's expected completion date is 15 weeks.
2. There is a 71.57% chance that the equipment will be in place within the 16-week deadline. PERT analysis can easily find the probability of finishing by any date Williams is interested in.
3. Five activities (A, C, E, G, and H) are on the critical path. If any one of these is delayed for any reason, the entire project will be delayed.
4. Three activities (B, D, F) are not critical and have some slack time built in. This means that Williams can borrow from their resources, and, if necessary, she may be able to speed up the whole project.
5. A detailed schedule of activity starting and ending dates, slack, and critical path activities has been made available (see Table 3.3 in Example 6).

Cost-Time Trade-Offs and Project Crashing

While managing a project, it is not uncommon for a project manager to be faced with either (or both) of the following situations: (1) the project is behind schedule, and (2) the scheduled project completion time has been moved forward. In either situation, some or all of the remaining activities need to be speeded up (usually by adding resources) to finish the project by the desired due date. The process by which we shorten the duration of a project in the cheapest manner possible is called project **crashing**.

Crashing
Shortening activity time in a network to reduce time on the critical path so total completion time is reduced.

CPM is a technique in which each activity has a *normal* or *standard* time that we use in our computations. Associated with this normal time is the *normal* cost of the activity. However, another time in project management is the *crash time*, which is defined as the shortest duration required to complete an activity. Associated with this crash time is the *crash cost* of the activity. Usually, we can shorten an activity by adding extra resources (e.g., equipment, people) to it. Hence, it is logical for the crash cost of an activity to be higher than its normal cost.

The amount by which an activity can be shortened (i.e., the difference between its normal time and crash time) depends on the activity in question. We may not be able to shorten some activities at all. For example, if a casting needs to be heat-treated in the furnace for 48 hours, adding more resources does not help shorten the time. In contrast, we may be able to shorten some activities significantly (e.g., frame a house in 3 days instead of 10 days by using three times as many workers).

Likewise, the cost of crashing (or shortening) an activity depends on the nature of the activity. Managers are usually interested in speeding up a project at the least additional cost. Hence, when choosing which activities to crash, and by how much, we need to ensure the following:

- The amount by which an activity is crashed is, in fact, permissible
- Taken together, the shortened activity durations will enable us to finish the project by the due date
- The total cost of crashing is as small as possible

Crashing a project involves four steps:

LO 3.6 *Crash* a project

STEP 1: Compute the crash cost per week (or other time period) for each activity in the network. If crash costs are linear over time, the following formula can be used:

$$\text{Crash cost per period} = \frac{(\text{Crash cost} - \text{Normal cost})}{(\text{Normal time} - \text{Crash time})} \qquad (3\text{-}11)$$

STEP 2: Using the current activity times, find the critical path(s) in the project network. Identify the critical activities.

STEP 3: If there is only one critical path, then select the activity on this critical path that (a) can still be crashed and (b) has the smallest crash cost per period. Crash this activity by one period.

If there is more than one critical path, then select one activity from each critical path such that (a) each selected activity can still be crashed and (b) the total crash cost per period of *all* selected activities is the smallest. Crash each activity by one period. Note that the same activity may be common to more than one critical path.

STEP 4: Update all activity times. If the desired due date has been reached, stop. If not, return to Step 2.

We illustrate project crashing in Example 12.

Example 12

PROJECT CRASHING TO MEET A DEADLINE AT MILWAUKEE PAPER

Suppose the plant manager at Milwaukee Paper Manufacturing has been given only 13 weeks (instead of 16 weeks) to install the new pollution control equipment. As you recall, the length of Julie Ann Williams's critical path was 15 weeks, but she must now complete the project in 13 weeks.

APPROACH ▶ Williams needs to determine which activities to crash, and by how much, to meet this 13-week due date. Naturally, Williams is interested in speeding up the project by 2 weeks, at the least additional cost.

SOLUTION ▶ The company's normal and crash times, and normal and crash costs, are shown in Table 3.5. Note, for example, that activity B's normal time is 3 weeks (the estimate used in computing the critical path), and its crash time is 1 week. This means that activity B can be shortened by up to 2 weeks if extra resources are provided. The cost of these additional resources is \$4,000 (= difference between the crash cost of \$34,000 and the normal cost of \$30,000). If we assume that the crashing cost is linear over time (i.e., the cost is the same each week), activity B's crash cost per week is \$2,000 (= \$4,000/2).

TABLE 3.5 Normal and Crash Data for Milwaukee Paper Manufacturing

ACTIVITY	TIME (WEEKS) NORMAL	TIME (WEEKS) CRASH	COST (\$) NORMAL	COST (\$) CRASH	CRASH COST PER WEEK (\$)	CRITICAL PATH?
A	2	1	22,000	22,750	750	Yes
B	3	1	30,000	34,000	2,000	No
C	2	1	26,000	27,000	1,000	Yes
D	4	3	48,000	49,000	1,000	No
E	4	2	56,000	58,000	1,000	Yes
F	3	2	30,000	30,500	500	No
G	5	2	80,000	84,500	1,500	Yes
H	2	1	16,000	19,000	3,000	Yes

This calculation for Activity B is shown in Figure 3.17. Crash costs for all other activities can be computed in a similar fashion.

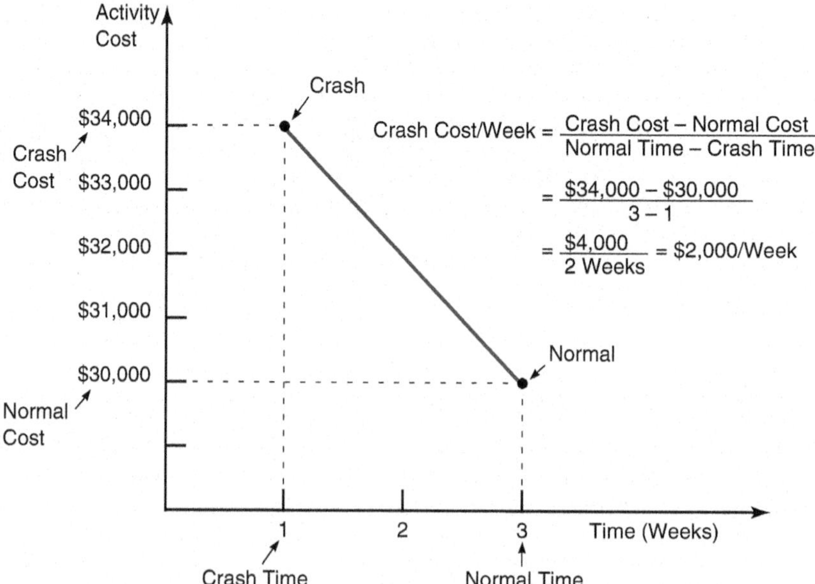

Figure 3.17 **Crash and Normal Times and Costs for Activity B**

Steps 2, 3, and 4 can now be applied to reduce Milwaukee Paper's project completion time at a minimum cost. We show the project network for Milwaukee Paper again in Figure 3.18.

The current critical path (using normal times) is Start–A–C–E–G–H, in which Start is just a dummy starting activity. Of these critical activities, activity A has the lowest crash cost per week of \$750. Julie Ann Williams should therefore crash activity A by 1 week to reduce the project completion time to

Figure 3.18

Critical Path and Slack Times for Milwaukee Paper

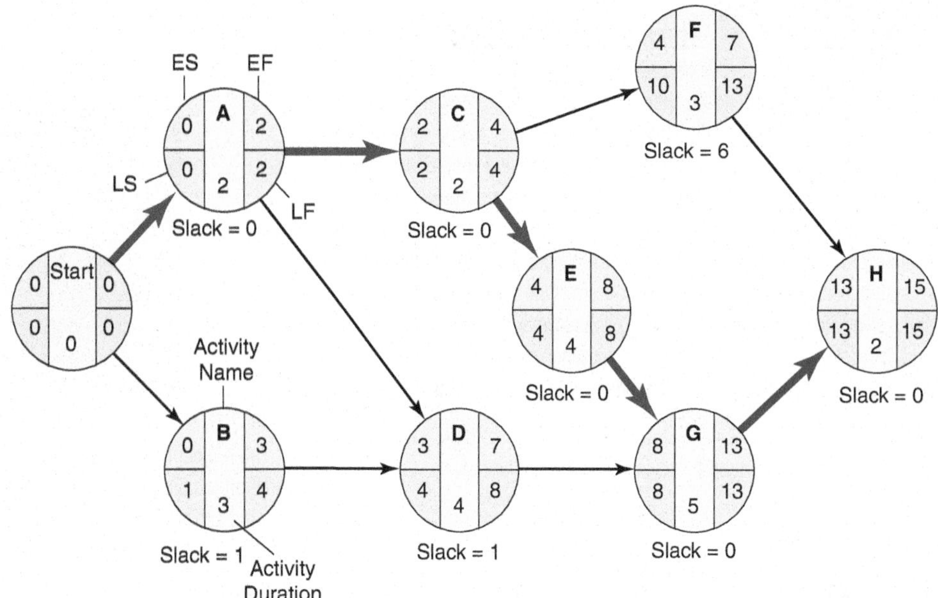

14 weeks. The cost is an additional $750. Note that activity A cannot be crashed any further, since it has reached its crash limit of 1 week.

At this stage, the original path Start–A–C–E–G–H remains critical with a completion time of 14 weeks. However, a new path Start–B–D–G–H is also critical now, with a completion time of 14 weeks. Hence, any further crashing must be done to both critical paths.

On each of these critical paths, we need to identify one activity that can still be crashed. We also want the total cost of crashing an activity on each path to be the smallest. We might be tempted to simply pick the activities with the smallest crash cost per period in each path. If we did this, we would select activity C from the first path and activity D from the second path. The total crash cost would then be $\$2,000\ (= \$1,000 + \$1,000)$.

But we spot that activity G is common to both paths. That is, by crashing activity G, we will simultaneously reduce the completion time of both paths. Even though the $1,500 crash cost for activity G is higher than that for activities C and D, we would still prefer crashing G because the total crashing cost will now be only $1,500 (compared with the $2,000 if we crash C and D).

INSIGHT ▶ To crash the project down to 13 weeks, Williams should crash activity A by 1 week and activity G by 1 week. The total additional cost will be $\$2,250\ (= \$750 + \$1,500)$. This is important because many contracts for projects include bonuses or penalties for early or late finishes.

LEARNING EXERCISE ▶ Say the crash cost for activity B is $31,000 instead of $34,000. How does this change the answer? [Answer: no change.]

RELATED PROBLEMS ▶ 3.28–3.33

EXCEL OM Data File **Ch03Ex12.xls** can be found online.

A Critique of PERT and CPM

As a critique of our discussions of PERT, here are some of its features about which operations managers need to be aware:

Advantages

1. Especially useful when scheduling and controlling large projects.
2. Straightforward concept and not mathematically complex.
3. Graphical networks help highlight relationships among project activities.
4. Critical path and slack time analyses help pinpoint activities that need to be closely watched.
5. Project documentation and graphs point out who is responsible for various activities.
6. Applicable to a wide variety of projects.
7. Useful in monitoring not only schedules but costs as well.

OM in Action Behind the Tour de France[3]

A Tour de France racing team needs the same large behind-the-scenes support that might be seen in the better-known Formula One or NBA playoff competitions. "A Tour de France team is like a large traveling circus," says the coach of the Belkin team. "The public only sees the riders, but they could not function without the unseen support staff."

The project management skills needed mean scheduling everyone from media relations staff to doctors to mechanics to the drivers of the team's huge truck, vans, and convoy of support cars and coach. Each move is made to guarantee that riders are in peak physical, nutritional, and psychological condition. Here are just some of the supplies the project management team for Belkin handles:

- 11 mattresses
- 36 aero suits, 45 bib shorts, 54 race jerseys, 250 podium caps
- 63 bikes
- 140 wheels, 220 tires
- 250 feeding bags, 3,000 water bottles
- 2,190 nutrition gels, 3,800 nutrition bars
- 10 jars of peanut butter, 10 boxes of chocolate sprinkles, 20 bags of wine gums, 20 jars of jam
- 80 kg of nuts, raisins, apricots, and figs, plus 50 kg of cereals

Marc Pagani Photography/Shutterstock

The project management behind a world-tour team is complex: These top teams often compete in two to three races simultaneously, in different countries and sometimes on different continents. Each team has 25–35 riders (9 compete in any single race), coming from different parts of the world, going to different races at different times, each with their own physique and strengths. They have customized bikes, uniforms, and food preferences. The support staff can include another 30 people.

Limitations

1. Project activities have to be clearly defined, independent, and stable in their relationships.
2. Precedence relationships must be specified and networked together.
3. Time estimates tend to be subjective and are subject to fudging by managers who fear the dangers of being overly optimistic or not pessimistic enough.
4. There is the inherent danger of placing too much emphasis on the longest, or critical, path. Near-critical paths need to be monitored closely as well.

Using Software to Manage Projects

The approaches discussed so far are effective for managing small projects. However, for large or complex projects, specialized project management software is much preferred. Commercial software is often tailored to specific types of projects, such as construction, information technology, or laboratory research. These commercial products are all extremely useful in drawing project networks, identifying the project schedule and the necessary resources, and tracking time and money. Additionally, various versions of project management software now provide document management and track allocation of multiple resources. Some even provide an opportunity to establish probabilities for the necessity of various activities. Indeed, no large projects are managed these days without project management software. There are many products available. These include Oracle Primavera, SAP, HP Project, and Microsoft Project.

In this section, we provide a brief introduction to one of the most popular examples of such specialized software, Microsoft Project. A trial version can be downloaded from Microsoft.[4]

Milwaukee Paper Co. Activities

ACTIVITY	TIME (WKS)	PREDECESSORS
A	2	—
B	3	—
C	2	A
D	4	A, B
E	4	C
F	3	C
G	5	D, E
H	2	F, G

Entering Data Let us again consider the Milwaukee Paper Manufacturing project. Recall that this project has eight activities (repeated in the margin). The first step is to define the activities and their precedence relationships. To do so, we select **File|New** to open a blank project. We type the project start date (as March 1), then enter all activity information (see Program 3.1). For each activity (or task, as Microsoft Project calls it), we fill in the name and duration. The description of the activity is also placed in the Task Name column in Program 3.1. As we enter activities and durations, the software automatically inserts start and finish dates.

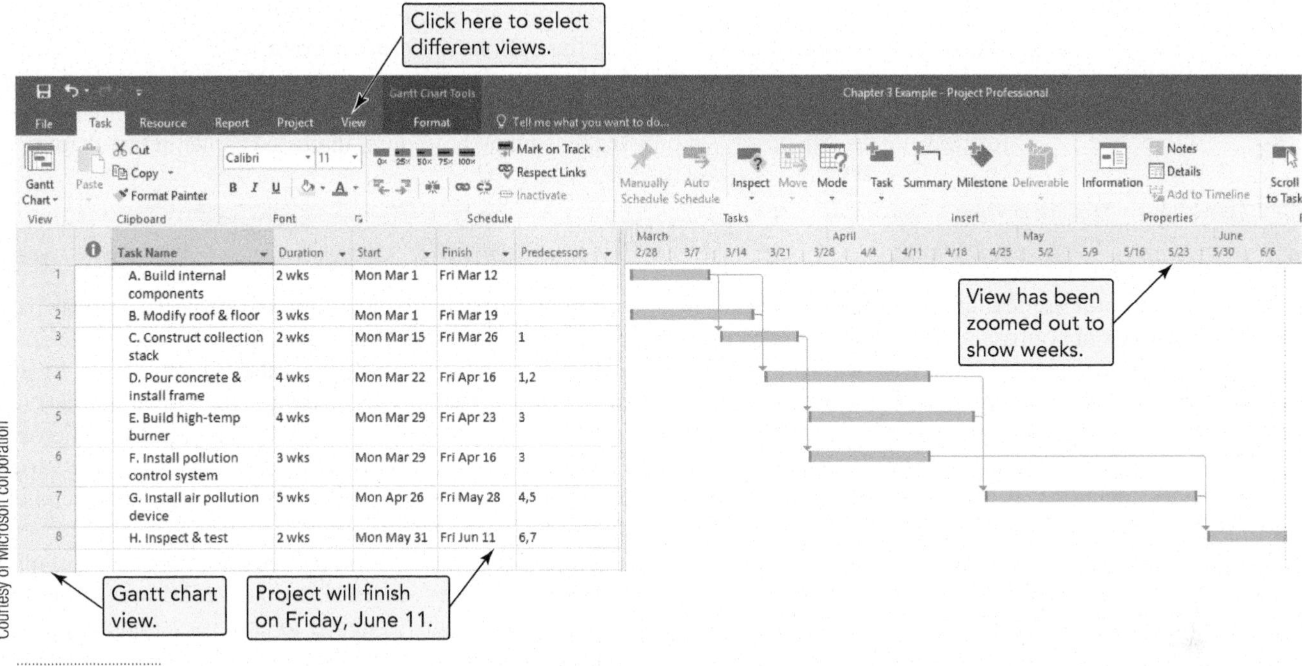

Program **3.1**

Gantt Chart in Microsoft Project for Milwaukee Paper Manufacturing

The next step is to define precedence relationships between these activities. To do so, we enter the relevant activity numbers (e.g., 1, 2) in the Predecessors column.

STUDENT TIP
Now that you understand the workings of PERT and CPM, you are ready to master this useful program. Knowing such software gives you an edge over others in the job market.

Viewing the Project Schedule When all links have been defined, the complete project schedule can be viewed as a Gantt chart. We can also select **View|Network Diagram** to view the schedule as a project network (shown in Program 3.2). The critical path is shown in red on the screen in the network diagram. We can click on any of the activities in the project network to view details of the activities. Likewise, we can easily add or remove activities from the project network. Each time we do so, Microsoft Project automatically updates all start dates, finish

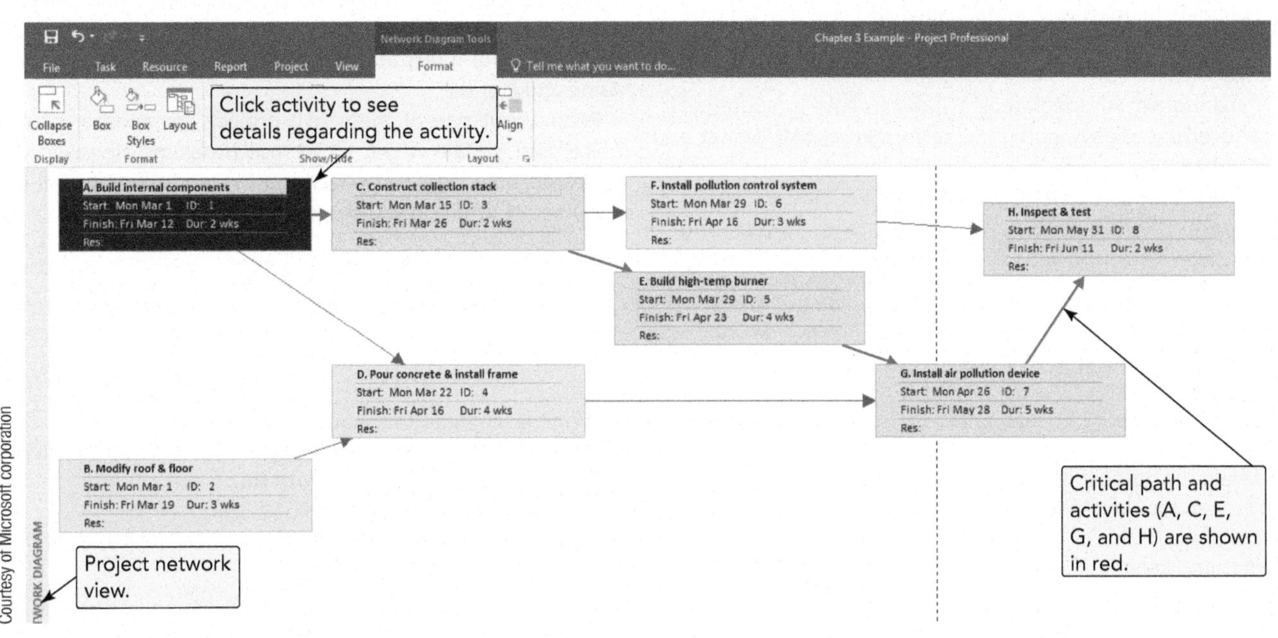

Program **3.2**

Project Network in Microsoft Project for Milwaukee Paper Manufacturing

dates, and the critical path(s). If desired, we can manually change the layout of the network (e.g., reposition activities) by changing the options in **Format|Layout**.

Programs 3.1 and 3.2 show that if Milwaukee Paper's project starts March 1, it can be finished on June 11. The start and finish dates for all activities are also clearly identified. Project management software, we see, can greatly simplify the scheduling procedures discussed earlier in this chapter.

Summary

PERT, CPM, and other scheduling techniques have proven to be valuable tools in controlling large and complex projects. Managers use such techniques to segment projects into discrete activities (work breakdown structures), identifying specific resources and time requirements for each. With PERT and CPM, managers can understand the status of each activity, including its earliest start, latest start, earliest finish, and latest finish (ES, LS, EF, and LF) times. By controlling the trade-off between ES and LS, managers can identify the activities that have slack and can address resource allocation, perhaps by smoothing resources. Effective project management also allows managers to focus on the activities that are critical to timely project completion. By understanding the project's critical path, they know where crashing makes the most economic sense.

Good project management also allows firms to efficiently create products and services for global markets and to respond effectively to global competition. Microsoft Project, illustrated in this chapter, is one of a wide variety of software packages available to help managers handle network modeling problems.

The models described in this chapter require good management practices, detailed work breakdown structures, clear responsibilities assigned to activities, and straightforward and timely reporting systems. All are critical parts of project management.

Key Terms

Project organization (p. 62)
Work breakdown structure (WBS) (p. 64)
Gantt charts (p. 65)
Waterfall projects (p. 67)
Agile projects (p. 67)
Program evaluation and review technique (PERT) (p. 67)
Critical path method (CPM) (p. 67)
Critical path (p. 67)
Activity-on-node (AON) (p. 68)
Activity-on-arrow (AOA) (p. 68)
Dummy activity (p. 70)
Critical path analysis (p. 71)
Forward pass (p. 72)
Backward pass (p. 74)
Slack time (p. 75)
Optimistic time (p. 78)
Pessimistic time (p. 78)
Most likely time (p. 78)
Crashing (p. 83)

Ethical Dilemma

Two examples of massively mismanaged projects are TAURUS and the "Big Dig." The first, formally called the London Stock Exchange Automation Project, cost $575 million before it was finally abandoned. Although most IT projects have a reputation for cost overruns, delays, and underperformance, TAURUS set a new standard.

But even TAURUS paled next to the biggest, most expensive public works project in U.S. history—Boston's 15-year-long Central Artery/Tunnel Project. Called the Big Dig, this was perhaps the poorest and most felonious case of project mismanagement in decades. From a starting $2 billion budget to a final price tag of $15 billion, the Big Dig cost more than the Panama Canal, Hoover Dam, or Interstate 95, the 1,919-mile highway between Maine and Florida.

Read about one of these two projects (or another of your choice) and explain why it faced such problems. How and why do project managers allow such massive endeavors to fall into such a state? What do you think are the causes?

Discussion Questions

1. Give an example of a situation in which project management is needed.
2. Explain the purpose of project organization.
3. What are the three phases involved in the management of a large project?
4. What are some of the questions that can be answered with PERT and CPM?
5. Define *work breakdown structure*. How is it used?
6. What is the use of Gantt charts in project management?
7. What is the difference between an activity-on-arrow (AOA) network and an activity-on-node (AON) network? Which is primarily used in this chapter?
8. What is the significance of the critical path?
9. What would a project manager have to do to crash an activity?
10. Describe how expected activity times and variances can be computed in a PERT network.
11. Define *earliest start, earliest finish, latest finish*, and *latest start* times.
12. Students are sometimes confused by the concept of critical path, and want to believe that it is the *shortest* path through a network. Convincingly explain why this is not so.
13. What are dummy activities? Why are they used in activity-on-arrow (AOA) project networks?
14. What are the three time estimates used with PERT?

15. Would a project manager ever consider crashing a noncritical activity in a project network? Explain convincingly.
16. How is the variance of the total project computed in PERT?
17. Describe the meaning of slack, and discuss how it can be determined.
18. How can we determine the probability that a project will be completed by a certain date? What assumptions are made in this computation?
19. Name some of the widely used project management software programs.
20. What is the difference between the *waterfall* approach and *agile* project management?

Using Software to Solve Project Management Problems

In addition to the Microsoft Project software illustrated earlier, both Excel OM and POM for Windows are available to readers of this text as project management tools and can be found online.

X USING EXCEL OM

Excel OM has a Project Scheduling module. Program 3.4 uses the data from the Milwaukee Paper Manufacturing example in this chapter (see Examples 4 and 5). The PERT/CPM analysis also handles activities with three time estimates.

Program **3.4**

Excel OM's Use of Milwaukee Paper Manufacturing's Data from Examples 4 and 5

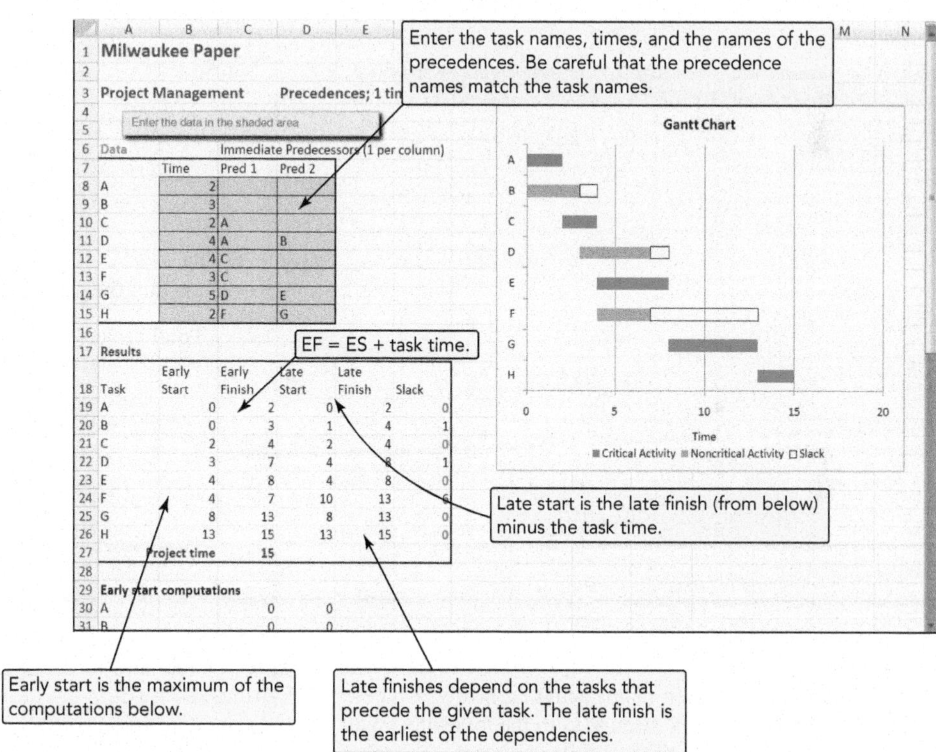

P USING POM FOR WINDOWS

POM for Window's Project Scheduling module can also find the expected project completion time for a CPM and PERT network with either one or three time estimates. POM for Windows also performs project crashing. For further details, refer to Appendix II.

Solved Problems

SOLVED PROBLEM 3.1

Construct an AON network based on the following:

ACTIVITY	IMMEDIATE PREDECESSOR(S)
A	—
B	—
C	—
D	A, B
E	C

SOLUTION

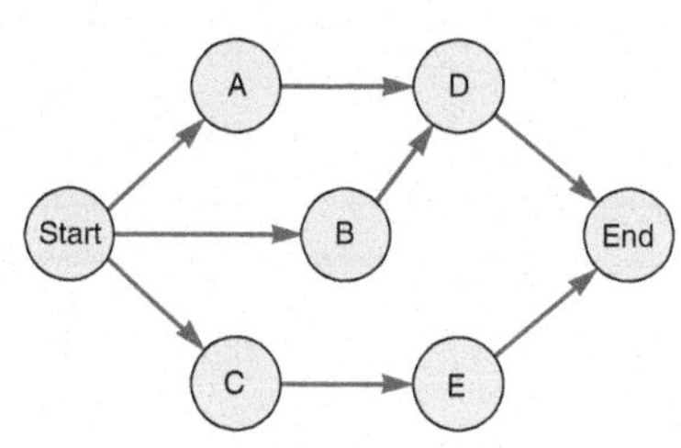

SOLVED PROBLEM 3.2

Insert a dummy activity and event to correct the following AOA network:

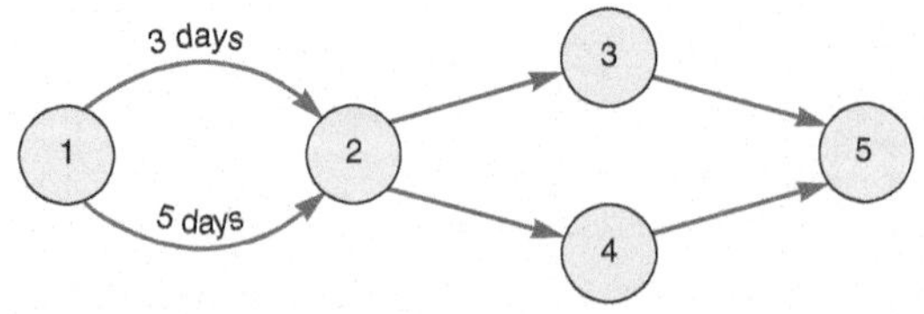

SOLUTION

Because we cannot have two activities starting and ending at the same node, we add the following dummy activity and dummy event to obtain the correct AOA network:

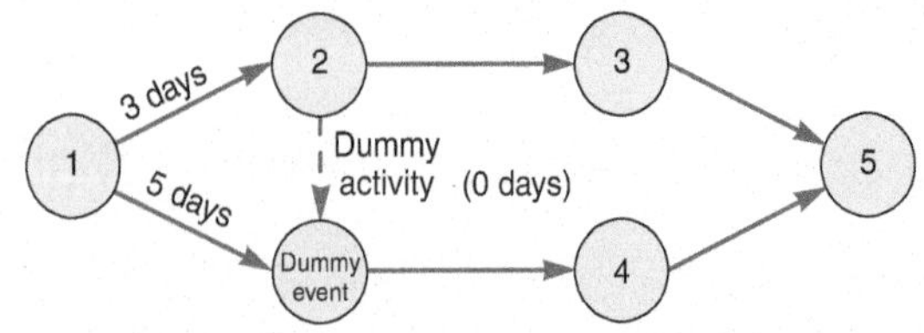

SOLVED PROBLEM 3.3

Calculate the critical path, project completion time T, and project variance σ_p^2, based on the following AON network information:

ACTIVITY	TIME	VARIANCE	ES	EF	LS	LF	SLACK
A	2	$\frac{2}{6}$	0	2	0	2	0
B	3	$\frac{2}{6}$	0	3	1	4	1
C	2	$\frac{4}{6}$	2	4	2	4	0
D	4	$\frac{4}{6}$	3	7	4	8	1
E	4	$\frac{2}{6}$	4	8	4	8	0
F	3	$\frac{1}{6}$	4	7	10	13	6
G	5	$\frac{1}{6}$	8	13	8	13	0

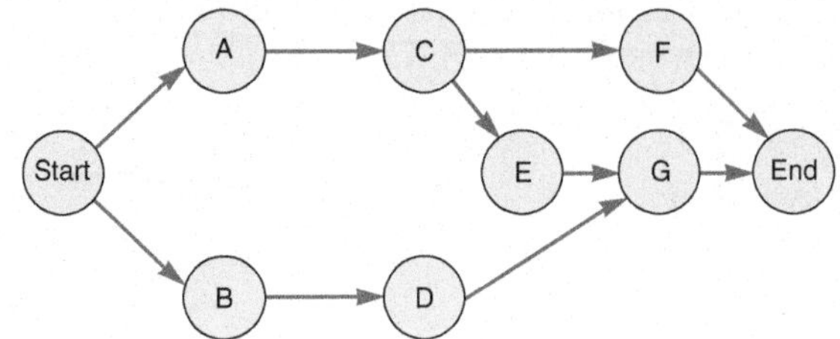

SOLUTION

We conclude that the critical path is Start–A–C–E–G–End:

$$\text{Total project time} = T = 2 + 2 + 4 + 5 = 13$$

and

$$\sigma_p^2 = \Sigma \text{ Variances on the critical path}$$

$$= \frac{2}{6} + \frac{4}{6} + \frac{2}{6} + \frac{1}{6} = \frac{9}{6} = 1.5$$

SOLVED PROBLEM 3.4

To complete the wing assembly for an experimental aircraft, Jim Gilbert has laid out the seven major activities involved. These activities have been labeled A through G in the following table, which also shows their estimated completion times (in weeks) and immediate predecessors. Determine the expected time and variance for each activity.

ACTIVITY	*a*	*m*	*b*	IMMEDIATE PREDECESSORS
A	1	2	3	—
B	2	3	4	—
C	4	5	6	A
D	8	9	10	B
E	2	5	8	C, D
F	4	5	6	D
G	1	2	3	E

SOLUTION

Expected times and variances can be computed using Equations (3–6) and (3–7). The results are summarized in the following table:

ACTIVITY	EXPECTED TIME (IN WEEKS)	VARIANCE
A	2	$\frac{1}{9}$
B	3	$\frac{1}{9}$
C	5	$\frac{1}{9}$
D	9	$\frac{1}{9}$
E	5	1
F	5	$\frac{1}{9}$
G	2	$\frac{1}{9}$

SOLVED PROBLEM 3.5

Referring to Solved Problem 3.4, now Jim Gilbert would like to determine the critical path for the entire wing assembly project as well as the expected completion time for the total project. In addition, he would like to determine the earliest and latest start and finish times for all activities.

SOLUTION

The AON network for Gilbert's project is shown in Figure 3.19. Note that this project has multiple activities (A and B) with no immediate predecessors, and multiple activities (F and G) with no successors. Hence, in addition to a unique starting activity (Start), we have included a unique finishing activity (End) for the project.

Figure 3.19 shows the earliest and latest times for all activities. The results are also summarized in the following table:

	ACTIVITY TIME				
ACTIVITY	ES	EF	LS	LF	SLACK
A	0	2	5	7	5
B	0	3	0	3	0
C	2	7	7	12	5
D	3	12	3	12	0
E	12	17	12	17	0
F	12	17	14	19	2
G	17	19	17	19	0

$$\text{Expected project length} = 19 \text{ weeks}$$

$$\text{Variance of the critical path} = 1.333$$

$$\text{Standard deviation of the critical path} = 1.155 \text{ weeks}$$

The activities along the critical path are B, D, E, and G. These activities have zero slack as shown in the table.

Figure **3.19**

Critical Path for Solved Problem 3.5

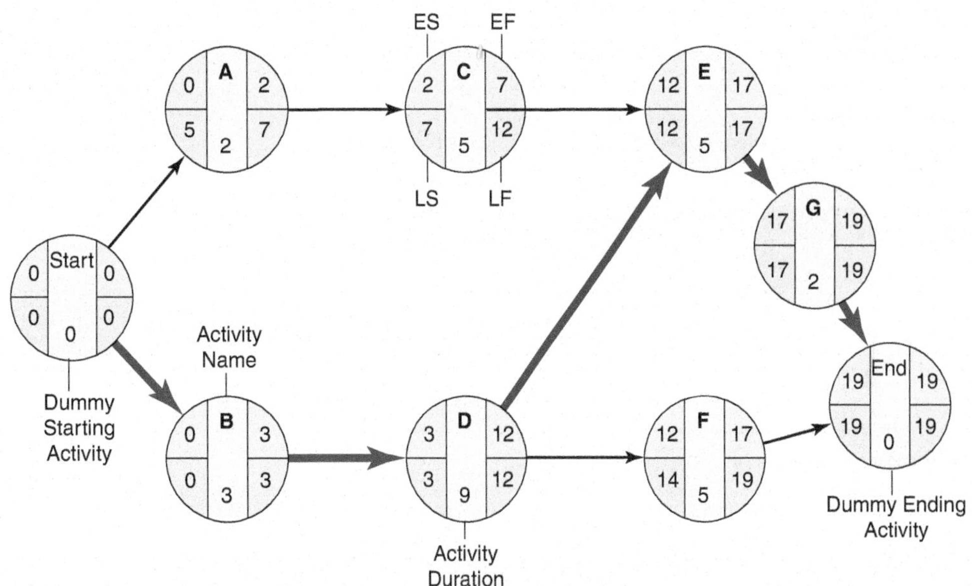

SOLVED PROBLEM 3.6

The following information has been computed from a project:

$$\text{Expected total project time } (T) = 62 \text{ weeks}$$

$$\text{Project variance } (\sigma_p^2) = 81$$

What is the probability that the project will be completed 18 weeks *before* its expected completion date?

SOLUTION

The desired completion date is 18 weeks before the expected completion date, 62 weeks. The desired completion date is 44 (or 62–18) weeks:

$$\sigma_p = \sqrt{\text{Project variance}}$$

$$Z = \frac{\text{Due date} - \text{Expected completion date}}{\sigma_p}$$

$$= \frac{44 - 62}{9} = \frac{-18}{9} = -2.0$$

The normal curve appears as follows:

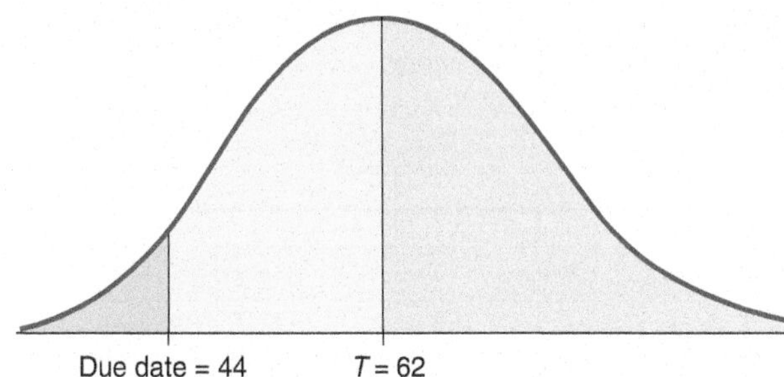

Because the normal curve is symmetrical and table values are calculated for positive values of Z, the area desired is equal to 1– (table value). For $Z = +2.0$ the area from the table is .97725. Thus, the area corresponding to a Z-value of –2.0 is .02275 (or 1 – .97725). Hence, the probability of completing the project 18 weeks before the expected completion date is approximately .023, or 2.3%.

SOLVED PROBLEM 3.7

Determine the least cost of reducing the project completion date by 3 months based on the following information:

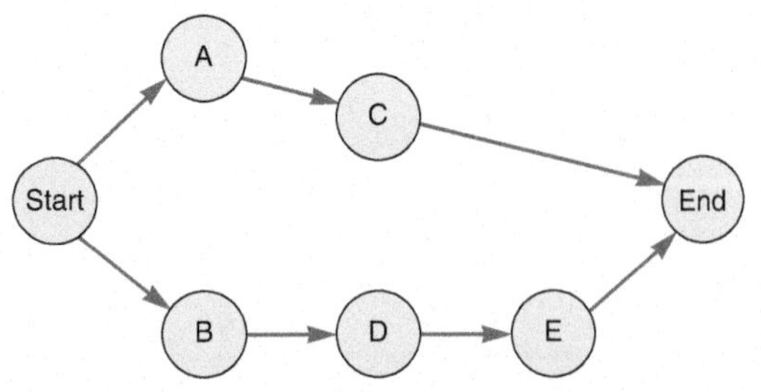

ACTIVITY	NORMAL TIME (MONTHS)	CRASH TIME (MONTHS)	NORMAL COST	CRASH COST
A	6	4	$2,000	$2,400
B	7	5	3,000	3,500
C	7	6	1,000	1,300
D	6	4	2,000	2,600
E	9	8	8,800	9,000

SOLUTION

The first step in this problem is to compute ES, EF, LS, LF, and slack for each activity.

ACTIVITY	ES	EF	LS	LF	SLACK
A	0	6	9	15	9
B	0	7	0	7	0
C	6	13	15	22	9
D	7	13	7	13	0
E	13	22	13	22	0

The critical path consists of activities B, D, and E.

Next, crash cost/month must be computed for each activity:

ACTIVITY	NORMAL TIME–CRASH TIME	CRASH COST–NORMAL COST	CRASH COST/MONTH	CRITICAL PATH?
A	2	$400	$200/month	No
B	2	500	250/month	Yes
C	1	300	300/month	No
D	2	600	300/month	Yes
E	1	200	200/month	Yes

Finally, we will select that activity on the critical path with the smallest crash cost/month. This is activity E. Thus, we can reduce the total project completion date by 1 month for an additional cost of $200. We still need to reduce the project completion date by 2 more months. This reduction can be achieved at least cost along the critical path by reducing activity B by 2 months for an additional cost of $500. Neither reduction has an effect on noncritical activities. This solution is summarized in the following table:

ACTIVITY	MONTHS REDUCED	COST
E	1	$200
B	2	500
		Total: $700

Problems

Note: **PX** means the problem may be solved with POM for Windows and/or Excel OM.

Problems 3.1–3.2 relate to Project Planning

• **3.1** The work breakdown structure (WBS) for building a house (levels 1 and 2) is shown below:

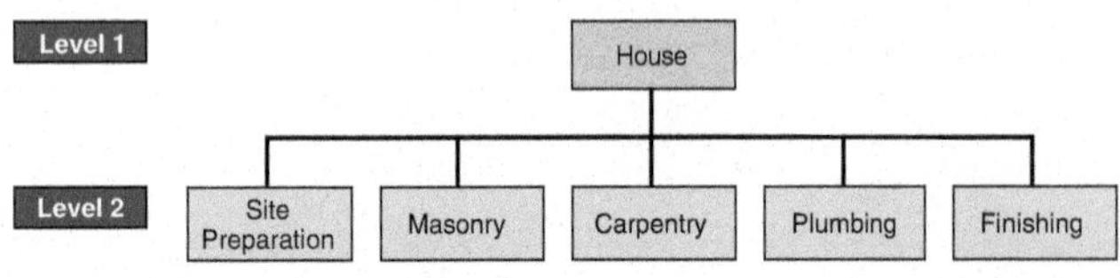

a) Add two level-3 activities to each of the level-2 activities to provide more detail to the WBS.

b) Select one of your level-3 activities and add two level-4 activities below it.

•• **3.2** James Lawson has decided to run for a seat as Congressman from the House of Representatives, District 34, in Florida. He views his 8-month campaign for office as a major project and wishes to create a work breakdown structure (WBS) to help control the detailed scheduling. So far, he has developed the following pieces of the WBS:

LEVEL	LEVEL ID NO.	ACTIVITY
1	1.0	Develop political campaign
2	1.1	Fund-raising plan
3	1.1.1	______
3	1.1.2	______
3	1.1.3	______
2	1.2	Develop a position on major issues
3	1.2.1	______
3	1.2.2	______
3	1.2.3	______
2	1.3	Staffing for campaign
3	1.3.1	______
3	1.3.2	______
3	1.3.3	______
3	1.3.4	______
2	1.4	Paperwork compliance for candidacy
3	1.4.1	______
3	1.4.2	______
2	1.5	Ethical plan/issues
3	1.5.1	______

Help Lawson by providing details where the blank lines appear. Are there any other major (level-2) activities to create? If so, add an ID no. 1.6 and insert them.

Problem 3.3 relates to Project Scheduling

•• **3.3** The City Commission of Nashville has decided to build a botanical garden and picnic area in the heart of the city for the recreation of its citizens. The precedence table for all the activities required to construct this area successfully is given. Draw the Gantt chart for the whole construction activity.

CODE	ACTIVITY	DESCRIPTION	TIME (IN HOURS)	IMMEDIATE PREDECESSOR(S)
A	Planning	Find location; determine resource requirements	20	None
B	Purchasing	Requisition of lumber and sand	60	Planning
C	Excavation	Dig and grade	100	Planning
D	Sawing	Saw lumber into appropriate sizes	30	Purchasing
E	Placement	Position lumber in correct locations	20	Sawing, excavation
F	Assembly	Nail lumber together	10	Placement
G	Infill	Put sand in and under the equipment	20	Assembly
H	Outfill	Put dirt around the equipment	10	Assembly
I	Decoration	Put grass all over the garden, landscape, paint	30	Infill, outfill

Problems 3.4–3.14 relate to Project Management Techniques

•• **3.4** Refer to the table in Problem 3.3.

a) Draw the AON network for the construction activity.
b) Draw the AOA network for the construction activity.

• **3.5** Draw the activity-on-node (AON) project network associated with the following activities for Carl Betterton's construction project. How long should it take Carl and his team to complete this project? What are the critical path activities?

ACTIVITY	IMMEDIATE PREDECESSOR(S)	TIME (DAYS)
A	—	3
B	A	4
C	A	6
D	B	6
E	B	4
F	C	4
G	D	6
H	E, F	8

• **3.6** Given the activities whose sequence is described by the following table, draw the appropriate activity-on-arrow (AOA) network diagram.

a) Which activities are on the critical path?
b) What is the length of the critical path?

ACTIVITY	IMMEDIATE PREDECESSOR(S)	TIME (DAYS)
A	—	5
B	A	2
C	A	4
D	B	5
E	B	5
F	C	5
G	E, F	2
H	D	3
I	G, H	5

• **3.7** Using AOA, diagram the network described below for Lillian Fok's construction project. Calculate its critical path and project completion time.

ACTIVITY	NODES	TIME (WEEKS)	ACTIVITY	NODES	TIME (WEEKS)
J	1–2	10	N	3–4	2
K	1–3	8	O	4–5	7
L	2–4	6	P	3–5	5
M	2–3	3			

•• **3.8** Roger Ginde is developing a program in supply chain management certification for managers. Ginde has listed a number of activities that must be completed before a training program of this nature could be conducted. The activities, immediate predecessors, and times appear in the accompanying table:

ACTIVITY	IMMEDIATE PREDECESSOR(S)	TIME (DAYS)
A	—	2
B	—	5
C	—	1
D	B	10
E	A, D	3
F	C	6
G	E, F	8

a) Develop an AON network for this problem.
b) What is the critical path?
c) What is the total project completion time?
d) What is the slack time for each individual activity? **PX**

•• **3.9** Task time estimates for the modification of an assembly line at Jim Goodale's Carbondale, Illinois, factory are as follows:

ACTIVITY	TIME (IN HOURS)	IMMEDIATE PREDECESSORS
A	6.0	—
B	7.2	—
C	5.0	A
D	6.0	B, C
E	4.5	B, C
F	7.7	D
G	4.0	E, F

a) Draw the project network using AON.
b) Identify the critical path.
c) What is the expected project completion time?
d) Draw a Gantt chart for the project. **PX**

• **3.10** The activities described by the following table are given for the J. C. Howard Corporation in Kansas:

ACTIVITY	IMMEDIATE PREDECESSOR(S)	TIME
A	—	9
B	A	7
C	A	3
D	B	6
E	B	9
F	C	4
G	E, F	6
H	D	5
I	G, H	3

a) Draw the appropriate AON PERT diagram for J. C. Howard's management team.
b) Find the critical path.
c) What is the project completion time? **PX**

•• **3.11** The following is a table of activities associated with a project at Rafay Ishfaq's software firm in Chicago, their durations, and what activities each must precede:

ACTIVITY	DURATION (WEEKS)	PRECEDES
A (start)	1	B, C
B	1	E
C	4	F
E	2	F
F (end)	2	—

a) Draw an AOA diagram of the project, including activity durations.
b) Define the critical path, listing all critical activities in chronological order.
c) What is the project completion time (in weeks)?
d) What is the slack (in weeks) associated with any and all noncritical paths through the project?

• **3.12** The activities needed to build a prototype laser scanning machine at Dave Fletcher Corp. are listed in the following table. Construct an AON network for these activities.

ACTIVITY	IMMEDIATE PREDECESSOR(S)	ACTIVITY	IMMEDIATE PREDECESSOR(S)
A	—	E	B
B	—	F	B
C	A	G	C, E
D	A	H	D, F

•• **3.13** The following represent activities in Marc Massoud's Construction Company project.

ACTIVITY	IMMEDIATE PREDECESSOR(S)	TIME (WEEKS)
A	—	1
B	—	4
C	A	1
D	B	5
E	B	2
F	C, E	7
G	D	2
H	F, G	3

a) Draw the network to represent this situation.
b) Which activities are on the critical path?
c) What is the length of the critical path? **PX**

•• **3.14** Aliyah White's political campaign must coordinate a number of necessary activities to be prepared for an upcoming election. The following table describes the relationships between these activities that need to be completed as well as estimated times.

ACTIVITY	IMMEDIATE PREDECESSOR(S)	TIME (WEEKS)
S	—	4
T	—	6
U	S	3
V	S	4
W	T, U	8
X	T	7
Y	V, W	2
Z	X	1

a) Develop a project network for this problem.
b) Determine the total project completion time and the critical path(s). **PX**

Problems 3.15–3.16 relate to Determining the Project Schedule

• **3.15** Dave Fletcher (see Problem 3.12) was able to determine the activity times for constructing his laser scanning machine. Fletcher would like to determine ES, EF, LS, LF, and slack for each activity. The total project completion time and the critical path should also be determined. Here are the activity times:

ACTIVITY	TIME (WEEKS)	ACTIVITY	TIME (WEEKS)
A	6	E	4
B	7	F	6
C	3	G	10
D	2	H	7

••• **3.16** The Rover 6 is a new custom-designed sports car. An analysis of the task of building the Rover 6 reveals the following list of relevant activities, their immediate predecessors, and their duration.

ACTIVITY	DESCRIPTION	TIME (DAYS)	IMMEDIATE PREDECESSOR(S)
A	Start	0	—
B	Design	8	A
C	Order special accessories	0.1	B
D	Build frame	1	B
E	Build doors	1	B
F	Attach axles, wheels, gas tank	1	D
G	Build body shell	2	B
H	Build transmission and drivetrain	3	B
I	Fit doors to body shell	1	G, E
J	Build engine	4	B
K	Bench-test engine	2	J
L	Assemble chassis	1	F, H, K
M	Road-test chassis	0.5	L
N	Paint body	2	I
O	Install wiring	1	N
P	Install interior	1.5	N
Q	Accept delivery of special accessories	5	C
R	Mount body and accessories on chassis	1	M, O, P, Q
S	Road test car	0.5	R
T	Attach exterior trim	1	S
U	Finish	0	T

a) Draw a network diagram for the project.
b) Mark the critical path and state its length.
c) If the Rover 6 had to be completed 2 days earlier, would it help to
   i) buy preassembled transmissions and drivetrains?
   ii) install robots to halve engine-building time?
   iii) speed delivery of special accessories by 3 days?
d) How might resources be borrowed from activities on the non-critical path to speed activities on the critical path? **PX**

Problems 3.17–3.27 relate to Variability in Activity Times

••• **3.17** Ross Hopkins, president of Hopkins Hospitality, has developed the tasks, durations, and predecessor relationships in the following table for building new motels. Draw the AON network and answer the questions that follow.

		TIME ESTIMATES (IN WEEKS)		
ACTIVITY	IMMEDIATE PREDECESSOR(S)	OPTIMISTIC	MOST LIKELY	PESSIMISTIC
A	—	4	8	10
B	A	2	8	24
C	A	8	12	16
D	A	4	6	10
E	B	1	2	3
F	E, C	6	8	20
G	E, C	2	3	4
H	F	2	2	2
I	F	6	6	6
J	D, G, H	4	6	12
K	I, J	2	2	3

a) What is the expected (estimated) time for activity C?
b) What is the variance for activity C?
c) Based on the calculation of estimated times, what is the critical path?
d) What is the estimated time of the critical path?
e) What is the activity variance along the critical path?
f) What is the probability of completion of the project before week 36? **PX**

• **3.18** A renovation of the gift shop at Orlando Amway Center has six activities (in hours). For the following estimates of *a*, *m*, and *b*, calculate the expected time and the standard deviation for each activity: **PX**

ACTIVITY	*a*	*m*	*b*
A	11	15	19
B	27	31	41
C	18	18	18
D	8	13	19
E	17	18	20
F	16	19	22

•• **3.19** Kelle Carpet and Trim installs carpet in commercial offices. Peter Kelle has been very concerned with the amount of time it took to complete several recent jobs. Some of his workers are very unreliable. A list of activities and their optimistic completion time, the most likely completion time, and the pessimistic completion time (all in days) for a new contract are given in the following table:

	TIME (DAYS)			
ACTIVITY	*a*	*m*	*b*	IMMEDIATE PREDECESSOR(S)
A	3	6	8	—
B	2	4	4	—
C	1	2	3	—
D	6	7	8	C
E	2	4	6	B, D
F	6	10	14	A, E
G	1	2	4	A, E
H	3	6	9	F
I	10	11	12	G
J	14	16	20	C
K	2	8	10	H, I

a) Determine the expected completion time and variance for each activity.
b) Determine the total project completion time and the critical path for the project.
c) Determine ES, EF, LS, LF, and slack for each activity.
d) What is the probability that Kelle Carpet and Trim will finish the project in 40 days or less? **PX**

••• **3.20** The estimated times and immediate predecessors for the activities in a project at George Kyparis's retinal scanning company are given in the following table. Assume that the activity times are independent.

		TIME (WEEKS)		
ACTIVITY	IMMEDIATE PREDECESSOR	*a*	*m*	*b*
A	—	9	10	11
B	—	4	10	16
C	A	9	10	11
D	B	5	8	11

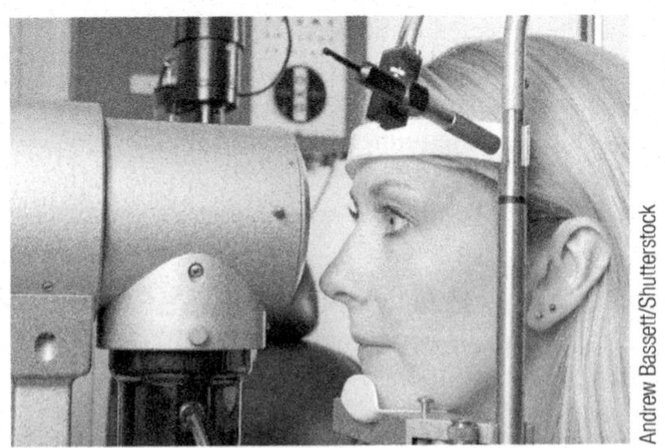
Andrew Bassett/Shutterstock

a) Calculate the expected time and variance for each activity.
b) What is the expected completion time of the critical path? What is the expected completion time of the other path in the network?
c) What is the variance of the critical path? What is the variance of the other path in the network?
d) If the time to complete path A–C is normally distributed, what is the probability that this path will be finished in 22 weeks or less?
e) If the time to complete path B–D is normally distributed, what is the probability that this path will be finished in 22 weeks or less?
f) Explain why the probability that the *critical path* will be finished in 22 weeks or less is not necessarily the probability that the *project* will be finished in 22 weeks or less. **PX**

••• **3.21** Coleman Rich Control Devices, Inc., produces custom-built relay devices for auto makers. The most recent project undertaken by Rich requires 14 different activities. Rich's managers would like to determine the total project completion time (in days) and those activities that lie along the critical path. The appropriate data are shown in the following table.

ACTIVITY	IMMEDIATE PREDECESSOR(S)	OPTIMISTIC TIME	MOST LIKELY TIME	PESSIMISTIC TIME
A	—	4	6	7
B	—	1	2	3
C	A	6	6	6
D	A	5	8	11
E	B, C	1	9	18
F	D	2	3	6
G	D	1	7	8
H	E, F	4	4	6
I	G, H	1	6	8
J	I	2	5	7
K	I	8	9	11
L	J	2	4	6
M	K	1	2	3
N	L, M	6	8	10

a) What is the probability of being done in 53 days?
b) What date results in a 99% probability of completion? **PX**

••• **3.22** Four Squares Productions, a firm hired to coordinate the release of the movie *Pirates of the Caribbean: Dead Men Tell No Tales* (starring Johnny Depp), identified 16 activities to be completed before the release of the film.

ACTIVITY	IMMEDIATE PREDECESSOR(S)	OPTIMISTIC TIME	MOST LIKELY TIME	PESSIMISTIC TIME
A	—	1	2	4
B	—	3	3.5	4
C	—	10	12	13
D	—	4	5	7
E	—	2	4	5
F	A	6	7	8
G	B	2	4	5.5
H	C	5	7.7	9
I	C	9.9	10	12
J	C	2	4	5
K	D	2	4	6
L	E	2	4	6
M	F, G, H	5	6	6.5
N	J, K, L	1	1.1	2
O	I, M	5	7	8
P	N	5	7	9

a) How many weeks in advance of the film release should Four Squares have started its marketing campaign? What is the critical path? The tasks (in time units of weeks) are as follows:
b) What is the probability of completing the marketing campaign in the time (in weeks) noted in part (a)?
c) If activities I and J were not necessary, what impact would this have on the critical path and the number of weeks needed to complete the marketing campaign? **PX**

•• **3.23** Using PERT, Adam Munson was able to determine that the expected project completion time for the construction of a pleasure yacht is 21 months, and the project variance is 4.
a) What is the probability that the project will be completed in 17 months?
b) What is the probability that the project will be completed in 20 months?
c) What is the probability that the project will be completed in 23 months?
d) What is the probability that the project will be completed in 25 months?
e) What is the due date that yields a 95% chance of completion? **PX**

•• **3.24** A small software development project at Krishna Dhir's firm has five major activities. The times are estimated and provided in the following table.

ACTIVITY	IMMEDIATE PREDECESSOR	*a*	*m*	*b*
A	—	2	5	8
B	—	3	6	9
C	A	4	7	10
D	B	2	5	14
E	C	3	3	3

a) Find the expected completion time for this project.
b) What variance value would be used to find probabilities of finishing by a certain time? **PX**

••• **3.25** Janice Eliasson must complete the activities in the following table to finish her consulting project.

ACTIVITY	EXPECTED TIME	STANDARD DEVIATION OF TIME ESTIMATE	IMMEDIATE PREDECESSOR(S)
A	7	2	—
B	3	1	A
C	9	3	A
D	4	1	B, C
E	5	1	B, C
F	8	2	E
G	8	1	D, F
H	6	2	G

a) Draw the appropriate PERT diagram.
b) Find the critical path and project completion time.
c) Find the probability that the project will take more than 49 time periods to complete. **PX**

••• **3.26** Shanice Brown's gaming software firm has developed the data for a new project, as given in the following table.

ACTIVITY	DAYS			IMMEDIATE PREDECESSOR(S)
	a	*m*	*b*	
A	8	10	12	—
B	1	6	23	—
C	9	12	15	A, B
D	2	5	8	C

a) What is the critical path? What is the expected completion time of the project?
b) What is the probability of all activities on the critical path being completed within 29 days?
c) What is the probability of all activities on the other (noncritical) path being completed within 29 days?
d) Is it safe to say that the probability of the project being completed within 29 days is equal to your answer from part (b)? **PX**

•• **3.27** Consider the data for a simple project at Carlos Lopez's real estate development company, as given in the following table.

ACTIVITY	DAYS			IMMEDIATE PREDECESSOR
	a	*m*	*b*	
Q	7	10	12	—
R	1	6	23	Q
S	10	12	16	R
T	1	5	7	S

a) Note that this project has only one path. What is the expected completion time of the project?
b) What is the probability of the project being completed within 38 days? **PX**

Problems 3.28–3.33 relate to Cost-Time Trade-Offs and Project Crashing

•• **3.28** Assume that the activities in Problem 3.11 have the following costs to shorten: A, $300/week; B, $100/week; C, $200/week; E, $100/week; and F, $400/week. Assume also that you can crash an activity down to 0 weeks in duration and that every week you can shorten the project is worth $250 to you. What activities would you crash? What is the total crashing cost?

••• **3.29** What is the minimum cost of crashing the following project that Roger Solano manages at Slippery Rock University by 4 days? **PX**

ACTIVITY	NORMAL TIME (DAYS)	CRASH TIME (DAYS)	NORMAL COST	CRASH COST	IMMEDIATE PREDECESSOR(S)
A	6	5	$ 900	$1,000	—
B	8	6	300	400	—
C	4	3	500	600	—
D	5	3	900	1,200	A
E	8	5	1,000	1,600	C

•• **3.30** Three activities are candidates for crashing on a project network for a large computer installation (all are, of course, critical). Activity details are in the following table:

ACTIVITY	PREDECESSOR	NORMAL TIME	NORMAL COST	CRASH TIME	CRASH COST
A	—	7 days	$6,000	6 days	$6,600
B	A	4 days	1,200	2 days	3,000
C	B	11 days	4,000	9 days	6,000

a) What action would you take to reduce the critical path by 1 day?
b) Assuming no other paths become critical, what action would you take to reduce the critical path 1 additional day?
c) What is the total cost of the 2-day reduction? **PX**

••• **3.31** Development of Version 2.0 of a particular accounting software product is being considered by Jose Noguera's technology firm in Baton Rouge. The activities necessary for the completion of this project are listed in the following table:

ACTIVITY	NORMAL TIME (WEEKS)	CRASH TIME (WEEKS)	NORMAL COST	CRASH COST	IMMEDIATE PREDECESSOR(S)
A	4	3	$2,000	$2,600	—
B	2	1	2,200	2,800	—
C	3	3	500	500	—
D	8	4	2,300	2,600	A
E	6	3	900	1,200	B
F	3	2	3,000	4,200	C
G	4	2	1,400	2,000	D, E

a) What is the project completion date?
b) What is the total cost required for completing this project on normal time?
c) If you wish to reduce the time required to complete this project by 1 week, which activity should be crashed, and how much will this increase the total cost?
d) What is the maximum time that can be crashed? How much would costs increase? **PX**

••• **3.32** Kimpel Products makes pizza ovens for commercial use. James Kimpel, CEO, is contemplating producing smaller ovens for use in high school and college kitchens. The activities necessary to build an experimental model and related data are given in the following table:

ACTIVITY	NORMAL TIME (WEEKS)	CRASH TIME (WEEKS)	NORMAL COST ($)	CRASH COST ($)	IMMEDIATE PREDECESSOR(S)
A	3	2	1,000	1,600	—
B	2	1	2,000	2,700	—
C	1	1	300	300	—
D	7	3	1,300	1,600	A
E	6	3	850	1,000	B
F	2	1	4,000	5,000	C
G	4	2	1,500	2,000	D, E

a) What is the project completion date?
b) Crash this project to 10 weeks at the least cost.
c) Crash this project to 7 weeks (which is the maximum it can be crashed) at the least cost. **PX**

•• **3.33** What is the minimum cost of crashing the following project at Sawaya Robotics by 4 days? **PX**

ACTIVITY	NORMAL TIME (DAYS)	CRASH TIME (DAYS)	NORMAL COST	CRASH COST	IMMEDIATE PREDECESSOR(S)
Design	6	5	900	1,000	—
Wiring	8	6	300	400	—
Chip install	4	3	500	600	—
Software	5	3	900	1,200	A, B
Testing	8	5	1,000	1,600	C

CASE STUDIES

Project Management at Arnold Palmer Hospital

Video Case

The equivalent of a new kindergarten class is born every day at Orlando's Arnold Palmer Hospital. With more than 13,000 births in a hospital that was designed for a capacity of 6,500 births a year, the newborn intensive care unit was stretched to the limit. Moreover, with continuing strong population growth in central Florida, the hospital was often full. It was clear that new facilities were needed. After much analysis, forecasting, and discussion, the management team decided to build a new 273-bed building across the street from the existing hospital. But the facility had to be built in accordance with the hospital's Guiding Principles and its uniqueness as a health center dedicated to the specialized needs of women and infants. Those Guiding Principles are: *Family-centered focus, a healing environment where privacy and dignity are respected, sanctuary of caring that includes warm, serene surroundings with natural lighting, sincere and dedicated staff providing the highest quality care, and patient-centered flow and function.*

The vice president of business development, Karl Hodges, wanted a hospital that was designed from the inside out by the people who understood the Guiding Principles, who knew most about the current system, and who were going to use the new system, namely, the doctors and nurses. Hodges and his staff spent 13 months discussing expansion needs with this group, as well as with patients and the community, before developing a proposal for the new facility. An administrative team created 35 user groups, which held over 1,000 planning meetings (lasting from 45 minutes to a whole day). They even created a "Supreme Court" to deal with conflicting views on the multifaceted issues facing the new hospital.

Funding and regulatory issues added substantial complexity to this major expansion, and Hodges was very concerned that the project stay on time and within budget. Tom Hyatt, director of facility development, was given the task of onsite manager of the $100 million project, in addition to overseeing ongoing renovations, expansions, and other projects. The activities in the multi-year project for the new building at Arnold Palmer are shown in Table 3.6 on the next page.

Discussion Questions*

1. Develop the network for planning and construction of the new hospital at Arnold Palmer.
2. What is the critical path, and how long is the project expected to take?
3. Why is the construction of this 11-story building any more complex than construction of an equivalent office building?
4. What percentage of the whole project duration was spent in planning that occurred prior to the proposal and reviews? Prior to the actual building construction? Why?

*You may wish to view the video accompanying this case before addressing these questions.

TABLE 3.6 Expansion Planning and Arnold Palmer Hospital Construction Activities and Times[a]

ACTIVITY	SCHEDULED TIME	PRECEDENCE ACTIVITY(IES)
1. Proposal and review	1 month	—
2. Establish master schedule	2 weeks	1
3. Architect selection process	5 weeks	1
4. Survey whole campus and its needs	1 month	1
5. Conceptual architect's plans	6 weeks	3
6. Cost estimating	2 months	2, 4, 5
7. Deliver plans to board for consideration/decision	1 month	6
8. Surveys/regulatory review	6 weeks	6
9. Construction manager selection	9 weeks	6
10. State review of need for more hospital beds ("Certificate of Need")	3.5 months	7, 8
11. Design drawings	4 months	10
12. Construction documents	5 months	9, 11
13. Site preparation/demolish existing building	9 weeks	11
14. Construction start/building pad	2 months	12, 13
15. Relocate utilities	6 weeks	12
16. Deep foundations	2 months	14
17. Building structure in place	9 months	16
18. Exterior skin/roofing	4 months	17
19. Interior buildout	12 months	17
20. Building inspections	5 weeks	15, 19
21. Occupancy	1 month	20

[a] This list of activities is abbreviated for purposes of this case study. For simplification, assume each week = .25 months (i.e., 2 weeks = .5 month, 6 weeks = 1.5 months, etc.).

Managing Hard Rock's Rockfest

Video Case

At the Hard Rock Cafe, like many organizations, project management is a key planning tool. With Hard Rock's constant growth in hotels and cafes, remodeling of existing cafes, scheduling for Hard Rock Live concert and event venues, and planning the annual Rockfest, managers rely on project management techniques and software to maintain schedule and budget performance.

"Without Microsoft Project," says Hard Rock Vice-President Chris Tomasso, "there is no way to keep so many people on the same page." Tomasso is in charge of the Rockfest event, which is attended by well over 100,000 enthusiastic fans. The challenge is pulling it off within a tight 9-month planning horizon. As the event approaches, Tomasso devotes greater energy to its activities. For the first 3 months, Tomasso updates his Microsoft Project charts monthly. Then at the 6-month mark, he updates his progress weekly. At the 9-month mark, he checks and corrects his schedule twice a week.

Early in the project management process, Tomasso identifies 10 major tasks (called level-2 activities in a work breakdown structure, or WBS):[†] talent booking, ticketing, marketing/PR, online promotion, television, show production, travel, sponsorships, operations, and merchandising. Using a WBS, each of these is further divided into a series of subtasks. Table 3.7 identifies 26 of the major activities and subactivities, their immediate predecessors, and time estimates. Tomasso enters all these into the Microsoft Project software.[‡] Tomasso alters the Microsoft Project document and the time line as the project progresses. "It's okay to change it as long as you keep on track," he states.

The day of the rock concert itself is not the end of the project planning. "It's nothing but surprises. A band not being able to get to the venue because of traffic jams is a surprise, but an 'anticipated' surprise. We had a helicopter on stand-by ready to fly the band in," says Tomasso.

On completion of Rockfest in July, Tomasso and his team have a 3-month reprieve before starting the project planning process again.

[†] The level-1 activity is the Rockfest concert itself.

[‡] There are actually 127 activities used by Tomasso; the list is abbreviated for this case study.

TABLE 3.7 Some of the Major Activities and Subactivities in the Rockfest Plan

ACTIVITY	DESCRIPTION	PREDECESSOR(S)	TIME (WEEKS)
A	Finalize site and building contracts	—	7
B	Select local promoter	A	3
C	Hire production manager	A	3
D	Design promotional Web site	B	5
E	Set TV deal	D	6
F	Hire director	E	4
G	Plan for TV camera placement	F	2
H	Target headline entertainers	B	4
I	Target support entertainers	H	4
J	Travel accommodations for talent	I	10
K	Set venue capacity	C	2
L	Ticketmaster contract	D, K	3
M	On-site ticketing	L	8
N	Sound and staging	C	6
O	Passes and stage credentials	G, R	7
P	Travel accommodations for staff	B	20
Q	Hire sponsor coordinator	B	4
R	Finalize sponsors	Q	4
S	Define/place signage for sponsors	R, X	3
T	Hire operations manager	A	4
U	Develop site plan	T	6
V	Hire security director	T	7
W	Set police/fire security plan	V	4
X	Power, plumbing, AC, toilet services	U	8
Y	Secure merchandise deals	B	6
Z	Online merchandise sales	Y	6

Discussion Questions§

1. Identify the critical path and its activities for Rockfest. How long does the project take?
2. Which activities have a slack time of 8 weeks or more?
3. Identify five major challenges a project manager faces in events such as this one.
4. Why is a work breakdown structure useful in a project such as this? Take the 26 activities and break them into what you think should be level-2, level-3, and level-4 tasks.

§You may wish to view the video accompanying this case before addressing these questions.

Endnotes

1. Sources: *Washington Post* (Sep. 13, 2017); *Knight Ridder Tribune Business News* (Jul. 16, 2005); and *New York Times* (Dec. 8, 2015).
2. This formula is based on the statistical concept that from one end of the beta distribution to the other is 6 standard deviations (±3 standard deviations from the mean). Because $(b - a)$ is 6 standard deviations, the variance is $[(b - a)/6]^2$.
3. Sources: **ComputerWeekly.com** (July, 2017); *BBC News* (July 6, 2014); and **ProjectManagement.com** (October 7, 2016).
4. A trial version of Microsoft Project can be downloaded from **www.microsoft.com/en-us/microsoft-365/project/project-management-software**.

Bibliography

Balakrishnan, R., B. Render, R. M. Stair, and C. Munson. *Managerial Decision Modeling: Business Analytics with Spreadsheets*, 4th ed. Boston: DeGruyter, 2017.

Birkinshaw, J. "What to Expect from Agile." *MIT Sloan Management Review* 59, no. 2 (Winter 2018): 39–42.

Gray, C. L., and E. W. Larson. *Project Management*, 6th ed. New York: McGraw-Hill/Irwin, 2014.

Kerzner, H. *Project Management: A Systems Approach to Planning, Scheduling, and Controlling*, 12th ed. New York: Wiley, 2017.

Matta, N. F., and R. N. Ashkenas. "Why Good Projects Fail Anyway." *Harvard Business Review* 85, no. 5 (September 2003): 109–114.

Render, B., R. M. Stair, M. Hanna, and T. Hale. *Quantitative Analysis for Management*, 13th ed. Boston: Pearson, 2018.

Verzuh, E. *The Fast Forward MBA in Project Management*. New York: Wiley, 2008.

Chapter 3 *Rapid* Review

Main Heading	Review Material	MyLab Operations Management
THE IMPORTANCE OF PROJECT MANAGEMENT	The management of projects involves three phases: 1. *Planning*—This phase includes goal setting, defining the project, and team organization. 2. *Scheduling*—This phase relates people, money, and supplies to specific activities and relates activities to each other. 3. *Controlling*—Here the firm monitors resources, costs, quality, and budgets. It also revises or changes plans and shifts resources to meet time and cost demands.	Concept Questions: 1.1–1.6 **VIDEO 3.1** Project Management at Hard Rock's Rockfest
PROJECT PLANNING	Projects can be defined as a series of related tasks directed toward a major output. ▪ **Project organization**—An organization formed to ensure that programs (projects) receive the proper management and attention. ▪ **Work breakdown structure (WBS)**—Defines a project by dividing it into more and more detailed components.	Concept Questions: 2.1–2.6 Problem 3.1–3.2
PROJECT SCHEDULING	▪ **Gantt charts**—Planning charts used to schedule resources and allocate time. Project scheduling serves several purposes: 4. It shows the relationship of each activity to others and to the whole project. 5. It identifies the precedence relationships among activities. 6. It encourages the setting of realistic time and cost estimates for each activity. 7. It helps make better use of people, money, and material resources by identifying critical bottlenecks in the project.	Concept Questions: 3.1–3.5 Problem 3.3
PROJECT CONTROLLING	Computerized programs produce a broad variety of PERT/CPM reports, including (1) detailed cost breakdowns for each task, (2) total program labor curves, (3) cost distribution tables, (4) functional cost and hour summaries, (5) raw material and expenditure forecasts, (6) variance reports, (7) time analysis reports, and (8) work status reports. ▪ **Waterfall projects**—Projects that progress smoothly in a step-by-step manner until completed. ▪ **Agile projects**—Ill-defined projects requiring collaboration and constant feedback to adjust to project unknowns.	Concept Questions: 4.1–4.4 **VIDEO 3.2** Project Management at Arnold Palmer Hospital
PROJECT MANAGEMENT TECHNIQUES: PERT AND CPM	▪ **Program evaluation and review technique (PERT)**—A project management technique that employs three time estimates for each activity. ▪ **Critical path method (CPM)**—A project management technique that uses only one estimate per activity. ▪ **Critical path**—The computed *longest* time path(s) through a network. PERT and CPM both follow six basic steps. The activities on the critical path will delay the entire project if they are not completed on time. ▪ **Activity-on-node (AON)**—A network diagram in which nodes designate activities. ▪ **Activity-on-arrow (AOA)**—A network diagram in which arrows designate activities. In an AOA network, the nodes represent the starting and finishing times of an activity and are also called *events*. ▪ **Dummy activity**—An activity having no time that is inserted into a network to maintain the logic of the network. A dummy ending activity can be added to the end of an AON diagram for a project that has multiple ending activities.	Concept Questions: 5.1–5.6 Problems: 3.4–3.14 Virtual Office Hours for Solved Problems: 3.1, 3.2
DETERMINING THE PROJECT SCHEDULE	▪ **Critical path analysis**—A process that helps determine a project schedule. To find the critical path, we calculate two distinct starting and ending times for each activity: ▪ *Earliest start (ES)* = Earliest time at which an activity can start, assuming that all predecessors have been completed ▪ *Earliest finish (EF)* = Earliest time at which an activity can be finished ▪ *Latest start (LS)* = Latest time at which an activity can start, without delaying the completion time of the entire project ▪ *Latest finish (LF)* = Latest time by which an activity has to finish so as to not delay the completion time of the entire project ▪ **Forward pass**—A process that identifies all the early start and early finish times. $ES = \text{Max}\{\text{EF of all immediate predecessors}\}$ (3-1) $EF = ES + \text{Activity time}$ (3-2) ▪ **Backward pass**—A process that identifies all the late start and late finish times. $LF = \text{Min}\{\text{LS of all immediate following activities}\}$ (3-3) $LS = LF - \text{Activity time}$ (3-4)	Concept Questions: 6.1–6.6 Problems: 3.15, 3.16

Main Heading	Review Material	MyLab Operations Management
	■ **Slack time**—Free time for an activity. Slack = LS − ES or Slack = LF − EF (3-5) The activities with zero slack are called *critical activities* and are said to be on the critical path. The critical path is a continuous path through the project network that starts at the first activity in the project, terminates at the last activity in the project, and includes only critical activities.	Virtual Office Hours for Solved Problem: 3.3 ACTIVE MODEL 3.1
VARIABILITY IN ACTIVITY TIMES	■ **Optimistic time** (a)—The "best" activity completion time that could be obtained in a PERT network. ■ **Pessimistic time** (b)—The "worst" activity time that could be expected in a PERT network. ■ **Most likely time** (m)—The most probable time to complete an activity in a PERT network. When using PERT, we often assume that activity time estimates follow the beta distribution. Expected activity time $t = (a + 4m + b)/6$ (3-6) Variance of activity completion time $= [(b - a)/6]^2$ (3-7) σ_p^2 = Project variance = Σ(variances of activities on critical path) (3-8) Z = (Due date − Expected date of completion) / σ_p (3-9) Due date = Expected completion time + $(Z \times \sigma_p)$ (3-10)	Concept Questions: 7.1–7.6 Problems 3.17–3.27 Virtual Office Hours for Solved Problems: 3.4, 3.5, 3.6
COST–TIME TRADE-OFFS AND PROJECT CRASHING	■ **Crashing**—Shortening activity time in a network to reduce time on the critical path so total completion time is reduced. $\text{Crash cost per period} = \frac{(\text{Crash cost} - \text{Normal cost})}{(\text{Normal time} - \text{Crash time})}$ (3-11)	Concept Questions: 8.1–8.6 Problems: 3.28–3.33 Virtual Office Hours for Solved Problem: 3.7
A CRITIQUE OF PERT AND CPM	As with every technique for problem solving, PERT and CPM have a number of advantages as well as several limitations.	Concept Questions: 9.1–9.6
USING SOFTWARE TO MANAGE PROJECTS	No large projects are managed these days without project management software. These commercial products are all extremely useful in drawing project networks, identifying the project schedule and the necessary resources, as well as tracking time and money.	Concept Questions: 10.1–10.6
ADDITIONAL MYLAB OPERATIONS MANAGEMENT RESOURCES	✔ Additional Case Studies (Shale Oil Company and Southwestern University (A)) ✔ Multiple Choice Case Questions (Southwestern University (A)) ✔ Recent Graduate Video: Kimberly Gersh, Project Manager, Little Green Software ✔ Project Management Simulation	

Self Test

LO 3.1 Which of the following statements regarding Gantt charts is true?
- **a)** Gantt charts give a timeline and precedence relationships for each activity of a project.
- **b)** Gantt charts use the four standard spines: Methods, Materials, Manpower, and Machinery.
- **c)** Gantt charts are visual devices that show the duration of activities in a project.
- **d)** Gantt charts are expensive.
- **e)** All of the above are true.

LO 3.2 Which of the following is true about AOA and AON networks?
- **a)** In AOA, arrows represent activities.
- **b)** In AON, nodes represent activities.
- **c)** Activities consume time and resources.
- **d)** Nodes are also called *events* in AOA.
- **e)** All of the above.

LO 3.3 Slack time equals:
- **a)** ES + t.
- **b)** LS − ES.
- **c)** zero.
- **d)** EF − ES.

LO 3.4 The critical path of a network is the:
- **a)** shortest-time path through the network.
- **b)** path with the fewest activities.
- **c)** path with the most activities.
- **d)** longest-time path through the network.

LO 3.5 PERT analysis computes the variance of the total project completion time as:
- **a)** the sum of the variances of all activities in the project.
- **b)** the sum of the variances of all activities on the critical path.
- **c)** the sum of the variances of all activities not on the critical path.
- **d)** the variance of the final activity of the project.

LO 3.6 The crash cost per period:
- **a)** is the difference in costs divided by the difference in times (crash and normal).
- **b)** is considered to be linear in the range between normal and crash.
- **c)** needs to be determined so that the smallest cost values on the critical path can be considered for time reduction first.
- **d)** all of the above.

Answers: LO 3.1. c; LO 3.2. e; LO 3.3. b; LO 3.4. d; LO 3.5. b; LO 3.6. d.

Forecasting

CHAPTER 4

CHAPTER OUTLINE

CityFiles/WireImage/Getty Images

Dick Loek/Toronto Star/Getty Images

Forecasters often have to deal with products that have highly seasonal demand. Yamaha, the manufacturer of this WaveRunner and snowmobile, produces products with complementary demands to address seasonal fluctuations.

CHAPTER 4

GLOBAL COMPANY PROFILE
Walt Disney Parks & Resorts

Forecasting Provides a Competitive Advantage for Disney

When it comes to the world's most respected global brands, Walt Disney Parks & Resorts is a visible leader. Although the monarch of this magic kingdom is no man but a mouse—Mickey Mouse—it's CEO Bob Chapek who daily manages the entertainment giant.

Disney's global portfolio includes Shanghai Disney (2016), Hong Kong Disneyland (2005), Disneyland Paris (1992), and Tokyo Disneyland (1983). But it is Walt Disney World Resort in Florida and Disneyland Resort in California that drive profits in this multibillion-dollar corporation, which is ranked in the top 100 in both the *Fortune* 500 and *Financial Times* Global 500.

Helen Sessions/Alamy Stock Photo

Mickey and Minnie Mouse provide the public image of Disney to the world. Forecasts drive the work schedules of 72,000 cast members working at Walt Disney World Resort near Orlando.

Revenues at Disney are all about people—how many visit the parks and how they spend money while there. When Chapek receives a daily report from his four theme parks and two water parks near Orlando, the report contains only two numbers: the *forecast* of yesterday's attendance at the parks (Magic Kingdom, Epcot, Disney's Animal Kingdom, Disney's Hollywood Studios, Typhoon Lagoon, and Blizzard Beach) and the *actual* attendance. An error close to zero is expected. Chapek takes his forecasts very seriously.

The forecasting team at Walt Disney World Resort doesn't just do a daily prediction, however, and Chapek is not its only customer. The team also provides daily, weekly, monthly, annual, and 5-year forecasts to the labor management, maintenance, operations, finance, and park scheduling departments. Forecasters use judgmental models, econometric models, moving-average models, and regression analysis.

The giant sphere is the symbol of Epcot, one of Disney's four Orlando parks, for which forecasts of meals, lodging, entertainment, and transportation must be made. This Disney monorail moves guests among parks and the 28 hotels on the massive 47-square-mile property (about the size of San Francisco and twice the size of Manhattan).

Nicolas Chan/Alamy Stock Photo

Kevin Fleming/The Image Bank Unreleased/Getty Images

A daily forecast of attendance is made by adjusting Disney's annual operating plan for weather forecasts, the previous day's crowds, conventions, and seasonal variations. One of the two water parks at Walt Disney World Resort, Typhoon Lagoon, is shown here.

Melvyn Longhurst/The Image Bank Unreleased/Getty Images

Cinderella's iconic castle is a focal point for meeting up with family and friends in the massive park. The statue of Walt Disney greets visitors to the open plaza.

With 20% of Walt Disney World Resort's customers coming from outside the United States, its economic model includes such variables as gross domestic product (GDP), cross-exchange rates, and arrivals into the U.S. Disney also uses 35 analysts and 70 field people to survey 1 million people each year. The surveys, administered to guests at the parks and its 20 hotels, to employees, and to travel industry professionals, examine future travel plans and experiences at the parks. This helps forecast not only attendance but also behavior at each ride (e.g., how long people will wait, how many times they will ride). Inputs to the monthly forecasting model include airline specials, speeches by the chair of the Federal Reserve, and Wall Street trends. Disney even monitors 3,000 school districts inside and outside the U.S. for holiday/vacation schedules. With this approach, Disney's 5-year attendance forecast yields just a 5% error on average. Its annual forecasts have a 0% to 3% error.

RSBPhoto1/Alamy Stock Photo

Forecasts are critical to making sure rides are not overcrowded. Disney is good at "managing demand" with techniques such as adding more street activities to reduce long lines for rides. On slow days, Disney calls fewer cast members to work.

Attendance forecasts for the parks drive a whole slew of management decisions. For example, capacity on any day can be increased by opening at 8 A.M. instead of the usual 9 A.M., by opening more shows or rides, by adding more food/beverage carts (9 million hamburgers and 50 million Cokes are sold per year!), and by bringing in more employees (called "cast members"). Cast members are scheduled in 15-minute intervals throughout the parks for flexibility. Demand can be managed by limiting the number of guests admitted to the parks, with the "FastPass+" reservation system, and by shifting crowds from rides to more street parades.

At Disney, forecasting is a key driver in the company's success and competitive advantage.

LEARNING OBJECTIVES

What Is Forecasting?

Every day, managers like those at Disney make decisions without knowing what will happen in the future. They order inventory without knowing what sales will be, purchase new equipment despite uncertainty about demand for products, and make investments without knowing what profits will be. Managers are always trying to make better estimates of what will happen in the future in the face of uncertainty. Making good estimates is the main purpose of forecasting.

STUDENT TIP An increasingly complex world economy makes forecasting challenging.

In this chapter, we examine different types of forecasts and present a variety of forecasting models. Our purpose is to show that there are many ways for managers to forecast. We also provide an overview of business sales forecasting and describe how to prepare, monitor, and judge the accuracy of a forecast. Good forecasts are an *essential* part of efficient service and manufacturing operations.

Forecasting
The art and science of predicting future events.

Forecasting is the art and science of predicting future events. Forecasting may involve taking historical data (such as past sales) and projecting them into the future with a mathematical model. It may be a subjective or an intuitive prediction (e.g., "this is a great new product and will sell 20% more than the old one"). It may be based on demand-driven data, such as customer plans to purchase, and projecting them into the future. Or the forecast may involve a combination of these, that is, a mathematical model adjusted by a manager's good judgment.

As we introduce different forecasting techniques in this chapter, you will see that there is seldom one superior method. Forecasts may be influenced by a product's position in its life cycle—whether sales are in an introduction, growth, maturity, or decline stage. Other products can be influenced by the demand for a related product; for example, navigation systems may track with new car sales. Because there are limits to what can be expected from forecasts, we develop error measures. Preparing and monitoring forecasts can also be costly and time consuming.

Few businesses, however, can afford to avoid the process of forecasting by just waiting to see what happens and then taking their chances. Effective planning in both the short run and long run depends on a forecast of demand for the company's products.

Forecasting Time Horizons

LO 4.1 *Understand* the three time horizons and which models apply for each

A forecast is usually classified by the *future time horizon* that it covers. Time horizons fall into three categories:

1. *Short-range forecast:* This forecast has a time span of up to 1 year but is generally less than 3 months. It is used for planning purchasing, job scheduling, workforce levels, job assignments, and production levels.
2. *Medium-range forecast:* A medium-range, or intermediate, forecast generally spans from 3 months to 3 years. It is useful in sales planning, production planning and budgeting, cash budgeting, and analysis of various operating plans.
3. *Long-range forecast:* Generally 3 years or more in time span, long-range forecasts are used in planning for new products, capital expenditures, facility location or expansion, and research and development.

Medium- and long-range forecasts are distinguished from short-range forecasts by three features:

1. First, intermediate and long-range forecasts *deal with more comprehensive issues* supporting management decisions regarding planning and products, plants, and processes. Implementing some facility decisions, such as GM's decision to open a new Brazilian manufacturing plant, can take 5 to 8 years from inception to completion.
2. Second, short-term forecasting usually *employs different methodologies* than longer-term forecasting. Mathematical techniques, such as moving averages, exponential smoothing, and trend extrapolation (all of which we shall examine shortly), are common to short-run projections. Broader, *less* quantitative methods are useful in predicting such issues as whether a new product, like the optical disk recorder, should be introduced into a company's product line.
3. Finally, as you would expect, short-range forecasts *tend to be more accurate* than longer-range forecasts. Factors that influence demand change every day. Thus, as the time horizon lengthens, it is likely that forecast accuracy will diminish. It almost goes without saying, then, that sales forecasts must be updated regularly to maintain their value and integrity. After each sales period, forecasts should be reviewed and revised.

Types of Forecasts

Organizations use three major types of forecasts in planning future operations:

1. **Economic forecasts** address the business cycle by predicting inflation rates, money supplies, housing starts, and other planning indicators.
2. **Technological forecasts** are concerned with rates of technological progress, which can result in the birth of exciting new products, requiring new plants and equipment.
3. **Demand forecasts** are projections of demand for a company's products or services. Forecasts drive decisions, so managers need immediate and accurate information about real demand. They need *demand-driven forecasts*, where the focus is on rapidly identifying and tracking customer desires. These forecasts may use recent point-of-sale (POS) data, retailer-generated reports of customer preferences, and any other information that will help to forecast with the most current data possible. Demand-driven forecasts drive a company's production, capacity, and scheduling systems and serve as inputs to financial, marketing, and personnel planning. In addition, the payoff in reduced inventory and obsolescence can be huge.

Economic forecasts
Planning indicators that are valuable in helping organizations prepare medium- to long-range forecasts.

Technological forecasts
Long-term forecasts concerned with the rates of technological progress.

Demand forecasts
Projections of a company's sales for each time period in the planning horizon.

Economic and technological forecasting are specialized techniques that may fall outside the role of the operations manager. The emphasis in this chapter will therefore be on demand forecasting.

The Strategic Importance of Forecasting

Good forecasts are of critical importance in all aspects of a business: *The forecast is the only estimate of demand until actual demand becomes known.* Forecasts of demand therefore drive decisions in many areas. Let's look at the impact of product demand forecast on three activities: (1) supply chain management, (2) human resources, and (3) capacity.

Supply Chain Management

Good supplier relations and the ensuing advantages in product innovation, cost, and speed to market depend on accurate forecasts. Here are just three examples:

- Apple has built an effective global system where it controls nearly every piece of the supply chain, from product design to retail store. With rapid communication and accurate data shared up and down the supply chain, innovation is enhanced, inventory costs are reduced, and speed to market is improved. Once a product goes on sale, Apple tracks demand by the hour for each store and adjusts production forecasts daily. At Apple, forecasts for its supply chain are a strategic weapon.

- Toyota develops sophisticated car forecasts with input from a variety of sources, including dealers. But forecasting the demand for accessories such as navigation systems, custom wheels, spoilers, and so on is particularly difficult. And there are more than 1,000 items that vary by model and color. As a result, Toyota not only reviews reams of data regarding vehicles that have been built and wholesaled but also looks in detail at vehicle forecasts before it makes judgments about the future accessory demand. When this is done correctly, the result is an efficient supply chain and satisfied customers.
- Walmart collaborates with suppliers such as Sara Lee and Procter & Gamble to make sure the right item is available at the right time in the right place and at the right price. For instance, in hurricane season, Walmart's ability to analyze 700 million store–item combinations means it can forecast that not only flashlights but also Pop-Tarts and beer sell at seven times the normal demand rate. These forecasting systems are known as *collaborative planning, forecasting, and replenishment* (CPFR). They combine the intelligence of multiple supply-chain partners. The goal of CPFR is to create significantly more accurate information that can power the supply chain to greater sales and profits.

Human Resources

Hiring, training, and laying off workers all depend on anticipated demand. If the human resources department must hire additional workers without warning, the amount of training declines, and the quality of the workforce suffers. A large Louisiana chemical firm almost lost its biggest customer when a quick expansion to around-the-clock shifts led to a total breakdown in quality control on the second and third shifts.

Capacity

When capacity is inadequate, the resulting shortages can lead to loss of customers and market share. This is exactly what happened to Nabisco when it underestimated the huge demand for its Snackwell Devil's Food Cookies. Even with production lines working overtime, Nabisco could not keep up with demand, and it lost customers. Procter & Gamble faced this problem during the COVID-19 pandemic with its Bounty paper towels, when demand vastly exceeded forecasts and hence capacity. On the other hand, when excess capacity exists, costs can skyrocket.

Seven Steps in the Forecasting System

Forecasting follows seven basic steps. We use Disney World, the focus of this chapter's *Global Company Profile*, as an example of each step:

1. *Determine the use of the forecast:* Disney uses park attendance forecasts to drive decisions about staffing, opening times, ride availability, and food supplies.
2. *Select the items to be forecasted:* For Disney World, there are six main parks. A forecast of daily attendance at each is the main number that determines labor, maintenance, and scheduling.
3. *Determine the time horizon of the forecast:* Is it short, medium, or long term? Disney develops daily, weekly, monthly, annual, and 5-year forecasts.
4. *Select the forecasting model(s):* Disney uses a variety of statistical models that we shall discuss, including moving averages, econometrics, and regression analysis. It also employs judgmental, or nonquantitative, models.
5. *Gather the data needed to make the forecast:* Disney's forecasting team employs 35 analysts and 70 field personnel to survey 1 million people/businesses every year. Disney also uses a firm called Global Insights for travel industry forecasts and gathers data on exchange rates, arrivals into the U.S., airline specials, Wall Street trends, and school vacation schedules.
6. *Make the forecast.*
7. *Validate and implement the results:* At Disney, forecasts are reviewed daily at the highest levels to make sure that the model, assumptions, and data are valid. Error measures are applied; then the forecasts are used to schedule personnel down to 15-minute intervals.

These seven steps present a systematic way of initiating, designing, and implementing a forecasting system. When the system is to be used to generate forecasts regularly over time, data must be routinely collected. Then actual computations are usually made by computer.

Regardless of the system that firms such as Disney use, each company faces several realities:

- Most forecasting techniques assume that there is some underlying stability in the system. Consequently, many firms automate their predictions using computerized forecasting software and then closely monitor only the product items whose demand is erratic.
- Both product family and aggregated forecasts are more accurate than individual product forecasts. Disney, for example, aggregates daily attendance forecasts by park. This approach helps balance the over- and underpredictions for each of the six attractions.
- Outside factors that we cannot predict or control often impact the forecast, and extreme events can wreak havoc on forecasting systems. During the COVID-19 pandemic, short-term demand for some products fell sharply (such as spare parts for airplanes, oil and gas, air travel, nail salons, and theater visits). Meanwhile, demand for some products rose sharply (such as face masks, vitamins, holiday lights, packaging, and dry ice). And the demand for other products had temporary spikes often due to hoarding (such as pasta, canned food, toilet paper, Clorox wipes, bleach, and hand sanitizer).

Forecasting Approaches

There are two general approaches to forecasting, just as there are two ways to tackle all decision modeling. One is a quantitative analysis; the other is a qualitative approach. **Quantitative forecasts** use a variety of mathematical models that rely on historical data and/or associative variables to forecast demand. Subjective or **qualitative forecasts** incorporate such factors as the decision maker's intuition, emotions, personal experiences, and value system in reaching a forecast. Some firms use one approach and some use the other. In practice, a combination of the two is usually most effective.

Quantitative forecasts
Forecasts that employ mathematical modeling to forecast demand.

Qualitative forecasts
Forecasts that incorporate such factors as the decision maker's intuition, emotions, personal experiences, and value system.

Overview of Qualitative Methods

In this section, we consider four different *qualitative* forecasting techniques:

1. **Jury of executive opinion**: Under this method, the opinions of a group of high-level experts or managers, often in combination with statistical models, are pooled to arrive at a group estimate of demand. Bristol-Myers Squibb Company, for example, uses 220 well-known research scientists as its jury of executive opinion to get a grasp on future trends in the world of medical research.
2. **Delphi method**: There are three different types of participants in the Delphi method: decision makers, staff personnel, and respondents. Decision makers usually consist of a group of 5 to 10 experts who will be making the actual forecast. Staff personnel assist decision makers by preparing, distributing, collecting, and summarizing a series of questionnaires and survey results. The respondents are a group of people, often located in different places, whose judgments are valued. This group provides inputs to the decision makers before the forecast is made.

   The state of Alaska, for example, has used the Delphi method to develop its long-range economic forecast. A large part of the state's budget is derived from the million-plus barrels of oil pumped daily through a pipeline at Prudhoe Bay. The large Delphi panel of experts had to represent all groups and opinions in the state and all geographic areas.
3. **Sales force composite**: In this approach, all salespeople estimate what sales will be in their region. These forecasts are then reviewed to ensure that they are realistic. Then they are combined at the district and national levels to reach an overall forecast. A variation of this approach occurs at Lexus, where every quarter Lexus dealers have a "make meeting." At this meeting, they talk about what is selling, in what colors, and with what options, so the factory knows what to build.
4. **Market survey**: This method solicits input from customers or potential customers regarding future purchasing plans. It can help not only in preparing a forecast but also in improving

Jury of executive opinion
A forecasting technique that uses the opinion of a small group of high-level managers to form a group estimate of demand.

Delphi method
A forecasting technique using a group process that allows experts to make forecasts.

LO 4.2 *Explain* when to use each of the four qualitative models

Sales force composite
A forecasting technique based on salespersons' estimates of expected sales.

Market survey
A forecasting method that solicits input from customers or potential customers regarding future purchasing plans.

product design and planning for new products. The consumer market survey and sales force composite methods can, however, suffer from overly optimistic forecasts that arise from customer input.

Overview of Quantitative Methods

Five quantitative forecasting methods,[1] all of which use historical data, are described in this chapter. They fall into two categories:

1. Naive approach
2. Moving averages
3. Exponential smoothing
4. Trend projection

} **Time-series models**

5. Linear regression } **Associative model**

Time series
A forecasting technique that uses a series of past data points to make a forecast.

Time-Series Models Time-series models predict on the assumption that the future is a function of the past. In other words, they look at what has happened over a period of time and use a series of past data to make a forecast. If we are predicting sales of lawn mowers, we use the past sales for lawn mowers to make the forecasts.

Associative Models Associative models, such as linear regression, incorporate the variables or factors that might influence the quantity being forecast. For example, an associative model for lawn mower sales might use factors such as new housing starts, advertising budget, and competitors' prices.

STUDENT TIP Here is the meat of this chapter. We now show you a wide variety of models that use time-series data.

Time-Series Forecasting

A time series is based on a sequence of evenly spaced (weekly, monthly, quarterly, and so on) data points. Examples include weekly sales of Nike Air Jordans, quarterly earnings reports of Microsoft stock, daily shipments of Coors beer, and annual consumer price indices. Forecasting time-series data implies that future values are predicted *only* from past values and that other variables, no matter how potentially valuable, may be ignored.

Decomposition of a Time Series

Analyzing time series means breaking down past data into components and then projecting them forward. A time series has four components:

1. *Trend* is the gradual upward or downward movement of the data over time. Changes in income, population, age distribution, or cultural views may account for movement in trend.
2. *Seasonality* is a data pattern that repeats itself after a period of days, weeks, months, or quarters. There are six common seasonality patterns:

STUDENT TIP The peak "seasons" for sales of Frito-Lay chips are the Super Bowl, Memorial Day, Labor Day, and the Fourth of July.

PERIOD LENGTH	"SEASON" LENGTH	NUMBER OF "SEASONS" IN PATTERN
Week	Day	7
Month	Week	$4–4\frac{1}{2}$
Month	Day	28–31
Year	Quarter	4
Year	Month	12
Year	Week	52

Restaurants and barbershops, for example, experience weekly seasons, with Saturday being the peak of business. See the *OM in Action* box, "Forecasting at Olive Garden." Beer distributors forecast yearly patterns, with monthly seasons. Three "seasons"—May, July, and September—each contain a big beer-drinking holiday.

OM in Action Forecasting at Olive Garden[2]

It's Friday night in the college town of Gainesville, Florida, and the local Olive Garden restaurant is humming. Customers may wait an average of 30 minutes for a table, but they can sample new wines and cheeses and admire scenic paintings of Italian villages on the Tuscan-style restaurant's walls. Then comes dinner with portions so huge that many people take home a doggie bag. The typical bill: under $17 per person.

Crowds flock to the Darden restaurant chain's Olive Garden, Seasons 52, and Bahama Breeze for value and consistency—*and* they get it.

Every night, Darden's computers crank out forecasts that tell store managers what demand to anticipate the next day. The forecasting software generates a total meal forecast and breaks that down into specific menu items. The system tells a manager, for instance, that if 625 meals will be served the next day, "you will serve these items in these quantities. So before you go home, pull 25 pounds of shrimp and 30 pounds of crab out, and tell your operations people to prepare 42 portion packs of chicken, 75 scampi dishes, 8 stuffed flounders, and so on." Managers often fine-tune the quantities based on local conditions, such as weather or a convention, but they know what their customers are going to order.

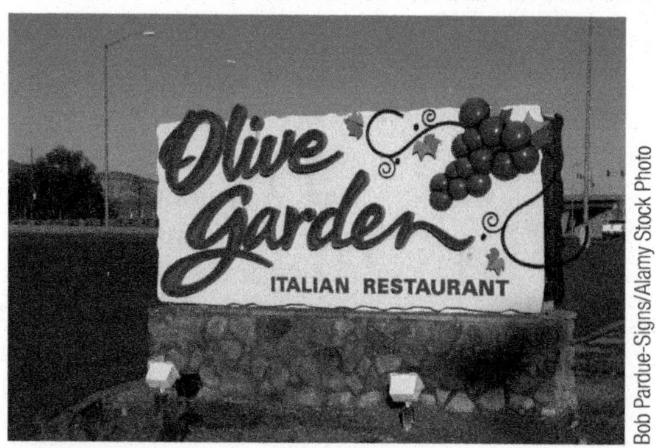

Bob Pardue-Signs/Alamy Stock Photo

By relying on demand history, the forecasting system has cut millions of dollars of waste out of the system. The forecast also reduces labor costs by providing the necessary information for improved scheduling. Labor costs decreased almost a full percent in the first year, translating into additional millions in savings for the Darden chain. In the low-margin restaurant business, every dollar counts.

3. *Cycles* are patterns in the data that occur every several years. They are usually tied into the business cycle and are of major importance in short-term business analysis and planning. Predicting business cycles is difficult because they may be affected by political events or by international turmoil.
4. *Random variations* are "blips" in the data caused by chance and unusual situations. They follow no discernible pattern, so they cannot be predicted.

Figure 4.1 illustrates a demand over a 4-year period. It shows the average, trend, seasonal components, and random variations around the demand curve. The average demand is the sum of the demand for each period divided by the number of data periods.

LO 4.3 *Apply* the naive, moving-average, exponential smoothing, and trend methods

Naive Approach

The simplest way to forecast is to assume that demand in the next period will be equal to demand in the most recent period. In other words, if sales of a product—say, Nokia cell phones—were 68 units in January, we can forecast that February's sales will also be 68 phones.

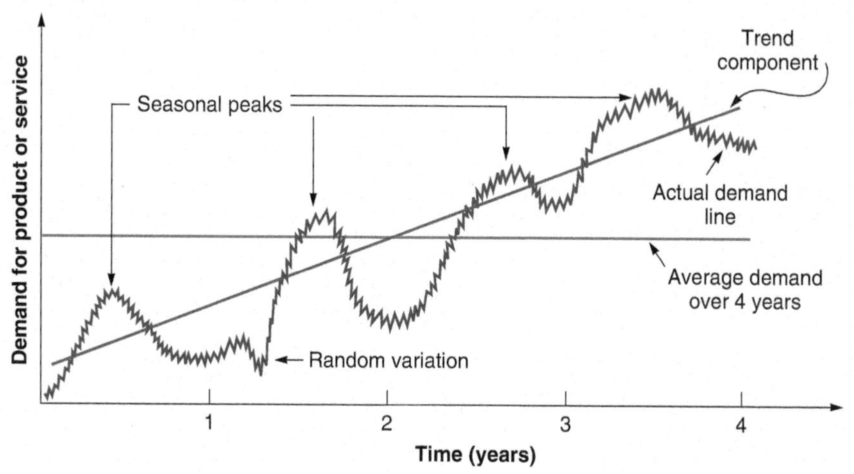

Figure 4.1 **Demand Charted over 4 Years, with a Growth Trend and Seasonality Indicated**

STUDENT TIP
Forecasting is easy when demand is stable. But with trend, seasonality, and cycles considered, the job is a lot more interesting.

Naive approach
A forecasting technique that assumes that demand in the next period is equal to demand in the most recent period.

Does this make any sense? It turns out that for some product lines, this naive approach is the most cost-effective and efficient objective forecasting model. At least it provides a starting point against which more sophisticated models that follow can be compared.

Moving Averages

Moving averages
A forecasting method that uses an average of the n most recent periods of data to forecast the next period.

A moving-average forecast uses a number of historical actual data values to generate a forecast. Moving averages are useful *if we can assume that market demands will stay fairly steady over time*. A 4-month moving average is found by simply summing the demand during the past 4 months and dividing by 4. With each passing month, the most recent month's data are added to the sum of the previous 3 months' data, and the earliest month is dropped. This practice tends to smooth out short-term irregularities in the data series.

Mathematically, the simple moving average (which serves as an estimate of the next period's demand) is expressed as:

$$\text{Moving average} = \frac{\sum \text{demand in previous } n \text{ periods}}{n} \qquad (4\text{-}1)$$

where n is the number of periods in the moving average—for example, 4, 5, or 6 months, respectively, for a 4-, 5-, or 6-period moving average.

Example 1 shows how moving averages are calculated.

Example 1

DETERMINING THE MOVING AVERAGE

Donna's Garden Supply wants a 3-month moving-average forecast, including a forecast for next January, for shed sales.

APPROACH ▶ Storage shed sales are shown in the middle column of the following table. A 3-month moving average appears on the right.

MONTH	ACTUAL SHED SALES	3-MONTH MOVING AVERAGE
January	10	
February	12	
March	13	
April	16	$(10 + 12 + 13)/3 = 11\frac{2}{3}$
May	19	$(12 + 13 + 16)/3 = 13\frac{2}{3}$
June	23	$(13 + 16 + 19)/3 = 16$
July	26	$(16 + 19 + 23)/3 = 19\frac{1}{3}$
August	30	$(19 + 23 + 26)/3 = 22\frac{2}{3}$
September	28	$(23 + 26 + 30)/3 = 26\frac{1}{3}$
October	18	$(26 + 30 + 28)/3 = 28$
November	16	$(30 + 28 + 18)/3 = 25\frac{1}{3}$
December	14	$(28 + 18 + 16)/3 = 20\frac{2}{3}$

SOLUTION ▶ The forecast for December is $20\frac{2}{3}$. To project the demand for sheds in the coming January, we sum the October, November, and December sales and divide by 3: January forecast $= (18 + 16 + 14)/3 = 16$.

INSIGHT ▶ Management now has a forecast that averages sales for the last 3 months. It is easy to use and understand.

LEARNING EXERCISE ▶ If actual sales in December were 18 (rather than 14), what is the new January forecast? [Answer: $17\frac{1}{3}$.]

RELATED PROBLEMS ▶ 4.1a, 4.2b, 4.5a, 4.6, 4.8a,b, 4.10a, 4.13b, 4.15, 4.33, 4.35, 4.38

EXCEL OM Data File **Ch04Ex1**.xls can be found online.

ACTIVE MODEL 4.1 This example is further illustrated in Active Model 4.1 found online.

When a detectable trend or pattern is present, *weights* can be used to place more emphasis on recent values. This practice makes forecasting techniques more responsive to changes because more recent periods may be more heavily weighted. Choice of weights is somewhat arbitrary because there is no set formula to determine them. Therefore, deciding which weights to use requires some experience. For example, if the latest month or period is weighted too heavily, the forecast may reflect a large unusual change in the demand or sales pattern too quickly.

A weighted moving average may be expressed mathematically as:

$$\text{Weighted moving average} = \frac{\Sigma((\text{Weight for period } n)(\text{Demand in period } n))}{\Sigma \text{ Weights}} \quad (4\text{-}2)$$

Example 2 shows how to calculate a weighted moving average.

Example 2

DETERMINING THE WEIGHTED MOVING AVERAGE

Donna's Garden Supply (see Example 1) wants to forecast storage shed sales by weighting the past 3 months, with more weight given to recent data to make them more significant.

APPROACH ▶ Assign more weight to recent data, as follows:

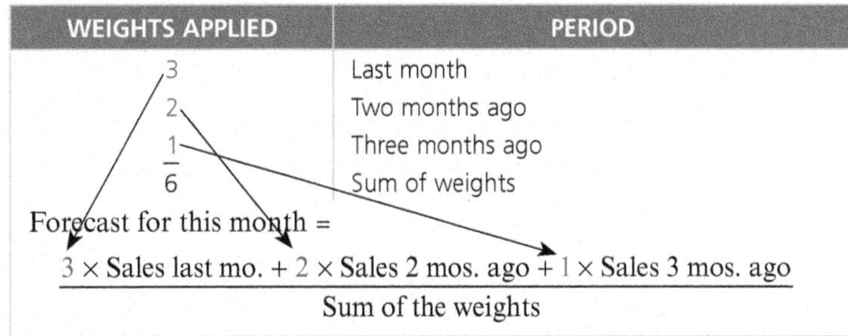

WEIGHTS APPLIED	PERIOD
3	Last month
2	Two months ago
1	Three months ago
6	Sum of weights

Forecast for this month =

$$\frac{3 \times \text{Sales last mo.} + 2 \times \text{Sales 2 mos. ago} + 1 \times \text{Sales 3 mos. ago}}{\text{Sum of the weights}}$$

SOLUTION ▶ The results of this weighted-average forecast are as follows:

MONTH	ACTUAL SHED SALES	3-MONTH WEIGHTED MOVING AVERAGE
January	10	
February	12	
March	13	
April	16	$[(3 \times 13) + (2 \times 12) + (10)]/6 = 12\frac{1}{6}$
May	19	$[(3 \times 16) + (2 \times 13) + (12)]/6 = 14\frac{1}{3}$
June	23	$[(3 \times 19) + (2 \times 16) + (13)]/6 = 17$
July	26	$[(3 \times 23) + (2 \times 19) + (16)]/6 = 20\frac{1}{2}$
August	30	$[(3 \times 26) + (2 \times 23) + (19)]/6 = 23\frac{5}{6}$
September	28	$[(3 \times 30) + (2 \times 26) + (23)]/6 = 27\frac{1}{2}$
October	18	$[(3 \times 28) + (2 \times 30) + (26)]/6 = 28\frac{1}{3}$
November	16	$[(3 \times 18) + (2 \times 28) + (30)]/6 = 23\frac{1}{3}$
December	14	$[(3 \times 16) + (2 \times 18) + (28)]/6 = 18\frac{2}{3}$

The forecast for January is $15\frac{1}{3}$. Do you see how this number is computed?

INSIGHT ▶ In this particular forecasting situation, you can see that more heavily weighting the latest month provides a more accurate projection.

LEARNING EXERCISE ▶ If the assigned weights were 0.50, 0.33, and 0.17 (instead of 3, 2, and 1), what is the forecast for January's weighted moving average? Why? [Answer: There is no change. These are the same *relative* weights. Note that Σ weights = 1 now, so there is no need for a denominator. When the weights sum to 1, calculations tend to be simpler.]

RELATED PROBLEMS ▶ 4.1b, 4.2c, 4.5c, 4.6, 4.7, 4.10b, 4.34, 4.38

EXCEL OM Data File Ch04Ex2.xls can be found online.

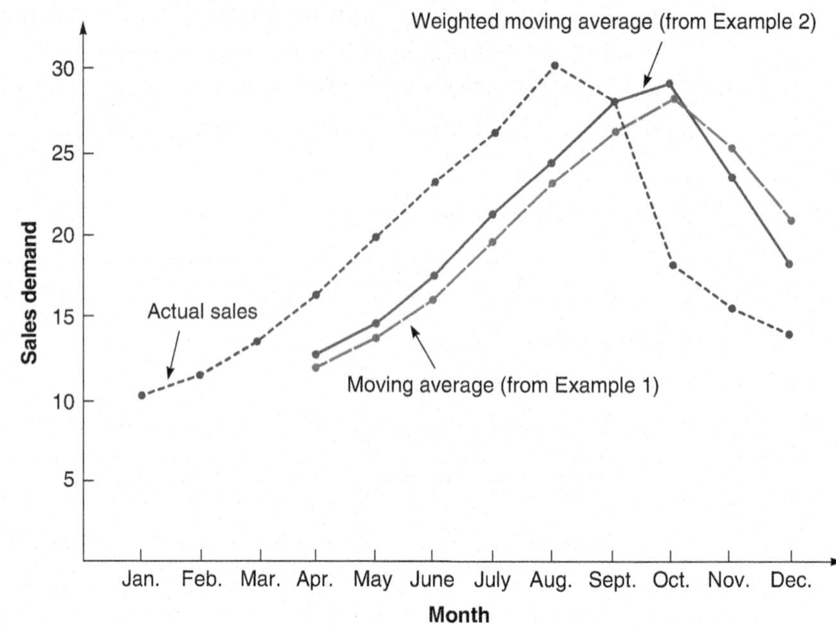

Figure 4.2

Actual Demand vs. Moving-Average and Weighted-Moving-Average Methods for Donna's Garden Supply

STUDENT TIP

Moving-average methods always lag behind when there is a trend present, as shown by the blue line (actual sales) for January through August.

Both simple and weighted moving averages are effective in smoothing out sudden fluctuations in the demand pattern to provide stable estimates. Moving averages do, however, present three problems:

1. Increasing the size of n (the number of periods averaged) does smooth out fluctuations better, but it makes the method less sensitive to changes in the data.
2. Moving averages cannot pick up trends very well. Because they are averages, they will always stay within past levels and will not predict changes to either higher or lower levels. That is, they *lag* the actual values.
3. Moving averages require extensive records of past data.

Figure 4.2, a plot of the data in Examples 1 and 2, illustrates the lag effect of the moving-average models. Note that both the moving-average and weighted-moving-average lines lag the actual demand. The weighted moving average, however, usually reacts more quickly to demand changes. Even in periods of downturn (see November and December), it more closely tracks the demand.

Exponential Smoothing

Exponential smoothing
A weighted-moving-average forecasting technique in which data points are weighted by an exponential function.

Exponential smoothing is another weighted-moving-average forecasting method. It involves very *little* record keeping of past data and is fairly easy to use. The basic exponential smoothing formula can be shown as follows:

$$\text{New forecast} = \text{Last period's forecast} + \alpha\,(\text{Last period's actual demand} - \text{Last period's forecast}) \quad (4\text{-}3)$$

Smoothing constant
The weighting factor used in an exponential smoothing forecast, a number greater than or equal to 0 and less than or equal to 1.

where α is a weight, or **smoothing constant**, chosen by the forecaster, that has a value greater than or equal to 0 and less than or equal to 1. Equation (4-3) can also be written mathematically as:

$$F_t = F_{t-1} + \alpha(A_{t-1} - F_{t-1}) \quad (4\text{-}4)$$

where
F_t = new forecast
F_{t-1} = previous period's forecast
α = smoothing (or weighting) constant ($0 \le \alpha \le 1$)
A_{t-1} = previous period's actual demand

The concept is not complex. The latest estimate of demand is equal to the old forecast adjusted by a fraction of the difference between the last period's actual demand and last period's forecast. Example 3 shows how to use exponential smoothing to derive a forecast.

Example 3

DETERMINING A FORECAST VIA EXPONENTIAL SMOOTHING

In January, a car dealer predicted February demand for 142 Ford Mustangs. Actual February demand was 153 autos. Using a smoothing constant chosen by management of $\alpha = .20$, the dealer wants to forecast March demand using the exponential smoothing model.

APPROACH ▶ The exponential smoothing model in Equations (4-3) and (4-4) can be applied.

SOLUTION ▶ Substituting the sample data into the formula, we obtain:

$$\begin{aligned}\text{New forecast (for March demand)} &= 142 + .2(153 - 142) = 142 + 2.2\\ &= 144.2\end{aligned}$$

Thus, the March demand forecast for Ford Mustangs is rounded to 144.

INSIGHT ▶ Using just two pieces of data, the forecast and the actual demand, plus a smoothing constant, we developed a forecast of 144 Ford Mustangs for March.

LEARNING EXERCISE ▶ If the smoothing constant is changed to .30, what is the new forecast? [Answer: 145.3]

RELATED PROBLEMS ▶ 4.1c, 4.3, 4.4, 4.5d, 4.6, 4.9d, 4.11, 4.12, 4.13a, 4.17, 4.18, 4.31, 4.33, 4.36, 4.61a

The *smoothing constant*, α, is generally in the range from .05 to .50 for business applications. It can be changed to give more weight to recent data (when α is high) or more weight to past data (when α is low). When α reaches the extreme of 1.0, then in Equation (4-4), $F_t = 1.0A_{t-1}$. All the older values drop out, and the forecast becomes identical to the naive model mentioned previously in this chapter. That is, the forecast for the next period is just the same as this period's demand.

The following table helps illustrate this concept. For example, when $\alpha = .5$, we can see that the new forecast is based almost entirely on demand in the last three or four periods. When $\alpha = .1$, the forecast places little weight on recent demand and takes many periods (about 19) of historical values into account.

	WEIGHT ASSIGNED TO				
SMOOTHING CONSTANT	MOST RECENT PERIOD (α)	2ND MOST RECENT PERIOD $\alpha(1-\alpha)$	3RD MOST RECENT PERIOD $\alpha(1-\alpha)^2$	4TH MOST RECENT PERIOD $\alpha(1-\alpha)^3$	5TH MOST RECENT PERIOD $\alpha(1-\alpha)^4$
$\alpha = .1$	.1	.09	.081	.073	.066
$\alpha = .5$	.5	.25	.125	.063	.031

Selecting the Smoothing Constant Exponential smoothing has been successfully applied in virtually every type of business. However, the appropriate value of the smoothing constant, α, can make the difference between an accurate forecast and an inaccurate forecast. High values of α are chosen when the underlying average is likely to change. Low values of α are used when the underlying average is fairly stable. In picking a value for the smoothing constant, the objective is to obtain the most accurate forecast.

STUDENT TIP
Forecasts tend to be more accurate as they become shorter. Therefore, forecast error also tends to drop with shorter forecasts.

Measuring Forecast Error

The overall accuracy of any forecasting model—moving average, exponential smoothing, or other—can be determined by comparing the forecasted values with the actual or observed values. If F_t denotes the forecast in period t, and A_t denotes the actual demand in period t, the *forecast error* (or deviation) is defined as:

$$\text{Forecast error} = \text{Actual demand} - \text{Forecast value}$$
$$= A_t - F_t$$

LO 4.4 *Compute* three measures of forecast accuracy

Several measures are used in practice to calculate the overall forecast error. These measures can be used to compare different forecasting models, as well as to monitor forecasts to ensure they are performing well. Three of the most popular measures are mean absolute deviation (MAD), mean squared error (MSE), and mean absolute percent error (MAPE). We now describe and give an example of each.

Mean absolute deviation (MAD)
A measure of the overall forecast error for a model.

Mean Absolute Deviation The first measure of the overall forecast error for a model is the mean absolute deviation (MAD). This value is computed by taking the sum of the absolute values of the individual forecast errors (deviations) and dividing by the number of periods of data (n):

$$\text{MAD} = \frac{\Sigma|\text{Actual} - \text{Forecast}|}{n} \tag{4-5}$$

Example 4 applies MAD, as a measure of overall forecast error, by testing two values of α.

Example 4

DETERMINING THE MEAN ABSOLUTE DEVIATION (MAD)

During the past 8 quarters, the Port of Baltimore has unloaded large quantities of grain from ships. The port's operations manager wants to test the use of exponential smoothing to see how well the technique works in predicting tonnage unloaded. The manager guesses that the forecast of grain unloaded in the first quarter was 175 tons. Two values of α are to be examined: $\alpha = .10$ and $\alpha = .50$.

APPROACH ▶ Compare the actual data with the data we forecast (using each of the two α values) and then find the absolute deviation and MADs.

SOLUTION ▶ The following table shows the *detailed* calculations for $\alpha = .10$ only:

QUARTER	ACTUAL TONNAGE UNLOADED	FORECAST WITH $\alpha = .10$	FORECAST WITH $\alpha = .50$
1	180	175	175
2	168	175.50 = 175.00 + .10(180 – 175)	177.50
3	159	174.75 = 175.50 + .10(168 – 175.50)	172.75
4	175	173.18 = 174.75 + .10(159 – 174.75)	165.88
5	190	173.36 = 173.18 + .10(175 – 173.18)	170.44
6	205	175.02 = 173.36 + .10(190 – 173.36)	180.22
7	180	178.02 = 175.02 + .10(205 – 175.02)	192.61
8	182	178.22 = 178.02 + .10(180 – 178.02)	186.30
9	?	178.59 = 178.22 + .10(182 – 178.22)	184.15

To evaluate the accuracy of each smoothing constant, we can compute forecast errors in terms of absolute deviations and MADs:

QUARTER	ACTUAL TONNAGE UNLOADED	FORECAST WITH $\alpha = .10$	ABSOLUTE DEVIATION FOR $\alpha = .10$	FORECAST WITH $\alpha = .50$	ABSOLUTE DEVIATION FOR $\alpha = .50$
1	180	175	5.00	175	5.00
2	168	175.50	7.50	177.50	9.50
3	159	174.75	15.75	172.75	13.75
4	175	173.18	1.82	165.88	9.12
5	190	173.36	16.64	170.44	19.56
6	205	175.02	29.98	180.22	24.78
7	180	178.02	1.98	192.61	12.61
8	182	178.22	3.78	186.30	4.30
Sum of absolute deviations:			82.45		98.62
$\text{MAD} = \frac{\Sigma\|\text{Deviations}\|}{n}$			10.31		12.33

INSIGHT ▶ On the basis of this comparison of the two MADs, a smoothing constant of $\alpha = .10$ is preferred to $\alpha = .50$ because its MAD is smaller.

LEARNING EXERCISE ▶ If the smoothing constant is changed from $\alpha = .10$ to $\alpha = .20$, what is the new MAD? [Answer: 10.21.]

RELATED PROBLEMS ▶ 4.5b, 4.8c, 4.9c, 4.14, 4.23, 4.35d, 4.37a, 4.38c, 4.59b, 4.61b

EXCEL OM Data File Ch04Ex4a.xls and **Ch04Ex4b.xls** can be found online.

ACTIVE MODEL 4.2 This example is further illustrated in Active Model 4.2 found online.

Most computerized forecasting software includes a feature that automatically finds the smoothing constant with the lowest forecast error. Some software modifies the α value if errors become larger than acceptable.

Mean Squared Error The mean squared error (MSE) is a second way of measuring overall forecast error. MSE is the average of the squared differences between the forecasted and observed values. Its formula is:

Mean squared error (MSE)
The average of the squared differences between the forecasted and observed values.

$$\text{MSE} = \frac{\Sigma(\text{Forecast errors})^2}{n} \tag{4-6}$$

Example 5 finds the MSE for the Port of Baltimore problem introduced in Example 4.

Example 5

DETERMINING THE MEAN SQUARED ERROR (MSE)

The operations manager for the Port of Baltimore now wants to compute MSE for $\alpha = .10$.

APPROACH ▶ Using the same forecast data for $\alpha = .10$ from Example 4, compute the MSE with Equation (4-6).

SOLUTION ▶

QUARTER	ACTUAL TONNAGE UNLOADED	FORECAST FOR $\alpha = .10$	$(\text{ERROR})^2$
1	180	175	$5^2 = 25$
2	168	175.50	$(-7.5)^2 = 56.25$
3	159	174.75	$(-15.75)^2 = 248.06$
4	175	173.18	$(1.82)^2 = 3.31$
5	190	173.36	$(16.64)^2 = 276.89$
6	205	175.02	$(29.98)^2 = 898.80$
7	180	178.02	$(1.98)^2 = 3.92$
8	182	178.22	$(3.78)^2 = 14.29$
			Sum of errors squared = 1,526.52

$$\text{MSE} = \frac{\Sigma(\text{Forecast errors})^2}{n} = 1{,}526.52 / 8 = 190.8$$

INSIGHT ▶ Is this MSE = 190.8 good or bad? It all depends on the MSEs for other forecasting approaches. A low MSE is better because we want to minimize MSE. MSE exaggerates errors because it squares them.

LEARNING EXERCISE ▶ Find the MSE for $\alpha = .50$. [Answer: MSE = 195.20. The result indicates that $\alpha = .10$ is a better choice because we seek a lower MSE. Coincidentally, this is the same conclusion we reached using MAD in Example 4.]

RELATED PROBLEMS ▶ 4.8d, 4.11c, 4.14, 4.15c, 4.16c, 4.20, 4.35d, 4.37b

The MSE tends to accentuate large deviations because of the squared term. For example, if the forecast error for period 1 is twice as large as the error for period 2, the squared error in period 1 is four times as large as that for period 2. Hence, using MSE as the measure of forecast error typically indicates that we prefer to have several smaller deviations rather than even one large deviation.

Mean absolute percent error (MAPE)
The average of the absolute differences between the forecast and actual values.

Mean Absolute Percent Error A problem with both the MAD and MSE is that their values depend on the magnitude of the item being forecast. If the forecast item is measured in thousands, the MAD and MSE values can be very large. To avoid this problem, we can use the mean absolute percent error (MAPE). This is computed as the average of the absolute difference between the forecasted and actual values, expressed as a percentage of the actual values. That is, if we have forecasted and actual values for n periods, the MAPE is calculated as:

$$\text{MAPE} = \frac{\sum_{i=1}^{n} 100|\text{Actual}_i - \text{Forecast}_i| / \text{Actual}_i}{n} \quad (4\text{-}7)$$

Example 6 illustrates the calculations using the data from Examples 4 and 5.

Example 6

DETERMINING THE MEAN ABSOLUTE PERCENT ERROR (MAPE)

The Port of Baltimore wants to now calculate the MAPE when $\alpha = .10$.

APPROACH ▶ Equation (4-7) is applied to the forecast data computed in Example 4.

SOLUTION ▶

QUARTER	ACTUAL TONNAGE UNLOADED	FORECAST FOR $\alpha = .10$	ABSOLUTE PERCENT ERROR 100 (|ERROR|/ACTUAL)
1	180	175.00	100(5/180) = 2.78%
2	168	175.50	100(7.5/168) = 4.46%
3	159	174.75	100(15.75/159) = 9.90%
4	175	173.18	100(1.82/175) = 1.05%
5	190	173.36	100(16.64/190) = 8.76%
6	205	175.02	100(29.98/205) = 14.62%
7	180	178.02	100(1.98/180) = 1.10%
8	182	178.22	100(3.78/182) = 2.08%
			Sum of % errors = 44.75%

$$\text{MAPE} = \frac{\sum \text{absolute percent error}}{n} = \frac{44.75\%}{8} = 5.59\%$$

INSIGHT ▶ MAPE expresses the error as a percentage of the actual values, undistorted by a single large value.

LEARNING EXERCISE ▶ What is MAPE when α is .50? [Answer: MAPE = 6.75%. As was the case with MAD and MSE, the $\alpha = .1$ was preferable for this series of data.]

RELATED PROBLEMS ▶ 4.8e, 4.29c

The MAPE is perhaps the easiest measure to interpret. For example, a result that the MAPE is 6% is a clear statement that is not dependent on issues such as the magnitude of the input data.

Table 4.1 summarizes how MAD, MSE, and MAPE differ.

Exponential Smoothing with Trend Adjustment

Simple exponential smoothing, the technique we just illustrated in Examples 3 to 6, is like any other moving-average technique: It fails to respond to trends. Other forecasting techniques that can deal with trends are certainly available. However, because exponential smoothing is such a popular modeling approach in business, let us look at it in more detail.

Table 4.1 Comparison of Measures of Forecast Error

MEASURE	MEANING	EQUATION	APPLICATION TO CHAPTER EXAMPLE
Mean absolute deviation (MAD)	How much the forecast missed the target	$\text{MAD} = \dfrac{\sum\lvert\text{Actual} - \text{Forecast}\rvert}{n}$ (4-5)	For $\alpha = .10$ in , the forecast for grain unloaded was off by an average of 10.31 tons.
Mean squared error (MSE)	The square of how much the forecast missed the target	$\text{MSE} = \dfrac{\sum(\text{Forecast errors})^2}{n}$ (4-6)	For $\alpha = .10$ in Example 5, the square of the forecast error was 190.8. This number does not have a physical meaning but is useful when compared to the MSE of another forecast.
Mean absolute percent error (MAPE)	The average percent error	$\text{MAPE} = \dfrac{\sum_{i=1}^{n} 100\lvert\text{Actual}_i - \text{Forecast}_i\rvert / \text{Actual}_i}{n}$ (4-7)	For $\alpha = .10$ in Example 6, the forecast is off by 5.59% on average. As in Examples 4 and 5, some forecasts were too high, and some were low.

Here is why exponential smoothing must be modified when a trend is present. Assume that demand for our product or service has been increasing by 100 units per month and that we have been forecasting with $\alpha = 0.4$ in our exponential smoothing model. The following table shows a severe lag in the second, third, fourth, and fifth months, even when our initial estimate for month 1 is perfect:

MONTH	ACTUAL DEMAND	FORECAST (F_t) FOR MONTHS 1–5
1	100	$F_1 = 100$ (given)
2	200	$F_2 = F_1 + \alpha(A_1 - F_1) = 100 + .4(100 - 100) = 100$
3	300	$F_3 = F_2 + \alpha(A_2 - F_2) = 100 + .4(200 - 100) = 140$
4	400	$F_4 = F_3 + \alpha(A_3 - F_3) = 140 + .4(300 - 140) = 204$
5	500	$F_5 = F_4 + \alpha(A_4 - F_4) = 204 + .4(400 - 204) = 282$

To improve our forecast, let us illustrate a more complex exponential smoothing model, one that adjusts for trend. The idea is to compute an exponentially smoothed average of the data and then adjust for positive or negative lag in trend. The new formula is:

$$\begin{aligned}\text{Forecast including trend}(FIT_t) &= \text{Exponentially smoothed forecast average}(F_t)\\ &\quad + \text{Exponentially smoothed trend}(T_t)\end{aligned} \tag{4-8}$$

With trend-adjusted exponential smoothing, estimates for both the average and the trend are smoothed. This procedure requires two smoothing constants: α for the average and β for the trend. We then compute the average and trend each period:

$$F_t = \alpha(\text{Actual demand last period}) + (1 - \alpha)(\text{Forecast last period} + \text{Trend estimate last period})$$

or:

$$F_t = \alpha(A_{t-1}) + (1 - \alpha)(F_{t-1} + T_{t-1}) \tag{4-9}$$

$$T_t = \beta(\text{Forecast this period} - \text{Forecast last period}) + (1 - \beta)(\text{Trend estimate last period})$$

or:

$$T_t = \beta(F_t - F_{t-1}) + (1 - \beta)T_{t-1} \tag{4-10}$$

where F_t = exponentially smoothed forecast average of the data series in period t
T_t = exponentially smoothed trend in period t
A_t = actual demand in period t
α = smoothing constant for the average = $(0 \le \alpha \le 1)$
β = smoothing constant for the trend $(0 \le \beta \le 1)$

So the three steps to compute a trend-adjusted forecast are:

STEP 1: Compute F_t, the exponentially smoothed forecast average for period t, using Equation (4-9).

STEP 2: Compute the smoothed trend, T_t, using Equation (4-10).

STEP 3: Calculate the forecast including trend, FIT_t, by the formula $FIT_t = F_t + T_t$ [from Equation (4-8)].

Example 7 shows how to use trend-adjusted exponential smoothing.

Example 7

COMPUTING A TREND-ADJUSTED EXPONENTIAL SMOOTHING FORECAST

A large Portland manufacturer wants to forecast demand for a piece of pollution-control equipment. A review of past sales, as shown below, indicates that an increasing trend is present:

MONTH (t)	ACTUAL DEMAND (A_t)	MONTH (t)	ACTUAL DEMAND (A_t)
1	12	6	21
2	17	7	31
3	20	8	28
4	19	9	36
5	24	10	?

Smoothing constants are assigned the values of $\alpha = .2$ and $\beta = .4$. The firm assumes the initial forecast average for month 1 (F_1) was 11 units and the trend over that period (T_1) was 2 units.

APPROACH ▶ A trend-adjusted exponential smoothing model, using Equations (4-9), (4-10), and (4-8) and the three steps above, is employed.

SOLUTION ▶

Step 1: Forecast average for month 2:

$$F_2 = \alpha A_1 + (1 - \alpha)(F_1 + T_1)$$
$$F_2 = (.2)(12) + (1 - .2)(11 + 2)$$
$$= 2.4 + (.8)(13) = 2.4 + 10.4 = 12.8 \text{ units}$$

Step 2: Compute the trend in period 2:

$$T_2 = \beta(F_2 - F_1) + (1 - \beta)T_1$$
$$= .4(12.8 - 11) + (1 - .4)(2)$$
$$= (.4)(1.8) + (.6)(2) = .72 + 1.2 = 1.92$$

Step 3: Compute the forecast including trend (FIT_t):

$$FIT_2 = F_2 + T_2$$
$$= 12.8 + 1.92$$
$$= 14.72 \text{ units}$$

We will also do the same calculations for the third month:

Step 1: $F_3 = \alpha A_2 + (1 - \alpha)(F_2 + T_2) = (.2)(17) + (1 - .2)(12.8 + 1.92)$
$= 3.4 + (.8)(14.72) = 3.4 + 11.78 = 15.18$

Step 2: $T_3 = \beta(F_3 - F_2) + (1 - \beta)T_2 = (.4)(15.18 - 12.8) + (1 - .4)(1.92)$
$= (.4)(2.38) + (.6)(1.92) = .952 + 1.152 = 2.10$

Step 3: $FIT_3 = F_3 + T_3$
$= 15.18 + 2.10 = 17.28.$

Table 4.2 completes the forecasts for the 10-month period.

TABLE 4.2 **Forecast with $\alpha = .2$ and $\beta = .4$**

MONTH	ACTUAL DEMAND	SMOOTHED FORECAST AVERAGE, F_t	SMOOTHED TREND, T_t	FORECAST INCLUDING TREND, FIT_t
1	12	11	2	13.00
2	17	12.80	1.92	14.72
3	20	15.18	2.10	17.28
4	19	17.82	2.32	20.14
5	24	19.91	2.23	22.14
6	21	22.51	2.38	24.89
7	31	24.11	2.07	26.18
8	28	27.14	2.45	29.59
9	36	29.28	2.32	31.60
10	—	32.48	2.68	35.16

INSIGHT ▶ Figure 4.3 compares actual demand (A_t) to an exponential smoothing forecast that includes trend (FIT_t). *FIT* picks up the trend in actual demand. A simple exponential smoothing model (as we saw in Examples 3 and 4) trails far behind.

LEARNING EXERCISE ▶ Using the data for actual demand for the 9 months, compute the exponentially smoothed forecast average *without* trend [using Equation (4-4) as we did earlier in Examples 3 and 4]. Apply $\alpha = .2$, and assume an initial forecast average for month 1 of 11 units. Then plot the months 2–10 forecast values on Figure 4.3. What do you notice? [Answer: Month 10 forecast $= 24.65$. All the points are below and lag the trend-adjusted forecast.]

RELATED PROBLEMS ▶ 4.19, 4.20, 4.21, 4.22, 4.32

EXCEL OM Data File **Ch04Ex7.xis** can be found online.

ACTIVE MODEL 4.3 This example is further illustrated in Active Model 4.3 found online.

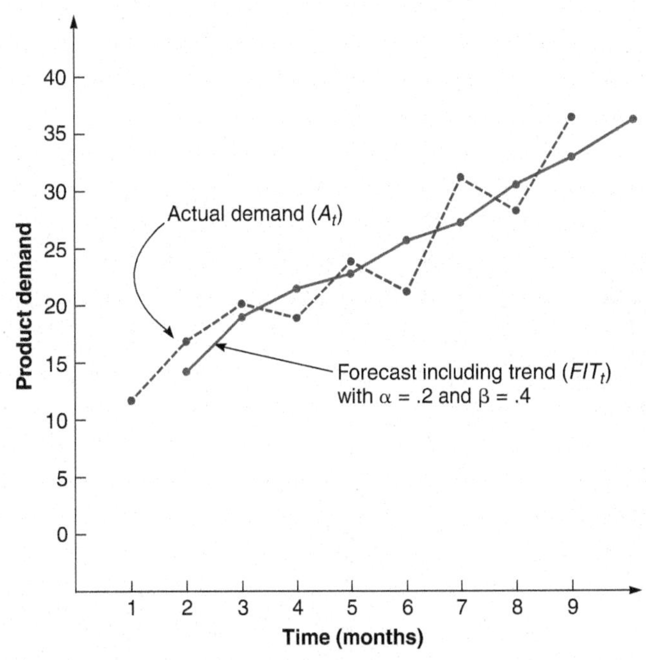

Figure **4.3**

Exponential Smoothing with Trend-Adjustment Forecasts Compared to Actual Demand Data

The value of the trend-smoothing constant, β, resembles the α constant because a high β is more responsive to recent changes in trend. A low β gives less weight to the most recent trends and tends to smooth out the present trend. Values of β can be found by the trial-and-error approach or by using sophisticated commercial forecasting software, with the MAD used as a measure of comparison.

Simple exponential smoothing is often referred to as *first-order smoothing*, and trend-adjusted smoothing is called *second-order smoothing* or *double smoothing*. Other advanced exponential-smoothing models are also used, including seasonal-adjusted and triple smoothing.

Trend Projections

Trend projection
A time-series forecasting method that fits a trend line to a series of historical data points and then projects the line into the future for forecasts.

The last time-series forecasting method we will discuss is **trend projection**. This technique fits a trend line to a series of historical data points and then projects the slope of the line into the future for medium- to long-range forecasts. Several mathematical trend equations can be developed (for example, exponential and quadratic), but in this section, we will look at *linear* (straight-line) trends only.

If we decide to develop a linear trend line by a precise statistical method, we can apply the *least-squares method.* This approach results in a straight line that minimizes the sum of the squares of the vertical differences or deviations from the line to each of the actual observations. Figure 4.4 illustrates the least-squares approach.

A least-squares line is described in terms of its y-intercept (the height at which it intercepts the y-axis) and its expected change (slope). If we can compute the y-intercept and slope, we can express the line with the following equation:

$$\hat{y} = a + bx \tag{4-11}$$

where $\hat{y}$ (called "y hat") = computed value of the variable to be predicted (called the *dependent variable*)
a = y-axis intercept
b = slope of the regression line (or the rate of change in y for given changes in x)
x = the independent variable (which in this case is *time*)

Statisticians have developed equations that we can use to find the values of a and b for any regression line. The slope b is found by:

$$b = \frac{\Sigma xy - n\bar{x}\,\bar{y}}{\Sigma x^2 - n\bar{x}^2} \tag{4-12}$$

Figure **4.4**

The Least-Squares Method for Finding the Best-Fitting Straight Line, Where the Asterisks Are the Locations of the Seven Actual Observations or Data Points

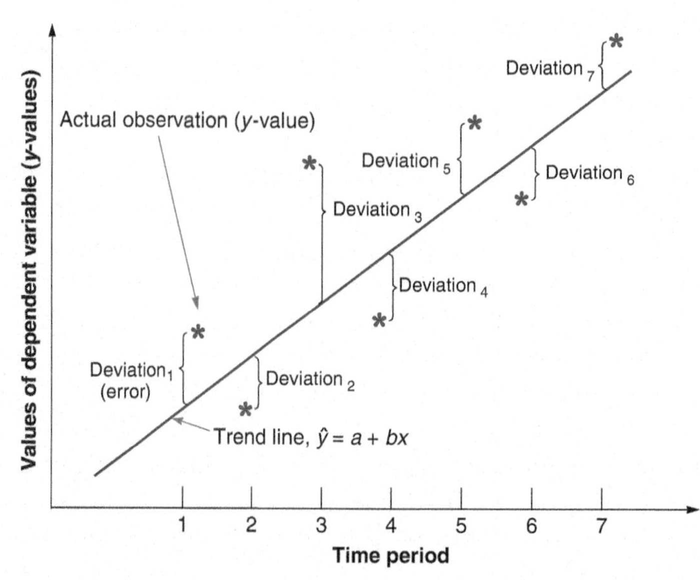

where b = slope of the regression line
Σ = summation sign
x = known values of the independent variable
y = known values of the dependent variable
$\bar{x}$ = average of the x-values
$\bar{y}$ = average of the y-values
n = number of data points or observations

We can compute the y-intercept a as follows:

$$a = \bar{y} - b\bar{x} \quad (4\text{-}13)$$

Example 8 shows how to apply these concepts.

Example 8

FORECASTING WITH LEAST SQUARES

The demand for electric power at N.Y. Edison over the past 7 years is shown in the following table, in megawatts. The firm wants to forecast next year's demand by fitting a straight-line trend to these data.

YEAR	ELECTRICAL POWER DEMAND	YEAR	ELECTRICAL POWER DEMAND
1	74	5	105
2	79	6	142
3	80	7	122
4	90		

APPROACH ▶ Equations (4-12) and (4-13) can be used to create the trend projection model.

SOLUTION ▶

YEAR (x)	ELECTRIC POWER DEMAND (y)	x^2	xy
1	74	1	74
2	79	4	158
3	80	9	240
4	90	16	360
5	105	25	525
6	142	36	852
7	122	49	854
$\Sigma x = 28$	$\Sigma y = 692$	$\Sigma x^2 = 140$	$\Sigma xy = 3{,}063$

$$\bar{x} = \frac{\Sigma x}{n} = \frac{28}{7} = 4 \quad \bar{y} = \frac{\Sigma y}{n} = \frac{692}{7} = 98.86$$

$$b = \frac{\Sigma xy - n\bar{x}\bar{y}}{\Sigma x^2 - n\bar{x}^2} = \frac{3{,}063 - (7)(4)(98.86)}{140 - (7)(4^2)} = \frac{295}{28} = 10.54$$

$$a = \bar{y} - b\bar{x} = 98.86 - 10.54(4) = 56.70$$

Thus, the least-squares trend equation is $\hat{y} = 56.70 + 10.54x$. To project demand next year, $x = 8$:

$$\begin{aligned}\text{Demand forecast in year 8} &= 56.70 + 10.54(8)\\ &= 141.02, \text{ or } 141 \text{ megawatts}\end{aligned}$$

INSIGHT ▶ To evaluate the model, we plot both the historical demand and the trend line in Figure 4.5. In this case, we may wish to be cautious and try to understand the swing in demand from year 6 to year 7.

Figure 4.5

Electrical Power and the Computed Trend Line

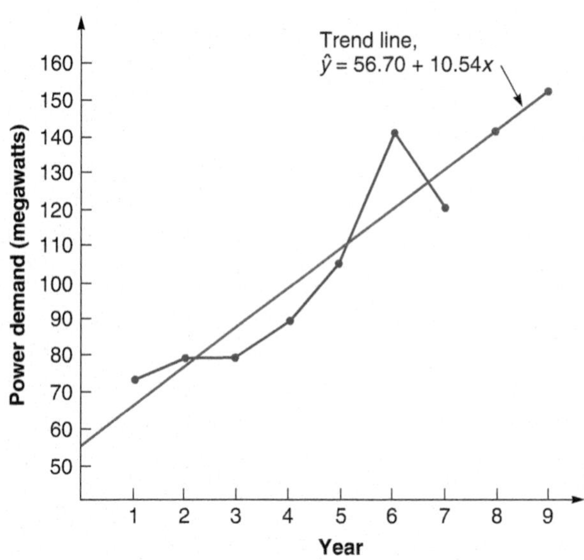

LEARNING EXERCISE ▶ Estimate demand for year 9. [Answer: 151.56, or 152 megawatts.]

RELATED PROBLEMS ▶ 4.6, 4.13c, 4.16, 4.24, 4.30, 4.39

EXCEL OM Data File **Ch04Ex8.xls** can be found online.

ACTIVE MODEL 4.4 This example is further illustrated in Active Model 4.4 found online.

Notes on the Use of the Least-Squares Method Using the least-squares method implies that we have met three requirements:

1. We always plot the data because least-squares data assume a linear relationship. If a curve appears to be present, curvilinear analysis is probably needed.
2. We do not predict time periods far beyond our given database. For example, if we have 20 months' worth of average prices of Microsoft stock, we can forecast only 3 or 4 months into the future. Forecasts beyond that have little statistical validity. Thus, you cannot take 5 years' worth of sales data and project 10 years into the future. The world is too uncertain.
3. Deviations around the least-squares line (see Figure 4.4) are assumed to be random and normally distributed, with most observations close to the line and only a smaller number farther out.

Seasonal Variations in Data

Seasonal variations
Regular upward or downward movements in a time series that tie to recurring events.

Seasonal variations in data are regular movements in a time series that relate to recurring events such as weather or holidays. Demand for coal and fuel oil, for example, peaks during cold winter months. Demand for golf clubs or sunscreen may be highest in summer.

Seasonality may be applied to hourly, daily, weekly, monthly, or other recurring patterns. Fast-food restaurants experience *daily* surges at noon and again at 5 P.M. Movie theaters see higher demand on Friday and Saturday evenings. The post office, The Christmas Store, and Hallmark Card Shops also exhibit seasonal variation in customer traffic and sales.

STUDENT TIP
John Deere understands seasonal variations: It has been able to obtain 70% of its orders in advance of seasonal use so it can smooth production.

Similarly, understanding seasonal variations is important for capacity planning in organizations that handle peak loads. These include electric power companies during extreme cold and warm periods, banks on Friday afternoons, and buses and subways during the morning and evening rush hours.

Time-series forecasts like those in Example 8 involve reviewing the trend of data over a series of time periods. The presence of seasonality makes adjustments in trend-line forecasts necessary. Seasonality is expressed in terms of the amount that actual values differ from average values in the time series. Analyzing data in monthly or quarterly terms usually makes it easy for a statistician to spot seasonal patterns. Seasonal indices can then be developed by several common methods.

In what is called a *multiplicative seasonal model*, seasonal factors are multiplied by an estimate of average demand to produce a seasonal forecast. Our assumption in this section is that trend has been removed from the data. Otherwise, the magnitude of the seasonal data will be distorted by the trend.

Here are the steps we will follow for a company that has "seasons" of 1 month:

1. Find the *average historical demand each season* (or month in this case) by summing the demand for that month in each year and dividing by the number of years of data available. For example, if, in January, we have seen sales of 8, 6, and 10 over the past 3 years, average January demand equals $(8 + 6 + 10)/3 = 8$ units.
2. Compute the *average demand over all months* by dividing the total average annual demand by the number of seasons. For example, if the total average demand for a year is 120 units and there are 12 seasons (each month), the average monthly demand is $120/12 = 10$ units.
3. Compute a *seasonal index* for each season by dividing that *month's* historical average demand (from Step 1) by the average demand over all months (from Step 2). For example, if the average historical January demand over the past 3 years is 8 units and the average demand over all months is 10 units, the seasonal index for January is $8/10 = .80$. Likewise, a seasonal index of 1.20 for February would mean that February's demand is 20% larger than the average demand over all months.
4. Estimate next year's total annual demand.
5. Divide this estimate of total annual demand by the number of seasons, then multiply it by the seasonal index for each month. This provides the *seasonal forecast*.

LO 4.5 *Develop* seasonal indices

Example 9 illustrates this procedure as it computes seasonal indices from historical data.

Example 9

DETERMINING SEASONAL INDICES

A Des Moines distributor of Lenova laptop computers wants to develop monthly indices for sales. Data from the past 3 years, by month, are available.

APPROACH ▶ Follow the five steps listed above.

SOLUTION ▶

	DEMAND					
MONTH	YEAR 1	YEAR 2	YEAR 3	AVERAGE PERIOD DEMAND	AVERAGE MONTHLY DEMAND[a]	SEASONAL INDEX[b]
Jan.	80	85	105	90	94	.957(= 90/94)
Feb.	70	85	85	80	94	.851(= 80/94)
Mar.	80	93	82	85	94	.904(= 85/94)
Apr.	90	95	115	100	94	1.064(= 100/94)
May	113	125	131	123	94	1.309(= 123/94)
June	110	115	120	115	94	1.223(= 115/94)
July	100	102	113	105	94	1.117(= 105/94)
Aug.	88	102	110	100	94	1.064(= 100/94)
Sept.	85	90	95	90	94	.957(= 90/94)
Oct.	77	78	85	80	94	.851(= 80/94)
Nov.	75	82	83	80	94	.851(= 80/94)
Dec.	82	78	80	80	94	.851(= 80/94)

Total average annual demand = 1,128

[a]Average monthly demand $= \frac{1{,}128}{12 \text{ months}} = 94.$ [b]Seasonal index $= \frac{\text{Average monthly demand for past 3 years}}{\text{Average monthly demand}}.$

If we expect the annual demand for computers to be 1,200 units next year, we would use these seasonal indices to forecast the monthly demand as follows:

MONTH	DEMAND	MONTH	DEMAND
Jan.	$\frac{1,200}{12} \times .957 = 96$	July	$\frac{1,200}{12} \times 1.117 = 112$
Feb.	$\frac{1,200}{12} \times .851 = 85$	Aug.	$\frac{1,200}{12} \times 1.064 = 106$
Mar.	$\frac{1,200}{12} \times .904 = 90$	Sept.	$\frac{1,200}{12} \times .957 = 96$
Apr.	$\frac{1,200}{12} \times 1.064 = 106$	Oct.	$\frac{1,200}{12} \times .851 = 85$
May	$\frac{1,200}{12} \times 1.309 = 131$	Nov.	$\frac{1,200}{12} \times .851 = 85$
June	$\frac{1,200}{12} \times 1.223 = 122$	Dec.	$\frac{1,200}{12} \times .851 = 85$

INSIGHT ▶ Think of these indices as percentages of average sales. The average sales (without seasonality) would be 94, but with seasonality, sales fluctuate from 85% to 131% of average.

LEARNING EXERCISE ▶ If next year's annual demand is 1,150 laptops (instead of 1,200), what will the January, February, and March forecasts be? [Answer: 91.7, 81.5, and 86.6, which can be rounded to 92, 82, and 87.]

RELATED PROBLEMS ▶ 4.26, 4.27, 4.40, 4.41a

EXCEL OM Data File **Ch04Ex9.xls** can be found online.

For simplicity, only 3 periods (years) are used for each monthly index in the preceding example. Example 10 illustrates how indices that have already been prepared can be applied to adjust trend-line forecasts for seasonality.

Example 10

APPLYING BOTH TREND AND SEASONAL INDICES

San Diego Hospital wants to improve its forecasting by applying both trend and seasonal indices to 66 months of data it has collected. It will then forecast "patient days" over the coming year.

APPROACH ▶ A trend line is created; then monthly seasonal indices are computed. Finally, a multiplicative seasonal model is used to forecast months 67 to 78.

SOLUTION ▶ Using 66 months of adult inpatient hospital days, the following equation was computed:

$$\hat{y} = 8,090 + 21.5x$$

where

$$\hat{y} = \text{patient days}$$
$$x = \text{time, in months}$$

Based on this model, which reflects only trend data, the hospital forecasts patient days for the next month (period 67) to be:

$$\text{Patient days} = 8{,}090 + (21.5)(67) = 9{,}530\,(\text{trend only})$$

While this model, as plotted in Figure 4.6, recognized the upward trend line in the demand for inpatient services, it ignored the seasonality that the administration knew to be present.

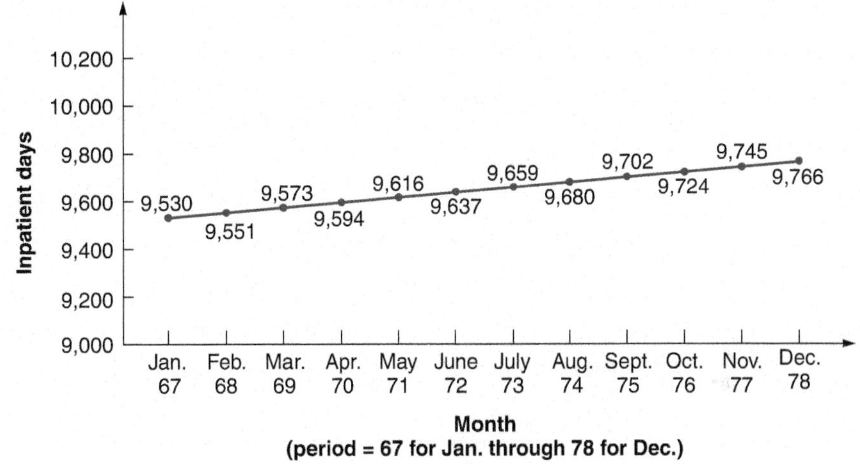

Figure 4.6

Trend Data for San Diego Hospital

Source: Based on "Modern Methods Improve Hospital Forecasting" by W. E. Sterk and E. G. Shryock from *Healthcare Financial Management* 41, no. 3, p. 97.

The following table provides seasonal indices based on the same 66 months. Such seasonal data, by the way, were found to be typical of hospitals nationwide.

Seasonality Indices for Adult Inpatient Days at San Diego Hospital

MONTH	SEASONALITY INDEX	MONTH	SEASONALITY INDEX
January	1.04	July	1.03
February	0.97	August	1.04
March	1.02	September	0.97
April	1.01	October	1.00
May	0.99	November	0.96
June	0.99	December	0.98

These seasonal indices are graphed in Figure 4.7. Note that January, March, July, and August seem to exhibit significantly higher patient days on average, while February, September, November, and December experience lower patient days.

However, neither the trend data nor the seasonal data alone provide a reasonable forecast for the hospital. Only when the hospital multiplied the trend-adjusted data by the appropriate seasonal index did it obtain good forecasts. Thus, for period 67 (January):

$$\text{Patient days} = (\text{Trend-adjusted forecast})(\text{Monthly seasonal index}) = (9{,}530)(1.04) = 9{,}911$$

Figure 4.7

Seasonal Index for San Diego Hospital

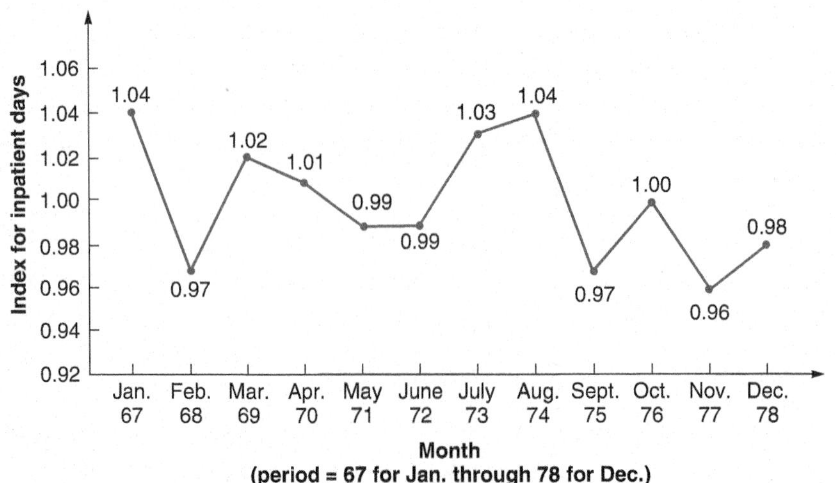

The patient days for each month are:

Period	67	68	69	70	71	72	73	74	75	76	77	78
Month	Jan.	Feb.	March	April	May	June	July	Aug.	Sept.	Oct.	Nov.	Dec.
Forecast with Trend & Seasonality	9,911	9,265	9,764	9,691	9,520	9,542	9,949	10,068	9,411	9,724	9,355	9,572

A graph showing the forecast that combines both trend and seasonality appears in Figure 4.8.

Figure 4.8

Combined Trend and Seasonal Forecast

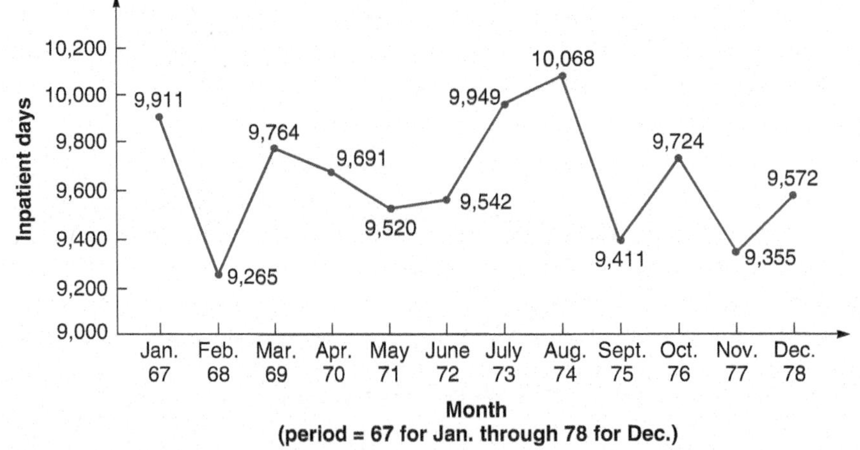

INSIGHT ▶ Notice that with trend only, the September forecast is 9,702, but with both trend and seasonal adjustments, the forecast is 9,411. By combining trend and seasonal data, the hospital was better able to forecast inpatient days and the related staffing and budgeting vital to effective operations.

LEARNING EXERCISE ▶ If the slope of the trend line for patient days is 22.0 (rather than 21.5) and the index for December is .99 (instead of .98), what is the new forecast for December inpatient days? [Answer: 9,708.]

RELATED PROBLEMS ▶ 4.25, 4.28, 4.41b

Example 11 further illustrates seasonality for quarterly data at a wholesaler.

Example 11

ADJUSTING TREND DATA WITH SEASONAL INDICES

Management at Jagoda Wholesalers, in Calgary, Canada, has used time-series regression based on point-of-sale data to forecast sales for the next 4 quarters. Sales estimates are \$100,000, \$120,000, \$140,000, and \$160,000 for the respective quarters. Seasonal indices for the 4 quarters have been found to be 1.30, .90, .70, and 1.10, respectively.

APPROACH ▶ To compute a seasonalized or adjusted sales forecast, we just multiply each seasonal index by the appropriate trend forecast:

$$\hat{y}_{\text{seasonal}} = \text{Index} \times \hat{y}_{\text{trend forecast}}$$

SOLUTION ▶

$$\begin{aligned}
\text{Quarter I:} \quad & \hat{y}_{I} = (1.30)(\$100{,}000) = \$130{,}000 \\
\text{Quarter II:} \quad & \hat{y}_{II} = (.90)(\$120{,}000) = \$108{,}000 \\
\text{Quarter III:} \quad & \hat{y}_{III} = (.70)(\$140{,}000) = \$98{,}000 \\
\text{Quarter IV:} \quad & \hat{y}_{IV} = (1.10)(\$160{,}000) = \$176{,}000
\end{aligned}$$

INSIGHT ▶ The straight-line trend forecast is now adjusted to reflect the seasonal changes.

LEARNING EXERCISE ▶ If the sales forecast for Quarter IV was $180,000 (rather than $160,000), what would be the seasonally adjusted forecast? [Answer: $198,000.]

RELATED PROBLEMS ▶ 4.25, 4.28, 4.41b

Cyclical Variations in Data

Cycles
Patterns in the data that occur every several years.

Cycles are like seasonal variations in data but occur every several *years*, not weeks, months, or quarters. Forecasting cyclical variations in a time series is difficult. This is because cycles include a wide variety of factors that cause the economy to go from recession to expansion to recession over a period of years. These factors include national or industrywide overexpansion in times of euphoria and contraction in times of concern. Forecasting demand for individual products can also be driven by product life cycles—the stages products go through from introduction through decline. Life cycles exist for virtually all products; striking examples include floppy disks, video recorders, and the original Game Boy. We leave cyclical analysis to forecasting texts.

Developing associative techniques of variables that affect one another is our next topic.

Associative Forecasting Methods: Regression and Correlation Analysis

STUDENT TIP
We now deal with the same mathematical model that we saw earlier, the least-squares method. But we use any potential "cause-and-effect" variable as *x*.

Unlike time-series forecasting, *associative forecasting* models usually consider *several* variables that are related to the quantity being predicted. Once these related variables have been found, a statistical model is built and used to forecast the item of interest. This approach is more powerful than the time-series methods that use only the historical values for the forecast variable.

Many factors can be considered in an associative analysis. For example, the sales of Dell PCs may be related to Dell's advertising budget, the company's prices, competitors' prices and promotional strategies, and even the nation's economy and unemployment rates. In this case, PC sales would be called the *dependent variable*, and the other variables would be called *independent variables*. The manager's job is to develop *the best statistical relationship between PC sales and the independent variables*. The most common quantitative associative forecasting model is linear regression analysis using one independent variable.

Linear regression analysis
A straight-line mathematical model to describe the functional relationships between independent and dependent variables.

Using Regression Analysis for Forecasting

We can use the same mathematical model that we employed in the least-squares method of trend projection to perform a linear regression analysis. The dependent variables that we want to forecast will still be $\hat{y}$. But now the independent variable, x, need no longer be time. We use the equation:

$$\hat{y} = a + bx$$

where
$\hat{y}$ = value of the dependent variable (in our example, sales)
a = y-axis intercept
b = slope of the regression line
x = independent variable

LO 4.6 *Conduct* a regression and correlation analysis

Example 12 shows how to use linear regression.

Example 12

COMPUTING A LINEAR REGRESSION EQUATION

Nodel Construction Company renovates old homes in West Bloomfield, Michigan. Over time, the company has found that its dollar volume of renovation work is dependent on the West Bloomfield area payroll. Management wants to establish a mathematical relationship to help predict sales.

APPROACH ▶ Nodel has prepared the following table, which lists company revenues and the amount of money earned by wage earners in West Bloomfield during the past 6 years:

NODEL'S SALES (IN \$ MILLIONS), y	AREA PAYROLL (IN \$ BILLIONS), x	NODEL'S SALES (IN \$ MILLIONS), y	AREA PAYROLL (IN \$ BILLIONS), x
2.0	1	2.0	2
3.0	3	2.0	1
2.5	4	3.5	7

The company needs to determine whether there is a straight-line (linear) relationship between area payroll and sales. It plots the known data on a scatter diagram:

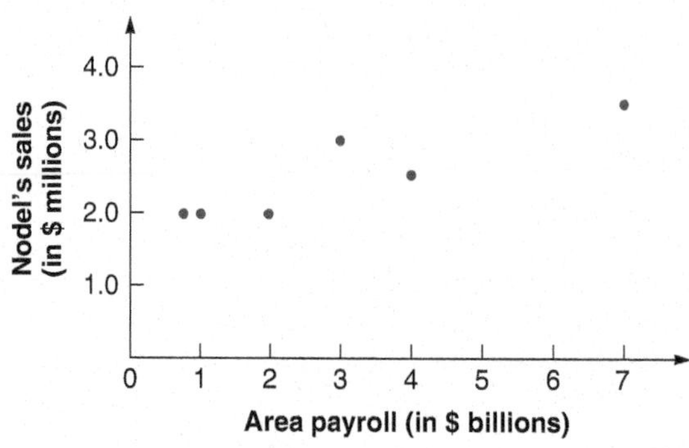

STUDENT TIP A scatter diagram is a powerful data analysis tool. It helps quickly size up the relationship between two variables.

From the six data points, there appears to be a slight positive relationship between the independent variable (payroll) and the dependent variable (sales): As payroll increases, Nodel's sales tend to be higher.

SOLUTION ▶ We can find a mathematical equation by using the least-squares regression approach:

VIDEO 4.1 Forecasting Ticket Revenue for Orlando Magic Basketball Games

SALES, y	PAYROLL, x	x^2	xy
2.0	1	1	2.0
3.0	3	9	9.0
2.5	4	16	10.0
2.0	2	4	4.0
2.0	1	1	2.0
3.5	7	49	24.5
$\sum y = 15.0$	$\sum x = 18$	$\sum x^2 = 80$	$\sum xy = 51.5$

$$\bar{x} = \frac{\sum x}{6} = \frac{18}{6} = 3$$

$$\bar{y} = \frac{\sum y}{6} = \frac{15}{6} = 2.5$$

$$b = \frac{\sum xy - n\bar{x}\,\bar{y}}{\sum x^2 - n\bar{x}^2} = \frac{51.5 - (6)(3)(2.5)}{80 - (6)(3^2)} = .25$$

$$a = \bar{y} - b\bar{x} = 2.5 - (.25)(3) = 1.75$$

The estimated regression equation, therefore, is:

$$\hat{y} = 1.75 + .25x$$

or:

$$\text{Sales} = 1.75 + .25(\text{payroll})$$

If the local chamber of commerce predicts that the West Bloomfield area payroll will be \$6 billion next year, we can estimate sales for Nodel with the regression equation:

$$\text{Sales (in \$ millions)} = 1.75 + .25(6) = 1.75 + 1.50 = 3.25$$

or:

$$\text{Sales} = \$3,250,000$$

INSIGHT ▶ Given our assumptions of a straight-line relationship between payroll and sales, we now have an indication of the slope of that relationship: on average, sales increase at the rate of $\frac{1}{4}$ million dollars for every billion dollars in the local area payroll. This is because $b = .25$.

LEARNING EXERCISE ▶ What are Nodel's sales when the local payroll is \$8 billion? [Answer: \$3.75 million.]

RELATED PROBLEMS ▶ 4.42–4.48, 4.50–4.54, 4.56a, 4.57, 4.58

EXCEL OM Data File **Ch04Ex12.xls** can be found online.

The final part of Example 12 shows a central weakness of associative forecasting methods like regression. Even when we have computed a regression equation, we must provide a forecast of the independent variable x—in this case, payroll—before estimating the dependent variable y for the next time period. Although this is not a problem for all forecasts, you can imagine the difficulty of determining future values of *some* common independent variables (e.g., unemployment rates, gross national product (GNP), price indices, and so on).

Standard Error of the Estimate

The forecast of \$3,250,000 for Nodel's sales in Example 12 is called a *point estimate* of y. The point estimate is really the *mean*, or *expected value*, of a distribution of possible values of sales. Figure 4.9 illustrates this concept.

To measure the accuracy of the regression estimates, we must compute the **standard error of the estimate**, $S_{y,x}$. This computation is called the *standard deviation of the regression:* It measures the error from the dependent variable, y, to the regression line, rather than to the mean. Equation (4-14) is a similar expression to that found in most statistics books for computing the standard deviation of an arithmetic mean:

Standard error of the estimate
A measure of variability around the regression line—its standard deviation.

$$S_{y,x} = \sqrt{\frac{\Sigma(y - y_c)^2}{n - 2}} \quad (4\text{-}14)$$

where y = y-value of each data point
y_c = computed value of the dependent variable, from the regression equation
n = number of data points

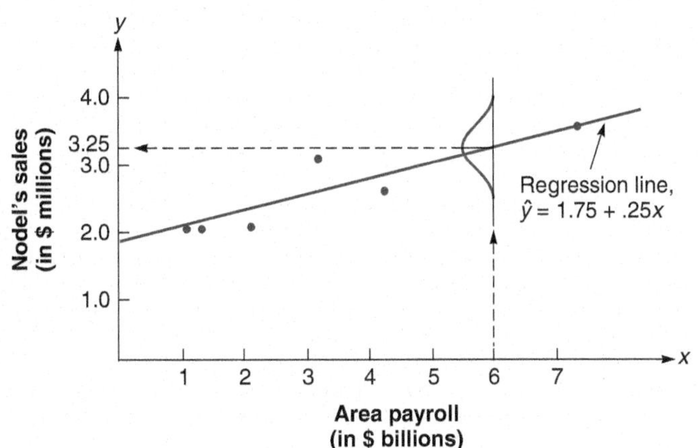

Figure **4.9**

Distribution About the Point Estimate of \$3.25 Million Sales

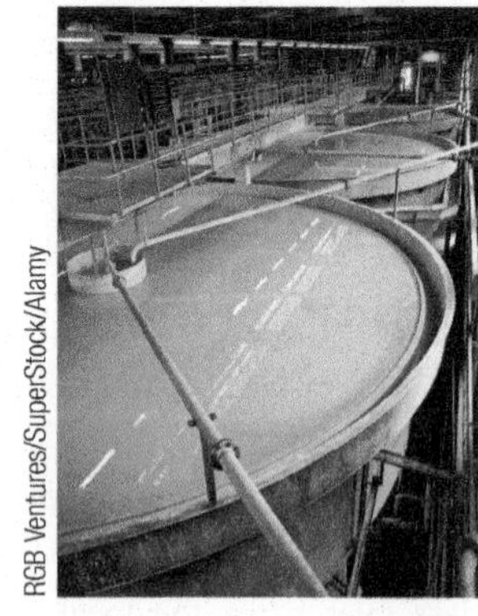
RGB Ventures/SuperStock/Alamy

Glidden Paints' assembly lines require thousands of gallons every hour. To predict demand, the firm uses associative forecasting methods such as linear regression, with independent variables such as disposable personal income and GNP. Although housing starts would be a natural variable, Glidden found that it correlated poorly with past sales. It turns out that most Glidden paint is sold through retailers to customers who already own homes or businesses.

Equation (4-15) may look more complex, but it is actually an easier-to-use version of Equation (4-14). Both formulas provide the same answer and can be used in setting up prediction intervals around the point estimate[3]:

$$S_{y,x} = \sqrt{\frac{\sum y^2 - a\sum y - b\sum xy}{n-2}} \tag{4-15}$$

Example 13 shows how we would calculate the standard error of the estimate in Example 12.

Example 13

COMPUTING THE STANDARD ERROR OF THE ESTIMATE

Nodel now wants to know the error associated with the regression line computed in Example 12.

APPROACH ▶ Compute the standard error of the estimate, $S_{y,x}$, using Equation (4-15).

SOLUTION ▶ The only number we need that is not available to solve for $S_{y,x}$ is $\sum y^2$. Some simple calculations reveal $\sum y^2 = 39.5$. Therefore:

$$S_{y,x} = \sqrt{\frac{\sum y^2 - a\sum y - b\sum xy}{n-2}}$$

$$= \sqrt{\frac{39.5 - 1.75(15.0) - .25(51.5)}{6-2}}$$

$$= \sqrt{.09375} = .306 \text{ (in \$ millions)}$$

The standard error of the estimate is then $306,000 in sales.

INSIGHT ▶ The interpretation of the standard error of the estimate is similar to the standard deviation; namely, ±1 standard deviation = .6827. So there is a 68.27% chance of sales being ±$306,000 from the point estimate of $3,250,000.

LEARNING EXERCISE ▶ What is the probability sales will exceed $3,556,000? [Answer: About 16%.]

RELATED PROBLEMS ▶ 4.52e, 4.54b, 4.56c, 4.57

Correlation Coefficients for Regression Lines

The regression equation is one way of expressing the nature of the relationship between two variables. Regression lines are not "cause-and-effect" relationships. They merely describe the relationships among variables. The regression equation shows how one variable relates to the value and changes in another variable.

Coefficient of correlation
A measure of the strength of the relationship between two variables.

Another way to evaluate the relationship between two variables is to compute the **coefficient of correlation**. This measure expresses the degree or strength of the linear relationship (but note

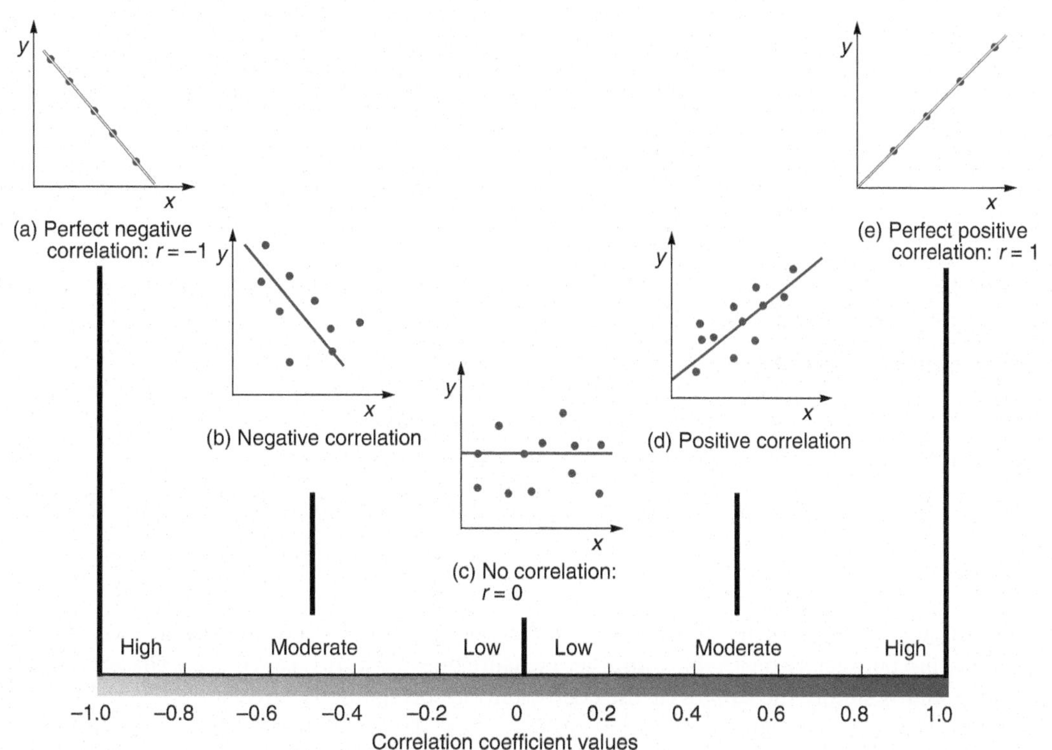

Figure **4.10**

Five Values of the Correlation Coefficient

that correlation does not necessarily imply causality). Usually identified as r, the coefficient of correlation can be any number between +1 and −1. Figure 4.10 illustrates what different values of r might look like.

To compute r, we use much of the same data needed earlier to calculate a and b for the regression line. The rather lengthy equation for r is:

$$r = \frac{n\sum xy - \sum x \sum y}{\sqrt{\left[n\sum x^2 - \left(\sum x\right)^2\right]\left[n\sum y^2 - \left(\sum y\right)^2\right]}} \tag{4-16}$$

Example 14 shows how to calculate the coefficient of correlation for the data given in Examples 12 and 13.

Example 14

DETERMINING THE COEFFICIENT OF CORRELATION

In Example 12, we looked at the relationship between Nodel Construction Company's renovation sales and payroll in its hometown of West Bloomfield. The company now wants to know the strength of the association between area payroll and sales.

APPROACH ▶ We compute the r value using Equation (4-16). We need to first add one more column of calculations—for y^2.

SOLUTION ▶ The data, including the column for y^2 and the calculations, are shown here:

y	x	x^2	xy	y^2
2.0	1	1	2.0	4.0
3.0	3	9	9.0	9.0
2.5	4	16	10.0	6.25
2.0	2	4	4.0	4.0
2.0	1	1	2.0	4.0
3.5	7	49	24.5	12.25
$\sum y = 15.0$	$\sum x = 18$	$\sum x^2 = 80$	$\sum xy = 51.5$	$\sum y^2 = 39.5$

$$r = \frac{(6)(51.5) - (18)(15.0)}{\sqrt{[(6)(80) - (18)^2][(6)(39.5) - (15.0)^2]}}$$

$$= \frac{309 - 270}{\sqrt{(156)(12)}} = \frac{39}{\sqrt{1,872}}$$

$$= \frac{39}{43.3} = .901$$

INSIGHT ▶ This r of .901 appears to be a significant correlation and helps confirm the closeness of the relationship between the two variables.

LEARNING EXERCISE ▶ If the coefficient of correlation was −.901 rather than +.901, what would this tell you? [Answer: The negative correlation would tell you that as payroll went up, Nodel's sales went down—a rather unlikely occurrence that would suggest you recheck your math.]

RELATED PROBLEMS ▶ 4.43d, 4.48d, 4.50c, 4.52f, 4.54b, 4.56b, 4.57

Coefficient of determination
A measure of the amount of variation in the dependent variable about its mean that is explained by the regression equation.

Although the coefficient of correlation is the measure most commonly used to describe the relationship between two variables, another measure does exist. It is called the **coefficient of determination** and is simply the square of the coefficient of correlation—namely, r^2. The value of r^2 will always be a positive number in the range $0 \le r^2 \le 1$. The coefficient of determination is the percent of variation in the dependent variable (y) that is explained by the regression equation. In Nodel's case, the value of r^2 is .81, indicating that 81% of the total variation is explained by the regression equation.

Multiple Regression Analysis

Multiple regression
An associative forecasting method with more than one independent variable.

Multiple regression is a practical extension of the linear regression model we just explored. It allows us to build a model with several independent variables instead of just one variable. For example, if Nodel Construction wanted to include average annual interest rates in its model for forecasting renovation sales, the proper equation would be:

$$\hat{y} = a + b_1x_1 + b_2x_2 \quad (4\text{-}17)$$

where
y = dependent variable, sales
a = a constant, the y intercept
x_1 and x_2 = values of the two independent variables, area payroll and interest rates, respectively
b_1 and b_2 = coefficients for the two independent variables

The mathematics of multiple regression becomes quite complex (and is usually tackled by computer), so we leave the formulas for a, b_1, and b_2 to statistics textbooks. However, Example 15 shows how to interpret Equation (4-17) in forecasting Nodel's sales.

Example 15

USING A MULTIPLE REGRESSION EQUATION

Nodel Construction wants to see the impact of a second independent variable, interest rates, on its sales.

APPROACH ▶ The new multiple regression line for Nodel Construction, calculated by computer software, is:

$$\hat{y} = 1.80 + .30x_1 - 5.0x_2$$

We also find that the new coefficient of correlation is .96, implying the inclusion of the variable x_2, interest rates, adds even more strength to the linear relationship.

SOLUTION ▶ We can now estimate Nodel's sales if we substitute values for next year's payroll and interest rate. If West Bloomfield's payroll will be \$6 billion and the interest rate will be .12 (12%), sales will be forecast as:

$$\text{Sales(\$ millions)} = 1.80 + .30(6) - 5.0(.12)$$
$$= 1.8 + 1.8 - .6$$
$$= 3.00$$

or:

$$\text{Sales} = \$3{,}000{,}000$$

INSIGHT ▶ By using both variables, payroll and interest rates, Nodel now has a sales forecast of \$3 million and a higher coefficient of correlation. This suggests a stronger relationship between the two variables and a more accurate estimate of sales. Adding a variable can be expected to increase r^2.

LEARNING EXERCISE ▶ If interest rates were only 6%, what would be the sales forecast? [Answer: 1.8 + 1.8 − 5.0(.06) = 3.3, or \$3,300,000.]

RELATED PROBLEMS ▶ 4.47, 4.49, 4.59

The *OM in Action* box, "NYC's Potholes and Regression Analysis," provides an interesting example of one city's use of regression and multiple regression.

OM in Action NYC's Potholes and Regression Analysis[4]

New York is famous for many things, but one it does not like to be known for is its large and numerous potholes. TV comedian David Letterman used to joke: "There is a pothole so big on 8th Avenue, it has its own Starbucks in it." When it comes to potholes, some years seem to be worse than others. A recent winter was exceptionally bad. City workers filled a record 300,000 potholes during the first 4 months of the year. That's an astounding accomplishment.

But potholes are to some extent a measure of municipal competence—and they are costly. NYC's poor streets cost the average motorist an estimated \$800 per year in repair work and new tires. There has been a steady and dramatic increase in potholes from around 70,000–80,000 in the 1990s to the devastatingly high 200,000–300,000 range in recent years. One theory is that bad weather causes the potholes. Using inches of snowfall as a measure of the severity of the winter, the graph below shows a plot of the number of potholes versus the inches of snow each winter.

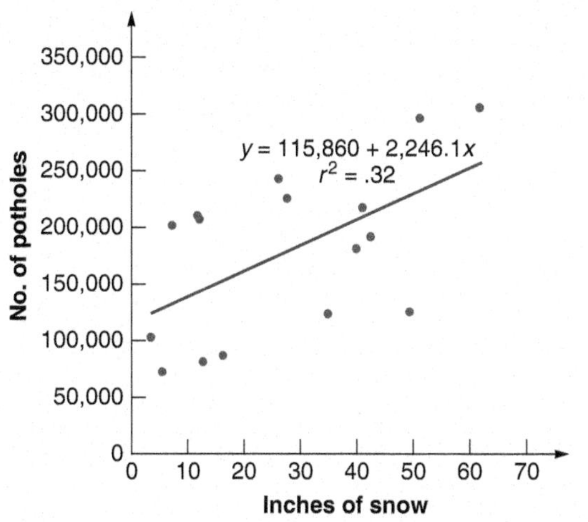

Research showed that the city would need to resurface at least 1,000 miles of roads per year just to stay even with road deterioration. Any amount below that would contribute to a "gap" or backlog of streets needing repair. The graph below shows the plot of potholes versus the gap. With an r^2 of .81, there is a very strong relationship between the increase in the "gap" and the number of potholes. It is obvious that the real reason for the steady and substantial increase in the number of potholes is because of the increasing gap in road resurfacing.

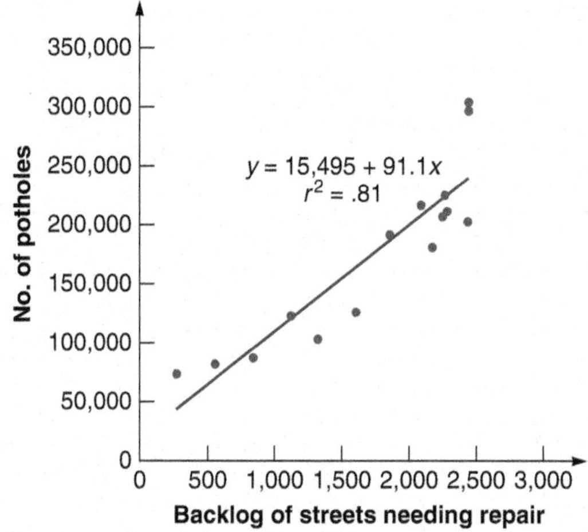

A third model performs a regression analysis using the resurfacing gap and inches of snow as two independent variables and number of potholes as the dependent variable. That regression model's r^2 is .91.

$$\text{Potholes} = 7{,}801.5 + 80.6 \times \text{Resurfacing gap} + 930.1 \times \text{Inches of snow}$$

Monitoring, Controlling, and Adapting Forecasts

Once a forecast has been completed, it should not be forgotten. No manager wants to be reminded that his or her forecast is horribly inaccurate, but a firm needs to determine why actual demand (or whatever variable is being examined) differed significantly from that projected. If the forecaster is accurate, that individual usually makes sure that everyone is aware of his or her talents. Very seldom does one read articles in *Fortune, Forbes*, or *The Wall Street Journal*, however, about money managers who are consistently off by 25% in their stock market forecasts.

Tracking signal
A measurement of how well a forecast is predicting actual values.

One way to monitor forecasts to ensure that they are performing well is to use a tracking signal. A **tracking signal** is a measurement of how well a forecast is predicting actual values. As forecasts are updated every week, month, or quarter, the newly available demand data are compared to the forecast values.

The tracking signal is computed as the cumulative error divided by the *mean absolute deviation (MAD)*:

$$\text{Tracking signal} = \frac{\text{Cumulative error}}{\text{MAD}} \tag{4-18}$$

$$= \frac{\Sigma(\text{Actual demand in period } i - \text{Forecast demand in period } i)}{\text{MAD}}$$

where

$$\text{MAD} = \frac{\Sigma|\text{Actual} - \text{Forecast}|}{n}$$

as seen earlier, in Equation (4-5).

STUDENT TIP Using a tracking signal is a good way to make sure the forecasting system is continuing to do a good job.

Positive tracking signals indicate that demand is *greater* than forecast. *Negative* signals mean that demand is *less* than forecast. A good tracking signal—that is, one with a low cumulative error—has about as much positive error as it has negative error. In other words, small deviations are okay, but positive and negative errors should balance one another so that the tracking signal centers closely around zero. A consistent tendency for forecasts to be greater or less than the actual values (that is, for a high absolute cumulative error) is called a **bias** error. Bias can occur if, for example, the wrong variables or trend line are used or if a seasonal index is misapplied.

Bias
A forecast that is consistently higher or consistently lower than actual values of a time series.

Once tracking signals are calculated, they are compared with predetermined control limits. When a tracking signal exceeds an upper or lower limit, there is a problem with the forecasting method, and management may want to reevaluate the way it forecasts demand. Figure 4.11 shows the graph of a tracking signal that is exceeding the range of acceptable variation. If the model being used is exponential smoothing, perhaps the smoothing constant needs to be readjusted.

LO 4.7 *Use* a tracking signal

How do firms decide what the upper and lower tracking limits should be? There is no single answer, but they try to find reasonable values—in other words, limits not so low as to be triggered with every small forecast error and not so high as to allow bad forecasts to be regularly overlooked. One MAD is equivalent to approximately .8 standard deviations, ± 2 MADs = ± 1.6 standard deviations, ± 3 MADs = ± 2.4 standard deviations, and ± 4 MADs = ± 3.2 standard deviations. This fact suggests that for a forecast to be "in control," 89% of the errors are expected to fall within ± 2 MADs, 98% within ± 3 MADs, or 99.9% within ± 4 MADs.[5]

Figure **4.11**

A Plot of Tracking Signals

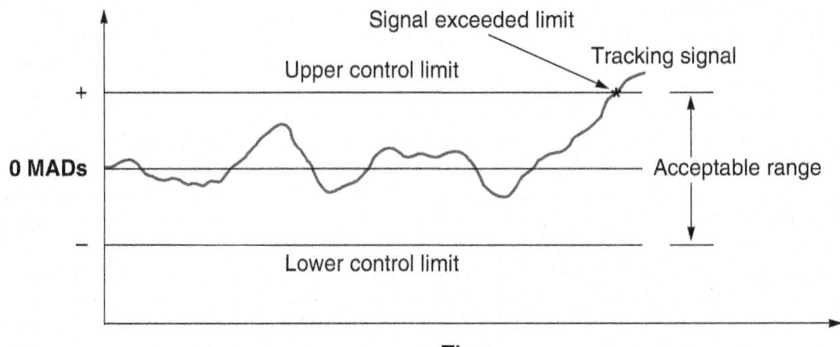

Example 16 shows how the tracking signal and cumulative error can be computed.

Example 16

COMPUTING THE TRACKING SIGNAL AT CARLSON'S BAKERY

Carlson's Bakery wants to evaluate performance of its croissant forecast.

APPROACH ▶ Develop a tracking signal for the forecast, and see if it stays within acceptable limits, which we define as ±4 MADs.

SOLUTION ▶ Using the forecast and demand data for the past 6 quarters for croissant sales, we develop a tracking signal in the following table:

QUARTER	ACTUAL DEMAND	FORECAST DEMAND	ERROR	CUMULATIVE ERROR	ABSOLUTE FORECAST ERROR	CUMULATIVE ABSOLUTE FORECAST ERROR	MAD	TRACKING SIGNAL (CUMULATIVE ERROR/MAD)
1	90	100	−10	−10	10	10	10.0	−10/10 = −1
2	95	100	−5	−15	5	15	7.5	−15/7.5 = −2
3	115	100	+15	0	15	30	10.0	0/10 = 0
4	100	110	−10	−10	10	40	10.0	−10/10 = −1
5	125	110	+15	+5	15	55	11.0	+5/11 = +0.5
6	140	110	+30	+35	30	85	14.2	+35/14.2 = +2.5

$$\text{At the end of quarter 6, MAD} = \frac{\Sigma|\text{Forecast errors}|}{n} = \frac{85}{6} = 14.2$$

$$\text{and Tracking signal} = \frac{\text{Cumulative error}}{\text{MAD}} = \frac{35}{14.2} = 2.5\text{ MADs}$$

INSIGHT ▶ Because the tracking signal drifted from −2 MAD to +2.5 MAD (between 1.6 and 2.0 standard deviations), we can conclude that it is within acceptable limits.

LEARNING EXERCISE ▶ If actual demand in quarter 6 was 130 (rather than 140), what would be the MAD and resulting tracking signal? [Answer: MAD for quarter 6 would be 12.5, and the tracking signal for period 6 would be 2 MADs.]

RELATED PROBLEMS ▶ 4.59, 4.60, 4.61c

Adaptive Smoothing

Adaptive forecasting refers to computer monitoring of tracking signals and self-adjustment if a signal passes a preset limit. For example, when applied to exponential smoothing, the α and β coefficients are first selected on the basis of values that minimize error forecasts and then adjusted accordingly whenever the computer notes an errant tracking signal. This process is called **adaptive smoothing**.

Adaptive smoothing
An approach to exponential smoothing forecasting in which the smoothing constant is automatically changed to keep errors to a minimum.

Focus Forecasting

Rather than adapt by choosing a smoothing constant, computers allow us to try a variety of forecasting models. Such an approach is called focus forecasting. **Focus forecasting** is based on two principles:

Focus forecasting
Forecasting that tries a variety of computer models and selects the best one for a particular application.

1. Sophisticated forecasting models are not always better than simple ones.
2. There is no single technique that should be used for all products or services.

Bernard Smith, inventory manager for American Hardware Supply, coined the term *focus forecasting*. Smith's job was to forecast quantities for 100,000 hardware products purchased by American's 21 buyers.[6] He found that buyers neither trusted nor understood the exponential smoothing model then in use. Instead, they used very simple approaches of their own. So Smith developed his new computerized system for selecting forecasting methods.

Smith chose to test seven forecasting methods. They ranged from the simple ones that buyers used (such as the naive approach) to statistical models. Every month, Smith applied the forecasts of all seven models to each item in stock. In these simulated trials, the forecast values were subtracted from the most recent actual demands, giving a simulated forecast error. The forecast method yielding the least error is selected by the computer, which then uses it to make next month's forecast. Although buyers still have an override capability, American Hardware finds that focus forecasting provides excellent results.

Forecasting Under a Pandemic or Major Disruption

The forecasting techniques presented so far assume some consistency in both the social and economic system. Consequently, forecasters may experience substantial challenges during unusual events such as the 2020 COVID-19 pandemic and the 2008 financial crisis. For example, as work-from-home exploded in March 2020, sales of toilet paper and hand sanitizer skyrocketed. Conversely, some crops had to be destroyed because the closing of restaurants decimated demand in the restaurant supply chain. Traditional techniques based on historical data may be of little use during such events or following disruptive natural disasters (e.g., hurricanes, tsunamis, or earthquakes). But managers can find easily prepared, readily discussed, and rapidly evaluated forecasts valuable not only for disruptions but also as an ongoing validating or auditing process for existing forecasting methods. Both Intel and Motorola use a forecasting technique called a **stagger chart** that provides such a tool.

Stagger chart
A short-term rolling forecast that is easily prepared, discussed, and evaluated.

Stagger charts emphasize current data, some of which may be intuitive or subjective, to provide a reoccurring fresh look at forecasts by using "rolling forecasts" each month for each of the next several months. The charts compare forecasts against a standard, such as a budget plan, average sales for this period in recent years, or booked sales.

One of the strengths of stagger charts is the discussion and evaluation that should follow each forecast. The evaluation might ask questions such as:

- "In what ways are the assumptions of the forecasts made in recent months different from those used when the annual budget plan was prepared?"
- "What do we know now that we did not know then?"
- "Is a competitor no longer in business?"
- "Have we had a significant change in demand due to the economy, interest rates, weather, and so on?"

Such questions may expose crucial insight necessary for adaptation during a disruptive period. Additionally, stagger charts provide a rapid and economical after-the-fact opportunity for evaluation, learning, and improvement of the forecasting process.

Forecasting in the Service Sector

STUDENT TIP
Forecasting at McDonald's, FedEx, and Walmart is as important and complex as it is for manufacturers such as Toyota and Dell.

Forecasting in the service sector presents some unusual challenges. A major technique in the retail sector is tracking demand by maintaining good short-term records. For instance, a barbershop catering to men expects peak flows on Fridays and Saturdays. Indeed, most barbershops are closed on Sunday and Monday, and many call in extra help on Friday and Saturday. A downtown restaurant, on the other hand, may need to track conventions and holidays for effective short-term forecasting.

Specialty Retail Shops Specialty retail facilities, such as flower shops, may have other unusual demand patterns, and those patterns will differ depending on the holiday. When Valentine's Day falls on a weekend, for example, flowers can't be delivered to offices, and those romantically inclined are likely to celebrate with outings rather than flowers. If a holiday falls on a Monday, some of the celebration may also take place on the weekend, reducing flower sales. However, when Valentine's Day falls in midweek, busy midweek schedules often make flowers the optimal way to celebrate. Because flowers for Mother's Day are to be delivered on Saturday or Sunday, this holiday forecast varies less. Due to special demand patterns, many service firms maintain records of sales, noting not only the day of the week but also unusual events, including the weather, so that patterns and correlations that influence demand can be developed.

VIDEO 4.2
Forecasting at Hard Rock Cafe

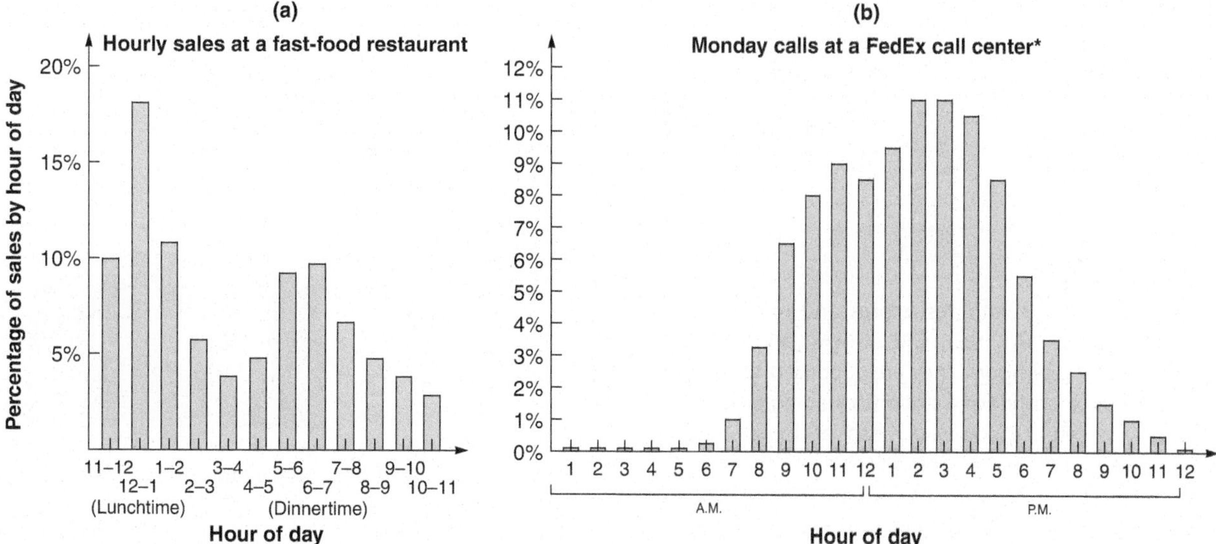

Figure **4.12**

Forecasts Are Unique: Note the Variations Between (a) Hourly Sales at a Fast-Food Restaurant and (b) Hourly Call Volume at FedEx

*Based on historical data: see *Journal of Business Forecasting* (Vol. 19, no. 4): 6–11.

Fast-Food Restaurants Fast-food restaurants are well aware of not only weekly, daily, and hourly but even 15-minute variations in demands that influence sales. Therefore, detailed forecasts of demand are needed. Figure 4.12(a) shows the hourly forecast for a typical fast-food restaurant. Note the lunchtime and dinnertime peaks. This contrasts to the mid-morning and mid-afternoon peaks at FedEx's call center in Figure 4.12(b).

Firms like Taco Bell now use point-of-sale computers that track sales every quarter hour. Taco Bell found that a 6-week moving average was the forecasting technique that minimized its mean squared error (MSE) of these quarter-hour forecasts. Building this forecasting methodology into each of Taco Bell's 7,500 U.S. stores' computers, the model makes weekly projections of customer transactions. These in turn are used by store managers to schedule staff, who begin in 15-minute increments, not 1-hour blocks as in other industries. The forecasting model has been so successful that Taco Bell has increased customer service while documenting more than $50 million in labor cost savings in 4 years of use.

Summary

Forecasts are a critical part of the operations manager's function. Demand forecasts drive a firm's production, capacity, and scheduling systems and affect the financial, marketing, and personnel planning functions.

There are a variety of qualitative and quantitative forecasting techniques. Qualitative approaches employ judgment, experience, intuition, and a host of other factors that are difficult to quantify. Quantitative forecasting uses historical data and causal, or associative, relations to project future demands. The Rapid Review for this chapter summarizes the formulas we introduced in quantitative forecasting. Forecast calculations are seldom performed by hand. Most operations managers turn to software packages such as Forecast PRO, NCSS, Minitab, Systat, Statgraphics, SAS, or SPSS.

No forecasting method is perfect under all conditions. And even once management has found a satisfactory approach, it must still monitor and control forecasts to make sure errors do not get out of hand. Forecasting can often be a very challenging, but rewarding, part of managing.

Key Terms

Forecasting (p. 106)
Economic forecasts (p. 107)
Technological forecasts (p. 107)
Demand forecasts (p. 107)
Quantitative forecasts (p. 109)
Qualitative forecasts (p. 109)
Jury of executive opinion (p. 109)
Delphi method (p. 109)
Sales force composite (p. 109)
Market survey (p. 109)
Time series (p. 110)
Naive approach (p. 112)
Moving averages (p. 112)
Exponential smoothing (p. 114)
Smoothing constant (p. 114)

Mean absolute deviation (MAD) (p. 116)
Mean squared error (MSE) (p. 117)
Mean absolute percent error (MAPE) (p. 118)
Trend projection (p. 122)
Seasonal variations (p. 124)
Cycles (p. 129)
Linear regression analysis (p. 129)
Standard error of the estimate (p. 131)
Coefficient of correlation (p. 132)
Coefficient of determination (p. 134)
Multiple regression (p. 134)
Tracking signal (p. 136)
Bias (p. 136)
Adaptive smoothing (p. 137)
Focus forecasting (p. 137)
Stagger chart (p. 138)

Ethical Dilemma

We live in a society obsessed with test scores and maximum performance. Think of the SAT, ACT, GRE, GMAT, and LSAT. Though they take only a few hours, they are supposed to give schools and companies a snapshot of a student's abiding talents.

But these tests are often spectacularly bad at forecasting performance in the real world. The SAT does a decent job (r^2 = .12) of predicting the grades of a college freshman. It is, however, less effective at predicting achievement *after* graduation. LSAT scores bear virtually no correlation to career success as measured by income, life satisfaction, or public service.

Cathy Yeulet/123RF

What does the r^2 mean in this context? Is it ethical for colleges to base admissions and financial aid decisions on scores alone? What role do these tests take at your own school?

Discussion Questions

1. What is a qualitative forecasting model, and when is its use appropriate?
2. Identify and briefly describe the two general forecasting approaches.
3. Identify the three forecasting time horizons. State an approximate duration for each.
4. Briefly describe the steps that are used to develop a forecasting system.
5. A skeptical manager asks what medium-range forecasts can be used for. Give the manager three possible uses/purposes.
6. Explain why such forecasting devices as moving averages, weighted moving averages, and exponential smoothing are not well-suited for data series that have trends.
7. What is the basic difference between a weighted moving average and exponential smoothing?
8. What three methods are used to determine the accuracy of any given forecasting method? How would you determine whether time-series regression or exponential smoothing is better in a specific application?
9. Research and briefly describe the Delphi technique. How would it be used by an employer you have worked for?
10. What is the primary difference between a time-series model and an associative model?
11. Define *time series*.
12. What effect does the value of the smoothing constant have on the weight given to the recent values?
13. Explain the value of seasonal indices in forecasting. How are seasonal patterns different from cyclical patterns?
14. Which forecasting technique can place the most emphasis on recent values? How does it do this?
15. In your own words, explain adaptive forecasting.
16. What is the purpose of a tracking signal?
17. Explain, in your own words, the meaning of the correlation coefficient. Discuss the meaning of a negative value of the correlation coefficient.
18. What is the difference between a dependent and an independent variable?
19. Give examples of industries that are affected by seasonality. Why would these businesses want to filter out seasonality?
20. Give examples of industries in which demand forecasting is dependent on the demand for other products.
21. What happens to the ability to forecast for periods farther into the future?
22. CEO John Goodale, at Southern Illinois Power and Light, has been collecting data on demand for electric power in its western subregion for only the past 2 years. Those data are shown in the table below.

    To plan for expansion and to arrange to borrow power from neighboring utilities during peak periods, Goodale needs to be able to forecast demand for each month next year. However, the standard forecasting models discussed in this chapter will not fit the data observed for the 2 years.
    a) What are the weaknesses of the standard forecasting techniques as applied to this set of data?
    b) Because known models are not appropriate here, propose your own approach to forecasting. Although there is no perfect solution to tackling data such as these (in other words, there are no 100% right or wrong answers), justify your model.
    c) Forecast demand for each month next year using the model you propose.

DEMAND IN MEGAWATTS		
MONTH	LAST YEAR	THIS YEAR
January	5	17
February	6	14
March	10	20
April	13	23
May	18	30
June	15	38
July	23	44
August	26	41
September	21	33
October	15	23
November	12	26
December	14	17

Using Software in Forecasting

This section presents three ways to solve forecasting problems with computer software. First, you can create your own Excel spreadsheets to develop forecasts. Second, you can use the Excel OM software available online. Third, POM for Windows is another program also available online.

CREATING YOUR OWN EXCEL SPREADSHEETS

Excel spreadsheets (and spreadsheets in general) are frequently used in forecasting. Exponential smoothing, trend analysis, and regression analysis (simple and multiple) are supported by built-in Excel functions.

Program 4.1 illustrates how to build an Excel forecast for the data in Example 8. The goal for N.Y. Edison is to create a trend analysis of the year 1 to year 7 data.

As an alternative, you may want to experiment with Excel's built-in regression analysis. To do so, under the **Data** menu bar selection choose **Data Analysis**, then **Regression**. Enter your *Y* and *X* data into two columns (say A and B). When the regression window appears, enter the *Y* and *X* ranges, then select **OK**. Excel offers several plots and tables to those interested in more rigorous analysis of regression problems.

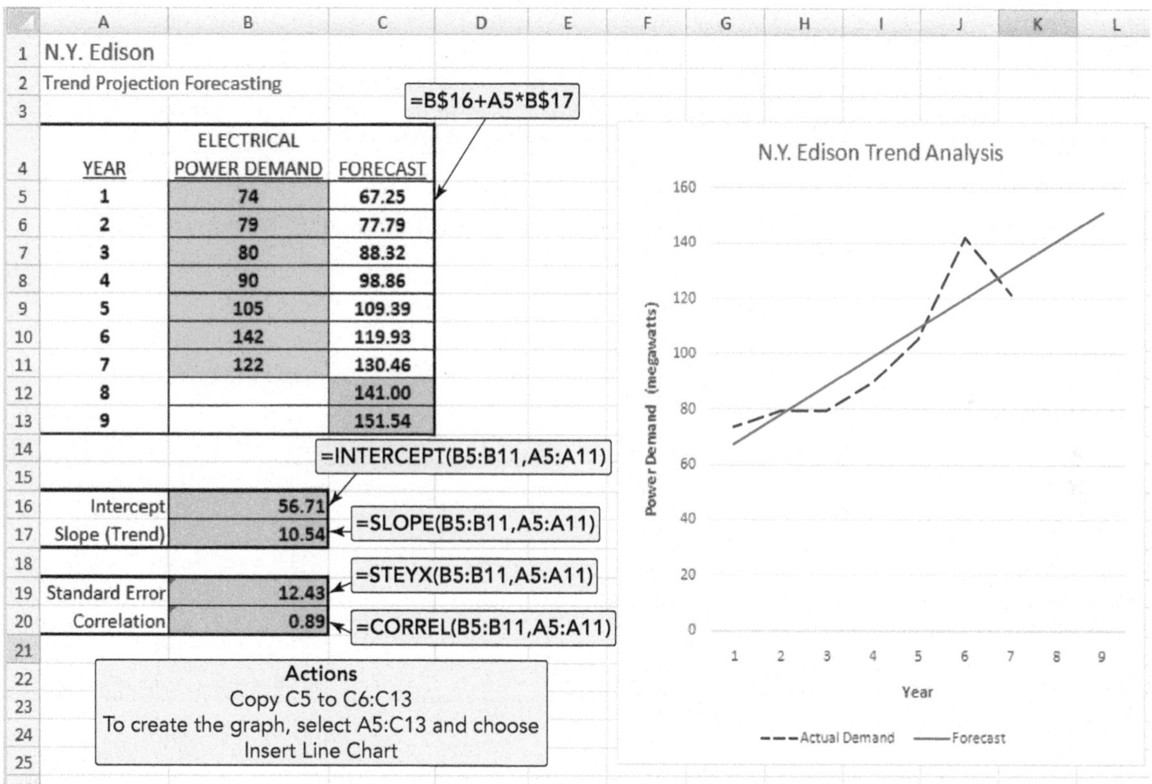

N.Y. Edison
Trend Projection Forecasting

YEAR	ELECTRICAL POWER DEMAND	FORECAST
1	74	67.25
2	79	77.79
3	80	88.32
4	90	98.86
5	105	109.39
6	142	119.93
7	122	130.46
8		141.00
9		151.54

Intercept	56.71
Slope (Trend)	10.54
Standard Error	12.43
Correlation	0.89

Program **4.1**

Using Excel to Develop Your Own Forecast, with Data from Example 8

USING EXCEL OM

Excel OM's forecasting module has five components: (1) moving averages, (2) weighted moving averages, (3) exponential smoothing, (4) regression (with one variable only), and (5) decomposition. Excel OM's error analysis is much more complete than that available with the Excel add-in.

Program 4.2 illustrates Excel OM's input and output, using Example 2's weighted-moving-average data.

USING POM FOR WINDOWS

POM for Windows can project moving averages (both simple and weighted), handle exponential smoothing (both simple and trend adjusted), forecast with least-squares trend projection, and solve linear regression (associative) models. A summary screen of error analysis and a graph of the data can also be generated. As a special example of exponential smoothing adaptive forecasting, when using an α of 0, POM for Windows will find the α value that yields the minimum MAD. Appendix II provides further details.

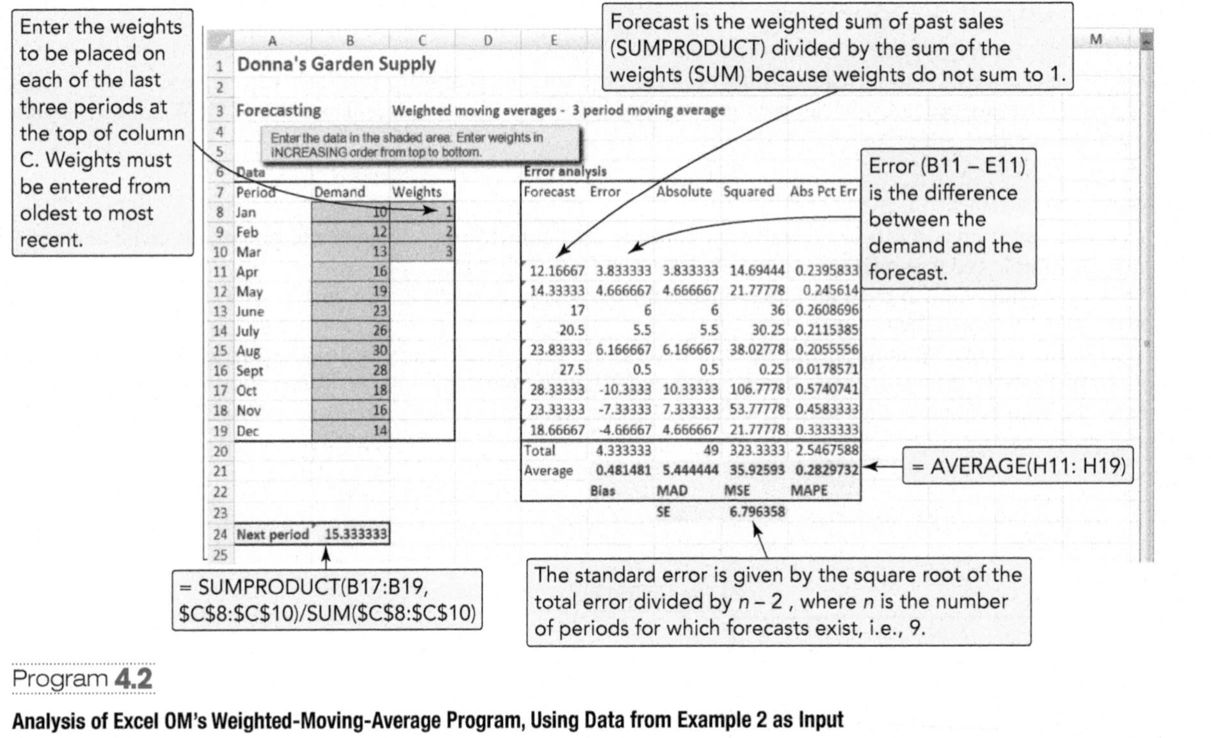

Enter the weights to be placed on each of the last three periods at the top of column C. Weights must be entered from oldest to most recent.
Forecast is the weighted sum of past sales (SUMPRODUCT) divided by the sum of the weights (SUM) because weights do not sum to 1.
Error (B11 – E11) is the difference between the demand and the forecast.
= AVERAGE(H11: H19)
= SUMPRODUCT(B17:B19, $C$8:$C$10)/SUM($C$8:$C$10)
The standard error is given by the square root of the total error divided by n – 2 , where n is the number of periods for which forecasts exist, i.e., 9.

Program **4.2**

Analysis of Excel OM's Weighted-Moving-Average Program, Using Data from Example 2 as Input

Solved Problems

SOLVED PROBLEM 4.1

Sales of Volkswagens have grown steadily at auto dealerships in Nevada during the past 5 years (see table below). The sales manager had predicted before the new model was introduced that first-year sales would be 410 VWs. Using exponential smoothing with a weight of $\alpha = .30$, develop forecasts for years 2 through 6.

YEAR	SALES	FORECAST
1	450	410
2	495	
3	518	
4	563	
5	584	
6	?	

SOLUTION

YEAR	FORECAST
1	410.0
2	422.0 = 410 + .3(450 – 410)
3	443.9 = 422 + .3(495 – 422)
4	466.1 = 443.9 + .3(518 – 443.9)
5	495.2 = 466.1 + .3(563 – 466.1)
6	521.8 = 495.2 + .3(584 – 495.2)

SOLVED PROBLEM 4.2

In Example 7, we applied trend-adjusted exponential smoothing to forecast demand for a piece of pollution-control equipment for months 2 and 3 (out of 9 months of data provided). Let us now continue this process for month 4. We want to confirm the forecast for month 4 shown in Table 4.2 and Figure 4.3.

For month 4, $A_4 = 19$, with $\alpha = .2$, and $\beta = .4$.

SOLUTION

$$\begin{aligned}
F_4 &= \alpha A_3 + (1-\alpha)(F_3 + T_3) \\
&= (.2)(20) + (1 - .2)(15.18 + 2.10) \\
&= 4.0 + (.8)(17.28) \\
&= 4.0 + 13.82 \\
&= 17.82 \\
T_4 &= \beta(F_4 - F_3) + (1-\beta)T_3 \\
&= (.4)(17.82 - 15.18) + (1 - .4)(2.10) \\
&= (.4)(2.64) + (.6)(2.10) \\
&= 1.056 + 1.26 \\
&= 2.32 \\
FIT_4 &= 17.82 + 2.32 \\
&= 20.14
\end{aligned}$$

SOLVED PROBLEM 4.3

Sales of hair dryers at the Walgreens stores in Youngstown, Ohio, over the past 4 months have been 100, 110, 120, and 130 units (with 130 being the most recent sales).

Develop a moving-average forecast for next month, using these three techniques:

a) 3-month moving average.
b) 4-month moving average.
c) Weighted 4-month moving average with the most recent month weighted 4, the preceding month 3, then 2, and the oldest month weighted 1.
d) If next month's sales turn out to be 140 units, forecast the following month's sales (months) using a 4-month moving average.

SOLUTION

a) 3-month moving average

$$= \frac{110 + 120 + 130}{3} = \frac{360}{3} = 120 \text{ dryers}$$

b) 4-month moving average

$$= \frac{100 + 110 + 120 + 130}{4} = \frac{460}{4} = 115 \text{ dryers}$$

c) Weighted moving average

$$= \frac{4(130) + 3(120) + 2(110) + 1(100)}{10}$$

$$= \frac{1{,}200}{10} = 120 \text{ dryers}$$

d) *Now* the four most recent sales are 110, 120, 130, and 140.

$$\text{4-month moving average} = \frac{110 + 120 + 130 + 140}{4}$$

$$= \frac{500}{4} = 125 \text{ dryers}$$

We note, of course, the lag in the forecasts, as the moving-average method does not immediately recognize trends.

SOLVED PROBLEM 4.4

Consider the following forecast and actual values.

PERIOD	FORECAST VALUES	ACTUAL VALUES
1	410	406
2	419	423
3	428	423
4	435	440

Compute the MAD and MSE.

SOLUTION

$$\text{MAD} = \frac{\Sigma|\text{Actual} - \text{Forecast}|}{n}$$

$$= \frac{|406 - 410| + |423 - 419| + |423 - 428| + |440 - 435|}{4}$$

$$= \frac{4 + 4 + 5 + 5}{4} = \frac{18}{4} = 4.5$$

$$\text{MSE} = \frac{\Sigma(\text{Forecast errors})^2}{n}$$

$$= \frac{(406 - 410)^2 + (423 - 419)^2 + (423 - 428)^2 + (440 - 435)^2}{4}$$

$$= \frac{4^2 + 4^2 + 5^2 + 5^2}{4} = \frac{16 + 16 + 25 + 25}{4} = 20.5$$

SOLVED PROBLEM 4.5

Room registrations in the Toronto Towers Plaza Hotel have been recorded for the past 9 years. To project future occupancy, management would like to determine the mathematical trend of guest registration. This estimate will help the hotel determine whether future expansion will be needed. Given the following time-series data, develop a regression equation relating registrations to time (e.g., a trend equation). Then forecast year 11 registrations. Room registrations are in the thousands:

Year 1: 17	Year 2: 16	Year 3: 16	Year 4: 21	Year 5: 20
Year 6: 20	Year 7: 23	Year 8: 25	Year 9: 24	

SOLUTION

YEAR	REGISTRANTS, y (IN THOUSANDS)	x^2	xy
1	17	1	17
2	16	4	32
3	16	9	48
4	21	16	84
5	20	25	100
6	20	36	120
7	23	49	161
8	25	64	200
9	24	81	216
$\Sigma x = 45$	$\Sigma y = 182$	$\Sigma x^2 = 285$	$\Sigma xy = 978$

$$b = \frac{\Sigma xy - n\bar{x}\bar{y}}{\Sigma x^2 - n\bar{x}^2} = \frac{978 - (9)(5)(20.22)}{285 - (9)(25)}$$

$$= \frac{978 - 909.9}{285 - 225} = \frac{68.1}{60} = 1.135$$

$$a = \bar{y} - b\bar{x} = 20.22 - (1.135)(5) = 20.22 - 5.675 = 14.545$$

$$\hat{y} = (\text{registrations}) = 14.545 + 1.135x$$

The projection of registrations in year 11 is:

$$\hat{y} = 14.545 + (1.135)(11) = 27.03 \text{ or } 27{,}030 \text{ guests in year 11.}$$

SOLVED PROBLEM 4.6

Quarterly demand for Ford F150 pickups at a New York auto dealer is forecast with the equation:

$$\hat{y} = 10 + 3x$$

where x = quarters, and:

Quarter I of year 1 = 0
Quarter II of year 1 = 1
Quarter III of year 1 = 2
Quarter IV of year 1 = 3
Quarter I of year 2 = 4
and so on

and:

$$\hat{y} = \text{quarterly demand}$$

The demand for trucks is seasonal, and the indices for Quarters I, II, III, and IV are 0.80, 1.00, 1.30, and 0.90, respectively. Forecast demand for each quarter of year 3. Then seasonalize each forecast to adjust for quarterly variations.

SOLUTION

Quarter II of year 2 is coded $x = 5$; Quarter III of year 2, $x = 6$; and Quarter IV of year 2, $x = 7$. Hence, Quarter I of year 3 is coded $x = 8$; Quarter II, $x = 9$; and so on.

$$\hat{y}(\text{Year 3 Quarter I}) = 10 + 3(8) = 34$$
$$\hat{y}(\text{Year 3 Quarter II}) = 10 + 3(9) = 37$$
$$\hat{y}(\text{Year 3 Quarter III}) = 10 + 3(10) = 40$$
$$\hat{y}(\text{Year 3 Quarter IV}) = 10 + 3(11) = 43$$

Adjusted forecast = (.80)(34) = 27.2
Adjusted forecast = (1.00)(37) = 37
Adjusted forecast = (1.30)(40) = 52
Adjusted forecast = (.90)(43) = 38.7

SOLVED PROBLEM 4.7

Cengiz Haksever runs an Istanbul high-end jewelry shop. He advertises weekly in local Turkish newspapers and is thinking of increasing his ad budget. Before doing so, he decides to evaluate the past effectiveness of these ads. Five weeks are sampled, and the data are shown in the table below:

SALES ($1,000s)	AD BUDGET THAT WEEK ($100s)
11	5
6	3
10	7
6	2
12	8

Develop a regression model to help Cengiz evaluate his advertising.

SOLUTION

We apply the least-squares regression model as we did in Example 12.

SALES, y	ADVERTISING, x	x^2	xy
11	5	25	55
6	3	9	18
10	7	49	70
6	2	4	12
12	8	64	96
$\Sigma y = 45$	$\Sigma x = 25$	$\Sigma x^2 = 151$	$\Sigma xy = 251$
$\bar{y} = \frac{45}{5} = 9$	$\bar{x} = \frac{25}{5} = 5$		

$$b = \frac{\Sigma xy - n\bar{x}\bar{y}}{\Sigma x^2 - n\bar{x}^2} = \frac{251 - (5)(5)(9)}{151 - (5)(5^2)}$$
$$= \frac{251 - 225}{151 - 125} = \frac{26}{26} = 1$$
$$a = \bar{y} - b\bar{x} = 9 - (1)(5) = 4$$

So the regression model is $\hat{y} = 4 + 1x$, or
Sales (in \$1,000s) = 4 + 1 (Ad budget in \$100s)
This means that for each 1-unit increase in x (or \$100 in ads), sales increase by 1 unit (or \$1,000).

SOLVED PROBLEM 4.8

Using the data in Solved Problem 4.7, find the coefficient of determination, r^2, for the model.

SOLUTION

To find r^2, we need to also compute Σy^2.

$$\Sigma y^2 = 11^2 + 6^2 + 10^2 + 6^2 + 12^2$$
$$= 121 + 36 + 100 + 36 + 144 = 437$$

The next step is to find the coefficient of correlation, r:

$$r = \frac{n\Sigma xy - \Sigma x \Sigma y}{\sqrt{[n\Sigma x^2 - (\Sigma x)^2][n\Sigma y^2 - (\Sigma y)^2]}}$$
$$= \frac{5(251) - (25)(45)}{\sqrt{[5(151) - (25)^2][5(437) - (45)^2]}}$$
$$= \frac{1{,}255 - 1{,}125}{\sqrt{(130)(160)}} = \frac{130}{\sqrt{20{,}800}} = \frac{130}{144.22}$$
$$= .9014$$

Thus, $r^2 = (.9014)^2 = .8125$, meaning that about 81% of the variability in sales can be explained by the regression model with advertising as the independent variable.

Problems

Note: **PX** means the problem may be solved with POM for Windows and/or Excel OM.

Problems 4.1–4.41 relate to Time-Series Forecasting

• **4.1** The following gives the number of pints of type B blood used at Woodlawn Hospital in the past 6 weeks:

WEEK OF	PINTS USED
August 31	360
September 7	389
September 14	410
September 21	381
September 28	368
October 5	374

a) Forecast the demand for the week of October 12 using a 3-week moving average.
b) Use a 3-week weighted moving average, with weights of .1, .3, and .6, using .6 for the most recent week. Forecast demand for the week of October 12.
c) Compute the forecast for the week of October 12 using exponential smoothing with a forecast for August 31 of 360 and $\alpha = .2$. **PX**

•• **4.2**

YEAR	1	2	3	4	5	6	7	8	9	10	11
DEMAND	7	9	5	9	13	8	12	13	9	11	7

a) Plot the above data on a graph. Do you observe any trend, cycles, or random variations?
b) Starting in year 4 and going to year 12, forecast demand using a 3-year moving average. Plot your forecast on the same graph as the original data.
c) Starting in year 4 and going to year 12, forecast demand using a 3-year moving average with weights of .1, .3, and .6, using .6 for the most recent year. Plot this forecast on the same graph.
d) As you compare forecasts with the original data, which seems to give the better results? **PX**

•• **4.3** Refer to Problem 4.2. Develop a forecast for years 2 through 12 using exponential smoothing with $\alpha = 4$ and a forecast for year 1 of 6. Plot your new forecast on a graph with the actual data and the naive forecast. Based on a visual inspection, which forecast is better? **PX**

• **4.4** A check-processing center uses exponential smoothing to forecast the number of incoming checks each month. The number of checks received in June was 40 million, while the forecast was 42 million. A smoothing constant of .2 is used.
a) What is the forecast for July?
b) If the center received 45 million checks in July, what would be the forecast for August?
c) Why might this be an inappropriate forecasting method for this situation? **PX**

•• **4.5** The Carbondale Hospital is considering the purchase of a new ambulance. The decision will rest partly on the anticipated mileage to be driven next year. The miles driven during the past 5 years are as follows:

YEAR	MILEAGE
1	3,000
2	4,000
3	3,400
4	3,800
5	3,700

a) Forecast the mileage for next year (6th year) using a 2-year moving average.
b) Find the MAD based on the 2-year moving average. (*Hint:* You will have only 3 years of matched data.)
c) Use a weighted 2-year moving average with weights of .4 and .6 to forecast next year's mileage. (The weight of .6 is for the most recent year.) What MAD results from using this approach to forecasting? (*Hint:* You will have only 3 years of matched data.)
d) Compute the forecast for year 6 using exponential smoothing, an initial forecast for year 1 of 3,000 miles, and $\alpha = .5$. **PX**

•• **4.6** The monthly sales for Yazici Batteries, Inc., were as follows:

MONTH	SALES
January	20
February	21
March	15
April	14
May	13
June	16
July	17
August	18
September	20
October	20
November	21
December	23

a) Plot the monthly sales data.
b) Forecast January sales using each of the following:
 i) Naive method.
 ii) A 3-month moving average.
 iii) A 6-month weighted average using .1, .1, .1, .2, .2, and .3, with the heaviest weights applied to the most recent months.
 iv) Exponential smoothing using an $\alpha = .3$ and a September forecast of 18.
 v) A trend projection.
c) With the data given, which method would allow you to forecast next March's sales? **PX**

•• **4.7** The actual demand for the patients at Providence Emergency Medical Clinic for the first 6 weeks of this year follows:

WEEK	ACTUAL NO. OF PATIENTS
1	65
2	62
3	70
4	48
5	63
6	52

Clinic administrator Dara Schniederjans wants you to forecast patient demand at the clinic for week 7 by using this data. You decide to use a weighted-moving-average method to find this forecast. Your method uses four actual demand levels, with weights of 0.333 on the present period, 0.25 one period ago, 0.25 two periods ago, and 0.167 three periods ago.

a) What is the value of your forecast? PX
b) If instead the weights were 20, 15, 15, and 10, respectively, how would the forecast change? Explain why.
c) What if the weights were 0.40, 0.30, 0.20, and 0.10, respectively? Now what is the forecast for week 7?

• **4.8** Daily high temperatures in St. Louis for the last week were as follows: 93, 94, 93, 95, 96, 88, 90 (yesterday).
a) Forecast the high temperature today, using a 3-day moving average.
b) Forecast the high temperature today, using a 2-day moving average.
c) Calculate the mean absolute deviation based on a 2-day moving average.
d) Compute the mean squared error (MSE) for the 2-day moving average.
e) Calculate the mean absolute percent error (MAPE) for the 2-day moving average. PX

••• **4.9** Lenovo uses the ZX-81 chip in some of its laptop computers. The prices for the chip during the past 12 months were as follows:

MONTH	PRICE PER CHIP ($)	MONTH	PRICE PER CHIP ($)
January	1.80	July	1.80
February	1.67	August	1.83
March	1.70	September	1.70
April	1.85	October	1.65
May	1.90	November	1.70
June	1.87	December	1.75

a) Use a 2-month moving average on all the data and plot the averages and the prices.
b) Use a 3-month moving average and add the 3-month plot to the graph created in part (a).
c) Which is better (using the mean absolute deviation): the 2-month average or the 3-month average?
d) Compute the forecasts for each month using exponential smoothing, with an initial forecast for January of $1.80. Use $\alpha = .1$, then $\alpha = .3$, and finally $\alpha = .5$. Using MAD, which a is the best? PX

•• **4.10** Data collected on the yearly registrations for a Six Sigma seminar at the Quality College are shown in the following table:

YEAR	1	2	3	4	5	6	7	8	9	10	11
REGISTRATIONS (000)	4	6	4	5	10	8	7	9	12	14	15

a) Develop a 3-year moving average to forecast registrations from year 4 to year 12.
b) Estimate demand again for years 4 to 12 with a 3-year weighted moving average in which registrations in the most recent year are given a weight of 2, and registrations in the other 2 years are each given a weight of 1.
c) Graph the original data and the two forecasts. Which of the two forecasting methods seems better? PX

• **4.11** Use exponential smoothing with a smoothing constant of 0.3 to forecast the registrations at the seminar given in Problem 4.10. To begin the procedure, assume that the forecast for year 1 was 5,000 people signing up.
a) What is the MAD? PX
b) What is the MSE?

•• **4.12** Consider the following actual and forecast demand levels for Big Mac hamburgers at a local McDonald's restaurant:

DAY	ACTUAL DEMAND	FORECAST DEMAND
Monday	88	88
Tuesday	72	88
Wednesday	68	84
Thursday	48	80
Friday		

The forecast for Monday was derived by observing Monday's demand level and setting Monday's forecast level equal to this demand level. Subsequent forecasts were derived by using exponential smoothing with a smoothing constant of 0.25. Using this exponential smoothing method, what is the forecast for Big Mac demand for Friday? PX

••• **4.13** As you can see in the following table, demand for heart transplant surgery at Washington General Hospital has increased steadily in the past few years:

YEAR	1	2	3	4	5	6
HEART TRANSPLANTS	45	50	52	56	58	?

The director of medical services predicted 6 years ago that demand in year 1 would be 41 surgeries.
a) Use exponential smoothing, first with a smoothing constant of .6 and then with one of .9, to develop forecasts for years 2 through 6.
b) Use a 3-year moving average to forecast demand in years 4, 5, and 6.
c) Use the trend projection (regression) model to forecast demand in years 1 through 6.
d) With MAD as the criterion, which of the four forecasting methods is best? PX

•• **4.14** Following are two weekly forecasts made by two different methods for the number of gallons of gasoline, in thousands, demanded at a local gasoline station. Also shown are actual demand levels, in thousands of gallons.

	FORECASTS		
WEEK	METHOD 1	METHOD 2	ACTUAL DEMAND
1	0.90	0.80	0.70
2	1.05	1.20	1.00
3	0.95	0.90	1.00
4	1.20	1.11	1.00

What are the MAD and MSE for each method?

• **4.15** Refer to Solved Problem 4.1.
a) Use a 3-year moving average to forecast the sales of Volkswagens in Nevada through year 6.
b) What is the MAD? PX
c) What is the MSE?

• **4.16** Refer to Solved Problem 4.1.
a) Using the trend-projection (regression) method, develop a forecast for the sales of Volkswagens in Nevada through year 6.
b) What is the MAD? PX
c) What is the MSE?

• **4.17** Refer to Solved Problem 4.1.Using smoothing constants of .6 and .9, develop forecasts for the sales of Volkswagens. What effect did the smoothing constant have on the forecast? Use MAD to determine which of the three smoothing constants (.3, .6, or .9) gives the most accurate forecast. **PX**

• • • **4.18** Consider the following actual (A_t) and forecast (F_t) demand levels for a commercial multiline telephone at Office Max:

TIME PERIOD, t	ACTUAL DEMAND, A_t	FORECAST DEMAND, F_t
1	50	50
2	42	50
3	56	48
4	46	50
5		

The first forecast, F_1, was derived by observing A_1 and setting F_1 equal to A_1. Subsequent forecast averages were derived by exponential smoothing. Using the exponential smoothing method, find the forecast for time period 5. (*Hint:* You need to first find the smoothing constant, α.)

• • • **4.19** Income at the architectural firm Spraggins and Yunes for the period February to July was as follows:

MONTH	FEBRUARY	MARCH	APRIL	MAY	JUNE	JULY
Income (in $ thousand)	70.0	68.5	64.8	71.7	71.3	72.8

Use trend-adjusted exponential smoothing to forecast the firm's August income. Assume that the initial forecast average for February is $65,000 and the initial trend adjustment is 0. The smoothing constants selected are $\alpha = .1$ and $\beta = .2$. **PX**

• • • **4.20** Solve Problem 4.19 again with $\alpha = .1$ and $\beta = .8$. Using MSE, determine which smoothing constants provide a better forecast. **PX**

• **4.21** Refer to the trend-adjusted exponential smoothing illustration in Example 7. Using $\alpha = .2$ and $\beta = .4$, we forecast sales for 9 months, showing the detailed calculations for months 2 and 3. In Solved Problem 4.2, we continued the process for month 4.

In this problem, show your calculations for months 5 and 6 for F_t, T_t, and FIT_t. **PX**

• **4.22** Refer to Problem 4.21. Complete the trend-adjusted exponential-smoothing forecast computations for periods 7, 8, and 9. Confirm that your numbers for F_t, T_t, and FIT_t match those in Table 4.2. **PX**

• • **4.23** Sales of quilt covers at Bud Banis' department store in Carbondale over the past year are shown below. Management prepared a forecast using a combination of exponential smoothing and its collective judgment for the 4 months (March, April, May, and June):

MONTH	UNIT SALES	MANAGEMENT'S FORECAST
July	100	
August	93	
September	96	
October	110	
November	124	
December	119	
January	92	
February	83	
March	101	120
April	96	114
May	89	110
June	108	108

a) Compute MAD and MAPE for management's technique.
b) Do management's results outperform (i.e., have smaller MAD and MAPE than) a naive forecast?
c) Which forecast do you recommend, based on lower forecast error? **PX**

• **4.24** The following gives the number of accidents that occurred on Florida State Highway 101 during the past 4 months:

MONTH	NUMBER OF ACCIDENTS
January	30
February	40
March	60
April	90

Forecast the number of accidents that will occur in May, using least-squares regression to derive a trend equation. **PX**

• **4.25** In the past, Peter Kelle's tire dealership in Baton Rouge sold an average of 1,000 radials each year. In the past 2 years, 200 and 250, respectively, were sold in fall, 350 and 300 in winter, 150 and 165 in spring, and 300 and 285 in summer. With a major expansion planned, Kelle projects sales next year to increase to 1,200 radials. What will be the demand during each season? **PX**

• • **4.26** George Kyparisis owns a company that manufactures sailboats. Actual demand for George's sailboats during each of the past four seasons was as follows:

	YEAR			
SEASON	1	2	3	4
Winter	1,400	1,200	1,000	900
Spring	1,500	1,400	1,600	1,500
Summer	1,000	2,100	2,000	1,900
Fall	600	750	650	500

George has forecasted that annual demand for his sailboats in year 5 will equal 5,600 sailboats. Based on this data and the multiplicative seasonal model, what will the demand level be for George's sailboats in the spring of year 5?

• • **4.27** Attendance at Orlando's newest Disney-like attraction, Minecraft World, has been as follows:

QUARTER	GUESTS (IN THOUSANDS)	QUARTER	GUESTS (IN THOUSANDS)
Winter Year 1	73	Summer Year 2	124
Spring Year 1	104	Fall Year 2	52
Summer Year 1	168	Winter Year 3	89
Fall Year 1	74	Spring Year 3	146
Winter Year 2	65	Summer Year 3	205
Spring Year 2	82	Fall Year 3	98

Compute seasonal indices using all of the data. **PX**

• **4.28** North Dakota Electric Company estimates its demand trend line (in millions of kilowatt hours) to be:

$$D = 77 + 0.43Q$$

where Q refers to the sequential quarter number and $Q = 1$ for winter of Year 1. In addition, the multiplicative seasonal factors are as follows:

QUARTER	FACTOR (INDEX)
Winter	.8
Spring	1.1
Summer	1.4
Fall	.7

Forecast energy use for the four quarters of year 26 (namely quarters 101 to 104), beginning with winter.

• **4.29** The number of disk drives (in millions) made at a plant in Taiwan during the past 5 years follows:

YEAR	DISK DRIVES
1	140
2	160
3	190
4	200
5	210

a) Forecast the number of disk drives to be made next year, using linear regression.
b) Compute the mean squared error (MSE) when using linear regression.
c) Compute the mean absolute percent error (MAPE). **PX**

•• **4.30** Dr. Lillian Fok, a New Orleans psychologist, specializes in treating patients who are agoraphobic (i.e., afraid to leave their homes). The following table indicates how many patients Dr. Fok has seen each year for the past 10 years. It also indicates what the robbery rate was in New Orleans during the same year:

YEAR	1	2	3	4	5	6	7	8	9	10
NUMBER OF PATIENTS	36	33	40	41	40	55	60	54	58	61
ROBBERY RATE PER 1,000 POPULATION	58.3	61.1	73.4	75.7	81.1	89.0	101.1	94.8	103.3	116.2

Using trend (linear regression) analysis, predict the number of patients Dr. Fok will see in years 11 and 12 as a function of time. How well does the model fit the data? **PX**

••• **4.31** Emergency calls to the 911 system of Durham, North Carolina, for the past 6 weeks are shown in the following table:

WEEK	1	2	3	4	5	6
CALLS	50	35	25	40	45	35

a) Compute the exponentially smoothed forecast of calls for each week. Assume an initial forecast of 50 calls in the first week, and use $\alpha = .20$. What is the forecast for week 7? (*Round each forecast to 2 decimal places.*)
b) Reforecast each period using $\alpha = .60$. (*Round each forecast to 2 decimal places.*)
c) Actual calls during week 7 were 33. Which smoothing constant provides a superior forecast based on MAD (covering weeks 1–6)? Which smoothing constant provides the closest forecast to the actual demand in week 7? **PX**

••• **4.32** Using the 911 call data in Problem 4.31, forecast calls for weeks 1 through 7 with a trend-adjusted exponential smoothing model. Assume an initial forecast for 50 calls for week 1 and an initial trend of zero. Use smoothing constants of $\alpha = .30$ and $\beta = .20$. Is this model better than that of Problem 4.31? What adjustment might be useful for further improvement? (Again, assume that actual calls in week 7 were 33.) **PX**

••• **4.33** Storrs Cycles has just started selling the new Cyclone mountain bike, with monthly sales as shown in the table. First, co-owner Bob Day wants to forecast by exponential smoothing by initially setting February's forecast equal to January's sales with $\alpha = .1$. Co-owner Sherry Snyder wants to use a 3-period moving average.

	SALES	BOB	SHERRY	BOB'S ERROR	SHERRY'S ERROR
JANUARY	400	—			
FEBRUARY	380	400			
MARCH	410				
APRIL	375				
MAY					

a) Is there a strong linear trend in sales over time?
b) Fill in the table with what Bob and Sherry each forecast for May and the earlier months, as relevant.
c) Assume that May's actual sales figure turns out to be 405. Complete the table's columns and then calculate the mean absolute deviation for both Bob's and Sherry's methods.
d) Based on these calculations, which method seems more accurate? **PX**

• **4.34** Given the following data, develop a demand forecast for period 7 using a 4-period weighted moving average, with weights of .1, .2, .3, and .4, where .4 is used for the most recent week. **PX**

PERIOD	1	2	3	4	5	6
DEMAND	7	9	5	9	13	8

•• **4.35** Registration numbers for an accounting seminar over the past 10 weeks are shown below:

WEEK	1	2	3	4	5	6	7	8	9	10
REGISTRATIONS	22	21	25	27	35	29	33	37	41	37

a) Starting with week 2 and ending with week 11, forecast registrations using the naive forecasting method.
b) Starting with week 3 and ending with week 11, forecast registration using a 2-week moving average.
c) Starting with week 5 and ending with week 11, forecast registrations using a 4-week moving average.
d) Plot the original data and the three forecasts on the same graph. Which forecast smooths the data the most? Which forecast responds to change the best? **PX**

• **4.36** Given the following data, use exponential smoothing ($\alpha = 0.2$) to develop a demand forecast. Assume the forecast for the initial period is 5. **PX**

PERIOD	1	2	3	4	5	6
DEMAND	7	9	5	9	13	8

• **4.37** Calculate (a) MAD and (b) MSE for the following forecast versus actual sales figures: **PX**

FORECAST	100	110	120	130
ACTUAL	95	108	123	130

••• **4.38** Sales of industrial vacuum cleaners at Jonquel Parker Supply Co. over the past 13 months are shown below:

MONTH	Jan.	Feb.	March	April	May	June	July
SALES (IN THOUSANDS)	11	14	16	10	15	17	11
MONTH	Aug.	Sept.	Oct.	Nov.	Dec.	Jan.	
SALES (IN THOUSANDS)	14	17	12	14	16	11	

a) Using a moving average with three periods, determine the demand for vacuum cleaners for next February.
b) Using a weighted moving average with three periods, determine the demand for vacuum cleaners for February. Use 3, 2, and 1 for the weights of the most recent, second most recent, and third most recent periods, respectively. For example, if you were forecasting the demand for February, November would have a weight of 1, December would have a weight of 2, and January would have a weight of 3.
c) Using MAD, determine which is the better forecast.
d) What other factors might Parker consider in forecasting sales? **PX**

• **4.39** Given the following data, use least-squares regression to derive a trend equation. What is your estimate of the demand in period 7? In period 12? **PX**

PERIOD	1	2	3	4	5	6
DEMAND	7	9	5	11	10	13

• **4.40** Thamer Almutairi, owner of Almutairi's Department Store, has used time-series extrapolation to forecast retail sales for the next 4 quarters. The sales estimates are $120,000, $140,000, $160,000, and $180,000 for the respective quarters. Seasonal indices for the 4 quarters have been found to be 1.25, .90, .75, and 1.10, respectively. Compute a seasonalized or adjusted sales forecast. **PX**

•• **4.41** The director of the Riley County, Kansas, library system would like to forecast evening patron usage for next week. Below are the data for the past 4 weeks:

	MON	TUE	WED	THU	FRI	SAT
WEEK 1	210	178	250	215	160	180
WEEK 2	215	180	250	213	165	185
WEEK 3	220	176	260	220	175	190
WEEK 4	225	178	260	225	176	190

a) Calculate a seasonal index for each day of the week.
b) If the trend equation for this problem is $\hat{y} = 201.74 + .18x$, what is the forecast for each day of week 5? Round your forecast to the nearest whole number. **PX**

Problems 4.42–4.58 relate to Associative Forecasting Methods

• **4.42** A careful analysis of the cost of operating an automobile was conducted by accounting manager Dia Bandaly. The following model was developed:

$$\hat{y} = 4{,}000 + 0.20x$$

Where $\hat{y}$ is the annual cost and x is the miles driven.
a) If the car is driven 15,000 miles this year, what is the forecasted cost of operating this automobile?
b) If the car is driven 25,000 miles this year, what is the forecasted cost of operating this automobile?

•• **4.43** Mark Gershon, owner of a musical instrument distributorship, thinks that demand for guitars may be related to the number of television appearances by the popular group Maroon 5 during the previous month. Mark has collected the data shown in the following table:

DEMAND FOR GUITARS	3	6	7	5	10	7
MAROON 5 TV APPEARANCES	3	4	7	6	8	5

a) Graph these data to see whether a linear equation might describe the relationship between the group's television shows and guitar sales.
b) Use the least-squares regression method to derive a forecasting equation.
c) What is your estimate for guitar sales if Maroon 5 performed on TV nine times last month?
d) What are the correlation coefficient (r) and the coefficient of determination (r^2) for this model, and what do they mean? **PX**

• **4.44** Lori Cook has developed the following forecasting model:

$$\hat{y} = 36 + 4.3x$$

where $\hat{y}$ = demand for Kool Air conditioners and
x = the outside temperature (°F)

a) Forecast demand for the Kool Air when the temperature is 70° F.
b) What is demand when the temperature is 80° F?
c) What is demand when the temperature is 90° F? **PX**

•• **4.45** Café Michigan's manager, Gary Stark, suspects that demand for mocha latte coffees depends on the price being charged. Based on historical observations, Gary has gathered the following data, which show the numbers of these coffees sold over six different price values:

PRICE	NUMBER SOLD
$2.70	760
$3.50	510
$2.00	980
$4.20	250
$3.10	320
$4.05	480

Using these data, how many mocha latte coffees would be forecast to be sold according to simple linear regression if the price per cup were $2.80? **PX**

• **4.46** The following data relate the sales figures of the bar in Mark Kaltenbach's small bed-and-breakfast inn in Portland to the number of guests registered that week:

WEEK	GUESTS	BAR SALES
1	16	$330
2	12	270
3	18	380
4	14	300

a) Perform a linear regression that relates bar sales to guests (not to time).
b) If the forecast is for 20 guests next week, what are the sales expected to be? **PX**

• **4.47** The number of auto accidents in Athens, Ohio, is related to the regional number of registered automobiles in thousands (X_1), alcoholic beverage sales in $10,000s ($X_2$), and rainfall in inches (X_3). Furthermore, the regression formula has been calculated as:

$$Y = a + b_1X_1 + b_2X_2 + b_3X_3$$

where

Y = number of automobile accidents
$a = 7.5$
$b_1 = 3.5$
$b_2 = 4.5$
$b_3 = 2.5$

Calculate the expected number of automobile accidents under conditions a, b, and c:

	X_1	X_2	X_3
(a)	2	3	0
(b)	3	5	1
(c)	4	7	2

•• **4.48** Janice Carrillo, a Gainesville, Florida, real estate developer, has devised a regression model to help determine residential housing prices in northeastern Florida. The model was developed using recent sales in a particular neighborhood. The price (Y) of the house is based on the size (square footage = X) of the house. The model is:

$$Y = 13{,}473 + 37.65X$$

The coefficient of correlation for the model is 0.63.

a) Use the model to predict the selling price of a house that is 1,860 square feet.
b) An 1,860-square-foot house recently sold for $95,000. Explain why this is not what the model predicted.
c) If you were going to use multiple regression to develop such a model, what other quantitative variables might you include?
d) What is the value of the coefficient of determination in this problem? **PX**

• **4.49** Accountants at the Tucson firm, Larry Youdelman, CPAs, believed that several traveling executives were submitting unusually high travel vouchers when they returned from business trips. First, they took a sample of 200 vouchers submitted from the past year. Then they developed the following multiple regression equation relating expected travel cost to number of days on the road (x_1) and distance traveled (x_2) in miles:

$$\hat{y} = \$90.00 + \$48.50x_1 + \$.40x_2$$

The coefficient of correlation computed was .68.

a) If Barbara Downey returns from a 300-mile trip that took her out of town for 5 days, what is the expected amount she should claim as expenses?
b) Downey submitted a reimbursement request for $685. What should the accountant do?
c) Should any other variables be included? Which ones? Why? **PX**

•• **4.50** City government has collected the following data on annual sales tax collections and new car registrations:

ANNUAL SALES TAX COLLECTIONS (IN MILLIONS)	1.0	1.4	1.9	2.0	1.8	2.1	2.3
NEW CAR REGISTRATIONS (IN THOUSANDS)	10	12	15	16	14	17	20

Determine the following:

a) The least-squares regression equation.
b) Using the results of part (a), find the estimated sales tax collections if new car registrations total 22,000.
c) The coefficients of correlation and determination. **PX**

•• **4.51** Using the data in Problem 4.30, apply linear regression to study the relationship between the robbery rate and Dr. Fok's patient load. If the robbery rate increases to 131.2 in year 11, how many phobic patients will Dr. Fok treat? If the robbery rate drops to 90.6, what is the patient projection? **PX**

••• **4.52** Bus and subway ridership for the summer months in London, England, is believed to be tied heavily to the number of tourists visiting the city. During the past 12 years, the following data have been obtained:

YEAR (SUMMER MONTHS)	NUMBER OF TOURISTS (IN MILLIONS)	RIDERSHIP (IN MILLIONS)
1	7	1.5
2	2	1.0
3	6	1.3
4	4	1.5
5	14	2.5
6	15	2.7
7	16	2.4
8	12	2.0
9	14	2.7
10	20	4.4
11	15	3.4
12	7	1.7

a) Plot these data and decide if a linear model is reasonable.
b) Develop a regression relationship.
c) What is expected ridership if 10 million tourists visit London in a year?
d) Explain the predicted ridership if there are no tourists at all.
e) What is the standard error of the estimate?
f) What is the model's correlation coefficient and coefficient of determination? **PX**

•• **4.53** Thirteen students entered the business program at Sante Fe College 2 years ago. The following table indicates what each student scored on the high school SAT math exam and their grade-point averages (GPAs) after students were in the Sante Fe program for 2 years:

STUDENT	A	B	C	D	E	F	G
SAT SCORE	421	377	585	690	608	390	415
GPA	2.90	2.93	3.00	3.45	3.66	2.88	2.15
STUDENT	H	I	J	K	L	M	
SAT SCORE	481	729	501	613	709	366	
GPA	2.53	3.22	1.99	2.75	3.90	1.60	

a) Is there a meaningful relationship between SAT math scores and grades?
b) If a student scores a 350, what do you think his or her GPA will be?
c) What about a student who scores 800?

•• **4.54** Dave Fletcher, the general manager of North Carolina Engineering Corporation (NCEC), thinks that his firm's engineering services contracted to highway construction firms are directly related to the volume of highway construction business contracted with companies in his geographic area. He wonders if this is really so, and if it is, can this information help him plan his operations better by forecasting the quantity of his engineering services required by construction firms in each quarter of the year? The following table presents the sales of his services and total amounts of contracts for highway construction over the past eight quarters:

QUARTER	1	2	3	4	5	6	7	8
Sales of NCEC Services (in $ thousands)	8	10	15	9	12	13	12	16
Construction Contracts (in $ thousands)	153	172	197	178	185	199	205	226

a) Using this data, develop a regression equation for predicting the level of demand of NCEC's services.
b) Determine the coefficient of correlation and the standard error of the estimate. **PX**

• **4.55** The following multiple regression model was developed to predict job performance as measured by a company job performance evaluation index based on a preemployment test score and college grade point average (GPA):

$$\hat{y} = 35 + 20x_1 + 50x_2$$

Where

$\hat{y}$ = job performance evaluation index
x_1 = preemployment test score
x_2 = college GPA

a) Forecast the job performance index for an applicant who had a 3.0 GPA and scored 80 on the preemployment test.
b) Forecast the job performance index for an applicant who had a 2.5 GPA and scored 70 on the preemployment test.

•• **4.56** A study to determine the correlation between bank deposits and consumer price indices in Birmingham, Alabama, revealed the following (which was based on $n = 5$ years of data):

- $\Sigma x = 15$
- $\Sigma x^2 = 55$
- $\Sigma xy = 70$
- $\Sigma y = 20$
- $\Sigma y^2 = 130$

a) What is the equation of the least-squares regression line?
b) Find the coefficient of correlation. What does it imply to you?
c) What is the standard error of the estimate?

• **4.57** The accountant at Bintong Chen Coal Distributors, Inc., in Newark, Delaware, notes that the demand for coal seems to be tied to an index of weather severity developed by the U.S. Weather Bureau. When weather was extremely cold in the U.S. over the past 5 years (and the index was thus high), coal sales were high. The accountant proposes that one good forecast of next year's coal demand could be made by developing a regression equation and then consulting the *Farmer's Almanac* to see how severe next year's winter would be. For the data in the following table, derive a least-squares regression and compute the coefficient of correlation of the data. Also compute the standard error of the estimate. **PX**

COAL SALES (IN MILLIONS OF TONS), y	4	1	4	6	5
WEATHER INDEX, x	2	1	4	5	3

•• **4.58** Given the following data, use least-squares regression to develop a relation between the number of rainy summer days and the number of games lost by the Boca Raton Cardinal baseball team. **PX**

YEARS	1	2	3	4	5	6	7	8	9	10
RAINY DAYS	15	25	10	10	30	20	20	15	10	25
GAMES LOST	25	20	10	15	20	15	20	10	5	20

Problems 4.59–4.61 relate to Monitoring, Controlling, and Adapting Forecasts

•• **4.59** Sales of tablet computers at Maria Gonzalez's electronics store in Washington, D.C., over the past 10 weeks are shown in the table below:

WEEK	DEMAND	WEEK	DEMAND
1	20	6	29
2	21	7	36
3	28	8	22
4	37	9	25
5	25	10	28

a) Forecast demand for each week, including week 10, using exponential smoothing with $\alpha = .5$ (initial forecast = 20).
b) Compute the MAD.
c) Compute the tracking signal. **PX**

••• **4.60** The following are monthly actual and forecast demand levels for May through December for units of a product manufactured by the Deborah Bishop Company in Des Moines:

MONTH	ACTUAL DEMAND	FORECAST DEMAND
May	100	100
June	80	104
July	110	99
August	115	101
September	105	104
October	110	104
November	125	105
December	120	109

What is the value of the tracking signal as of the end of December?

••• **4.61** Passenger miles flown on Northeast Airlines, a commuter firm serving the Boston hub, are shown for the past 12 weeks:

WEEK	1	2	3	4	5	6	7	8	9	10	11	12
ACTUAL PASSENGER MILES (IN THOUSANDS)	17	21	19	23	18	16	20	18	22	20	15	22

a) Assuming an initial forecast for week 1 of 17,000 miles, use exponential smoothing to compute miles for weeks 2 through 12. Use $\alpha = .2$
b) What is the MAD for this model?
c) Compute the Cumulative Forecast Errors and tracking signals. Are they within acceptable limits? **PX**

CASE STUDIES

Southwestern University (B)

Southwestern University (SWU), a large state college in Stephenville, Texas, enrolls close to 20,000 students. The school is a dominant force in the small city, with more students during fall and spring than permanent residents.

Always a football powerhouse, SWU is usually in the top 20 in college football rankings. Since the legendary Phil Flamm was hired as its head coach in 2015 (in hopes of reaching the elusive number-one ranking), attendance at the five Saturday home games each year increased. Prior to Flamm's arrival, attendance generally averaged 25,000 to 29,000 per game. Season ticket sales bumped up by 10,000 just with the announcement of the new coach's arrival. Stephenville and SWU were ready to move to the big time!

The immediate issue facing SWU, however, was not NCAA ranking. It was capacity. The existing SWU stadium, built in 1953, has seating for 54,000 fans. The table indicates attendance at each game for the past 6 years.

One of Flamm's demands upon joining SWU had been a stadium expansion, or possibly even a new stadium. With attendance increasing, SWU administrators began to face the issue head-on. Flamm had wanted dormitories solely for his athletes in the stadium as an additional feature of any expansion.

Southwestern University Football Game Attendance, 2016–2021

	2016		2017		2018	
GAME	ATTENDEES	OPPONENT	ATTENDEES	OPPONENT	ATTENDEES	OPPONENT
1	34,200	Rice	36,100	Miami	35,900	USC
2[a]	39,800	Texas	40,200	Nebraska	46,500	Texas Tech
3	38,200	Duke	39,100	Ohio State	43,100	Alaska
4[b]	26,900	Arkansas	25,300	Nevada	27,900	Arizona
5	35,100	TCU	36,200	Boise State	39,200	Baylor

	2019		2020		2021	
GAME	ATTENDEES	OPPONENT	ATTENDEES	OPPONENT	ATTENDEES	OPPONENT
1	41,900	Arkansas	42,500	Indiana	46,900	LSU
2[a]	46,100	Missouri	48,200	North Texas	50,100	Texas
3	43,900	Florida	44,200	Texas A&M	45,900	South Florida
4[b]	30,100	Central Florida	33,900	Southern	36,300	Montana
5	40,500	LSU	47,800	Oklahoma	49,900	Arizona State

[a] Homecoming games.

[b] During the fourth week of each season, Stephenville hosted a hugely popular southwestern crafts festival. This event brought tens of thousands of tourists to the town, especially on weekends, and had an obvious negative impact on game attendance.

SWU's president, Dr. Joel Wisner, decided it was time for his vice president of development to forecast when the existing stadium would "max out." The expansion was, in his mind, a given. But Wisner needed to know how long he could wait. He also sought a revenue projection, assuming an average ticket price of $50 in 2022 and a 5% increase each year in future prices.

Discussion Questions

1. Develop a forecasting model, justifying its selection over other techniques, and project attendance through 2023.
2. What revenues are to be expected in 2022 and 2023?
3. Discuss the school's options.

Forecasting Ticket Revenue for Orlando Magic Basketball Games

Video Case

For its first two decades of existence, the NBA's Orlando Magic basketball team set seat prices for its 41-game home schedule the same for each game. If a lower-deck seat sold for $150, that was the price charged, regardless of the opponent, day of the week, or time of the season. If an upper-deck seat sold for $10 in the first game of the year, it likewise sold for $10 for every game.

But when Anthony Perez, director of business strategy, finished his MBA at the University of Florida, he developed a valuable database of ticket sales. Analysis of the data led him to build a forecasting model he hoped would increase ticket revenue. Perez hypothesized that selling a ticket for similar seats should differ based on demand.

Studying individual sales of Magic tickets on the open Stub Hub marketplace during the prior season, Perez determined the additional potential sales revenue the Magic could have made had they charged prices the fans had proven they were willing to pay on Stub Hub. This became his dependent variable, y, in a multiple regression model.

Table 4.3 Data for Last Year's Magic Ticket Sales Pricing Model

TEAM	DATE	DAY OF WEEK*	TIME OF YEAR	RATING OF OPPONENT	ADDITIONAL SALES POTENTIAL
Phoenix Suns	November 4	Wednesday	0	0	$12,331
Detroit Pistons	November 6	Friday	0	1	$29,004
Cleveland Cavaliers	November 11	Wednesday	0	6	$109,412
Miami Heat	November 25	Wednesday	0	3	$75,783
Houston Rockets	December 23	Wednesday	3	2	$42,557
Boston Celtics	January 28	Thursday	1	4	$120,212
New Orleans Pelicans	February 3	Monday	1	1	$20,459
L. A. Lakers	March 7	Sunday	2	8	$231,020
San Antonio Spurs	March 17	Wednesday	2	1	$28,455
Denver Nuggets	March 23	Sunday	2	1	$110,561
NY Knicks	April 9	Friday	3	0	$44,971
Philadelphia 76ers	April 14	Wednesday	3	1	$30,257

*Day of week rated as 1 = Monday, 2 = Tuesday, 3 = Wednesday, 4 = Thursday, 5 = Friday, 6 = Saturday, 5 = Sunday, 3 = holiday.

He also found that three variables would help him build the "true market" seat price for every game. With his model, it was possible that the same seat in the arena would have as many as seven different prices created at season onset—sometimes higher than expected on average and sometimes lower.

The major factors he found to be statistically significant in determining how high the demand for a game ticket, and hence, its price, would be were:

- The day of the week (x_1)
- A rating of how popular the opponent was (x_2)
- The time of the year (x_3)

For the day of the week, Perez found that Mondays were the least-favored game days (and he assigned them a value of 1). The rest of the weekdays increased in popularity, up to a Saturday game, which he rated a 6. Sundays and Fridays received 5 ratings, and holidays a 3 (refer to the footnote in Table 4.3).

His ratings of opponents, done just before the start of the season, were subjective and range from a low of 0 to a high of 8. A very high-rated team in that particular season may have had one or more superstars on its roster, or have won the NBA finals the prior season, making it a popular fan draw.

Finally, Perez believed that the NBA season could be divided into four periods in popularity:

- Early games (which he assigned 0 scores)
- Games during the Christmas season (assigned a 3)
- Games until the All-Star break (given a 2)
- Games leading into the playoffs (scored with a 3)

The first year Perez built his multiple regression model, the dependent variable y, which was a "potential premium revenue score," yielded an $r^2 = .86$ with this equation:

$$y = 14{,}996 + 10{,}801x_1 + 23{,}397x_2 + 10{,}784x_3$$

Table 4.3 illustrates, for brevity in this case study, a sample of 12 games that year (out of the total 41 home game regular season), including the potential extra revenue per game (y) to be expected using the variable pricing model.

A leader in NBA variable pricing, the Orlando Magic have learned that regression analysis is indeed a profitable forecasting tool.

Discussion Questions*

1. Use the data in Table 4.3 to build a regression model with day of the week as the only independent variable.
2. Use the data to build a model with rating of the opponent as the sole independent variable.
3. Using Perez's multiple regression model, what would be the additional sales potential of a Thursday Miami Heat game played during the Christmas holiday?
4. What additional independent variables might you suggest to include in Perez's model?

*You may wish to view the video that accompanies this case before answering these questions.

Forecasting at Hard Rock Cafe

Video Case 

With the growth of Hard Rock Cafe—from one pub in London in 1971 to more than 185 restaurants in 74 countries today—came a corporatewide demand for better forecasting. Hard Rock uses long-range forecasting in setting a capacity plan and intermediate-term forecasting for locking in contracts for leather goods (used in jackets) and for such food items as beef, chicken, and pork. Its short-term sales forecasts are conducted each month, by cafe, and then aggregated for a headquarters view.

The heart of the sales forecasting system is the point-of-sale (POS) system, which, in effect, captures transaction data on nearly every person who walks through a cafe's door. The sale of each entrée represents one customer; the entrée sales data are transmitted daily to the Orlando corporate headquarters' database. There, the financial team, headed by Todd Lindsey, begins the forecast process. Lindsey forecasts monthly guest counts, retail sales, banquet sales, and concert sales (if applicable) at each cafe. The general managers of individual cafes tap into the same database to prepare a daily forecast for their sites. A cafe manager pulls up prior years' sales for that day, adding information from the local Chamber of Commerce or Tourist Board on upcoming events such as a major convention, sporting event, or concert in the city where the cafe is located. The daily forecast is further broken into hourly sales, which drives employee scheduling. An hourly forecast of $5,500 in sales translates into 19 workstations, which are further broken down into a specific number of waitstaff, hosts, bartenders, and kitchen staff. Computerized scheduling software plugs in people based on their availability. Variances between forecast and actual sales are then examined to see why errors occurred.

Hard Rock doesn't limit its use of forecasting tools to sales. To evaluate managers and set bonuses, a 3-year weighted moving average is applied to cafe sales. If cafe general managers exceed their targets, a bonus is computed. Todd Lindsey, at corporate headquarters, applies weights of 40% to the most recent year's sales, 40% to the year before, and 20% to sales 2 years ago in reaching his moving average.

An even more sophisticated application of statistics is found in Hard Rock's menu planning. Using multiple regression, managers can compute the impact on demand of other menu items if the price of one item is changed. For example, if the price of a cheeseburger increases from $14.99 to $15.99, Hard Rock can predict the effect this will have on sales of chicken sandwiches, pork sandwiches, and salads. Managers do the same analysis on menu placement, with the center section driving higher sales volumes. When an item such as a hamburger is moved off the center to one of the side flaps, the corresponding effect on related items, say french fries, is determined.

Discussion Questions*

1. Describe three different forecasting applications at Hard Rock. Name three other areas in which you think Hard Rock could use forecasting models.
2. What is the role of the POS system in forecasting at Hard Rock?
3. Justify the use of the weighting system used for evaluating managers for annual bonuses.
4. Name several variables besides those mentioned in the case that could be used as good predictors of daily sales in each cafe.
5. At Hard Rock's Moscow restaurant, the manager is trying to evaluate how a new advertising campaign affects guest counts. Using data for the past 10 months (see the table), develop a least-squares regression relationship and then forecast the expected guest count when advertising is $65,000.

HARD ROCK'S MOSCOW CAFE[a]

MONTH	1	2	3	4	5	6	7	8	9	10
Guest count (in thousands)	21	24	27	32	29	37	43	43	54	66
Advertising (in $ thousand)	14	17	25	25	35	35	45	50	60	60

[a] These figures are used for purposes of this case study.

*You may wish to view the video that accompanies this case before answering these questions.

Endnotes

1. For a good review of statistical terms, refer to Tutorial 1, "Statistical Review for Managers," found online.
2. Sources: *InformationWeek* (April 1, 2014); and *USA Today* (Oct. 13, 2014).
3. When the sample size is large ($n > 30$), the prediction interval value of y can be computed using normal tables. When the number of observations is small, the t-distribution is appropriate. See D. Groebner et al., *Business Statistics: A Decision-Making Approach*, 11th ed. (Boston: Pearson, 2021).
4. Sources: *OR/MS Today* (June 2014); *New York Daily News* (March 5, 2014); and *New York Post* (March 19, 2015).
5. To prove these three percentages to yourself, just set up a normal curve for ± 1.6 standard deviations (z-values). Using the normal table in Appendix I, you find that the area under the curve is .89. This represents ± 2 MADs. Likewise, ± 3 MADs = ± 2.4 standard deviations encompass 98% of the area, and so on for ± 4 MADs.
6. B. T. Smith, *Focus Forecasting: Computer Techniques for Inventory Control* (Boston: CBI Publishing, 1978).

Bibliography

Balakrishnan, R., B. Render, R. M. Stair, and C. Munson. *Managerial Decision Modeling: Business Analytics with Spreadsheets*, 4th ed. Boston: DeGruyter, 2017.

Berenson, M., T. Krehbiel, and D. Levine. *Basic Business Statistics*, 14th ed. Boston: Pearson, 2019.

Campbell, O. "Forecasting in Direct Selling Business: Tupperware's Experience." *The Journal of Business Forecasting* 27, no. 2 (Summer 2008): 18–19.

Georgoff, D. M., and R. G. Murdick. "Manager's Guide to Forecasting." *Harvard Business Review* 64 (January–February 1986): 110–120.

Gilliland, M., and M. Leonard. "Forecasting Software—The Past and the Future." *The Journal of Business Forecasting* 25, no. 1 (Spring 2006): 33–36.

Heizer, J. H. "Evaluate Your Forecast with Stagger Charts." *SCM Now Magazine* (March 8, 2021): 1–8.

Jain, C. L. "Benchmarking Forecasting Software and Systems." *The Journal of Business Forecasting* 26, no. 4 (Winter 2007/2008): 30–34.

Onkal, D., M. S. Gonul, and M. Lawrence. "Judgmental Adjustments of Previously Adjusted Forecasts." *Decision Sciences* 39, no. 2 (May 2008): 213–238.

Render, B., R. M. Stair, M. Hanna, and T. Hale. *Quantitative Analysis for Management*, 13th ed. Boston: Pearson, 2018.

Shah, P. "Techniques to Support Better Forecasting." *APICS Magazine* (November/December 2008): 49–50.

Urs, R. "How to Use a Demand Planning System for Best Forecasting and Planning Results." *The Journal of Business Forecasting* 27, no. 2 (Summer 2008): 22–25.

Yurklewicz, J. "Forecasting an Upward Trend." *OR/MS Today* (June 2012): 52–61.

Chapter 4 *Rapid* Review

Main Heading	Review Material	MyLab Operations Management
WHAT IS FORECASTING?	▪ **Forecasting**—The art and science of predicting future events. ▪ **Economic forecasts**—Planning indicators that are valuable in helping organizations prepare medium- to long-range forecasts. ▪ **Technological forecasts**—Long-term forecasts concerned with the rates of technological progress. ▪ **Demand forecasts**—Projections of a company's sales for each time period in the planning horizon.	Concept Questions: 1.1–1.6
THE STRATEGIC IMPORTANCE OF FORECASTING	*The forecast is the only estimate of demand until actual demand becomes known.* Forecasts of demand drive decisions in many areas, including: human resources, capacity, and supply chain management.	Concept Questions: 2.1–2.3
SEVEN STEPS IN THE FORECASTING SYSTEM	▪ Forecasting follows seven basic steps: (1) Determine the use of the forecast; (2) Select the items to be forecasted; (3) Determine the time horizon of the forecast; (4) Select the forecasting model(s); (5) Gather the data needed to make the forecast; (6) Make the forecast; (7) Validate and implement the results.	Concept Questions: 3.1–3.6
FORECASTING APPROACHES	▪ **Quantitative forecasts**—Forecasts that employ mathematical modeling to forecast demand. ▪ **Qualitative forecast**—Forecasts that incorporate such factors as the decision maker's intuition, emotions, personal experiences, and value system. ▪ **Jury of executive opinion**—Takes the opinion of a small group of high-level managers and results in a group estimate of demand. ▪ **Delphi method**—Uses an interactive group process that allows experts to make forecasts. ▪ **Sales force composite**—Based on salespersons' estimates of expected sales. ▪ **Market survey**—Solicits input from customers or potential customers regarding future purchasing plans. ▪ **Time series**—Uses a series of past data points to make a forecast.	Concept Questions: 4.1–4.6
TIME-SERIES FORECASTING	▪ **Naive approach**—Assumes that demand in the next period is equal to demand in the most recent period. ▪ **Moving average**—Uses an average of the n most recent periods of data to forecast the next period. $$\text{Moving average} = \frac{\Sigma \text{ demand in previous } n \text{ periods}}{n} \quad (4\text{-}1)$$ $$\text{Weighted moving average} = \frac{\Sigma((\text{Weight for period } n)(\text{Demand in period } n))}{\Sigma\text{Weights}} \quad (4\text{-}2)$$ ▪ **Exponential smoothing**—A weighted-moving-average forecasting technique in which data points are weighted by an exponential function. ▪ **Smoothing constant**—The weighting factor, α, used in an exponential smoothing forecast, a number between 0 and 1. Exponential smoothing formula: $$F_t = F_{t-1} + \alpha(A_{t-1} - F_{t-1}) \quad (4\text{-}4)$$ ▪ **Mean absolute deviation (MAD)**—A measure of the overall forecast error for a model. $$\text{MAD} = \frac{\Sigma\lvert\text{Actual} - \text{Forecast}\rvert}{n} \quad (4\text{-}5)$$ ▪ **Mean squared error (MSE)**—The average of the squared differences between the forecast and observed values. $$\text{MSE} = \frac{\Sigma(\text{Forecast errors})^2}{n} \quad (4\text{-}6)$$ ▪ **Mean absolute percent error (MAPE)**—The average of the absolute differences between the forecast and actual values, expressed as a percentage of actual values. $$\text{MAPE} = \frac{\sum_{i=1}^{n} 100\lvert\text{Actual}_i - \text{Forecast}_i\rvert / \text{Actual}_i}{n} \quad (4\text{-}7)$$ Exponential smoothing with trend adjustment $$\text{Forecast including trend } (FIT_t) = \text{Exponentially smoothed forecast average } (F_t) + \text{Exponentially smoothed trend } (T_t) \quad (4\text{-}8)$$	Concept Questions: 5.1–5.6 Problems: 4.1–4.41 Shorter (brief) versions of Problems: 4.2, 4.6, 4.9, 4.17, 4.19, 4.25 Virtual Office Hours for Solved Problems: 4.1–4.4 ACTIVE MODELS 4.1–4.4

Main Heading	Review Material	MyLab Operations Management
	■ **Trend projection**—A time-series forecasting method that fits a trend line to a series of historical data points and then projects the line into the future for forecasts. Trend projection and regression analysis $\hat{y} = a + bx$, where $b = \frac{\sum xy - n\bar{x}\,\bar{y}}{\sum x^2 - n\bar{x}^2}$ and $a = \bar{y} - b\bar{x}$ (4-11), (4-12), (4-13) ■ **Seasonal variations**—Regular upward or downward movements in a time series that tie to recurring events. ■ **Cycles**—Patterns in the data that occur every several years.	Virtual Office Hours for Solved Problems: 4.5–4.6
ASSOCIATIVE FORECASTING METHODS: REGRESSION AND CORRELATION ANALYSIS	■ **Linear regression analysis**—A straight-line mathematical model to describe the functional relationships between independent and dependent variables. ■ **Standard error of the estimate**—A measure of variability around the regression line. ■ **Coefficient of correlation**—A measure of the strength of the relationship between two variables. ■ **Coefficient of determination**—A measure of the amount of variation in the dependent variable about its mean that is explained by the regression equation. ■ **Multiple regression**—An associative forecasting method with > 1 independent variable. Multiple regression forecast: $\hat{y} = a + b_1x_1 + b_2x_2$ (4-17)	Concept Questions: 6.1–6.6 Problems: 4.42-4.58 **VIDEO 4.1** Forecasting Ticket Revenue for Orlando Magic Basketball Games Virtual Office Hours for Solved Problems: 4.7–4.8
MONITORING, CONTROLLING, AND ADAPTING FORECASTS	■ **Tracking signal**—A measurement of how well the forecast is predicting actual values. $\text{Tracking signal} = \frac{\sum(\text{Actual demand in period } i - \text{Forecast demand in period } i)}{\text{MAD}}$ (4-18) ■ **Bias**—A forecast that is consistently higher or lower than actual values of a time series. ■ **Adaptive smoothing**—An approach to exponential smoothing forecasting in which the smoothing constant is automatically changed to keep errors to a minimum. ■ **Focus forecasting**—Forecasting that tries a variety of computer models and selects the best one for a particular application. ■ **Stagger chart**—A short-term rolling forecast that is easily prepared, discussed, and evaluated.	Concept Questions: 7.1–7.6 Problems: 4.59–4.61
FORECASTING IN THE SERVICE SECTOR	Service-sector forecasting may require good short-term demand records, even per 15-minute intervals. Demand during holidays or specific weather events may also need to be tracked.	Concept Question: 8.1 **VIDEO 4.2** Forecasting at Hard Rock Cafe
ADDITIONAL MYLAB OPERATIONS MANAGEMENT RESOURCES	✔ Video for Creating Your Own Excel Spreadsheets (Example 8) ✔ Additional Case Studies (North-South Airlines and Digital Cell Phone, Inc.) ✔ Southwestern University Case Studies are integrated in Chapters 3, 4, 6, 8, 12, and 13 and in Supplement 7 ✔ Multiple Choice Case Questions (Southwestern University (B)) ✔ Recent Graduate Video: Kenzie Schmitt, Forecasting Analyst, Carel USA ✔ Forecasting Simulation	

Self Test

LO 4.1 Forecasting time horizons include:
- **a)** long range.
- **b)** medium range.
- **c)** short range.
- **d)** all of the choices.

LO 4.2 Qualitative methods of forecasting include:
- **a)** sales force composite.
- **b)** jury of executive opinion.
- **c)** consumer market survey.
- **d)** exponential smoothing.
- **e)** all except (d).

LO 4.3 The difference between a *moving-average* model and an *exponential smoothing* model is that ________.

LO 4.4 Three popular measures of forecast accuracy are:
- **a)** total error, average error, and mean error.
- **b)** average error, median error, and maximum error.
- **c)** median error, minimum error, and maximum absolute error.
- **d)** mean absolute deviation, mean squared error, and mean absolute percent error.

LO 4.5 Average demand for iPods in the Apple store in Rome, Italy, is 800 units per month. The May monthly index is 1.25. What is the seasonally adjusted sales forecast for May?
- **a)** 640 units
- **b)** 798.75 units
- **c)** 800 units
- **d)** 1,000 units
- **e)** cannot be calculated with the information given

LO 4.6 The main difference between simple and multiple regression is ________.

LO 4.7 The tracking signal is the:
- **a)** standard error of the estimate.
- **b)** cumulative error.
- **c)** mean absolute deviation (MAD).
- **d)** ratio of the cumulative error to MAD.
- **e)** mean absolute percent error (MAPE).

Answers: LO 4.1. d; LO 4.2. e; LO 4.3. exponential smoothing is a weighted moving-average model in which all prior values are weighted with a set of exponentially declining weights; LO 4.4. d; LO 4.5. d; LO 4.6. simple regression has only one independent variable; LO 4.7. d.

Design of Goods and Services

CHAPTER 5

CHAPTER OUTLINE

Felix Choo/Alamy Stock Photo

Amid a backlash against the fast-food industry for contributing to America's obesity problem, McDonald's reduced the size of its French fry servings in Happy Meals by more than half and added sliced apples. Later, the chain stopped promoting soft drinks as an option in children's meals and instead started including juice, low-fat milk, and water. The changes are the latest steps in the evolution of the Happy Meal toward healthier options. Product design is an ongoing process at McDonald's, just as it is at Apple, Honda, and other industry leaders.

10 OM STRATEGY DECISIONS

- *Design of Goods and Services*
- Managing Quality
- Process Strategies
- Location Strategies
- Layout Strategies
- Human Resources
- Supply Chain Management
- Inventory Management
- Scheduling
- Maintenance

GLOBAL COMPANY PROFILE
Nautique Boats

Product Design Provides Competitive Advantage at Nautique Boats

Designing and defining a product has implications for all subsequent operations decisions. As such, product design is the first of the 10 critical decisions of OM around which this textbook is structured.

At Nautique Boat Company, famous for its waterskiing, wakeboarding, and wake-surfing products, design is at the heart of corporate strategy. "We create the standards that others are judged by," says President Greg Meloon. "We have to be innovative in building boats. And this involves product design."

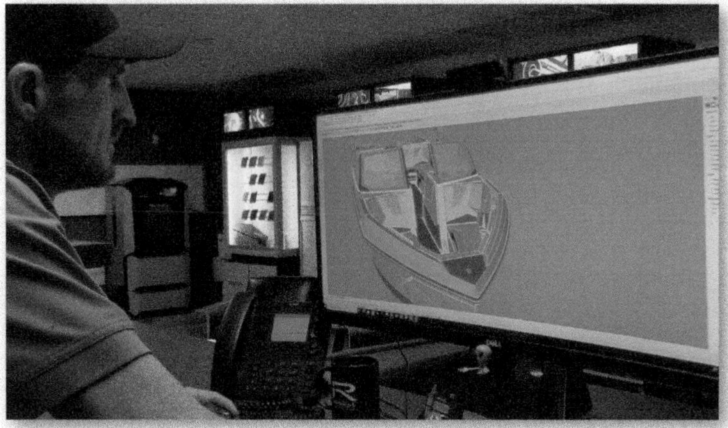

Beverly Amer/Aspenleaf Productions

It all begins with concepts developed via computer-aided design (CAD). Note the 3D printer in the background, which is used to create parts that help model new design features for the production team.

With changing consumer tastes, compounded by material changes and ever-improving marine engineering, the design function is under constant pressure. Added to these pressures are the ongoing issue of cost competitiveness and the need to provide good value for customers.

Consequently, Nautique Boats is a heavy user of computer-aided design (CAD). Its sophisticated CAD software system not only has reduced product development time and cost but also has improved tooling and reduced production issues, resulting in a superior product.

Chief Designer Steve Carlton creates a clay model of his vision for the next prototype. A full-scale wooden mock-up follows, with upholstery and technology in place.

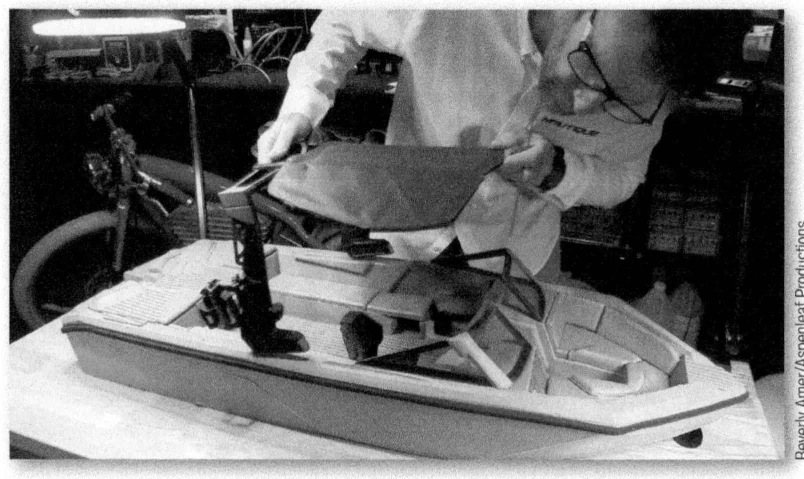

Beverly Amer/Aspenleaf Productions

Courtesy of Correct Craft Holding Company, LLC

Hulls and decks are separately hand-produced by spraying preformed molds with three to five layers of fiberglass laminate. The hulls and decks harden and become the upper and lower structure of the boat.

Beverly Amer/Aspenleaf Productions

On the five assembly lines, each boat moves from station to station in accordance with changing production schedules. Here the "deck," suspended from ceiling cranes, is being lowered to join the "hull."

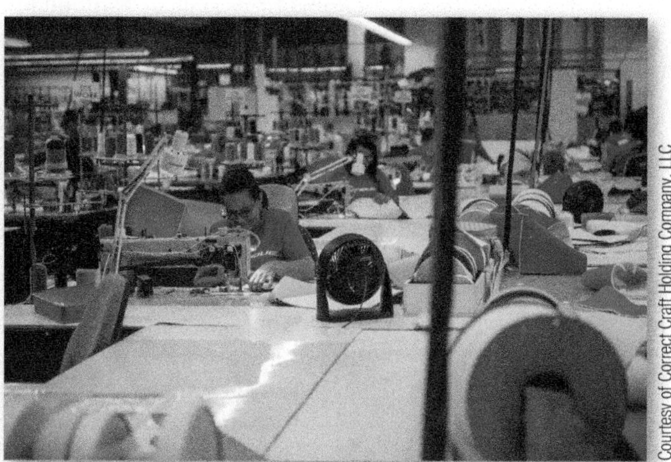

Courtesy of Correct Craft Holding Company, LLC

The upholstery department custom-manufactures the interiors for each boat on a JIT basis. These subassemblies are moved to the assembly line on specially designed carts just in time to be installed that day.

Courtesy of Correct Craft Holding Company, LLC

Once completed, each boat is put through a series of 44-minute extreme tests in a nearby lake owned by Nautique Boats. Testers are told to "drive the boats hard."

All of Nautique's products, from its $80,000 Ski Nautique to the nearly $300,000 Paragon G-23, follow a similar production process. Hulls and decks are produced separately and then joined on the assembly line, where components such as engines, towers, and upholstery are added at individual workstations.

With a focus on design and production details, Nautique Boats superbly manages the product from design to water testing. That is how this 100-year-old company has stayed as number one in its competitive market.

LEARNING OBJECTIVES

Goods and Services Selection

STUDENT TIP Product strategy is critical to achieving competitive advantage.

Global firms such as Nautique Boats know that the basis for an organization's existence is the good or service it provides society. Great products are the keys to success. Anything less than an excellent product strategy can be devastating to a firm. To maximize the potential for success, many companies focus on only a few products and then concentrate on those products. For instance, Honda's focus, its core competency, is engines. Virtually all of Honda's sales (autos, motorcycles, generators, lawn mowers) are based on its outstanding engine technology. Likewise, Intel's focus is on microprocessors, and Michelin's is on tires.

VIDEO 5.1 Product Design at Nautique Boat Company

However, because most products have a limited and even predictable life cycle, companies must constantly be looking for new products to design, develop, and take to market. Operations managers insist on strong communication among customer, product, processes, and suppliers that results in a high success rate for their new products. 3M's goal is to produce 30% of its profit from products introduced in the past 4 years. Apple generates almost 60% of its revenue from products launched in the past 4 years. Benchmarks, of course, vary by industry—Rubbermaid introduces a new product each day!

The importance of successfully integrating talent, capital, and technology to build new products cannot be overestimated. As Figure 5.1 shows, leading companies generate a substantial portion of their sales from products less than 5 years old. The need for new products is why Gillette developed its multiblade razors, in spite of continuing high sales of its phenomenally successful Sensor razor, and why Disney continues to innovate with new rides, new parks, and creative new multimedia innovations even though it is already the world's leading family entertainment company.

Despite constant efforts to introduce viable new products, many new products do not succeed. Product selection, definition, and design occur frequently—perhaps hundreds of times

Figure **5.1**

Innovation and New Products

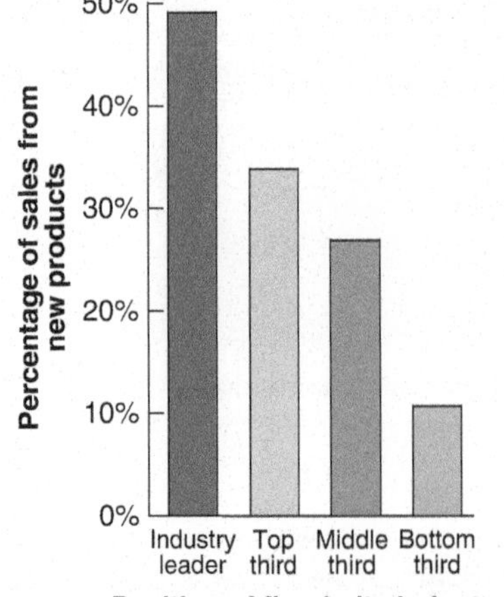

The higher the percentage of sales from the last 5 years, the more likely the firm is to be a leader.

for each financially successful product. DuPont estimates that it takes 250 ideas to yield one *marketable* product. Operations managers and their organizations build cultures that accept this risk and tolerate failure. They learn to accommodate a high volume of new product ideas while maintaining the production activities to which they are already committed.

STUDENT TIP
Motorola went through 3,000 working models before it developed its first pocket cell phone.

Although the term *products* often refers to tangible goods, it also refers to offerings by service organizations. For instance, when Allstate Insurance offers a new homeowner's policy, it is referred to as a new "product." Similarly, when Citicorp opens a mortgage department, it offers a number of new mortgage "products."

An effective product strategy links product decisions with investment, market share, and product life cycle, and defines the breadth of the product line. The *objective of the* **product decision** *is to develop and implement a product strategy that meets the demands of the marketplace with a competitive advantage.* As one of the 10 decisions of OM, product strategy may focus on developing a competitive advantage via differentiation, low cost, rapid response, or a combination of these.

Product decision
The selection, definition, and design of products.

Product Strategy Options Support Competitive Advantage

A world of options exists in the selection, definition, and design of products. Product selection is choosing the good or service to provide customers or clients. For instance, hospitals specialize in various types of patients and medical procedures. A hospital's management may decide to operate a general-purpose hospital or a maternity hospital or, as in the case of the Canadian hospital Shouldice, to specialize in hernias. Hospitals select their products when they decide what kind of hospital to be. Numerous other options exist for hospitals, just as they exist for Taco Bell and Toyota.

Service organizations like Shouldice Hospital *differentiate* themselves through their product. Shouldice differentiates itself by offering a distinctly unique and high-quality product. Its world-renowned specialization in hernia-repair service is so effective it allows patients to return to normal living in 8 days as opposed to the average 2 weeks—and with very few complications. The entire production system is designed for this one product. Local anesthetics are used; patients enter and leave the operating room on their own; meals are served in a common dining room, encouraging patients to get out of bed for meals and join fellow patients in the lounge. As Shouldice demonstrates, product selection affects the entire production system.

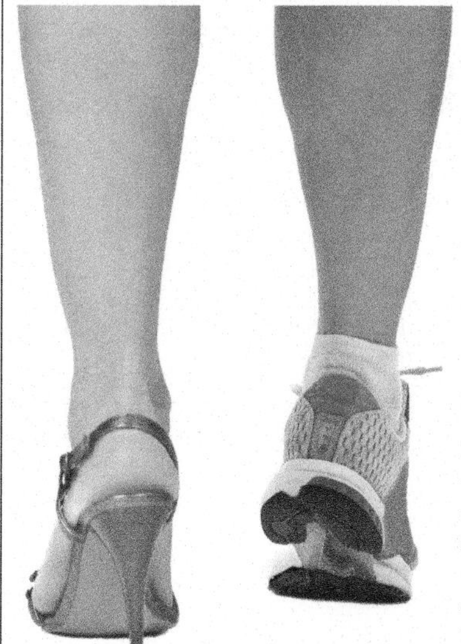

Radu Razvan/Shutterstock

(a) Markets: In its creative way, the market has moved athletic shoes from utilitarian footwear into fashionable accessories.

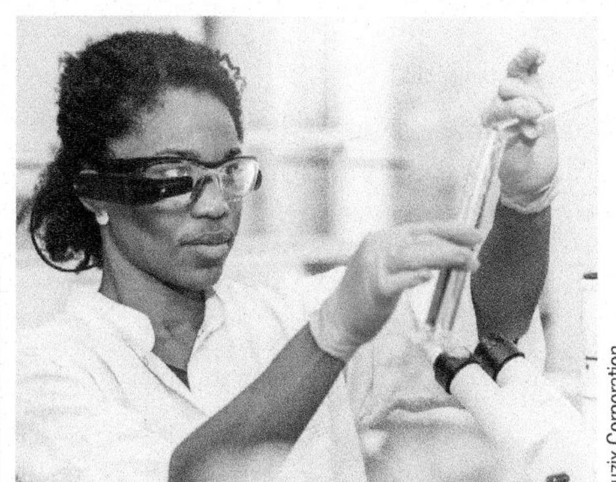

Vuzix Corporation

(b) Technology: Vuzix's wearable augmented reality smart glasses allow interactive feedback while working.

P&G

(c) Packaging: Tide has cut down the plastic in packaging by 60%. And the boxed detergent doesn't need to be packed in another box—online retailers just slap an address on it.

Product Innovation Can Be Driven by Markets, Technology, and Packaging. Whether it is design focused on changes in the market (a), the application of technology at Vuzix (b), or a new detergent container for Tide (c), operations managers need to remind themselves that the creative process is ongoing with major production implications.

Taco Bell has developed and executed a *low-cost* strategy through product design. By designing a product (its menu) that can be produced with a minimum of labor in small kitchens, Taco Bell has developed a product line that is both low cost and high value. Successful product design has allowed Taco Bell to increase the food content of its products from 27¢ to 45¢ of each sales dollar.

Toyota's strategy is *rapid response* to changing consumer demand. By executing the fastest automobile design in the industry, Toyota has driven the speed of product development down to well under 2 years in an industry whose standard is still more than 2 years. The shorter design time allows Toyota to get a car to market before consumer tastes change and to do so with the latest technology and innovations.

Product decisions are fundamental to an organization's strategy and have major implications throughout the operations function. For instance, GM's steering columns are a good example of the strong role product design plays in both quality and efficiency. The redesigned steering column is simpler, with about 30% fewer parts than its predecessor. The result: Assembly time is one-third that of the older column, and the new column's quality is about seven times higher. As an added bonus, machinery on the new line costs a third less than that on the old line. Similarly, Tesla's Model 3 undercarriage has 70 parts, each with its own drawings, scheduling, tooling, and inventory. However, the new Model Y has been redesigned to have just two parts replacing all of the manufacturing demands of the other 68. Now, with the world's largest die-casting machine installed in its Fremont, California, plant, Tesla plans to ultimately cast a single component that will form the rear underbody of its Model Y electric crossover vehicle.

Product Life Cycles

Products are born. They live and they die. They are cast aside by a changing society. It may be helpful to think of a product's life as divided into four phases. Those phases are introduction, growth, maturity, and decline.

Product life cycles may be a matter of a few days (a concert t-shirt), months (seasonal fashions), years (Madden NFL football video game), or decades (Boeing 737). Regardless of the length of the cycle, the task for the operations manager is the same: to design a system that helps introduce new products successfully. If the operations function cannot perform effectively at this stage, the firm may be saddled with losers—products that cannot be produced efficiently and perhaps not at all.

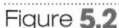

LO 5.1 *Define* product life cycle

Figure 5.2 shows the four life cycle stages and the relationship of product sales, cash flow, and profit over the life cycle of a product. Note that typically a firm has a negative cash flow while it develops a product. When the product is successful, those losses may be recovered. Eventually, the successful product may yield a profit prior to its decline. However, the profit is fleeting—hence, the constant demand for new products.

Life Cycle and Strategy

Just as operations managers must be prepared to develop new products, they must also be prepared to develop *strategies* for new and *existing* products. Periodic examination of products is appropriate because *strategies change as products move through their life cycle*. Successful product strategies require determining the best strategy for each product based

Figure **5.2**

Product Life Cycle, Sales, Cost, Profit, and Loss

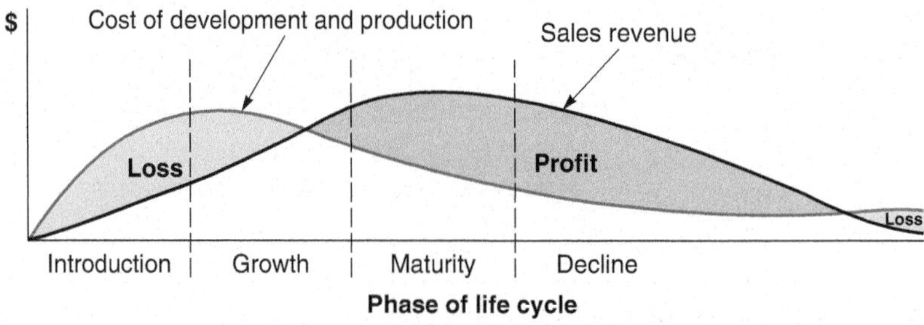

on its position in its life cycle. A firm, therefore, identifies products or families of products and their position in the life cycle. Let us review some strategy options as products move through their life cycles.

Introductory Phase Because products in the introductory phase are still being "fine-tuned" for the market, as are their production techniques, they may warrant unusual expenditures for (1) research, (2) product development, (3) process modification and enhancement, and (4) supplier development. For example, when the iPhone was first introduced, the features desired by the public were still being determined. At the same time, operations managers were still groping for the best manufacturing techniques.

Growth Phase In the growth phase, product design has begun to stabilize, and effective forecasting of capacity requirements is necessary. Adding capacity or enhancing existing capacity to accommodate the increase in product demand may be necessary.

Maturity Phase By the time a product is mature, competitors are established. So high-volume, innovative production may be appropriate. Improved cost control, reduction in options, and a paring down of the product line may be effective or necessary for profitability and market share.

Decline Phase Management may need to be ruthless with those products whose life cycle is at an end. Dying products are typically poor products in which to invest resources and managerial talent. Unless dying products make some unique contribution to the firm's reputation or its product line or can be sold with an unusually high contribution, their production should be terminated.[1]

Product-by-Value Analysis

The effective operations manager selects items that show the greatest promise. This is the Pareto principle applied to product mix: Resources are to be invested in the critical few and not the trivial many. **Product-by-value analysis** lists products in descending order of their *individual dollar contribution* to the firm. It also lists the *total annual dollar contribution* of the product. Low contribution on a per-unit basis by a particular product may look substantially different if it represents a large portion of the company's sales.

Product-by-value analysis
A list of products, in descending order of their individual dollar contribution to the firm, as well as the *total annual dollar contribution* of the product.

A product-by-value report allows management to evaluate possible strategies for each product. These may include increasing cash flow (e.g., increasing contribution by raising selling price or lowering cost), increasing market penetration (improving quality and/or reducing cost or price), or reducing costs (improving the production process). The report may also tell management which product offerings should be eliminated and which fail to justify further investment in research and development or capital equipment. Product-by-value analysis focuses attention on the strategic direction for each product.

Generating New Products

STUDENT TIP
Societies reward those who supply new products that reflect their needs.

Managers face a dilemma: The very things they focus on to manage an ongoing firm—building core competencies, listening to existing customers, and investing in current technologies—moves them away from recognizing new disruptive innovations. But products die and existing products must be weeded out and replaced. Moreover, because firms generate most of their revenue and profit from new products, product selection, definition, and design must take place on a continuing basis. Consider recent product changes: DVDs to video streaming, coffee shops to Starbucks lifestyle coffee, traveling circuses to Cirque du Soleil, landlines to cell phones, cell phones to smartphones and Alexa, and an Internet of digital information to an Internet of "things" that connects you and your smartphone to your home, car, and doctor. And the list goes on. Knowing how to successfully find and develop new products is a requirement.

Aggressive new product development requires that organizations build structures internally that have open communication with customers, innovative product development cultures, aggressive R&D, strong leadership, formal incentives, and training. Only then can a firm profitably and energetically focus on specific opportunities such as the following:

1. *Understanding the customer* is the premier issue in new-product development. Many commercially important products are initially thought of and even prototyped by users rather than producers. Such products tend to be developed by "lead users"—companies, organizations, or individuals that are well ahead of market trends and have needs that go far beyond those of average users. The operations manager must be "tuned in" to the market and particularly these innovative lead users.
2. *Economic change* brings increasing levels of affluence in the long run but economic cycles and price changes in the short run. In the long run, for instance, more and more people can afford automobiles, but in the short run, a recession may weaken the demand for automobiles.
3. *Sociological and demographic change* may appear in such factors as decreasing family size. This trend alters the size preference for homes, apartments, and automobiles.
4. *Technological change* makes possible everything from smartphones to iPads to artificial hearts.
5. *Political and legal change* brings about new trade agreements, tariffs, and government requirements.
6. Other changes may be brought about through *market practice, professional standards, suppliers*, and *distributors*.

Operations managers must be aware of these dynamics and be able to anticipate and manage changes in product opportunities, the products themselves, product volume, and product mix.

Product Development

Product Development System

LO 5.2 *Describe* a product development system

An effective product strategy links product decisions with other business functions, such as R&D, engineering, marketing, and finance. A firm requires cash for product development, an understanding of the marketplace, and the necessary human talents. The product development system may well determine not only product success but also the firm's future. Figure 5.3 shows the stages of product development. In this system, product options go through a series of steps, each having its own screening and evaluation criteria, but providing a continuing flow of information to prior steps.

Optimum product development depends not only on support from other parts of the firm but also on the successful integration of all 10 of the OM decisions, from product design to maintenance. Identifying products that appear likely to capture market share, be cost-effective, and be profitable but are, in fact, very difficult to produce may lead to failure rather than success.

Quality function deployment (QFD)
A process for determining customer requirements (customer "wants") and translating them into the attributes (the "hows") that each functional area can understand and act on.

House of quality
A part of the quality function deployment process that utilizes a planning matrix to relate customer "wants" to "how" the firm is going to meet those "wants."

Quality Function Deployment (QFD)

Quality function deployment (QFD) refers to both (1) determining what will satisfy the customer and (2) translating those customer desires into the target design. The idea is to capture a rich understanding of customer wants and to identify alternative process solutions. This information is then integrated into the evolving product design. QFD is used early in the design process to help determine *what will satisfy the customer* and *where to deploy quality efforts*.

One of the tools of QFD is the **house of quality**, a graphic technique for defining the relationship between customer desires and product (or service). Only by defining this relationship in a

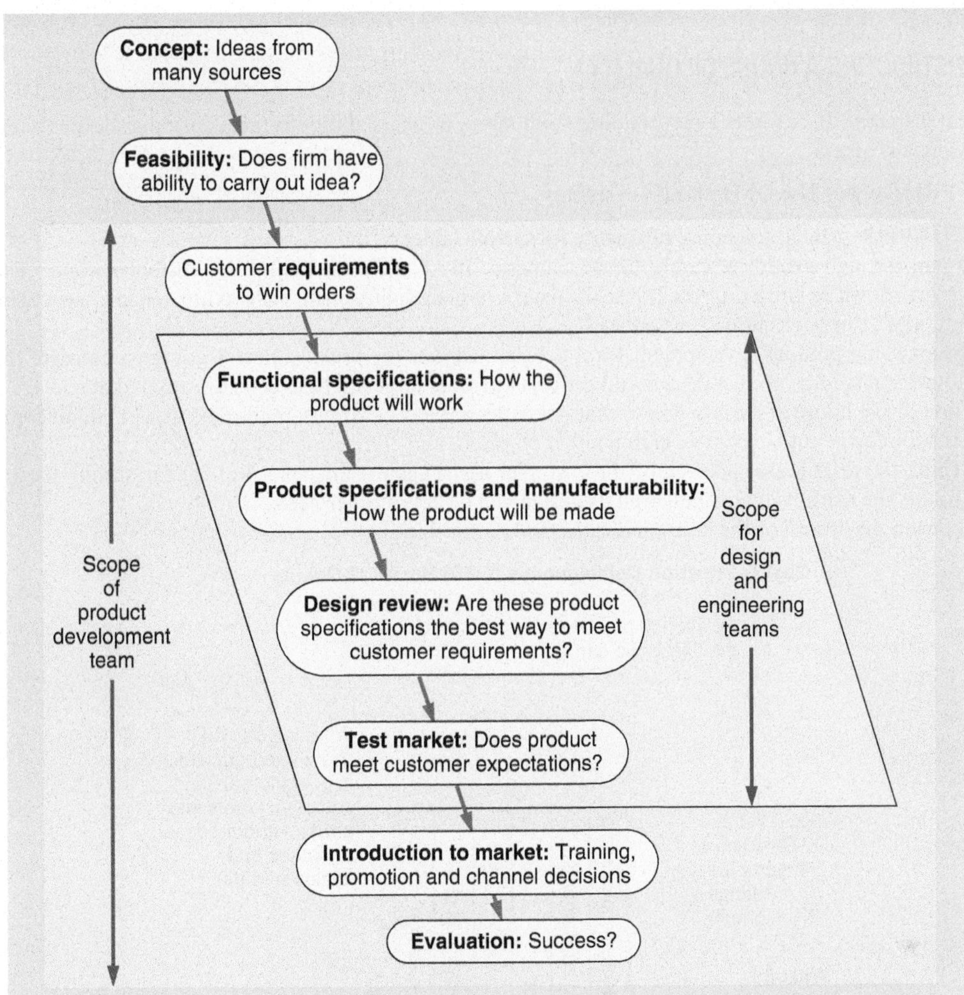

Figure 5.3

Product Development Stages

Product concepts are developed from a variety of sources, both external and internal to the firm. Concepts that survive the product idea stage progress through various stages, with nearly constant review, feedback, and evaluation in a highly participative environment to minimize failure.

rigorous way can managers design products and processes with features desired by customers. Defining this relationship is the first step in building a world-class production system. To build the house of quality, we perform seven basic steps:

1. Identify customer *wants*. (What do customers want in this product?)
2. Identify *how* the good/service will satisfy customer wants. (Identify specific product characteristics, features, or attributes and show how they will satisfy customer *wants*.)
3. Relate customer *wants* to product *hows*. (Build a matrix, as in Example 1, that shows this relationship.)
4. Identify relationships between the firm's *hows*. (How do our *hows* tie together? For instance, in the following example, there is a high relationship between low electricity requirements and auto focus, auto exposure, and number of pixels because they all require electricity. This relationship is shown in the "roof" of the house in Example 1.)
5. Develop importance ratings. (Using the *customer's* importance ratings and weights for the relationships shown in the matrix, compute *our* importance ratings, as in Example 1.)
6. Evaluate competing products. (How well do competing products meet customer wants? Such an evaluation, as shown in the two columns on the right of the figure in Example 1, would be based on market research.)
7. Determine the desirable technical attributes, your performance, and the competitor's performance against these attributes. (This is done at the bottom of the figure in Example 1.)

LO 5.3 *Build* a house of quality

Example 1

CONSTRUCTING A HOUSE OF QUALITY

Great Cameras, Inc., wants a methodology that strengthens its ability to meet customer desires with its new digital camera.

APPROACH ▶ Use QFD's house of quality.

SOLUTION ▶ Build the house of quality for Great Cameras, Inc.

First, through market research, Great Cameras, Inc., determined what the customer *wants*. Those *wants* are shown on the left of the house of quality, namely, lightweight, easy to use, reliable, easy to hold steady, and color correction.

Second, the product development team determined *how* the organization is going to translate those customer *wants* into product design and process attribute targets. These *hows* are entered across the top portion of the house of quality. These characteristics are low electricity requirements, aluminum components, auto focus, auto exposure, high number of pixels, and ergonomic design.

Third, the team evaluated each of the customer *wants* against the *hows*. In the relationship matrix of the house, the team evaluated how well its design will meet customer needs.

Fourth, the "roof" of the house indicates the relationships between the attributes.

STUDENT TIP QFD Capture Software is a management aid for prioritizing choices for better products and services. A free evaluation version is available at **https://qfdcapture-professional-edition.software.informer.com**.

Quality Function Deployment's (QFD) House of Quality

Relationship between the things **we** can do

What we can **do** (**how** the organization is going to translate customer wants into product and process attributes and design targets)

Customer importance ratings (5 = highest)

What the customer **wants**

G = good
F = fair
P = poor

What the customer wants	Importance	Low electricity requirements	Aluminum components	Auto focus	Auto exposure	High number of pixels	Ergonomic design	Company A	Company B
Lightweight	3	•	○				•	G	P
Easy to use	4	•		○	○	○	○	G	P
Reliable	5	○		○	○	○		F	G
Easy to hold steady	2						◉	G	P
Color correction	1					◉		P	P
Our importance ratings		22	9	27	27	32	25		
Target values (technical attributes)		0.5A	75%	2' to ∞	2 circuits	Failure 1 per 10,000	Panel ranking		
Technical evaluation		0.7	60%	Yes	1	ok	G		
		0.6	50%	Yes	2	ok	F		
		0.5	75%	Yes	2	ok	G		

Competitive assessment

How well what we do meets the customer's wants (relationship matrix)

Weighted rating of 25 = (1 × 3) + (3 × 4) + (5 × 2)

◉ High relationship (5) ○ Medium relationship (3) • Low relationship (1)

Fifth, the team developed importance ratings for its design attributes on the bottom row of the table. This was done by assigning values (5 for high, 3 for medium, and 1 for low) to each entry in the relationship matrix and then multiplying each of these values by the associated customer importance rating. These values in the "Our importance ratings" row provide a ranking of how to proceed with product and process design, with the highest values being the most critical to a successful product.

Sixth, the house of quality is also used for the evaluation of *competitors*. How well do *competitors* meet customer demand? The two columns on the right indicate how market research thought that competitors, A and B, satisfy customer wants (**G**ood, **F**air, or **P**oor). So company A does a good job on "lightweight," "easy to use," and "easy to hold steady;" a fair job on "reliable;" and a poor job on "color correction." Company B does a good job with "reliable" but poorly on the other attributes. Products from other firms and even the proposed product can be added next to company B.

Seventh, the team identified the technical attributes that are critical to the camera and evaluated how well both Great Cameras, Inc., and competitors address these attributes. Here the team decided on the noted technical attributes.

INSIGHT ▶ QFD provides an analytical tool that structures design features and technical issues, as well as providing importance rankings and competitor comparison.

LEARNING EXERCISE ▶ If the market research for another country indicates that "lightweight" has the most important customer ranking (5) and reliability has a 3, what is the new total importance ranking for low-electricity requirements, aluminum components, and ergonomic design? [Answer: 18, 15, 27, respectively.]

RELATED PROBLEMS ▶ 5.4, 5.5, 5.6, 5.7, 5.8

Another use of quality function deployment (QFD) is to show how the quality effort will be *deployed*. As Figure 5.4 shows, *design characteristics* of House 1 become the inputs to House 2, which are satisfied by *specific components* of the product. Similarly, the concept is carried to House 3, where the specific components are to be satisfied through particular *production processes*. Once those production processes are defined, they become requirements of House 4 to be satisfied by a *quality plan* that will ensure conformance of those processes. The quality plan is a set of specific tolerances, procedures, methods, and sampling techniques that will ensure that the production process meets the customer requirements.

The QFD effort is devoted to meeting customer requirements. The *sequence* of houses is a very effective way of identifying, communicating, and deploying production resources. In this way we produce quality products, meet customer requirements, and win orders.

Organizing for Product Development

Let's look at four approaches to organizing for product development. *First*, the traditional U.S. approach to product development is an organization with distinct departments: a research and development department to do the necessary research; an engineering department to design the product; a manufacturing engineering department to design a product that can be produced;

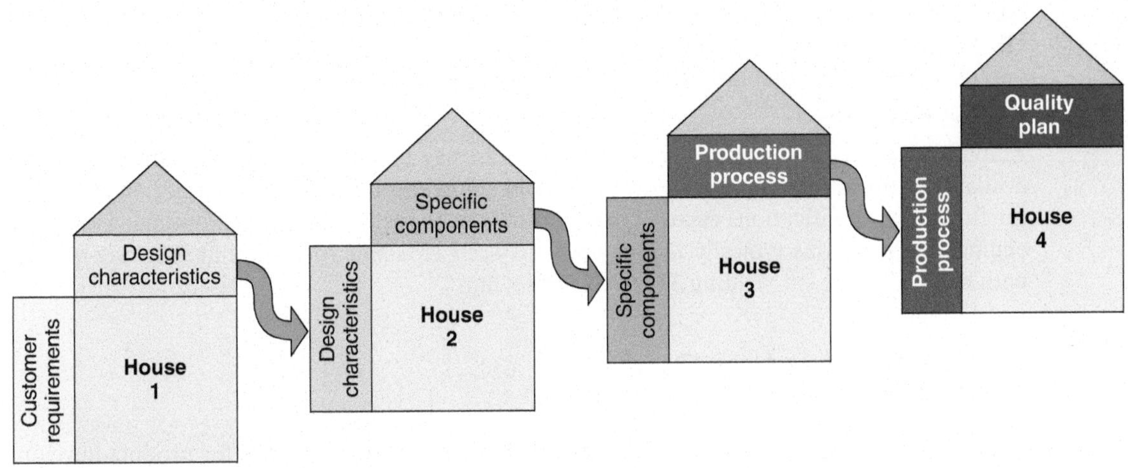

Figure 5.4

House of Quality Sequence Indicates How to Deploy Resources to Achieve Customer Requirements

and a production department that produces the product. The distinct advantage of this approach is that fixed duties and responsibilities exist. The distinct disadvantage is lack of forward thinking: How will downstream departments in the process deal with the concepts, ideas, and designs presented to them, and ultimately what will the customer think of the product?

A *second* and popular approach is to assign a product manager to "champion" the product through the product development system and related organizations. However, a *third*, and perhaps the best, product development approach used in the U.S. seems to be the use of teams. Such teams are known variously as *product development teams, design for manufacturability teams*, and *value engineering teams*.

The Japanese use a *fourth* approach. They bypass the team issue by not subdividing organizations into research and development, engineering, production, and so forth. Consistent with the Japanese style of group effort and teamwork, these activities are all in one organization. Japanese culture and management style are more collegial and the organization less structured than in most Western countries. Therefore, the Japanese find it unnecessary to have "teams" provide the necessary communication and coordination. However, the typical Western style, and the conventional wisdom, is to use teams.

Product development teams
Teams charged with moving from market requirements for a product to achieving product success.

Product development teams are charged with the responsibility of moving from market requirements for a product to achieving a product success (refer to Figure 5.3). Such teams often include representatives from marketing, manufacturing, purchasing, quality assurance, and field service personnel. Many teams also include representatives from vendors. Regardless of the formal nature of the product development effort, research suggests that success is more likely in an open, highly participative environment where those with potential contributions are allowed to make them. The objective of a product development team is to make the good or service a success. This includes marketability, manufacturability, and serviceability.

Concurrent engineering
Simultaneous performance of the various stages of product development.

Concurrent engineering implies speedier product development by bringing product, process, and quality engineers, as well as suppliers and marketing personnel, together to achieve simultaneous performance of the various stages of product development (as we saw earlier in Figure 5.3). Often the concept is expanded to include all elements of a product's life cycle, from customer requirements to disposal and recycling. Concurrent engineering is facilitated by teams representing all affected areas (known as *cross-functional* teams).

Manufacturability and Value Engineering

Manufacturability and value engineering
Activities that help improve a product's design, production, maintainability, and use.

Manufacturability and value engineering activities are concerned with improvement of design and specifications at the research, development, design, and preproduction stages of product development. In addition to immediate, obvious cost reduction, design for manufacturability and value engineering may produce other benefits. These include:

1. Reduced complexity of the product.
2. Reduction of environmental impact.
3. Additional standardization of components.
4. Improvement of functional aspects of the product.
5. Improved job design and job safety.
6. Improved maintainability (serviceability) of the product.
7. Robust design.

Manufacturability and value engineering activities may be the best cost-avoidance technique available to operations management. They yield value improvement by focusing on achieving the functional specifications necessary to meet customer requirements in an optimal way. Value engineering programs typically reduce costs between 15% and 70% without reducing quality, with every dollar spent yielding \$10 to \$25 in savings.

Issues for Product Design

In addition to developing an effective system and organization structure for product development, several considerations are important to the design of a product. We will now review six of these: (1) robust design, (2) modular design, (3) computer-aided design/computer-aided manufacturing (CAD/CAM), (4) virtual reality technology, (5) value analysis, and (6) sustainability/life cycle assessment (LCA).

Robust Design

Robust design means that the product is designed so that small variations in production or assembly do not adversely affect the product. For instance, Nokia developed an integrated circuit that could be used in many products to amplify voice signals. As originally designed, the circuit had to be manufactured very expensively to avoid variations in the strength of the signal. But after testing and analyzing the design, Nokia engineers realized that if the resistance of the circuit was reduced—a minor change with no associated costs—the circuit would be far less sensitive to manufacturing variations. The result was a 40% improvement in quality.

Robust design
A design that can be produced to requirements even with unfavorable conditions in the production process.

Modular Design

Products designed in easily segmented components are known as **modular designs**. Modular designs offer flexibility to both production and marketing. Operations managers find modularity helpful because it makes product development, production, and subsequent changes easier. Marketing may like modularity because it adds flexibility to the ways customers can be satisfied. For instance, virtually all premium high-fidelity sound systems are produced and sold this way. The customization provided by modularity allows customers to mix and match to their own taste. This is also the approach taken by Harley-Davidson, where relatively few different engines, chassis, gas tanks, and suspension systems are mixed to produce a huge variety of motorcycles. It has been estimated that many automobile manufacturers can, by mixing the available modules, never make two cars alike. This same concept of modularity is carried over to many industries, from airframe manufacturers to fast-food restaurants. Airbus uses the same wing modules on several planes, just as McDonald's and Burger King use relatively few modules (cheese, lettuce, buns, sauces, pickles, meat patties, french fries, etc.) to make a variety of meals.

Modular design
A design in which parts or components of a product are subdivided into modules that are easily interchanged or replaced.

Computer-Aided Design (CAD) and Computer-Aided Manufacturing (CAM)

Computer-aided design (CAD) is the use of computers to interactively design products and prepare engineering documentation. CAD uses three-dimensional (3D) drawing to save time and money by shortening development cycles for virtually all products (see the 3D design photo in the Nautique Boat Global Company Profile that opens this chapter). The speed and ease with which sophisticated designs can be manipulated, analyzed, and modified with CAD make review of numerous options possible before final commitments are made. Faster development, better products, and accurate flow of information to other departments all contribute to a tremendous payoff for CAD. The payoff is particularly significant because most product costs are determined at the design stage.

Computer-aided design (CAD)
Interactive use of a computer to develop and document a product.

STUDENT TIP
Siemens' CAD software webpage contains a short video demonstration: **https://trials.sw.siemens.com/nx-coredesigner/**.

One extension of CAD is **design for manufacture and assembly (DFMA)** software, which focuses on the effect of design on assembly. For instance, DFMA allows Ford to build new vehicles in a virtual factory where designers examine how to put a transmission in a car on the production line, even while both the transmission and the car are still in the design stage.

Design for manufacture and assembly (DFMA)
Software that allows designers to look at the effect of design on manufacturing of the product.

CAD systems have moved to the Internet through e-commerce, where they link computerized design with purchasing, outsourcing, manufacturing, and long-term maintenance. This move also speeds up design collaboration, as staff around the world can work on their unique work schedules. Rapid product change also supports the trend toward "mass customization" and, when carried to an extreme, allows customers to enter a supplier's design libraries and make changes. The result is faster and less expensive customized products. As product life cycles shorten, designs become more complex, and global collaboration has grown, the European Community (EU) has developed a **standard for the exchange of product data (STEP**; ISO 10303). STEP permits 3D product information to be expressed in a standard format so it can be exchanged internationally.

Standard for the exchange of product data (STEP)
A standard that provides a format allowing the electronic transmission of three-dimensional data.

Computer-aided manufacturing (CAM) refers to the use of specialized computer programs to direct and control manufacturing equipment. When CAD information is translated into instructions for CAM, the result of these two technologies is CAD/CAM. The combination is a powerful tool for manufacturing efficiency. Fewer defective units are produced, translating into less rework and lower inventory. More precise scheduling also contributes to less inventory and more efficient use of personnel.

Computer-aided manufacturing (CAM)
The use of information technology to control machinery.

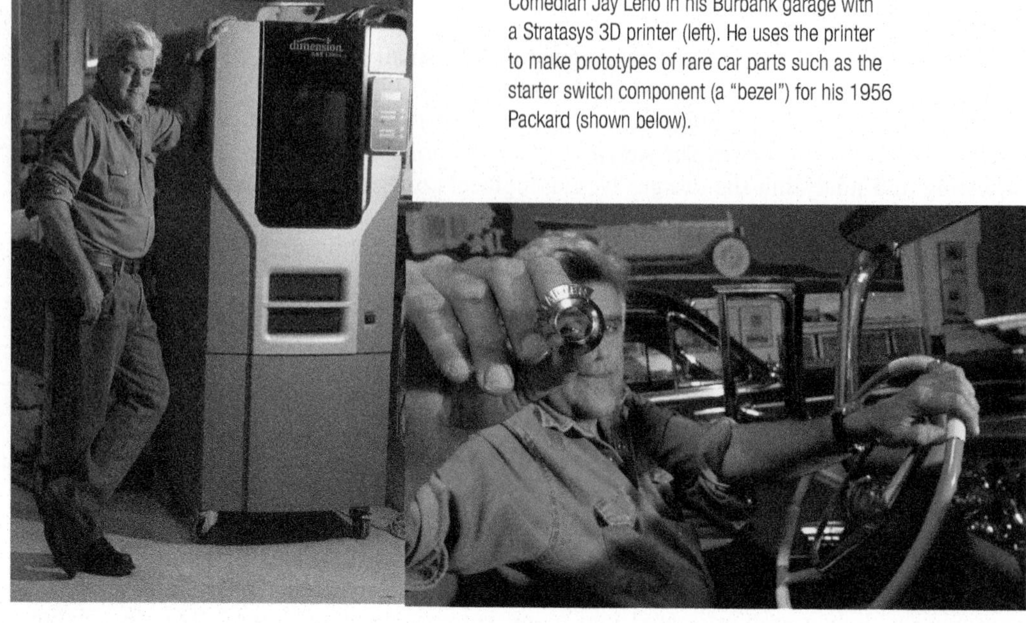

Comedian Jay Leno in his Burbank garage with a Stratasys 3D printer (left). He uses the printer to make prototypes of rare car parts such as the starter switch component (a "bezel") for his 1956 Packard (shown below).

Gary Friedman/Los Angeles Times/Getty Images

Additive manufacturing
An extension of CAD that builds products by adding material layer upon layer.

Additive manufacturing (sometimes called 3D printing) uses CAD technology to produce products by adding material layer upon layer. The raw material can be plastic, metal, ceramic, glass, or even human tissues, and the output can be everything from jet engine fan blades to human organs to printed steaks (see the *OM in Action* box, "Israeli Startups Roll Out 3D Printed Steaks"). Traditional manufacturing removes material, while additive manufacturing adds material. (See Figure 5.5.)

OM in Action Israeli Startups Roll Out 3D Printed Steaks[2]

The walls of Redefined Meat's lab in Rehovot, Israel, are plastered with posters of cuts of beef, including sirloins, T-bones, and rib eyes. But the startup isn't selling the perfect cut of beef. Instead, it is using plant-based alternatives. The company is using 3D bioprinting technology to produce a printed steak that's so fatty, juicy, and perfectly meaty that even the most dedicated carnivore won't know the difference. The 3D printing technology controls what's happening inside the mass to improve the texture and to improve the flavor.

Competing Israeli startup Aleph Farms says that 3D printing promises to give diners the same savory experience as eating a real T-bone or rump roast. The technology involves developing a design that can then be printed countless times. First, proprietary computer software creates a detailed model of a steak, including the muscle, fat, and blood, based on whichever cut it's emulating. That blueprint is then transmitted to a printer loaded with plant-based "inks." Hit the start button and out comes a "steak."

While ground-meat replacements are widely available, mimicking an actual cut of meat has proved far more challenging. That's because replicating the mouthfeel and visual appeal of a juicy sirloin is a lot tougher than cranking out something that's going to be slapped in a bun. Aleph Farms CEO Didier Toubia says the product mirrors the sensory quality, texture, flavor, and fatty marbling of a traditionally produced rib eye. "A beefsteak is the holy grail of plant-based meat," says one executive.

The faux-meat category has already reached an estimated $14 billion in annual sales worldwide and is estimated to grow to $140 billion in 2029. Redefined Meat and Aleph Farms expect to supply customers, including restaurants, meat distributors, and retailers, with both the printers and cartridges. Redefined Meat's printer can now deliver 35 seven-ounce steaks in an hour. That means the equivalent of a cow's worth of steak a day.

Dmytro Zinkevych/Shutterstock

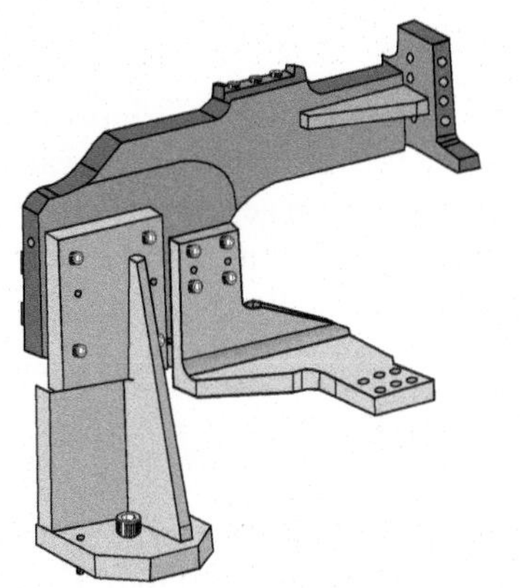
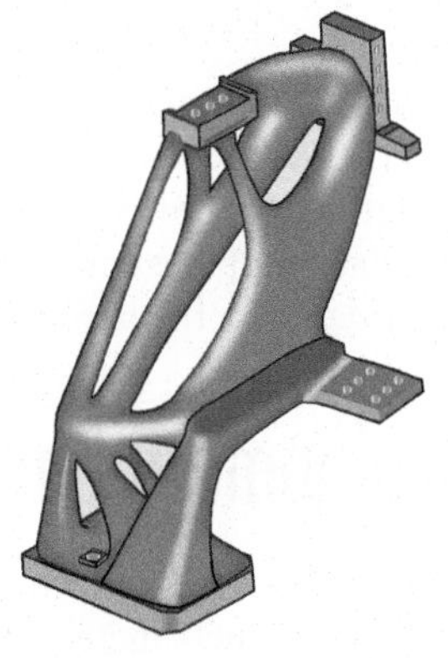

Figure 5.5

Cost Reduction via Additive Manufacturing

Additive manufacturing, sometimes called 3D printing, uses CAD technology to place (add) material layer upon layer to build and replace this multi-piece clamp assembly (left) with the one-piece clamp (right). Production is often faster and more economical, particularly for low-volume production.

Virtual Reality Technology

Virtual reality is a visual form of communication in which images substitute for the real thing but still allow the user to respond interactively. The roots of virtual reality technology in operations are in CAD. Once design information is in a CAD system, it is also in electronic digital form for other uses, such as developing 3D layouts of everything from retail stores and restaurant layouts to amusement parks. Procter & Gamble, for instance, builds walk-in virtual stores, and Celebrity Cruises builds virtual ships, to rapidly generate and test ideas.

Virtual reality
A visual form of communication in which images substitute for reality and typically allow the user to respond interactively.

Augmented reality is an extension of virtual reality where the real world is enhanced by the use of technology. Digital information or images are superimposed on an existing image being viewed through "smart" glasses, goggles, or a device such as a smartphone. The use of the yellow line for first downs on televised football games is one such application. So is your GPS location on a smartphone map. Augmented reality assists assembly workers by superimposing digital images of the proper tool or specifications on the task at hand.

Augmented reality
The integration of digital information with the user's environment in real time.

The advances made with CAD/CAM, virtual reality, and augmented reality are indicative of the tremendous challenges and opportunities available to innovators as we accelerate into the digital world. Factories are automated. Digital instructions control automated warehouses. Airplanes constantly download maintenance data. Farm tractors go driverless. Music is available from your handheld device wherever you may be.

Value Analysis

Although value engineering focuses on *preproduction* design and manufacturing issues, value analysis, a related technique, takes place *during* the production process, when it is clear that a new product is a success. **Value analysis** seeks improvements that lead to a better product, a product made more economically, or a product with less environmental impact. The techniques and advantages for value analysis are the same as for value engineering, although minor changes in implementation may be necessary because value analysis is taking place while the product is being produced.

Value analysis
A review of successful products that takes place during the production process.

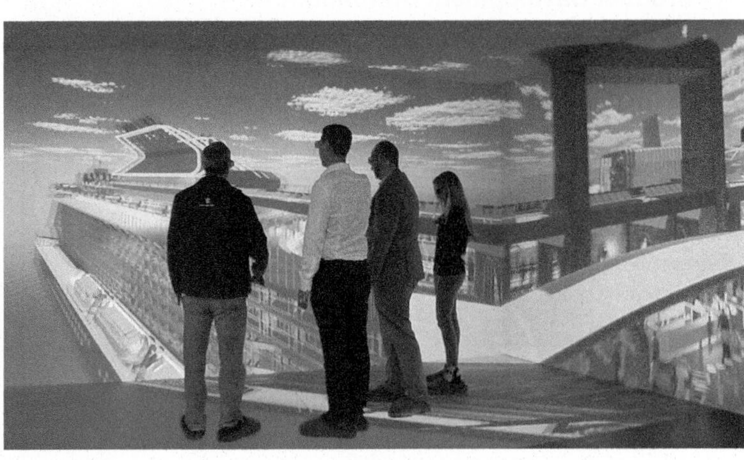

Beverly Amer/Aspenleaf Productions

Celebrity Cruises' Innovation Lab in Miami, Florida, includes the world's largest virtual reality facility. This significant investment allows management to evaluate design features and creative opportunities as design progresses.

Sustainability and Life Cycle Assessment (LCA)

Product design requires that managers evaluate product options. Addressing sustainability and life cycle assessment (LCA) are two ways of doing this. *Sustainability* means meeting the needs of the present without compromising the ability of future generations to meet their needs. An LCA is a formal evaluation of the environmental impact of a product. Both sustainability and LCA are discussed in depth in the supplement to this chapter.

STUDENT TIP Fast communication, rapid technological change, and short product life cycles push product development.

Product Development Continuum

As product life cycles shorten, the need for faster product development increases. And as technological sophistication of products increases, so do the expense and risk. For instance, drug firms invest an average of 12 to 15 years and billions of dollars before receiving regulatory approval for a new drug. And even then, only 1 of 5 will actually be a success. Those operations managers who master this art of product development continually gain on slower product developers. To the swift goes the competitive advantage. This concept is called **time-based competition**.

Time-based competition
Competition based on time; rapidly developing products and moving them to market.

Often, the first company into production may have its product adopted for use in a variety of applications that will generate sales for years. It may become the "standard." Consequently, there is often more concern with getting the product to market than with optimum product design or process efficiency. Even so, rapid introduction to the market may be good management because until competition begins to introduce copies or improved versions, the product can sometimes be priced high enough to justify somewhat inefficient production design and methods.

LO 5.4 *Explain* how time-based competition is implemented by OM

Because time-based competition is so important, instead of developing new products from scratch (which has been the focus thus far in this chapter), a number of other strategies can be used. Figure 5.6 shows a continuum that goes from new, internally developed products (on the lower left) to "alliances." *Enhancements* and *migrations* use the organization's existing product strengths for innovation and therefore are typically faster while at the same time being less risky than developing entirely new products.

Enhancements may include changes in color, size, weight, taste, or features, such as are taking place in fast-food menu items (see the McDonald's description at the beginning of this chapter) or even changes in commercial aircraft. Boeing's enhancements of the 737 since its introduction in 1967 have made the 737 the largest-selling commercial aircraft in history.

Boeing also uses its engineering prowess in air frames to *migrate* from one model to the next. This allows Boeing to speed development while reducing both cost and risk for new designs. This approach is also referred to as building on *product platforms*. Similarly, Volkswagen is using a versatile automobile platform (the MQB chassis) for small to midsize front-wheel-drive cars. This includes VW's Polo, Golf, Passat, Tiguan, and Skoda Octavia, and it may eventually include 44 different vehicles. The advantages are downward pressure on cost as well as faster development. Hewlett-Packard has done the same in the printer business. Enhancements and platform migrations are a way of building on existing expertise, speeding product development, and extending a product's life cycle.

Figure **5.6**

Product Development Continuum

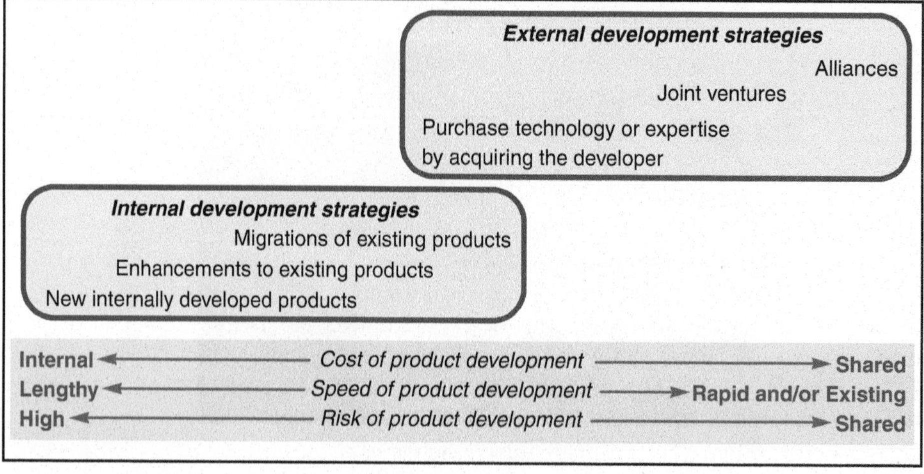

STUDENT TIP Managers seek a variety of approaches to obtain speed to market. As the president of one U.S. firm said: "If I miss one product cycle, I'm dead."

The product development strategies on the lower left of Figure 5.6 are *internal* development strategies, while the three approaches we now introduce can be thought of as *external* development strategies. Firms use both. The external strategies are (1) purchase the technology, (2) establish joint ventures, and (3) develop alliances.

Purchasing Technology by Acquiring a Firm

Microsoft and Cisco Systems are examples of companies on the cutting edge of technology that often speed development by *acquiring entrepreneurial firms* that have already developed the technology that fits their mission. The issue then becomes fitting the purchased organization, its technology, its product lines, and its culture into the buying firm, rather than a product development issue.

Joint Ventures

In an effort to reduce the weight of new cars, GM is in a joint venture with Tokyo-based Teijin Ltd. to bring lightweight carbon fiber to GM's customers. **Joint ventures** such as this are combined ownership, usually between just two firms, to form a new entity. Ownership can be 50–50, or one owner can assume a larger portion to ensure tighter control. Joint ventures are often appropriate for exploiting specific product opportunities that may not be central to the firm's mission. Such ventures are more likely to work when the risks are known and can be equitably shared.

Joint ventures
Firms establishing joint ownership to pursue new products or markets.

Alliances

When new products are central to the mission, but substantial resources are required and sizable risk is present, then alliances may be a good strategy for product development. **Alliances** are cooperative agreements that allow firms to remain independent but use complementing strengths to pursue strategies consistent with their individual missions. Alliances are particularly beneficial when the products to be developed also have technologies that are in ferment. For example, Microsoft is pursuing alliances with a variety of companies to deal with the convergence of computing, the Internet, and television broadcasting. Alliances in this case are appropriate because the technological unknowns, capital demands, and risks are significant. Similarly, three firms, Mercedes-Benz, Ford Motor, and Ballard Power Systems, have formed an alliance to develop "green" cars powered by fuel cells. Alliances are much more difficult to achieve and maintain than joint ventures because of the ambiguities associated with them. It may be helpful to think of an alliance as an incomplete contract between the firms. The firms remain separate.

Alliances
Cooperative agreements that allow firms to remain independent, but pursue strategies consistent with their individual missions.

Enhancements, migration, acquisitions, joint ventures, and alliances are all strategies for speeding product development. Moreover, they typically reduce the risk associated with product development while enhancing the human and capital resources available.

Defining a Product

STUDENT TIP
Before anything can be produced, a product's functions and attributes must be defined.

Once new goods or services are selected for introduction, they must be defined. First, a good or service is defined in terms of its *functions*—that is, what it is to *do*. The product is then designed, and the firm determines how the functions are to be achieved. Management typically has a variety of options as to how a product should achieve its functional purpose. For instance, when an alarm clock is produced, aspects of design such as the color, size, or location of buttons may make substantial differences in ease of manufacture, quality, and market acceptance.

LO 5.5 *Describe* how goods and services are defined by OM

Rigorous specifications of a product are necessary to ensure efficient production. Equipment, layout, and human resources cannot be determined until the product is defined, designed, and documented. Therefore, every organization needs documents to define its products. This is true of everything from meat patties, to cheese, to computers, to a medical procedure. In the case of cheese, a written specification is typical. Indeed, written specifications or standard grades exist and provide the definition for many products. For instance, Monterey Jack cheese has a written description that specifies the characteristics necessary for each Department of Agriculture grade. A portion of the Department of Agriculture grade for Monterey Jack Grade AA is shown in Figure 5.7. Similarly, McDonald's Corp. has 60 specifications for potatoes that are to be made into french fries.

Figure 5.7

Monterey Jack

A portion of the general requirements for the U.S. grades of Monterey Jack cheese is shown here.

Source: Based on 58.2469 Specifications for U.S. grades of Monterey (Monterey Jack) cheese, (May 10, 1996).

§ 58.2469 Specifications for U.S. grades of Monterey (Monterey Jack) cheese

(a) *U.S. grade AA.* Monterey Cheese shall conform to the following requirements:

(1) *Flavor.* Is fine and highly pleasing, free from undesirable flavors and odors. May possess a very slight acid or feed flavor.

(2) *Body and texture.* A plug drawn from the cheese shall be reasonably firm. It shall have numerous small mechanical openings evenly distributed throughout the plug. It shall not possess sweet holes, yeast holes, or other gas holes.

(3) *Color.* Shall have a natural, uniform, bright, attractive appearance.

(4) *Finish and appearance—bandaged and paraffin-dipped.* The rind shall be sound, firm, and smooth, providing a good protection to the cheese.

Code of Federal Regulation, Parts 53 to 109, General Service Administration.

Philip Dowell/Dorling Kindersley ltd/ Alamy Stock Photo

Engineering drawing
A drawing that shows the dimensions, tolerances, materials, and finishes of a component.

Bill of material (BOM)
A list of the hierarchy of components, their description, and the quantity of each required to make one unit of a product.

Most manufactured items, as well as their components, are defined by a drawing, usually referred to as an engineering drawing. An **engineering drawing** shows the dimensions, tolerances, materials, and finishes of a component. The engineering drawing will be an item on a bill of material. An engineering drawing is shown in Figure 5.8. The **bill of material (BOM)** lists the hierarchy of components, their description, and the quantity of each required to make one unit of a product. A bill of material for a manufactured item is shown in Figure 5.9(a). Note that subassemblies and components (lower-level items) are indented at each level to indicate their subordinate position. An engineering drawing shows how to make one item on the bill of material.

In the food-service industry, bills of material manifest themselves in *portion-control standards*. The portion-control standard for Hard Rock Cafe's hickory BBQ bacon cheeseburger is shown in Figure 5.9(b). In a more complex product, a bill of material is referenced on other bills of material of which they are a part. In this manner, subunits (subassemblies) are part of the next higher unit (their parent bill of material) that ultimately makes a final product. In addition to being defined by written specifications, portion-control documents, or bills of material, products can be defined in other ways. For example, products such as chemicals, paints, and petroleums may be defined by formulas or proportions that describe how they are to be made. Movies are defined by storyboards and scripts, and insurance coverage by legal documents known as *policies*.

Make-or-Buy Decisions

Make-or-buy decision
The choice between producing a component or a service and purchasing it from an outside source.

For many components of products, firms have the option of producing the components themselves or purchasing them from outside sources. Choosing between these options is known as the *make-or-buy decision*. The **make-or-buy decision** distinguishes between what the firm wants

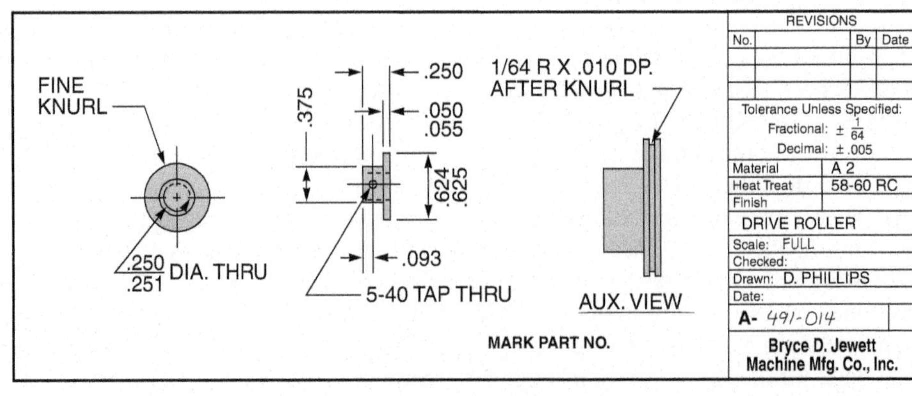

Figure 5.8

Engineering Drawings Such as This One Show Dimensions, Tolerances, Materials, and Finishes

(a) **Bill of Material for a Panel Weldment**

NUMBER	DESCRIPTION	QTY
A 60-71	PANEL WELDM'T	1
A 60-7	LOWER ROLLER ASSM.	1
R 60-17	ROLLER	1
R 60-428	PIN	1
P 60-2	LOCKNUT	1
A 60-72	GUIDE ASSM. REAR	1
R 60-57-1	SUPPORT ANGLE	1
A 60-4	ROLLER ASSEM.	1
02-50-1150	BOLT	1
A 60-73	GUIDE ASSM. FRONT	1
A 60-74	SUPPORT WELDM'T	1
R 60-99	WEAR PLATE	1
02-50-1150	BOLT	1

(b) **Hard Rock Cafe's Hickory BBQ Bacon Cheeseburger**

DESCRIPTION	QTY
Bun	1
Hamburger patty	8 oz.
Cheddar cheese	2 slices
Bacon	2 strips
BBQ onions	1/2 cup
Hickory BBQ sauce	1 oz.
Burger set	
Lettuce	1 leaf
Tomato	1 slice
Red onion	4 rings
Pickle	1 slice
French fries	5 oz.
Seasoned salt	1 tsp.
11-inch plate	1
HRC flag	1

Figure 5.9

Bills of Material Take Different Forms in (a) Manufacturing Plants and (b) Restaurants, but in Both Cases, the Product Must Be Defined

STUDENT TIP

Hard Rock's recipe here serves the same purpose as a bill of material in a factory: It defines the product for production.

to *produce* and what it wants to *purchase*. Because of variations in quality, cost, and delivery schedules, the make-or-buy decision is critical to product definition. Many items can be purchased as a "standard item" produced by someone else. Examples are the standard bolts listed twice on the bill of material shown in Figure 5.9(a), for which there will be Society of Automotive Engineers (SAE) specifications. Therefore, there typically is no need for the firm to duplicate this specification in another document.

Group Technology

Engineering drawings may also include codes to facilitate group technology. **Group technology** identifies components by a coding scheme that specifies size, shape, and the type of processing (such as drilling). This facilitates standardization of materials, components, and processes as well as the identification of families of parts. As families of parts are identified, activities and machines can be grouped to minimize setups, routings, and material handling. An example of how families of parts may be grouped is shown in Figure 5.10. Group technology provides a systematic way to review a family of components to see if an existing component might suffice on a new project. Using existing or standard components eliminates all the costs connected with the design and development of the new part, which is a major cost reduction.

Group technology
A product and component coding system that specifies the size, shape, and type of processing; it allows similar products to be grouped.

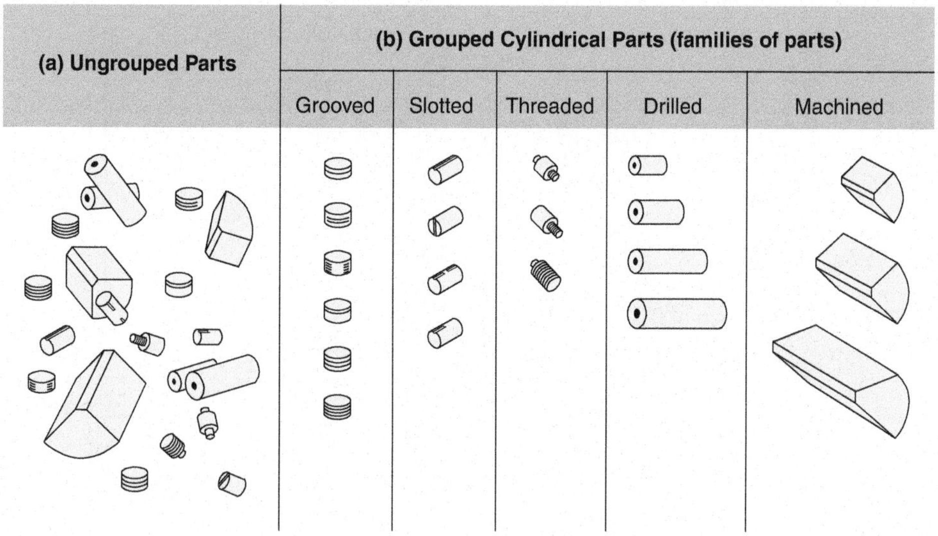

(a) Ungrouped Parts	(b) Grouped Cylindrical Parts (families of parts)				
	Grooved	Slotted	Threaded	Drilled	Machined

Figure 5.10

A Variety of Group Technology Coding Schemes Move Manufactured Components from (a) Ungrouped to (b) Grouped (Families of Parts)

STUDENT TIP Production personnel need clear, specific documents to help them make the product.

Documents for Production

Once a product is selected, designed, and ready for production, production is assisted by a variety of documents. We will briefly review some of these.

Assembly drawing
An exploded view of the product.

An **assembly drawing** simply shows an exploded view of the product. An assembly drawing is usually a three-dimensional drawing, known as an *isometric drawing*; the relative locations of components are drawn in relation to each other to show how to assemble the unit [see Figure 5.11(a)].

Assembly chart
A graphic means of identifying how components flow into subassemblies and final products.

The **assembly chart** shows in schematic form how a product is assembled. Manufactured components, purchased components, or a combination of both may be shown on an assembly chart. The assembly chart identifies the point of production at which components flow into subassemblies and ultimately into a final product. An example of an assembly chart is shown in Figure 5.11(b).

Route sheet
A listing of the operations necessary to produce a component with the material specified in the bill of material.

The **route sheet** lists the operations necessary to produce the component with the material specified in the bill of material. The route sheet for an item will have one entry for each operation to be performed on the item. When route sheets include specific methods of operation and labor standards, they are often known as *process sheets*.

Work order
An instruction to make a given quantity of a particular item.

The **work order** is an instruction to make a given quantity of a particular item, usually to a given schedule. The ticket that a server writes in your favorite restaurant is a work order. In a hospital or factory, the work order is a more formal document that provides authorization to draw items from inventory, to perform various functions, and to assign personnel to perform those functions.

Engineering change notice (ECN)
A correction or modification of an engineering drawing or bill of material.

Configuration management
A system by which a product's planned and changing components are accurately identified.

Engineering change notices (ECNs) change some aspect of the product's definition or documentation, such as an engineering drawing or a bill of material. For a complex product that has a long manufacturing cycle, such as a Boeing 777, the changes may be so numerous that no two 777s are built exactly alike—which is indeed the case. Such dynamic design change has fostered the development of a discipline known as *configuration management*, which is concerned with product identification, control, and documentation. **Configuration management** is the system by which a product's planned and changing configurations are accurately identified and for which control and accountability of change are maintained.

Product Life-Cycle Management (PLM)

Product life-cycle management (PLM)
Software programs that tie together many phases of product design and manufacture.

Product life-cycle management (PLM) is an umbrella of software programs that attempts to bring together phases of product design and manufacture—tying together many of the techniques discussed in the prior two sections, *Defining a Product* and *Documents for Production*.

Figure 5.11
Assembly Drawing and Assembly Chart
Source: Assembly drawing and assembly chart produced by author.

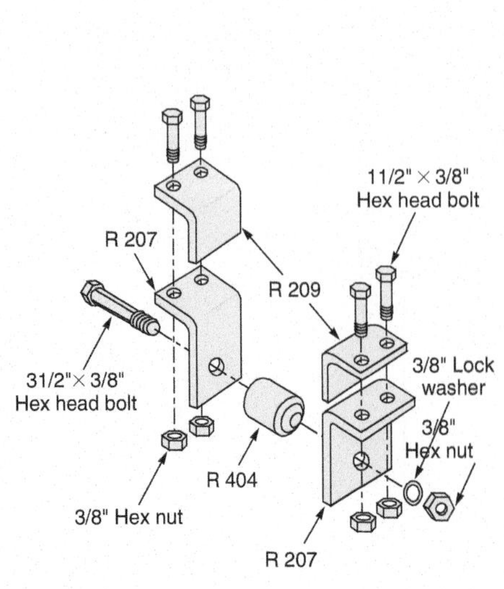

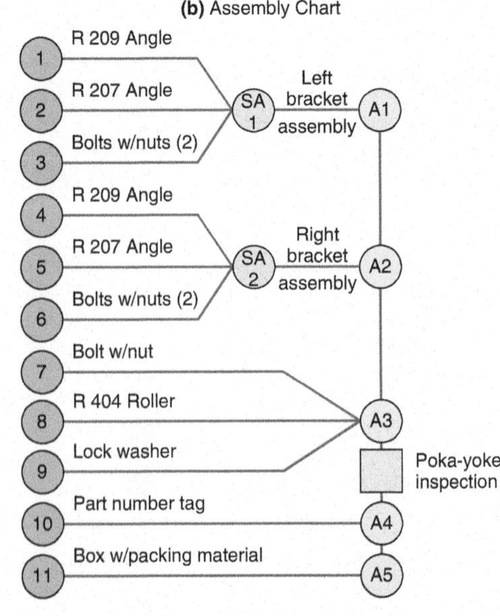

David R. Frazier Photolibrary, Inc. /Alamy Stock Photo

David R. Frazier Photolibrary, Inc. /Alamy Stock Photo

Each year the JR Simplot potato-processing facility in Caldwell, Idaho, produces billions of french fries for quick-service restaurant chains and many other customers, both domestically and overseas (left photo). Sixty specifications (including a special blend of frying oil, a unique steaming process, and exact time and temperature for prefrying and drying) define how these potatoes become french fries. Further, 40% of all french fries must be 2 to 3 inches long, 40% must be more than 3 inches long, and a few stubby ones constitute the final 20%. Quality control personnel use a Vernier caliper to measure the fries (right photo).

The idea behind PLM software is that product design and manufacture decisions can be performed more creatively, faster, and more economically when the data are integrated and consistent.

Although there is not one standard, PLM products often start with product design (CAD/CAM); move on to design for manufacture and assembly (DFMA); and then into product routing, materials, layout, assembly, maintenance, and even environmental issues. Integration of these tasks makes sense because many of these decision areas require overlapping pieces of data. PLM software is now a tool of many large organizations, including Lockheed Martin, GE, Procter & Gamble, Toyota, and Boeing. PLM is now finding its way into medium and small manufacture as well.

LO 5.6 *Describe* the documents needed for production

Shorter life cycles, more technologically challenging products, more regulations regarding materials and manufacturing processes, and more environmental issues all make PLM an appealing tool for operations managers. Major vendors of PLM software include SAP PLM (**www.mySAP.com**), Parametric Technology Corp. (**www.ptc.com**), Siemens (**www.plm.automation.siemens.com**), and Proplanner (**www.proplanner.com**).

Service Design

STUDENT TIP

Services also need to be defined and documented.

Much of our discussion so far has focused on what we call *tangible products*—that is, goods. On the other side of the product coin are, of course, services. Service industries include banking, insurance, health care, transportation, hospitality, entertainment, and communication. Note that many aspects of the service product are intangible. The products offered by service firms range from a medical procedure that leaves only the tiniest scar after an appendectomy, to the joy of a trip to Disneyland, or a delightful vacation with Celebrity Cruises. The success of the appendectomy may depend on how the product is defined in the operating procedure, the training and skill of the surgeon, and the OR staff. Disney's product is the selection and availability of exciting rides, cleanliness of the park, entertainment, and so forth. And the vacation with Celebrity Cruises depends on how its product is defined by such things as ship amenities, room options, food quality, and port-of-call selection (see the case at the end of this chapter and the video, *Celebrity Cruises Designs a New Ship*).

VIDEO 5.2
Celebrity Cruises Designs a New Ship

Designing successful services can be challenging because they often have a strong element of customer interaction. This suggests some creative ways of making services more efficient.

Designing More Efficient Services

Service productivity is notoriously low, in part because of customer involvement in the *design* or *delivery* of the service, or both. This complicates the product design challenge. We will now discuss a number of ways to increase service efficiency and, among these, several ways to limit this interaction.

Limit the Options Because customers may participate in the design of the service (e.g., for a funeral or a hairstyle), design specifications may take the form of everything from a menu (in a restaurant), to a list of options (for a funeral), to a verbal description (a hairstyle). However, by providing a list of options (in the case of the funeral) or a series of photographs (in the case of the hairstyle), ambiguity may be reduced. An early resolution of the product's definition can aid efficiency as well as aid in meeting customer expectations.

Delay Customization Design the product so that *customization is delayed* as late in the process as possible. This is the way a hair salon operates. Although shampoo and condition are done in a standard way with lower-cost labor, the color and styling (customizing) are done last. It is also the way most restaurants operate: How would you like that cooked? Which dressing would you prefer with your salad?

Modularization *Modularize* the service so that customization takes the form of changing modules. This strategy allows for "custom" services to be designed as standard modular entities. Just as modular design allows you to buy a high-fidelity sound system with just the features you want, modular flexibility also lets you buy meals, clothes, and insurance on a mix-and-match (modular) basis. Investments (portfolios of stocks and bonds) and education (college curricula) are examples of how the modular approach can be used to customize a service.

Automation Divide the service into small parts, and identify those parts that lend themselves to automation. For instance, by isolating check-cashing activity via ATM, banks have been very effective at designing a product that both increases customer service and reduces costs. Similarly, airlines have moved to ticketless service via kiosks. A technique such as kiosks reduces both costs and lines at airports—thereby increasing customer

OM in Action Amazon Pushes Product Design[3]

Bring your Amazon smartphone app with you to get through the entry stands at the new Amazon Go Store. The digital-driven store is a radical innovation... no checkout counters, no lines, and no sales clerks. Once inside the new 1,800-square-foot cashier-less convenience store, customers become a digital 3D object with hundreds of cameras scanning the shelves and tracking their interaction with items.

The shopping experience is unique. Shoppers twist, turn, and bend to examine items—a real challenge for designers. Then to complicate matters more, customers tend to cover unique aspects of the product with their hand, another challenge. Nevertheless, designers developed the computer-vision and machine-learning algorithms necessary to track shoppers and automatically charge their Amazon account for the items selected.

There are no shopping carts or baskets inside the Amazon Go Store. Instead, customers put items directly into a shopping bag. Every time customers grab an item off a shelf, the product is automatically put into the shopping cart of their online account. If customers put the item back on the shelf, Amazon removes it from their virtual cart. No cashiers or clunky self-checkout kiosks to slow the process.

Paul Christian Gordon/Alamy Stock Photo

Amazon Go Store's opening was delayed almost a year as the technology proved more difficult to master than expected. Although a designer's challenge, it is a technological marvel pushing the frontier of the new digital world.

satisfaction—and providing a win–win "product" design. The Amazon Go Store is another such innovation in the service sector (see the *OM in Action* box, "Amazon Pushes Product Design").

Moment of Truth High customer interaction means that in the service industry there is a *moment of truth* when the relationship between the provider and the customer is crucial. At that moment, the customer's satisfaction with the service is defined. The moment of truth is the moment that exemplifies, enhances, or detracts from the customer's expectations. That moment may be as simple as a smile from a Starbucks barista or having the checkout clerk focus on you rather than talking over his shoulder to the clerk at the next counter. Moments of truth can occur when you order at McDonald's, get a haircut, or register for college courses. The operations manager's task is to identify moments of truth and design operations that meet or exceed the customer's expectations.

Documents for Services

Because of the high customer interaction of most services, the documents for moving the product to production often take the form of explicit *job instructions* or *script*. For instance, regardless of how good a bank's products may be in terms of checking, savings, trusts, loans, mortgages, and so forth, if the interaction between participants is not done well, the product may be poorly received. Example 2 shows the kind of documentation a bank may use to move a product (drive-up window banking) to "production." Similarly, a telemarketing service has the product design communicated to production personnel in the form of a *telephone script*, while a *manuscript* is used for books, and a *storyboard* is used for movie and TV production.

Example 2

SERVICE DOCUMENTATION FOR PRODUCTION

First Bank Corp. wants to ensure effective delivery of service to its drive-up customers.

APPROACH ▶ Develop a "production" document for the tellers at the drive-up window that provides the information necessary to do an effective job.

SOLUTION ▶

Documentation for Tellers at Drive-up Windows

Tellers face communication challenges with drive-up customers because of the physical distance and machinery between them. Tellers should follow these guidelines:

- Say "please" and "thank you" in all conversations.
- Be discrete when speaking into the microphone, as others may hear the conversation.
- Give customers written instructions if they need to fill out forms you give them.
- Look directly at customers if there is a line of sight.
- If the customer needs to park and enter the bank, apologize for the inconvenience.

Source: Ideas adapted from Teller Operations, The Institute of Financial Education, Chicago.

INSIGHT ▶ By providing documentation in the form of a script/guideline for tellers, the likelihood of effective communication and a good product/service is improved.

LEARNING EXERCISE ▶ Modify the guidelines above to show how they would be different for a drive-through restaurant. [Answer: Written instructions, marking lines to be completed, or coming into the store are seldom necessary, but techniques for making change and proper transfer of the order should be included.]

RELATED PROBLEM ▶ 5.11

Application of Decision Trees to Product Design

STUDENT TIP A decision tree is a great tool for thinking through a problem.

Decision trees can be used for new-product decisions as well as for a wide variety of other management problems when uncertainty is present. They are particularly helpful when there are a series of decisions and various outcomes that lead to *subsequent* decisions followed by other outcomes. (Module A provides a complete introduction to the topic.)

To form a decision tree, we use the following procedure:

LO 5.7 *Apply* decision trees to product issues

1. Be sure that all possible alternatives and states of nature (beginning on the left and moving right) are included in the tree. This includes an alternative of "doing nothing."
2. Payoffs are entered at the end of the appropriate branch. This is the place to develop the payoff of achieving this branch.
3. The objective is to determine the expected monetary value (EMV) of each course of action. We accomplish this by starting at the end of the tree (the right-hand side) and working toward the beginning of the tree (the left), calculating values at each step and "pruning" alternatives that are not as good as others from the same node.

Example 3 shows the use of a decision tree applied to product design.

Example 3

DECISION TREE APPLIED TO PRODUCT DESIGN

Silicon, Inc., a semiconductor manufacturer, is investigating the possibility of producing and marketing a microprocessor. Undertaking this project will require either purchasing a sophisticated CAD system or hiring and training several additional engineers. The market for the product could be either favorable or unfavorable. Silicon, Inc., of course, has the option of not developing the new product at all.

With favorable acceptance by the market, sales would be 25,000 processors selling for \$100 each. With unfavorable acceptance, sales would be only 8,000 processors selling for \$100 each. The cost of CAD equipment is \$500,000, but that of hiring and training three new engineers is only \$375,000. However, manufacturing costs should drop from \$50 each when manufacturing without CAD to \$40 each when manufacturing with CAD.

The probability of favorable acceptance of the new microprocessor is .40; the probability of unfavorable acceptance is .60.

APPROACH ▶ Use of a decision tree seems appropriate as Silicon, Inc., has the basic ingredients: a choice of decisions, probabilities, and payoffs.

SOLUTION ▶ In Figure 5.12 we draw a decision tree with a branch for each of the three decisions, assign the respective probabilities and payoff for each branch, and then compute the respective EMVs. The EMVs have been circled at each step of the decision tree. For the top branch:

$$\begin{aligned}\text{EMV (Purchase CAD system)} &= (.4)(\$1{,}000{,}000) + (.6)(-\$20{,}000)\\ &= \$388{,}000\end{aligned}$$

This figure represents the results that will occur if Silicon, Inc., purchases CAD.

The expected value of hiring and training engineers is the second series of branches:

$$\begin{aligned}\text{EMV (Hire / train engineers)} &= (.4)(\$875{,}000) + (.6)(\$25{,}000)\\ &= \$365{,}000\end{aligned}$$

The EMV of doing nothing is \$0.

Because the top branch has the highest expected monetary value (an EMV of \$388,000 vs. \$365,000 vs. \$0), it represents the best decision. Management should purchase the CAD system.

INSIGHT ▶ Use of the decision tree provides both objectivity and structure to our analysis of the Silicon, Inc., decision.

LEARNING EXERCISE ▶ If Silicon, Inc., thinks the probabilities of high sales and low sales may be equal, at .5 each, what is the best decision? [Answer: Purchase CAD remains the best decision, but with an EMV of \$490,000.]

RELATED PROBLEMS ▶ 5.18–5.27

ACTIVE MODEL 5.1 This example is further illustrated in Active Model 5.1 found online.

Figure 5.12
Decision Tree for Development of a New Product

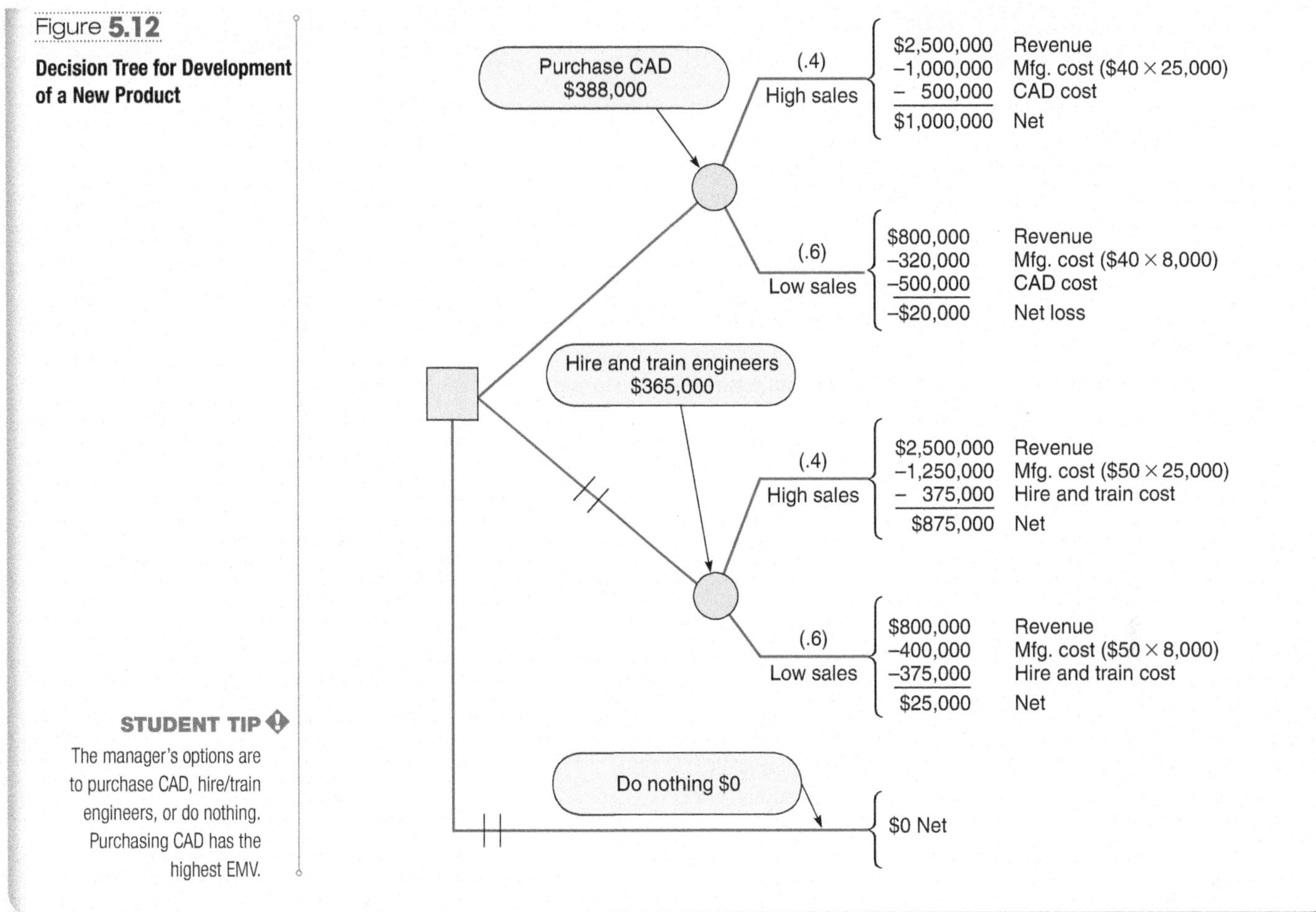

STUDENT TIP The manager's options are to purchase CAD, hire/train engineers, or do nothing. Purchasing CAD has the highest EMV.

Transition to Production

STUDENT TIP One of the arts of management is knowing when a product should move from development to production.

Eventually, a product, whether a good or service, has been selected, designed, and defined. It has progressed from an idea to a functional definition, and then perhaps to a design. Now, management must make a decision as to further development and production or termination of the product idea. One of the arts of management is knowing when to move a product from development to production; this move is known as *transition to production*. The product development staff is always interested in making improvements in a product. Because this staff tends to see product development as evolutionary, they may never have a completed product, but as we noted earlier, the cost of late product introduction is high. Although these conflicting pressures exist, management must make a decision—more development or production.

Once this decision is made, there is usually a period of trial production to ensure that the design is indeed producible. This is the manufacturability test. This trial also gives the operations staff the opportunity to develop proper tooling, quality control procedures, and training of personnel to ensure that production can be initiated successfully. Finally, when the product is deemed both marketable and producible, line management will assume responsibility.

To ensure that the transition from development to production is successful, some companies appoint a *project manager*; others use *product development teams*. Both approaches allow a wide range of resources and talents to be brought to bear to ensure satisfactory production of a product that is still in flux. A third approach is *integration of the product development and manufacturing organizations*. This approach allows for easy shifting of resources between the two organizations as needs change. The operations manager's job is to make the transition from R&D to production seamless.

Summary

Effective product strategy requires selecting, designing, and defining a product and then transitioning that product to production. Only when this strategy is carried out effectively can the production function contribute its maximum to the organization. The operations manager must build a product development system that has the ability to conceive, design, and produce products that will yield a competitive advantage for the firm. As products move through their life cycle (introduction, growth, maturity, and decline), the options that the operations manager should pursue change. Both manufactured and service products have a variety of techniques available to aid in performing this activity efficiently.

Written specifications, bills of material, and engineering drawings aid in defining products. Similarly, assembly drawings, assembly charts, route sheets, and work orders are often used to assist in the actual production of the product. Once a product is in production, value analysis is appropriate to ensure maximum product value. Engineering change notices and configuration management provide product documentation.

Key Terms

Product decision (p. 161)
Product-by-value analysis (p. 163)
Quality function deployment (QFD) (p. 164)
House of quality (p. 164)
Product development teams (p. 168)
Concurrent engineering (p. 168)
Manufacturability and value engineering (p. 168)
Robust design (p. 169)
Modular design (p. 169)
Computer-aided design (CAD) (p. 169)
Design for manufacture and assembly (DFMA) (p. 169)
Standard for the exchange of product data (STEP) (p. 169)
Computer-aided manufacturing (CAM) (p. 169)
Additive manufacturing (p. 170)
Virtual reality (p. 171)
Augmented reality (p. 171)
Value analysis (p. 171)
Time-based competition (p. 172)
Joint ventures (p. 173)
Alliances (p. 173)
Engineering drawing (p. 174)
Bill of material (BOM) (p. 174)
Make-or-buy decision (p. 174)
Group technology (p. 175)
Assembly drawing (p. 176)
Assembly chart (p. 176)
Route sheet (p. 176)
Work order (p. 176)
Engineering change notice (ECN) (p. 176)
Configuration management (p. 176)
Product life-cycle management (PLM) (p. 176)

Ethical Dilemma

Monica Lopez, president of Lopez Toy Company, Inc., in Oregon, has just reviewed the design of a new pull-toy locomotive for 1- to 3-year-olds. Monica's design and marketing staff are very enthusiastic about the market for the product and the potential of follow-on circus train cars. The sales manager is looking forward to a very good reception at the annual toy show in Dallas next month. Monica, too, is delighted, as she is faced with a layoff if orders do not improve.

Monica's production people have worked out the manufacturing issues and produced a successful pilot run. However, the quality assessment staff suggests that under certain conditions, a hook to attach cars to the locomotive and the crank for the bell can be broken off. This is an issue because children can choke on small parts such as these. In the quality test, 1- to 3-year-olds were unable to break off these parts; there were *no* failures. But when the test simulated the force of an adult tossing the locomotive into a toy box or a 5-year-old throwing it on the floor, there were failures. The estimate is that one of the two parts can be broken off 4 times out of 100,000 throws. Neither the design nor the material people know how to make the toy safer and still perform as designed. The failure rate is low and certainly normal for this type of toy, but not at the Six Sigma level that Monica's firm strives for. And, of course, someone, someday may sue. A child choking on the broken part is a serious matter. Also, Monica was recently reminded in a discussion with legal counsel that U.S. case law suggests that new products may not be produced if there is "actual or foreseeable knowledge of a problem" with the product.

The design of successful, ethically produced new products, as suggested in this chapter, is a complex task. What should Monica do?

Discussion Questions

1. Why is it necessary to document a product explicitly?
2. What techniques do we use to define a product?
3. In what ways is product strategy linked to product decisions?
4. Once a product is defined, what documents are used to assist production personnel in its manufacture?
5. What is time-based competition?
6. Describe the differences between joint ventures and alliances.
7. Describe four organizational approaches to product development. Which of these is generally thought to be best?
8. Explain what is meant by robust design.
9. What are three specific ways in which computer-aided design (CAD) benefits the design engineer?
10. What information is contained in a bill of material?
11. What information is contained in an engineering drawing?
12. What information is contained in an assembly chart? In a process sheet?
13. Explain what is meant in service design by the "moment of truth."
14. Explain how the house of quality translates customer desires into product/service attributes.
15. What strategic advantages does computer-aided design provide?
16. Describe several applications of virtual reality in product design.
17. What is the difference between virtual reality and augmented reality?
18. Why are documents for service useful? Provide examples of four types.

Solved Problem

SOLVED PROBLEM 5.1

Sarah King, president of King Electronics, Inc., has two design options for her new line of high-resolution monitors for CAD workstations. The life cycle for this model is 100,000 units.

Design option A has a .90 probability of yielding 60 good monitors per 100 and a .10 probability of yielding 65 good monitors per 100. This development will cost $1,000,000.

Design option B has a .80 probability of yielding 64 good units per 100 and a .20 probability of yielding 59 good units per 100. This development will cost $1,350,000.

Good or bad, each monitor will cost $75. Each good monitor will sell for $150. Bad monitors are destroyed and have no salvage value. We ignore any disposal costs in this problem.

SOLUTION

We draw the decision tree to reflect the two decisions and the probabilities associated with each decision. We then determine the payoff associated with each branch. The resulting tree is shown in Figure 5.13.

For design A:

$$\text{EMV(design A)} = (.9)(\$500,000) + (.1)(\$1,250,000) = \$575,000$$

For design B:

$$\text{EMV(design B)} = (.8)(\$750,000) + (.2)(\$0) = \$600,000$$

The highest payoff is design option B, at $600,000.

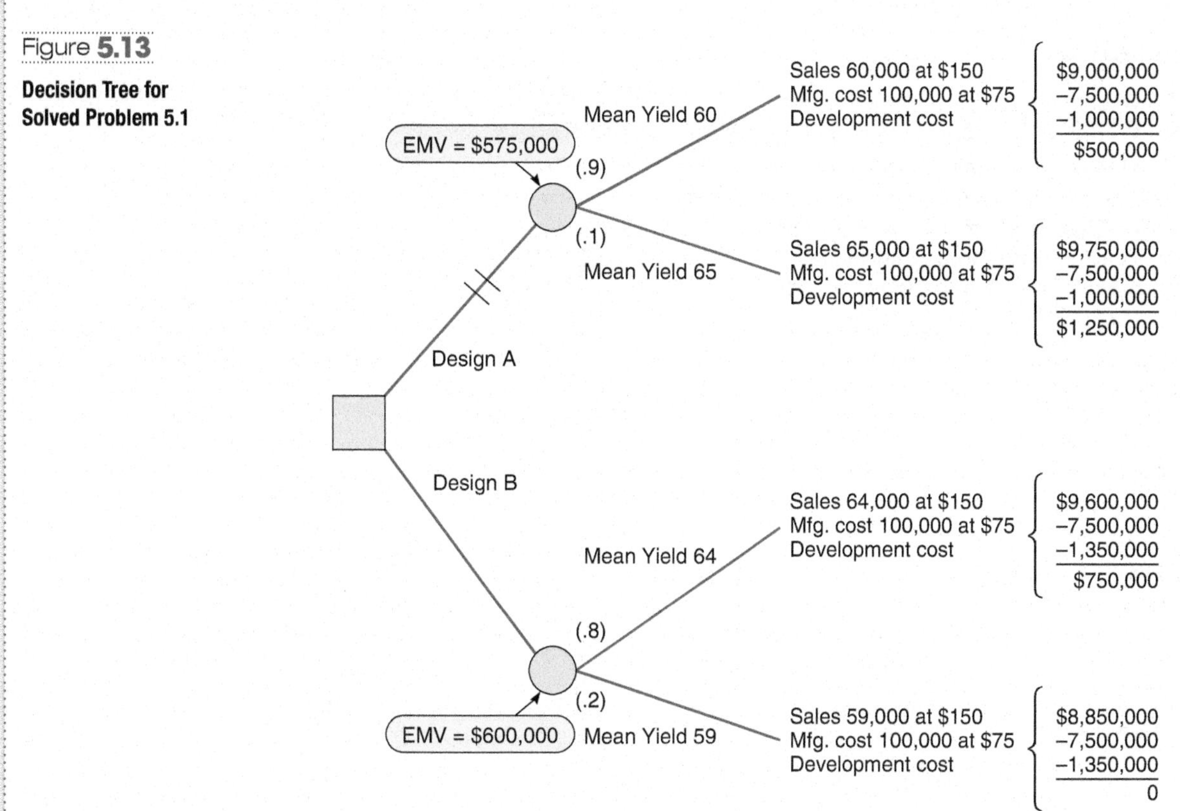

Figure 5.13 **Decision Tree for Solved Problem 5.1**

Problems

Note: **PX** means the problem may be solved with POM for Windows and/or Excel OM.

Problems 5.1–5.3 relate to Goods and Services Selection

••• **5.1** Prepare a product-by-value analysis for the following products, and given the position in its life cycle, identify the issues likely to confront the operations manager and his or her possible actions. Product Alpha has annual sales of 1,000 units and a contribution of $2,500; it is in the introductory stage. Product Bravo has annual sales of 1,500 units and a contribution of $3,000; it is in the growth stage. Product Charlie has annual sales of 3,500 units and a contribution of $1,750; it is in the decline stage.

•• **5.2** Given the contribution made on each of the three products in the following table and their position in the life cycle, identify a reasonable operations strategy for each:

PRODUCT	PRODUCT CONTRIBUTION (% OF SELLING PRICE)	COMPANY CONTRIBUTION (%: TOTAL ANNUAL CONTRIBUTION DIVIDED BY TOTAL ANNUAL SALES)	POSITION IN LIFE CYCLE
Smart watch	30	40	Introduction
Tablet	30	50	Growth
Hand calculator	50	10	Decline

• **5.3** Perform a "product-by-value" analysis on products A, B, C, D, and E.

	A	B	C	D	E
Unit Contribution	$0.75	$0.33	$1.25	$0.85	$0.75
Total Contribution	$63,000	$82,000	$95,000	$115,000	$57,000

Problems 5.4–5.8 relate to Product Development

•• **5.4** Construct a house of quality matrix for a wristwatch. Be sure to indicate specific customer wants that you think the general public desires. Then complete the matrix to show how an operations manager might identify specific attributes that can be measured and controlled to meet those customer desires.

•• **5.5** Using the house of quality, pick a real product (a good or service) and analyze how an existing organization satisfies customer requirements.

•• **5.6** Prepare a house of quality for a mousetrap.

•• **5.7** Conduct an interview with a prospective purchaser of a new bicycle and translate the customer's *wants* into the specific *hows* of the firm.

•••• **5.8** Using the house of quality sequence, as described in Figure 5.4, determine how you might deploy resources to achieve the desired quality for a product or service whose production process you understand.

Problems 5.9–5.17 relate to Defining a Product

•• **5.9** Prepare a bill of material for (a) a pair of eyeglasses and its case or (b) a fast-food sandwich (visit a local sandwich shop like Subway, McDonald's, Blimpie, Quizno's; perhaps a clerk or the manager will provide you with details on the quantity or weight of various ingredients—otherwise, estimate the quantities).

•• **5.10** Draw an assembly chart for a pair of eyeglasses and its case.

•• **5.11** Prepare a script for telephone callers at the university's annual "phone-a-thon" fundraiser.

•• **5.12** Prepare an assembly chart for a table lamp.

•• **5.13** Prepare a bill of material and an assembly chart for a salad of your own choosing.

•• **5.14** Prepare a bill of material for a wooden pencil, complete with eraser.

•• **5.15** Prepare a bill of material for the table illustrated:

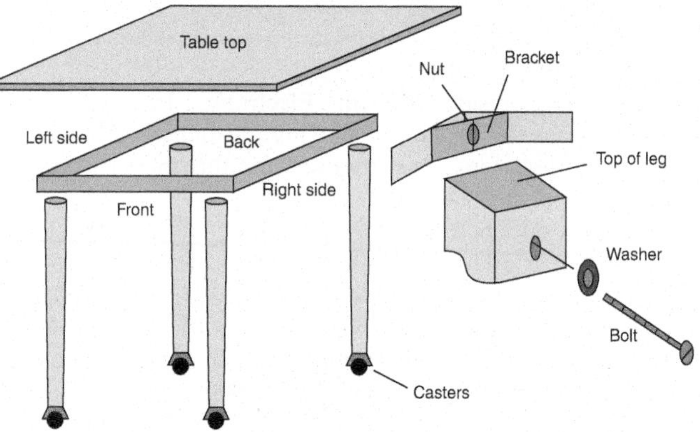

•• **5.16** Prepare a bill of material for a computer mouse.

•• **5.17** Prepare a bill of material for a mechanical pencil.

Problems 5.18–5.27 relate to the Application of Decision Trees to Product Design

•• **5.18** The product design group of Iyengar Electric Supplies, Inc., has determined that it needs to design a new series of switches. It must decide on one of three design strategies. The market forecast is for 200,000 units. The better and more sophisticated the design strategy and the more time spent on value engineering, the less will be the variable cost. The chief of engineering design, Dr. W. L. Berry, has decided that the following costs are a good estimate of the initial and variable costs connected with each of the three strategies:

a) *Low-tech:* A low-technology, low-cost process consisting of hiring several new junior engineers. This option has a fixed cost of $45,000 and variable-cost probabilities of .3 for $.55 each, .4 for $.50, and .3 for $.45.
b) *Subcontract:* A medium-cost approach using a good outside design staff. This approach would have a fixed cost of $65,000 and variable-cost probabilities of .7 of $.45, .2 of $.40, and .1 of $.35.
c) *High-tech:* A high-technology approach using the very best of the inside staff and the latest computer-aided design technology. This approach has a fixed cost of $75,000 and variable-cost probabilities of .9 of $.40 and .1 of $.35.

What is the best decision based on an expected monetary value (EMV) criterion? (*Note:* We want the lowest EMV, as we are dealing with costs in this problem.) **PX**

• • **5.19** MacDonald Products, Inc., of Clarkson, New York, has the option of (a) proceeding immediately with production of a new top-of-the-line stereo TV that has just completed prototype testing or (b) having the value analysis team complete a study. If Tyrone Martin, VP for operations, proceeds with the existing prototype (option a), the firm can expect sales to be 100,000 units at $550 each, with a probability of .6, and a .4 probability of 75,000 at $550. If, however, he uses the value analysis team (option b), the firm expects sales of 75,000 units at $750, with a probability of .7, and a .3 probability of 70,000 units at $750. Value analysis, at a cost of $100,000, is only used in option b. Which option has the highest expected monetary value (EMV)? **PX**

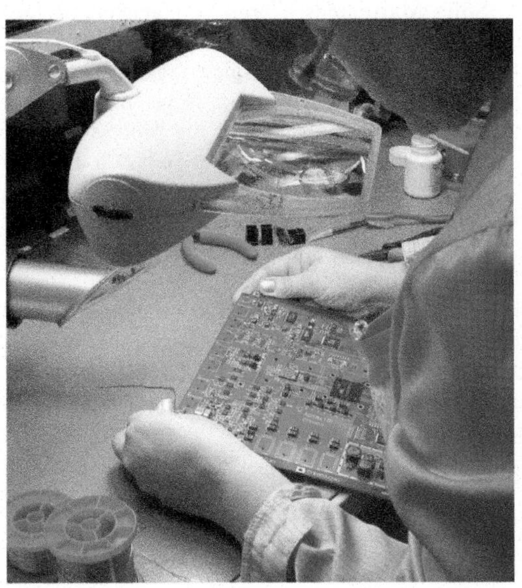

Romanchuck Dimitry/Shutterstock

• • **5.20** Residents of Mill River have fond memories of ice skating at a local park. An artist, Shaquille Smith, has captured the experience in a drawing and is hoping to reproduce it and sell framed copies to current and former residents. He thinks that if the market is good he can sell 400 copies of the elegant version at $125 each. If the market is not good, he will sell only 300 at $90 each. He can make a deluxe version of the same drawing instead. He feels that if the market is good he can sell 500 copies of the deluxe version at $100 each. If the market is not good, he will sell only 400 copies at $70 each. In either case, production costs will be approximately $35,000. He can also choose to do nothing. If he believes there is a 50% probability of a good market, what should he do? Why? **PX**

• • **5.21** Ritz Products' materials manager, Tej Dhakar, must determine whether to make or buy a new semiconductor for the wrist TV that the firm is about to produce. One million units are expected to be produced over the life cycle. If the product is made, start-up and production costs of the *make* decision total $1 million, with a probability of .4 that the product will be satisfactory and a .6 probability that it will not. If the product is not satisfactory, the firm will have to reevaluate the decision. If the decision is reevaluated, the choice will be whether to spend another $1 million to redesign the semiconductor or to purchase. Likelihood of success the second time that the make decision is made is .9. If the second *make* decision also fails, the firm must purchase. Regardless of when the purchase takes place, Dhakar's best judgment of cost is that Ritz will pay $.50 for each purchased semiconductor plus $1 million in vendor development cost.

a) Assuming that Ritz must have the semiconductor (stopping or doing without is not a viable option), what is the best decision?
b) What criteria did you use to make this decision?
c) What is the worst that can happen to Ritz as a result of this particular decision? What is the best that can happen? **PX**

• • **5.22** De León Engineering designs and constructs air conditioning and heating systems for hospitals and clinics. Currently, the company's staff is overloaded with design work. There is a major design project due in 8 weeks. The penalty for completing the design late is $14,000 per week because any delay will cause the facility to open later than anticipated and cost the client significant revenue. If the company uses its inside engineers to complete the design, it will have to pay them overtime for all work. Owner Roberta De León has estimated that it will cost $12,000 per week (wages and overhead), including late weeks, to have company engineers complete the design. Roberta is also considering having an outside engineering firm do the design. A bid of $90,000 has been received for the completed design. Yet another option for completing the design is to conduct a joint design by having a third engineering company complete all electromechanical components of the design at a cost of $50,000. De León Engineering would then complete the rest of the design and control systems at an estimated cost of $30,000.

Roberta has estimated the following probabilities of completing the project within various time frames when using each of the three options. Those estimates are shown in the following table:

	PROBABILITY OF COMPLETING THE DESIGN			
OPTION	ON TIME	1 WEEK LATE	2 WEEKS LATE	3 WEEKS LATE
Internal Engineers	.4	.5	.1	—
External Engineers	.2	.4	.3	.1
Joint Design	.1	.3	.4	.2

What is the best decision based on an expected monetary value (EMV) criterion? (*Note:* You want the lowest EMV because we are dealing with costs in this problem.) **PX**

• • • **5.23** Use the data in Solved Problem 5.1 to examine what happens to the decision if Sarah King can increase all of Design B yields from 59,000 to 64,000 by applying an expensive phosphorus to the screen at an added manufacturing cost of $250,000. Prepare the modified decision tree. What are the payoffs, and which branch has the greatest expected monetary value (EMV)? **PX**

• • • • **5.24** McBurger, Inc., wants to redesign its kitchens to improve productivity and quality. Three designs, called designs K1, K2, and K3, are under consideration. No matter which design is used, daily production of sandwiches at a typical McBurger restaurant is for 500 sandwiches. A sandwich costs

$1.30 to produce. Nondefective sandwiches sell, on the average, for $2.50 per sandwich. Defective sandwiches cannot be sold and are scrapped. The goal is to choose a design that maximizes the expected profit at a typical restaurant over a 300-day period. Designs K1, K2, and K3 cost $100,000, $130,000, and $180,000, respectively. Under design K1, there is a .80 chance that 90 out of each 100 sandwiches are non-defective and a .20 chance that 70 out of each 100 sandwiches are non-defective. Under design K2, there is a .85 chance that 90 out of each 100 sandwiches are non-defective and a .15 chance that 75 out of each 100 sandwiches are non-defective. Under design K3, there is a .90 chance that 95 out of each 100 sandwiches are non-defective and a .10 chance that 80 out of each 100 sandwiches are non-defective. What is the expected profit level of the design that achieves the maximum expected 300-day profit level? **PX**

••• **5.25** Ling Li of Li Windows, Inc. is considering making a change in the material the firm uses for panes in its residential window line. The new material has a slight mirror attribute that assists in reflecting ultraviolet light and restricts the transmission of heat. The new material will raise the cost of a standard window by $3.76. The current window is in the mature stage of the life cycle, and, with no modifications, Li has estimated that sales of the window line will be 240,000 units per year with a probability of 0.3 or 180,000 units per year with a probability of 0.7. The standard price of a window unit is $45. With the new glass material, the price per unit can be increased to $50. However, Li estimates that the demand for the newly designed window will be 210,000 units with a probability of 0.6, and the there will be a 0.4 probability of sales of 150,000 units. Which option will allow the company to maximize its expected monetary value (EMV)? **PX**

•• **5.26** Shamrock Oil owns a parcel of land that has the potential to be an underground oil field. It will cost $500,000 to drill for oil. If oil does exist on the land, Shamrock will realize a payoff of $4,000,000 (not including drilling costs). With current information, Shamrock estimates that there is a 0.2 probability that oil is present on the site. Shamrock also has the option of selling the land as is for $400,000 without further information about the likelihood of oil being present. A third option is to perform geological tests at the site, which would cost $100,000. There is a 30% chance that the test results will be positive, after which Shamrock can sell the land for $650,000 or drill the land, with a 0.65 probability that oil exists. If the test results are negative, Shamrock can sell the land for $50,000 or drill the land, with a 0.05 probability that oil exists. Using a decision tree, recommend a course of action for Shamrock Oil. **PX**

•• **5.27** Sofia Martinez is a product manager for Diamond Chemical. The firm is considering whether to launch a new product line that will require building a new facility. The technology required to produce the new product is yet untested. If Sofia decides to build the new facility and the process is successful, Diamond Chemical will realize a profit of $675,000. If the process does not succeed, the company will lose $825,000. Sofia estimates that there is a 0.6 probability that the process will succeed.

Sofia can also decide to build a pilot plant for $60,000 to test the new process before deciding to build the full-scale facility. If the pilot plant succeeds, Sofia feels the chance of the full-scale facility succeeding is 85%. If the pilot plant fails, Sofia feels the chance of the full-scale facility succeeding is only 20%. The probability that the pilot plant will succeed is estimated at 0.6. Structure this problem with a decision tree and advise Sofia what to do. **PX**

CASE STUDIES

De Mar's Product Strategy

De Mar, a plumbing, heating, and air-conditioning company located in Fresno, California, has a simple but powerful product strategy: *Solve the customer's problem no matter what, solve the problem when the customer needs it solved, and make sure the customer feels good when you leave.* De Mar offers guaranteed, same-day service for customers requiring it. The company provides 24-hour-a-day, 7-day-a-week service at no extra charge for customers whose air conditioning dies on a hot summer Sunday or whose toilet overflows at 2:30 A.M. As assistant service coordinator Janie Walter puts it: "We will be there to fix your A/C on the fourth of July, and it's not a penny extra. When our competitors won't get out of bed, we'll be there!"

De Mar guarantees the price of a job to the penny before the work begins. Whereas most competitors guarantee their work for 30 days, De Mar guarantees all parts and labor for one year. The company assesses no travel charge because "it's not fair to charge customers for driving out." Owner Larry Harmon says: "We are in an industry that doesn't have the best reputation. If we start making money our main goal, we are in trouble. So I stress customer satisfaction; money is the by-product."

De Mar uses selective hiring, ongoing training and education, performance measures, and compensation that incorporate customer satisfaction, strong teamwork, peer pressure, empowerment, and aggressive promotion to implement its strategy. Says credit manager Anne Semrick: "The person who wants a nine-to-five job needs to go somewhere else."

De Mar is a premium pricer. Yet customers respond because De Mar delivers value—that is, benefits for costs. In 8 years, annual sales increased from about $200,000 to more than $3.3 million.

Discussion Questions

1. What is De Mar's product? Identify the tangible parts of this product and its service components.
2. How should other areas of De Mar (marketing, finance, personnel) support its product strategy?
3. Even though De Mar's product is primarily a service product, how should each of the 10 strategic OM decisions in the text be managed to ensure that the product is successful?

Source: Reprinted with permission: *On Great Service: A Framework for Action*, by Leonard L. Berry.

Product Design at Nautique Boat Company

Video Case

For nearly 100 years, Florida-based Nautique Boat Company has built innovative boats in a very competitive market. Nautique, as the premier boat for waterskiing, wakeboarding, and wake surfing, is on the cutting edge of style, customer satisfaction, and performance. There is continuing and rapid change in this industry, which sees substantial input from imaginative, experimenting customers. Success means integrating customer feedback, technological change, and creative engineering talent into a dynamic, but ongoing, product line. As number one in its market, Nautique is a vivid example of what it takes to be a creative leader.

From the introduction of waterskiing in the 1920s to barefoot skiing to wake boarding in the 1990s and wake surfing in the 2000s, Nautique has led. While these new sport expectations were placed on boat performance, changes in marine engineering and technology were also taking place. Ski boats were initially made of wood, changing to fiberglass in the late 1950s, with tracking fins added in the 1960s. The 1990s brought longer and wider boats with hull changes, slopping transoms, spray relief, and flight control towers. This was followed by *Total Wake Control*, which allows users to instantly switch from wakeboarding to wake surfing. Indeed, the wake can now be customized, providing ramp-style and vertical-style wakes behind the same boat. Simultaneously, carpet, trim, and color options and a variety of powerful engines, as well as major electronic innovations such as sonar, GPS, and sophisticated sound systems with strategically placed speakers, have been introduced. Nautique's design team is now developing the boat of the future—the electric boat—with the prototype in testing.

To maintain Nautique's innovative prowess, Chief Designer Steve Carlton (shown in photo) has organized his 40-person department into three teams: model integration, design, and engineering. In addition to creating, developing, and designing new innovations, Nautique also plans on two to four new models each year, with remodeling or updating every 2 years.

To facilitate efficient development and design integrity, Carlton's design and development team uses sophisticated CAD software from Rhino and SolidWorks. NX software is also available for tooling, and StrataSys 3D printers facilitate prototyping small parts. Design typically begins with sketches and clay models, progressing to CAD drawings and then to a full-scale wooden model, followed by a full-size boat for testing in Nautique's private lake.

Each design innovation requires ongoing discussion with the in-house engineering staff to provide design integrity. Similarly, the team sends staff to the factory floor to work with production personnel to ensure production capability and with external suppliers to ensure raw material and component quality and delivery. This coordination with external suppliers and internal production by the design group has the goal of facilitating a smooth transition to production.

Beverly Amer/Aspenleaf Productions

Discussion Questions*

1. How does the concept of product life cycle apply to the Nautique Boat Company?
2. How would you define Nautique's strategy, and how does its product design fit into that strategy?
3. What kinds of efficiency does the use of CAD bring to Nautique?
4. How does Nautique handle the "transition" to production?

* The Global Company Profile featuring Nautique Boat Company (which opens Chapter 5) provides further background on Nautique's operation as does the video that accompanies this case. You may wish to review both prior to answering these questions.

Celebrity Cruises Designs a New Ship

Video Case

Cruising is a growth industry, expanding by 6% to 7% annually for the past 30 years. Recognizing this growth opportunity, Celebrity Cruises, Inc., has enlarged its offerings in the premium "upscale experience at an intelligent price" segment of the industry by continuously growing its fleet. As CEO Lisa Lutoff-Perlo notes, "Celebrity Cruises has always been recognized as an innovative, trend-setting brand, and the Celebrity ships are just another instance in which we keep raising the bar in modern luxury travel." To meet this opportunity and related challenges, Celebrity has committed the resources necessary for a five-ship expansion. But that was the easy part. Now Celebrity must determine the exact specifications of the new ships.

Like other products, ships are designed for specific tasks and markets. For a cruise line, the design options are huge. Consider Celebrity Cruises' new ships. First, think of a hotel for 3,000 passengers and a staff of over 2,000. Then put the hotel on a huge power plant that can move the hotel around the world at 22 knots and supply all of the power necessary for heating, air-conditioning, laundry, kitchen, and all the other amenities expected on a luxury vacation. Then add the massive support necessary for 5,000 people. This means warehousing and refrigeration for all the food to feed 5,000 people at least three meals every day (although the average passenger on a cruise consumes close to 4½ meals per day); fresh water for cooking, bathing, and multiple swimming pools; plus storage tanks for fuel to be used at the rate of 3,500 gallons per day.

In the cruise business, it helps to think of the ship as a small city where everyone in town is on holiday. The ship size, room accommodations, restaurants, and sundry onboard features from theater to library to retail space must all be designed to meet unique customer demands for specific market segments. Even the type of power plant (turbine versus diesel) impacts cost and environmental issues. Additionally, the selection of ports that accommodate both the ship and passenger preferences must be evaluated as part of the "product design."

With a capital investment of over $400 million per ship, Celebrity needs to get the product design right. To develop new ships, Celebrity has put together a "building committee" made up of key executives including the Senior VP of Global Marine Operations, the VP of Hotel Operations, the VP of Sales, and the VP of Revenue Management. Additionally, in a uniquely creative way, Celebrity has also invested in a special "Innovation Center" featuring the largest virtual-reality (VR) facility in the world (as shown earlier in this chapter) so it can design a "state-of-the art" experience on board the ship. The VR capability allows management personnel, as well as design personnel, to visualize various areas (restaurants and bars), features (theater, library, card room), and activities (spa, gym, putting green) on the ship early in the design stage. This is particularly helpful by providing insight into dynamic aspects of design such as the flow of passengers and supplies, as well as a deeper understanding of tasks such as the loading and unloading of luggage.

Discussion Questions*

1. Identify the product design options that Celebrity Cruises has available as it designs a new ship.
2. Visit the websites of Celebrity's sister companies, Royal Caribbean Cruises Ltd. and Azamara Club Cruises. Explain how Celebrity's product strategy differs from the strategies of these two cruise lines.
3. How might Figure 5.3 be modified to reflect the product development cycle at Celebrity Cruises?
4. How does Celebrity's product development process compare with the product development organizations suggested in your text?

*You may wish to view the video accompanying this case before addressing these questions.

Endnotes

1. *Contribution* is defined as the difference between direct cost and selling price. Direct costs are directly attributable to the product, namely labor and material that go into the product.
2. Sources: *The Washington Post* (February 9, 2021); and *Businessweek* (November 25, 2019).
3. Sources: *USA Today* (February 25, 2020); and *Forbes* (January 13, 2019).

Bibliography

Brown, B., and S. D. Anthony. "How P&G Tripled Its Innovation Success Rate." *Harvard Business Review* 89, no. 6 (June 2011): 64–72.

Buckley, P. J., and R. Strange. "The Governance of the Global Factory: Location and Control of World Economic Activity." *Academy of Management Perspectives* 29, no. 2 (May 2015): 237–249.

Gwynne, P. "Measuring R&D Productivity." *Research Technology Management* 58, no. 1 (Jan/Feb. 2015): 19–22.

Haines, S. *Managing Product Management*. New York: McGraw-Hill, 2012.

Madu, C. N. *The House of Quality in a Minute: A Guide to Quality Function Deployment*, 3rd ed. Charlotte, NC: Information Age Publishing, 2020.

Oliver A., P. Criscuelo, and A. Salter. "Managing Unsolicited Ideas for R&D." *California Management Review* 54, no. 3 (Spring 2012): 116–139.

Reeves, M., and M. Deimler. "Adaptability: The New Competitive Advantage." *Harvard Business Review* 90, no. 7 (July–August 2011): 135–141.

Seider, W. D., et al. *Product and Process Design Principles: Synthesis, Analysis and Evaluation*, 4th ed. New York: Wiley, 2017.

Ulrich, K., and S. Eppinger. *Product Design and Development*, 6th ed. New York: McGraw-Hill, 2016.

Chapter 5 ***Rapid*** Review

Main Heading	Review Material	MyLab Operations Management
GOODS AND SERVICES SELECTION	Although the term *products* may often refer to tangible goods, it also refers to offerings by service organizations. *The objective of the product decision is to develop and implement a product strategy that meets the demands of the marketplace with a competitive advantage.* ■ **Product decision**—The selection, definition, and design of products. The four phases of the product life cycle are introduction, growth, maturity, and decline. ■ **Product-by-value analysis**—A list of products, in descending order of their individual dollar contribution to the firm, as well as the *total annual dollar* contribution of the product.	Concept Questions: 1.1–1.6 Problems: 5.1–5.3 **VIDEO 5.1** Product Design at Nautique Boat Company
GENERATING NEW PRODUCTS	Product selection, definition, and design take place on a continuing basis. Changes in product opportunities, the products themselves, product volume, and product mix may arise due to understanding the customer, economic change, sociological and demographic change, technological change, political/legal change, market practice, professional standards, suppliers, or distributors.	Concept Question: 2.1
PRODUCT DEVELOPMENT	■ **Quality function deployment (QFD)**—A process for determining customer requirements (customer "wants") and translating them into attributes (the "hows") that each functional area can understand and act on. ■ **House of quality**—A part of the quality function deployment process that utilizes a planning matrix to relate customer wants to how the firm is going to meet those wants. ■ **Product development teams**—Teams charged with moving from market requirements for a product to achieving product success. ■ **Concurrent engineering**—Simultaneous performance of the various stages of product development. ■ **Manufacturability and value engineering**—Activities that help improve a product's design, production, maintainability, and use.	Concept Questions: 3.1–3.6
ISSUES FOR PRODUCT DESIGN	■ **Robust design**—A design that can be produced to requirements even with unfavorable conditions in the production process. ■ **Modular design**—A design in which parts or components of a product are subdivided into modules that are easily interchanged or replaced. ■ **Computer-aided design (CAD)**—Interactive use of a computer to develop and document a product. ■ **Design for manufacture and assembly (DFMA)**—Software that allows designers to look at the effect of design on manufacturing of a product. ■ **Standard for the exchange of product data (STEP)**—A standard that provides a format allowing the electronic transmission of three-dimensional data. ■ **Computer-aided manufacturing (CAM)**—The use of information technology to control machinery. ■ **Additive manufacturing (also called 3D printing)**—An extension of CAD that builds products by adding material layer upon layer. ■ **Virtual reality**—A visual form of communication in which images substitute for reality and typically allow the user to respond interactively. ■ **Augmented reality**—The integration of digital information with the user's environment in real time. ■ **Value analysis**—A review of successful products that takes place during the production process. Sustainability is meeting the needs of the present without compromising the ability of future generations to meet their needs. Life cycle assessment (LCA) is part of ISO 14000; it assesses the environmental impact of a product from material and energy inputs to disposal and environmental releases. Both sustainability and LCA are discussed in depth in Supplement 5.	Concept Questions: 4.1–4.6
PRODUCT DEVELOPMENT CONTINUUM	■ **Time-based competition**—Competition based on time; rapidly developing products and moving them to market. *Internal development strategies* include (1) new internally developed products, (2) enhancements to existing products, and (3) migrations of existing products. *External development strategies* include (1) purchase the technology or expertise by acquiring the developer, (2) establish joint ventures, and (3) develop alliances. ■ **Joint ventures**—Firms establishing joint ownership to pursue new products or markets. ■ **Alliances**—Cooperative agreements that allow firms to remain independent but pursue strategies consistent with their individual missions.	Concept Questions: 5.1–5.5

Main Heading	Review Material	MyLab Operations Management
DEFINING A PRODUCT	■ **Engineering drawing**—A drawing that shows the dimensions, tolerances, materials, and finishes of a component. ■ **Bill of material (BOM)**—A list of the components, their description, and the quantity of each required to make one unit of a product. ■ **Make-or-buy decision**—The choice between producing a component or a service and purchasing it from an outside source. ■ **Group technology**—A product and component coding system that specifies the size, shape, and type of processing; it allows similar products to be grouped.	Concept Questions: 6.1–6.5 Problems: 5.9, 5.10, 5.12–5.17
DOCUMENTS FOR PRODUCTION	■ **Assembly drawing**—An exploded view of a product. ■ **Assembly chart**—A graphic means of identifying how components flow into subassemblies and final products. ■ **Route sheet**—A list of the operations necessary to produce a component with the material specified in the bill of material. ■ **Work order**—An instruction to make a given quantity of a particular item. ■ **Engineering change notice (ECN)**—A correction or modification of an engineering drawing or bill of material. ■ **Configuration management**—A system by which a product's planned and changing components are accurately identified. ■ **Product life cycle management (PLM)**—Software programs that tie together many phases of product design and manufacture.	Concept Questions: 7.1–7.6
SERVICE DESIGN	To enhance service efficiency, companies: (1) limit options, (2) delay customization, (3) modularize, (4) automate, and (5) design for the "moment of truth."	Concept Questions: 8.1–8.4 **VIDEO 5.2** Celebrity Cruises Designs a New Ship
APPLICATION OF DECISION TREES TO PRODUCT DESIGN	To form a decision tree, (1) include all possible alternatives (including "do nothing") and states of nature; (2) enter payoffs at the end of the appropriate branch; and (3) determine the expected value of each course of action by starting at the end of the tree and working toward the beginning, calculating values at each step and "pruning" inferior alternatives.	Concept Questions: 9.1–9.2 Problems: 5.18–5.27 Virtual Office Hours for Solved Problem: 5.1 **ACTIVE MODEL 5.1**
TRANSITION TO PRODUCTION	One of the arts of management is knowing when to move a product from development to production; this move is known as *transition to production*.	Concept Questions: 10.1–10.2
ADDITIONAL MYLAB OPERATIONS MANAGEMENT RESOURCES	✔ Multiple Choice Case Questions (De Mar's Product Strategy) ✔ Recent Graduate Video: Ari Davis, Product Designer and Entrepreneur, The Soil Co.	

***Self* Test**

LO 5.1 A product's life cycle is divided into four stages, including:
- **a)** introduction.
- **b)** growth.
- **c)** maturity.
- **d)** all of the above.

LO 5.2 Product development systems include:
- **a)** bills of material.
- **b)** routing charts.
- **c)** functional specifications.
- **d)** product-by-values analysis.
- **e)** configuration management.

LO 5.3 A house of quality is:
- **a)** a matrix relating customer "wants" to the firm's "hows."
- **b)** a schematic showing how a product is put together.
- **c)** a list of the operations necessary to produce a component.
- **d)** an instruction to make a given quantity of a particular item.
- **e)** a set of detailed instructions about how to perform a task.

LO 5.4 Time-based competition focuses on:
- **a)** moving new products to market more quickly.
- **b)** reducing the life cycle of a product.
- **c)** linking QFD to PLM.
- **d)** design database availability.
- **e)** value engineering.

LO 5.5 Products are defined by:
- **a)** value analysis.
- **b)** value engineering.
- **c)** routing sheets.
- **d)** assembly charts.
- **e)** engineering drawings.

LO 5.6 A route sheet:
- **a)** lists the operations necessary to produce a component.
- **b)** is an instruction to make a given quantity of a particular item.
- **c)** is a schematic showing how a product is assembled.
- **d)** is a document showing the flow of product components.
- **e)** all of the above.

LO 5.7 Decision trees use:
- **a)** probabilities.
- **b)** payoffs.
- **c)** logic.
- **d)** options.
- **e)** all of the above.

Answers: LO 5.1. d; LO 5.2. c; LO 5.3. a; LO 5.4. a; LO 5.5. e; LO 5.6. a; LO 5.7. e.

Sustainability in the Supply Chain

SUPPLEMENT 5

SUPPLEMENT OUTLINE

LEARNING OBJECTIVES

LO S5.1 ***Describe*** corporate social responsibility 192

LO S5.2 ***Describe*** sustainability 192

LO S5.3 ***Explain*** the circular economy 195

LO S5.4 ***Calculate*** design for disassembly 197

LO S5.5 ***Explain*** the impact of sustainable regulations on operations 200

Airlines from around the world, including Air China, Virgin Atlantic Airways, KLM, Alaska, Air New Zealand, and Japan Airlines, are experimenting with alternative fuels to power their jets in an effort to reduce greenhouse gas emissions and to reduce their dependence on traditional petroleum-based jet fuel. Alternative biofuels are being developed from recycled cooking oil, sewage sludge, municipal waste, coconuts, sugar cane, and genetically modified algae that feed on plant waste.

DING CHEN/CHINE NOUVELLE/SIPA/Newscom

Corporate Social Responsibility[1]

LO S5.1 *Describe* corporate social responsibility

Managers must consider how the products and services they provide affect both people and the environment. Certainly, firms must provide products and services that are innovative and attractive to buyers. But today's technologies allow consumers, communities, public interest groups, and regulators to be well informed about all aspects of an organization's performance. As a result, stakeholders can have strong views about firms that fail to respect the environment or that engage in unethical conduct. Firms need to consider all the implications of a product—from design to disposal.

Many companies realize that "doing what's right" and doing it properly can be beneficial to all stakeholders. Companies that practice **corporate social responsibility (CSR)** introduce policies that consider environmental, societal, and financial impacts in their decision making. As managers consider approaches to CSR, they find it helpful to consider the concept of creating **shared value**. *Shared value* suggests finding policies and practices that enhance the organization's competitiveness while simultaneously advancing the economic and social conditions in the communities in which it operates. For instance, note how automakers Tesla, Toyota, and Nissan find shared value in low-emission vehicles... vehicles that enhance their competiveness in a global market while meeting society's interest in low-emission vehicles. Similarly, Dow Chemical finds social benefits and profit in Nexera canola and sunflower seeds. These seeds yield twice as much cooking oil as soybeans, enhancing profitability to the grower. They also have a longer shelf life, which reduces operating costs throughout the supply chain. As an added bonus, the oils have lower levels of saturated fat than traditional products and contain no trans fats. A win–win for Dow and society.

Corporate social responsibility (CSR)
Managerial decision making that considers environmental, societal, and financial impacts.

Shared value
Developing policies and practices that enhance the competitiveness of an organization while advancing the economic and social conditions in the communities in which it operates.

Operations functions—from supply chain management to product design to production to packaging and logistics—provide an opportunity for finding shared value and meeting CSR goals.[2]

Sustainability

Sustainability
Meeting the needs of the present without compromising the ability of future generations to meet their needs.

Sustainability is often associated with corporate social responsibility. The term **sustainability** refers to meeting the needs of the present without compromising the ability of future generations to meet their needs. Many people who hear of sustainability for the first time think of green products or "going green"—recycling, global warming, and saving rainforests. This is certainly part of it. However, it is more than this. True sustainability involves thinking not only about environmental resources but also about employees, customers, community, and the company's reputation. Three concepts may be helpful as managers consider sustainability decisions: a *systems* view, the *commons*, and the *triple bottom line*.

LO S5.2 *Describe* sustainability

Systems View

Managers may find that their decisions regarding sustainability improve when they take a *systems* view. This means looking at a product's life from design to disposal, including all the resources required. Recognizing that both raw materials and human resources are subsystems of any production process may provide a helpful perspective. Similarly, the product or service itself is a small part of much larger social, economic, and environmental systems. Indeed, managers need to understand the inputs and interfaces between the interacting systems and identify how changes in one system affect others. For example, hiring or laying off employees can be expected to have morale implications for internal systems (within an organization), as well as socioeconomic implications for external systems. Similarly, dumping chemicals down the drain has implications on systems beyond the firm. Once managers understand that the systems immediately under their control have interactions with systems below them and above them, more informed judgments regarding sustainability can be made.

VIDEO S5.1
Building Sustainability at the Orlando Magic's Amway Center

Commons

Many inputs to a production system have market prices, but others do not. Those that do not often have unidentified or ambiguous ownership (sunshine, air, airspace) or are held by the public. Such resources are often referred to as being in the *common*. Resources held in the *common* are often misallocated. Examples include depletion of fish in international waters and polluted air and waterways. The attitude seems to be that just a little more fishing or a little more pollution will not matter, or the adverse results may be perceived as someone else's problem. Society is still groping for solutions for use of those resources in the *common*. The answer is slowly being found in a number of ways: (1) moving some of the *common* to private property (e.g., selling radio frequency spectrum), (2) allocation of rights (e.g., establishing fishing boundaries), and (3) allocation of yield (e.g., only a given quantity of fish can be harvested). As managers understand the issues of the *commons*, they have further insight about sustainability and the obligation of caring for the *commons*.

Triple Bottom Line

STUDENT TIP
Profit is now just one of the three *Ps*: people, planet, and profit.

Firms that do not consider the impact of their decisions on all their stakeholders see reduced sales and profits. Profit maximization is not the only measure of success. A one-dimensional bottom line, profit, will not suffice; the larger socioeconomic systems beyond the firm demand more. One way to think of sustainability is to consider the systems necessary to support the triple bottom line of the three *Ps*: *people, planet*, and *profit* (see Figure S5.1), which we will now discuss.

People Companies are becoming more aware of how their decisions affect people—not only their employees and customers but also those who live in the communities in which they operate. Most employers want to pay fair wages, offer educational opportunities, and provide a safe and healthy workplace. So do their suppliers. But globalization and the reliance on outsourcing to suppliers around the world complicate the task. This means companies must create policies that guide supplier selection and performance. Sustainability suggests that supplier selection and performance criteria evaluate safety in the work environment, whether living wages are paid, if child labor is used, and whether work hours are excessive. Apple, GE, Procter & Gamble, and Walmart are examples of companies that conduct supplier audits to uncover any harmful or exploitative business practices that are counter to their sustainability goals and objectives.

STUDENT TIP
Walmart has become a global leader in sustainability. Read *Force of Nature: The Unlikely Story of Wal-Mart's Green Revolution*, by Edward Humes.

Accordingly, operations managers must consider the working conditions in which they place their own employees. This includes training and safety orientations, before-shift exercises, safety equipment, and rest breaks to reduce the possibility of worker fatigue and injury. Operations managers must also make decisions regarding the disposal of material and chemical waste, including hazardous materials to avoid harm to employees and the community.

Planet When discussing the subject of sustainability, our planet's environment is the first thing that comes to mind, so it understandably gets substantial attention from managers. Operations managers look for ways to reduce the environmental impact of their operations, whether from raw material selection, process innovation, alternative product delivery methods,

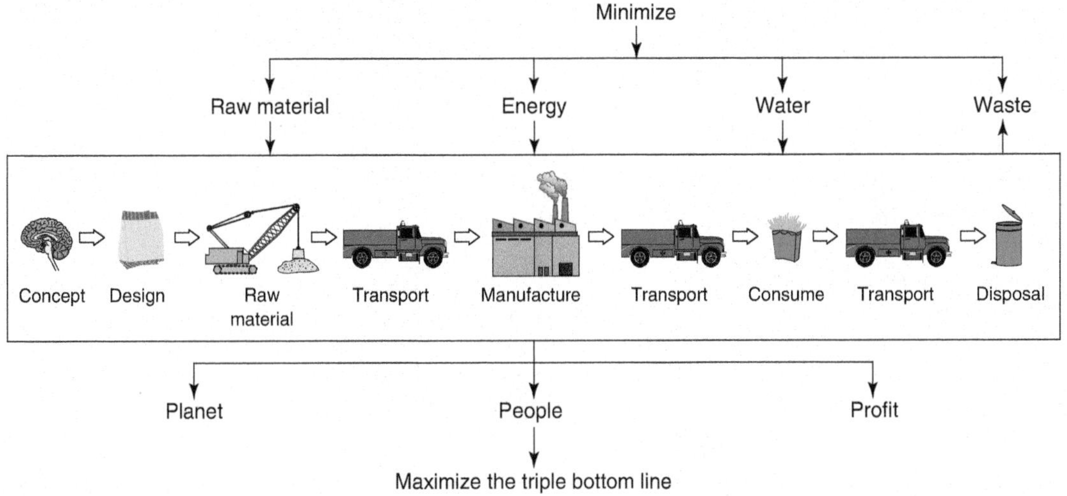

Figure **S5.1**

Improving the Triple Bottom Line with Sustainability

or disposal of products at their end of life. The overarching objective for operations managers is to conserve scarce resources, thereby reducing the negative impact on the environment. For example, Levi Strauss & Co. has started a campaign to save water in the creation of jeans, as seen in the OM in Action box, "Blue Jeans and Sustainability."

Carbon footprint
A measure of total greenhouse gas emissions caused directly or indirectly by an organization, a product, an event, or a person.

To gauge their environmental impact on the planet, many companies are measuring their carbon footprint. **Carbon footprint** is a measure of the total greenhouse gas (GHG) emissions caused directly and indirectly by an organization, a product, an event, or a person. A substantial portion of GHGs are released naturally by farming, cattle, and decaying forests and, to a lesser degree, by manufacturing and services. The most common GHG produced by human activities is carbon dioxide, primarily from burning fossil fuels for electricity generation, heating, and transport. Operations managers are being asked to do their part to reduce GHG emissions.

VIDEO S5.2
Green Manufacturing and Sustainability at Frito-Lay

Industry leaders such as Frito-Lay have been able to break down the carbon emissions from various stages in the production process. For instance, in potato chip production, a 34.5-gram (1.2-ounce) bag of chips is responsible for about twice its weight in emissions—75 grams per bag (see Figure S5.2).

Economic sustainability
Appropriately allocating scarce resources to make a profit.

Profit Social and environmental sustainability do not exist without economic sustainability. **Economic sustainability** refers to how companies remain in business. Staying in business requires making investments, and investments require making profits. Though profits may be relatively easy to determine, other measures can also be used to gauge economic sustainability. The alternative measures that point to a successful business include risk profile, intellectual property, employee morale, and company valuation. To support economic sustainability, firms may supplement standard financial accounting and reporting with some version of social accounting. Social accounting can include accounting for assets and liabilities related to brand equity, management talent, human capital development and benefits, research and development, productivity, philanthropy, and taxes paid.

OM in Action Blue Jeans and Sustainability[3]

The recent drought in California is hurting more than just farmers. It is also having a significant impact on the fashion industry and spurring changes in how jeans are made and how they should be laundered. Southern California is estimated to be the world's largest supplier of so-called premium denim, the $100 to $200-plus-a-pair of designer jeans. Water is a key component in the various steps of the processing and repeated washing with stones, or bleaching and dyeing that create that "distressed" vintage look. Southern California produces 75% of the high-end denim in the U.S. that is sold worldwide. The area employs about 200,000 people, making it the largest U.S. fashion manufacturing hub.

Now that water conservation is a global priority, major denim brands are working to cut water use. Levi Strauss & Co., with sales of $5 billion, is using ozone machines to replace the bleach traditionally used to lighten denim. It is also reducing the number of times it washes jeans. The company has saved more than a billion liters of water since 2011 with its Levi's Water Less campaign. Eighty percent of Levi's brand products are now made using the Water Less process, up from about 25% a decade ago.

Traditionally, about 34 liters of water are used in the cutting, sewing, and finishing process to make a pair of Levi's signature 501 jeans. Nearly 3,800 liters of water are used throughout the lifetime of a pair of Levi's 501. A study found cotton cultivation represents 68% of that and consumer washing another

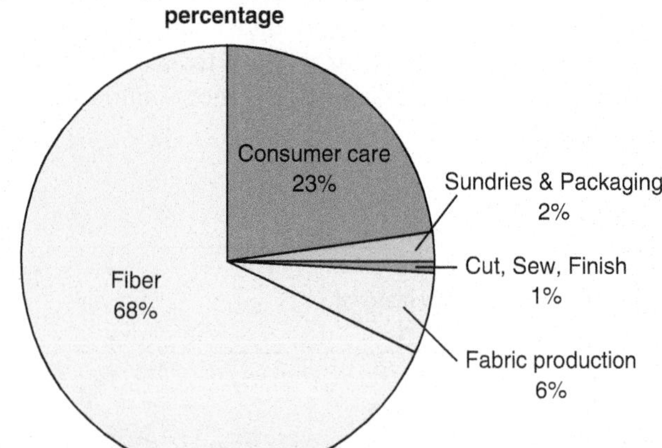

23%. So Levi is promoting the idea that jeans only need washing after 10 *wears*. (The average American consumer washes after 2 *wears*.) Levi's CEO recently urged people to stop washing their jeans, saying he hadn't washed his one-year-old jeans at the time. "You can air dry and spot clean instead," he said.

Figure **S5.2**

Carbon Footprint of a 34.5-gram Bag of Frito-Lay Chips

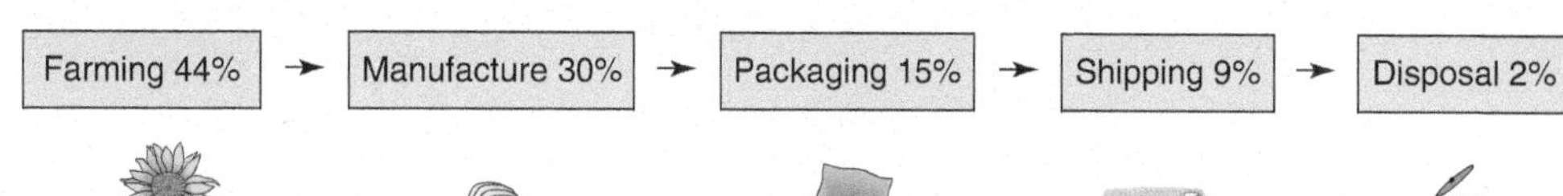

Design and Production for Sustainability

The operations manager's greatest opportunity to make substantial contributions to the company's environmental objectives occurs during product life cycle assessment. **Life cycle assessment** evaluates the environmental impact of a product, from raw material and energy inputs all the way to the disposal of the product at its end of life. The goal is to make decisions that help reduce the environmental impact of a product throughout its entire life.

Life cycle assessment
Analysis of environmental impacts of products from the design stage through end of life.

This brings us to the idea of the **circular economy**. The circular economy is an alternative to the traditional linear economy of make–use–dispose. In the circular economy, we keep and use resources for as long as possible and then recover and regenerate to the maximum possible value at the end of service life. We now address how managers can implement the circular economy by looking at product design, production processes, logistics, and end of life.

Circular economy
Keeps and uses resources for as long as possible, recovering and regenerating the maximum possible value at end of life.

LO S5.3 *Explain* the circular economy

Product Design

Product design is the most critical phase in product life cycle assessment. The decisions that are made during this phase greatly affect materials, quality, cost, processes, related packaging and logistics, and ultimately how the product will be processed when discarded. During design, one of the goals is to incorporate a systems view in the product or service design that lowers the environmental impact. Such an approach reduces waste and energy costs at the supplier, in the logistics system, and for the end user. For instance, by taking a systems view, Procter &

Dan Bates/The Herald/AP Images

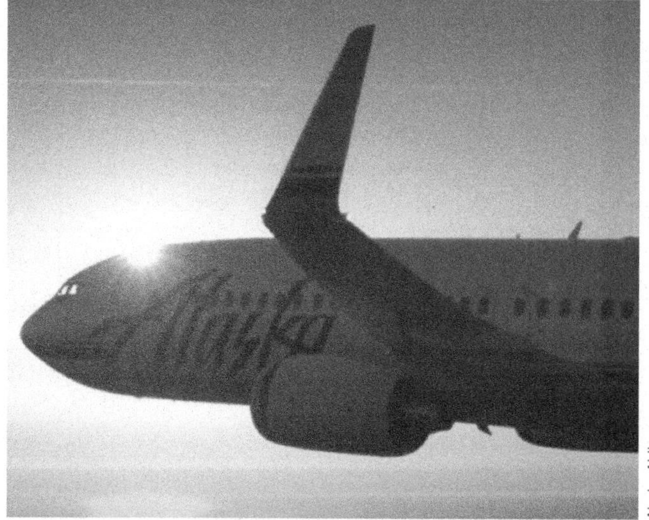

Alaska Airlines

An excellent place for operations managers to begin the sustainability challenge is with good product design. Here Tom Malone, CEO, of MicroGreen Polymers, discusses the company's new ultra-light cup with production personnel (left). The cup can be recycled over and over and never go to a landfill. Another recent design is the "winglet" (right). These wing tip extensions increase climb speed, reduce noise by 6.5%, cut carbon dioxide emissions by 5%, and save 6% in fuel costs. Alaska Air has retrofitted its entire 737 fleet with winglets, saving $20 million annually.

Gamble developed *Tide Coldwater*, a detergent that gets clothes clean with cold water, saving the consumer about three-fourths of the energy used in a typical wash.

Other successful design efforts include:

- Boston's Park Plaza Hotel eliminated bars of soap and bottles of shampoo by installing pump dispensers in its bathrooms, saving the need for 1 million plastic containers a year.
- UPS reduced the amount of materials it needs for its envelopes by developing its *reusable express envelopes,* which are made from 100% recycled fiber. These envelopes are designed to be used twice, and after the second use, the envelope can be recycled.
- Coca-Cola's redesigned Dasani bottle reduced the amount of plastic needed and is now 30% lighter than when it was introduced.

Product design teams also look for *alternative* materials from which to make their products. Innovating with alternative materials can be expensive, but it may make autos, trucks, and aircraft more environmentally friendly while improving payload and fuel efficiency. Aircraft and auto makers, for example, constantly seek lighter materials to use in their products. Lighter materials translate into better fuel economy, fewer carbon emissions, and reduced operating cost. For instance:

- Mercedes is building some car exteriors from a banana fiber that is both biodegradable and lightweight.
- Boeing is using carbon fiber, epoxy composites, and titanium graphite laminate to reduce weight in its 787 Dreamliner.
- Celebrity Cruises builds its ships with energy efficiency as a major criterion, as seen in its numerous green efforts in Figure S5.3.

VIDEO S5.3
"Saving the Waves" at Celebrity Cruises

Product designers often must decide between two or more environmentally friendly design alternatives. Example S1 deals with a *design for disassembly* cost–benefit analysis. The design team analyzes the amount of revenue that might be reclaimed against the cost of disposing of the product at its end of life.

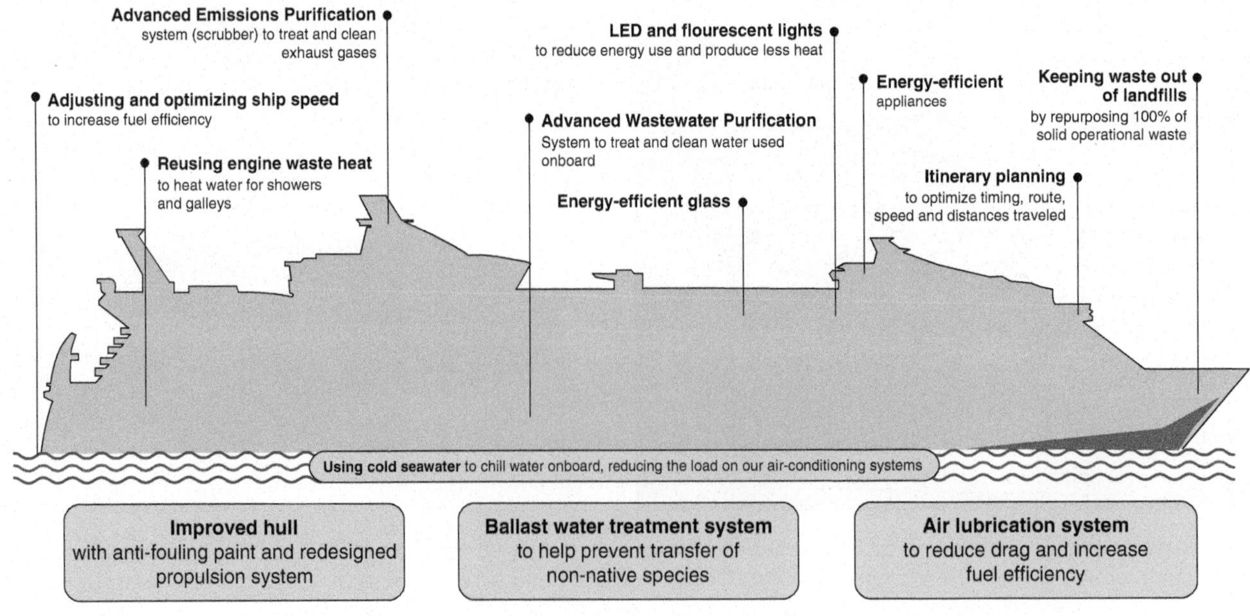

Figure **S5.3**

Celebrity Cruises' Green Efforts

See the case study at the end of this supplement for details on how Celebrity "saves the waves."

Example S1

DESIGN FOR DISASSEMBLY

Sound Barrier, Inc., needs to decide which of two speaker designs is better environmentally.

APPROACH ▶ The design team collected the following information for two audio speaker designs, the *Harmonizer* and the *Rocker*:

1. Resale value of the components minus the cost of transportation to the disassembly facility
2. Revenue collected from recycling
3. Processing costs, which include disassembly, sorting, cleaning, and packaging
4. Disposal costs, including transportation, fees, taxes, and processing time

SOLUTION ▶ The design team developed the following revenue and cost information for the two speaker design alternatives:

LO S5.4 *Calculate* design for disassembly

Harmonizer

PART	RESALE REVENUE PER UNIT	RECYCLING REVENUE PER UNIT	PROCESSING COST PER UNIT	DISPOSAL COST PER UNIT
Printed circuit board	$5.93	$1.54	$3.46	$0.00
Laminate back	0.00	0.00	4.53	1.74
Coil	8.56	5.65	6.22	0.00
Processor	9.17	2.65	3.12	0.00
Frame	0.00	0.00	2.02	1.23
Aluminum case	11.83	2.10	2.98	0.00
Total	$35.49	$11.94	$22.33	$2.97

Rocker

PART	RESALE REVENUE PER UNIT	RECYCLING REVENUE PER UNIT	PROCESSING COST PER UNIT	DISPOSAL COST PER UNIT
Printed circuit board	$7.88	$3.54	$2.12	$0.00
Coil	6.67	4.56	3.32	0.00
Frame	0.00	0.00	4.87	1.97
Processor	8.45	4.65	3.43	0.00
Plastic case	0.00	0.00	4.65	3.98
Total	$23.00	$12.75	$18.39	$5.95

Using the Equation (S5-1), the design team can compare the two design alternatives:

$$\text{Revenue retrieval} = \text{Total resale revenue} + \text{Total recycling revenue} - \text{Total processing cost} - \text{Total disposal cost} \quad \text{(S5-1)}$$

$$\text{Revenue retrieval for Harmonizer} = \$35.49 + \$11.94 - \$22.33 - \$2.97 = \$22.13$$

$$\text{Revenue retrieval for Rocker} = \$23.00 + \$12.75 - \$18.39 - \$5.95 = \$11.41$$

INSIGHT ▶ After analyzing both environmental revenue and cost components of each speaker design, the design team finds that the Harmonizer is the better environmental design alternative because it achieves a higher revenue retrieval opportunity. Note that the team is assuming that both products have the same market acceptance, profitability, and environmental impact.

LEARNING EXERCISE ▶ What would happen if there was a change in the supply chain that caused the processing and disposal costs to triple for the laminate back part of the Harmonizer? [Answer: The revenue retrieval from the Harmonizer is $35.49 + $11.94 − $31.39 − $6.45 = $9.59. This is less than the Rocker's revenue retrieval of $11.41, so the Rocker becomes the better environmental design alternative, as it achieves a higher revenue retrieval opportunity.]

RELATED PROBLEMS ▶ S5.1, S5.2, S5.3, S5.9, S5.12, S5.13, S5.14

Production Process

Manufacturers look for ways to reduce the amount of resources in the production process. Opportunities to reduce environmental impact during production typically revolve around the themes of energy, water, and environmental contamination. Conservation of energy and improving energy efficiency come from the use of alternative energy and more energy-efficient machinery. For example:

- S.C. Johnson built its own power plant that runs on natural gas and methane piped in from a nearby landfill, cutting back its reliance on coal-fired power.
- PepsiCo developed *Resource Conservation (ReCon)*, a diagnostic tool for understanding and reducing in-plant water and energy usage. In its first 2 years, *ReCon* helped sites across the world identify 2.2 billion liters of water savings, with a corresponding cost savings of nearly $2.7 million.
- Frito-Lay decided to extract water from potatoes, which are 80% water. Each year, a single factory processes 350,000 tons of potatoes, and as those potatoes are processed, the company reuses the extracted water for that factory's daily production.

STUDENT TIP Las Vegas, always facing a water shortage, pays residents $40,000 an acre to take out lawns and replace them with rocks and native plants.

These and similar successes in the production process reduce both costs and environmental concerns. Less energy is consumed, and less material is going to landfills.

Logistics

As products move along in the supply chain, managers strive to achieve efficient route and delivery networks, just as they seek to drive down operating costs. Doing so reduces environmental impact. Management analytics (such as linear programming, queuing, and vehicle routing software) help firms worldwide optimize elaborate supply-chain and distribution networks. Networks of container ships, airplanes, trains, and trucks are being analyzed to reduce the number of miles traveled or the number of hours required to make deliveries. For example:

- UPS has found that making left turns increases the time it takes to make deliveries. This in turn increases fuel usage and carbon emissions. So UPS plans its delivery truck routes with the fewest possible left turns. Likewise, airplanes fly at different altitudes and routes to take advantage of favorable wind conditions in an effort to reduce fuel use and carbon emissions.
- Food distribution companies now have trucks with three temperature zones (frozen, cool, and nonrefrigerated) instead of using three different types of trucks.
- Whirlpool radically revised its packaging to reduce "dings and dents" of appliances during delivery, generating huge savings in transportation and warranty costs.

To further enhance logistic efficiency, operations managers also evaluate equipment alternatives, taking into account cost, payback period, and the firm's stated environmental objectives. Example S2 deals with decision making that takes into account life cycle ownership costs. A firm must decide whether to pay *more* up front for vehicles to further its sustainability goals or to pay *less* up front for vehicles that do not.

Example S2

LIFE CYCLE OWNERSHIP AND CROSSOVER ANALYSIS

Blue Star is starting a new distribution service that delivers auto parts to the service departments of auto dealerships in the local area. Blue Star has found two light-duty trucks that would do the job well, so now it needs to pick one to perform this new service. The Ford TriVan costs $28,000 to buy and uses regular unleaded gasoline, with an average fuel efficiency of 24 miles per gallon. The TriVan has an operating cost of $.20 per mile. The Honda CityVan, a hybrid truck, costs $32,000 to buy and uses regular unleaded gasoline and battery power; it gets an average of 37 miles per gallon. The CityVan has an operating cost of $.22 per mile. The distance traveled annually is estimated to be 22,000 miles, with the life of either truck expected to be 8 years. The average gas price is $4.25 per gallon.

APPROACH ▶ Blue Star applies Equation (S5-2) to evaluate total life cycle cost for each vehicle:

$$\text{Total life cycle cost} = \text{Cost of vehicle} + \text{Life cycle cost of fuel} + \text{Life cycle operating cost} \qquad \text{(S5-2)}$$

a) Based on life cycle cost, which model truck is the best choice?
b) How many miles does Blue Star need to put on a truck for the costs to be equal?
c) What is the crossover point in years?

SOLUTION ▶

a) Ford TriVan:

$$\text{Total life-cycle cost} = \$28{,}000 + \left[\frac{22{,}000\,\dfrac{miles}{year}}{24\,\dfrac{miles}{gallon}}\right](\$4.25\ /\ gallon)(8\ years) + \left(22{,}000\,\frac{miles}{year}\right)(\$.20\ /\ mile)(8\ years)$$

$$= \$28{,}000 + \$31{,}167 + \$35{,}200 = \$94{,}367$$

Honda CityVan:

$$\text{Total life-cycle cost} = \$32{,}000 + \left[\frac{22{,}000\,\dfrac{miles}{year}}{37\,\dfrac{miles}{gallon}}\right](\$4.25\ /\ gallon)(8\ years) + \left(22{,}000\,\frac{miles}{year}\right)(\$.22\ /\ mile)(8\ years)$$

$$= \$32{,}000 + \$20{,}216 + \$38{,}720 = \$90{,}936$$

b) Blue Star lets M be the crossover (break-even) point in miles, sets the two life cycle cost equations equal to each other, and solves for M:

Total cost for Ford TriVan = Total cost for Honda CityVan

$$\$28{,}000 + \left[\frac{4.25\,\dfrac{\$}{gallon}}{24\,\dfrac{miles}{gallon}} + .20\,\frac{\$}{mile}\right](M\ miles) = \$32{,}000 + \left[\frac{4.25\,\dfrac{\$}{gallon}}{37\,\dfrac{miles}{gallon}} + .22\,\frac{\$}{mile}\right](M\ miles)$$

or,

$$\$28{,}000 + \left(.3770\,\frac{\$}{mile}\right)(M) = \$32{,}000 + \left(.3349\,\frac{\$}{mile}\right)(M)$$

or,

$$\left(.0421\,\frac{\$}{mile}\right)(M) = \$4{,}000$$

$$M = \frac{\$4{,}000}{.0421\,\dfrac{\$}{mile}} = 95{,}012\ miles$$

c) The crossover point in years is:

$$\text{Crossover point} = \frac{95{,}012\ miles}{22{,}000\,\dfrac{miles}{year}} = 4.32\ years$$

INSIGHTS ▶

a) Honda CityVan is the best choice, even though the initial fixed cost and variable operating cost per mile are higher. The savings comes from the better fuel mileage (more miles per gallon) for the Honda CityVan.
b) The crossover (break-even) point is at 95,012 miles, which indicates that at this mileage point, the cost for either truck is the same.

c) It will take 4.32 years to recoup the cost of purchasing and operating either vehicle. This is the point of indifference between the two vehicles. However, it will cost Blue Star approximately \$.03 per mile less to operate the Honda CityVan than the Ford TriVan over the 8-year expected life.

LEARNING EXERCISE ▶ If the cost of gasoline drops to \$3.25, what will be the total life-cycle cost of each van, the break-even point in miles, and the crossover point in years? [Answer: The cost of the Ford TriVan is \$87,033; the Honda CityVan costs \$86,179; the crossover point in miles is 144,927; and the crossover point in years is 6.59.]

RELATED PROBLEMS ▶ S5.4, S5.5, S5.6, S5.10, S5.11, S5.15, S5.16, S5.17, S5.18, S5.19

End-of-Life Phase

We noted earlier that during product design, managers need to consider what happens to a product or its materials after the product reaches its end-of-life stage. Products with less material, with recycled material, or with recyclable materials all contribute to sustainability efforts, reducing the need for the "burn or bury" decision and conserving scarce natural resources.

Closed-loop supply chains
Supply chains that consider forward and reverse product flows over the entire life cycle.

Innovative and sustainability-conscious companies are now designing **closed-loop supply chains**, also called *reverse logistics*. Firms can no longer sell a product and then forget about it. They need to design and implement end-of-life systems for the physical return of products that facilitate recycling or reuse.

Caterpillar, through its expertise in remanufacturing technology and processes, has devised *Cat Reman,* a remanufacturing initiative, in an effort to show its commitment to sustainability. Caterpillar remanufactures parts and components that provide same-as-new performance and reliability at a fraction of new cost, while reducing the impact on the environment. The remanufacturing program is based on an exchange system where customers return a used component in exchange for a remanufactured product. The result is lower operating costs for the customer, reduced material waste, and less need for raw material to make new products. In a 1-year period, Caterpillar took back 2.1 million end-of-life units and remanufactured more than 130 million pounds of material from recycled iron.

The *OM in Action* box, "Designing for End of Life," describes Apple's design philosophy to facilitate the disassembly, recycling, and reuse of its iPhones that have reached their end of life.

Regulations and Industry Standards

LO S5.5 *Explain* the impact of sustainable regulations on operations

Government, industry standards, and company policies are all important factors in operational decisions. Failure to recognize these constraints can be costly. Over the last 100 years, we have seen development of regulations, standards, and policies to guide managers in product design, manufacturing/assembly, and disassembly/disposal.

To guide decisions in *product design*, U.S. laws and regulations, such as those of the Food and Drug Administration, Consumer Product Safety Commission, and National Highway Safety Administration, provide guidance and often explicit regulations.

Manufacturing and assembly activities have their own set of regulatory agencies providing guidance and standards of operations. These include the Occupational Safety and Health Administration (OSHA), Environmental Protection Agency (EPA), and many state and local agencies that regulate workers' rights and employment standards.

STUDENT TIP
A group of 100 apparel brands and retailers have created the Eco Index to display an eco-value on a tag, like the Energy Star rating does for appliances.

U.S. agencies that govern the *disassembly and disposal of hazardous products* include the EPA and the Department of Transportation. As product life spans shorten due to ever-changing trends and innovation, product designers are under added pressure to *design for disassembly*. This encourages designers to create products that can be disassembled and whose components can be recovered, minimizing impact on the environment.

Organizations are obliged by society and regulators to reduce harm to consumers, employees, and the environment. The result is a proliferation of community, state, federal, and even international laws that often complicate compliance. The lack of coordination of regulations and reporting requirements between jurisdictions adds not just complexity but cost.

OM in Action Designing for End of Life[4]

Apple just introduced a piece of technology that will likely never be used by any consumer. Instead, it kind of cleans up after consumers: a robot that breaks down 1.2 million iPhones a year for recycling. The arrival of Liam—a 29-armed robot—addresses a big challenge facing tech manufacturers today. Even as they entice consumers to ditch their existing devices for the next new thing, companies must figure out what to do with the growing number of devices that are destined for the scrapheap. "We think as much now about the recycling and end of life of products as the design of products itself," says Apple's VP of environment, policy and social initiatives.

Stephan Lam/REUTERS Pictures

Other electronics makers take a different recycling approach, designing products that simplify disassembly by replacing glue and screws with parts that snap together, for instance. Some also have reduced the variety of plastics used and avoid mercury and other hazardous materials that can complicate disposal. Samsung designed a recent model of its 55-inch curved television for easier disassembly, eliminating 30 of 38 screws and replacing them with snap closures. Now the TVs can be dismantled in less than 10 minutes.

The constant stream of new devices has contributed to an increase in electronic waste, with its inherent health and environmental risks. A product's design choices for the types of materials and varieties of components are an important part of a successful circular economy.

From the following examples it is apparent that nearly all industries must abide by regulations in some form or another:

- Commercial homebuilders are required not just to manage water runoff but to have a pollution prevention plan for each site.
- Public drinking water systems must comply with the federal Safe Drinking Water Act's arsenic standard, even for existing facilities.
- Hospitals are required to meet the terms of the Resource Conservation and Recovery Act, which governs the storage and handling of hazardous material.

The consequences of ignoring regulations can be disastrous and even criminal. The EPA investigates environmental crimes in which companies and individuals are held accountable. Prison time and expensive fines can be handed down. (British Petroleum paid billions of dollars in fines in the past few years for breaking U.S. environmental and safety laws.) Even if a crime has not been committed, the financial impacts and customer upheaval can be disastrous to companies that do not comply with regulations. Due to lack of supplier oversight, Mattel, Inc., the largest U.S. toymaker, has recalled more than 10 million toys in recent years because of consumer health hazards such as lead paint.

International Environmental Policies and Standards

Organizations such as the U.N. Framework Convention on Climate Change (UNFCCC), International Organization for Standardization (ISO), and governments around the globe are guiding businesses to reduce environmental impacts from disposal of materials to reductions in greenhouse gas (GHG) emissions. Some governments are implementing laws that mandate the outright reduction of GHG emissions by forcing companies to pay taxes based on the amount of GHG emissions that are emitted. We now provide an overview of some of the international standards that apply to how businesses operate, manufacture, and distribute goods and services.

European Union Emissions Trading System The European Union (EU) has developed and implemented the EU Emissions Trading System (EUETS) to combat climate change. This is the key tool for reducing industrial greenhouse gas emissions in the EU. The EUETS works on the "cap-and-trade" principle. This means there is a cap, or limit, on the total amount of certain GHGs that can be emitted by factories, power plants, and airlines in EU airspace. Within this cap, companies receive emission allowances, which they can sell to, or buy from, one another as needed.

ISO 14001 and 50001 The International Organization for Standardization (ISO) is widely known for its contributions in ISO 9000 quality assurance standards (discussed in Chapter 6). The ISO 14001 and ISO 50001 family grew out of the ISO's commitment to support U.N. and government objectives of sustainable development. ISO 14001 is a series of environmental management standards that contain five core elements: (1) environmental management, (2) auditing, (3) performance evaluation, (4) labeling, and (5) life cycle assessment. The more recently introduced ISO 50001 sets out the criteria for improving energy use, as well as the efficiency and security of energy consumption. These standards are primarily aimed at working toward the reduction of greenhouse gas (carbon dioxide) emissions, while at the same time helping businesses reduce their energy usage and costs. ISO 50001 is designed to work alongside ISO 14001. Its main objective is to help businesses and individuals maintain constant awareness of the need for good energy reduction practices and identify potential energy savings. The ISO standards have several advantages:

ISO 14001
A series of environmental management standards established by the International Organization for Standardization (ISO).

ISO 50001
Environmental standards for improving energy performance.

- Positive public image and reduced exposure to liability
- Good systematic approach to pollution prevention through minimization of ecological impact of products and activities
- Compliance with regulatory requirements and opportunities for competitive advantage

ISO sustainability standards have been implemented by more than 200,000 organizations in 155 countries. Companies that have done so report environmental and economic benefits such as reduced raw material/resource use, reduced energy consumption, lower distribution costs, improved corporate image, improved process efficiency, reduced waste generation and disposal costs, and better utilization of recoverable resources.

ISO 14001 and ISO 50001 give guidance to companies to minimize harmful effects on the environment caused by their activities. The *OM in Action* box, "Subaru's Clean, Green Set of Wheels with ISO 14001 and ISO 50001," illustrates the growing application of the ISO environmental standards.

OM in Action — Subaru's Clean, Green Set of Wheels with ISO 14001 and ISO 50001[5]

"Going green" had humble beginnings. First, it was newspapers, soda cans and bottles, and corrugated packaging—the things you typically throw into your own recycling bins. Similarly, at Subaru's Lafayette, Indiana, plant, the process of becoming the first completely waste-free auto plant in North America began with employees dropping these items in containers throughout the plant. Then came employee empowerment. "We had 268 suggestions for different things to improve our recycling efforts," said Denise Coogan, plant ISO environmental compliance leader.

Some ideas were easy to handle. "With plastic shrink wrap, we found some (recyclers) wouldn't take colored shrink wrap. So we went back to our vendors and asked for only clear shrink wrap," Coogan said. Some suggestions were a lot dirtier. "We went dumpster diving to see what we were throwing away and see what we could do with it."

The last load of waste generated by Subaru made its way to a landfill 10 years ago. Since then, everything that enters the plant eventually exits as a usable product. Coogan adds, "We didn't redefine 'zero.' Zero means zero. Nothing from our manufacturing process goes to the landfill."

Last year alone, the Subaru plant recycled 13,142 tons of steel, 1,448 tons of paper products, 194 tons of plastics, 10 tons of solvent-soaked rags, and 4 tons of light bulbs. Doing so conserved 29,200 trees, 670,000 gallons of oil, 34,700 gallons of gas, 10 million gallons of water, and 53,000 million watts of electricity. "Going green" isn't easy, but it can be done!

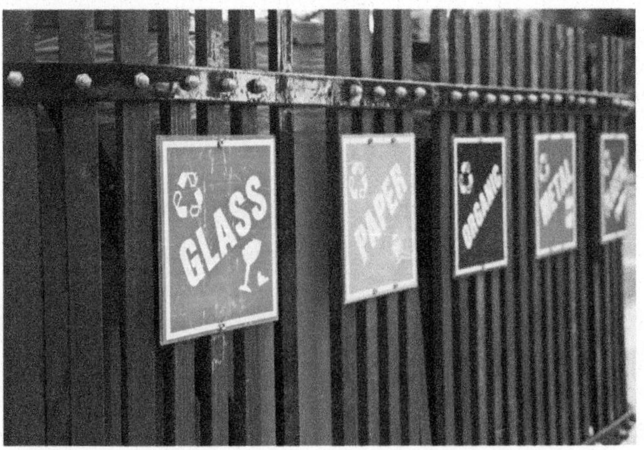

Cnky Photography/Fotolia

Summary

If a firm wants to be viable and competitive, it must have a strategy for corporate social responsibility and sustainability. Operations and supply chain managers understand that they have a critical role in a firm's sustainability objectives. Their actions impact all the stakeholders. They must continually seek new and innovative ways to design, produce, deliver, and dispose of profitable, customer-satisfying products while adhering to many environmental regulations. Without the expertise and commitment of operations and supply chain managers, firms are unable to meet their sustainability obligations.

Key Terms

Corporate social responsibility (CSR) (p. 192)
Shared value (p. 192)
Sustainability (p. 192)
Carbon footprint (p. 194)
Economic sustainability (p. 194)
Life cycle assessment (p. 195)
Circular economy (p. 195)
Closed-loop supply chains (p. 200)
ISO 14001 (p. 202)
ISO 50001 (p. 202)

Discussion Questions

1. Why must companies practice corporate social responsibility?
2. Find statements of sustainability for a well-known company online and analyze that firm's policy.
3. Explain sustainability.
4. Discuss the circular economy.
5. Explain closed-loop supply chains.
6. How would you classify a company as green?
7. Why are sustainable business practices important?

Solved Problems

SOLVED PROBLEM S5.1

The design team for Superior Electronics is creating a mobile audio player and must choose between two design alternatives. Which is the better environmental design alternative, based on achieving a higher revenue retrieval opportunity?

SOLUTION

Collecting the resale revenue per unit, recycling revenue per unit, processing cost per unit, and the disposal cost per unit, the design team computes the revenue retrieval for each design:

Design 1

PART	RESALE REVENUE PER UNIT	RECYCLING REVENUE PER UNIT	PROCESSING COST PER UNIT	DISPOSAL COST PER UNIT
Tuner	$4.93	$2.08	$2.98	$0.56
Speaker	0.00	0.00	4.12	1.23
Case	6.43	7.87	4.73	0.00
Total	$11.36	$9.95	$11.83	$1.79

Design 2

PART	RESALE REVENUE PER UNIT	RECYCLING REVENUE PER UNIT	PROCESSING COST PER UNIT	DISPOSAL COST PER UNIT
Tuner	$6.91	$4.92	$3.41	$2.13
Case	5.83	3.23	2.32	1.57
Amplifier	1.67	2.34	4.87	0.00
Speaker	0.00	0.00	3.43	1.97
Total	$14.41	$10.49	$14.03	$5.67

Using the following formula [Equation (S5-1)], compare the two design alternatives:

$$\text{Revenue retrieval} = \text{Total resale revenue} + \text{Total recycling recycling revenue} - \text{Total processing cost} - \text{Total disposal cost}$$

$$\text{Revenue retrieval Design 1} = \$11.36 + \$9.95 - \$11.83 - \$1.79 = \$7.69$$

$$\text{Revenue retrieval Design 2} = \$14.41 + \$10.49 - \$14.03 - \$5.67 = \$5.20$$

Design 1 brings in the most revenue from its design when the product has reached its end of life.

SOLVED PROBLEM S5.2

The City of High Point is buying new school buses for the local school system. High Point has found two models of school buses that it is interested in. Eagle Mover costs $80,000 to buy and uses diesel fuel, with an average fuel efficiency of 10 miles per gallon. Eagle Mover has an operating cost of $.28 per mile. Yellow Transport, a hybrid bus, costs $105,000 to buy and uses diesel fuel and battery power, getting an average of 22 miles per gallon. Yellow Transport has an operating cost of $.32 per mile. The distance traveled annually is determined to be 25,000 miles, with the expected life of either bus to be 10 years. The average diesel price is $3.50 per gallon.

SOLUTION

a) Based on life cycle cost, which bus is the better choice?

Eagle Mover:

$$\$80{,}000 + \left[\frac{25{,}000 \dfrac{miles}{year}}{10 \dfrac{miles}{gallon}}\right](\$3.50\ /\ gallon)(10\ years) + \left(25{,}000 \frac{miles}{year}\right)(\$.28\ /\ mile)(10\ years)$$

$$= \$80{,}000 + \$87{,}500 + \$70{,}000 = \$237{,}500$$

Yellow Transport:

$$\$105{,}000 + \left[\frac{25{,}000\,\frac{miles}{year}}{22\,\frac{miles}{gallon}}(\$3.50\ /\ gallon)(10\ years) + \left(25{,}000\,\frac{miles}{year}\right)(\$.32\ /\ mile)(10\ years)\right]$$

$$= \$105{,}000 + \$39{,}773 + \$80{,}000 = \$224{,}773$$

Yellow Transport is the better choice.

b) How many miles does the school district need to put on a bus for costs to be equal?

Let M be the break-even point in miles, set the equations equal to each other, and solve for M:

$$\text{Total cost for Eagle Mover} = \text{Total cost for Yellow Transport}$$

$$\$80{,}000 + \left[\frac{3.50\,\frac{\$}{gallon}}{10\,\frac{miles}{gallon}} + .28\,\frac{\$}{mile}\right](M\ miles) = \$105{,}000 + \left[\frac{3.50\,\frac{\$}{gallon}}{22\,\frac{miles}{gallon}} + .32\,\frac{\$}{mile}\right](M\ miles)$$

$$\$80{,}000 + \left(.630\,\frac{\$}{mile}\right)(M) = \$105{,}000 + \left(.479\,\frac{\$}{mile}\right)(M)$$

$$\left(.151\,\frac{\$}{mile}\right)(M) = \$25{,}000$$

$$M = \frac{\$25{,}000}{.151\,\frac{\$}{mile}} = 165{,}563\ miles$$

c) What is the crossover point in years?

$$\text{Crossover point} = \frac{165{,}563\ miles}{25{,}000\,\frac{miles}{year}} = 6.62\ years$$

Problems

Problems S5.1–S5.19 relate to Design and Production for Sustainability

•• **S5.1** The Brew House needs to decide which of two coffee maker designs is better environmentally. Using the following tables, determine which model is the better design alternative.

Brew Master

PART	RESALE REVENUE PER UNIT	RECYCLING REVENUE PER UNIT	PROCESSING COST PER UNIT	DISPOSAL COST PER UNIT
Metal frame	$1.65	$2.87	$1.25	$0.75
Timer	0.50	0.00	1.53	1.45
Plug/cord	4.25	5.65	6.22	0.00
Coffee pot	2.50	2.54	2.10	1.35

Brew Mini

PART	RESALE REVENUE PER UNIT	RECYCLING REVENUE PER UNIT	PROCESSING COST PER UNIT	DISPOSAL COST PER UNIT
Plastic frame	$1.32	$3.23	$0.95	$0.95
Plug/cord	3.95	4.35	5.22	0.00
Coffee pot	2.25	2.85	2.05	1.25

•• **S5.2** Using the information in Problem S5.1, which design alternative is the better environmental choice if the Brew House decided to add a timer to the Brew Mini model? The timer revenue and costs are identical to those of the Brew Master.

•• **S5.3** Using the information in Problem S5.1, which design alternative is the better environmental choice if the Brew House decided to remove the timer from the Brew Master model?

•• **S5.4** What is the total vehicle life cycle cost of this hybrid car, given the information provided in the following table?

VEHICLE PURCHASE COST	$17,000
VEHICLE OPERATING COST PER MILE	$0.12
USEFUL LIFE OF VEHICLE	15 years
MILES PER YEAR	14,000
MILES PER GALLON	32
AVERAGE FUEL PRICE PER GALLON	$3.75

•• **S5.5** What is the crossover point in miles between the hybrid vehicle in Problem S5.4 and this alternative vehicle from a competing auto manufacturer?

VEHICLE PURCHASE COST	$19,000
VEHICLE OPERATING COST PER MILE	$0.09
USEFUL LIFE OF VEHICLE	15 years
MILES PER YEAR	14,000
MILES PER GALLON	35
AVERAGE FUEL PRICE PER GALLON	$3.75

•• **S5.6** Given the crossover mileage in Problem S5.5, what is the crossover point in years?

•• **S5.7** In Problem S5.5, if gas prices rose to $4.00 per gallon, what would be the new crossover point in miles?

•• **S5.8** Using the new crossover mileage in Problem S5.7, what is the crossover point in years?

•• **S5.9** Mercedes is assessing which of two windshield suppliers provides a better environmental design for disassembly. Using the tables below, select between PG Glass and Glass Unlimited.

PG Glass

PART	RESALE REVENUE PER UNIT	RECYCLING REVENUE PER UNIT	PROCESSING COST PER UNIT	DISPOSAL COST PER UNIT
Glass	$12	$10	$6	$2
Steel frame	2	1	1	1
Rubber insulation	1	2	1	1

Glass Unlimited

PART	RESALE REVENUE PER UNIT	RECYCLING REVENUE PER UNIT	PROCESSING COST PER UNIT	DISPOSAL COST PER UNIT
Reflective glass	$15	$12	$7	$3
Aluminum frame	4	3	2	2
Rubber insulation	2	2	1	1

•• **S5.10** Environmentally conscious Susan has been told that a new electric car will only generate 6 ounces of greenhouse gases (GHGs) per mile, but that a standard internal combustion car is double that at 12 ounces per mile. However, the nature of electric cars is such that the new technology and electric batteries generate 30,000 lbs. of GHGs to manufacture and another 10,000 lbs. to recycle. A standard car generates only 14,000 lbs. of GHGs to manufacture, and recycling with established technology is only 1,000 lbs. Susan is interested in taking a systems approach that considers the life-cycle impact of her decision. How many miles must she drive the electric car for it to be the preferable decision in terms of reducing greenhouse gases?

••• **S5.11** A Southern Georgia school district is considering ordering 53 propane-fueled school buses. "They're healthier, they're cleaner burning, and they're much quieter than the diesel option," said a school administrator. Propane-powered buses also reduce greenhouse gases by 22% compared to gasoline-powered buses and 6% compared to diesel ones. But they come at a premium—$103,000 for a propane model, $15,000 more than the diesel equivalent.

The propane bus operating cost (above and beyond fuel cost) is 30 cents/mile, compared to 40 cents for the diesel. Diesel fuel costs about $2/gallon in Georgia, about $1 more than propane. Bus mileage is 12 mpg for the propane model vs. 10 mpg for diesel. The life of a school bus in the district averages 9 years, and each bus travels an average of 30,000 miles per year because the district is so large and rural.

Which bus is the better choice based on a life cycle analysis?

•• **S5.12** Green Forever, a manufacturer of lawn equipment, has preliminary drawings for two grass trimmer designs. Charla Fraley's job is to determine which is better environmentally. Specifically, she is to use the following data to help the company determine:

a) The revenue retrieval for the GF Deluxe
b) The revenue retrieval for the Premium Mate
c) Which model is the better design alternative based on revenue retrieval

GF Deluxe

PART	RESALE REVENUE PER UNIT	RECYCLING REVENUE PER UNIT	PROCESSING COST PER UNIT	DISPOSAL COST PER UNIT
Metal drive	$3.27	$4.78	$1.05	$0.85
Battery	0.00	3.68	6.18	3.05
Motor housing	3.93	2.95	2.05	1.25
Trimmer head	1.25	0.75	1.00	0.65

Premium Mate

PART	RESALE REVENUE PER UNIT	RECYCLING REVENUE PER UNIT	PROCESSING COST PER UNIT	DISPOSAL COST PER UNIT
Metal drive	$3.18	$3.95	$1.15	$0.65
Battery	0.00	2.58	4.98	2.90
Motor housing	4.05	3.45	2.45	1.90
Trimmer head	1.05	0.85	1.10	0.75

•• **S5.13** Green Forever (see Problem S5.12) has decided to add an automatic string feeder system with cost and revenue estimates as shown below to the GF Deluxe model.

a) What is the new revenue retrieval value for each model?
b) Which model is the better environmental design alternative?

PART	RESALE REVENUE PER UNIT	RECYCLING REVENUE PER UNIT	PROCESSING COST PER UNIT	DISPOSAL COST PER UNIT
String feeder system	$1.05	$1.25	$1.50	$1.40

•• **S5.14** Green Forever's challenge (see Problem S5.12) is to determine which design alternative is the better environmental choice if it uses a different battery for the Premium Mate. The alternate battery revenue and costs are as follows:

PART	RESALE REVENUE PER UNIT	RECYCLING REVENUE PER UNIT	PROCESSING COST PER UNIT	DISPOSAL COST PER UNIT
Battery	$0.00	$3.68	$4.15	$3.00

a) What is the revenue retrieval for the GF Deluxe?
b) What is the revenue retrieval for the Premium Mate?
c) Which is the better environmental design alternative?

•• **S5.15** Hartley Auto Supply delivers parts to area auto service centers and is replacing its fleet of delivery vehicles. What is the total vehicle life cycle cost of this gasoline engine truck given the information provided in the following table?

VEHICLE PURCHASE COST	$25,000
VEHICLE OPERATING COST PER MILE	$0.13
USEFUL LIFE OF VEHICLE	10 years
MILES PER YEAR	18,000
MILES PER GALLON	25
AVERAGE FUEL PRICE PER GALLON	$2.55

• • **S5.16** Given the data in Problem S5.15 and an alternative hybrid vehicle with the specifications shown below:

a) What is the crossover point in miles?
b) Which vehicle is has the lowest cost until the crossover point is reached?

VEHICLE PURCHASE COST	$29,000
VEHICLE OPERATING COST PER MILE	$0.08
USEFUL LIFE OF VEHICLE	10 years
MILES PER YEAR	18,000
MILES PER GALLON	40
AVERAGE FUEL PRICE PER GALLON	$2.55

• **S5.17** Based on the crossover point in miles found in Problem S5.16, what is this point in years?

• • **S5.18** Using the data from Problem S5.16, if gas prices rose to $3.00 per gallon, what would be the new crossover point in miles?

• **S5.19** Using the new crossover point in Problem S5.18, how many years does it take to reach that point?

CASE STUDIES

Building Sustainability at the Orlando Magic's Amway Center

Video Case 

When the Amway Center opened in Orlando in 2011, it became the first Leadership in Energy and Environmental Design (LEED) gold-certified professional basketball arena in the country. It took 10 years for Orlando Magic's management to develop a plan for the new state-of-the-art sports and entertainment center. The community received not only an entertainment center but an environmentally sustainable building to showcase in its revitalized downtown location. "We wanted to make sure we brought the most sustainable measures to the construction, so in operation we can be a good partner to our community and our environment," states CEO Alex Martins. The new 875,000-square-foot facility—almost triple the size of the Amway Arena it replaced—is now the benchmark for other sports facilities.

Here are a few of the elements in the Amway Center project that helped earn the LEED certification:

- The roof of the building is designed to minimize daytime heat gain by using reflective and insulated materials.
- Rainwater and air-conditioning condensation are captured and used for irrigation.
- There is 40% less water usage than in similar arenas (saving 800,000 gallons per year), mostly through use of high-efficiency restrooms, including low-flow, dual-flush toilets.
- There is 20% energy savings (about $750,000 per year) with the use of high-efficiency heating and cooling systems.
- The center used environmentally friendly building materials and recycled 83% of the wood, steel, and concrete construction waste that would have ended up in a landfill.
- There is preferred parking for hybrids and other energy-efficient cars.
- The center is maintained using green-friendly cleaning products.

LEED certification means five environmental measures and one design measure must be met when a facility is graded by the U.S. Green Building Council, which is a nationally accepted benchmark program. The categories are sustainability of site, water efficiency, energy, materials/resources, indoor environmental quality, and design innovation.

Other Amway Center design features include efficient receiving docks, food storage layouts, and venue change-over systems. Massive LED electronic signage controlled from a central control room also contributes to lower operating costs. From an operations management perspective, combining these savings with the significant ongoing savings from reduced water and energy usage will yield a major reduction in annual operating expenses. "We think the LEED certification is not only great for the environment but good business overall," says Martins.

Discussion Questions*

1. Find a LEED-certified building in your area and compare its features to those of the Amway Center.
2. What does a facility need to do to earn the gold LEED rating? What other ratings exist?
3. Why did the Orlando Magic decide to "go green" in its new building?

*You may wish to view the video that accompanies this case before answering these questions.

Fernando Medina/NBAE/Getty Images

Green Manufacturing and Sustainability at Frito-Lay

Video Case

Frito-Lay, the multi-billion-dollar snack food giant, requires vast amounts of water, electricity, natural gas, and fuel to produce its 41 well-known brands. In keeping with growing environmental concerns, Frito-Lay has initiated ambitious plans to produce environmentally friendly snacks. But even environmentally friendly snacks require resources. Recognizing the environmental impact, the firm is an aggressive "green manufacturer," with major initiatives in resource reduction and sustainability.

For instance, the company's energy management program includes a variety of elements designed to engage employees in reducing energy consumption. These elements include scorecards and customized action plans that empower employees and recognize their achievements.

At Frito-Lay's factory in Casa Grande, Arizona, more than 500,000 pounds of potatoes arrive every day to be washed, sliced, fried, seasoned, and portioned into bags of Lay's and Ruffles chips. The process consumes enormous amounts of energy and creates vast amounts of wastewater, starch, and potato peelings. Frito-Lay plans to take the plant off the power grid and run it almost entirely on renewable fuels and recycled water. The managers at the Casa Grande plant have also installed skylights in conference rooms, offices, and a finished goods warehouse to reduce the need for artificial light. More fuel-efficient ovens recapture heat from exhaust stacks. Vacuum hoses that pull moisture from potato slices to recapture the water and to reduce the amount of heat needed to cook the potato chips are also being used.

Frito-Lay has also built more than 50 acres of solar concentrators behind its Modesto, California, plant to generate solar power. The solar power is being converted into heat and used to cook Sun Chips. A biomass boiler, which will burn agricultural waste, is also planned to provide additional renewable fuel.

Frito-Lay is installing high-tech filters that recycle most of the water used to rinse and wash potatoes. It also recycles corn by-products to make Doritos and other snacks; starch is reclaimed and sold, primarily as animal feed, and leftover sludge is burned to create methane gas to run the plant boiler.

There are benefits besides the potential energy savings. Like many other large corporations, Frito-Lay is striving to establish its green credentials as consumers become more focused on environmental issues. There are marketing opportunities, too. The company, for example, advertises that its popular Sun Chips snacks are made using solar energy.

At Frito-Lay's Florida plant, only 3.5% of the waste goes to landfills, but that is still 1.5 million pounds annually. The goal is zero waste to landfills. The snack food maker earned its spot in the National Environmental Performance Task Program by maintaining a sustained environmental compliance record and making new commitments to reduce, reuse, and recycle at this facility.

Substantial resource reductions have been made in the production process, with an energy reduction of 21% across Frito-Lay's 34 U.S. plants. But the continuing battle for resource reduction continues. The company is also moving toward biodegradable packaging and seasoning bags and cans and bottles. While these multiyear initiatives are expensive, they have the backing at the highest levels of Frito-Lay as well as corporate executives at PepsiCo, the parent company.

Discussion Questions*

1. Why do Frito-Lay's stakeholders have an interest in reducing the firm's environmental footprint?
2. Identify the specific techniques that Frito-Lay is using to become a "green manufacturer."
3. Select another company and compare its green policies to those of Frito-Lay.
4. What other changes could Frito-Lay make to increase sustainability in its processes and physical facilities?

*You may wish to view the video that accompanies this case before answering these questions.

"Saving the Waves" at Celebrity Cruises

Video Case

Celebrity Cruises, with its reputation as a leader in the "premium cruise" industry, decided in 2013 to become an industry leader in environmental sustainability as well. Celebrity's "Save The Waves" program was initially set up to increase compliance with the government's "reduce, reuse, and recycle" policies. But Celebrity's "Above and Beyond Compliance" (ABC) philosophy has committed the company to creating improvements over and above those imposed by law, often years ahead of federal mandatory requirements. Celebrity has an environmental officer aboard each ship.

How does Celebrity reach its sustainability targets? "We have learned over the years that what you measure gets better, so these goals are measurable, but ambitious," says Senior V. P. Patrik Dahlgren. "First, we aim to reduce waste whenever possible. We work with our suppliers to reduce packaging materials and use more sustainable resources. Next, we reuse materials—participating in container return programs and establishing a donation database for our fleet, just as we work with passengers to reduce the laundering of towels and bed linen. Last year we donated tens of thousands of mattresses, towels, and furniture items that would have otherwise gone to landfills. Finally, we recycle. All trash onboard our ships is hand-sorted by our crew members to determine what can be recycled."

Celebrity's ships are now able to repurpose 100% of their operational waste with a policy of zero discharge into the oceans. And with Celebrity's goal of zero waste, its vendors are expected to supply food products and cleaning agents in reusable and returnable containers. Total waste recycled has increased every year since 2008, from 2 million pounds to over 30 million pounds recently.

By 2012, all of Celebrity's ships had already installed advanced water purification (AWP) systems (which add 1–1.5% to a ship's construction price). Their job is to treat "gray water" (the runoff produced by showers, laundry, pools, spas, and kitchens) and "black water" (drainage from toilets and medical facilities). The AWP systems produce output cleaner than water from most cities and cleaner than that required by law.

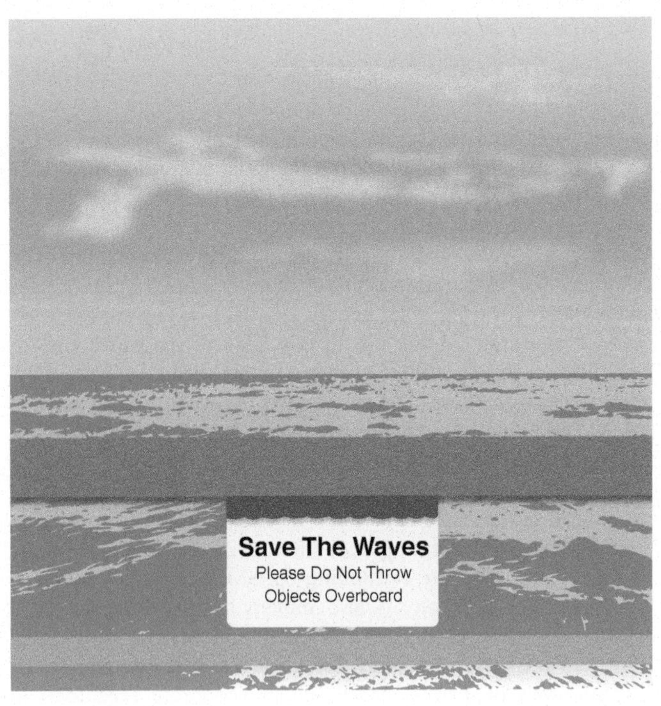

Save The Waves
Please Do Not Throw
Objects Overboard

Finally, Celebrity has worked with its architects and shipbuilders to find new ways to save energy and achieve investments in energy efficiency that yield 3-year paybacks. To encourage innovation from ship builders, Celebrity allocates a new ship's full first-year fuel savings due to innovation back to the builders as a financial reward. Celebrity's goal is a 20% fuel reduction with each new class of ships. Its "scrubbers," an advanced emission purification system, decrease sulfur dioxide emissions by 98%. New speed management software that optimizes engine use was rolled out in 2013. Figure S5.3, shown earlier in this chapter, illustrates the variety of "Save The Waves" initiatives underway.

Discussion Questions*

1. How are cruise line sustainability programs similar to and different from those of hotels on shore?
2. What environmental events in the cruise industry precipitated green efforts?
3. Compare Celebrity to other cruise lines in terms of sustainability.
4. What is the purpose of the "scrubbing" system?

*You may wish to view the video that accompanies this case before addressing these questions.

Endnotes

1. The authors wish to thank Dr. Steve Leon, Appalachian State University, for his contributions to this supplement.
2. See related discussions in M. E. Porter and M. R. Kramer, "Creating Shared Value," *Harvard Business Review* (Jan.–Feb. 2011) and M. Pfitzer, V. Bockstette, and M. Stamp, "Innovating for Shared Values," *Harvard Business Review* (Sept. 2013).
3. Sources: *The Wall Street Journal* (April 10, 2015); *Forbes* (April 17, 2016); and *The New York Times* (March 31, 2015).
4. Sources: *The Wall Street Journal* (June 8, 2016); *Fortune* (March 27, 2016); and *PC World* (March 29, 2016).
5. Sources: **Chase.com/news** (November 10, 2017); *IndyStar* (May 10, 2014); and *Businessweek* (June 6, 2011).

Bibliography

Banerjee, S. B. "Embedding Sustainability Across the Organization: A Critical Perspective." *Academy of Management, Learning & Education* 10, no. 4 (December 2011): 719–731.

Brinkmann, R. *Introduction to Sustainability*. Hoboken, NJ: Wiley-Blackwell, 2016.

Cerny, A. *Sustainability: Global Issues, Global Perspectives*. San Diego, CA: Cognella Publishing, 2016.

Haugh, H. M., and A. Talwar. "How Do Corporations Embed Sustainability Across the Organization?" *Academy of Management, Learning & Education* 9, no. 3 (September 2010): 384–396.

Laszlo, C., and N. Zhexembayeva. *Embedded Sustainability*. Stanford, CA: Stanford University Press, 2011.

Munson, C. (ed.), *The Supply Chain Management Casebook: Comprehensive Coverage and Best Practices in SCM*. Upper Saddle River, NJ: FT Press, 2013.

Robertson, M. *Sustainability Principles and Practice*, 2nd ed. Abingdon, Oxon: Routledge, 2017.

Supplement 5 ***Rapid* Review**

Main Heading	Review Material	MyLab Operations Management
CORPORATE SOCIAL RESPONSIBILITY	Managers must consider how the products and services they make affect people and the environment in which they operate. ■ **Corporate social responsibility (CSR)**—Managerial decision making that considers environmental, societal, and financial impacts. ■ **Shared value**—Developing policies and practices that enhance the competitiveness of an organization, while advancing the economic and social conditions in the communities in which it operates.	Concept Questions 1.1–1.2
SUSTAINABILITY	■ **Sustainability**—Meeting the needs of the present without compromising the ability of future generations to meet their needs. Systems view—Looking at a product's life from design to disposal, including all of the resources required. The commons—Inputs or resources for a production system that are held by the public. Triple bottom line—Systems needed to support the three *Ps: people, planet*, and *profit*. To support their *people,* many companies evaluate safety in the work environment, the wages paid, work hours/week. Apple, GE, P&G, and Walmart conduct audits of their suppliers to make sure sustainability goals are met. To support the *planet,* operation managers look for ways to reduce the environmental impact of their operations. ■ **Carbon footprint**—A measure of the total greenhouse gas emissions caused directly and indirectly by an organization, product, event or person. ■ **Economic sustainability**—Appropriately allocating scarce resources to make a profit. To support their *profits,* company investments must be sustainable economically. Firms may supplement standard accounting with social accounting.	Concept Questions: 2.1–2.6 **VIDEO S5.1** Building Sustainability at the Orlando Magic's Amway Center **VIDEO S5.2** Green Manufacturing and Sustainability at Frito-Lay
DESIGN AND PRODUCTION FOR SUSTAINABILITY	■ **Life cycle assessment**—Analysis of environmental impacts of products from the design stage through end of life. ■ **Circular economy**—Keeps and uses resources for as long as possible, recovering and regenerating the maximum possible value at end of life. Product design is the most critical phase in the product life cycle assessment. Design for disassembly focuses on reuse and recycle. Revenue retrieval = Total resale revenue + Total recycling revenue − Total processing cost − Total disposal cost (S5-1) Manufacturers also look for ways to reduce the amount of scarce resources in the production process. As products move along the supply chain, logistics managers strive to achieve efficient route and delivery networks, which reduce environmental impact. Vehicles are also evaluated on a life cycle ownership cost basis. A firm must decide whether to pay more up front for sustainable vehicles or pay less up front for vehicles that may be less sustainable. Total life cycle cost = Cost of vehicle + Life cycle cost of fuel + Life cycle operating cost (S5-2) ■ **Closed-loop supply chains**, also called *reverse logistics*—Supply chains that consider the product or its materials after the product reaches its end-of-life stage. This includes forward and reverse product flows. Green disassembly lines help take cars apart so that parts can be recycled. Recycling is the 16th-largest industry in the U.S.	Concept Questions: 3.1–3.4 Problems: S5.1–S5.19 Virtual Office Hours for Solved Problems S5.1–S5.2 **VIDEO S5.3** "Saving the Waves" at Celebrity Cruises

Main Heading	Review Material	MyLab Operations Management
REGULATIONS AND INDUSTRY STANDARDS	To guide *product design* decisions, U.S. laws and regulations often provide explicit regulations. *Manufacturing and assembly activities* are guided by OSHA, EPA, and many state and local agencies. There are also U.S. agencies that govern the *disassembly and disposal of hazardous products.* International environmental policies and standards come from the U.N., ISO, the EU, and governments around the globe. The EU has implemented the Emissions Trading System to help reduce greenhouse gas emissions. It works on a "cap-and-trade" principle. ▪ **ISO 14001**—The International Organization of Standardization's family of guidelines for sustainable development. It addresses environmental management systems. ▪ **ISO 50001**—The International Organization of Standardization's guidelines for improving energy management. They work alongside ISO 14001.	Concept Questions: 4.1–4.5
ADDITIONAL MYLAB OPERATIONS MANAGEMENT RESOURCES	✔ Additional Case Study (Environmental Sustainability at Walmart) ✔ Multiple Choice Case Questions (Green Manufacturing and Sustainability at Frito-Lay)	

Self Test

LO S5.1 Corporate social responsibility includes:
- **a)** doing what's right.
- **b)** having policies that consider environmental, societal, and financial impact.
- **c)** considering a product from design to disposal.
- **d)** all of the above.
- **e)** a and b only.

LO S5.2 Sustainability deals:
- **a)** solely with green products, recycling, global warming, and rain forests.
- **b)** with keeping products that are not recyclable.
- **c)** with meeting the needs of present and future generations.
- **d)** with three views—systems, commons, and defects.
- **e)** with not laying off older workers.

LO S5.3 The circular economy means:
- **a)** keep and use resources as long as possible, then recover and regenerate to the maximum possible value at end of service life.
- **b)** meet the needs of the present without compromising the ability of future generations to meet their needs.
- **c)** appropriately allocating scarce resources to make a profit.
- **d)** globalization has led to a "borderless" economy.
- **e)** focus on production design, production process, logistics, and end-of-life value.

LO S5.4 Design for disassembly is:
- **a)** cost–benefit analysis for old parts.
- **b)** analysis of the amount of revenue that might be reclaimed versus the cost of disposing of a product.
- **c)** a means of recycling plastic parts in autos.
- **d)** the use of lightweight materials in products.

LO S5.5 U.S. and international agencies provide policies and regulations to guide managers in product design, manufacturing/assembly, and disassembly/disposal. They include:
- **a)** U.N. Commission on Resettlement.
- **b)** World Health Organization (WHO).
- **c)** OSHA, FDA, EPA, and NHSA.
- **d)** EPA, ISO, and British High Commission.
- **e)** GHG Commission, UN, and ISO.

Answers: LO S5.1. d; LO S5.2. c; LO S5.3. a; LO S5.4. b; LO S5.5. c.

Managing Quality

CHAPTER 6

CHAPTER OUTLINE

GLOBAL COMPANY PROFILE: *Arnold Palmer Hospital*

Alaska Airlines

Like many service organizations, Alaska Airlines sets quality standards in areas such as courtesy, appearance, and time. Shown here are some of Alaska Airlines' 50 quality checkpoints, based on a timeline for each departure.

10 OM STRATEGY DECISIONS

- Design of Goods and Services
- *Managing Quality*
- Process Strategies
- Location Strategies
- Layout Strategies
- Human Resources
- Supply Chain Management
- Inventory Management
- Scheduling
- Maintenance

GLOBAL COMPANY PROFILE
Arnold Palmer Hospital

Managing Quality Provides a Competitive Advantage at Arnold Palmer Hospital

Since 1989, Arnold Palmer Hospital, named after its famous golfing benefactor, has touched the lives of over 7 million children and women and their families. Its patients come not only from its Orlando location but also from all 50 states and around the world. More than 14,000 babies are delivered every year at Arnold Palmer, and its huge neonatal intensive care unit boasts one of the highest survival rates in the U.S.

Every hospital professes quality health care, but at Arnold Palmer quality is the mantra—practiced in a fashion like the Ritz-Carlton practices it in the hotel industry. The hospital typically scores in the top 10% of national benchmark studies in terms of patient satisfaction. And its managers follow patient questionnaire results daily. If anything is amiss, corrective action takes place immediately.

Virtually every quality management technique we present in this chapter is employed at Arnold Palmer Hospital:

- *Continuous improvement:* The hospital constantly seeks new ways to lower infection rates, readmission rates, deaths, costs, and hospital stay times.

Courtesy Arnold Palmer Medical Center

The lobby of Arnold Palmer Hospital, with its 20-foot-high Genie, is clearly intended as a warm and friendly place for children.

Arnold Palmer Hospital for Children

The Storkboard is a visible chart of the status of each baby about to be delivered, so all nurses and doctors are kept up to date at a glance.

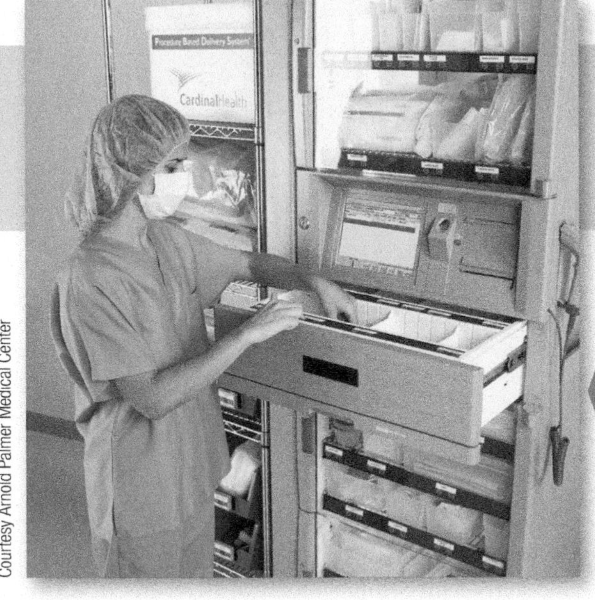

Courtesy Arnold Palmer Medical Center

This PYXIS inventory station gives nurses quick access to medicines and supplies needed in their departments. When the nurse removes an item for patient use, the item is automatically billed to that account, and usage is noted at the main supply area.

The hospital has redesigned its neonatal rooms. In the old system, there were 16 neonatal beds in an often noisy and large room. The new rooms are semiprivate, with a quiet simulated-night atmosphere. These rooms have proven to help babies develop and improve more quickly.

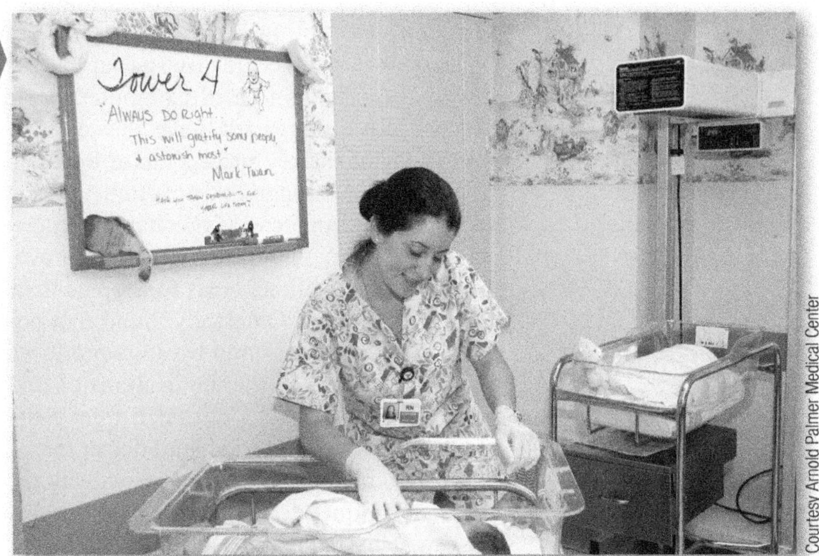

Courtesy Arnold Palmer Medical Center

- *Employee empowerment:* When employees see a problem, they are trained to take care of it; staff are empowered to give gifts to patients displeased with some aspect of service.

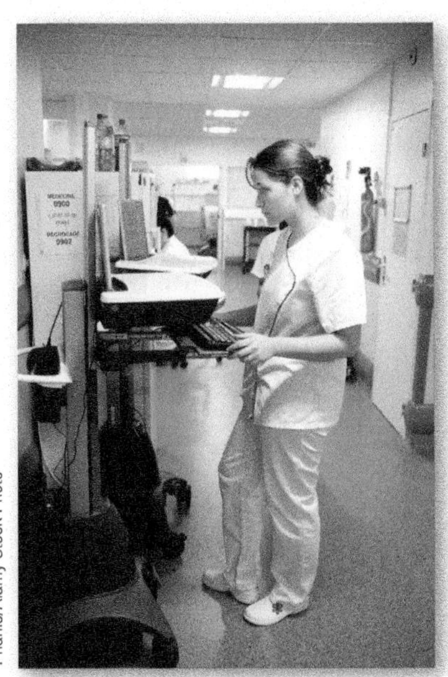

Phanie/Alamy Stock Photo

Technology is moving from a central computer at the nursing station to computers on carts and even in each room. This saves much walking time by nurses, helping them better care for patients.

- *Benchmarking:* The hospital belongs to a 2,000-member organization that monitors standards in many areas and provides monthly feedback to the hospital.
- *Just-in-time (JIT):* Supplies are delivered to Arnold Palmer on a JIT basis. This keeps inventory costs low and keeps quality problems from hiding.
- *Tools such as Pareto charts and flowcharts:* These tools monitor processes and help the staff graphically spot problem areas and suggest ways they can be improved.

From their first day of orientation, employees from janitors to nurses learn that the patient comes first. Staff standing in hallways will never be heard discussing their personal lives or commenting on confidential issues of health care. This culture of quality at Arnold Palmer Hospital makes a hospital visit, often traumatic to children and their parents, a warmer and more comforting experience.

LEARNING OBJECTIVES

Quality and Strategy

VIDEO 6.1
The Culture of Quality at Arnold Palmer Hospital

As Arnold Palmer Hospital and many other organizations have found, quality is a wonderful tonic for improving operations. Managing quality helps build successful strategies of *differentiation, low cost*, and *response*. For instance, defining customer quality expectations has helped Bose Corp. successfully *differentiate* its stereo speakers as among the best in the world. Nucor has learned to produce quality steel at *low cost* by developing efficient processes that produce consistent quality. And Dell Computers rapidly *responds* to customer orders because quality systems, with little rework, have allowed it to achieve rapid throughput in its plants. Indeed, quality may be the key success factor for these firms, just as it is at Arnold Palmer Hospital.

STUDENT TIP
High-quality products and services are the most profitable.

As Figure 6.1 suggests, improvements in quality help firms increase sales and reduce costs, both of which can increase profitability. Increases in sales often occur as firms speed response, increase or lower selling prices, and improve their reputation for quality products. Similarly, improved quality allows costs to drop as firms increase productivity and lower rework, scrap, and warranty costs. One study found that companies with the highest quality were five times as productive (as measured by units produced per labor-hour) as companies with the poorest quality. Indeed, when the implications of an organization's long-term costs and the potential for increased sales are considered, total costs may well be at a minimum when 100% of the goods or services are perfect and defect free.

Quality, or the lack of quality, affects the entire organization from supplier to customer and from product design to maintenance. Perhaps more important, *building* an organization that can achieve quality is a demanding task. Figure 6.2 lays out the flow of activities for an organization to use to achieve total quality management (TQM). A successful quality strategy begins with an organizational culture that fosters quality, followed by an understanding of the principles of quality, and then engaging employees in the necessary activities to implement quality. When these things are done well, the organization typically satisfies its customers and obtains a competitive advantage. The ultimate goal is to win customers. Because quality causes so many other good things to happen, it is a great place to start.

Figure **6.1**
Ways Quality Improves Profitability

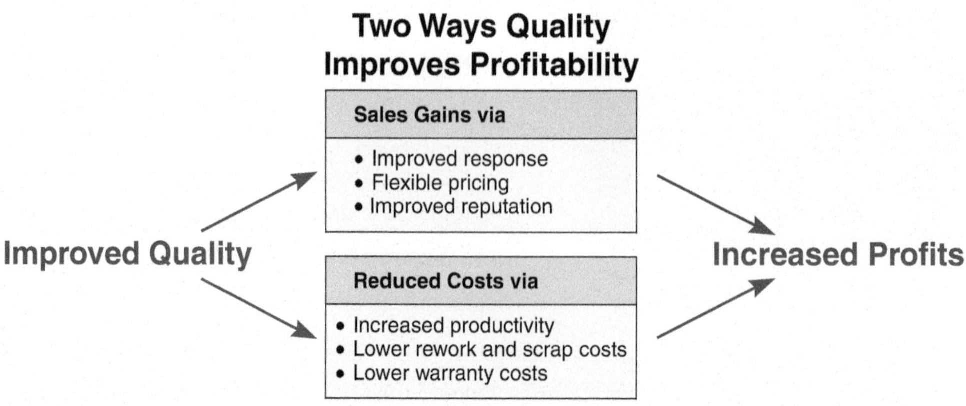

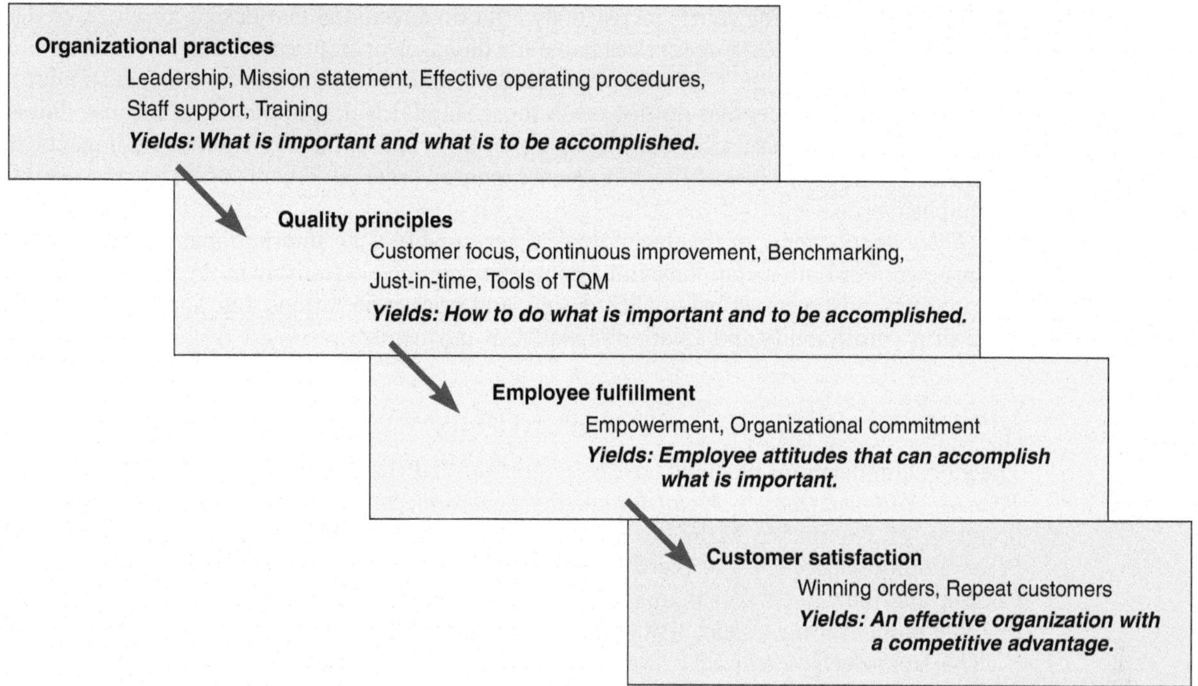

Figure **6.2**

The Flow of Activities Necessary to Achieve Total Quality Management

Defining Quality

The operations manager's objective is to build a total quality management system that identifies and satisfies customer needs. Total quality management takes care of the customer. Consequently, we accept the definition of **quality** as adopted by the American Society for Quality (ASQ; **www.asq.org**): "The totality of features and characteristics of a product or service that bears on its ability to satisfy stated or implied needs."

Quality
The ability of a product or service to meet customer needs.

Others, however, believe that definitions of quality fall into several categories. Some definitions are *user based*. They propose that quality "lies in the eyes of the beholder." Marketing people like this approach and so do customers. To them, higher quality means better performance, nicer features, and other (sometimes costly) improvements. To production managers, quality is *manufacturing based*. They believe that quality means conforming to standards and "making it right the first time." Yet a third approach is *product based*, which views quality as a precise and measurable variable. In this view, for example, really good ice cream has high butterfat levels.

LO 6.1 *Define* quality and TQM

This text develops approaches and techniques to address all three categories of quality. The characteristics that connote quality must first be identified through research (a user-based approach to quality). These characteristics are then translated into specific product attributes (a product-based approach to quality). Then, the manufacturing process is organized to ensure that products are made precisely to specifications (a manufacturing-based approach to quality). A process that ignores any one of these steps will not result in a quality product.

STUDENT TIP
To create a quality good or service, operations managers need to know what the customer expects.

Implications of Quality

In addition to being a critical element in operations, quality has other implications. Here are three other reasons why quality is important:

1. *Company reputation:* An organization can expect its reputation for quality—be it good or bad—to follow it. Quality will show up in perceptions about the firm's new products, employment practices, and supplier relations. Self-promotion is not a substitute for quality products.

2. *Product liability:* The courts increasingly hold organizations that design, produce, or distribute faulty products or services liable for damages or injuries resulting from their use. Legislation such as the Consumer Product Safety Act sets and enforces product standards by banning products that do not reach those standards. Impure foods that cause illness, nightgowns that burn, tires that fall apart, or auto fuel tanks that explode on impact can all lead to injury or loss of life, huge legal expenses, large settlements or losses, and terrible publicity.
3. *Global implications:* In this technological age, quality is an international, as well as OM, concern. For both a company and a country to compete effectively in the global economy, products must meet global quality, design, and price expectations. Inferior products harm a firm's profitability and a nation's balance of payments.

Malcolm Baldrige National Quality Award

The global implications of quality are so important that the U.S. has established the *Malcolm Baldrige National Quality Award* for quality achievement. The award is named for former Secretary of Commerce Malcolm Baldrige. Recent winners include such organizations as PricewaterhouseCoopers, Lockheed Martin, Nestle Purina, Honeywell, Howard Community College, and the City of Germantown, Tennessee. (For details about the Baldrige Award and its 1,000-point scoring system, visit **www.nist.gov/baldrige/**.)

The Japanese have a similar award, the Deming Prize, named after an American, Dr. W. Edwards Deming.

ISO 9000 International Quality Standards

ISO 9000
A set of quality standards developed by the International Organization for Standardization (ISO).

LO 6.2 *Describe* the ISO international quality standards

The move toward global supply chains has placed so much emphasis on quality that the world has united around a single quality standard, **ISO 9000**. ISO 9000 is *the* quality standard with international recognition. Its focus is to enhance success through eight quality management principles: (1) top management leadership, (2) customer satisfaction, (3) continual improvement, (4) involvement of people, (5) process analysis, (6) use of data-driven decision making, (7) a systems approach to management, and (8) mutually beneficial supplier relationships.

The ISO standard encourages establishment of quality management procedures, detailed documentation, work instructions, and recordkeeping. Like the Baldrige Awards, the assessment includes self-appraisal and problem identification. Unlike the Baldrige, ISO certified organizations must be reaudited every 3 years.

The latest modification of the standard, ISO 9004: 2018, gives guidelines for enhancing an organization's ability to achieve sustained success.

STUDENT TIP
International quality standards grow in prominence every year. See **www.iso.ch**.

Over 1.6 million certifications have been awarded to firms in 201 countries, including about 30,000 in the U.S. To do business globally, it is critical for a firm to be certified and listed in the ISO directory.

Cost of Quality (COQ)

Cost of quality (COQ)
The cost of doing things wrong—that is, the price of nonconformance.

Four major categories of costs are associated with quality. Called the **cost of quality (COQ)**, they are:

- *Prevention costs:* costs associated with reducing the potential for defective parts or services (e.g., training, quality improvement programs).
- *Appraisal costs:* costs related to evaluating products, processes, parts, and services (e.g., testing, labs, inspectors).
- *Internal failure costs:* costs that result from production of defective parts or services before delivery to customers (e.g., rework, scrap, downtime).
- *External failure costs:* costs that occur after delivery of defective parts or services (e.g., rework, returned goods, liabilities, lost goodwill, costs to society).

The first three costs can be reasonably estimated, but external costs are very hard to quantify. When GE had to recall 3.1 million dishwashers (because of a defective switch alleged to

TABLE 6.1 Pioneers in the Field of Quality Management

LEADER	PHILOSOPHY/CONTRIBUTION
W. Edwards Deming	Deming insisted management accept responsibility for building good systems. The employee cannot produce products that on average exceed the quality of what the process is capable of producing. His 14 points for implementing quality improvement are presented in this chapter.
Joseph M. Juran	A pioneer in teaching the Japanese how to improve quality, Juran believed strongly in top-management commitment, support, and involvement in the quality effort. He was also a believer in teams that continually seek to raise quality standards. Juran varies from Deming somewhat in focusing on the customer and defining quality as fitness for use, not necessarily the written specifications.
Armand Feigenbaum	His 1961 book *Total Quality Control* laid out 40 steps to quality improvement processes. He viewed quality not as a set of tools but as a total field that integrated the processes of a company. His work in how people learn from each other's successes led to the field of cross-functional teamwork.
Philip B. Crosby	*Quality Is Free* was Crosby's attention-getting book published in 1979. Crosby believed that in the traditional trade-off between the cost of improving quality and the cost of poor quality, the cost of poor quality is understated. The cost of poor quality should include all of the things that are involved in not doing the job right the first time. Crosby coined the term *zero defects* and stated, "There is absolutely no reason for having errors or defects in any product or service." *Source*: Based on *Quality Is Free* by Philip B. Crosby (New York, McGraw-Hill, 1979) p. 58.

have started seven fires), the cost of repairs exceeded the value of all the machines. This leads to the belief by many experts that the cost of poor quality is consistently underestimated.

Observers of quality management believe that, on balance, the cost of quality products is only a fraction of the benefits. They think the real losers are organizations that fail to work aggressively at quality. For instance, Philip Crosby stated that quality is free. "What costs money are the unquality things—all the actions that involve not doing it right the first time."[1]

巧

Takumi is a Japanese character that symbolizes a broader dimension than quality, a deeper process than education, and a more perfect method than persistence.

Pioneers in Quality Besides Crosby there are several other giants in the field of quality management, including Deming, Feigenbaum, and Juran. Table 6.1 summarizes their philosophies and contributions.

Ethics and Quality Management

For operations managers, one of the most important jobs is to deliver healthy, safe, and quality products and services to customers. The development of poor-quality products, because of inadequate design and production processes, not only results in higher production costs but also leads to injuries, lawsuits, and increased government regulation.

If a firm believes that it has introduced a questionable product, ethical conduct must dictate the responsible action. This may be a worldwide recall, as conducted by both Johnson & Johnson (for Tylenol) and Perrier (for sparkling water), when each of these products was found to be contaminated. A manufacturer must accept responsibility for any poor-quality product released to the public.

There are many stakeholders involved in the production and marketing of poor-quality products, including stockholders, employees, customers, suppliers, distributors, and creditors. As a matter of ethics, management must ask if any of these stakeholders are being wronged. Every company needs to develop core values that become day-to-day guidelines for everyone from the CEO to production-line employees.

Total Quality Management

Total quality management (TQM) refers to a quality emphasis that encompasses the entire organization, from supplier to customer. TQM stresses a commitment by management to have a continuing companywide drive toward excellence in all aspects of products and services that are important to the customer. Each of the 10 decisions made by operations managers deals with some aspect of identifying and meeting customer expectations. Meeting those expectations requires an emphasis on TQM if a firm is to compete as a leader in world markets.

Total quality management (TQM)
Management of an entire organization so that it excels in all aspects of products and services that are important to the customer.

TABLE 6.2 Deming's 14 Points for Implementing Quality Improvement[2]

1. Create consistency of purpose.
2. Lead to promote change.
3. Build quality into the product; stop depending on inspections to catch problems.
4. Build long-term relationships based on performance instead of awarding business on the basis of price.
5. Continuously improve product, quality, and service.
6. Start training.
7. Emphasize leadership.
8. Drive out fear.
9. Break down barriers between departments.
10. Stop haranguing workers.
11. Support, help, and improve.
12. Remove barriers to pride in work.
13. Institute a vigorous program of education and self-improvement.
14. Put everybody in the company to work on the transformation.

STUDENT TIP Here are seven concepts that make up the heart of an effective TQM program.

Quality expert W. Edwards Deming used 14 points (see Table 6.2) to indicate how he implemented TQM. We incorporate these into seven concepts for an effective TQM program: (1) continuous improvement, (2) Six Sigma, (3) employee empowerment, (4) benchmarking, (5) just-in-time (JIT), (6) Taguchi concepts, and (7) knowledge of TQM tools.

Continuous Improvement

TQM requires a never-ending process of continuous improvement that covers people, equipment, suppliers, materials, and procedures. The basis of the philosophy is that every aspect of an operation can be improved. The end goal is perfection, which is never achieved but always sought.

PDCA
A continuous improvement model of plan, do, check, act.

Plan-Do-Check-Act Walter Shewhart, another pioneer in quality management, developed a circular model known as PDCA (plan, do, check, act) as his version of continuous improvement. Deming later took this concept to Japan during his work there after World War II. The PDCA cycle (also called a Deming circle or a Shewhart circle) is shown in Figure 6.3 as a circle to stress the continuous nature of the improvement process.

The Japanese use the word *kaizen* to describe this ongoing process of unending improvement—the setting and achieving of ever-higher goals. In the U.S., *TQM* and *zero defects* are also used to describe continuous improvement efforts. But whether it's PDCA, kaizen, TQM, or zero defects, the operations manager is a key player in building a work culture that endorses continuous improvement.

Figure 6.3
PDCA Cycle

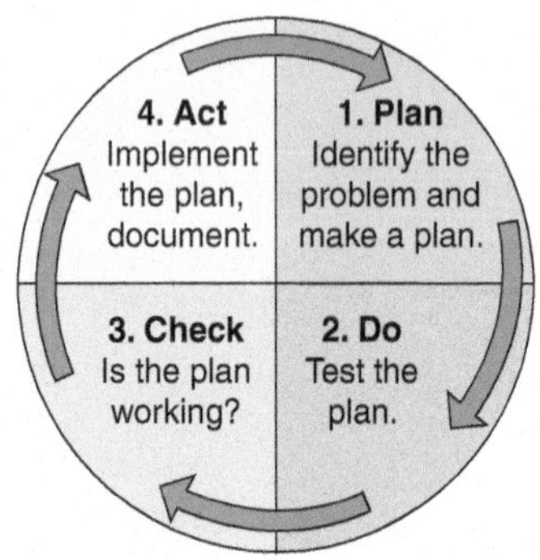

Six Sigma

The term Six Sigma, popularized by Motorola, Honeywell, and General Electric (GE), has two meanings in TQM. In a *statistical* sense, it describes a process, product, or service with an extremely high capability.

The second TQM definition of Six Sigma is a *program* designed to reduce defects to help lower costs, save time, and improve customer satisfaction. Six Sigma is a comprehensive system—a strategy, a discipline, and a set of tools—for achieving and sustaining business success:

Six Sigma
A program to save time, improve quality, and lower costs.

- It is a *strategy* because it focuses on total customer satisfaction.
- It is a *discipline* because it follows the formal Six Sigma Improvement Model known as **DMAIC**. This five-step process improvement model (1) ***Defines*** the project's purpose, scope, and outputs and then identifies the required process information, keeping in mind the

customer's definition of quality; (2) ***Measures*** the process and collects data; (3) ***Analyzes*** the data, ensuring repeatability (the results can be duplicated) and reproducibility (others get the same result); (4) ***Improves***, by modifying or redesigning, existing processes and procedures; and (5) ***Controls*** the new process to make sure performance levels are maintained.

- It is a *set of seven tools* that we introduce shortly in this chapter: check sheets, scatter diagrams, cause-and-effect diagrams, Pareto charts, flowcharts, histograms, and statistical process control.

LO 6.3 *Explain* Six Sigma

Motorola developed Six Sigma in the 1980s, in response to customer complaints about its products and in response to stiff competition. The company first set a goal of reducing defects by 90%. Within 1 year, it had achieved such impressive results—through benchmarking competitors, soliciting new ideas from employees, changing reward plans, adding training, and revamping critical processes—that it documented the procedures into what it called Six Sigma. Although the concept was rooted in manufacturing, GE later expanded Six Sigma into services, including human resources, sales, customer services, and financial/credit services. The concept of wiping out defects turns out to be the same in both manufacturing and services.

Implementing Six Sigma Implementing Six Sigma is a big commitment. Indeed, successful Six Sigma programs in every firm, from GE to Motorola to DuPont to Texas Instruments, require a major time commitment, especially from top management. These leaders have to formulate the plan, communicate their buy-in and the firm's objectives, and take a visible role in setting the example for others.

Successful Six Sigma projects are clearly related to the strategic direction of an organization. It is a management-directed, team-based, and expert-led approach.

Employee Empowerment

Employee empowerment means involving employees in every step of the production process. Consistently, research suggests that some 85% of quality problems have to do with materials and processes, not with employee performance. Therefore, the task is to design equipment and processes that produce the desired quality. This is best done with a high degree of involvement by those who understand the shortcomings of the system. Those dealing with the system on a daily basis understand it better than anyone else. One study indicated that TQM programs that delegate responsibility for quality to shop-floor employees tend to be twice as likely to succeed as those implemented with "top-down" directives.

Employee empowerment
Enlarging employee jobs so that the added responsibility and authority is moved to the lowest level possible in the organization.

When nonconformance occurs, the worker is seldom at fault. Either the product was designed wrong, the process that makes the product was designed wrong, or the employee was improperly trained. Although the employee may be able to help solve the problem, the employee rarely causes it.

Techniques for building employee empowerment include (1) building communication networks that include employees; (2) developing open, supportive supervisors; (3) moving responsibility from both managers and staff to production employees; (4) building high-morale organizations; and (5) creating such formal organization structures as teams and quality circles.

Teams can be built to address a variety of issues. One popular focus of teams is quality. Such teams are often known as quality circles. A **quality circle** is a group of employees who meet regularly to solve work-related problems. The members receive training in group planning, problem solving, and statistical quality control. They generally meet once a week (usually after work but sometimes on company time). Although the members are not rewarded financially, they do receive recognition from the firm. A specially trained team member, called the *facilitator*, usually helps train the members and keeps the meetings running smoothly. Teams with a quality focus have proven to be a cost-effective way to increase productivity as well as quality.

Quality circle
A group of employees meeting regularly with a facilitator to solve work-related problems in their work area.

Benchmarking

Benchmarking is another ingredient in an organization's TQM program. **Benchmarking** involves selecting a demonstrated standard of products, services, costs, or practices that represent the very best performance for processes or activities very similar to your own. The idea is to

Benchmarking
Selecting a demonstrated standard of performance that represents the very best performance for a process or an activity.

develop a target at which to shoot and then to develop a standard or benchmark against which to compare your performance. The steps for developing benchmarks are:

1. Determine what to benchmark.
2. Form a benchmark team.
3. Identify benchmarking partners.
4. Collect and analyze benchmarking information.
5. Take action to match or exceed the benchmark.

LO 6.4 *Explain* how benchmarking is used in TQM

Typical performance measures used in benchmarking include percentage of defects, cost per unit or per order, processing time per unit, service response time, return on investment, customer satisfaction rates, and customer retention rates.

In the ideal situation, you find one or more similar organizations that are leaders in the particular areas you want to study. Then you compare yourself (benchmark yourself) against them. The company need not be in your industry. Indeed, to establish world-class standards, it may be best to look outside your industry. If one industry has learned how to compete via rapid product development while yours has not, it does no good to study your industry.

This is exactly what Xerox and Mercedes-Benz did when they went to L.L. Bean for order-filling and warehousing benchmarks. Xerox noticed that L.L. Bean was able to "pick" orders three times faster. After benchmarking, Xerox was immediately able to pare warehouse costs by 10%. Mercedes-Benz observed that L.L. Bean warehouse employees used flowcharts to spot wasted motions. The auto giant followed suit and now relies more on problem solving at the worker level.

Benchmarks often take the form of "best practices" found in other firms or in other divisions. Table 6.3 illustrates best practices for resolving customer complaints.

Likewise, Britain's Great Ormond Street Hospital benchmarked the Ferrari Racing Team's pit stops to improve one aspect of medical care. (See the *OM in Action* box, "A Hospital Benchmarks Against the Ferrari Racing Team?")

Internal Benchmarking When an organization is large enough to have many divisions or business units, a natural approach is the internal benchmark. Data are usually much more accessible than from outside firms. Typically, one internal unit has superior performance worth learning from.

Xerox's almost religious belief in benchmarking has paid off not only by looking outward to L.L. Bean but by examining the operations of its various country divisions. For example, Xerox Europe, a $6 billion subsidiary of Xerox Corp., formed teams to see how better sales could result through internal benchmarking. Somehow, France sold five times as many color copiers as did other divisions in Europe. By copying France's approach, namely, better sales training and use of dealer channels to supplement direct sales, Norway increased sales by 152%, Holland by 300%, and Switzerland by 328%!

Benchmarks can and should be established in a variety of areas. Total quality management requires no less.

TABLE 6.3 Best Practices for Resolving Customer Complaints

BEST PRACTICE	JUSTIFICATION
Make it easy for clients to complain.	It is free market research.
Respond quickly to complaints.	It adds customers and loyalty.
Resolve complaints on the first contact.	It reduces cost.
Use computers to manage complaints.	Discover trends, share them, and align your services.
Recruit the best for customer service jobs.	It should be part of formal training and career advancement.

Source: Based on Canadian Government Guide on Complaint Mechanism.

Just-in-Time (JIT)

The philosophy behind just-in-time (JIT) is one of continuous improvement and reinforced problem solving. JIT systems are designed to produce or deliver goods just as they are needed. JIT is related to quality in three ways:

- *JIT cuts the cost of quality:* This occurs because scrap, rework, inventory investment, and damage costs are directly related to inventory on hand. Because there is less inventory on hand with JIT, costs are lower. In addition, inventory hides bad quality, whereas JIT immediately *exposes* bad quality.
- *JIT improves quality:* As JIT shrinks lead time, it keeps evidence of errors fresh and limits the number of potential sources of error. JIT creates, in effect, an early warning system for quality problems, both within the firm and with vendors.
- *Better quality means less inventory and a better, easier-to-employ JIT system:* Often the purpose of keeping inventory is to protect against poor production performance resulting from unreliable quality. If consistent quality exists, JIT allows firms to reduce all the costs associated with inventory.

Taguchi Concepts

Most quality problems are the result of poor product and process design. Genichi Taguchi has provided us with three concepts aimed at improving both product and process quality: *quality robustness, target-oriented quality, and the quality loss function.*

Quality robust products are products that can be produced uniformly and consistently in adverse manufacturing and environmental conditions. Taguchi's idea is to remove the *effects* of adverse conditions instead of removing the causes. Taguchi suggests that removing the effects is often cheaper than removing the causes and more effective in producing a robust product. In this way, small variations in materials and process do not destroy product quality.

Quality robust
Products that are consistently built to meet customer needs despite adverse conditions in the production process.

OM in Action A Hospital Benchmarks Against the Ferrari Racing Team?[3]

After surgeons successfully completed a 6-hour operation to fix a hole in a 3-year-old boy's heart, Dr. Angus McEwan supervised one of the most dangerous phases of the procedure: the boy's transfer from surgery to the intensive care unit.

Thousands of such "handoffs" occur in hospitals every day, and devastating mistakes can happen during them. In fact, at least 35% of preventable hospital mishaps take place because of handoff problems. Risks come from many sources: using temporary nursing staff, frequent shift changes for interns, surgeons working in larger teams, and an ever-growing tangle of wires and tubes connected to patients.

Using an unlikely benchmark, Britain's largest children's hospital turned to Italy's Formula One Ferrari racing team for help in revamping patient handoff techniques. Armed with videos and slides, the racing team described how they analyze pit crew performance. It also explained how its system for recording errors stressed the small ones that go unnoticed in pit-stop handoffs.

To move forward, Ferrari invited a team of doctors to attend practice sessions at the British Grand Prix to get closer looks at pit stops. Ferrari's technical director, Nigel Stepney, then watched a video of a hospital handoff. Stepney was not impressed. "In fact, he was amazed at how clumsy, chaotic, and informal the process appeared," said one hospital official. At that meeting, Stepney described how each Ferrari crew member is required to do a specific job, in a specific sequence, and in silence. The hospital handoff, in contrast, had several conversations going on at once, while different members of its team disconnected or reconnected patient equipment, but in no particular order.

Results of the benchmarking process: handoff errors fell more than 40%, with a bonus of faster handoff time.

Oliver Multhaup/AP Images

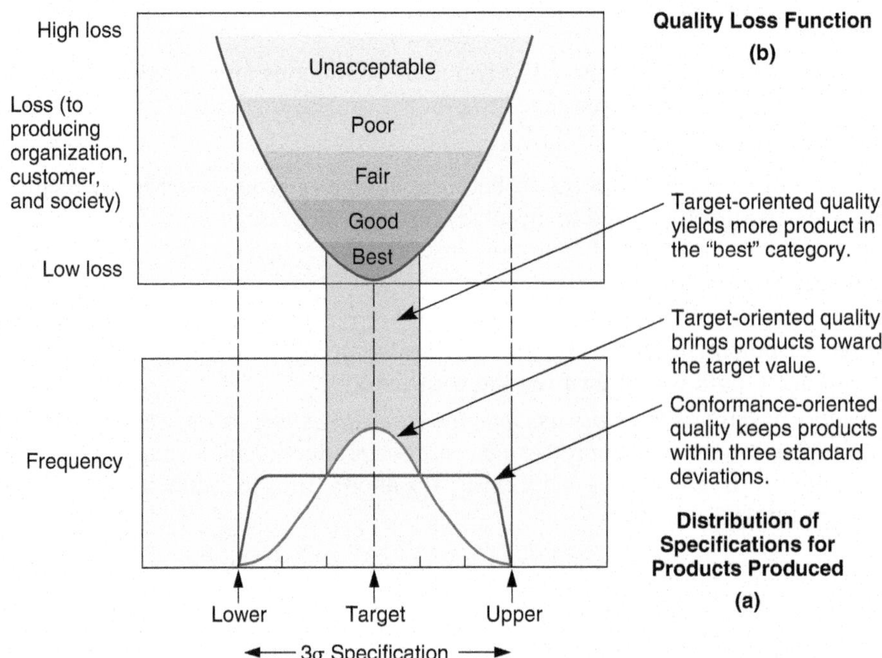

Figure **6.4**

(a) Distribution of Products Produced and (b) Quality Loss Function

Taguchi aims for the target because products produced near the upper and lower acceptable specifications result in a higher quality loss.

LO 6.5 *Explain* quality robust products and Taguchi concepts

A study found that U.S. consumers preferred Sony TVs made in Japan to Sony TVs made in the U.S., even though both factories used the exact same designs and specifications. The difference in approaches to quality generated the difference in consumer preferences. In particular, the U.S. factory was *conformance-oriented*, accepting all components that were produced within specification limits (the blue curve in Figure 6.4(a)). On the other hand, the Japanese factory strove to produce as many components as close to the actual target as possible (the red curve in Figure 6.4(a)).

This suggests that even though components made close to the boundaries of the specification limits may technically be acceptable, they may still create problems. For example, TV screens produced near their diameter's lower spec limit may provide a loose fit with screen frames produced near their upper spec limit, and vice versa. This implies that a final product containing many parts produced near their specification boundaries may contain numerous loose and tight fits, which could cause assembly, performance, or aesthetic concerns. Customers may be dissatisfied, resulting in possible returns, service work, or decreased future demand.

Target-oriented quality
A philosophy of continuous improvement to bring a product exactly on target.

Quality loss function (QLF)
A mathematical function that identifies all costs connected with poor quality and shows how these costs increase as output moves away from the target value.

Taguchi introduced the concept of **target-oriented quality** as a philosophy of continuous improvement to bring the product exactly on target. As a measure, Taguchi's **quality loss function (QLF)** attempts to estimate the cost of deviating from the target value. Even though the item is produced within specification limits, the variation in quality can be expected to increase costs as the item output moves away from its target value. (These quality-related costs are estimates of the average cost over many such units produced.)

The QLF is an excellent way to estimate quality costs of different processes. A process that produces closer to the actual target value may be more expensive, but it may yield a more valuable product. The QLF is the tool that helps the manager determine if this added cost is worthwhile. The QLF takes the general form of a simple quadratic equation (see Figure 6.4(b)).

Knowledge of TQM Tools

To empower employees and implement TQM as a continuing effort, everyone in the organization must be trained in the techniques of TQM. In the following section, we focus on some of the diverse and expanding tools that are used in the TQM crusade.

Tools for Generating Ideas

(a) *Check Sheet:* An organized method of recording data

	Hour							
Defect	1	2	3	4	5	6	7	8
A	///	/		/	/	/	///	/
B	//	/	/	/			//	///
C	/	//					//	////

(b) *Scatter Diagram:* A graph of the value of one variable vs. another variable

(c) *Cause-and-Effect Diagram:* A tool that identifies process elements (causes) that may affect an outcome

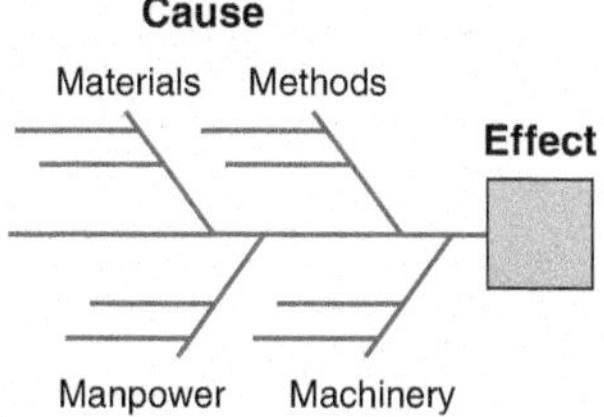

Tools for Organizing the Data

(d) *Pareto Chart:* A graph that identifies and plots problems or defects in descending order of frequency

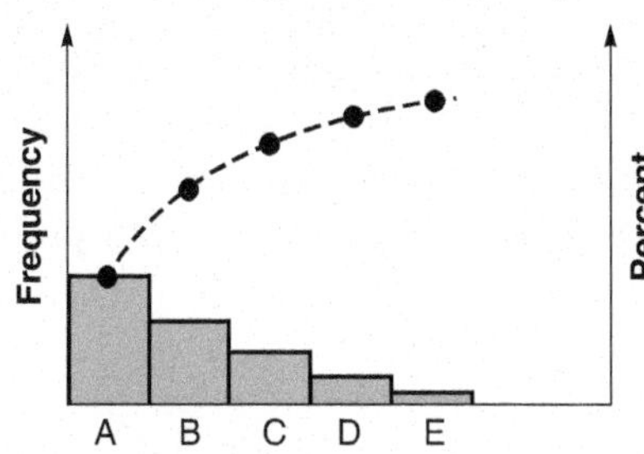

(e) *Flowchart (Process Diagram):* A chart that describes the steps in a process

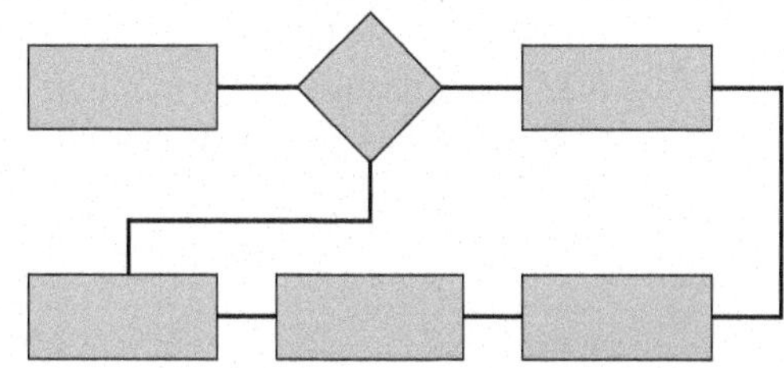

Tools for Identifying Problems

(f) *Histogram:* A distribution that shows the frequency of occurrences of a variable

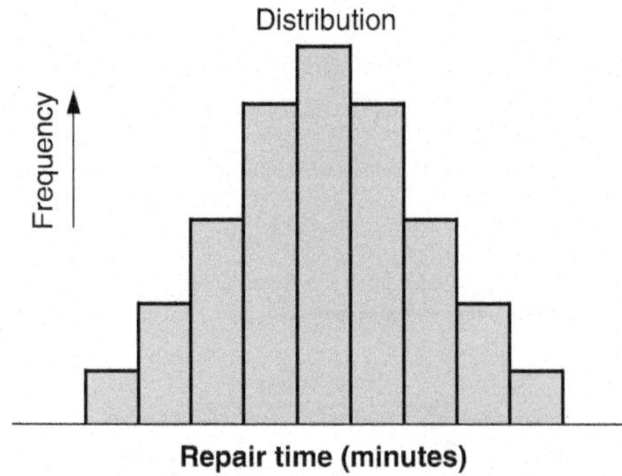

(g) *Statistical Process Control Chart:* A chart with time on the horizontal axis for plotting values of a statistic

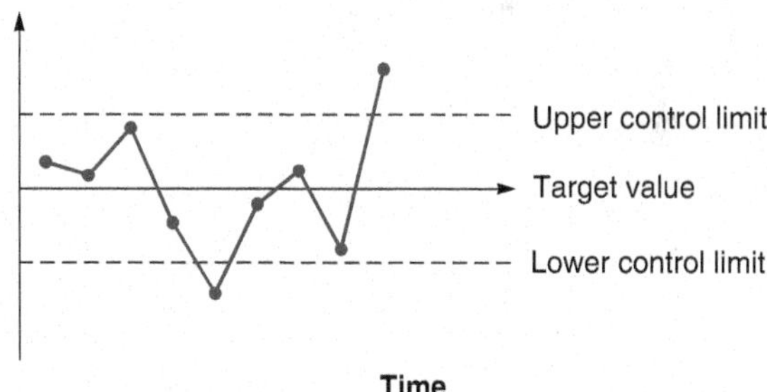

Figure **6.5**

Seven Tools of TQM

Tools of TQM

STUDENT TIP
These tools will prove useful in many of your courses and throughout your career.

Seven tools that are particularly helpful in the TQM effort are shown in Figure 6.5. We will now introduce these tools.

Check Sheets

LO 6.6 *Use* the seven tools of TQM

A check sheet is any kind of a form that is designed for recording data. In many cases, the recording is done so the patterns are easily seen while the data are being taken [see Figure 6.5(a)]. Check sheets help analysts find the facts or patterns that may aid subsequent analysis. An example might be a drawing that shows a tally of the areas where defects are occurring or a check sheet showing the type of customer complaints.

Scatter Diagrams

Scatter diagrams show the relationship between two measurements. An example is the positive relationship between length of a service call and the number of trips a repair person makes back to the truck for parts. Another example might be a plot of productivity and absenteeism, as shown in Figure 6.5(b). If the two items are closely related, the data points will form a tight band. If a random pattern results, the items are unrelated.

Cause-and-Effect Diagrams

Cause-and-effect diagram
A schematic technique used to discover possible locations of quality problems.

Another tool for identifying quality issues and inspection points is the **cause-and-effect diagram**, also known as an **Ishikawa diagram** or a **fish-bone chart**. Figure 6.6 illustrates a chart (note the shape resembling the bones of a fish) for a basketball quality control problem—missed free throws. Each "bone" represents a possible source of error.

The operations manager starts with four categories: material, machinery/equipment, manpower/people, and method. These four *M*s are the "causes." They provide a good checklist for initial analysis. Individual causes associated with each category are tied in as separate bones along that branch, often through a brainstorming process. For example, the method branch in Figure 6.6 has problems caused by hand position, follow-through, aiming point, bent knees, and balance. When a fish-bone chart is systematically developed, possible quality problems and inspection points are highlighted.

Pareto Charts

Pareto charts
A graphic way of classifying problems by their level of importance, often referred to as the 80–20 rule.

Pareto charts are a method of organizing errors, problems, or defects to help focus on problem-solving efforts. They are based on the work of Vilfredo Pareto, a 19th-century economist. Joseph M. Juran popularized Pareto's work when he suggested that 80% of a firm's problems are a result of only 20% of the causes.

Example 1 indicates that of the five types of complaints identified, the vast majority were of one type—poor room service.

Pareto analysis indicates which problems may yield the greatest payoff. Pacific Bell discovered this when it tried to find a way to reduce damage to buried phone cable, the number-one

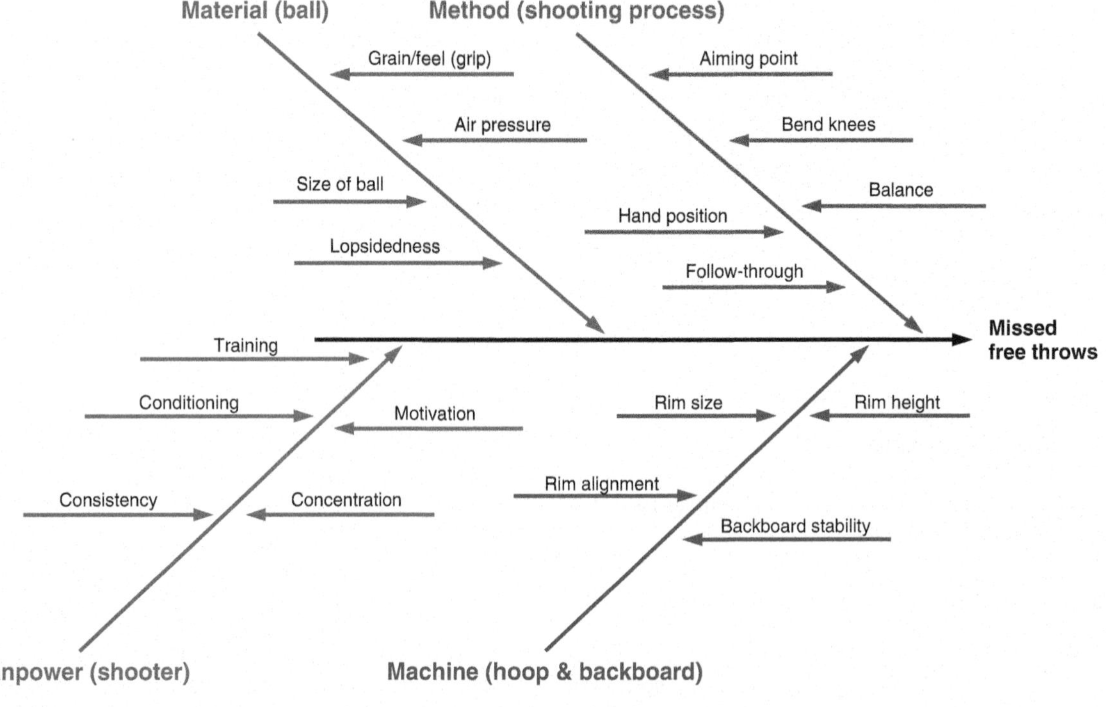

Figure **6.6**

Fish-Bone Chart (or Cause-and-Effect Diagram) for Problems with Missed Free Throws

Source: Courtesy of MoreSteam.com, LLC.

Example 1

A PARETO CHART AT THE HARD ROCK HOTEL

The Hard Rock Hotel in Bali has just collected the data from 75 complaint calls to the general manager during the month of October. The manager wants to prepare an analysis of the complaints. The data provided are room service, 54; check-in delays, 12; hours the pool is open, 4; minibar prices, 3; and miscellaneous, 2.

APPROACH ▶ A Pareto chart is an excellent choice for this analysis. The chart sums the errors, problems, or defects and notes them in decreasing order along the horizontal axis of a table.

SOLUTION ▶ The Pareto chart shown below indicates that 72% of the calls were the result of one cause: room service. The majority of complaints will be eliminated when this one cause is corrected.

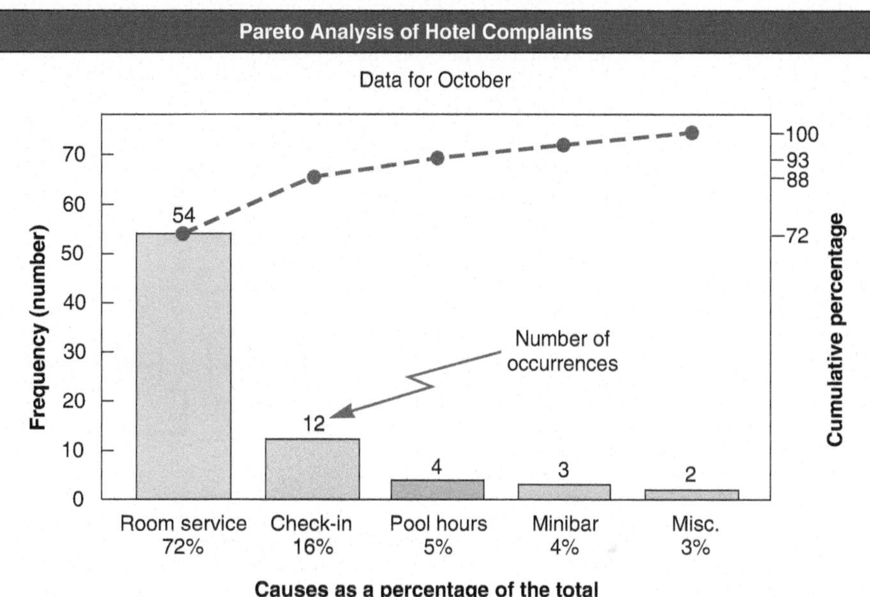

INSIGHT ▶ This visual means of summarizing data is very helpful—particularly with large amounts of data, as in the Southwestern University case study at the end of this chapter. We can immediately spot the top problems and prepare a plan to address them.

LEARNING EXERCISE ▶ Hard Rock's bar manager decides to conduct a similar analysis on complaints collected over the past year: too expensive, 22; weak drinks, 15; slow service, 65; short hours, 8; unfriendly bartender, 12. Prepare a Pareto chart. [Answer: slow service, 53%; expensive, 18%; drinks, 12%; bartender, 10%; hours, 7%.]

RELATED PROBLEMS ▶ 6.1, 6.3, 6.7b, 6.12, 6.13, 6.16c, 6.17b

ACTIVE MODEL 6.1 This example is further illustrated in Active Model 6.1 found online.

cause of phone outages. Pareto analysis showed that 41% of cable damage was caused by construction work. Armed with this information, Pacific Bell was able to devise a plan to reduce cable cuts by 24% in 1 year, saving $6 million.

Likewise, Japan's Ricoh Corp., a copier maker, used the Pareto principle to tackle the "callback" problem. Callbacks meant the job was not done right the first time and that a second visit, at Ricoh's expense, was needed. Identifying and retraining only the 11% of the customer engineers with the most callbacks resulted in a 19% drop in return visits.

Flowcharts

Flowcharts graphically present a process or system using annotated boxes and interconnected lines [see Figure 6.5(e)]. They are a simple but great tool for trying to make sense of a process or explain a process. Example 2 uses a flowchart to show the process of completing an MRI at a hospital.

Flowcharts
Block diagrams that graphically describe a process or system.

Example 2

A FLOWCHART FOR HOSPITAL MRI SERVICE

Arnold Palmer Hospital has undertaken a series of process improvement initiatives. One of these is to make the MRI service efficient for patient, doctor, and hospital. The first step, the administrator believes, is to develop a flowchart for this process.

APPROACH ▶ A process improvement staffer observed a number of patients and followed them (and information flow) from start to end. Here are the 11 steps:

1. Physician schedules MRI after examining patient (START).
2. Patient taken from the examination room to the MRI lab with test order and copy of medical records.
3. Patient signs in, completes required paperwork.
4. Patient is prepped by technician for scan.
5. Technician carries out the MRI scan.
6. Technician inspects film for clarity.
7. If MRI not satisfactory (20% of time), Steps 5 and 6 are repeated.
8. Patient taken back to hospital room.
9. MRI is read by radiologist and report is prepared.
10. MRI and report are transferred electronically to physician.
11. Patient and physician discuss report (END).

SOLUTION ▶ Here is the flowchart:

STUDENT TIP Flowcharting any process is an excellent way to understand and then try to improve that process.

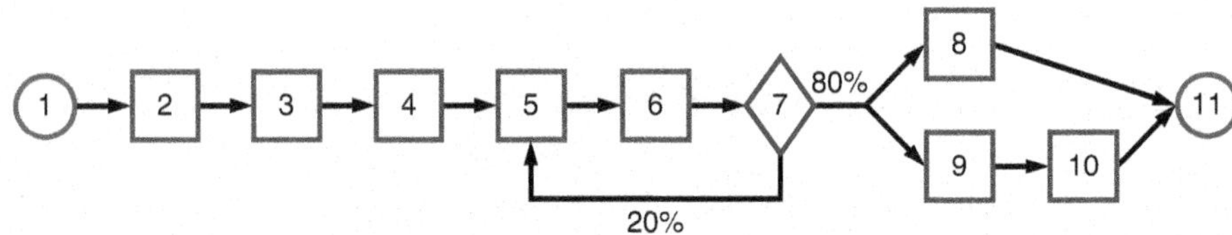

INSIGHT ▶ With the flowchart in hand, the hospital can analyze each step and identify value-added activities and activities that can be improved or eliminated.

LEARNING EXERCISE ▶ A new procedure requires that if the patient's blood pressure is over 200/120 when being prepped for the MRI, the patient is taken back to the hospital room for 2 hours and the process returns to Step 2. How does the flowchart change? Answer:

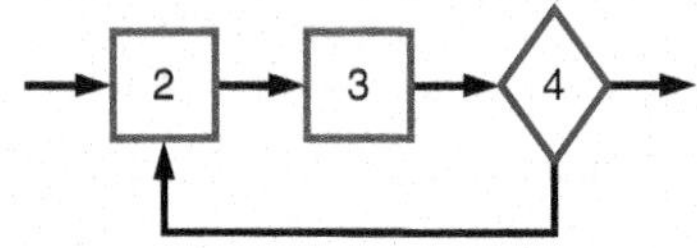

RELATED PROBLEMS ▶ 6.6, 6.15

Histograms

Histograms show the range of values of a measurement and the frequency with which each value occurs [see Figure 6.5(f)]. They show the most frequently occurring readings as well as the variations in the measurements. Descriptive statistics, such as the average and standard deviation, may be calculated to describe the distribution. However, the data should always be plotted so the shape of the distribution can be "seen." A visual presentation of the distribution may also provide insight into the cause of the variation.

Statistical process control (SPC)
A process used to monitor standards, make measurements, and take corrective action as a product or service is being produced.

Control charts
Graphic presentations of process data over time, with predetermined control limits.

Statistical Process Control (SPC)

Statistical process control (SPC) monitors standards, makes measurements, and takes corrective action as a product or service is being produced. Samples of process outputs are examined; if they are within acceptable limits, the process is permitted to continue. If they fall outside certain specific ranges, the process is stopped and, typically, the assignable cause located and removed.

Control charts are graphic presentations of data over time that show upper and lower limits for the process we want to control [see Figure 6.5(g)]. Control charts are constructed in such a way that new

Figure 6.7

Control Chart for Percentage of Free Throws Missed in the Team's First Nine Games of the New Season

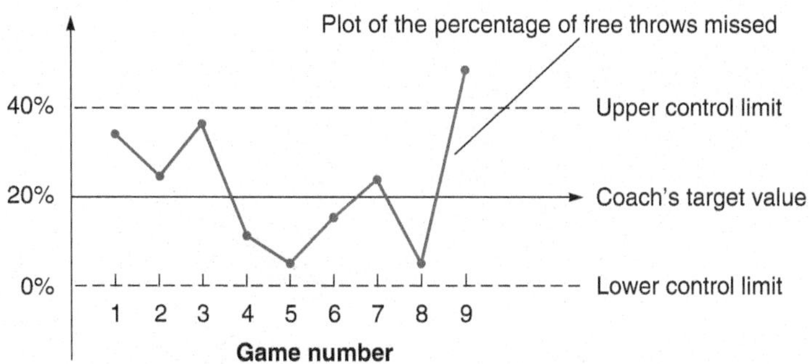

Stephen M. Dowell/Orlando Sentinel/McClatchy/Tribune Content LLC/Alamy Stock Photo

data can be quickly compared with past performance data. We take samples of the process output and plot the average of each of these samples on a chart that has the limits on it. The upper and lower limits in a control chart can be in units of temperature, pressure, weight, length, and so on.

Figure 6.7 shows the plot of sample averages in a control chart. When the samples fall within the upper and lower control limits and no discernible pattern is present, the process is said to be in control with only natural variation present. Otherwise, the process is out of control or out of adjustment.

The supplement to this chapter details how control charts of different types are developed. It also deals with the statistical foundation underlying the use of this important tool.

The Role of Inspection

To make sure a system is producing as expected, control of the process is needed. The best processes have little variation from the standard expected. In fact, if variation were completely eliminated, there would be no need for inspection because there would be no defects. The operations manager's challenge is to build such systems. However, inspection must often be performed to ensure that processes are performing to standard. This **inspection** can involve measurement, tasting, touching, weighing, or testing of the product (sometimes even destroying it when doing so). Its goal is to detect a bad process immediately. Inspection does not correct deficiencies in the system or defects in the products, nor does it change a product or increase its value. Inspection only finds deficiencies and defects. Moreover, inspections are expensive and do not add value to the product.

Inspection
A means of ensuring that an operation is producing at the quality level expected.

Inspection should be thought of as a vehicle for improving the system. Operations managers need to know critical points in the system: (1) *when to inspect* and (2) *where to inspect*.

When and Where to Inspect

Deciding when and where to inspect depends on the type of process and the value added at each stage. Notice, in Table 6.4, Samsung's pervasive testing processes. Inspections can take place at any of the following points:

1. At your supplier's plant while the supplier is producing.
2. At your facility upon receipt of goods from your supplier.

TABLE 6.4 How Samsung Tests Its Smartphones[4]

Durability	Stress testing with nail punctures, extreme temperatures and overcharging
Visual inspection	Comparing the battery with standardized models
X-ray	Looking for internal abnormalities
Charge and discharge	Power up and down the completed phone
Organic compounds (TVOC)	Looking for battery leakage
Disassembling	Opening the battery cell to inspect tab welding and insulation tape conditions
Accelerated usage	Simulated 2 weeks of real-life use in 5 days
Voltage volatility (OCV)	Checking for change in voltage throughout the manufacturing process

OM in Action — Inspecting the Boeing 787[5]

Imagine you're buying a $270 million car. You'd want to kick the tires pretty hard. That's what airlines do with new airplanes. Delivering one widebody airplane is a big deal—each plane has a list price roughly equivalent to the cost of a high-rise hotel.

Carriers such as American Airlines station their own engineers at Boeing factories to watch their flying machines get built and check parts as they arrive. Then they send flight attendants, mechanics, and pilots for what are called shakedown inspections.

"The rubber meets the road here," says an American manager, as he begins checking a brand new Boeing 787. "It's inspected, and it's inspected, and it's inspected. And yet we still find things." Boeing does its own testing, but buyers do their own extra inspection—and note an average of 140 items on a plane's punch list. (The paint on the tail in the photo is carefully checked because it has 13 different colors).

Five flight attendants, a couple of mechanical experts, and an American test pilot attack the 285-passenger plane. All the doors and panels are opened for inspection. Flight attendants shake each seat violently, grab the headrest, and pull it up and jerk the cord on each entertainment controller. They test power ports, USB ports, audio jacks, and the entertainment system. They open all tray tables and turn all lights on and off. They recline each seat with knee-knocking force. They flush all the toilets, blow fake smoke into smoke alarms, and make sure all prerecorded emergency messages sound when required.

Travis Dove

Inside the cockpit, an American test pilot flies the jet to its limits, making sure alarms sound when increasing air speed or slowing the plane down to stall speed. The pilot turns it sharply until "bank angle" warnings sound. Each engine gets shut down and restarted in the air. Every backup and emergency system is activated to make sure it works.

3. Before costly or irreversible processes.
4. During the step-by-step production process.
5. When production or service is complete.
6. Before delivery to your customer.
7. At the point of customer contact.

As we see in the Boeing 787 *OM in Action* box, with large, expensive purchases, many companies perform a comprehensive and demanding customer acceptance review.

The seven tools of TQM discussed in the previous section aid in this "when and where to inspect" decision. However, inspection is not a substitute for a robust product produced by well-trained employees in a good process. In one well-known experiment conducted by an independent research firm, 100 defective pieces were added to a "perfect" lot of items and then subjected to 100% inspection. The inspectors found only 68 of the defective pieces in their first inspection. It took another three passes by the inspectors to find the next 30 defects. The last two defects were never found. So the bottom line is that there is variability in the inspection process. In addition,

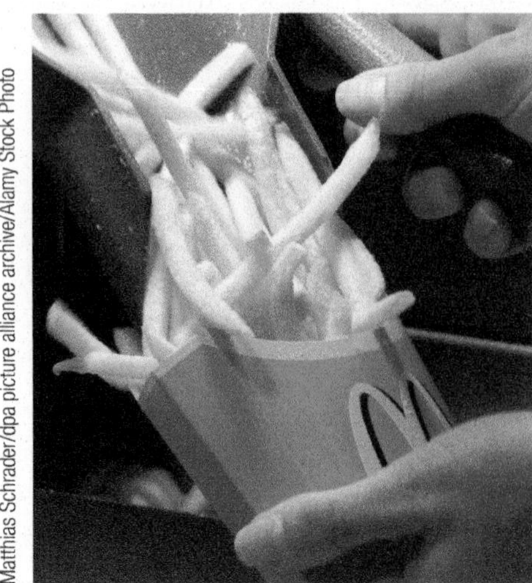
Matthias Schrader/dpa picture alliance archive/Alamy Stock Photo

Good methods analysis and the proper tools can result in poka-yokes that improve both quality and speed. Here, two poka-yokes are demonstrated. First, the aluminum scoop automatically positions the french fries vertically, and second, the properly sized container ensures that the portion served is correct. McDonald's thrives by bringing rigor and consistency to the restaurant business.

inspectors are only human: They become bored, they become tired, and the inspection equipment itself has variability. Even with 100% inspection, inspectors cannot guarantee perfection. Therefore, good processes, employee empowerment, and source control are a better solution than trying to find defects by inspection. You cannot inspect quality into the product.

STUDENT TIP
One of our themes of quality is that "quality cannot be inspected into a product."

Source Inspection

The best inspection can be thought of as no inspection at all; this "inspection" is always done at the source—it is just doing the job properly with the operator ensuring that this is so. This may be called **source inspection** (or source control) and is consistent with the concept of employee empowerment, where individual employees self-check their own work. The idea is that each supplier, process, and employee *treats the next step in the process as the customer*, ensuring perfect product to the next "customer." This inspection may be assisted by the use of checklists and controls such as a fail-safe device called a *poka-yoke*, a name borrowed from the Japanese.

Source inspection
Controlling or monitoring at the point of production or purchase—at the source.

A **poka-yoke** is a foolproof device or technique that ensures production of good units every time. These special devices avoid errors and provide quick feedback of problems. A simple example of a poka-yoke device is the diesel gas pump nozzle that will not fit into the "unleaded" gas tank opening on your car. In McDonald's, the french fry scoop and standard-size container used to measure the correct quantity are poka-yokes. Similarly, in a hospital, the prepackaged surgical coverings that contain exactly the items needed for a medical procedure are poka-yokes.

Poka-yoke
Literally translated, "mistake proofing"; it has come to mean a device or technique that ensures the production of a good unit every time.

Checklists are a type of poka-yoke to help ensure consistency and completeness in carrying out a task. A basic example is a to-do list. This tool may take the form of preflight checklists used by airplane pilots, surgical safety checklists used by doctors, or software quality assurance lists used by programmers. The *OM in Action* box, "Safe Patients, Smart Hospitals," illustrates the important role checklists have in hospital quality.

Checklist
A type of poka-yoke that lists the steps needed to ensure consistency and completeness in a task.

The idea of source inspection, poka-yokes, and checklists is to guarantee 100% good product or service at each step of a process.

Service Industry Inspection

In *service*-oriented organizations, inspection points can be assigned at a wide range of locations, as illustrated in Table 6.5. Again, the operations manager must decide where inspections are justified and may find the seven tools of TQM useful when making these judgments.

OM in Action — Safe Patients, Smart Hospitals[6]

Simple and avoidable errors are made in hospitals each day, causing patients to die. Inspired by two tragic medical mistakes—his father's misdiagnosed cancer and sloppiness that killed an 18-month-old child at Johns Hopkins—Dr. Peter Pronovost has made it his mission, often swimming upstream against the medical culture, to improve patient safety and prevent deaths.

He began by developing a basic 5-step checklist to reduce catheter infections. Inserted into veins in the groin, neck, or chest to administer fluids and medicines, catheters can save lives. But every year, 80,000 Americans get infections from *central venous catheters* (or lines), and more than 30,000 of these patients die. Pronovost's checklist has dropped infection rates at hospitals that use it down to zero, saving thousands of lives and tens of millions of dollars.

His steps for doctors and nurses are simple: (1) wash your hands; (2) use sterile gloves, masks, and drapes; (3) use antiseptic on the area being opened for the catheter; (4) avoid veins in the arms and legs; and (5) take the catheter out as soon as possible. He also created a special cart, where all supplies needed are stored.

Dr. Pronovost believes that many hospital errors are due to lack of standardization, poor communications, and a noncollaborative culture that is "antiquated and toxic." He points out that checklists in the airline industry are a science, and *every* crew member works as part of the safety team. Pronovost's book has shown that one person, with small changes, can make a huge difference.

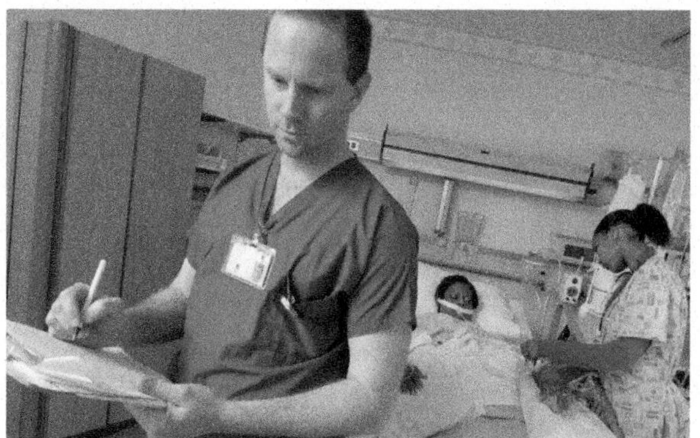
David Joel/Photographer's Choice RF/Getty Images

VIDEO 6.2
Quality Counts at Alaska Airlines

TABLE 6.5 Examples of Inspection in Services

ORGANIZATION	WHAT IS INSPECTED	STANDARD
Alaska Airlines	Last bag on carousel	Less than 20 minutes after arrival at the gate
	Airplane door opened	Less than 2 minutes after arrival at the gate
Jones Law Offices	Receptionist performance	Phone answered by the second ring
	Billing	Accurate, timely, and correct format
	Attorney	Promptness in returning calls
Hard Rock Hotel	Reception desk	Use customer's name
	Doorkeeper	Greet guest in less than 30 seconds
	Room	All lights working, spotless bathroom
	Minibar	Restocked and charges accurately posted to bill
Arnold Palmer Hospital	Billing	Accurate, timely, and correct format
	Pharmacy	Prescription accuracy, inventory accuracy
	Lab	Audit for lab-test accuracy
	Nurses	Charts immediately updated
	Admissions	Data entered correctly and completely
Olive Garden Restaurant	Busser	Serves water and bread within one minute
	Busser	Clears all entrée items and crumbs prior to dessert
	Server	Knows and suggests specials, desserts
Nordstrom Department Store	Display areas	Attractive, well organized, stocked, good lighting
	Stockrooms	Rotation of goods, organized, clean
	Salesclerks	Neat, courteous, very knowledgeable

Inspection of Attributes versus Variables

Attribute inspection
An inspection that classifies items as being either good or defective.

Variable inspection
Classifications of inspected items as falling on a continuum scale, such as dimension or strength.

When inspections take place, quality characteristics may be measured as either *attributes* or *variables*. **Attribute inspection** classifies items as being either good or defective. It does not address the *degree* of failure. For example, the lightbulb burns or it does not. **Variable inspection** measures such dimensions as weight, speed, size, or strength to see if an item falls within an acceptable range. If a piece of electrical wire is supposed to be 0.01 inch in diameter, a micrometer can be used to see if the product is close enough to pass inspection.

Knowing whether attributes or variables are being inspected helps us decide which statistical quality control approach to take, as we will see in the supplement to this chapter.

TQM in Services

VIDEO 6.3
Celebrity Cruises: A Premium Experience

The personal component of services is more difficult to measure than the quality of the tangible component. Generally, the user of a service, like the user of a good, has features in mind that form a basis for comparison among alternatives. Lack of any one feature may eliminate the service from further consideration. Quality also may be perceived as a bundle of attributes in which many lesser characteristics are superior to those of competitors. This approach to product comparison differs little between goods and services. However, what is very different about the selection of services is the poor definition of the (1) *intangible differences between products* and (2) *the intangible expectations customers have of those products*. Indeed, the intangible attributes may not be defined at all. They are often unspoken images in the purchaser's mind. This is why all of those marketing issues such as advertising, image, and promotion can make a difference.

TABLE 6.6 Determinants of Service Quality[7]

Reliability involves consistency of performance and dependability. It means that the firm performs the service right the first time and that the firm honors its promises.
Responsiveness concerns the willingness or readiness of employees to provide service. It involves timeliness of service.
Competence means possession of the required skills and knowledge to perform the service.
Access involves approachability and ease of contact.
Courtesy involves politeness, respect, consideration, and friendliness of contact personnel (including receptionists, telephone operators, etc.).
Communication means keeping customers informed in language they can understand and listening to them. It may mean that the company has to adjust its language for different consumers—increasing the level of sophistication with a well-educated customer and speaking simply and plainly with a novice.
Credibility involves trustworthiness, believability, and honesty. It involves having the customer's best interests at heart.
Security is the freedom from danger, risk, or doubt.
Understanding/knowing the customer involves making the effort to understand the customer's needs.
Tangibles include the physical evidence of the service.

The operations manager plays a significant role in addressing several major aspects of service quality. First, the *tangible component of many services is important*. How well the service is designed and produced does make a difference. This might be how accurate, clear, and complete your checkout bill at the hotel is, how warm the food is at Taco Bell, or how well your car runs after you pick it up at the repair shop.

Second, another aspect of service and service quality is the process. Notice in Table 6.6 that 9 out of 10 of the determinants of service quality are related to *the service process*. Such things as reliability and courtesy are part of the process. An operations manager can *design processes that have these attributes* and can ensure their quality through the TQM techniques discussed in this chapter.

Third, the operations manager should realize that the customer's expectations are the standard against which the service is judged. Customers' perceptions of service quality result from a comparison of their "before-service expectations" with their "actual-service experience." In other words, service quality is judged on the basis of whether it meets expectations. *The manager may be able to influence both the quality of the service and the expectation.* Don't promise more than you can deliver.

Fourth, the manager must expect exceptions. There is a standard quality level at which the regular service is delivered, such as the bank teller's handling of a transaction. However, there are "exceptions" or "problems" initiated by the customer or by less-than-optimal operating conditions (e.g., the computer "crashed"). This implies that the quality control system must recognize and *have a set of alternative plans for less-than-optimal operating conditions*.

Service recovery
Training and empowering frontline workers to solve a problem immediately.

Well-run companies have **service recovery** strategies. This means they train and empower frontline employees to immediately solve a problem. For instance, staff at Marriott Hotels are drilled in the **LEARN** routine—**L**isten, **E**mpathize, **A**pologize, **R**eact, **N**otify—with the final step ensuring that the complaint is fed back into the system. And at the Ritz-Carlton, staff members are trained not to say merely "sorry" but "please accept my apology." The Ritz gives them a budget for reimbursing upset guests. Similarly, employees at Alaska Airlines are empowered to soothe irritated travelers by drawing from a "toolkit" of options at their disposal.

SERVQUAL
A popular measurement scale for service quality that compares service expectations with service performance.

Managers of service firms may find **SERVQUAL** useful when evaluating performance. SERVQUAL is a widely used instrument that provides direct comparisons between customer service expectations and the actual service provided. SERVQUAL focuses on the *gaps* between the customer service expectations and the service provided on 10 service quality determinants.

The most common version of the scale collapses the 10 service quality determinants shown in Table 6.6 into five factors for measurement: reliability, assurance, tangibles, empathy, and responsiveness.

Designing the product, managing the service process, matching customer expectations to the product, and preparing for the exceptions are keys to quality services. The *OM in Action* box, "Richey International's Spies," provides another glimpse of how OM managers improve quality in services.

OM in Action Richey International's Spies[8]

How do luxury hotels maintain quality? They inspect. But when the product is one-on-one service, largely dependent on personal behavior, how do you inspect? You hire spies!

Richey International is the spy. Preferred Hotels and Resorts Worldwide and Intercontinental Hotels have both hired Richey to do quality evaluations via spying. Richey employees posing as customers perform the inspections. However, even then management must have established what the customer expects and specific services that yield customer satisfaction. Only then do managers know where and how to inspect. Aggressive training and objective inspections reinforce behavior that will meet those customer expectations.

The hotels use Richey's undercover inspectors to ensure performance to exacting standards. The hotels do not know when the evaluators will arrive. Nor what aliases they will use. More than 50 different standards are evaluated before the inspectors even check in at a luxury hotel. Over the next 24 hours, using checklists, tape recordings, and photos, written reports are prepared. The reports include evaluation of standards such as:

- Does the doorkeeper greet each guest in less than 30 seconds?
- Does the front-desk clerk use the guest's name during check-in?
- Are the bathroom tub and shower spotlessly clean?
- How many minutes does it take to get coffee after the guest sits down for breakfast?
- Did the server make eye contact?
- Were minibar charges posted correctly on the bill?

Established standards, aggressive training, and inspections are part of the TQM effort at these hotels. Quality does not happen by accident.

Summary

Quality is a term that means different things to different people. We define quality as "the totality of features and characteristics of a product or service that bears on its ability to satisfy stated or implied needs." Defining quality expectations is critical to effective and efficient operations.

Quality requires building a total quality management (TQM) environment because quality cannot be inspected into a product. The chapter also addresses seven TQM concepts: continuous improvement, Six Sigma, employee empowerment, benchmarking, just-in-time, Taguchi concepts, and knowledge of TQM tools. The seven TQM tools introduced in this chapter are check sheets, scatter diagrams, cause-and-effect diagrams, Pareto charts, flowcharts, histograms, and statistical process control (SPC).

Key Terms

Quality (p. 215)
ISO 9000 (p. 216)
Cost of quality (COQ) (p. 216)
Total quality management (TQM) (p. 217)
PDCA (p. 218)
Six Sigma (p. 218)
Employee empowerment (p. 219)
Quality circle (p. 219)
Benchmarking (p. 219)
Quality robust (p. 221)
Target-oriented quality (p. 222)
Quality loss function (QLF) (p. 222)
Cause-and-effect diagram, Ishikawa diagram, or fish-bone chart (p. 224)
Pareto charts (p. 224)
Flowcharts (p. 225)
Statistical process control (SPC) (p. 226)
Control charts (p. 226)
Inspection (p. 227)
Source inspection (p. 229)
Poka-yoke (p. 229)
Checklist (p. 229)
Attribute inspection (p. 230)
Variable inspection (p. 230)
Service recovery (p. 231)
SERVQUAL (p. 231)

Ethical Dilemma

A lawsuit a few years ago made headlines worldwide when a McDonald's drive-through customer spilled a cup of scalding hot coffee on herself. Claiming the coffee was too hot to be safely consumed in a car, the badly burned 80-year-old woman won $2.9 million in court. (The judge later reduced the award to $640,000.) McDonald's claimed the product was served to the correct specifications and was of proper quality. Further, the cup read "Caution—Contents May Be Hot." McDonald's coffee, at 180°, is substantially hotter (by corporate rule) than typical restaurant coffee, despite hundreds of coffee-scalding complaints in the past 10 years. Similar court cases, incidentally, resulted in smaller verdicts, but again in favor of the plaintiffs. For example, Motor City Bagel Shop was sued for a spilled cup of coffee by a drive-through patron, and Starbucks by a customer who spilled coffee on her own ankle.

Are McDonald's, Motor City, and Starbucks at fault in situations such as these? How do quality and ethics enter into these cases?

Discussion Questions

1. Explain how improving quality can lead to reduced costs.
2. As an Internet exercise, determine the Baldrige Award criteria. See the Web site **www.nist.gov/baldrige/**.
3. Which 3 of Deming's 14 points do you think are most critical to the success of a TQM program? Why?
4. List the seven concepts that are necessary for an effective TQM program. How are these related to Deming's 14 points?
5. Name three of the important people associated with the quality concepts of this chapter. In each case, write a sentence about each one summarizing his primary contribution to the field of quality management.
6. What are seven tools of TQM?
7. How does fear in the workplace (and in the classroom) inhibit learning?
8. How can a university control the quality of its output (that is, its graduates)?
9. Philip Crosby said that quality is free. Why?
10. List the three concepts central to Taguchi's approach.
11. What is the purpose of using a Pareto chart for a given problem?
12. What are the four broad categories of "causes" to help initially structure an Ishikawa diagram or cause-and-effect diagram?
13. Of the several points where inspection may be necessary, which apply especially well to manufacturing?
14. What roles do operations managers play in addressing the major aspects of service quality?
15. Explain, in your own words, what is meant by *source inspection*.
16. What are 10 determinants of service quality?
17. Name several products that do not require high quality.
18. In this chapter, we have suggested that building quality into a process and its people is difficult. Inspections are also difficult. To indicate just how difficult inspections are, count the number of *E*s (both capital *E* and lowercase *e*) in the *OM in Action* box, "Inspecting the Boeing 787" (including the title but not the photo credit). How many did you find? If each student does this individually, you are very likely to find a distribution rather than a single number!

Solved Problems

SOLVED PROBLEM 6.1

Northern Airlines' frequent flyer complaints about redeeming miles for free, discounted, and upgraded travel are summarized below, in 5 categories, from 600 letters received this year.

COMPLAINT	FREQUENCY
Could not get through to customer service to make requests	125
Seats not available on date requested	270
Had to pay fees to get "free" seats	62
Seats were available but only on flights at odd hours	110
Rules kept changing whenever customer called	33

Develop a Pareto chart for the data.

SOLUTION

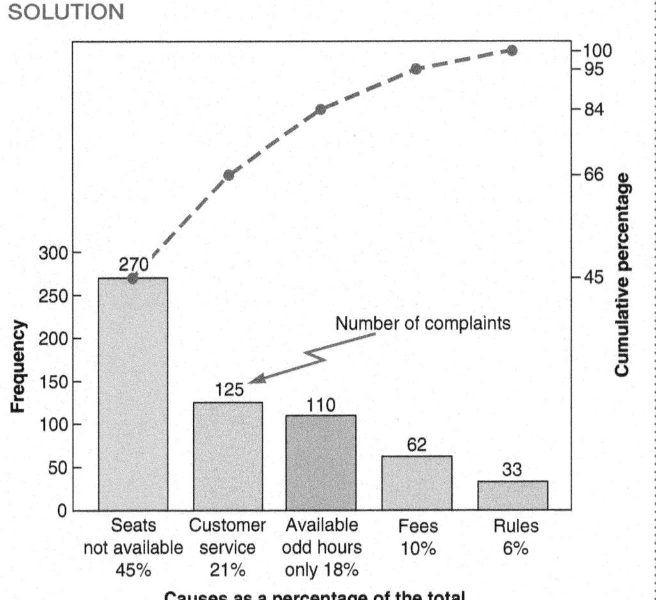

Problems

Problems 6.1–6.19 relate to Tools of TQM

• **6.1** An avant-garde clothing manufacturer runs a series of high-profile, risqué ads on a billboard on Highway 101 and regularly collects protest calls from people who are offended by them. The company has no idea how many people in total see the ads, but it has been collecting statistics on the number of phone calls from irate viewers:

TYPE	DESCRIPTION	NUMBER OF COMPLAINTS
R	Offensive racially/ethnically	10
M	Demeaning to men	4
W	Demeaning to women	14
I	Ad is incomprehensible	6
O	Other	2

a) Depict this data with a Pareto chart. Also depict the cumulative complaint line.
b) What percentage of the total complaints can be attributed to the most prevalent complaint?

• **6.2** Develop a scatter diagram for two variables of interest [say pages in the newspaper by day of the week; see the example in Figure 6.5(b)].

• **6.3** Develop a Pareto chart of the following causes of poor grades on an exam:

REASON FOR POOR GRADE	FREQUENCY
Insufficient time to complete	15
Late arrival to exam	7
Difficulty understanding material	25
Insufficient preparation time	2
Studied wrong material	2
Distractions in exam room	9
Calculator batteries died during exam	1
Forgot exam was scheduled	3
Felt ill during exam	4

• **6.4** Develop a histogram of the time it took for you or your friends to receive six recent orders at a fast-food restaurant.

•• **6.5** Kathleen McFadden's restaurant in Boston has recorded the following data for eight recent customers:

CUSTOMER NUMBER, i	MINUTES FROM TIME FOOD ORDERED UNTIL FOOD ARRIVED (y_i)	NO. OF TRIPS TO KITCHEN BY WAITRESS (x_i)
1	10.50	4
2	12.75	5
3	9.25	3
4	8.00	2
5	9.75	3
6	11.00	4
7	14.00	6
8	10.75	5

a) McFadden wants you to graph the eight points (x_i, y_i), $i = 1, 2, \ldots 8$. She has been concerned because customers have been waiting too long for their food, and this graph is intended to help her find possible causes of the problem.
b) This is an example of what type of graph?

•• **6.6** Develop a flowchart [as in Figure 6.5(e) and Example 2] showing all the steps involved in planning a party.

•• **6.7** Consider the types of poor driving habits that might occur at a traffic light. Make a list of the 10 you consider most likely to happen. Add the category of "other" to that list.
a) Compose a check sheet [like that in Figure 6.5(a)] to collect the frequency of occurrence of these habits. Using your check sheet, visit a busy traffic light intersection at four different times of the day, with two of these times being during high-traffic periods (rush hour, lunch hour). For 15 to 20 minutes each visit, observe the frequency with which the habits you listed occurred.
b) Construct a Pareto chart showing the relative frequency of occurrence of each habit.

•• **6.8** Draw a fish-bone chart detailing reasons why an airline customer might be dissatisfied.

•• **6.9** Consider the everyday task of getting to work on time or arriving at your first class on time in the morning. Draw a fish-bone chart showing reasons why you might arrive late in the morning.

•• **6.10** Construct a cause-and-effect diagram to reflect "student dissatisfied with university registration process." Use the "four *M*s" or create your own organizing scheme. Include at least 12 causes.

•• **6.11** Draw a fish-bone chart depicting the reasons that might give rise to an incorrect fee statement at the time you go to pay for your registration at school.

••• **6.12** Mary Beth Marrs, the manager of an apartment complex, feels overwhelmed by the number of complaints she is receiving. Below is the check sheet she has kept for the past 12 weeks. Develop a Pareto chart using this information. What recommendations would you make?

WEEK	GROUNDS	PARKING/ DRIVES	POOL	TENANT ISSUES	ELECTRICAL/ PLUMBING
1	✓✓✓	✓✓	✓	✓✓✓	
2	✓	✓✓✓	✓✓	✓✓	✓
3	✓✓✓	✓✓✓	✓✓	✓	
4	✓	✓✓✓✓	✓	✓	✓✓
5	✓✓	✓✓✓	✓✓✓✓	✓✓	
6	✓	✓✓✓✓	✓✓		
7		✓✓✓	✓✓	✓✓	
8	✓	✓✓✓✓	✓✓	✓✓✓	✓
9	✓	✓✓	✓		
10	✓	✓✓✓✓	✓✓	✓✓	
11		✓✓✓	✓✓	✓	
12	✓✓	✓✓✓	✓✓✓	✓	

• **6.13** Use Pareto analysis to investigate the following data collected on a printed-circuit-board assembly line:

DEFECT	NUMBER OF DEFECT OCCURRENCES
Components not adhering	143
Excess adhesive	71
Misplaced transistors	601
Defective board dimension	146
Mounting holes improperly positioned	12
Circuitry problems on final test	90
Wrong component	212

a) Prepare a graph of the data.
b) What conclusions do you reach?

•• **6.14** A list of 16 issues that led to incorrect formulations in Tuncey Bayrak's jam manufacturing unit in New England is provided below:

List of Issues

1. Incorrect measurement	9. Variability in scale accuracy
2. Antiquated scales	10. Equipment in disrepair
3. Lack of clear instructions	11. Technician calculation off
4. Damaged raw material	12. Jars mislabeled
5. Operator misreads display	13. Temperature controls off
6. Inadequate cleanup	14. Incorrect weights
7. Incorrect maintenance	15. Priority miscommunication
8. Inadequate flow controls	16. Inadequate instructions

Create a fish-bone diagram and categorize each of these issues correctly, using the "four *M*s" method.

•• **6.15** Develop a flowchart for one of the following:
a) Filling up with gasoline at a self-serve station.
b) Determining your account balance and making a withdrawal at an ATM.
c) Getting a cone of yogurt or ice cream from an ice cream store.

•••• **6.16** Boston Electric Generators has been getting many complaints from its major customer, Home Station, about the quality of its shipments of home generators. Daniel Shimshak, the plant manager, is alarmed that a customer is providing him with the only information the company has on shipment quality. He decides to collect information on defective shipments through a form he has asked his drivers to complete on arrival at customers' stores. The forms for the first 279 shipments have been turned in. They show the following over the past 8 weeks:

			REASON FOR DEFECTIVE SHIPMENT			
WEEK	NO. OF SHIPMENTS	NO. OF SHIPMENTS WITH DEFECTS	INCORRECT BILL OF LADING	INCORRECT TRUCKLOAD	DAMAGED PRODUCT	TRUCKS LATE
1	23	5	2	2	1	
2	31	8	1	4	1	2
3	28	6	2	3	1	
4	37	11	4	4	1	2
5	35	10	3	4	2	1
6	40	14	5	6	3	
7	41	12	3	5	3	1
8	44	15	4	7	2	2

Even though Daniel increased his capacity by adding more workers to his normal contingent of 30, he knew that for many weeks he exceeded his regular output of 30 shipments per week. A review of his turnover over the past 8 weeks shows the following:

WEEK	NO. OF NEW HIRES	NO. OF TERMINATIONS	TOTAL NO. OF WORKERS
1	1	0	30
2	2	1	31
3	3	2	32
4	2	0	34
5	2	2	34
6	2	4	32
7	4	1	35
8	3	2	36

a) Develop a scatter diagram using total number of shipments and number of defective shipments. Does there appear to be any relationship?
b) Develop a scatter diagram using the variable "turnover" (number of new hires plus number of terminations) and the number of defective shipments. Does the diagram depict a relationship between the two variables?
c) Develop a Pareto chart for the type of defects that have occurred.
d) Draw a fish-bone chart showing the possible causes of the defective shipments.

••• **6.17** A recent Gallup poll of 519 adults who flew in the past year found the following number of complaints about flying: cramped seats (45), cost (16), dislike or fear of flying (57), security measures (119), poor service (12), connecting flight problems (8), overcrowded planes (42), late planes/waits (57), food (7), lost luggage (7), and other (51).

a) What percentage of those surveyed found nothing they disliked?
b) Draw a Pareto chart summarizing these responses. Include the "no complaints" group.
c) Use the "four *M*s" method to create a fish-bone diagram for the 10 specific categories of dislikes (exclude "other" and "no complaints").

•• **6.18** Prepare a fish-bone chart explaining how a pizza delivery can arrive late on a Friday or Saturday.

••• **6.19** Draw a fish-bone chart showing why an assistant you paid to prepare a spreadsheet produced one with numerous errors.

Problems 6.20-6.21 relate to The Role of Inspection

•• **6.20** Choose criteria for a checklist to identify the quality problems (possible defect issues) for manufacturing shirts and blouses.

•• **6.21** Identify and explain six quality inspections in a service with which you have had some experience.

CASE STUDIES

Southwestern University: (C)

The popularity of Southwestern University's football program under its new coach Phil Flamm surged in each of the 5 years since his arrival at the Stephenville, Texas, college. (See Southwestern University (B) in Chapter 4.) With a football stadium close to maxing out at 54,000 seats and a vocal coach pushing for a new stadium, SWU president Joel Wisner faced some difficult decisions. After a phenomenal upset victory over its archrival, the University of Texas, at the homecoming game in the fall, Dr. Wisner was not as happy as one would think. Instead of ecstatic alumni, students, and faculty, all Wisner heard were complaints. "The lines at the concession stands were too long"; "Parking was harder to find and farther away than in the old days" (that is, before the team won regularly); "Seats weren't comfortable"; "Traffic was backed up halfway to Dallas"; and on and on. "A college president just can't win," muttered Wisner to himself.

At his staff meeting the following Monday, Wisner turned to his VP of administration, Leslie Gardner. "I wish you would take care of these football complaints, Leslie," he said. "See what the *real* problems are and let me know how you've resolved them." Gardner wasn't surprised at the request. "I've already got a handle on it, Joel," she replied. "We've been randomly surveying 50 fans per game for the past year to see what's on their minds. It's all part of my campus-wide TQM effort. Let me tally things up and I'll get back to you in a week."

When she returned to her office, Gardner pulled out the file her assistant had compiled (see Table 6.7). "There's a lot of information here," she thought.

TABLE 6.7 Fan Satisfaction Survey Results (N = 250)

		OVERALL GRADE				
		A	B	C	D	F
Game Day	A. Parking	90	105	45	5	5
	B. Traffic	50	85	48	52	15
	C. Seating	45	30	115	35	25
	D. Entertainment	160	35	26	10	19
	E. Printed Program	66	34	98	22	30
Tickets	A. Pricing	105	104	16	15	10
	B. Season Ticket Plans	75	80	54	41	0
Concessions	A. Prices	16	116	58	58	2
	B. Selection of Foods	155	60	24	11	0
	C. Speed of Service	35	45	46	48	76
Respondents						
Alumnus	113					
Student	83					
Faculty/Staff	16					
None of the above	38					

Open-Ended Comments on Survey Cards:

Parking a mess Add a skybox Get better cheerleaders Double the parking attendants Everything is okay Too crowded Seats too narrow Great food Phil F. for President! I smelled drugs being smoked Stadium is ancient Seats are like rocks Not enough cops for traffic Game starts too late Hire more traffic cops Need new band Great!	More hot dog stands Seats are all metal Need skyboxes Seats stink Go SWU! Lines are awful Seats are uncomfortable I will pay more for better view Get a new stadium Student dress code needed I want cushioned seats Not enough police Students too rowdy Parking terrible Toilets weren't clean Not enough handicap spots in lot Well done, SWU	Put in bigger seats Friendly ushers Need better seats Expand parking lots Hate the bleacher seats Hot dogs cold $3 for a coffee? No way! Get some skyboxes Love the new uniforms Took an hour to park Coach is terrific More water fountains Better seats Seats not comfy Bigger parking lot I'm too old for bench seats Cold coffee served at game	My company will buy a skybox—build it! Programs overpriced Want softer seats Beat those Longhorns! I'll pay for a skybox Seats too small Band was terrific Love Phil Flamm Everything is great Build new stadium Move games to Dallas No complaints Dirty bathroom

Discussion Questions

1. Using at least two different quality tools, analyze the data and present your conclusions.
2. How could the survey have been more useful?
3. What is the next step?

The Culture of Quality at Arnold Palmer Hospital

Video Case

Founded in 1989, Arnold Palmer Hospital is one of the largest hospitals for women and children in the U.S., with 431 beds in two facilities totaling 676,000 square feet. Located in downtown Orlando, Florida, and named after its famed golf benefactor, the hospital, with more than 2,000 employees, serves an 18-county area in central Florida and is the only Level 1 trauma center for children in that region. Arnold Palmer Hospital provides a broad range of medical services including neonatal and pediatric intensive care, pediatric oncology and cardiology, care for high-risk pregnancies, and maternal intensive care.

The Issue of Assessing Quality Health Care

Quality health care is a goal all hospitals profess, but Arnold Palmer Hospital has actually developed comprehensive and scientific means of asking customers to judge the quality of care they receive. Participating in a national benchmark comparison against other hospitals, Arnold Palmer Hospital consistently scores in the top 10% in overall patient satisfaction. Executive Director Kathy Swanson states, "Hospitals in this area will be distinguished largely on the basis of their customer satisfaction. We must have accurate information about how our patients and their families judge the quality of our care, so I follow the questionnaire results daily. The in-depth survey helps me and others on my team to gain quick knowledge from patient feedback." Arnold Palmer Hospital employees are empowered to provide gifts in value up to $200 to patients who find reason to complain about any hospital service such as food, courtesy, responsiveness, or cleanliness.

Swanson doesn't focus just on the customer surveys, which are mailed to patients one week after discharge, but also on a variety of internal measures. These measures usually start at the grassroots level, where the staff sees a problem and develops ways to track performance. The hospital's longstanding philosophy supports the concept that each patient is important and respected as a person. That patient has the right to comprehensive, compassionate family-centered health care provided by a knowledgeable physician-directed team.

Some of the measures Swanson carefully monitors for continuous improvement are morbidity, infection rates, readmission rates, costs per case, and length of stays. The tools she uses daily include Pareto charts, flowcharts, and process charts, in addition to benchmarking against hospitals both nationally and in the southeast region.

The result of all of these efforts has been a quality culture as manifested in Arnold Palmer's high ranking in patient satisfaction and one of the highest survival rates of critically ill babies.

Discussion Questions*

1. Why is it important for Arnold Palmer Hospital to obtain a patient's assessment of health care quality? Do patients have the expertise to judge the health care they receive?
2. How would you build a culture of quality in an organization such as Arnold Palmer Hospital?
3. What techniques does Arnold Palmer Hospital practice in its drive for quality and continuous improvement?
4. Develop a fish-bone diagram illustrating the quality variables for a patient who just gave birth at Arnold Palmer Hospital (or any other hospital).

*You may wish to view the video that accompanies this case before addressing these questions.

Quality Counts at Alaska Airlines

Video Case

Alaska Airlines, with nearly 100 destinations, including regular service to Alaska, Hawaii, Canada, and Mexico, is the seventh-largest U.S. carrier. Alaska Airlines has won the J. D. Power and Associates Award for highest customer satisfaction in the industry for 12 years in a row while being the number one on-time airline for 5 years in a row.

Management's unwavering commitment to quality has driven much of the firm's success and generated an extremely loyal customer base. Executive VP Ben Minicucci exclaims, "We have rewritten our DNA." Building an organization that can achieve quality is a demanding task, and the management at Alaska Airlines accepted the challenge. This is a highly participative quality culture, reinforced by leadership training, constant process improvement, comprehensive metrics, and frequent review of those metrics. The usual training of flight crews and pilots is supplemented with classroom training in areas such as Six Sigma. Over 200 managers have obtained Six Sigma Green Belt certification.

Alaska Air lines

ELEMENTS	WEIGHTING	PERFORMANCE	SCORE	BONUS POINTS	TOTAL	GRADE
Process Compliance	20		15		15	B
Staffing	15		15	5	20	A+
MAP Rate (for bags)	20		15		15	B
Delays	10		9		9	A
Time to Carousel (total weight = 10)		98.7%	10		10	A
Percentage of flights scanned	2	70.9%				
Percentage of bags scanned	2	92.5%				
20 Minutes all bags dropped (% compliance)	4	2				
Outliers (>25mins)	2					
Safety Compliance	15		15	5	20	A+
Quality Compliance	10		10		10	A
Total - 100%	**100**		**89**	**10**	**99**	**A+**

Time to Carousel

Points	2	1.5	1	0
Percentage of flights scanned	95%–100%	90%–94.9%	89.9%–85%	< 84.9%

Points	2	0
Percentage of bags scanned	60% or above	≤ 59.9%

Points	0	4
Last bag percent compliance	Below 89.9%	90%–100%

Points	0	1	1.5	2
Last Bag >25 min. (Outliers)	20	15	10	5

Alaska collects more than 100 quality and performance metrics every day. For example, the accompanying picture tells the crew that it has 6 minutes to close the door and back away from the gate to meet the "time to pushback" target. Operations personnel review each airport hub's performance scorecard daily and the overall operations scorecard weekly. As Director of System Operations Control, Wayne Newton proclaims, "If it is not measured, it is not managed." The focus is on identifying problem areas or trends, determining causes, and working on preventive measures.

Within the operations function there are numerous detailed input metrics for station operations (such as the percentage of time that hoses are free of twists, the ground power cord is stowed, and no vehicles are parked in prohibited zones). Management operates under the assumption that if all the detailed input metrics are acceptable, the major key performance indicators, such as Alaska's on-time performance and 20-minute luggage guarantee, will automatically score well.

The accompanying table displays a sample monthly scorecard for Alaska's ground crew provider in Seattle. The major evaluation categories include process compliance, staffing (degree that crew members are available when needed), MAP rate (minimum acceptable performance for mishandled bags), delays, time to carousel, safety compliance, and quality compliance. The quality compliance category alone tracks 64 detailed input metrics using approximately 30,000 monthly observations. Each of the major categories on the scorecard has an importance weight, and the provider is assigned a weighted average score at the end of each month. The contract with the supplier provides for up to a 3.7% bonus for outstanding performance and as much as a 5.0% penalty for poor performance. The provider's line workers receive a portion of the bonus when top scores are achieved.

As a company known for outstanding customer service, service recovery efforts represent a necessary area of emphasis. When things go wrong, employees mobilize to first communicate with, and in many cases compensate, affected customers. "It doesn't matter if it's not our fault," says Minicucci. Front-line workers are empowered with a "toolkit" of options to offer to inconvenienced customers, including the ability to provide up to 5,000 frequent flyer miles and/or vouchers for meals, hotels, luggage, and tickets. When an Alaska flight had to make an emergency landing in Eugene, Oregon, due to a malfunctioning oven, passengers were immediately texted with information about what happened and why, and they were told that a replacement plane would be arriving within 1 hour. Within that hour, an apology letter along with a $450 ticket voucher were already in the mail to each passenger's home. No customer complaints subsequently appeared on Twitter or Facebook. It's no wonder why Alaska's customers return again and again.

Discussion Questions*

1. What are some ways that Alaska can ensure that quality and performance metric standards are met when the company outsources its ground operations to a contract provider?
2. Identify several quality metrics, in addition to those identified earlier, that you think Alaska tracks or should be tracking.
3. Think about a previous problem that you had when flying, for example, a late flight, a missed connection, or lost luggage. How, if at all, did the airline respond? Did the airline adequately address your situation? If not, what else should they have done? Did your experience affect your desire (positively or negative) to fly with that airline in the future? (If you

have not had such an airline experience, feel free to comment about some other service experience you had where a problem occurred.)

4. See the accompanying table. The contractor received a perfect Time to Carousel score of 10 total points, even though its performance was not "perfect." How many total overall points would the contractor have received with the following performance scores: 93.2% of flights scanned, 63.5% of bags scanned, 89.6% of all bags dropped within 20 minutes, and 15 bags arriving longer than 25 minutes? (The contractor's current total score is 99.)

*You may wish to view the video that accompanies this case before answering these questions.

Celebrity Cruises: A Premium Experience

Video Case 

Twenty-five years ago, Celebrity Cruises, Inc., decided to make a name for itself in the premium market by offering an "upscale experience at an intelligent price." Evoking images of luxury similar to the Ritz-Carlton brand, this "hotel on the water" treats quality as if it is the heartbeat of the company. Consequently, Celebrity has consistently been awarded the "Best Premium Cruise Line."

In the cruise and hotel industries, quality can be hard to quantify. Traveling guests are buying an experience—not just a tangible product. So creating the right combination of elements to make the experience stand out is the goal of every employee, from cabin attendants to galley staff to maintenance to entertainers. The captain even has an important social role, often hosting dinners for a dozen guests a night.

"Our target audience consists of savvy, discerning guests who know what they want from this cruise," says Brian Abel, Associate VP for Hotel Operations. "We meet their needs by being best in the competitive class of modern luxury ships."

Crew-to-guest ratios at Celebrity, and other premium lines, are 1 crew member for every 2 guests. Employees are expected to greet guests with a formal style, to say "good morning" instead of "hi" and "with pleasure" instead of "no problem." With crew members from 70 countries, such preferred phrases, dress codes, and many other manners of dealing with customers are detailed in employee training manuals.

Employees sign 4- to 9-month contracts and then typically take 6–8 weeks off. They have difficult jobs, often working 7 days a week, but even with intense schedules, most Celebrity staffers remain on the job for 5 to 7 years.

Food is a very important part of the cruise experience. "Food is the number-one reason people rebook a cruise," says Abel. So everything served aboard a Celebrity ship is prepared from scratch. About 200 people work in the galley in a structured, well-planned operation, using years of historical data, to forecast demand for each component of each meal.

Guest surveys are provided online at the end of each cruise, and guests complete them at a high rate (about 85%), having been strongly encouraged to do so by their cabin attendants. The surveys serve not only as a measure of overall satisfaction, but two other purposes as well: (1) they are used as a brand marketing tool comparing Celebrity to other cruise lines, hotels, and competing entertainment venues, and (2) they provide management with specific feedback, down to the individual employee level in some cases. Abel personally reviews results of each completed cruise within 48 hours and takes action if defects are found in any aspect of the experience. The initial part of the questionnaire appears in Figure 6.8.

Celebrity's main quality feedback tool is called the net provider score (NPS), which tallies the guests' answers to a wide series of questions about their experience. The question: "How likely would you be to recommend Celebrity Cruises to a friend, family member, or colleague?" is critical and is the measure used to compare ships within Celebrity's fleet as well as with competing cruise lines. Scores of 9–10 on this question label the customers as "advocates." A 7–8 is "neutral," and a score of 6 or below is a "detractor."

The NPS computation is simple: The percentage of detractors is subtracted from the percentage of advocates. For example, if 70% of the guests score the cruise a 9–10, 17% score it a 7–8, and 13% give a 6 or less, the NPS = 70 – 13 = 57. An elite line tries to attain a score over 60 on each cruise. Celebrity averages a 65.

Figure 6.8

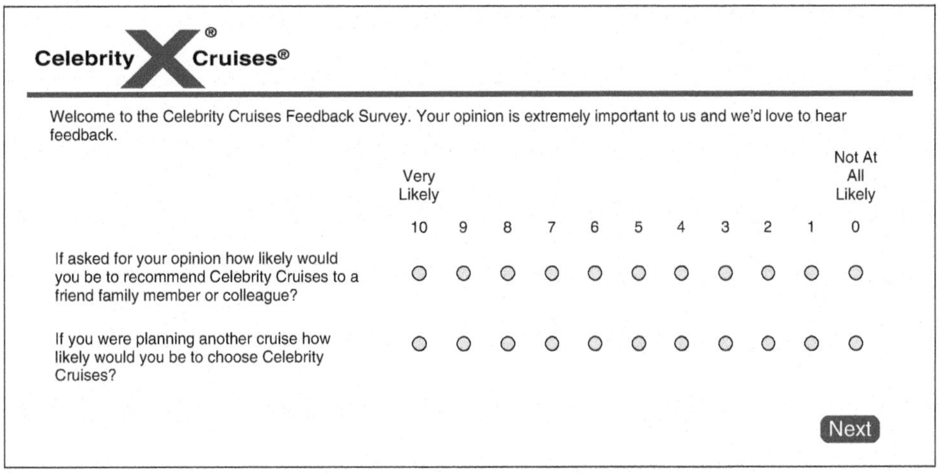

Celebrity X® Cruises®

Welcome to the Celebrity Cruises Feedback Survey. Your opinion is extremely important to us and we'd love to hear feedback.

	Very Likely										Not At All Likely
	10	9	8	7	6	5	4	3	2	1	0
If asked for your opinion how likely would you be to recommend Celebrity Cruises to a friend family member or colleague?	O	O	O	O	O	O	O	O	O	O	O
If you were planning another cruise how likely would you be to choose Celebrity Cruises?	O	O	O	O	O	O	O	O	O	O	O

Next

Discussion Questions*

1. What unique aspects of the cruise industry make quality service more difficult to attain? What aspects help raise quality?
2. How does the cruise operation differ from that at a land-based hotel?
3. How could control charts, Pareto diagrams, and cause-and-effect diagrams be used to identify quality problems at Celebrity?
4. Suppose that on two successive cruises of the same ship, the cruise line receives NPS scores of: (Trip 1) 78% "advocates," 4% "neutrals," and 18% "detractors" and (Trip 2) 70% "advocates," 20% "neutrals," and 10% "detractors." Which would be preferable and why?
5. List a dozen quality indicators (besides NPS) that Celebrity also measures. (There are 35 on its guest evaluation form.)

*You may wish to view the video that accompanies this case before addressing these questions.

Endnotes

1. Philip B. Crosby, *Quality Is Free* (New York: McGraw-Hill, 1979). Further, J. M. Juran states, in his book *Juran on Quality by Design* (The Free Press 1992, p. 119), that costs of poor quality "are huge, but the amounts are not known with precision. In most companies the accounting system provides only a minority of the information needed to quantify this cost of poor quality. It takes a great deal of time and effort to extend the accounting system so as to provide full coverage."
2. Source: Based on Deming, W. Edwards. *Out of the Crisis*, pp. 23–24, © 2000 W. Edwards Deming Institute, published by The MIT Press.
3. Sources: **www.asq.org** (August, 2008); and *The Wall Street Journal* (December 3, 2007) and (November 14, 2006).
4. Sources: Based on Samsung Electronics Co. and *The Wall Street Journal* (January 23, 2017:B4).
5. Sources: *The Wall Street Journal* (August 31, 2017) and (December 14, 2020).
6. Sources: **FierceHealthcare.com** (July 21, 2019); and **Cleveland.com** (February 9, 2020).
7. Sources: Adapted from A. Parasuranam, Valarie A. Zeithaml, and Leonard L. Berry, "A Conceptual Model of Service Quality and Its Implications for Future Research," *Journal of Marketing* (1985): 49. Copyright © 1985 by the American Marketing Association. Reprinted with permission.
8. Sources: *Daily Mail* (October 7, 2015) and (July 14, 2012); and *Travel Weekly* (July 14, 2015).

Bibliography

Besterfield, D. H. *Quality Control*, 9th ed. Boston: Pearson, 2013.

Brown, M. G. *Baldrige Award Winning Quality*, 19th ed. University Park, IL: Productivity Press, 2010.

Crosby, P. B. *Quality Is Still Free*. New York: McGraw-Hill, 1996.

Feigenbaum, A. V. "Raising the Bar." *Quality Progress* 41, no. 7 (July 2008): 22–28.

Foster, T. *Managing Quality*, 6th ed. Boston: Pearson, 2017.

Gitlow, H. S. *A Guide to Lean Six Sigma Management Skills*. University Park, IL: Productivity Press, 2009.

Gryna, F. M., R. C. H. Chua, and J. A. DeFeo. *Juran's Quality Planning and Analysis for Enterprise Quality*, 5th ed. New York: McGraw-Hill, 2007.

Luthra, S., D. Garg, A. Agarwal, and S. K. Mangla. *Total Quality Management (TQM): Principles, Methods, and Applications*. Boca Raton, FL: CRC Press, 2021.

Mitra, A. *Fundamentals of Quality Control and Improvement*, 4th ed. New York: Wiley, 2016.

Schroeder, R. G., et al. "Six Sigma: Definition and Underlying Theory." *Journal of Operations Management* 26, no. 4 (2008): 536–554.

Summers, D. *Quality*, 6th ed. Boston: Pearson, 2018.

Chapter 6 *Rapid* Review

Main Heading	Review Material	MyLab Operations Management
QUALITY AND STRATEGY	Managing quality helps build successful strategies of differentiation, low cost, and *response.* Two ways that quality improves profitability are: ▪ *Sales gains* via improved response, price flexibility, increased market share, and/or improved reputation ▪ *Reduced costs* via increased productivity, lower rework and scrap costs, and/or lower warranty costs	Concept Questions: 1.1–1.5 **VIDEO 6.1** The Culture and Quality at Arnold Palmer Hospital
DEFINING QUALITY	An operations manager's objective is to build a total quality management system that identifies and satisfies customer needs. ▪ **Quality**—The ability of a product or service to meet customer needs. The American Society for Quality (ASQ) defines quality as "the totality of features and characteristics of a product or service that bears on its ability to satisfy stated or implied needs." The two most well-known quality awards are: ▪ *U.S.*: Malcolm Baldrige National Quality Award, named after a former secretary of commerce ▪ *Japan*: Deming Prize, named after an American, Dr. W. Edwards Deming ▪ **ISO 9000**—A set of quality standards developed by the International Organization for Standardization (ISO). ISO 9000 is the only quality standard with international recognition. To do business globally, being listed in the ISO directory is critical. ▪ **Cost of quality (COQ)**—The cost of doing things wrong; that is, the price of nonconformance. The four major categories of costs associated with quality are *prevention costs, appraisal costs, internal failure costs,* and *external failure costs.* Four leaders in the field of quality management are W. Edwards Deming, Joseph M. Juran, Armand Feigenbaum, and Philip B. Crosby.	Concept Questions: 2.1–2.6
TOTAL QUALITY MANAGEMENT	▪ **Total quality management (TQM)**—Management of an entire organization so that it excels in all aspects of products and services that are important to the customer. Seven concepts for an effective TQM program are (1) continuous improvement, (2) Six Sigma, (3) employee empowerment, (4) benchmarking, (5) just-in-time (JIT), (6) Taguchi concepts, and (7) knowledge of TQM tools. ▪ **PDCA**—A continuous improvement model that involves four stages: plan, do, check, and act. The Japanese use the word *kaizen* to describe the ongoing process of unending improvement—the setting and achieving of ever-higher goals. ▪ **Six Sigma**—A program to save time, improve quality, and lower costs. In a statistical sense, Six Sigma describes a process, product, or service with an extremely high capability—99.9997% accuracy, or 3.4 defects per million. ▪ **Employee empowerment**—Enlarging employee jobs so that the added responsibility and authority are moved to the lowest level possible in the organization. Business literature suggests that some 85% of quality problems have to do with materials and processes, not with employee performance. ▪ **Quality circle**—A group of employees meeting regularly with a facilitator to solve work-related problems in their work area. ▪ **Benchmarking**—Selecting a demonstrated standard of performance that represents the very best performance for a process or an activity. The philosophy behind just-in-time (JIT) involves continuing improvement and enforced problem solving. JIT systems are designed to produce or deliver goods just as they are needed. ▪ **Quality robust**—Products that are consistently built to meet customer needs, despite adverse conditions in the production process. ▪ **Target-oriented quality**—A philosophy of continuous improvement to bring the product exactly on target. ▪ **Quality loss function (QLF)**—A mathematical function that identifies all costs connected with poor quality and shows how these costs increase as output moves away from the target value.	Concept Questions: 3.1–3.6

Main Heading	Review Material	MyLab Operations Management
TOOLS OF TQM	TQM tools that generate ideas include the *check sheet* (organized method of recording data), *scatter diagram* (graph of the value of one variable vs. another variable), and *cause-and-effect diagram*. Tools for organizing the data are the *Pareto chart* and *flowchart*. Tools for identifying problems are the *histogram* (distribution showing the frequency of occurrences of a variable) and *statistical process control chart*. ▪ **Cause-and-effect diagram**—A schematic technique used to discover possible locations of quality problems. (Also called an Ishikawa diagram or a fish-bone chart.) The 4 *M*s (material, machinery/equipment, manpower, and methods) may be broad "causes." ▪ **Pareto chart**—A graphic that identifies the few critical items as opposed to many less important ones. ▪ **Flowchart**—A block diagram that graphically describes a process or system. ▪ **Statistical process control (SPC)**—A process used to monitor standards, make measurements, and take corrective action as a product or service is being produced. ▪ **Control chart**—A graphic presentation of process data over time, with predetermined control limits.	Concept Questions: 4.1–4.6 Problems: 6.1, 6.3, 6.5, 6.8–6.14, 6.16–6.19 **ACTIVE MODEL 6.1** Virtual Office Hours for Solved Problem: 6.1
THE ROLE OF INSPECTION	▪ **Inspection**—A means of ensuring that an operation is producing at the quality level expected. ▪ **Source inspection**—Controlling or monitoring at the point of production or purchase: at the source. ▪ **Poka-yoke**—Literally translated, "mistake proofing"; it has come to mean a device or technique that ensures the production of a good unit every time. ▪ **Checklist**—A type of poka-yoke that lists the steps needed to ensure consistency and completeness in a task. ▪ **Attribute inspection**—An inspection that classifies items as being either good or defective. ▪ **Variable inspection**—Classifications of inspected items as falling on a continuum scale, such as dimension, size, or strength.	Concept Questions: 5.1–5.6 Problems: 6.20–6.21 **VIDEO 6.2** Quality Counts at Alaska Airlines
TQM IN SERVICES	Determinants of service quality: reliability, responsiveness, competence, access, courtesy, communication, credibility, security, understanding/knowing the customer, and tangibles. ▪ **Service recovery**—Training and empowering frontline workers to solve a problem immediately. ▪ **SERVQUAL**—A popular measurement scale for service quality that compares service expectations with service performance.	Concept Questions: 6.1–6.6 **VIDEO 6.3** Celebrity Cruises: A Premium Experience
ADDITIONAL MYLAB OPERATIONS MANAGEMENT RESOURCES	✔ Additional Case Studies (Westover Electrical, Inc. and Quality at the Ritz-Carlton Hotel Company) ✔ Southwestern University Case Studies are integrated in Chapters 3, 4, 6, 8, 12, and 13 and in Supplement 7 ✔ Multiple Choice Case Questions (Southwestern University (C) and Westover Electrical, Inc.) ✔ Quality Management Simulation	

Self Test

LO 6.1 In this chapter, *quality* is defined as:
- **a)** the degree of excellence at an acceptable price and the control of variability at an acceptable cost.
- **b)** how well a product fits patterns of consumer preferences.
- **c)** the totality of features and characteristics of a product or service that bears on its ability to satisfy stated or implied needs.
- **d)** being impossible to define, but you know what it is.

LO 6.2 ISO 9000 is an international standard that addresses _____.

LO 6.3 A Six Sigma program:
- **a)** is a process that has a very high level of capability.
- **b)** is a program that focuses on customer satisfaction.
- **c)** uses a set of tools such as histograms and flowcharts.
- **d)** uses the DMAIC model.
- **e)** All of the above are features of a Six Sigma program.

LO 6.4 The process of identifying other organizations that are best at some facet of your operations and then modeling your organization after them is known as:
- **a)** continuous improvement.
- **b)** employee empowerment.
- **c)** benchmarking.
- **d)** copycatting.
- **e)** patent infringement.

LO 6.5 The Taguchi method includes all except which of the following major concepts?
- **a)** Employee involvement
- **b)** Remove the effects of adverse conditions
- **c)** Quality loss function
- **d)** Target specifications

LO 6.6 The seven tools of total quality management are _______, _______, _______, _______, _______, _______, and _______.

Answers: LO 6.1. c; LO 6.2. quality management systems; LO 6.3. e; LO 6.4. c; LO 6.5. a; LO 6.6. check sheets, scatter diagrams, cause-and-effect diagrams, Pareto charts, flowcharts, histograms, SPC charts.

Statistical Process Control

SUPPLEMENT 6

SUPPLEMENT OUTLINE

LEARNING OBJECTIVES

LO S6.1 ***Explain*** the purpose of a control chart 245

LO S6.2 ***Explain*** the role of the central limit theorem in SPC 246

LO S6.3 ***Build*** $\bar{x}$-charts and R-charts 247

LO S6.4 ***List*** the five steps involved in building control charts 251

LO S6.5 ***Build*** p-charts and c-charts 253

LO S6.6 ***Explain*** process capability and compute C_p and C_{pk} 257

LO S6.7 ***Explain*** acceptance sampling 260

As part of its statistical process control system, Flowers Bakery, in Georgia, uses a digital camera to inspect just-baked sandwich buns as they move along the production line. Items that don't measure up in terms of color, shape, seed distribution, or size are identified and removed automatically from the conveyor.

Courtesy of Georgia Institute of Technology

Statistical Process Control (SPC)

Statistical process control (SPC)
A process used to monitor standards by taking measurements and corrective action as a product or service is being produced.

Control chart
A graphical presentation of process data over time.

In this supplement, we address statistical process control—the same techniques used at Flowers Bakery, Arnold Palmer Hospital, GE, and Southwest Airlines to achieve quality standards. **Statistical process control (SPC)** is the application of statistical techniques to ensure that processes meet standards. All processes are subject to a certain degree of variability. While studying process data in the 1920s, Walter Shewhart of Bell Laboratories made the distinction between the common (natural) and special (assignable) causes of variation. He developed a simple but powerful tool to separate the two—the **control chart**.

A process is said to be operating *in statistical control* when the only source of variation is common (natural) causes. The process must first be brought into statistical control by detecting and eliminating special (assignable) causes of variation.[1] Then its performance is predictable, and its ability to meet customer expectations can be assessed. The *objective* of a process control system is to *provide a statistical signal when assignable causes of variation are present*. Such a signal can quicken appropriate action to eliminate assignable causes.

Natural variations
Variability that affects every production process to some degree and is to be expected; also known as common cause.

Assignable variation
Variation in a production process that can be traced to specific causes.

Natural Variations Natural variations affect almost every process and are to be expected. **Natural variations** are the many sources of variation that occur within a process, even one that is in statistical control. Natural variations form a pattern that can be described as a *distribution*.

As long as the distribution (output measurements) remains within specified limits, the process is said to be "in control," and natural variations are tolerated.

Assignable Variations **Assignable variation** in a process can be traced to a specific reason. Factors such as machine wear, misadjusted equipment, fatigued or untrained workers, or new batches of raw material are all potential sources of assignable variations.

Natural and assignable variations distinguish two tasks for the operations manager. The first is to *ensure that the process is capable* of operating under control with only natural variation. The second is, of course, to *identify and eliminate assignable variations* so that the processes will remain under control.

Figure **S6.1**

Natural and Assignable Variation

(a) Samples of the product, say five boxes of cereal taken off the filling machine line, vary from one another in weight.

(b) After enough sample means are taken from a stable process, they form a pattern called a *distribution*.

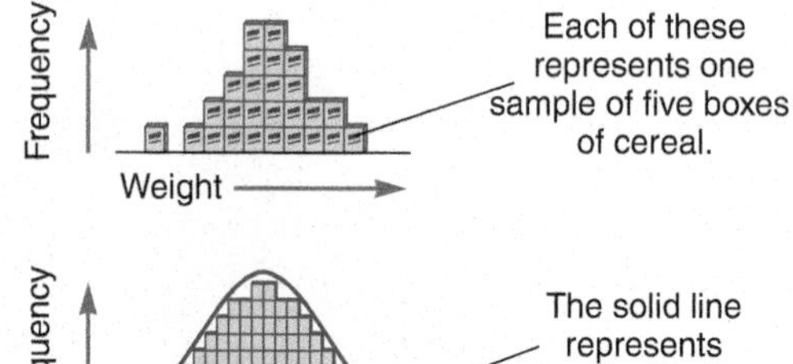

(c) There are many types of distributions, including the normal (bell-shaped) distribution, but distributions do differ in terms of central tendency (mean), standard deviation or variance, and shape.

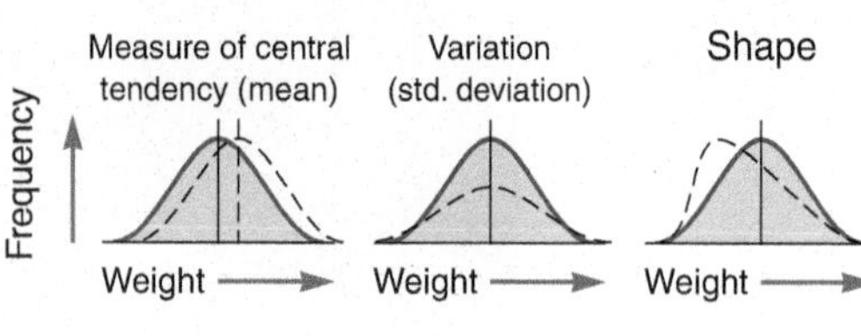

(d) If only natural causes of variation are present, the output of a process forms a distribution that is stable over time and is predictable.

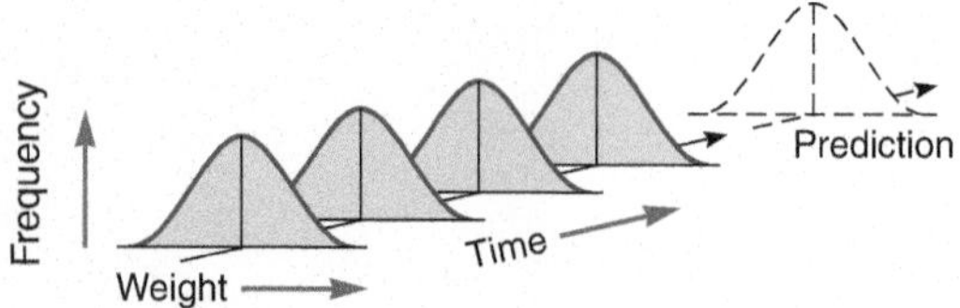

(e) If assignable causes of variation are present, the process output is not stable over time and is not predictable. That is, when causes that are not an expected part of the process occur, the samples will yield unexpected distributions that vary by central tendency, standard deviation, and shape.

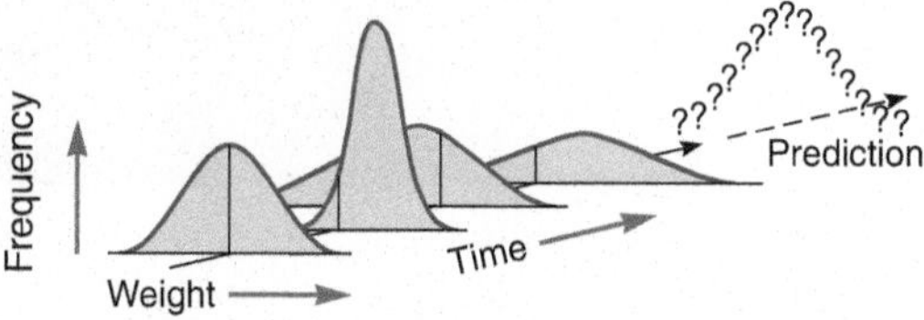

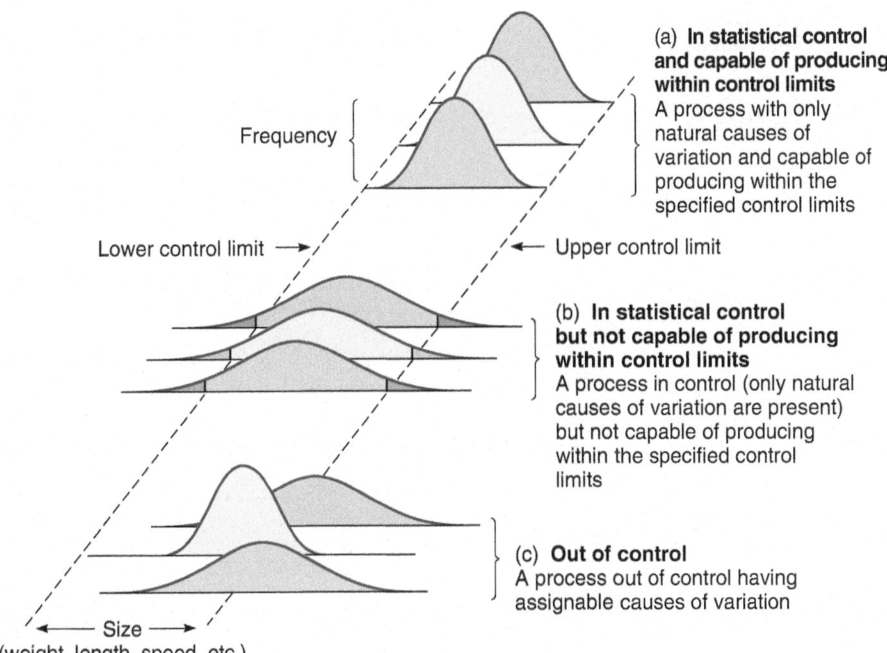

(a) In statistical control and capable of producing within control limits
A process with only natural causes of variation and capable of producing within the specified control limits
Frequency
Lower control limit
Upper control limit
(b) In statistical control but not capable of producing within control limits
A process in control (only natural causes of variation are present) but not capable of producing within the specified control limits
(c) Out of control
A process out of control having assignable causes of variation
Size
(weight, length, speed, etc.)

Figure **S6.2**

Process Control: Three Types of Process Outputs

Samples Because of natural and assignable variation, statistical process control uses averages of small samples (often of four to eight items) as opposed to data on individual parts. Individual pieces tend to be too erratic to make trends quickly visible.

Figure S6.1 provides a detailed look at the important steps in determining process variation. The horizontal scale can be weight (as in the number of ounces in boxes of cereal) or length (as in fence posts) or any physical measure. The vertical scale is frequency. The samples of five boxes of cereal in Figure S6.1 **(a)** are weighed, **(b)** form a distribution, and **(c)** can vary. The distributions formed in **(b)** and **(c)** will fall in a predictable pattern **(d)** if only natural variation is present. If assignable causes of variation are present, then we can expect either the mean to vary or the dispersion to vary, as is the case in **(e)**.

LO S6.1 *Explain* the purpose of a control chart

Control Charts The process of building control charts is based on the concepts presented in Figure S6.2. This figure shows three distributions that are the result of outputs from three types of processes. We plot small samples and then examine characteristics of the resulting data to see if the process is within "control limits." The purpose of control charts is to help distinguish between natural variations and variations due to assignable causes. As seen in Figure S6.2, a process is **(a)** in control *and the process is capable of producing within established control limits*, **(b)** in control *but the process is not capable of producing within established limits*, or **(c)** out of control. We now look at ways to build control charts that help the operations manager keep a process under control.

Control Charts for Variables

The variables of interest here are those that have continuous dimensions. They have an infinite number of possibilities. Examples are weight, speed, length, or strength. Control charts for the mean, $\bar{x}$ or *x*-bar, and the range, *R*, are used to monitor processes that have continuous dimensions. The $\bar{x}$-chart tells us whether changes have occurred in the central tendency (the mean, in this case) of a process. These changes might be due to such factors as tool wear, a gradual increase in temperature, a different method used on the second shift, or new and stronger materials. The *R*-chart values indicate that a gain or loss in dispersion has occurred. Such a change may be due to worn bearings, a loose tool, an erratic flow of lubricants to a machine, or to sloppiness on the part of a machine operator. The two types of charts go hand in hand when monitoring variables because they measure the two critical parameters: central tendency and dispersion.

$\bar{x}$-chart
A quality control chart for variables that indicates when changes occur in the central tendency of a production process.

***R*-chart**
A control chart that tracks the "range" within a sample; it indicates that a gain or loss in uniformity has occurred in dispersion of a production process.

The Central Limit Theorem

The theoretical foundation for $\bar{x}$-charts is the central limit theorem. This theorem states that regardless of the distribution of the population, the distribution of $\bar{x}$ s (each of which is a mean of a sample drawn from the population) will tend to follow a normal curve as the number of samples increases. Fortunately, even if each sample (*n*) is fairly small (say, 4 or 5), the distributions of the

Central limit theorem
The theoretical foundation for $\bar{x}$-charts, which states that regardless of the distribution of the population of all parts or services, the distribution of $\bar{x}$ s tends to follow a normal curve as the number of samples increases.

Figure S6.3

The Relationship Between Population and Sampling Distributions

Even though the population distributions will differ (e.g., normal, beta, uniform), each with its own mean (μ) and standard deviation (σ), the distribution of sample means always approaches a normal distribution.

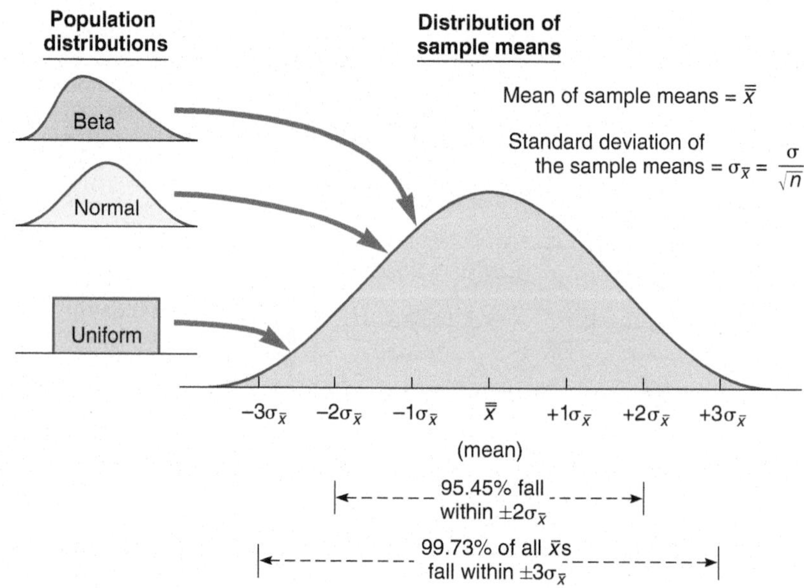

averages will still roughly follow a normal curve. The theorem also states that: (1) the mean of the distribution of the $\bar{x}$ s (called $\bar{\bar{x}}$) will equal the mean of the overall population (called μ); and (2) the standard deviation of the *sampling distribution*, $\sigma_{\bar{x}}$, will be the *population (process) standard deviation*, divided by the square root of the sample size, n. In other words:[2]

LO S6.2 *Explain* the role of the central limit theorem in SPC

$$\bar{\bar{x}} = \mu \quad \text{(S6-1)}$$

and

$$\sigma_{\bar{x}} = \frac{\sigma}{\sqrt{n}} \quad \text{(S6-2)}$$

Figure S6.3 shows three possible population distributions, each with its own mean, μ and standard deviation, σ. If a series of random samples ($\bar{x}_1$, $\bar{x}_2$, $\bar{x}_3$, $\bar{x}_4$, and so on), each of size n, is drawn from any population distribution (which could be normal, beta, uniform, and so on), the resulting distribution of $\bar{x}_i$s will approximate a normal distribution (see Figure S6.3).

Moreover, the sampling distribution, as is shown in Figure S6.4(a), will have less variability than the process distribution. Because the sampling distribution is normal, we can state that:

- 95.45% of the time, the sample averages will fall within $\pm 2\sigma_{\bar{x}}$ if the process has only natural variations.
- 99.73% of the time, the sample averages will fall within $\pm 3\sigma_{\bar{x}}$ if the process has only natural variations.

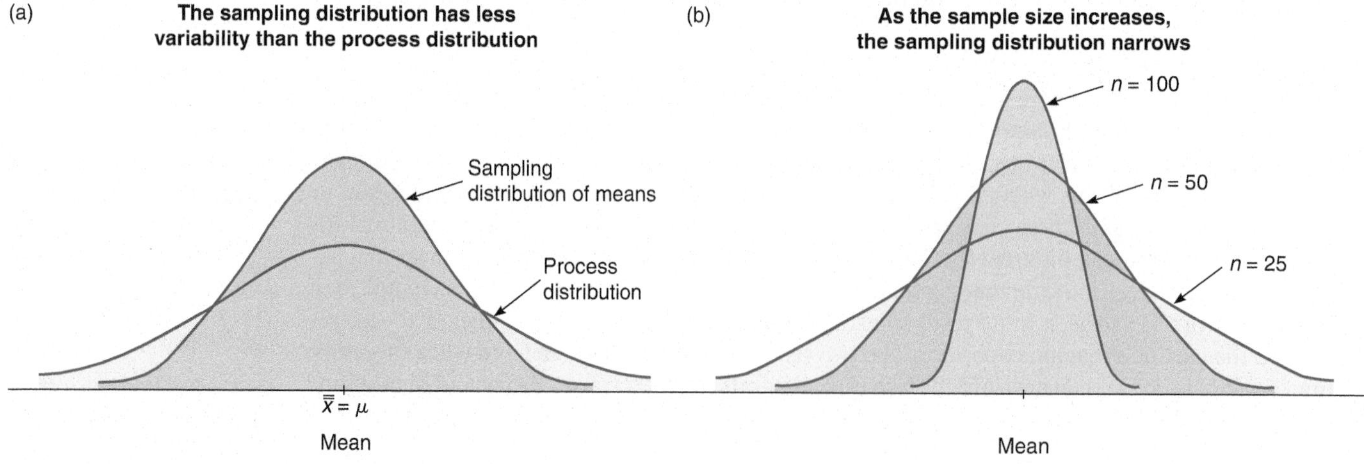

Figure S6.4

The Sampling Distribution of Means Is Normal

The process distribution from which the sample was drawn was also normal, but it could have been any distribution.

If a point on the control chart falls outside of the $\pm 3\sigma_{\bar{x}}$ control limits, then we are 99.73% sure the process has changed. Figure S6.4(b) shows that as the sample size increases, the sampling distribution becomes narrower. So the sample statistic is closer to the true value of the population for larger sample sizes. This is the theory behind control charts.

Setting Mean Chart Limits ($\bar{X}$-Charts)

If we know, through past data, the standard deviation of the population (process), σ, we can set upper and lower control limits[3] by using these formulas:

$$\text{Upper control limit (UCL)} = \bar{\bar{x}} + z\sigma_{\bar{x}} \qquad \text{(S6-3)}$$

$$\text{Lower control limit (LCL)} = \bar{\bar{x}} - z\sigma_{\bar{x}} \qquad \text{(S6-4)}$$

where

$\bar{\bar{x}}$ = mean of the sample means or a target value set for the process
z = number of normal standard deviations (2 for 95.45% confidence, 3 for 99.73%)
$\sigma_{\bar{x}}$ = standard deviation of the sample means = $\sigma/\sqrt{n}$
σ = population (process) standard deviation
n = sample size

LO S6.3 *Build $\bar{x}$-charts and R-charts*

Example S1 shows how to set control limits for sample means using standard deviations.

Example S1

SETTING CONTROL LIMITS USING SAMPLES

The weights of boxes of Oat Flakes within a large production lot are sampled each hour. Managers want to set control limits that include 99.73% of the sample means.

APPROACH ▶ Randomly select and weigh nine ($n = 9$) boxes each hour. Then find the overall mean and use Equations (S6-3) and (S6-4) to compute the control limits. Here are the nine boxes chosen for Hour 1:

Oat Flakes	Oat Flakes	Oat Flakes	Oat Flakes	Oat Flakes	Oat Flakes	Oat Flakes	Oat Flakes	Oat Flakes
17 oz.	13 oz.	16 oz.	18 oz.	17 oz.	16 oz.	15 oz.	17 oz.	16 oz.

STUDENT TIP
If you want to see an example of such variability in your supermarket, go to the soft drink section and line up a few 2-liter bottles of Coke or Pepsi.

SOLUTION ▶

$$\text{The average weight in the first hourly sample} = \frac{17+13+16+18+17+16+15+17+16}{9} = 16.1 \text{ ounces.}$$

Also, the *population (process)* standard deviation (σ) is known to be 1 ounce. We do not show each of the boxes randomly selected in hours 2 through 12, but here are all 12 hourly samples:

WEIGHT OF SAMPLE		WEIGHT OF SAMPLE		WEIGHT OF SAMPLE	
HOUR	(AVG. OF 9 BOXES)	HOUR	(AVG. OF 9 BOXES)	HOUR	(AVG. OF 9 BOXES)
1	16.1	5	16.5	9	16.3
2	16.8	6	16.4	10	14.8
3	15.5	7	15.2	11	14.2
4	16.5	8	16.4	12	17.3

The average mean $\bar{\bar{x}}$ of the 12 samples is calculated to be exactly 16 ounces $\left[\bar{\bar{x}} = \frac{\sum_{i=1}^{12}(\text{Avg. of 9 Boxes})}{12}\right]$.

We therefore have $\bar{\bar{x}} = 16$ ounces, $\sigma = 1$ ounce, $n = 9$, and $z = 3$. The control limits are:

$$UCL_{\bar{x}} = \bar{\bar{x}} + z\sigma_{\bar{x}} = 16 + 3\left(\frac{1}{\sqrt{9}}\right) = 16 + 3\left(\frac{1}{3}\right) = 17 \text{ ounces}$$

$$LCL_{\bar{x}} = \bar{\bar{x}} - z\sigma_{\bar{x}} = 16 - 3\left(\frac{1}{\sqrt{9}}\right) = 16 - 3\left(\frac{1}{3}\right) = 15 \text{ ounces}$$

The 12 samples are then plotted on the following control chart:

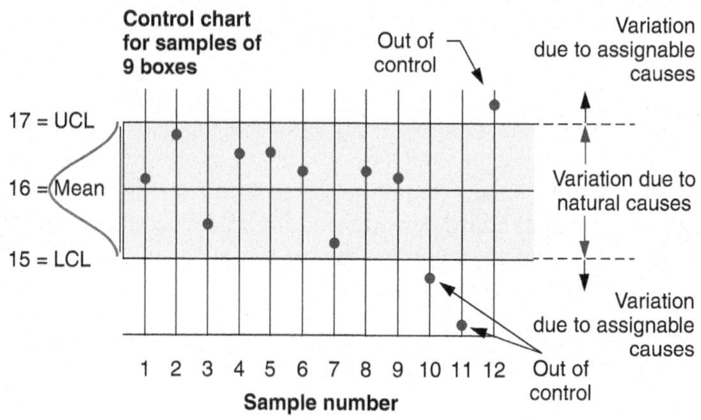

INSIGHT ▶ Because the means of recent sample averages fall outside the upper and lower control limits of 17 and 15, we can conclude that the process is becoming erratic and is *not* in control.

LEARNING EXERCISE ▶ If Oat Flakes' population standard deviation (σ) is 2 (instead of 1), what is your conclusion? [Answer: LCL = 14, UCL = 18; the process would be in control.]

RELATED PROBLEMS ▶ S6.1, S6.2, S6.4, S6.8, S6.10a,b, S6.26, S6.27, S6.28

ACTIVE MODEL S6.1 This example is further illustrated in Active Model S6.1 found online.

EXCEL OM Data File **CH06ExS1.xls** can be found online.

Because process standard deviations are often not available, we usually calculate control limits based on the average *range* values rather than on standard deviations. Table S6.1 provides the necessary conversion for us to do so. The *range* (R_i) is defined as the difference between the largest and smallest items in one sample. For example, the heaviest box of Oat Flakes in Hour 1 of Example S1 was 18 ounces and the lightest was 13 ounces, so the range for that hour is 5 ounces. We use Table S6.1 and the equations:

$$UCL_{\bar{x}} = \bar{\bar{x}} + A_2\bar{R} \qquad \text{(S6-5)}$$

and:

$$LCL_{\bar{x}} = \bar{\bar{x}} - A_2\bar{R} \qquad \text{(S6-6)}$$

where $\bar{R} = \dfrac{\sum_{i=1}^{k} R_i}{k}$ = average range of all the samples; R_i = range for sample i

A_2 = value found in Table S6.1 $\qquad$ k = total number of samples

$\bar{\bar{x}}$ = mean of the sample means

Example S2 shows how to set control limits for sample means by using Table S6.1 and the average range.

TABLE S6.1 Factors for Computing Control Chart Limits (3 sigma)

SAMPLE SIZE, n	MEAN FACTOR, A_2	UPPER RANGE, D_4	LOWER RANGE, D_3
2	1.880	3.268	0
3	1.023	2.574	0
4	.729	2.282	0
5	.577	2.115	0
6	.483	2.004	0
7	.419	1.924	0.076
8	.373	1.864	0.136
9	.337	1.816	0.184
10	.308	1.777	0.223
12	.266	1.716	0.284

Source: Based on *Special Technical Publication 15–C,* "Quality Control of Materials," pp. 63 and 72. Copyright ASTM INTERNATIONAL.

Example S2

SETTING MEAN LIMITS USING TABLE VALUES

Super Cola bottles soft drinks labeled "net weight 12 ounces." Indeed, an overall process average of 12 ounces has been found by taking 10 samples, in which each sample contained 5 bottles. The OM team wants to determine the upper and lower control limits for averages in this process.

APPROACH ▶ Super Cola first examines the 10 samples to compute the average range of the process. Here are the data and calculations:

SAMPLE	WEIGHT OF LIGHTEST BOTTLE IN SAMPLE OF $n = 5$	WEIGHT OF HEAVIEST BOTTLE IN SAMPLE OF $n = 5$	RANGE (R_i) = DIFFERENCE BETWEEN THESE TWO
1	11.50	11.72	.22
2	11.97	12.00	.03
3	11.55	12.05	.50
4	12.00	12.20	.20
5	11.95	12.00	.05
6	10.55	10.75	.20
7	12.50	12.75	.25
8	11.00	11.25	.25
9	10.60	11.00	.40
10	11.70	12.10	.40
			$\sum R_i = 2.50$ ounces

$$\text{Average Range} = \frac{2.50}{10 \text{ samples}} = .25 \text{ ounces}$$

Now Super Cola applies Equations (S6-5) and (S6-6) and uses the A_2 column of Table S6.1.

SOLUTION ▶ Looking in Table S6.1 for a sample size of 5 in the mean factor A_2 column, we find the value .577. Thus, the upper and lower control chart limits are:

$$\begin{aligned} \text{UCL}_{\bar{x}} &= \bar{\bar{x}} + A_2\bar{R} \\ &= 12 + (.577)(.25) \\ &= 12 + .144 \\ &= 12.144 \text{ ounces} \end{aligned}$$

$$\begin{aligned} \text{LCL}_{\bar{x}} &= \bar{\bar{x}} - A_2\bar{R} \\ &= 12 - .144 \\ &= 11.856 \text{ ounces} \end{aligned}$$

INSIGHT ▶ The advantage of using this range approach, instead of the standard deviation, is that it is easy to apply and may be less confusing.

LEARNING EXERCISE ▶ If the sample size was $n = 4$ and the average range = .20 ounces, what are the revised $UCL_{\bar{x}}$ and $LCL_{\bar{x}}$? [Answer: 12.146, 11.854.]

RELATED PROBLEMS ▶ S6.3a, S6.5, S6.6, S6.7, S6.9, S6.10b,c,d, S6.11, S6.26–S6.27, S6.29a, S6.30a, S6.31, S6.32a, S6.33a

EXCEL OM Data File **CH06ExS2.xls** can be found online.

Setting Range Chart Limits (*R*-Charts)

In Examples S1 and S2, we determined the upper and lower control limits for the process *average*. In addition to being concerned with the process average, operations managers are interested in the process *dispersion*, or *range*. Even though the process average is under control, the dispersion of the process may not be. For example, something may have worked itself loose in a piece of equipment that fills boxes of Oat Flakes. As a result, the average of the samples may remain the same, but the variation within the samples could be entirely too large. For this reason, operations managers use control charts for ranges to monitor the process variability, as well as control charts for averages, which monitor the process central tendency. The theory behind the control charts for ranges is the same as that for process average control charts. Limits are established that contain ±3 standard deviations of the distribution for the average range $\bar{R}$. We can use the following equations to set the upper and lower control limits for ranges:

$$UCL_R = D_4\bar{R} \tag{S6-7}$$

$$LCL_R = D_3\bar{R} \tag{S6-8}$$

where

UCL_R = upper control chart limit for the range

LCL_R = lower control chart limit for the range

D_4 and D_3 = values from Table S6.1

VIDEO S6.1
Farm to Fork: Quality of Darden Restaurants

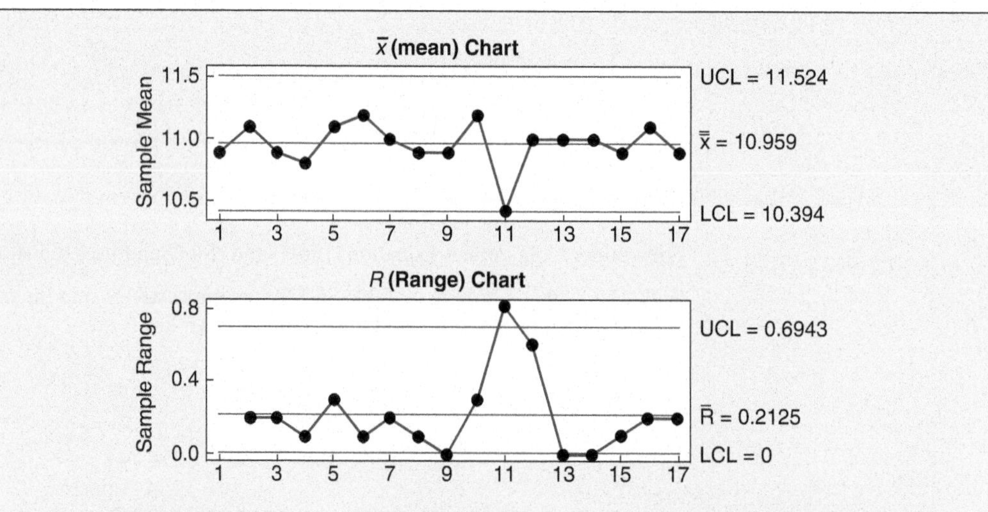

Salmon filets are monitored by Darden Restaurant's SPC software, which includes $\bar{x}$-(mean) charts and *R*-(range) charts. Darden uses average weight as a measure of central tendency for salmon filets. The range is the difference between the heaviest and the lightest filets in each sample. The video case study "Farm to Fork," at the end of this supplement, asks you to interpret these figures.

Example S3 shows how to set control limits for sample ranges using Table S6.1 and the average range.

Example S3 SETTING RANGE LIMITS USING TABLE VALUES

Esmail Mohebbi's mail-ordering business wants to measure the response time of its operators in taking customer orders over the phone. Esmail lists below the time recorded (in minutes) from five different samples of the ordering process with four customer orders per sample. He wants to determine the upper and lower range control chart limits.

APPROACH ▶ Looking in Table S6.1 for a sample size of 4, he finds that $D_4 = 2.282$ and $D_3 = 0$.

SOLUTION ▶

SAMPLE	OBSERVATIONS (MINUTES)	SAMPLE RANGE (R_i)
1	5, 3, 6, 10	10 − 3 = 7
2	7, 5, 3, 5	7 − 3 = 4
3	1, 8, 3, 12	12 − 1 = 11
4	7, 6, 2, 1	7 − 1 = 6
5	3, 15, 6, 12	15 − 3 = 12
		$\Sigma R_i = 40$

$$\overline{R} = \frac{40}{5} = 8$$

$$\text{UCL}_R = 2.282(8) = 18.256 \text{ minutes}$$

$$\text{LCL}_R = 0(8) = 0 \text{ minutes}$$

INSIGHT ▶ Computing ranges with Table S6.1 is straightforward and an easy way to evaluate dispersion. No sample ranges are out of control.

LEARNING EXERCISE ▶ Esmail decides to increase the sample size to $n = 6$ (with no change in average range, $\overline{R}$). What are the new UCL_R and LCL_R values? [Answer: 16.032, 0.]

RELATED PROBLEMS ▶ S6.3b, S6.5, S6.6, S6.7, S6.9, S6.10c, S6.11, S6.12, S6.26–S6.27, S6.29b, S6.30b, S6.31, S6.32b, S6.33b

Using Mean and Range Charts

The normal distribution is defined by two parameters, the *mean* and *standard deviation*. The $\overline{x}$ (mean)-chart and the *R*-chart mimic these two parameters. The $\overline{x}$-chart is sensitive to shifts in the process mean, whereas the *R*-chart is sensitive to shifts in the process standard deviation. Consequently, by using both charts we can track changes in the process distribution.

For instance, the samples and the resulting $\overline{x}$-chart in Figure S6.5(a) show the shift in the process mean, but because the dispersion is constant, no change is detected by the *R*-chart. Conversely, the samples and the $\overline{x}$-chart in Figure S6.5(b) detect no shift (because none is present), but the *R*-chart does detect the shift in the dispersion. Both charts are required to track the process accurately.

LO S6.4 *List* the five steps involved in building control charts

Steps to Follow When Building Control Charts There are five steps that are generally followed in building $\overline{x}$- and *R*-charts:

1. Collect 20 to 25 samples, often of $n = 4$ or $n = 5$ observations each, from a stable process, and compute the mean and range of each.
2. Compute the overall means ($\overline{\overline{x}}$ and $\overline{R}$), set appropriate control limits, usually at the 99.73% level, and calculate the preliminary upper and lower control limits. Refer to Table S6.2 for other control limits. *If the process is not currently stable and in control*, use the desired mean, μ, instead of $\overline{\overline{x}}$ to calculate limits.

Figure **S6.5**

Mean and Range Charts Complement Each Other by Showing the Mean and Dispersion of the Normal Distribution

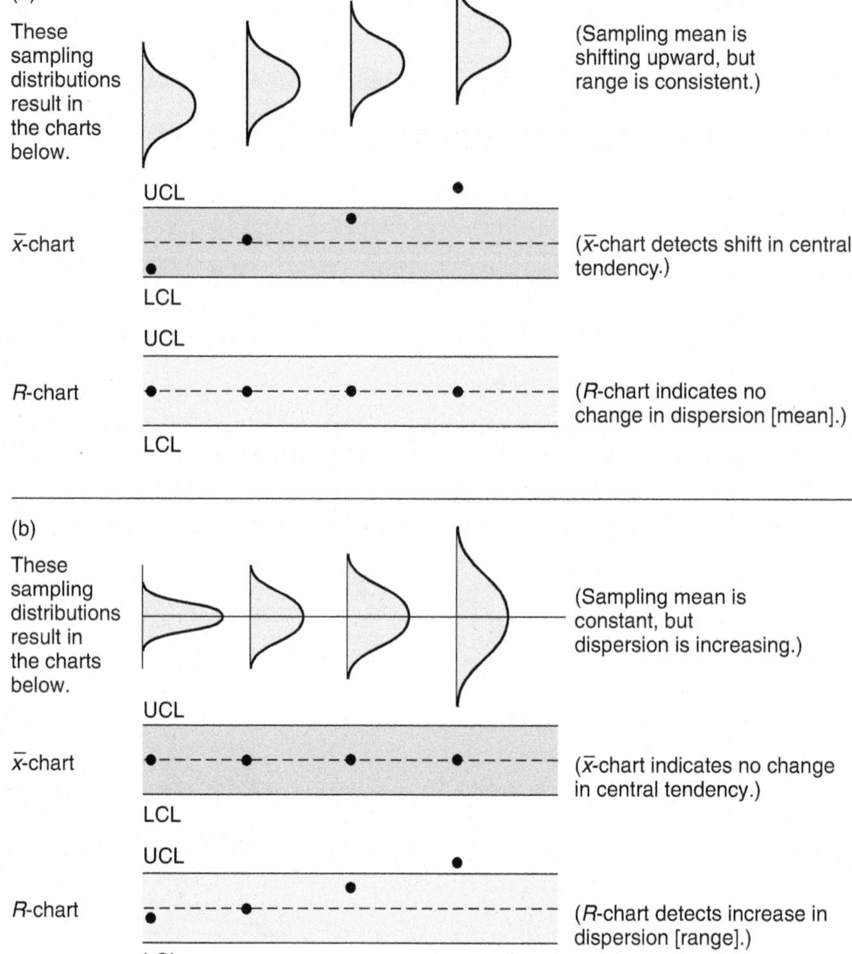

(a)
These sampling distributions result in the charts below.
(Sampling mean is shifting upward, but range is consistent.)
UCL
$\bar{x}$-chart
($\bar{x}$-chart detects shift in central tendency.)
LCL
UCL
R-chart
(R-chart indicates no change in dispersion [mean].)
LCL
(b)
These sampling distributions result in the charts below.
(Sampling mean is constant, but dispersion is increasing.)
UCL
$\bar{x}$-chart
($\bar{x}$-chart indicates no change in central tendency.)
LCL
UCL
R-chart
(R-chart detects increase in dispersion [range].)
LCL

STUDENT TIP

Mean ($\bar{X}$) charts are a measure of *central tendency*, while range (*R*) charts are a measure of *dispersion*. SPC requires both charts for a complete assessment because a sample mean could be out of control while the range is in control and vice versa.

TABLE S6.2
Common z Values

DESIRED CONTROL LIMIT (%)	Z-VALUE (STANDARD DEVIATION REQUIRED FOR DESIRED LEVEL OF CONFIDENCE)
90.0	1.65
95.0	1.96
95.45	2.00
99.0	2.58
99.73	3.00

3. Graph the sample means and ranges on their respective control charts, and determine whether they fall outside the acceptable limits.
4. Investigate points or patterns that indicate the process is out of control. Try to assign causes for the variation, address the causes, and then resume the process.
5. Collect additional samples and, if necessary, revalidate the control limits using the new data.

Donna Mcwilliam/AP/Shutterstock

FritoLay

WEIGHTED SINGLE SAMPLE CENTERLINING CHART

FL-5885 (Rev. 11/03)

SHIFT	NUMBER OF CHECKS	MAX. TIME GAP BETWEEN CHECKS
1st SHIFT	10 Checks	40 min
2nd SHIFT		
3rd SHIFT		

INSTRUCTIONS

1. CURRENT VALUE MUST BE WITHIN SPECIFICATION LIMITS. IF NOT, FOLLOW PRESCRIBED PROCEDURES (ABORT, HOLD, ETC.). RECORD CORRECTIVE ACTION AT BOTTOM OF CHART.
2. PREDICTED VALUE = AVERAGE OF CURRENT VALUE AND PREVIOUS PREDICTED VALUE. THE PREDICTED VALUE IS ALWAYS PLOTTED ON THE CHART.
3. DECISION RULE: ANY PREDICTED VALUE IN THE YELLOW (RED TAKE ACTION) INDICATES IMMEDIATE CORRECTIVE ACTION IS REQUIRED. RECORD CORRECTIVE ACTION AT BOTTOM OF CHART.
4. AT STARTUP OR AFTER MAJOR PROCESS ADJUSTMENTS THE CURRENT VALUE WILL ALSO BE USED AS THE PREDICTED VALUE.

RED ABORT SPECIFICATION LIMITS 1.70--2.50

TIME:	710	750	830	910	950	1000	1035	1115	1155	1235	115
CURRENT VALUE	218	182	210	215	194	Break	1.99	193	193	205	191
PREDICTED VALUE (PLOT ON CHART)	218	200	205	210	202	↓	201	197	193	199	195

ADJUST
UCL = 2.22
216
AIM = 2.10
204
LCL = 1.98
ADJUST

218, 200, 205, 210, 202, 201, 197, 193, 199, 195

Frito-Lay uses $\bar{x}$ charts to control production quality at critical points in the process. About every 40 minutes, three batches of chips are taken from the conveyor (on the left) and analyzed electronically to get an average salt content, which is plotted on an $\bar{x}$-chart (on the right). Points plotted in the green zone are "in control," while those in the yellow zone are "out of control." The SPC chart is displayed where all production employees can monitor process stability.

Control Charts for Attributes

Control charts for $\bar{x}$ and R do not apply when we are sampling *attributes*, which are typically classified as *defective* or *nondefective*. Measuring defectives involves counting them (for example, number of bad lightbulbs in a given lot, or number of letters or data entry records typed with errors), whereas *variables* are usually measured for length or weight. There are two kinds of attribute control charts: (1) those that measure the *percent* defective in a sample—called *p*-charts—and (2) those that count the *number* of defects—called *c*-charts.

LO S6.5 *Build p*-charts and *c*-charts

p-Charts Using *p*-charts is the chief way to control attributes. Although attributes that are either good or bad follow the binomial distribution, the normal distribution can be used to calculate *p*-chart limits when sample sizes are large. The size of each sample should be large enough to include some defects. The procedure resembles the $\bar{x}$-chart approach, which is also based on the central limit theorem.

p-chart
A quality control chart that is used to control attributes.

The formulas for *p*-chart upper and lower control limits follow:

VIDEO S6.2
Frito-Lay's Quality-Controlled Potato Chips

$$\text{UCL}_p = \bar{p} + z\sigma_p \quad \text{(S6-9)}$$

$$\text{LCL}_p = \bar{p} - z\sigma_p \quad \text{(S6-10)}$$

where $\bar{p}$ = mean fraction (percent) defective in the samples = $\dfrac{\text{total number of defects}}{\text{sample size} \times \text{number of samples}}$

z = number of standard deviations (z = 2 for 95.45% limits; z = 3 for 99.73% limits)

σ_p = standard deviation of the sampling distribution

σ_p is estimated by the formula:

$$\hat{\sigma}_p = \sqrt{\frac{\bar{p}(1-\bar{p})}{n}} \quad \text{(S6-11)}$$

where n = number of observations in *each* sample[4]

Example S4 shows how to set control limits for *p*-charts for these standard deviations.

Example S4

SETTING CONTROL LIMITS FOR PERCENT DEFECTIVE

Clerks at Mosier Data Systems key in thousands of insurance records each day for a variety of client firms. CEO Donna Mosier wants to set control limits to include 99.73% of the random variation in the data entry process when it is in control.

APPROACH ▶ Samples of the work of 20 clerks are gathered (and shown in the table). Mosier carefully examines 100 records entered by each clerk and counts the number of errors. She also computes the fraction (percent) defective in each sample. Equations (S6-9), (S6-10), and (S6-11) are then used to set the control limits.

SAMPLE NUMBER	NUMBER OF ERRORS	FRACTION DEFECTIVE	SAMPLE NUMBER	NUMBER OF ERRORS	FRACTION DEFECTIVE
1	6	.06	11	6	.06
2	5	.05	12	1	.01
3	1	.01	13	8	.08
4	1	.01	14	5	.05
5	4	.04	15	5	.05
6	2	.02	16	4	.04
7	5	.05	17	11	.11
8	3	.03	18	3	.03
9	3	.03	19	1	.01
10	2	.02	20	4	.04
				80	

SOLUTION ▶

$$\bar{p} = \frac{\text{Total number of errors}}{\text{Total number of records examined}} = \frac{80}{(100)(20)} = .04$$

$$\hat{\sigma}_p = \sqrt{\frac{(.04)(1-.04)}{100}} = .02 \text{ (rounded up from .0196)}$$

(*Note:* 100 is the size of *each* sample = n.)

$$UCL_p = \bar{p} + z\hat{\sigma}_p = .04 + 3(.02) = .10$$

$$LCL_p = \bar{p} - z\hat{\sigma}_p = .04 - 3(.02) = 0$$

(because we cannot have a negative percent defective)

INSIGHT ▶ When we plot the control limits and the sample fraction defectives, we find that only one data-entry clerk (number 17) is out of control. The firm may wish to examine that individual's work a bit more closely to see if a serious problem exists (see Figure S6.6).

Figure **S6.6**

***p*-Chart for Data Entry for Example S4**

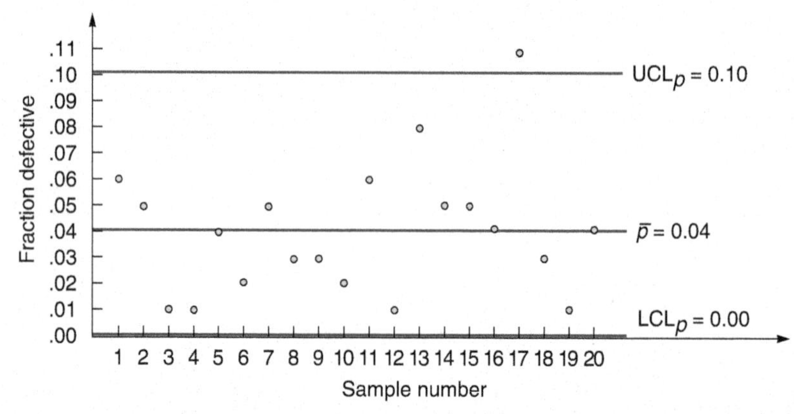

STUDENT TIP

We are always pleased to be at zero or below the center line in a *p*-chart.

LEARNING EXERCISE ▶ Mosier decides to set control limits at 95.45% instead. What are the new UCL_p and LCL_p? [Answer: 0.08, 0]

RELATED PROBLEMS ▶ S6.13–S6.20, S6.25, S6.36–S6.39

ACTIVE MODEL S6.2 This example is further illustrated in Active Model S6.2 found online.

EXCEL OM Data File **CH06ExS4.xls** can be found online.

Again, note that as in the example, when the LCL is negative, it is rounded up to zero. A negative percent defective is not possible.

The *OM in Action* box "Trying to Land a Seat with Frequent Flyer Miles" provides a real-world follow-up to Example S4.

***c*-chart**

A quality control chart used to control the number of defects per unit of output.

***c*-Charts** In Example S4, we counted the number of defective records entered. A defective record was one that was not exactly correct because it contained at least one defect. However, a bad record may contain more than one defect. We use *c*-charts to control the *number* of defects per unit of output (or per insurance record, in the preceding case).

Control charts for defects are helpful for monitoring processes in which a large number of potential errors can occur, but the actual number that do occur is relatively small. Defects may be errors in newspaper words, bad circuits in a microchip, blemishes on a table, or missing pickles on a fast-food hamburger.

The Poisson probability distribution,[5] which has a variance equal to its mean, is the basis for *c*-charts. Because $\bar{c}$ is the mean number of defects per unit, the standard deviation is equal to $\sqrt{\bar{c}}$. To compute 99.73% control limits for $\bar{c}$, we use the formula:

$$\text{Control limits} = \bar{c} \pm 3\sqrt{\bar{c}} \qquad \text{(S6-12)}$$

Example S5 shows how to set control limits for a $\bar{c}$-chart.

OM in Action — Trying to Land a Seat with Frequent Flyer Miles[6]

How hard is it to redeem your 25,000 frequent flyer points for your airline tickets? That depends on the airline. (It also depends on the city. Don't try to get in or out of San Francisco!) When the consulting firm Idea Works made 280 requests for a standard mileage award to each of 25 airlines' Web sites (a total of 7,000 requests), the success rates ranged from a low of 45%, 53.6%, and 56.4% (at LAN, Avianca, and American, respectively) to a high of 100% at Southwest and Air Berlin.

The overall average of 76.6% for the 25 carriers provides the centerline in a *p*-chart. With 3-sigma upper and lower control limits of 84.2% and 69.0%, respectively, the other top and other bottom performers are easily spotted. "Out of control" (but in a positive *outperforming* way) are Virgin Australia (94.3%), Jet Blue (92.9%), Lufthansa (92.9%), Turkish (92.1%), Air Canada (90.7%), Qantas (89.3%), and Korean (84.3%).

Others *out of control on the negative side* are Scandinavian (57.9%), Air Asia (60%), Alitalia (61.4%), Air France (67.1%), and Delta (68.6%).

Control charts can help airlines see where they stand relative to competitors in such customer activities as lost bags, on-time rates, and ease of redeeming mileage points. "I think airlines are getting the message that availability is important. Overall, I think the consumer is better served than the year before," says the president of Idea Works.

Christophe Testi/Shutterstock

Example S5 — SETTING CONTROL LIMITS FOR NUMBER OF DEFECTS

Red Top Cab Company receives several complaints per day about the behavior of its drivers. Over a 9-day period (where days are the units of measure), the owner, Gordon Hoft, received the following numbers of calls from irate passengers: 3, 0, 8, 9, 6, 7, 4, 9, 8, for a total of 54 complaints. Hoft wants to compute 99.73% control limits.

APPROACH ▶ He applies Equation (S6–12).

SOLUTION ▶ $\bar{c} = \frac{54}{9} = 6$ complaints per day

Thus:

$$UCL_c = \bar{c} + 3\sqrt{\bar{c}} = 6 + 3\sqrt{6} = 6 + 3(2.45) = 13.35, \text{ or } 13$$

$$LCL_c = \bar{c} - 3\sqrt{\bar{c}} = 6 - 3\sqrt{6} = 6 - 3(2.45) = 0 \leftarrow \text{(since it cannot be negative)}$$

INSIGHT ▶ After Hoft plotted a control chart summarizing these data and posted it prominently in the drivers' locker room, the number of calls received dropped to an average of three per day. Can you explain why this occurred?

LEARNING EXERCISE ▶ Hoft collects 3 more days' worth of complaints (10, 12, and 8 complaints) and wants to combine them with the original 9 days to compute updated control limits. What are the revised UCL_c and LCL_c? [Answer: 14.94, 0.]

RELATED PROBLEMS ▶ S6.14, S6.21, S6.22, S6.23, S6.24, S6.35

EXCEL OM Data File **CH06SExS5.xls** can be found online.

Note that *c*-charts, like *p*-charts, cannot have a negative LCL, so negatives are always rounded up to zero.

Sampling wine from these wooden barrels, to make sure it is aging properly, uses both SPC (for alcohol content and acidity) and subjective measures (for taste).

Managerial Issues and Control Charts

In an ideal world, there is no need for control charts. Quality is uniform and so high that employees need not waste time and money sampling and monitoring variables and attributes. But because most processes have not reached perfection, managers must make three major decisions regarding control charts.

First, managers must select the points in their process that need SPC. They may ask "Which parts of the job are critical to success?" or "Which parts of the job have a tendency to become out of control?"

Second, managers need to decide if variable charts (i.e., $\bar{x}$ and R) or attribute charts (i.e., p and c) are appropriate. Variable charts monitor weights or dimensions. Attribute charts are more of a "yes–no" or "go–no go" gauge and tend to be less costly to implement. Table S6.3 can help you understand when to use each of these types of control charts.

Third, the company must set clear and specific SPC policies for employees to follow. For example, should the data-entry process be halted if a trend is appearing in percent defective records being keyed? Should an assembly line be stopped if the average length of five successive samples is above the centerline? Figure S6.7 illustrates some of the patterns to look for over time in a process.

Run test
A test used to examine the points in a control chart to see if nonrandom variation is present.

A tool called a run test is available to help identify the kind of abnormalities in a process that we see in Figure S6.7. In general, a run of 5 points above or below the target or centerline may suggest that an assignable, or nonrandom, variation is present. When this occurs, even though all the points may fall inside the control limits, a flag has been raised. This means the process may not be statistically in control. A variety of run tests are described in books on the subject of quality methods.

STUDENT TIP
This is a really useful table. When you are not sure which control chart to use, turn here for clarification.

TABLE S6.3 Helping You Decide Which Control Chart to Use

VARIABLE DATA USING AN $\bar{x}$-CHART AND AN R-CHART
1. Observations are *variables*, which are usually products measured for size or weight. Examples are the width or length of a wire and the weight of a can of Campbell's soup. 2. Collect 20 to 25 samples, usually of $n = 4$, $n = 5$, or more, each from a stable process, and compute the means for an $\bar{x}$-chart and the ranges for an R-chart. 3. We track samples of n observations each, as in Example S1.
ATTRIBUTE DATA USING A p-CHART
1. Observations are *attributes* that can be categorized as good or bad (or pass–fail, or functional–broken); that is, in two states. 2. We deal with fraction, proportion, or percent defectives. 3. There are several samples, with many observations in each. For example, 20 samples of $n = 100$ observations in each, as in Example S4.
ATTRIBUTE DATA USING A c-CHART
1. Observations are *attributes* whose defects per unit of output can be counted. 2. We deal with the number counted, which is a small part of the possible occurrences. 3. Defects may be: number of blemishes on a desk; flaws in a bolt of cloth; crimes in a year; broken seats in a stadium; typos in a chapter of this text; or complaints in a day, as is shown in Example S5.

Upper control limit

Target

Lower control limit

Normal behavior. Process is "in control."

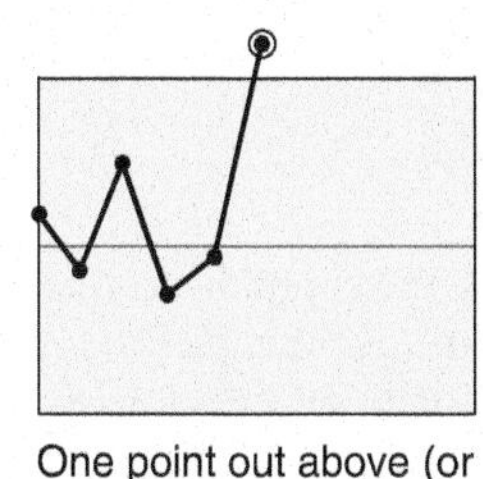

One point out above (or below). Investigate for cause. Process is "out of control."

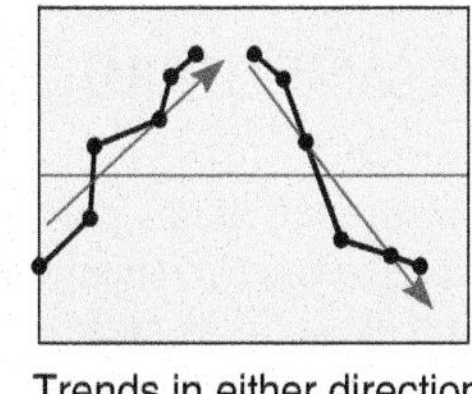

Trends in either direction, 5 points. Investigate for cause of progressive change. This could be the result of gradual tool wear.

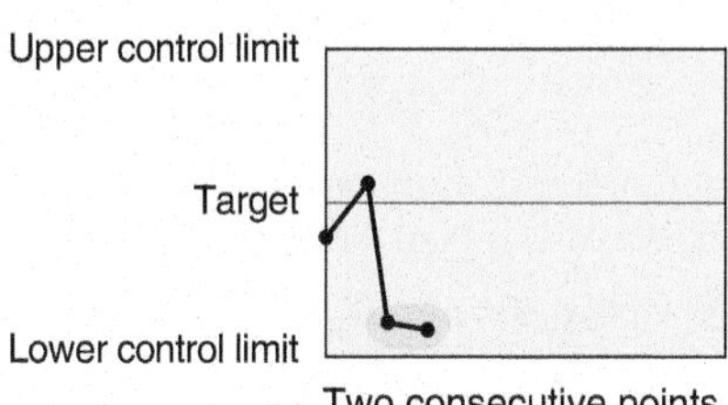

Two consecutive points very near lower (or upper) control. Investigate for cause.

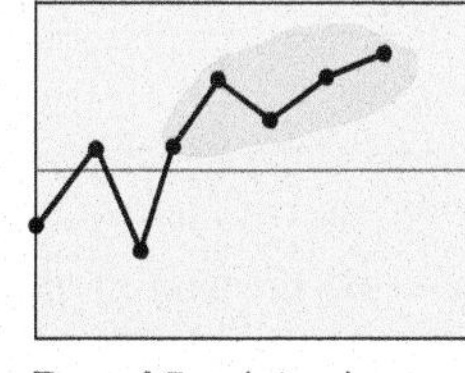

Run of 5 points above (or below) central line. Investigate for cause.

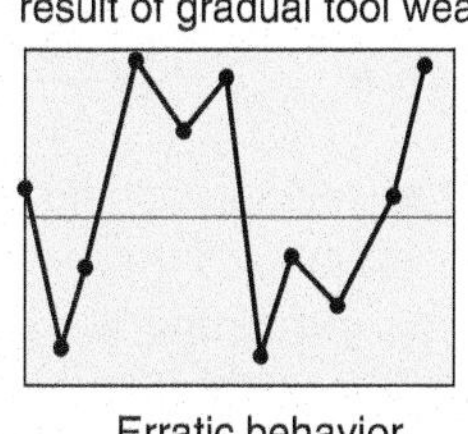

Erratic behavior. Investigate.

Figure **S6.7**

Patterns to Look for on Control Charts

Source: Based on Bertrand L. Hansen, *Quality Control: Theory and Applications* (1991): 65.

STUDENT TIP

Workers in companies such as Frito-Lay are trained to follow rules like these.

Process Capability

STUDENT TIP

Here we deal with whether a process meets the specification it was *designed* to yield.

Statistical process control means keeping a process in control. This means that the natural variation of the process must be stable. However, a process that is in statistical control may not yield goods or services that meet their *design specifications* (tolerances). In other words, the variation should be small enough to produce consistent output within specifications. The ability of a process to meet specifications, which are set by engineering design or customer requirements, is called **process capability**. Even though that process may be statistically in control (stable), the output of that process may not conform to specifications.

Process capability
The ability to meet specifications.

For example, let's say the time a customer expects to wait for the completion of a lube job at Quik Lube is 12 minutes, with an acceptable tolerance of ±2 minutes. This tolerance gives an upper specification of 14 minutes and a lower specification of 10 minutes. The lube process has to be capable of operating within these specifications—if not, some customers will not have their requirements met. As a manufacturing example, the tolerances for Harley-Davidson cam gears are extremely low, only 0.0005 inch—and a process must be developed that is capable of achieving this tolerance.

There are two popular measures for quantitatively determining if a process is capable: process capability ratio (C_p) and process capability index (C_{pk}).

LO S6.6 *Explain* process capability and compute C_p and C_{pk}

Process Capability Ratio (C_p)

For a process to be capable, its values must fall within upper and lower specifications. This typically means the process capability is within ±3 standard deviations from the process mean. Because this range of values is 6 standard deviations, a capable process tolerance, which is the difference between the upper and lower specifications, must be greater than or equal to 6.

If the C_p is less than 1.0, the process yields products or services that are outside their allowable tolerance. A capable process has a C_p of at least 1.0. With a C_p of 1.0, 2.7 parts in 1,000 can be expected to be "out of spec." The higher the process capability ratio, the greater the likelihood the process will be within specifications. Many firms have chosen a C_p of 1.33 (a 4-sigma standard) as a desirable target for process capability. A C_p of 1.33 means that only 64 parts per million can be expected to be out of specification.

C_p
A ratio for determining whether a process meets specifications; a ratio of the specification to the process variation.

The process capability ratio, **C_p**, is computed as:

$$C_p = \frac{\text{Upper specification} - \text{Lower specification}}{6\sigma} \quad \text{(S6-13)}$$

Example S6 shows the computation of C_p.

Example S6

PROCESS CAPABILITY RATIO (C_p)

In a GE insurance claims process, the company strives to process claims in 210 minutes $\left(\bar{X} = 210\right)$ and experiences a standard deviation of .516 minute $(\sigma = .516)$.

The specification to meet customer expectations is 210 ± 3 minutes. So the Upper Specification is 213 minutes and the lower specification is 207 minutes. The OM manager wants to compute the process capability ratio.

APPROACH ▶ GE applies Equation (S6-13).

SOLUTION ▶ $C_p = \dfrac{\text{Upper specification} - \text{Lower specification}}{6\sigma} = \dfrac{213 - 207}{6(.516)} = 1.938$

INSIGHT ▶ Because a ratio of 1.00 means that 99.73% of a process's outputs are within specifications, this ratio suggests a very capable process, with nonconformance of less than 4 claims per million.

LEARNING EXERCISE ▶ If $\sigma = .60$ (instead of .516), what is the new C_p? [Answer: 1.667, a very capable process still.]

RELATED PROBLEMS ▶ S6.40, S6.41, S6.50

ACTIVE MODEL S6.3 This example is further illustrated in Active Model S6.3 found online.

EXCEL OM Data File **CH06SExS6.xls** can be found online.

C_p relates to the spread (dispersion) of the process output. However, because the target may intentionally be set closer to one of the specification limits than the other or because many processes are not stable, we may want to use a different measure. When the process mean is not exactly centered between the specification limits, we use the process capability index (C_{pk}).

Process Capability Index (C_{pk})

C_{pk}
A proportion of variation (3σ) between the center of the process and the nearest specification limit.

The process capability index, C_{pk}, is a ratio for determining whether a process meets specifications based on the distance between the process mean and the nearest specification limit.

The formula for C_{pk} is:

$$C_{pk} = \text{Minimum of}\left[\frac{\text{Upper specification limit} - \bar{X}}{3\sigma}, \frac{\bar{X} - \text{Lower specification limit}}{3\sigma}\right] \tag{S6-14}$$

Where $\bar{X}$ = process mean
σ = standard deviation of the process population

When the C_{pk} index for both the upper and lower specification limits equals 1.0, the process variation is centered and the process is capable of producing within ± 3 standard deviations (fewer than 2,700 defects per million). When the mean is exactly on target and $C_{pk} = 2.0$, only 2 defects per billion would be expected. For C_{pk} to exceed 1, σ must be less than $\frac{1}{3}$ of the difference between each specification limit and the process mean ($\bar{X}$). Figure S6.8 shows the meaning of various measures of C_{pk}. Figure S6.8(c) illustrates a case where the process is capable on the high side but not the low side, hence not capable overall (shown in Example S7).

Example S7

PROCESS CAPABILITY INDEX (C_{pk})

You are the process improvement manager and have developed a new machine to cut insoles for the company's top-of-the-line running shoes. You are excited because the company's goal is no more than 64 defects per million, and this machine may be the innovation you need. The insoles cannot be smaller than .248" or larger than .252".

Because insoles that are too thick can be trimmed but insoles that are too thin must be scrapped, the process mean is set at .251". The standard deviation of the new process = σ = .0005 inch.

APPROACH ▶ You decide to determine the C_{pk}, using Equation (S6-14), for the new machine and make a decision on that basis.

SOLUTION ▶

$$C_{pk} = \text{Minimum of}\left[\frac{\text{Upper specification limit} - \bar{X}}{3\sigma}, \frac{\bar{X} - \text{Lower specification limit}}{3\sigma}\right]$$

$$C_{pk} = \text{Minimum of}\left[\frac{.252 - .251}{(3).0005}, \frac{.251 - .248}{(3).0005}\right] = \text{Minimum of}\left[\frac{.001}{.0015}, \frac{.003}{.0015}\right]$$

$$= \text{Minimum of } (.67, 2) = .67$$

INSIGHT ▶ Because the new machine has a C_{pk} of only 0.67, the new machine *is not* capable. It will not meet the company's target. Furthermore, the C_p ratio for this machine equals (.252 − .248)/[6(.0005)] = 1.33. So, the C_p measure would have provided an *incorrect* conclusion of capability because C_p assumes the process is centered exactly halfway between the specification limits.

LEARNING EXERCISE ▶ If the process standard deviation can be reduced from .0005 to .00025, what is the new C_{pk}, and will the new machine meet the desired standards? [Answer: The new C_{pk} is 1.33. With the reduced variance, the new machine is now capable and meets the standard of 1.33.]

RELATED PROBLEMS ▶ S6.41–S6.49

ACTIVE MODEL S6.2 This example is further illustrated in Active Model S6.2 found online.

EXCEL OM Data File **CH06SExS7.xls** can be found online.

C_p and C_{pk} will be the same when the process is centered. However, if the mean of the process is not exactly halfway between the specification limits, then the minimum distance to the upper or lower specification limit determines process capability. C_{pk} is the standard criterion used to express process performance. Another application of C_{pk} is shown in Solved Problem S6.4.

Acceptance Sampling[7]

Acceptance sampling is a form of testing that involves taking random samples of "lots," or batches, of finished products and measuring them against predetermined standards. Sampling is more economical than 100% inspection. The quality of the sample is used to judge the quality of all

Acceptance sampling
A method of measuring random samples of lots or batches of products against predetermined standards.

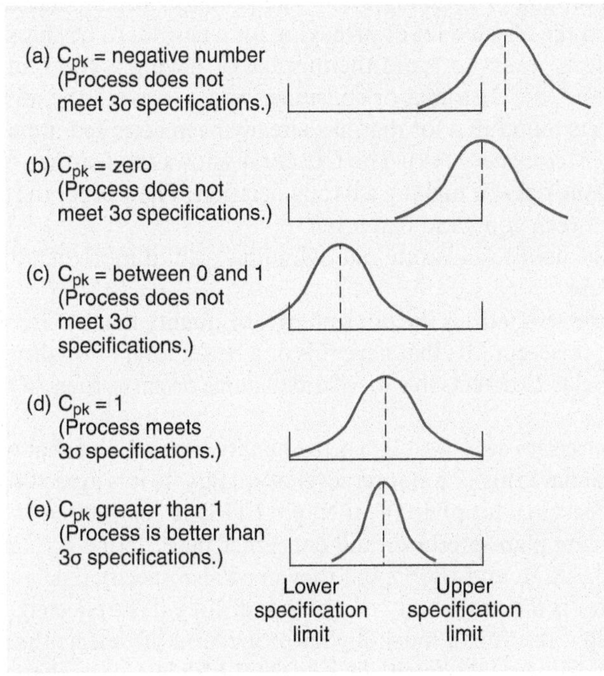
(a) C_{pk} = negative number (Process does not meet 3σ specifications.)
(b) C_{pk} = zero (Process does not meet 3σ specifications.)
(c) C_{pk} = between 0 and 1 (Process does not meet 3σ specifications.)
(d) C_{pk} = 1 (Process meets 3σ specifications.)
(e) C_{pk} greater than 1 (Process is better than 3σ specifications.)
Lower specification limit
Upper specification limit

Figure **S6.8**

Meanings of C_{pk} Measures
A C_{pk} index of 1.0 for both the upper and lower specification limits indicates that the process variation is within the upper and lower 3σ specification limits. As the C_{pk} index goes above 1.0, the process becomes increasingly target oriented, with fewer defects. If the C_{pk} is less than 1.0, the process will not produce within the specified 3σ tolerance. Because a process may not be centered, or may "drift," a C_{pk} above 1 is desired.

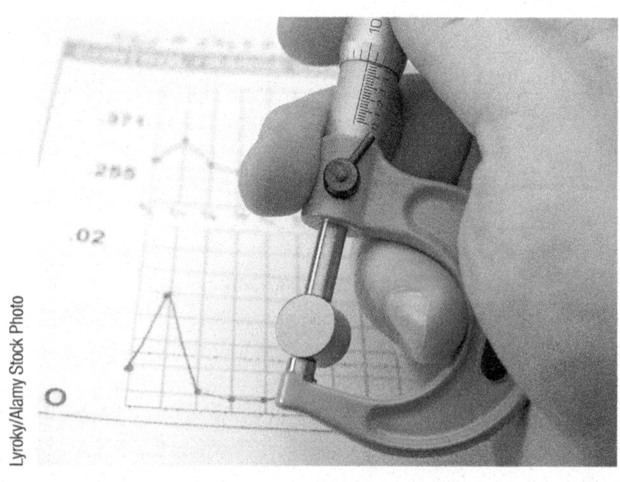
Lyroky/Alamy Stock Photo

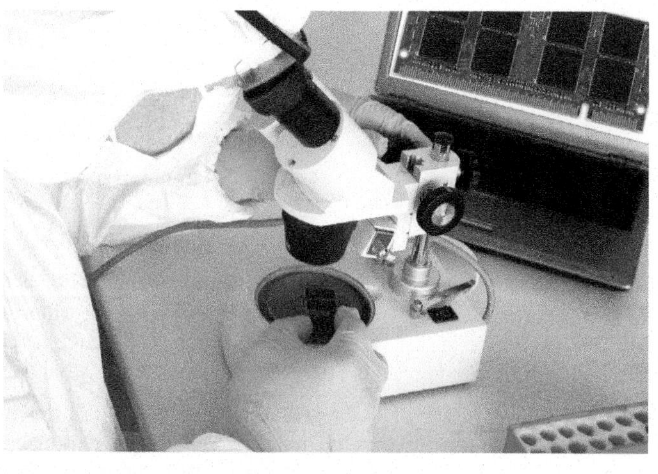
Mikeledray/Shutterstock

Raw data for statistical process control is collected in a wide variety of ways. Here physical measures using a micrometer (on the left) and a microscope (on the right) are being made.

items in the lot. Although both attributes and variables can be inspected by acceptance sampling, attribute inspection is more commonly used, as illustrated in this section.

LO S6.7 *Explain acceptance sampling*

Acceptance sampling can be applied either when materials arrive at a plant or at final inspection, but it is usually used to control incoming lots of purchased products. A lot of items rejected, based on an unacceptable level of defects found in the sample, can (1) be returned to the supplier or (2) be 100% inspected to cull out all defects, with the cost of this screening usually billed to the supplier. However, acceptance sampling is not a substitute for adequate process controls. In fact, the current approach is to build statistical quality controls at suppliers so that acceptance sampling can be eliminated.

Operating Characteristic Curve

Operating characteristic (OC) curve
A graph that describes how well an acceptance plan discriminates between good and bad lots.

The operating characteristic (OC) curve describes how well an acceptance plan discriminates between good and bad lots. A curve pertains to a specific plan—that is, to a combination of n (sample size) and c (acceptance level). It is intended to show the probability that the plan will accept lots of various quality levels.

With acceptance sampling, two parties are usually involved: the producer of the product and the consumer of the product. In specifying a sampling plan, each party wants to avoid costly mistakes in accepting or rejecting a lot. The producer usually has the responsibility of replacing all defects in the rejected lot or of paying for a new lot to be shipped to the customer. The producer, therefore, wants to avoid the mistake of having a good lot rejected (producer's risk). On the other hand, the customer or consumer wants to avoid the mistake of accepting a bad lot because defects found in a lot that has already been accepted are usually the responsibility of the customer (consumer's risk). The OC curve shows the features of a particular sampling plan, including the risks of making a wrong decision. The steeper the curve, the better the plan distinguishes between good and bad lots.[8]

Producer's risk
The mistake of having a producer's good lot rejected through sampling.

Consumer's risk
The mistake of a customer's acceptance of a bad lot overlooked through sampling.

Figure S6.9 can be used to illustrate one sampling plan in more detail. Four concepts are illustrated in this figure.

Acceptable quality level (AQL)
The quality level of a lot considered good.

The acceptable quality level (AQL) is the poorest level of quality that we are willing to accept. In other words, we wish to accept lots that have this or a better level of quality, but no worse. If an acceptable quality level is 20 defects in a lot of 1,000 items or parts, then AQL is $20/1{,}000 = 2\%$ defectives.

Lot tolerance percentage defective (LTPD)
The quality level of a lot considered bad.

The lot tolerance percentage defective (LTPD) is the quality level of a lot that we consider bad. We wish to reject lots that have this or a poorer level of quality. If it is agreed that an unacceptable quality level is 70 defects in a lot of 1,000, then the LTPD is $70/1{,}000 = 7\%$ defective.

To derive a sampling plan, producer and consumer must define not only "good lots" and "bad lots" through the AQL and LTPD, but they must also specify risk levels.

Producer's risk (α) is the probability that a "good" lot will be rejected. This is the risk that a random sample might result in a much higher proportion of defects than the population of all items. A lot with an acceptable quality level of AQL still has an α chance of being rejected. Sampling plans are often designed to have the producer's risk set at $\alpha = .05$, or 5%.

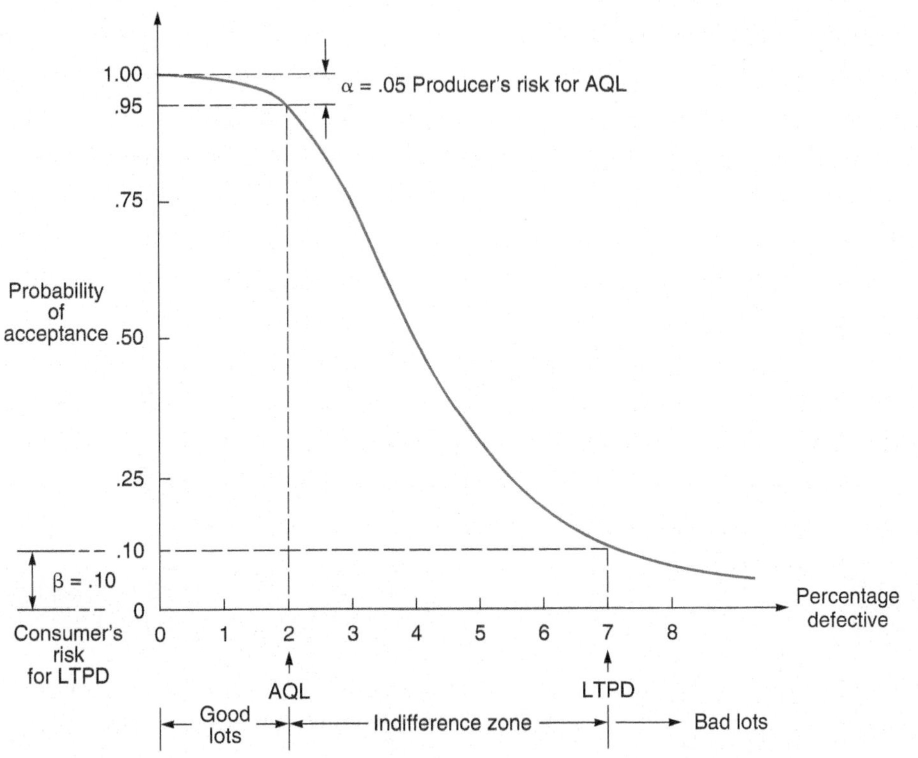

Figure S6.9

An Operating Characteristic (OC) Curve Showing Producer's and Consumer's Risks

A good lot for this particular acceptance plan has less than or equal to 2% defectives. A bad lot has 7% or more defectives.

RELATED PROBLEMS

▶ S6.51–S6.55

Consumer's risk (β) is the probability that a "bad" lot will be accepted. This is the risk that a random sample may result in a lower proportion of defects than the overall population of items. A common value for consumer's risk in sampling plans is $\beta = .10$, or 10%.

The probability of rejecting a good lot is called a type I error. The probability of accepting a bad lot is a type II error.

Type I error
Statistically, the probability of rejecting a good lot.

Type II error
Statistically, the probability of accepting a bad lot.

Sampling plans and OC curves may be developed by computer (as seen in the Excel OM and POM for Windows software available with this text), by published tables, or by calculation, using binomial or Poisson distributions.

Average Outgoing Quality

In most sampling plans, when a lot is rejected, the entire lot is inspected and all defective items replaced. Use of this replacement technique improves the average outgoing quality in terms of percent defective. In fact, given (1) any sampling plan that replaces all defective items encountered and (2) the true incoming percent defective for the lot, it is possible to determine the average outgoing quality (AOQ) in percentage defective. The equation for AOQ is:

Average outgoing quality (AOQ)
The percentage defective in an average lot of goods inspected through acceptance sampling.

$$\text{AOQ} = \frac{(P_d)(P_a)(N - n)}{N} \tag{S6-15}$$

Where

P_d = true percentage defective of the lot

P_a = probability of accepting the lot for a given sample size and quantity defective

N = number of items in the lot

n = number of items in the sample

The maximum value of AOQ corresponds to the highest average percentage defective or the lowest average quality for the sampling plan. It is called the *average outgoing quality limit (AOQL)*.

Acceptance sampling is useful for screening incoming lots. When the defective parts are replaced with good parts, acceptance sampling helps to increase the quality of the lots by reducing the outgoing percent defective.

Figure S6.10 compares acceptance sampling, SPC, and C_{pk}. As the figure shows, (a) acceptance sampling by definition accepts some bad units, (b) control charts try to keep the process in control, but (c) the C_{pk} index places the focus on improving the process. As operations managers, that is what we want to do—improve the process.

Figure S6.10

The Application of Statistical Process Control Techniques Contributes to the Identification and Systematic Reduction of Process Variability

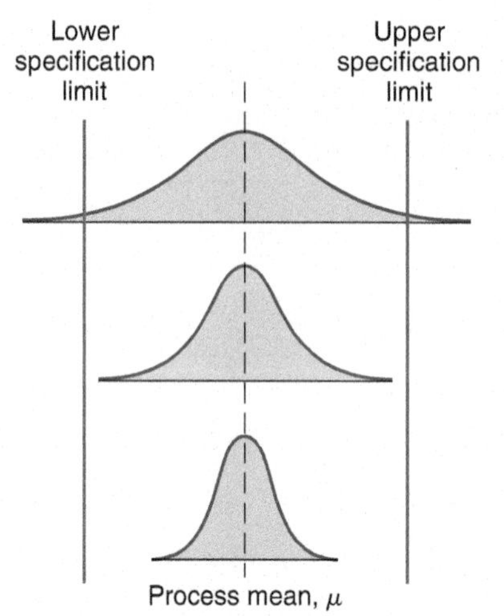

Lower specification limit
Upper specification limit
(a) Acceptance sampling (Some bad units accepted; the "lot" is good or bad.)
(b) Statistical process control (Keep the process "in control.")
(c) C_{pk} > 1 (Design a process that is within specification.)
Process mean, μ

Summary

Statistical process control is a major statistical tool of quality control. Control charts for SPC help operations managers distinguish between natural and assignable variations. The $\bar{x}$-chart and the R-chart are used for variable sampling, and the p-chart and the c-chart for attribute sampling. The C_{pk} index is a way to express process capability. Operating characteristic (OC) curves facilitate acceptance sampling and provide the manager with tools to evaluate the quality of a production run or shipment.

Key Terms

Statistical process control (SPC) (p. 244)
Control chart (p. 244)
Natural variations (p. 244)
Assignable variation (p. 244)
$\bar{x}$-chart (p. 245)
R-chart (p. 245)
Central limit theorem (p. 245)
p-chart (p. 253)
c-chart (p. 254)
Run test (p. 256)
Process capability (p. 257)
C_p (p. 257)
C_{pk} (p. 258)
Acceptance sampling (p. 259)
Operating characteristic (OC) curve (p. 260)
Producer's risk (p. 260)
Consumer's risk (p. 260)
Acceptable quality level (AQL) (p. 260)
Lot tolerance percentage defective (LTPD) (p. 260)
Type I error (p. 261)
Type II error (p. 261)
Average outgoing quality (AOQ) (p. 261)

Discussion Questions

1. List Shewhart's two types of variation. What are they also called?
2. Define "in statistical control."
3. Explain briefly what an $\bar{x}$-chart and an *R*-chart do.
4. What might cause a process to be out of control?
5. List five steps in developing and using $\bar{x}$-charts and *R*-charts.
6. List some possible causes of assignable variation.
7. Explain how a person using 2-sigma control charts will more easily find samples "out of bounds" than 3-sigma control charts. What are some possible consequences of this fact?
8. When is the desired mean, μ, used in establishing the centerline of a control chart instead of $\bar{\bar{x}}$?
9. Can a production process be labeled as "out of control" because it is too good? Explain.
10. In a control chart, what would be the effect on the control limits if the sample size varied from one sample to the next?
11. Define C_{pk} and explain what a C_{pk} of 1.0 means. What is C_p?
12. What does a run of 5 points above or below the centerline in a control chart imply?
13. What are the acceptable quality level (AQL) and the lot tolerance percentage defective (LTPD)? How are they used?
14. What is a run test, and when is it used?
15. Discuss the managerial issues regarding the use of control charts.
16. What is an OC curve?
17. What is the purpose of acceptance sampling?
18. What two risks are present when acceptance sampling is used?
19. Is a *capable* process a *perfect* process? That is, does a capable process generate only output that meets specifications? Explain.

Using Software for SPC

Excel, Excel OM, and POM for Windows may be used to develop control charts for most of the problems in this chapter.

CREATING YOUR OWN EXCEL SPREADSHEETS TO DETERMINE CONTROL LIMITS FOR A C-CHART

Excel and other spreadsheets are extensively used in industry to maintain control charts. Program S6.1 is an example of how to use Excel to determine the control limits for a *c*-chart. A *c*-chart is used when the number of defects per unit of output is known. The data from Example S5 are used here. In this example, 54 complaints occurred over 9 days. Excel also contains a built-in graphing ability with Chart Wizard.

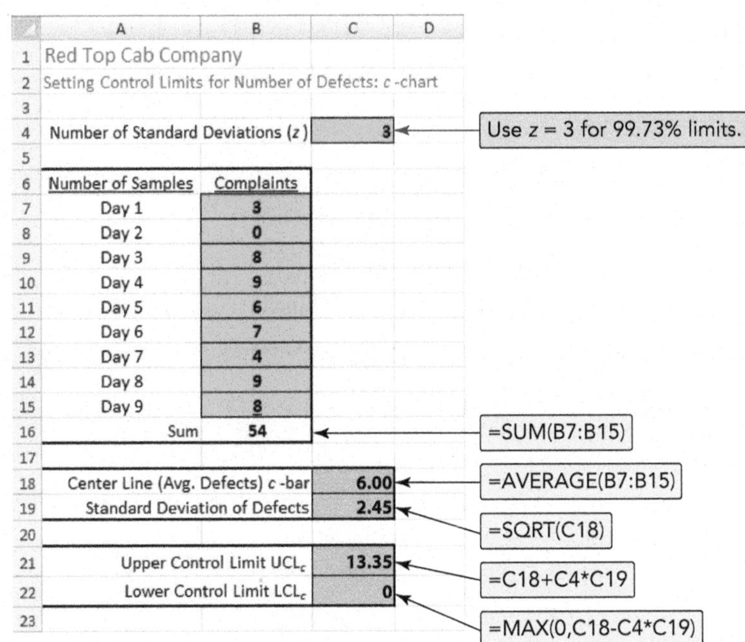

Program **S6.1**

An Excel Spreadsheet for Creating a *c*-Chart for Example S5

USING EXCEL OM

Excel OM's Quality Control module has the ability to develop $\bar{x}$-charts, *p*-charts, and *c*-charts. It also handles OC curves, acceptance sampling, and process capability. Program S6.2 illustrates Excel OM's spreadsheet approach to computing the $\bar{x}$ control limits for the Oat Flakes company in Example S1.

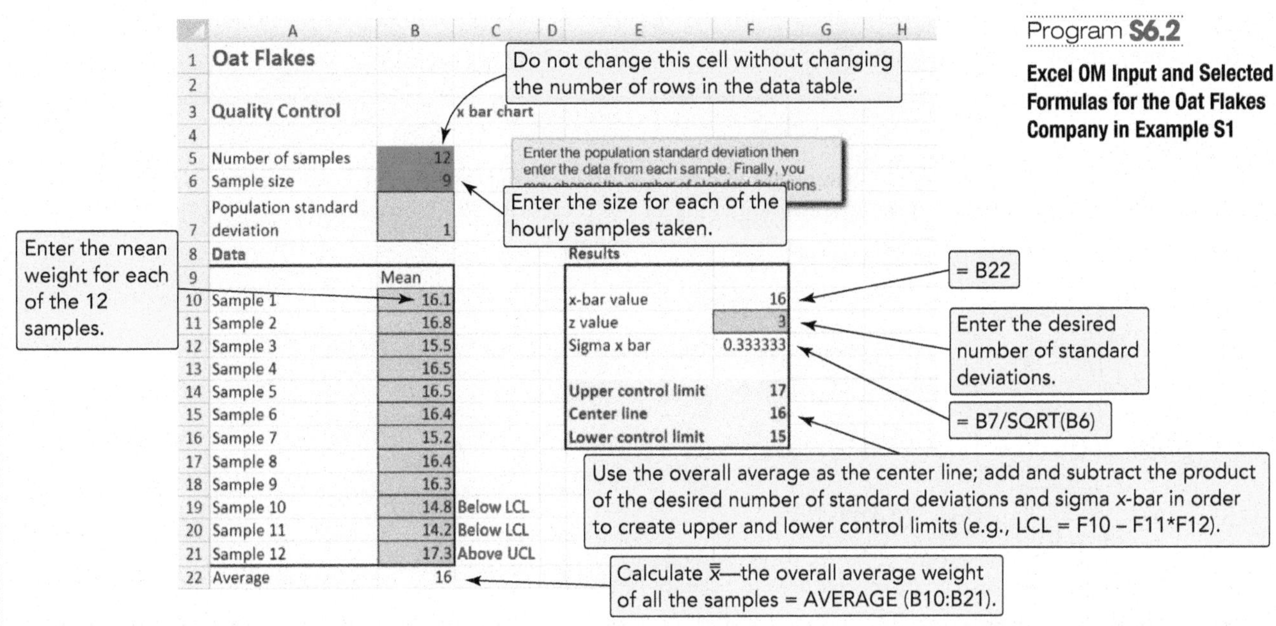

Program **S6.2**

Excel OM Input and Selected Formulas for the Oat Flakes Company in Example S1

USING POM FOR WINDOWS

The POM for Windows Quality Control module has the ability to compute all the SPC control charts we introduced in this supplement, as well as OC curves, acceptance sampling, and process capability. See Appendix II for further details.

Solved Problems

SOLVED PROBLEM S6.1

A manufacturer of precision machine parts produces round shafts for use in the construction of drill presses. The average diameter of a shaft is .56 inch. Inspection samples contain 6 shafts each. The average range of these samples is .006 inch. Determine the upper and lower $\bar{x}$ control chart limits.

SOLUTION

The mean factor A_2 from Table S6.1, where the sample size is 6, is seen to be .483. With this factor, you can obtain the upper and lower control limits:

$$\begin{aligned} \text{UCL}_{\bar{x}} &= .56 + (.483)\,(.006) \\ &= .56 + .0029 \\ &= .5629 \text{ inch} \\ \text{LCL}_{\bar{x}} &= .56 - .0029 \\ &= .5571 \text{ inch} \end{aligned}$$

SOLVED PROBLEM S6.2

Nocaf Drinks, Inc., a producer of decaffeinated coffee, bottles Nocaf. Each bottle should have a net weight of 4 ounces. The machine that fills the bottles with coffee is new, and the operations manager wants to make sure that it is properly adjusted. Bonnie Crutcher, the operations manager, randomly selects and weighs $n = 8$ bottles and records the average and range in ounces for each sample. The data for several samples is given in the following table. Note that every sample consists of 8 bottles.

SAMPLE	SAMPLE RANGE	SAMPLE AVERAGE	SAMPLE	SAMPLE RANGE	SAMPLE AVERAGE
A	.41	4.00	E	.56	4.17
B	.55	4.16	F	.62	3.93
C	.44	3.99	G	.54	3.98
D	.48	4.00	H	.44	4.01

Is the machine properly adjusted and in control?

SOLUTION

We first find that $\bar{\bar{x}} = 4.03$ and $\bar{R} = .505$. Then, using Table S6.1, we find:

$$\begin{aligned} \text{UCL}_{\bar{x}} &= \bar{\bar{x}} + A_2\bar{R} = 4.03 + (.373)\,(.505) = 4.22 \\ \text{LCL}_{\bar{x}} &= \bar{\bar{x}} - A_2\bar{R} = 4.03 - (.373)\,(.505) = 3.84 \\ \text{UCL}_R &= D_4\bar{R} = (1.864)\,(.505) = .94 \\ \text{LCL}_R &= D_3\bar{R} = (.136)\,(.505) = .07 \end{aligned}$$

It appears that the process average and range are both in statistical control.

The operations manager needs to determine if a process with a mean (4.03) slightly above the desired mean of 4.00 is satisfactory; if it is not, the process will need to be changed.

SOLVED PROBLEM S6.3

Altman Distributors, Inc., fills catalog orders. Samples of size $n = 100$ orders have been taken each day over the past 6 weeks. The average defect rate was .05. Determine the upper and lower limits for this process for 99.73% confidence.

SOLUTION

$z = 3$, $\bar{p} = .05$. Using Equations (S6-9), (S6-10), and (S6-11):

$$\text{UCL}_p = \bar{p} + 3\sqrt{\frac{\bar{p}(1-\bar{p})}{n}} = .05 + 3\sqrt{\frac{(.05)\,(1-.05)}{100}}$$

$$= .05 + 3(0.0218) = .1154$$

$$\text{LCL}_p = \bar{p} - 3\sqrt{\frac{\bar{p}(1-\bar{p})}{n}} = .05 - 3(0.0218)$$

$$= .05 - .0654 = 0 \text{ (because percentage defective cannot be negative)}$$

SOLVED PROBLEM S6.4

Ettlie Engineering has a new catalyst injection system for your countertop production line. Your process engineering department has conducted experiments and determined that the mean is 8.01 grams with a standard deviation of .03. Your specifications are: $\mu = 8.0$ and $\sigma = .04$, which means an upper specification limit of 8.12 [= 8.0 + 3(.04)] and a lower specification limit of 7.88 [= 8.0 − 3(.04)].

What is the C_{pk} performance of the injection system?

SOLUTION

Using Equation (S6-14):

$$C_{pk} = \text{Minimum of}\left[\frac{\text{Upper specification limit} - \bar{X}}{3\sigma}, \frac{\bar{X} - \text{Lower specification limit}}{3\sigma}\right]$$

where

$\bar{X}$ = process mean
σ = standard deviation of the process population

$$C_{pk} = \text{Minimum of}\left[\frac{8.12 - 8.01}{(3)\,(.03)}, \frac{8.01 - 7.88}{(3)\,(.03)}\right]$$

$$\left[\frac{.11}{.09} = 1.22, \frac{.13}{.09} = 1.44\right]$$

The minimum is 1.22, so the C_{pk} is within specifications and has an implied error rate of less than 2,700 defects per million.

SOLVED PROBLEM S6.5

Airlines lose thousands of checked bags every day, and America South Airlines is no exception to the industry rule. Over the past 6 weeks, the number of bags "misplaced" on America South flights has been 18, 10, 4, 6, 12, and 10. The head of customer service wants to develop a *c*-chart at 99.73% levels.

SOLUTION

She first computes $\bar{c} = \dfrac{18 + 10 + 4 + 6 + 12 + 10}{6} = \dfrac{60}{6} = 10$ bags/week

Then, using Equation (S6-12):

$$UCL_c = \bar{c} + 3\sqrt{\bar{c}} = 10 + 3\sqrt{10} = 10 + 3(3.16) = 19.48 \text{ bags}$$

$$LCL_c = \bar{c} - 3\sqrt{\bar{c}} = 10 - 3\sqrt{10} = 10 - 3(3.16) = .52 \text{ bag}$$

Problems

Note: **PX** means the problem may be solved with POM for Windows and/or Excel OM/Excel.

Problems S6.1–S6.39 relate to Statistical Process Control (SPC)

• **S6.1** Boxes of Honey-Nut Oatmeal are produced to contain 14 ounces, with a standard deviation of .1 ounce. Set up the 3-sigma $\bar{x}$-chart for a sample size of 36 boxes. **PX**

• **S6.2** The overall average on a process you are attempting to monitor is 50 units. The process population standard deviation is 1.72. Determine the upper and lower control limits for a mean chart, if you choose to use a sample size of 5. **PX**

a) Set $z = 3$.
b) Now set $z = 2$. How do the control limits change?

• **S6.3** Thirty-five samples of size 7 each were taken from a fertilizer-bag-filling machine at Panos Kouvelis Lifelong Lawn Ltd. The results were overall mean = 57.75 lb; average range = 1.78 lb. **PX**

a) Determine the upper and lower control limits of the $\bar{x}$-chart, where $\sigma = 3$.
b) Determine the upper and lower control limits of the *R*-chart, where $\sigma = 3$. **PX**

• **S6.4** Rosters Chicken advertises "lite" chicken with 30% fewer calories than standard chicken. When the process for "lite" chicken breast production is in control, the average chicken breast contains 420 calories, and the standard deviation in caloric content of the chicken breast population is 25 calories.

Rosters wants to design an $\bar{x}$-chart to monitor the caloric content of chicken breasts, where 25 chicken breasts would be chosen at random to form each sample.

a) What are the lower and upper control limits for this chart if these limits are chosen to be *four* standard deviations from the target?
b) What are the limits with three standard deviations from the target? **PX**

• **S6.5** Ross Hopkins is attempting to monitor a filling process that has an overall average of 705 mL. The average range is 6 mL. If you use a sample size of 10, what are the upper and lower control limits for the mean and range?

•• **S6.6** Sampling four pieces of precision-cut wire (to be used in computer assembly) every hour for the past 24 hours has produced the following results:

HOUR	$\bar{X}$	R	HOUR	$\bar{X}$	R
1	3.25"	.71"	13	3.11"	.85"
2	3.10	1.18	14	2.83	1.31
3	3.22	1.43	15	3.12	1.06
4	3.39	1.26	16	2.84	.50
5	3.07	1.17	17	2.86	1.43
6	2.86	.32	18	2.74	1.29
7	3.05	.53	19	3.41	1.61
8	2.65	1.13	20	2.89	1.09
9	3.02	.71	21	2.65	1.08
10	2.85	1.33	22	3.28	.46
11	2.83	1.17	23	2.94	1.58
12	2.97	.40	24	2.64	.97

Develop appropriate control charts and determine whether there is any cause for concern in the cutting process. Plot the information and look for patterns. **PX**

•• **S6.7** Auto pistons at Wenming Chung's plant in Shanghai are produced in a forging process, and the diameter is a critical factor that must be controlled. From sample sizes of 10 pistons produced each day, the mean and the range of this diameter have been as follows:

DAY	MEAN (MM)	RANGE (MM)
1	156.9	4.2
2	153.2	4.6
3	153.6	4.1
4	155.5	5.0
5	156.6	4.5

a) What is the value of $\bar{\bar{x}}$?
b) What is the value of $\bar{R}$?
c) What are the $UCL_{\bar{x}}$ and $LCL_{\bar{x}}$, using 3σ? Plot the data.
d) What are the UCL_R and LCL_R, using 3σ? Plot the data.
e) If the true diameter mean should be 155 mm and you want this as your center (nominal) line, what are the new $UCL_{\bar{x}}$ and $LCL_{\bar{x}}$? **PX**

•• **S6.8** A. Choudhury's bowling ball factory in Illinois makes bowling balls of adult size and weight only. The standard deviation in the weight of a bowling ball produced at the factory is known to be 0.12 pounds. Each day for 24 days, the average weight, in pounds, of nine of the bowling balls produced that day has been assessed as follows:

DAY	AVERAGE (LB)	DAY	AVERAGE (LB)	DAY	AVERAGE (LB)
1	16.3	9	15.9	17	16.1
2	15.9	10	16.2	18	15.9
3	15.8	11	15.9	19	16.2
4	15.5	12	15.9	20	15.9
5	16.3	13	16.3	21	15.9
6	16.2	14	15.9	22	16.0
7	16.0	15	16.3	23	15.5
8	16.1	16	16.2	24	15.8

a) Establish a control chart for monitoring the average weights of the bowling balls in which the upper and lower control limits are each two standard deviations from the mean. What are the values of the control limits?
b) If three standard deviations are used in the chart, how do these values change? Why? **PX**

•• **S6.9** Organic Grains LLC uses statistical process control to ensure that its health-conscious, low-fat, multigrain sandwich loaves have the proper weight. Based on a previously stable and in-control process, the control limits of the $\bar{x}$- and R-charts are $UCL_{\bar{x}} = 6.56$. $LCL_{\bar{x}} = 5.84$, $UCL_R = 1.141$, $LCL_R = 0$. Over the past few days, they have taken five random samples of four loaves each and have found the following:

SAMPLE	NET WEIGHT			
	LOAF #3	LOAF #2	LOAF #3	LOAF #4
1	6.3	6.0	5.9	5.9
2	6.0	6.0	6.3	5.9
3	6.3	4.8	5.6	5.2
4	6.2	6.0	6.2	5.9
5	6.5	6.6	6.5	6.9

Is the process still in control? Explain why or why not. **PX**

••• **S6.10** A process that is considered to be in control measures an ingredient in ounces. Below are the last 10 samples (each of size $n = 5$) taken. The population process standard deviation, σ, is 1.36.

SAMPLES									
1	2	3	4	5	6	7	8	9	10
10	9	13	10	12	10	10	13	8	10
9	9	9	10	10	10	11	10	8	12
10	11	10	11	9	8	10	8	12	9
9	11	10	10	11	12	8	10	12	8
12	10	9	10	10	9	9	8	9	12

a) What is $\sigma_{\bar{x}}$?
b) If $z = 3$, what are the control limits for the mean chart?
c) What are the control limits for the range chart?
d) Is the process in control? **PX**

••• **S6.11** Twelve samples, each containing five parts, were taken from a process that produces steel rods at Emmanuel Kodzi's factory. The length of each rod in the samples was determined. The results were tabulated and sample means and ranges were computed. The results were:

SAMPLE	SAMPLE MEAN (IN.)	RANGE (IN.)
1	10.002	0.011
2	10.002	0.014
3	9.991	0.007
4	10.006	0.022
5	9.997	0.013
6	9.999	0.012
7	10.001	0.008
8	10.005	0.013
9	9.995	0.004
10	10.001	0.011
11	10.001	0.014
12	10.006	0.009

a) Determine the upper and lower control limits and the overall means for $\bar{x}$-charts and R-charts.
b) Draw the charts and plot the values of the sample means and ranges.
c) Do the data indicate a process that is in control?
d) Why or why not? **PX**

•• **S6.12** Eagletrons are all-electric automobiles produced by Mogul Motors, Inc. One of the concerns of Mogul Motors is that the Eagletrons be capable of achieving appropriate maximum speeds. To monitor this, Mogul executives take samples of 8 Eagletrons at a time. For each sample, they determine the average maximum speed and the range of the maximum speeds within the sample. They repeat this with 35 samples to obtain 35 sample means and 35 ranges. They find that the average sample mean is 88.50 miles per hour, and the average range is 3.25 miles per hour. Using these results, the executives decide to establish an R chart. They would like this chart to be established so that when it shows that the range of a sample is not within the control limits, there is only approximately a 0.0027 probability that this is due to natural variation. What will be the upper control limit (UCL) and the lower control limit (LCL) in this chart? **PX**

•• **S6.13** The defect rate for data entry of insurance claims at Sadegh Kazemi Insurance Co. has historically been about 1.5%.
a) What are the upper and lower control chart limits if you wish to use a sample size of 100 and 3-sigma limits?
b) What if the sample size used were 50, with 3σ?
c) What if the sample size used were 100, with 2σ?
d) What if the sample size used were 50, with 2σ?
e) What happens to $\hat{\sigma}_p$ when the sample size is larger?
f) Explain why the lower control limit cannot be less than 0. **PX**

• **S6.14** Sue Abdinnour has been hand-painting wooden Christmas ornaments for several years. Recently, she has hired some friends to help her increase the volume of her business. In checking the quality of the work, she notices that some slight blemishes occasionally are apparent. A sample of 20 pieces of work resulted in the following number of blemishes on each piece: 0, 2, 1, 0, 0, 3, 2, 0, 4, 1, 2, 0, 0, 1, 2, 1, 0, 0, 0, 1. Develop upper and lower control limits for the number of blemishes on each piece. **PX**

•• **S6.15** The results of an inspection of DNA samples taken over the past 10 days are given below. Sample size is 100.

DAY	1	2	3	4	5	6	7	8	9	10
DEFECTIVES	7	6	6	9	5	6	1	7	9	1

a) Construct a 3-sigma p-chart using this information.
b) Using the control chart in part (a), and finding that the number of defectives on the next three days are 12, 5, and 13, is the process in control? **PX**

• **S6.16** In the past, the defective rate for your product has been 1.5%. What are the upper and lower control chart limits if you wish to use a sample size of 500 and $z = 3$? **PX**

• **S6.17** Refer to Problem S6.16. If the defective rate was 3.5% instead of 1.5%, what would be the control limits ($z = 3$)? **PX**

•• **S6.18** Five data entry operators work at the data processing department of the Birmingham Bank. Each day for 30 days, the number of defective records in a sample of 250 records typed by these operators has been noted, as follows:

SAMPLE NO.	NO. DEFECTIVE	SAMPLE NO.	NO. DEFECTIVE	SAMPLE NO.	NO. DEFECTIVE
1	7	11	18	21	17
2	5	12	5	22	12
3	19	13	16	23	6
4	10	14	4	24	7
5	11	15	11	25	13
6	8	16	8	26	10
7	12	17	12	27	14
8	9	18	4	28	6
9	6	19	6	29	12
10	13	20	16	30	3

a) Establish 3σ upper and lower control limits.
b) Why can the lower control limit not be a negative number?
c) The industry standards for the upper and lower control limits are 0.10 and 0.01, respectively. What does this imply about Birmingham Bank's own standards? **PX**

•• **S6.19** Houston North Hospital is trying to improve its image by providing a positive experience for its patients and their relatives. Part of the "image" program involves providing tasty, inviting patient meals that are also healthful. A questionnaire accompanies each meal served, asking the patient, among other things, whether he or she is satisfied or unsatisfied with the meal. A 100-patient sample of the survey results over the past 7 days yielded the following data:

DAY	NO. OF UNSATISFIED PATIENTS	SAMPLE SIZE
1	24	100
2	22	100
3	8	100
4	15	100
5	10	100
6	26	100
7	17	100

Construct a p-chart that plots the percentage of patients unsatisfied with their meals. Set the control limits to include 99.73% of the random variation in meal satisfaction. Comment on your results. **PX**

•• **S6.20** Jamison Kovach Supply Company manufactures paper clips and other office products. Although inexpensive, paper clips have provided the firm with a high margin of profitability. Sample size is 200. Results are given for the last 10 samples:

SAMPLE	1	2	3	4	5	6	7	8	9	10
DEFECTIVES	5	7	4	4	6	3	5	6	2	8

a) Establish upper and lower control limits for the control chart and graph the data.
b) Has the process been in control?
c) If the sample size were 100 instead, how would your limits and conclusions change? **PX**

• **S6.21** Peter Ittig's department store, Ittig Brothers, is Amherst's largest independent clothier. The store receives an average of six returns per day. Using $z = 3$, would 9 returns in a day warrant action? **PX**

•• **S6.22** An ad agency tracks the complaints, by week received, about the billboards in its city:

WEEK	1	2	3	4	5	6
NO. OF COMPLAINTS	4	5	4	11	3	9

a) What type of control chart would you use to monitor this process? Why?
b) What are the 3-sigma control limits for this process? Assume that the historical complaint rate is unknown.
c) Is the process in control, according to the control limits? Why or why not?
d) Assume now that the historical complaint rate has been 4 calls a week. What would the 3-sigma control limits for this process be now? Has the process been in control according to the control limits? **PX**

•• **S6.23** The school board is trying to evaluate a new math program introduced to second-graders in five elementary schools across the county this year. A sample of the student scores on standardized math tests in each elementary school yielded the following data:

SCHOOL	A	B	C	D	E
NO. OF TEST ERRORS	52	27	35	44	55

Construct a *c*-chart for test errors, and set the control limits to contain 99.73% of the random variation in test scores. What does the chart tell you? Has the new math program been effective? **PX**

•• **S6.24** Telephone inquiries of 100 IRS "customers" are monitored daily at random by Xiangling Hu. Incidents of incorrect information or other nonconformities (such as impoliteness to customers) are recorded. The data for last week follow:

DAY	1	2	3	4	5
NO. OF NONCONFORMITIES	5	10	23	20	15

a) Construct a 3-standard deviation *c*-chart of nonconformities.
b) What does the control chart tell you about the IRS telephone operators? **PX**

••• **S6.25** The accounts receivable department at Rick Wing Manufacturing has been having difficulty getting customers to pay the full amount of their bills. Many customers complain that the bills are not correct and do not reflect the materials that arrived at their receiving docks. The department has decided to implement SPC in its billing process. To set up control charts, 10 samples of 50 bills each were taken over a month's time and the items on the bills checked against the bill of lading sent by the company's shipping department to determine the number of bills that were not correct. The results were:

SAMPLE NO.	NO. OF INCORRECT BILLS	SAMPLE NO.	NO. OF INCORRECT BILLS
1	6	6	4
2	5	7	3
3	11	8	4
4	4	9	7
5	1	10	2

a) Determine the value of *p*-bar, the mean fraction defective. Then determine the control limits for the *p*-chart using a 99.73% confidence level (3 standard deviations). Has this process been in control? If not, which samples were out of control?
b) How might you use the quality tools discussed in Chapter 6 to determine the source of the billing defects and where you might start your improvement efforts to eliminate the causes? **PX**

• **S6.26** Nathan Kunz Storage Technologies produces refrigeration units for food producers and retail food establishments. The overall average temperature that these units maintain is 46 degrees Fahrenheit. The average range is 2 degrees Fahrenheit. Samples of six are taken to monitor the process. Determine the upper and lower control chart limits for the mean and range for these refrigeration units. **PX**

• **S6.27** Susan Martonosi Products, Inc., produces granola cereal, granola bars, and other natural food products. Its natural granola cereal is sampled to ensure proper weight. Each sample contains eight boxes of cereal. The overall average for the samples is 17 ounces. The average range is only 0.5 ounce. Determine the upper and lower control-chart limits for the mean and range for the boxes of cereal. **PX**

• **S6.28** The overall average of a process you are attempting to monitor at Gihan Edirisinghe Motors is 75 units. The process standard deviation is 1.95, and the sample size is $n = 10$. What would be the upper and lower control limits for a 3-sigma control chart? **PX**

• **S6.29** The overall average of a process you are attempting to monitor at Feryal Erhun Enterprises is 50 units. The average range is 4 units. The sample size you are using is $n = 5$.
a) What are the upper and lower control limits of the 3-sigma mean chart?
b) What are the upper and lower control limits of the 3-sigma range chart? **PX**

•• **S6.30** Autopitch devices are made for both major- and minor-league teams to help them improve their batting averages. When set at the standard position, Autopitch can throw hardballs toward a batter at an average speed of 60 mph. To monitor these devices and to maintain the highest quality, Autopitch executive Neil Geismar takes samples of 10 devices at a time. The average range is 3 mph. Using this information, construct control limits for:
a) $\bar{x}$ chart.
b) *R* chart. **PX**

•• **S6.31** For the past two months, Muge Yayla-Kullu has been concerned about machine number 5 at the West Factory. To ensure that the machine is operating correctly, samples are taken, and the average and range for each sample are computed. Each sample consists of 12 items produced from the machine. Recently, 12 samples were taken, and for each, the sample range and average were computed. The sample range and sample average were 1.1 and 46 for the first sample, 1.31 and 45 for the second sample, 0.91 and 46 for the third sample, and 1.1 and 47 for the fourth sample. After the fourth sample, the sample averages increased. For the fifth sample, the range was 1.21, and the average was 48; for sample number 6, it was 0.82 and 47; for sample number 7, it was 0.86 and 50; and for the eighth sample, it was 1.11 and 49. After the eighth sample, the sample average continued to increase, never getting below 50. For sample number 9, the range and average were 1.12 and 51; for sample number 10, they were 0.99 and 52; for sample number 11, they were 0.86 and 50; and for sample number 12, they were 1.2 and 52.

Although Muge's boss wasn't overly concerned about the process, Muge was. It was Muge's feeling that something was definitely wrong with machine number 5. Do you agree? Why or why not? **PX**

••• **S6.32** A process at Amit Eynan Bottling Company that is considered in control measures liquid in ounces. Below are the last 12 samples taken. The sample size = 4.

SAMPLES											
1	2	3	4	5	6	7	8	9	10	11	12
20.2	19.9	19.8	20.1	19.8	19.7	20.2	20.1	20.0	19.9	20.1	19.9
20.1	19.9	19.9	19.9	20.1	19.6	20.1	20.1	19.9	19.8	20.1	19.4
19.8	19.7	20.0	19.9	20.1	19.7	19.9	19.8	20.1	20.1	19.8	19.7
19.9	20.1	20.1	19.8	19.9	19.4	19.9	19.7	19.8	19.9	19.9	19.8

a) What are the control limits for the mean chart?
b) What are the control limits for the range chart?
c) Graph the data. Is the process in control? **PX**

•• **S6.33** Your supervisor, Lisa Lehmann, has asked that you report on the output of a machine on the factory floor. This machine is supposed to be producing optical lenses with a mean weight of 50 grams and a range of 3.5 grams. The following table contains the data for a sample size of $n = 6$ taken during the past 3 hours:

SAMPLE NUMBER	1	2	3	4	5	6	7	8	9	10
SAMPLE AVERAGE	55	47	49	50	52	57	55	48	51	56
SAMPLE RANGE	3	1	5	3	2	6	3	2	2	3

a) What are the $\bar{x}$-chart control limits when the machine is working properly?
b) What are the R-chart control limits when the machine is working properly?
c) What seems to be happening? (*Hint:* Graph the data points. Run charts may be helpful.) **PX**

•• **S6.34** A part at Belleh Fontem Manufacturing requires holes of size .5 ± 0.01 inches. The drill bit has a stated size of .5 ± 0.005 inches and wears at the rate of .00017 inches per hole drilled. Give the most and least number of holes that can be drilled with one drill bit before it has to be replaced.

•• **S6.35** Cybersecurity is an area of increasing concern. The National Security Agency (NSA) monitors the number of hits at sensitive websites. When the number of hits is much larger than normal, there is cause for concern, and further investigation is warranted. For each of the past 12 months, the number of hits at one such website has been 181, 162, 172, 169, 185, 212, 190, 168, 190, 191, 197, and 204. Determine the upper and lower control limits (99.7%) for the associated c-chart. Using the upper control limit as your reference, at what point should the NSA take action? **PX**

• **S6.36** The smallest defect in a computer chip will render the entire chip worthless. Therefore, tight quality control measures must be established to monitor these chips. In the past, the percent defective at Chieh Lee's Computer Chips has been 1.1%. The sample size is 1,000. Determine upper and lower control chart limits for these computer chips. Use $z = 3$. **PX**

••• **S6.37** Daily samples of 100 power drills are removed from the assembly line at Drills by Stergios Fotopoulos and inspected for defects. Over the past 21 days, the following information has been gathered. Develop a 3 standard deviation (99.73% confidence) p-chart and graph the samples. Is the process in control?

DAY	1	2	3	4	5	6	7	8	9	10	11
DEFECTIVES	6	5	6	4	3	4	5	3	6	3	7

DAY	12	13	14	15	16	17	18	19	20	21	
DEFECTIVES	5	4	3	4	5	6	5	4	3	7	

PX

•• **S6.38** The new president at Big State University, Dr. Misty Blessley, has made student satisfaction with the enrollment and registration process one of her highest priorities. Students must see an advisor, sign up for classes, obtain a parking permit, pay tuition and fees, and buy textbooks and other supplies. During one registration period, students were sampled and asked about satisfaction with each of these areas. Twelve different groups of 100 students each were sampled, and the number in each group who had a least one complaint was as follows: 3, 2, 1, 2, 3, 1, 3, 2, 1, 2, 2, 3.

Develop upper and lower control limits (99.7%) for the proportion of students with complaints. **PX**

•• **S6.39** In order to monitor the allocation of patrol cars and other police resources, the local police chief, Timothy Kraft, collects data on the incidence of crime by city sector. The city is divided into ten sectors of 1000 residents each. The number of crime incidents reported last month in each sector is as follows:

SECTOR	1	2	3	4	5	6	7	8	9	10
CRIME INCIDENCE	6	25	5	11	20	17	10	22	7	33

Construct a p-chart that plots the rate of crime by sector. Set the control limits to include 99.73 percent of the random variation in crime. Is the crime rate in any sector out of control? Do you have any suggestions for the reallocation of police resources? **PX**

Problems S6.40–S6.50 relate to Process Capability

• **S6.40** The difference between the upper specification and the lower specification for a process is 0.6". The standard deviation is 0.1". What is the process capability ratio, C_p? Interpret this number. **PX**

•• **S6.41** Meena Chavan Corp.'s computer chip production process yields DRAM chips with an average life of 1,800 hours and $\sigma = 100$ hours. The tolerance upper and lower specification limits are 2,400 hours and 1,600 hours, respectively. Is this process capable of producing DRAM chips to specification? **PX**

• • **S6.42** Linda Boardman, Inc., an equipment manufacturer in Boston, has submitted a sample cutoff valve to improve your manufacturing process. Your process engineering department has conducted experiments and found that the valve has a mean (μ) of 8.00 and a standard deviation (σ) of .04. Your desired performance is $\mu = 8.0 \pm 3\sigma$, where $\sigma = .045$. What is the C_{pk} of the Boardman valve? **PX**

• • **S6.43** The specifications for a plastic liner for concrete highway projects calls for a thickness of 3.0 mm ± .1 mm. The standard deviation of the process is estimated to be .02 mm. What are the upper and lower specification limits for this product? The process is known to operate at a mean thickness of 3.0 mm. What is the C_{pk} for this process? About what percentage of all units of this liner will meet specifications? **PX**

• • **S6.44** Frank Pianki, the manager of an organic yogurt processing plant, desires a quality specification with a mean of 16 ounces, an upper specification limit of 16.5, and a lower specification limit of 15.5. The process has a mean of 16 ounces and a standard deviation of 1 ounce. Determine the C_{pk} of the process. **PX**

• • **S6.45** A process filling small bottles with baby formula has a target of 3 ounces ± 0.150 ounce. Two hundred bottles from the process were sampled. The results showed the average amount of formula placed in the bottles to be 3.042 ounces. The standard deviation of the amounts was 0.034 ounce. Determine the value of C_{pk}. Roughly what proportion of bottles meet the specifications? **PX**

• • **S6.46** A new process has just been established on Jianli Hu's assembly line. The process is supposed to place 4 grams of a deep red coloring in each bottle of nail polish with an Upper Specification Limit of 4.1 grams and a Lower Specification Limit of 3.9 grams. Although the mean is 4 grams, the standard deviation is 0.1 gram. What is the C_{pk} of the new process? **PX**

• • • **S6.47** NutraFlakes cereal is to be packaged with a mean of 10 ounces plus or minus 0.1 ounce. Using the data provided in the following table, determine the standard deviation of the 20 weights and then determine the C_{pk} of the process. (*Hint:* The standard deviation of the 20 cereal boxes can be computed using Endnote 2 of this supplement.)

TIME	BOX 1	BOX 2	BOX 3	BOX 4
9 A.M.	9.8	10.4	9.9	10.3
10 A.M.	10.1	10.2	9.9	9.8
11 A.M.	9.9	10.5	10.3	10.1
12 P.M.	9.7	9.8	10.3	10.2
1 P.M.	9.7	10.1	9.9	9.9

• • **S6.48** Two machines are currently in use in a process at the Dennis Kira Mfg. Co. The standards for this process are LSL = .400" and USL = .403". Machine One is currently producing with mean = .401" and standard deviation .0004". Machine Two is currently producing with mean .4015" and standard deviation .0005". Which machine has the higher capability index? **PX**

• • **S6.49** Product specifications call for a part at Vaidy Jayaraman's Metalworks to have a length of 1.000" ± .040". Currently, the process is performing at a grand average of 1.000" with a standard deviation of 0.010". Calculate the capability index of this process. Is the process "capable"? **PX**

• • **S6.50** The upper and lower design specification limits for a medical procedure performed by Dr. Roberta Rodriguez are 15 minutes and 13 minutes, respectively. The process $\bar{x} = 14.5$ minutes and $\sigma = .2$ minutes. Compute the C_p and interpret its meaning. **PX**

Problems S6.51–S6.55 relate to Acceptance Sampling

• • **S6.51** As the supervisor in charge of shipping and receiving, you need to determine *the average outgoing quality* in a plant where the known incoming lots from your assembly line have an average defective rate of 3%. Your plan is to sample 80 units of every 1,000 in a lot. The number of defects in the sample is not to exceed 3. Such a plan provides you with a probability of acceptance of each lot of .79 (79%). What is your average outgoing quality? **PX**

• • **S6.52** An acceptance sampling plan has lots of 500 pieces and a sample size of 60. The number of defects in the sample may not exceed 2. This plan, based on an OC curve, has a probability of .57 of accepting lots when the incoming lots have a defective rate of 4%, which is the historical average for this process. What do you tell your customer the average outgoing quality is? **PX**

• • **S6.53** The percent defective from an incoming lot is 3%. An OC curve showed the probability of acceptance to be 0.55. Given a lot size of 2,000 and a sample of 100, determine the average outgoing quality in percent defective.

• • **S6.54** In an acceptance sampling plan developed for lots containing 1,000 units, the sample size n is 85. The percent defective of the incoming lots is 2%, and the probability of acceptance is 0.64. What is the average outgoing quality?

• • **S6.55** We want to determine the AOQ for an acceptance sampling plan when the quality of the incoming lots in percent defective is 1.5%, and then again when the incoming percent defective is 5%. The sample size is 80 units for a lot size of 550 units. Furthermore, P_a at 1.5% defective levels is 0.95. At 5% incoming defective levels, the P_a is found to be 0.5. Determine the average outgoing quality for both incoming percent defective levels. **PX**

CASE STUDIES

Alabama Airlines' On-Time Schedule

Alabama Airlines opened its doors in 2020 as a commuter service with its headquarters and only hub located in Birmingham. A product of airline deregulation, Alabama Air joined the growing number of short-haul, point-to-point airlines, including Lone Star, Comair, Atlantic Southeast, and Skywest.

Alabama Air was started and managed by two former pilots, Sheila Hawkins (who had been with United Airlines) and Leilani Parker (formerly with Delta Airlines). It acquired a fleet of 12 used prop-jet planes and the airport gates vacated by Delta Airlines in 2020 when it curtailed flights due to the pandemic.

One of Alabama Air's top competitive priorities is on-time arrivals. The airline defines "on-time" to mean any arrival that is within 20 minutes of the scheduled time.

Sheila Hawkins decided to personally monitor Alabama Air's performance. Each week for the past 30 weeks, she checked a random sample of 100 flight arrivals for on-time performance. The accompanying table contains the number of flights that did not meet Alabama Air's definition of "on time."

SAMPLE (WEEK)	LATE FLIGHTS	SAMPLE (WEEK)	LATE FLIGHTS
1	2	16	2
2	4	17	3
3	10	18	7
4	4	19	3
5	1	20	2
6	1	21	3
7	13	22	5
8	9	23	4
9	11	24	3
10	1	25	2
11	3	26	2
12	4	27	1
13	2	28	1
14	2	29	3
15	8	30	4

Discussion Questions

1. Using a 95% confidence interval, plot the overall percentage of late flights (p) and the upper and lower control limits on a control chart.
2. Assume that the airline industry's upper and lower control limits for flights that are not on time are .1000 and .0400, respectively. Draw them on your control chart.
3. Plot the percentage of late flights in each sample. Do all samples fall within Alabama Airlines' control limits? When one falls outside the control limits, what should be done?
4. What can Sheila Hawkins report about the quality of service?

Frito-Lay's Quality-Controlled Potato Chips

Video Case 

Frito-Lay, the multi-billion-dollar snack food giant, produces billions of pounds of product every year at its dozens of U.S. and Canadian plants. From the farming of potatoes—in Florida, North Carolina, and Michigan—to factory and to retail stores, the ingredients and final product of Lay's chips, for example, are inspected at least 11 times: in the field, before unloading at the plant, after washing and peeling, at the sizing station, at the fryer, after seasoning, when bagged (for weight), at carton filling, in the warehouse, and as they are placed on the store shelf by Frito-Lay personnel. Similar inspections take place for its other famous products, including Cheetos, Fritos, Ruffles, and Tostitos.

In addition to these employee inspections, the firm uses proprietary vision systems to look for defective potato chips. Chips are pulled off the high-speed line and checked twice if the vision system senses them to be too brown.

The company follows the very strict standards of the American Institute of Baking (AIB), standards that are much tougher than those of the U.S. Food and Drug Administration. Two unannounced AIB site visits per year keep Frito-Lay's plants on their toes. Scores, consistently in the "excellent" range, are posted, and every employee knows exactly how the plant is doing.

There are two key metrics in Frito-Lay's continuous improvement quality program: (1) total customer complaints (measured on a complaints per million bag basis) and (2) hourly or daily statistical process control scores (for oil, moisture, seasoning, and salt content, for chip thickness, for fryer temperature, and for weight).

In the Florida plant, Angela McCormack, who holds engineering and MBA degrees, oversees a 15-member quality assurance staff. They watch all aspects of quality, including training employees on the factory floor, monitoring automated processing equipment, and developing and updating statistical process control (SPC) charts. The upper and lower control limits for one checkpoint, salt content in Lay's chips, are 2.22% and 1.98%, respectively. To see exactly how these limits are created using SPC, watch the video that accompanies this case.

Discussion Questions*

1. Angela is now going to evaluate a new salt process delivery system and wants to know if the upper and lower control limits at 3 standard deviations for the new system will meet the upper and lower control specifications noted earlier.

   The data (in percents) from the initial trial samples are:

   Sample 1: 1.98, 2.11, 2.15, 2.06
   Sample 2: 1.99, 2.0, 2.08, 1.99
   Sample 3: 2.20, 2.10. 2.20, 2.05
   Sample 4: 2.18, 2.01, 2.23, 1.98
   Sample 5: 2.01, 2.08, 2.14, 2.16

   Provide the report to Angela.
2. What are the advantages and disadvantages of Frito-Lay drivers stocking their customers' shelves?
3. Why is quality a critical function at Frito-Lay?

*You may wish to view the video that accompanies this case before answering these questions.

Farm to Fork: Quality at Darden Restaurants

Video Case 

Darden Restaurants, the $8 billion owner of such popular brands as Olive Garden, Seasons 52, Cheddar's, Longhorn Steakhouse, Capital Grille, and Bahama Breeze, serves hundreds of millions of meals annually in its 2,100 restaurants across the U.S. and Canada. Before any one of these meals is placed before a guest, the ingredients for each recipe must pass quality control inspections at the source, ranging from measurement and weighing to tasting, touching, or lab testing. Darden has differentiated itself from its restaurant peers by developing the gold standard in continuous improvement.

To assure both customers and the company that quality expectations are met, Darden uses a rigorous inspection process, employing statistical process control (SPC) as part of its "Farm to Fork" program. More than 50 food scientists, microbiologists, and public health professionals report to Ana Hooper, vice president of quality assurance.

As part of Darden's Point Source program, Hooper's team, based in Southeast Asia (in China, Thailand, and Singapore) and Latin America (in Ecuador, Honduras, and Chile), approves and inspects—and works with Darden buyers to purchase—more than 50 million pounds of seafood each year for restaurant use. Darden used to build quality in at the end by inspecting shipments as they reached U.S. distribution centers. Now, thanks to coaching and partnering with vendors abroad, Darden needs but a few domestic inspection labs to verify compliance to its exacting standards. Food vendors in source countries know that when supplying Darden, they are subject to regular audits that are stricter than U.S. Food and Drug Administration (FDA) standards.

Two Quality Success Stories

Quality specialists' jobs include raising the bar and improving quality and safety at all plants in their geographic area. The Thai quality representative, for example, worked closely with several of Darden's largest shrimp vendors to convert them to a production-line-integrated quality assurance program. The vendors were able to improve the quality of shrimp supplied and reduce the percentage of defects by 19%.

Likewise, when the Darden quality teams visited fields of growers/shippers in Mexico recently, it identified challenges such as low employee hygiene standards, field food safety problems, lack of portable toilets, child labor, and poor working conditions. Darden addressed these concerns and hired third-party independent food safety verification firms to ensure continued compliance to standards.

SPC Charts

SPC charts, such as the ones shown in this supplement, are particularly important. These charts document precooked food weights; meat, seafood and poultry temperatures; blemishes on produce; and bacteria counts on shrimp—just to name a few. Quality assurance is part of a much bigger process that is key to Darden's success—its supply chain. (See Chapters 2 and 11 for discussion and case studies on this topic.) That's because quality comes from the source and flows through distribution to the restaurant and guests.

Discussion Questions*

1. How does Darden build quality into the supply chain?
2. Select two potential problems—one in the Darden supply chain and one in a restaurant—that can be analyzed with a fish-bone chart. Draw a complete chart to deal with each problem.
3. Darden applies SPC in many product attributes. Identify where these are probably used.
4. The SPC charts appearing earlier in this supplement illustrate Darden's use of control charts to monitor the weight of salmon filets. Given these data, what conclusion do you, as a Darden quality control inspector, draw? What report do you issue to your supervisor? How do you respond to the salmon vendor?

*You might want to view the video that accompanies this case before answering these questions.

Endnotes

1. Removing assignable causes is work. Quality expert W. Edwards Deming observed that a state of statistical control is not a natural state for a manufacturing process. Deming instead viewed it as an achievement, arrived at by elimination, one by one, by determined effort, of special causes of excessive variation.
2. The standard deviation is easily calculated as

$$\sigma = \sqrt{\frac{\sum_{i=1}^{n}(x_i - \bar{x})^2}{n-1}}$$. For a good review of this and other statistical terms, refer to Tutorial 1, "Statistical Review for Managers," found online.
3. Lower control limits cannot take negative values in control charts. So the LCL = max $(0, \bar{\bar{x}} - z\sigma_{\bar{x}})$.
4. If the sample sizes are not the same, other techniques must be used.
5. A Poisson probability distribution is a discrete distribution commonly used when the items of interest (in this case, defects) are infrequent or occur in time or space.
6. Sources: *The Wall Street Journal* (May 11, 2016); and **ideaworkscompany.com.**
7. Refer to Tutorial 2 found online for an extended discussion of acceptance sampling.
8. Note that sampling always runs the danger of leading to an erroneous conclusion. Let us say that in one company the total population under scrutiny is a load of 1,000 computer chips, of which in reality only 30 (or 3%) are defective. This means that we would want to accept the shipment of chips, because for this particular firm 4% is the allowable defect rate. However, if a random sample of $n = 50$ chips was drawn, we could conceivably end up with 0 defects and accept that shipment (that is, it is okay), or we could find all 30 defects in the sample. If the latter happened, we could wrongly conclude that the whole population was 60% defective and reject them all.

Bibliography

Besterfield, D. H. *Quality Control*, 9th ed. Boston: Pearson, 2013.

Elg, M., J. Olsson, and J. J. Dahlgaard. "Implementing Statistical Process Control." *The International Journal of Quality and Reliability Management* 25, no. 6 (2008): 545.

Goetsch, D. L., and S. B. Davis. *Quality Management*, 8th ed. Boston: Pearson, 2016.

Gryna, F. M., R. C. H. Chua, and J. A. DeFeo. *Juran's Quality Planning and Analysis*, 5th ed. New York: McGraw-Hill, 2007.

Mitra, A. *Fundamentals of Quality Control and Improvement*, 4th ed. New York: Wiley, 2016.

Montgomery, D. C. *Introduction to Statistical Quality Control*, 8th ed. New York: Wiley, 2019.

Montgomery, D. C., C. L. Jennings, and M. E. Pfund. *Managing, Controlling, and Improving Quality.* New York: Wiley, 2011.

Sower, V. E. *Essentials of Quality.* New York: Wiley, 2012.

Summers, D. *Quality*, 6th ed. Boston: Pearson, 2018.

Supplement 6 *Rapid* Review

Main Heading	Review Material	MyLab Operations Management
STATISTICAL PROCESS CONTROL (SPC)	■ **Statistical process control (SPC)**—A process used to monitor standards by taking measurements and corrective action as a product or service is being produced. ■ **Control chart**—A graphical presentation of process data over time. A process is said to be operating *in statistical control* when the only source of variation is common (natural) causes. The process must first be brought into statistical control by detecting and eliminating special (assignable) causes of variation. *The objective of a process control system is to provide a statistical signal when assignable causes of variation are present.* ■ **Natural variations**—The variability that affects every production process to some degree and is to be expected; also known as common cause. When natural variations form a *normal distribution,* they are characterized by two parameters: ■ Mean, μ (the measure of central tendency—in this case, the average value) ■ Standard deviation, σ (the measure of dispersion) As long as the distribution (output measurements) remains within specified limits, the process is said to be "in control," and natural variations are tolerated. ■ **Assignable variation**—Variation in a production process that can be traced to specific causes. Control charts for the mean, $\bar{x}$, and the range, R, are used to monitor *variables* (outputs with continuous dimensions), such as weight, speed, length, or strength. ■ **$\bar{x}$-chart**—A quality control chart for variables that indicates when changes occur in the central tendency of a production process. ■ **R-chart**—A control chart that tracks the range within a sample; it indicates that a gain or loss in uniformity has occurred in dispersion of a production process. ■ **Central limit theorem**—The theoretical foundation for $\bar{x}$-charts, which states that regardless of the distribution of the population of all parts or services, the $\bar{x}$ distribution will tend to follow a normal curve as the number of samples increases: $$\bar{\bar{x}} = \mu \quad \text{(S6-1)}$$ $$\sigma_{\bar{x}} = \frac{\sigma}{\sqrt{n}} \quad \text{(S6-2)}$$ The $\bar{x}$-chart limits, if we know the true standard deviation σ of the process population, are: $$\text{Upper control limit (UCL)} = \bar{\bar{x}} + z\sigma_{\bar{x}} \quad \text{(S6-3)}$$ $$\text{Lower control limit (LCL)} = \bar{\bar{x}} - z\sigma_{\bar{x}} \quad \text{(S6-4)}$$ where z = confidence level selected (e.g., $z = 3$ is 99.73% confidence). The *range, R,* of a sample is defined as the difference between the largest and smallest items. If we do not know the true standard deviation, σ, of the population, the $\bar{x}$-chart limits are: $$\text{UCL}_{\bar{x}} = \bar{\bar{x}} + A_2\bar{R} \quad \text{(S6-5)}$$ $$\text{LCL}_{\bar{x}} = \bar{\bar{x}} - A_2\bar{R} \quad \text{(S6-6)}$$ In addition to being concerned with the process average, operations managers are interested in the process dispersion, or range. The R-chart control limits for the range of a process are: $$\text{UCL}_R = D_4\bar{R} \quad \text{(S6-7)}$$ $$\text{LCL}_R = D_3\bar{R} \quad \text{(S6-8)}$$ Attributes are typically classified as *defective* or *nondefective*. The two attribute charts are (1) p-charts (which measure the *percent* defective in a sample), and (2) c-charts (which *count* the number of defects in a sample). ■ **p-chart**—A quality control chart that is used to control attributes: $$\text{UCL}_p = \bar{p} + z\sigma_p \quad \text{(S6-9)}$$ $$\text{LCL}_p = \bar{p} - z\sigma_p \quad \text{(S6-10)}$$ $$\hat{\sigma}_p = \sqrt{\frac{\bar{p}(1-\bar{p})}{n}} \quad \text{(S6-11)}$$ ■ **c-chart**—A quality control chart used to control the number of defects per unit of output. The Poisson distribution is the basis for c-charts, whose 99.73% limits are computed as: $$\text{Control limits} = \bar{c} \pm 3\sqrt{\bar{c}} \quad \text{(S6-12)}$$ ■ **Run test**—A test used to examine the points in a control chart to determine whether nonrandom variation is present.	Concept Questions: 1.1–1.6 Problems: S6.1–S6.39 **VIDEO S6.1** Farm to Fork: Quality at Darden Restaurants Virtual Office Hours for Solved Problems: S6.1–S6.3 **ACTIVE MODELS S6.1 and S6.2** **VIDEO S6.2** Frito-Lay's Quality-Controlled Potato Chips Virtual Office Hours for Solved Problem: S6.5

Main Heading	Review Material	MyLab Operations Management
PROCESS CAPABILITY	■ **Process capability**—The ability to meet specifications. ■ C_p—A ratio for determining whether a process meets design specifications. $$C_p = \frac{(\text{Upper specification} - \text{Lower specification})}{6\sigma} \quad \text{(S6-13)}$$ ■ C_{pk}—A proportion of variation (3σ) between the center of the process and the nearest specification limit: $$C_{pk} = \text{Minimum of}\left[\frac{\text{Upper spec limit} - \bar{X}}{3\sigma}, \frac{\bar{X} - \text{Lower spec limit}}{3\sigma}\right] \quad \text{(S6-14)}$$	Concept Questions: 2.1–2.6 Problems: S6.40–S6.50 Virtual Office Hours for Solved Problems: S6.4 ACTIVE MODEL S6.3
ACCEPTANCE SAMPLING	■ **Acceptance sampling**—A method of measuring random samples of lots or batches of products against predetermined standards. ■ **Operating characteristic (OC) curve**—A graph that describes how well an acceptance plan discriminates between good and bad lots. ■ **Producer's risk**—The mistake of having a producer's good lot rejected through sampling. ■ **Consumer's risk**—The mistake of a customer's acceptance of a bad lot overlooked through sampling. ■ **Acceptable quality level (AQL)**—The quality level of a lot considered good. ■ **Lot tolerance percent defective (LTPD)**—The quality level of a lot considered bad. ■ **Type I error**—Statistically, the probability of rejecting a good lot. ■ **Type II error**—Statistically, the probability of accepting a bad lot. ■ **Average outgoing quality (AOQ)**—The percent defective in an average lot of goods inspected through acceptance sampling: $$AOQ = \frac{(P_d)(P_a)(N - n)}{N} \quad \text{(S6-15)}$$	Concept Questions: 3.1–3.6 Problems: S6.51–S6.55
ADDITIONAL MYLAB OPERATIONS MANAGEMENT RESOURCES	✔ Video for Creating Your Own Excel Spreadsheets (Example S5) ✔ Additional Case Studies (PEI Potato Purveyors and Green River Chemical Company) ✔ Multiple Choice Case Questions (PEI Potato Purveyors) ✔ Recent Graduate Video: Kylie Bertoncello, Design Engineer, Siemens Energy	

Self Test

LO S6.1 If the mean of a particular sample is within control limits and the range of that sample is not within control limits:
- **a)** the process is in control, with only assignable causes of variation.
- **b)** the process is not producing within the established control limits.
- **c)** the process is producing within the established control limits, with only natural causes of variation.
- **d)** the process has both natural and assignable causes of variation.

LO S6.2 The central limit theorem:
- **a)** is the theoretical foundation of the *c*-chart.
- **b)** states that the average of assignable variations is zero.
- **c)** allows managers to use the normal distribution as the basis for building some control charts.
- **d)** states that the average range can be used as a proxy for the standard deviation.
- **e)** controls the steepness of an operating characteristic curve.

LO S6.3 The type of chart used to control the central tendency of variables with continuous dimensions is:
- **a)** $\bar{x}$-chart.
- **b)** *R*-chart.
- **c)** *p*-chart.
- **d)** *c*-chart.
- **e)** none of the above.

LO S6.4 If parts in a sample are measured and the mean of the sample measurement is outside the control limits:
- **a)** the process is out of control, and the cause should be established.
- **b)** the process is in control but not capable of producing within the established control limits.
- **c)** the process is within the established control limits, with only natural causes of variation.
- **d)** all of the above are true.

LO S6.5 Control charts for attributes are:
- **a)** *p*-charts.
- **b)** *c*-charts.
- **c)** *R*-charts.
- **d)** $\bar{x}$-charts.
- **e)** both a and b.

LO S6.6 The ability of a process to meet design specifications is called:
- **a)** Taguchi.
- **b)** process capability.
- **c)** capability index.
- **d)** acceptance sampling.
- **e)** average outgoing quality.

LO S6.7 The ______ risk is the probability that a lot will be rejected despite the quality level exceeding or meeting the ______.

Answers: LO S6.1. b; LO S6.2. c; LO S6.3. a; LO S6.4. a; LO S6.5. e; LO S6.6. b; LO S6.7. producer's, AQL

Process Strategies

CHAPTER 7

CHAPTER OUTLINE

This new Austrian steel mill is operated by just 14 people who make 500,000 tons of steel wire per year. The technicians sitting in what's called the "pulpit"—high above the plant floor—mostly watch for warning signs such as spikes in temperature or pressure. As new control processes and innovations in the past 20 years have improved productivity, the number of worker hours to make a ton of steel has fallen from 700 to 250.

Bloomberg/Getty Images

10 OM STRATEGY DECISIONS

- Design of Goods and Services
- Managing Quality
- *Process Strategies*
- Location Strategies
- Layout Strategies
- Human Resources
- Supply Chain Management
- Inventory Management
- Scheduling
- Maintenance

GLOBAL COMPANY PROFILE
Harley-Davidson

Repetitive Manufacturing Works at Harley-Davidson

Since Harley-Davidson's founding in Milwaukee in 1903, it has competed with hundreds of manufacturers, foreign and domestic. The competition has been tough. Recent competitive battles have been with the Japanese, and previous battles were with the German, English, and Italian manufacturers. But after over a century, Harley is the only major U.S. motorcycle company. The company now has three U.S. facilities and an assembly plant in Brazil. The Sportster powertrain is manufactured in Wauwatosa, Wisconsin, and the sidecars, saddlebags, windshields, and other specialty items are produced in Tomahawk, Wisconsin. The Touring and Softail, as well as the Sportster, Dyna, and VRSC models, are assembled in York, Pennsylvania.

As a part of management's lean manufacturing effort, Harley groups production of parts that require similar processes together. The result is work cells. Using the latest technology, work cells perform in one location all the operations necessary for production of a specific module. Raw materials are moved to the work cells for module production. The modules then proceed

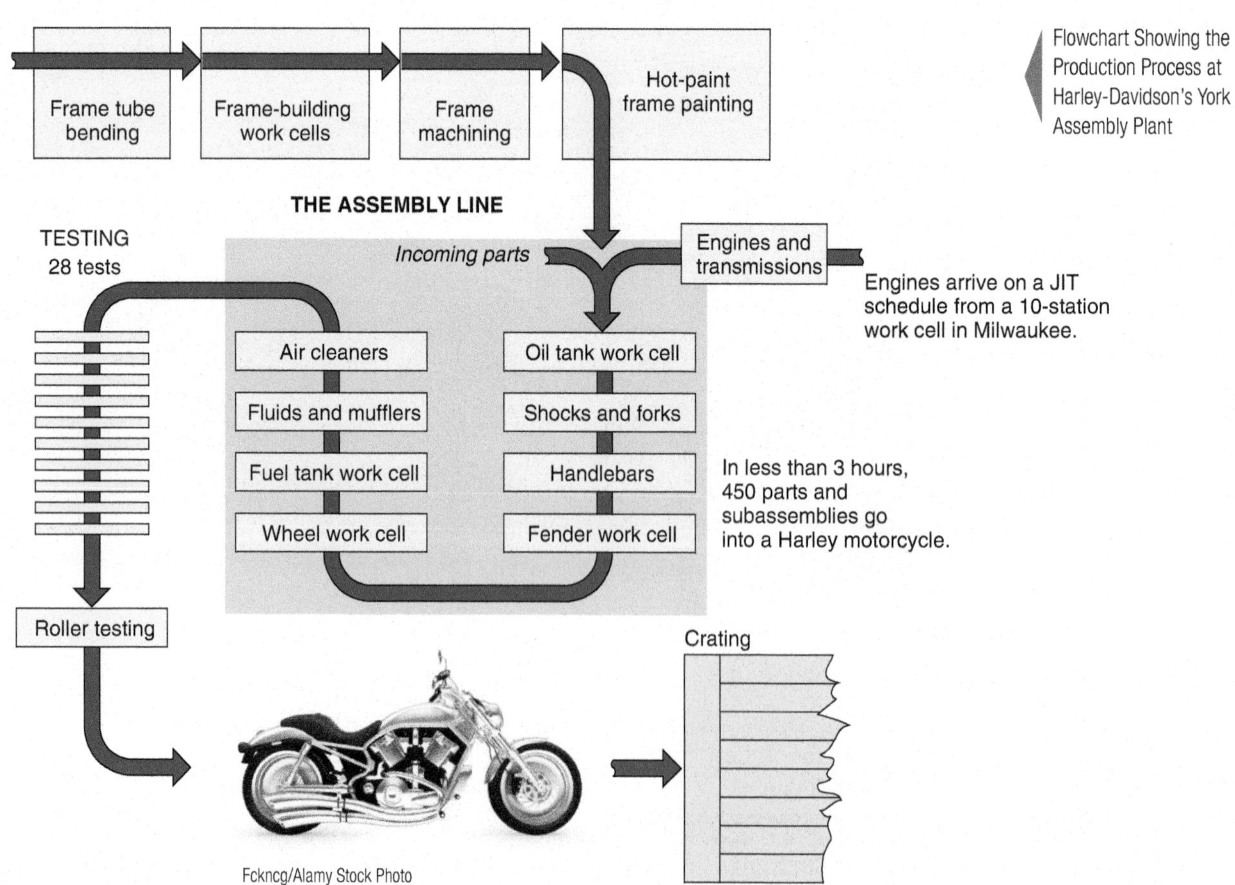

Fckncg/Alamy Stock Photo

Flowchart Showing the Production Process at Harley-Davidson's York Assembly Plant

Rick Friedman/Corbis Historical/Getty Images

Wheel assembly modules are prepared in a work cell for JIT delivery to the assembly line.

Rick Friedman/Corbis Historical/Getty Images

For manufacturers like Harley-Davidson, which produces a large number of end products from a relatively small number of options, modular bills of material provide an effective solution.

to the assembly line. As a double check on quality, Harley has also installed "light curtain" technology, which uses an infrared sensor to verify the bin from which an operator is taking parts. Materials go to the assembly line on a just-in-time basis, or as Harley calls it, using a Materials as Needed (MAN) system.

The 12.5-million-square-foot York facility includes manufacturing cells that perform tube bending, frame building, machining, painting, and polishing. Innovative manufacturing techniques use robots to load machines and highly automated production to reduce machining time. Automation and precision sensors play a key role in maintaining tolerances and producing a quality product. Each day the York facility produces up to 600 heavy-duty factory-custom motorcycles. Bikes are assembled with different engine displacements, multiple wheel options, colors, and accessories. The result is a huge number of variations in the motorcycles available, which allows customers to individualize their purchase. (See **www.Harley-Davidson.com** for an example of modular customization.) The Harley-Davidson production system works because high-quality modules are brought together on a tightly scheduled repetitive production line.

Nuccio DiNuzzo/KRT/Newscom

Engines are assembled in Menomonee Falls, Wisconsin, and placed in their own protective containers for shipment to the York facility. Upon arrival in York, engines are placed on an overhead conveyor for movement directly to the assembly line.

Rick Friedman/Corbis Historical/Getty Images

It all comes together on the line. Any employee who spots a problem has the authority to stop the line until the problem is corrected. The multicolored "andon" light above the line signals the severity of the problem.

LEARNING OBJECTIVES

LO 7.1 ***Describe*** four process strategies 278

LO 7.2 ***Compute*** crossover points for different processes 282

LO 7.3 ***Use*** the tools of process analysis 284

LO 7.4 ***Describe*** customer interaction in service processes 288

LO 7.5 ***Identify*** recent advances in production technology 290

Four Process Strategies

In Chapter 5, we examined the need for the selection, definition, and design of goods and services. Our purpose was to create environmentally friendly goods and services that could be delivered in an ethical, sustainable manner. We now turn to their production. A major decision for an operations manager is finding the best way to produce so as not to waste our planet's resources. Let's look at ways to help managers design a process for achieving this goal.

Process strategy
An organization's approach to transforming resources into goods and services.

A **process strategy** is an organization's approach to transforming resources into goods and services. *The objective is to create a process that can produce offerings that meet customer requirements within cost and other managerial constraints.* The process selected will have a long-term effect on efficiency and flexibility of production, as well as on cost and quality of the goods produced.

LO 7.1 *Describe* four process strategies

Virtually every good or service is made by using some variation of one of four process strategies: (1) process focus, (2) repetitive focus, (3) product focus, and (4) mass customization. The relationship of these four strategies to volume and variety is shown in Figure 7.1. We examine *Arnold Palmer Hospital* as an example of a process-focused firm, *Harley-Davidson* as a repetitive producer, *Frito-Lay* as a product-focused operation, and *Dell* as a mass customizer.

Process Focus

The vast majority of global production is devoted to making *low-volume, high-variety* products in places called "job shops." Such facilities are organized around specific activities or processes. In a factory, these processes might be departments devoted to welding, grinding, and painting. In an office, the processes might be accounts payable, sales, and payroll. In a restaurant, they might be bar, grill, and bakery. Such facilities are **process focused** in terms of equipment, layout, and supervision. They provide a high degree of product flexibility as products move between the specialized processes. Each process is designed to perform a variety of activities and handle frequent changes. Consequently, they are also called *intermittent processes*.

Process focus
A production facility organized around processes to facilitate low-volume, high-variety production.

Figure **7.1**

Process Selected Must Fit with Volume and Variety

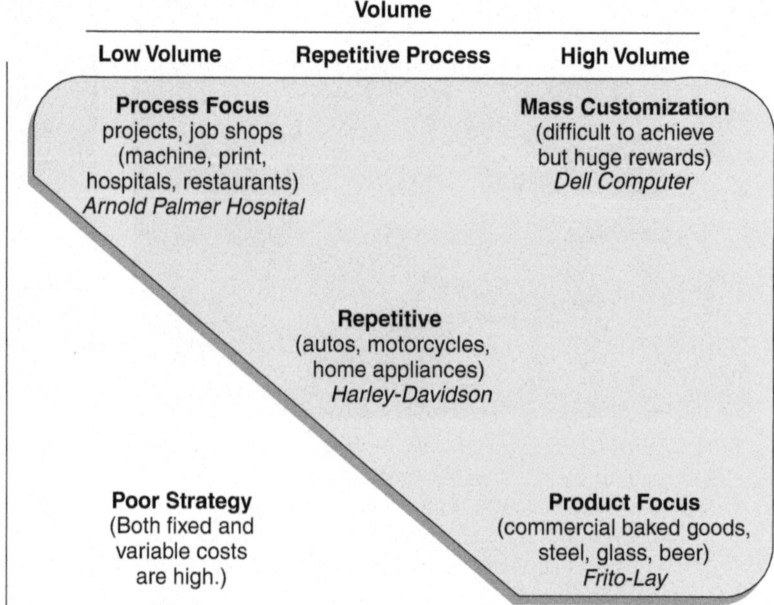

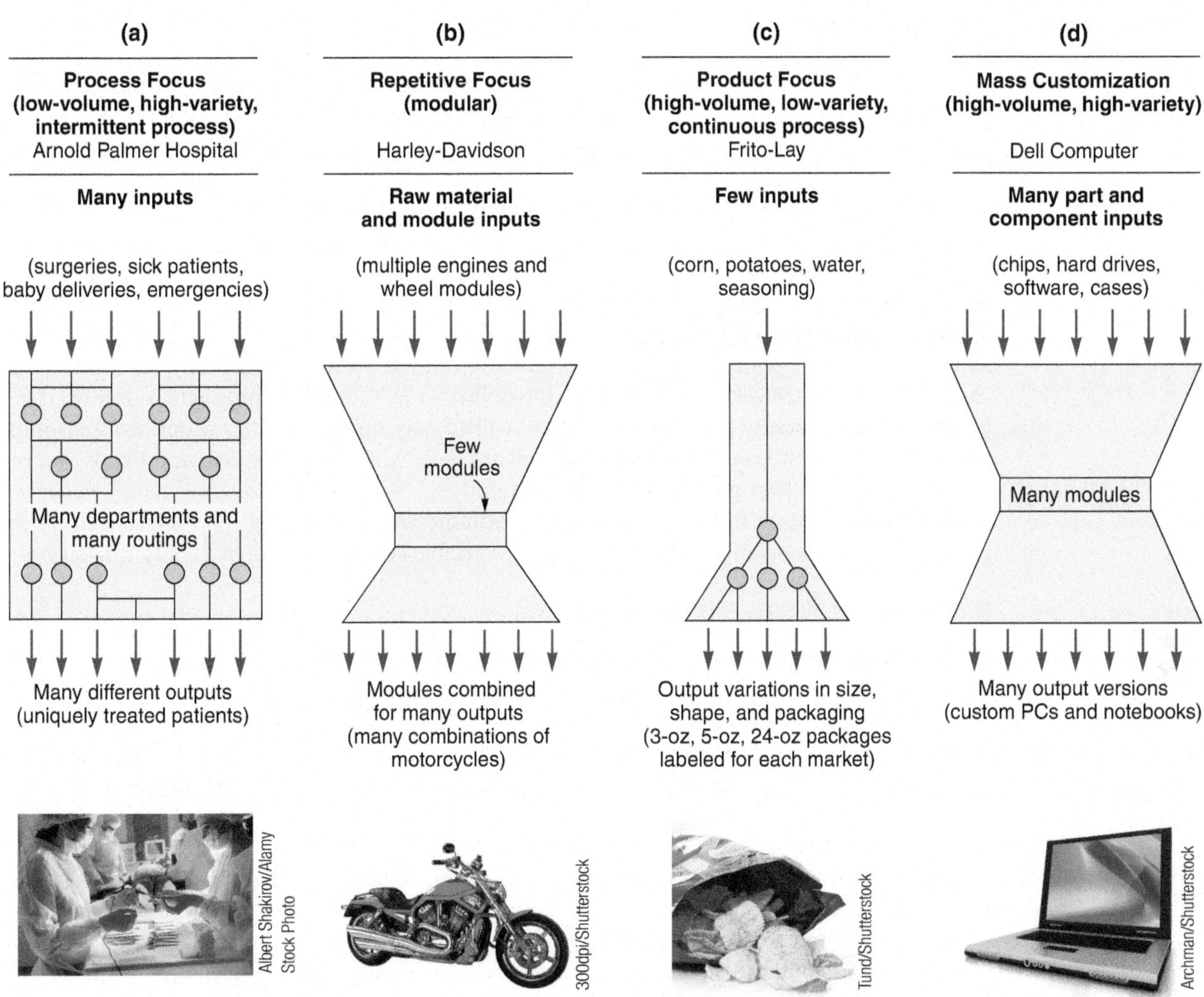

Figure **7.2**

Four Process Options with an Example of Each

Referring to Figure 7.2(a), imagine a diverse group of patients entering Arnold Palmer Hospital, a process-focused facility, to be routed to specialized departments, treated in a distinct way, and then exiting as uniquely cared-for individuals.

Process-focused facilities have high variable costs with extremely low utilization of facilities, as low as 5%. This is the case for many restaurants, hospitals, and machine shops. However, facilities that lend themselves to electronic controls can do somewhat better.

VIDEO 7.1
Process Strategy at Wheeled Coach Ambulance

Repetitive Focus

Repetitive processes, as we saw in the Global Company Profile on Harley-Davidson, use modules (see Figure 7.2b). **Modules** are parts or components previously prepared, often in a product-focused (continuous) process.

The **repetitive process** is the classic assembly line. Widely used in the assembly of virtually all automobiles and household appliances, it has more structure and consequently less flexibility than a process-focused facility.

Fast-food firms are another example of a repetitive process using modules. This type of production allows more customizing than a product-focused facility; modules (for example, meat, cheese, sauce, tomatoes, onions) are assembled to get a quasi-custom product, a cheeseburger. In this manner, the firm obtains both the economic advantages of the product-focused model (where many of the modules are prepared) and the custom advantage of the low-volume, high-variety model.

Modules
Parts or components of a product previously prepared, often in a continuous process.

Repetitive process
A product-oriented production process that uses modules.

Product Focus

Product focus
A facility organized around products; a product-oriented, high-volume, low-variety process.

High-volume, low-variety processes are **product focused**. Such facilities are organized around *products*. They are also called *continuous processes* because they have very long, continuous production runs. Products such as glass, paper, tin sheets, lightbulbs, beer, and potato chips are made via a continuous process. Some products, such as lightbulbs, are discrete; others, such as rolls of paper, are made in a continuous flow. Still others, such as repaired hernias at Canada's famous Shouldice Hospital, are services. It is only with standardization and effective quality control that firms have established product-focused facilities. An organization producing the same lightbulb or hot dog bun day after day can organize around a product. These organizations have an inherent ability to set standards and maintain a given quality, as opposed to an organization that is producing unique products every day, such as a print shop or general-purpose hospital. For example, Frito-Lay's family of products is produced in a product-focused facility [see Figure 7.2(c)]. At Frito-Lay, corn, potatoes, water, and seasoning are the relatively few inputs, but outputs (such as Cheetos, Ruffles, Tostitos, and Fritos) vary in seasoning and packaging within the product family.

A product-focused facility produces high volume and low variety. The specialized nature of the facility requires high fixed cost, but low variable costs reward high facility utilization.

Mass Customization Focus

Mass customization
Rapid, low-cost production that caters to constantly changing unique customer desires.

Consumers in our increasingly wealthy and sophisticated world request individualized goods and services. The result is an explosion of variety in both goods and services. Operations managers use *mass customization* to produce this vast array of goods and services. **Mass customization** is the rapid, low-cost production of goods and services that fulfill unique customer desires. But mass customization (see the upper-right section of Figure 7.1) is not just about variety; it is about making precisely *what* the customer wants *when* the customer wants it economically.

Mass customization brings us the variety of products traditionally provided by low-volume manufacture (a process focus) at the low cost of standardized high-volume (product-focused) production. However, achieving mass customization is a challenge that requires sophisticated operational capabilities. Building agile processes that rapidly and economically produce custom products requires that the link between sales, design, production, supply chain, and logistics be complete and rapid.

Dell Computer [see Figure 7.2(d)] has demonstrated that the payoff for mass customization can be substantial. More traditional manufacturers include Toyota, which recently announced delivery of custom-ordered cars in 5 days. Similarly, digitalization allows designers in the textile industry to rapidly revamp their lines and respond to changes.

The service industry is also moving toward mass customization. For instance, insurance companies are adding and tailoring new products with shortened development times to meet the unique needs of their customers. And firms such as Apple Music, Spotify, Napster, Amazon, and eMusic maintain a music inventory on the Internet that allows customers to select an almost unlimited number of songs of their choosing and have them made into a custom playlist. Mass customization places significant demands on operations managers who must create and align the processes that provide this expanding variety of goods and services.

Build-to-order (BTO)
Produce to customer order rather than to a forecast.

Making Mass Customization Work Mass customization suggests a high-volume system in which products are built-to-order. **Build-to-order (BTO)** means producing to customer orders, not forecasts. High-volume build-to-order is difficult, but resourceful managers address challenges in a number of ways, including:

Postponement
The delay of any modifications or customization to a product as long as possible in the production process.

- *Product design* must be imaginative, often with a limited product line and using product modules. Ping Inc., a premier golf club manufacturer, uses different combinations of club heads, grips, shafts, and angles to make 20,000 variations of its golf clubs.
- *Process design* must be flexible and able to accommodate changes in both design and technology. For instance, **postponement** allows for customization late in the production process.

Toyota installs unique interior modules very late in production for its popular Scion, a process also typical with customized vans. Postponement is further discussed in Chapter 11.

- *Inventory management* requires tight control. To be successful in this mass customization and build-to-order environment, a firm must avoid being stuck with unpopular or obsolete components. With very little parts inventory, Dell puts custom computers together in less than a day.
- *Digitized communication* tracks orders and material from design through delivery to help maintain tight schedules. Align Technology, a well-known name in orthodontics, figured out how to achieve competitive advantage by delivering custom-made clear plastic aligners within 3 weeks of the first visit to the dentist's office (see photo).
- *Responsive partners* in the supply chain must be effective collaborators. Suppliers of men's dress shirts in Hong Kong track sales, provide forecasts, manage inventory, and place orders for American retailers.

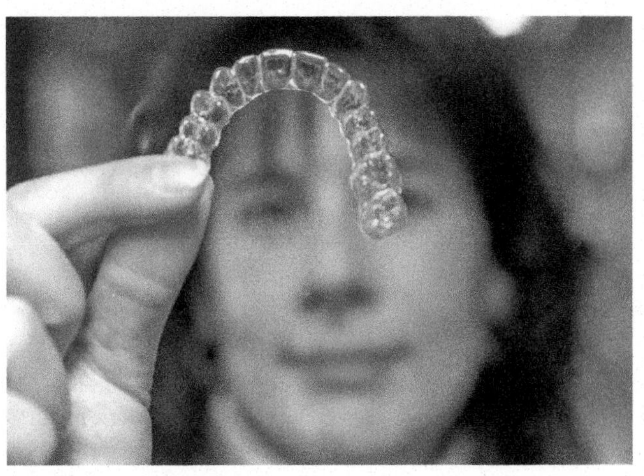

HughGrannum/KRT/Newscom

Align Technology of Santa Clara, California, is a mass customizer for orthodontic treatments, printing more than 433,000 individualized aligners a day. And each patient is *very* custom, from dental impressions to customized teeth aligners.

Mass customization/build-to-order is the new imperative for operations.

Process Comparison

The characteristics of the four processes are shown in Table 7.1 and Figure 7.2, and each may provide a strategic advantage. For instance, unit costs will be less in the product (continuous) or repetitive case when high volume (and high utilization) exists. However, a low-volume differentiated product is likely to be produced more economically under process focus. And mass customization requires exceptional competence in product and process design, scheduling, supply chain, and inventory management. Proper evaluation and selection of process strategies are critical.

TABLE 7.1 Comparison of the Characteristics of Four Types of Processes

PROCESS FOCUS (LOW VOLUME, HIGH VARIETY; e.g., ARNOLD PALMER HOSPITAL)	REPETITIVE FOCUS (MODULAR; e.g., HARLEY-DAVIDSON)	PRODUCT FOCUS (HIGH VOLUME, LOW VARIETY; e.g., FRITO-LAY)	MASS CUSTOMIZATION (HIGH VOLUME, HIGH VARIETY; e.g., DELL COMPUTER)
1. Small quantity and large variety of products	1. Long runs, a standardized product from modules	1. Large quantity and small variety of products	1. Large quantity and large variety of products
2. Broadly skilled operators	2. Moderately trained employees	2. Less broadly skilled operators	2. Flexible operators
3. Instructions for each job	3. Few changes in job instructions	3. Standardized job instructions	3. Custom orders requiring many job instructions
4. High inventory	4. Low inventory	4. Low inventory	4. Low inventory relative to the value of the product
5. Finished goods are made to order and not stored	5. Finished goods are made to frequent forecasts	5. Finished goods are made to a forecast and stored	5. Finished goods are build-to-order (BTO)
6. Scheduling is complex	6. Scheduling is routine	6. Scheduling is routine	6. Sophisticated scheduling accommodates custom orders
7. Fixed costs are low and variable costs high	7. Fixed costs are dependent on flexibility of the facility	7. Fixed costs are high, and variable costs low	7. Fixed costs tend to be high and variable costs low

Crossover Charts The comparison of processes can be further enhanced by looking at the point where the total cost of the processes changes. For instance, Figure 7.3 shows three alternative processes compared on a single chart. Such a chart is sometimes called a **crossover chart**. Process A has the lowest cost for volumes below V_1, process B has the lowest cost between V_1 and V_2, and process C has the lowest cost at volumes above V_2.

Crossover chart
A chart of costs at the possible volumes for more than one process.

Example 1 illustrates how to determine the exact volume where one process becomes more expensive than another.

Example 1

CROSSOVER CHART

Kleber Enterprises would like to evaluate three accounting software products (A, B, and C) to support changes in its internal accounting processes. The resulting processes will have cost structures similar to those shown in Figure 7.3. The costs of the software for these processes are:

	TOTAL FIXED COST	DOLLARS REQUIRED PER ACCOUNTING REPORT
Software A	$200,000	$60
Software B	$300,000	$25
Software C	$400,000	$10

LO 7.2 Compute crossover points for different processes

APPROACH ▶ Solve for the crossover point for software A and B and then the crossover point for software B and C.

SOLUTION ▶ Software A yields a process that is most economical up to V_1, but to exactly what number of reports (volume)? To determine the volume at V_1, we set the cost of software A equal to the cost of software B. V_1 is the unknown volume:

$$\begin{aligned} 200{,}000 + (60)V_1 &= 300{,}000 + (25)V_1 \\ 35V_1 &= 100{,}000 \\ V_1 &= 2{,}857 \end{aligned}$$

This means that software A is most economical from 0 reports to 2,857 reports (V_1).

Similarly, to determine the crossover point for V_2, we set the cost of software B equal to the cost of software C:

$$\begin{aligned} 300{,}000 + (25)V_2 &= 400{,}000 + (10)V_2 \\ 15V_2 &= 100{,}000 \\ V_2 &= 6{,}666 \end{aligned}$$

This means that software B is most economical if the number of reports is between 2,857 (V_1) and 6,666 (V_2) and that software C is most economical if reports exceed 6,666 (V_2).

INSIGHT ▶ As you can see, the software and related process chosen is highly dependent on the forecasted volume.

LEARNING EXERCISE ▶ If the vendor of software A reduces the fixed cost to $150,000, what is the new crossover point between A and B? [Answer: 4,286.]

RELATED PROBLEMS ▶ 7.1, 7.2, 7.3, 7.4, 7.5, 7.6, 7.7, 7.8, 7.9, 7.10, 7.11, 7.12

ACTIVE **MODEL 7.1** This example is further illustrated in Active Model 7.1 found online.

EXCEL **OM** Data File **Ch07Ex1.xls** can be found online.

Focused Processes In an ongoing quest for efficiency, industrialized societies continue to move toward specialization. The focus that comes with specialization contributes to efficiency. Managers who focus on a limited number of activities, products, and technologies do better. As the variety of products in a facility increases, overhead costs increase even faster. Similarly, as the variety of products, customers, and technology increases, so does complexity. The resources necessary to cope with the complexity expand disproportionately. A focus on depth of product line as opposed to breadth is typical of outstanding firms, of which Intel, L.M. Ericsson, and Bosch are world-class examples. *Focus*, defined here as specialization,

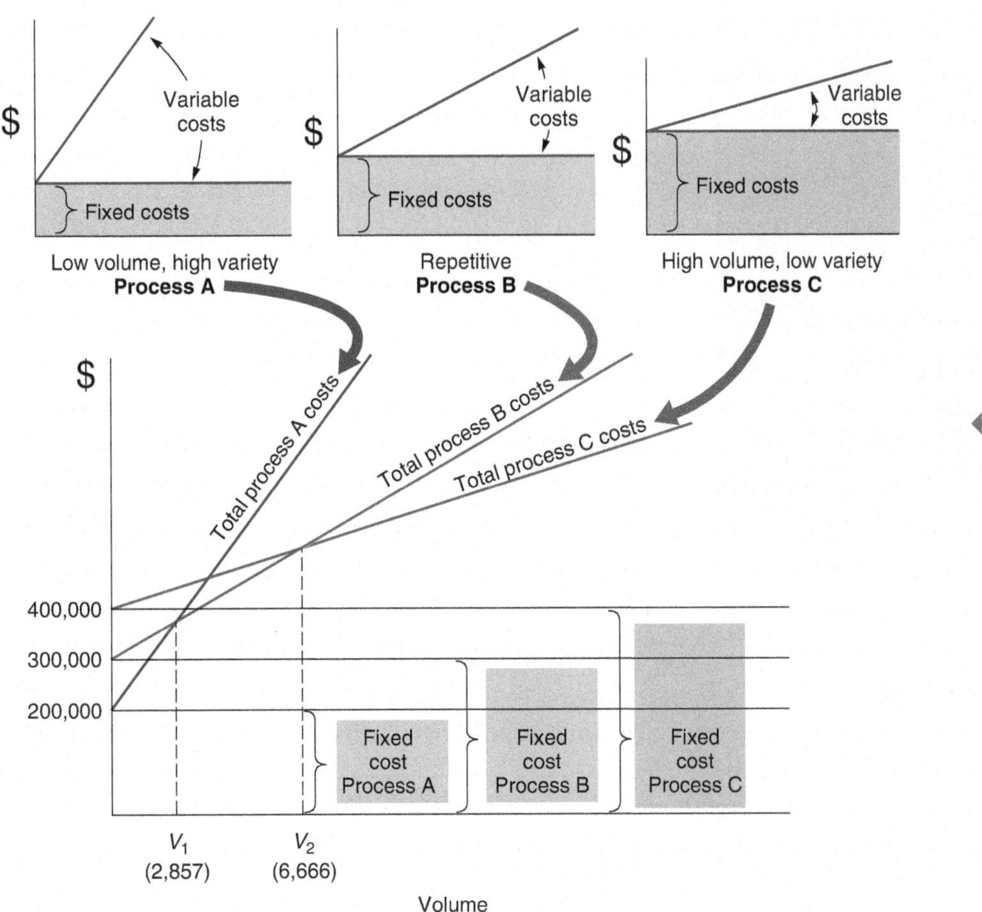

Figure 7.3
Crossover Charts

STUDENT TIP
Different processes can be expected to have different costs. However, at any given volume, only one will have the lowest cost.

simplification, and concentration, yields efficiency. Focus also contributes to building a core competence that fosters market and financial success. The focus can be:

- *Customers* (such as Winterhalter Gastronom, a German company that focuses on dishwashers for hotels and restaurants for whom spotless glasses and dishes are critical)
- *Products* with similar attributes (such as Nucor Steel's Crawford, Ohio, plant, which processes only high-quality sheet steels, and Gallagher, a New Zealand company, which has 45% of the world market in electric fences)
- *Service* (such as Orlando's Arnold Palmer Hospital, with a focus on children and women; and Shouldice Hospital, in Canada, with a focus on hernia repair)
- *Technology* (such as Texas Instruments, with a focus on only certain specialized kinds of semiconductors; and SAP, which despite a world of opportunities, remains focused on software)

The key for the operations manager is to move continuously toward specialization, focusing on the core competence necessary to excel at that specialty.

Selection of Equipment

STUDENT TIP
A process that is going to win orders often depends on the selection of the proper equipment.

Ultimately, selection of a particular process strategy requires decisions about equipment and technology. These decisions can be complex, as alternative methods of production are present in virtually all operations functions, from hospitals, to restaurants, to manufacturing facilities. Picking the best equipment requires understanding the specific industry and available processes and technology. The choice of equipment, be it an X-ray machine for a hospital, a computer-controlled lathe for a factory, or a new computer for an office, requires considering cost, cash flow, market stability, quality, capacity, and flexibility. To make this decision, operations managers develop documentation that indicates the capacity, size, tolerances, and maintenance requirements of each option.

Courtesy of Yum China

Yum Brands is testing self-serve kiosks at a concept store where customers choose menu items on a video screen. Facial-recognition software and an app automatically generate a charge.

In this age of rapid technological change and short product life cycles, adding flexibility to the production process can be a major competitive advantage. **Flexibility** is the ability to respond with little penalty in time, cost, or customer value. This may mean modular, movable, or digitally controlled equipment. Honda's process flexibility, for example, has allowed it to become the industry leader at responding to market dynamics by modifying production volume and product mix.

Building flexibility into a production process can be difficult and expensive, but if it is not present, change may mean starting over. Consider what would be required for a rather simple change—such as McDonald's adding the flexibility necessary to serve you a charbroiled hamburger. What appears to be rather straightforward would require changes in many of the 10 OM decisions. For instance, changes may be necessary in (1) purchasing (a different quality of meat, perhaps with more fat content, and supplies such as charcoal), (2) quality standards (how long and at what temperature the patty will cook), (3) equipment (the charbroiler), (4) layout (space for the new process and for new exhaust vents), (5) training, and (6) maintenance. You may want to consider the implications of another change, such as a change from paper menus and cash to choosing items on a video screen, then paying by facial-recognition. (See the Yum Brands photo.)

Flexibility
The ability to respond with little penalty in time, cost, or customer value.

Changing processes or equipment can be difficult and expensive. It is best to get this critical decision right the first time.

STUDENT TIP Here we look at five tools that help understand processes.

Process Analysis and Design

LO 7.3 *Use* the tools of process analysis

When analyzing and designing processes, we ask questions such as the following:

- Is the process designed to achieve competitive advantage in terms of differentiation, response, or low cost?
- Does the process eliminate steps that do not add value?
- Does the process maximize customer value as perceived by the customer?
- Will the process win orders?

Process analysis and design addresses not only these issues but also related OM issues such as throughput, cost, and quality. Process is key. Examine the process; then continuously improve the process.

VIDEO 7.2 Alaska Airlines: 20-minute Baggage Process—Guaranteed!

The following tools help us understand the complexities of process design and redesign. They are simply ways of making sense of what happens or must happen in a process. We now look at: flowcharts, time-function mapping, process charts, value-stream mapping, and service blueprinting.

Flowchart

Flowchart
A drawing used to analyze movement of people or material.

The first tool is the **flowchart**, which is a schematic or drawing of the movement of material, product, or people. For instance, the flowchart in the *Global Company Profile* for this chapter shows the assembly processes for Harley-Davidson. Such charts can help understanding, analysis, and communication of a process.

Time-Function Mapping

Time-function mapping (or process mapping)
A flowchart with time added on the horizontal axis.

A second tool for process analysis and design is a modified flowchart with time added on the horizontal axis. Such charts are sometimes called **time-function mapping**, or **process mapping**. With time-function mapping, nodes indicate the activities, and the arrows indicate the flow direction, with time on the horizontal axis. This type of analysis allows users to identify and eliminate waste such as extra steps, duplication, and delay. Figure 7.4 shows the use of process mapping before and after process improvement at American National Can Company. In this example, substantial reduction in waiting time and process improvement in order processing contributed to a savings of 46 days.

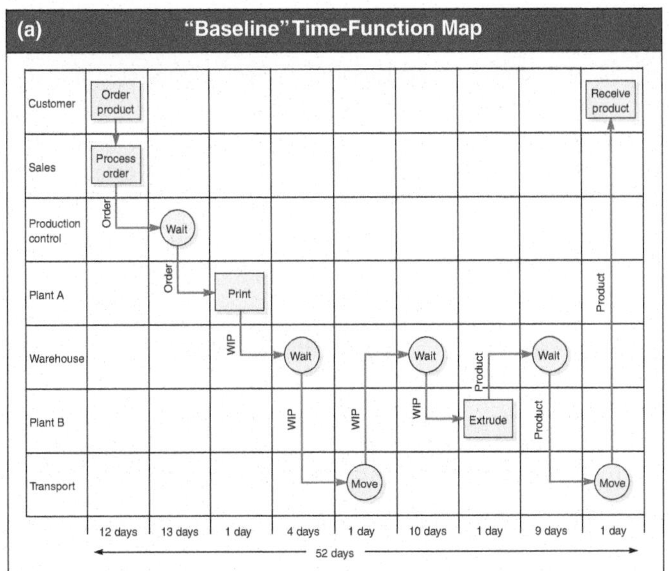

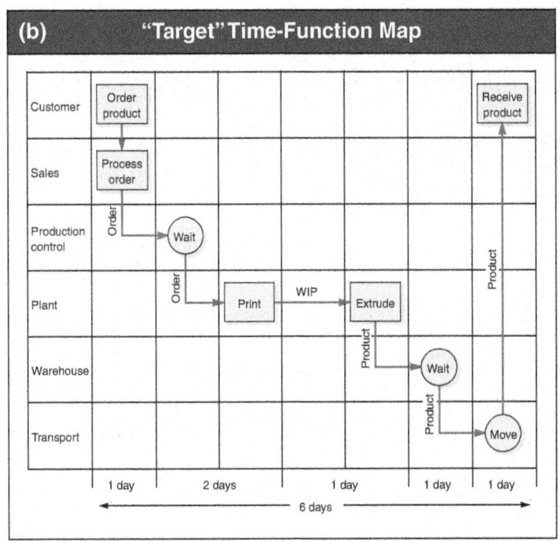

Figure 7.4

Time-Function Mapping (Process Mapping) for a Product Requiring Printing and Extruding Operations at American National Can Company[1]

Process Charts

The third tool is the *process chart*. **Process charts** use symbols, time, and distance to provide an objective and structured way to analyze and record the activities that make up a process.[2] They allow us to focus on value-added activities. For instance, the process chart shown in Figure 7.5, which includes the present method of hamburger assembly at a fast-food restaurant, includes a value-added line to help us distinguish between value-added activities and waste. Identifying all value-added operations (as opposed to inspection, storage, delay, and transportation, which add no value) allows us to determine the percent of value added to total activities.[3] We can see from the computation at the bottom of Figure 7.5 that the percentage of value added in this case is 85.7%. The operations manager's job is to reduce waste and increase the percent of value added. The non-value-added items are a waste; they are resources lost to the firm and to society forever.

Process charts
Charts that use symbols to analyze the movement of people or material.

VIDEO 7.3
Process Analysis at Arnold Palmer Hospital

Present Method [X] PROCESS CHART Proposed Method [ ]

SUBJECT CHARTED *Hamburger Assembly Process* DATE *12 / 1 / 21*

DEPARTMENT ______ CHART BY *KH* SHEET NO. *1* OF *1*

DIST. IN FEET	TIME IN MINS.	CHART SYMBOLS	PROCESS DESCRIPTION
	—	○ ⇨ □ D ▽	*Meat Patty in Storage*
1.5	*.05*	○ ⇨ □ D ▽	*Transfer to Broiler*
	2.50	○ ⇨ □ D ▽	*Broiler*
	.05	○ ⇨ □ D ▽	*Visual Inspection*
1.0	*.05*	○ ⇨ □ D ▽	*Transfer to Rack*
	.15	○ ⇨ □ D ▽	*Temporary Storage*
.5	*.10*	○ ⇨ □ D ▽	*Obtain Buns, Lettuce, etc.*
	.20	○ ⇨ □ D ▽	*Assemble Order*
.5	*.05*	○ ⇨ □ D ▽	*Place in Finish Rack*
		○ ⇨ □ D ▽	
3.5	*3.15*	*2 4 1 – 2*	TOTALS

Value-added time = Operation time/Total time = (2.50+.20)/3.15 = 85.7%

○ = operation; ⇨ = transport; □ = inspect; D = delay; ▽ = storage.

Figure 7.5

Process Chart Showing a Hamburger Assembly Process at a Fast-Food Restaurant

Value-Stream Mapping

Value-stream mapping (VSM)
A process that helps managers understand how to add value in the flow of material and information through the entire production process.

A variation of time-function mapping is **value-stream mapping (VSM)**; however, value-stream mapping takes an expanded look at where value is added (and not added) in the entire production process, including the supply chain. As with time-function mapping, the idea is to start with the customer and understand the production process, but value-stream mapping extends the analysis back to suppliers.

Value-stream mapping takes into account not only the process but also, as shown in Example 2, the management decisions and information systems that support the process.

Example 2

VALUE-STREAM MAPPING

GPS Tech, Inc. has received an order for 11,000 GPS modules per month and wants to understand how the order will be processed through manufacturing.

APPROACH ▶ To fully understand the process from customer to supplier, GPS Tech prepares a value-stream map.

SOLUTION ▶ Although value-stream maps appear complex, their construction is easy. Here are the steps needed to complete the value-stream map shown in Figure 7.6.

1. Begin with symbols for customer, supplier, and production to ensure the big picture.
2. Enter customer order requirements.
3. Calculate the daily production requirements.
4. Enter the outbound shipping requirements and delivery frequency.
5. Determine inbound shipping method and delivery frequency.
6. Add the process steps (i.e., machine, assemble) in sequence, left to right.
7. Add communication methods, add their frequency, and show the direction with arrows.
8. Add inventory quantities (shown with △I) between every step of the entire flow.
9. Determine total working time (value-added time) and delay (non-value-added time).

Figure 7.6
Value-Stream Mapping (VSM)

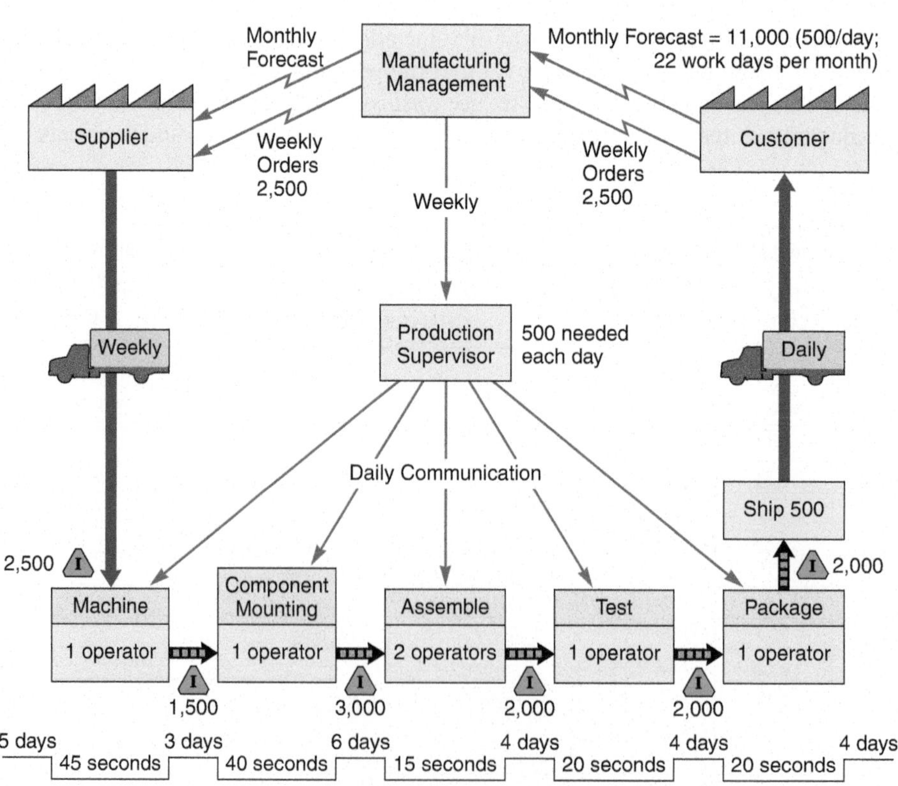

INSIGHT ▶ From Figure 7.6 we note that large inventories exist in incoming raw material and between processing steps, and that the value-added time is low as a proportion of the entire process.

LEARNING EXERCISE ▶ How might raw material inventory be reduced? [Answer: Have deliveries twice per week rather than once per week.]

RELATED PROBLEM ▶ 7.17

Service Blueprinting

Products with a high service content may warrant use of yet a fifth process technique. **Service blueprinting** is a process analysis technique that focuses on the customer and the provider's interaction with the customer. For instance, the activities at level one of Figure 7.7 are under the control of the customer. In the second level are activities of the service provider interacting with

Service blueprinting
A process analysis technique that lends itself to a focus on the customer and the provider's interaction with the customer.

Figure **7.7**

Service Blueprint for Service at Speedy Lube, Inc.

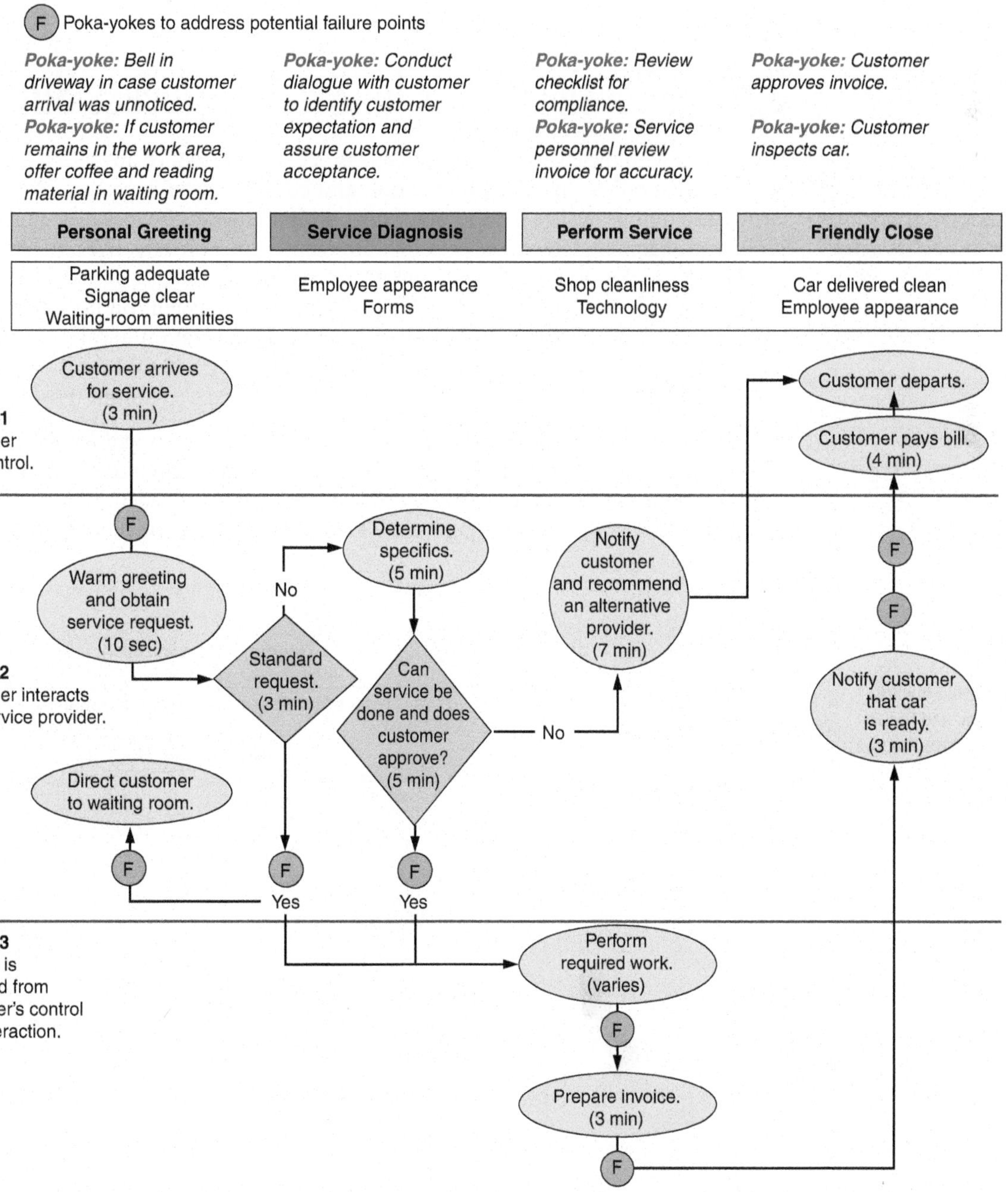

(F) Poka-yokes to address potential failure points

Personal Greeting	Service Diagnosis	Perform Service	Friendly Close
Poka-yoke: Bell in driveway in case customer arrival was unnoticed. *Poka-yoke: If customer remains in the work area, offer coffee and reading material in waiting room.*	*Poka-yoke: Conduct dialogue with customer to identify customer expectation and assure customer acceptance.*	*Poka-yoke: Review checklist for compliance.* *Poka-yoke: Service personnel review invoice for accuracy.*	*Poka-yoke: Customer approves invoice.* *Poka-yoke: Customer inspects car.*
Physical Attributes to Support Service: Parking adequate, Signage clear, Waiting-room amenities	Employee appearance, Forms	Shop cleanliness, Technology	Car delivered clean, Employee appearance

Level #1
Customer is in control.

Customer arrives for service. (3 min)
Customer pays bill. (4 min)
Customer departs.

Level #2
Customer interacts with service provider.

Warm greeting and obtain service request. (10 sec)
Standard request. (3 min) — No → Determine specifics. (5 min); Yes
Can service be done and does customer approve? (5 min) — No → Notify customer and recommend an alternative provider. (7 min); Yes
Direct customer to waiting room.
Notify customer that car is ready. (3 min)

Level #3
Service is removed from customer's control and interaction.

Perform required work. (varies)
Prepare invoice. (3 min)

STUDENT TIP Service blueprinting helps evaluate the impact of customer interaction with the process.

the customer. The third level includes those activities that are performed away from, and not immediately visible to, the customer. Each level suggests different management issues. For instance, the top level may suggest educating the customer or modifying expectations, whereas the second level may require a focus on personnel selection and training. Finally, the third level lends itself to more typical process innovations. The service blueprint shown in Figure 7.7 also notes potential failure points and shows how poka-yoke techniques can be added to improve quality. The consequences of these failure points can be greatly reduced if identified at the design stage when modifications or appropriate poka-yokes can be included. A time dimension is included in Figure 7.7 to aid understanding, extend insight, and provide a focus on customer service.

Each of these five process analysis tools has strengths and variations. Flowcharts provide a quick way to view the big picture and try to make sense of the entire system. Time-function mapping adds some rigor and a time element to the macro analysis. Value-stream mapping extends beyond the immediate organization to customers and suppliers. Process charts are designed to provide a much more detailed view of the process, adding items such as value-added time, delay, distance, storage, and so forth. Service blueprinting, on the other hand, is designed to help us focus on the customer interaction part of the process. Because customer interaction is often an important variable in process design, we now examine some additional aspects of service process design.

Special Considerations for Service Process Strategies

STUDENT TIP Customer interaction within service processes increases the design challenge.

Interaction with the customer often affects process performance adversely. But a service, by its very nature, implies that some interaction and customization is needed. Recognizing that the customer's unique desires tend to play havoc with a process, the more the manager designs the process to accommodate these special requirements, the more effective and efficient the process will be. The trick is to find the right combination.

The four quadrants of Figure 7.8 provide additional insight on how operations managers modify service processes to find the best level of specialization and focus while maintaining the necessary customer interaction and customization. The 10 OM decisions we introduced in Chapters 1 and 2 are used with a different emphasis in each quadrant. For instance:

LO 7.4 *Describe* customer interaction in service processes

- In the upper sections (quadrants) of *mass service* and *professional service*, where *labor content is high*, we expect the manager to focus extensively on human resources. This is often done with personalized services, requiring high labor involvement and therefore significant personnel selection and training issues. This is particularly true in the professional service quadrant.

Figure **7.8**

Services Moving toward Specialization and Focus within the Service Process Matrix[4]

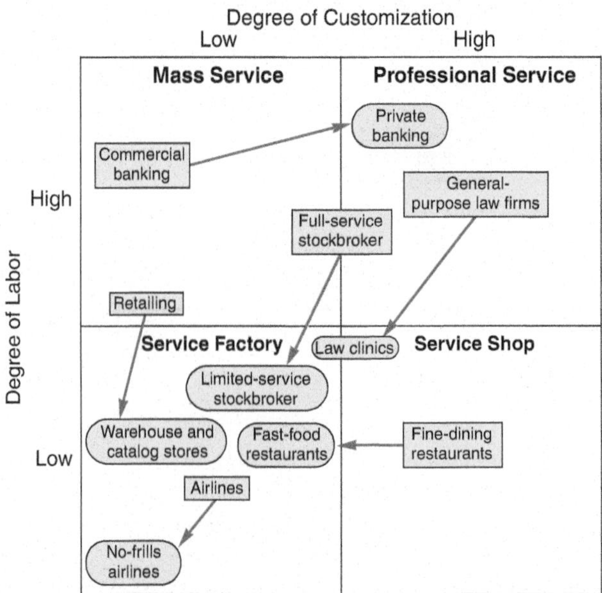

STUDENT TIP Notice how services find a competitive opportunity by moving from the rectangles to the ovals.

TABLE 7.2 Techniques for Improving Service Productivity

STRATEGY	TECHNIQUE	EXAMPLE
Separation	*Structuring service* so customers must go where the service is offered	Bank customers go to a manager to open a new account, to loan officers for loans, and to tellers for deposits.
Self-service	*Self-service* so customers examine, compare, and check out at their own pace	Supermarkets and department stores. Internet ordering.
Postponement	*Customizing at delivery*	Customizing vans at delivery rather than at production.
Focus	*Restricting* the offerings	Limited-menu restaurant.
Modules	*Modular* selection of service *Modular* production	Investment and insurance selection. Prepackaged food modules in restaurants.
Automation	*Separating services* that may lend themselves to some type of automation	Automated teller machines. Ordering via an app.
Scheduling	Precise personnel *scheduling*	Scheduling airline ticket-counter personnel at 15-minute intervals.
Training	*Clarifying* the service options *Explaining* how to avoid problems	Investment counselor, funeral directors. After-sale maintenance personnel.

- The quadrants with *low customization* tend to (1) standardize or restrict some offerings, as do fast-food restaurants, (2) automate, as have airlines with ticket-vending machines, or (3) remove some services, such as seat assignments, as has Southwest Airlines. Offloading some aspect of the service through automation may require innovations in process design. Such is the case with airline ticket vending, self-checkout at Home Depot, and bank ATMs. This move to standardization and automation may also require changes in other areas, such as added capital expenditure and new OM skills for the purchase and maintenance of equipment. A reduction in a customization capability will require added strength in other areas.
- Because customer feedback is lower in the quadrants with *low customization*, tight control may be required to maintain quality standards.
- Operations with *low labor intensity* may lend themselves particularly well to innovations in process technology and scheduling.

Table 7.2 shows some additional techniques for innovative process design in services. Managers focus on designing innovative processes that enhance the service. For instance, supermarket *self-service* reduces cost while it allows customers to check for the specific features they want, such as freshness or color. Dell Computer provides another version of self-service by allowing customers to design their own product on the Web. Customers seem to like this, and it is cheaper and faster for Dell.

Production Technology

The future of OM is increasingly tied to the collection of data from critical points in the production system for instantaneous processing. Technology is shrinking the cost of digital sensors, coding, and computing. Sensors are being embedded in machinery and processes of every conceivable type. Huge streams of data are captured, transmitted, cleaned, and interpreted to improve decision making. Digitalization is superseding traditional human data collection. These technological advances that enhance production and productivity are changing how things are designed, made, and serviced around the world.

In this section, we introduce nine areas of technology: (1) machine technology, (2) automatic identification systems (AIS), (3) process control, (4) vision systems, (5) robots, (6) automated storage and retrieval systems (ASRSs), (7) automated guided vehicles (AGVs), (8) flexible manufacturing systems (FMSs), and (9) computer-integrated manufacturing (CIM). Consider the impact on operations managers as we digitally link these technologies within the firm. Then consider the implications when they are combined and linked

STUDENT TIP
Here are nine technologies that can improve employee safety, product quality, and productivity.

globally in a seamless chain that can immediately respond to changing consumer demands, supplier dynamics, and producer innovations. The implications for the world economy and OM are huge.

LO 7.5 *Identify* recent advances in production technology

Machine Technology

Much of the world's machinery performs operations by *removing material,* performing operations such as cutting, drilling, boring, and milling. This technology is undergoing tremendous progress in both precision and control. Machinery now turns out metal components that vary less than a micron—1/76 the width of a human hair. They can accelerate water to three times the speed of sound to cut titanium for surgical tools. Such machinery is often five times more productive than that of previous generations while being smaller and using less power. And continuing advances in lubricants now allow the use of water-based lubricants rather than oil-based. Water-based lubricants enhance sustainability by eliminating hazardous waste and allowing shavings to be easily recovered and recycled.

Computer intelligence often controls this new machinery, allowing more complex and precise items to be made faster. Such machinery, with its own computer and memory, is referred to as having **computer numerical controls (CNC)**. Electronic controls increase throughput by cutting changeover time, reducing waste (because of fewer mistakes), and enhancing flexibility.

Computer numerical control (CNC)
Machinery with its own computer and memory.

Advanced versions of such technology are used on Pratt & Whitney's turbine blade plant in Connecticut. The machinery has improved the loading and alignment task so much that Pratt has cut the total time for the grinding process of a turbine blade from 10 days to 2 hours. The new machinery has also contributed to process improvements that mean the blades now travel just 1,800 feet in the plant, down from 8,100 feet, cutting throughput time from 22 days to 7 days.

Additive manufacturing
The production of physical items by adding layer upon layer, much in the same way an inkjet printer lays down ink.

New advances in machinery suggest that rather than *removing* material as has traditionally been done, *adding* material may in many cases be more efficient. **Additive manufacturing** or, as it is often called, *3D printing*, is frequently used for design testing, prototypes, and custom products. The technology continues to advance and now supports innovative product design (variety and complexity), minimal custom tooling (little tooling is needed), minimal assembly (integrated assemblies can be "printed"), low inventory (make-to-order systems), and reduced time to market. As a result, additive manufacturing is being increasingly used to enhance production efficiency for high-volume products. The convergence of software advances, digitalization, worldwide communication, and additive manufacturing is bringing enormous changes to operations.

With RFID, a cashier could scan the entire contents of a shopping cart in seconds.

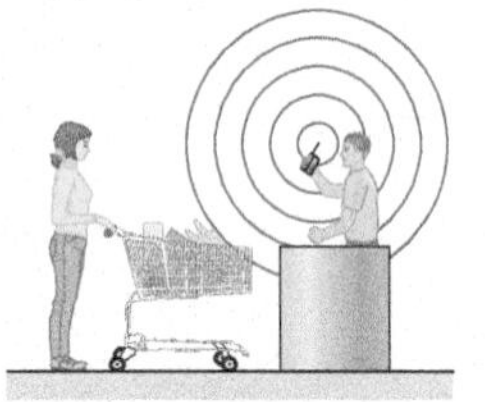

Automatic Identification Systems (AISs) and RFID

New equipment, from numerically controlled manufacturing machinery to ATMs, is controlled by digital electronic signals. Electrons are a great vehicle for transmitting information, but they have a major limitation—most OM data does not start out in bits and bytes. Therefore, operations managers must get the data into an electronic form. Making data digital is done via computer keyboards, bar and QR codes, radio frequencies, optical characters, and so forth. These **automatic identification systems (AISs)** help us move data into electronic form, where it is easily manipulated.

Automatic identification system (AIS)
A system for transforming data into electronic form, for example, bar codes.

Because of its decreasing cost and increasing pervasiveness, **radio frequency identification (RFID)** warrants special note. RFID is integrated circuitry with its own tiny antennas that use radio waves to send signals a limited range—usually a matter of yards. These RFID tags provide unique identification that enables the tracking and monitoring of parts, pallets, people, and pets—virtually everything that moves. RFID requires no line of sight between tag and reader.

Radio frequency identification (RFID)
A wireless system in which integrated circuits with antennas send radio waves.

Process Control

Process control
The use of information technology to control a physical process.

Process control is the use of information technology to monitor and control a physical process. For instance, process control is used to measure the moisture content and thickness of paper as it travels over a paper machine at thousands of feet per minute. Note the process technology at work in the Donawitz rolling mill seen in the opening photo of this chapter. Process control is also used to determine and control temperatures, pressures, and quantities in petroleum refineries, petrochemical processes, cement plants, nuclear reactors, and other product-focused facilities.

Process control systems operate in a number of ways, but the following are typical:

- Sensors collect data, which is read on some periodic basis, perhaps once a minute or second.
- Measurements are translated into digital signals, which are transmitted to a computer.
- Computer programs read the file and analyze the data.
- The resulting output may take numerous forms. These include warning lights or horns, messages on computer consoles, printouts, and statistical process control charts. Increasingly, such outputs are sending signals directly to other parts of the process to help control them.

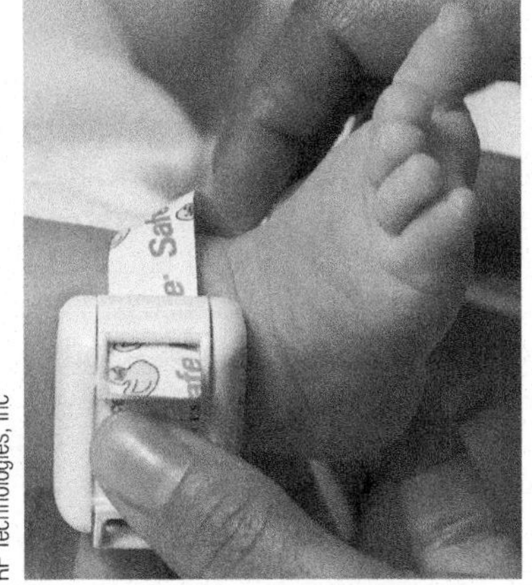
RF Technologies, Inc

The Safe Place® Infant Security Solution from RF Technologies® monitors infants with small, lightweight transmitters and soft, comfortable banding. When a protected infant approaches a monitored exit, the transmitter triggers the exit's lock and notifies staff to ensure a fast response.

Vision Systems

Vision systems combine video cameras and computer technology and are often used in inspection roles. Visual inspection is an important task in most food-processing and manufacturing organizations. Moreover, in many applications, visual inspection performed by humans is tedious, mind-numbing, and error prone. Thus, vision systems are widely used when the items being inspected are very similar. For instance, vision systems are used to inspect Frito-Lay's potato chips so that imperfections can be identified as the chips proceed down the production line. BMW also uses vision systems to ensure that auto chassis bolts are properly secured. And, as we see in the *OM in Action* box, "Tyson's Computer Vision Technology Improves Process Performance," Tyson's computer vision systems track chickens packed into trays en route to supermarkets.

Vision systems
Systems that use video cameras and computer technology in inspection roles.

Vision systems are consistently accurate, do not become bored, and are of modest cost. These systems are vastly superior to individuals trying to perform these tasks.

OM in Action — Tyson's Computer Vision Technology Improves Process Performance[5]

Tyson is rolling out a computer-vision-enabled tracking system at facilities where it packs chicken into trays for grocery stores. The system can read SKU information and weight, replacing what Tyson described as communication by hand gestures followed by manual entry. Tyson's automated tracking technology combines computer vision, machine learning, and edge computing to expand its speed and processing capability.

Automated tracking using computer vision led to a double-digit increase in inventory accuracy in the three facilities currently using the technology. The company plans to expand the program to all 10 of its poultry plants.

Though cold, wet storage environments make implementing new electronic technologies difficult, the payoff of real-time accurate information is already evident for Tyson. The company recently opened the Tyson Manufacturing Automation Center, where it works with manufacturers and suppliers to develop new technologies and trains employees to use it. The company has spent $215 million on new technologies in the last 5 years.

Precise, real-time visibility increases the frequency of Tyson's on-time orders in the best of times. But process performance is particularly key in times of uncertainty, and Tyson has been dealing with plenty of uncertainty. The disruptive forces of the pandemic on its workforce, of shifting global trade policies, and of a fire at an important Tyson facility all distorted usual supply and demand patterns, making a relevant forecast next to impossible at various times.

Anton Mislawsky/Shutterstock

Indeed, the entire meat industry is moving toward similar monitoring technologies and automation that goes beyond production processing. For instance, Cargill is starting to use computer vision to track animal health in dairy operations. But a more process-directed application inspired Tyson's work: the technology used in Amazon's *cashier-less* stores.

Robots

Robot
A flexible machine with the ability to hold, move, or grab items. It functions through electronic impulses that activate motors and switches.

When a machine is flexible and has the ability to hold, move, and perhaps "grab" items, we tend to use the word *robot*. **Robots** are mechanical devices that use electronic impulses to activate motors and switches. Robots may be used effectively to perform tasks that are especially monotonous or dangerous or those that can be improved by the substitution of mechanical for human effort. Such is the case when consistency, accuracy, speed, strength, or power can be enhanced by the substitution of machines for people. The automobile industry, for example, uses robots to do virtually all the welding and painting on automobiles. And a new, more sophisticated, generation of robots is fitted with sensors and cameras that provide enough dexterity to assemble, test, and pack (or in the case of Apple's iPhone, disassemble) small parts.

Automated Storage and Retrieval Systems (ASRSs)

Automated storage and retrieval system (ASRS)
Computer-controlled warehouses that provide for the automatic placement of parts into and from designated places in a warehouse.

Because of the tremendous labor involved in error-prone warehousing, computer-controlled warehouses have been developed. These systems, known as **automated storage and retrieval systems (ASRSs)**, provide for the automatic placement and withdrawal of parts and products into and from designated places in a warehouse. Such systems are commonly used in distribution facilities of retailers such as Walmart, Tupperware, and Benetton. These systems are also found in inventory and test areas of manufacturing firms.

Automated Guided Vehicles (AGVs)

Automated guided vehicle (AGV)
Electronically guided and controlled cart used to move materials.

Automated material handling can take the form of monorails, conveyors, robots, or automated guided vehicles. **Automated guided vehicles (AGVs)** are electronically guided and controlled carts used in manufacturing and warehousing to move parts and equipment. They are also used in agriculture to distribute feed, in offices to move mail, and in hospitals and jails to deliver supplies and meals.

Flexible Manufacturing Systems (FMSs)

Flexible manufacturing system (FMS)
A system that uses electronic signals from a centralized computer to automate production and material flow.

When a central computer provides instructions to each workstation *and* to the material-handling equipment such as robots, ASRSs, and AGVs (as just noted), the system is known as an automated work cell or, more commonly, a **flexible manufacturing system (FMS)**. An FMS is flexible because both the material-handling devices and the machines themselves are controlled by easily changed electronic signals (computer programs). Operators simply load new programs, as necessary, to produce different products. The result is a system that can economically produce low volume but high variety. For example, the Lockheed Martin facility, near Dallas, efficiently builds one-of-a-kind spare parts for military aircraft. The costs associated with changeover and low utilization have been reduced substantially. FMSs bridge the gap between product-focused and process-focused facilities.

Computer-Integrated Manufacturing (CIM)

Computer-integrated manufacturing (CIM)
A manufacturing system in which CAD, FMS, inventory control, warehousing, and shipping are integrated.

Flexible manufacturing systems can be extended backward electronically into the engineering and inventory control departments and forward to the warehousing and shipping departments. In this way, computer-aided design (CAD) generates the necessary electronic instructions to run a numerically controlled machine. In a computer-integrated manufacturing environment, a design change initiated at a CAD terminal can result in that change being made in the part produced on the shop floor in a matter of minutes. When this capability is integrated with inventory control, warehousing, and shipping as a part of a flexible manufacturing system, the entire system is called **computer-integrated manufacturing (CIM)** (Figure 7.9).

Flexible manufacturing systems and computer-integrated manufacturing are reducing the distinction between low-volume/high-variety and high-volume/low-variety production. Information technology is allowing FMS and CIM to handle increasing variety while expanding to include a growing range of volumes.

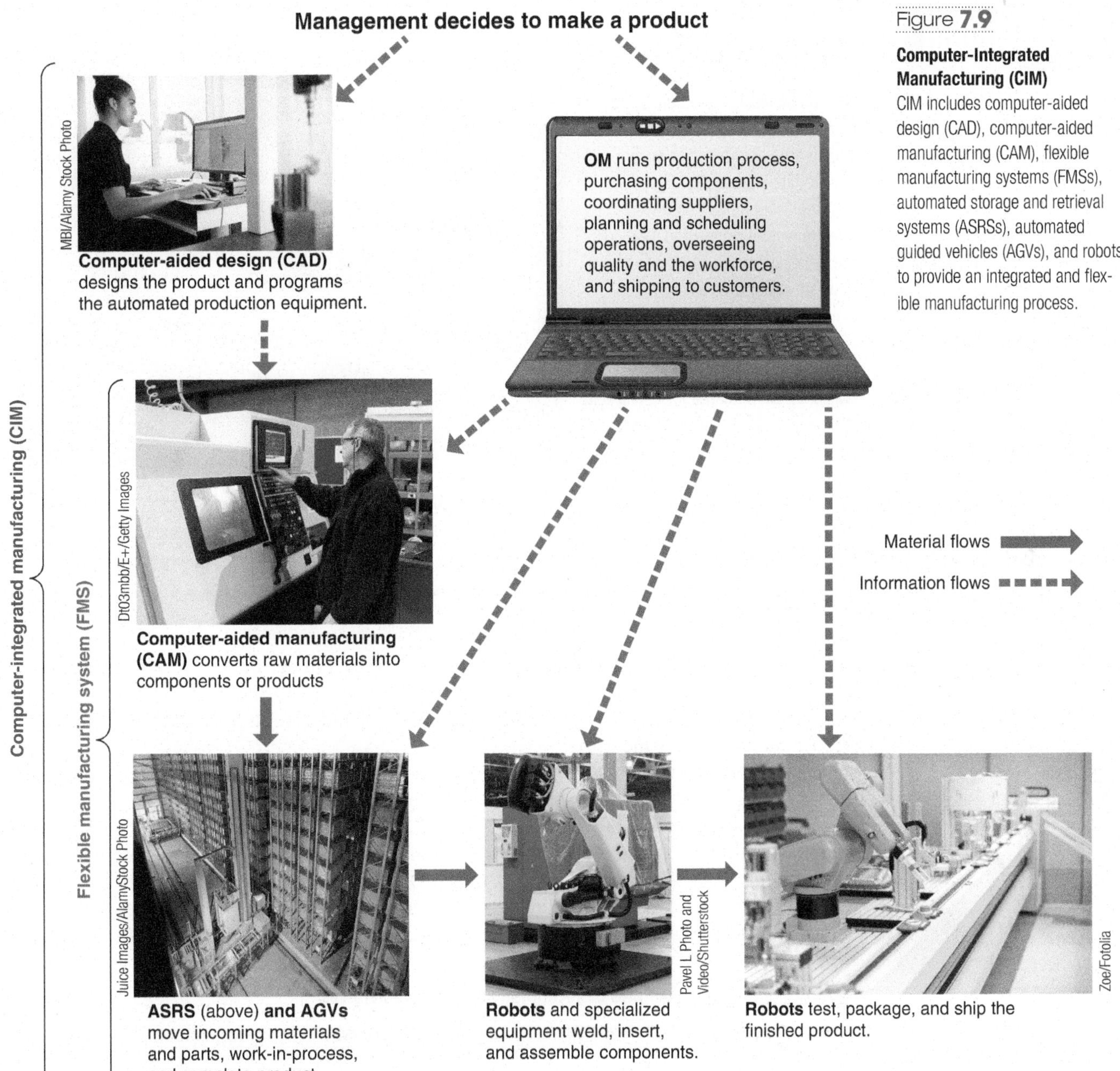
Management decides to make a product
MBI/Alamy Stock Photo
Computer-aided design (CAD) designs the product and programs the automated production equipment.
OM runs production process, purchasing components, coordinating suppliers, planning and scheduling operations, overseeing quality and the workforce, and shipping to customers.
Computer-integrated manufacturing (CIM)
Flexible manufacturing system (FMS)
Dt03mbb/E+/Getty Images
Computer-aided manufacturing (CAM) converts raw materials into components or products
Material flows
Information flows
Juice Images/AlamyStock Photo
ASRS (above) and AGVs move incoming materials and parts, work-in-process, and complete product.
Pavel L Photo and Video/Shutterstock
Robots and specialized equipment weld, insert, and assemble components.
Zoe/Fotolia
Robots test, package, and ship the finished product.

Figure **7.9**

Computer-Integrated Manufacturing (CIM)

CIM includes computer-aided design (CAD), computer-aided manufacturing (CAM), flexible manufacturing systems (FMSs), automated storage and retrieval systems (ASRSs), automated guided vehicles (AGVs), and robots to provide an integrated and flexible manufacturing process.

Technology in Services

Just as we have seen rapid advances in technology in the manufacturing sector, so we also find dramatic changes in the service sector. These range from electronic diagnostic equipment at auto repair shops, to blood- and urine-testing equipment in hospitals, to retinal security scanners at airports, to the Genki Sushi restaurant chain's automated service (see the *OM in Action* box, "The Automated Sushi Restaurant)." The fast-food industry has introduced kiosks and cashless payments. The labor savings when ordering and speedier checkout service provide valuable productivity increases for both the restaurant and the customer.

In the retail sector, self-service and point-of-sale (POS) technology immediately reflect changing costs, market conditions, and sales prices, tracking sales in 15-minute segments to aid personnel scheduling. With tools such as Google Analytics and custom algorithms, the retail sector uses market research, store sales, returns, blog posts, search engines, and loyalty card data to develop product-specific forecasts.

OM in Action The Automated Sushi Restaurant[6]

As with workers in manufacturing, fast-food employees are beginning to see automation enter into their places of employment. Along with kiosks popping up at McDonald's, diners in Japan are now treated to sushi-making robots, and in California, robots are flipping burgers. Fast-food chains are thinking more seriously about replacing human staff with technology as a way to combat increased minimum wages and food costs.

Genki Sushi, a Japanese sushi chain expanding throughout Asia, is known for its conveyer belt style of food service, where customers can grab a roll passing by their table. That's not entirely new.

But lately it has rolled out restaurants in which customers order customized dishes on a tablet. The food is delivered by an automated train that comes straight to the diner's seat or booth. Upon being seated, the customer has everything from sushi to noodle dishes and even cheesecake delivered by the train. There's no human involvement at all or even someone to explain the process. Twenty-four sets of tracks that crisscross the restaurant have the capacity to serve up to 158 patrons at once. When customers are ready to leave, they simply pay for their meal on a self-service tablet. They also clear their own tables by simply dropping the plates into a slot that leads to a hidden water-driven conveyor belt.

richardernestyap/ Shutterstock

Genki Sushi automatically tracks what every single customer eats and at what time of day. This large database of customer demand produces highly accurate forecasts that minimize food waste in the conveyor system.

Drug companies, such as Purdue Pharma LP, track critical medications with radio frequency identification (RFID) tags to reduce counterfeiting and theft. Facial recognition has been adopted at airports, stadiums, traffic intersections, and schools and is being used in retailing to enhance security measures.

Heat map
A graphical representation of data where values are depicted by color.

Heat mapping technology is helping retailers track customers' behavior electronically in real time. Security camera images generate a heat map of customer traffic superimposed on a store's layout, helping managers allocate personnel optimally. Heat mapping is so specific it can tell which items on display get the most attention. This exciting topic is explored in more detail in Module G: "Applying Analytics to Big Data in Operations Management."

Table 7.3 provides a broader glimpse of the impact of technology on services. Operations managers in services, as in manufacturing, must be able to evaluate the impact of technology

TABLE 7.3 Examples of Technology's Impact on Services

SERVICE INDUSTRY	EXAMPLE
Financial services	Debit cards, electronic funds transfer, automatic teller machines, Internet stock trading, online banking via cell phone, Apple pay
Education	Online journals and textbooks; interactive assignments via Zoom, Blackboard, and smartphones
Utilities and government	Automated one-person garbage trucks, optical mail scanners, flood-warning systems, meters that allow homeowners to control energy usage and costs
Restaurants and foods	Wireless orders from servers to the kitchen, robot butchering, transponders monitor traffic at drive-throughs
Communications	Interactive TV, e-books, speech recognition
Hotels	Electronic check-in/check-out, electronic key/lock systems, mobile Web bookings
Wholesale/retail trade	Point-of-sale (POS) technology; e-commerce; electronic communication between store, supplier, and customer; bar-coded data; RFID; self-checkout; pay via facial recognition
Transportation	Automatic toll booths, satellite-directed navigation systems, autonomous automobiles
Health care	Online patient-monitoring systems, online medical information systems, robotic surgery, artificial intelligence for medical diagnoses
Airlines	Ticketless travel, scheduling, Internet purchases, boarding passes downloaded as codes on smartphones

on their firm. This ability requires particular skill when evaluating reliability, investment analysis, human-resource requirements, and maintenance/service.

Process Redesign

Often a firm finds that the initial assumptions of its process are no longer valid. The world is a dynamic place, and customer desires, product technology, and product mix change. Consequently, processes are redesigned. **Process redesign** (sometimes called *process reengineering*) is the fundamental rethinking of business processes to bring about dramatic improvements in performance. Effective process redesign relies on reevaluating the purpose of the process and questioning both purpose and underlying assumptions. It works only if the basic process and its objectives are reexamined.

Process redesign
The fundamental rethinking of business processes to bring about dramatic improvements in performance.

Process redesign also focuses on those activities that cross functional lines. Because managers are often in charge of specific "functions" or specialized areas of responsibility, those activities (processes) that cross from one function or specialty to another may be neglected. Redesign casts aside all notions of how the process is currently being done and focuses on dramatic improvements in cost, time, and customer value. Any process is a candidate for radical redesign. The process can be a factory layout, a purchasing procedure, a new way of processing credit applications, or a new order-fulfillment process.

Shell Lubricants, for example, reinvented its order-fulfillment process by replacing a group of people who handled different parts of an order with one individual who does it all. As a result, Shell has cut the cycle time of turning an order into cash by 75%, reduced operating expenses by 45%, and boosted customer satisfaction 105%—all by introducing a new way of handling orders. Time, cost, and customer satisfaction—the dimension of performance shaped by operations—get major boosts from operational innovation.

Summary

Effective operations managers understand how to use process strategy as a competitive weapon. They select a production process with the necessary quality, flexibility, and cost structure to meet product and volume requirements. They also seek creative ways to combine the low unit cost of high-volume, low-variety manufacturing with the customization available through low-volume, high-variety facilities. Managers use the techniques of lean production and employee participation to encourage the development of efficient equipment and processes. They design their equipment and processes to have capabilities beyond the tolerance required by their customers, while ensuring the flexibility needed for adjustments in technology, features, and volumes.

Key Terms

Process strategy (p. 278)
Process focus (p. 278)
Modules (p. 279)
Repetitive process (p. 279)
Product focus (p. 280)
Mass customization (p. 280)
Build-to-order (BTO) (p. 280)
Postponement (p. 280)
Crossover chart (p. 282)
Flexibility (p. 284)
Flowchart (p. 284)
Time-function mapping (or process mapping) (p. 284)
Process charts (p. 285)
Value-stream mapping (VSM) (p. 286)
Service blueprinting (p. 287)
Computer numerical control (CNC) (p. 290)
Additive manufacturing (p. 290)
Automatic identification system (AIS) (p. 290)
Radio frequency identification (RFID) (p. 290)
Process control (p. 290)
Vision systems (p. 291)
Robot (p. 292)
Automated storage and retrieval system (ASRS) (p. 292)
Automated guided vehicle (AGV) (p. 292)
Flexible manufacturing system (FMS) (p. 292)
Computer-integrated manufacturing (CIM) (p. 292)
Heat map (p. 294)
Process redesign (p. 295)

Ethical Dilemma

For the sake of efficiency and lower costs, Premium Standard Farms of Princeton, Missouri, has turned pig production into a standardized product-focused process. Slaughterhouses have done this for a hundred years—but after the animal was dead. Doing it while the animal is alive is a relatively recent innovation. Here is how it works.

Impregnated female sows wait for 40 days in metal stalls so small that they cannot turn around. After an ultrasound test, they wait 67 days in a similar stall until they give birth. Two weeks after delivering 10 or 11 piglets, the sows are moved back to breeding rooms for another cycle. After 3 years, the sow is slaughtered. Animal-welfare advocates say such confinement drives pigs crazy. Premium Standard replies that its hogs are in fact comfortable, arguing that only 1% die before Premium Standard wants them to and that their system helps reduce the cost of pork products.

Glenda/Shutterstock

Discuss the productivity and ethical implications of this industry and these two divergent opinions.

Discussion Questions

1. What is process strategy?
2. What type of process is used for making each of the following products?
   a) beer
   b) wedding invitations
   c) automobiles
   d) paper
   e) Big Macs
   f) custom homes
   g) motorcycles
3. What is service blueprinting?
4. What is process redesign?
5. What are the techniques for improving service productivity?
6. Name the four quadrants of the service process matrix. Discuss how the matrix is used to classify services into categories.
7. What is CIM?
8. What do we mean by a process-control system, and what are the typical elements in such systems?
9. Identify *manufacturing* firms that compete on each of the four processes shown in Figure 7.1.
10. Identify the competitive advantage of each of the four firms identified in Discussion Question 9.
11. Identify *service* firms that compete on each of the four processes shown in Figure 7.1.
12. Identify the competitive advantage of each of the four firms identified in Discussion Question 11.
13. What are numerically controlled machines?
14. Describe briefly what an automatic identification system (AIS) is and how service organizations could use AIS to increase productivity and at the same time increase the variety of services offered.
15. Name some of the advances being made in technology that enhance production and productivity.
16. Explain what a flexible manufacturing system (FMS) is.
17. In what ways do CAD and FMS connect?
18. What is additive manufacturing?
19. Discuss the advantages and disadvantages of 3D printing.

Solved Problem

SOLVED PROBLEM 7.1

Bagot Copy Shop has a volume of 125,000 black-and-white copies per month. Two salespeople have made presentations to Gordon Bagot for machines of equal quality and reliability. The *Print Shop 5* has a cost of $2,000 per month and a variable cost of $.03. The other machine (a *Speed Copy 100*) will cost only $1,500 per month, but the toner is more expensive, driving the cost per copy up to $.035. If cost and volume are the only considerations, which machine should Bagot purchase?

SOLUTION

$$2{,}000 + .03X = 1{,}500 + .035X$$
$$2{,}000 - 1{,}500 = .035X - .03X$$
$$500 = .005X$$
$$100{,}000 = X$$

Because Bagot expects his volume to exceed 100,000 units, he should choose the *Print Shop 5*.

Problems

Note: **PX** means the problem may be solved with POM for Windows and/or Excel OM.

Problems 7.1–7.12 relate to Four Process Strategies

• **7.1** Borges Machine Shop, Inc., has a 1-year contract for the production of 200,000 gear housings for a new off-road vehicle. Owner Luis Borges hopes the contract will be extended and the volume increased next year. Borges has developed costs for three alternatives. They are general-purpose equipment (GPE), flexible manufacturing system (FMS), and expensive, but efficient, dedicated machine (DM). The cost data follow:

	GENERAL-PURPOSE EQUIPMENT (GPE)	FLEXIBLE MANUFACTURING SYSTEM (FMS)	DEDICATED MACHINE (DM)
Annual contracted units	200,000	200,000	200,000
Annual fixed cost	$100,000	$200,000	$500,000
Per unit variable cost	$ 15.00	$ 14.00	$ 13.00

Which process is best for this contract? **PX**

• **7.2** Using the data in Problem 7.1, determine the most economical volume for each process. **PX**

• **7.3** Using the data in Problem 7.1, determine the best process for each of the following volumes: (1) 75,000, (2) 275,000, and (3) 375,000.

• **7.4** Refer to Problem 7.1. If a contract for the second and third years is pending, what are the implications for process selection?

•• **7.5** Stan Fawcett's company is considering producing a gear assembly that it now purchases from Salt Lake Supply, Inc. Salt Lake Supply charges $4 per unit, with a minimum order of 3,000 units. Stan estimates that it will cost $15,000 to set up the process and then $1.82 per unit for labor and materials.

a) Draw a graph illustrating the crossover (or indifference) point.

b) Determine the number of units where either choice has the same cost. **PX**

•• **7.6** Ski Boards, Inc., wants to enter the market quickly with a new finish on its ski boards. It has three choices: (a) Refurbish the old equipment at a cost of $800, (b) make major modifications at a cost of $1,100, or (c) purchase new equipment at a net cost of $1,800. If the firm chooses to refurbish the equipment, materials and labor will be $1.10 per board. If it chooses to make modifications, materials and labor will be $0.70 per board. If it buys new equipment, variable costs are estimated to be $0.40 per board.

a) Graph the three total cost lines on the same chart.
b) Which alternative should Ski Boards, Inc., choose if it thinks it can sell more than 3,000 boards?
c) Which alternative should the firm use if it thinks the market for boards will be between 1,000 and 2,000? **PX**

Eric Limon/Shutterstock

•• **7.7** Tim Urban, owner/manager of Urban's Motor Court in Key West, is considering outsourcing the daily room cleanup for his motel to Duffy's Maid Service. Tim rents an average of 50 rooms for each of 365 nights (365 × 50 equals the total rooms rented for the year). Tim's cost to clean a room is $12.50. The Duffy's Maid Service quote is $18.50 per room plus a fixed cost of $25,000 for sundry items such as uniforms with the motel's name. Tim's annual fixed cost for space, equipment, and supplies is $61,000. Which is the preferred process for Tim, and why? **PX**

•• **7.8** Matthew Bailey, as manager of Designs by Bailey, is upgrading his CAD software. The high-performance (HP) software rents for $3,000 per month per workstation. The standard-performance (SP) software rents for $2,000 per month per workstation. The productivity figures that he has available suggest that the HP software is faster for his kind of design. Therefore, with the HP software he will need five engineers and with the SP software he will need six. This translates into a variable cost of $200 per drawing for the HP system and $240 per drawing for the SP system. At his projected volume of 80 drawings per month, which system should he rent? **PX**

••• **7.9** Makayla Metters Cabinets, Inc., needs to choose a production method for its new office shelf, the Maxistand. To help accomplish this, the firm has gathered the following production cost data:

PROCESS TYPE	ANNUALIZED FIXED COST OF PLANT & EQUIP.	*VARIABLE COSTS (PER UNIT) ($)* LABOR	MATERIAL	ENERGY
Mass Customization	$1,260,000	30	18	12
Intermittent	$1,000,000	24	26	20
Repetitive	$1,625,000	28	15	12
Continuous	$1,960,000	25	15	10

Makayla Metters Cabinets projects an annual demand of 24,000 units for the Maxistand. The Maxistand will sell for $120 per unit.

a) Which process type will maximize the annual profit from producing the Maxistand?
b) What is the value of this annual profit? **PX**

•• **7.10** California Gardens, Inc., prewashes, shreds, and distributes a variety of salad mixes in 2-pound bags. Doug Voss, Operations VP, is considering a new Hi-Speed shredder to replace the old machine, referred to in the shop as "Clunker." Hi-Speed will have a fixed cost of $85,000 per month and a variable cost of $1.25 per bag. Clunker has a fixed cost of only $44,000 per month, but a variable cost of $1.75. Selling price is $2.50 per bag.

a) What is the crossover point in units (point of indifference) for the processes?
b) What is the monthly profit or loss if the company changes to the Hi-Speed shredder and sells 60,000 bags per month?
c) What is the monthly profit or loss if the company stays with Clunker and sells 60,000 bags per month?

•• **7.11** Guadalupe Electric, Inc. must replace a robotic MIG welder and is evaluating two alternatives. Machine A has a fixed cost for the first year of $75,000 and a variable cost of $16, with a capacity of 18,000 units per year. Machine B is slower, with a speed of one-half of A's, but the fixed cost is only $60,000. The variable cost will be higher, at $20 per unit. Each unit is expected to sell for $28.

a) What is the crossover point (point of indifference) in units for the two machines?
b) What is the range of units for which machine A is preferable?
c) What is the range of units for which machine B is preferable?

•• **7.12** Arike Ogwumike Manufacturing intends to increase capacity through the addition of new equipment. Two vendors have presented proposals. The fixed cost for proposal A is $65,000, and for proposal B, $34,000. The variable cost for A is $10, and for B, $14. The revenue generated by each unit is $18.

a) What is the crossover point in units for the two options?
b) At an expected volume of 8,300 units, which alternative should be chosen?

Problems 7.13–7.17 relate to Process Analysis and Design

• **7.13** Prepare a flowchart for the process at the local car wash.

• **7.14** Prepare a process chart for a shoeshine.

•• **7.15** Prepare a time-function map for a shoeshine.

•• **7.16** Prepare a service blueprint for the process at an automatic car wash.

•• **7.17** Using Figure 7.6 in the discussion of value-stream mapping as a starting point, analyze an opportunity for improvement in a process with which you are familiar and develop an improved process.

CASE STUDIES

Rochester Manufacturing's Process Decision

Rochester Manufacturing Corporation (RMC) is considering moving some of its production from traditional numerically controlled machines to a flexible manufacturing system (FMS). Its computer numerical control machines have been operating in a high-variety, low-volume manner. Machine utilization, as near as it can determine, is hovering around 10%. The machine tool salespeople and a consulting firm want to put the machines together in an FMS. They believe that a $3 million expenditure on machinery and the transfer machines will handle about 30% of RMC's work. There will, of course, be transition and startup costs in addition to this.

The firm has not yet entered all its parts into a comprehensive group technology system, but believes that the 30% is a good estimate of products suitable for the FMS. This 30% should fit very nicely into a "family." A reduction, because of higher utilization, should take place in the number of pieces of machinery. The firm should be able to go from 15 to about 4 machines, and personnel should go from 15 to perhaps as low as 3. Similarly, floor space reduction will go from 20,000 square feet to about 6,000. Throughput of orders should also improve with processing of this family of parts in 1 to 2 days rather than 7 to 10. Inventory reduction is estimated to yield a one-time $750,000 savings, and annual labor savings should be in the neighborhood of $300,000.

Although the projections all look very positive, an analysis of the project's return on investment showed it to be between 10% and 15% per year. The company has traditionally had an expectation that projects should yield well over 15% and have payback periods of substantially less than 5 years.

Discussion Questions

1. As a production manager for RMC, what do you recommend? Why?
2. Prepare a case by a conservative plant manager for maintaining the status quo until the returns are more obvious.
3. Prepare the case for an optimistic sales manager that you should move ahead with the FMS now.

Process Strategy at Wheeled Coach

Video Case

Wheeled Coach, based in Winter Park, Florida, is the world's largest manufacturer of ambulances. Working four 10-hour days each week, 350 employees make only custom-made ambulances; virtually every vehicle is unique. Wheeled Coach accommodates the marketplace by providing a wide variety of options and an engineering staff accustomed to innovation and custom design. Continuing growth, which now requires that more than 20 ambulances roll off the assembly line each week, makes process design a continuing challenge. Wheeled Coach's response has been to build a focused factory: Wheeled Coach builds nothing but ambulances. Within the focused factory, Wheeled Coach established work cells for every major module feeding an assembly line, including aluminum bodies, electrical wiring harnesses, interior cabinets, windows, painting, and upholstery.

Labor standards drive the schedule so that every work cell feeds the assembly line on schedule, just-in-time for installations. The chassis, usually that of a Ford truck, moves to a station at which the aluminum body is mounted. Then the vehicle is moved to painting. Following a custom paint job, it moves to the assembly line, where it will spend 7 days. During each of these 7 workdays, each work cell delivers its respective module to the appropriate position on the assembly line. During the first day, electrical wiring is installed; on the second day, the unit moves forward to the station at which cabinetry is delivered and installed, then to a window and lighting station, on to upholstery, to fit and finish, to further customizing, and finally to inspection and road testing. The *Global Company Profile* featuring Wheeled Coach, which opens Chapter 14, provides further details about this process.

Discussion Questions*

1. Why do you think major auto manufacturers do not build ambulances?
2. What is an alternative process strategy to the assembly line that Wheeled Coach currently uses?
3. Why is it more efficient for the work cells to prepare "modules" and deliver them to the assembly line than it would be to produce the component (e.g., interior upholstery) on the line?
4. How does Wheeled Coach manage the tasks to be performed at each workstation?

*You may wish to view the video that accompanies this case before addressing these questions.

Alaska Airlines: 20-Minute Baggage Process—Guaranteed!

Video Case 

Alaska Airlines is unique among the nine major U.S. carriers not only for its extensive flight coverage of remote towns throughout Alaska (it also covers the U.S., Hawaii, and Mexico from its primary hub in Seattle). It is also one of the smallest independent airlines, with 10,300 employees, including 3,000 flight attendants and 1,500 pilots. What makes it really unique, though, is its ability to build state-of-the-art processes, using the latest technology, that yield high customer satisfaction. Indeed, J. D. Power and Associates has ranked Alaska Airlines highest in North America for 7 years in a row for customer satisfaction.

Alaska Airlines was the first to sell tickets via the Internet, first to offer Web check-in and print boarding passes online, and first with kiosk check-in. As Wayne Newton, Director of System Operation Control, states, "We are passionate about our processes. If it's not measured, it's not managed."

One of the processes Alaska is most proud of is its baggage-handling system. Passengers can check in at kiosks, tag their own bags with bar code stickers, and deliver them to a customer service agent at the carousel, which carries the bags through the vast underground system that eventually delivers the bags to a baggage handler. En route, each bag passes through TSA automated screening and is manually opened or inspected if it appears suspicious. With the help of bar code readers, conveyer belts automatically sort and transfer bags to their location (called a "pier") at the tarmac level. A baggage handler then loads the bags onto a cart and takes it to the plane for loading by the ramp team waiting inside the cargo hold. There are different procedures for "hot bags" (bags that have less than 30 minutes between transfer) and for "cold bags" (bags with over 60 minutes between plane transfers). Hot bags are delivered directly from one plane to another (called "tail-to-tail"). Cold bags are sent back into the normal conveyer system.

The process continues on the destination side with Alaska's unique guarantee that customer luggage will be delivered to the terminal's carousel within 20 minutes of the plane's arrival at the gate. If not, Alaska grants each passenger a 2,000 frequent-flier mile bonus!

The airline's use of technology includes bar code scanners to check in the bag when a passenger arrives, and again before it is placed on the cart to the plane. Similarly, on arrival, the time the passenger door opens is electronically noted and bags are again scanned as they are placed on the baggage carousel at the destination—tracking this metric means that the "time to carousel" (TTC) deadline is seldom missed. And the process almost guarantees that the lost bag rate approaches zero. On a recent day, only one out of 100 flights missed the TTC mark. The baggage process relies not just on technology, though. There are detailed, documented procedures to ensure that bags hit the 20-minute timeframe. Within 1 minute of the plane door opening at the gate, baggage handlers must begin the unloading. The first bag must be out of the plane within 3 minutes of parking the plane. This means the ground crew must be in the proper location—with their trucks and ramps in place and ready to go.

Courtesy of Alaska Airlines

Largely because of technology, flying on Alaska Airlines is remarkably reliable—even in the dead of an Alaska winter with only 2 hours of daylight, 50 mph winds, slippery runways, and low visibility. Alaska Airlines has had the industry's best on-time performance, with 87% if its flights landing on time.

Discussion Questions*

1. Prepare a flowchart of the process a passenger's bag follows from kiosk to destination carousel. (See Example 2 in Chapter 6 for a sample flowchart.) Include the exception process for the TSA opening of selected bags.
2. What other processes can an airline examine? Why is each important?
3. How does the kiosk alter the check-in process?
4. What metrics (quantifiable measures) are needed to track baggage?
5. What is the role of bar code scanners in the baggage process?

*You may wish to view the video that accompanies this case before addressing these questions.

Process Analysis at Arnold Palmer Hospital

Video Case 

The Arnold Palmer Hospital (APH) in Orlando, Florida, is one of the busiest and most respected hospitals for the medical treatment of children and women in the U.S. Since its opening on golfing legend Arnold Palmer's birthday September 10, 1989, more than 1.6 million children and women have passed through its doors. It is the fourth-busiest labor and delivery hospital in the U.S. and one of the largest neonatal intensive care units in the Southeast. APH ranks in the top 10% of hospitals nationwide in patient satisfaction.

"Part of the reason for APH's success," says Executive Director Kathy Swanson, "is our continuous improvement process. Our goal is 100% patient satisfaction. But getting there means constantly examining and reexamining everything we do, from patient flow, to cleanliness, to layout space, to a work-friendly environment, to speed of medication delivery from the pharmacy to a patient. Continuous improvement is a huge and never-ending task."

One of the tools the hospital uses consistently is process charts [like those in Figures 7.4 to 7.7 in this chapter and Figure 6.6(e) in Chapter 6]. Staffer Diane Bowles, who carries the title "clinical practice improvement consultant," charts scores of processes. Bowles's flowcharts help study ways to improve the turnaround of a vacated room (especially important in a hospital that has pushed capacity for years), speed up the admission process, and deliver warm meals warm.

Lately, APH has been examining the flow of maternity patients (and their paperwork) from the moment they enter the hospital until they are discharged, hopefully with their healthy baby, a day or two later. The flow of maternity patients follows these steps:

1. Enter APH's Labor & Delivery (L&D) check-in desk entrance.
2. If the baby is born en route or if birth is imminent, the mother and baby are taken directly to Labor & Delivery on the second floor and registered and admitted directly at the bedside. If there are no complications, the mother and baby go to Step 6.
3. If the baby is *not* yet born, the front desk asks if the mother is pre-registered. (Most do preregister at the 28- to 30-week pregnancy mark.) If she is not, she goes to the registration office on the first floor.
4. The pregnant woman is then taken to L&D Triage on the 8th floor for assessment. If she is in active labor, she is taken to an L&D room on the 2nd floor until the baby is born. If she is not ready, she goes to Step 5.
5. Pregnant women not ready to deliver (i.e., no contractions or false alarms) are either sent home to return on a later date and reenter the system at that time, or if contractions are not yet close enough, they are sent to walk around the hospital grounds (to encourage progress) and then return to L&D Triage at a prescribed time.
6. When the baby is born, if there are no complications, after 2 hours the mother and baby are transferred to a "mother–baby care unit" room on floors 3, 4, or 5 for an average of 40–44 hours.
7. If there *are* complications with the mother, she goes to an operating room and/or intensive care unit. From there, she goes back to a mother–baby care room upon stabilization—or is discharged at another time if not stabilized. Complications for the baby may result in a stay in the neonatal intensive care unit (NICU) before transfer to the baby nursery near the mother's room. If the baby is not stable enough for discharge with the mother, the baby is discharged later.
8. Mother and/or baby, when ready, are discharged and taken by wheelchair to the discharge exit for pickup to travel home.

Discussion Questions*

1. As Diane's new assistant, you need to flowchart this process. Explain how the process might be improved once you have completed the chart.
2. If a mother is scheduled for a Caesarean-section birth (i.e., the baby is removed from the womb surgically), how would this flowchart change?
3. If *all* mothers were electronically (or manually) preregistered, how would the flowchart change? Redraw the chart to show your changes.
4. Describe in detail a process that the hospital could analyze, besides the ones mentioned in this case.

*You may wish to view the video that accompanies this case before addressing these questions.

Endnotes

1. Source: Excerpted from Elaine J. Labach, "Faster, Better, and Cheaper," *Target* no. 5:43 with permission of the Association for Manufacturing Excellence, 380 West Palatine Road, Wheeling, IL 60090-5863, 847/520-3282. **www.ame.org**. Reprinted with permission of *Target Magazine*.
2. An additional example of a process chart is shown in Chapter 10.
3. Waste includes *inspection* (if the task is done properly, then inspection is unnecessary); *transportation* (movement of material within a process may be a necessary evil, but it adds no value); *delay* (an asset sitting idle and taking up space is waste); and *storage* (unless part of a "curing" process, storage is waste).
4. See related discussions in Gary J. Salegna and Farzanch Fazel, "An Integrative Approach for Classifying Services," *Journal of Global Business Management* (vol. 9, no. 1), 2013; and Roger Schmenner, "Services Moving toward Specialization and Focus with the Service Matrix," *MIT Sloan Management Review*, 1986.
5. Sources: *Supply Chain Dive* (February 11, 2020); and *The Wall Street Journal* (February 10, 2020).
6. Sources: *QSR* (November–December 2020); **CNBC.com** (July 23, 2017); and **People.com** (December 3, 2020).

Bibliography

Davenport, T. H. "The Coming Commoditization of Processes." *Harvard Business Review* 83, no. 6 (June 2005): 101–108.

Hall, J. M., and M. E. Johnson. "When Should a Process Be Art, Not Science?" *Harvard Business Review* 87, no. 3 (March 2009): 58–65.

Hegde, V. G., et al. "Customization: Impact on Product and Process Performance." *Production and Operations Management* 14, no. 4 (Winter 2005): 388–399.

Inderfurth, K., and I. M. Langella. "An Approach for Solving Disassembly-to-order Problems under Stochastic Yields." In *Logistik Management*. Heidelberg: Physica, 2004: 309–331.

Lee, E. K. et al. "Machine Learning for Predicting Vaccine Immunogenicity." *Interfaces* 46, no. 5 (Sept.–Oct. 2016): 368–390.

Moultrie, J., and A. M. Maier. "A Simplified Approach to Design for Assembly." *Journal of Engineering Design* 25, no. 1–3 (March 2014): 44–63.

Rugtusanatham, M. J., and F. Salvador. "From Mass Production to Mass Customization." *Production and Operations Management* 17, no. 3 (May–June 2008): 385–396.

Seider, W. D. et al. *Product and Process Design Principles: Synthesis, Analysis and Evaluation*, 4th ed. New York: Wiley, 2017.

Zhang, M., Y. Qi, and X. Zhao. "The Impact of Mass Customization Practices on Performance. "*International Journal of Mass Customization* 4, no. 1–2 (2011): 44–46.

Chapter 7 *Rapid* Review

Main Heading	Review Material	MyLab Operations Management
FOUR PROCESS STRATEGIES	■ **Process strategy**—An organization's approach to transforming resources into goods and services. *The objective of a process strategy is to build a production process that meets customer requirements and product specifications within cost and other managerial constraints.* Virtually every good or service is made by using some variation of one of four process strategies. ■ **Process focus**—A facility organized around processes to facilitate low-volume, high-variety production. The vast majority of global production is devoted to making low-volume, high-variety products in process-focused facilities, also known as job shops or *intermittent process* facilities. Process-focused facilities have high variable costs with extremely low utilization (5% to 25%) of facilities. ■ **Modules**—Parts or components of a product previously prepared, often in a continuous process. ■ **Repetitive process**—A product-oriented production process that uses modules. The repetitive process is the classic assembly line. It allows the firm to use modules and combine the economic advantages of the product-focused model with the customization advantages of the process-focus model. ■ **Product focus**—A facility organized around products; a product-oriented, high-volume, low-variety process. Product-focused facilities are also called *continuous processes* because they have very long, continuous production runs. The specialized nature of a product-focused facility requires high fixed cost; however, low variable costs reward high facility utilization. ■ **Mass customization**—Rapid, low-cost production that caters to constantly changing unique customer desires. ■ **Build-to-order (BTO)**—Produce to customer order rather than to a forecast. Major challenges of a build-to-order system include: *Product design, Process design, Inventory management, Tight schedules*, and *Responsive partners.* ■ **Postponement**—The delay of any modifications or customization to a product as long as possible in the production process. ■ **Crossover chart**—A chart of costs at the possible volumes for more than one process.	Concept Questions: 1.1–1.6 Problems: 7.1–7.12 ACTIVE MODEL 7.1 **VIDEO 7.1** Process Strategy at Wheeled Coach Ambulance Virtual Office Hours for Solved Problem: 7.1
SELECTION OF EQUIPMENT	Picking the best equipment involves understanding the specific industry and available processes and technology. The choice requires considering cost, quality, capacity, and flexibility. ■ **Flexibility**—The ability to respond with little penalty in time, cost, or customer value.	Concept Questions: 2.1–2.3
PROCESS ANALYSIS AND DESIGN	Five tools of process analysis are (1) flowcharts, (2) time-function mapping, (3) process charts, (4) value-stream mapping, and (5) service blueprinting. ■ **Flowchart**—A drawing used to analyze movement of people or materials. ■ **Time-function mapping (or process mapping)**—A flowchart with time added on the horizontal axis. ■ **Process charts**—Charts that use symbols to analyze the movement of people or material. Process charts allow managers to focus on value-added activities and to compute the percentage of value-added time (= operation time/total time). ■ **Value-stream mapping (VSM)**—A tool that helps managers understand how to add value in the flow of material and information through the entire production process. ■ **Service blueprinting**—A process analysis technique that lends itself to a focus on the customer and the provider's interaction with the customer.	Concept Questions: 3.1–3.6 Problems: 7.13–7.16 **VIDEO 7.2** Alaska Airlines 20-Minute Baggage Process–Guaranteed! **VIDEO 7.3** Process Analysis at Arnold Palmer Hospital

Main Heading	Review Material	MyLab Operations Management
SPECIAL CONSIDERATIONS FOR SERVICE PROCESS STRATEGIES	Services can be classified into one of four quadrants, based on relative degrees of labor and customization: 1. *Service factory* 2. *Service shop* 3. *Mass service* 4. *Professional service* Techniques for improving service productivity include: ▪ *Separation*—Structuring service so customers must go where the service is offered ▪ *Self-service*—Customers examining, comparing, and evaluating at their own pace ▪ *Postponement*—Customizing at delivery ▪ *Focus*—Restricting the offerings ▪ *Modules*—Modular selection of service; modular production ▪ *Automation*—Separating services that may lend themselves to a type of automation ▪ *Scheduling*—Precise personnel scheduling ▪ *Training*—Clarifying the service options; explaining how to avoid problems	Concept Questions: 4.1–4.6
PRODUCTION TECHNOLOGY	▪ **Computer numerical control (CNC)**—Machinery with its own computer and memory. ▪ **Additive manufacturing**—The production of physical items by adding layer upon layer, much in the same way an ink jet printer lays down ink; often referred to as 3D printing. ▪ **Automatic identification system (AIS)**—A system for transforming data into electronic form (e.g., bar codes). ▪ **Radio frequency identification (RFID)**—A wireless system in which integrated circuits with antennas send radio waves. ▪ **Process control**—The use of information technology to control a physical process. ▪ **Vision systems**—Systems that use video cameras and computer technology in inspection roles. ▪ **Robot**—A flexible machine with the ability to hold, move, or grab items. ▪ **Automated storage and retrieval systems (ASRS)**—Computer-controlled warehouses that provide for the automatic placement of parts into and from designated places within a warehouse. ▪ **Automated guided vehicle (AGV)**—Electronically guided and controlled cart used to move materials. ▪ **Flexible manufacturing system (FMS)**—Automated work cell controlled by electronic signals from a common centralized computer facility. ▪ **Computer-integrated manufacturing (CIM)**—A manufacturing system in which CAD, FMS, inventory control, warehousing, and shipping are integrated.	Concept Questions: 5.1–5.6
TECHNOLOGY IN SERVICES	Many rapid technological developments have occurred in the service sector, ranging from self-checkout and RFID to e-books and speech recognition. ▪ **Heat map**—A graphical representation of data where values are depicted by color.	Concept Questions: 6.1–6.4
PROCESS REDESIGN	▪ **Process redesign**—The fundamental rethinking of business processes to bring about dramatic improvements in performance. ▪ Process redesign often focuses on activities that cross functional lines.	Concept Questions: 7.1–7.3
ADDITIONAL MYLAB OPERATIONS MANAGEMENT RESOURCES	✔ Additional Case Study (Matthew Yachts, Inc.) ✔ Multiple Choice Case Questions (Rochester Manufacturing's Process Decision) ✔ Recent Graduate Video: Cameron Tinney, Process Improvement Analyst, Bimbo Bakeries	

Self Test

LO 7.1 Low-volume, high-variety processes are also known as:
- **a)** continuous processes.
- **b)** process focused.
- **c)** repetitive processes.
- **d)** product focused.

LO 7.2 A crossover chart for process selection focuses on:
- **a)** labor costs.
- **b)** material cost.
- **c)** both labor and material costs.
- **d)** fixed and variable costs.
- **e)** fixed costs.

LO 7.3 Tools for process analysis include all of the following except:
- **a)** flowchart.
- **b)** vision systems.
- **c)** service blueprinting.
- **d)** time-function mapping.
- **e)** value-stream mapping.

LO 7.4 Customer feedback in process design is lower as:
- **a)** the degree of customization is increased.
- **b)** the degree of labor is increased.
- **c)** the degree of customization is lowered.
- **d)** both a and b.
- **e)** both b and c.

LO 7.5 Computer-integrated manufacturing (CIM) includes manufacturing systems that have:
- **a)** computer-aided design, direct numerical control machines, and material-handling equipment controlled by automation.
- **b)** transaction processing, a management information system, and decision support systems.
- **c)** automated guided vehicles, robots, and process control.
- **d)** robots, automated guided vehicles, and transfer equipment.

Answers: LO 7.1. b; LO 7.2. d; LO 7.3. b; LO 7.4. c; LO 7.5. a.

Capacity and Constraint Management

SUPPLEMENT 7

SUPPLEMENT OUTLINE

LEARNING OBJECTIVES

LO S7.1 ***Define*** capacity 304

LO S7.2 ***Determine*** design capacity, effective capacity, and utilization 305

LO S7.3 ***Perform*** bottleneck analysis 311

LO S7.4 ***Compute*** break-even 314

LO S7.5 ***Determine*** expected monetary value of a capacity decision 318

LO S7.6 ***Compute*** net present value 319

Klaus Lang/All Canada Photos/Alamy Stock Photo

When designing facilities, including a concert hall, management hopes that the forecasted capacity (the product mix—opera, symphony, and special events—and the technology needed for these events) is accurate and adequate for operation above the break-even point. However, in many concert halls, even when operating at full capacity, break-even is not achieved, and supplemental funding must be obtained.

Capacity

LO S7.1 *Define* capacity

What should be the seating capacity of a concert hall? How many customers per day should an Olive Garden or a Hard Rock Cafe be able to serve? How large should a Frito-Lay plant be to produce 75,000 bags of Ruffles in an 8-hour shift? In this supplement we look at tools that help a manager make these decisions.

Capacity
The "throughput," or number of units a facility can hold, receive, store, or produce in a period of time.

After selection of a production process (Chapter 7), managers need to determine capacity. **Capacity** is the "throughput," or the number of units a facility can hold, receive, store, or produce in a given time. Capacity decisions often determine capital requirements and therefore a large portion of fixed cost. Capacity also determines whether demand will be satisfied or whether facilities will be idle. If a facility is too large, portions of it will sit unused and add cost to existing production. If a facility is too small, customers—and perhaps entire markets—will be lost. Determining facility size, with an objective of achieving high levels of utilization and a high return on investment, is critical.

STUDENT TIP
Too little capacity loses customers and too much capacity is expensive. Like Goldilocks's porridge, capacity needs to be *just* right.

Capacity planning can be viewed in three time horizons. In Figure S7.1 we note that long-range capacity (generally greater than 3 years) is a function of adding facilities and equipment that have a long lead time. In the intermediate range (usually 3 to 36 months), we can add equipment, personnel, and shifts; we can subcontract; and we can build or use inventory. This is the "aggregate planning" task. In the short run (usually up to 3 months), we are primarily concerned with scheduling jobs and people, as well as allocating machinery. Modifying capacity in the short run is difficult, as we are usually constrained by existing capacity.

Design and Effective Capacity

Design capacity
The theoretical maximum output of a system in a given period under ideal conditions.

Design capacity is the maximum theoretical output of a system in a given period under ideal conditions. It is normally expressed as a rate, such as the number of tons of steel that can be produced per week, per month, or per year. For many companies, measuring capacity can be straightforward: it is the maximum number of units the company is capable of producing in a specific time. However, for some organizations, determining capacity can be more difficult. Capacity can be measured in terms of beds (a hospital), active members (a place of worship), or billable hours (a CPA firm). Other organizations use total work time available as a measure of overall capacity.

Most organizations operate their facilities at a rate less than the design capacity. They do so because they have found that they can operate more efficiently when their resources are not stretched to the limit. For example, Ian's Bistro has tables set with 2 or 4 chairs seating a total of 270 guests. But the tables are never filled that way. Some tables will have 1 or 3 guests; tables can be pulled together for parties of 6 or 8. There are always unused chairs. *Design capacity* is 270, but *effective capacity* is often closer to 220, which is 81% of design capacity.

Effective capacity
The capacity a firm can expect to achieve, given its product mix, methods of scheduling, maintenance, and standards of quality.

Effective capacity is the capacity a firm *expects* to achieve given the current operating constraints. Effective capacity is often lower than design capacity because the facility may have

Figure **S7.1**
Time Horizons and Capacity Options

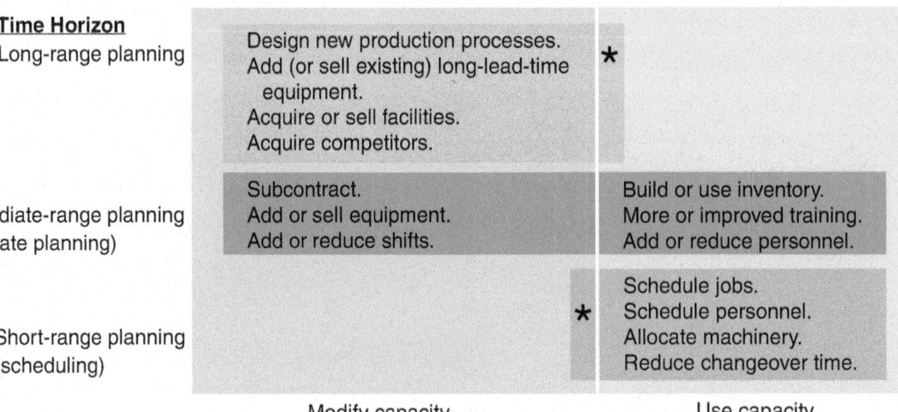

TABLE S7.1 Capacity Measurements

MEASURE	DEFINITION	EXAMPLE
Design capacity Andrey Eremin/123rf	Ideal conditions exist during the time that the system is available	If machines at Frito-Lay are designed to produce 1,000 bags of chips/hr., and the plant operates 16 hrs./day: $\textbf{Design Capacity} = \textbf{1,000 bags/hr.} \times \textbf{16 hrs.}$ $= \textbf{16,000 bags/day}$
Effective capacity Julie Jacobson/AP/Shutterstock	Design capacity minus lost output because of *planned* resource unavailability (e.g., preventive maintenance, machine setups/changeovers, changes in product mix, scheduled breaks)	If Frito-Lay loses 3 hours of output per day (namely 0.5 hr./day on preventive maintenance + 1 hr./day on employee breaks + 1.5 hrs./day setting up machines for different products): $\textbf{Effective Capacity} = \textbf{16,000 bags/day} - (\textbf{1,000 bags/hr.})(\textbf{3 hrs./day})$ $= \textbf{16,000 bags/day} - \textbf{3,000 bags/day}$ $= \textbf{13,000 bags/day}$
Actual output Hongqi Zhang/123rf	Effective capacity minus lost output during *unplanned* resource idleness (e.g., absenteeism, machine breakdowns, unavailable parts, quality problems)	On average, if machines at Frito-Lay are not running 0.25 hr./day due to late parts and machine breakdowns: $\textbf{Actual Output} = \textbf{13,000 bags/day} - (\textbf{1,000 bags/hr.})(\textbf{0.25 hrs./day})$ $= \textbf{13,000 bags/day} - \textbf{250 bags/day}$ $= \textbf{12,750 bags/day}$

been designed for an earlier version of the product or a different product mix than is currently being produced. Table S7.1 further illustrates the relationship between design capacity, effective capacity, and *actual output*.

Two measures of system performance are particularly useful: utilization and efficiency. **Utilization** is simply the percentage of *design capacity* actually achieved. **Efficiency** is the percentage of *effective capacity* actually achieved. Depending on how facilities are used and managed, it may be difficult or impossible to reach 100% efficiency. Operations managers tend to be evaluated on efficiency. The key to improving efficiency is often found in correcting quality problems and in effective scheduling, training, and maintenance. Utilization and efficiency are computed below:

Utilization
Actual output as a percentage of design capacity.

Efficiency
Actual output as a percentage of effective capacity.

$$\text{Utilization} = \text{Actual output/Design capacity} \qquad \text{(S7-1)}$$

$$\text{Efficiency} = \text{Actual output/Effective capacity} \qquad \text{(S7-2)}$$

LO S7.2 *Determine* design capacity, effective capacity, and utilization

In Example S1 we determine these values.

Example S1

DETERMINING CAPACITY UTILIZATION AND EFFICIENCY

Sara James Bakery has a plant for processing *Deluxe* breakfast rolls and wants to better understand its capability. Last week the facility produced 148,000 rolls. The effective capacity is 175,000 rolls. The production line operates 7 days per week, with three 8-hour shifts per day. The line was designed to process the nut-filled, cinnamon-flavored *Deluxe* roll at a rate of 1,200 per hour. Determine the design capacity, utilization, and efficiency for this plant when producing this *Deluxe* roll.

APPROACH ▶ First compute the design capacity and then use Equation (S7-1) to determine utilization and Equation (S7-2) to determine efficiency.

SOLUTION ▶

$$\text{Design capacity} = (7 \text{ days} \times 3 \text{ shifts} \times 8 \text{ hours}) \times (1{,}200 \text{ rolls per hour}) = 201{,}600 \text{ rolls}$$

$$\text{Utilization} = \text{Actual output/Design capacity} = 148{,}000 \;/\; 201{,}600 = 73.4\%$$

$$\text{Efficiency} = \text{Actual output/Effective capacity} = 148{,}000 \;/\; 175{,}000 = 84.6\%$$

INSIGHT ▶ The bakery now has the information necessary to evaluate efficiency.

LEARNING EXERCISE ▶ If the actual output is 150,000, what is the efficiency? [Answer: 85.7%.]

RELATED PROBLEMS ▶ S7.1, S7.2, S7.3, S7.4, S7.5, S7.6, S7.7, S7.8

ACTIVE MODEL S7.1 This example is further illustrated in Active Model S7.1 found online.

In Example S2 we see how the effectiveness of new capacity additions depends on how well management can perform on the utilization and efficiency of those additions.

Example S2

EXPANDING CAPACITY

The manager of Sara James Bakery (see Example S1) now needs to increase production of the increasingly popular *Deluxe* roll. To meet this demand, she will be adding a second production line. The second line has the same design capacity (201,600) and effective capacity (175,000) as the first line; however, new workers will be operating the second line. Quality problems and other inefficiencies stemming from the inexperienced workers are expected to reduce output on the second line to 130,000 (compared to 148,000 on the first). The utilization and efficiency were 73.4% and 84.6%, respectively, on the first line. Determine the new utilization and efficiency for the *Deluxe* roll operation after adding the second line.

APPROACH ▶ First, determine the new design capacity, effective capacity, and actual output after adding the second line. Then, use Equation (S7-1) to determine utilization and Equation (S7-2) to determine efficiency.

SOLUTION ▶ Design capacity = 201,600 × 2 = 403,200 rolls

Effective capacity = 175,000 × 2 = 350,000 rolls

Actual output = 148,000 + 130,000 = 278,000 rolls

Utilization = Actual output/Design capacity = 278,000 / 403,200 = 68.95%

Efficiency = Actual output/Effective capacity = 278,000 / 350,000 = 79.43%

INSIGHT ▶ Although adding equipment increases capacity, that equipment may not be operated as efficiently with new employees as might be the case with experienced employees. For Sara James Bakery, a doubling of equipment investment did not result in a doubling of output; other variables drove both utilization and efficiency lower.

LEARNING EXERCISE ▶ Suppose that Sara James reduces changeover time (setup time) by 3 hours per week on each production line. Assuming that the efficiency on each line remains constant, actual output will increase to 150,643 and 132,321 on lines 1 and 2, respectively. What will be the new overall utilization of the bakery? [Answer: utilization increases to 70.18%]

RELATED PROBLEMS ▶ S7.1, S7.2, S7.3, S7.4, S7.5, S7.6, S7.7, S7.8

Actual output, as used in Equation (S7-2), represents current conditions. Alternatively, with a knowledge of effective capacity and a current or target value for efficiency, the future *expected output* can be computed by reversing Equation (S7-2):

$$\text{Expected output} = \text{Effective capacity} \times \text{Efficiency}$$

If the expected output is inadequate, additional capacity may be needed. Much of the remainder of this supplement addresses how to effectively and efficiently add that capacity.

Capacity and Strategy

Sustained profits come from building competitive advantage, not just from a good financial return on a specific process. Capacity decisions must be integrated into the organization's mission and strategy. Investments are not to be made as isolated expenditures, but as part of a coordinated plan that will place the firm in an advantageous position. The questions to be asked are, "Will these investments eventually win profitable customers?" and "What competitive advantage (such as process flexibility, speed of delivery, improved quality, and so on) do we obtain?"

All 10 OM decisions we discuss in this text, as well as other organizational elements such as marketing and finance, are affected by changes in capacity. Change in capacity will have sales and cash flow implications, just as capacity changes have quality, supply chain, human resource, and maintenance implications. All must be considered.

Capacity Considerations

In addition to tight integration of strategy and investments, there are four special considerations for a good capacity decision:

1. *Forecast demand accurately:* Product additions and deletions, competition actions, product life cycle, and unknown sales volumes all add challenge to accurate forecasting.
2. *Match technology increments and sales volume:* Capacity options are often constrained by technology. Some capacity increments may be large (e.g., steel mills, paper mills, or power plants), while others may be small (hand-crafted Louis Vuitton handbags). Large capacity increments complicate the difficult but necessary job of matching capacity to sales.
3. *Find the optimum operating size (volume):* Economies and diseconomies of scale often dictate an optimal size for a facility. *Economies of scale* exist when average cost declines as size increases, whereas *diseconomies of scale* occur when a larger size raises the average cost. As Figure S7.2 suggests, most businesses have an optimal size—at least until someone comes along with a new business model. For decades, very large integrated steel mills were considered optimal. Then along came Nucor, CMC, and other minimills, with a new process and a new business model that radically reduced the optimum size of a steel mill.
4. *Build for change:* Managers build flexibility into facilities and equipment; changes will occur in processes, as well as products, product volume, and product mix.

STUDENT TIP
Each industry and technology has an optimum size.

Next, we note that rather than strategically manage capacity, managers may tactically manage demand.

Managing Demand

Even with good forecasting and facilities built to accommodate that forecast, there may be a poor match between the actual demand that occurs and available capacity. A poor match may mean demand exceeds capacity or capacity exceeds demand. However, in both cases, firms have options.

Demand Exceeds Capacity When *demand exceeds capacity*, the firm may be able to curtail demand simply by raising prices, scheduling long lead times (which may be inevitable), and discouraging marginally profitable business. However, because inadequate facilities reduce revenue below what is possible, the long-term solution is usually to increase capacity.

Capacity Considerations for Krispy Kreme Stores

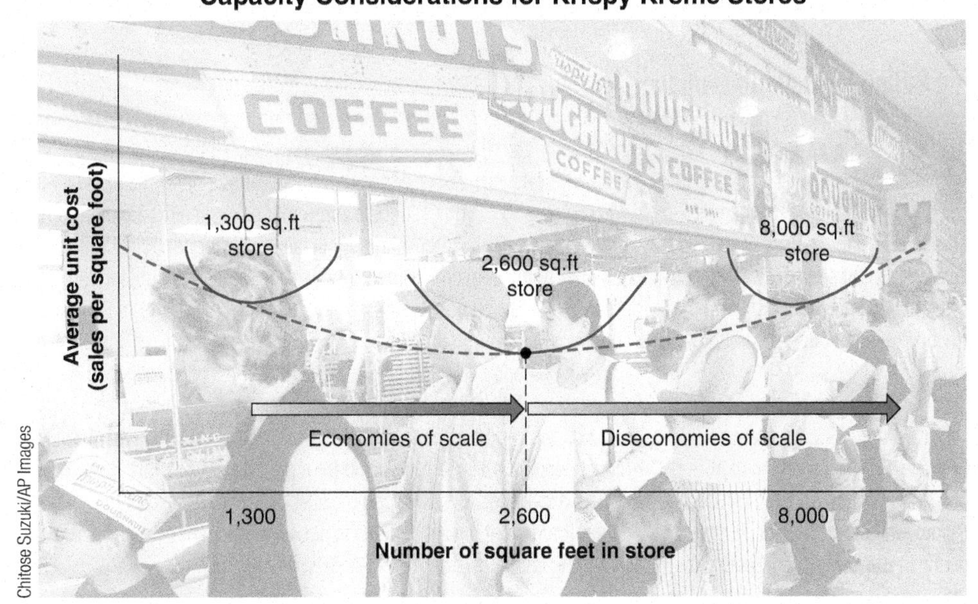

Figure **S7.2**

Economies and Diseconomies of Scale

Krispy Kreme originally had 8,000-square-foot stores but found them too large and too expensive for many markets. Then they tried tiny 1,300-square-foot stores, which required less investment, but such stores were too small to provide the mystique of seeing and smelling Krispy Kreme doughnuts being made. Krispy Kreme finally got it right with a 2,600-foot-store.

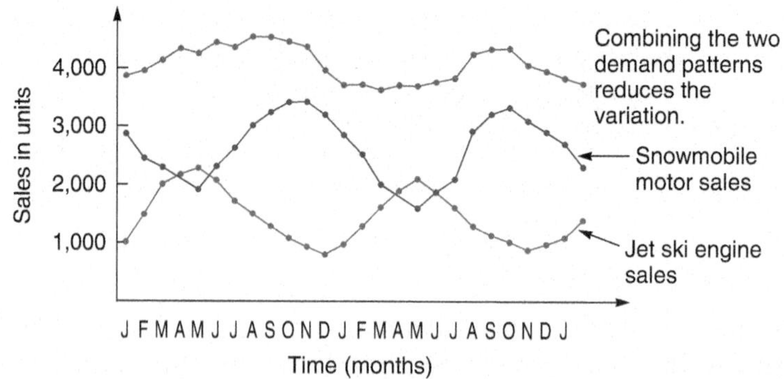

Figure **S7.3**

By Combining Products That Have Complementary Seasonal Patterns, Capacity Can Be Better Utilized

Capacity Exceeds Demand When capacity exceeds demand, the firm may want to stimulate demand through price reductions or aggressive marketing, or it may accommodate the market through product changes. When decreasing customer demand is combined with old and inflexible processes, layoffs and plant closings may be necessary to bring capacity in line with demand.

Adjusting to Seasonal Demands A seasonal or cyclical pattern of demand is another capacity challenge. In such cases, management may find it helpful to offer products with complementary demand patterns—that is, products for which the demand is high for one when low for the other. For example, in Figure S7.3 the firm is adding a line of snowmobile motors to its line of jet skis to smooth demand. With appropriate complementing of products, perhaps the utilization of facility, equipment, and personnel can be smoothed (as we see in the OM in Action box, "Matching Airline Capacity to Demand").

Tactics for Matching Capacity to Demand Various tactics for adjusting capacity to demand include:

1. Making staffing changes (increasing or decreasing the number of employees or shifts)
2. Adjusting equipment (purchasing additional machinery or selling or leasing out existing equipment)
3. Improving processes to increase throughput (e.g., reducing setup times at M2 Global Technology added the equivalent of 17 shifts of capacity)
4. Redesigning products to facilitate more throughput
5. Adding process flexibility to better meet changing product preferences
6. Closing facilities

The foregoing tactics can be used to adjust demand to existing facilities. The strategic issue is, of course, how to have a facility of the correct size.

OM in Action Matching Airline Capacity to Demand[1]

Airlines constantly struggle to control their capital expenditures and to adapt to unstable demand patterns.

Southwest and Lufthansa have each taken their own approach to increasing capacity while holding down capital investment. To manage capacity constraints on the cheap, Southwest squeezes seven flight segments out of its typical plane schedule per day—one more than most competitors. Its operations personnel find that quick ground turnaround, long a Southwest strength, is a key to this capital-saving technique.

Lufthansa has cut hundreds of millions of dollars in new jet purchases by squashing rows of seats 2 inches closer together. On the A320, for example, Lufthansa added two rows of seats, giving the plane 174 seats instead of 162. For its European fleet, this is the equivalent of having 12 more Airbus A320 jets.

Unstable demands in the airline industry provide another capacity challenge. Seasonal patterns (e.g., fewer people fly in the winter), compounded by spikes in demand during major holidays and summer vacations, play havoc with efficient use of capacity. Airlines attack costly seasonality in several ways. First, they schedule more planes for maintenance and renovations during slow winter months, curtailing winter capacity; second, they seek out contra-seasonal routes. And when capacity is substantially above demand, placing planes in storage (as shown in the photo) may be the most economical answer.

While demand struggled during the 2020–2021 pandemic (passenger traffic was down 20%), there was a huge increase in demand for cargo capacity. This occurred because many supplies for the fight against COVID-19 in North America were purchased in Asia. So for a period in May 2020, Anchorage, Alaska's airport became the world's busiest, surpassing even Chicago and Atlanta. It turns out that Anchorage is positioned to perfect geographical advantage, being 9.5 flying hours to 90% of the industrial world.

Joe McNally/Getty Images

The global pandemic (2020–2021), recessions (e.g., 2008–2010), and terrorist attacks (e.g., September 11, 2001) can make even the best capacity decision for an airline look bad. And excess capacity for an airline can be very expensive, with storage costs running as high as $60,000 per month per aircraft. Here, as a testimonial to recent excess capacity, aircraft sit idle in the Mojave Desert.

Service-Sector Demand and Capacity Management

In the service sector, scheduling customers is *demand management*, and scheduling the workforce is *capacity management*.

Demand Management When demand and capacity are fairly well matched, demand management can often be handled with appointments, reservations, or a first-come, first-served rule. In some businesses, such as doctors' and lawyers' offices, an *appointment system* is the schedule and is adequate. *Reservations systems* work well in rental car agencies, hotels, and some restaurants as a means of minimizing customer waiting time and avoiding disappointment over unfilled service. In retail shops, a post office, or a fast-food restaurant, a *first-come, first-served* rule for serving customers may suffice. Each industry develops its own approaches to matching demand and capacity. Other more aggressive approaches to demand management include many variations of discounts: "early bird" specials in restaurants, discounts for matinee performances or for seats at odd hours on an airline, and cheap weekend hotel rooms.

Capacity Management When managing demand is not feasible, then managing capacity through changes in full-time, temporary, or part-time staff may be an option. This is the approach in many services. For instance, hospitals may find capacity limited by a shortage of board-certified radiologists willing to cover the overnight shifts. Getting fast and reliable radiology readings can be the difference between life and death for an emergency room patient. As the photo illustrates, when an overnight reading is required (and 40% of CT scans are done between 8 P.M. and 8 A.M.), the image can be sent by e-mail to a doctor in Europe or Australia for immediate analysis.

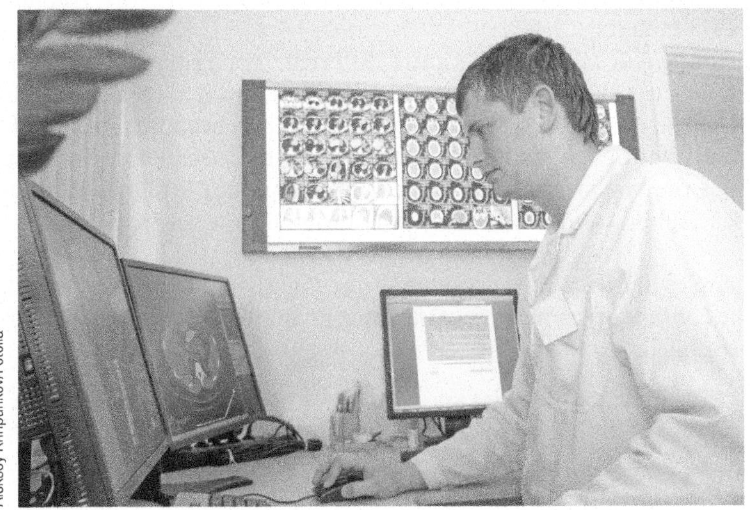

Aleksey Khripunkov/Fotolia

Many U.S. hospitals use services abroad to manage capacity for radiologists during night shifts. Night Hawk, an Idaho-based service with 50 radiologists in Zurich and Sydney, contracts with 900 facilities (20% of all U.S. hospitals). These trained experts, wide awake and alert in their daylight hours, usually return a diagnosis in 10 to 20 minutes, with a guarantee of 30 minutes.

Bottleneck Analysis and the Theory of Constraints

As managers seek to match capacity to demand, decisions must be made about the size of specific operations or work areas in the larger system. Each of the interdependent work areas can be expected to have its own unique capacity. **Capacity analysis** involves determining the throughput capacity of workstations in a system and ultimately the capacity of the entire system.

Capacity analysis
A means of determining throughput capacity of workstations or an entire production system.

A key concept in capacity analysis is the role of a constraint or **bottleneck**. A bottleneck is an operation that is the limiting factor or constraint. The term *bottleneck* refers to the literal neck of a bottle that constrains flow or, in the case of a production system, constrains throughput. A bottleneck has the lowest effective capacity of any operation in the system and thus limits the system's output. Bottlenecks occur in all facets of life—from job shops where a machine is constraining the work flow to highway traffic where two lanes converge into one inadequate lane, resulting in traffic congestion.

Bottleneck
The limiting factor or constraint in a system.

We define the **process time** of a station as the time to produce a unit (or a specified batch size of units) at that workstation. For example, if 16 customers can be checked out in a supermarket line every 60 minutes, then the process time at that station is 3.75 minutes per customer (= 60/16). (Process time is simply the inverse of capacity, which in this case is 60 minutes per hour/3.75 minutes per customer = 16 customers per hour.)

Process time
The time to produce a unit (or specified batch of units) at a workstation.

To determine the bottleneck in a production system, simply identify the station with the slowest process time. The **bottleneck time** is the process time of the slowest workstation (the one that takes the longest) in a production system. For example, the flowchart in Figure S7.4 shows a simple assembly line. Individual station process times are 2, 4, and 3 minutes, respectively. The bottleneck time is 4 minutes. This is because station B is the slowest station. Even if we were to speed up station A, the entire production process would not be faster. Inventory would simply pile up in front of station B even more than now. Likewise, if station C could work faster, we could not tap its excess capacity because station B will not be able to feed products to it any faster than 1 every 4 minutes.

Bottleneck time
The process time of the longest (slowest) process, i.e., the bottleneck.

Figure **S7.4**

Three-Station Assembly Line
A box represents an operation, a triangle represents inventory, and arrows represent precedence relationships

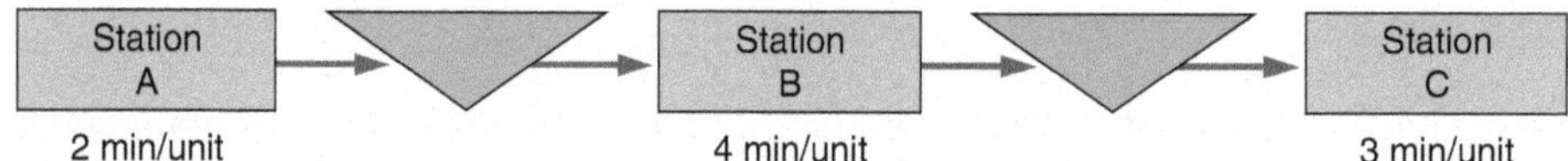

The **throughput time**, on the other hand, is the time it takes a unit to go through production from start to end, *with no waiting*. (Throughput time describes the behavior in an empty system. In contrast, flow time describes the time to go through a production process from beginning to end, including idle time waiting for stations to finish working on other units.) The throughput time to produce a new completed unit in Figure S7.4 is 9 minutes (= 2 minutes + 4 minutes + 3 minutes).

Throughput time
The time it takes for a product to go through the production process *with no waiting*. It is the time of the longest path through the system.

Bottleneck time and *throughput time* may be quite different. For example, a Ford assembly line may roll out a new car every minute (bottleneck time), but it may take 25 hours to actually make a car from start to finish (throughput time). This is because the assembly line has many workstations, with each station contributing to the completed car. Thus, bottleneck time determines the system's capacity (one car per minute), while its throughput time determines potential ability to produce a newly ordered product from scratch in 25 hours.

The following two examples illustrate capacity analysis for slightly more complex systems. Example S3 introduces the concept of parallel processes, and Example S4 introduces the concept of simultaneous processing.

Example S3

CAPACITY ANALYSIS WITH PARALLEL PROCESSES

Howard Kraye's sandwich shop provides healthy sandwiches for customers. Howard has two identical sandwich assembly lines. A customer first places an order, which takes 30 seconds. The order is then sent to one of the two assembly lines. Each assembly line has two workers and three operations: (1) assembly worker 1 retrieves and cuts the bread (15 seconds/sandwich), (2) assembly worker 2 adds ingredients and places the sandwich onto the toaster conveyor belt (20 seconds/sandwich), and (3) the toaster heats the sandwich (40 seconds/sandwich). Finally, another employee wraps the heated sandwich coming out of the toaster and delivers it to the customer (37.5 seconds/sandwich). A flowchart of the process is shown below. Howard wants to determine the bottleneck time and throughput time of this process.

LO S7.3 *Perform* bottleneck analysis

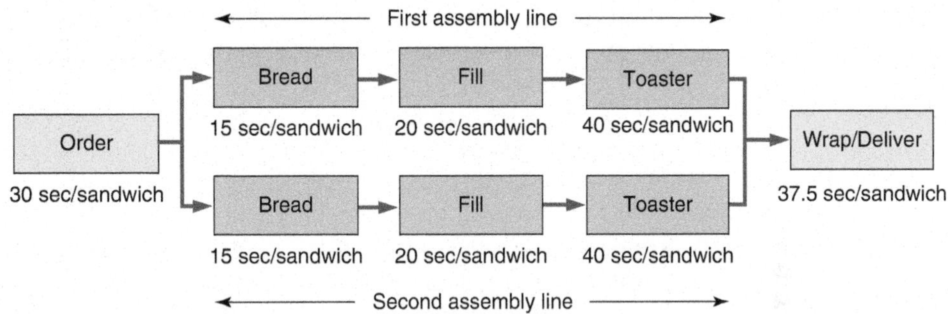

APPROACH ▶ Clearly the toaster is the single-slowest resource in the five-step process, but is it the bottleneck? Howard should first determine the bottleneck time of each of the two assembly lines separately, then the bottleneck time of the combined assembly lines, and finally the bottleneck time of the entire operation. For throughput time, each assembly line is identical, so Howard should just sum the process times for all five operations.

SOLUTION ▶ Because each of the three assembly-line operations uses a separate resource (worker or machine), different partially completed sandwiches can be worked on simultaneously at each station. Thus, the bottleneck time of each assembly line is the longest process time of each of the three operations. In this case, the 40-second toasting time represents the bottleneck time of each assembly line. Next, the bottleneck time of the *combined* assembly line operations is 40 seconds per *two* sandwiches, or 20 seconds per sandwich. Therefore, the wrapping and delivering operation, with a process time of 37.5 seconds, appears to be the bottleneck for the entire operation. The capacity per hour equals 3,600 seconds per hour/37.5 seconds per sandwich = 96 sandwiches per hour. The throughput time equals 30 + 15 + 20 + 40 + 37.5 = 142.5 seconds (or 2 minutes and 22.5 seconds), assuming no wait time in line to begin with.

INSIGHT ▶ Doubling the resources at a workstation effectively cuts the time at that station in half. (If n parallel [redundant] operations are added, the process time of the combined workstation operation will equal $1/n$ times the original process time.)

LEARNING EXERCISE ▶ If Howard hires an additional wrapper, what will be the new hourly capacity? [Answer: The new bottleneck is now the order-taking station: Capacity = 3,600 seconds per hour/30 seconds per sandwich = 120 sandwiches per hour]

RELATED PROBLEMS ▶ S7.9, S7.10, S7.11, S7.12, S7.13

In Example S3, how could we claim that the process time of the toaster was 20 seconds per sandwich when it takes 40 seconds to toast a sandwich? The reason is that we had two toasters; thus, two sandwiches could be toasted every 40 seconds, for an average of one sandwich every 20 seconds. And that time for a toaster can actually be achieved if the start times for the two are *staggered* (i.e., a new sandwich is placed in a toaster every 20 seconds). In that case, even though each sandwich will sit in the toaster for 40 seconds, a sandwich could emerge from one of the two toasters every 20 seconds. As we see, doubling the number of resources effectively cuts the process time at that station in half, resulting in a doubling of the capacity of those resources.

Example S4

CAPACITY ANALYSIS WITH SIMULTANEOUS PROCESSES

Dr. Cynthia Knott's dentistry practice has been cleaning customers' teeth for decades. The process for a basic dental cleaning is relatively straightforward: (1) the customer checks in (2 minutes); (2) a lab technician takes and develops X-rays (2 and 4 minutes, respectively); (3) the dentist processes and examines the X-rays (5 minutes) *while* the hygienist cleans the teeth (24 minutes); (4) the dentist meets with the patient to poke at a few teeth, explain the X-ray results, and tell the patient to floss more often (8 minutes); and (5) the customer pays and books her next appointment (6 minutes). A flowchart of the customer visit is shown below. Dr. Knott wants to determine the bottleneck time and throughput time of this process.

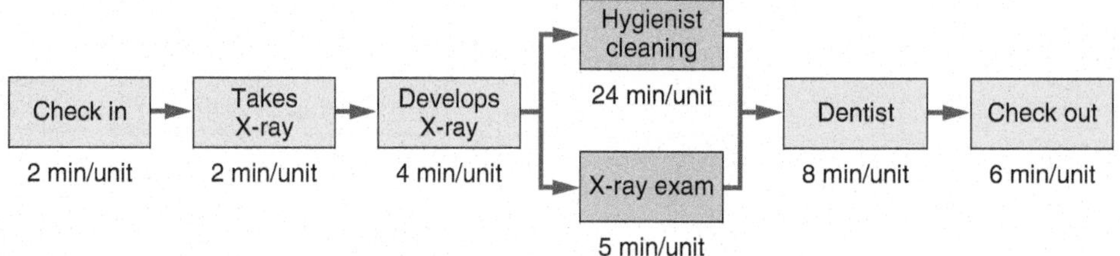

APPROACH ▶ With simultaneous processes, an order or a product is essentially *split* into different paths to be rejoined later on. To find the bottleneck time, each operation is treated separately, just as though all operations were on a sequential path. To find the throughput time, the time over *all* paths must be computed, and the throughput time is the time of the *longest* path.

SOLUTION ▶ The bottleneck in this system is the hygienist cleaning operation at 24 minutes per patient, resulting in an hourly system capacity of 60 minutes/24 minutes per patient = 2.5 patients. The throughput time is the maximum of the two paths through the system. The path through the X-ray exam is $2 + 2 + 4 + 5 + 8 + 6 = 27$ minutes, while the path through the hygienist cleaning operation is $2 + 2 + 4 + 24 + 8 + 6 = 46$ minutes. Thus, a patient should be out the door after 46 minutes (i.e., the maximum of 27 and 46).

INSIGHT ▶ With simultaneous processing, all operation times in the entire system are not simply added together to compute throughput time because some operations are occurring simultaneously. Instead, the time of the longest path through the system is deemed the throughput time.

LEARNING EXERCISE ▶ Suppose that the same technician now has the hygienist start immediately after the X-rays are taken (allowing the hygienist to start 4 minutes sooner). The technician then develops the X-rays while the hygienist is cleaning teeth. The dentist still examines the X-rays while the teeth cleaning is occurring. What would be the new system capacity and throughput time? [Answer: The X-ray development operation is now on the parallel path with cleaning and X-ray exam, reducing the total patient visit duration by 4 minutes, for a throughput time of 42 minutes (the maximum of 27 and 42). However, the hygienist cleaning operation is still the bottleneck, so the capacity remains 2.5 patients per hour.]

RELATED PROBLEMS ▶ S7.14, S7.15

To summarize: (1) the *bottleneck* is the operation with the longest (slowest) process time, after dividing by the number of parallel (redundant) operations, (2) the *system capacity* is the inverse of the *bottleneck time*, and (3) the *throughput time* is the total time through the longest path in the system, assuming no waiting.

Theory of Constraints

Theory of constraints (TOC)
A body of knowledge that deals with anything that limits an organization's ability to achieve its goals.

The **theory of constraints (TOC)** is a body of knowledge that deals with anything that limits or constrains an organization's ability to achieve its goals. Constraints can be physical (e.g., process or personnel availability, raw materials, or supplies) or nonphysical (e.g., procedures, morale, and training). Recognizing and managing these limitations through a five-step process is the basis of TOC.

STEP 1: Identify the constraints.

STEP 2: Develop a plan for overcoming the identified constraints.

STEP 3: Focus resources on accomplishing Step 2.

STEP 4: Reduce the effects of the constraints by offloading work or by expanding capability. Make sure that the constraints are recognized by all those who can have an impact on them.

STEP 5: When one set of constraints is overcome, go back to Step 1 and identify new constraints.

Bottleneck Management

A crucial constraint in any system is the bottleneck, and managers must focus significant attention on it. We present four principles of bottleneck management:

1. *Release work orders to the system at the pace set by the bottleneck's capacity:* The theory of constraints utilizes the concept of *drum, buffer, rope* to aid in the implementation of bottleneck and nonbottleneck scheduling. In brief, the *drum* is the beat of the system. It provides the schedule—the pace of production. The *buffer* is the resource, usually inventory, which may be helpful to keep the bottleneck operating at the pace of the drum. Finally, the *rope* provides the synchronization or communication necessary to pull units through the system. The rope can be thought of as signals between workstations.
2. *Lost time at the bottleneck represents lost capacity for the whole system:* This principle implies that the bottleneck should always be kept busy with work. Well-trained and cross-trained employees and inspections prior to the bottleneck can reduce lost capacity at a bottleneck.
3. *Increasing the capacity of a nonbottleneck station is a mirage:* Increasing the capacity of *nonbottleneck* stations has no impact on the system's overall capacity. Working faster on a nonbottleneck station may just create extra inventory, with all of its adverse effects. This implies that nonbottlenecks should have planned idle time. Extra work or setups at nonbottleneck stations will not cause delay, which allows for smaller batch sizes and more frequent product changeovers at nonbottleneck stations.
4. *Increasing the capacity of the bottleneck increases capacity for the whole system:* Managers should focus improvement efforts on the bottleneck. Bottleneck capacity may be improved by various means, including offloading some of the bottleneck operations to another workstation (e.g., let the beer foam settle next to the tap at the bar, not under it, so the next beer can be poured), increasing capacity of the bottleneck (adding resources, working longer or working faster), subcontracting, developing alternative routings, and reducing setup times.

STUDENT TIP
There are always bottlenecks; a manager must identify and manage them.

Even when managers have process and quality variability under control, changing technology, personnel, products, product mixes, and volumes can create multiple and shifting bottlenecks. Identifying and managing bottlenecks is a required operations task, but by definition, bottlenecks cannot be "eliminated." A system will always have at least one.

Break-Even Analysis

Break-even analysis is the critical tool for determining the capacity a facility must have to achieve profitability. The objective of **break-even analysis** is to find the point, in dollars and units, at which costs equal revenue. This point is the break-even point. Firms must operate above this level to achieve profitability. As shown in Figure S7.5, break-even analysis requires an estimation of fixed costs, variable costs, and revenue.

Break-even analysis
A means of finding the point, in dollars and units, at which costs equal revenues.

Fixed costs are costs that continue even if no units are produced. Examples include depreciation, taxes, debt, and mortgage payments. *Variable costs* are those that vary with the volume of units produced. The major components of variable costs are labor and materials. However, other costs, such as the portion of the utilities that varies with volume, are also variable costs. The difference between selling price and variable cost is *contribution*. Only when total contribution exceeds total fixed cost will there be profit.

Another element in break-even analysis is the *revenue function*. In Figure S7.5, revenue begins at the origin and proceeds upward to the right, increasing by the selling price of each unit. Where the revenue function crosses the total cost line (the sum of fixed and variable costs) is the break-even point, with a profit corridor to the right and a loss corridor to the left.

Figure **S7.5**

Basic Break-Even Point

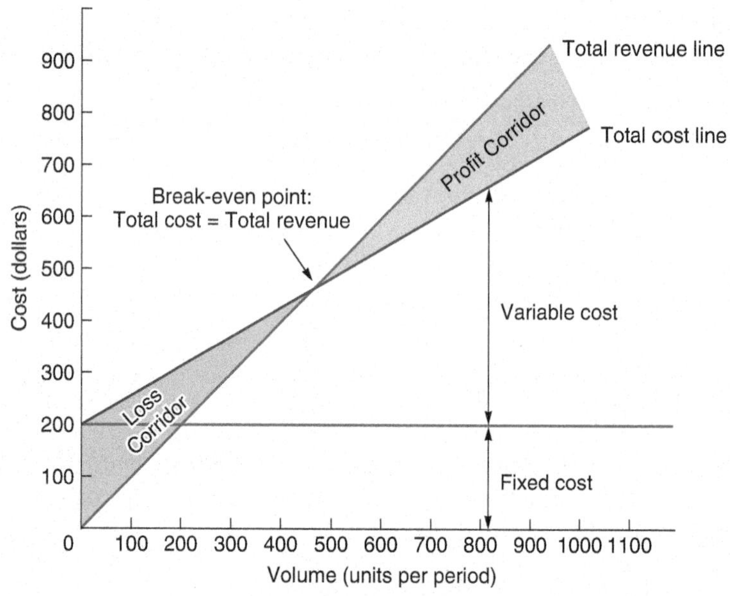

Assumptions A number of assumptions underlie the basic break-even model. Notably, costs and revenue are shown as straight lines. They are shown to increase linearly—that is, in direct proportion to the volume of units being produced. However, neither fixed costs nor variable costs (nor, for that matter, the revenue function) need be a straight line. For example, fixed costs change as more capital equipment or warehouse space is used; labor costs change with overtime or as marginally skilled workers are employed; the revenue function may change with such factors as volume discounts.

Single-Product Case

LO S7.4 *Compute* break-even

The formulas for the break-even point in units and dollars for a single product are shown below. Let:

BEP_x = break-even point in units TR = total revenue = Px

$BEP_\$$ = break-even point in dollars F = fixed costs

P = price per unit (after all discounts) V = variable costs per unit

x = number of units produced TC = total costs = $F + Vx$

The break-even point occurs where total revenue equals total costs. Therefore:

$$TR = TC \quad \text{or} \quad Px = F + Vx$$

Solving for x, we get:

$$\text{Break-even point in units } (BEP_x) = \frac{F}{P - V}$$

and:

$$\text{Break-even point in dollars } (BEP_\$) = BEP_x P = \frac{F}{P - V} P = \frac{F}{1 - V/P}$$

$$\text{Profit} = TR - TC = Px - (F + Vx) = Px - F - Vx = (P - V)x - F$$

Using these equations, we can solve directly for break-even point and profitability. The two break-even formulas of particular interest are:

$$\text{Break-even in units } (BEP_x) = \frac{\text{Total fixed cost}}{\text{Price} - \text{Variable cost}} = \frac{F}{P - V} \tag{S7-3}$$

$$\text{Break-even in dollars } (BEP_\$) = \frac{\text{Total fixed cost}}{1 - \dfrac{\text{Variable cost}}{\text{Price}}} = \frac{F}{1 - \dfrac{V}{P}} \tag{S7-4}$$

In Example S5, we determine the break-even point in dollars and units for one product.

Example S5

SINGLE-PRODUCT BREAK-EVEN ANALYSIS

Stephens, Inc., wants to determine the minimum dollar volume and unit volume needed at its new facility to break even.

APPROACH ▶ The firm first determines that it has fixed costs of \$10,000 this period. Direct labor is \$1.50 per unit, and material is \$.75 per unit. The selling price is \$4.00 per unit.

SOLUTION ▶ The break-even point in dollars is computed as follows:

$$BEP_\$ = \frac{F}{1-(V/P)} = \frac{\$10,000}{1-[(\$1.50+\$0.75)/(\$4.00)]} = \frac{\$10,000}{.4375} = \$22,857.14$$

The break-even point in units is:

$$BEP_x = \frac{F}{P-V} = \frac{\$10,000}{\$4.00-(\$1.50+\$0.75)} = 5,714$$

Note that we use total variable costs (that is, both labor and material).

INSIGHT ▶ The management of Stevens, Inc., now has an estimate in both units and dollars of the volume necessary for the new facility.

LEARNING EXERCISE ▶ If Stevens finds that fixed cost will increase to \$12,000, what happens to the break-even in units and dollars? [Answer: The break-even in units increases to 6,857, and break-even in dollars increases to \$27,428.57.]

RELATED PROBLEMS ▶ S7.16–S7.25, S7.28–S7.31

EXCEL OM Data File **Ch07SExS3.xls** can be found online.

ACTIVE MODEL S7.2 This example is further illustrated in Active Model S7.2 found online.

Multiproduct Case

Most firms, from manufacturers to restaurants, have a variety of offerings. We can expect each offering to have a different selling price and variable cost. Using break-even analysis, we modify Equation (S7-4) to reflect the proportion of sales for each product. We do this by "weighting" each product's contribution by its proportion of sales. The formula is then:

$$\text{Break-even point in dollars } (BEP_\$) = \frac{F}{\sum\left[\left(1-\frac{V_i}{P_i}\right)\times(W_i)\right]} \qquad \text{(S7-5)}$$

where F = fixed cost
V_i = variable cost per unit for product i
P_i = price per unit for product i
W_i = percentage of total dollar sales for product i

Image Ideas/Stockbyte/Getty Images

Paper machines such as the one shown here require a high capital investment. This investment results in a high fixed cost but allows production of paper at a very low variable cost. The production manager's job is to maintain utilization above the break-even point to achieve profitability.

Example S6 shows how to determine the break-even point for the multiproduct case at the Le Bistro restaurant.

Example S6

MULTIPRODUCT BREAK-EVEN ANALYSIS

Le Bistro, like most other restaurants, makes more than one product and would like to know its break-even point in dollars. Information for Le Bistro follows. Fixed costs are $3,000 per month.

ITEM	ANNUAL FORECASTED SALES UNITS	PRICE	COST
Sandwich	9,000	$5.00	$3.00
Drinks	9,000	1.50	0.50
Baked potato	7,000	2.00	1.00

APPROACH ▶ With a variety of offerings, we proceed with break-even analysis just as in a single-product case, except that we weight each of the products by its proportion of total sales using Equation (S7-5).

SOLUTION ▶ Multiproduct Break-Even: Determining Contribution

1	2	3	4	5	6	7	8	9
ITEM (i)	ANNUAL FORECASTED SALES UNITS	SELLING PRICE (P_i)	VARIABLE COST (V_i)	(V_i/P_i)	CONTRIBUTION $1-(V_i/P_i)$	ANNUAL FORECASTED SALES $	% OF SALES (W_i)	WEIGHTED CONTRIBUTION (COL. 6 × COL. 8)
Sandwich	9,000	$5.00	$3.00	.60	.40	$45,000	.621	.248
Drinks	9,000	1.50	0.50	.33	.67	13,500	.186	.125
Baked potato	7,000	2.00	1.00	.50	.50	14,000	.193	.097
						$72,500	1.000	.470

Note: Revenue for sandwiches is $45,000 (= 5.00 × 9,000), which is 62.1% of the total revenue of $72,500. Therefore, the contribution for sandwiches is "weighted" by .621. The weighted contribution is .621 × .40 = .248. In this manner, its *relative* contribution is properly reflected.

Using this approach for each product, we find that the total weighted contribution is .47 for each dollar of sales, and the break-even point in dollars is $76,596:

$$BEP_\$ = \frac{F}{\sum\left[\left(1-\frac{V_i}{P_i}\right)\times(W_i)\right]} = \frac{\$3{,}000\times 12}{.470} = \frac{\$36{,}000}{.470} = \$76{,}596$$

The information given in this example implies total daily sales (52 weeks at 6 days each) of:

$$\frac{\$76{,}596}{312\text{ days}} = \$245.50$$

INSIGHT ▶ The management of Le Bistro now knows that it must generate average sales of $245.50 each day to break even. Management also knows that if the forecasted sales of $72,500 are correct, Le Bistro will lose money, as break-even is $76,596.

LEARNING EXERCISE ▶ If the manager of Le Bistro wants to make an additional $1,000 per month in salary, and considers this a fixed cost, what is the new break-even point in average sales per day? [Answer: $327.33.]

RELATED PROBLEMS ▶ S7.26, S7.27

Break-even figures by product provide managers with added insight as to the realism of their sales forecast. These break-even amounts indicate exactly what must be sold each day, as we illustrate in Example S7.

Example S7

UNIT SALES AT BREAK-EVEN

Le Bistro also wants to know the break-even for the number of sandwiches that must be sold every day.

APPROACH ▶ Using the data in Example S6, we take the forecast sandwich sales of 62.1% times the daily break-even of $245.50 divided by the selling price of each sandwich ($5.00).

SOLUTION ▶ At break-even, sandwich sales must then be:

$$\frac{.621 \times \$245.50}{\$5.00} = \text{Number of sandwiches} = 30.5 \approx 31 \text{ sandwiches each day}$$

INSIGHT ▶ With knowledge of individual product sales, the manager has a basis for determining material and labor requirements.

LEARNING EXERCISE ▶ At a dollar break-even of $327.33 per day, how many sandwiches must Le Bistro sell each day? [Answer: ≈ 41.]

RELATED PROBLEMS ▶ S7.26b, S7.27b

Once break-even analysis has been prepared, analyzed, and judged to be reasonable, decisions can be made about the type and capacity of equipment needed. Indeed, a better judgment of the likelihood of success of the enterprise can now be made.

Reducing Risk with Incremental Changes

When demand for goods and services can be forecast with a reasonable degree of precision, determining a break-even point and capacity requirements can be rather straightforward. But, more likely, determining the capacity and how to achieve it will be complicated, as many factors are difficult to measure and quantify. Factors such as technology, competitors, building restrictions, cost of capital, human-resource options, and regulations make the decision interesting. To complicate matters further, demand growth is usually in small units, while capacity additions are likely to be both instantaneous and in large units. This contradiction adds to the capacity decision risk. To reduce risk, incremental changes that hedge demand forecasts may be a good option. Figure S7.6 illustrates four approaches to new capacity.

STUDENT TIP
Capacity decisions require matching capacity to forecasts, which is always difficult.

Alternative Figure S7.6(a) *leads* capacity—that is, acquires capacity to stay ahead of demand, with new capacity being acquired at the beginning of period 1. This capacity handles increased demand until the beginning of period 2. At the beginning of period 2, new capacity is again acquired, allowing the organization to stay ahead of demand until the beginning of period 3. This process can be continued indefinitely into the future. Here capacity is acquired *incrementally*—at the beginning of period 1 *and* at the beginning of period 2.

VIDEO S7.1
Capacity Planning at Arnold Palmer Hospital

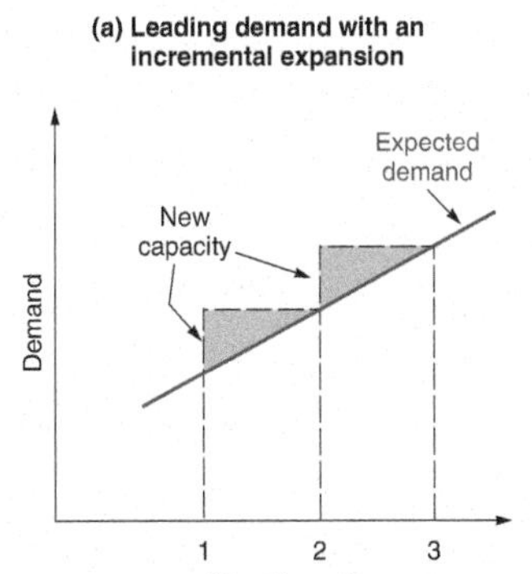

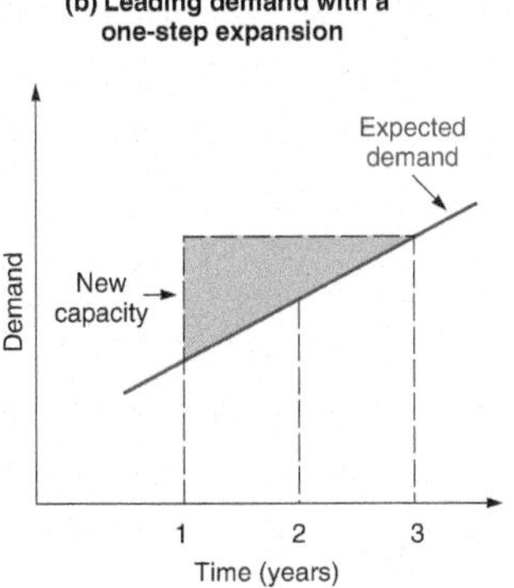

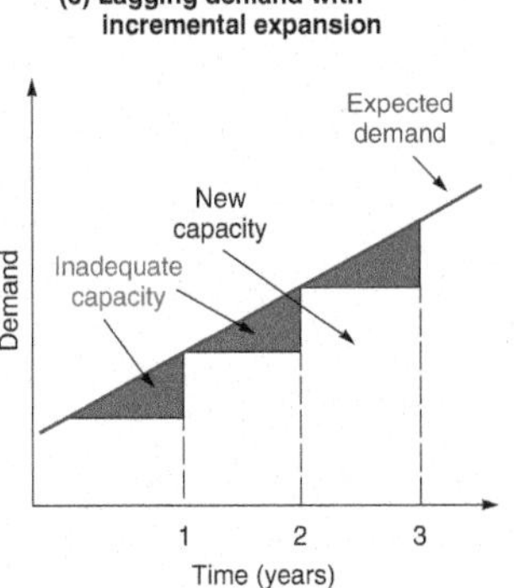

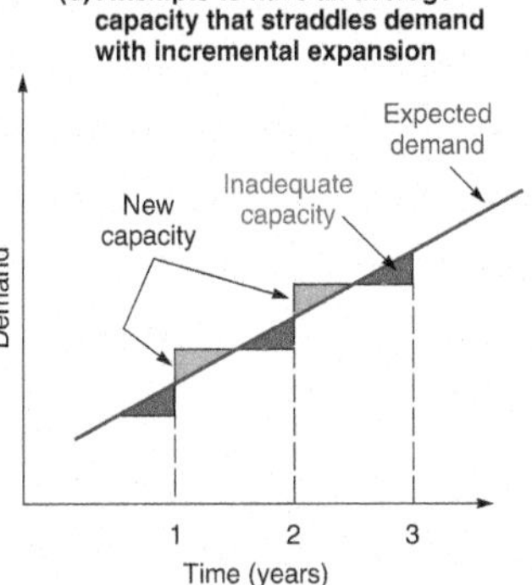

Figure **S7.6**

Four Approaches to Capacity Expansion

But managers can also elect to make a larger increase at the beginning of period 1 [Figure S7.6(b)]—an increase that may satisfy expected demand until the beginning of period 3. Excess capacity gives operations managers flexibility. For instance, in the hotel industry, added (extra) capacity in the form of rooms can allow a wider variety of room options and perhaps flexibility in room cleanup schedules. In manufacturing, excess capacity can be used to do more setups, shorten production runs, and drive down inventory costs.

Figure S7.6(c) shows an option that *lags* capacity, perhaps using overtime or subcontracting to accommodate excess demand. Finally, Figure S7.6(d) *straddles* demand by building capacity that is "average," sometimes lagging demand and sometimes leading it. Both the lag and straddle option have the advantage of delaying capital expenditure.

In cases where the business climate is stable, deciding between alternatives can be relatively easy. The total cost of each alternative can be computed, and the alternative with the least total cost can be selected. However, when capacity requirements are subject to significant unknowns, "probabilistic" models may be appropriate. One technique for making successful capacity planning decisions with an uncertain demand is decision theory, including the use of expected monetary value.

STUDENT TIP Uncertainty in capacity decisions makes EMV a helpful tool.

Applying Expected Monetary Value (EMV) to Capacity Decisions

LO S7.5 *Determine* expected monetary value of a capacity decision

Determining expected monetary value (EMV) requires specifying alternatives and various states of nature. For capacity-planning situations, the state of nature usually is future demand or market favorability. By assigning probability values to the various states of nature, we can make decisions that maximize the expected value of the alternatives. Example S8 shows how to apply EMV to a capacity decision.

Example S8

EMV APPLIED TO CAPACITY DECISIONS

Southern Hospital Supplies, a company that makes hospital gowns, is considering capacity expansion.

APPROACH: ▶ Southern's major alternatives are to do nothing, build a small plant, build a medium plant, or build a large plant. The new facility would produce a new type of gown, and currently the potential or marketability for this product is unknown. If a large plant is built and a favorable market exists, a profit of $100,000 could be realized. An unfavorable market would yield a $90,000 loss. However, a medium plant would earn a $60,000 profit with a favorable market. A $10,000 loss would result from an unfavorable market. A small plant, on the other hand, would return $40,000 with favorable market conditions and lose only $5,000 in an unfavorable market. Of course, there is always the option of doing nothing.

Recent market research indicates that there is a .4 probability of a favorable market, which means that there is also a .6 probability of an unfavorable market. With this information, the alternative that will result in the highest expected monetary value (EMV) can be selected.

SOLUTION ▶ Compute the EMV for each alternative:

$$\begin{aligned} \text{EMV (large plant)} &= (.4)(\$100,000) + (.6)(-\$90,000) = -\$14,000 \\ \text{EMV (medium plant)} &= (.4)(\$60,000) + (.6)(-\$10,000) = +\$18,000 \\ \text{EMV (small plant)} &= (.4)(\$40,000) + (.6)(-\$5,000) = +\$13,000 \\ \text{EMV (do nothing)} &= \$0 \end{aligned}$$

Based on EMV criteria, Southern should build a medium plant.

INSIGHT ▶ If Southern makes many decisions like this, then determining the EMV for each alternative and selecting the highest EMV is a good decision criterion.

LEARNING EXERCISE ▶ If a new estimate of the loss from a medium plant in an unfavorable market increases to –$20,000, what is the new EMV for this alternative? [Answer: $12,000, which changes the decision because the small plant EMV is now higher.]

RELATED PROBLEMS ▶ S7.32, S7.33

Applying Investment Analysis to Strategy-Driven Investments

Once the strategy implications of potential investments have been considered, traditional investment analysis is appropriate. We introduce the investment aspects of capacity next.

STUDENT TIP
An operations manager may be held responsible for return on investment (ROI).

Investment, Variable Cost, and Cash Flow

Because capacity and process alternatives exist, so do options regarding capital investment and variable cost. Managers must choose from among different financial options as well as capacity and process alternatives. Analysis should show the capital investment, variable cost, and cash flows as well as net present value for each alternative.

Net Present Value

Determining the discount value of a series of future cash receipts is known as the **net present value** technique. By way of introduction, let us consider the time value of money. Say you invest \$100.00 in a bank at 5% for 1 year. Your investment will be worth \$100.00 + (\$100.00)(.05) = \$105.00. If you invest the \$105.00 for a second year, it will be worth \$105.00 + (\$105.00)(.05) = \$110.25 at the end of the second year. Of course, we could calculate the future value of \$100.00 at 5% for as many years as we wanted by simply extending this analysis. However, there is an easier way to express this relationship mathematically. For the first year:

Net present value
A means of determining the discounted value of a series of future cash receipts.

$$\$105 = \$100(1 + .05)$$

For the second year:

$$\$110.25 = \$105(1 + .05) = \$100(1 + .05)^2$$

In general:

$$F = P(1 + i)^N \qquad \text{(S7-6)}$$

Where F = future value (such as \$110.25 or \$105)
P = present value (such as \$100.00)
i = **discount rate** (such as .05)
N = number of years (such as 1 year or 2 years)

Discount rate
The rate of return offered by investment alternatives that is used to discount expected future payoffs. (It may also be referred to as *cost of capital* or *interest rate*.)

In most investment decisions, however, we are interested in calculating the present value of a series of future cash receipts. Solving for P, we get:

$$P = \frac{F}{(1 + i)^N} \qquad \text{(S7-7)}$$

When the number of years is not too large, the preceding equation is effective. However, when the number of years, N, is large, the formula is cumbersome. For 20 years, you would have to compute $(1 + i)^{20}$. Discount-rate tables, such as Table S7.2, can help. We restate the present value equation:

LO S7.6 *Compute* net present value

$$P = \frac{F}{(1 + i)^N} = FX \qquad \text{(S7-8)}$$

where X = a factor from Table S7.2 defined as = $1/(1 + i)^N$ and F = future value

Thus, all we have to do is find the factor X and multiply it by F to calculate the present value, P. The factors, of course, are a function of the discount rate, i, and the number of years, N. Table S7.2 lists some of these factors.

Equations (S7-7) and (S7-8) are used to determine the present value of one future cash amount, but there are situations in which an investment generates a series of uniform and

TABLE S7.2 Present Value of $1

YEAR	5%	6%	7%	8%	9%	10%	12%	14%
1	.952	.943	.935	.926	.917	.909	.893	.877
2	.907	.890	.873	.857	.842	.826	.797	.769
3	.864	.840	.816	.794	.772	.751	.712	.675
4	.823	.792	.763	.735	.708	.683	.636	.592
5	.784	.747	.713	.681	.650	.621	.567	.519
6	.746	.705	.666	.630	.596	.564	.507	.456
7	.711	.665	.623	.583	.547	.513	.452	.400
8	.677	.627	.582	.540	.502	.467	.404	.351
9	.645	.592	.544	.500	.460	.424	.361	.308
10	.614	.558	.508	.463	.422	.386	.322	.270
15	.481	.417	.362	.315	.275	.239	.183	.140
20	.377	.312	.258	.215	.178	.149	.104	.073

equal cash amounts. This type of investment is called an *annuity*. For example, an investment might yield $300 per year for 3 years. Easy-to-use factors have been developed for the present value of annuities. These factors are shown in Table S7.3. The basic relationship is:

$$S = RX$$

where X = factor from Table S7.3
S = present value of a series of uniform annual receipts
R = receipts that are received every year for the life of the investment (the annuity)

The present value of a uniform annual series of amounts is an extension of the present value of a single amount, and thus Table S7.3 can be directly developed from Table S7.2. The factors for any given discount rate in Table S7.3 are the cumulative sum of the values in Table S7.2. In Table S7.2, for example, .943, .890, and .840 are the factors for years 1, 2, and 3 when the discount rate is 6%. The cumulative sum of these factors is 2.673. Now look at the point in Table S7.3 where the discount rate is 6% and the number of years is 3. The factor for the present value of an annuity is 2.673, as you would expect. Alternatively, the PV formula in Microsoft Excel can be used: =–PV(discount rate,year,1), e.g., =–PV(.06,3,1) = 2.673.

TABLE S7.3 Present Value of an Annuity of $1

YEAR	5%	6%	7%	8%	9%	10%	12%	14%
1	.952	.943	.935	.926	.917	.909	.893	.877
2	1.859	1.833	1.808	1.783	1.759	1.736	1.690	1.647
3	2.723	2.673	2.624	2.577	2.531	2.487	2.402	2.322
4	3.546	3.465	3.387	3.312	3.240	3.170	3.037	2.914
5	4.329	4.212	4.100	3.993	3.890	3.791	3.605	3.433
6	5.076	4.917	4.767	4.623	4.486	4.355	4.111	3.889
7	5.786	5.582	5.389	5.206	5.033	4.868	4.564	4.288
8	6.463	6.210	5.971	5.747	5.535	5.335	4.968	4.639
9	7.108	6.802	6.515	6.247	5.995	5.759	5.328	4.946
10	7.722	7.360	7.024	6.710	6.418	6.145	5.650	5.216
15	10.380	9.712	9.108	8.559	8.061	7.606	6.811	6.142
20	12.462	11.470	10.594	9.818	9.129	8.514	7.469	6.623

Example S9 shows how to determine the present value of an annuity.

Example S9

DETERMINING NET PRESENT VALUE OF FUTURE RECEIPTS OF EQUAL VALUE

River Road Medical Clinic is thinking of investing in a sophisticated new piece of medical equipment. It will generate \$7,000 per year in receipts for 5 years.

APPROACH ▶ Determine the present value of this cash flow; assume a discount rate of 6%.

SOLUTION ▶ The factor from Table S7.3 (4.212) is obtained by finding that value when the discount rate is 6% and the number of years is 5 (alternatively using the Excel formula = –PV (.06,5,1)):

$$S = RX = \$7{,}000(4.212) = \$29{,}484$$

INSIGHT ▶ There is another way of looking at this example. If you went to a bank and took a loan for \$29,484 today, your payments would be \$7,000 per year for 5 years if the bank used an interest rate of 6% compounded yearly. Thus, \$29,484 is the present value.

LEARNING EXERCISE ▶ If the discount rate is 8%, what is the present value? [Answer: \$27,951.]

RELATED PROBLEMS ▶ S7.34–S7.45

EXCEL OM Data File **Ch07SExS9.xls** can be found online.

The net present value method is straightforward: You simply compute the present value of all cash flows for each investment alternative. When deciding among investment alternatives, you pick the investment with the highest net present value. Similarly, when making several investments, those with higher net present values are preferable to investments with lower net present values.

Solved Problem S7.4 shows how to use the net present value to choose between investment alternatives.

Although net present value is one of the best approaches to evaluating investment alternatives, it does have its faults. Limitations of the net present value approach include the following:

1. Investments with the same net present value may have significantly different projected lives and different salvage values.
2. Investments with the same net present value may have different cash flows. Different cash flows may make substantial differences in the company's ability to pay its bills.
3. The assumption is that we know future rates of return, which we do not.
4. Payments are always made at the end of the period (week, month, or year), which is not always the case.

Summary

Managers tie equipment selection and capacity decisions to the organization's missions and strategy. Four additional considerations are critical: (1) accurately forecasting demand; (2) understanding the equipment, processes, and capacity increments; (3) finding the optimum operating size; and (4) ensuring the flexibility needed for adjustments in technology, product features and mix, and volumes.

Techniques that are particularly useful to operations managers when making capacity decisions include good forecasting, bottleneck analysis, break-even analysis, expected monetary value, cash flow, and net present value.

The single most important criterion for investment decisions is the contribution to the overall strategic plan and the winning of profitable orders. Successful firms select the correct process and capacity.

Key Terms

Capacity (p. 304)
Design capacity (p. 304)
Effective capacity (p. 304)
Utilization (p. 305)
Efficiency (p. 305)
Capacity analysis (p. 310)
Bottleneck (p. 310)
Process time (p. 310)
Bottleneck time (p. 310)
Throughput time (p. 310)
Theory of constraints (TOC) (p. 312)
Break-even analysis (p. 313)
Net present value (p. 319)
Discount rate (p. 319)

Discussion Questions

1. Distinguish between design capacity and effective capacity.
2. What is effective capacity?
3. What is efficiency?
4. Distinguish between effective capacity and actual output.
5. Explain why doubling the capacity of a bottleneck may not double the system capacity.
6. Distinguish between bottleneck time and throughput time.
7. What is the theory of constraints?
8. What are the assumptions of break-even analysis?
9. What keeps plotted revenue data from falling on a straight line in a break-even analysis?
10. Under what conditions would a firm want its capacity to lag demand? to lead demand?
11. Explain how net present value is an appropriate tool for comparing investments.
12. Describe the five-step process that serves as the basis of the theory of constraints.
13. What are the techniques available to operations managers to deal with a bottleneck operation? Which of these does not decrease throughput time?

Using Software for Break-Even Analysis

Excel, Excel OM, and POM for Windows all handle break-even and cost–volume analysis problems.

CREATING YOUR OWN EXCEL SPREADSHEETS

It is a straightforward task to develop the formulas to conduct a single-product break-even analysis in Excel. Although we do not demonstrate the basics here, Active Model S7.2 provides a working example. Program S7.1 illustrates how you can make an Excel model to solve Example S6, which is a multiproduct break-even analysis.

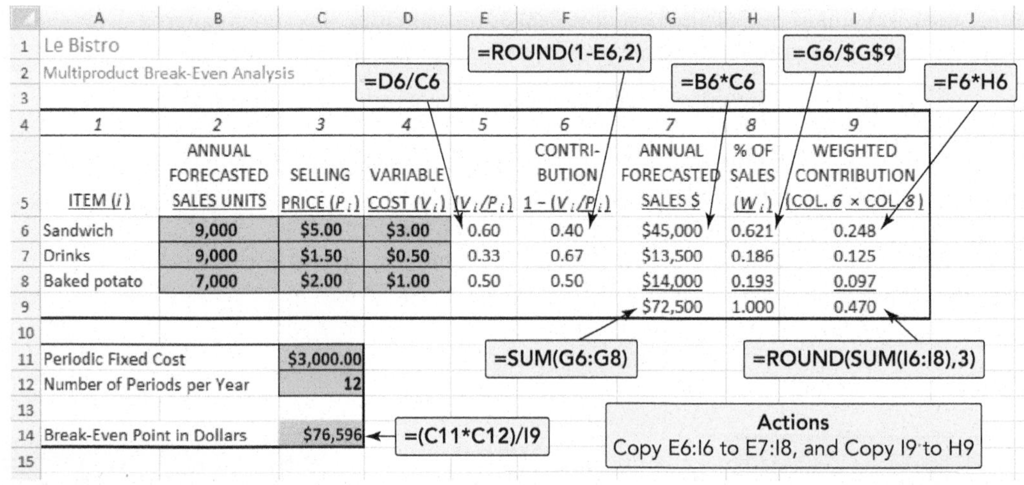

Le Bistro
Multiproduct Break-Even Analysis

1	2	3	4	5	6	7	8	9
ITEM (i)	ANNUAL FORECASTED SALES UNITS	SELLING PRICE (P_i)	VARIABLE COST (V_i)	(V_i/P_i)	CONTRIBUTION $1-(V_i/P_i)$	ANNUAL FORECASTED SALES $	% OF SALES (W_i)	WEIGHTED CONTRIBUTION (COL. 6 × COL. 8)
Sandwich	9,000	$5.00	$3.00	0.60	0.40	$45,000	0.621	0.248
Drinks	9,000	$1.50	$0.50	0.33	0.67	$13,500	0.186	0.125
Baked potato	7,000	$2.00	$1.00	0.50	0.50	$14,000	0.193	0.097
						$72,500	1.000	0.470

Periodic Fixed Cost	$3,000.00
Number of Periods per Year	12
Break-Even Point in Dollars	$76,596

Program **S7.1**

An Excel Spreadsheet for Performing Break-Even Analysis for Example S6

USING EXCEL OM

Excel OM's Break-Even Analysis module provides the Excel formulas needed to compute the break-even points, and the solution and graphical output.

USING POM FOR WINDOWS

Similar to Excel OM, POM for Windows also contains a break-even/cost–volume analysis module. Please refer to Appendix II for further details.

Solved Problems

SOLVED PROBLEM S7.1

Sara James Bakery, described in Examples S1 and S2, has decided to increase its facilities by adding one additional process line. The firm will have two process lines, each working 7 days a week, 3 shifts per day, 8 hours per shift, with effective capacity of 300,000 rolls. This addition, however, will reduce overall system efficiency to 85%. Compute the expected production with this new effective capacity.

SOLUTION

$$\begin{aligned}\text{Expected production} &= (\text{Effective capacity})(\text{Efficiency})\\ &= 300{,}000(.85)\\ &= 255{,}000 \text{ rolls per week}\end{aligned}$$

SOLVED PROBLEM S7.2

Marty McDonald has a business packaging software in Wisconsin. His annual fixed cost is $10,000, direct labor is $3.50 per package, and material is $4.50 per package. The selling price will be $12.50 per package. What is the break-even point in dollars? What is break-even in units?

SOLUTION

$$BEP_\$ = \frac{F}{1-(V/P)} = \frac{\$10{,}000}{1-(\$8.00/\$12.50)} = \frac{\$10{,}000}{.36} = \$27{,}777$$

$$BEP_x = \frac{F}{P-V} = \frac{\$10{,}000}{\$12.50-\$8.00} = \frac{\$10{,}000}{\$4.50} = 2{,}222 \text{ units}$$

SOLVED PROBLEM S7.3

John has been asked to determine whether the $22.50 cost of tickets for the community dinner theater will allow the group to achieve break-even and whether the 175 seating capacity is adequate. The cost for each performance of a 10-performance run is $2,500. The facility rental cost for the entire 10 performances is $10,000. Drinks and parking are extra charges and have their own price and variable costs, as shown below:

1	2	3	4	5	6	7	8	9
	SELLING PRICE (P)	VARIABLE COST (V)	PERCENTAGE VARIABLE COST (V/P)	CONTRIBUTION 1 − (V/P)	ESTIMATED QUANTITY OF SALES UNITS (SALES)	DOLLAR SALES (SALES × P)	PERCENTAGE OF SALES	CONTRIBUTION WEIGHTED BY PERCENTAGE OF SALES (COL.5 × COL. 8)
Tickets with dinner	$22.50	$10.50	0.467	0.533	175	$3,938	0.741	0.395
Drinks	$ 5.00	$ 1.75	0.350	0.650	175	$ 875	0.165	0.107
Parking	$ 5.00	$ 2.00	0.400	0.600	100	$ 500	0.094	0.056
					450	$5,313	1.000	0.558

SOLUTION

$$BEP_\$ = \frac{F}{\sum\left[\left(1-\frac{V_i}{P_i}\right)\times(W_i)\right]} = \frac{\$(10\times 2{,}500)+\$10{,}000}{0.558} = \frac{\$35{,}000}{0.558} = \$62{,}724$$

Revenue for each performance (from column 7) = $5,313
Total forecasted revenue for the 10 performances = (10 × $5,313) = $53,130
Forecasted revenue with this mix of sales shows a break-even of $62,724

Thus, given this mix of costs, sales, and capacity John determines that the theater will not break even.

SOLVED PROBLEM S7.4

Your boss has told you to evaluate the cost of two machines. After some questioning, you are assured that they have the costs shown at the right. Assume:

a) The life of each machine is 3 years.
b) The company thinks it knows how to make 14% on investments no riskier than this one.

Determine via the present value method which machine to purchase.

	MACHINE A	MACHINE B
Original cost	$13,000	$20,000
Labor cost per year	2,000	3,000
Floor space per year	500	600
Energy (electricity) per year	1,000	900
Maintenance per year	2,500	500
Total annual cost	$ 6,000	$ 5,000
Salvage value	$ 2,000	$ 7,000

SOLUTION

		MACHINE A			MACHINE B		
		COLUMN 1	COLUMN 2	COLUMN 3	COLUMN 4	COLUMN 5	COLUMN 6
Now	Expense	1.000	$13,000	$13,000	1.000	$20,000	$20,000
1 yr.	Expense	.877	6,000	5,262	.877	5,000	4,385
2 yr.	Expense	.769	6,000	4,614	.769	5,000	3,845
3 yr.	Expense	.675	6,000	4,050	.675	5,000	3,375
				$26,926			$31,605
3 yr.	Salvage revenue	.675	$ 2,000	−1,350	.675	$ 7,000	−4,725
				$25,576			$26,880

We use 1.0 for payments with no discount applied against them (that is, when payments are made now, there is no need for a discount). The other values in columns 1 and 4 are from the 14% column and the respective year in Table S7.2 (for example, the intersection of 14% and 1 year is .877, etc.). Columns 3 and 6 are the products of the present value figures times the combined costs. This computation is made for each year and for the salvage value.

The calculation for machine A for the first year is:

$$.877 \times (\$2,000 + \$500 + \$1,000 + \$2,500) = \$5,262$$

The salvage value of the product is subtracted from the summed costs, because it is a receipt of cash. Because the sum of the net costs for machine B is larger than the sum of the net costs for machine A, machine A is the low-cost purchase, and your boss should be so informed.

SOLVED PROBLEM S7.5

T. Smunt Manufacturing Corp. has the process displayed below. The drilling operation occurs separately from and simultaneously with the sawing and sanding operations. The product only needs to go through one of the three assembly operations (the assembly operations are "parallel").

a) Which operation is the bottleneck?
b) What is the throughput time for the overall system?
c) If the firm operates 8 hours per day, 22 days per month, what is the monthly capacity of the manufacturing process?
d) Suppose that a second drilling machine is added, and it takes the same time as the original drilling machine. What is the new bottleneck time of the system?
e) Suppose that a second drilling machine is added, and it takes the same time as the original drilling machine. What is the new throughput time?

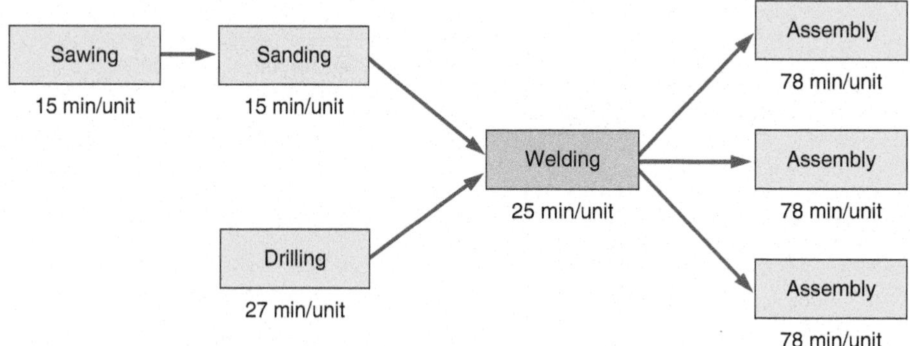

SOLUTION

a) The time for *assembly* is 78 minutes/3 operators = 26 minutes per unit, so the station that takes the longest time, hence the bottleneck, is *drilling*, at 27 minutes.
b) System throughput time is the maximum of [(15 + 15 + 25 + 78), (27 + 25 + 78)] = maximum of (133, 130) = 133 minutes.
c) Monthly capacity = (60 minutes)(8 hours)(22 days)/27 minutes per unit = 10,560 minutes per month/27 minutes per unit = 391.11 units/month.
d) The bottleneck shifts to *Assembly*, with a time of 26 minutes per unit.
e) Redundancy does not affect throughput time. It is still 133 minutes.

Problems

Note: **PX** means the problem may be solved with POM for Windows and/or Excel OM.

Problems S7.1–S7.8 relate to Capacity

• **S7.1** Amy Xia's plant was designed to produce 7,000 hammers per day but is limited to making 6,000 hammers per day because of the time needed to change equipment between styles of hammers. What is the utilization?

• **S7.2** For the past month, the plant in Problem S7.1, which has an effective capacity of 6,500, has made only 4,500 hammers per day because of material delay, employee absences, and other problems. What is its efficiency?

•• **S7.3** If a plant has an effective capacity of 6,500 and an efficiency of 88%, what is the actual (planned) output?

• **S7.4** A plant has an effective capacity of 900 units per day and produces 800 units per day with its product mix. What is its efficiency?

• **S7.5** Material delays have routinely limited production of household sinks to 400 units per day. If the plant efficiency is 80%, what is the effective capacity?

•• **S7.6** The effective capacity and efficiency for the next quarter at MMU Mfg. in Waco, Texas, for each of three departments are shown:

DEPARTMENT	EFFECTIVE CAPACITY	RECENT EFFICIENCY
Design	93,600	.95
Fabrication	156,000	1.03
Finishing	62,400	1.05

Compute the expected production for next quarter for each department.

•• **S7.7** Southeastern Oklahoma State University's business program has the facilities and faculty to handle an enrollment of 2,000 new students per semester. However, in an effort to limit class sizes to a "reasonable" level (under 200, generally), Southeastern's dean, Holly Lutze, placed a ceiling on enrollment of 1,500 new students. Although there was ample demand for business courses last semester, conflicting schedules allowed only 1,450 new students to take business courses. What are the utilization and efficiency of this system?

•• **S7.8** Under ideal conditions, a service bay at a Fast Lube can serve 6 cars per hour. The effective capacity and efficiency of a Fast Lube service bay are known to be 5.5 and 0.880, respectively. What is the minimum number of service bays Fast Lube needs to achieve an anticipated servicing of 200 cars per 8-hour day?

Problems S7.9–S7.15 relate to Bottleneck Analysis and the Theory of Constraints

• **S7.9** A production line at V. J. Sugumaran's machine shop has three stations. The first station can process a unit in 10 minutes. The second station has two identical machines, each of which can process a unit in 12 minutes. (Each unit only needs to be processed on one of the two machines.) The third station can process a unit in 8 minutes. Which station is the bottleneck station?

•• **S7.10** A work cell at Chris Ellis Commercial Laundry has a workstation with two machines, and each unit produced at the station needs to be processed by both of the machines. (The same unit cannot be worked on by both machines simultaneously.) Each machine has a production capacity of 4 units per hour. What is the throughput time of the work cell?

•• **S7.11** The three-station work cell illustrated in Figure S7.7 has a product that must go through one of the two machines at station 1 (they are parallel) before proceeding to station 2.

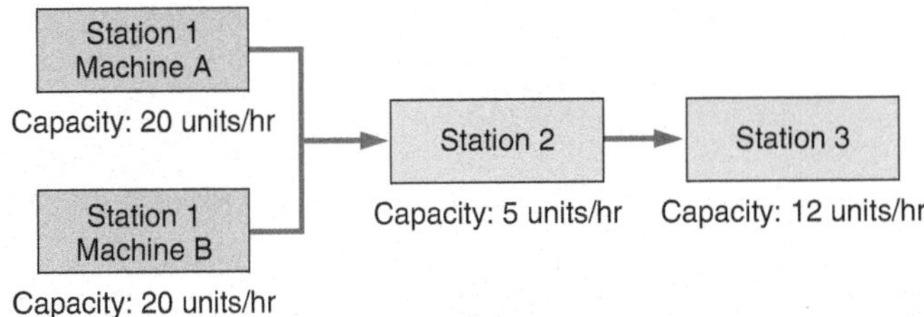

Figure **S7.7**

a) What is the bottleneck time of the system?
b) What is the bottleneck station of this work cell?
c) What is the throughput time?
d) If the firm operates 10 hours per day, 5 days per week, what is the weekly capacity of this work cell?

•• **S7.12** The three-station work cell at Pullman Mfg., Inc. is illustrated in Figure S7.8. It has two machines at station 1 in parallel (i.e., the product needs to go through only one of the two machines before proceeding to station 2).

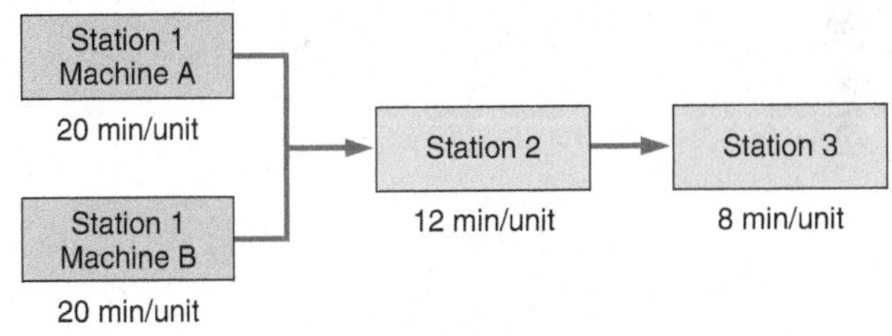

Figure **S7.8**

a) What is the throughput time of this work cell?
b) What is the bottleneck time of this work cell?
c) What is the bottleneck station?
d) If the firm operates 8 hours per day, 6 days per week, what is the weekly capacity of this work cell?

•• **S7.13** The Pullman Mfg., Inc., three-station work cell illustrated in Figure S7.8 has two machines at station 1 in parallel. (The product needs to go through only one of the two machines before proceeding to station 2.) The manager, Seimone Hartley, has asked you to evaluate the system if she adds a parallel machine at station 2.

a) What is the throughput time of the new work cell?
b) What is the bottleneck time of the new work cell?
c) If the firm operates 8 hours per day, 6 days per week, what is the weekly capacity of this work cell?
d) How did the addition of the second machine at workstation 2 affect the performance of the work cell from Problem S7.12?

• **S7.14** Klassen Toy Company, Inc., assembles two parts (parts 1 and 2): Part 1 is first processed at workstation A for 15 minutes per unit and then processed at workstation B for 10

minutes per unit. Part 2 is simultaneously processed at workstation C for 20 minutes per unit. Workstations B and C feed the parts to an assembler at workstation D, where the two parts are assembled. The time at workstation D is 15 minutes.
a) What is the bottleneck of this process?
b) What is the hourly capacity of the process?

•• **S7.15** A production process at Brianna Bryant Manufacturing is shown in Figure S7.9. The drilling operation occurs separately from, and simultaneously with, sawing and sanding, which are independent and sequential operations. A product needs to go through only one of the three assembly operations (the operations are in parallel).
a) Which operation is the bottleneck?
b) What is the bottleneck time?
c) What is the throughput time of the overall system?
d) If the firm operates 8 hours per day, 20 days per month, what is the monthly capacity of the manufacturing process?

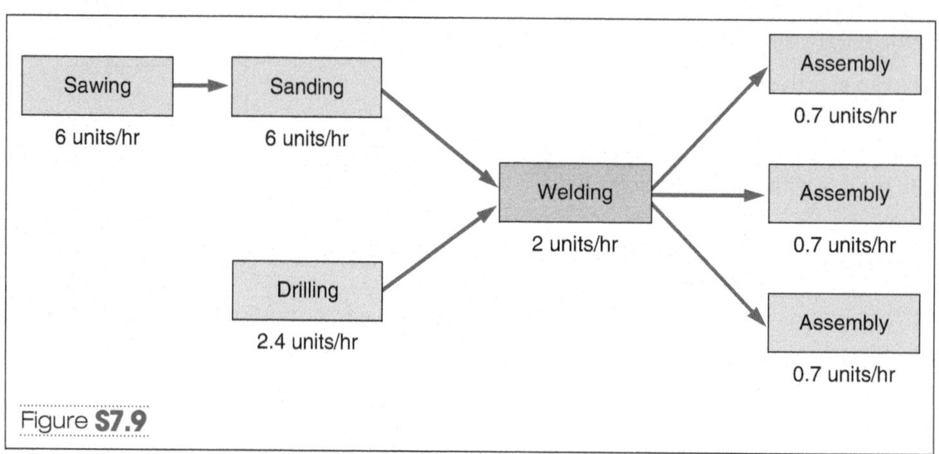

Figure **S7.9**

Problems S7.16–S7.31 relate to Break-Even Analysis

• **S7.16** Achonwa Cutting is opening a new line of scissors for supermarket distribution. It estimates its fixed cost to be $500.00 and its variable cost to be $0.50 per unit. Selling price is expected to average $0.75 per unit.
a) What is Achonwa's break-even point in units?
b) What is the break-even point in dollars? **PX**

• **S7.17** Cortez Manufacturing intends to increase capacity by overcoming a bottleneck operation by adding new equipment. Two vendors have presented proposals. The fixed costs for proposal A are $50,000, and for proposal B, $70,000. The variable cost for A is $12.00, and for B, $10.00. The revenue generated by each unit is $20.00.
a) What is the break-even point in units for proposal A?
b) What is the break-even point in units for proposal B? **PX**

• **S7.18** Using the data in Problem S7.17:
a) What is the break-even point in dollars for proposal A if you add $10,000 installation to the fixed cost?
b) What is the break-even point in dollars for proposal B if you add $10,000 installation to the fixed cost? **PX**

• **S7.19** Given the data in Problem S7.17, at what volume (units) of output would the two alternatives yield the same profit?

•• **S7.20** Janelle Heinke, the owner of Ha'Peppas!, is considering a new oven in which to bake the firm's signature dish, vegetarian pizza. Oven type A can handle 20 pizzas an hour. The fixed costs associated with oven A are $20,000 and the variable costs are $2.00 per pizza. Oven B is larger and can handle 40 pizzas an hour. The fixed costs associated with oven B are $30,000 and the variable costs are $1.25 per pizza. The pizzas sell for $14 each.
a) What is the break-even point for each oven?
b) If the owner expects to sell 9,000 pizzas, which oven should she purchase?
c) If the owner expects to sell 12,000 pizzas, which oven should she purchase?
d) At what volume should Janelle switch ovens? **PX**

Rocha Ribeiro/Shutterstock

•• **S7.21** Given the following data, calculate a) BEP_x; b) $BEP_\$$; and c) the profit at 100,000 units:

$$P = \$8/\text{unit} \quad V = \$4/\text{unit} \quad F = \$50,000$$ **PX**

•• **S7.22** You are considering opening a copy service in the student union. You estimate your fixed cost at $15,000 and the variable cost of each copy sold at $.01. You expect the selling price to average $.05.
a) What is the break-even point in dollars?
b) What is the break-even point in units? **PX**

•• **S7.23** An electronics firm is currently manufacturing an item that has a variable cost of $.50 per unit and a selling price of $1.00

per unit. Fixed costs are $14,000. Current volume is 30,000 units. The firm can substantially improve the product quality by adding a new piece of equipment at an additional fixed cost of $6,000. Variable cost would increase to $.60, but volume should jump to 50,000 units due to a higher-quality product. Should the company buy the new equipment? **PX**

•• **S7.24** The electronics firm in Problem S7.23 is now considering the new equipment and increasing the selling price to $1.10 per unit. With the higher-quality product, the new volume is expected to be 45,000 units. Under these circumstances, should the company purchase the new equipment and increase the selling price? **PX**

•••• **S7.25** Zania Azlett and Angela Zesiger have joined forces to start A&Z Lettuce Products, a processor of packaged shredded lettuce for institutional use. Zania has years of food-processing experience, and Angela has extensive commercial food-preparation experience. The process will consist of opening crates of lettuce and then sorting, washing, slicing, preserving, and finally packaging the prepared lettuce. Together, with help from vendors, they think they can adequately estimate demand, fixed costs, revenues, and variable cost per 5-pound bag of lettuce. They think a largely manual process will have monthly fixed costs of $37,500 and variable costs of $1.75 per bag. A more mechanized process will have fixed costs of $75,000 per month with variable costs of $1.25 per 5-pound bag. They expect to sell the shredded lettuce for $2.50 per 5-pound bag.

a) What is the break-even quantity for the manual process?
b) What is the revenue at the break-even quantity for the manual process?
c) What is the break-even quantity for the mechanized process?
d) What is the revenue at the break-even quantity for the mechanized process?
e) What is the monthly profit or loss of the *manual* process if they expect to sell 60,000 bags of lettuce per month?
f) What is the monthly profit or loss of the *mechanized* process if they expect to sell 60,000 bags of lettuce per month?
g) At what quantity would Zania and Angela be indifferent to the process selected?
h) Over what range of demand would the *manual* process be preferred over the mechanized process? Over what range of demand would the *mechanized* process be preferred over the manual process? **PX**

•••• **S7.26** As a prospective owner of a club known as the Red Rose, you are interested in determining the sales dollars necessary for the coming month to reach the break-even point. You have decided to break down the sales for the club into four categories, the first category being beer. Your estimate of the beer sales is that 30,000 drinks will be served. The selling price for each unit will average $1.50; the cost is $.75. The second major category is meals, which you expect to be 10,000 units with an average price of $10.00 and a cost of $5.00. The third major category is desserts and wine, of which you also expect to sell 10,000 units, but with an average price of $2.50 per unit sold and a cost of $1.00 per unit. The final category is lunches and inexpensive sandwiches, which you expect to total 20,000 units at an average price of $6.25 with a food cost of $3.25. Your fixed cost (i.e., rent, utilities, and so on) is $1,800 per month plus $2,000 per month for entertainment.

a) What is your break-even point in dollars per month?
b) What is the expected number of meals each day if you are open 20 days a month?

••• **S7.27** As manager of the St. Cloud Theatre Company, you have decided that concession sales will support themselves. The following table provides the information you have been able to put together thus far:

ITEM	SELLING PRICE	VARIABLE COST	% OF REVENUE
Soft drink	$2.00	$.65	25
Wine	2.75	.95	25
Coffee	2.00	.75	30
Candy	1.00	.30	20

Last year's manager, Scott Ellis, has advised you to be sure to add 10% of variable cost as a waste allowance for all categories.

You estimate labor cost to be $250.00 (5 booths with 2 people each). Even if nothing is sold, your labor cost will be $250.00, so you decide to consider this a fixed cost. Booth rental, which is a contractual cost at $50.00 for *each* booth per night, is also a fixed cost.

a) What is the break-even volume per evening performance?
b) How much wine would you expect to sell each evening at the break-even point?

•• **S7.28** Wolfgang Kersten Mfg. intends to increase capacity through the addition of new equipment. Two vendors have presented proposals. The fixed costs for proposal X are $150,000, and for proposal Y, $170,000. The variable cost for X is $120.00, and for Y, $100.00. The revenue generated by each unit is $200.00.

a) What is the break-even point in units for proposal X?
b) What is the break-even point in units for proposal Y? **PX**

•• **S7.29** You are given the data in Problem S7.28:

a) What is the break-even point in dollars for proposal X?
b) What is the break-even point in dollars for proposal Y? **PX**

•• **S7.30** Given the data in Problem S7.28, at what volume (units) of output would the two alternatives yield the same profit (loss)? **PX**

•• **S7.31** Use the same data in Problem S7.28:

a) If the expected volume is 28,500 units, which alternative should be chosen?
b) If the expected volume is 15,000 units, which alternative should be chosen? **PX**

Problems S7.32–S7.33 relate to Applying Expected Monetary Value (EMV) to Capacity Decisions

•• **S7.32** Jasmine Dantas' Bed and Breakfast, in a small historic Mississippi town, must decide how to subdivide (remodel) the large old home that will become its inn. There are three alternatives: Option A would modernize all baths and combine rooms, leaving the inn with four suites, each suitable for two to four adults. Option B would modernize only the second floor; the results would be six suites, four for two to four adults, two for two adults only. Option C (the status quo option) leaves all walls intact. In this case, there are eight rooms available, but only two are suitable for four adults, and four rooms will not have private baths. The details of profit and demand patterns that will accompany each option are:

	ANNUAL PROFIT UNDER VARIOUS DEMAND PATTERNS			
ALTERNATIVES	HIGH	P	AVERAGE	P
A (modernize all)	$90,000	.5	$25,000	.5
B (modernize 2nd)	$80,000	.4	$70,000	.6
C (status quo)	$60,000	.3	$55,000	.7

Which option has the highest expected monetary value? **PX**

•••• **S7.33** As operations manager of Holz Furniture, you must make a decision about adding a line of rustic furniture. In discussing the possibilities with your sales manager, Steve Gilbert, you decide that there will definitely be a market and that your firm should enter that market. However, because rustic furniture has a different finish than your standard offering, you decide you need another process line. There is no doubt in your mind about the decision, and you are sure that you should have a second process. But you do question how large to make it. A large process line is going to cost $400,000; a small process line will cost $300,000. The question, therefore, is the demand for rustic furniture. After extensive discussion with Mr. Gilbert and Rosalita Ferrera of Ferrera Market Research, Inc., you determine that the best estimate you can make is that there is a two-out-of-three chance of profit from sales as large as $600,000 and a one-out-of-three chance as low as $300,000. With a large process line, you could handle the high figure of $600,000. However, with a small process line you could not and would be forced to expand (at a cost of $150,000), after which time your profit from sales would be $500,000 rather than the $600,000 because of the lost time in expanding the process. If you do not expand the small process, your profit from sales would be held to $400,000. If you build a small process and the demand is low, you can handle all of the demand.

Should you open a large or small process line?

Problems S7.34–S7.45 relate to Applying Investment Analysis to Strategy-Driven Investments

•• **S7.34** What is the net present value of an investment that costs $75,000 and has a salvage value of $45,000? The annual profit from the investment is $15,000 each year for 5 years. The discount rate at this risk level is 12%. **PX**

• **S7.35** The initial cost of an investment is $65,000 and the discount rate is 10%. The return is $16,000 per year for 8 years. What is the net present value? **PX**

• **S7.36** What is the present value of $5,600 when the discount rate is 8% and the return of $5,600 will not be received for 15 years? **PX**

•• **S7.37** Tim Smunt has been asked to evaluate two machines. After some investigation, he determines that they have the costs shown in the following table. He is told to assume that:

1. The life of each machine is 3 years.
2. The company thinks it knows how to make 12% on investments no riskier than this one.
3. Labor and maintenance are paid at the end of the year.

	MACHINE A	MACHINE B
Original cost	$10,000	$20,000
Labor per year	2,000	4,000
Maintenance per year	4,000	1,000
Salvage value	2,000	7,000

Determine, via the present value method, which machine Tim should recommend.

•••• **S7.38** Your boss has told you to evaluate two ovens for Tink-the-Tinkers, a gourmet sandwich shop. After some questioning of vendors and receipt of specifications, you are assured that the ovens have the attributes and costs shown in the following table. The following two assumptions are appropriate:

1. The life of each machine is 5 years.
2. The company thinks it knows how to make 14% on investments no riskier than this one.

a) Determine via the present value method which machine to tell your boss to purchase.
b) What assumption are you making about the ovens?
c) What assumptions are you making in your methodology?

	THREE SMALL OVENS AT $1,250 EACH	TWO LARGE OVENS AT $2,500 EACH
Original cost	$3,750	$5,000
Labor per year in excess of larger models	$ 750 (total)	
Cleaning/ maintenance	$ 750 ($250 each)	$ 400 ($200 each)
Salvage value	$ 750 ($250 each)	$1,000 ($500 each)

•••• **S7.39** Bold's Gym, a health club chain, is considering expanding into a new location: the initial investment would be $1 million in equipment, renovation, and a 6-year lease, and its annual upkeep and expenses would be $75,000 (paid at the beginning of the year). Its planning horizon is 6 years out, and at the end, it can sell the equipment for $50,000. Club capacity is 500 members who would pay an annual fee of $600. Bold's expects to have no problems filling membership slots. Assume that the discount rate is 10%. (See Table S7.2.)

a) What is the present value profit/loss of the deal?
b) The club is considering offering a special deal to the members in the first year. For $3,000 upfront they get a full 6-year membership (i.e., 1 year free). Would it make financial sense to offer this deal?

•• **S7.40** Kurt Hozak, VP of Operations at Monterrey Manufacturing, has to make a decision between two investment alternatives. Investment A has an initial cost of $61,000, and investment B has an initial cost of $74,000. The useful life of investment A is 6 years; the useful life of investment B is 7 years. Given a discount rate of 9% and the following cash flows for each alternative, determine the most desirable investment alternative according to the net present value criterion. **PX**

	YEAR 1	YEAR 2	YEAR 3	YEAR 4	YEAR 5	YEAR 6	YEAR 7
Investment A's Cash Flow	$19,000	$19,000	$19,000	$19,000	$19,000	$19,000	–
Investment B's Cash Flow	19,000	20,000	21,000	22,000	21,000	20,000	11,000

•• **S7.41** Evaluate the following capital investments according to net present value. Each alternative requires an initial investment of $20,000. Assume a 10% discount rate. Which is the preferred investment? **PX**

YEAR	CASH FLOW FROM INVESTMENT 1	CASH FLOW FROM INVESTMENT 2	CASH FLOW FROM INVESTMENT 3
1	$1,000	$ 7,000	$ 10,000
2	1,000	6,000	5,000
3	3,000	5,000	3,000
4	15,000	4,000	2,000
5	3,000	4,000	1,000
6	1,000	4,000	1,000
7	–	4,000	1,000
8	1,000	2,000	–
9	–	–	1,000

•• **S7.42** Tony and Steve are considering whether to purchase a new "bending brake." This machine puts precise bends in a material used in their vinyl siding business. The machine will cost \$70,000. Tony and Steve estimate that the machine will generate profits as follows: \$20,000 in its first year; \$15,000 in years 2, 3, and 4; and \$10,000 in years 5 and 6. They believe the machine will have no value after year 6.

a) Should they purchase the machine if they believe they can make 11% on their money in other investments of similar risk?
b) Should they purchase the machine if they believe they can make only 4% on their money in other investments of similar risk? **PX**

•• **S7.43** Justin Bateh is considering a new machine for his shop. The machine is expected to generate receipts as follows: \$50,000 in year 1, \$30,000 in year 2, nothing in the next year, and \$20,000 in year 4. At a discount rate of 6%, what is the present value of these receipts? **PX**

•• **S7.44** Cheryl Druehl needs to purchase a new milling machine. She is considering two different competing machines. Milling Machine A will cost \$300,000 and will return \$80,000 per year for 6 years, with no salvage value. Milling machine B will cost \$220,000 and will return \$60,000 for 5 years, with a salvage value of \$30,000. The firm is currently using 7% as the discount rate. Using net present value as the criterion, which machine should be purchased?

•• **S7.45** Portland Savings and Loan is considering new computer software, which, because of installation and training cost, will have an unusual pattern of net receipts. The expected receipts are: \$20,000 in year 1, nothing in the next year, \$30,000 in year 3, and \$50,000 in year 4. At a discount rate of 6% what is the present value of these receipts? **PX**

CASE STUDY

Capacity Planning at Arnold Palmer Hospital

Video Case

Since opening day, Arnold Palmer Hospital has experienced an explosive growth in demand for its services. One of only six hospitals in the U.S. to specialize in health care for women and children, Arnold Palmer Hospital has cared for more than 1,500,000 patients who came to the Orlando facility from all 50 states and more than 100 other countries. With patient satisfaction scores in the top 10% of U.S. hospitals surveyed (over 95% of patients would recommend the hospital to others), one of Arnold Palmer Hospital's main focuses is delivery of babies. Originally built with 281 beds and a capacity for 6,500 births per year, the hospital steadily approached and then passed 10,000 births. Looking at Table S7.4, Executive Director Kathy Swanson knew an expansion was necessary.

With continuing population growth in its market area serving 18 central Florida counties, Arnold Palmer Hospital was delivering the equivalent of a kindergarten class of babies every day and still not meeting demand. Supported with substantial additional demographic analysis, the hospital was ready to move ahead with a capacity expansion plan and a new 11-story hospital building across the street from the existing facility.

Thirty-five planning teams were established to study such issues as (1) specific forecasts, (2) services that would transfer to the new facility, (3) services that would remain in the existing facility, (4) staffing needs, (5) capital equipment, (6) pro forma accounting data, and (7) regulatory requirements. Ultimately, Arnold Palmer Hospital was ready to move ahead with a budget of \$100 million and a commitment to an additional 150 beds. But given the growth of the central Florida region, Swanson decided to expand the hospital in stages: the top two floors would be empty interiors ("shell") to be completed at a later date, and the fourth-floor operating room could be doubled in size when needed. "With the new facility in place, we are now able to handle up to 16,000 births per year," says Swanson.

Discussion Questions*

1. Given the capacity planning discussion in the text (see Figure S7.6), what approach is being taken by Arnold Palmer Hospital toward matching capacity to demand?
2. What kind of major changes could take place in Arnold Palmer Hospital's demand forecast that would leave the hospital with an underutilized facility (namely, what are the risks connected with this capacity decision)?
3. Use regression analysis to forecast the point at which Swanson needs to "build out" the top two floors of the new building, namely, when demand will exceed 16,000 births.

TABLE S7.4 Births at Arnold Palmer Hospital

YEAR	BIRTHS
2000	8,655
2001	9,536
2002	9,825
2003	10,253
2004	10,555
2005	12,316
2006	13,070
2007	14,028
2008	14,241
2009	13,050
2010	12,571
2011	12,978
2012	13,529
2013	13,576
2014	13,994
2015	14,898
2016	14,600
2017	14,111
2018	13,228
2019	13,535
2020	13,990

*You may wish to view the video that accompanies the case before addressing these questions.

Endnote

1. Sources: *International Business Times* (March 3, 2017); *Forbes* (May 1, 2020); and **USnews.com** (November 3, 2020).

Bibliography

Anderson, E. J., and S.-J. S. Yang. "The Timing of Capacity Investment with Lead Times: When Do Firms Act in Unison?" *Production and Operations Management* 24, no. 1 (January 2015): 21–41.

Anupindi, R., S. Chopra, S. D. Deshmukh, J. A. Van Mieghem, and E. Zemel. *Managing Business Process Flows: Principles of Operations Management,* 3rd ed. Boston: Pearson (2011).

Goldratt, E, and J. Cox. *The Goal: A Process of Ongoing Improvement*, 4th rev. ed. Great Barrington, MA: North River Press (2014).

Jackson, J. E., and C. L. Munson. "Shared Resource Capacity Expansion Decisions for Multiple Products with Quantity Discounts." *European Journal of Operational Research* 253 (2016): 602–613.

Li, S., and L. Wang. "Outsourcing and Capacity Planning in an Uncertain Global Environment." *European Journal of Operational Research* 207, no. 1 (November 2010): 131–141.

Roy, A. "Strategic Positioning and Capacity Utilization: Factors in Planning for Profitable Growth in Banking." *Journal of Performance Management* 23, no. 3 (November 2010): 23.

Schuz, J. D. "Managing Capacity, Key to Profitability." *Logistics Management* 54, no. 7 (July 2015): 28–30.

Watson, K. J., J. H. Blackstone, and S. C. Gardiner. "The Evolution of a Management Philosophy: The Theory of Constraints." *Journal of Operations Management* 25, no. 2 (March 2007): 387–402.

Supplement 7 *Rapid* Review

Main Heading	Review Material	MyLab Operations Management
CAPACITY	■ **Capacity**—The "throughput," or number of units a facility can hold, receive, store, or produce in a period of time. Capacity decisions often determine capital requirements and therefore a large portion of fixed cost. Capacity also determines whether demand will be satisfied or whether facilities will be idle. *Determining facility size, with an objective of achieving high levels of utilization and a high return on investment, is critical.* Capacity planning can be viewed in three time horizons: 1. *Long-range* (>1 year)—Adding facilities and long lead-time equipment 2. *Intermediate-range* (3–18 months)—"Aggregate planning" tasks, including adding equipment, personnel, and shifts; subcontracting; and building or using inventory 3. *Short-range* (<3 months)—Scheduling jobs and people, and allocating machinery ■ **Design capacity**—The theoretical maximum output of a system in a given period, under ideal conditions. Most organizations operate their facilities at a rate less than the design capacity. ■ **Effective capacity**—The capacity a firm can expect to achieve, given its product mix, methods of scheduling, maintenance, and standards of quality. ■ **Utilization**—Actual output as a percent of design capacity. ■ **Efficiency**—Actual output as a percent of effective capacity. Utilization = Actual output/Design capacity (S7-1) Efficiency = Actual output/Effective capacity (S7-2) When demand exceeds capacity, a firm may be able to curtail demand simply by raising prices, increasing lead times (which may be inevitable), and discouraging marginally profitable business. When capacity exceeds demand, a firm may want to stimulate demand through price reductions or aggressive marketing, or it may accommodate the market via product changes. In the service sector, scheduling customers is *demand management,* and scheduling the workforce is *capacity management.* When demand and capacity are fairly well matched, demand management in services can often be handled with appointments, reservations, or a first-come, first-served rule. Otherwise, discounts based on time of day may be used (e.g., "early bird" specials, matinee pricing). When managing demand in services is not feasible, managing capacity through changes in full-time, temporary, or part-time staff may be an option.	Concept Questions: 1.1–1.6 Problems: S7.1–S7.8 Virtual Office Hours for Solved Problem: S7.1 ACTIVE MODEL S7.1
BOTTLENECK ANALYSIS AND THE THEORY OF CONSTRAINTS	■ **Capacity analysis**—Determining throughput capacity of workstations or an entire production system. ■ **Bottleneck**—The limiting factor or constraint in a system. ■ **Process time**—The time to produce a unit (or batch) at a workstation. ■ **Bottleneck time**—The process time of the longest (slowest) process. ■ **Throughput time**—The time it takes for a product to go through the production process *with no waiting*, i.e., the time of the longest path through the system. If n parallel (redundant) operations are added, the process time of the combined operations will equal $1/n$ times the process time of the original. With simultaneous processing, an order or product is essentially *split* into different paths to be rejoined later on. The longest path through the system is deemed the throughput time. ■ **Theory of constraints (TOC)**—A body of knowledge that deals with anything limiting an organization's ability to achieve its goals.	Concept Questions: 2.1–2.6 Problems: S7.9–S7.15 Virtual Office Hours for Solved Problem: S7.5
BREAK-EVEN ANALYSIS	■ **Break-even analysis**—A means of finding the point, in dollars and units, at which costs equal revenues. *Fixed costs* are costs that exist even if no units are produced. Variable costs are those that vary with the volume of units produced. In the break-even model, costs and revenue are assumed to increase linearly.	Concept Questions: 3.1–3.5 Problems: S7.16–S7.31

Main Heading	Review Material	MyLab Operations Management
	$$\text{Break-even in units} = \frac{\text{Total Fixed cost}}{\text{Price} - \text{Variable cost}} = \frac{F}{P - V} \quad \text{(S7-3)}$$ $$\text{Break-even in dollars} = \frac{\text{Total Fixed cost}}{1 - \frac{\text{Variable cost}}{\text{Price}}} = \frac{F}{1 - \left(\frac{V}{P}\right)} \quad \text{(S7-4)}$$ $$\text{Multiproduct break-even in dollars} = BEP_\$ = \frac{F}{\sum\left[\left(1 - \frac{V_i}{P_i}\right) \times (W_i)\right]} \quad \text{(S7-5)}$$	Virtual Office Hours for Solved Problem: S7.3 **ACTIVE MODEL S7.2**
REDUCING RISK WITH INCREMENTAL CHANGES	Demand growth is usually in small units, while capacity additions are likely to be both instantaneous and in large units. To reduce risk, incremental changes that hedge demand forecasts may be a good option. Four approaches to capacity expansion are (1) *leading* strategy, with incremental expansion, (2) *leading* strategy with one step expansion, (3) *lag* strategy, and (4) *straddle* strategy. Both lag strategy and straddle strategy delay capital expenditure.	Concept Questions: 4.1–4.5 **VIDEO S7.1** Capacity Planning at Arnold Palmer Hospital
APPLYING EXPECTED MONETARY VALUE (EMV) TO CAPACITY DECISIONS	Determining expected monetary value requires specifying alternatives and various states of nature (e.g., demand or market favorability). By assigning probability values to the various states of nature, we can make decisions that maximize the expected value of the alternatives.	Concept Questions: 5.1–5.4 Problems: S7.32–S7.33
APPLYING INVESTMENT ANALYSIS TO STRATEGY-DRIVEN INVESTMENTS	■ **Net present value**—A means of determining the discounted value of a series of future cash receipts. ■ **Discount rate**—The rate of return offered by investment alternatives that is used to discount expected future payoffs. (It may also be referred to as *cost of capital* or *interest rate*.) $$F = P(1 + i)^N \quad \text{(S7-6)}$$ $$P = \frac{F}{(1 + i)^N} \quad \text{(S7-7)}$$ $$P = \frac{F}{(1 + i)^N} = FX \quad \text{(S7-8)}$$ When making several investments, those with higher net present values are preferable to investments with lower net present values.	Concept Questions: 6.1–6.4 Problems: S7.34–S7.45 Virtual Office Hours for Solved Problem: S7.4
ADDITIONAL MYLAB OPERATIONS MANAGEMENT RESOURCES	✔ Video for Creating Your Own Excel Spreadsheets (Example S6) ✔ Additional Case Studies (Southwestern University (D) & Global Chemical's COVID-19 Capacity Decision) ✔ Southwestern University Case Studies are integrated in Chapters 3, 4, 6, 8, 12, and 13 and in Supplement 7 ✔ Multiple Choice Case Questions (Southwestern University (D) & Global Chemical's COVID-19 Capacity Decision) ✔ Recent Graduate Video: Gabrielle Sliwinski, Controller, Student Transportation of America	

Self Test

LO S7.1 Capacity decisions should be made on the basis of:
- **a)** building sustained competitive advantage.
- **b)** good financial returns.
- **c)** a coordinated plan.
- **d)** integration into the company's strategy.
- **e)** all of the above.

LO S7.2 Effective capacity is:
- **a)** the capacity a firm expects to achieve, given the current operating constraints.
- **b)** the percentage of design capacity actually achieved.
- **c)** the percentage of capacity actually achieved.
- **d)** actual output.
- **e)** efficiency.

LO S7.3 System capacity is based on:
- **a)** the bottleneck.
- **b)** throughput time.
- **c)** time of the fastest station.
- **d)** throughput time plus waiting time.
- **e)** none of the above.

LO S7.4 The break-even point is:
- **a)** adding processes to meet the point of changing product demands.
- **b)** improving processes to increase throughput.
- **c)** the point in dollars or units at which cost equals revenue.
- **d)** adding or removing capacity to meet demand.
- **e)** the total cost of a process alternative.

LO S7.5 Expected monetary value is most appropriate:
- **a)** when the payoffs are equal.
- **b)** when the probability of each decision alternative is known.
- **c)** when probabilities are the same.
- **d)** when both revenue and cost are known.
- **e)** when probabilities of each state of nature are known.

LO S7.6 Net present value:
- **a)** is greater if cash receipts occur later rather than earlier.
- **b)** is greater if cash receipts occur earlier rather than later.
- **c)** is revenue minus fixed cost.
- **d)** is preferred over break-even analysis.
- **e)** is greater if \$100 monthly payments are received in a lump sum (\$1,200) at the end of the year.

Answers: LO S7.1. e; LO S7.2. a; LO S7.3. a; LO S7.4. c; LO S7.5. b; LO S7.6. b.

Location Strategies

CHAPTER 8

CHAPTER OUTLINE

Geographic information systems (GISs) are used by a variety of firms to identify target markets by income, ethnicity, product use, age, etc. Here, data from MapInfo helps with competitive analysis for a retailer adding a new location. Three concentric blue rings, each representing various mile radii, were drawn around the competitor's store. The heavy red line indicates the "drive" time to the firm's own central store (the red dot).

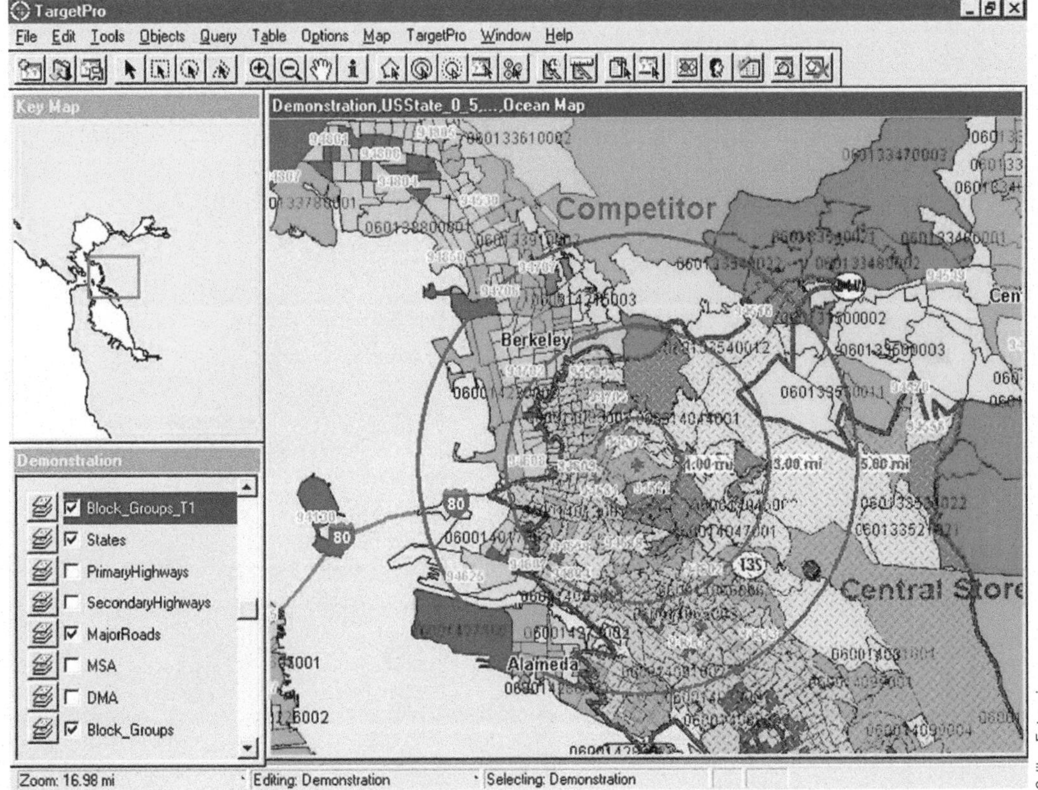

Galilee Enterprise

10 OM STRATEGY DECISIONS

- Design of Goods and Services
- Managing Quality
- Process Strategies
- *Location Strategies*
- Layout Strategies
- Human Resources
- Supply Chain Management
- Inventory Management
- Scheduling
- Maintenance

GLOBAL COMPANY PROFILE
FedEx

Location Provides Competitive Advantage for FedEx

Overnight-delivery powerhouse FedEx has believed in the hub concept for its 50-plus year existence. Even though Fred Smith, founder and CEO, got a C on his college paper proposing a hub for small-package delivery, the idea has proven extremely successful. Starting with one central location in Memphis, Tennessee (now called its *superhub*), the $69 billion firm has added a European hub in Paris, an Asian hub in Guangzhou, China, a Latin American hub in Miami, and a Canadian hub in Toronto. FedEx's fleet of over 680 planes flies into 375 airports worldwide, then delivers to the door with more than 100,000 vans and trucks.

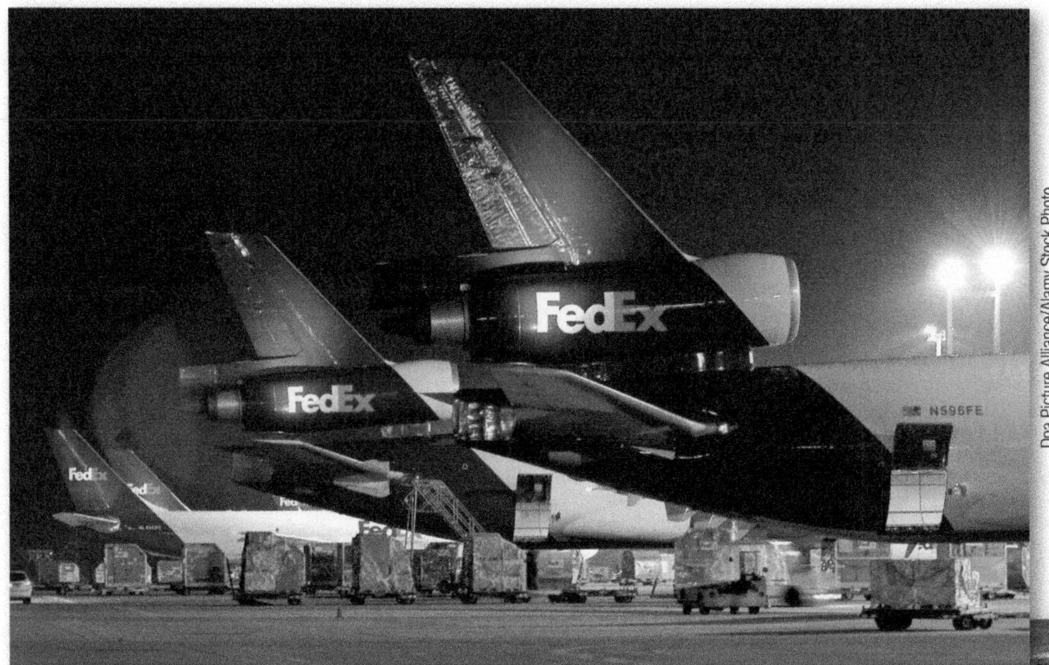

Dpa Picture Alliance/Alamy Stock Photo

At the FedEx hub in Memphis, Tennessee, approximately 100 FedEx aircraft converge each night around midnight with more than 5 million documents and packages.

Ben Margot/AP/Shutterstock

Packages are sorted by priority, air, ground, etc., and then the air freight packages are re-sorted to a container that goes to the plane. The Memphis facility covers 2.4 million square feet; it is big enough to hold 50 football fields. Packages are sorted and exchanged until 4 A.M.

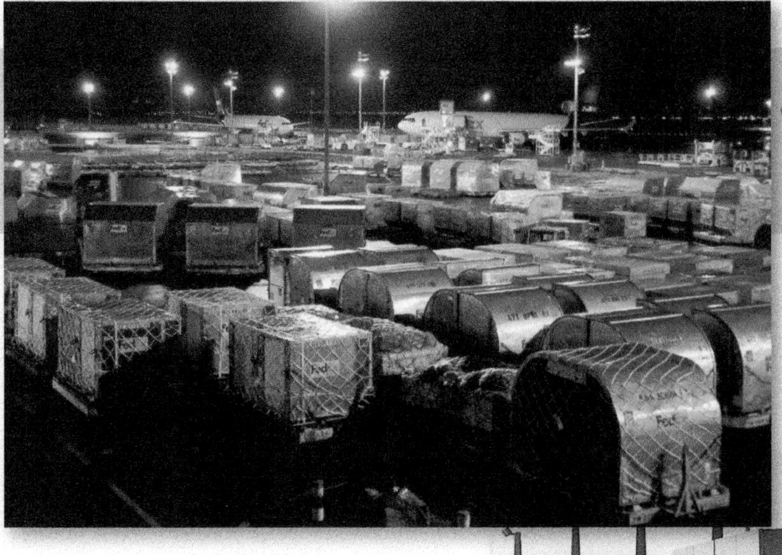

Gilles ROLLE/REA/Redux

Packages and documents that have already gone through the sorting process are checked by city, state, and zip code. They are then placed in containers that are loaded onto aircraft for delivery to their final destinations in 236 countries.

FedEx's fleet of over 680 planes makes it the largest airline in the world. More than 100,000 trucks complete the delivery process.

Matt York/AP Images

The $150 million hub opened in Guangzhou lies in the heart of one of China's fastest-growing manufacturing districts. FedEx controls 39% of the China-to-U.S. air express market.

testing/Shutterstock

Why was Memphis picked as FedEx's central location? (1) It is located in the middle of the U.S. (2) It has very few hours of bad weather closures, perhaps contributing to the firm's excellent flight-safety record. (3) It provided FedEx with generous tax incentives.

Each night, except Sunday, FedEx brings to Memphis packages from throughout the world that are going to cities for which FedEx does not have direct flights. The central hub permits service to a far greater number of points with fewer aircraft than the traditional City-A-to-City-B system. It also allows FedEx to match aircraft flights with package loads each night and to reroute flights when load volume requires it, a major cost savings. Moreover, FedEx also believes that the central hub system helps reduce mishandling and delay in transit because there is total control over the packages from pickup point through delivery.

LEARNING OBJECTIVES

- **LO 8.1** *Identify* and explain seven major factors that affect location decisions 338
- **LO 8.2** *Compute* labor productivity 338
- **LO 8.3** *Apply* the factor-rating method 341
- **LO 8.4** *Complete* a locational cost–volume analysis graphically and mathematically 343
- **LO 8.5** *Use* the center-of-gravity method 344
- **LO 8.6** *Understand* the differences between service- and industrial-sector location analysis 347

The Strategic Importance of Location

VIDEO 8.1
Hard Rock's Location Selection

World markets continue to expand, and the global nature of business is accelerating. Indeed, one of the most important strategic decisions made by many companies, including FedEx, Amazon, and Hard Rock, is where to locate their operations. When FedEx opened its Asian hub in Guangzhou, China, it set the stage for "round-the-world" flights linking its Paris and Memphis package hubs to Asia. When Seattle-based Amazon announced its plan to build a second major headquarters (called HQ2) in Northern Virginia, it had completed a 14-month competition with proposals from 230 cities in Canada, Mexico, and the U.S. When Hard Rock Cafe opened in Moscow, it ended 3 years of advance preparation of a Russian food-supply chain. The strategic impact, cost, and international aspect of these decisions indicate how significant location decisions are.

Firms throughout the world are using the concepts and techniques of this chapter to address the location decision because location greatly affects both fixed and variable costs. Location has a major impact on the overall risk and profit of the company. For instance, depending on the product and type of production or service taking place, transportation costs alone can total as much as 25% of the product's selling price. That is, one-fourth of a firm's total revenue may be needed just to cover logistics expenses of the raw materials coming in and finished products going out. Other costs that may be influenced by location include taxes, wages, raw material costs, and rents. When all costs are considered, location may alter total operating expenses as much as 50%.

The economics of transportation are so significant that companies—and even cities—have coalesced around a transportation advantage. For centuries, rivers and ports, and more recently rail hubs and then interstate highways, were a major ingredient in the location decision. Today airports are often the deciding factor, providing fast, low-cost transportation of goods and people.

Companies make location decisions relatively infrequently, usually because demand has outgrown the current plant's capacity or because of changes in labor productivity, exchange rates, costs, or local attitudes. Companies may also relocate their manufacturing or service facilities because of shifts in demographics and customer demand.

Location options include (1) expanding an existing facility instead of moving, (2) maintaining current sites while adding another facility elsewhere, or (3) closing the existing facility and moving to another location.

The location decision often depends on the type of business. For industrial location decisions, the strategy is usually minimizing costs, although locations that foster innovation and creativity may also be critical. For retail and professional service organizations, the strategy focuses on maximizing revenue. Warehouse location strategy, however, may be driven by a combination of cost and speed of delivery. *The objective of location strategy is to maximize the benefit of location to the firm.*

Location and Costs Because location is such a significant cost and revenue driver, location often has the power to make (or break) a company's business strategy. Key multinationals in every major industry, from automobiles to cellular phones, now have or are planning a presence in each of their major markets. Location decisions to support a low-cost strategy require particularly careful consideration.

Once management is committed to a specific location, many costs are firmly in place and difficult to reduce. For instance, if a new factory location is in a region with high energy costs, even good management with an outstanding energy strategy is starting at a disadvantage. Management is in a similar bind with its human resource strategy if labor in the selected location is expensive or ill-trained or has a poor work ethic. Consequently, hard work to determine an optimal facility location is a good investment.

TABLE 8.1

Ease of Doing Business Ranking in 190 Countries

RANK	COUNTRY	SCORE
1	New Zealand	86.8
2	Singapore	86.2
3	Hong Kong, Denmark (tie)	85.3
5	South Korea, U.S. (tie)	84.0
	⋮	
22	Germany	79.7
23	Canada	79.6
	⋮	
28	Russia	78.2
29	Japan	78.0
	⋮	
31	China	77.9
	⋮	
35	Israel	76.7
	⋮	
60	Mexico	72.4
	⋮	
63	India	71.0
	⋮	
124	Brazil	59.1
	⋮	
190	Somalia	20.0

Source: World Bank, 2020.

Factors That Affect Location Decisions

Selecting a facility location is becoming much more complex with globalization. As we saw in Chapter 2, globalization has taken place because of the development of (1) market economics; (2) better international communications; (3) more rapid, reliable travel and shipping; (4) ease of capital flow between countries; and (5) high differences in labor costs. Many firms now consider opening new offices, factories, retail stores, or banks outside their home country. Location decisions transcend national borders. In fact, as Figure 8.1 shows, the sequence of location decisions often begins with choosing a country in which to operate.

One approach to selecting a country is to identify what the parent organization believes are key success factors (KSFs) needed to achieve competitive advantage. Six possible country KSFs are listed at the top of Figure 8.1. Using such factors (including some negative ones, such as crime) the World Bank annually ranks the ease of doing business in 190 countries (see Table 8.1).

Once a firm decides which country is best for its location, it focuses on a region of the chosen country and a community. The final step in the location decision process is choosing a specific site within a community. The company must pick the one location that is best suited for shipping and receiving, zoning, utilities, size, and cost. Again, Figure 8.1 summarizes this series of decisions and the factors that affect them.

Country Decision

Key Success Factors

1. Political risks, government rules, attitudes, incentives
2. Cultural and economic issues
3. Location of markets
4. Labor talent, attitudes, productivity, costs
5. Availability of supplies, communications, energy
6. Exchange rates and currency risk

Region/Community Decision

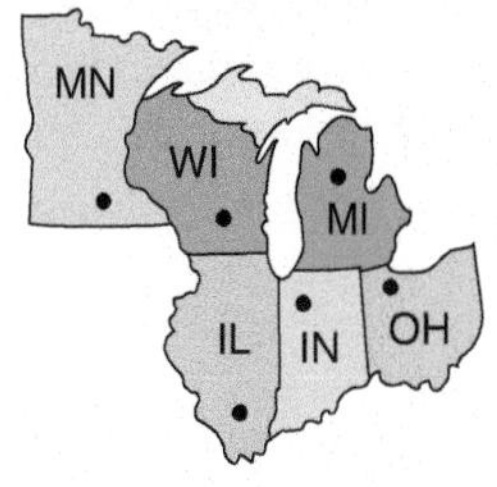

1. Corporate desires
2. Attractiveness of region (culture, taxes, climate, etc.)
3. Labor availability, costs, attitudes toward unions
4. Cost and availability of utilities
5. Environmental regulations of state and town
6. Government incentives and fiscal policies
7. Proximity to raw materials and customers
8. Land/construction costs

Site Decision

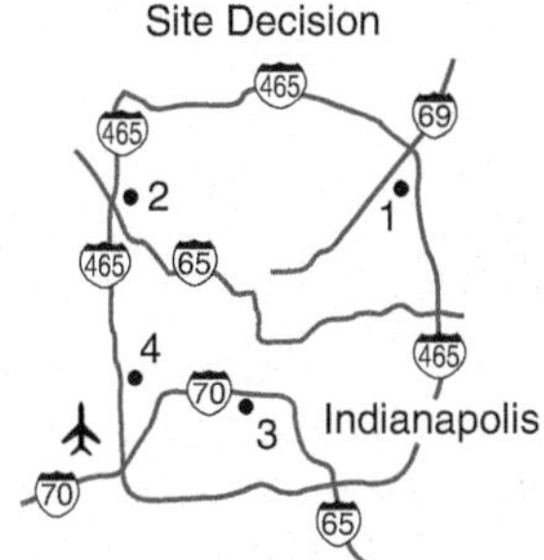

1. Site size and cost
2. Air, rail, highway, and waterway systems
3. Zoning restrictions
4. Proximity of services/supplies needed
5. Environmental impact issues
6. Customer density and demographics

Figure **8.1**

Some Considerations and Factors That Affect Location Decisions

LO 8.1 *Identify* and explain seven major factors that affect location decisions

Besides globalization, a number of other factors affect the location decision. Among these are labor productivity, foreign exchange, culture, changing attitudes toward the industry, and proximity to markets, suppliers, and competitors.

Labor Productivity

LO 8.2 *Compute* labor productivity

When deciding on a location, management may be tempted by an area's relatively lower wage rates. However, wage rates cannot be considered by themselves, as Otis Elevator discovered when it opened a plant in Latin America. Otis subsequently found a move to an automated plant in South Carolina more advantageous. Management must also consider productivity.

As discussed in Chapter 1, differences exist in productivity in various countries. What management is really interested in is the combination of production and the wage rate. For example, if Otis Elevator pays $70 per day with 60 units produced per day in South Carolina, it will spend less on labor than at a Latin American plant that pays $25 per day with production of 20 units per day:

$$\frac{\text{Labor cost per day}}{\text{Production (units per day)}} = \text{Labor cost per unit}$$

1. Case 1: South Carolina plant:

$$\frac{\$70 \text{ Wages per day}}{60 \text{ Units produced per day}} = \frac{\$70}{60} = \$1.17 \text{ per unit}$$

2. Case 2: Latin American plant:

$$\frac{\$25 \text{ Wages per day}}{20 \text{ Units produced per day}} = \frac{\$25}{20} = \$1.25 \text{ per unit}$$

STUDENT TIP Final cost is the critical factor, and lower productivity can negate lower wages.

When opportunities for training and access to education are limited, employee productivity often declines. In addition, if the infrastructure is inadequate, employees' punctuality and attendance rates will be impacted. Management may find that low productivity is not a good buy even at bargain wages. (Labor cost per unit is sometimes called the *labor content* of the product.)

Exchange Rates and Currency Risk

Although wage rates and productivity may make a country seem economical, unfavorable exchange rates may negate any savings. Sometimes, though, firms can take advantage of a particularly favorable exchange rate by relocating or exporting to a foreign country. However, the values of foreign currencies continually rise and fall in most countries. Such changes could well make what was a good location in 2022 a disastrous one in 2026. *Operational hedging* describes the situation where firms have excess capacity in multiple countries and then shift production levels from location to location as exchange rates change.

Costs

Tangible costs
Readily identifiable costs that can be measured with some precision.

We can divide location costs into two categories, tangible and intangible. **Tangible costs** are those costs that are readily identifiable and precisely measured. They include electricity (as we see in the *OM in Action* box, "Bitcoin Goes to Where the Power Is Cheap"), other utilities, labor, material, taxes, depreciation, and other costs that the accounting department and management can identify. In addition, such costs as transportation of raw materials, transportation of finished goods, and site construction are all factored into the overall cost of a location.

Intangible costs
A category of location costs that cannot be easily quantified, such as quality of life and government.

Intangible costs are less easily quantified. They include quality of education, public transportation facilities, community attitudes toward the industry and the company, and quality and attitude of prospective employees. They also include quality-of-life variables, such as climate and sports teams, that may influence personnel recruiting.

OM in Action Bitcoin Goes to Where the Power Is Cheap[1]

Home to hydroelectric dams that harness the flow of the Columbia River, north central Washington has some of the cheapest power in the U.S. That has made the largely rural area best known for its apple orchards a magnet for bitcoin miners, who use powerful specialized computers to generate new units of cryptocurrencies—a process that requires vast amounts of electricity to run and cool thousands of machines. "If you ask the guys at UPS or FedEx what they're delivering to Wenatchee, I think they'd tell you it's a whole bunch of bitcoin mining machines," says that town's mayor.

Mining operations can squeeze into small spaces. Shoebox-size computer servers that suck up as much power as 1,000 homes can be packed into a 25-by-25-foot room. Miners have popped up in unexpected places in the area: an old laundromat, a former warehouse, and apartments. There are already at least 30 *known* cryptocurrency-mining operations in north central Washington.

These aren't the first businesses to come to the region for its cheap power. Aluminum smelters once flocked here. In more recent years, companies including Microsoft and Dell have built data-storage centers. Electricity in the region costs 2 to 4 cents per kwh compared with more than 10 cents nationwide. Some residents and officials hope that mining will be the first step toward transforming the area into a business hub for *blockchain* technology, bringing new jobs.

PHOTOCREO Michal Bednarek/Shutterstock

Others worry these miners will drain the area of the surplus power that helps keep rates low. Here is why: Comparative power usage rates (per sq. ft. per year): school—10, home—12, hotel—18, hospital—32, grocery store—40, computer data center—2,100!

Political Risk, Values, and Culture

The political risk associated with national, state, and local governments' attitudes toward private and intellectual property, zoning, pollution, and employment stability may be in flux. Governmental positions at the time a location decision is made may not be lasting ones. However, management may find that these attitudes can be influenced by their own leadership.

Worker values may also differ from country to country, region to region, and small town to city. Worker views regarding turnover, unions, and absenteeism are all relevant factors. In turn, these values can affect a company's decision whether to make offers to current workers if the firm relocates to a new location. The case study at the end of this chapter, "National Assembly Services," describes a Cincinnati firm that actively chose *not to relocate* any of its union workers when it moved to Mississippi.

One of the greatest challenges in a global operations decision is dealing with another country's culture. Cultural variations in punctuality by employees and suppliers make a marked difference in production and delivery schedules. Bribery and other forms of corruption also create substantial economic inefficiency, as well as ethical and legal problems in the global arena. As a result, operations managers face significant challenges when building effective supply chains across cultures. Table 8.2 provides one ranking of corruption in countries around the world.

TABLE 8.2

Ranking Corruption in Selected Countries (Score of 100 Represents a Corruption-Free Country

RANK	SCORE
1 New Zealand	88
2 Denmark	88
3 Finland	85
⋮	
11 Canada	77
⋮	
25 U.S., Chile	67 (tie)
⋮	
35 Israel	60
⋮	
78 China	42
⋮	
129 Russia	30
⋮	
170 North Korea	18
⋮	
179 Somalia	12

Source: Transparency International's 2020 survey, at **www.transparency.org**. Used with permission of Transparency International.

Proximity to Markets

For many firms, locating near customers is extremely important. Particularly, service organizations, like drugstores, restaurants, post offices, or barbers, find that demographics and proximity to market are *the* primary location factors. Manufacturing firms find it useful to be close to customers when transporting finished goods is expensive or difficult (perhaps because they are bulky, heavy, or fragile). To be near U.S. markets, foreign-owned auto giants such as Mercedes, Honda, Toyota, and Hyundai are building millions of cars each year in the U.S.

In addition, with just-in-time production, suppliers want to locate near users. For a firm like Coca-Cola, whose product's primary ingredient is water, it makes sense to have bottling plants in many cities rather than shipping heavy (and sometimes fragile glass) containers cross country.

Proximity to Suppliers

Firms locate near their raw materials and suppliers because of (1) perishability, (2) transportation costs, or (3) bulk. Bakeries, dairy plants, and frozen seafood processors deal with *perishable* raw materials, so they often locate close to suppliers. Companies dependent on inputs of heavy or bulky raw materials (such as steel producers using coal and iron ore) face expensive inbound *transportation costs*, so transportation costs become a major factor. And goods for which there is a *reduction in bulk* during production (such as a sawmill cutting trees to lumber) typically need facilities near the raw material.

Proximity to Competitors (Clustering)

Clustering
The location of competing companies near each other, often because of a critical mass of information, talent, venture capital, or natural resources.

Both manufacturing and service organizations also like to locate, somewhat surprisingly, near competitors. This tendency, called **clustering**, often occurs when a major resource is found in that region. Such resources include natural resources, information resources, venture capital resources, and talent resources. Table 8.3 presents nine examples of industries that exhibit clustering, and the reasons why.

Italy may be the true leader when it comes to clustering, however, with northern zones of that country holding world leadership in such specialties as ceramic tile (Modena), gold jewelry (Vicenza), machine tools (Busto Arsizio), cashmere and wool (Biella), designer eyeglasses (Belluma), and pasta machines (Parma). When it comes to clusters for innovations in slaughtering, however (see the *OM in Action* box), Denmark is the leader.

Methods of Evaluating Location Alternatives

Four major methods are used for solving location problems: the factor-rating method, locational cost–volume analysis, the center-of-gravity method, and the transportation model. This section describes these approaches.

TABLE 8.3 Clustering of Companies

INDUSTRY	LOCATIONS	REASON FOR CLUSTERING
Wine making	Napa Valley (U.S.), Bordeaux region (France)	Natural resources of land and climate
Software firms	Silicon Valley, Boston, Bangalore, Israel	Talent resources of bright graduates in scientific/technical areas, venture capitalists nearby
Clean energy	Colorado	Critical mass of talent and information, with 1,000 companies
Theme parks (e.g., Disney World, Universal Studios, and Sea World)	Orlando, Florida	A hot spot for entertainment, warm weather, tourists, and inexpensive labor
Electronics firms (e.g., Sony, IBM, HP, Motorola, and Panasonic)	Northern Mexico	North American trade agreements, duty-free export to U.S. (24% of all TVs are built here)
Computer hardware manufacturing	Singapore, Taiwan	High technological penetration rates and per capita GDP, skilled/educated workforce with large pool of engineers
Fast-food chains (e.g., Wendy's, McDonald's, Burger King, Pizza Hut)	Sites within 1 mile of one another	Stimulate food sales, high traffic flows
General aviation aircraft (e.g., Cessna, Learjet, Boeing, Raytheon)	Wichita, Kansas	Mass of aviation skills (60–70% of world's small planes/jets are built here)
Athletic footwear, outdoor wear	Portland, Oregon	300 companies, many spawned by Nike, deep talent pool and outdoor culture

OM in Action Denmark's Meat Cluster[2]

Every day, 20,000 pigs are delivered to the Danish Crown company's slaughterhouse in central Denmark. The pigs trot into the stunning room, guided by workers armed with giant fly swats. The animals are killed, hung upside down, divided in two, shaved, and scalded clean. A machine cuts them into pieces, which are then cooled, boned, and packed.

The slaughterhouse is enormous: 10 football fields long with 7 miles of conveyor belts. Its managers attend to the tiniest detail. The workers wear green rather than white because this puts the pigs in a better mood. The cutting machine photographs a carcass before adjusting its blades to the exact carcass contours. The company calibrates not only how to carve the flesh, but also where the various parts will fetch the highest prices.

Denmark is a tiny country, with 5.8 million people and wallet-draining labor costs. But it is an agricultural giant, home to 30 million pigs and numerous global brands. Farm products make up 25% of its goods exports—and the value of these exports exceeds $20 billion.

How is this meat-processing cluster still thriving? It is because clustering can be applied to ancient industries like slaughtering as well as to new ones. The cluster includes several big companies: Danish Crown, Arla, Rose Poultry, and DuPont Danisco, as well as plenty of smaller firms, which act as indicators of nascent trends and incubators of new ideas. Other firms are contributing information technology tools for the cluster. Among these are LetFarm for fields, Bovisoft for stables, Agrosoft for pigs, Webstech for grain, and InOMEGA for food.

Thuwanan Krueabudda/Shutterstock

The cluster also has a collection of productivity-spurring institutions (the Cattle Research Center, for example, creates ways to boost pork productivity through robotics) and Danish Tech University, where 1,500 people work on food-related subjects.

The Factor-Rating Method

There are many factors, both qualitative and quantitative, to consider in choosing a location. Some of these factors are more important than others, so managers can use weightings to make the decision process more objective. The **factor-rating method** is popular because a wide variety of factors, from education to recreation to labor skills, can be objectively included. Figure 8.1 listed a few of the many factors that affect location decisions.

Factor-rating method
A location method that instills objectivity into the process of identifying hard-to-evaluate costs.

The factor-rating method (which we introduced in Chapter 2) has six steps:

1. Develop a list of relevant factors called *key success factors* (such as those in Figure 8.1).
2. Assign a weight to each factor to reflect its relative importance in the company's objectives.
3. Develop a scale for each factor (for example, 1 to 10 or 1 to 100 points).
4. Have management score each location for each factor, using the scale in Step 3.
5. Multiply the score by the weights for each factor and total the score for each location.
6. Make a recommendation based on the maximum point score, considering the results of other quantitative approaches as well.

Example 1 FACTOR-RATING METHOD FOR AN EXPANDING THEME PARK

LO 8.3 *Apply* the factor-rating method

Five Flags over Florida, a U.S. chain of 10 family-oriented theme parks, has decided to expand overseas by opening its first park in Europe. It wishes to select between France and Denmark.

APPROACH ▶ The ratings sheet in Table 8.4 lists key success factors that management has decided are important; their weightings and their rating for two possible sites—Dijon, France, and Copenhagen, Denmark—are shown.

STUDENT TIP These weights do not need to be on a 0–1 scale or total to 1. We can use a 1–10 scale, 1–100 scale, or any other scale we prefer.

TABLE 8.4 Weights, Scores, and Solution

KEY SUCCESS FACTOR	WEIGHT	SCORES (OUT OF 100)		WEIGHTED SCORES	
		FRANCE	DENMARK	FRANCE	DENMARK
Labor availability and attitude	.25	70	60	(.25)(70) = 17.50	(.25)(60) = 15.00
People-to-car ratio	.05	50	60	(.05)(50) = 2.50	(.05)(60) = 3.00
Per capita income	.10	85	80	(.10)(85) = 8.50	(.10)(80) = 8.00
Tax structure	.39	75	70	(.39)(75) = 29.25	(.39)(70) = 27.30
Education and health	.21	60	70	(.21)(60) = 12.60	(.21)(70) = 14.70
Totals	1.00			70.35	68.00

SOLUTION ▶ Table 8.4 uses weights and scores to evaluate alternative site locations. Given the option of 100 points assigned to each factor, the French location is preferable.

INSIGHT ▶ By changing the points or weights slightly for those factors about which there is some doubt, we can analyze the sensitivity of the decision. For instance, we can see that changing the scores for "labor availability and attitude" by 10 points can change the decision. The numbers used in factor weighting can be subjective, and the model's results are not "exact" even though this is a quantitative approach.

LEARNING EXERCISE ▶ If the weight for "tax structure" drops to .20 and the weight for "education and health" increases to .40, what is the new result? [Answer: Denmark is now chosen, with a 68.0 vs. a 67.5 score for France.]

RELATED PROBLEMS ▶ 8.5–8.15, 8.24–8.28, 8.33–8.34

EXCEL OM Data File **Ch08Ex1.xls** can be found online.

When a decision is sensitive to minor changes, further analysis of the weighting and the points assigned may be appropriate. Alternatively, management may conclude that these intangible factors are not the proper criteria on which to base a location decision. Managers therefore place primary weight on the more quantitative aspects of the decision.

Locational Cost–Volume Analysis

Locational cost–volume analysis A method of making an economic comparison of location alternatives.

Locational cost–volume analysis is a technique for making an economic comparison of location alternatives. By identifying fixed and variable costs and graphing them for each location, we can determine which one provides the lowest cost. Locational cost–volume analysis can be done mathematically or graphically. The graphic approach has the advantage of providing the range of volume over which each location is preferable.

The three steps to locational cost–volume analysis are as follows:

1. Determine the fixed and variable cost for each location.
2. Plot the costs for each location, with costs on the vertical axis of the graph and annual volume on the horizontal axis.
3. Select the location that has the lowest total cost for the expected production volume.

Example 2

LOCATIONAL COST–VOLUME ANALYSIS FOR A PARTS MANUFACTURER

Esmail Mohebbi, owner of European Ignitions Manufacturing, needs to expand his capacity. He is considering three locations—Athens, Brussels, and Lisbon—for a new plant. The company wishes to find the most economical location for an expected volume of 2,000 units per year.

APPROACH ▶ Mohebbi conducts locational cost–volume analysis. To do so, he determines that fixed costs per year at the sites are $30,000, $60,000, and $110,000, respectively; and variable costs are $75 per unit, $45 per unit, and $25 per unit, respectively. The expected selling price of each ignition system produced is $120.

SOLUTION ▶ For each of the three locations, Mohebbi can plot the fixed costs (those at a volume of zero units) and the total cost (fixed costs + variable costs) at the expected volume of output. These lines have been plotted in Figure 8.2.

Figure 8.2

Crossover Chart for Locational Cost–Volume Analysis

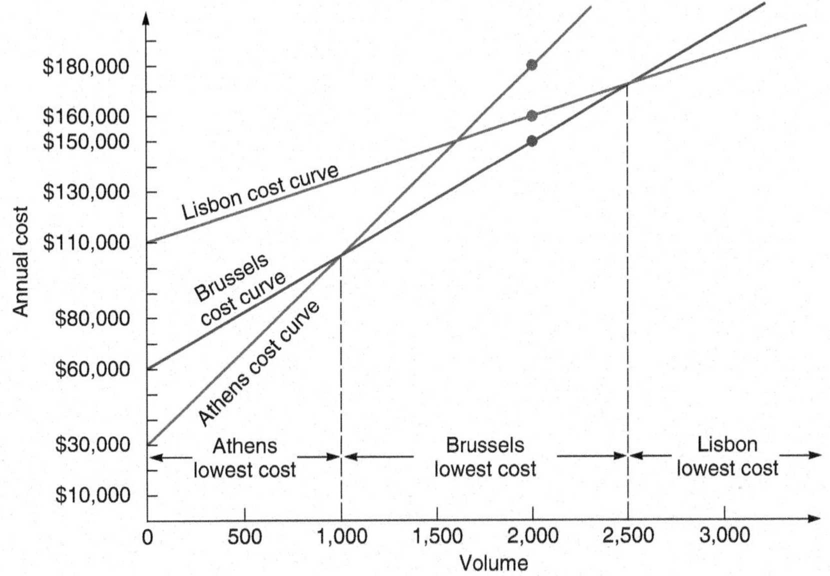

For Athens:

$$\text{Total cost} = \$30,000 + \$75(2,000) = \$180,000$$

For Brussels:

$$\text{Total cost} = \$60,000 + \$45(2,000) = \$150,000$$

For Lisbon:

$$\text{Total cost} = \$110,000 + \$25(2,000) = \$160,000$$

LO 8.4 *Complete* a locational cost–volume analysis graphically and mathematically

With an expected volume of 2,000 units per year, Brussels provides the lowest cost location. The expected profit is:

$$\text{Total revenue} - \text{Total cost} = \$120(2,000) - \$150,000 = \$90,000 \text{ per year}$$

The crossover point for Athens and Brussels is:

$$30,000 + 75(x) = 60,000 + 45(x)$$
$$30(x) = 30,000$$
$$x = 1,000$$

and the crossover point for Brussels and Lisbon is:

$$60,000 + 45(x) = 110,000 + 25(x)$$
$$20(x) = 50,000$$
$$x = 2,500$$

INSIGHT ▶ As with every other OM model, locational cost–volume analysis can be sensitive to input data. For example, for a volume of less than 1,000, Athens would be preferred. For a volume greater than 2,500, Lisbon would yield the greatest profit.

LEARNING EXERCISE ▶ The variable cost for Lisbon is now expected to be $22 per unit. What is the new crossover point between Brussels and Lisbon? [Answer: 2,174 units.]

RELATED PROBLEMS ▶ 8.16–8.19, 8.29, 8.30

EXCEL OM Data File **Ch08Ex2.xls** can be found online.

Center-of-Gravity Method

Center-of-gravity method A mathematical technique used for finding the best location for a single distribution point that services several stores or areas.

The **center-of-gravity method** is a mathematical technique used for finding the location of a distribution center that will minimize distribution costs. The method takes into account the location of markets, the volume of goods shipped to those markets, and shipping costs in finding the best location for a distribution center.

The first step in the center-of-gravity method is to place the locations on a coordinate system. This will be illustrated in Example 3. The origin of the coordinate system and the scale used are arbitrary, just as long as the relative distances are correctly represented. This can be done easily by placing a grid over an ordinary map. The center of gravity is determined using Equations (8-1) and (8-2):

$$x\text{-coordinate of the center of gravity} = \frac{\sum_i x_i Q_i}{\sum_i Q_i} \tag{8-1}$$

$$y\text{-coordinate of the center of gravity} = \frac{\sum_i y_i Q_i}{\sum_i Q_i} \tag{8-2}$$

where x_i = x-coordinate of location i
y_i = y-coordinate of location i
Q_i = Quantity of goods moved to or from location i

LO 8.5 *Use* the center-of-gravity method

Note that Equations (8-1) and (8-2) include the term Q_i, the quantity of supplies transferred to or from location i.

Because the number of containers shipped each month affects cost, distance alone should not be the principal criterion. The center-of-gravity method assumes that cost is directly proportional to both distance and volume shipped. The ideal location is that which minimizes the weighted distance between sources and destinations, where the distance is weighted by the number of containers shipped.[3]

Example 3

CENTER OF GRAVITY

Quain's Discount Department Stores, a chain of four large Target-type outlets, has store locations in Chicago, Pittsburgh, New York, and Atlanta; they are currently being supplied out of an old and inadequate warehouse in Pittsburgh, the site of the chain's first store. The firm wants to find some "central" location in which to build a new warehouse.

APPROACH ▶ Quain's will apply the center-of-gravity method. It gathers data on demand rates at each outlet (see Table 8.5).

TABLE 8.5 Demand for Quain's Discount Department Stores

STORE LOCATION	NUMBER OF CONTAINERS SHIPPED PER MONTH
Chicago	2,000
Pittsburgh	1,000
New York	1,000
Atlanta	2,000

Its current store locations are shown in Figure 8.3. For example, location 1 is Chicago, and from Table 8.5 and Figure 8.3, we have:

$$x_1 = 30$$
$$y_1 = 120$$
$$Q_1 = 2{,}000$$

Figure 8.3

Coordinate Locations of Four Quain's Department Stores and Center of Gravity

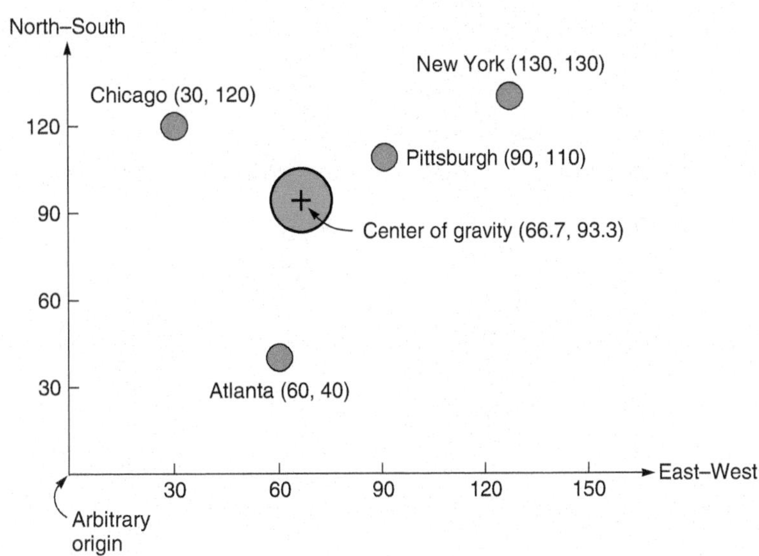

SOLUTION ▶ Using the data in Table 8.5 and Figure 8.3 for each of the other cities, and Equations (8-1) and (8-2), we find:

x-coordinate of the center of gravity:

$$= \frac{(30)(2000) + (90)(1000) + (130)(1000) + (60)(2000)}{2000 + 1000 + 1000 + 2000} = \frac{400,000}{6,000}$$

$$= 66.7$$

y-coordinate of the center of gravity:

$$= \frac{(120)(2000) + (110)(1000) + (130)(1000) + (40)(2000)}{2000 + 1000 + 1000 + 2000} = \frac{560,000}{6,000}$$

$$= 93.3$$

This location (66.7, 93.3) is shown by the crosshairs in Figure 8.3.

INSIGHT ▶ By overlaying a U.S. map on Figure 8.3, we find this location (66.7, 93.3) is near central Ohio. The firm may well wish to consider Columbus, Ohio, or a nearby city as an appropriate location. But it is important to have both north–south and east–west interstate highways near the city selected to make delivery times quicker.

LEARNING EXERCISE ▶ The number of containers shipped per month to Atlanta is expected to grow quickly to 3,000. How does this change the center of gravity, and where should the new warehouse be located? [Answer: (65.7, 85.7), which is closer to Cincinnati, Ohio.]

RELATED PROBLEMS ▶ 8.20–8.23, 8.31, 8.32

EXCEL OM Data File **Ch08Ex3.xls** can be found online.

ACTIVE MODEL 8.1 This example is further illustrated in Active Model 8.1 found online.

Transportation Model

The objective of the transportation model is to determine the best pattern of shipments from several points of supply (sources) to several points of demand (destinations) so as to minimize total production and transportation costs. Every firm with a network of supply-and-demand points faces such a problem. The complex Volkswagen supply network (shown in Figure 8.4) provides one such illustration. We note in Figure 8.4, for example, that VW of Mexico ships vehicles for assembly and parts to VW of Nigeria, sends assemblies to VW of Brasil, and receives parts and assemblies from headquarters in Germany.

Transportation model
A technique for solving a class of linear programming problems.

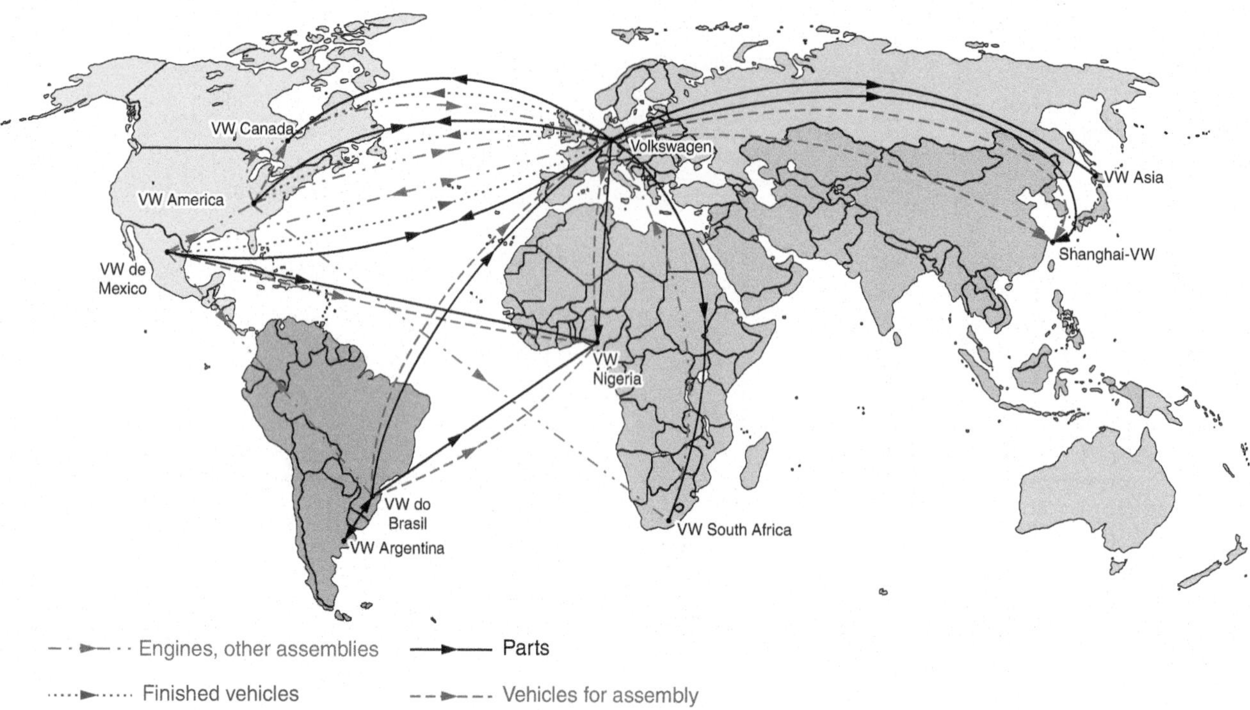

Figure 8.4

Volkswagen, the Largest Automaker in the World, Finds It Advantageous to Locate Its Plants throughout the World

This graphic shows a portion of VW's supply network. There are 61 plants in Europe, along with nine countries in the Americas, Asia, and Africa.

Although the linear programming (LP) technique (see Module B) can be used to solve this type of problem, more efficient, special-purpose algorithms have been developed for the transportation application. The transportation model (see Module C) finds an initial feasible solution and then makes step-by-step improvement until an optimal solution is reached.

Service Location Strategy

While the focus in industrial-sector location analysis is on *minimizing cost,* the focus in the service sector is on *maximizing revenue.* This is because manufacturing firms find that costs tend to vary substantially among locations, while service firms find that location often has more impact on revenue than cost. Therefore, the location focus for service firms should be on determining the volume of customers and revenue.

STUDENT TIP Retail stores often attract more shoppers when competitors are close.

There are eight major determinants of volume and revenue for the service firm:

1. Purchasing power of the customer-drawing area
2. Service and image compatibility with demographics of the customer-drawing area
3. Competition in the area
4. Quality of the competition
5. Uniqueness of the firm's and competitors' locations
6. Physical qualities of facilities and neighboring businesses
7. Operating policies of the firm
8. Quality of management

Realistic analysis of these factors can provide a reasonable picture of the revenue expected. The techniques used in the service sector include regression analysis (see the *OM in Action* box, "How La Quinta Selects Profitable Hotel Sites"), traffic counts, demographic analysis, purchasing power analysis, the factor-rating method, the center-of-gravity method, and geographic information systems. Table 8.6 provides a summary of location strategies for both service and goods-producing organizations.

OM in Action: How La Quinta Selects Profitable Hotel Sites[4]

One of the most important decisions a lodging chain makes is location. Those that pick good sites more accurately and quickly than competitors have a distinct advantage. La Quinta Inns, headquartered in San Antonio, Texas, is a moderately priced chain of 800 inns. To model motel selection behavior and predict success of a site, La Quinta turned to regression analysis.

The hotel started by testing 35 independent variables, trying to find which of them would have the highest correlation with predicted profitability, the dependent variable. Variables included: the number of hotel rooms in the vicinity and their average room rates; local attractions such as office buildings and hospitals that drew potential customers to a 4-mile-radius trade area; local population and unemployment rate; the number of inns in a region; and physical characteristics of the site, such as ease of access or sign visibility.

In the end, the regression model chosen, with an r^2 of 51%, included four predictive variables: (1) the price of the inn, (2) median income levels, (3) the state population per inn, and (4) the location of nearby colleges (which serves as a proxy for other demand generators). La Quinta then used the regression model to predict profitability and developed a cutoff that gave the best results for predicting success or failure of a site. A spreadsheet is now used to implement the model, which applies the decision rule and suggests "build" or "don't build." The CEO likes the model so much that he no longer feels obliged to personally select new sites.

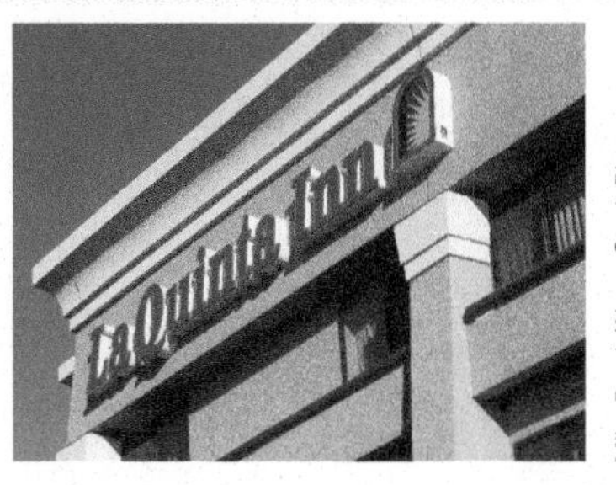

Mike Booth/Alamy Stock Photo

TABLE 8.6 Location Strategies—Service vs. Goods-Producing Organizations

SERVICE/RETAIL/PROFESSIONAL	*GOODS-PRODUCING*
REVENUE FOCUS	COST FOCUS
Volume/revenue Drawing area; purchasing power Competition; advertising/pricing **Physical quality** Parking/access; security/lighting Appearance/image **Cost determinants** Rent Management caliber Operation policies (hours, wage rates)	**Tangible costs** Transportation cost of raw material Shipment cost of finished goods Energy and utility cost; labor; raw material; taxes, and so on **Intangible and future costs** Attitude toward union Quality of life Education expenditures by state Quality of state and local government
TECHNIQUES	TECHNIQUES
Regression models to determine importance of various factors Factor-rating method Traffic counts Demographic analysis of drawing area Purchasing power analysis of area Center-of-gravity method Geographic information systems	Transportation method Factor-rating method Locational cost–volume analysis Crossover charts
ASSUMPTIONS	ASSUMPTIONS
Location is a major determinant of revenue High customer-interaction issues are critical Costs are relatively constant for a given area; therefore, the revenue function is critical	Location is a major determinant of cost Most major costs can be identified explicitly for each site Low customer contact allows focus on the identifiable costs Intangible costs can be evaluated

STUDENT TIP
This table helps differentiate between service- and manufacturing-sector decisions.

LO 8.6 *Understand* the differences between service- and industrial-sector location analysis

Geographic Information Systems

Given that every potential location has differing demographics and infrastructure, geographic information systems are an important tool to help firms make successful, analytical decisions with regard to location. A **geographic information system (GIS)** stores, accesses, displays, and links demographic information to a geographical location. For instance, retailers, banks, food

Geographic information system (GIS)
A system that stores and displays information that can be linked to a geographic location.

chains, gas stations, and print shop franchises can all use geographically coded files from a GIS to conduct demographic analyses. By combining population, age, income, traffic flow, and density figures with geography, a retailer can pinpoint the best location for a new store or restaurant. (See this chapter's opening photo.)

Here are some of the geographic databases available in many GISs:

- Census data by block, tract, city, county, congressional district, metropolitan area, state, and zip code
- Maps of every street, highway, bridge, and tunnel in the U.S.
- Utilities such as electrical, water, and gas lines
- All rivers, mountains, lakes, and forests
- All major airports, colleges, and hospitals

For example, airlines use GISs to identify airports where ground services are the most effective. This information is then used to help schedule and to decide where to purchase fuel, meals, and other services.

Commercial office building developers use GISs in the selection of cities for future construction. Building new office space takes several years; therefore, developers value the database approach that a GIS can offer. GIS is used to analyze factors that influence the location decisions by addressing five elements for each city: (1) residential areas, (2) retail shops, (3) cultural and entertainment centers, (4) crime incidence, and (5) transportation options.

Here are five examples of how location-scouting GIS software is turning commercial real estate into a science.

- *Carvel Ice Cream:* This 95-year-old chain of ice cream shops uses GIS to create a demographic profile of what a typically successful neighborhood for a Carvel looks like—mostly in terms of income and ages.
- *Saber Roofing:* Rather than send workers out to estimate the costs for reroofing jobs, this Redwood City, California, firm pulls up aerial shots of the building via Google Earth. The owner can measure roofs, evaluate their condition, and e-mail the client an estimate, saving hundreds of miles of driving daily. In one case, while on the phone, a potential client was told her roof was too steep for the company to tackle after the Saber employee quickly looked up the home on Google Earth.
- *Arby's:* As this fast-food chain learned, specific products can affect behavior. Using MapInfo, Arby's discovered that diners drove up to 20% farther for their roast beef sandwich (which they consider a "destination" product) than for its chicken sandwich.
- *Home Depot:* Wanting a store in New York City, even though Home Depot demographics are usually for customers who own big homes, the company opened in Queens when GIS software predicted it would do well. Although most people there live in apartments and very small homes, the store has become one of the chain's highest-volume outlets. Similarly, Home Depot thought it had saturated Atlanta two decades ago, but GIS analysis suggested expansion. There are now over 40 Home Depots in that area.
- *Jo-Ann Stores:* This fabric and craft retailer's 70 superstores were doing well a few years ago, but managers were afraid more big-box stores could not justify building expenses. So Jo-Ann used its GIS to create an ideal customer profile—female homeowners with families—and mapped it against demographics. The firm found it could build 700 superstores, which in turn increased the sales from \$105 to \$150 per square foot.

Other packages similar to MapInfo are Atlas GIS (from Strategic Mapping), ArcGIS (by Esri), SAS/GIS (by SAS Institute), Maptitude (by Caliper), and GeoMedia (by Intergraph).

These GISs can be extensive, including comprehensive sets of map and demographic data. The maps have millions of miles of streets and points of interest to allow users to locate restaurants, airports, hotels, gas stations, ATMs, museums, campgrounds, and freeway exits. Demographic data include statistics for population, age, income, education, and housing. These data can be mapped by state, county, city, zip code, or census tract.

VIDEO 8.2
Locating the Next Red Lobster Restaurant

The *Video Case Study* "Locating the Next Red Lobster Restaurant" that appears at the end of this chapter describes how that chain uses its GIS to define trade areas based on market size and population density.

Summary

Location may determine up to 50% of operating expense. Location is also a critical element in determining revenue for the service, retail, or professional firm. Industrial firms need to consider both tangible and intangible costs. Industrial location problems are typically addressed via a factor-rating method, locational cost–volume analysis, the center-of-gravity method, and the transportation method of linear programming.

For service, retail, and professional organizations, analysis is typically made of a variety of variables including purchasing power of a drawing area, competition, advertising and promotion, physical qualities of the location, and operating policies of the organization.

Key Terms

Tangible costs (p. 338)
Intangible costs (p. 338)
Clustering (p. 340)
Factor-rating method (p. 341)
Locational cost–volume analysis (p. 342)
Center-of-gravity method (p. 344)
Transportation model (p. 345)
Geographic information system (GIS) (p. 347)

Ethical Dilemma

In this chapter, we have discussed a number of location decisions. Consider another: United Airlines announced its competition to select a town for a new billion-dollar aircraft-repair base. The bidding for the prize of 7,500 high-paying jobs was fast and furious, with Orlando offering $154 million in incentives and Denver more than twice that amount. Kentucky's governor angrily rescinded Louisville's offer of $300 million, likening the bidding to "squeezing every drop of blood out of a turnip."

When United finally selected from among the 93 cities bidding on the base, the winner was Indianapolis and its $320 million offer of taxpayers' money.

But a few years later, with United near bankruptcy, and having fulfilled its legal obligation, the company walked away from the massive center. This left the city and state governments out all that money, with no new tenant in sight. The city now even owns the tools, neatly arranged in each of the 12 elaborately equipped hangar bays. United outsourced its maintenance to mechanics at a southern firm (which pays one-third of what United paid in salary and benefits in Indianapolis).

What are the ethical, legal, and economic implications of such location bidding wars? Who pays for such giveaways? Are local citizens allowed to vote on offers made by their cities, counties, or states? Should there be limits on these incentives?

Leonard Zhukovsky/Shutterstock

Discussion Questions

1. How is FedEx's location a competitive advantage? Discuss.
2. Why do so many U.S. firms build facilities in other countries?
3. Why do so many foreign companies build facilities in the U.S.?
4. What is clustering?
5. How does factor weighting incorporate personal preference in location choices?
6. What are the advantages and disadvantages of a qualitative (as opposed to a quantitative) approach to location decision making?
7. Provide two examples of clustering in the service sector.
8. What are the major factors that firms consider when choosing a country in which to locate?
9. What factors affect region/community location decisions?
10. Although most organizations may make the location decision infrequently, there are some organizations that make the decision quite regularly and often. Provide one or two examples. How might their approach to the location decision differ from the norm?

11. List factors, other than globalization, that affect the location decision.
12. Explain the assumptions behind the center-of-gravity method. How can the model be used in a service facility location?
13. What are the three steps to locational cost–volume analysis?
14. "Manufacturers locate near their resources, retailers locate near their customers." Discuss this statement, with reference to the proximity-to-markets arguments covered in the text. Can you think of a counter-example in each case? Support your choices.
15. Why shouldn't low wage rates alone be sufficient to select a location?
16. List the techniques used by service organizations to select locations.
17. Contrast the location of a food distributor and a supermarket. (The distributor sends truckloads of food, meat, produce, etc., to the supermarket.) Show the relevant considerations (factors) they share; show those where they differ.
18. Elmer's Fudge Factory is planning to open 10 retail outlets in Oregon over the next 2 years. Identify (and weight) those factors relevant to the decision. Provide this list of factors and weights.

Using Software to Solve Location Problems

This section presents three ways to solve location problems with computer software. First, you can create your own spreadsheets to compute factor ratings, the center of gravity, and locational cost–volume analysis. Second, Excel OM (available online) is programmed to solve all three models. Third, POM for Windows is also found online and can solve all problems labeled with a **P**.

CREATING YOUR OWN EXCEL SPREADSHEETS

Excel spreadsheets are easily developed to solve most of the problems in this chapter. Consider the Quain's Department Store center-of-gravity analysis in Example 3. You can see from Program 8.1 how the formulas are created.

X USING EXCEL OM

Excel OM may be used to solve Example 1 (with the Factor Rating module), Example 2 (with the Cost–Volume Analysis module), and Example 3 (with the Center-of-Gravity module), as well as other location problems. The factor-rating method was illustrated in Chapter 2.

P USING POM FOR WINDOWS

POM for Windows also includes three different facility location models: the factor-rating method, the center-of-gravity model, and locational cost–volume analysis. Please refer to Appendix II for further details.

	A	B	C	D
1	Quain's Discount Department Stores			
2	Center-of-Gravity Method			
3				
4	STORE LOCATION	NUMBER OF CONTAINERS SHIPPED PER MONTH	x-coordinate x_i	y-coordinate y_i
5	Chicago	2,000	30	120
6	Pittsburgh	1,000	90	110
7	New York	1,000	130	130
8	Atlanta	2,000	60	40
9	Sum	6,000		
10				
11	Center of Gravity		66.7	93.3

=SUM(B5:B8)

Action
Copy D11 to C11

=SUMPRODUCT(D5:D8,$B5:$B8)/$B9

Program **8.1**

An Excel Spreadsheet for Creating a Center-of-Gravity Analysis for Example 3, Quain's Discount Department Stores

Solved Problems

SOLVED PROBLEM 8.1

Just as cities and communities can be compared for location selection by the weighted approach model, as we saw earlier in this chapter, so can actual site decisions within those cities. Table 8.7 illustrates four factors of importance to Washington, DC, and the health officials charged with opening that city's first public drug treatment clinic. Of primary concern (and given a weight of 5) was location of the clinic so it would be as accessible as possible to the largest number of patients. Due to a tight budget, the annual lease cost was also of some concern. A suite in the city hall, at 14th and U Streets, was highly rated because its rent would be free. An old office building near the downtown bus station received a much lower rating because of its cost. Equally important as lease cost was the need for confidentiality of patients and, therefore, for a relatively inconspicuous clinic. Finally, because so many of the staff at the clinic would be donating their time, the safety, parking, and accessibility of each site were of concern as well.

Using the factor-rating method, which site is preferred?

SOLUTION

From the three rightmost columns in Table 8.7, the weighted scores are summed. The bus terminal area has a low score and can be excluded from further consideration. The other two sites are virtually identical in total score. The city may now want to consider other factors, including political ones, in selecting between the two remaining sites.

TABLE 8.7 Potential Clinic Sites in Washington, DC

		*POTENTIAL LOCATIONS**			*WEIGHTED SCORES*		
FACTOR	IMPORTANCE WEIGHT	HOMELESS SHELTER (2nd AND D, SE)	CITY HALL (14th AND U, NW)	BUS TERMINAL AREA (7th AND H, NW)	HOMELESS SHELTER	CITY HALL	BUS TERMINAL AREA
Accessibility for addicts	5	9	7	7	45	35	35
Annual lease cost	3	6	10	3	18	30	9
Inconspicuous	3	5	2	7	15	6	21
Accessibility for health staff	2	3	6	2	6	12	4
					Total scores: 84	83	69

*All sites are rated on a 1 to 10 basis, with 10 as the highest score and 1 as the lowest.

SOLVED PROBLEM 8.2

Ching-Chang Kuo is considering opening a new foundry in Denton, Texas; Edwardsville, Illinois; or Fayetteville, Arkansas, to produce high-quality rifle sights. He has assembled the following fixed-cost and variable-cost data:

		PER-UNIT COSTS		
LOCATION	FIXED COST PER YEAR	MATERIAL	VARIABLE LABOR	OVERHEAD
Denton	$200,000	$ .20	$ .40	$ .40
Edwardsville	$180,000	$ .25	$ .75	$ .75
Fayetteville	$170,000	$1.00	$1.00	$1.00

a) Graph the total cost lines.
b) Over what range of annual volume is each facility going to have a competitive advantage?
c) What is the volume at the intersection of the Edwardsville and Fayetteville cost lines?

SOLUTION

a) A graph of the total cost lines is shown in Figure 8.5.
b) Below 8,000 units, the Fayetteville facility will have a competitive advantage (lowest cost); between 8,000 units and 26,666 units, Edwardsville has an advantage; and above 26,666, Denton has the advantage. (We have made the assumption in this problem that other costs—that is, delivery and intangible factors—are constant regardless of the decision.)
c) From Figure 8.5, we see that the cost line for Fayetteville and the cost line for Edwardsville cross at about 8,000. We can also determine this point with a little algebra:

$$\$180,000 + 1.75Q = \$170,000 + 3.00Q$$
$$\$10,000 = 1.25Q$$
$$8,000 = Q$$

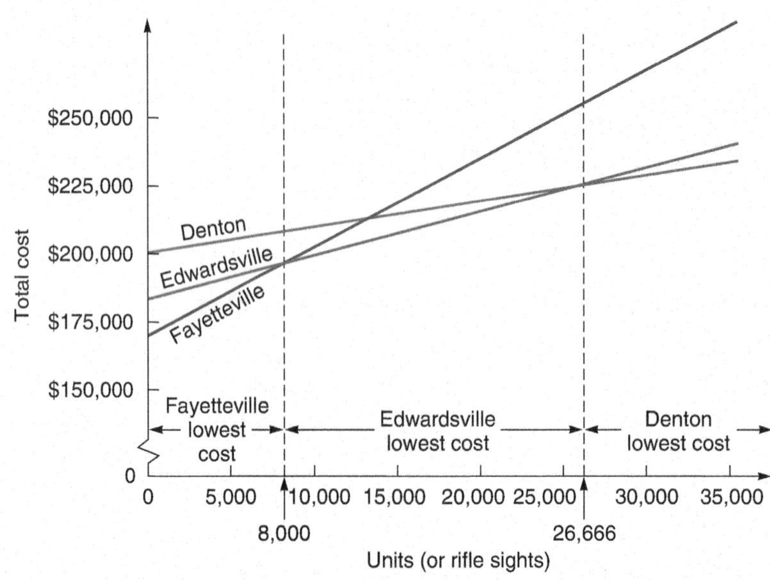

Figure **8.5**

Graph of Total Cost Lines for Ching-Chang Kuo

SOLVED PROBLEM 8.3

The Metropolis Public Library plans to expand with its first major branch library in the city's growing north side. The branch will serve six census tracts. Here are the coordinates of each tract and the population within it:

CENSUS TRACT	CENTER OF TRACT	POPULATION IN TRACT
503—Logan Square	(3, 4)	45,000
519—Albany Park	(4, 5)	25,000
522—Rogers Park	(3, 6)	62,000
538—Kentwood	(4, 7)	51,000
540—Roosevelt	(2, 3)	32,000
561—Western	(5, 2)	29,000

Using the center-of-gravity method, what should be the coordinate location of the branch library?

SOLUTION

$$x\text{-coordinate} = \frac{\sum_i x_i Q_i}{\sum_i Q_i} = \frac{3(45{,}000) + 4(25{,}000) + 3(62{,}000) + 4(51{,}000) + 2(32{,}000) + 5(29{,}000)}{244{,}000} = 3.42$$

$$y\text{-coordinate} = \frac{\sum_i y_i Q_i}{\sum_i Q_i} = \frac{4(45{,}000) + 5(25{,}000) + 6(62{,}000) + 7(51{,}000) + 3(32{,}000) + 2(29{,}000)}{244{,}000} = 4.87$$

The new branch library will sit just west of Logan Square and Rogers Park, at the (3.42, 4.87) tract location.

Problems

Note: **PX** means the problem may be solved with POM for Windows and/or Excel OM.

Problems 8.1–8.4 relate to Factors That Affect Location Decisions

• **8.1** In Myanmar (formerly Burma), 6 laborers, each making the equivalent of $6 per day, can produce 40 units per day. In rural China, 10 laborers, each making the equivalent of $4 per day, can produce 45 units. In Billings, Montana, 2 laborers, each making $120 per day, can make 100 units. Based on labor costs only, which location would be most economical to produce the item?

• **8.2** Refer to Problem 8.1. Shipping cost from Myanmar to Denver, Colorado, the final destination, is $3.00 per unit. Shipping cost from China to Denver is $2.00 per unit, while the shipping cost from Billings to Denver is $0.50 per unit. Considering both labor and transportation costs, which is the most favorable production location?

•• **8.3** Purchasing manager Susie Robinson has been asked to analyze the bids for 200 polished disks used in solar panels.

These bids have been submitted by three suppliers: Thailand Polishing, India Shine, and Sacramento Glow. Thailand Polishing has submitted a bid of 2,000 baht. India Shine has submitted a bid of 2,000 rupees. Sacramento Glow has submitted a bid of \$200. Susie checks with the local bank and finds that \$1 = 10 baht and \$1 = 8 rupees. Which company should Susie choose?

• **8.4** Refer to Problem 8.3. If the final destination is New Delhi, India, and there is a 30% import tax, which firm should be chosen?

Problems 8.5–8.34 relate to Methods of Evaluating Location Alternatives

• • **8.5** Subway, with more than 27,000 outlets in the U.S., is planning for a new restaurant in Buffalo, New York. Three locations are being considered. The following table gives the factors for each site.

FACTOR	WEIGHT	MAITLAND	BAPTIST CHURCH	NORTHSIDE MALL
Space	.30	60	70	80
Costs	.25	40	80	30
Traffic density	.20	50	80	60
Neighborhood income	.15	50	70	40
Zoning laws	.10	80	20	90

a) At which site should Subway open the new restaurant?
b) If the weights for Space and Traffic density are reversed, how would this affect the decision? **PX**

• **8.6** Ken Gilbert owns the Knoxville Warriors, a minor league baseball team in Tennessee. He wishes to move the Warriors south, to either Mobile (Alabama) or Jackson (Mississippi). The following table gives the factors that Gilbert thinks are important, their weights, and the scores for Mobile and Jackson.

FACTOR	WEIGHT	MOBILE	JACKSON
Incentive	.4	80	60
Player satisfaction	.3	20	50
Sports interest	.2	40	90
Size of city	.1	70	30

a) Which site should he select?
b) Jackson just raised its incentive package, and the new score is 75. Why doesn't this impact your decision in part (a)? **PX**

Andrea Catenaro/Shutterstock

• • **8.7** Northeastern Insurance Company is considering opening an office in the U.S. The two cities under consideration are Philadelphia and New York. The factor ratings (higher scores are better) for the two cities are given in the following table. In which city should Northeastern locate?

FACTOR	WEIGHT	PHILADELPHIA	NEW YORK
Customer convenience	.25	70	80
Bank accessibility	.20	40	90
Computer support	.20	85	75
Rental costs	.15	90	55
Labor costs	.10	80	50
Taxes	.10	90	50

• • **8.8** Marilyn Helm Retailers is attempting to decide on a location for a new retail outlet. At the moment, the firm has three alternatives—stay where it is but enlarge the facility; locate along the main street in nearby Newbury; or locate in a new shopping mall in Hyde Park. The company has selected the four factors listed in the following table as the basis for evaluation and has assigned weights as shown:

FACTOR	FACTOR DESCRIPTION	WEIGHT
1	Average community income	.30
2	Community growth potential	.15
3	Availability of public transportation	.20
4	Labor availability, attitude, and cost	.35

Helm has rated each location for each factor, on a 100-point basis. These ratings are given below:

	LOCATION		
FACTOR	PRESENT LOCATION	NEWBURY	HYDE PARK
1	40	60	50
2	20	20	80
3	30	60	50
4	80	50	50

a) What should Helm do?
b) A new subway station is scheduled to open across the street from the present location in about a month, so its third factor score should be raised to 40. How does this change your answer? **PX**

• • **8.9** A location analysis for Cook Controls, a small manufacturer of parts for high-technology cable systems, has been narrowed down to four locations. Cook will need to train assemblers, testers, and robotics maintainers in local training centers. Lori Cook, the president, has asked each potential site to offer training programs, tax breaks, and other industrial incentives. The critical factors, their weights, and the ratings for each location are shown in the following table. High scores represent favorable values.

		LOCATION			
FACTOR	WEIGHT	AKRON, OH	BILOXI, MS	CARTHAGE, TX	DENVER, CO
Labor availability	.15	90	80	90	80
Technical school quality	.10	95	75	65	85
Operating cost	.30	80	85	95	85
Land and construction cost	.15	60	80	90	70
Industrial incentives	.20	90	75	85	60
Labor cost	.10	75	80	85	75

a) Compute the composite (weighted average) rating for each location.
b) Which site would you choose?
c) Would you reach the same conclusion if the weights for operating cost and labor cost were reversed? Recompute as necessary and explain. **PX**

••• **8.10** Pan American Refineries, headquartered in Houston, must decide among three sites for the construction of a new oil-processing center. The firm has selected the six factors listed below as a basis for evaluation and has assigned rating weights from 1 to 5 on each factor:

FACTOR	FACTOR NAME	RATING WEIGHT
1	Proximity to port facilities	5
2	Power-source availability and cost	3
3	Workforce attitude and cost	4
4	Distance from Houston	2
5	Community desirability	2
6	Equipment suppliers in area	3

Subhajit Chakraborty, the CEO, has rated each location for each factor on a 1- to 100-point basis.

FACTOR	LOCATION A	LOCATION B	LOCATION C
1	100	80	80
2	80	70	100
3	30	60	70
4	10	80	60
5	90	60	80
6	50	60	90

a) Which site will be recommended based on *total* weighted scores?
b) If location B's score for Proximity to port facilities was reset at 90, how would the result change?
c) What score would location B need on Proximity to port facilities to change its ranking? **PX**

•• **8.11** A company is planning on expanding and building a new plant in one of three Southeast Asian countries. Chris Ellis, the manager charged with making the decision, has determined that five key success factors can be used to evaluate the prospective countries. Ellis used a rating system of 1 (least desirable country) to 5 (most desirable) to evaluate each factor.

KEY SUCCESS FACTOR		*CANDIDATE COUNTRY RATINGS*		
	WEIGHT	TAIWAN	THAILAND	SINGAPORE
Technology	0.2	4	5	1
Level of education	0.1	4	1	5
Political and legal aspects	0.4	1	3	3
Social and cultural aspects	0.1	4	2	3
Economic factors	0.2	3	3	2

a) Which country should be selected for the new plant?
b) Political unrest in Thailand results in a lower score, 2, for Political and legal aspects. Does your conclusion change?
c) What if Thailand's score drops even further, to a 1, for Political and legal aspects? **PX**

•• **8.12** Harden College is contemplating opening a European campus where students from the main campus could go to take courses for 1 of the 4 college years. At the moment, it is considering five countries: The Netherlands, Great Britain, Italy, Belgium, and Greece. The college wishes to consider eight factors in its decision. The first two factors are given weights of 0.2, while the rest are assigned weights of 0.1. The following table illustrates its assessment of each factor for each country (5 is best).

FACTOR	FACTOR DESCRIPTION	THE NETHERLANDS	GREAT BRITAIN	ITALY	BELGIUM	GREECE
1	Stability of government	5	5	3	5	4
2	Degree to which the population can converse in English	4	5	3	4	3
3	Stability of the monetary system	5	4	3	4	3
4	Communications infrastructure	4	5	3	4	3
5	Transportation infrastructure	5	5	3	5	3
6	Availability of historic/cultural sites	3	4	5	3	5
7	Import restrictions	4	4	3	4	4
8	Availability of suitable quarters	4	4	3	4	3

a) In which country should Harden College choose to set up its European campus?
b) How would the decision change if the "degree to which the population can converse in English" was not an issue? **PX**

•• **8.13** Daniel Tracy, owner of Martin Manufacturing, must expand by building a new factory. The search for a location for this factory has been narrowed to four sites: A, B, C, or D. The following table shows the results thus far obtained by Tracy by using the factor-rating method to analyze the problem. The scale used for each factor scoring is 1 through 5.

		SITE SCORES			
FACTOR	WEIGHT	A	B	C	D
Quality of labor	10	5	4	4	5
Construction cost	8	2	3	4	1
Transportation costs	8	3	4	3	2
Proximity to markets	7	5	3	4	4
Taxes	6	2	3	3	4
Weather	6	2	5	5	4
Energy costs	5	5	4	3	3

a) Which site should Tracy choose?
b) If site D's score for Energy costs increases from a 3 to a 5, do results change?
c) If site A's Weather score is adjusted to a 4, what is the impact? What should Tracy do at this point? **PX**

••• **8.14** An American consulting firm is planning to expand globally by opening a new office in one of four countries: Germany, Italy, Spain, or Greece. The chief partner entrusted with the decision, L. Wayne Shell, has identified eight key success factors that he views as essential for the success of any consultancy. He used a rating system of 1 (least desirable country) to 5 (most desirable) to evaluate each factor.

KEY SUCCESS FACTOR	WEIGHT	CANDIDATE COUNTRY RATINGS			
		GERMANY	ITALY	SPAIN	GREECE
Level of education					
Number of consultants	.05	5	5	5	2
National literacy rate	.05	4	2	1	1
Political aspects					
Stability of government	0.2	5	5	5	2
Product liability laws	0.2	5	2	3	5
Environmental regulations	0.2	1	4	1	3
Social and cultural aspects					
Similarity in language	0.1	4	2	1	1
Acceptability of consultants	0.1	1	4	4	3
Economic factors					
Incentives	0.1	2	3	1	5

a) Which country should be selected for the new office?
b) If Spain's score were lowered in the Stability of government factor, to a 4, how would its overall score change? On this factor, at what score for Spain *would* the rankings change? **PX**

•• **8.15** A British hospital chain wishes to make its first entry into the U.S. market by building a medical facility in the Midwest, a region with which its director, Doug Moodie, is comfortable because he got his medical degree at Northwestern University. After a preliminary analysis, four cities are chosen for further consideration. They are rated and weighted according to the factors shown below:

FACTOR	WEIGHT	CITY			
		CHICAGO	MILWAUKEE	MADISON	DETROIT
Costs	2.0	8	5	6	7
Need for a facility	1.5	4	9	8	4
Staff availability	1.0	7	6	4	7
Local incentives	0.5	8	6	5	9

a) Which city should Moodie select?
b) Assume a minimum score of 5 is now required for all factors. Which city should be chosen? **PX**

•• **8.16** The fixed and variable costs for three potential manufacturing plant sites for a rattan chair weaver are shown:

SITE	FIXED COST PER YEAR	VARIABLE COST PER UNIT
1	$ 500	$11
2	1,000	7
3	1,700	4

a) Over what range of production is each location optimal?
b) For a production of 200 units, which site is best? **PX**

• **8.17** Peter Billington Stereo, Inc., supplies car radios to auto manufacturers and is going to open a new plant. The company is undecided between Detroit and Dallas as the site. The fixed costs in Dallas are lower due to cheaper land costs, but the variable costs in Dallas are higher because shipping distances would increase. Given the following costs:

COST	DALLAS	DETROIT
Fixed costs	$600,000	$800,000
Variable costs	$28/radio	$22/radio

a) Perform an analysis of the volume over which each location is preferable.
b) How does your answer change if Dallas's fixed costs increase by 10%? **PX**

••• **8.18** Hyundai Motors is considering three sites—A, B, and C—at which to locate a factory to build its new electric car batteries. The goal is to locate at a minimum-cost site, where cost is measured by the annual fixed plus variable costs of production. Hyundai Motors has gathered the following data:

SITE	ANNUALIZED FIXED COST	VARIABLE COST PER BATTERY PRODUCED
A	$10,000,000	$2,500
B	$20,000,000	$2,000
C	$25,000,000	$1,000

The firm knows it will produce between 0 and 60,000 batteries at the new plant each year, but, thus far, that is the extent of its knowledge about production plans.

a) For what values of volume, V, of production, if any, is site C a recommended site?
b) What volume indicates site A is optimal?
c) Over what range of volume is site B optimal? Why? **PX**

•• **8.19** Peggy Lane Corp., a producer of machine tools, wants to move to a larger site. Two alternative locations have been identified: Bonham and McKinney. Bonham would have fixed costs of $800,000 per year and variable costs of $14,000 per standard unit produced. McKinney would have annual fixed costs of $920,000 and variable costs of $13,000 per standard unit. The finished items sell for $29,000 each.

a) At what volume of output would the two locations have the same profit?
b) For what range of output would Bonham be superior (have higher profits)?
c) For what range would McKinney be superior?
d) What is the relevance of break-even points for these cities? **PX**

•• **8.20** The following table gives the map coordinates and the shipping loads for a set of cities that we wish to connect through a central hub.

CITY	MAP COORDINATE (x, y)	SHIPPING LOAD
A	(5, 10)	5
B	(6, 8)	10
C	(4, 9)	15
D	(9, 5)	5
E	(7, 9)	15
F	(3, 2)	10
G	(2, 6)	5

a) Near which map coordinates should the hub be located?
b) If the shipments from city A triple, how does this change the coordinates? **PX**

•• **8.21** A chain of home health care firms in Louisiana needs to locate a central office from which to conduct internal audits and other periodic reviews of its facilities. These facilities are scattered throughout the state, as detailed in the following table. Each site, except for Houma, will be visited three times each year by a team of workers, who will drive from the central office to the site. Houma will be visited five times a year. Which coordinates represent a good central location for this office? What other factors might influence the office location decision? Where would you place this office? Explain. **PX**

	MAP COORDINATES	
CITY	x	y
Covington	9.2	3.5
Donaldsonville	7.3	2.5
Houma	7.8	1.4
Monroe	5.0	8.4
Natchitoches	2.8	6.5
New Iberia	5.5	2.4
Opelousas	5.0	3.6
Ruston	3.8	8.5

•• **8.22** A small rural county has experienced unprecedented growth over the past 6 years, and as a result, the local school district built the new 500-student North Park Elementary School. The district has three older and smaller elementary schools: Washington, Jefferson, and Lincoln. Now the growth pressure is being felt at the secondary level. The school district would like to build a centrally located middle school to accommodate students and reduce busing costs. The older middle school is adjacent to the high school and will become part of the high school campus.

a) What are the coordinates of the central location?
b) What other factors should be considered before building a school? **PX**

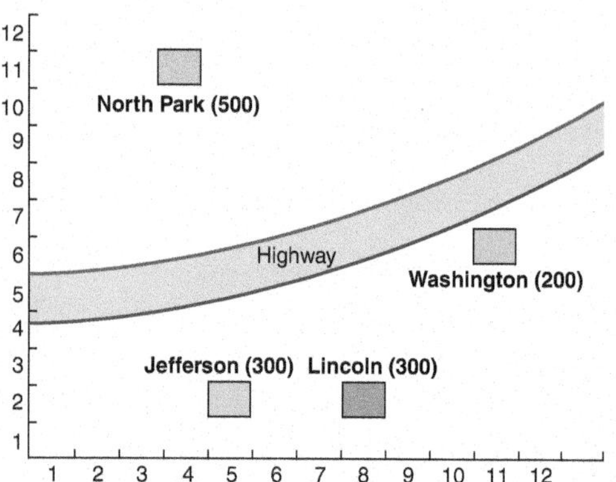

•• **8.23** Todd's Direct, a major TV sales chain headquartered in New Orleans, is about to open its first outlet in Mobile, Alabama, and wants to select a site that will place the new outlet in the center of Mobile's population base. Todd examines the seven census tracts in Mobile, plots the coordinates of the center of each from a map, and looks up the population base in each to use as a weighting. The information gathered appears in the following table.

CENSUS TRACT	POPULATION IN CENSUS TRACT	(x, y) MAP COORDINATES
101	2,000	(25, 45)
102	5,000	(25, 25)
103	10,000	(55, 45)
104	7,000	(50, 20)
105	10,000	(80, 50)
106	20,000	(70, 20)
107	14,000	(90, 25)

a) At what center-of-gravity coordinates should the new store be located?
b) Census tracts 103 and 105 are each projected to grow by 20% in the next year. How will this influence the new store's coordinates? **PX**

•••• **8.24** Eagle Electronics must expand by building a second facility. The search has been narrowed down to locating the new facility in one of four cities: Atlanta (A), Baltimore (B), Chicago (C), or Dallas (D). The factors, scores, and weights follow:

			SCORES BY SITE			
I	FACTOR	WEIGHT (W_i)	A	B	C	D
1	Labor quality	20	5	4	4	5
2	Quality of life	16	2	3	4	1
3	Transportation	16	3	4	3	2
4	Proximity to markets	14	5	3	4	4
5	Proximity to suppliers	12	2	3	3	4
6	Taxes	12	2	5	5	4
7	Energy supplies	10	5	4	3	3

a) Using the factor-rating method, what is the recommended site for Eagle Electronics' new facility?
b) For what range of values for the weight (currently $w_7 = 10$) does the site given as the answer to part (a) remain a recommended site? **PX**

•• **8.25** With ambitious expansion plans, Lufthansa Airlines has decided it needs a second service hub in Germany to complement its large Frankfurt repair facility. The location selection is critical, and with the potential for 1,000 new skilled blue-collar jobs on the line, virtually every city in Germany is actively bidding for the new business. After initial investigations by Andrew Johnson, head of the Operations Department, the company has narrowed the list to three cities. Each is then rated on five factors, as shown in the following table.

			CITY	
FACTOR	WEIGHT	MUNICH	BONN	BERLIN
Financial incentives	80	6	6	8
Skilled labor pool	85	8	8	9
Existing facility	70	9	10	3
Wage rates	70	4	4	4
Competition for jobs	70	3	7	8

a) What is the best location for the aircraft servicing facility?
b) After further investigation, Johnson decides that an existing set of hangar facilities for repairs (the third factor) is not nearly as important as earlier thought. If he lowers the weight of that factor to 50, what is the best location for the aircraft servicing facility? **PX**

• • **8.26** Katie Reynolds owns two exclusive women's clothing stores in Miami. In her plan to expand to a third location, she has narrowed her decision to three sites—one in a downtown office building, one in a shopping mall, and one in an old Victorian house in the suburban area of Coral Gables. She feels that rent is absolutely the most important factor to be considered, although walk-in traffic is 90% as important as rent. Further, the more distant the new store is from her two existing stores, the better. She weights this factor to be 80% as important as walk-in traffic. Katie developed the following table, in which she graded each site from A to D, with A being best. Which site is preferable? **PX**

	DOWNTOWN	SHOPPING MALL	CORAL GABLES HOUSE
Rent	D	C	A
Walk-in traffic	B	A	D
Distance from existing stores	B	A	C

• • **8.27** When placing a new medical clinic, the county health commissioner, Vicky Luo, wishes to consider three sites. The pertinent data are given in the following table. Which is the best site (assuming a higher weight is more desirable than a lower one)? **PX**

		SCORES		
LOCATION FACTOR	WEIGHT	DOWNTOWN	SUBURB A	SUBURB B
Facility utilization	9	9	7	6
Average time per emergency trip	8	6	6	8
Employee preferences	5	2	5	6
Accessibility to major roadways	5	8	4	5
Land costs	4	2	9	6

• • **8.28** A Detroit seafood restaurant is considering opening a second facility in the suburb of West Bloomfield. The following table shows its ratings of five factors at each of four potential sites. Which site should the owner, Lucy Lyu, select? **PX**

		SITE			
FACTOR	WEIGHT	1	2	3	4
Affluence of local population	10	70	60	85	90
Construction and land cost	10	85	90	80	60
Traffic flow	25	70	60	85	90
Parking availability	20	80	90	90	80
Growth potential	15	90	80	90	75

• • **8.29** The fixed and variable costs (in U.S. dollars) for the new recycling plant that Napheesa Griner is opening in Australia are computed as follows:

COST	VICTORIA	PERTH
Fixed	$1,000,000	$800,000
Variable	$73/unit	$112/unit

a) Over what range is each city preferable?
b) For a recycling of 5,000 units, which site is best? **PX**

• • **8.30** The fixed and variable costs for four potential plant sites for Brent Snyder's Ski Supplies are shown here:

SITE	FIXED COST PER YEAR	VARIABLE COST PER UNIT
Atlanta	$125,000	$6
Burlington	75,000	5
Cleveland	100,000	4
Denver	50,000	12

a) Graph the total-cost lines for the four potential sites.
b) Over what range of annual volume is each location the preferable one (that with lowest expected cost)?
c) If expected volume of the ski equipment is 5,000 units, which location would you recommend? **PX**

• • **8.31** Xun Xu Health and Fitness has several production facilities scattered around the central and northern part of Iowa. It would be desirable to have a distribution center from which to send product to retailers such as Target, Costco, Walmart, and large sporting equipment retailers. Using the data in the following table, calculate the center of gravity location for the proposed new facility. **PX**

CITY	MAP COORDINATES (*x*, *y*)	TRUCK ROUND TRIPS PER DAY
A	(2, 1)	20
B	(2, 13)	10
C	(4, 17)	5
D	(7, 7)	20
E	(8, 18)	15
F	(12, 16)	10
G	(17, 4)	20
H	(18, 18)	20

• • **8.32** The main post office in Tampa, Florida, is due to be replaced with a much larger, more modern facility that can handle the tremendous flow of mail that has followed the city's growth since 1970. Since all mail, incoming or outgoing, travels from the seven regional post offices in Tampa through the main post office, its site selection can mean a big difference in overall delivery and movement efficiency. Using the data in the following table, help Postmaster Purushottam Meena calculate the center of gravity location for the proposed new facility. **PX**

REGIONAL POST OFFICE	MAP COORDINATES (*x*, *y*)	TRUCK ROUND TRIPS PER DAY
Ybor City	(10, 5)	3
Davis Island	(3, 8)	3
Dale-Mabry	(4, 7)	2
Palma Ceia	(15, 10)	6
Bayshore	(13, 3)	5
Temple Terrace	(1, 12)	3
Hyde Park	(5, 5)	10

•• **8.33** Global Publishing is going to add two additional distribution centers. One center will be located in each of two countries selected from the set of four candidates. The four potential countries and three KSF ratings are given in the following table. The KSFs were evaluated on the basis of a 1-to-5 rating system, where 1 means least desirable country and 5 means most desirable.

KEY SUCCESS FACTOR (KSF)	SPAIN	ENGLAND	ITALY	POLAND
Rate of technology change	1	3	5	2
Innovations in process design	3	5	5	4
Tax rates	1	5	3	5

On the assumption that each of the three KSFs is equally important, which two countries would be the best choices for locating the two new distribution centers? Why? (Assume equal weights of 1.) **PX**

•• **8.34** Amazing Electronics Distributors wishes to select a new location for its first European warehouse. The location will be selected from the set of four potential candidate countries. The five KSF ratings were evaluated on the basis of a 1-to-5 rating system, where 1 means least desirable.

KEY SUCCESS FACTOR (KSF)	SPAIN	ENGLAND	ITALY	POLAND
Stability of government	5	1	1	1
Product liability laws	1	2	1	2
Export restrictions	4	2	1	5
Inflation	5	3	5	2
Availability of raw materials	5	5	5	5

The first two KSF ratings are each given a mathematical weight of 0.1, and the remaining three KSFs are equally weighted at 0.05. What is the best country for the new warehouse? What is the second-best choice, and why? **PX**

CASE STUDIES

National Assembly Services

Starting from humble beginnings in 2004, Shirin Shahsavand successfully built a mid-sized business assembling small electronic components in Cincinnati, Ohio. Her company, National Assembly Services, gained a reputation among those manufacturers that outsourced to her as a fast, reliable, and flexible assembler. But in 2020, the economy was changing, and so too were the expectations placed upon her firm. Quicker turnaround, faster delivery, and lower costs were all in demand. Competition was fierce, and National found itself with diminishing market share and revenues, along with increasing costs. In order to save the company, Shirin and her management team were forced to make some difficult decisions.

It was clear to Shirin that she could operate from a somewhat distant location because her customers are spread all over the southern U.S. Cincinnati was becoming too expensive: Facility and overhead costs were high, as were wages in the immediate area. Thus, the hunt began for a new location, and contact was made with the economic development offices in certain targeted cities throughout the South in order to ascertain the types of incentives that might be available.

Ultimately, National Assembly Services decided upon Hattiesburg, Mississippi. Hattiesburg offered the firm property tax breaks for 5 years, reductions in utility costs, financial support for training costs when hiring residents, and other considerations including minor relief for moving expenses. In addition, the new facility costs would be substantially less in Hattiesburg versus Cincinnati, as would be the wage rate for comparable work. Overall, the reduction in both capital and operating costs plus the municipal incentives contributed to making the decision to move to Hattiesburg justifiable.

On July 20, 2021, Shirin advised her staff of the decision. She provided each of them the following letter:

To: Employees of National Assembly Services

From: Shirin Shahsavand, President

Thank you for your dedication and service over the years. I sincerely appreciate your hard work and loyalty. It is with regret that I announce that National Assembly Services will discontinue operations at the Cincinnati location after November 1, 2021. Reduced revenues coupled with increased costs and unreasonable demands on the part of the union have forced us to make this decision. I extend to you best wishes for your future endeavors.

Discussion Questions

1. What are the ethical implications of such municipal incentives?
2. What are the challenges in relocating National's senior management team to a small town such as Hattiesburg as opposed to staying in Cincinnati (a much larger city)?
3. Assess the notification made by Shirin to her employees. Are there aspects of this letter that are inappropriate? Would you rewrite this notice? If so, how?
4. Does a company in National's position have a legal, moral, or ethical responsibility to its employees?

Locating the Next Red Lobster Restaurant

Video Case 

From its first Red Lobster in 1968, the chain has grown to 675 locations, with over $1.6 billion in U.S. sales annually. The casual dining market may be crowded, with competitors such as Chili's, Ruby Tuesday, Applebee's, TGI Friday's, and Outback, but Red Lobster's continuing success means the chain thinks there is still plenty of room to grow. Robert Reiner, director of market development, is charged with identifying the sites that will maximize new store sales without cannibalizing sales at the existing Red Lobster locations.

Characteristics for identifying a good site have not changed in 40 years; they still include real estate prices, customer age, competition, ethnicity, income, family size, population density, nearby hotels, and buying behavior, to name just a few. What *has* changed is the powerful software that allows Reiner to analyze a new site in 5 minutes, as opposed to the 8 hours he spent just a few years ago.

Red Lobster has partnered with MapInfo Corp., whose geographic information system (GIS) contains a powerful module for analyzing a trade area (see the discussion of GIS in the chapter). With the U.S. geo-coded down to the individual block, MapInfo allows Reiner to create a psychographic profile of existing and potential Red Lobster trade areas. "We can now target areas with greatest sales potential," says Reiner.

The U.S. is segmented into 72 "clusters" of customer profiles by MapInfo. If, for example, cluster #7, Equestrian Heights (see MapInfo description below), represents 1.7% of a household base within a Red Lobster trade area, but this segment also accounts for 2.4% of sales, Reiner computes that this segment is effectively spending 1.39 times more than average (Index = 2.4/1.7) and adjusts his analysis of a new site to reflect this added weight.

CLUSTER	PSYTE 2003	SNAP SHOT DESCRIPTION
7	Equestrian Heights	They may not have a stallion in the barn, but they likely pass a corral on the way home. These families with teens live in older, larger homes adjacent to, or between, suburbs but not usually tract housing. Most are married with teenagers, but 40% are empty nesters. They use their graduate and professional school education—56% are dual earners. Over 90% are white, non-Hispanic. Their mean family income is $99,000, and they live within commuting distance of central cities. They have white-collar jobs during the week but require a riding lawn mower to keep the place up on weekends.

When Reiner maps the U.S., a state, or a region for a new site, he wants one that is at least 3 miles from the nearest Red Lobster and won't negatively impact its sales by more than 8%; MapInfo pinpoints the best spot. The software also recognizes the nearness of non-Red Lobster competition and assigns a probability of success (as measured by reaching sales potential).

The specific spot selected depends on Red Lobster's seven real estate brokers, whose list of considerations includes proximity to a vibrant retail area, proximity to a freeway, road visibility, nearby hotels, and a corner location at a primary intersection.

"Picking a new Red Lobster location is one of the most critical functions we can do," says Reiner. "And the software we use serves as an independent voice in assessing the quality of an existing or proposed location."

Discussion Questions*

1. Visit the Web site for PSTYE 2003 (**www.gemapping.com/downloads/targetpro_brochure.pdf**). Describe the psychological profiling (PSYTE) clustering system. Select an industry, other than restaurants, and explain how the software can be used for that industry.
2. What are the major differences in site location for a restaurant versus a retail store versus a manufacturing plant?
3. Red Lobster also defines its trade areas based on market size and population density. Here are its seven density classes:

DENSITY CLASS	DESCRIPTION	HOUSEHOLDS PER SQ. MILE
1	Super Urban	8,000+
2	Urban	4,000–7,999
3	Light Urban	2,000–3,999
4	First Tier Suburban	1,000–1,999
5	Second Tier Suburban	600–999
6	Exurban/Small	100–599
7	Rural	0–99

Note: Density classes are based on the households and land area within 3 miles of the geography (e.g., census tract) using population-weighted centroids.

The majority (92%) of the Red Lobster restaurants fall into three of these classes. Which three classes do you think the chain has the most restaurants in? Why?

*You may wish to view the video that accompanies this case before answering the questions.

Where to Place the Hard Rock Café

Video Case

Some people would say that Oliver Munday, Hard Rock's vice president for cafe development, has the best job in the world. Travel the world to pick a country for Hard Rock's next cafe, select a city, and find the ideal site. It's true that selecting a site involves lots of incognito walking around, visiting nice restaurants, and drinking in bars. But that is not where Mr. Munday's work begins, nor where it ends. At the front end, selecting the country and city first involves a great deal of research. At the back end, Munday not only picks the final site and negotiates the deal but then works with architects and planners and stays with the project through the opening and first year's sales.

Munday is currently looking heavily into global expansion in Europe, Latin America, and Asia. "We've got to look at political risk, currency, and social norms—how does our brand fit into the country," he says. Once the country is selected, Munday focuses on the region and city. His research checklist is extensive, as seen in the accompanying table.

Site location now tends to focus on the tremendous resurgence of "city centers," where nightlife tends to concentrate. That's what Munday selected in Moscow and Bogota, although in both locations he chose to find a local partner and franchise the operation. In these two political environments, "Hard Rock wouldn't dream of operating by ourselves," says Munday. The location decision also is at least a 10- to 15-year commitment by Hard Rock, which employs tools such as locational cost–volume analysis to help decide whether to purchase land and build, or to remodel an existing facility.

Currently, Munday is considering four European cities for Hard Rock's next expansion. Although he could not provide the names, for competitive reasons, the following is known:

	EUROPEAN CITY UNDER CONSIDERATION				IMPORTANCE OF THIS FACTOR AT THIS TIME
FACTOR	A	B	C	D	
A. Demographics	70	70	60	90	20
B. Visitor market	80	60	90	75	20
C. Transportation	100	50	75	90	20
D. Restaurants/ nightclubs	80	90	65	65	10
E. Low political risk	90	60	50	70	10
F. Real estate market	65	75	85	70	10
G. Comparable market analysis	70	60	65	80	10

Discussion Questions*

1. From Munday's Standard Market Report checklist, select any other four categories, such as population (A1), hotels (B2), or restaurants/nightclubs (D), and provide three subcategories that should be evaluated. (See item C1 [airport] for a guide.)

Hard Rock's Standard Market Report (for offshore sites)

A. Demographics (local, city, region, SMSA), with trend analysis
   1. Population of area
   2. Economic indicators

B. Visitor market, with trend analysis
   1. Tourists/business visitors
   2. Hotels
   3. Convention center
   4. Entertainment
   5. Sports
   6. Retail

C. Transportation
   1. Airport
   2. Rail
   3. Road
   4. Sea/river

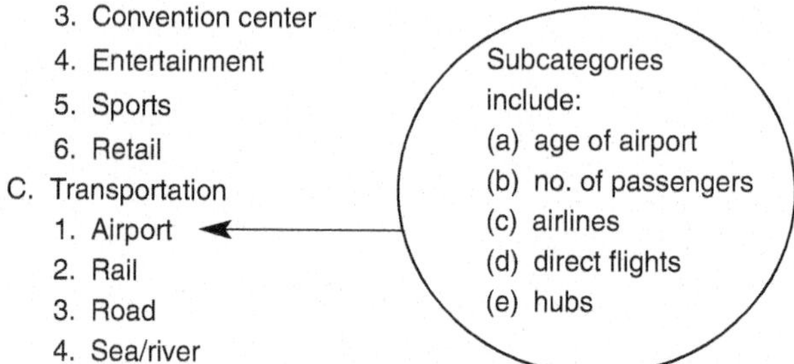

Subcategories include:
(a) age of airport
(b) no. of passengers
(c) airlines
(d) direct flights
(e) hubs

D. Restaurants and nightclubs (a selection in key target market areas)

E. Political risk

F. Real estate market

G. Hard Rock Cafe comparable market analysis

2. Which is the highest rated of the four European cities under consideration, using the table?
3. Why does Hard Rock put such serious effort into its location analysis?
4. Under what conditions do you think Hard Rock prefers to franchise a cafe?

*You may wish to view the video case before answering the questions.

Endnotes

1. Sources: *The Wall Street Journal* (February 12, 2018); and *The Seattle Times* (January 9, 2021).
2. Sources: *The Guardian* (November 30, 2019); and **foodnation-denmark.com**.
3. Equations (8-1) and (8-2) compute a center of gravity (COG) under "squared Euclidean" distances and may actually result in transportation costs slightly (less than 2%) higher than an *optimal* COG computed using "Euclidean" (straight-line) distances. The latter, however, is a more complex and involved procedure mathematically, so the formulas we present are generally used as an attractive substitute. See C. Kuo and R. E. White, "A Note on the Treatment of the Center-of-Gravity Method in Operations Management Textbooks," *Decision Sciences Journal of Innovative Education* 2: 219–227.
4. Sources: S. Kimes and J. Fitzsimmons, *Interfaces* 20, no. 2; and G. Keller, *Statistics for Management and Economics*, 11th ed. Cincinnati: Cengage, 2018.

Bibliography

Bartness, A. D. "The Plant Location Puzzle." *Harvard Business Review* 72, no. 2 (March–April 1994).

Church, R. L., and A. T. Murray. *Business Site Selection, Location Analysis, and GIS*. New York: Wiley, 2009.

Denton, B. "Decision Analysis, Location Models, and Scheduling Problems." *Interfaces* 30, no. 3 (May–June 2005): 262–263.

Kang, K.-T. *Introduction to Geographic Information Systems*, 9th ed. New York: McGraw-Hill, 2019.

Laporte, G., S. Nickel, and F. S. da Gama, eds. *Location Science*, 2nd ed. Cham, Switzerland: Springer Nature Switzerland, 2019.

Mentzer, J. T. "Seven Keys to Facility Location." *Supply Chain Management Review* 12, no. 5 (May 2008): 25.

Porter, M. E., and S. Stern. "Innovation: Location Matters." *MIT Sloan Management Review* (Summer 2001): 28–36.

Render, B., R. M. Stair, M. Hanna, and T. Hale. *Quantitative Analysis for Management*, 13th ed. Boston: Pearson, 2018.

White, G. "Location, Location, Location." *Nation's Restaurant News* 42, no. 27 (July 14, 2008): S10–S11.

Chapter 8 *Rapid* Review

Main Heading	Review Material	MyLab Operations Management
THE STRATEGIC IMPORTANCE OF LOCATION	Location has a major impact on the overall risk and profit of the company. Transportation costs alone can total as much as 25% of the product's selling price. When all costs are considered, location may alter total operating expenses as much as 50%. Companies make location decisions relatively infrequently, usually because demand has outgrown the current plant's capacity or because of changes in labor productivity, exchange rates, costs, or local attitudes. Companies may also relocate their manufacturing or service facilities because of shifts in demographics and customer demand. Location options include (1) expanding an existing facility instead of moving, (2) maintaining current sites while adding another facility elsewhere, and (3) closing the existing facility and moving to another location. For industrial location decisions, the location strategy is usually minimizing costs. For retail and professional service organizations, the strategy focuses on maximizing revenue. Warehouse location strategy may be driven by a combination of cost and speed of delivery. *The objective of location strategy is to maximize the benefit of location to the firm.* When innovation is the focus, overall competitiveness and innovation are affected by (1) the presence of high-quality and specialized inputs such as scientific and technical talent, (2) an environment that encourages investment and intense local rivalry, (3) pressure and insight gained from a sophisticated local market, and (4) local presence of related and supporting industries.	Concept Questions: 1.1–1.4 **VIDEO 8.1** Hard Rock's Location Selection
FACTORS THAT AFFECT LOCATION DECISIONS	Globalization has taken place because of the development of (1) market economics; (2) better international communications; (3) more rapid, reliable travel and shipping; (4) ease of capital flow between countries; and (5) large differences in labor costs. Labor cost per unit is sometimes called the *labor content* of the product: Labor cost per unit = Labor cost per day ÷ Production (that is, units per day) Sometimes firms can take advantage of a particularly favorable exchange rate by relocating or exporting to (or importing from) a foreign country. ■ **Tangible costs**—Readily identifiable costs that can be measured with some precision. ■ **Intangible costs**—A category of location costs that cannot be easily quantified, such as quality of life and government. Many service organizations find that proximity to market is *the* primary location factor. Firms locate near their raw materials and suppliers because of (1) perishability, (2) transportation costs, or (3) bulk. ■ **Clustering**—Location of competing companies near each other, often because of a critical mass of information, talent, venture capital, or natural resources.	Concept Questions: 2.1–2.6 Problems: 8.1–8.4
METHODS OF EVALUATING LOCATION ALTERNATIVES	■ **Factor-rating method**—A location method that instills objectivity into the process of identifying hard-to-evaluate costs. The six steps of the factor-rating method are: 1. Develop a list of relevant factors called *key success factors.* 2. Assign a weight to each factor to reflect its relative importance in the company's objectives. 3. Develop a scale for each factor (for example, 1 to 10 or 1 to 100 points). 4. Have management score each location for each factor, using the scale in step 3. 5. Multiply the score by the weight for each factor and total the score for each location. 6. Make a recommendation based on the maximum point score, considering the results of other quantitative approaches as well. ■ **Locational cost–volume analysis**—A method used to make an economic comparison of location alternatives. The three steps to locational cost–volume analysis are: 1. Determine the fixed and variable cost for each location. 2. Plot the costs for each location, with costs on the vertical axis of the graph and annual volume on the horizontal axis. 3. Select the location that has the lowest total cost for the expected production volume.	Concept Questions: 3.1–3.5 Problems: 8.5–8.34 Virtual Office Hours for Solved Problems: 8.1, 8.2 ACTIVE MODEL 8.1

Main Heading	Review Material	MyLab Operations Management
	■ **Center-of-gravity method**—A mathematical technique used for finding the best location for a single distribution point that services several stores or areas. The center-of-gravity method chooses the ideal location that minimizes the *weighted* distance between itself and the locations it serves, where the distance is weighted by the number of containers shipped, Q_i: $x\text{-coordinate of the center of gravity} = \sum_i x_i Q_i \div \sum_i Q_i$ (8-1) $y\text{-coordinate of the center of gravity} = \sum_i y_i Q_i \div \sum_i Q_i$ (8-2) ■ **Transportation model**—A technique for solving a class of linear programming problems. The transportation model determines the best pattern of shipments from several points of supply to several points of demand to minimize total production and transportation costs.	Virtual Office Hours for Solved Problem: 8.3
SERVICE LOCATION STRATEGY	The eight major determinants of volume and revenue for the service firm are: 1. Purchasing power of the customer-drawing area 2. Service and image compatibility with demographics of the customer-drawing area 3. Competition in the area 4. Quality of the competition 5. Uniqueness of the firm's and competitors' locations 6. Physical qualities of facilities and neighboring businesses 7. Operating policies of the firm 8. Quality of management	Concept Questions: 4.1–4.6
GEOGRAPHIC INFORMATION SYSTEMS	■ **Geographic information system (GIS)**—A system that stores and displays information that can be linked to a geographic location. Some of the geographic databases available in many GISs include (1) census data by block, tract, city, county, congressional district, metropolitan area, state, and zip code; (2) maps of every street, highway, bridge, and tunnel in the U.S.; (3) utilities such as electrical, water, and gas lines; (4) all rivers, mountains, lakes, and forests; and (5) all major airports, colleges, and hospitals.	Concept Questions: 5.1–5.4 **VIDEO 8.2** Locating the Next Red Lobster Restaurant
ADDITIONAL MYLAB OPERATIONS MANAGEMENT RESOURCES	✔ Video for Creating Your Own Excel Spreadsheets (Example 3) ✔ Additional Case Study (Southwestern University (E)) ✔ Southwestern University Case Studies are integrated in Chapters 3, 4, 6, 8, 12, and 13 and in Supplement 7 ✔ Multiple Choice Case Questions (National Assembly Services) ✔ Recent Graduate Video: Greg Friedman, Real Estate Acquisitions Consultant, JCR Property Management	

Self **Test**

LO 8.1 The factors involved in location decisions include
- **a)** foreign exchange.
- **b)** attitudes.
- **c)** labor productivity.
- **d)** all of the above.

LO 8.2 If Fender Guitar pays \$30 per day to a worker in its Ensenada, Mexico, plant, and the employee completes four instruments per 8-hour day, the labor cost/unit is
- **a)** \$30.00.
- **b)** \$3.75.
- **c)** \$7.50.
- **d)** \$4.00.
- **e)** \$8.00.

LO 8.3 Evaluating location alternatives by comparing their composite (weighted-average) scores involves
- **a)** factor-rating analysis.
- **b)** cost–volume analysis.
- **c)** transportation model analysis.
- **d)** linear regression analysis.
- **e)** crossover analysis.

LO 8.4 On the cost–volume analysis chart where the costs of two or more location alternatives have been plotted, the quantity at which two cost curves cross is the quantity at which:
- **a)** fixed costs are equal for two alternative locations.
- **b)** variable costs are equal for two alternative locations.
- **c)** total costs are equal for all alternative locations.
- **d)** fixed costs equal variable costs for one location.
- **e)** total costs are equal for two alternative locations.

LO 8.5 A regional bookstore chain is about to build a distribution center that is centrally located for its eight retail outlets. It will most likely employ which of the following tools of analysis?
- **a)** Assembly-line balancing
- **b)** Load–distance analysis
- **c)** Center-of-gravity model
- **d)** Linear programming
- **e)** All of the above

LO 8.6 What is the major difference in focus between location decisions in the service sector and in the manufacturing sector?
- **a)** There is no difference in focus.
- **b)** The focus in manufacturing is revenue maximization, while the focus in service is cost minimization.
- **c)** The focus in service is revenue maximization, while the focus in manufacturing is cost minimization.
- **d)** The focus in manufacturing is on raw materials, while the focus in service is on labor.

Answers: LO 8.1. d; LO 8.2. c; LO 8.3. a; LO 8.4. e; LO 8.5. c; LO 8.6. c.

Layout Strategies

9 CHAPTER

CHAPTER OUTLINE

A critical element contributing to the bottom line at Hard Rock Cafe is the layout of each cafe's retail shop space. The retail space, from 600 to 1,300 square feet in size, is laid out in conjunction with the restaurant area to create the maximum traffic flow before and after eating. The payoffs for cafes like this one in London are huge. Almost half of a cafe's annual sales are generated from these small shops, which have very high retail sales per square foot.

Michael Weber/imageBROKER/Alamy Stock Photo

10 OM STRATEGY DECISIONS

- Design of Goods and Services
- Managing Quality
- Process Strategies
- Location Strategies
- *Layout Strategies*
- Human Resources
- Supply Chain Management
- Inventory Management
- Scheduling
- Maintenance

GLOBAL COMPANY PROFILE
McDonald's

McDonald's Looks for Competitive Advantage through Layout

In its over 80 years of existence, McDonald's has revolutionized the restaurant industry by inventing the limited-menu fast-food restaurant. It has also made seven major innovations. The first, the introduction of *indoor seating* (1950s), was a layout issue, as was the second, *drive-through windows* (1970s). The third, adding *breakfasts* to the menu (1980s), was a product strategy. The fourth, *adding play areas* (late 1980s), was again a layout decision.

In the 1990s, McDonald's completed its fifth innovation, a radically new *redesign of the kitchens* in its 14,000 North American outlets to facilitate a mass-customization process. Dubbed the "Made by You" kitchen system, sandwiches were assembled to order with the revamped layout.

In 2004, the chain began the rollout of its sixth innovation, a new food ordering layout: the *self-service kiosk*. Self-service kiosks have been infiltrating the service sector. Alaska Airlines was the first airline to provide self-service airport check-in, in 1996. Most passengers of the major airlines now check themselves in for flights. Kiosks take up less space than an employee and reduce waiting line time.

Now, McDonald's is working on its seventh innovation, and not surprisingly, it also deals with restaurant layout. The company, on an unprecedented scale, is redesigning all 30,000 eateries around the globe to take on a contemporary look. The dining area will be separated into three sections with distinct personalities: (1) the "linger" zone focuses on young adults and offers

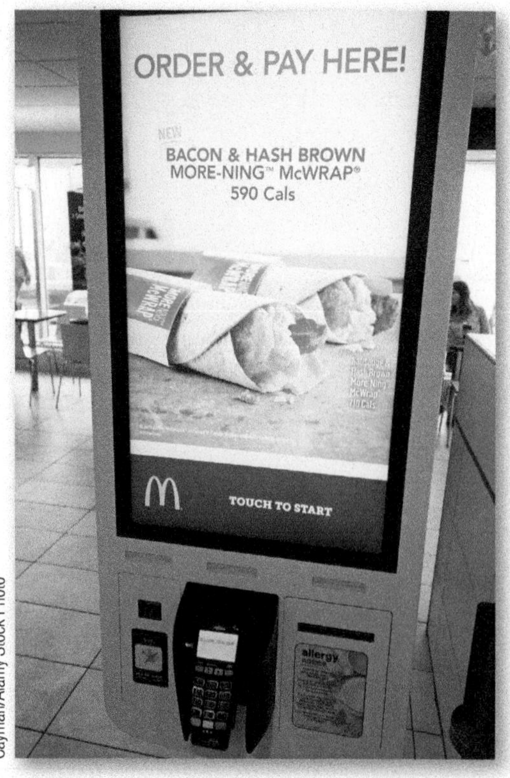

Cayman/Alamy Stock Photo

McDonald's finds that kiosks reduce both space requirements and waiting; order taking is faster. An added benefit is that customers like them. Also, kiosks are reliable—they don't call in sick. And, most important, sales are up 10%–15% (an average of $1) when a customer orders from a kiosk, which consistently recommends the larger size and other extras.

Nancy Siesel/The New York Times/Redux

The redesigned kitchen of a McDonald's in Manhattan. The more efficient layout requires less labor, reduces waste, and provides faster service. A graphic of this "assembly line" is shown in Figure 9.11.

comfortable furniture and Wi-Fi connections; (2) the "grab and go" zone features tall counters, bar stools, and flat-screen TVs; and (3) the "flexible" zone has colorful family booths, flexible seating, and kid-oriented music. The cost per outlet: a whopping $300,000–$400,000 renovation fee.

As McDonald's has discovered, facility layout is indeed a source of competitive advantage.

ZUMA Press, Inc./Alamy

Flexible Zone

This area is geared for family and larger groups, with movable tables and chairs.

McDonald's Corporation

Grab & Go Zone

This section has tall counters with bar stools for customers who eat alone. Flat-screen TVs keep them company.

McDonald's Corporation

Linger Zone

Cozy booths, plus Wi-Fi connections, make these areas attractive to those who want to hang out and socialize.

LEARNING OBJECTIVES

LO 9.1 ***Discuss*** important issues in office layout 368
LO 9.2 ***Define*** the objectives of retail layout 370
LO 9.3 ***Discuss*** modern warehouse management and terms such as ASRS, cross-docking, and random stocking 371
LO 9.4 ***Identify*** when fixed-position layouts are appropriate 373
LO 9.5 ***Explain*** how to achieve a good process-oriented facility layout 374
LO 9.6 ***Define*** work cell and the requirements of a work cell 379
LO 9.7 ***Define*** product-oriented layout 381
LO 9.8 ***Explain*** how to balance production flow in a repetitive or product-oriented facility 382

The Strategic Importance of Layout Decisions

Layout is one of the key decisions that determines the long-run efficiency of operations. Layout has strategic implications because it establishes an organization's competitive priorities in regard to capacity, processes, flexibility, and cost, as well as quality of work life, customer contact, and image. An effective layout can help an organization achieve a strategy that supports differentiation, low cost, or response. Benetton, for example, supports a *differentiation* strategy by heavy investment in warehouse layouts that contribute to fast, accurate sorting and shipping to its 5,000 outlets. Walmart store layouts support a strategy of *low cost*, as do its warehouse layouts. Hallmark's office layouts, where many professionals operate with open communication in work cells, support *rapid development* of greeting cards. *The objective of layout strategy is to develop an effective and efficient layout that will meet the firm's competitive requirements*. These firms have done so.

In all cases, layout design must consider how to achieve the following:

- Higher utilization of space, equipment, and people
- Improved flow of information, materials, and people
- Improved employee morale and safer working conditions
- Improved customer/client interaction
- Flexibility (whatever the layout is now, it will need to change)

In our increasingly short-life-cycle, mass-customized world, layout designs need to be viewed as dynamic. This means considering small, movable, and flexible equipment. Store displays need to be movable, office desks and partitions modular, and warehouse racks prefabricated. To make quick and easy changes in product models and in production rates, operations managers must design flexibility into layouts. To obtain flexibility in layout, managers cross-train their workers, maintain equipment, keep investments low, place workstations close together, and use small, movable equipment. In some cases, equipment on wheels is appropriate, in anticipation of the next change in product, process, or volume.

The digital world is providing additional resources for layout design. Infrared technology and computer vision are now used to provide insight about how facilities are used. This technology counts the number of people in a room, tracks their location, and measures how far apart they are. For instance, offices, conference centers, and restaurants can now automatically determine and track occupancy and seating preference, allowing management to modify layouts and provide appropriate staffing.

Types of Layout

Layout decisions include the best placement of machines (in production settings), offices and desks (in office settings), or service centers (in settings such as hospitals or department stores). An effective layout facilitates the flow of materials, people, and information within and between areas. To achieve these objectives, a variety of approaches has been developed. We will discuss seven of them in this chapter:

1. *Office layout:* Positions workers, their equipment, and spaces/offices to facilitate collaboration.
2. *Retail layout:* Allocates display space and responds to customer behavior.

TABLE 9.1 Layout Strategies

	OBJECTIVE	EXAMPLES
Office	Balance physical and social proximity, privacy, and permission	Allstate Insurance Microsoft Corp.
Retail	Maximize profitability per square foot of floor space or linear foot of shelf space	Kroger's Supermarket Walgreens Bloomingdale's
Warehouse (storage)	Balance low-cost storage with low-cost material handling	Federal-Mogul's warehouse The Gap's distribution center
Project (fixed position)	Move material to and from the limited storage areas around the site	Ingall Ship Building Corp. Trump Plaza Pittsburgh Airport
Job shop (process oriented)	Manage varied material flow for each product	Arnold Palmer Hospital Hard Rock Cafe Olive Garden
Work cell (product families)	Identify a product family, build teams, cross-train team members	Hallmark Cards Wheeled Coach Ambulances
Repetitive/continuous (product oriented)	Equalize the task time at each workstation	Sony's TV assembly line Toyota Sequoia

3. *Warehouse layout:* Addresses trade-offs between space and material handling.
4. *Fixed-position layout:* Addresses the layout requirements of large, bulky projects such as ships and buildings.
5. *Process-oriented layout:* Deals with low-volume, high-variety production (also called "job shop," or intermittent production).
6. *Work-cell layout:* Arranges machinery and equipment to focus on production of a single product or group of related products.
7. *Product-oriented layout:* Seeks the best personnel and machine utilization in repetitive or continuous production.

Examples for each of these classes of layouts are noted in Table 9.1.

Because only a few of these seven classes can be modeled mathematically, layout and design of physical facilities are still something of an art. However, we do know that a good layout requires determining the following:

- *Material-handling equipment:* Managers must decide about equipment to be used, including conveyors, cranes, automated storage and retrieval systems, and robot-driven carriers to deliver and store material.
- *Capacity and space requirements:* Only when personnel, machines, and equipment requirements are known can managers proceed with layout and provide space for each component. In the case of office work, operations managers must make judgments about the space requirements for each employee. They must also consider allowances for requirements that address safety, noise, dust, fumes, temperature, and space around equipment and machines.
- *Environment and aesthetics:* Layout concerns often require decisions about windows, planters, and height of partitions to facilitate airflow, reduce noise, and provide privacy.
- *Flows of information:* Communication is important to any organization and must be facilitated by the layout. This issue may require decisions about proximity, as well as decisions about open spaces versus half-height dividers versus private offices.
- *Cost of moving between various work areas:* There may be unique considerations related to moving materials or to the importance of having certain areas next to each other. For example, moving molten steel is more difficult than moving cold steel.

Office Layout

Office layouts require the grouping of workers, their equipment, and spaces to provide for comfort, safety, and movement of information. Office layouts are in constant flux as the technological changes sweeping society alter the way offices function.

Office layout
The grouping of workers, their equipment, and spaces/offices to provide for comfort, safety, and movement of information.

Here are five versions of the office layout.

Everett Collection/Shutterstock

Managers and architects have pondered how to design an office to encourage productivity for more than 100 years. In the early 20th century, large offices resembled factories, where clerical workers sat in long rows, often performing repetitive tasks.

Starting in the 1960s, layouts changed to foster teamwork where managers and support staff sat together, and groupings were geared toward specific tasks.

Courtesy of Herman Miller, Inc.

Chad McDermott/Shutterstock

With computers, more individual work was possible and the "Cube Farm" era became ubiquitous through the '80s and '90s. An office full of high-walled cubicles offered both an open environment and personal office space.

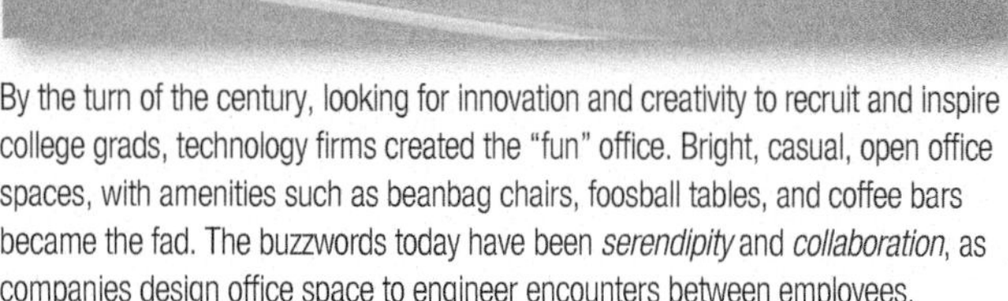

By the turn of the century, looking for innovation and creativity to recruit and inspire college grads, technology firms created the "fun" office. Bright, casual, open office spaces, with amenities such as beanbag chairs, foosball tables, and coffee bars became the fad. The buzzwords today have been *serendipity* and *collaboration*, as companies design office space to engineer encounters between employees.

Monkey Business Images/Shutterstock

Andrey_Popov/Shutterstock

Virtual meetings have now become commonplace as it is no longer unusual for employees to work remotely.

Some layout considerations are universal and may apply to factories as well as to offices. Workspace can inspire informal and productive encounters if it balances three physical and social aspects:

LO 9.1 *Discuss important issues in office layout*

- *Proximity:* Spaces should allow people to collaborate.
- *Privacy:* People must be able to control access to their conversations.
- *Permission:* The culture should signal that nonwork interactions are encouraged.

We note two major trends in office layout. First, technology (such as smartphones, scanners, the Internet, laptop computers, and tablets) allows increasing layout flexibility by moving information electronically and allowing employees substantial flexibility in location. The COVID-19 pandemic accelerated this flexibility and brought a potentially permanent change in the number of people who work from home. Second, even before the pandemic, firms have been creating dynamic needs for space and services.

Here are two examples:

- When Deloitte & Touche found that 30% to 40% of desks were empty at any given time, the firm developed its "hoteling programs." Consultants lost their permanent offices; anyone who plans to be in the building (rather than out with clients) books an office through a "concierge," who hangs that consultant's name on the door for the day and stocks the space with requested supplies.
- Cisco Systems cut rent and workplace service costs by 37% and saw productivity benefits of $2.4 billion per year by reducing square footage, reconfiguring space, creating movable, everything-on-wheels offices, and designing "get away from it all" innovation areas.

Retail Layout

Retail layouts are based on the idea that sales and profitability vary directly with customer exposure to products. Thus, most retail operations managers try to expose customers to as many products as possible. Studies do show that greater product exposure generates sales and a higher return on investment (at least until the customer is overwhelmed by options). The operations manager can change exposure with store arrangement and the allocation of space to various products within that arrangement.

Retail layout
An approach that addresses flow, allocates space, and responds to customer behavior.

Five ideas are helpful for determining the overall arrangement of many stores:

1. Locate the high-draw items around the periphery of the store. Thus, we tend to find dairy products on one side of a supermarket and bread and bakery products on another. An example of this tactic is shown in Figure 9.1.

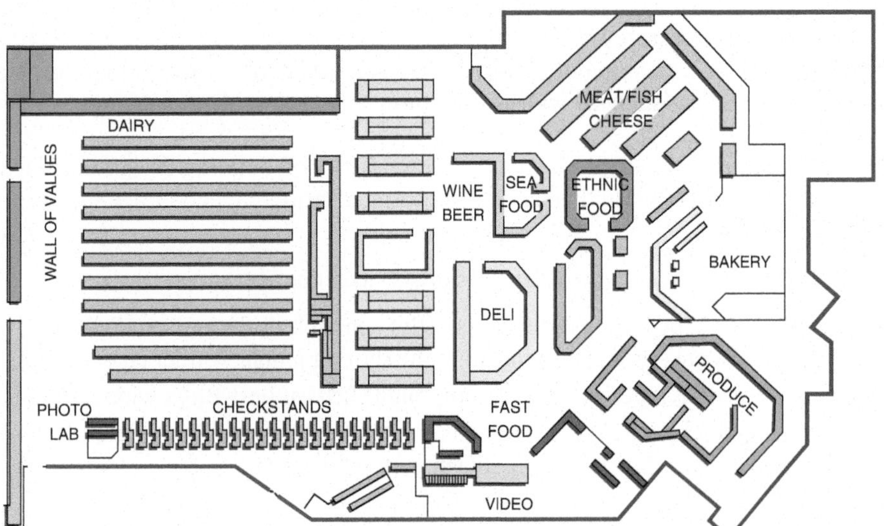

Figure 9.1
Store Layout with Dairy and Bakery, High-Draw Items, in Different Areas of the Store

2. Use prominent locations for high-impulse and high-margin items. Best Buy puts fast-growing, high-margin digital goods—such as printers—in the front and center of its stores.
3. Distribute what are known in the trade as "power items"—items that may dominate a purchasing trip—to both sides of an aisle, and disperse them to increase the viewing of other items.
4. Use end-of-aisle and checkout counter locations for high-margin and impulse items because of their very high exposure rates. Checkout counter locations are so valuable that the space is sometimes sold by the inch.
5. Convey the mission of the store by carefully selecting the position of the lead-off department. For instance, if prepared foods are part of a supermarket's mission, position the bakery and deli up front to appeal to convenience-oriented customers. Walmart's push to increase sales of clothes means those departments are in broad view upon entering a store.

STUDENT TIP The goal in a retail layout is to maximize profit per square foot of store space.

LO 9.2 *Define* the objectives of retail layout

Once the overall layout of a retail store has been decided, products need to be arranged for sale. Many considerations go into this arrangement. However, the main *objective of retail layout is to maximize profitability per square foot of floor space* (or, in some stores, on linear foot of shelf space). Big-ticket, or expensive, items may yield greater dollar sales, but the profit per square foot may be lower. Many firms use computer programs to assist managers in evaluating the profitability of various merchandising plans for hundreds of categories—this technique is known as *category management.*

Slotting fees
Fees manufacturers pay to get shelf space for their products.

An additional, and somewhat controversial, issue in retail layout is called *slotting.* **Slotting fees** are fees manufacturers pay to get their goods on the shelf in a retail store or supermarket chain. The result of a massive flood of new products, retailers can now demand up to $50,000 to place an item in their chain. Marketplace economics and consolidations, as well as improved supply chain management and inventory control systems, have provided retailers with this leverage. Many small firms question the ethics of slotting fees, claiming the fees stifle new products, limit their ability to expand, and cost consumers money. Walmart is one of the few major retailers that does not demand slotting fees, removing a barrier to entry. (See the *Ethical Dilemma* at the end of this chapter.)

Servicescapes

Servicescape
The physical surroundings in which a service takes place, and how they affect customers and employees.

Although a major goal of retail layout is to maximize profit through product exposure, there are other aspects of the service that managers consider. The term **servicescape** describes the physical surroundings in which the service is delivered and how the surroundings have a humanistic effect on customers and employees. To provide a good service layout, a firm considers three elements:

1. *Ambient conditions*, which are background characteristics such as lighting, sound, smell, and temperature. All these affect workers *and* customers and can affect how much is spent and how long a person stays in the building.
2. *Spatial layout and functionality*, which involve customer circulation path planning, aisle characteristics (such as width, direction, angle, and shelf spacing), and product grouping.
3. *Signs, symbols, and artifacts*, which are characteristics of building design that carry social significance (such as carpeted areas of a department store that encourage shoppers to slow down and browse).

Examples of each of these three elements of servicescape are:

- *Ambient conditions:* Fine-dining restaurants with linen tablecloths and candlelit atmosphere; Mrs. Field's Cookie bakery smells permeating the shopping mall; leather chairs at Starbucks.
- *Layout/functionality:* Kroger's long aisles and high shelves; Best Buy's wide center aisle.
- *Signs, symbols, and artifacts:* Walmart's greeter at the door; Hard Rock Cafe's wall of guitars; Disneyland's entrance looking like hometown heaven.

Warehouse and Storage Layouts

The objective of **warehouse layout** *is to find the optimum trade-off between handling cost and costs associated with warehouse space.* Consequently, management's task is to maximize the utilization of the total "cube" of the warehouse—that is, utilize its full volume while maintaining low material-handling costs. We define *material-handling costs* as all the costs related to the transaction. This consists of incoming transport, storage, and outgoing transport of the materials to be warehoused. These costs include equipment, people, material, supervision, insurance, and depreciation. Effective warehouse layouts do, of course, also minimize the damage and spoilage of material within the warehouse.

Warehouse layout
A design that attempts to minimize total cost by addressing trade-offs between space and material handling.

Management minimizes the sum of the resources spent on finding and moving material plus the deterioration and damage to the material itself. The variety of items stored and the number of items "picked" has direct bearing on the optimum layout. A warehouse storing a few unique items lends itself to higher density than a warehouse storing a variety of items. Modern warehouse management is, in many instances, an automated procedure using *automated storage and retrieval systems* (ASRSs).

LO 9.3 *Discuss* modern warehouse management and terms such as ASRS, cross-docking, and random stocking

The Stop & Shop grocery chain, with 350 supermarkets in New England, has recently completed the largest ASRS in the world. The 1.3-million-square-foot distribution center in Freetown, Massachusetts, employs 77 rotating-fork automated storage and retrieval machines. These 77 ASRS machines each access 11,500 pick slots on 90 aisles—a total of 64,000 pallets of food. The *OM in Action* box, "Amazon Warehouses Are Full of Robots," shows another way that technology can help minimize warehouse costs.

An important component of warehouse layout is the relationship between the receiving/unloading area and the shipping/loading area. Facility design depends on the type of supplies unloaded, what they are unloaded from (trucks, rail cars, barges, and so on), and where they are unloaded. In some companies, the receiving and shipping facilities, or *docks*, as they are called, are even in the same area; sometimes they are receiving docks in the morning and shipping docks in the afternoon. With aggressive warehouse management, Costco has driven the average time an item is in the warehouse down to 6 hours.

STUDENT TIP
In warehouse layout, we want to maximize use of the whole building—from floor to ceiling.

OM in Action Amazon Warehouses Are Full of Robots[1]

An Amazon warehouse is a flurry of activity. Workers, in a cavernous warehouse, are diligently plopping items into yellow and black crates while towering hydraulic arms lift heavy boxes toward the rafters. And hundreds of stubby orange robots slide along the floor like giant hockey pucks, piled high with towers of consumer products.

Those are Kiva robots, once the marvel of warehouses everywhere. When Amazon threw down $775 million to purchase Kiva in 2012, it effectively gave the firm command of an entire industry. And it decided to use the robots for Amazon and Amazon alone, ending the sale of Kiva's products to competitors that had come to rely on them. The acquisition set off an arms race among robot makers and retailers like Walmart and Target that scurried to keep up with the e-commerce giant.

For decades, warehouse operators were focused on the task of loading pallets and shipping them to retailers, who broke up the shipments and routed them to retail locations. Fulfilling online orders, on the other hand, requires shippers to pack boxes with a diverse set of individual items and route them on to customers' homes. That shift has given way to *collaborative robotics*, in which a human warehouse worker toils alongside an autonomous machine.

Noah Berger/Reuters

It has taken several years, but competing firms are ready to take on Kiva and equip the world's warehouses with new robotics. Meanwhile, Amazon has more than 200,000 Kiva robots at its warehouses across the globe (and is adding 15,000 per year), which has reduced operating expenses by about 20%. Adding them to one new warehouse saves $22 million in fulfillment expenses. Bringing the Kivas to the 100 or so distribution centers that still haven't implemented them will save Amazon an additional $2.5 billion.

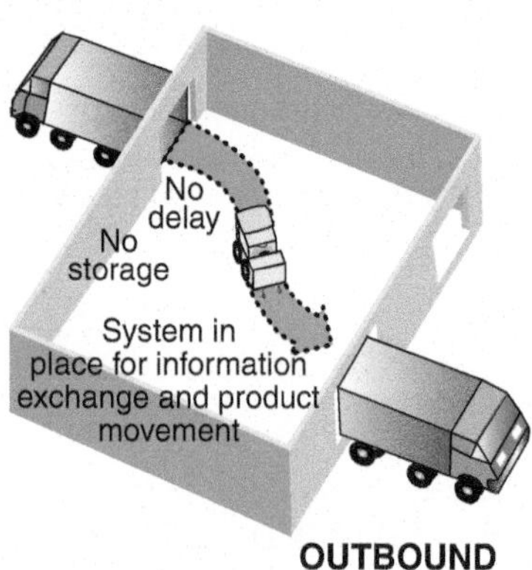

Cross-Docking

Cross-docking means to avoid placing materials or supplies in storage by processing them as they are received. In a manufacturing facility, product is received directly by the assembly line. In a distribution center, labeled and presorted loads arrive at the shipping dock for immediate rerouting, thereby avoiding formal receiving, stocking/storing, and order-selection activities. Because these activities add no value to the product, their elimination is 100% cost savings. Walmart, an early advocate of cross-docking, uses the technique as a major component of its continuing low-cost strategy. With cross-docking, Walmart reduces distribution costs and speeds restocking of stores, thereby improving customer service. Although cross-docking reduces product handling, inventory, and facility costs, it requires both (1) tight scheduling and (2) accurate inbound product identification.

Cross-docking
Avoiding the placement of materials or supplies in storage by processing them as they are received for shipment.

Random stocking
Used in warehousing to locate stock wherever there is an open location.

Customizing
Using warehousing to add value to a product through component modification, repair, labeling, and packaging.

Random Stocking

Automatic identification systems (AISs), usually in the form of bar codes, allow accurate and rapid item identification. When automatic identification systems are combined with effective management information systems, operations managers know the quantity and location of every unit. This information can be used with human operators or with automatic storage and retrieval systems to load units anywhere in the warehouse—randomly. Accurate inventory quantities and locations mean the potential utilization of the whole facility because space does not need to be reserved for certain stock-keeping units (SKUs) or part families. Computerized random stocking systems often include the following tasks:

1. Maintaining a list of "open" locations
2. Maintaining accurate records of existing inventory and its locations
3. Sequencing items to minimize the travel time required to "pick" orders
4. Combining orders to reduce picking time
5. Assigning certain items or classes of items, such as high-usage items, to particular warehouse areas so that the total distance traveled within the warehouse is minimized

Andrew Hetherington/Redux

At IKEA's distribution center in Almhult, Sweden, pallets are stacked and retrieved through a fully automated process.

Random stocking systems can increase facility utilization and decrease labor cost, but they require comprehensive product information and accurate records.

Customizing

Although we expect warehouses to store as little product as possible and hold it for as short a time as possible, we are now asking warehouses to customize products. Warehouses can be places where value is added through customizing. Warehouse customization is a particularly useful way to generate competitive advantage in markets where products have multiple configurations. For instance, a warehouse can be a place where computer components are put together, software loaded, and repairs made. Warehouses may also provide customized labeling and packaging for retailers so items arrive ready for display.

This type of work often goes on adjacent to major airports, in facilities such as the FedEx terminal in Memphis. Adding value at warehouses adjacent to major airports also facilitates overnight delivery. For example, if your computer has failed, the replacement may be sent to you from such a warehouse for delivery the next morning. When your old machine arrives back at the warehouse, it is repaired and sent to someone else. These value-added activities at "quasi-warehouses" contribute to strategies of differentiation, low cost, and rapid response.

Fixed-Position Layout

In a **fixed-position layout**, the project remains in one place, and workers and equipment come to that one work area. Examples of this type of project are a ship, a highway, a bridge, a house, and an operating table in a hospital operating room.

Fixed-position layout
A system that addresses the layout requirements of stationary projects.

The techniques for addressing the fixed-position layout are complicated by three factors. First, there is limited space at virtually all sites. Second, at different stages of a project, different materials are needed; therefore, different items become critical as the project develops. Third, the volume of materials needed is dynamic. For example, the rate of use of steel panels for the hull of a ship changes as the project progresses.

LO 9.4 *Identify* when fixed-position layouts are appropriate

Because problems with fixed-position layouts are so difficult to solve well on site, an alternative strategy is to complete as much of the project as possible off site. This approach is used in the shipbuilding industry when standard units—say, pipe-holding brackets—are assembled on a nearby assembly line (a product-oriented facility). In an attempt to add efficiency to shipbuilding, Ingall Ship Building Corporation has moved toward product-oriented production when sections of a ship (modules) are similar or when it has a contract to build the same section of several similar ships. Also, as the photo of the house under construction shows, many home builders are moving from a fixed-position layout strategy to one that is more product oriented. About one-third of all new homes in the U.S. are built this way. In addition, many houses that are built on site (fixed position) have the majority of components such as doors, windows, fixtures, trusses, stairs, and wallboard built as modules in more efficient off-site processes.

Here are three versions of the fixed-position layout.

Craig Ruttle/AP Images

A house built via traditional fixed-position layout would be constructed onsite, with equipment, materials, and workers brought to the site. Then a "meeting of the trades" would assign space for various time periods. However, the home pictured here can be built at a much lower cost. The house is built in two movable modules in a factory. Scaffolding and hoists make the job easier, quicker, and cheaper, and the indoor work environment aids labor productivity.

A service example of a fixed-position layout is an operating room; the patient remains stationary on the table, and medical personnel and equipment are brought to the site.

Arctic-Images/Stone/Getty Images

U.S. Navy/Hulton Archive/Getty Images

In shipbuilding, there is limited space next to the fixed-position layout. Shipyards call these loading areas platens, and they are assigned for various time periods to each contractor.

Process-Oriented Layout

Process-oriented layout
A layout that deals with low-volume, high-variety production in which like machines and equipment are grouped together.

A **process-oriented layout** can simultaneously handle a wide variety of products or services. This is the traditional way for operations to support a differentiation strategy. It is most efficient when making products with different requirements or when handling customers, patients, or clients with different needs. A process-oriented layout is typically the low-volume, high-variety strategy discussed in Chapter 7. In this job-shop environment, each product or each small group of products undergoes a different sequence of operations. A product or small order is produced by moving it from one department to another in the sequence required for that product. A good example of the process-oriented layout is a hospital or clinic. Figure 9.2 illustrates the process for two patients, A and B, at an emergency clinic in Chicago. An inflow of patients, each with his or her own needs, requires routing through admissions, laboratories, operating rooms, radiology, pharmacies, nursing beds, and so on. Equipment, skills, and supervision are organized around these processes.

VIDEO 9.1
Laying Out Arnold Palmer Hospital's New Facility

A big advantage of process-oriented layout is its flexibility in equipment and labor assignments. The breakdown of one machine, for example, need not halt an entire process; work can be transferred to other machines in the department. Process-oriented layout is also especially good for handling the manufacture of parts in small batches, or **job lots**, and for the production of a wide variety of parts in different sizes or forms.

Job lots
Groups or batches of parts processed together.

The disadvantages of process-oriented layout come from the general-purpose use of the equipment. Orders take more time to move through the system because of difficult scheduling, changing setups, and unique material handling. In addition, general-purpose equipment requires high labor skills, and work-in-process inventories are higher because of imbalances in the production process. High labor-skill needs also increase the required level of training and experience, and high work-in-process levels increase capital investment.

When designing a process layout, the most common tactic is to arrange departments or work centers so as to minimize the costs of material handling. In other words, departments with large flows of parts or people between them should be placed next to one another. Material-handling costs in this approach depend on (1) the number of loads (or people) to be moved between two departments during some period of time and (2) the distance-related costs of moving loads (or people) between departments. Cost is assumed to be a function of distance between departments. The objective can be expressed as follows:

LO 9.5 *Explain* how to achieve a good process-oriented facility layout

$$\text{Minimize cost} = \sum_{i=1}^{n}\sum_{j=1}^{n} X_{ij}C_{ij} \qquad (9\text{-}1)$$

where n = total number of work centers or departments
i, j = individual departments
X_{ij} = number of loads moved from department i to department j
C_{ij} = cost to move a load between department i and department j

Figure 9.2
An Emergency Room Process Layout Showing the Routing of Two Patients

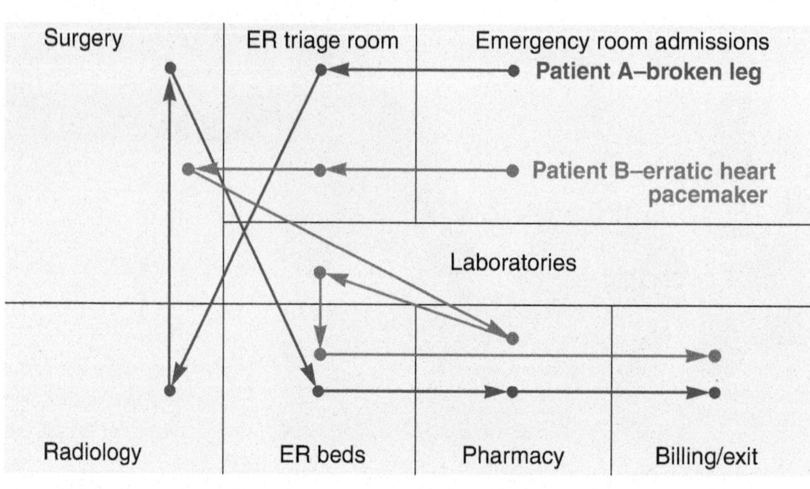

STUDENT TIP
Patient A (broken leg) proceeds (blue arrow) to ER triage, to radiology, to surgery, to a bed, to pharmacy, to billing. Patient B (pacemaker problem) moves (red arrow) to ER triage, to surgery, to pharmacy, to lab, to a bed, to billing.

Process-oriented facilities (and fixed-position layouts as well) try to minimize loads, or trips, multiplied by distance-related costs. The term C_{ij} combines distance and other costs into one factor. We thereby assume not only that the difficulty of movement is equal but also that the pickup and set-down costs are constant. Although they are not always constant, for simplicity's sake we summarize these data (that is, distance, difficulty, and pickup and setdown costs) in this one variable, cost. The best way to understand the steps involved in designing a process layout is to look at an example.

Example 1

DESIGNING A PROCESS LAYOUT

Walters Company management wants to arrange the six departments of its factory in a way that will minimize interdepartmental material-handling costs. They make an initial assumption (to simplify the problem) that each department is 20 × 20 feet and that the building is 60 feet long and 40 feet wide.

APPROACH AND SOLUTION ▶ The process layout procedure that they follow involves six steps:

Step 1: *Construct a "from–to matrix"* showing the flow of parts or materials from department to department (see Figure 9.3).

Figure **9.3**

Interdepartmental Flow of Parts

STUDENT TIP

The high flows between 1 and 3 and between 3 and 6 are immediately apparent. Departments 1, 3, and 6, therefore, should be close together.

Number of loads per week

Department	Assembly (1)	Painting (2)	Machine Shop (3)	Receiving (4)	Shipping (5)	Testing (6)
Assembly (1)		50	100	0	0	20
Painting (2)			30	50	10	0
Machine Shop (3)				20	0	100
Receiving (4)					50	0
Shipping (5)						0
Testing (6)						

Step 2: *Determine the space requirements* for each department. (Figure 9.4 shows available plant space.)

Figure **9.4**

Building Dimensions and One Possible Department Layout

STUDENT TIP

Think of this as a starting, initial, layout. Our goal is to improve it, if possible.

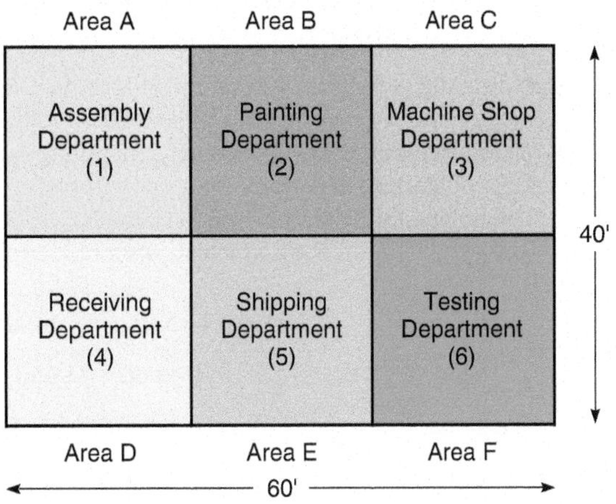

Step 3: *Develop an initial schematic diagram* showing the sequence of departments through which parts must move. Try to place departments with a heavy flow of materials or parts next to one another. (See Figure 9.5.)

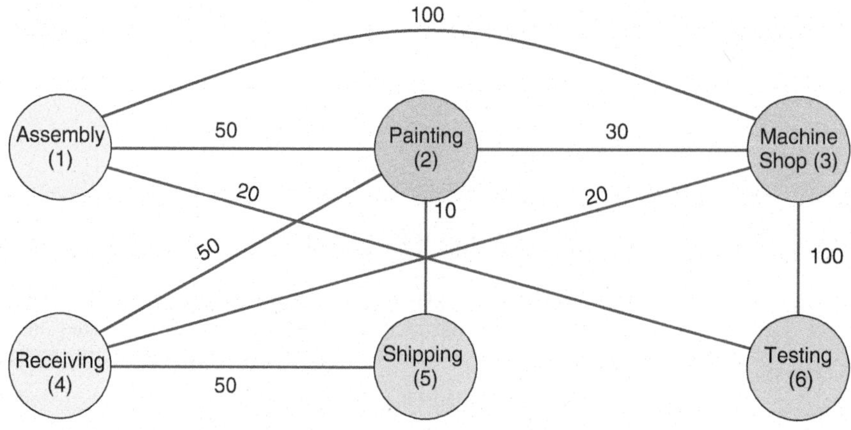

Figure **9.5**

Interdepartmental Flow Graph Showing Number of Weekly Loads

STUDENT TIP

This shows that 100 loads also move weekly between Assembly and the Machine Shop. We will probably want to move these two departments closer to one another to minimize the flow of parts through the factory.

Step 4: *Determine the cost of this layout* by using the material-handling cost equation:

$$\text{Cost} = \sum_{i=1}^{n}\sum_{j=1}^{n} X_{ij}C_{ij}$$

For this problem, Walters Company assumes that a forklift carries all interdepartmental loads. The cost of moving one load between adjacent departments is estimated to be \$1. Moving a load between nonadjacent departments costs \$2. Looking at Figures 9.3 and 9.4, we thus see that the handling cost between departments 1 and 2 is \$50 (\$1 × 50 loads), \$200 between departments 1 and 3 (\$2 × 100 loads), \$40 between departments and 6 (\$2 × 20 loads), and so on. Work areas that are diagonal to one another, such as 2 and 4, are treated as adjacent. The total cost for the layout shown in Figure 9.5 is:

Cost = \$50 + \$200 + \$40 + \$30 + \$50
(1 and 2) (1 and 3) (1 and 6) (2 and 3) (2 and 4)
+ \$10 + \$40 + \$100 + \$50
(2 and 5) (3 and 4) (3 and 6) (4 and 5)
= \$570

Step 5: By trial and error (or by a more sophisticated computer program approach that we discuss shortly), *try to improve the layout* pictured in Figure 9.4 to establish a better arrangement of departments.

By looking at both the flow graph (Figure 9.5) and the cost calculations, we see that placing departments 1 and 3 closer together appears desirable. They currently are nonadjacent, and the high volume of flow between them causes a large handling expense. Looking the situation over, we need to check the effect of shifting departments and possibly raising, instead of lowering, overall costs.

One possibility is to switch departments 1 and 2. This exchange produces a second departmental flow graph (Figure 9.6), which shows a reduction in cost to \$480, a savings in material handling of \$90:

Cost = \$50 + \$100 + \$20 + \$60 + \$50
(1 and 2) (1 and 3) (1 and 6) (2 and 3) (2 and 4)
+ \$10 + \$40 + \$100 + \$50
(2 and 5) (3 and 4) (3 and 6) (4 and 5)
= \$480

Figure **9.6**

Second Interdepartmental Flow Graph

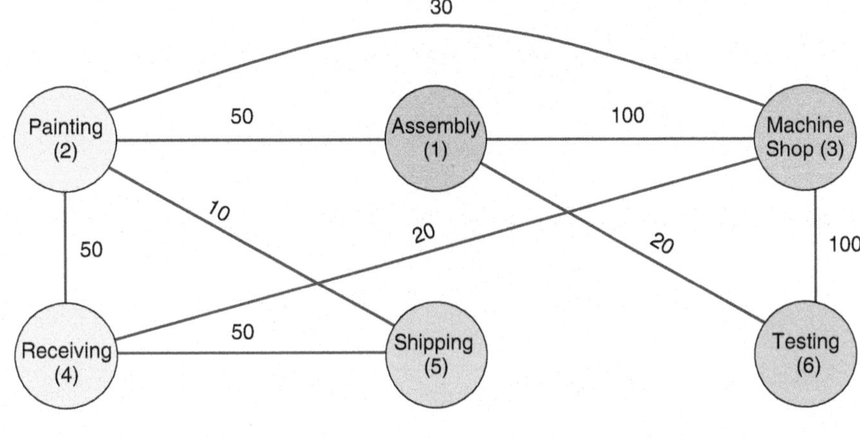

STUDENT TIP
Notice how Assembly and Machine Shop are now adjacent. Testing stayed close to the Machine Shop also.

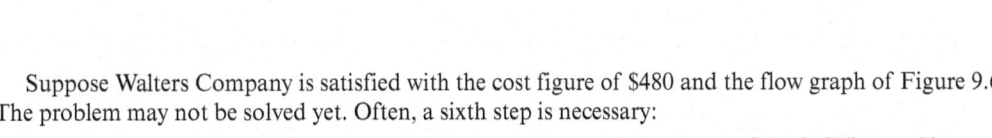

Suppose Walters Company is satisfied with the cost figure of $480 and the flow graph of Figure 9.6. The problem may not be solved yet. Often, a sixth step is necessary:

Step 6: *Prepare a detailed plan* arranging the departments to fit the shape of the building and its nonmovable areas (such as the loading dock, washrooms, and stairways). Often this step involves ensuring that the final plan can be accommodated by the electrical system, floor loads, aesthetics, and other factors.

In the case of Walters Company, space requirements are a simple matter (see Figure 9.7).

Figure **9.7**

A Feasible Layout for Walters Company

Area A	Area B	Area C
Painting Department (2)	Assembly Department (1)	Machine Shop Department (3)
Receiving Department (4)	Shipping Department (5)	Testing Department (6)
Area D	Area E	Area F

STUDENT TIP
Here we see the departments moved to areas A–F to try to improve the flow.

INSIGHT ▶ This switch of departments is only one of a large number of possible changes. For a six-department problem, there are actually 720 (or $6! = 6 \times 5 \times 4 \times 3 \times 2 \times 1$) potential arrangements! In layout problems, we may not find the optimal solution and may have to be satisfied with a "reasonable" one.

LEARNING EXERCISE ▶ Can you improve on the layout in Figures 9.6 and 9.7? [Answer: Yes, it can be lowered to $430 by placing Shipping in area A, Painting in area B, Assembly in area C, Receiving in area D (no change), Machine Shop in area E, and Testing in area F (no change).]

RELATED PROBLEMS ▶ 9.1–9.11

EXCEL OM Data File **Ch09Ex1.xls** can be found online.

ACTIVE MODEL 9.1 Example 1 is further illustrated in Active Model 9.1 found online.

Figure **9.8**

Muther's Grid for Process-Oriented Layouts

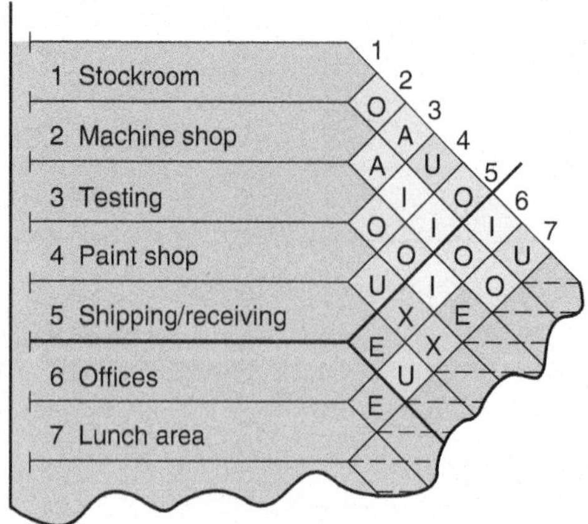

Code	CLOSENESS
A	Absolutely necessary
E	Especially important
I	Important
O	Ordinary OK
U	Unimportant
X	Not desirable

The solution procedure in Example 1 works well when load data are available. But when quantitative data are not readily obtained or when subjective or behavioral location issues are to be considered, *Muther's Grid* is a useful tool. Also called a "relationship chart," it uses the A, E, I, O, U, X coding system, as shown in Figure 9.8.

Relationships between departments are prioritized based on many factors, including the need for interdepartmental communication, use of common equipment or tools, safety issues, workflow, etc. In the figure, we see that it is especially important (E) for the office and lunch areas to be nearby, absolutely important (A) for testing and the machine shop to be close, but not desirable (X) for the paint shop (which is quite pungent) to be near either the office or lunch areas.

Computer Software for Process-Oriented Layouts

The graphic approach in Example 1 is fine for small problems. It does not, however, suffice for larger problems. When 20 departments are involved in a layout problem, more than 600 *trillion* different department configurations are possible. Fortunately, computer programs have been written to handle large layouts. These programs (see the Flow Path Calculator graphic) often add sophistication with flowcharts, multiple-story capability, storage and container placement, material volumes, time analysis, and cost comparisons. These programs tend to be interactive—that is, require participation by the user. And most only claim to provide "good," not "optimal," solutions.

Siemens Corp. software such as this allows operations managers to quickly place factory equipment for a full three-dimensional view of the layout. Such presentations provide added insight into the issues of facility layout in terms of process, material handling, efficiency, and safety. (Images created with Tecnomatix Plant Simulation software, courtesy of Siemens PLM Software)

Julian Stratenschulte/picture-alliance/dpa/AP Images

Proplanner Software for Process-Oriented Layouts

Working with computer-aided design software, analysts with the click of a mouse can use Proplanner's Flow Path Calculator to generate material flow diagrams and calculate material-handling distances, time, and cost. Variable-width flow lines indicating volume, color-coded by product, part, or material-handling method, allow users to identify how layouts should be arranged and where to eliminate excessive material handling.

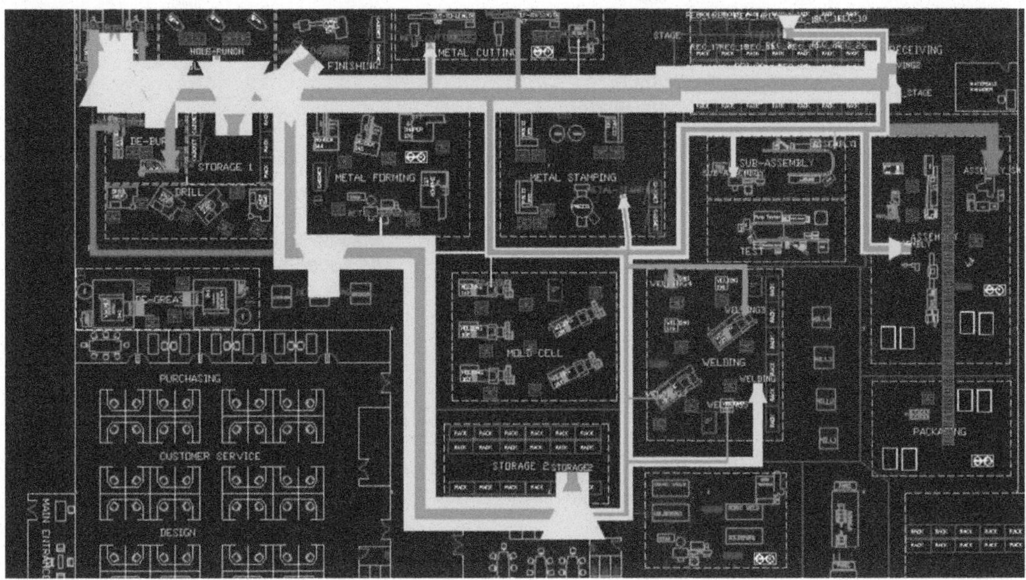

Focused Facilities

Focused facilities are appropriate when a firm has identified a family of similar products that have a reasonably stable demand. Focused facilities are designed to meet the needs of the customer by matching facilities to both customer product requirements and customer demand. When designed well, extra capacity and wasted resources do not exist. Such facilities are created and managed to run lean and in sync with the customer. This process requires a takt time. As shown in the equation below, **takt time** is the total work time available divided by units required to satisfy customer demand. The customer demand can be virtually anything, from parts in a factory, to patients in an emergency room, to Big Macs at McDonald's.

Takt time
Pace of production to meet the customer's demand.

$$\text{Takt time} = \text{Total work time available} \;/\; \text{Units required to satisfy customer demand}$$

For example, if there are 7½ working hours available at Illinois Tool Works and the customer's daily demand is for 600 parts, then

$$\text{Takt time} = \left(7\tfrac{1}{2} \text{ hours} \times 60 \text{ minutes}\right) / 600 \text{ units} = 450 \text{ minutes} / 600 \text{ units} = .75 \text{ minutes} = 45 \text{ seconds}$$

Therefore, the customer requirement (and the takt time) is one part every 45 seconds. Takt time sets the pace of the production facility to match the rate of the customer demand. Equipment, personnel, and suppliers are all managed to adhere to the takt time. Takt time becomes the *drumbeat* of the focused facility to which all facets of the organization march.

We will now discuss three types of focused facilities: The *work cell*, the *focused work center*, and the *focused factory*.

LO 9.6 *Define* work cell and the requirements of a work cell

Work Cells

A **work cell**, the smallest of the three focused facilities, reorganizes people and machines that would ordinarily be dispersed in various departments into a group so that they can focus on making a single product or a group of related products (see Figures 9.9 and 9.10). Cellular work arrangements are used when volume warrants a special arrangement of machinery and equipment. Work cells are reconfigured as takt time and product designs change.

Work cell
An arrangement of machines and personnel that focuses on making a single product or family of related products with similar routings.

Figure 9.9

Improve the Process by Moving from Individual Work Areas to a Work Cell

Note in both Figures 9.9 and 9.10 that U-shaped work cells can reduce material and employee movement. The U shape may also reduce space requirements, enhance communication, cut the number of workers, and make inspection easier.

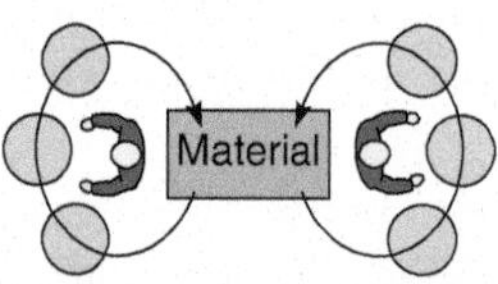

Current layout—workers are in small individual areas.

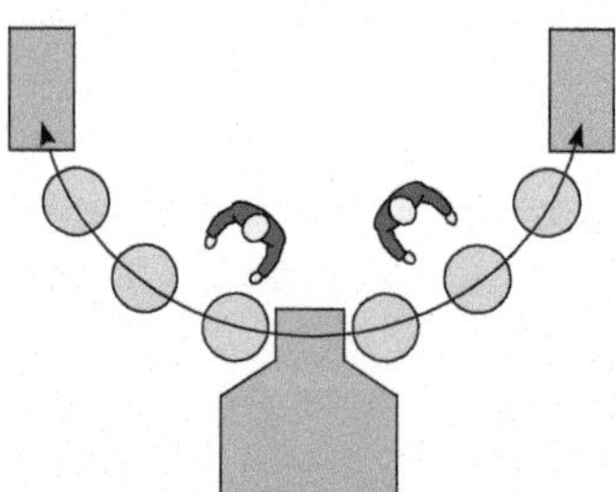

Improved layout—cross-trained workers can assist each other. May be able to add a third worker as added output is needed.

Figure 9.10

Improve the Process by Moving from a Straight Line to a Work Cell

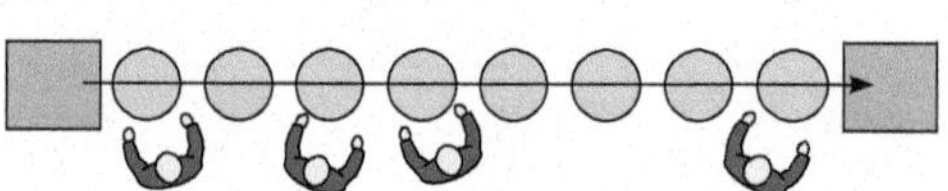

Current layout—straight lines make it hard to balance tasks because work may not be divided evenly.

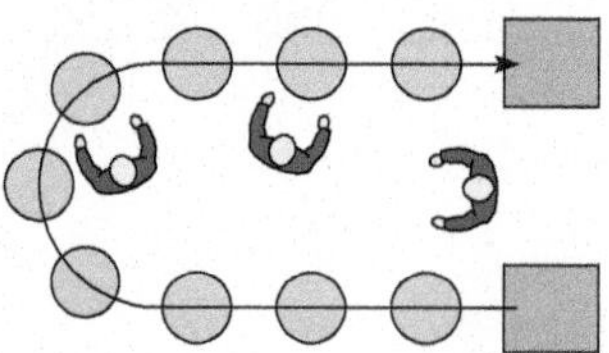

Improved layout—in U shape, workers have better access. Four cross-trained workers were reduced to three.

Requirements of Work Cells The requirements of work cells include:

- Identification of families of products, often using group technology codes or equivalents
- A high level of training, flexibility, and empowerment of employees
- Self-contained equipment and resources
- Testing (poka-yoke) at each station in the cell

Advantages of Work Cells The advantages of work cells are:

1. *Reduced work-in-process inventory* because the work cell is set up to provide one-piece flow from machine to machine
2. *Less floor space* required because less space is needed between machines to accommodate work-in-process inventory
3. *Reduced raw material and finished goods inventories* because less work-in-process allows more rapid movement of materials through the work cell
4. *Reduced direct labor cost* because of improved communication among employees, better material flow, and improved scheduling
5. *Heightened sense of employee participation* in the organization and the product: employees accept the added responsibility of product quality because it is directly associated with them and their work cell
6. *Increased equipment and machinery utilization* because of better scheduling and faster material flow
7. *Reduced investment in machinery and equipment* because good utilization reduces the number of machines and the amount of equipment and tooling

STUDENT TIP Using work cells is a big step toward manufacturing efficiency. They can make jobs more interesting, save space, and cut inventory.

Work cells are often superior to assembly lines and process-oriented facilities because: (1) tasks are grouped, meaning inspection can often be immediate; (2) fewer workers are needed; (3) workers can reach more of the work area; (4) work can be more efficiently balanced; and (5) communication is enhanced. Work cells are sometimes organized in a U shape, as shown on the right side of Figures 9.9 and 9.10. The shape of the cell, however, is secondary to process flow. The focus should be on a flow that optimizes people, material, and communication.

In a work cell, as in any production facility, we must consider staffing and potential bottlenecks because bottlenecks constrain flow through the cell. However, imbalance in a work cell is seldom an issue if the operation is manual, as cell members are, by definition, part of a cross-trained team. Consequently, the inherent flexibility of work cells typically overcomes modest imbalance issues. However, if the imbalance is a machine constraint, then an adjustment in machinery, process, or operations may be necessary. In such situations, the use of traditional assembly-line balancing analysis, the topic of our next section, may be helpful.

The success of work cells is not limited to manufacturing. Kansas City's Hallmark, which has more than half of the U.S. greeting-card market and produces some 40,000 different cards, has modified the offices into a cellular design. In the past, its 700 creative professionals would take up to 2 years to develop a new card. Hallmark's decision to create work cells consisting of artists, writers, lithographers, merchandisers, and accountants, all located in the same area, has resulted in card preparation in a fraction of the time that the old layout required. Similarly, work cells have yielded higher performance and better service for the American Red Cross blood donation process.

Focused Work Center

When the similar products are of significant volume and the demand stable, the cell may be organized as a focused work center. A focused work center (also called a *plant within a plant*) moves production to a large cellular work center that remains part of the present facility. For example, bumpers and dashboards in Toyota's Texas plant are produced in a focused work center with takt times that respond to vehicle, model, and color changes on the assembly line.

Focused Factory

If the focused work center is in a separate facility, it is often called a **focused factory**. For example, the separate plants that produce seat belts, fuel tanks, and exhaust systems for Toyota are focused factories. A fast-food restaurant is also a focused factory—most are easily reconfigured for adjustments to product mix and volume. Burger King changes the number of personnel and task assignments, rather than moving machines and equipment, to balance the process to meet changing takt times. In effect, the layout changes numerous times each day.

Focused factory
A plant established to focus the entire manufacturing system on a limited set of products defined by the company's competitive strategy.

The term *focused factories* may also refer to facilities that are focused in ways other than by product line or layout. For instance, facilities may focus on their core competence, such as low cost, quality, new product introductions, or flexibility.

Focused facilities in both manufacturing and services are, by design, very good at establishing takt time. Takt time sets a basis that enables companies to stay in tune with their customers, to produce quality products, and to operate at higher margins. This is true whether they are auto manufacturers such as Toyota, restaurants such as McDonald's, or hospitals such as Arnold Palmer Hospital in Orlando.

Repetitive and Product-Oriented Layout

Product-oriented layouts are organized around products or families of similar high-volume, low-variety products. Repetitive production and continuous production, which are discussed in Chapter 7, use product layouts. The assumptions are that:

1. Volume is adequate for high equipment utilization
2. Product demand is stable enough to justify high investment in specialized equipment
3. Product is standardized or approaching a phase of its life cycle that justifies investment in specialized equipment
4. Supplies of raw materials and components are adequate and of uniform quality (adequately standardized) to ensure that they will work with the specialized equipment

LO 9.7 *Define* product-oriented layout

Two types of a product-oriented layout are fabrication and assembly lines. The **fabrication line** builds components, such as automobile tires or metal parts for a refrigerator, on a series of machines, while an **assembly line** puts the fabricated parts together at a series of workstations. However, both are repetitive processes, and in both cases, the line must be "balanced"; that is, the time spent to perform work on one machine must equal or "balance" the time spent to perform work on the next machine in the fabrication line, just as the time spent at one workstation by one assembly-line employee must "balance" the time spent at the next workstation by the next employee. The same issues arise when designing the "disassembly lines" of slaughterhouses and automobile recyclers.

Fabrication line
A machine-paced, product-oriented facility for building components.

Assembly line
An approach that puts fabricated parts together at a series of workstations; used in repetitive processes.

A well-balanced assembly line has the advantage of high personnel and facility utilization and equity among employees' workloads. Some union contracts require that workloads be nearly equal among those on the same assembly line. The term most often used to describe this process is **assembly-line balancing**. Indeed, the *objective of the product-oriented layout is to minimize imbalance in the fabrication or assembly line*.

Assembly-line balancing
Obtaining output at each workstation on a production line so delay is minimized.

The main advantages of product-oriented layout are:

1. The low variable cost per unit usually associated with high-volume, standardized products
2. Low material-handling costs
3. Reduced work-in-process inventories
4. Easier training and supervision
5. Rapid throughput

The disadvantages of product layout are:

1. The high volume required because of the large investment needed to establish the process
2. Work stoppage at any one point can tie up the whole operation
3. The process flexibility necessary for a variety of products and production rates can be a challenge

VIDEO 9.2
Facility Layout at Wheeled Coach Ambulances

Because the problems of fabrication lines and assembly lines are similar, we focus our discussion on assembly lines. On an assembly line, the product typically moves via automated means, such as a conveyor, through a series of workstations until completed. This is the way fast-food hamburgers are made (see Figure 9.11), automobiles and some planes (see the photo of the Boeing 737) are assembled, and television sets and ovens are produced. Product-oriented layouts use more automated and specially designed equipment than do process layouts.

Assembly-Line Balancing

LO 9.8 *Explain* how to balance production flow in a repetitive or product-oriented facility

Line balancing is usually undertaken to minimize imbalance between machines or personnel while meeting a required output from the line. To produce at a specified rate, management must know the tools, equipment, and work methods used. Then the time requirements for each assembly task (e.g., drilling a hole, tightening a nut, or spray-painting a part) must be determined. Management also needs to know the *precedence relationship* among the activities—that is, the sequence in which various tasks must be performed. Example 2 shows how to turn these task data into a precedence diagram.

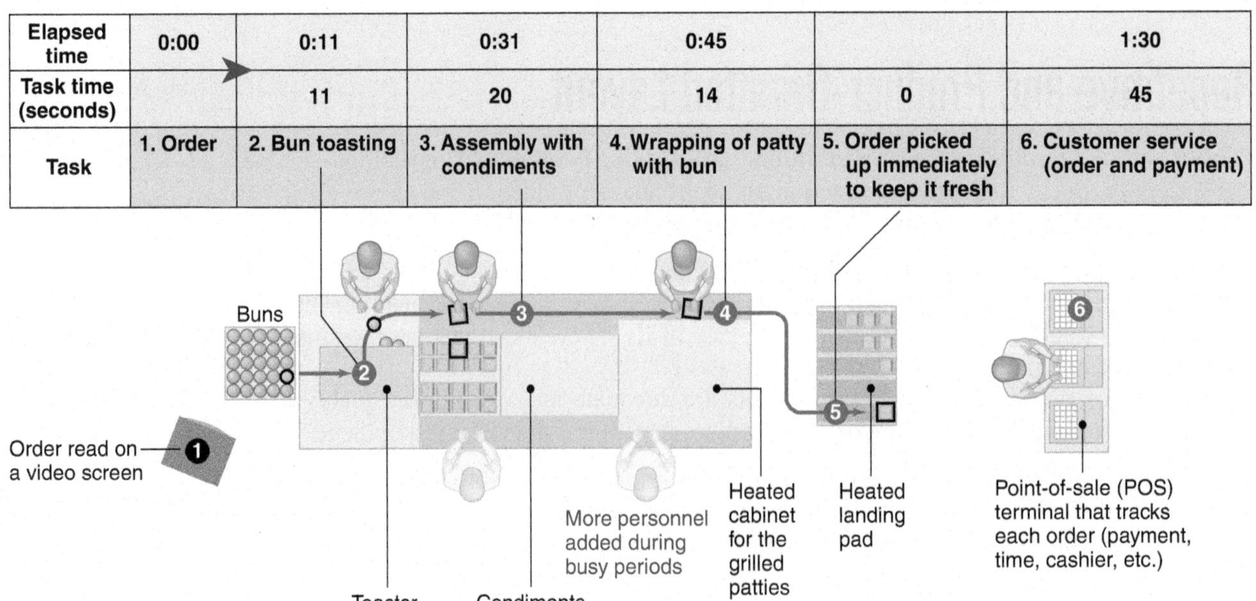

Figure **9.11**
McDonald's Hamburger Assembly Line

The Boeing 737, the world's most popular commercial airplane, is produced on a moving production line, traveling at 2 inches a minute through the final assembly process. The moving line, one of several lean manufacturing innovations at the Renton, Washington, facility, has enhanced quality, reduced flow time, slashed inventory levels, and cut space requirements. Final assembly is only 8 days—a time savings of 70%—and inventory is down more than 55%.

Example 2

DEVELOPING A PRECEDENCE DIAGRAM FOR AN ASSEMBLY LINE

Boeing wants to develop a precedence diagram for an electrostatic wing component that requires a total assembly time of 65 minutes.

APPROACH ▶ Staff gather tasks, assembly times, and sequence requirements for the component in Table 9.2.

TABLE 9.2 Precedence Data for Wing Component

TASK	ASSEMBLY TIME (MINUTES)	TASK MUST FOLLOW TASK LISTED BELOW	
A	10	—	This means that tasks B and E cannot be done until task A has been completed
B	11	A	
C	5	B	
D	4	B	
E	11	A	
F	3	C, D	
G	7	F	
H	11	E	
I	3	G, H	
	Total time 65		

SOLUTION ▶ Figure 9.12 shows the precedence diagram.

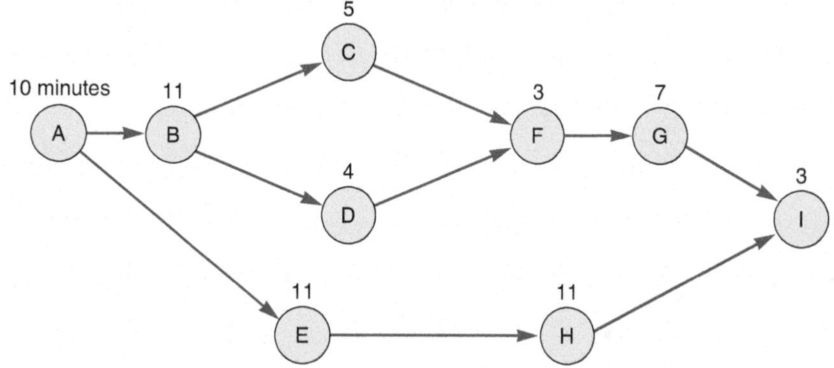

Figure **9.12**

Precedence Diagram

INSIGHT ▶ The diagram helps structure an assembly line and workstations, and it makes it easier to visualize the sequence of tasks.

LEARNING EXERCISE ▶ If task D had a second preceding task (C), how would Figure 9.12 change? [Answer: There would also be an arrow pointing from C to D.]

RELATED PROBLEMS ▶ 9.13a, 9.15a, 9.16a, 9.17a, 9.20a, 9.25a,d, 9.26a, 9.27

Once we have constructed a precedence chart summarizing the sequences and performance times, we turn to the job of grouping tasks into job stations so that we can meet the specified production rate. This process involves three steps:

Cycle time
The maximum time that a product is allowed at each workstation.

1. Take the units required (demand or production rate) per day and divide them into the productive time available per day (in minutes or seconds). This operation gives us what is called the cycle time—namely, the maximum time allowed at each workstation if the production rate is to be achieved:

$$\text{Cycle time} = \frac{\text{Production time available per day}}{\text{Units required per day}} \tag{9-2}$$

2. Calculate the theoretical minimum number of workstations. This is the total task-duration time (the time it takes to make the product) divided by the cycle time. Fractions are rounded to the next higher whole number:

$$\text{Minimum number of workstations} = \frac{\sum_{i=1}^{n} \text{Time for task } i}{\text{Cycle time}} \tag{9-3}$$

where n is the number of assembly tasks.

3. Balance the line by assigning specific assembly tasks to each workstation. An efficient balance is one that will complete the required assembly, follow the specified sequence, and keep the idle time at each workstation to a minimum. A formal procedure for doing this is the following:
   a. Identify a master list of tasks.
   b. Eliminate those tasks that have been assigned.
   c. Eliminate those tasks whose precedence relationship has not been satisfied.
   d. Eliminate those tasks for which inadequate time is available at the workstation.
   e. Use one of the line-balancing "heuristics" described in Table 9.3. The five choices are (1) longest task time, (2) most following tasks, (3) ranked positional weight, (4) shortest task time, and (5) least number of following tasks. You may wish to test several of these heuristics to see which generates the "best" solution—that is, the smallest number of workstations and highest efficiency. Remember, however, that although heuristics provide solutions, they do not guarantee an optimal solution.

Heuristic
Problem solving using procedures and rules rather than mathematical optimization.

TABLE 9.3 Layout Heuristics That May Be Used to Assign Tasks to Workstations in Assembly-Line Balancing

1. *Longest task (operation) time*	From the available tasks, choose the task with the largest (longest) time.
2. *Most following tasks*	From the available tasks, choose the task with the largest number of following tasks.
3. *Ranked positional weight*	From the available tasks, choose the task for which the sum of the times for each following task is longest. (In Example 3 we see that the ranked positional weight of task C = 5(C) + 3(F) + 7(G) + 3(I) = 18, whereas the ranked positional weight of task D = 4(D) + 3(F) + 7(G) + 3(I) = 17; therefore, C would be chosen first, using this heuristic.)
4. *Shortest task (operations) time*	From the available tasks, choose the task with the shortest task time.
5. *Least number of following tasks*	From the available tasks, choose the task with the least number of subsequent tasks.

Example 3 illustrates a simple line-balancing procedure.

Example 3

BALANCING THE ASSEMBLY LINE

On the basis of the precedence diagram and activity times given in Example 2, Boeing determines that there are 480 productive minutes of work available per day. Furthermore, the production schedule requires that 40 units of the wing component be completed as output from the assembly line each day. It now wants to group the tasks into workstations.

APPROACH ▶ Following the three steps above, we compute the cycle time using Equation (9-2) and minimum number of workstations using Equation (9-3), and we assign tasks to workstations—in this case using the *most following tasks* heuristic.

SOLUTION ▶

$$\text{Cycle time (in minutes)} = \frac{480 \text{ minutes}}{40 \text{ units}}$$
$$= 12 \text{ minutes/unit}$$

$$\text{Minimum number of workstations} = \frac{\text{Total task time}}{\text{Cycle time}} = \frac{65}{12}$$
$$= 5.42, \text{ or } 6 \text{ stations}$$

Figure 9.13 shows one solution that does not violate the sequence requirements and that groups tasks into 6 one-person stations. To obtain this solution, activities with the most following tasks were moved into workstations to use as much of the available cycle time of 12 minutes as possible. The first workstation consumes 10 minutes and has an idle time of 2 minutes.

Figure 9.13

A Six-Station Solution to the Line-Balancing Problem

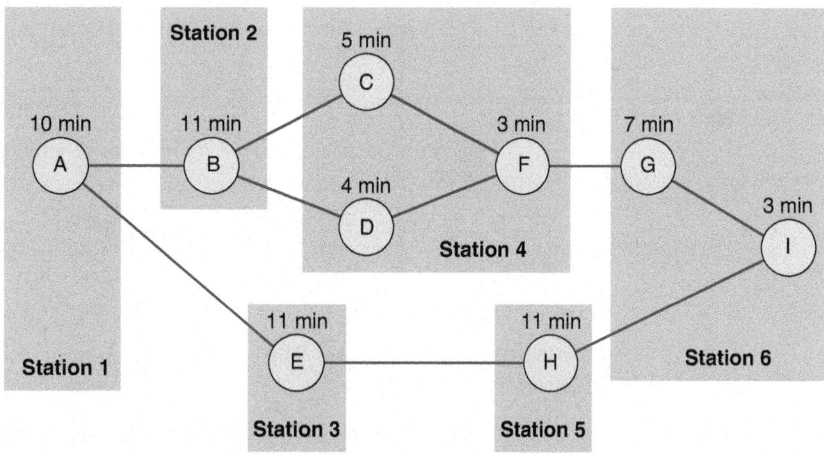

STUDENT TIP

Tasks C, D, and F can be grouped together in one workstation, provided that the physical facilities and skill levels meet the work requirements.

INSIGHT ▶ This is a reasonably well-balanced assembly line. The second and third workstations use 11 minutes. The fourth workstation groups three small tasks and balances perfectly at 12 minutes. The fifth has 1 minute of idle time, and the sixth (consisting of tasks G and I) has 2 minutes of idle time per cycle. Total idle time for this solution is 7 minutes per cycle.

LEARNING EXERCISE ▶ If task I required 6 minutes (instead of 3 minutes), how would this change the solution? [Answer: the cycle time would not change, and the *theoretical* minimum number of workstations would still be 6 (rounded up from 5.67), but it would take 7 stations to balance the line.]

RELATED PROBLEMS ▶ 9.12–9.27

EXCEL OM Data File **Ch09Ex4.xls** can be found online.

There are two measures of effectiveness of a balance assignment. The first measure computes the *efficiency* of a line balance by dividing the total task times by the product of the number of workstations required times the assigned (actual) cycle time of the *longest* workstation:

$$\text{Efficiency} = \frac{\Sigma \text{ Task times}}{(\textit{Actual} \text{ number of workstations}) \times (\text{Largest assigned cycle time})} \quad (9\text{-}4)$$

Operations managers compare different levels of efficiency for various numbers of workstations. In this way, a firm can determine the sensitivity of the line to changes in the production rate and workstation assignments.

The second measure computes the *idle time* for the line.

$$\text{Idle Time} = (\text{Actual number of workstations} \times \text{Largest assigned cycle time}) - \Sigma\text{Task times} \quad (9\text{-}5)$$

Example 4

DETERMINING LINE EFFICIENCY

Boeing wishes to calculate the efficiency and idle time for Example 3.

APPROACH ▶ Equation (9-4) and (9-5) are applied.

SOLUTION ▶

$$\text{Efficiency} = \frac{65 \text{ minutes}}{(6 \text{ stations}) \times (12 \text{ minutes})} = \frac{65}{72} = 90.3\%$$

$$\text{Idle time} = (6 \text{ stations} \times 12 \text{ minutes}) - 65 \text{ minutes} = 7 \text{ minutes}$$

Note that opening a seventh workstation, for whatever reason, would decrease the efficiency of the balance to 77.4% (assuming that at least one of the workstations still required 12 minutes):

$$\text{Efficiency} = \frac{65 \text{ minutes}}{(7 \text{ stations}) \times (12 \text{ minutes})} = 77.4\%$$

$$\text{And now the idle time} = (7 \text{ stations} \times 12 \text{ minutes}) - 65 \text{ minutes} = 19 \text{ minutes}$$

INSIGHT ▶ Increasing efficiency may require that some tasks be divided into smaller elements and reassigned to other tasks. This facilitates a better balance between workstations and means higher efficiency. Note that we can also compute efficiency as 1 – (% Idle time), that is, [1 – (Idle time)/(Total time in workstations)].

LEARNING EXERCISE ▶ What is the efficiency if an eighth workstation is opened? [Answer: Efficiency = 67.7%.]

RELATED PROBLEMS ▶ 9.13f, 9.14c, 9.15f, 9.17c, 9.18b, 9.19b, 9.20e,g, 9.25e, 9.26c, 9.27

Large-scale line-balancing problems, like large process-layout problems, are often solved by computers. Computer programs such as Assembly Line Pro, Proplanner, Timer Pro, Flexible Line Balancing, and Promodel are available to handle the assignment of workstations on assembly lines with numerous work activities. Such software evaluates the thousands, or even millions, of possible workstation combinations much more efficiently than could ever be done by hand.

Summary

Layouts make a substantial difference in operating efficiency. The seven layout situations discussed in this chapter are (1) office, (2) retail, (3) warehouse, (4) fixed position, (5) process oriented, (6) work cells, and (7) product oriented. A variety of techniques have been developed to solve these layout problems. Office layouts often seek to maximize information flows, retail firms focus on product exposure, and warehouses attempt to optimize the trade-off between storage space and material-handling cost.

The fixed-position layout problem attempts to minimize material-handling costs within the constraint of limited space at the site. Process layouts minimize travel distances times the number of trips. Product layouts focus on reducing waste and the imbalance in an assembly line. Work cells are the result of identifying a family of products that justify a special configuration of machinery and equipment that reduces material travel and adjusts imbalances with cross-trained personnel.

Often, the issues in a layout problem are so wide-ranging and dynamic that finding an optimal solution may not be possible. For this reason, layout decisions, although the subject of substantial research effort, remain something of an art.

Key Terms

Office layout (p. 367)
Retail layout (p. 369)
Slotting fees (p. 370)
Servicescape (p. 370)
Warehouse layout (p. 371)
Cross-docking (p. 372)
Random stocking (p. 372)
Customizing (p. 372)
Fixed-position layout (p. 373)
Process-oriented layout (p. 374)
Job lots (p. 374)
Takt time (p. 379)
Work cell (p. 379)
Focused factory (p. 381)
Fabrication line (p. 381)
Assembly line (p. 381)
Assembly-line balancing (p. 382)
Cycle time (p. 384)
Heuristic (p. 384)

Ethical Dilemma

Although buried by mass customization and a proliferation of new products of numerous sizes and variations, grocery chains continue to seek to maximize payoff from their layout. Their layout includes a marketable commodity—shelf space—and they charge for it. This charge is known as a *slotting fee*. Recent estimates are that food manufacturers now spend some 13% of sales on trade promotions, which is paid to grocers to get them to promote and discount the manufacturer's products. A portion of these fees is for slotting, but slotting fees drive up the manufacturer's cost. They also put the small company with a new product at a disadvantage because small companies with limited resources may be squeezed out of the marketplace. Slotting fees may also mean that customers may no longer be able to find the special local brand. How ethical are slotting fees?

Image Source/Alamy Stock Photo

Discussion Questions

1. What are the seven layout strategies presented in this chapter?
2. What are the three factors that complicate a fixed-position layout?
3. What are the advantages and disadvantages of process layout?
4. How would an analyst obtain data and determine the number of trips in:
   (a) a hospital?
   (b) a machine shop?
   (c) an auto-repair shop?
5. What are the advantages and disadvantages of product layout?
6. What are the four assumptions (or preconditions) of establishing layout for high-volume, low-variety products?
7. What are the alternative forms of work cells discussed in this textbook?
8. What are the advantages and disadvantages of work cells?
9. What are the requirements for a focused work center or focused factory to be appropriate?
10. What are the major trends influencing office layout?
11. How could Muther's Grid be used in an office layout?
12. What layout innovations have you noticed recently in retail establishments?

13. What are the variables that a manager can manipulate in a retail layout?
14. Visit a local supermarket and sketch its layout. What are your observations regarding departments and their locations?
15. What is random stocking?
16. What information is necessary for random stocking to work?
17. Explain the concept of cross-docking.
18. What is a heuristic? Name several that can be used in assembly-line balancing.
19. Create Muther's Grid (see Figure 9.8) for a company that has three departments: Tooling, Warehousing, and Painting. It is "important" that Tooling be near Painting, "not desirable" that Painting be near Warehousing, and "OK" that Tooling is near Warehousing.

Using Software to Solve Layout Problems

In addition to the many commercial software packages available for addressing layout problems, Excel OM and POM for Windows, both of which accompany this text and can be found online, contain modules for the process problem and the assembly-line-balancing problem.

X USING EXCEL OM

Excel OM can assist in evaluating a series of department work assignments like the one we saw for the Walters Company in Example 1. The layout module can generate an optimal solution by enumeration or by computing the "total movement" cost for each layout you wish to examine. As such, it provides a speedy calculator for each flow–distance pairing.

Program 9.1 illustrates our inputs in the top two tables. We first enter department flows and then provide distances between work areas. Entering area assignments on a trial-and-error basis in the upper left of the top table generates movement computations at the bottom of the screen. Total movement is recalculated each time we try a new area assignment. It turns out that the assignment shown is optimal at 430 feet of movement.

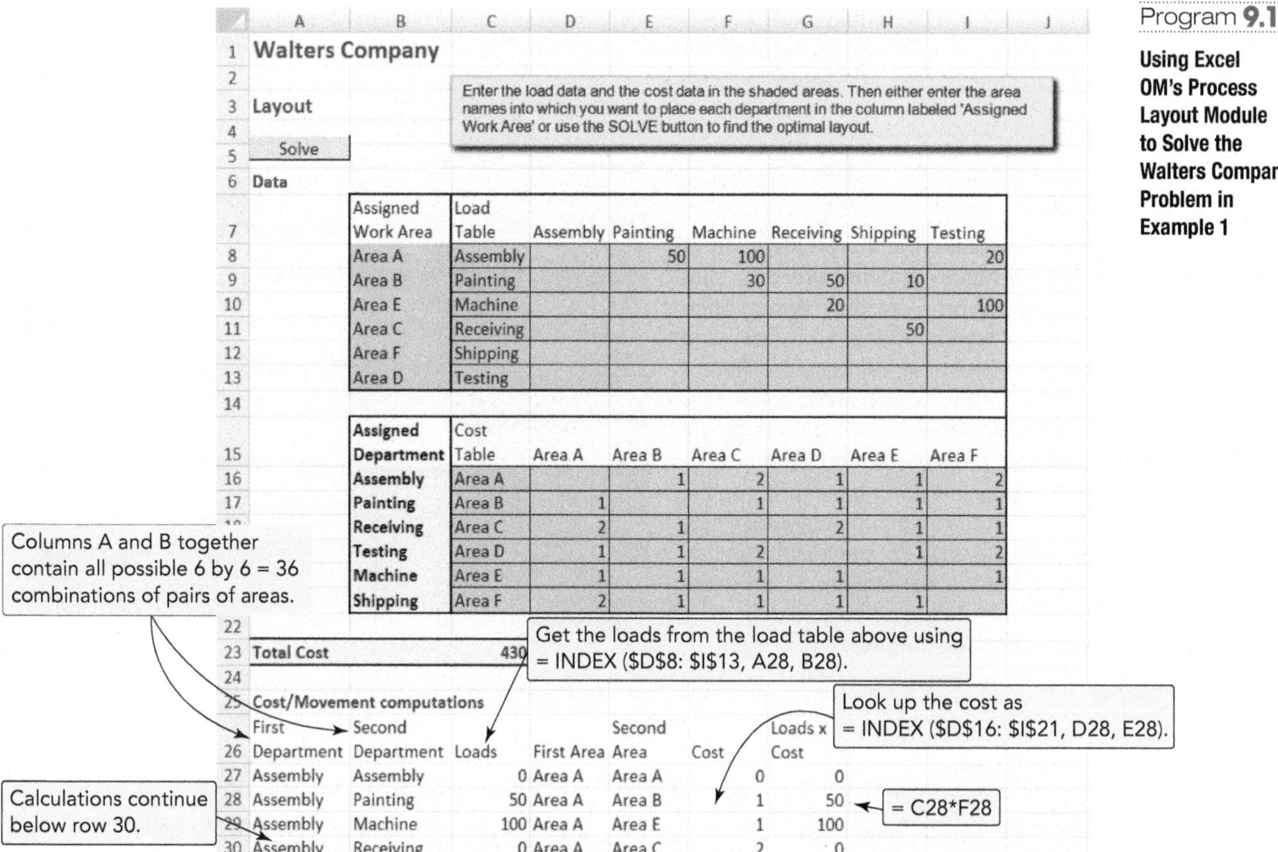

Walters Company

Layout

Solve

Enter the load data and the cost data in the shaded areas. Then either enter the area names into which you want to place each department in the column labeled 'Assigned Work Area' or use the SOLVE button to find the optimal layout.

Data

Assigned Work Area	Load Table	Assembly	Painting	Machine	Receiving	Shipping	Testing
Area A	Assembly		50	100			20
Area B	Painting			30	50	10	
Area E	Machine				20		100
Area C	Receiving					50	
Area F	Shipping						
Area D	Testing						

Assigned Department	Cost Table	Area A	Area B	Area C	Area D	Area E	Area F
Assembly	Area A		1	2	1	1	2
Painting	Area B	1		1	1	1	1
Receiving	Area C	2	1		2	1	1
Testing	Area D	1	1	2		1	2
Machine	Area E	1	1	1	1		1
Shipping	Area F	2	1	1	1	1	

Total Cost 430

Cost/Movement computations

First Department	Second Department	Loads	First Area	Second Area	Cost	Loads x Cost
Assembly	Assembly	0	Area A	Area A	0	0
Assembly	Painting	50	Area A	Area B	1	50
Assembly	Machine	100	Area A	Area E	1	100
Assembly	Receiving	0	Area A	Area C	2	0

Program 9.1

Using Excel OM's Process Layout Module to Solve the Walters Company Problem in Example 1

P USING POM FOR WINDOWS

The POM for Windows facility layout module can be used to place up to 10 departments in 10 rooms to minimize the total distance traveled as a function of the distances between the rooms and the flow between departments. The program exchanges departments until no exchange will reduce the total amount of movement, meaning an optimal solution has been reached.

The POM for Windows and Excel OM modules for line balancing can handle a line with up to 99 tasks, each with up to six immediate predecessors. In this program, cycle time can be entered as either (1) *given*, if known, or (2) the *demand* rate can be entered with time available as shown. All five "heuristic rules" are used: (1) longest operation (task) time, (2) most following tasks, (3) ranked positional weight, (4) shortest operation (task) time, and (5) least number of following tasks. No one rule can guarantee an optimal solution, but POM for Windows displays the number of stations needed for each rule.

Please refer to Appendix II for further details.

Solved Problems

SOLVED PROBLEM 9.1

Aero Maintenance is a small aircraft engine maintenance facility located in Wichita, Kansas. Its new administrator, Ann Daniel, decides to improve material flow in the facility, using the process layout method she studied at Wichita State University. The current layout of Aero Maintenance's eight departments is shown in Figure 9.14.

The only physical restriction perceived by Daniel is the need to keep the entrance in its current location. All other departments can be moved to a different work area (each 10 feet square) if layout analysis indicates a move would be beneficial.

First, Daniel analyzes records to determine the number of material movements among departments in an average month. These data are shown in Figure 9.15. Her objective, Daniel decides, is to lay out the departments so as to minimize the total movement (distance traveled) of material in the facility. She writes her objective as:

$$\text{Minimize material movement} = \sum_{i=1}^{8}\sum_{j=1}^{8} X_{ij}C_{ij}$$

where X_{ij} = number of material movements per month (loads or trips) moving from department i to department j

C_{ij} = distance in feet between departments i and j (which, in this case, is the equivalent of cost per load to move between departments)

Note that this is only a slight modification of the cost-objective equation shown earlier in the chapter.

Daniel assumes that adjacent departments, such as entrance (now in work area A) and receiving (now in work area B), have a walking distance of 10 feet. Diagonal departments are also considered adjacent and assigned a distance of 10 feet. Nonadjacent departments, such as the entrance and parts (now in area C) or the entrance and inspection (area G) are 20 feet apart, and nonadjacent rooms, such as entrance and metallurgy (area D), are 30 feet apart. (Hence, 10 feet is considered 10 units of cost, 20 feet is 20 units of cost, and 30 feet is 30 units of cost.)

Given the above information, redesign Aero Maintenance's layout to improve its material flow efficiency.

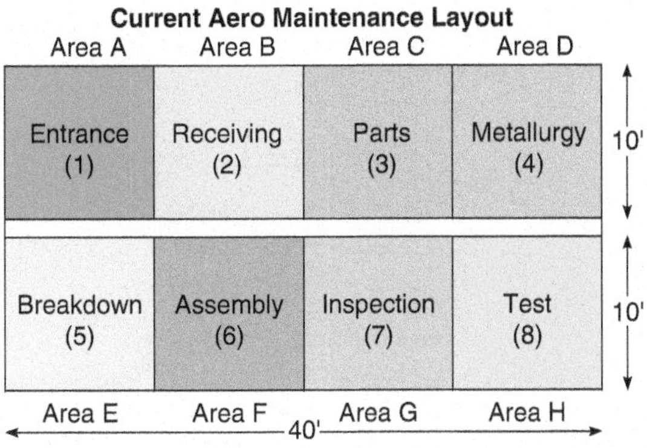

Current Aero Maintenance Layout

Area A	Area B	Area C	Area D	
Entrance (1)	Receiving (2)	Parts (3)	Metallurgy (4)	10'
Breakdown (5)	Assembly (6)	Inspection (7)	Test (8)	10'
Area E	Area F	Area G	Area H	

40'

Figure **9.14**

Aero Maintenance Layout

Figure **9.15**

Number of Material Movements (Loads) between Departments in 1 Month

Department	Entrance (1)	Receiving (2)	Parts (3)	Metallurgy (4)	Breakdown (5)	Assembly (6)	Inspection (7)	Test (8)
Entrance (1)		100	100	0	0	0	0	0
Receiving (2)			0	50	20	0	0	0
Parts (3)				30	30	0	0	0
Metallurgy (4)					20	0	0	20
Breakdown (5)						20	0	10
Assembly (6)							30	0
Inspection (7)								0
Test (8)								

SOLUTION

First, establish Aero Maintenance's current layout, as shown in Figure 9.16. Then, by analyzing the current layout, compute material movement:

$$
\begin{aligned}
\text{Total movement} &= \underset{1\text{ to }2}{(100 \times 10')} + \underset{1\text{ to }3}{(100 \times 20')} + \underset{2\text{ to }4}{(50 \times 20')} + \underset{2\text{ to }5}{(20 \times 10')} \\
&\quad + \underset{3\text{ to }4}{(30 \times 10')} + \underset{3\text{ to }5}{(30 \times 20')} + \underset{4\text{ to }5}{(20 \times 30')} + \underset{4\text{ to }8}{(20 \times 10')} \\
&\quad + \underset{5\text{ to }6}{(20 \times 10')} + \underset{5\text{ to }8}{(10 \times 30')} + \underset{5\text{ to }7}{(30 \times 10')} \\
&= 1{,}000 + 2{,}000 + 1{,}000 + 200 + 300 + 600 + 600 \\
&\quad +200 + 200 + 300 + 300 \\
&= 6{,}700 \text{ feet}
\end{aligned}
$$

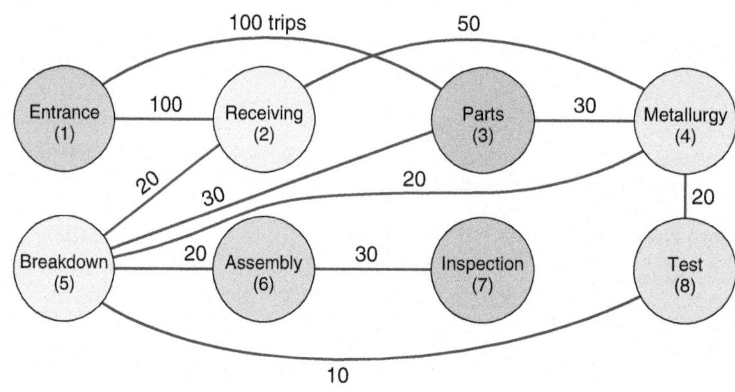

Figure 9.16

Current Material Flow

Propose a new layout that will reduce the current figure of 6,700 feet. Two useful changes, for example, are to switch departments 3 and 5 and to interchange departments 4 and 6. This change would result in the schematic shown in Figure 9.17:

$$
\begin{aligned}
\text{Total movement} &= \underset{1\text{ to }2}{(100 \times 10')} + \underset{1\text{ to }3}{(100 \times 10')} + \underset{2\text{ to }4}{(50 \times 10')} + \underset{2\text{ to }5}{(20 \times 10')} \\
&\quad + \underset{3\text{ to }4}{(30 \times 10')} + \underset{3\text{ to }5}{(30 \times 20')} + \underset{4\text{ to }5}{(20 \times 10')} + \underset{4\text{ to }8}{(20 \times 20')} \\
&\quad + \underset{5\text{ to }6}{(20 \times 10')} + \underset{5\text{ to }8}{(10 \times 10')} + \underset{6\text{ to }7}{(30 \times 10')} \\
&= 1{,}000 + 1{,}000 + 500 + 200 + 300 + 600 + 200 \\
&\quad + 400 + 200 + 100 + 300 \\
&= 4{,}800 \text{ feet}
\end{aligned}
$$

Do you see any room for further improvement?

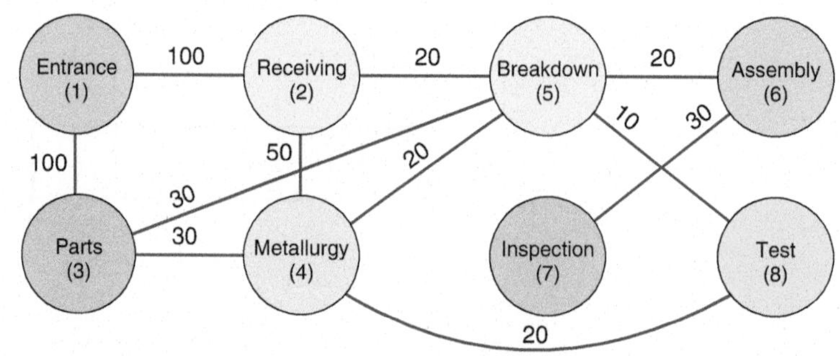

Figure 9.17

Improved Layout

SOLVED PROBLEM 9.2

The assembly line whose activities are shown in Figure 9.18 has an 8-minute cycle time. Draw the precedence graph, and find the minimum possible number of one-person workstations. Then arrange the work activities into workstations so as to balance the line. What is the efficiency of your line balance?

TASK	PERFORMANCE TIME (MINUTES)	TASK MUST FOLLOW THIS TASK
A	5	—
B	3	A
C	4	B
D	3	B
E	6	C
F	1	C
G	4	D, E, F
H	2	G
	28	

Figure **9.18**

Four-Station Solution to the Line-Balancing Problem

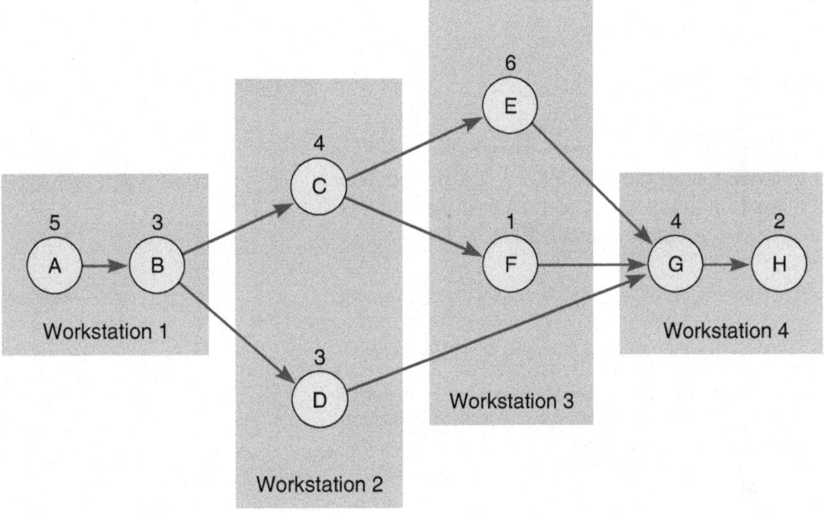

SOLUTION

The theoretical minimum number of workstations is:

$$\frac{\Sigma t_i}{\text{Cycle time}} = \frac{28 \text{ minutes}}{8 \text{ minutes}} = 3.5, \text{ or } 4 \text{ stations}$$

The precedence graph and one good layout are shown in Figure 9.18:

$$\text{Efficiency} = \frac{\text{Total task time}}{(\text{Actual number of workstations}) \times (\text{Largest assigned cycle time})} = \frac{28}{(4)(8)} = 87.5\%$$

Problems

Note: **PX** means the problem may be solved with POM for Windows and/or Excel OM.

Problems 9.1–9.11 relate to Process-Oriented Layout

•• **9.1** Gordon Miller's job shop has four work areas, A, B, C, and D. Distances in feet between centers of the work areas are:

	A	B	C	D
A	—	4	9	7
B	—	—	6	8
C	—	—	—	10
D	—	—	—	—

Workpieces moved, in hundreds of workpieces per week, between pairs of work areas, are:

	A	B	C	D
A	—	8	7	4
B	—	—	3	2
C	—	—	—	6
D	—	—	—	—

It costs Gordon $1 to move 1 work piece 1 foot. What is the weekly total material handling cost of the layout? **PX**

•• **9.2** A Missouri job shop has four departments—machining (M), dipping in a chemical bath (D), finishing (F), and plating (P)—assigned to four work areas. The operations manager, Mary Marrs, has gathered the following data for this job shop as it is currently laid out (Plan A).

100s of Workpieces Moved between Work Areas Each Year Plan A

	M	D	F	P
M	—	6	18	2
D	—	—	4	2
F	—	—	—	18
P	—	—	—	—

Distances between Work Areas (Departments) in Feet

	M	D	F	P
M	—	20	12	8
D	—	—	6	10
F	—	—	—	4
P	—	—	—	—

It costs $0.50 to move 1 workpiece 1 foot in the job shop. Marrs' goal is to find a layout that has the lowest material-handling cost.

a) Determine cost of the current layout, Plan A, from the given data.
b) One alternative is to switch those departments with the high loads, namely, finishing (F) and plating (P), which alters the distance between them and machining (M) and dipping (D), as follows:

Distances between Work Areas (Departments) in Feet Plan B

	M	D	F	P
M	—	20	8	12
D	—	—	10	6
F	—	—	—	4
P	—	—	—	—

What is the cost of *this* layout?

c) Marrs now wants you to evaluate Plan C, which also switches milling (M) and drilling (D), below.

Distance between Work Areas (Departments) in Feet Plan C

	M	D	F	P
M	—	20	10	6
D	—	—	8	12
F	—	—	—	4
P	—	—	—	—

What is the cost of *this* layout?

d) Which layout is best from a cost perspective? **PX**

• **9.3** Three departments—milling (M), drilling (D), and sawing (S)—are assigned to three work areas in Victor Berardis' machine shop in Kent, Ohio. The number of workpieces moved per day and the distances between the centers of the work areas, in feet, follow.

Pieces Moved between Work Areas Each Day

	M	D	S
M	—	23	32
D	—	—	20
S	—	—	—

Distances between Centers of Work Areas (Departments) in Feet

	M	D	S
M	—	10	5
D	—	—	8
S	—	—	—

It costs $2 to move 1 workpiece 1 foot. What is the cost?

•• **9.4** Roy Creasey Enterprises, a machine shop, is planning to move to a new, larger location. The new building will be 60 feet long by 40 feet wide. Creasey envisions the building as having six distinct production areas, roughly equal in size. He feels strongly about safety and intends to have marked pathways throughout the building to facilitate the movement of people and materials. See the following building schematic.

Building Schematic (with work areas 1–6)

1	2	3
4	5	6

His foreman has completed a month-long study of the number of loads of material that have moved from one process to another in the current building. This information is contained in the following flow matrix.

Flow Matrix between Production Processes

FROM \ TO	MATERIALS	WELDING	DRILLS	LATHES	GRINDERS	BENDERS
Materials	0	100	50	0	0	50
Welding	25	0	0	50	0	0
Drills	25	0	0	0	50	0
Lathes	0	25	0	0	20	0
Grinders	50	0	100	0	0	0
Benders	10	0	20	0	0	0

Finally, Creasey has developed the following matrix to indicate distances between the work areas shown in the building schematic.

Distance between Work Areas

	1	2	3	4	5	6
1		20	40	20	40	60
2			20	40	20	40
3				60	40	20
4					20	40
5						20
6						

What is the appropriate layout of the new building? **PX**

•• **9.5** Adam Munson Manufacturing, in Gainesville, Florida, wants to arrange its four work centers so as to minimize interdepartmental parts handling costs. The flows and existing facility layout are shown in Figure 9.19. For example, to move a part from Work Center A to Work Center C is a 60-foot movement distance. It is 90 feet from A to D.

Parts Moved between Work Centers

	A	B	C	D
A	—	450	550	50
B	350	—	200	0
C	0	0	—	750
D	0	0	0	—

Existing Layout

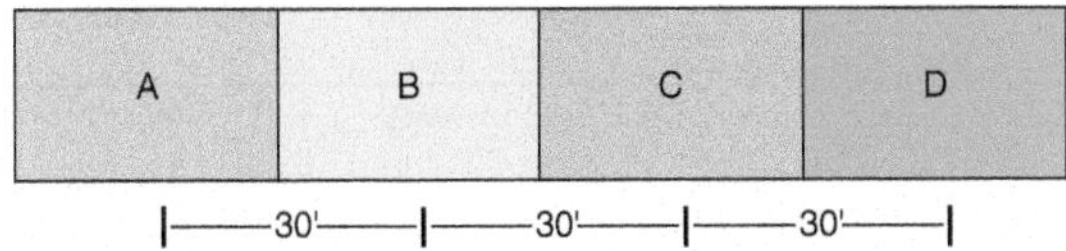

Figure **9.19**

Munson Manufacturing

a) What is the "load × distance," or "movement cost," of the layout shown?
b) Provide an improved layout and compute its movement cost. **PX**

••• **9.6** You have just been hired as the director of operations for Reid Chocolates, a purveyor of exceptionally fine candies. Reid Chocolates has two kitchen layouts under consideration for its recipe making and testing department. The strategy is to provide the best kitchen layout possible so that food scientists can devote their time and energy to product improvement, not wasted effort in the kitchen. You have been asked to evaluate these two kitchen layouts and to prepare a recommendation for your boss, Mr. Reid, so that he can proceed to place the contract for building the kitchens. [See Figure 9.20(a), and Figure 9.20(b).] **PX**

Number of trips between work centers:

From: \ To:		Refrigerator 1	Counter 2	Sink 3	Storage 4	Stove 5
Refrig.	1	0	8	13	0	0
Counter	2	5	0	3	3	8
Sink	3	3	12	0	4	0
Storage	4	3	0	0	0	5
Stove	5	0	8	4	10	0

Figure **9.20(a)**

Layout Options

Kitchen layout #1

Walking distance in feet

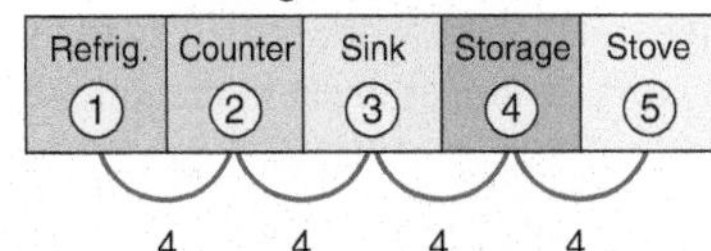

Kitchen layout #2

Walking distance in feet

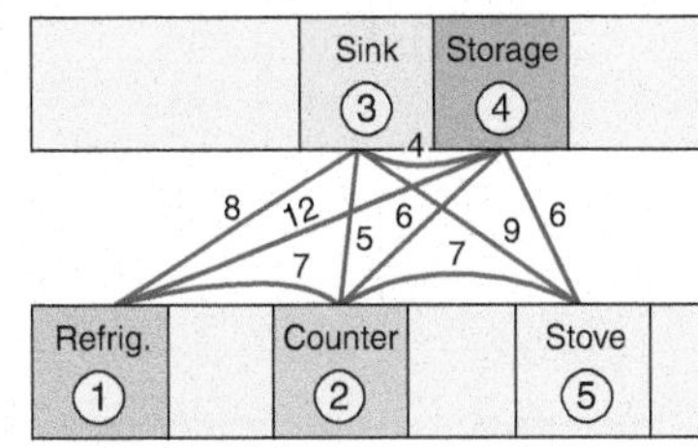

Figure **9.20(b)**

••• **9.7** Reid Chocolates (see Problem 9.6) is considering a third layout, as shown below. Evaluate its effectiveness in trip-distance feet. **PX**

Kitchen layout #3

Walking distance in feet

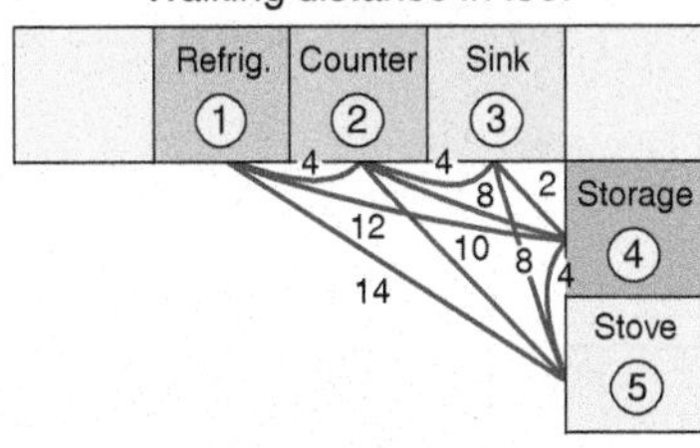

••• **9.8** Reid Chocolates (see Problem 9.6) has another layout to consider. Layout 4 is shown below. What is the total trip distance? **PX**

Kitchen layout #4

Walking distance in feet

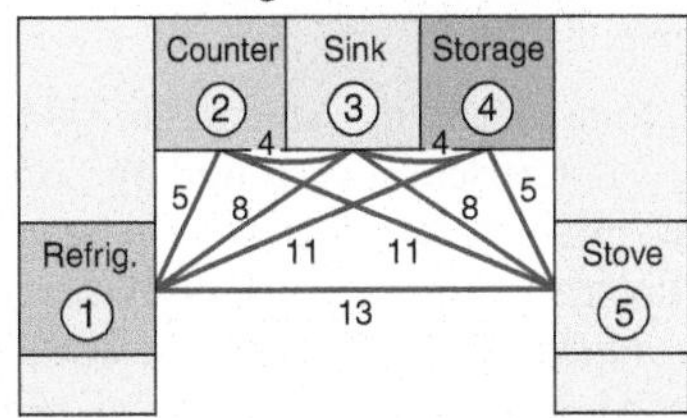

••• **9.9** Reid Chocolates (see Problem 9.6) has yet another layout it wishes to consider. Layout 5, which follows, has what total trip distance? **PX**

Kitchen layout #5

Walking distance in feet

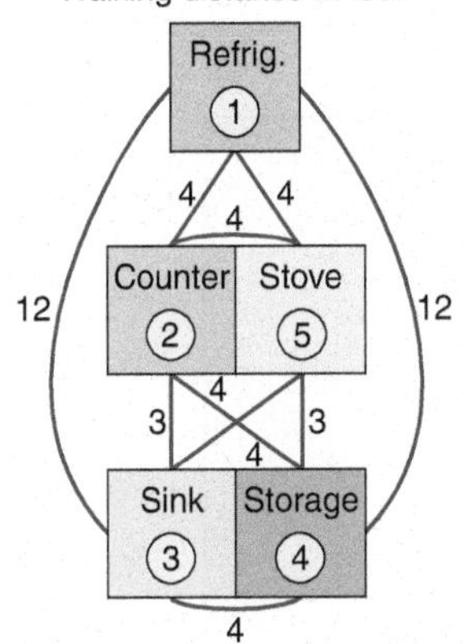

•• **9.10** Six processes are to be laid out in six areas along a long corridor at Rita Gibson Accounting Services in Daytona Beach. The distance between adjacent work centers is 40 feet. The number of trips between work centers is given in the following table:

	TRIPS BETWEEN PROCESSES					
	TO					
FROM	A	B	C	D	E	F
A		18	25	73	12	54
B			96	23	31	45
C				41	22	20
D					19	57
E						48
F						

a) Assign the processes to the work areas in a way that minimizes the total flow, using a method that places processes with highest flow adjacent to each other.
b) What assignment minimizes the total traffic flow? **PX**

••• **9.11** Snow-Bird Hospital is a small emergency-oriented facility located in a popular ski resort in Northern Michigan. Its new administrator, Mary Lord, decides to reorganize the hospital using the process-layout method she studied in business school.

The current layout is:

Area A	Area B	Area C	Area D
Entrance (1)	Exam room 1 (2)	Exam room 2 (3)	X-ray (4)
Lab test/EKG (5)	Operating room (6)	Recovery room (7)	Cast-setting room (8)
Area E	Area F	Area G	Area H

With this layout, total patient movement is 6,700 feet.

The only physical restriction perceived by Lord is the need to keep the entrance/initial processing room in its current location. All other departments or rooms (each of 10 feet square) can be moved if layout analysis indicates a move would be beneficial. Her objective, Lord decides, is to lay out the departments so as to minimize the total movement (distance traveled) of patients in the hospital.

First, Lord analyzes records to determine the number of patient movements among departments in an average month. These data are:

(1)	(2)	(3)	(4)	(5)	(6)	(7)	(8)	
	100	100	0	0	0	0	0	Entrance/Initial processing (1)
		0	50	20	0	0	0	Exam room 1 (2)
			30	30	0	0	0	Exam room 2 (3)
				20	0	0	20	X-ray (4)
					20	0	10	Lab test/EKG (5)
						30	0	Operating room (6)
							0	Recovery room (7)
								Cast-setting room (8)

Lord assumes that adjacent departments, such as entrance (now in work area A) and exam room 1 (now in work area B), have a walking distance of 10 feet. Diagonal departments are also considered adjacent and assigned a distance of 10 feet. Nonadjacent departments, such as the entrance and exam room 2 (now in area C) or the entrance and recovery room (area G) are 20 feet apart, and nonadjacent rooms, such as entrance and X-ray (area D), are 30 feet apart. (Hence, 10 feet is considered 10 units of cost, 20 feet is 20 units of cost, and 30 feet is 30 units of cost.)

Lord determined that two useful changes are to switch departments 3 and 5 and to interchange departments 4 and 6. This change would result in the schematic shown the following figure:

Area A	Area B	Area C	Area D
Entrance (1)	Exam room 1 (2)	Lab test/EKG (5)	Operating room (6)
Exam room 2 (3)	X-ray (4)	Recovery room (7)	Cast-setting room (8)
Area E	Area F	Area G	Area H

With this layout, total patient movement is 4,800 feet.

Is further improvement possible? If so, what is it? **PX**

Problems 9.12–9.27 relate to Repetitive and Product-Oriented Layout

•• **9.12** Amelio Rodriguez Computing wants to establish an assembly line for producing a new product, the Personal Digital Assistant (PDA). The tasks, task times, and immediate predecessors for the tasks are as follows:

TASK	TIME (sec)	IMMEDIATE PREDECESSORS
A	12	—
B	15	A
C	8	A
D	5	B, C
E	20	D

Amelio's goal is to produce 180 PDAs per hour.

a) What is the cycle time?
b) What is the theoretical minimum for the number of workstations that Amelio can achieve in this assembly line?
c) Can the theoretical minimum actually be reached when workstations are assigned? **PX**

••• **9.13** Illinois Furniture, Inc., produces all types of office furniture. The "Executive Secretary" is a chair that has been designed using ergonomics to provide comfort during long work hours. The chair sells for $130. There are 480 minutes available during the day, and the average daily demand has been 50 chairs. There are eight tasks:

TASK	PERFORMANCE TIME (min)	TASK MUST FOLLOW TASK LISTED BELOW
A	4	—
B	7	—
C	6	A, B
D	5	C
E	6	D
F	7	E
G	8	E
H	6	F, G

a) Draw a precedence diagram of this operation.
b) What is the cycle time for this operation?
c) What is the *theoretical* minimum number of workstations?
d) Assign tasks to workstations.
e) What is the idle time per cycle?
f) How much total idle time is present in an 8-hour shift?
g) What is the efficiency of the assembly line, given your answer in (d)? **PX**

•• **9.14** Sue Helms Appliances wants to establish an assembly line to manufacture its new product, the Micro Popcorn Popper. The goal is to produce five poppers per hour. The tasks, task times, and immediate predecessors for producing one Micro Popcorn Popper are as follows:

TASK	TIME (min)	IMMEDIATE PREDECESSORS
A	10	—
B	12	A
C	8	A, B
D	6	B, C
E	6	C
F	6	D, E

a) What is the *theoretical* minimum for the smallest number of workstations that Helms can achieve in this assembly line?
b) Graph the assembly line, and assign workers to workstations. Can you assign them with the theoretical minimum?
c) What is the efficiency of *your* assignment? **PX**

•• **9.15** The Action Toy Company has decided to manufacture a new train set, the production of which is broken into six steps. The demand for the train is 4,800 units per 40-hour workweek:

TASK	PERFORMANCE TIME (sec)	PREDECESSORS
A	20	None
B	30	A
C	15	A
D	15	A
E	10	B, C
F	30	D, E

a) Draw a precedence diagram of this operation.
b) Given the demand, what is the cycle time for this operation?
c) What is the *theoretical* minimum number of workstations?
d) Assign tasks to workstations.
e) How much total idle time is present each cycle?
f) What is the efficiency of the assembly line with five stations? With six stations? **PX**

•• **9.16** The following table details the tasks required for Indiana-based Frank Pianki Industries to manufacture a fully portable industrial vacuum cleaner. The times in the table are in minutes. Demand forecasts indicate a need to operate with a cycle time of 10 minutes.

ACTIVITY	ACTIVITY DESCRIPTION	IMMEDIATE PREDECESSORS	TIME
A	Attach wheels to tub	—	5
B	Attach motor to lid	—	1.5
C	Attach battery pack	B	3
D	Attach safety cutoff	C	4
E	Attach filters	B	3
F	Attach lid to tub	A, E	2
G	Assemble attachments	—	3
H	Function test	D, F, G	3.5
I	Final inspection	H	2
J	Packing	I	2

a) Draw the appropriate precedence diagram for this production line.
b) Assign tasks to workstations and determine how much idle time is present each cycle.
c) Discuss how this balance could be improved to 100%.
d) What is the *theoretical* minimum number of workstations? **PX**

•• **9.17** Tailwind, Inc., produces high-quality but expensive training shoes for runners. The Tailwind shoe, which sells for $210, contains both gas- and liquid-filled compartments to provide more stability and better protection against knee, foot, and back injuries. Manufacturing the shoes requires 10 separate tasks. There are 400 minutes available for manufacturing the shoes in the plant each day. Daily demand is 60. The information for the tasks is as follows:

TASK	PERFORMANCE TIME (min)	TASK MUST FOLLOW TASK LISTED BELOW
A	1	—
B	3	A
C	2	B
D	4	B
E	1	C, D
F	3	A
G	2	F
H	5	G
I	1	E, H
J	3	I

a) Draw the precedence diagram.
b) Assign tasks to the minimum feasible number of workstations according to the "ranked positioned weight" decision rule.
c) What is the efficiency of the process you completed in (b)?
d) What is the idle time per cycle? **PX**

•• **9.18** The Mach 10 is a one-person sailboat manufactured by Creative Leisure. The final assembly plant is in Cupertino, California. The assembly area is available for production of the Mach 10 for 200 minutes per day. (The rest of the time it is busy making other products.) The daily demand is 60 boats. Given the information in the following table,

a) Draw the precedence diagram and assign tasks using five workstations.
b) What is the efficiency of the assembly line, using your answer to (a)?
c) What is the *theoretical* minimum number of workstations?
d) What is the idle time per boat produced? **PX**

TASK	PERFORMANCE TIME (min)	TASK MUST FOLLOW TASK LISTED BELOW
A	1	—
B	1	A
C	2	A
D	1	C
E	3	C
F	1	C
G	1	D, E, F
H	2	B
I	1	G, H

Ivan Smuk/Shutterstock

•• **9.19** Because of the expected high demand for Mach 10, Creative Leisure has decided to increase manufacturing time available to produce the Mach 10 (see Problem 9.18).

a) If demand remained the same but 300 minutes were available each day on the assembly line, how many workstations would be needed?
b) What would be the efficiency of the new system, using the actual number of workstations from (a)?
c) What would be the impact on the system if 400 minutes were available? **PX**

••• **9.20** Dr. Lori Baker, operations manager at Nesa Electronics, prides herself on excellent assembly-line balancing. She has been told that the firm needs to complete 96 instruments per 24-hour day. The assembly-line activities are:

TASK	TIME (min)	PREDECESSORS
A	3	—
B	6	—
C	7	A
D	5	A, B
E	2	B
F	4	C
G	5	F
H	7	D, E
I	1	H
J	6	E
K	4	G, I, J
	50	

a) Draw the precedence diagram.
b) If the daily (24-hour) production rate is 96 units, what is the highest allowable cycle time?
c) If the cycle time after allowances is given as 10 minutes, what is the daily (24-hour) production rate?
d) With a 10-minute cycle time, what is the theoretical minimum number of stations with which the line can be balanced?
e) With a 10-minute cycle time and six workstations, what is the efficiency?
f) What is the total idle time per cycle with a 10-minute cycle time and six workstations?
g) What is the best workstation assignment you can make without exceeding a 10-minute cycle time, and what is its efficiency? **PX**

•• **9.21** Suppose production requirements in Solved Problem 9.2 increase and require a reduction in cycle time from 8 minutes to 7 minutes. Balance the line once again, using the new cycle time. Note that it is not possible to combine task times so as to group tasks into the minimum number of workstations. This condition occurs in actual balancing problems fairly often. **PX**

•• **9.22** The preinduction physical examination given by the U.S. Army involves the following seven activities:

ACTIVITY	AVERAGE TIME (min)
Medical history	10
Blood tests	8
Eye examination	5
Measurements (e.g., weight, height, blood pressure)	7
Medical examination	16
Psychological interview	12
Exit medical evaluation	10

These activities can be performed in any order, with two exceptions: Medical history must be taken first, and Exit medical evaluation is last. At present, there are three paramedics and two physicians on duty during each shift. Only physicians can perform exit evaluations and conduct psychological interviews. Other activities can be carried out by either physicians or paramedics.

a) Develop a layout and balance the line.
b) How many people can be processed per hour?
c) Which activity accounts for the current bottleneck?
d) What is the total idle time per cycle?
e) If one more physician and one more paramedic can be placed on duty, how would you redraw the layout? What is the new throughput?

••• **9.23** Carlos Torres' company wants to establish an assembly line to manufacture its new product, the iStar phone. Carlos' goal is to produce 60 iStars per hour. Tasks, task times, and immediate predecessors are as follows:

TASK	TIME (sec)	IMMEDIATE PREDECESSORS	TASK	TIME (sec)	IMMEDIATE PREDECESSORS
A	40	—	F	25	C
B	30	A	G	15	C
C	50	A	H	20	D, E
D	40	B	I	18	F, G
E	6	B	J	30	H, I

a) What is the theoretical minimum for the number of workstations that Carlos can achieve in this assembly line?
b) Use the *most following tasks* heuristic to balance an assembly line for the iStar phone.
c) How many workstations are in your answer to (b)?
d) What is the efficiency of your answer to (b)? **PX**

•• **9.24** Advanced Ergonomics has developed a new automated desktop that raises and lowers electronically (so a person using a computer can work standing or sitting). Its assembly line must produce 80 desktop devices in a standard 8-hour workday. The five tasks, their performance times, and their predecessors are as follows:

TASK	TIME (min)	IMMEDIATE PREDECESSORS
A	5	—
B	6	A
C	4	B
D	2	B
E	6	C, D

a) What is the cycle time?
b) What is the minimum number of workstations?
c) Draw the precedence diagram.

••• **9.25** An assembly plant for the Blood Pressure RX, a portable blood pressure device, has 400 minutes available daily in the plant for the device, and the average demand is 80 units per day. Final assembly requires 6 separate tasks. Information concerning these tasks is given in the following table.

TASK	PERFORMANCE TIME (minutes)	TASK MUST FOLLOW TASK LISTED BELOW
A	1	—
B	1	—
C	4	A, B
D	1	C
E	2	D
F	4	E

a) Draw a precedence diagram of this operation
b) Given the demand, what is the cycle time for this operation?
c) What is the theoretical minimum number of workstations?
d) Assign tasks to workstations.
e) What is the overall actual efficiency of the assembly line? **PX**

•• **9.26** Given the following task, times, and sequence, develop a balanced line capable of operating with a 10-minute cycle time at Dave Visser's company.

TASK ELEMENT	TIME (minutes)	ELEMENT PREDECESSOR
A	3	—
B	5	A
C	7	B
D	5	—
E	3	C
F	3	B, D
G	5	D
H	6	G

a) Draw the precedence diagram.
b) Assign tasks to the minimum feasible number of workstations according to the greatest time remaining (ranked positional weight) decision rule.
c) What is the efficiency of the process? **PX**

•• **9.27** Marilyn Hart, operations manager at Hart Electronics, faces assembly-line balancing problems. She has been told that the firm needs to complete 1,400 electronic relays per day. There are 420 minutes of productive time in each working day (which is equivalent to 25,200 seconds). Group the assembly-line activities in the following table into appropriate workstations and calculate the efficiency of the balance. **PX**

TASK	TIME (seconds)	MUST FOLLOW TASK	TASK	TIME (seconds)	MUST FOLLOW TASK
A	13	—	G	5	E
B	4	A	H	6	F, G
C	10	B	I	7	H
D	10	—	J	5	H
E	6	D	K	4	I, J
F	12	E	L	15	C, K

CASE STUDIES

State Automobile License Renewals[2]

LaToya Clarendon, the manager of a metropolitan branch office of the state department of motor vehicles, attempted to analyze the driver's license-renewal operations. She had to perform several steps. After examining the license-renewal process, she identified those steps and associated times required to perform each step, as shown in the following table:

State Automobile License Renewal Process Times

STEP	AVERAGE TIME TO PERFORM (sec)
1. Review renewal application for correctness	15
2. Process and record payment	30
3. Check file for violations and restrictions	60
4. Conduct eye test	40
5. Photograph applicant	20
6. Issue temporary license	30

LaToya found that each step was assigned to a different person. Each application was a separate process in the sequence shown. She determined that her office should be prepared to accommodate a maximum demand of processing 120 renewal applicants per hour.

She observed that work was unevenly divided among clerks and that the clerk responsible for checking violations tended to shortcut her task to keep up with the others. Long lines built up during the maximum-demand periods.

LaToya also found that Steps 1 to 4 were handled by general clerks who were each paid $24 per hour. Step 5 was performed by a photographer paid $32 per hour. (Branch offices were charged $20 per hour for each camera to perform photography.) Step 6, issuing temporary licenses, was required by state policy to be handled by uniformed motor vehicle officers. Officers were paid $36 per hour but could be assigned to any job except photography.

A review of the jobs indicated that Step 1, reviewing applications for correctness, had to be performed before any other step could be taken. Similarly, Step 6, issuing temporary licenses, could not be performed until all the other steps were completed.

LaToya Clarendon was under severe pressure to increase productivity and reduce costs, but she was also told by the regional director that she must accommodate the demand for renewals. Otherwise, "heads would roll."

Discussion Questions

1. What is the maximum number of applications per hour that can be handled by the present configuration of the process?
2. How many applications can be processed per hour if a second clerk is added to check for violations?
3. Using the original personnel, is there a way to combine stations to increase the maximum number of applications the process can handle? What is the new configuration, and what is the new output per hour?
4. How would you suggest modifying the process to accommodate 120 applications per hour? (This may include hiring more staff. And if a second photographer is needed, a second camera would be needed as well.) What is the cost per application of this new configuration?

Laying Out Arnold Palmer Hospital's New Facility

Video Case

When Orlando's Arnold Palmer Hospital began plans to create a new 273-bed, 11-story hospital across the street from its existing facility, which was bursting at the seams in terms of capacity, a massive planning process began. The $100 million building, opened in 2006, was long overdue, according to Executive Director Kathy Swanson: "We started Arnold Palmer Hospital in 1989, with a mission to provide quality services for children and women in a comforting, family-friendly environment. Since then we have served well over 1.5 million women and children and now deliver more than 12,000 babies a year. By 2001, we simply ran out of room, and it was time for us to grow."

The new hospital's unique, circular pod design provides a maximally efficient layout in all areas of the hospital, creating a patient-centered environment. *Servicescape* design features include a serene environment created through the use of warm colors, private rooms with pull-down Murphy beds for family members, 14-foot ceilings, and natural lighting with oversized windows in patient rooms. But these radical new features did not come easily. "This pod concept with a central nursing area and pie-shaped rooms resulted from over 1,000 planning meetings of 35 user groups, extensive motion and time studies, and computer simulations of the daily movements of nurses," says Swanson.

In a traditional linear hospital layout, called the *racetrack* design, patient rooms line long hallways, and a nurse might walk 2.7 miles per day serving patient needs at Arnold Palmer. "Some nurses spent 30% of their time simply walking. With the nursing shortage and the high cost of healthcare professionals, efficiency is a major concern," added Swanson. With the nursing station in the center of 10- or 12-bed circular pods, no patient room is more than 14 feet from a station. The time savings are in the 20% range. Swanson pointed to Figures 9.21 and 9.22 as examples of the old and new walking and trip distances.*

"We have also totally redesigned our neonatal rooms," says Swanson. "In the old system, there were 16 neonatal beds in a large and often noisy rectangular room. The new building features semiprivate rooms for these tiny babies. The rooms are much improved, with added privacy and a quiet, simulated night atmosphere, in addition to pull-down beds for parents to use. Our research shows that babies improve and develop much more quickly with this layout design. Layout and environment indeed impact patient care!"

*Layout and walking distances, including some of the numbers in Figures 9.21 and 9.22, have been simplified for purposes of this case.

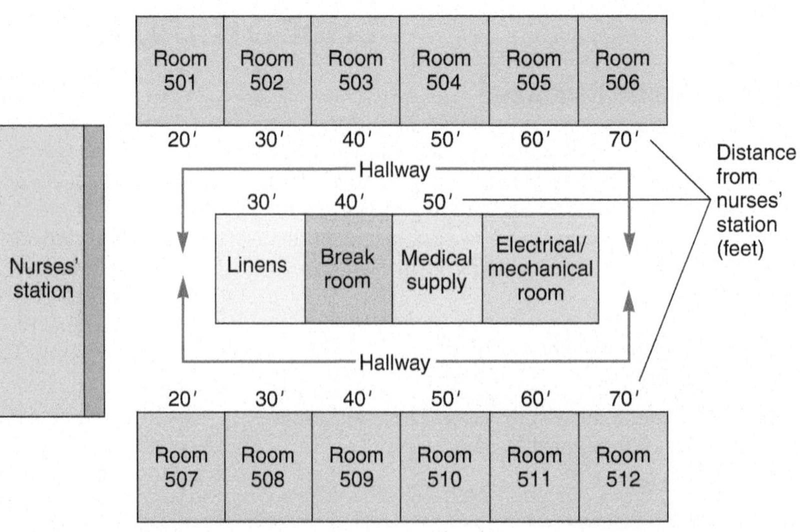

Figure 9.21

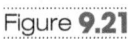

Traditional Hospital Layout

Patient rooms are on two linear hallways with exterior windows. Supply rooms are on interior corridors. This layout is called a "racetrack" design.

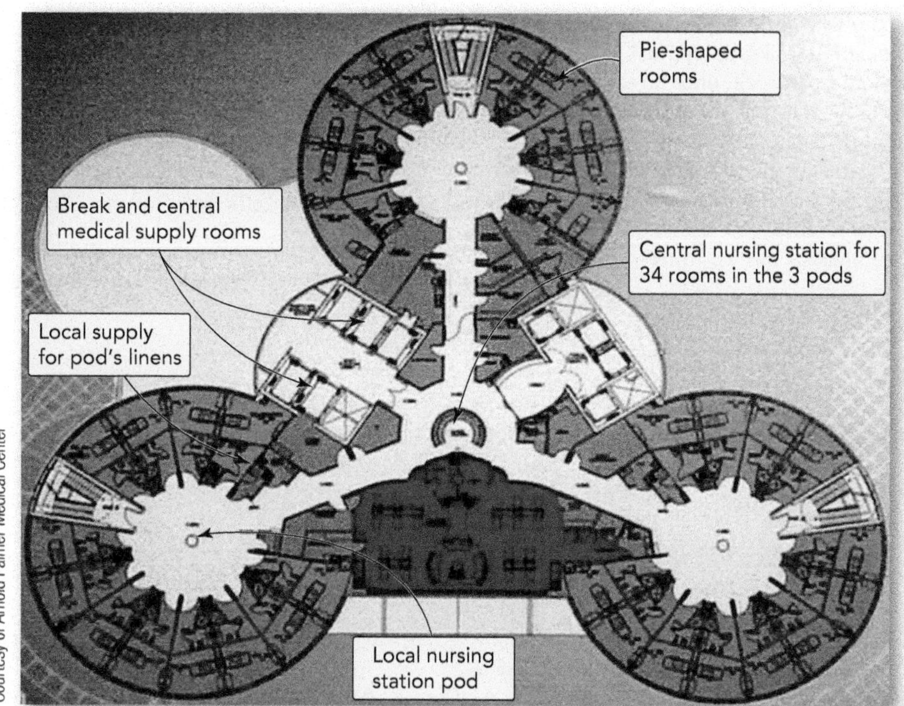

Figure **9.22**

New Pod Design for Hospital Layout

Note that each room is 14 feet from the pod's local nursing station. The break rooms and the central medical station are each about 60 feet from the local nursing pod. Pod linen supply rooms are also 14 feet from the local nursing station.

Discussion Questions*

1. Identify the many variables that a hospital needs to consider in layout design.
2. What are the advantages of the circular pod design over the traditional linear hallway layout found in most hospitals?
3. Figure 9.21 illustrates a sample linear hallway layout. During a period of random observation, nurse Thomas Smith's day includes 6 trips from the nursing station to each of the 12 patient rooms (back and forth), 20 trips to the medical supply room, 5 trips to the break room, and 12 trips to the linen supply room. What is his total distance traveled in miles?
4. Figure 9.22 illustrates an architect's drawing of Arnold Palmer Hospital's new circular pod system. If nurse Susan Jones's day includes 7 trips from the nursing pod to each of the 12 rooms (back and forth), 20 trips to central medical supply, 6 trips to the break room, and 12 trips to the pod linen supply, how many miles does she walk during her shift? What are the differences in the travel times between the two nurses for this random day?
5. The concept of *servicescapes* is discussed in this chapter. Describe why this is so important at Arnold Palmer Hospital, and give examples of its use in layout design.
6. As technology and costs change, hospitals continue to innovate. The reduced cost of computers means some hospitals have moved from a central computer at the nurse's station to computers in the room or on carts (see photo in the Global Company Profile for Chapter 6). What changes in overall hospital layout would these innovations suggest?

*You may wish to view the video that accompanies this case before addressing these questions.

Facility Layout at Wheeled Coach

Video Case  

When President Bob Collins began his career at Wheeled Coach, the world's largest manufacturer of ambulances, there were only a handful of employees. Now the firm's Florida plant has a workforce of 350. The physical plant has also expanded, with offices, R&D, final assembly, and wiring, cabinetry, and upholstery work cells in one large building. Growth has forced the painting work cell into a separate building, aluminum fabrication and body installation into another, inspection and shipping into a fourth, and warehousing into yet another.

Like many other growing companies, Wheeled Coach was not able to design its facility from scratch. And although management realizes that material-handling costs are a little higher than an ideal layout would provide, Collins is pleased with the way the facility has evolved and employees have adapted. The aluminum cutting work cell lies adjacent to body fabrication, which, in turn, is located next to the body-installation work cell. And while the vehicle must be driven across a street to one building for painting and then to another for final assembly, at least the ambulance is on wheels. Collins is also satisfied with the flexibility shown in the design of the work cells. Cell construction is flexible and can accommodate changes in product mix and volume. In addition, work cells are typically small and movable, with many work benches and staging racks borne on wheels so that they can be easily rearranged and products transported to the assembly line.

Assembly-line balancing is one key problem facing Wheeled Coach and every other repetitive manufacturer. Produced on a schedule calling for four 10-hour workdays per week, once an ambulance is on one of the six final assembly lines, it *must* move forward each day to the next workstation. Balancing just enough workers and tasks at each of the seven workstations is a never-ending challenge. Too many workers end up running into each other; too few can't finish an ambulance in seven days. Constant shifting of design and mix and improved analysis has led to frequent changes.

Discussion Questions*

1. What analytical techniques are available to help a company like Wheeled Coach deal with layout problems?
2. What suggestions would you make to Bob Collins about his layout?
3. How would you measure the "efficiency" of this layout?

*You may wish to view the video that accompanies this case and read the Global Company Profile opening Chapter 14 before addressing these questions.

Endnotes

1. Sources: *Robotics and Automation* (January 21, 2020); and **TechHQ.com** (September 21, 2020).
2. Source: Modified from a case by W. Earl Sasser, Paul R. Olson, and D. Daryl Wyckoff, *Management of Services Operations: Text, Cases, and Readings* (Boston: Allyn & Bacon).

Bibliography

Aghazadeh, S. M., S. Hafeznezami, L. Najjar, and Z. Hug. "The Influence of Work-Cells and Facility Layout on Manufacturing Efficiency." *Journal of Facilities Management* 9, no. 3 (2011): 213–224.

Fayurd, A. L., and J. Weeks. "Who Moved My Cube?" *Harvard Business Review* 89, no. 7/8 (July–August 2011): 102–110.

Francis, R. L., L. F. McGinnis, and J. A. White. *Facility Layout and Location*, 3rd ed. Upper Saddle River, NJ: Prentice Hall, 1998.

Garcia-Diaz, A., and J. MacGregor Smith. *Facilities Planning and Design*. Essex, England: Pearson Education Ltd., 2014.

Khazanchi, S., T. A. Sprinkle, S. S. Masterson, and N. Tong. "A Spatial Model of Work Relationships: The Relationship-Building and Relationship-Straining Effects of Workspace Design." *Academy of Management Review* 43, no. 4 (2018): 590–609.

Ortiz, C. A. *The Cell Manufacturing Playbook*. Boca Raton, FL: CRC Press, 2016.

Quillien, J. *Clever Digs: How Workspaces Can Enable Thought*. Ames, Iowa: Culicidae Press, LLC., 2012.

Roodbergen, K. J., and I. F. A. Vis. "A Model for Warehouse Layout." *IIE Transactions* 38, no. 10 (October 2006): 799–811.

Tompkins, J. A., J. A. White, Y. A. Bozer, and J. M. A. Tanchoco. *Facilities Planning*, 4th ed. New York: Wiley, 2010.

Upton, D. "What Really Makes Factories Flexible?" *Harvard Business Review* 73, no. 4 (July–August 1995): 74–84.

Chapter 9 *Rapid* Review

Main Heading	Review Material	MyLab Operations Management
THE STRATEGIC IMPORTANCE OF LAYOUT DECISIONS	Layout has numerous strategic implications because it establishes an organization's competitive priorities in regard to capacity, processes, flexibility, and cost, as well as quality of work life, customer contact, and image. *The objective of layout strategy is to develop an effective and efficient layout that will meet the firm's competitive requirements.*	Concept Questions: 1.1–1.4
TYPES OF LAYOUT	Types of layout and examples of their typical objectives include: 1. *Office layout*: Locate workers requiring frequent contact close to one another. 2. *Retail layout*: Expose customers to high-margin items. 3. *Warehouse layout*: Balance low-cost storage with low-cost material handling. 4. *Fixed-position layout*: Move material to the limited storage areas around the site. 5. *Process-oriented layout*: Manage varied material flow for each product. 6. *Work-cell layout*: Identify a product family, build teams, and cross-train team members. 7. *Product-oriented layout*: Equalize the task time at each workstation.	Concept Questions: 2.1–2.6
OFFICE LAYOUT	■ **Office layout**—The grouping of workers, their equipment, and spaces/offices to provide for comfort, safety, and movement of information.	Concept Questions: 3.1–3.3
RETAIL LAYOUT	■ **Retail layout**—An approach that addresses flow, allocates space, and responds to customer behavior. Retail layouts are based on the idea that sales and profitability vary directly with customer exposure to products. The main *objective of retail layout is to maximize profitability per square foot of floor space* (or, in some stores, per linear foot of shelf space). ■ **Slotting fees**—Fees manufacturers pay to get shelf space for their products. ■ **Servicescape**—The physical surroundings in which a service takes place and how they affect customers and employees.	Concept Questions: 4.1–4.6
WAREHOUSE AND STORAGE LAYOUTS	■ **Warehouse layout**—A design that attempts to minimize total cost by addressing trade-offs between space and material handling. The variety of items stored and the number of items "picked" has direct bearing on the optimal layout. Modern warehouse management is often an automated procedure using *automated storage and retrieval systems* (ASRSs). ■ **Cross-docking**—Avoiding the placement of materials or supplies in storage by processing them as they are received for shipment. Cross-docking requires both tight scheduling and accurate inbound product identification. ■ **Random stocking**—Used in warehousing to locate stock wherever there is an open location. ■ **Customizing**—Using warehousing to add value to a product through component modification, repair, labeling, and packaging.	Concept Questions: 5.1–5.6
FIXED-POSITION LAYOUT	■ **Fixed-position layout**—A system that addresses the layout requirements of stationary projects. Fixed-position layouts involve three complications: (1) there is limited space at virtually all sites, (2) different materials are needed at different stages of a project, and (3) the volume of materials needed is dynamic.	Concept Questions: 6.1–6.4
PROCESS-ORIENTED LAYOUT	■ **Process-oriented layout**—A layout that deals with low-volume, high-variety production in which like machines and equipment are grouped together. ■ **Job lots**—Groups or batches of parts processed together. $\text{Minimize cost} = \sum_{i=1}^{n} \sum_{j=1}^{n} X_{ij} C_{ij}$ (9-1) *Muther's Grid* (a relationship chart) displays a "closeness value" between each department.	Concept Questions: 7.1–7.8 Problems: 9.1–9.11 Virtual Office Hours for Solved Problem: 9.1 **VIDEO 9.1** Laying Out Arnold Palmer Hospital's New Facility **ACTIVE MODEL 9.1**
FOCUSED FACILITIES	■ **Takt time**—Pace of production to meet customer demands. Takt time = Total work time available/ Units required to satisfy customer demand ■ **Work cell**—An arrangement of machines and personnel that focuses on making a single product or family of related products with similar routings. ■ **Focused factory**—A plant established to focus the entire manufacturing system on a limited set of products defined by the company's competitive strategy.	Concept Questions: 8.1–8.6

Main Heading	Review Material	MyLab Operations Management
REPETITIVE AND PRODUCT-ORIENTED LAYOUT	■ **Fabrication line**—A machine-paced, product-oriented facility for building components. ■ **Assembly line**—An approach that puts fabricated parts together at a series of workstations; a repetitive process. ■ **Assembly-line balancing**—Obtaining output at each workstation on a production line in order to minimize delay. ■ **Cycle time**—The maximum time that a product is allowed at each workstation. Cycle time = Production time available per day ÷ Units required per day (9-2) $\text{Minimum number of workstations} = \sum_{i=1}^{n} \text{Time for task } i \,/\, (\text{Cycle time})$ (9-3) ■ **Heuristic**—Problem solving using procedures and rules rather than mathematical optimization. Line-balancing heuristics include *longest task (operation) time, most following tasks, ranked positional weight, shortest task (operation) time,* and *least number of following tasks.* $\text{Efficiency} = \dfrac{\Sigma\,\text{Task times}}{(\textit{Actual}\text{ number of workstations}) \times (\text{Largest assigned cycle time})}$ (9-4) Idle time = (Actual number of workstations × Largest assigned cycle time) − Σ Task times (9-5)	Concept Questions: 9.1–9.6 Problems: 9.12–9.27 **VIDEO 9.2** Facility Layout at Wheeled Coach Ambulances Virtual Office Hours for Solved Problem: 9.2
ADDITIONAL MYLAB OPERATIONS MANAGEMENT RESOURCES	✔ Additional Case Study (Microfix, Inc.) ✔ Multiple Choice Case Questions (State Automobile License Renewals) ✔ Recent Graduate Video: Megan Jones, Production Line Operations Leader, Textron Specialized Vehicles	

Self Test

LO 9.1 Which of the following statements best describes *office layout*?
- **a)** Groups workers, their equipment, and spaces/offices to provide for movement of information.
- **b)** Addresses the layout requirements of large, bulky projects such as ships and buildings.
- **c)** Seeks the best personnel and machine utilization in repetitive or continuous production.
- **d)** Allocates shelf space and responds to customer behavior.
- **e)** Deals with low-volume, high-variety production.

LO 9.2 Which of the following does *not* support the retail layout objective of maximizing customer exposure to products?
- **a)** Locate high-draw items around the periphery of the store.
- **b)** Use prominent locations for high-impulse and high-margin items.
- **c)** Maximize exposure to expensive items.
- **d)** Use end-aisle locations.
- **e)** Convey the store's mission with the careful positioning of the lead-off department.

LO 9.3 The major problem addressed by the warehouse layout strategy is:
- **a)** minimizing difficulties caused by material flow varying with each product.
- **b)** requiring frequent contact close to one another.
- **c)** addressing trade-offs between space and material handling.
- **d)** balancing product flow from one workstation to the next.
- **e)** none of the above.

LO 9.4 A fixed-position layout:
- **a)** groups workers to provide for movement of information.
- **b)** addresses the layout requirements of large, bulky projects such as ships and buildings.
- **c)** seeks the best machine utilization in continuous production.
- **d)** allocates shelf space based on customer behavior.
- **e)** deals with low-volume, high-variety production.

LO 9.5 A process-oriented layout:
- **a)** groups workers to provide for movement of information.
- **b)** addresses the layout requirements of large, bulky projects such as ships and buildings.
- **c)** seeks the best machine utilization in continuous production.
- **d)** allocates shelf space based on customer behavior.
- **e)** deals with low-volume, high-variety production.

LO 9.6 For a focused work center or focused factory to be appropriate, the following three factors are required:
- **a)** ____________________
- **b)** ____________________
- **c)** ____________________

LO 9.7 Before considering a product-oriented layout, it is important to be certain of:
- **a)** ____________________
- **b)** ____________________
- **c)** ____________________
- **d)** ____________________

LO 9.8 An assembly line is to be designed for a product whose completion requires 21 minutes of work. The factory works 400 minutes per day. Can a production line with five workstations make 100 units per day?
- **a)** Yes, with exactly 100 minutes to spare.
- **b)** No, but four workstations would be sufficient.
- **c)** No, it will fall short even with a perfectly balanced line.
- **d)** Yes, but the line's efficiency is very low.
- **e)** Cannot be determined from the information given.

Answers: LO 9.1. a; LO 9.2. c; LO 9.3. c; LO 9.4. b; LO 9.5. e; LO 9.6. family of products, stable forecast (demand), volume; LO 9.7. adequate volume, stable demand, standardized product, adequate/quality supplies; LO 9.8. c.

Human Resources, Job Design, and Work Measurement

CHAPTER 10

CHAPTER OUTLINE

At Alaska Airlines, as an integral part of job design, flight attendants and pilots receive extensive training in the many safety aspects of air travel. This includes using emergency exit slides and oxygen masks, along with how to treat passengers in need of CPR. Here we see a new class practicing CPR under a supervisor's watch.

Courtesy of Alaska Airlines

10 OM STRATEGY DECISIONS

- Design of Goods and Services
- Managing Quality
- Process Strategies
- Location Strategies
- Layout Strategies
- *Human Resources*
- Supply Chain Management
- Inventory Management
- Scheduling
- Maintenance

GLOBAL COMPANY PROFILE
NASCAR's Racing Teams

High-Performance Teamwork Makes the Difference between Winning and Losing

A new century brought new popularity to NASCAR (National Association for Stock Car Auto Racing). Hundreds of millions of TV and sponsorship dollars poured into the sport. With more money, competition increased, as did the rewards for winning on Sunday. The teams, headed by such names as Rusty Wallace, Jeff Gordon, Dale Earnhardt, Jr., and Juan Pablo Montoya, are as famous as the New York Yankees, Atlanta Hawks, or Chicago Bears.

Grindstone Media Group/Shutterstock

After 40 laps, all four Goodyear tires are changed and gas added in less than 12 seconds.

The race car drivers may be famous, but it's the pit crews who often determine the outcome of a race. Years ago, crews were auto mechanics during the week who simply did double duty on Sundays in the pits. They did pretty well to change four tires in less than 30 seconds. Today, because NASCAR teams find competitive advantage wherever they can, taking more than 12 seconds can be disastrous. A botched pit stop is the equivalent of ramming your car against the wall—crushing all hopes for the day.

On all the top NASCAR squads, the crew members who go "over the wall" are now athletes, usually ex-college football or basketball players with proven agility and strength. The Evernham team, for example, includes a former defensive back from Fairleigh Dickinson (who is now a professional tire carrier) and a 300-pound lineman from East Carolina University (who handles the jack). The Chip Ganassi racing team includes baseball players from Wake Forest, football players from University of Kentucky and North Carolina, and a hockey player from Dartmouth.

Grindstone Media Group/Shutterstock

At lap 91, another near-record-setting tire change and more gas.

Tire changers—the crew members who wrench lug nuts off and on—are a scarce human resource and average $100,000 a year in salary. Jeff Gordon was reminded of the importance of coordinated teamwork when five members of his "over-the-wall" crew jumped to Dale Jarrett's organization a few years ago; it was believed to be a $500,000 per year deal.

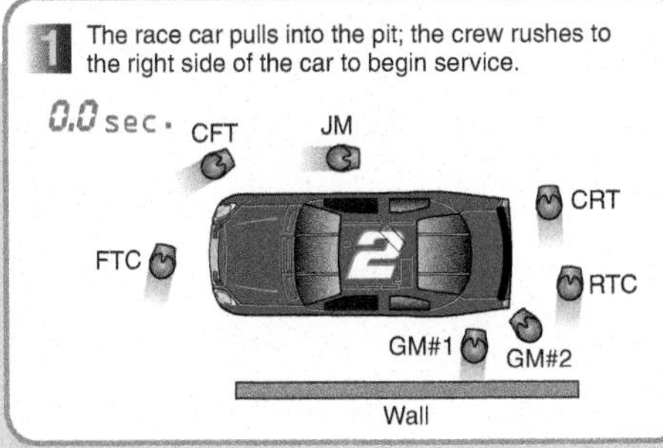
1
The race car pulls into the pit; the crew rushes to the right side of the car to begin service.
0.0 sec.
CFT
JM
CRT
FTC
RTC
GM#1
GM#2
Wall

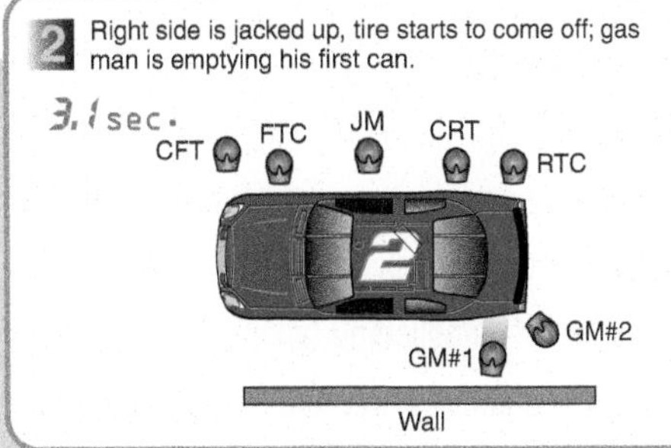
2
Right side is jacked up, tire starts to come off; gas man is emptying his first can.
3.1 sec.
CFT
FTC
JM
CRT
RTC
GM#2
GM#1
Wall

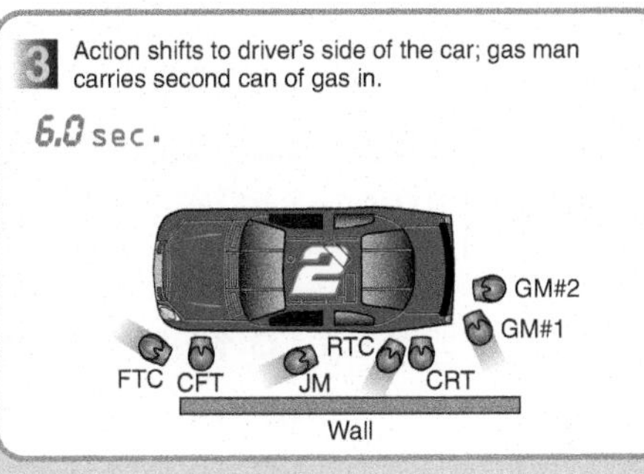
3
Action shifts to driver's side of the car; gas man carries second can of gas in.
6.0 sec.
GM#2
GM#1
RTC
FTC
CFT
JM
CRT
Wall

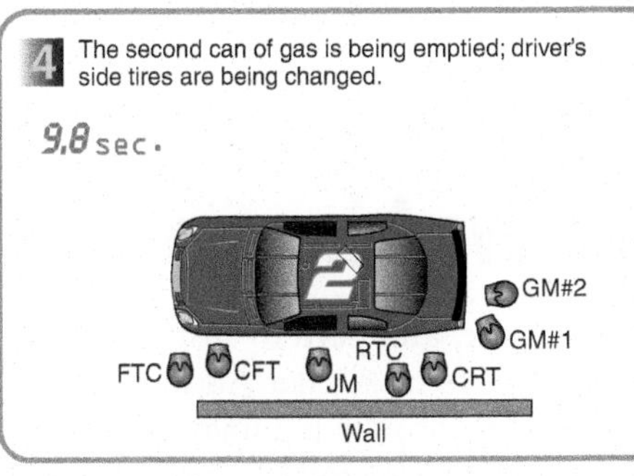
4
The second can of gas is being emptied; driver's side tires are being changed.
9.8 sec.
GM#2
GM#1
RTC
FTC
CFT
JM
CRT
Wall

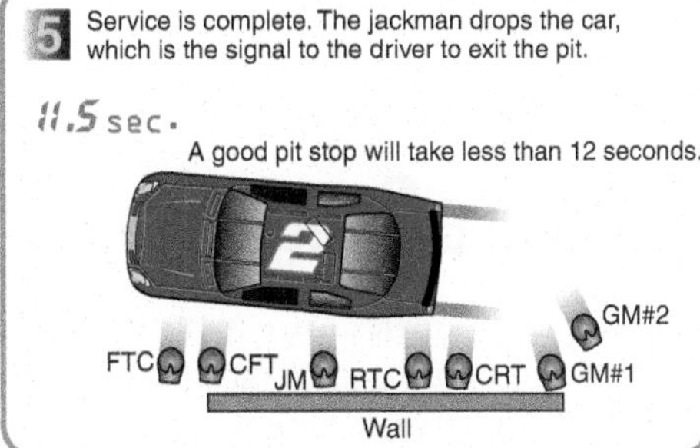
5
Service is complete. The jackman drops the car, which is the signal to the driver to exit the pit.
11.5 sec.
A good pit stop will take less than 12 seconds.
GM#2
FTC
CFT
JM
RTC
CRT
GM#1
Wall

Movement of the pit crew members who go over the wall...

JM = Jackman
FTC = Front tire carrier
CFT = Changer front tire
RTC = Rear tire carrier
CRT = Changer rear tire
GM#1 = Gas man #1
GM#2 = Gas man #2

JM (Jackman) The jackman carries the hydraulic jack from the pit wall to raise the car's right side. After new tires are bolted on, he drops the car to the ground and repeats the process on the left side. His timing is crucial during this left side change, because when he drops the car again, it's the signal for the driver to go. The jackman has the most dangerous job of all the crew members; during the right-side change, he is exposed to oncoming traffic down pit row. **FTC (Front tire carrier)** Each tire carrier hauls a new 75-pound tire to the car's right side, places it on the wheel studs, and removes the old tire after the tire change. They repeat this process on the left side of the car with a new tire rolled to them by crew members behind the pit wall. **CFT (Changer front tire)** Tire changers run to the car's right side and, using an air impact wrench, they remove five lug nuts off the old tire and bolt on a new tire. They repeat the process on the left side. **RTC (Rear tire carrier)** Same as front tire carrier, except RTC may also adjust the rear jack bolt to alter the car's handling. **CRT (Changer rear tire)** Same as FT but on two rear tires. **Gas man #1** This gas man is usually the biggest and strongest person on the team. He goes over the wall carrying a 75-pound, 11-gallon "dump can" whose nozzle he jams into the car's fuel cell receptacle. He is then handed (or tossed) another can, and the process is repeated. **Gas man #2** Gets second gas can to Gas man #1 and catches excess fuel that spills out.

A pit crew consists of seven members: a front-tire changer, a rear-tire changer, front- and rear-tire carriers, a person who jacks the car up, and two gas men with an 11-gallon can.

Every sport has its core competencies and key metrics—for example, the speed of a pitcher's fastball and a running back's time on the 40-yard dash. In NASCAR, a tire changer should get 5 lug nuts off in 1.2 seconds. The jackman should haul a 25-pound aluminum jack from the car's right side to left in 3.8 seconds. For tire carriers, it should take .7 seconds to get a tire from the ground to mounted on the car.

The seven crew members who go over the wall are coached and orchestrated. Coaches use the tools of OM and watch "game tape" of pit stops and make intricate adjustments to the choreography.

"There's a lot of pressure," says D. J. Richardson, a team tire changer—and one of the best in the business. Richardson trains daily with the rest of the crew in the shop of the team owner. They focus on cardiovascular work and two muscle groups daily. Twice a week, they simulate pit stops—there can be from 12 to 14 variations—to work on their timing.

In a recent race in Michigan, Richardson and the rest of his team, with ergonomically designed gas cans, tools, and special safety gear, were ready. On lap 43, the split-second frenzy began, with Richardson—air gun in hand—jumping over a 2-foot white wall and sprinting to the right side of the team's Dodge. A teammate grabbed the tire and set it in place while Richardson secured it to the car. The process was repeated on the left side while the front crew followed the same procedure. Coupled with refueling, the pit stop took 11.734 seconds.

After catching their breath for a minute, Richardson and the other pit crew members reviewed a videotape, looking for split-second flaws.

LEARNING OBJECTIVES	**LO 10.1**	***Describe*** labor-planning policies 407
	LO 10.2	***Identify*** the major issues in job design 408
	LO 10.3	***Identify*** major ergonomic and work environment issues 412
	LO 10.4	***Use*** the tools of methods analysis 414
	LO 10.5	***Identify*** four ways of establishing labor standards 417
	LO 10.6	***Compute*** the normal and standard times in a time study 419
	LO 10.7	***Find*** the proper sample size for a time study 420

Human Resource Strategy for Competitive Advantage

VIDEO 10.1
The "People" Focus: Human Resources at Alaska Airlines

Good human resource strategies are expensive, difficult to achieve, and hard to sustain. But, like a NASCAR team, many organizations, from Hard Rock Cafe to Alaska Airlines, have demonstrated that sustainable competitive advantage can be built through a human resource strategy. The payoff can be significant and difficult for others to duplicate. Indeed, as London's Four Seasons Hotel notes, "We've identified that our key *competitive difference is our people*." In this chapter, we will examine some of the tools available to operations managers for achieving competitive advantage via human resource management.

The objective of a human resource strategy is to manage personnel and design jobs so people are effectively and efficiently utilized. As we focus on a human resource strategy, we want to ensure that people:

VIDEO 10.2
Human Resources at Hard Rock Cafe

1. Are efficiently utilized within the constraints of other operations management decisions.
2. Have a reasonable quality of work life in an atmosphere of mutual commitment and trust.

By reasonable *quality of work life* we mean a job that is not only reasonably safe and for which the pay is equitable but that also achieves an appropriate level of both physical and psychological requirements. *Mutual commitment* means that both management and employee strive to meet common objectives. *Mutual trust* is reflected in reasonable, documented employment policies that are honestly and equitably implemented to the satisfaction of both management and employee. When management has a genuine respect for its employees and their contributions to the firm, establishing a reasonable quality of work life and mutual trust is not particularly difficult.

Constraints on Human Resource Strategy

STUDENT TIP
An operations manager knows how to build an effective human resource strategy.

As Figure 10.1 suggests, many decisions made about people are constrained by other decisions. First, the product mix may determine seasonality and stability of employment. Second, technology, equipment, and processes may have implications for safety and job content. Third, the location decision may have an impact on the ambient environment in which the employees work. Finally, layout decisions, such as assembly line versus work cell, influence job content.

Technology decisions impose substantial constraints. For instance, some of the jobs in foundries are dirty, noisy, and dangerous; slaughterhouse jobs may be stressful and subject workers to stomach-crunching stench; assembly-line jobs are often boring and mind numbing; and high capital investments such as those required for manufacturing semiconductor chips may require 24-hour, 7-day-a-week operation in restrictive clothing.

We are not going to change these jobs without making changes in our other strategic decisions, so the trade-offs necessary to reach a tolerable quality of work life are difficult. Effective managers consider such decisions simultaneously. The result: a system in which both individual and team performance are enhanced through optimum job design.

We now look at three distinct decision areas of human resource strategy: *labor planning, job design*, and *labor standards*.

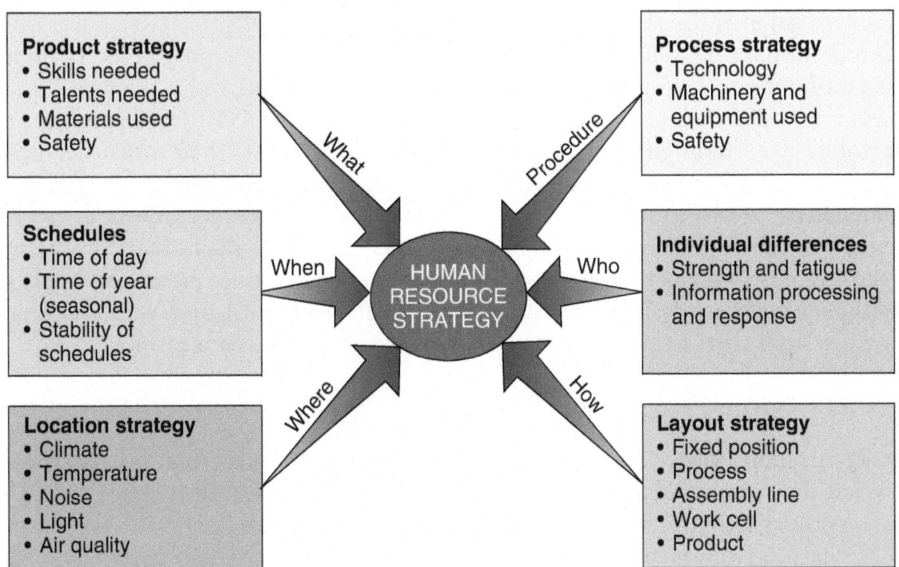

Figure **10.1**

Constraints on Human Resource Strategy

Labor Planning

Labor planning is determining staffing policies that deal with (1) employment stability, (2) work schedules, and (3) work rules.

Labor planning
A means of determining staffing policies dealing with employment stability, work schedules, and work rules.

Employment-Stability Policies

Employment stability deals with the number of employees maintained by an organization at any given time. There are two very basic policies for dealing with stability:

1. *Follow demand exactly:* Following demand exactly keeps direct labor costs tied to production but incurs other costs. These other costs include (a) hiring and layoff costs, (b) unemployment insurance, and (c) premium wages to entice personnel to accept unstable employment. This policy tends to treat labor as a variable cost.
2. *Hold employment constant:* Holding employment levels constant maintains a trained workforce and keeps hiring, layoff, and unemployment costs to a minimum. However, with employment held constant, employees may not be utilized fully when demand is low, and the firm may not have the human resources it needs when demand is high. This policy tends to treat labor as a fixed cost.

These policies are only two of many that can be efficient *and* provide a reasonable quality of work life. Firms must determine policies about employment stability.

LO 10.1 *Describe* labor-planning policies

Work Schedules

Although the standard work schedule in the U.S. is still five 8-hour days, many variations exist. A popular variation is a work schedule called flextime. *Flextime* allows employees, within limits, to determine their own schedules. A flextime policy might allow an employee (with proper notification) to be at work at 8 A.M. plus or minus 2 hours. This policy allows more autonomy and independence on the part of the employee. Some firms have found flextime a low-cost fringe benefit that enhances job satisfaction. The problem from the OM perspective is that much production work requires full staffing for efficient operations. A machine that requires three people cannot run at all if only two show up. Having a server show up to serve lunch at 1:30 P.M. rather than 11:30 A.M. is not much help either.

Similarly, some industries find that their process strategies severely constrain their human resource scheduling options. For instance, paper manufacturing, petroleum refining, and power stations require around-the-clock staffing except for maintenance and repair shutdown.

Another option is the *flexible workweek*. This plan often calls for fewer but longer days, such as four 10-hour days or, as in the case of less physically demanding tasks, 12-hour shifts. Working 12-hour shifts usually means working 3 days one week and 4 the next. Such shifts are sometimes called *compressed workweeks*. More radically, the COVID-19 pandemic accelerated a remote-work trend offering office workers the ability to set their own schedules. These schedules are viable for many operations functions—as long as suppliers and customers can be accommodated.

Another option is shorter days rather than longer days. This plan often moves employees to *part-time status*. Such an option is particularly attractive in service industries, where staffing for peak loads is necessary. Banks and restaurants often hire part-time workers. Also, many firms reduce labor costs by reducing fringe benefits for part-time employees.

Job Classifications and Work Rules

Many organizations have strict job classifications and work rules that specify who can do what, when they can do it, and under what conditions they can do it, often as a result of union pressure. These job classifications and work rules restrict employee flexibility on the job, which in turn reduces the flexibility of the operations function. Yet part of an operations manager's task is to manage the unexpected. Therefore, the more flexibility a firm has when staffing and establishing work schedules, the more efficient and responsive it *can* be. This is particularly true in service organizations, where extra capacity often resides in extra or flexible staff. Building morale and meeting staffing requirements that result in an efficient, responsive operation are easier if managers have fewer job classifications and work-rule constraints. If the strategy is to achieve a competitive advantage by responding rapidly to the customer, a flexible workforce may be a prerequisite.

Job Design

Job design
An approach that specifies the tasks that constitute a job for an individual or a group.

Job design specifies the tasks that constitute a job for an individual or a group. We examine five components of job design: (1) job specialization, (2) job expansion, (3) psychological components, (4) self-directed teams, and (5) motivation and incentive systems.

Labor Specialization

Labor specialization (or job specialization)
The division of labor into unique ("special") tasks.

The importance of job design as a management variable is credited to the 18th-century economist Adam Smith. Smith suggested that a division of labor, also known as **labor specialization** (or job specialization), would assist in reducing labor costs of multiskilled artisans. This is accomplished in several ways:

1. *Development of dexterity* and faster learning by the employee because of repetition
2. *Less loss of time* because the employee would not be changing jobs or tools
3. *Development of specialized tools* and the reduction of investment because each employee has only a few tools needed for a particular task

LO 10.2 *Identify* the major issues in job design

The 19th-century British mathematician Charles Babbage determined that a fourth consideration was also important for labor efficiency. Because pay tends to follow skill with a rather high correlation, Babbage suggested *paying exactly the wage needed for the particular skill required*. If the entire job consists of only one skill, then we would pay for only that skill. Otherwise, we would tend to pay for the highest skill contributed by the employee. These four advantages of labor specialization are still valid today.

A classic example of labor specialization is the assembly line. Such a system is often very efficient, although it may require employees to do short, repetitive, mind-numbing jobs. The wage rate for many of these jobs, however, is good. Given the relatively high wage rate for the modest skills required in many of these jobs, there is often a large pool of employees from which to choose.

From the manager's point of view, a major limitation of specialized jobs is their failure to bring the whole person to the job. Job specialization tends to bring only the employee's manual skills to work. In an increasingly sophisticated knowledge-based society, managers want employees to bring their mind to work as well.

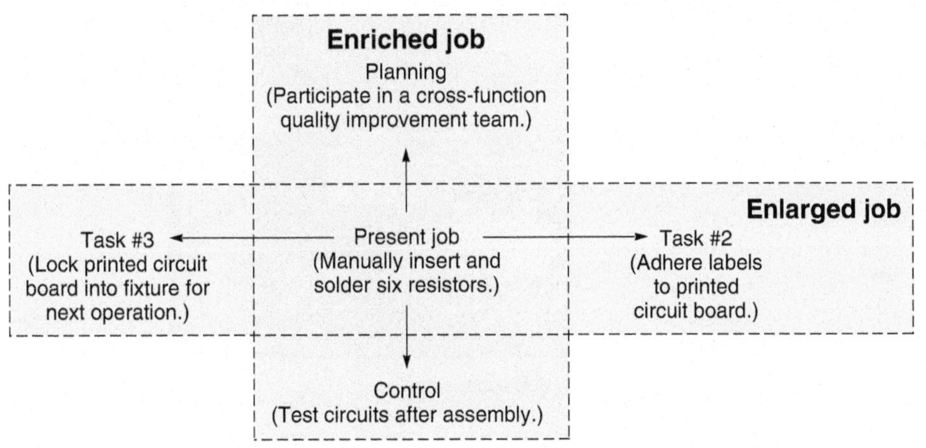

Figure 10.2

An Example of Job Enlargement (horizontal job expansion) and Job Enrichment (vertical job expansion)

Job Expansion

Moving from labor specialization toward more varied job design may improve the quality of work life. The theory is that variety makes the job "better" and that the employee therefore enjoys a higher quality of work life. This flexibility thus benefits the employee and the organization.

We modify jobs in a variety of ways. The first approach is **job enlargement**, which occurs when we add tasks requiring similar skill to an existing job. **Job rotation** is a version of job enlargement that occurs when the employee is allowed to move from one specialized job to another. Variety has been added to the employee's perspective of the job. Another approach is **job enrichment**, which adds planning and control to the job. An example is to have department store salespeople responsible for ordering, as well as selling, their goods. Job enrichment can be thought of as *vertical expansion*, as opposed to job enlargement, which is *horizontal*. These ideas are shown in Figure 10.2.

A popular extension of job enrichment, **employee empowerment** is the practice of enriching jobs so employees accept responsibility for a variety of decisions normally associated with staff specialists. Empowering employees helps them take "ownership" of their jobs so they have a personal interest in improving performance.

Job enlargement
The grouping of a variety of tasks about the same skill level; horizontal enlargement.

Job rotation
A system in which an employee is moved from one specialized job to another.

Job enrichment
A method of giving an employee more responsibility that includes some of the planning and control necessary for job accomplishment; vertical expansion.

Employee empowerment
Enlarging employee jobs so that the added responsibility and authority are moved to the lowest level possible.

Psychological Components of Job Design

An effective human resources strategy also requires consideration of the psychological components of job design. These components focus on how to design jobs that meet some minimum psychological requirements.

Hawthorne Studies The Hawthorne studies introduced psychology to the workplace. They were conducted in the 1920s at Western Electric's Hawthorne plant near Chicago. These studies were initiated to determine the impact of lighting on productivity. Instead, they found the dynamic social system and distinct roles played by employees to be more important than the intensity of the lighting. The researchers also found that individual differences may be dominant in what employees expect from the job and what they think their contributions to the job should be.

Core Job Characteristics Substantial research regarding the psychological components of job design has taken place since the Hawthorne studies. J. R. Hackman and G. R. Oldham have incorporated much of that work into five desirable characteristics of job design. They suggest that jobs should include the following characteristics:

1. **Skill variety**, requiring the worker to use a variety of skills and talents
2. **Job identity**, allowing the worker to perceive the job as a whole and recognize a start and a finish
3. **Job significance**, providing a sense that the job has an impact on the organization and society
4. **Autonomy**, offering freedom, independence, and discretion
5. **Feedback**, providing clear, timely information about performance

Figure **10.3**

Job Design Continuum

An increasing reliance on the employee's contribution can increase the responsibility accepted by the employee.

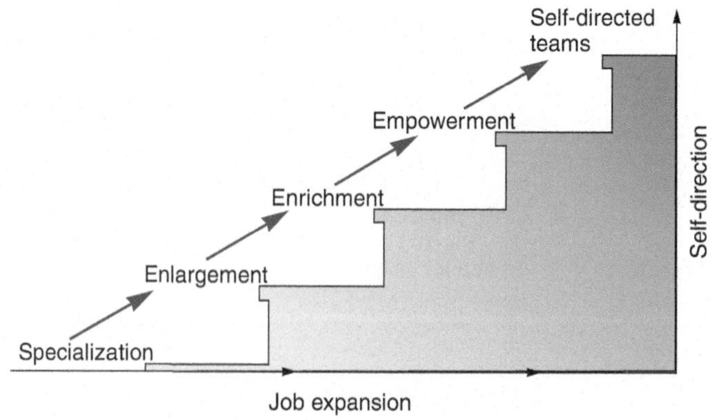

Including these five ingredients in job design is consistent with job enlargement, job enrichment, and employee empowerment. We now want to look at some of the ways in which teams can be used to expand jobs and achieve these five job characteristics.

Self-Directed Teams

Self-directed team
A group of empowered individuals working together to reach a common goal.

Many world-class organizations have adopted teams to foster mutual trust and commitment, and provide the core job characteristics. One team concept of particular note is the **self-directed team**: a group of empowered individuals working together to reach a common goal. These teams may be organized for long- or short-term objectives. Teams are effective primarily because they can easily provide employee empowerment, ensure core job characteristics, and satisfy many of the psychological needs of individual team members. A job design continuum is shown in Figure 10.3.

Limitations of Job Expansion If job designs that enlarge, enrich, empower, and use teams are so good, why are they not universally used? Mostly it is because of costs. Here are a few limitations of expanded job designs:

- *Higher capital cost:* Job expansion may require additional equipment and facilities.
- *Individual differences:* Some employees opt for the less-complex jobs.
- *Higher wage rates:* Expanded jobs may well require a higher average wage.
- *Smaller labor pool:* Because expanded jobs require more skill and acceptance of more responsibility, job requirements have increased.
- *Higher training costs:* Job expansion requires training and cross-training. Therefore, training budgets need to increase.

Despite these limitations, firms are finding a substantial payoff in job expansion.

Jim West/Alamy Stock Photo

Bloomberg/Getty Images

Southwest Airlines—consistently near the top of the airline industry in travel surveys, fewest lost bags and complaints, and highest profits—hires people with enthusiasm and empowers them to excel. Co-founder and former chairman Herb Kelleher (right photo) said, "I've tried to create a culture of caring for people in the totality of their lives, not just at work. Someone can go out and buy airplanes and ticket counters, but they can't buy our culture, our *esprit de corps.*"

OM in Action Using Incentives to Unsnarl Traffic Jams in the OR[1]

Hospitals have long offered surgeons a precious perk: scheduling the bulk of their elective surgeries in the middle of the week so they can attend conferences, teach, or relax during long weekends. But at Boston Medical Center, St. John's Health Center (in Missouri), and Elliot Health System (in New Hampshire), this practice, one of the biggest impediments to a smooth-running hospital, is changing. "Block scheduling" jams up operating rooms, overloads nurses at peak times, and bumps scheduled patients for hours and even days.

Boston Medical Center's delays and cancellations of elective surgeries were nearly eliminated after surgeons agreed to stop block scheduling and to dedicate one OR for emergency cases. Cancellations dropped to 3, from 334, in just one 6-month period. In general, hospitals changing to the new system of spreading out elective surgeries during the week increase their surgery capacity by 10%, move patients through the operating room faster, and reduce nursing overtime.

To get doctors on board at St. John's, the hospital offered a carrot and two sticks: Doctors who were more than 10 minutes late 10% of the time lost their coveted 7:30 A.M. start times *and* were fined a portion of their fee—with proceeds going to a kitty that rewarded the best on-time performers. Surgeons' late start times quickly dropped from 16% to 5% and then to less than 1% within a year.

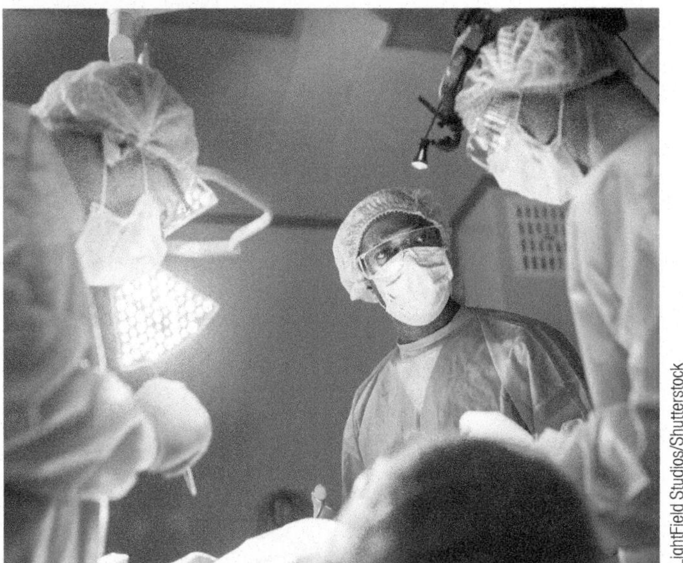

LightField Studios/Shutterstock

Motivation and Incentive Systems

Our discussion of the psychological components of job design provides insight into the factors that contribute to job satisfaction and motivation. In addition to these psychological factors, there are monetary factors. Money often serves as a psychological as well as a financial motivator. Monetary rewards take the form of bonuses, profit and gain sharing, and incentive systems.

Bonuses, in cash, stock ownership, or stock options, are often used to reward employees. Almost half of U.S. employees have one or more forms of profit sharing that distributes part of the profit to employees. A variation of profit sharing is gain sharing, which rewards employees for improvements made in an organization's performance. The most popular of these is the Scanlon plan, in which any reduction in the cost of labor is shared between management and labor.

Incentive systems based on individual or group productivity are used throughout the world in a wide variety of applications, including nearly half of the manufacturing firms in the U.S. Production incentives often require employees or crews to produce at or above a predetermined standard. The standard can be based on a "standard time" per task or number of pieces made. Both systems typically guarantee the employee at least a base rate. Incentives, of course, need not be monetary. Awards, recognition, and other kinds of preferences such as a preferred work schedule can be effective. (See the *OM in Action* box, "Using Incentives to Unsnarl Traffic Jams in the OR.") Hard Rock Cafe has successfully reduced its turnover by giving every employee—from the CEO to the bussers—a $10,000 gold Rolex watch on their 10th anniversary with the firm.

With the increasing use of teams, various forms of team-based pay are also being developed. Many are based on traditional pay systems supplemented with some form of bonus or incentive system. However, because many team environments require cross-training, *knowledge-based* pay systems have also been developed. Under knowledge-based (or skill-based) pay systems, a portion of the employee's pay depends on demonstrated knowledge or skills. At Wisconsin's Johnsonville Sausage Co., employees receive pay raises *only* by mastering new skills such as scheduling, budgeting, and quality control.

Ergonomics and the Work Environment

With the foundation provided by Frederick W. Taylor, the father of the era of scientific management, we have developed a body of knowledge about people's capabilities and limitations. This knowledge is necessary because humans are hand/eye tool-using animals possessing

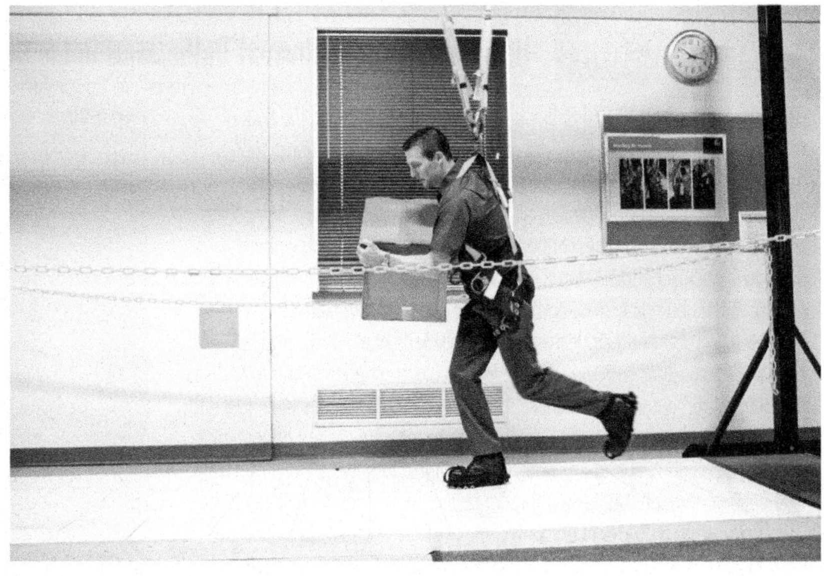

With a commitment to efficiency and an understanding of ergonomics, UPS trains drivers in the company's "340 methods" that save seconds and improve safety. Here a UPS driver is learning to walk on "ice" with the help of a "slip and fall" simulator.

Stephen Voss

exceptional capabilities and some limitations. Because managers must design jobs that can be done, we now introduce a few of the issues related to people's capabilities and limitations.

Ergonomics

The study of the human interface with the environment and machines.

Ergonomics The operations manager is interested in building a good interface between humans, the environment, and machines. Studies of this interface are known as **ergonomics**. Ergonomics means "the study of work." (*Ergon* is the Greek word for "work.") The term *human factors* is often substituted for the word *ergonomics*. Understanding ergonomic issues helps to improve human performance.

Male and female adults come in limited configurations and abilities. Therefore, design of tools and the workplace depends on the study of people to determine what they can and cannot do. Substantial data have been collected that provide basic strength and measurement data needed to design tools and the workplace. The design of the workplace can make the job easier or impossible. In addition, we now have the ability, through the use of computer modeling, to analyze human motions and efforts. The *OM in Action* box, "The Rise of the Exoskeleton," shows how new technological advances can help worker performance.

LO 10.3 *Identify* major ergonomic and work environment issues

Operator Input to Machines Operator response to machines, be they hand tools, pedals, levers, or buttons, needs to be evaluated. Operations managers need to be sure that operators have the strength, reflexes, perception, and mental capacity to provide necessary control. Such problems as *carpal tunnel syndrome* may result when a tool as simple as a keyboard is poorly designed. The photo of the race car steering wheel shows one innovative approach to critical operator input.

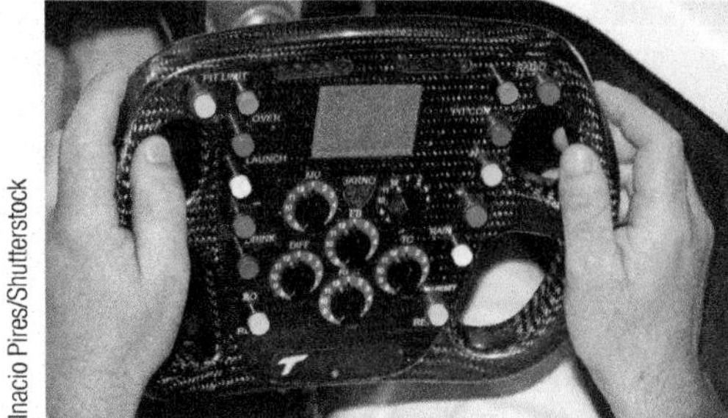

Inacio Pires/Shutterstock

Drivers of race cars have no time to grasp for controls or to look for small hidden gauges. Controls and instrumentation for modern race cars have migrated to the steering wheel itself—the critical interface between man and machine.

Beverly Amer/Aspenleaf Productions

Airline pilots now have modern "glass cockpits" to comprehensively display real-time information and realistic 3D depictions.

OM in Action The Rise of the Exoskeleton[2]

In the welding shop of Toyota's huge Ontario plant, workers inspect the steel frame of a RAV4. The workers move ultrasonic wands over metal to test the integrity of dozens of welds. Until recently, this task was performed by seated workers wielding hammers and chisels. But the latest RAV4 uses a lighter, stronger steel that requires ultrasonic testing. A new frame arrives every 60 seconds. The prolonged reaching is shoulder-breaking work, the kind that can lead to debilitating injuries and decreased productivity.

But now employees are assisted by wearable *exoskeletons.* The exoskeletons use a system of springs, cables, and pulleys to transfer weight from the arms and backs to the outside of the hips, easing the strain of overhead or bending work. When a worker raises his or her arms, the exoskeleton provides a counterweight that makes the arms feel buoyant, as if the upper body is suspended in water. The system gradually releases as the limbs are lowered, allowing the arms to hang unassisted.

Exoskeletons are becoming commonplace on factory floors, construction sites, and film sets. Toyota is the first large manufacturer to require the use of exoskeletons. Meanwhile, Ford uses about 100 exoskeletons across 16 plants in 8 countries. BMW has 66 in use at its Spartanburg, South Carolina, plant, and Boeing deploys hundreds.

There are upper-body, lower-body, and full-body models (as seen in the photo). Most range in price from $4,000 to $6,000, weigh 5 to 10 pounds, and require a one-time adjustment to the user's frame. Factory workers who have tried exoskeletons report less back and shoulder pain, and they return home at night more active and relaxed. Ultimately, the hope is that the devices will reduce work-related musculoskeletal disorders, which currently cost employers about $50 billion annually.

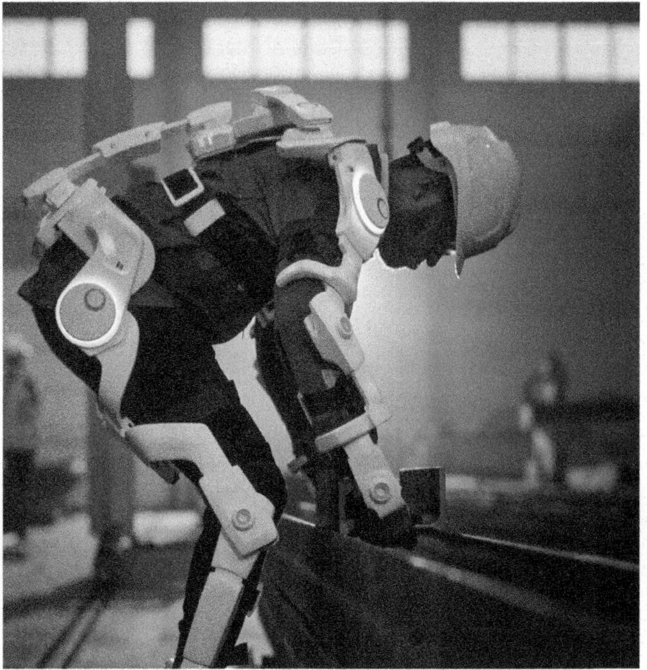
Gorodenkoff/Shutterstock

Feedback to Operators Feedback to operators is provided by sight, sound, and feel; it should not be left to chance. The mishap at the Three Mile Island nuclear facility, America's worst nuclear experience, was in large part the result of poor feedback to the operators about reactor performance. Nonfunctional groups of large, unclear instruments and inaccessible controls, combined with hundreds of confusing warning lights, contributed to that failure. Such relatively simple issues make a difference in operator response and, therefore, performance.

An important human factor/ergonomic issue in the aircraft industry is cockpit design. Newer "glass cockpits" (see photo) display information in more concise form than the traditional rows of round analog dials and gauges. These displays reduce the chance of human error, which is a factor in about two-thirds of commercial air accidents.

The Work Environment The physical environment in which employees work affects their performance, safety, and quality of work life. Illumination, noise and vibration, temperature, humidity, and air quality are work-environment factors under the control of the organization and the operations manager. The manager must approach them as controllable.

Illumination is necessary, but the proper level depends on the work being performed. Figure 10.4(a) provides some guidelines. However, other lighting factors are important. These include reflective ability, contrast of the work surface with surroundings, glare, and shadows.

Noise of some form is usually present in the work area, and most employees seem to adjust well. However, high levels of sound will damage hearing. Figure 10.4(b) provides indications of the sound generated by various activities. Extended periods of exposure to decibel levels above 85 dB are permanently damaging. The Occupational Safety and Health Administration (OSHA) requires ear protection above this level if exposure equals or exceeds 8 hours. Even at low levels, noise and vibration can be distracting and can raise a person's blood pressure, so managers make substantial effort to reduce noise and vibration through good machine design, enclosures, or insulation.

Temperature and humidity parameters have also been well established. Managers with activities operating outside the established comfort zone should expect adverse effect on performance.

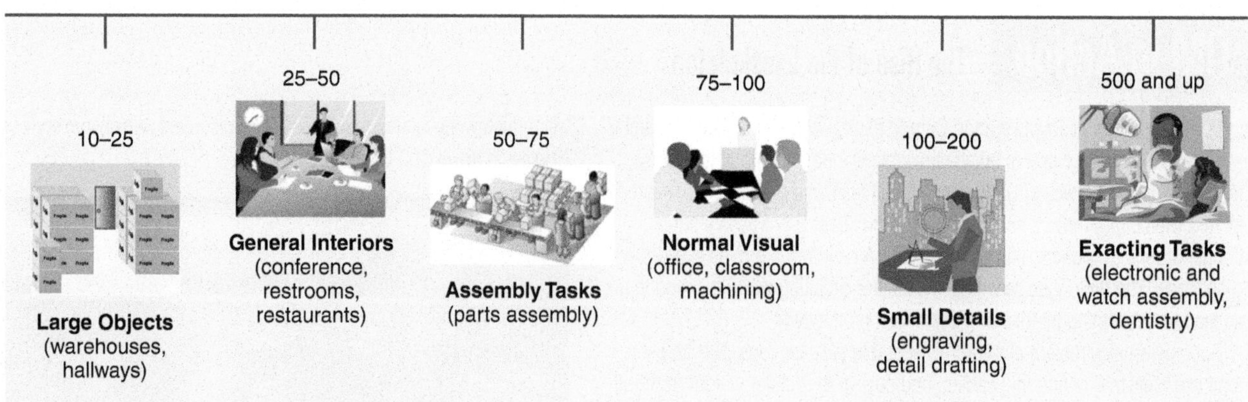

Figure **10.4(a)**

Recommended Levels of Illumination (using foot-candles (ft-c) as the measure of illumination)

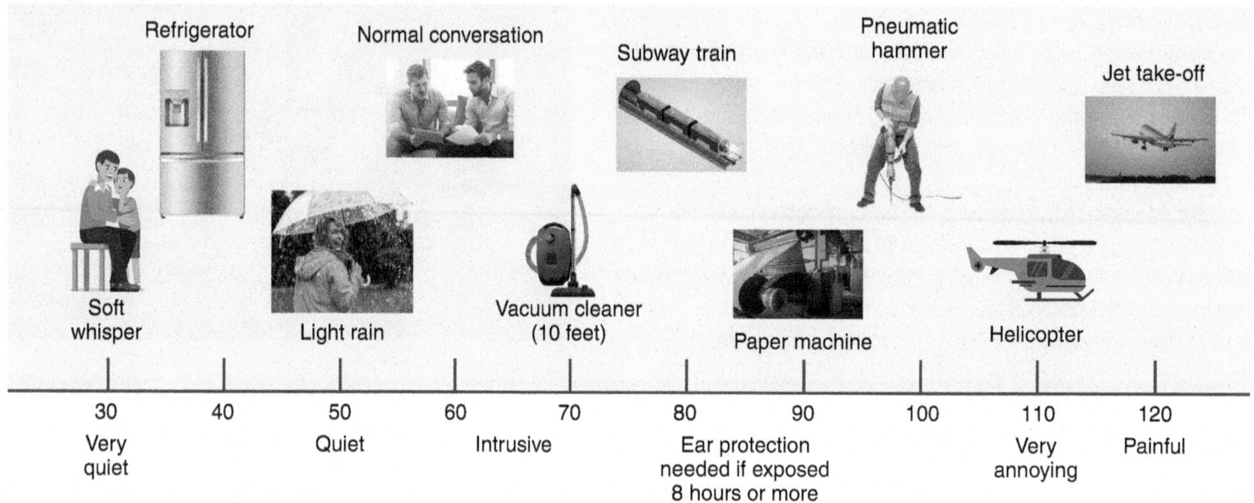

Figure **10.4(b)**

Decibel (dB) Levels for Various Sounds

Image credits (from left to right): Iconic Bestiary/Shutterstock; ApoGapo/Shutterstock; Olena Yakobchuk/Shutterstock; marvent/Shutterstock; YAY Media AS/Alamy Stock Vector; Golden Sikorka/Shutterstock; K-STUDIO/Shutterstock; Tom Wang/Shutterstock; Jemastock/Shutterstock; belkos/Shutterstock

Methods Analysis

Methods analysis
A system that involves developing work procedures that are safe and produce quality products efficiently.

Methods analysis focuses on *how* a task is accomplished. Whether controlling a machine or making or assembling components, how a task is done makes a difference in performance, safety, and quality. Using knowledge from ergonomics and methods analysis, methods engineers are charged with ensuring that quality and quantity standards are achieved efficiently and safely. Methods analysis and related techniques are useful in office environments as well as in the factory. Methods techniques are used to analyze:

1. **Movement of individuals or material.** The analysis is performed using *flow diagrams* and *process charts* with varying amounts of detail.
2. **Activity of human and machine and crew activity.** This analysis is performed using *activity charts* (also known as *man–machine charts* and *crew charts*).
3. **Body movement** (primarily arms and hands). This analysis is performed using *operations charts.*

LO 10.4 *Use* the tools of methods analysis

Flow diagram
A drawing used to analyze movement of people or material.

Process chart
Graphic representations that depict a sequence of steps for a process.

Flow diagrams are schematics (drawings) used to investigate long-cycle repetitive movement of people or material. Britain's Paddy Hopkirk Factory, which manufactures auto parts, demonstrates one version of a flow diagram in Figure 10.5. Hopkirk's old work flow is shown in Figure 10.5(a), and a new method, with improved work flow and requiring less storage and space, is shown in Figure 10.5(b). **Process charts** use symbols, as in Figure 10.5(c), to help us understand the movement of people or material. In this way nonvalue-added activities can

Figure **10.5**

Flow Diagrams and Process Chart of Axle-Stand Production at Paddy Hopkirk Factory

(a) Old method; (b) new method; (c) process chart of axle-stand production using Paddy Hopkirk's new method (shown in (b)).

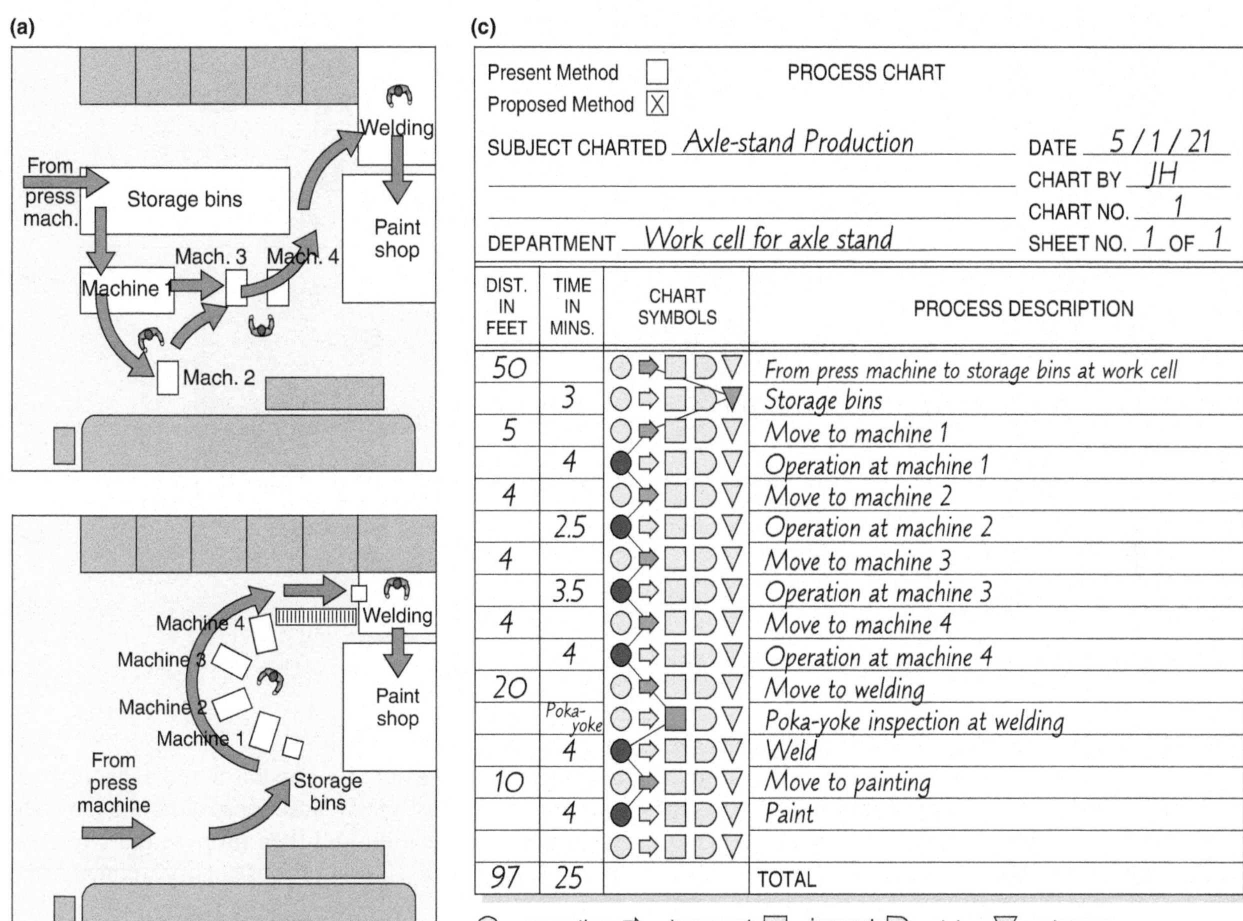

(c)

PROCESS CHART

Present Method ☐
Proposed Method ☒

SUBJECT CHARTED *Axle-stand Production* DATE *5 / 1 / 21*
CHART BY *JH*
CHART NO. *1*
DEPARTMENT *Work cell for axle stand* SHEET NO. *1* OF *1*

DIST. IN FEET	TIME IN MINS.	CHART SYMBOLS	PROCESS DESCRIPTION
50		○ ⇨ □ D ▽	*From press machine to storage bins at work cell*
	3	○ ⇨ □ D ▽	*Storage bins*
5		○ ⇨ □ D ▽	*Move to machine 1*
	4	○ ⇨ □ D ▽	*Operation at machine 1*
4		○ ⇨ □ D ▽	*Move to machine 2*
	2.5	○ ⇨ □ D ▽	*Operation at machine 2*
4		○ ⇨ □ D ▽	*Move to machine 3*
	3.5	○ ⇨ □ D ▽	*Operation at machine 3*
4		○ ⇨ □ D ▽	*Move to machine 4*
	4	○ ⇨ □ D ▽	*Operation at machine 4*
20		○ ⇨ □ D ▽	*Move to welding*
	Poka-yoke	○ ⇨ □ D ▽	*Poka-yoke inspection at welding*
	4	○ ⇨ □ D ▽	*Weld*
10		○ ⇨ □ D ▽	*Move to painting*
	4	○ ⇨ □ D ▽	*Paint*
		○ ⇨ □ D ▽	
97	25		TOTAL

○ = operation; ⇨ = transport; □ = inspect; D = delay; ▽ = storage

ACTIVITY CHART

	OPERATOR #1 TIME	OPERATOR #1 %	OPERATOR #2 TIME	OPERATOR #2 %
WORK	12	100	12	100
IDLE	0	0	0	0

OPERATION: Oil change & fluid check
EQUIPMENT: One bay/pit
OPERATOR: Two-person crew
STUDY NO.: ____ ANALYST: BR

SUBJECT *Quick Car Lube* DATE *5-1-21*
PRESENT / (PROPOSED) DEPT. SHEET *1* OF *1* CHART BY *LSA*

	TIME	Operator #1	TIME	Operator #2	TIME
		Take order		Move car to pit	
	2	Vacuum car		Drain oil	
	4	Clean windows		Check transmission	
	6	Check under hood		Change oil filter	
	8	Fill with oil		Replace oil plug	
	10	Complete bill		Move car to front for customer	
	12	Greet next customer		Move next car to pit	
Repeat cycle	14	Vacuum car		Drain oil	
	16	Clean windows		Check transmission	
	18				

Figure **10.6**

Activity Chart for Two-Person Crew Doing an Oil Change in 12 Minutes at Quick Car Lube

OPERATIONS CHART

SYMBOLS	PRESENT LH	PRESENT RH	PROPOSED LH	PROPOSED RH
○ OPERATION	2	3		
⇨ TRANSPORT.	1	1		
□ INSPECTION				
D DELAY	4	3		
▽ STORAGE				

PROCESS: Scooping Ice for Coffee
EQUIPMENT: Scoop
OPERATOR: Starbucks
STUDY NO: ____ ANALYST: CM
DATE: 5 /1 /21 SHEET NO. 1 of 2
METHOD (PRESENT/PROPOSED)
REMARKS: Partial Study

LEFT-HAND ACTIVITY Present METHOD	DIST.	SYMBOLS	SYMBOLS	DIST.	RIGHT-HAND ACTIVITY Present METHOD
1 Reach for cup		○⇨□D▽	○⇨□D▽		Idle
2 Grasp cup		○⇨□D▽	○⇨□D▽		Idle
3 Move cup	6"	○⇨□D▽	○⇨□D▽		Idle
4 Hold cup		○⇨□D▽	○⇨□D▽		Reach for scoop
5 Hold cup		○⇨□D▽	○⇨□D▽		Grasp scoop
6 Hold cup		○⇨□D▽	○⇨□D▽	8"	Move scoop to ice
7 Hold cup		○⇨□D▽	○⇨□D▽		Scoop ice

Figure **10.7**

Operations Chart (right-hand/left-hand chart) for Scooping Ice to Coffee Cup

Visual utensil holder encourages housekeeping.

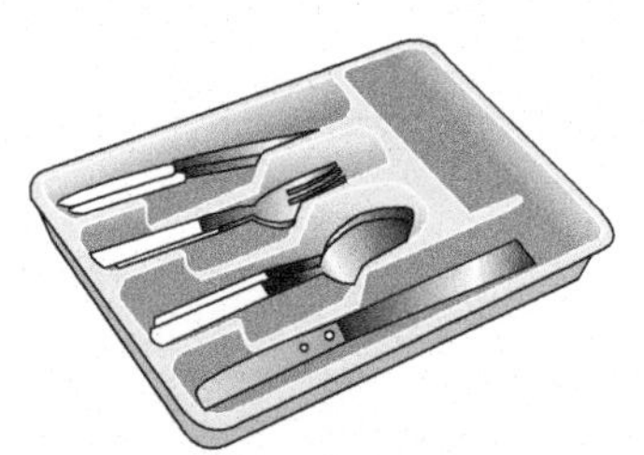

A "3-minute service" clock reminds employees of the goal.

Visual signals at the machine notify support personnel.

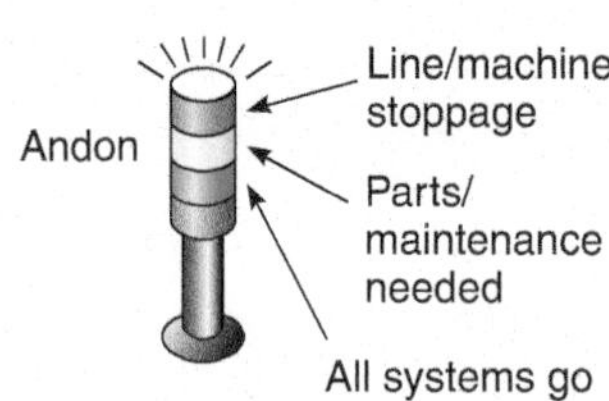

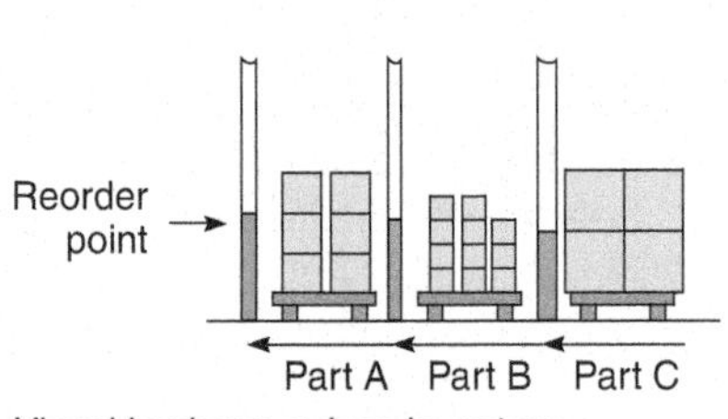

Visual kanbans reduce inventory and foster just-in-time (JIT) flows.

Quantities in bins indicate ongoing daily requirements, and clipboards provide information on schedule changes.

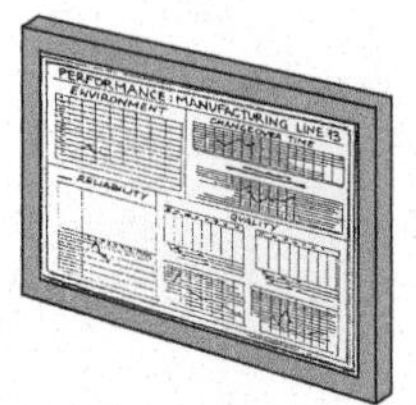

Company data, process specifications, and operating procedures are posted in each work area.

Figure **10.8**

The Visual Workplace

be recognized and operations made more efficient. Figure 10.5(c) is a process chart used to supplement the flow diagrams shown in Figure 10.5(b).

Activity chart
A way of improving utilization of an operator and a machine or some combination of operators (a crew) and machines.

Activity charts are used to study and improve the utilization of an operator and a machine or some combination of operators (a "crew") and machines. The typical approach is for the analyst to record the present method through direct observation and then propose the improvement on a second chart. Figure 10.6 is an activity chart to show a proposed improvement for a two-person crew at Quick Car Lube.

Operations chart
A chart depicting right- and left-hand motions.

Body movement is analyzed by an **operations chart**. It is designed to show economy of motion by pointing out wasted motion and idle time (delay). An operations chart for Starbucks (also known as a *right-hand/left-hand chart*) is shown in Figure 10.7.

The Visual Workplace

Visual workplace
Uses a variety of visual communication techniques to rapidly communicate information to stakeholders.

A **visual workplace** uses low-cost visual devices to share information quickly and accurately. Well-designed displays and graphs root out confusion and replace difficult-to-understand printouts and paperwork. Because workplace data change quickly and often, operations managers need to share accurate and up-to-date information. Changing customer requirements, specifications, schedules, and other details must be rapidly communicated to those who can make things happen.

Visuals are more meaningful when metrics have been set and are known. The visual workplace can eliminate nonvalue-added activities by making standards, problems, and abnormalities visual (see Figure 10.8). And supervision is reduced because employees understand the standard, see the results, and know what to do.

Labor Standards

Labor standards
The amount of time required to perform a job or part of a job.

So far in this chapter, we have discussed labor planning and job design. The third requirement of an effective human resource strategy is the establishment of labor standards. **Labor standards** are the amount of time required to perform a job or part of a job, and they exist, formally or informally, for all jobs. Effective workforce planning is dependent on a knowledge of the labor required.

Modern labor standards originated with the works of Frederick W. Taylor and Frank and Lillian Gilbreth at the beginning of the 20th century. At that time, a large proportion of work was manual, and the resulting labor content of products was high. Little was known about what constituted a fair day's work, so managers initiated studies to improve work methods and understand human effort. These efforts continue to this day. Although labor costs are often less than 10% of sales, labor standards remain important and continue to play a major role in both service and manufacturing organizations. They are often a beginning point for determining staffing requirements. With over half of the manufacturing plants in the U.S. using some form of labor incentive system, good labor standards are a requirement.

Effective operations management requires meaningful standards that help a firm determine:

1. Labor content of items produced (the labor cost)
2. Staffing needs (how many people it will take to meet required production)
3. Cost and time estimates prior to production (to assist in a variety of decisions, from cost estimates to make-or-buy decisions)
4. Crew size and work balance (who does what in a group activity or on an assembly line)
5. Expected production (so that both manager and worker know what constitutes a fair day's work)
6. Basis of wage-incentive plans (what provides a reasonable incentive)
7. Efficiency of employees and supervision (a standard is necessary against which to determine efficiency)

LO 10.5 *Identify* four ways of establishing labor standards

Properly set labor standards represent the amount of time that it should take an average employee to perform specific job activities under normal working conditions. Labor standards are set in four ways:

1. **Historical experience**
2. **Time studies**
3. **Predetermined time standards**
4. **Work sampling**

Historical Experience

Labor standards can be estimated based on *historical experience*—that is, how many labor-hours were required to do a task the last time it was performed. Historical standards have the advantage of being relatively easy and inexpensive to obtain. They are usually available from employee time cards or production records. However, they are not objective, and we do not know their accuracy, whether they represent a reasonable or a poor work pace, and whether unusual occurrences are included. Because these variables are unknown, their use is not recommended. Instead, time studies, predetermined time standards, and work sampling are preferred.

Time Studies

Time study
Timing a sample of a worker's performance and using it as a basis for setting a standard time.

The classical stopwatch study, or **time study**, originally proposed by Frederick W. Taylor in 1881, involves timing a sample of a worker's performance and using it to set a standard. Stopwatch studies are the most widely used labor standard method. A trained and experienced person can establish a standard by following these eight steps:

1. Define the task to be studied (after methods analysis has been conducted).
2. Divide the task into precise elements (parts of a task that often take no more than a few seconds).
3. Decide how many times to measure the task (the number of job cycles or samples needed).
4. Time and record element times and ratings of performance.
5. Compute the average observed (actual) time. The **average observed time** is the arithmetic mean of the times for *each* element measured, adjusted for unusual influence for each element:

Average observed time
The arithmetic mean of the times for each element measured, adjusted for unusual influence for each element.

$$\text{Average observed time} = \frac{\text{Sum of the times recorded to perform each element}}{\text{Number of observations}} \quad (10\text{-}1)$$

Normal time
The average observed time, adjusted for pace.

6. Determine performance rating (work pace) and then compute the **normal time** for each element.

$$\text{Normal time} = \text{Average observed time} \times \text{Performance rating factor} \qquad (10\text{-}2)$$

The performance rating adjusts the average observed time to what a trained worker could expect to accomplish working at a normal pace. For example, workers should be able to walk 3 miles per hour. They should also be able to deal a deck of 52 cards into 4 equal piles in 30 seconds. A performance rating of 1.05 would indicate that the observed worker performs the task slightly *faster* than average. Numerous videos specify work pace on which professionals agree, and benchmarks have been established by the Society for the Advancement of Management. Performance rating, however, is still something of an art.

7. Add the normal times for each element to develop a total normal time for the task.
8. Compute the **standard time**. This adjustment to the total normal time provides for allowances such as *personal* needs, unavoidable work *delays*, and worker *fatigue*:

Standard time
An adjustment to the total normal time; the adjustment provides allowances for personal needs, unavoidable work delays, and fatigue.

$$\text{Standard time} = \frac{\text{Total normal time}}{1 - \text{Allowance factor}} \qquad (10\text{-}3)$$

Personal time allowances are often established in the range of 4% to 7% of total time, depending on nearness to restrooms, water fountains, and other facilities. *Delay allowances* are often set as a result of the actual studies of the delay that occurs. *Fatigue allowances* are based on our growing knowledge of human energy expenditure under various physical and environmental conditions. A sample set of personal and fatigue allowances is shown in Table 10.1.

TABLE 10.1 Allowance Factors (in percentage) for Various Classes of Work[3]

Allowance	%
1. Constant allowances:	
(A) Personal allowance	5
(B) Basic fatigue allowance	4
2. Variable allowances:	
(A) Standing allowance	2
(B) Abnormal position allowance:	
(i) Awkward (bending)	2
(ii) Very awkward (lying, stretching)	7
(C) Use of force or muscular energy in lifting, pulling, pushing	
Weight lifted (pounds):	
20	3
40	9
60	17
(D) Bad light:	
(i) Well below recommended	2
(ii) Quite inadequate	5
(E) Noise level:	
(i) Intermittent—loud	2
(ii) Intermittent—very loud or high pitched	5

Example 1 illustrates the computation of standard time.

Example 1

DETERMINING NORMAL AND STANDARD TIME

The time study of a work operation at a Red Lobster restaurant yielded an average observed time of 4.0 minutes. The analyst rated the observed worker at 85%. This means the worker performed at 85% of normal when the study was made. The firm uses a 13% allowance factor. Red Lobster wants to compute the normal time and the standard time for this operation.

APPROACH ▶ The firm needs to apply Equations (10-2) and (10-3).

SOLUTION ▶

$$\text{Average observed time} = 4.0 \text{ min}$$

$$\text{Normal time} = (\text{Average observed time}) \times (\text{Performance rating factor})$$
$$= (4.0)(0.85)$$
$$= 3.4 \text{ min}$$

LO 10.6 *Compute* the normal and standard times in a time study

$$\text{Standard time} = \frac{\text{Normal time}}{1 - \text{Allowance factor}} = \frac{3.4}{1-0.13} = \frac{3.4}{0.87} = 3.9 \text{ min}$$

INSIGHT ▶ Because the observed worker was rated at 85% (slower than average), the normal time is less than the worker's 4.0-minute average time.

LEARNING EXERCISE ▶ If the observed worker is rated at 115% (faster than average), what are the new normal and standard times? [Answer: 4.6 min, 5.287 min.]

RELATED PROBLEMS ▶ 10.13–10.21, 10.33, 10.38–10.40

EXCEL OM Data File **Ch10Ex1.xls** can be found online.

Example 2 uses a series of actual stopwatch times for each element.

Example 2

USING TIME STUDIES TO COMPUTE STANDARD TIME

Management Science Associates offers management development seminars. A time study has been conducted for the registration process for seminar attendees. On the basis of the following observations, Management Science Associates wants to develop a time standard for this process. The firm's personal, delay, and fatigue allowance factor is 15%.

	OBSERVATIONS (MINUTES)					
JOB ELEMENT	1	2	3	4	5	PERFORMANCE RATING
(A) Register participant	8	10	9	21*	11	120%
(B) Explain schedule and answer questions	2	3	2	1	3	105%
(C) Provide badges, readings, and meal tickets	2	1	5*	2	1	110%

APPROACH ▶ Once the data have been collected, the procedure is to:

1. Delete unusual or nonrecurring observations.
2. Compute the *average time* for each element, using Equation (10-1).
3. Compute the *normal time* for each element, using Equation (10-2).
4. Find the total normal time.
5. Compute the *standard time*, using Equation (10-3).

SOLUTION ▶

1. Delete observations such as those marked with an asterisk (*). (These may be due to business interruptions, conferences with the boss, or mistakes of an unusual nature; they are not part of the job element but may be personal or delay time.)
2. Average time for each job element:

$$\text{Average time for A} = \frac{8 + 10 + 9 + 11}{4} = 9.5 \text{ min}$$

$$\text{Average time for B} = \frac{2 + 3 + 2 + 1 + 3}{5} = 2.2 \text{ min}$$

$$\text{Average time for C} = \frac{2 + 1 + 2 + 1}{4} = 1.5 \text{ min}$$

3. Normal time for each job element:

$$\text{Normal time for A} = (\text{Average observed time}) \times (\text{Performance rating})$$
$$= (9.5)(1.2) = 11.4 \text{ min}$$
$$\text{Normal time for B} = (2.2)(1.05) = 2.31 \text{ min}$$
$$\text{Normal time for C} = (1.5)(1.10) = 1.65 \text{ min}$$

Note: Normal times are computed for each element because the performance rating factor (work pace) may vary for each element, as it did in this case.

4. Add the normal times for each element to find the total normal time (the normal time for the whole job):

$$\text{Total normal time} = 11.40 + 2.31 + 1.65 = 15.36 \text{ min}$$

5. Standard time for the job:

$$\text{Standard time} = \frac{\text{Total normal time}}{1 - \text{Allowance factor}} = \frac{15.36}{1 - 0.15} = 18.07 \text{ min}$$

Thus, 18.07 minutes is the time standard for this job.

INSIGHT ▶ When observed times are not consistent, they need to be reviewed. Abnormally short times may be the result of an observational error and are usually discarded. Abnormally long times need to be analyzed to determine if they, too, are an error. However, they may *include* a seldom occurring but legitimate activity for the element (such as a machine adjustment) or may be personal, delay, or fatigue time.

LEARNING EXERCISE ▶ If the two observations marked with an asterisk were *not* deleted, what would be the total normal time and the standard time? [Answer: 18.89 min, 22.22 min.]

RELATED PROBLEMS ▶ 10.22–10.25, 10.28a, b, 10.29a, 10.30a, 10.41–10.43

Time study requires a sampling process, so the question of sampling error in the average observed time naturally arises. In statistics, error varies inversely with sample size. Thus, to determine just how many "cycles" we should time, we must consider the variability of each element in the study.

LO 10.7 *Find* the proper sample size for a time study

To determine an adequate sample size, three items must be considered:

1. How accurate we want to be (e.g., is ±5% of observed time close enough?).
2. The desired level of confidence (e.g., the z-value; is 95% adequate or is 99% required?).
3. How much variation exists within the job elements (e.g., if the variation is large, a larger sample will be required).

TABLE 10.2

Common z-Values

DESIRED CONFIDENCE (%)	Z-VALUE (STANDARD DEVIATION REQUIRED FOR DESIRED LEVEL OF CONFIDENCE)
90.0	1.65
95.0	1.96
95.45	2.00
99.0	2.58
99.73	3.00

The formula for finding the appropriate sample size, given these three variables, is:

$$\text{Required sample size} = n = \left(\frac{zs}{h\overline{x}}\right)^2 \quad (10\text{-}4)$$

where
h = accuracy level (acceptable error) desired in percent of the job element, expressed as a decimal ($5\% = .05$)
z = number of standard deviations required for desired level of confidence (90% confidence = 1.65; see Table 10.2 or Appendix I for more z-values)
s = standard deviation of the initial sample
$\overline{x}$ = mean of the initial sample
n = required sample size

We demonstrate with Example 3.

Example 3

COMPUTING SAMPLE SIZE

Andy Johnson Mfg., in Orlando, has asked you to check a labor standard prepared by a recently terminated analyst. Your first task is to determine the correct sample size. Your accuracy is to be within ±5% and your confidence level at 95%. The standard deviation of the sample is 1.0 and the mean 3.00.

APPROACH ▶ You apply Equation (10-4).

SOLUTION ▶

$$h = 0.05 \quad \bar{x} = 3.00 \quad s = 1.0$$

$z = 1.96$ (from Table 10.2 or Appendix I)

$$n = \left(\frac{zs}{h\bar{x}}\right)^2$$

$$n = \left(\frac{1.96 \times 1.0}{0.05 \times 3}\right)^2 = 170.74 \approx 171$$

Therefore, you recommend a sample size of 171.

INSIGHT ▶ Notice that as the confidence level required increases, the sample size also increases. Similarly, as the desired accuracy level increases (say, from 5% to 1%), the sample size increases.

LEARNING EXERCISE ▶ The confidence level for Andy Johnson Mfg. can be set lower, at 90%, while retaining the same ±5% accuracy levels. What sample size is needed now? [Answer: $n = 121$.]

RELATED PROBLEMS ▶ 10.26, 10.27, 10.28c, 10.29b, 10.30b, 10.44–10.46

EXCEL OM Data File **Ch10Ex3.xls** can be found online.

Now let's look at two variations of Example 3.

First, if h, the desired accuracy, is expressed as an absolute amount of error (say, ±1 minute of error is acceptable), then substitute e for $h\bar{x}$, and the appropriate formula is:

$$n = \left(\frac{zs}{e}\right)^2 \tag{10-5}$$

where e is the absolute time amount of acceptable error.

Second, for those cases when s, the standard deviation of the sample, is not provided (which is typically the case outside the classroom), it must be computed. The formula for doing so is given in Equation (10-6):

$$s = \sqrt{\frac{\sum (x_i - \bar{x})^2}{n-1}} = \sqrt{\frac{\sum (\text{Each sample observation} - \bar{x})^2}{\text{Number in sample} - 1}} \tag{10-6}$$

where x_i = value of each observation
$\bar{x}$ = mean of the observations
n = number of observations in the sample

An example of this computation is provided in Solved Problem 10.4.

With the development of handheld computers, job elements, time, performance rates, and statistical confidence intervals can be easily created, logged, edited, and managed. Although time studies provide accuracy in setting labor standards (see the *OM in Action* box, "UPS: The Tightest Ship in the Shipping Business"), they have two disadvantages. First, they require a trained staff of analysts. Second, these standards cannot be set before tasks are actually performed. This leads us to two alternative work-measurement techniques that we discuss next.

Predetermined Time Standards

In addition to historical experience and time studies, we can set production standards by using predetermined time standards. **Predetermined time standards** divide manual work into small basic elements that already have established times (based on very large samples of workers). To estimate the time for a particular task, the time factors for each basic element of that task are

Predetermined time standards
A division of manual work into small basic elements that have established and widely accepted times.

OM in Action UPS: The Tightest Ship in the Shipping Business[4]

Akerri/Shutterstock

United Parcel Service (UPS) employs 543,000 people and delivers more than 21 million packages a day to locations throughout the U.S. and 220 other countries. To achieve its claim of "running the tightest ship in the shipping business," UPS methodically trains its delivery drivers in how to do their jobs as efficiently as possible.

Industrial engineers at UPS have time-studied each driver's route and set standards for each delivery, stop, and pickup. These engineers have recorded every second taken up by stoplights, traffic volume, detours, doorbells, walkways, stairways, and coffee breaks. Even bathroom stops are factored into the standards. All this information is then fed into company computers to provide detailed time standards for every driver, every day.

To meet their objective of 200 deliveries and pickups each day (versus 100 at FedEx), UPS drivers must follow procedures exactly. As they approach a delivery stop, drivers unbuckle their seat belts, honk their horns, and cut their engines. Ignition keys have been dispensed with and replaced by a digital remote fob that turns off the engine and unlocks the bulkhead door that leads to the packages. In one seamless motion, drivers are required to yank up their emergency brakes and push their gearshifts into first. Then they slide to the ground with their electronic clipboards under their right arm and their packages in their left hand. They walk to the customer's door at the prescribed 3 feet per second and knock first to avoid lost seconds searching for the doorbell. After making the delivery, they do the paperwork on the way back to the truck.

Productivity experts describe UPS as one of the most efficient companies anywhere in applying effective labor standards.

added together. Developing a comprehensive system of predetermined time standards would be prohibitively expensive for any given firm. Consequently, a number of systems are commercially available. The most common predetermined time standard is *methods time measurement* (MTM), which is a product of the MTM Association.

Therbligs
Basic physical elements of motion.

Time measurement units (TMUs)
Units for very basic micromotions in which 1 TMU = .0006 min, or 100,000 TMUs = 1 hr.

Predetermined time standards are an outgrowth of basic motions called therbligs. The term *therblig* was coined by Frank Gilbreth (*Gilbreth* spelled backward, with the *t* and *h* reversed). **Therbligs** include such activities as select, grasp, position, assemble, reach, hold, rest, and inspect. These activities are stated in terms of **time measurement units (TMUs)**, which are equal to only .00001 hour, or .0006 minute each. MTM values for various therbligs are specified in very detailed tables. Figure 10.9, for example, provides the set of time standards for the motion GET and PLACE. To use GET and PLACE, one must know what is "gotten," its approximate weight, and where and how far it is supposed to be placed.

Figure **10.9**

Sample MTM Table for GET and PLACE Motion

Time values are in TMUs. MTM is a process language to describe and time labor processes and to set work standards. The original MTM systems are available at MTM Association Standards and Research.

Source: Copyrighted by the MTM Association e. V. (mtm.org). No reprint permission without written consent from the MTM Association e. V. Elbchaussee 352; 22609, Hamburg.

GET and PLACE			MOTION LENGTH in cm	≤ 20	> 20 to ≤ 50	> 50 to ≤ 80
WEIGHT	CONDITIONS OF GET	PLACE ACCURACY	MTM CODE	1	2	3
≤ 1 kg	EASY	APPROXIMATE	AA	20	35	50
		LOOSE	AB	30	45	60
		TIGHT	AC	40	55	70
	DIFFICULT	APPROXIMATE	AD	20	45	60
		LOOSE	AE	30	55	70
		TIGHT	AF	40	65	80
	HANDFUL	APPROXIMATE	AG	40	65	80
> 1 kg to ≤ 8 kg		APPROXIMATE	AH	25	45	55
		LOOSE	AJ	40	65	75
		TIGHT	AK	50	75	85
> 8 kg to ≤ 22 kg		APPROXIMATE	AL	90	105	115
		LOOSE	AM	95	120	130
		TIGHT	AN	120	145	160

Example 4 shows a use of predetermined time standards in setting service labor standards.

Example 4

USING PREDETERMINED TIME (MTM ANALYSIS) TO DETERMINE STANDARD TIME*

General Hospital wants to set the standard time for lab technicians to pour a tube specimen using MTM.

APPROACH ▶ This is a repetitive task for which the MTM data in Table 10.3 may be used to develop standard times. The sample tube is in a rack and the centrifuge tubes in a nearby box. A technician removes the sample tube from the rack, uncaps it, gets the centrifuge tube, pours, and places both tubes in the rack.

TABLE 10.3 MTM-UAS Analysis: Pouring Tube Specimen

ELEMENT DESCRIPTION	ELEMENT	TIME
Get tube from rack	AA2	35
Uncap, place on counter	AA2	35
Get centrifuge tube, place at sample tube	AD2	45
Pour (3 sec)	PT	83
Place tubes in rack (simo)	PC2	40
		Total TMU 238
.0006 × 238 = Total standard minutes = .143 or about 8.6 seconds		

*Permission for the example received from the MTM Association e. V. (mtm.org). No reprint permission without written consent from the MTM Association e. V., Elbchaussee 352, 22609, Hamburg, Germany.

SOLUTION ▶ The first work element involves getting the tube from the rack. The conditions for GETTING the tube and PLACING it in front of the technician are:

- *Weight:* (less than 1 kilogram)
- *Conditions of GET:* (easy)
- *Place accuracy:* (approximate)
- *Distance range:* (20 to 50 centimeters)

Then the MTM element for this activity is AA2 (as seen in Figure 10.9). The rest of Table 10.3 is developed from similar MTM tables.

INSIGHT ▶ Most MTM calculations are computerized, so the user need only key in the appropriate MTM codes, such as AA2 in this example.

LEARNING EXERCISE ▶ General Hospital decides that the first step in this process really involves a distance range of 10 centimeters (getting the tube from the rack). The other work elements are unchanged. What is the new standard time? [Answer: .134 minutes, or just over 8 seconds]

RELATED PROBLEM ▶ 10.36

Predetermined time standards have several advantages over direct time studies. First, they may be established in a laboratory environment, where the procedure will not upset actual production activities (which time studies tend to do). Second, because the standard can be set *before* a task is actually performed, it can be used for planning. Third, no performance ratings are necessary. Fourth, unions tend to accept this method as a fair means of setting standards. Finally, predetermined time standards are particularly effective in firms that do substantial numbers of studies of similar tasks. To ensure accurate labor standards, some firms use both time studies and predetermined time standards.

STUDENT TIP Families of predetermined time standards have been developed for many occupations.

Work Sampling

The fourth method of developing labor or production standards, work sampling, was developed in England by L. Tippet in the 1930s. **Work sampling** estimates the percentage of time that a worker spends on various tasks. Random observations are used to record the activity that a worker is performing. The results are primarily used to determine how employees allocate their time among various activities. Knowledge of this allocation may lead to staffing changes, reassignment of duties, estimates of activity cost, and the setting of delay allowances for labor standards. When work sampling is performed to establish delay allowances, it is sometimes called a *ratio delay study*.

Work sampling
An estimate, via sampling, of the percentage of time that a worker spends on various tasks.

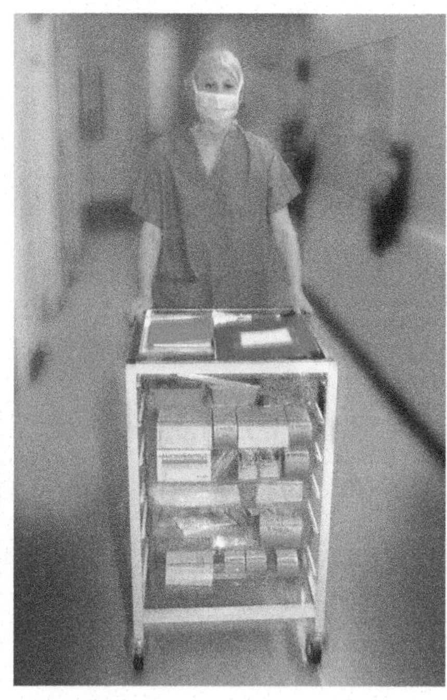
Samuel Ashfield/Science Source

Using the techniques of this chapter to develop labor standards, operations managers at Orlando's Arnold Palmer Hospital determined that nurses walked an average of 2.7 miles per day. This constitutes up to 30% of the nurse's time, a terrible waste of critical talent. Analysis resulted in a new layout design that has reduced walking distances by 20%.

The work-sampling procedure can be summarized in five steps:

1. Take a preliminary sample to obtain an estimate of the parameter value (e.g., percentage of time a worker is busy).
2. Compute the sample size required.
3. Prepare a schedule for observing the worker at appropriate times. The concept of random numbers is used to provide for random observation. For example, let's say we draw the following five random numbers from a table: 10, 24, 03, 32, and 23 (1st 5 numbers from the 7th column of Table F.4 in Module F). These can then be used to create an observation schedule of 9:03 A.M., 9:10, 9:23, 9:24, and 9:32.
4. Observe and record worker activities.
5. Determine how workers spend their time (usually as a percentage).

To determine the number of observations required, management must decide on the desired confidence level and accuracy. First, however, the analyst must select a preliminary value for the parameter under study (Step 1 above). The choice is usually based on a small sample of perhaps 50 observations. The following formula then gives the sample size for a desired confidence and accuracy:

$$n = \frac{z^2 p(1-p)}{h^2} \quad (10\text{-}7)$$

where n = required sample size
z = number of standard deviations for the desired confidence level ($z = 1$ for 68.27% confidence, $z = 2$ for 95.45% confidence, and $z = 3$ for 99.73% confidence—these values are obtained from Table 10.2 or the normal table in Appendix I)
p = estimated value of sample proportion (of time worker is observed busy or idle)
n = acceptable error level, in percent (as a decimal)

Example 5 shows how to apply this formula.

Example 5

DETERMINING THE NUMBER OF WORK SAMPLE OBSERVATIONS NEEDED

The manager of Michigan County's Social Security Office, Dana Johnson, estimates that her employees are idle 25% of the time. She would like to take a work sample that is accurate within ±3% and wants to have 95.45% confidence in the results.

APPROACH ▶ Dana applies Equation (10-7) to determine how many observations should be taken.

SOLUTION ▶ Dana computes n:

$$n = \frac{z^2 p(1-p)}{h^2}$$

where n = required sample size
z = confidence level (2 for 95.45% confidence)
p = estimate of idle proportion= 25%= .25
h = acceptable error of 3% = .03

She finds that

$$n = \frac{(2)^2(.25)(.75)}{(.03)^2} = 833 \text{ observations}$$

INSIGHT ▶ Thus, 833 observations should be taken. If the percentage of idle time observed is not close to 25% as the study progresses, then the number of observations may have to be recalculated and increased or decreased as appropriate.

LEARNING EXERCISE ▶ If the confidence level increases to 99.73%, how does the sample size change? [Answer: n = 1,875.]

RELATED PROBLEMS ▶ 10.31, 10.32, 10.35, 10.37

ACTIVE MODEL 10.1 This example is further illustrated in Active Model 10.1 found online.

The focus of work sampling is to determine how workers allocate their time among various activities. This is accomplished by establishing the percentage of time individuals spend on these activities rather than the exact amount of time spent on specific tasks. The analyst simply records in a random, nonbiased way the occurrence of each activity. Example 6 shows the procedure for evaluating employees at the state Social Security Office introduced in Example 5.

Example 6

DETERMINING EMPLOYEE TIME ALLOCATION WITH WORK SAMPLING

Dana Johnson, the manager of Michigan County's Social Security Office, wants to be sure her employees have adequate time to provide prompt, helpful service. She believes that service to clients who phone or walk in without an appointment deteriorates rapidly when employees are busy more than 75% of the time. Consequently, she does not want her employees to be occupied with client service activities more than 75% of the time.

APPROACH ▶ The study requires several things: First, based on the calculations in Example 5, 833 observations are needed. Second, observations are to be made in a random, nonbiased way over a period of 2 weeks to ensure a true sample. Third, the analyst must define the activities that are "work." In this case, work is defined as all the activities necessary to take care of the client (filing, meetings, data entry, discussions with the supervisor, etc.). Fourth, personal time is to be included in the 25% of nonwork time. Fifth, the observations are made in a nonintrusive way so as not to distort the normal work patterns. At the end of the 2 weeks, the 833 observations yield the following results:

NO. OF OBSERVATIONS	ACTIVITY
485	On the phone or meeting with a client
126	Idle
62	Personal time
23	Discussions with supervisor
137	Filing, meeting, and computer data entry
833	

SOLUTION ▶ The analyst concludes that all but 188 observations (126 idle and 62 personal) are work related. Because 22.6% (= 188/833) is less idle time than Dana believes necessary to ensure a high client service level, she needs to find a way to reduce current workloads. This could be done through a reassignment of duties or the hiring of additional personnel.

INSIGHT ▶ Work sampling is particularly helpful when determining staffing needs or the reallocation of duties (see Figure 10.10).

LEARNING EXERCISE ▶ The analyst working for Dana recategorizes several observations. There are now 450 "on the phone/meeting with client" observations, 156 "idle," and 67 "personal time" observations. The last two categories saw no changes. Do the conclusions change? [Answer: Yes; now about 27% of employee time is not work related—over the 25% Dana desires.]

RELATED PROBLEM ▶ 10.34

The results of similar studies of salespeople and assembly-line employees are shown in Figure 10.10.

Work sampling offers several advantages over time-study methods. First, because a single observer can observe several workers simultaneously, it is less expensive. Second, observers usually do not require much training, and no timing devices are needed. Third, the study can be temporarily delayed at any time with little impact on the results. Fourth, because work sampling uses instantaneous observations over a long period, the worker has little chance of affecting the study's outcome. Fifth, the procedure is less intrusive and therefore less likely to generate objections.

The disadvantages of work sampling are (1) it does not divide work elements as completely as time studies, (2) it can yield biased or incorrect results if the observer does not follow random routes of travel and observation, and (3) because it is less intrusive, it tends to be less accurate; this is particularly true when job content times are short.

Digital Monitoring Techniques

With more work moving from manual tasks to the office, new tools for evaluating "desk job" productivity have been developed. The transition is particularly evident with computer-oriented

Figure **10.10**

Work-Sampling Time Studies

These two work-sampling time studies were done to determine what salespeople do at a wholesale electronics distributor (left) and a composite of several auto assembly-line employees (right).

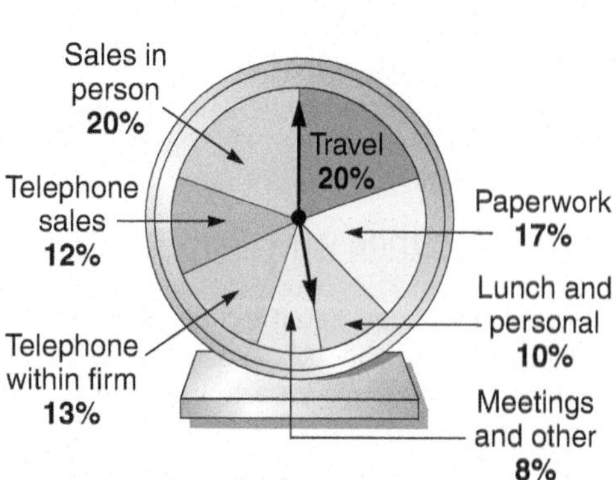

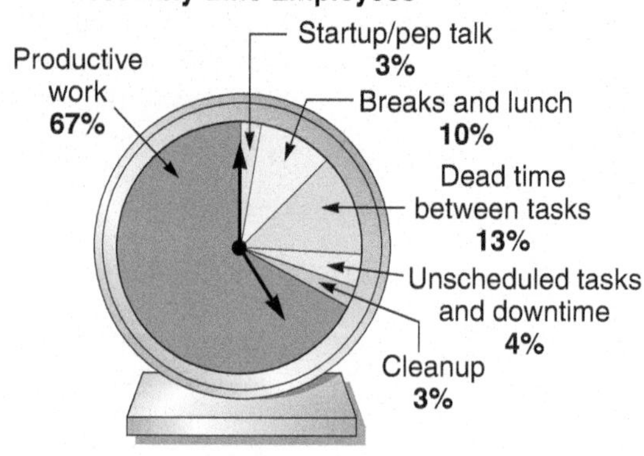

tasks. These stationary tasks that are part of the new digital world require new approaches for establishing performance standards, monitoring employees, and evaluating worker performance. For instance, computer-oriented jobs are now tracked by software that recognizes every little thing that happens on an employee's computer and for how long. Such software identifies and tracks company-oriented activity vs. surfing the Internet or watching YouTube or ESPN.

Similarly, for many physical tasks, from the factory floor to warehouse employees, labor standards are now built into software that tracks, monitors, and rates individual performance. Additionally, the digital world has hospitals installing sensors to detect everything from the location of nurses to handwashing practices. And restaurant software continues to be refined to track table occupancy and number of customers served, time of service, waitstaff upselling (for instance, sales of wine and desserts), and total sales—all in real time. For other jobs, employee location, if not precise activities, both within and outside the facility can be tracked.

STUDENT TIP Mutual trust and commitment cannot be achieved without ethical behavior.

Ethics

Ethics in the workplace presents some interesting challenges. As we have suggested in this chapter, many constraints influence job design. The issues of fairness, equity, and ethics are pervasive. Whether the issue is equal opportunity or safe working conditions, an operations manager is often the one responsible. Managers do have some guidelines. By knowing the law, working with OSHA, state agencies, unions, trade associations, insurers, and employees, managers can often determine the parameters of their decisions. Human resource and legal departments are also available for help and guidance through the labyrinth of laws and regulations.

Management's role is to educate employees; specify the necessary equipment, work rules, and work environment; and then enforce those requirements, even when employees think it is not necessary to wear safety equipment. We began this chapter with a discussion of mutual trust and commitment, and that is the environment that managers should foster. Ethical management requires no less.

Summary

Outstanding firms know that their human resource strategy can yield a competitive advantage. Often a large percentage of employees and a large part of labor costs are under the direction of OM. Consequently, an operations manager usually has a major role to play in achieving human resource objectives. A requirement is to build an environment with mutual respect and commitment and a reasonable quality of work life. Successful organizations have designed jobs that use both the mental and physical capabilities of their employees. Regardless of the strategy chosen, the skill with which a firm manages its human resources ultimately determines its success.

Labor standards are required for an efficient operations system. They are needed for production planning, labor planning, costing, and evaluating performance. They are used throughout industry—from the factory to finance, sales, and office. They can also be used as a basis for incentive systems. Standards may be established via historical data, time studies, predetermined time standards, and work sampling.

Key Terms

Labor planning (p. 407)
Job design (p. 408)
Labor specialization (or job specialization) (p. 408)
Job enlargement (p. 409)
Job rotation (p. 409)
Job enrichment (p. 409)
Employee empowerment (p. 409)
Self-directed team (p. 410)
Ergonomics (p. 412)
Methods analysis (p. 414)
Flow diagram (p. 414)
Process chart (p. 414)
Activity chart (p. 415)
Operations chart (p. 415)
Visual workplace (p. 415)
Labor standards (p. 416)
Time study (p. 417)
Average observed time (p. 417)
Normal time (p. 418)
Standard time (p. 418)
Predetermined time standards (p. 421)
Therbligs (p. 422)
Time measurement units (TMUs) (p. 422)
Work sampling (p. 423)

Ethical Dilemma

Johnstown Foundry, Inc., with several major plants, is one of the largest makers of cast-iron water and sewer pipes in the U.S. In one of the nation's most dangerous industries, Johnstown is perhaps one of the most unsafe, with 4 times the injury rate of its 6 competitors combined. Its worker death rate is 6 times the industry average. In a recent 7-year period, Johnstown's plants were also found to be in violation of pollution and emission limits 450 times.

Workers who protest dangerous work conditions claim they are "bull's-eyed"—marked for termination. Supervisors have bullied injured workers and intimidated union leaders. Line workers who fail to make daily quotas get disciplinary actions. Managers have put up safety signs *after* a worker was injured to make it appear that the worker ignored posted policies. They doctor safety records and alter machines to cover up hazards. When the government investigated one worker's death recently, inspectors found the Johnstown policy "was not to correct anything until OSHA found it."

Johnstown plants have also been repeatedly fined for failing to stop production to repair broken pollution controls. Three plants have been designated "high-priority" violators by the EPA. Inside the plants, workers have repeatedly complained of blurred vision, severe headaches, and respiratory problems after being exposed, without training or protection, to chemicals used in the production process. Near one Pennsylvania plant, school crossing guards have had to wear gas masks; that location alone has averaged over a violation every month for 7 years. Johnstown's "standard procedure," according to a former plant manager, is to illegally dump industrial contaminants into local rivers and creeks. Workers wait for night or heavy rainstorms before flushing thousands of gallons from their sump pumps.

Given the following scenarios, what is your position, and what action should you take?

a) On your spouse's recent move to the area, you accepted a job, perhaps somewhat naively, as a company nurse in one of the Johnstown plants. After 2 weeks on the job, you became aware of the unsafe work environment.
b) You are a contractor who has traditionally used Johnstown's products, which meet specifications. Johnstown is consistently the low bidder. Your customers are happy with the product.
c) You are Johnstown's banker.
d) You are a supplier to Johnstown.

Mark Wintrey/Shutterstock

Discussion Questions

1. How would you define a good quality of work life?
2. What are some of the worst jobs you know about? Why are they bad jobs? Why do people want these jobs?
3. If you were redesigning the jobs described in Question 2, what changes would you make? Are your changes realistic? Would they improve productivity (not just *production* but *productivity*)?
4. Can you think of any jobs that push the human–machine interface to the limits of human capabilities?
5. What are the five core characteristics of a good job design?
6. What are the differences among job enrichment, job enlargement, job rotation, job specialization, and employee empowerment?
7. Define ergonomics. Discuss the role of ergonomics in job design.
8. List the techniques available for carrying out methods analysis.
9. Identify four ways in which labor standards are set.
10. What are some of the uses to which labor standards are put?
11. How would you classify the following job elements? Are they personal, fatigue, or delay?
    a) The operator stops to talk to you.
    b) The operator lights up a cigarette.
    c) The operator opens her lunch pail (it is not lunchtime), removes an apple, and takes an occasional bite.
12. How do you classify the time for a drill press operator who is idle for a few minutes at the beginning of every job waiting for the setup person to complete the setup? Some of the setup time is used in going for stock, but the operator typically returns with stock before the setup person is finished with the setup.
13. How do you classify the time for a machine operator who, between every job and sometimes in the middle of jobs, turns off the machine and goes for stock?
14. The operator drops a part, which you pick up and hand back. Does this make any difference in a time study? If so, how?

Solved Problems

SOLVED PROBLEM 10.1

As pit crew manager for the NASCAR team (see the *Global Company Profile* that opens this chapter), you would like to evaluate how your "Jackman" (JM) and "Gas Man #1" (GM #1) are utilized. Recent stopwatch studies have verified the following times:

PIT CREW	ACTIVITY	TIME (SECONDS)
JM	Move to right side of car and raise car	4.0
GM #1	Move to rear gas filler	2.5
JM	Wait for tire	1.0
JM	Move to left side of car and raise car	3.8
GM #1	Load fuel (per gallon)	0.5
JM	Wait for tire	1.2
JM	Move back over wall from left side	2.5
GM #1	Move back over the wall from gas filler	2.5

Use an activity chart similar to the one in Figure 10.6 as an aid.

SOLUTION

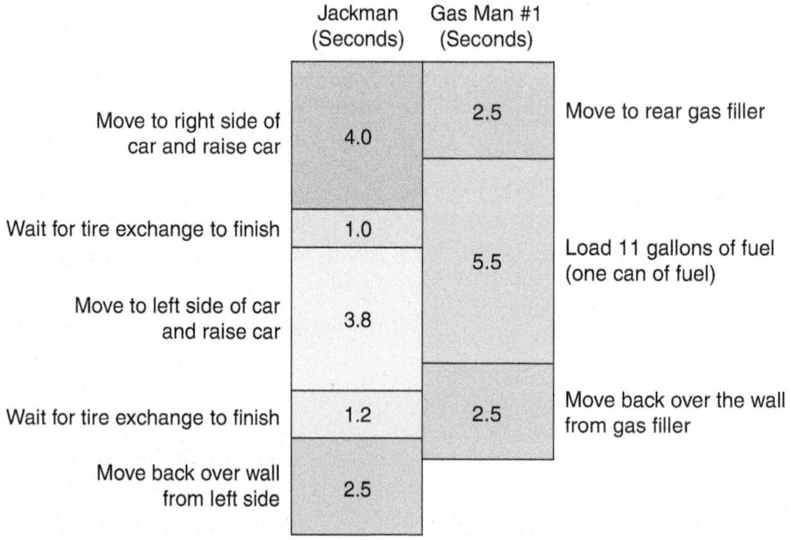

Jackman activity	Jackman (Seconds)	Gas Man #1 (Seconds)	Gas Man #1 activity
Move to right side of car and raise car	4.0	2.5	Move to rear gas filler
Wait for tire exchange to finish	1.0	5.5	Load 11 gallons of fuel (one can of fuel)
Move to left side of car and raise car	3.8		
Wait for tire exchange to finish	1.2	2.5	Move back over the wall from gas filler
Move back over wall from left side	2.5		

SOLVED PROBLEM 10.2

A work operation consisting of three elements has been subjected to a stopwatch time study. The recorded observations are shown in the following table. By union contract, the allowance time for the operation is personal time 5%, delay 5%, and fatigue 10%. Determine the standard time for the work operation.

JOB ELEMENT	OBSERVATIONS (MINUTES) 1	2	3	4	5	6	PERFORMANCE RATING (%)
A	.1	.3	.2	.9	.2	.1	90
B	.8	.6	.8	.5	3.2	.7	110
C	.5	.5	.4	.5	.6	.5	80

SOLUTION

First, delete the two observations that appear to be very unusual (.9 minute for job element A and 3.2 minutes for job element B). Then:

$$\text{A's average observed time} = \frac{.1+.3+.2+.2+.1}{5} = 0.18 \text{ min}$$

$$\text{B's average observed time} = \frac{.8+.6+.8+.5+.7}{5} = 0.68 \text{ min}$$

$$\text{C's average observed time} = \frac{.5+.5+.4+.5+.6+.5}{6} = 0.50 \text{ min}$$

$$\text{A's normal time} = (0.18)(0.90) = 0.16 \text{ min}$$

$$\text{B's normal time} = (0.68)(1.10) = 0.75 \text{ min}$$

$$\text{C's normal time} = (0.50)(0.80) = 0.40 \text{ min}$$

$$\text{Normal time for job} = 0.16 + 0.75 + 0.40 = 1.31 \text{ min}$$

Note, the total allowance factor = 0.05 + 0.05 + 0.10 = 0.20

$$\text{Then: Standard time} = \frac{1.31}{1 - 0.20} = 1.64 \text{ min}$$

SOLVED PROBLEM 10.3

The preliminary work sample of an operation indicates the following:

Number of times operator working	60
Number of times operator idle	40
Total number of preliminary observations	100

What is the required sample size for a 99.73% confidence level with ±4% precision?

SOLUTION

$z = 3$ for 99.73 confidence; $p = \frac{60}{100} = 0.6$; $h = 0.04$

So:

$$n = \frac{z^2 p(1-p)}{h^2} = \frac{(3)^2(0.6)(0.4)}{(0.04)^2} = 1{,}350 \text{ sample size}$$

SOLVED PROBLEM 10.4

Amor Manufacturing Co. of Geneva, Switzerland, has just observed a job in its laboratory in anticipation of releasing the job to the factory for production. The firm wants rather good accuracy for costing and labor forecasting. Specifically, it wants to provide a 99% confidence level and a cycle time that is within 3% of the true value. How many observations should it make? The data collected so far are as follows:

OBSERVATION	TIME
1	1.7
2	1.6
3	1.4
4	1.4
5	1.4

SOLUTION

First, solve for the mean, $\bar{x}$, and the sample standard deviation, s:

$$s = \sqrt{\frac{\sum(\text{Each sample observation} - \bar{x})^2}{\text{Number in sample} - 1}}$$

OBSERVATION	x_i	$\bar{x}$	$x_i - \bar{x}$	$(x_i - \bar{x})^2$
1	1.7	1.5	.2	0.04
2	1.6	1.5	.1	0.01
3	1.4	1.5	−.1	0.01
4	1.4	1.5	−.1	0.01
5	1.4	1.5	−.1	0.01
	$\bar{x} = 1.5$			$0.08 = \Sigma(x_i - \bar{x})^2$

$$s = \sqrt{\frac{0.08}{n-1}} = \sqrt{\frac{0.08}{4}} = 0.141$$

$$\text{Then, solve for } n = \left(\frac{zs}{h\bar{x}}\right)^2 = \left[\frac{(2.58)(0.141)}{(0.03)(1.5)}\right]^2 = 65.3$$

where $\bar{x} = 1.5$
$s = 0.141$
$z = 2.58$ (from Table 10.2)
$h = 0.03$

Finally, round up to 66 observations.

SOLVED PROBLEM 10.5

At Maggard Micro Manufacturing, Inc., workers press semiconductors into predrilled slots on printed circuit boards. The elemental motions for normal time used by the company are as follows:

Reach 6 inches for semiconductors	40 TMU
Grasp the semiconductor	10 TMU
Move semiconductor to printed circuit board	30 TMU
Position semiconductor	35 TMU
Press semiconductor into slots	65 TMU
Move board aside	20 TMU

(Each time measurement unit is equal to .0006 min.) Determine the normal time for this operation in minutes and in seconds.

SOLUTION

Add the time measurement units:

$$40 + 10 + 30 + 35 + 65 + 20 = 200$$
$$\text{Time in minutes} = (200)(.0006 \text{ min.}) = 0.12 \text{ min}$$
$$\text{Time in seconds} = (0.12)(60 \text{ sec}) = 7.2 \text{ sec}$$

SOLVED PROBLEM 10.6

To obtain the estimate of time a worker is busy for a work sampling study, a manager divides a typical workday into 480 minutes. Using a random-number table to decide what time to go to an area to sample work occurrences, the manager records observations on a tally sheet like the following:

STATUS	TALLY
Productively working	~~IIII~~ ~~IIII~~ ~~IIII~~ I
Idle	IIII

SOLUTION

In this case, the supervisor made 20 observations and found that employees were working 80% of the time. So, out of 480 minutes in an office workday, 20%, or 96 minutes, was idle time, and 384 minutes were productive. Note that this procedure describes that a worker is busy, not necessarily doing what the worker *should* be doing.

Problems

Note: **PX** means the problem may be solved with POM for Windows and/or Excel.

Problem 10.1 relates to Job Design

• **10.1** Rate a job (or volunteer or club experience) you have had using Hackman and Oldham's core job characteristics on a scale from 1 to 10. What is your total score? What about the job could have been changed to make you give it a higher score?

Problems 10.2–10.12 relate to Methods Analysis

• **10.2** Make a process chart for changing the right rear tire on an automobile.

• **10.3** Draw an activity chart for a machine operator with the following operation. The relevant times are as follows:

Prepare mill for loading (cleaning, oiling, and so on)	.50 min
Load mill	1.75 min
Mill operating (cutting material)	2.25 min
Unload mill	.75 min

••• **10.4** Draw an activity chart (a crew chart similar to Figure 10.6) for a concert (for example, Tim McGraw, Taylor Swift, Beyoncé, or Bruce Springsteen) and determine how to put together the concert so the star has reasonable breaks. For instance, at what point is there an instrumental number, a visual effect, a duet, a dance moment, that allows the star to pause and rest physically or at least rest their voice? Do other members of the show have moments of pause or rest?

Hurricanehank/Shutterstock

•• **10.5** Make an operations chart for putting paper into a printer.

• **10.6** Develop a process chart for installing a new memory board in your personal computer.

•• **10.7** Using the data in Solved Problem 10.1, prepare an activity chart like the one in the Solved Problem, but a second Gas Man also delivers 11 gallons.

•• **10.8** Prepare a process chart for the Jackman in Solved Problem 10.1.

•• **10.9** Draw an activity chart for changing the right rear tire on an automobile with:
a) Only one person working
b) Two people working

••• **10.10** Draw an activity chart for washing the dishes in a double-sided sink. Two people participate, one washing, the other rinsing and drying. The rinser dries a batch of dishes from the drip rack as the washer fills the right sink with clean but unrinsed dishes. Then the rinser rinses the clean batch and places them on the drip rack. All dishes are stacked before being placed in the cabinets.

••• **10.11** Your campus club is hosting a car wash. Due to demand, three people are going to be scheduled per wash line. (Three people have to wash each vehicle.) Design an activity chart for washing and drying a typical sedan. You must wash the wheels but ignore the cleaning of the interior because this part of the operation will be done at a separate vacuum station.

•••• **10.12** Design a process chart for printing a short document on a laser printer at an office. Unknown to you, the printer in the hallway is out of paper. The paper is located in a supply room at the other end of the hall. You wish to make five stapled copies of the document once it is printed. The copier, located next to the printer, has a sorter but no stapler. How could you make the task more efficient with the existing equipment?

Problems 10.13–10.46 relate to Labor Standards

• **10.13** If Charlene Brewster has times of 8.4, 8.6, 8.3, 8.5, 8.7, and 8.5 and a performance rating of 110%, what is the normal time for this operation? Is she faster or slower than normal? **PX**

• **10.14** If Charlene, the worker in Problem 10.13, has a performance rating of 90%, what is the normal time for the operation? Is she faster or slower than normal? **PX**

•• **10.15** Refer to Problem 10.13.
a) If the allowance factor is 15%, what is the standard time for this operation?
b) If the allowance factor is 18% and the performance rating is now 90%, what is the standard time for this operation? **PX**

•• **10.16** Claudine Soosay recorded the following times assembling a watch. Determine (a) the average time, (b) the normal time, and (c) the standard time taken by her, using a performance rating of 95% and a personal allowance of 8%.

Assembly Times Recorded

OBSERVATION NO.	TIME (MINUTES)	OBSERVATION NO.	TIME (MINUTES)
1	0.11	9	0.12
2	0.10	10	0.09
3	0.11	11	0.12
4	0.10	12	0.11
5	0.14	13	0.10
6	0.10	14	0.12
7	0.10	15	0.14
8	0.09	16	0.09

• **10.17** A Northeast Airlines gate agent, Betnijah Gray, gives out seat assignments to ticketed passengers. She takes an average of 50 seconds per passenger and is rated 110% in performance. How long should a *typical* agent be expected to take to make seat assignments? **PX**

• **10.18** After being observed many times, Beverly Demarr, a hospital lab analyst, had an average observed time for blood tests of 12 minutes. Beverly's performance rating is 105%. The hospital has a personal, fatigue, and delay allowance of 16%.
a) Find the normal time for this process.
b) Find the standard time for this blood test. **PX**

• **10.19** Jell Lee Beans is famous for its boxed candies, which are sold primarily to businesses. One operator had the following observed times for gift wrapping in minutes: 2.2, 2.6, 2.3, 2.5, 2.4. The operator has a performance rating of 105% and an allowance factor of 10%. What is the standard time for gift wrapping? **PX**

• **10.20** After training, Mary Fernandez, a computer technician, had an average observed time for memory-chip tests of 12 seconds. Mary's performance rating is 100%. The firm has a personal fatigue and delay allowance of 15%.
a) Find the normal time for this process.
b) Find the standard time for this process. **PX**

•• **10.21** Susan Cottenden clocked the observed time for welding a part onto truck doors at 5.3 minutes. The performance rating of the worker timed was estimated at 105%. Find the normal time for this operation.

Note: According to the local union contract, each welder is allowed 3 minutes of personal time per hour and 2 minutes of fatigue time per hour. Further, there should be an average delay allowance of 1 minute per hour. Compute the allowance factor and then find the standard time for the welding activity. **PX**

•• **10.22** A hotel housekeeper, Alison Harvey, was observed five times on each of four task elements, as shown in the following table. On the basis of these observations, find the standard time for the process. Assume a 10% allowance factor.

		OBSERVATIONS (MINUTES PER CYCLE)				
ELEMENT	PERFORMANCE RATING (%)	1	2	3	4	5
Check minibar	100	1.5	1.6	1.4	1.5	1.5
Make one bed	90	2.3	2.5	2.1	2.2	2.4
Vacuum floor	120	1.7	1.9	1.9	1.4	1.6
Clean bath	100	3.5	3.6	3.6	3.6	3.2

•• **10.23** Virginia College promotes a wide variety of executive-training courses for firms in the Arlington, Virginia, region. Director Wendy Tate believes that individually written letters add a personal touch to marketing. To prepare letters for mailing, she conducts a time study of her secretaries. On the basis of the observations shown in the following table, she wishes to develop a time standard for the whole job.

The college uses a total allowance factor of 12%. Tate decides to delete all unusual observations from the time study. What is the standard time?

	OBSERVATIONS (MINUTES)						
ELEMENT	1	2	3	4	5	6	PERFORMANCE RATING (%)
Customizing letter	2.5	3.5	2.8	2.1	2.6	3.3	85
Labeling envelope	.8	.8	.6	.8	3.1[a]	.7	100
Stuffing envelope	.4	.5	1.9[a]	.3	.6	.5	95
Sealing, sorting	1.0	2.9[b]	.9	1.0	4.4[b]	.9	125

[a]Disregard—secretary stopped to answer the phone.
[b]Disregard—interruption by supervisor. **PX**

• **10.24** The results of a time study to perform a quality control test are shown in the following table. On the basis of these observations, determine the normal and standard time for the test, assuming a 23% allowance factor. **PX**

		OBSERVATIONS (MINUTES)				
TASK ELEMENT	PERFORMANCE RATING (%)	1	2	3	4	5
1	97	1.5	1.8	2.0	1.7	1.5
2	105	.6	.4	.7	3.7[a]	.5
3	86	.5	.4	.6	.4	.4
4	90	.6	.8	.7	.6	.7

[a]Disregard—employee is smoking a cigarette (included in personal time).

•• **10.25** Latisha Simmons, a loan processor at Wentworth Bank, has been timed performing four work elements, with the results shown in the following table. The allowances for tasks such as this are personal, 7%; fatigue, 10%; and delay, 3%.

		OBSERVATIONS (MINUTES)				
TASK ELEMENT	PERFORMANCE RATING (%)	1	2	3	4	5
1	110	.5	.4	.6	.4	.4
2	95	.6	.8	.7	.6	.7
3	90	.6	.4	.7	.5	.5
4	85	1.5	1.8	2.0	1.7	1.5

a) What is the normal time?
b) What is the standard time? **PX**

•• **10.26** Each year, Nordstrom sets up a gift-wrapping station to assist its customers with holiday shopping. Preliminary observations of one worker at the station produced the following sample time (in minutes per package): 3.5, 3.2, 4.1, 3.6, 3.9. Based on this small sample, what number of observations would be necessary to determine the true cycle time with a 95% confidence level and an accuracy of ±5%? **PX**

•• **10.27** A time study of a factory worker has revealed an average observed time of 3.20 minutes, with a standard deviation of 1.28 minutes. These figures were based on a sample of 45 observations. Is this sample adequate in size for the firm to be 99% confident that the standard time is within ±5% of the true value? If not, what should be the proper number of observations? **PX**

•• **10.28** Based on a careful work study in the Hofstetter Corp., the results shown in the following table have been observed:

	OBSERVATIONS (MINUTES)					
ELEMENT	1	2	3	4	5	PERFORMANCE RATING (%)
Prepare daily reports	35	40	33	42	39	120
Photocopy results	12	10	36[a]	15	13	110
Label and package reports	3	3	5	5	4	90
Distribute reports	15	18	21	17	45[b]	85

[a]Photocopying machine broken; included as delay in the allowance factor.
[b]Power outage; included as delay in the allowance factor.

a) Compute the normal time for each work element.
b) If the allowance for this type of work is 15%, what is the standard time?
c) How many observations are needed for a 95% confidence level within ±5% accuracy? (*Hint:* Calculate the sample size of each element.)

•• **10.29** The Dubuque Cement Company packs 80-pound bags of concrete mix. Time-study data for the filling activity are shown in the following table. Because of the high physical demands of the job, the company's policy is a 23% allowance for workers.
a) Compute the standard time for the bag-packing task.
b) How many observations are necessary for 99% confidence, within ±5% accuracy?

	OBSERVATIONS (SECONDS)					
ELEMENT	1	2	3	4	5	PERFORMANCE RATING (%)
Grasp and place bag	8	9	8	11	7	110
Fill bag	36	41	39	35	112[a]	85
Seal bag	15	17	13	20	18	105
Place bag on conveyor	8	6	9	30[b]	35[b]	90

[a]Bag breaks open; included as delay in the allowance factor.
[b]Conveyor jams; included as delay in the allowance factor.

•• **10.30** Installing mufflers at the O'Sullivan Garage in Golden, Colorado, involves five work elements. Jill O'Sullivan has timed workers performing these tasks seven times, with the results shown in the following table:

JOB ELEMENT	OBSERVATIONS (MINUTES) 1	2	3	4	5	6	7	PERFORMANCE RATING (%)
Select correct mufflers	4	5	4	6	4	15[a]	4	110
Remove old muffler	6	8	7	6	7	6	7	90
Weld/install new muffler	15	14	14	12	15	16	13	105
Check/inspect work	3	4	24[a]	5	4	3	18[a]	100
Complete paperwork	5	6	8	—	7	6	7	130

[a]Employee has lengthy conversations with boss (not job related).

By agreement with her workers, Jill allows a 10% fatigue factor and a 10% personal-time factor, but no time for delay. To compute standard time for the work operation, Jill excludes all observations that appear to be unusual or nonrecurring. She does not want an error of more than ±5%.

a) What is the standard time for the task?
b) How many observations are needed to assure a 95% confidence level? **PX**

•• **10.31** Bank manager Art Hill wants to determine the percentage of time that tellers are working and idle. He decides to use work sampling, and his initial estimate is that the tellers are idle 15% of the time. How many observations should Hill take to be 95.45% confident that the results will not be more than ±4% from the true result? **PX**

•• **10.32** Supervisor Kenneth Peterson wants to determine the percentage of time a machine in his area is idle. He decides to use work sampling, and his initial estimate is that the machine is idle 20% of the time. How many observations should Peterson take to be 98% confident that the results will be less than 5% from the true results?

••• **10.33** Tim Nelson's job as an inspector for La-Z-Boy is to inspect 130 chairs per day.

a) If he works an 8-hour day, how many minutes is he allowed for each inspection (i.e., what is his "standard time")?
b) If he is allowed a 6% fatigue allowance, a 6% delay allowance, and 6% for personal time, what is the normal time that he is assumed to take to perform each inspection?

••• **10.34** A random work sample of operators taken over a 160-hour work month at Tele-Marketing, Inc., has produced the following results. What is the percentage of time spent working?

On phone with customer	858
Idle time	220
Personal time	85

•• **10.35** A total of 300 observations of Bob Ramos, an assembly-line worker, were made over a 40-hour workweek. The sample also showed that Bob was busy working (assembling the parts) during 250 observations.

a) Find the percentage of time Bob was working.
b) If you want a confidence level of 95%, and if ±3% is an acceptable error, what size should the sample be?
c) Was the sample size adequate? **PX**

• **10.36** Sharpening your pencil is an operation that may be divided into eight small elemental motions. In MTM terms, each element may be assigned a certain number of TMUs:

Reach 10 centimeters for the pencil	6 TMU
Grasp the pencil	2 TMU
Move the pencil 15 centimeters	10 TMU
Position the pencil	20 TMU
Insert the pencil into the sharpener	4 TMU
Sharpen the pencil	120 TMU
Disengage the pencil	10 TMU
Move the pencil 15 centimeters	10 TMU

What is the total normal time for sharpening one pencil? Convert your answer into minutes and seconds.

•• **10.37** Supervisor Tom Choi at Tempe Equipment Company is concerned that material is not arriving as promptly as needed at work cells. A new kanban system has been installed, but there seems to be some delay in getting the material moved to the work cells so that the job can begin promptly. Choi is interested in determining how much delay there is on the part of his highly paid machinists. Ideally, the delay would be close to zero. He has asked his assistant to determine the delay factor among his 10 work cells. The assistant collects the data on a random basis over the next 2 weeks and determines that of the 1,200 observations, 105 were made while the operators were waiting for materials. Use a 95% confidence level and a ±3% acceptable error. What report does he give to Choi? **PX**

•••• **10.38** The Miami Central Hotel has 400 rooms. Every day, the housekeepers clean any room that was occupied the night before. If a guest is checking out of the hotel, the housekeepers give the room a thorough cleaning to get it ready for the next guest. This takes 30 minutes. If a guest is staying another night, the housekeeper only "refreshes" the room, which takes 15 minutes. Each day, each housekeeper reports for a 6-hour shift and then prepares a cart. The housekeeper pushes the cart to the assigned floor and begins work. Usually having to restock the cart once per day, the housekeeper pushes it back to the storeroom at the end of the day and delivers dirty laundry, etc. Here is a timetable:

1. Arrive at work and stock cart (0.10 hr.)
2. Push cart to floor (0.10 hr.)
3. Take morning break (0.33 hr.)
4. Stop for lunch (0.50 hr.)
5. Restock cart (0.30 hr.)
6. Take afternoon break (0.33 hr.)
7. Push cart back to laundry and store items (0.33 hr.)

Last night, the hotel was full (all 400 rooms were occupied). People are checking out of 200 rooms. Their rooms will need to be thoroughly cleaned. The other 200 rooms will need to be refreshed.

a) What is the number of minutes per day that each housekeeper can perform actual room cleaning?
b) Based on the given occupancy and the guests checking out, how many total minutes of room cleaning are required by the Miami Central Hotel today?
c) What is the number of housekeepers that are needed today to perform the room cleaning?
d) What is the number of housekeepers that would be needed by the hotel for performing a thorough cleaning of all the rooms (assuming that all the guests check out this morning)?

• **10.39** An assembly-line worker at Joseph Milner's Fabrication Shop inserts Part A into Part B. A time study was conducted and produced the following results in seconds: 8.4, 8.6, 8.3, 8.5, 8.7, 8.5. What is the average cycle time? **PX**

• **10.40** The cycle time for performing a certain task has been clocked at 10 minutes. The performance rating of Richard Muszynski III was estimated at 110%. Common practice in this department is to allow 5 minutes of personal time and 3 minutes of fatigue time per hour. In addition, there should be an extra allowance of 2 minutes per hour for inspection.

a) Find the normal time for the operation.
b) Compute the allowance factor and the standard time. **PX**

•• **10.41** The data in the following table represent time-study observations for an assembly process at Nagesh Murthy's Toy Factory. On the basis of these observations, find the normal time for each element and the standard time for the process. Assume a 10% allowance factor. PX

ELEMENT	PERFORMANCE RATING (%)	OBSERVATIONS (MINUTES)				
		1	2	3	4	5
1	100	1.5	1.6	1.4	0.1*	1.5
2	90	2.3	2.5	2.1	2.2	2.4
3	120	1.7	1.9	1.9	1.4	1.6
4	100	3.5	3.6	3.6	3.6	3.2

*Disregard—possible error.

•• **10.42** Arunachalam Narayanan has been clocked performing three work elements in the office, with the results shown in the following table. The allowance for tasks such as this is 15%.

ELEMENT	OBSERVATIONS (SECONDS)						PERFORMANCE RATING (%)
	1	2	3	4	5	6	
1	13	11	14	16	51*	15	100
2	3*	21	25	73*	26	23	110
3	3	3.3	3.1	2.9	3.4	2.8	100

*Disregard—possible error.

a) Find the normal time.
b) Find the standard time. PX

••• **10.43** The data in the following table represent time-study observations for a metalworking process. On the basis of these observations, find the normal time for each element and the standard time for the process, assuming a 25% allowance factor.

ELEMENT	PERFORMANCE RATING (%)	OBSERVATIONS (MINUTES)						
		1	2	3	4	5	6	7
1	90	1.80	1.70	1.66	1.91	1.85	1.77	1.60
2	100	6.90	7.30	6.80	7.10	15.30*	7.00	6.40
3	115	3.00	9.00*	9.50*	3.80	2.90	3.10	3.20
4	90	10.10	11.10	12.30	9.90	12.00	11.90	12.00

*Disregard—employee is smoking a cigarette (included in personal time). PX

•• **10.44** A time study has revealed an average observed time of 5 minutes, with a standard deviation of 1.25 minutes. These figures are based on a sample of 75 cycles. Is this sample large enough for analyst Anand Paul to be 99.73% confident that the standard time is within 5% of the true value?

•• **10.45** The following data represent observations for a task in an assembly process. How many observations would be necessary for the observer to be 99% confident that the average task time is within 5% of the true value? PX

OBSERVATIONS (MINUTES)				
1	2	3	4	5
1.5	1.6	1.4	1.5	1.5

•• **10.46** Martha's Sweets is famous for its boxed candies, which are sold primarily to businesses. About 30% of the candies are gift wrapped. One operator had the following observed times for gift wrapping in minutes: 2.2, 2.6, 2.3, 2.5, and 2.4. What is the correct sample size if your accuracy is to be within 10% and your confidence level is 95.45%? If you desire an accuracy of 5%, what would the sample size need to be? PX

CASE STUDIES

Jackson Manufacturing Company

Kathleen McFadden, vice president of operations at Jackson Manufacturing Company, has just received a request for quote (RFQ) from DeKalb Electric Supply for 400 units per week of a motor armature. The components are standard and either easy to work into the existing production schedule or readily available from established suppliers on a just-in-time (JIT) basis. But there is some difference in assembly. Ms. McFadden has identified eight tasks that Jackson must perform to assemble the armature. Seven of these tasks are very similar to ones performed by Jackson in the past; therefore, the average time and resulting labor standard of those tasks is known.

The eighth task, an *overload* test, requires performing a task that is very different from any performed previously, however. Kathleen has asked you to conduct a time study on the task to determine the standard time. Then an estimate can be made of the cost to assemble the armature. This information, combined with other cost data, will allow the firm to put together the information needed for the RFQ.

To determine a standard time for the task, an employee from an existing assembly station was trained in the new assembly process. Once proficient, the employee was then asked to perform the task 17 times so a standard could be determined. The actual times observed (in minutes) were as follows:

1	2	3	4	5	6	7	8	9	10	11	12	13	14	15	16	17
2.05	1.92	2.01	1.89	1.77	1.80	1.86	1.83	1.93	1.96	1.95	2.05	1.79	1.82	1.85	1.85	1.99

The worker had a 115% performance rating. The task can be performed in a sitting position at a well-designed ergonomic work-station in an air-conditioned facility. Although the armature itself weighs 10.5 pounds, there is a carrier that holds it so that the operator need only rotate the armature. But the detail work remains high; therefore, the fatigue allowance should be 8%. The company has an established personal allowance of 6%. Delay should be very low. Previous studies of delay in this department average 2%. This standard is to use the same figure.

The workday is 7.5 hours, but operators are paid for 8 hours at an average of $12.50 per hour.

Discussion Questions

In your report to Ms. McFadden, you realize you will want to address several factors:

1. How big should the sample be for a statistically accurate standard (at, say, the 99.73% confidence level and accuracy of ±5%)?
2. Is the sample size adequate?
3. How many units should be produced at this workstation per day?
4. What is the cost per unit for this task in direct labor cost?

Source: Professor Hank Maddux, Sam Houston State University

The "People" Focus: Human Resources at Alaska Airlines

Video Case

With thousands of employees spread across nearly 100 locations in the United States, Mexico, and Canada, building a committed and cohesive workforce is a challenge. Yet Alaska Airlines is making it work. The company's "people" focus states:

> *While airplanes and technology enable us to do what we do, we recognize this is fundamentally a people business, and our future depends on how we work together to win in this extremely competitive environment. As we grow, we want to strengthen our small company feel… We will succeed where others fail because of our pride and passion, and because of the way we treat our customers, our suppliers and partners, and each other.*

Managerial excellence requires a committed workforce. Alaska Airlines' pledge of respect for people is one of the key elements of a world-class operation.

Effective organizations require talented, committed, and trained personnel. Alaska Airlines conducts comprehensive training at all levels. Its "Flight Path" leadership training for all 10,000 employees is now being followed by "Gear Up" training for 800 front-line managers. In addition, training programs have been developed for Lean and Six Sigma as well as for the unique requirements for pilots, flight attendants, baggage, and ramp personnel. Because the company only hires pilots into first officer positions—the right seat in the cockpit, it offers a program called the "Fourth Stripe" to train for promotion into the captain's seat on the left side, along with all the additional responsibility that entails.

Customer service agents receive specific training on the company's "Empowerment Toolkit." Like the Ritz-Carlton's famous customer service philosophy, agents have the option of awarding customers hotel and meal vouchers or frequent flier miles when the customer has experienced a service problem.

Because many managers are cross-trained in operational duties outside the scope of their daily positions, they have the ability to pitch in to ensure that customer-oriented processes go smoothly. Even John Ladner, Director of Seattle Airport Operations, who is a fully licensed pilot, has left his desk to cover a flight at the last minute for a sick colleague.

Along with providing development and training at all levels, managers recognize that inherent personal traits can make a huge difference. For example, when flight attendants are hired, the ones who are still engaged, smiling, and fresh at the end of a very long interview day are the ones Alaska wants on the team. Why? The job requires these behaviors and attitudes to fit with the Alaska Airlines team—and smiling and friendly flight attendants are particularly important at the end of a long flight.

Visual workplace tools also complement and close the loop that matches training to performance. Alaska Airlines makes full use of color-coded graphs and charts to report performance against key metrics to employees. Twenty top managers gather weekly in an operations leadership meeting, run by Executive VP of Operations, Ben Minicucci, to review activity consolidated into visual summaries. Key metrics are color-coded and posted prominently in every work area.

Alaska's training approach results in empowered employees who are willing to assume added responsibility and accept the unknowns that come with that added responsibility.

Discussion Questions*

1. Summarize Alaska Airlines' human resources focus in your own words.
2. Why is employee empowerment useful to companies such as Alaska Airlines?
3. What tools discussed in the chapter might be employed to enhance the company's training and performance efforts? Why?

*Before answering these questions, you may wish to view the video that accompanies this case.

Hard Rock's Human Resource Strategy

Video Case 

Everyone—managers and hourly employees alike—who goes to work for Hard Rock Cafe takes Rock 101, an initial 2-day training class. The Hard Rock value system is to bring a fun, healthy, nurturing environment into the Hard Rock Cafe culture. This initial course and many other courses help employees develop both personally and professionally. The human resource department plays a critical role in any service organization, but at Hard Rock, with its "experience strategy," the human resource department takes on added importance.

Long before Jim Knight, manager of corporate training, begins the class, the human resource strategy of Hard Rock has had an impact. Hard Rock's strategic plan includes building a culture that allows for acceptance of substantial diversity and individuality. From a human resource perspective, this has the benefit of enlarging the pool of applicants as well as contributing to the Hard Rock culture.

Creating a work environment above and beyond a paycheck is a unique challenge. Outstanding pay and benefits are a start, but the key is to provide an environment that works for the employees. This includes benefits that start for part-timers who work at least 19 hours per week (while others in the industry start at 35 hours per week); a unique respect for individuality; continuing training; and a high level of internal promotions—some 60% of the managers are promoted from hourly employee ranks. The company's training is very specific, with job-oriented interactive DVDs covering kitchen, retail, and front-of-the-house service. Outside volunteer work is especially encouraged to foster a bond between the workers, their community, and issues of importance to them.

Applicants also are screened on their interest in music and their ability to tell a story. Hard Rock builds on a hiring criterion of bright, positive-attitude, self-motivated individuals with an employee bill of rights and substantial employee empowerment. The result is a unique culture and work environment, which no doubt contributes to the low turnover of hourly people—one-half the industry average.

The layout, memorabilia, music, and videos are important elements in the Hard Rock "experience," but it falls on the waiters and waitresses to make the experience come alive. They are particularly focused on providing an authentic and memorable dining experience. Like Alaska Airlines, Hard Rock is looking for people with a cause—people who like to serve. By succeeding with its human resource strategy, Hard Rock obtains a competitive advantage.

Discussion Questions*

1. What has Hard Rock done to lower employee turnover to half the industry average?
2. How does Hard Rock's human resource department support the company's overall strategy?
3. How would Hard Rock's value system work for automobile assembly line workers? (*Hint:* Consider Hackman and Oldham's core job characteristics.)
4. How might you adjust a traditional assembly line to address more "core job characteristics"?

*Before answering these questions, you may wish to view the video that accompanies this case.

Endnotes

1. Sources: *Science Direct* (April 2018); *JAMA Surgery* (September 2014); and *Ocular Surgery News* (March 25, 2017).
2. Sources: *The Wall Street Journal* (January 19–20, 2019); and *New Equipment Digest* (January 15, 2020).
3. Sources: George Kanawaty (ed.), *Introduction to Work Study*, International Labour Office, Geneva, 1992; B. W. Niebel, *Motion and Time Study*, 8th ed. (Homewood, IL: Richard D. Irwin), 1988; and Stephan Konz, *Work Design* (Columbus, Ohio: Grid Publishing, Inc.), 1979.
4. Sources: **Quora.com** (October 5, 2020); and **VentureBeat.com** (January 29, 2020).

Bibliography

Barnes, R. M. *Motion and Time Study, Design and Measurement of Work*, 7th ed. New York: Wiley, 1980.

Bridger, R. S. *Introduction to Human Factors and Ergonomics*, 4th ed. New York: CRC Press, 2017.

Freivalds, A., and B. W. Niebel. *Methods, Standards, and Work Design*, 13th ed. New York: Irwin/McGraw-Hill, 2014.

Gurses, A. P., et al. "Overcoming COVID-19: What Can Human Factors and Ergonomics Offer?" *Journal of Patient Safety and Risk Management* 25, no. 2 (2020): 49–54.

Hackman, J. R., and G. R. Oldham. *Work Redesign*. Reading, MA: Addison-Wesley, 1980.

Mital, A., A. Desai, and A. Mital. *Fundamentals of Work Measurement: What Every Engineer Should Know*. New York: CRC Press, 2016.

Salvendy, G., ed. *Handbook of Human Factors and Ergonomics*, 4th ed. New York: Wiley, 2012.

Shahrokh, S., and N. A. Bakar. "Relationship between Human Factors Engineering and Productivity." *Interdisciplinary Journal of Contemporary Research in Business* 3, no. 3 (July 2011): 706–712.

Village, J., and P. W. Neumann. "Designing for Human Factors." *Industrial Engineer: IE* 46, no. 11 (November 2014): 36–41.

Chapter 10 *Rapid* Review

Main Heading	Review Material	MyLab Operations Management
HUMAN RESOURCE STRATEGY FOR COMPETITIVE ADVANTAGE	*The objective of a human resource strategy is to manage labor and design jobs so people are effectively and efficiently utilized.* *Quality of work life* refers to a job that is not only reasonably safe with equitable pay but that also achieves an appropriate level of both physical and psychological requirements. *Mutual commitment* means that both management and employees strive to meet common objectives. *Mutual trust* is reflected in reasonable, documented employment policies that are honestly and equitably implemented to the satisfaction of both management and employees.	Concept Questions: 1.1–1.4 **VIDEO 10.1** The "People" Focus: Human Resources at Alaska Airlines **VIDEO 10.2** Human Resources at Hard Rock Cafe
LABOR PLANNING	▪ **Labor planning**—A means of determining staffing policies dealing with employment stability, work schedules, and work rules. *Flextime* allows employees, within limits, to determine their own schedules. *Flexible* (or *compressed*) *workweeks* often call for fewer but longer workdays. *Part-time status* is particularly attractive in service industries with fluctuating demand loads.	Concept Questions: 2.1–2.6
JOB DESIGN	▪ **Job design**—Specifies the tasks that constitute a job for an individual or group. ▪ **Labor specialization** (or **job specialization**)—The division of labor into unique ("special") tasks. ▪ **Job enlargement**—The grouping of a variety of tasks about the same skill level; horizontal enlargement. ▪ **Job rotation**—A system in which an employee is moved from one specialized job to another. ▪ **Job enrichment**—A method of giving an employee more responsibility that includes some of the planning and control necessary for job accomplishment; vertical expansion. ▪ **Employee empowerment**—Enlarging employee jobs so that the added responsibility and authority are moved to the lowest level possible. ▪ **Self-directed team**—A group of empowered individuals working together to reach a common goal.	Concept Questions: 3.1–3.6
ERGONOMICS AND THE WORK ENVIRONMENT	▪ **Ergonomics**—The study of the human interface with the environment and machines. The physical environment affects performance, safety, and quality of work life. Illumination, noise and vibration, temperature, humidity, and air quality are controllable by management.	Concept Questions: 4.1–4.4
METHODS ANALYSIS	▪ **Methods analysis**—A system that involves developing work procedures that are safe and produce quality products efficiently. ▪ **Flow diagram**—A drawing used to analyze movement of people or material. ▪ **Process chart**—A graphic representation that depicts a sequence of steps for a process. ▪ **Activity chart**—A way of improving utilization of an operator and a machine or some combination of operators (a crew) and machines. ▪ **Operations chart**—A chart depicting right- and left-hand motions.	Concept Questions: 5.1–5.6 Problems: 10.2, 10.6, 10.8 Virtual Office Hours for Solved Problem: 10.1
THE VISUAL WORKPLACE	▪ **Visual workplace**—Uses a variety of visual communication techniques to rapidly communicate information to stakeholders.	Concept Questions: 6.1–6.5
LABOR STANDARDS	▪ **Labor standards**—The amount of time required to perform a job or part of a job. Labor standards are set in four ways: (1) historical experience, (2) time studies, (3) predetermined time standards, and (4) work sampling. ▪ **Time study**—Timing a sample of a worker's performance and using it as a basis for setting a standard time. ▪ **Average observed time**—The arithmetic mean of the times for each element measured, adjusted for unusual influence for each element. $\text{Average observed time} = \dfrac{\text{Sum of the times recorded to perform each element}}{\text{Number of observations}}$ (10-1) ▪ **Normal time**—The average observed time, adjusted for pace: $\text{Normal time} = (\text{Average observed time}) \times (\text{Performance rating factor})$ (10-2)	Concept Questions: 7.1–7.6 Problems: 10.13–10.46 Virtual Office Hours for Solved Problems: 10.2–10.6

Main Heading	Review Material	MyLab Operations Management
	■ **Standard time**—An adjustment to the total normal time; the adjustment provides allowances for personal needs, unavoidable work delays, and fatigue: $$\text{Standard time} = \frac{\text{Total normal time}}{1 - \text{Allowance factor}} \quad (10\text{-}3)$$ *Personal time allowances* are often established in the range of 4% to 7% of total time. $$\text{Required sample size} = n = \left(\frac{zs}{h\bar{x}}\right)^2 \quad (10\text{-}4)$$ $$n = \left(\frac{zs}{e}\right)^2 \quad (10\text{-}5)$$ $$s = \sqrt{\frac{\sum (x_i - \bar{x})^2}{n-1}} = \sqrt{\frac{\sum (\text{Each sample observation} - \bar{x})^2}{\text{Number in sample} - 1}} \quad (10\text{-}6)$$ ■ **Predetermined time standards**—A division of manual work into small basic elements that have established and widely accepted times. The most common predetermined time standard is *methods time measurement* (MTM). ■ **Therbligs**—Basic physical elements of motion. ■ **Time measurement units (TMUs)**—Units for very basic micromotions in which 1 TMU = 0.0006 min or 100,000 TMUs= 1 hr. ■ **Work sampling**—An estimate, via sampling, of the percentage of time that a worker spends on various tasks. Work sampling sample size for a desired confidence and accuracy: $$n = \frac{z^2 p(1-p)}{h^2} \quad (10\text{-}7)$$	ACTIVE MODEL 10.1
ETHICS	Management's role is to educate the employee; specify the necessary equipment, work rules, and work environment; and then enforce those requirements.	Concept Questions: 8.1–8.2
ADDITIONAL MYLAB OPERATIONS MANAGEMENT RESOURCES	✔ Additional Case Studies (Chicago Southern Hospital and The Fleet That Wanders) ✔ Multiple Choice Case Questions (Jackson Manufacturing Company)	

Self Test

LO 10.1 When product demand fluctuates and yet you maintain a constant level of employment, some of your cost savings might include:
- **a)** reduction in hiring costs.
- **b)** reduction in layoff costs and unemployment insurance costs.
- **c)** lack of need to pay a premium wage to get workers to accept unstable employment.
- **d)** having a trained workforce rather than having to retrain new employees each time you hire for an upswing in demand.
- **e)** all of the above.

LO 10.2 The difference between *job enrichment* and *job enlargement* is that:
- **a)** enlarged jobs contain a larger number of similar tasks, while enriched jobs include some of the planning and control necessary for job accomplishment.
- **b)** enriched jobs contain a larger number of similar tasks, while enlarged jobs include some of the planning and control necessary for job accomplishment.
- **c)** enriched jobs enable an employee to do a number of boring jobs instead of just one.
- **d)** all of the above.

LO 10.3 The work environment includes these factors:
- **a)** Lighting, noise, temperature, and air quality
- **b)** Illumination, carpeting, and high ceilings
- **c)** Enough space for meetings and videoconferencing
- **d)** Noise, humidity, and number of coworkers
- **e)** Job enlargement and space analysis

LO 10.4 *Methods analysis* focuses on:
- **a)** the design of the machines used to perform a task.
- **b)** how a task is accomplished.
- **c)** the raw materials that are consumed in performing a task.
- **d)** reducing the number of steps required to perform a task.

LO 10.5 The least preferred method of establishing labor standards is:
- **a)** time studies.
- **b)** work sampling.
- **c)** historical experience.
- **d)** predetermined time standards.

LO 10.6 The allowance factor in a time study:
- **a)** adjusts normal time for errors and rework.
- **b)** adjusts standard time for lunch breaks.
- **c)** adjusts normal time for personal needs, unavoidable delays, and fatigue.
- **d)** allows workers to rest every 20 minutes.

LO 10.7 To set the required sample size in a time study, you must know:
- **a)** the number of employees.
- **b)** the number of parts produced per day.
- **c)** the desired accuracy and confidence levels.
- **d)** management's philosophy toward sampling.

Answers: LO 10.1. e; LO 10.2. a; LO 10.3. a; LO 10.4. b; LO 10.5. c; LO 10.6. c; LO 10.7. c.

Supply Chain Management

CHAPTER 11

CHAPTER OUTLINE

GLOBAL COMPANY PROFILE: *Red Lobster*

Simone Hogan/Alamy Stock Photo

Millions of people panicked during the early days of the global pandemic. Store shelves emptied. Social media burst with pleas for rolls of toilet paper, whose sales skyrocketed a whopping 845% in March 2020. But toilet paper flows from paper mills to retail stores through a tight, efficient supply chain. It is bulky and not very profitable, so retailers don't keep a lot of inventory on hand; they just get frequent shipments and restock their shelves. The amount of toilet paper the average American uses hasn't changed; it's around 141 rolls per year. But even small changes in buying habits can throw everything into disarray in many supply chains.

10 OM STRATEGY DECISIONS

- Design of Goods and Services
- Managing Quality
- Process Strategies
- Location Strategies
- Layout Strategies
- Human Resources
- *Supply Chain Management*
- Inventory Management
- Scheduling
- Maintenance

GLOBAL COMPANY PROFILE
Red Lobster

Red Lobster's Supply Chain Yields a Competitive Advantage

Red Lobster is the world's largest seafood restaurant company. Headquartered in Orlando, Florida, Red Lobster serves over 140 million meals annually from more than 700 restaurants in the U.S. and Canada, and it has a growing international footprint. With more than 55,000 employees, the firm focuses on delivering the highest quality freshly prepared seafood at reasonable prices.

In the restaurant business, a winning operations management strategy requires a winning supply chain. Nothing is more important to Red Lobster than sourcing and delivering healthy, high-quality food. And there are very few other industries where supplier performance is so closely tied to the customer.

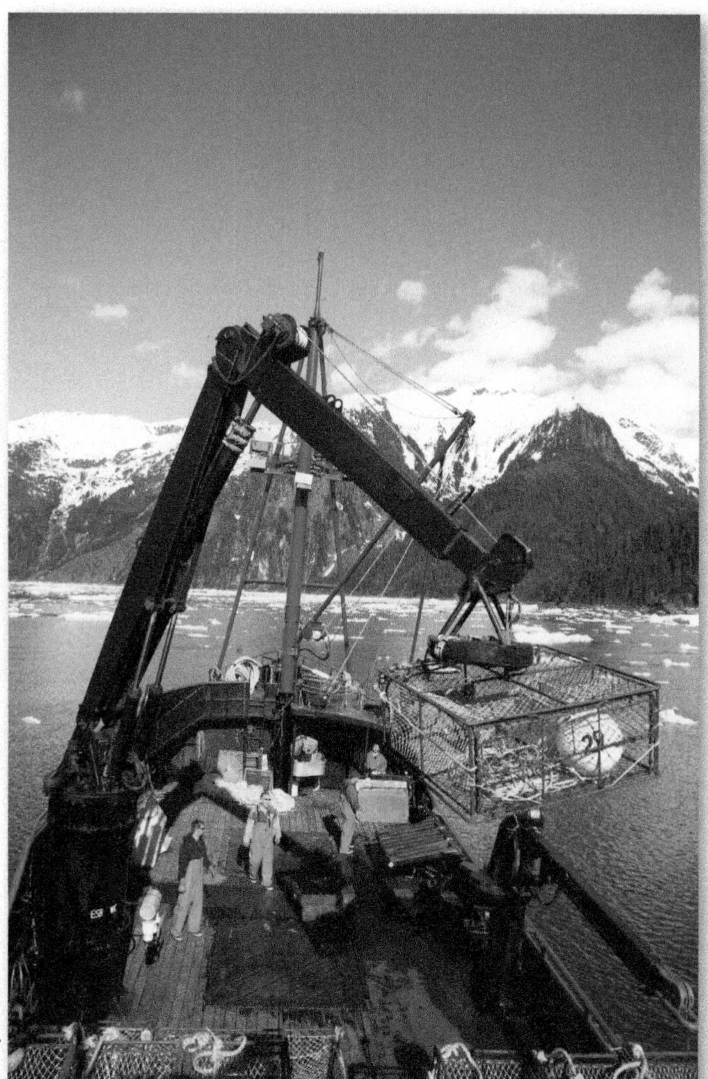

Courtesy of Red Lobster

Red Lobster has also been committed to seafood sustainability since it opened its doors in 1968. "We helped establish guidelines for best practices and aquaculture certifications, and we spent decades building relationships with suppliers who share our values. We believe it's our responsibility to protect and preserve our oceans and marine life for generations to come," says Executive VP Horace Dawson.

Qualifying Worldwide Sources: Part of Red Lobster's supply chain begins with a crab harvest in the frigid waters off the coast of Alaska. But long before a supplier is qualified to sell to Red Lobster, a total quality team is appointed. The team provides guidance, assistance, support, and training to the suppliers to ensure that overall objectives are understood and desired results accomplished.

Courtesy of Red Lobster

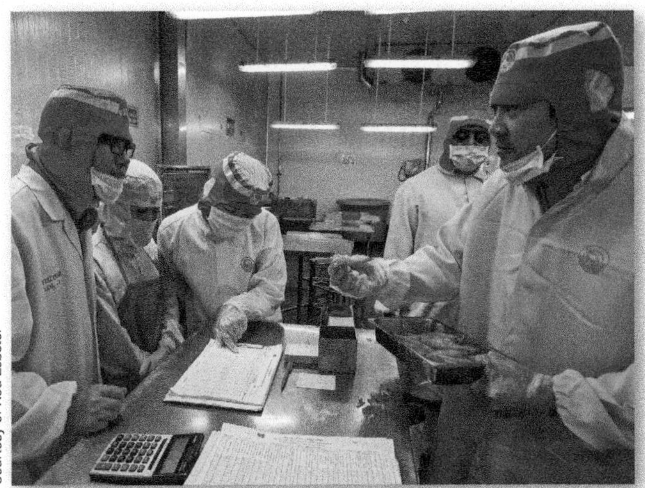

Courtesy of Red Lobster

Aquaculture Certification: Shrimp in this Asian plant are certified to ensure traceability. The focus is on quality control certified by the Global Aquaculture Alliance, of which Red Lobster was a founding member in 1997. Farming and inspection practices yield safe and wholesome shrimp. Joe Zhou, Senior Director of Purchasing, appears at the far left, and Nelson Griffin, Executive VP for Supply Chains/Purchasing, is second from the right.

Managing for Control: Executive Vice President Horace Dawson works with Jessica Wix, General Manager of the Red Lobster in Altamonte Springs, Florida, and her staff to ensure that quality and temperature tracking are maintained.

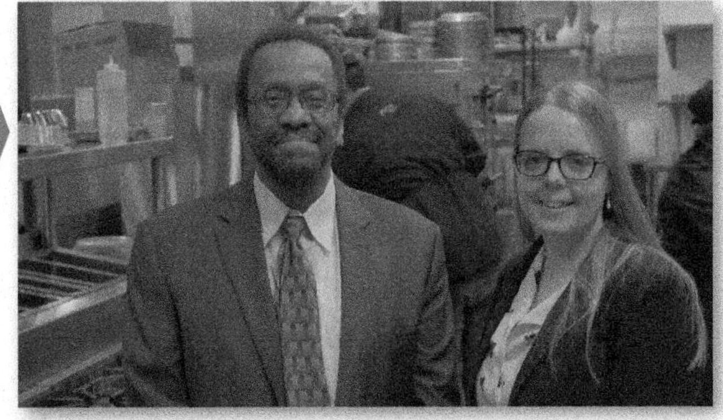

Courtesy of Red Lobster

Red Lobster sources its food from five continents and thousands of suppliers. Its supply channels all require *supplier qualification*, have *product tracking*, are subject to *independent audits*, and employ *just-in-time delivery*. With best-in-class techniques and processes, Red Lobster's supply chain alliances are rapid, transparent, and efficient. "We've got the best seafood supply chain in the industry," states Nelson Griffin, Executive VP for Supply Chains/Purchasing.

Courtesy of Red Lobster

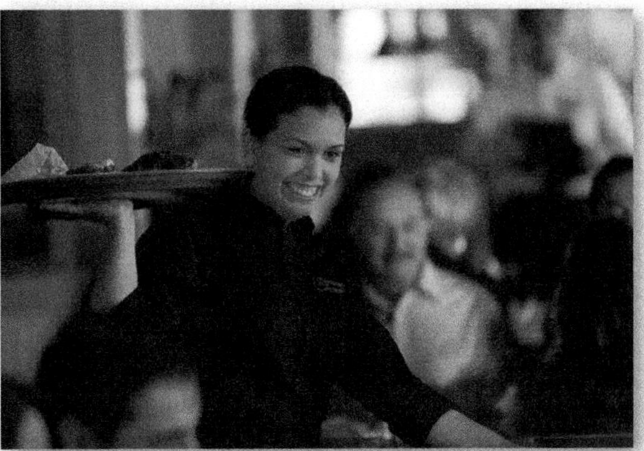

Courtesy of Red Lobster

JIT Delivery: For many products, temperature monitoring begins immediately and is tracked through the entire supply chain, to the kitchen at each of Red Lobster's 700+ restaurants, and ultimately to the guest.

LEARNING OBJECTIVES

STUDENT TIP Competition today is not between companies; it is between supply chains.

The Supply Chain's Strategic Importance

Like Red Lobster, most firms spend a huge portion of their sales dollars on purchases. Because an increasing percentage of an organization's costs are determined by purchasing, relationships with suppliers are increasingly integrated and long term. Combined efforts that improve innovation, speed design, and reduce costs are common. Such efforts, when part of a corporate-wide strategy, can dramatically improve all partners' competitiveness. This integrated focus places added emphasis on managing supplier relationships.

Supply chain management
The coordination of all supply chain activities involved in enhancing customer value.

Supply chain management describes the coordination of all supply chain activities, starting with raw materials and ending with a satisfied customer. Thus, a supply chain includes suppliers; manufacturers and/or service providers; and distributors, wholesalers, and/or retailers who deliver the product and/or service to the final customer. Figure 11.1 provides an example of the breadth of links and activities that a supply chain may cover.

The objective of supply chain management is to structure the supply chain to maximize its competitive advantage and benefits to the ultimate consumer. Just as with championship teams, a central feature of successful supply chains is members acting in ways that benefit the team (the supply chain).

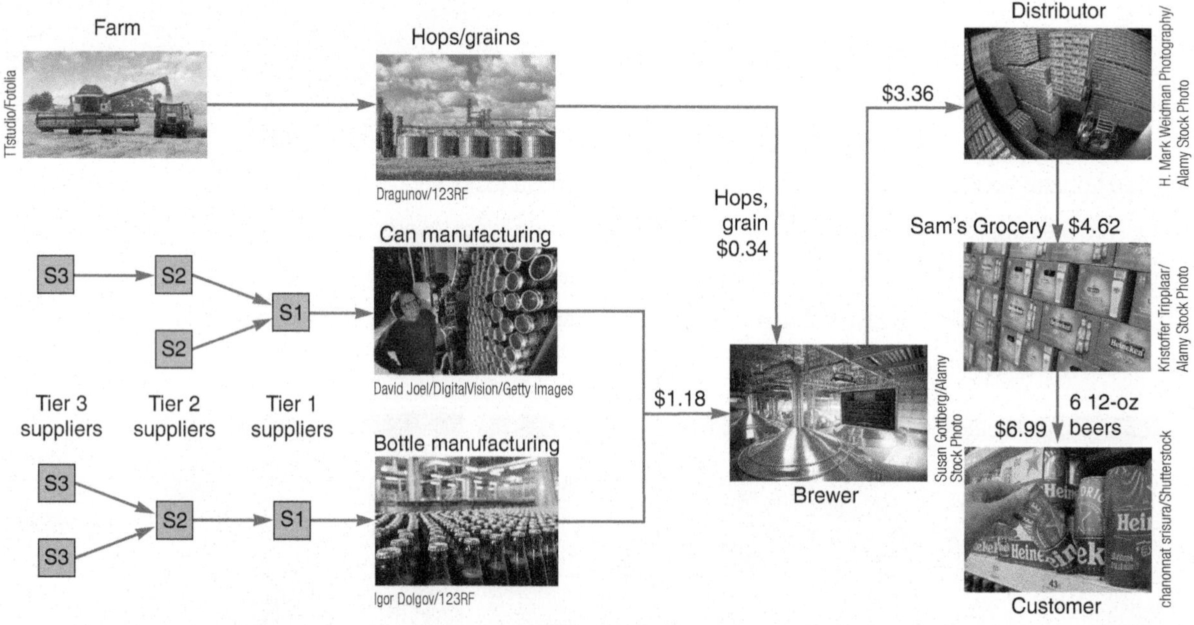

Figure **11.1**

A Supply Chain for Beer

The supply chain includes all the interactions among suppliers, manufacturers, distributors, and customers. A well-functioning supply chain has information flowing between all partners. The chain includes transportation, scheduling information, cash and credit transfers, as well as ideas, designs, and material transfers. Even can and bottle manufacturers have their own tiers of suppliers providing components such as lids, labels, packing containers, etc. (Costs are approximate and include substantial taxes.)

With collaboration, costs for both buyers and suppliers can drop. For example, when both parties are willing to share sales and cost information, profit can increase for both. Examples of supply chain coordination include:

- Walmart cooperates with its top 200 supplier factories in China to reach the goal of 20% energy efficiency improvement.
- Mercury Marine, the large boat-engine producer, uses the Internet to enhance design with boat builders and engine dealers as it fights off competition from Honda, Yamaha, and Volvo.
- Unifi, the leading U.S. maker of synthetic yarn, shares daily production-scheduling and quality-control information with raw materials supplier DuPont.
- Amazon, to reduce logistics costs, has moved its fulfillment activities for Procter & Gamble products directly into Procter & Gamble's warehouse.

As Table 11.1 indicates, a huge part of a firm's revenue is typically spent on purchases, so supply chains are a good place to look for savings. Example 1 further illustrates the amount of leverage available to the operations manager through the supply chain. These percentages indicate the strong role that supply chains play in potential profitability. Effective cost cutting may help a firm reach its profit goals more easily than would an increased sales effort.

TABLE 11.1

Supply Chain Costs as a Percentage of Sales

INDUSTRY	% PURCHASED
Automobile	67
Beverages	52
Chemical	62
Food	60
Lumber	61
Metals	65
Paper	55
Petroleum	79
Restaurants	35
Transportation	62

Example 1

SUPPLY CHAIN STRATEGY VS. SALES STRATEGY TO ACHIEVE A TARGET PROFIT

Hau Lee Furniture, Inc., spends 60% of its sales dollars in the supply chain and has a current gross profit of \$10,000. Hau wishes to increase gross profit by \$5,000 (50%). He would like to compare two strategies: reducing material costs vs. increasing sales.

APPROACH ▶ Use the following table to make the analysis.

SOLUTION ▶ The current material costs and production costs are 60% and 20%, respectively, of sales dollars, with fixed cost at a constant \$10,000. Analysis indicates that an improvement in the supply chain that would *reduce material costs by 8.3%* (\$5,000/\$60,000) would produce a 50% net profit gain for Hau, whereas a *much larger 25% increase in sales* (\$25,000/\$100,000) would be required to produce the same result.

	CURRENT SITUATION	SUPPLY CHAIN STRATEGY	SALES STRATEGY
Sales	\$100,000	\$100,000	\$125,000
Cost of materials	\$60,000 (60%)	\$55,000 (55%)	\$75,000 (60%)
Production costs	\$20,000 (20%)	\$20,000 (20%)	\$25,000 (20%)
Fixed costs	\$10,000 (10%)	\$10,000 (10%)	\$10,000 (8%)
Profit	\$10,000 (10%)	\$15,000 (15%)	\$15,000 (12%)

INSIGHT ▶ Supply chain savings flow directly to the bottom line. In general, supply chain costs need to shrink by a much lower percentage than sales revenue needs to increase to attain a profit goal. Effective management of the supply chain can generate substantial benefits.

LEARNING EXERCISE ▶ If Hau wants to double the original gross profits (from \$10,000 to \$20,000), what would be required of the supply chain and sales strategies? [Answer: Supply chain strategy = 16.7% *reduction* in material costs; sales strategy = 50% *increase* in sales.]

RELATED PROBLEMS ▶ 11.2, 11.3

As firms strive to increase their competitiveness via product customization, high quality, cost reductions, and speed to market, added emphasis is placed on the supply chain. Through long-term strategic relationships, suppliers become "partners" as they contribute to competitive advantage.

LO 11.1 *Explain* the strategic importance of the supply chain

To ensure that the supply chain supports a firm's strategy, managers need to consider the supply chain issues shown in Table 11.2. Activities of supply chain managers cut across the accounting, finance, marketing, and operations disciplines. Just as the OM function supports the firm's overall strategy, the supply chain must support the OM strategy. Strategies of low cost or

OM in Action A Rose Is a Rose, but Only If It Is Fresh[1]

Supply chains for food and flowers must be fast, and they must be good. When the food supply chain has a problem, the best that can happen is the customer does not get fed on time; the worst that happens is the customer gets food poisoning and dies. In the floral industry, the timing and temperature are also critical. Indeed, flowers are the most perishable agricultural item—even more so than fish. Flowers not only need to move fast, but they must also be kept cool, at a constant temperature of 33 to 37 degrees Fahrenheit. And they must be provided preservative-treated water while in transit. Roses are especially delicate, fragile, and perishable.

Eighty percent of the roses sold in the U.S. market arrive by air from rural Colombia and Ecuador. Roses move through this supply chain via an intricate but fast transportation network. This network stretches from growers who cut, grade, bundle, pack, and ship; to importers who make the deal; to the U.S. Department of Agriculture personnel who quarantine and inspect for insects, diseases, and parasites; to U.S. Customs agents who inspect and approve; to facilitators who provide clearance and labeling; to wholesalers who distribute; to retailers who arrange and sell; and finally to the customer. Each and every minute the product is deteriorating. The time and temperature sensitivity of perishables like roses requires sophistication and refined standards in the supply chain. Success yields quality and low losses. After all, when it's Valentine's Day, what good is a shipment of roses that arrives wilted or late? This is a difficult supply chain; only an excellent one will get the job done.

Africa Studio/Fotolia

rapid response demand different things from a supply chain than a strategy of differentiation. For instance, a low-cost strategy, as Table 11.2 indicates, requires suppliers be selected based primarily on cost. Such suppliers should have the ability to design low-cost products that meet the functional requirements, minimize inventory, and drive down lead times. However, if you want roses that are fresh, build a supply chain that focuses on response (see the *OM in Action* box, "A Rose Is a Rose, but Only If It Is Fresh").

Firms must achieve integration of strategy up and down the supply chain. And they must expect that strategy to be different for different products and to change as products move through their life cycle.

Sourcing Issues: Make-or-Buy and Outsourcing

As suggested in Table 11.2, a firm needs to determine strategically how to design the supply chain. However, prior to embarking on supply chain design, operations managers must first consider the "make-or-buy" and outsourcing decisions.

TABLE 11.2 How Corporate Strategy Impacts Supply Chain Decisions[2]

	LOW-COST STRATEGY	RESPONSE STRATEGY	DIFFERENTIATION STRATEGY
Primary supplier selection criteria	• Cost	• Capacity • Speed • Flexibility	• Product development skills • Willing to share information • Jointly and rapidly develop products
Supply chain inventory	• Minimize inventory to hold down costs	• Use buffer stocks to ensure speedy supply	• Minimize inventory to avoid product obsolescence
Distribution network	• Inexpensive transportation • Sell through discount distributors/retailers	• Fast transportation • Provide premium customer service	• Gather and communicate market research data • Knowledgeable sales staff
Product design characteristics	• Maximize performance • Minimize cost	• Low setup time • Rapid production ramp-up	• Modular design to aid product differentiation

Make-or-Buy Decisions

A wholesaler or retailer buys everything that it sells; a manufacturing operation hardly ever does. Manufacturers, restaurants, and assemblers of products buy components and subassemblies that go into final products. As we saw in Chapter 5, choosing products and services that can be advantageously obtained *externally* as opposed to produced *internally* is known as the **make-or-buy decision**. Supply chain personnel evaluate alternative suppliers and provide current, accurate, and complete data relevant to the buy alternative.

Make-or-buy decision
A choice between producing a component or service in-house or purchasing it from an outside source.

Outsourcing

Outsourcing transfers some of what are traditional internal activities and resources of a firm to outside vendors, making it slightly different from the traditional make-or-buy decision. Outsourcing, discussed in Chapter 2, is part of the continuing trend toward using the efficiency that comes with specialization. The vendor performing the outsourced service is an expert in that particular specialty. This leaves the outsourcing firm to focus on its key success factors and its core competencies.

Outsourcing
Transferring a firm's activities that have traditionally been internal to external suppliers.

Six Sourcing Strategies

LO 11.2 *Identify* six sourcing strategies

Having decided *what* to outsource, managers have six strategies to consider.

Many Suppliers

With the many-suppliers strategy, a supplier responds to the demands and specifications of a "request for quotation," with the order usually going to the low bidder. This is a common strategy when products are commodities. This strategy plays one supplier against another and places the burden of meeting the buyer's demands on the supplier. Suppliers aggressively compete with one another. This approach holds the supplier responsible for maintaining the necessary technology, expertise, and forecasting abilities, as well as cost, quality, and delivery competencies. Long-term "partnering" relationships are not the goal.

STUDENT TIP
Supply chain strategies come in many varieties; choosing the correct one is the trick.

Few Suppliers

A strategy of few suppliers implies that rather than looking for short-term attributes, such as low cost, a buyer is better off forming a long-term relationship with a few dedicated suppliers. Long-term suppliers are more likely to understand the broad objectives of the procuring firm and the end customer. Using few suppliers can create value by allowing suppliers to have economies of scale and a learning curve that yields both lower transaction costs and lower production costs. This strategy also encourages those suppliers to provide design innovations and technological expertise.

VIDEO 11.1
Darden's Global Supply Chain

Ford chooses suppliers even before parts are designed. Motorola evaluates suppliers on rigorous criteria, but in many instances has eliminated traditional supplier bidding, placing added emphasis on quality and reliability. On occasion these relationships yield contracts that extend through the product's life cycle. The British retailer Marks & Spencer finds that cooperation with its suppliers yields new products that win customers for the supplier and themselves. The move toward tight integration of the suppliers and purchasers is occurring in both manufacturing and services.

As with all other strategies, a downside exists. With few suppliers, the cost of changing partners is huge, so both buyer and supplier run the risk of becoming captives of the other. (See the *OM in Action box*, "The Complex Supply Chain for Apple and Samsung.") Poor supplier performance is only one risk the purchaser faces. The purchaser must also be concerned about trade secrets and suppliers that make other alliances or venture out on their own. This happened when the U.S. Schwinn Bicycle Co., needing additional capacity, taught Taiwan's Giant Manufacturing Company to make and sell bicycles. Giant Manufacturing is now the largest bicycle manufacturer in the world, and Schwinn was acquired out of bankruptcy by Pacific Cycle LLC.

OM in Action The Complex Supply Chain for Apple and Samsung[3]

Apple rival Samsung has good reason to hope the iPhone is a roaring success. The South Korean company makes $110 from each $1,000 iPhone that Apple sells. The fact reflects a love-hate dynamic between the phone makers that is one of the more unusual supply chain relationships. While each company vies to get consumers to buy its gadgets, Samsung stands to make billions of dollars supplying screens and memory chips for the iPhone—parts that Apple relies on for its most important product. "These are two of the largest companies on the planet deeply tied at the hip and directly competitive," says Harvard Professor David Yoffie.

Apple and Samsung are two of the world's most-profitable companies—and they depend on each other. Apple needs Samsung's parts to make the iPhones that account for about two-thirds of the company's revenue. Samsung needs Apple's orders to fuel a component business that delivers 35% of the firm's total revenue.

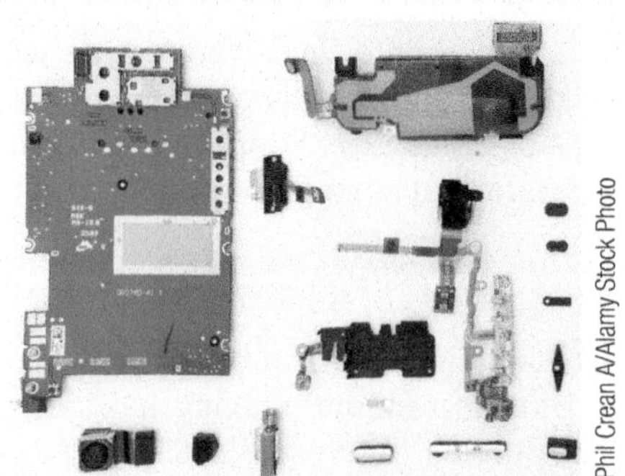
Phil Crean A/Alamy Stock Photo

The relationship grew after Apple moved into selling smartphones. Apple's immense demand for parts—it sells more than 200 million iPhones a year—limits the field of possible suppliers. Samsung is one of a handful of semiconductor makers that can produce a specialized chip crammed with extra memory capacity. And it was the only significant manufacturer of the organic light-emitting diode displays Apple has adopted to create the iPhone screen.

Supply chains can always be a challenge. But when demand is huge and the number of suppliers limited, the challenges increase.

Vertical Integration

Vertical integration
Developing the ability to produce goods or services previously purchased or actually buying a supplier or a distributor.

Purchasing can be extended to take the form of vertical integration. By **vertical integration**, we mean developing the ability to produce goods or services previously purchased or to actually buy a supplier or a distributor. As shown in Figure 11.2, vertical integration can take the form of *forward* or *backward integration*.

Backward integration suggests a firm purchase its suppliers, as in the case of Apple deciding to manufacture its own semiconductors. Apple also uses forward integration by establishing its own revolutionary retail stores.

Vertical integration can offer a strategic opportunity for the operations manager. For firms with the capital, managerial talent, and required demand, vertical integration may provide substantial opportunities for cost reduction, higher quality, timely delivery, and inventory reduction. Vertical integration appears to work best when the organization has a large market share and the management talent to operate an acquired vendor successfully.

The relentless march of specialization continues, meaning that a model of "doing everything" or "vertical integration" is increasingly difficult. Backward integration may be particularly dangerous for firms in industries undergoing technological change if management cannot keep abreast of those changes or invest the financial resources necessary for the next wave of technology. Research and development costs are too high and technology changes too rapid for one company to sustain leadership in every aspect of their product. Most organizations are better served concentrating on their own specialty and leveraging suppliers' contributions.

Figure **11.2**

Vertical Integration Can Be Forward or Backward

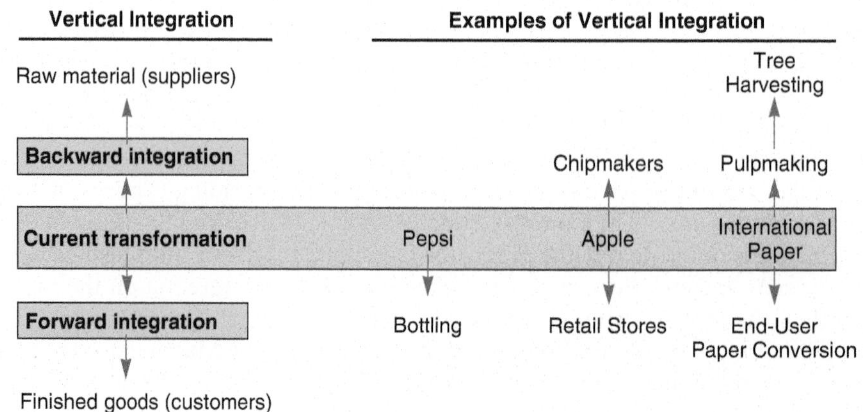

Joint Ventures

Because vertical integration is so dangerous, firms may opt for some form of formal collaboration. As we noted in Chapter 5, firms may engage in collaboration to enhance their new product prowess or technological skills. But firms also engage in collaboration to secure supply or reduce costs. One version of a joint venture is the current Daimler–BMW effort to develop and produce standard automobile components. Given the global consolidation of the auto industry, these two rivals in the luxury segment of the automobile market are at a disadvantage in volume. Their relatively low volume means fewer units over which to spread fixed costs, hence the interest in consolidating to cut development and production costs. As in all other such collaborations, the trick is to cooperate without diluting the brand or conceding a competitive advantage.

Keiretsu Networks

Many large Japanese manufacturers have found another strategy: it is part collaboration, part purchasing from few suppliers, and part vertical integration. These manufacturers are often financial supporters of suppliers through ownership or loans. The supplier becomes part of a company coalition known as a **keiretsu**. Members of the *keiretsu* are assured long-term relationships and are therefore expected to collaborate as partners, providing technical expertise and stable quality production to the manufacturer. Members of the *keiretsu* can also have second- and even third-tier suppliers as part of the coalition. Similar networks exist in other countries—in South Korea, they are more family-controlled than *keiretsu* networks and known as *chaebols*.

Keiretsu
A Japanese term that describes suppliers who become part of a company coalition.

Virtual Companies

Virtual companies rely on a variety of good, stable supplier relationships to provide services on demand. Suppliers may provide a variety of services that include doing the payroll, hiring personnel, designing products, providing consulting services, manufacturing components, conducting tests, or distributing products. The relationships may be short- or long-term and may include true partners, collaborators, or simply able suppliers and subcontractors. Whatever the formal relationship, the result can be exceptionally lean performance. The advantages of virtual companies include specialized management expertise, low capital investment, flexibility, and speed. The result is efficiency.

Virtual companies
Companies that rely on a variety of supplier relationships to provide services on demand. Also known as hollow corporations or network companies.

The apparel business provides a *traditional* example of virtual organizations. The designers of clothes seldom manufacture their designs; rather, they license the manufacture. The manufacturer may then rent space, lease sewing machines, and contract for labor. The result is an organization that has low overhead, remains flexible, and can respond rapidly to the market.

A *contemporary* example is exemplified by Vizio, Inc., a California-based producer of flat-screen TVs that has about 400 employees but huge sales. Vizio uses modules to assemble its own brand of TVs. Because the key components of TVs are now readily available and sold almost as commodities, innovative firms such as Vizio can specify the components, hire a contract manufacturer, and market the TVs with very little startup cost. In a virtual company, the supply chain is the company. Managing it is dynamic and demanding.

Supply Chain Risk

In this age of increasing specialization, low communication cost, and fast transportation, companies are making less and buying more. This means more reliance on supply chains and more risk. Managing integrated supply chains is a strategic challenge. Having fewer suppliers makes the supplier and customer more dependent on each other, increasing risk for both. This risk is compounded by globalization and logistical complexity. In any supply chain, vendor reliability and quality may be challenging. But the new model of a tight, fast, low-inventory supply chain, operating across political and cultural boundaries, adds a new dimension to risk. As organizations go global, shipping time (lead time) may increase, logistics may be less reliable, and tariffs and quotas may block companies from doing business. In addition, international supply chains complicate information flows and increase political/currency risks.

STUDENT TIP
The environment, controls, and process performance all affect supply chain risk.

Jack Sullivan/Alamy Stock Photo

Supply chain risks arise in many ways. As this mishap illustrates, expected shipments can literally sink into the ocean.

Risks and Mitigation Tactics

Supply chain risks arise in numerous ways, and you cannot outsource risk! Table 11.3 identifies major categories of risks and tactics to help manage them. The development of a successful strategic plan for supply chain management requires careful research, a thorough assessment of the risks involved, and innovative planning. Companies need to focus not only on reducing potential disruptions but also on how to prepare for responses to inevitable negative events. Flexible, secure supply chains and sufficient insurance against a variety of disruptions are a start. They may also choose to diversify their supplier base by using multiple sources for critical components. **Cross-sourcing** represents a hybrid technique where two suppliers each provide a different

Cross-sourcing
Using one supplier for a component and a second supplier for another component, where each supplier acts as a backup for the other.

TABLE 11.3 Supply Chain Risks and Tactics

RISK	RISK REDUCTION TACTICS	EXAMPLE
Supplier failure to deliver	Use multiple suppliers; effective contracts with penalties; subcontractors on retainer; preplanning	**McDonald's** planned its supply chain 6 years before its opening in Russia. Every plant—bakery, meat, chicken, fish, and lettuce—is closely monitored to ensure strong links.
Supplier quality failures	Careful supplier selection, training, certification, and monitoring	**Darden Restaurants** has placed extensive controls, including third-party audits, on supplier processes and logistics to ensure constant monitoring and reduction of risk.
Outsourcing	Take over production; provide or perform the service yourself	**Tyson** took over chicken farm production in China to mitigate product quality and safety concerns related to using independent farmers.
Logistics delays or damage	Multiple/redundant transportation modes and warehouses; secure packaging; effective contracts with penalties	**Walmart,** with its own trucking fleet and numerous distribution centers located throughout the U.S., finds alternative origins and delivery routes bypassing problem areas.
Distribution	Careful selection, monitoring, and effective contracts with penalties	**Toyota** trains its dealers around the world, invoking principles of the Toyota Production System to help dealers improve customer service, used-car logistics, and body and paint operations.
Information loss or distortion	Redundant databases; secure IT systems; training of supply chain partners on the proper interpretations and uses of information	**Boeing** utilizes a state-of-the-art international communication system that transmits engineering, scheduling, and logistics data to Boeing facilities and suppliers worldwide.
Political	Political risk insurance; cross-country diversification; franchising and licensing	**Hard Rock Cafe** reduces political risk by franchising and licensing, rather than owning, when the political and cultural barriers seem significant.
Economic	Hedging to combat exchange rate risk; purchasing contracts that address price fluctuations	**Honda and Nissan** are moving more manufacturing out of Japan as the exchange rate for the yen makes Japanese-made autos more expensive.
Natural catastrophes	Insurance; alternate sourcing; cross-country diversification	**Toyota,** after its experience with fires, earthquakes, and tsunamis, now attempts to have at least two suppliers, each in a different geographical region, for each component.
Theft, vandalism, and terrorism	Insurance; patent protection; security measures including RFID and GPS; diversification	**Domestic Port Radiation Initiative:** The U.S. government has set up radiation portal monitors that scan nearly all imported containers for radiation.

component, but they have the capability of producing each other's component—that is, each acting as a backup source. Another option is to create excess capacity that can be used in response to problems in the supply chain. Such contingency plans can reduce risk.

Richard Cummins/Robertharding/Alamy Stock Photo

As this photo of the port of Charleston suggests, with over 25 million containers entering the U.S. annually, tracking location, content, and condition of trucks and containers is a challenge. But new technology may improve both security and JIT shipments.

Security and JIT

There is probably no society more open than the U.S. This includes its borders and ports—but they are swamped. Millions of containers enter U.S. ports each year, along with thousands of planes, cars, and trucks each day. Even under the best of conditions, some 5% of the container movements are misrouted, stolen, damaged, or excessively delayed.

Since the September 11, 2001, terrorist attacks, supply chains have become more complex. However, technological innovations in the supply chain are improving both security and inventory management, making logistics more reliable. Technology is now capable of knowing truck and container location, content, and condition. New devices can even detect broken container seals. Motion detectors can also be installed inside containers. Other sensors record interior data including temperature, shock, radioactivity, and whether a container is moving. Tracking lost containers, identifying delays, or just reminding individuals in the supply chain that a shipment is on its way will help expedite shipments.

Managing the Integrated Supply Chain

As managers move toward integration of the supply chain, substantial efficiencies are possible. The cycle of materials—as they flow from suppliers, to production, to warehousing, to distribution, to the customer—takes place among separate and often very independent organizations. It can lead to actions that may not optimize the entire chain. On the other hand, the supply chain is full of opportunities to reduce waste and enhance value. We now look at some of the significant *issues* and *opportunities*.

VIDEO 11.2
Arnold Palmer Hospital's Supply Chain

Issues in Managing the Integrated Supply Chain

LO 11.3 *Explain* issues and opportunities in the supply chain

Three issues complicate development of an efficient, integrated supply chain: local optimization, incentives, and large lots.

Local Optimization Members of the chain are inclined to focus on maximizing local profit or minimizing immediate cost based on their limited knowledge. Slight upturns in demand are overcompensated for because no one wants to be caught short. Similarly, slight downturns are overcompensated for because no one wants to be caught holding excess inventory. So fluctuations are magnified. For instance, a pasta distributor does not want to run out of pasta for its retail customers; the natural response to an extra large order from the retailer is to compensate with an even larger order to the manufacturer on the assumption that retail sales are picking up. Neither the distributor nor the manufacturer knows that the retailer had a major one-time promotion that moved a lot of pasta. This is exactly the issue that complicated the implementation of efficient distribution at the Italian pasta maker Barilla.

Incentives (Sales Incentives, Quantity Discounts, Quotas, and Promotions) Incentives push merchandise into the chain for sales that have not occurred. This generates fluctuations that are ultimately expensive to all members of the chain.

Large Lots There is often a bias toward large lots because large lots tend to reduce unit costs. A logistics manager wants to ship large lots, preferably in full trucks, and a production manager wants long production runs. Both actions drive down unit shipping and production costs, but they increase holding costs and fail to reflect actual sales.

These three common occurrences—local optimization, incentives, and large lots—contribute to distortions of information about what is really occurring in the supply chain. A well-running supply system needs to be based on accurate information about how many products are truly being pulled through the chain. The inaccurate information is unintentional, but it results in distortions and fluctuations, causing what is known as the bullwhip effect.

Bullwhip effect
The increasing fluctuation in orders that often occurs as orders move through the supply chain.

The **bullwhip effect** occurs as orders are relayed from retailers, to distributors, to wholesalers, to manufacturers, with fluctuations increasing at each step in the sequence. The "bullwhip" fluctuations in the supply chain increase the costs associated with inventory, transportation, shipping, and receiving, while decreasing customer service and profitability. A number of specific opportunities exist for reducing the bullwhip effect and improving supply chain performance. The bullwhip effect is discussed more thoroughly in the supplement to this chapter.

Opportunities in Managing the Integrated Supply Chain

Opportunities for effective management in the supply chain are numerous. Here we discuss 12 of them.

Pull data
Accurate sales data that initiate transactions to "pull" product through the supply chain.

Accurate "Pull" Data Accurate **pull data** are generated by sharing (1) point-of-sales (POS) information so that each member of the chain can schedule effectively and (2) computer-assisted ordering (CAO). This implies using POS systems that collect sales data and then adjusting that data for market factors, inventory on hand, and outstanding orders. Then a net order is sent directly to the supplier, who is responsible for maintaining the finished-goods inventory.

Lot Size Reduction Lot sizes are reduced through aggressive management. This may include (1) developing economical shipments of less than truckload lots; (2) providing discounts based on total annual volume rather than size of individual shipments; and (3) reducing the cost of ordering through techniques such as standing orders and various forms of electronic purchasing.

Single-stage control of replenishment
Fixing responsibility for monitoring and managing inventory for the retailer.

Single-Stage Control of Replenishment **Single-stage control of replenishment** means designating a member in the chain as responsible for monitoring and managing inventory in the supply chain based on the "pull" from the end user. This approach removes distorted information and multiple forecasts that create the bullwhip effect. Control may be in the hands of:

- A sophisticated retailer who understands demand patterns. Walmart does this for some of its inventory with radio frequency ID (RFID) tags.
- A distributor who manages the inventory for a particular distribution area. Distributors who handle grocery items, beer, and soft drinks may do this. Anheuser-Busch manages beer inventory and delivery for many of its customers.
- A manufacturer who has a well-managed forecasting, manufacturing, and distribution system. Hong Kong's TAL Apparel Ltd. does this for retailers worldwide.

Vendor-managed inventory (VMI)
A system in which a supplier maintains material for the buyer, often delivering directly to the buyer's using department.

Vendor-Managed Inventory **Vendor-managed inventory (VMI)** means the use of a local supplier (usually a distributor) to maintain inventory for the manufacturer or retailer. The supplier delivers directly to the purchaser's using department rather than to a receiving dock or stockroom. If the supplier can maintain the stock of inventory for a variety of customers who use the same product or whose differences are very minor (say, at the packaging stage), then there should be a net savings. These systems work without the immediate direction of the purchaser.

Collaborative planning, forecasting, and replenishment (CPFR)
A system in which members of a supply chain share information in a joint effort to reduce supply chain costs.

Collaborative Planning, Forecasting, and Replenishment (CPFR) As with single-stage control and vendor-managed inventory, **collaborative planning, forecasting, and replenishment (CPFR)** is another effort to manage inventory in the supply chain. With CPFR, members of the supply chain share planning, demand, forecasting, and inventory information. Partners in a CPFR effort begin with collaboration on product definition and a joint marketing plan. Promotion, advertising, forecasts, joint order commitments, and timing of shipments are all included in the plan in a concerted effort to drive down inventory and related costs. CPFR can help to significantly reduce the bullwhip effect.

Blanket order
A long-term purchase commitment to a supplier for items that are to be delivered against short-term releases to ship.

Blanket Orders Blanket orders are unfilled orders with a vendor and are also called *open orders* or *incomplete orders*. A **blanket order** is a contract to purchase certain items from a vendor. It is not an authorization to ship anything. Shipment is made only on receipt of an agreed-on document, perhaps a shipping requisition or shipment release. Often such agreements include a discount for reaching an agreed-upon volume.

Standardization The purchasing department should make special efforts to increase levels of standardization. That is, rather than obtaining a variety of similar components with labeling, coloring, packaging, or perhaps even slightly different engineering specifications, the purchasing agent should try to have those components standardized.

Postponement Postponement withholds any modification or customization to the product (keeping it generic) as long as possible. The concept is to minimize internal variety while maximizing external variety. For instance, after analyzing the supply chain for its printers, Hewlett-Packard (HP) determined that if the printer's power supply was moved out of the printer itself and into a power cord, HP could ship the basic printer anywhere in the world. HP modified the printer, its power cord, its packaging, and its documentation so that only the power cord and documentation needed to be added at the final distribution point. This modification allowed the firm to manufacture and hold centralized inventories of the generic printer for shipment as demand changed. Only the unique power system and documentation had to be held in each country. This understanding of the entire supply chain reduced both risk and inventory investment. Similarly, Benetton leaves a portion of each style of its sweaters white so that they can be dyed the color the market is demanding at the last possible moment.

Postponement
Delaying any modifications or customization to a product as long as possible in the production process.

Electronic Ordering and Funds Transfer Electronic ordering and bank transfers are traditional approaches to speeding transactions and reducing paperwork. Transactions between firms often use electronic data interchange (EDI), which is a standardized data-transmittal format for computerized communications between organizations. EDI also provides for the use of advanced shipping notice (ASN), which notifies the purchaser that the vendor is ready to ship. Although many firms use EDI and ASN, the Internet's ease of use and low cost are proving popular.

Omnichannel Strategy, Drop Shipping, and Special Packaging Huge advances in logistics, from inventory accuracy to order tracking to speed of shipment, are taking place. These advances have opened the door to communicating with and delivering to customers the way they want to be served. This can be thought of as an omnichannel approach, which means multiple integrated channels of both communication and shipping. Orders can be filled directly from retail stores, routed through distribution centers, or drop shipped. Drop shipping means the supplier will ship directly to the end consumer rather than to the seller, saving both time and reshipping costs.

Omnichannel
An integrated multichannel strategy for communication and delivery that improves the user experience.

Drop shipping
Shipping directly from the supplier to the end consumer rather than from the seller, saving both time and reshipping costs.

Cost-saving measures within the distribution system include the use of special packaging and labels along with optimal placement of labels and bar codes on containers. The final location, down to the department and number of units in each shipping container, can also be indicated. These techniques can yield substantial savings and be of particular benefit to wholesalers and retailers by reducing shrinkage (lost, damaged, or stolen merchandise) and handling cost.

Blockchain Blockchain seems poised to play a significant role in logistics, where it could change how supply chains work. Blockchains allow participants to add "blocks" of information after each firm validates transactions as they progress along the production, shipping, and delivery phases of a supply chain. The blocks are time-stamped and added to the chain at each transaction.

Blockchain
A technology that tracks, documents, and verifies by adding "blocks" of information after each transaction.

Let's take the French grocery chain Carrefour as an example. Did the chicken you just bought at the supermarket have a nice life, roam free, and eat healthy grains? Every chicken at Carrefour comes complete with its very own life story, thanks to blockchain software. All you need to do is scan the label with your smartphone to get all the details. Carrefour wants to do whatever it can to ensure its products aren't tainted, part of a broader industry trend that buys into the concept that blockchain can improve food safety. Figure 11.3 shows how a chicken is traced with blockchain.

Digitalization and the Internet-of-Things (IoT) All of the forgoing opportunities are being enhanced by digitalization. As a part of the IoT, digitalization is permeating supply chain processes—just as it is many aspects of our lives. With its inherent precision and speed, digitalization enables supply chains to manage a deluge of information from multiple channels. The result is improved visibility throughout the supply chain. The digital world enables managers to make decisions in real time based on accurate current data. Airlines, for example, determine aircraft allocation, manage air traffic control, track and schedule maintenance, make crew assignments, determine gate designations, and coordinate takeoffs and landings—all at 40-second intervals.

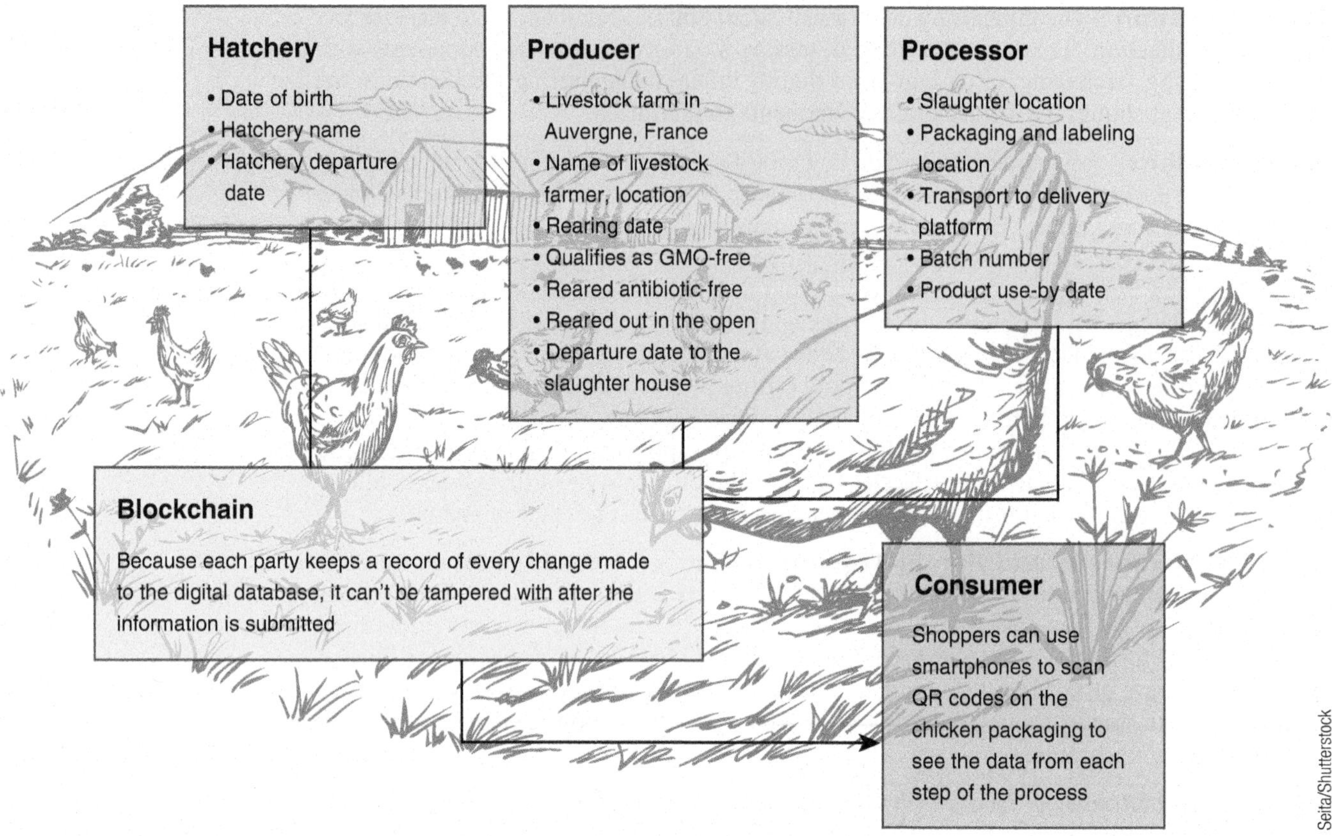

Hatchery
• Date of birth
• Hatchery name
• Hatchery departure date
Producer
• Livestock farm in Auvergne, France
• Name of livestock farmer, location
• Rearing date
• Qualifies as GMO-free
• Reared antibiotic-free
• Reared out in the open
• Departure date to the slaughter house
Processor
• Slaughter location
• Packaging and labeling location
• Transport to delivery platform
• Batch number
• Product use-by date
Blockchain
Because each party keeps a record of every change made to the digital database, it can't be tampered with after the information is submitted
Consumer
Shoppers can use smartphones to scan QR codes on the chicken packaging to see the data from each step of the process
Seita/Shutterstock

Figure **11.3**

Tracing Food via Blockchain

Agriculture companies are using blockchain software as a way to establish their products' documented history. Each party provides details related to its link in the supply chain. Here are the data points for a single Auvergne chicken sold by French supermarket chain Carrefour.

Building the Supply Base

LO 11.4 *Describe* the steps in supplier selection

For those goods and services a firm buys, suppliers, also known as *vendors*, must be selected and actively managed. Supplier selection considers numerous factors, such as strategic fit, supplier competence, delivery, and quality performance. Because a firm may have some competence in all areas and may have exceptional competence in only a few, selection can be challenging. Procurement policies also need to be established. Those might address issues such as percentage of business done with any one supplier or with minority businesses. We now examine supplier selection as a four-stage process: (1) supplier evaluation, (2) supplier development, (3) negotiations, and (4) contracting.

Supplier Evaluation

The first stage of supplier selection, *supplier evaluation*, involves finding potential suppliers and determining the likelihood of their becoming *good* suppliers. If good suppliers are not selected, then all other supply chain efforts are wasted. As firms move toward long-term suppliers, the issues of financial strength, quality, management, research, technical ability, and potential for a close, long-term relationship play an increasingly important role. Evaluation criteria critical to the firm might include these categories as well as production process capability, location, and information systems. The supplement to this chapter provides an example of the commonly used *factor weighting* approach to supplier evaluation.

Supplier Certification International quality certifications such as ISO 9000 and ISO 14000 are designed to provide an external verification that a firm follows sound quality management and environmental management standards. Buying firms can use such certifications to pre-qualify potential suppliers. Despite the existence of the ISO standards, firms often create their own supplier-certification programs. Buyers audit potential suppliers and award a certified status to those that meet the specified qualification. A certification process often involves three steps: (1) qualification, (2) education, and (3) the certification performance process. Once certified, the supplier may be awarded special treatment and priority, allowing the buying firm to reduce or eliminate incoming inspection of materials. Such an arrangement may facilitate JIT production for the buying firm. Most large companies use some sort of supplier certification program.

Supplier Development

The second stage of supplier selection is *supplier development*. Assuming that a firm wants to proceed with a particular supplier, how does it integrate this supplier into its system? The buyer makes sure the supplier has an appreciation of quality requirements, product specifications, schedules and delivery, and procurement policies. Supplier development may include everything from training, to engineering and production help, to procedures for information transfer.

Negotiations

Although the prices that consumers pay are often inflexible (printed on the price tag, listed in the catalog, etc.), a significant number of final prices paid in business-to-business transactions are negotiated. In addition to the price itself, several other aspects of the full product "package" must be determined. These may include credit and delivery terms, quality standards, and cooperative advertising agreements. In fact, negotiation represents a significant element in a purchasing manager's job, and well-honed negotiation skills are highly valued.

Here are three classic types of negotiation strategies: the cost-based model, the market-based price model, and competitive bidding.

Cost-Based Price Model The cost-based price model requires that the supplier open its books to the purchaser. The contract price is then based on time and materials or on a fixed cost with an escalation clause to accommodate changes in the vendor's labor and materials cost.

Market-Based Price Model In the market-based price model, price is based on a published, auction, or index price. Many commodities (agricultural products, paper, metal, etc.) are priced this way. Paperboard prices, for instance, are available via the *Official Board Markets* weekly publication (**www.advanstar.com**).

Competitive Bidding When suppliers are not willing to discuss costs or where near-perfect markets do not exist, competitive bidding is often appropriate. Competitive bidding is the typical policy in many firms for the majority of their purchases. Bidding policies usually require that the purchasing agent have several potential suppliers and quotations from each. The major disadvantage of this method, as mentioned earlier, is that the development of long-term relations between buyer and seller is hindered. It may also make difficult the communication and performance that are vital for engineering changes, quality, and delivery.

Yet a fourth approach is *to combine one or more* of the preceding negotiation techniques. The supplier and purchaser may agree to review cost data, accept some form of market-based cost, or agree that the supplier will "remain competitive."

Contracting

Supply chain partners often develop contracts to spell out terms of the relationship. Contracts are designed to share risks, share benefits, and create incentive structures to encourage supply chain members to adopt policies that are optimal for the entire chain. The idea is to make the total pie (of supply chain profits) bigger and then divide the bigger pie among all participants. The goal is collaboration. Some common features of contracts include *quantity discounts* (lower prices for larger orders), *buybacks* (common in the magazine and book business where there is a buyback of unsold units), and *revenue sharing* (where both partners share the risk of uncertainty by sharing revenue).

Centralized Purchasing

Companies with multiple facilities (e.g., multiple manufacturing plants or multiple retail outlets) must determine which items to purchase centrally and which to allow local sites to purchase for themselves. Unmonitored decentralized purchasing can create havoc. For example, different plants for Nestle USA's brands used to pay 29 different prices for its vanilla ingredient *to the same supplier*! Important cost, efficiency, and "single-voice" benefits often accrue from a centralized purchasing function. Typical benefits include:

- Leverage purchase volume for better pricing
- Develop specialized staff expertise
- Develop stronger supplier relationships
- Maintain professional control over the purchasing process
- Devote more resources to the supplier selection and negotiation process
- Reduce the duplication of tasks
- Promote standardization

However, local managers enjoy having their own purchasing control, and decentralized purchasing can offer certain inventory control, transportation cost, or lead-time benefits. Often firms use a hybrid approach—using centralized purchasing for some items and/or sites while allowing local purchasing for others.

Electronic Procurement

Electronic procurement speeds purchasing, reduces costs, and integrates the supply chain. It reduces the traditional barrage of paperwork and, at the same time, provides purchasing personnel with an extensive database of supplier, delivery, and quality data.

Online Catalogs and Exchanges Purchase of standard items is often accomplished via online catalogs. Such catalogs support cost comparisons and incorporate voice and video clips, making the process efficient for both buyers and sellers.

Online exchanges are typically industry-specific Internet sites that bring buyers and sellers together. Marriott and Hyatt created one of the first, Avendra (**www.avendra.com**), which facilitates economic purchasing of the huge range of goods needed by the 5,000 hospitality industry customers now in the exchange. Online catalogs and exchanges can help move companies from a multitude of individual phone calls, faxes, and emails to a centralized system and drive billions of dollars of waste out of the supply chain.

Online Auctions In addition to catalogs, some suppliers and buyers have established online auction sites. Operations managers find online auctions a fertile area for disposing of excess raw material and discontinued or excess inventory. Online auctions lower entry barriers, encourage sellers to join, and simultaneously increase the potential number of buyers. The key for intermediaries is to find and build a huge base of potential bidders, improve client buying procedures, and qualify new suppliers.

In a traditional auction, a seller offers a product or service and generates competition between bidders—bidding the price up. In contrast, buyers often utilize online *reverse auctions* (or *Dutch auctions*). In reverse auctions, a buyer initiates the process by submitting a description of the desired product or service. Potential suppliers then submit bids, which may include price and other delivery information. Thus, price competition occurs on the selling side of the transaction—bidding the price down. Note that, as with traditional supplier selection decisions, price is important but may not be the only factor in winning the bid.

Logistics Management

Mikbiz/Shutterstock

Four critical points in the logistics process are *point of origin, warehousing, transit,* and *destination*—each with a unique set of challenges that impact the efficiency of supply chains. Things that go wrong in the logistics process include pilfering, asset misplacement, and physical damage due to improper handling and storage, as well as unexpected events. Sensors, as part of a digital network, can alert managers to problems. Sensors can monitor environmental conditions, prevent misplacement, identify damages, avoid accidents, ensure compliance, track location, and reveal real-time conditions.

Procurement activities may be combined with various shipping, warehousing, and inventory activities to form a logistics system. The purpose of **logistics management** is to obtain efficiency of operations through the integration of all material acquisition, movement, and storage activities. When transportation and inventory costs are substantial on both the input and output sides of the production process, an emphasis on logistics may be appropriate. Many firms opt for outsourcing the logistics function, as logistics specialists can often bring expertise not available in-house. For instance, logistics companies often have tracking technology that reduces transportation losses and supports delivery schedules that adhere to precise delivery windows. The potential for competitive advantage is found via both reduced costs and improved customer service.

Logistics management
An approach that seeks efficiency of operations through the integration of all material acquisition, movement, and storage activities.

Shipping Systems

LO 11.5 *Explain* major issues in logistics management

Firms recognize that the transportation of goods to and from their facilities can represent as much as 25% of the cost of products. Because of this high cost, firms constantly evaluate their means of shipping. Six major means of shipping are trucking, railroads, airfreight, waterways, pipelines, and multimodal.

STUDENT TIP
Logistics represents a substantial part of the economy, as logistics cost comprises 11.3% of the U.S. gross domestic product.

Trucking The vast majority of manufactured goods moves by truck. The flexibility of shipping by truck is only one of its many advantages. Companies that have adopted JIT programs in recent years have put increased pressure on truckers to pick up and deliver on time, with no damage, with paperwork in order, and at low cost. (See the *OM in Action* box, "New York City Chokes on Deliveries.") Trucking firms are using computers to monitor weather, find the most effective route, reduce fuel cost, and analyze the most efficient way to unload. And Uber's freight division has created a potentially significant advancement in logistics efficiency. With technology similar to its ride-share system, Uber matches trucking companies with shipping requests. As though ordering a ride, users download the app, search for a shipping request, and click to book the shipment.

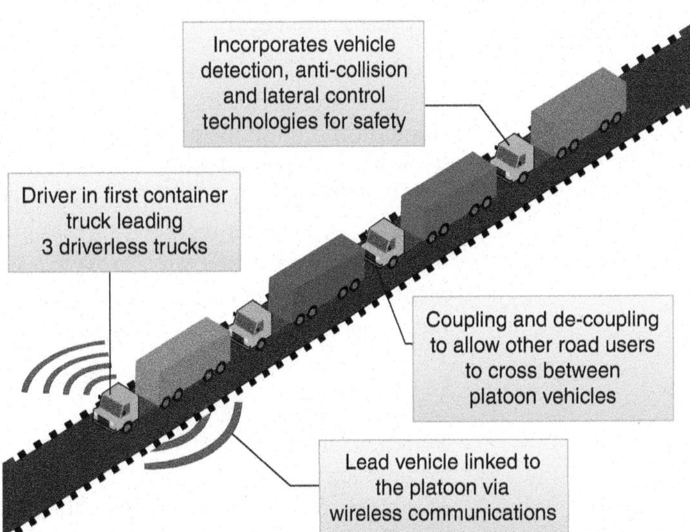
Incorporates vehicle detection, anti-collision and lateral control technologies for safety
Driver in first container truck leading 3 driverless trucks
Coupling and de-coupling to allow other road users to cross between platoon vehicles
Lead vehicle linked to the platoon via wireless communications

New autonomous vehicle technology may facilitate a *platooning* system for trucks. Platooning involves a number of trucks following each other very closely so they can take advantage of the aerodynamics and save fuel. This technique could generate a 4.5% fuel savings for the lead truck and 10% for those following.

Railroads Railroads in the U.S. employ 220,000 people and ship 40% of the ton-miles of all commodities, including 93% of coal, 57% of cereal grains, and 52% of basic chemicals. Containerization has made shipping of truck trailers on railroad flat cars a popular means of distribution. The equivalent of 47 million trailer loads are moved in the U.S. each year by rail.

Airfreight Airfreight represents less than 1% of tonnage shipped in the U.S. However, the proliferation of airfreight carriers such as FedEx, UPS, and DHL makes it a fast-growing mode of shipping.

OM in Action New York City Chokes on Deliveries[4]

An Amazon order starts with a tap of a finger. A day or two later—or even in a matter of hours—the package arrives. It seems simple enough. But to deliver Amazon orders and countless others from businesses that sell over the Internet, the very fabric of major urban areas around the world is being transformed. And New York City (NYC), where more than 2.4 million packages are delivered daily, shows the impact that this push for convenience is having on roadway safety, gridlock, and pollution.

The main entryway for packages into NYC, leading to the George Washington Bridge from New Jersey, has become the most congested interchange in the country. Officials are racing to keep track of the numerous warehouses sprouting up to create more zones for trucks to unload and to encourage some deliveries to be made by boat or at night as the city struggles to cope with a booming online economy.

The average number of daily deliveries to households in NYC quadrupled to more than 1.9 million shipments this past decade (and over 90,000 of these are stolen or disappear daily). With households receiving more shipments than businesses, trucks are pushed into neighborhoods where they had rarely ventured. Over 15% of NYC households receive a package every day. Delivery trucks double-park on streets and block bus and bike lanes. UPS and FedEx alone racked up more than 471,000 parking violations last year (see photo). Delivery firms must now build the substantial cost of traffic violations into their budgets.

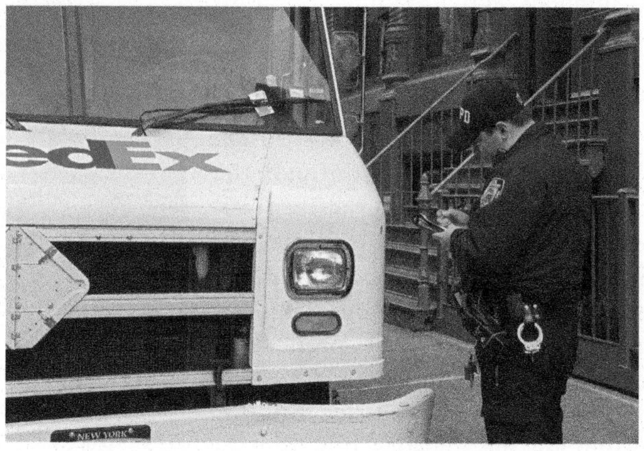

Ira Berger/Alamy Stock Photo

Clearly, for national and international movement of lightweight items, such as medical and emergency supplies, flowers, fruits, and electronic components, airfreight offers speed and reliability.

Waterways Waterways are one of the nation's oldest means of freight transportation, dating back to construction of the Erie Canal in 1817. Included in U.S. waterways are the nation's rivers, canals, the Great Lakes, coastlines, and oceans connecting to other countries. The usual cargo on internal waterways is bulky, low-value cargo such as iron ore, grains, cement, coal, chemicals, limestone, and petroleum products. Internationally, millions of containers holding all sorts of industrial and consumer goods are shipped at very low cost via huge oceangoing ships each year. Water transportation is often preferred when cost is more important than speed.

Pipelines Pipelines are an important form of transporting crude oil, natural gas, and other petroleum and chemical products.

Multimodal Multimodal shipping combines shipping methods and is a common means of getting a product to its final destination, particularly for international shipments. The use of standardized containers facilitates easy transport from truck to rail to ship and back again, without having to unload products from the containers until the very end.

While freight rates are often based on very complicated pricing systems, in general, clients pay for speed. Faster methods such as airfreight tend to be much more expensive, while slower methods, such as waterways, provide a much cheaper shipping rate per unit. The size of shipments follows a similar pattern. The faster methods tend to involve smaller shipment sizes, while the slower methods involve very large shipment sizes.

Warehousing

Warehousing often adds 8–10% to the cost of a product, making warehousing a significant expense for many firms. Warehouses come in all shapes and sizes, from tiny rooms in the back of a store to enormous facilities that could fit multiple football fields. Warehouses may be extremely expensive to operate, but the alternatives (e.g., either no storage at all or storage at local operating facilities, along with the related logistics issues) may be much more costly.

The fundamental purpose of a warehouse is to store goods. However, some warehouses also provide other crucial functions. For example, a warehouse can serve as a *consolidation point*, gathering shipments from multiple sources to send outbound in one cheaper, fully loaded truck. Alternatively, a warehouse can provide a *break-bulk* function by accepting a cheaper full truckload inbound shipment and then dividing it for distribution to individual sites. Further, similar to a major airport hub, a warehouse can serve simply as a *cross-docking* facility—accepting shipments from a variety of sources and recombining them for distribution to a variety of destinations, often without actually storing any goods during the transition. Finally, a warehouse can serve as a point of *postponement* in the process, providing final customer-specific value-added processing to the product before final shipment.

Channel assembly represents one way to implement postponement. **Channel assembly** sends individual components and modules, rather than finished products, to the distributor. The distributor then assembles, tests, and ships. Channel assembly treats distributors more as manufacturing partners than as distributors. This technique has proven successful in industries where products are undergoing rapid change, such as PCs. With this strategy, finished-goods inventory is reduced because units are built to a shorter, more accurate forecast. Consequently, market response is better, with lower investment—a nice combination.

Channel assembly
Postpones final assembly of a product so the distribution channel can assemble it.

Third-Party Logistics (3PL)

Third-party logistics, as is the case with most specialization, tends to bring added innovation and expertise to the logistics system. Consequently, supply chain managers outsource logistics to meet three goals: (1) drive down inventory investment, (2) lower delivery costs, and (3) improve delivery reliability and speed. Specialized logistics firms support these goals by creatively coordinating the supplier's inventory system with the service capabilities of the delivery firm. FedEx, UPS, and DHL play a core role in other firms' logistics processes. For instance, UPS works with Nike at a shipping hub in Louisville, Kentucky, to store and immediately expedite shipments.

Distribution Management

Management of the supply chain focuses on incoming materials, but just as important, *distribution management* focuses on the outbound flow of products. Designing distribution networks to meet customer expectations suggests three criteria: (1) *rapid response*, (2) *product choice*, and (3) *service*.

Staples, Inc., for example, addresses these customer concerns by having several office supply stores in a town for convenience and quick response time. But it also offers an online shopping presence to accommodate customers requiring a much larger selection of products. It may even offer delivery directly to large customers. These varying customer expectations suggest both different distribution channels and multiple outlets.

So how many stores should Staples open in a town? As Figure 11.4(a) indicates, an increase in the number of facilities generally implies a quicker response and increased customer satisfaction. On the cost side, three logistics-related costs [see Figure 11.4(b)] are shown: *inventory*

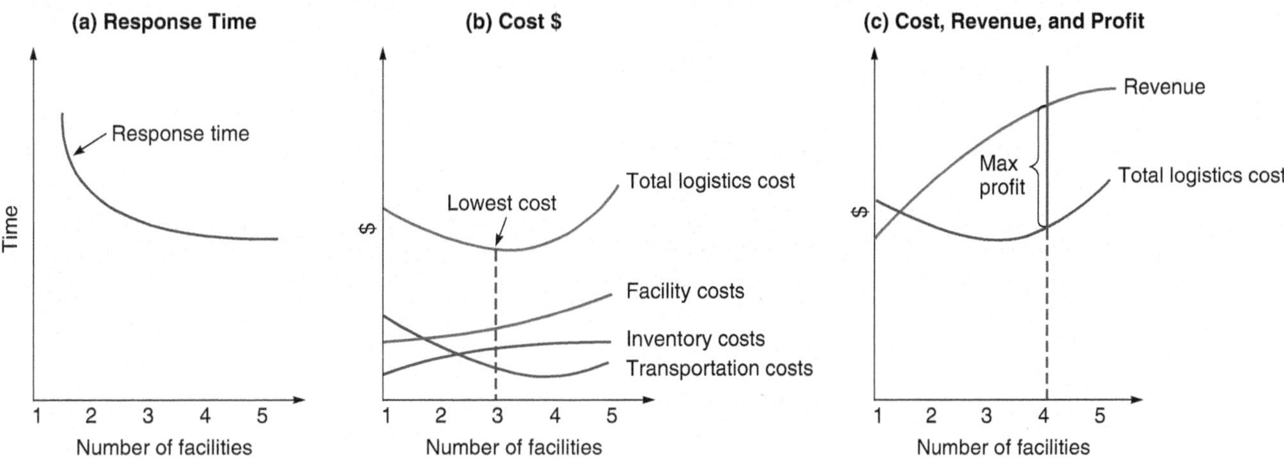

Figure 11.4

Number of Facilities in a Distribution Network

The focus should be on profit maximization (c) rather than cost minimization (b).

costs, transportation costs, and *facility costs*. Taken together, *total logistics costs* tend to follow the top curve, first declining, and then rising. For this particular example, it appears that total logistics costs are minimized with three facilities. However, when revenue is considered [see Figure 11.4(c)], we note that profit is maximized with four facilities.

Whether creating a network of warehouses or retail outlets, finding the optimal number of facilities represents a critical and often dynamic decision. For instance, barely a year after adding 3 million square feet of warehouse capacity, market dynamics caused **Amazon.com** to close three of its U.S. distribution centers.

Just as firms need an effective *supplier management* program, an effective *distribution management* program may make the difference between supply chain success and failure. For example, in addition to facilities, packaging and logistics are necessary for the network to perform well. Packaging and logistics are also important distribution decisions because the manufacturer is usually held responsible for breakages and serviceability. Further, selection and development of dealers or retailers are necessary to ensure ethical and enthusiastic representation of the firm's products. Top-notch supply chain performance requires good *downstream* (distributors and retailers) management, just as it does good *upstream* (suppliers) management.

Ethics and Sustainable Supply Chain Management

Let's look at two issues that OM managers must address every day when dealing with supply chains: ethics and sustainability.

Supply Chain Management Ethics

STUDENT TIP Because so much money passes through the supply chain, the opportunity for ethical lapses is significant.

We consider three aspects of ethics: personal ethics, ethics within the supply chain, and ethical behavior regarding the environment. As the supply chain becomes increasingly international, each of these becomes even more significant.

Personal Ethics Ethical decisions are critical to the long-term success of any organization. However, the supply chain is particularly susceptible to ethical lapses. With sales personnel anxious to sell and purchasing agents spending huge sums, temptations abound. Salespeople become friends with customers, do favors for them, take them to lunch, or present small (or large) gifts. Determining when tokens of friendship become bribes can be challenging. Many companies have strict rules and codes of conduct that limit what is acceptable.

Recognizing these issues, the Institute for Supply Management has developed the following principles and standards to be used as guidelines for ethical behavior:

- *Promote and uphold* responsibilities to one's employer; positive supplier and customer relationships; sustainability and social responsibility; protection of confidential and proprietary information; applicable laws, regulations, and trade agreements; and development of professional competence.
- *Avoid* perceived impropriety; conflicts of interest; behaviors that negatively influence supply chain decisions; and improper reciprocal agreements.

Ethics within the Supply Chain In this age of hyper-specialization, much of any organization's resources are purchased, putting great stress on ethics in the supply chain. Managers may be tempted to ignore ethical lapses by suppliers or offload pollution to suppliers. But firms must establish standards for their suppliers, just as they have established standards for themselves. Society expects ethical performance throughout the supply chain. For instance, Gap, Inc., reported that of its 3,000-plus factories worldwide, about 90% failed their initial evaluation. Gap found that 10% to 25% of its Chinese factories engaged in psychological or verbal abuse, and more than 50% of the factories in sub-Saharan Africa operated without proper safety devices. The challenge of enforcing ethical standards is significant, but responsible firms such as Gap are finding ways to deal with this difficult issue.

Ethical Behavior Regarding the Environment While ethics on both a personal basis and in the supply chain are important, so is ethical behavior in regard to the environment. Good ethics extends to doing business in a way that supports conservation and renewal of

TABLE 11.4 Management Challenges of Reverse Logistics

ISSUE	FORWARD LOGISTICS	REVERSE LOGISTICS
Forecasting	Relatively straightforward	More uncertain
Product quality	Uniform	Not uniform
Product packaging	Uniform	Often damaged
Pricing	Relatively uniform	Dependent on many factors
Speed	Often very important	Often not a priority
Distribution costs	Easily visible	Less directly visible
Inventory management	Consistent	Not consistent

Source: Based on Reverse Logistics Executive Council (**www.rlec.org**).

resources. This requires evaluation of the entire environmental impact, from raw material, to manufacture, through use and final disposal. For instance, Red Lobster and Walmart both require their shrimp and fish suppliers in Southeast Asia to abide by the standards of the Global Aquaculture Alliance. These standards must be met if suppliers want to maintain the business relationship. Operations managers also ensure that sustainability is reflected in the performance of second- and third-tier suppliers. Enforcement can be done by in-house inspectors, third-party auditors, governmental agencies, or nongovernmental watchdog organizations. All four approaches are used.

Establishing Sustainability in Supply Chains

The incoming supply chain garners most of the attention, but it is only part of the challenge of sustainability. The "return" supply chain is also significant. **Reverse logistics** involves the processes for sending returned products back up the supply chain for resale, repair, reuse, remanufacture, recycling, or disposal. The operations manager's goal should be to limit burning or burying of returned products and instead strive for reuse. Reverse logistics initiates a new set of challenges, as shown in Table 11.4.

Reverse logistics
The process of sending returned products back up the supply chain for value recovery or disposal.

Although sometimes used as a synonym for reverse logistics, a **closed-loop supply chain** refers more to the proactive design of a supply chain that tries to optimize all forward and reverse flows. A closed-loop supply chain prepares for returns prior to product introduction. For instance, IBM has recognized that components often have much longer life cycles than the original products that they go into. So the company has established a systematic method for dismantling returns and used equipment to extract components that still have value, such as boards, cards, and hard-disk assemblies. IBM has realized millions of dollars of savings in procurement costs by exploiting its "dismantling channel" of used parts.

Closed-loop supply chain
A supply chain designed to optimize both forward and reverse flows.

Measuring Supply Chain Performance

STUDENT TIP
If you can't measure it, you can't control it.

Like all other managers, supply chain managers require standards (or *metrics*, as they are often called) to evaluate performance. For example, the large grocery chain HEB tracks metrics such as total freight cost per $1 million of sales, errors and returns in distribution, and lead-time compliance. Lancers, a beverage dispenser manufacturer, tracks metrics such as on-time delivery percentage, defects per million, and lead time. We now introduce several financial-based inventory metrics.

Assets Committed to Inventory

LO 11.6 *Compute* the percentage of assets committed to inventory and inventory turnover

Supply chain managers make scheduling and quantity decisions that determine the assets committed to inventory. Three specific measures can be helpful here. The first is the amount of money invested in inventory, usually expressed as a percentage of assets, as shown in Equation (11-1) and Example 2:

$$\text{Percentage invested in inventory} = (\text{Average inventory investment}/\text{Total assets}) \times 100 \qquad (11\text{-}1)$$

Example 2

TRACKING HOME DEPOT'S INVENTORY INVESTMENT

Home Depot's management wishes to track its investment in inventory as one of its performance measures. Recently, Home Depot had \$12.5 billion invested in inventory and total assets of \$42.9 billion.

APPROACH ▶ Determine the investment in inventory and total assets and then use Equation (11-1).

SOLUTION ▶ Percentage invested in inventory $= (12.5/42.9) \times 100 = 29.1\%$

INSIGHT ▶ Over one-fourth of Home Depot assets are committed to inventory.

LEARNING EXERCISE ▶ If Home Depot can drive its inventory investment down to 20% of assets, how much money will it free up for other uses? [Answer: $12.5 - (42.9 \times 0.2) = \$3.92$ billion.]

RELATED PROBLEMS ▶ 11.5b, 11.6b

TABLE 11.5

Inventory as Percentage of Total Assets (with examples of exceptional performance)

Manufacturer (Toyota 5%)	15%
Wholesale (Coca-Cola 2.9%)	34%
Restaurants (McDonald's .05%)	2.9%
Retail (Home Depot 25.7%)	27%

Inventory turnover
Cost of goods sold divided by average inventory.

Specific comparisons with competitors may assist evaluation. Total assets committed to inventory in manufacturing approach 15%, in wholesale 34%, and retail 27%—with wide variations, depending on the specific business model, the business cycle, and management (see Table 11.5).

The second common measure of supply chain performance is *inventory turnover* (see Table 11.6). Its reciprocal, *weeks of supply*, is the third. **Inventory turnover** is computed on an annual basis, using Equation (11-2):

$$\text{Inventory turnover} = \text{Cost of goods sold}/\text{Average inventory investment} \quad (11\text{-}2)$$

Cost of goods sold is the cost to produce the goods or services sold for a given period. Inventory investment is the average inventory value for the same period. This may be the average of several periods of inventory or beginning and ending inventory added together and divided by 2. Often, average inventory investment is based on nothing more than the inventory investment at the end of the period—typically at year-end.

In Example 3, we look at inventory turnover applied to PepsiCo.

Example 3

INVENTORY TURNOVER AT PEPSICO, INC.

PepsiCo, Inc., manufacturer and distributor of drinks, Frito-Lay, and Quaker Foods, provides the following in a recent annual report (shown here in \$ billions). Determine PepsiCo's turnover.

Net revenue		\$63.50
Cost of goods sold		\$28.70
Inventory:		
Raw material inventory	\$1.32	
Work-in-process inventory	\$0.15	
Finished goods inventory	\$1.26	
Total average inventory investment		\$ 2.73

APPROACH ▶ Use the inventory turnover computation in Equation (11-2) to measure inventory performance. Cost of goods sold is \$28.7 billion. Total inventory is the sum of raw material at \$1.32 billion, work-in-process at \$.15 billion, and finished goods at \$1.26 billion, for total average inventory investment of \$2.73 billion.

SOLUTION ▶

$$\begin{aligned}\text{Inventory turnover} &= \text{Cost of goods sold}/\text{Average inventory investment}\\ &= 28.7/2.73\\ &= 10.5\end{aligned}$$

INSIGHT ▶ We now have a standard, popular measure by which to evaluate performance.

LEARNING EXERCISE ▶ If Coca-Cola's cost of goods sold is \$13.2 billion and inventory investment is \$2.6 billion, what is its inventory turnover? [Answer: 5.1.]

RELATED PROBLEMS ▶ 11.5a, 11.6c, 11.7

Weeks of supply, as shown in Example 4, may have more meaning in the wholesale and retail portions of the service sector than in manufacturing. It is computed as the reciprocal of inventory turnover:

$$\text{Weeks of supply} = \text{Average inventory investment}/(\text{Annual cost of goods sold}/52 \text{ weeks}) \quad (11\text{-}3)$$

Example 4

DETERMINING WEEKS OF SUPPLY AT PEPSICO

Using the PepsiCo data in Example 3, management wants to know the weeks of supply.

APPROACH ▶ We know that inventory investment is \$2.73 billion and that weekly sales equal annual cost of goods sold (\$28.7 billion) divided by 52 = \$28.7/52 = \$.55 billion.

SOLUTION ▶ Using Equation (11-3), we compute weeks of supply as:

$$\text{Weeks of supply} = (\text{Average inventory investment}/\text{Average weekly cost of goods sold})$$

$$= 2.73/.55 = 4.96 \text{ weeks}$$

INSIGHT ▶ We now have a standard measurement by which to evaluate a company's continuing performance or by which to compare companies.

LEARNING EXERCISE ▶ If Coca-Cola's average inventory investment is \$2.6 billion and its average weekly cost of goods sold is \$.25 billion, what is the firm's weeks of supply? [Answer: 10.4 weeks.]

RELATED PROBLEMS ▶ 11.6a, 11.8

Supply chain management is critical in driving down inventory investment. The rapid movement of goods is key. Walmart, for example, has set the pace in the retailing sector with its world-renowned supply chain management. By doing so, it has established a competitive advantage. With its own truck fleet, distribution centers, and a state-of-the-art communication system, Walmart (with the help of its suppliers) replenishes store shelves an average of twice per week. Competitors resupply every other week. Economical and speedy resupply means both rapid response to product changes and customer preferences, as well as lower inventory investment. Similarly, while many manufacturers struggle to move inventory turnover up to 10 times per year, Dell Computer has inventory turns exceeding 90 and supply measured in *days*—not weeks. Supply chain management provides a competitive advantage when firms effectively respond to the demands of global markets and global sources.

TABLE 11.6

Examples of Annual Inventory Turnover

FOOD, BEVERAGE, RETAIL	
Anheuser Busch	15
Coca-Cola	5
Home Depot	5
McDonald's	112
MANUFACTURING	
Dell Computer	90
Johnson Controls	22
Toyota (overall)	13
Nissan (assembly)	150

Benchmarking the Supply Chain

While metric values convey their own meaning and are useful when compared to past data, another important use compares these values to those of benchmark firms. Several organizations and websites allow companies to submit their own data and receive reports on how they stack up against other firms in their own industry or against world-class firms from any industry. Table 11.7 provides a few examples of metric values for typical firms and for benchmark firms in the consumer packaged goods industry. World-class benchmarks are the result of well-managed supply chains that drive down costs, lead times, late deliveries, and shortages while improving service levels.

TABLE 11.7 **Supply Chain Metrics in the Consumer Packaged Goods Industry**

	TYPICAL FIRMS	BENCHMARK FIRMS
Order fill rate	71%	98%
Order fulfillment lead time (days)	7	3
Cash-to-cash cycle time (days)	100	30
Inventory days of supply	50	20

Source: Based on Institute for Industrial Engineers.

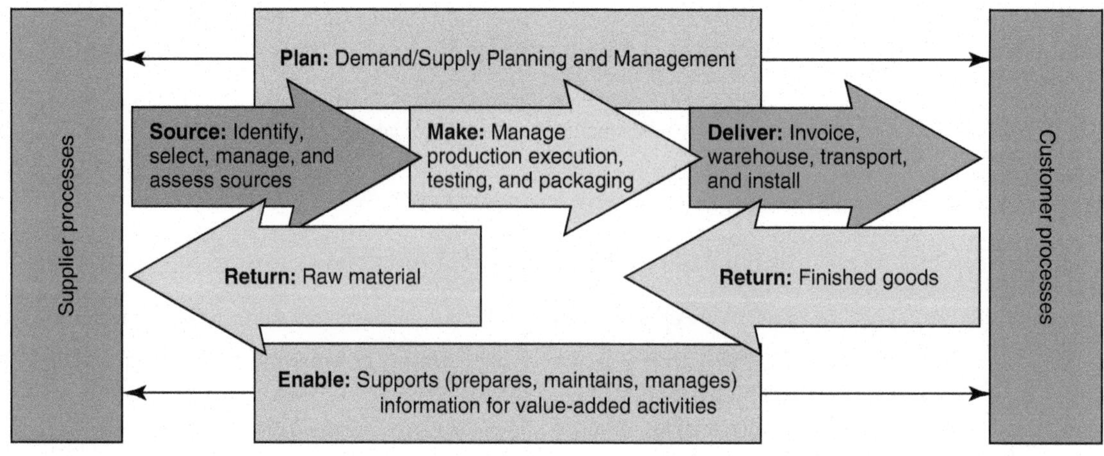

Figure 11.5
The Supply Chain Operations Reference (SCOR) Model

The SCOR Model

Supply Chain Operations Reference (SCOR) model
A set of processes, metrics, and best practices developed by the APICS Supply Chain Council.

Perhaps the best-known benchmarking system is the six-part **Supply Chain Operations Reference (SCOR) model**. As shown in Figure 11.5, the six parts are Plan (planning activities for supply and demand), Source (purchasing activities), Make (production activities), Deliver (distribution activities), Return (closed-loop supply chain activities), and Enable (support for value-added activities beyond the supply chain). The system is maintained by the APICS Supply Chain Council (**www.apics.org/sites/apics-supply-chain-council**). Firms use SCOR to identify, measure, reorganize, and improve supply chain processes.

The SCOR model defines processes, metrics, and best practices. The best practices describe the techniques used by benchmark firms that have scored very well on the metrics. SCOR combines these metrics with "Performance Attributes" (see Table 11.8) to facilitate comparisons of companies that compete by using different strategies (for example, low cost vs. responsiveness).

Benchmarking can be very useful, but it is not always adequate for excellence in the supply chain. Audits based on continuing communication, understanding, trust, performance, and corporate strategy are necessary. The relationships should manifest themselves in the mutual belief that "we are in this together" and go well beyond written agreements.

TABLE 11.8 SCOR Model Metrics Help Firms Benchmark Performance Against the Industry

SUPPLY CHAIN PERFORMANCE ATTRIBUTE	SAMPLE METRIC	CALCULATION
Reliability	Perfect order fulfillment	(Total perfect orders)/(Total number of orders)
Responsiveness	Average order fulfillment cycle time	(Sum of actual cycle times for all orders delivered)/ (Total number of orders delivered)
Agility	Upside supply chain flexibility	Time required to achieve an unplanned 20% increase in delivered quantities
Costs	Supply chain management cost	Cost to plan + Cost to source + Cost to deliver + Cost to return
Asset management	Cash-to-cash cycle time	Inventory days of supply + Days of receivables outstanding – Days of payables outstanding

Summary

Competition is no longer between companies but between supply chains. The key to success is to collaborate with members on both the supply side and the distribution side of the supply chain to make decisions that will benefit the whole channel. For many firms, the supply chain determines a substantial portion of product cost and quality, as well as opportunities for responsiveness and differentiation. The challenge of building a great supply chain is significant, but with good sourcing tactics, a thoughtful logistics plan, and active management of the distribution network, each link in the chain can be firmly forged. A number of metrics are available to help managers evaluate their supply chain performance and benchmark against the industry. Skillful supply chain management provides a great strategic opportunity for competitive advantage.

Key Terms

Supply chain management (p. 442)
Make-or-buy decision (p. 445)
Outsourcing (p. 445)
Vertical integration (p. 446)
Keiretsu (p. 447)
Virtual companies (p. 447)
Cross-sourcing (p. 448)
Bullwhip effect (p. 450)
Pull data (p. 450)
Single-stage control of replenishment (p. 450)
Vendor-managed inventory (VMI) (p. 450)
Collaborative planning, forecasting, and replenishment (CPFR) (p. 450)
Blanket order (p. 450)
Postponement (p. 451)
Omnichannel (p. 451)
Drop shipping (p. 451)
Blockchain (p. 451)
Logistics management (p. 455)
Channel assembly (p. 456)
Reverse logistics (p. 459)
Closed-loop supply chain (p. 459)
Inventory turnover (p. 460)
Supply Chain Operations Reference (SCOR) model (p. 462)

Ethical Dilemma

As a buyer for a discount retail chain, you find yourself caught in a maelstrom. Just last month, your chain began selling an economy-priced line of clothing endorsed by a famous movie star. To be price competitive, you have followed the rest of the industry and sourced the clothing from a low-wage region of Asia. Initial sales have been brisk; however, the movie star has recently called you screaming and crying because an investigative news outlet has reported that the clothes with her name on them are being made by children.

Outraged, you fly to the outsourcing manufacturing facility only to find that conditions are not quite as clear-cut as the news had reported. As you ride through the town, you see signs of poverty everywhere and children begging in the streets. When you enter the plant, you observe a very clean facility. The completely female workforce appears to be very industrious, but many of them do appear to be young. You confront the plant manager and explain your firm's strict international sourcing policies. You demand to know why these girls aren't in school. The manager provides the following response: "The truth is that some of these workers may be underage. We check IDs, but the use of falsified records is commonplace in this country. Plus, you don't understand the alternatives. If you shut this plant down, you will literally take food off the table for these families. There are no other opportunities in this town at this time, and there's no comprehensive welfare system in our country. As for the young women, school is not an option. In this town, only boys receive an education past the sixth grade. If you shut us down, these girls will be out on the street, begging or stealing. Your business offers them a better existence. Please don't take that away!"

What do you say to your company, the movie star, the media, and the protestors picketing your stores? Is the best option to shut down and try someplace else?

Discussion Questions

1. Define *supply chain management*.
2. What are the objectives of supply chain management?
3. What is the objective of logistics management?
4. How do we distinguish between the types of risk in the supply chain?
5. What is vertical integration? Give examples of backward and forward integration.
6. What are three basic approaches to negotiations?
7. How does a traditional adversarial relationship with suppliers change when a firm makes a decision to move to a few suppliers?
8. What is the difference between postponement and channel assembly?
9. What is CPFR?
10. What is the value of online auctions in e-commerce?
11. Explain how FedEx uses the Internet to meet requirements for quick and accurate delivery.
12. How does Walmart use drop shipping?
13. What are blanket orders? How do they differ from invoiceless purchasing?
14. What can purchasing do to implement just-in-time deliveries?
15. What is e-procurement?
16. How does Red Lobster, described in the *Global Company Profile*, find competitive advantage in its supply chain?
17. What is SCOR, and what purpose does it serve?

Solved Problem

SOLVED PROBLEM 11.1

Jack's Pottery Outlet has total end-of-year assets of \$5 million. The first-of-the-year inventory was \$375,000, with a year-end inventory of \$325,000. The annual cost of goods sold was \$7 million. The owner, Eric Jack, wants to evaluate his supply chain performance by measuring his percentage of assets in inventory, his inventory turnover, and his weeks of supply. We use Equations (11-1), (11-2), and (11-3) to provide these measures.

SOLUTION

First, determine *average inventory*:

$$(\$375{,}000 + \$325{,}000)/2 = \$350{,}000$$

Then, use Equation (11-1) to determine percentage invested in inventory:

$$\begin{aligned}\text{Percentage invested in inventory} &= (\text{Average inventory investment}/\text{Total assets}) \times 100\\ &= (350{,}000/5{,}000{,}000) \times 100\\ &= 7\%\end{aligned}$$

Third, determine inventory turnover, using Equation (11-2):

$$\begin{aligned}\text{Inventory turnover} &= \text{Cost of goods sold}/\text{Average inventory investment}\\ &= 7{,}000{,}000/350{,}000\\ &= 20\end{aligned}$$

Finally, to determine weeks of inventory, use Equation (11-3), adjusted to weeks:

$$\begin{aligned}\text{Weeks of inventory} &= \text{Average inventory investment}/\text{Weekly cost of goods sold}\\ &= 350{,}000/(7{,}000{,}000/52)\\ &= 350{,}000/134{,}615\\ &= 2.6\end{aligned}$$

We conclude that Jack's Pottery Outlet has 7% of its assets invested in inventory, that the inventory turnover is 20, and that weeks of supply is 2.6.

Problems

Problems 11.1–11.3 relate to The Supply Chain's Strategic Importance

•• **11.1** Choose a local establishment that is a member of a relatively large chain. From interviews with workers and information from the Internet, identify the elements of the supply chain. Determine whether the supply chain supports a low-cost, rapid response, or differentiation strategy (refer to Chapter 2). Are the supply chain characteristics significantly different from one product to another?

••• **11.2** Hau Lee Furniture, Inc., described in Example 1 of this chapter, finds its current profit of \$10,000 inadequate. The bank is insisting on an improved profit picture prior to approval of a loan for some new equipment. Hau would like to improve the profit line to \$25,000 so he can obtain the bank's approval for the loan.

a) What percentage improvement is needed in the *supply chain strategy* for profit to improve to \$25,000? What is the cost of material with a \$25,000 profit?
b) What percentage improvement is needed in the *sales strategy* for profit to improve to \$25,000? What must sales be for profit to improve to \$25,000?

•••• **11.3** Kamal Fatehl, production manager of Kennesaw Manufacturing, finds his profit at \$15,000 (as shown in the following statement)—inadequate for expanding his business. The bank is insisting on an improved profit picture prior to approval of a loan for some new equipment. Kamal would like to improve the profit line to \$25,000 so he can obtain the bank's approval for the loan.

		% OF SALES
Sales	\$250,000	100%
Cost of supply chain purchases	175,000	70%
Other production costs	30,000	12%
Fixed costs	30,000	12%
Profit	15,000	6%

a) What percentage improvement is needed in a *supply chain strategy* for profit to improve to \$25,000? What is the cost of material with a \$25,000 profit?
b) What percentage improvement is needed in a *sales strategy* for profit to improve to \$25,000? What must sales be for profit to improve to \$25,000? (*Hint:* See Example 1.)

Problem 11.4 relates to Six Sourcing Strategies

•• **11.4** Using sources from the Internet, identify some of the problems faced by a company of your choosing as it moves toward, or operates as, a virtual organization. Does its operating as a virtual organization simply exacerbate old problems, or does it create new ones?

Problems 11.5–11.11 relate to Measuring Supply Chain Performance

•• **11.5** Jenna Baker, CEO of Baker Mfg. Inc., wishes to compare her company's inventory turnover (see Table 11.9) to that of industry leaders, who have turnover of about 13 times per year and 8% of their assets invested in inventory.

a) What is Baker's inventory turnover?
b) What is Baker's percentage of assets committed to inventory?
c) How does Baker's performance compare to the industry leaders?

TABLE 11.9 For Problems 11.5 and 11.6

ARROW DISTRIBUTING CORP.	
Net revenue	$16,500
Cost of sales	$13,500
Average inventory	$ 1,000
Total assets	$ 8,600
BAKER MFG. INC.	
Net revenue	$27,500
Cost of sales	$21,500
Average inventory	$ 1,250
Total assets	$16,600

•• **11.6** Dara Jones, operations manager of Arrow Distributing Corp. (see Table 11.9), likes to track inventory by using weeks of supply as well as by inventory turnover.

a) What is Arrow's weeks of supply?
b) What percentage of Arrow's assets are committed to inventory?
c) What is Arrow's inventory turnover?
d) Is Arrow's supply chain performance, as measured by these inventory metrics, better than that of Baker Mfg. in Problem 11.5?

• **11.7** The grocery industry has an annual inventory turnover of about 14 times. Organic Grocers, Inc. had a cost of goods sold last year of $10.5 million; its average inventory was $1.0 million. What was Organic Grocers' inventory turnover, and how does that performance compare with that of the industry?

Tyler Olson/Shutterstock

•• **11.8** Mattress Wholesalers, Inc., is constantly trying to reduce inventory in its supply chain. Last year, cost of goods sold was $7.5 million and inventory was $1.5 million. This year, cost of goods sold is $8.6 million and inventory investment is $1.6 million.

a) What were the weeks of supply last year?
b) What are the weeks of supply this year?
c) Is Mattress Wholesalers making progress in its inventory-reduction effort?

•• **11.9** York Technologies makes Aircraft Navigation Systems. The plant manager Murat Kristal expects you, as the new OM analyst, to provide some insight for performance of the plant. High on his list is an understanding of his inventory turnover based on the financial data in Table 11.10. He expects you to provide him with the following:

a) Based on total inventory, what is the inventory turnover for last year?
b) Based on total inventory, what is the inventory turnover for this year?
c) Has inventory turnover improved this year?

•• **11.10** Based on the data in Table 11.10, you have been asked to determine:

a) The company's percentage of assets committed to inventory last year.
b) The company's percentage of assets committed to inventory this year.
c) The change in the percentage of assets committed to inventory.

•• **11.11** Based on the data in Table 11.10 determine:

a) How many weeks of finished goods were on hand at the end of last year.
b) How many weeks of finished goods were on hand at the end of this year.

TABLE 11.10 For Problems 11.9, 11.10, and 11.11

	LAST YEAR	THIS YEAR
Sales	$ 215,000	$ 228,000
Cost of goods sold	121,000	131,000
Gross margin	94,000	97,000
Other expenses	49,000	49,000
Net income	45,000	48,000
Finished goods inventory	2,000	5,000
Work-in-process inventory	10,000	12,000
Raw material inventory	5,000	6,000
Total inventory (average for year)	17,000	23,000
Other current assets	91,000	110,000
Other assets	213,000	242,000
Total assets	321,000	375,000

CASE STUDIES

Premier Bicycle's COVID Problem

Premier Bicycle Company, located in Knoxville, Tennessee, knows that the strength of its extensive supply chain is a key factor in its production success. The 90-year-old company is the nation's premier manufacturer of off-the-road mountain bikes, with prices ranging from $2,800 to $8,000.

Exceptional product quality is essential to a loyal and growing customer base along with enduring profitability and reputation of the firm. And its longstanding and loyal suppliers have made Wendy Cohen's job of purchasing manager a relatively smooth one—that is, until the 2020 COVID-19 pandemic.

With over 40 suppliers providing more than 1,000 SKUs ranging from common items such as handlebars (in 18 configurations) to wheels (15 options including the $1,800 carbon enduro 29-inch boost wheel) to seats—and to components most laypersons can't even name, such as headset caps, cogalicious cogs, cable guide chucks, and lock-on bar plugs, Wendy knew that any missing link could damage her sensitive JIT logistics system and customer loyalty.

When COVID-19 shuttered her plant in April–June 2020, 90% of Premier's 180 workers were placed on furlough, and production ground to a halt. With new safety measures (distancing, masks, sanitizers, etc.) in place as the economy began to reopen in July 2020, Premier geared up for normal production.

But an amazing change occurred during the summer of 2020. Instead of demand drying up during the shutdown, Americans decided that not working, or staying home to work, meant less need for their cars and SUVs. By the hundreds of thousands, they instead took to the road on bicycles. At which point, retailers ranging from Walmart to local bike shops found their inventories sold out and the supply chain that provided bikes depleted or at a standstill. At one point in August, it became almost impossible to find a specialty bike of any sort.

This exploding demand portended good news to Premier and its many competitors of lesser-priced bicycles. Wendy's boss ordered the plant to resume full production capacity to meet the surprising backlog. Wendy contacted each supplier and immediately prepared to revise economic order quantities based on the newest forecasts. Unfortunately, to her dismay, three of her suppliers had gone out of business—unable to cover their costs with no revenues during the economic shutdown. Others were slowly gearing up, trying to return furloughed workers back to the production line even though many remained fearful of catching the virus at work.

The supply chains for specialty bikes such as Premier's are thin. In some cases, there are only one or two producers of small, but critical parts.

Working 16-hour days, Wendy and her staff scoured the world for parts. Some were air-expressed from new European suppliers. Others required more novel and desperate approaches. In one case, Wendy even formed an alliance with the purchasing group at Dynamic Bicycle Co., a small and lower-end producer also located in Knoxville. Her counterpart could indeed share several SKUs that Wendy could then modify for Premier's products—and vice versa.

Nonetheless, because of the disruption, sales dropped nearly 20% in 2020 as the supply chain very slowly mended. Wendy and her firm needed a new road map.

Discussion Questions

1. What can Wendy do to prevent such disruption in the future?
2. What mistakes did Premier make with respect to its suppliers?
3. What will Premier need to do when forecasting sales in the next few years?
4. What are the disadvantages of an alliance, such as that made with Dynamic?

Darden's Global Supply Chains

Video Case

Darden Restaurants, owner of popular brands such as Olive Garden, Bahama Breeze, Seasons 52, and LongHorn Steakhouse, requires unique supply chains to serve hundreds of millions of meals annually. Darden's strategy is operations excellence, and Senior VP Jim Lawrence's task is to ensure competitive advantage via Darden's supply chains. For a firm with purchases exceeding $1.8 billion from 1,500 suppliers, managing the supply chains is a complex and challenging task.

Darden, like other casual dining restaurants, has unique supply chains that reflect its menu options. Darden's supply chains are rather shallow, often having just one tier of suppliers. But it has four distinct supply chains.

First, "smallware" is a restaurant industry term for items such as linens, dishes, tableware and kitchenware, and silverware. These are purchased, with Darden taking title as they are received at the Darden Direct Distribution (DDD) warehouse in Orlando, Florida. From this single warehouse, smallware items are shipped via common carrier (trucking companies) to Olive Garden, Bahama Breeze, and Seasons 52 restaurants.

Second, frozen, dry, and canned food products are handled economically by Darden's 11 distribution centers in North America, which are managed by major U.S. food distributors, such as MBM, Maines, and Sygma. This is Darden's second supply line.

Third, the fresh food supply chain (not frozen and not canned), where product life is measured in days, includes dairy products, produce, and meat. This supply chain is B2B, where restaurant managers directly place orders with a preselected group of independent suppliers.

Fourth, Darden's worldwide seafood supply chain is the final link. Here Darden has developed independent suppliers of salmon, shrimp, tilapia, scallops, and other fresh fish that are source inspected by Darden's overseas representatives to ensure quality. These fresh products are flown to the U.S. and shipped to 16 distributors, with 22 locations, for quick delivery to the restaurants. With suppliers in 35 countries, Darden must be on the cutting edge when it comes to collaboration, partnering, communication, and food safety. It does this with heavy travel schedules for purchasing and quality control personnel, native-speaking employees on site, and aggressive communication. Communication is a critical element; Darden tries to develop as much forecasting transparency as possible. "Point of sale (POS) terminals," says Lawrence, "feed actual sales every night to suppliers."

Discussion Questions*

1. What are the advantages of each of Darden's four supply chains?
2. What are the complications of having four supply chains?
3. Where would you expect ownership/title to change in each of Darden's four supply chains?
4. How do Darden's four supply chains compare with those of other firms, such as Dell or an automobile manufacturer? Why do the differences exist, and how are they addressed?

*You may wish to view the video that accompanies this case before answering these questions.

Arnold Palmer Hospital's Supply Chain

Arnold Palmer Hospital, one of the nation's top hospitals dedicated to serving women and children, is a large business with more than 2,000 employees working in a 431-bed facility totaling 676,000 square feet in Orlando, Florida. Like many other hospitals, and other companies, Arnold Palmer Hospital had been a long-time member of a large buying group, one servicing 900 members. But the group did have a few limitations. For example, it might change suppliers for a particular product every year (based on a new lower-cost bidder) or stock only a product that was not familiar to the physicians at Arnold Palmer Hospital. The buying group was also not able to negotiate contracts with local manufacturers to secure the best pricing.

So in 2003, Arnold Palmer Hospital, together with seven other partner hospitals in central Florida, formed its own much smaller, but still powerful (with $200 million in annual purchases) Healthcare Purchasing Alliance (HPA) corporation. The new alliance saved the HPA members $7 million in its first year with two main changes. First, it was structured and staffed to ensure that the bulk of the savings associated with its contracting efforts went to its eight members. Second, it struck even better deals with vendors by guaranteeing a *committed* volume and signing not 1-year deals but 3- to 5-year contracts. "Even with a new internal cost of $400,000 to run HPA, the savings and ability to contract for what our member hospitals really want makes the deal a winner," says George DeLong, head of HPA.

Effective supply chain management in manufacturing often focuses on development of new product innovations and efficiency through buyer–vendor collaboration. However, the approach in a service industry has a slightly different emphasis. At Arnold Palmer Hospital, supply chain opportunities often manifest themselves through the Medical Economic Outcomes Committee. This committee (and its subcommittees) consists of users (including the medical and nursing staff) who evaluate purchase options with a goal of better medicine while achieving economic targets. For instance, the heart pacemaker negotiation by the cardiology subcommittee allowed for the standardization to two manufacturers, with annual savings of $2 million for just this one product.

Arnold Palmer Hospital is also able to develop custom products that require collaboration down to the third tier of the supply chain. This is the case with custom packs that are used in the operating room. The custom packs are delivered by a distributor, McKesson General Medical, but assembled by a pack company that uses materials the hospital wanted purchased from specific manufacturers. The HPA allows Arnold Palmer Hospital to be creative in this way. With major cost savings, standardization, blanket purchase orders, long-term contracts, and more control of product development, the benefits to the hospital are substantial.

Discussion Questions*

1. How does this supply chain differ from that in a manufacturing firm?
2. What are the constraints on making decisions based on economics alone at Arnold Palmer Hospital?
3. What role do doctors and nurses play in supply chain decisions in a hospital? How is this participation handled at Arnold Palmer Hospital?
4. Doctor Smith just returned from the Annual Physician's Orthopedic Conference, where she saw a new hip joint replacement demonstrated. She decides she wants to start using the replacement joint at Arnold Palmer Hospital. What process will Dr. Smith have to go through at the hospital to introduce this new product into the supply chain for future surgical use?

*You may wish to view the video that accompanies this case before answering the questions.

Endnotes

1 Sources: *The Wall Street Journal* (February 15, 2018); *Supply Chain Dive* (February 14, 2017); and **EconLife.com** (February 12, 2018).

2 See related table and discussion in Marshall L. Fisher, "What Is the Right Supply Chain for Your Product?" *Harvard Business Review* (March–April 1997): 105.

3 Sources: *Forbes* (March 13, 2020); *Engineering & Technology* (July 10, 2020); and *Financial Times* (May 7, 2020).

4 Sources: *The New York Times* (October 27, 2019) and (March 4, 2021).

Bibliography

Cetinkaya, S., H. Uster, G. Easwaran, and B. B. Keskin. "An Integrated Outbound Logistics Model for Frito-Lay: Coordinating Aggregate-Level Production and Distribution Decisions." *Interfaces* 39, no. 5 (Sep./Oct. 2009): 460–475.

Chaman, L. J. "Building Resilient Supply Chains During Times of Extreme Uncertainty," *Journal of Business Forecasting* 39, no. 4 (Winter 2020-2021): 4–9.

Chopra, S. *Supply Chain Management: Strategy, Planning, and Operation*, 7th ed. Boston: Pearson, 2019.

Crandall, R. E. "The Next Wave of Sourcing." *APICS* 24, no. 3 (May/June 2014): 26–28.

Hu, J., and C. L. Munson. "Speed versus Reliability Trade-offs in Supplier Selection." *International Journal of Procurement Management* 1, no. 1/2 (2007): 238–259.

Kersten, W., and T. Blecker (eds.). *Managing Risk in Supply Chains.* Berlin: Erich Schmidt Verlag GmbH & Co., 2006.

Kersten, W., M. Seiter, B. von See, N. Hackius, and T. Maurer. *Trends and Strategies in Logistics and Supply Chain Management*. Hamburg: DVV Media Group, 2017.

Kumar, G., R. N. Banerjee, P. L. Meena, and K. Ganguly. "Collaborative Culture and Relationship Strength Roles in Collaborative Relationships: A Supply Chain Perspective." *Journal of Business & Industrial Marketing*, 31, no. 5 (2016): 587–599.

Monczka, R. M., R. B. Handfield, L. C. Gianipero, and J. L. Patterson. *Purchasing and Supply Chain Management*, 6th ed. Mason, OH: Cengage, 2016.

Munson, C. (ed.). *The Supply Chain Management Casebook: Comprehensive Coverage and Best Practices in SCM.* Upper Saddle River, NJ: FT Press, 2013.

Wisner, J., K. Tan, and G. K. Leong. *Principles of Supply Chain Management: A Balanced Approach*, 5th ed., Mason, OH: Cengage, 2019.

Yang, S, C. L. Munson, B. Chen, and C. Shi. "Coordinated Contracts for Supply Chains that Market with Mail-In Rebates and Retailer Promotions." *Journal of the Operational Research Society* 66, no. 12 (2015), 2025–2036.

Chapter 11 *Rapid* Review

Main Heading	Review Material	MyLab Operations Management
THE SUPPLY CHAIN'S STRATEGIC IMPORTANCE	Most firms spend a huge portion of their sales dollars on purchases. ■ **Supply chain management**—Management of activities related to procuring materials and services, transforming them into intermediate goods and final products, and delivering them through a distribution system. *The objective is to build a chain of suppliers that focuses on maximizing value to the ultimate customer*. Competition is no longer between companies; it is between supply chains.	Concept Questions: 1.1–1.6 Problems: 11.2–11.3
SOURCING ISSUES: MAKE-OR-BUY AND OUTSOURCING	■ **Make-or-buy decision**—A choice between producing a component or service within the firm or purchasing it from an outside source. ■ **Outsourcing**—Transferring to external suppliers a firm's activities that have traditionally been internal.	Concept Questions: 2.1–2.2
SIX SOURCING STRATEGIES	Six supply chain strategies for goods and services to be obtained from outside sources are: 1. Negotiating with many suppliers and playing one supplier against another 2. Developing long-term partnering relationships with a few suppliers 3. Vertical integration 4. Joint ventures 5. Developing *keiretsu* networks 6. Developing virtual companies that use suppliers on an as-needed basis. ■ **Vertical integration**—Developing the ability to produce goods or services previously purchased or actually buying a supplier or a distributor. ■ ***Keiretsu***—A Japanese term that describes suppliers who become part of a company coalition. ■ **Virtual companies**—Companies that rely on a variety of supplier relationships to provide services on demand. Also known as hollow corporations or network companies.	Concept Questions: 3.1–3.6 **VIDEO 11.1** Darden's Global Supply Chain
SUPPLY CHAIN RISK	The development of a supply chain plan requires a thorough assessment of the risks involved. ■ **Cross-sourcing**—Using one supplier for a component and a second supplier for another component, where each supplier acts as a backup for the other.	Concept Questions: 4.1–4.6
MANAGING THE INTEGRATED SUPPLY CHAIN	Supply chain integration success begins with mutual agreement on goals, followed by mutual trust, and continues with compatible organizational cultures. Three issues complicate the development of an efficient, integrated supply chain: local optimization, incentives, and large lots. ■ **Bullwhip effect**—Increasing fluctuation in orders or cancellations that often occurs as orders move through the supply chain. ■ **Pull data**—Accurate sales data that initiate transactions to "pull" product through the supply chain. ■ **Single-stage control of replenishment**—Fixing responsibility for monitoring and managing inventory for the retailer. ■ **Vendor-managed inventory (VMI)**—A system in which a supplier maintains material for the buyer, often delivering directly to the buyer's using department. ■ **Collaborative planning, forecasting, and replenishment (CPFR)**—A system in which members of a supply chain share information in a joint effort to reduce supply chain costs. ■ **Blanket order**—A long-term purchase commitment to a supplier for items that are to be delivered against short-term releases to ship. The purchasing department should make special efforts to increase levels of standardization. ■ **Postponement**—Delaying any modifications or customization to a product as long as possible in the production process. Postponement strives to minimize internal variety while maximizing external variety. ■ **Omnichannel**—An integrated multichannel strategy for communication and delivery that improves the user experience. ■ **Drop shipping**—Shipping directly from the supplier to the end consumer rather than from the seller, saving both time and reshipping costs. ■ **Blockchain**—A technology that tracks, documents, and verifies by adding "blocks" of information after each transaction. Online catalogs move companies from a multitude of individual phone calls, faxes, and e-mails to a centralized online system and drive billions of dollars of waste out of the supply chain.	Concept Questions: 5.1–5.6 **VIDEO 11.2** Arnold Palmer Hospital's Supply Chain
BUILDING THE SUPPLY BASE	Supplier selection is a four-stage process: (1) supplier evaluation, (2) supplier development, (3) negotiations, and (4) contracting. *Supplier evaluation* involves finding potential vendors and determining the likelihood of their becoming good suppliers. *Supplier development* may include everything from training, to engineering and production help, to procedures for information transfer.	Concept Questions: 6.1–6.6

Chapter 11 ***Rapid*** **Review** *continued*

Main Heading	Review Material	MyLab Operations Management
	Negotiations involve approaches taken by supply chain personnel to set prices. Three classic types of negotiation strategies are (1) the cost-based price model, (2) the market-based price model, and (3) competitive bidding. *Contracting* involves a design to share risks, share benefits, and create incentives so as to optimize the whole supply chain.	
LOGISTICS MANAGEMENT	■ **Logistics management**—An approach that seeks efficiency of operations through the integration of all material acquisition, movement, and storage activities. Six major means of distribution are trucking, railroads, airfreight, waterways, pipelines, and multimodal. The vast majority of manufactured goods move by truck. Third-party logistics involves the outsourcing of the logistics function. ■ **Channel assembly**—A system that postpones final assembly of a product so the distribution channel can assemble it.	Concept Questions: 7.1–7.6
DISTRIBUTION MANAGEMENT	Distribution management focused on the outbound flow of final products. Total logistics costs are the sum of facility costs, inventory costs, and transportation costs (Figure 11.3). The optimal number of distribution facilities focuses on maximizing profit.	Concept Questions: 8.1–8.4
ETHICS AND SUSTAINABLE SUPPLY CHAIN MANAGEMENT	Ethics includes personal ethics, ethics within the supply chain, and ethical behavior regarding the environment. The Institute for Supply Management has developed a set of Principles and Standards for ethical conduct. ■ **Reverse logistics**—The process of sending returned products back up the supply chain for value recovery or disposal. ■ **Closed-loop supply chain**—A supply chain designed to optimize all forward and reverse flows.	Concept Questions: 9.1–9.5
MEASURING SUPPLY CHAIN PERFORMANCE	Typical supply chain benchmark metrics include lead time, time spent placing an order, percentage of late deliveries, percentage of rejected material, and number of shortages per year: Percentage invested in inventory = (Average inventory investment/Total assets) × 100 (11-1) ■ **Inventory turnover**—Cost of goods sold divided by average inventory: Inventory turnover = Cost of goods sold ÷ Average inventory investment (11-2) Weeks of supply = Avg. Inv. investment ÷ (Annual cost of goods sold/52 weeks) (11-3) ■ **Supply Chain Operations Reference (SCOR) model**—A set of processes, metrics, and best practices developed by the APICS Supply Chain Council. The six parts of the SCOR model are Plan, Source, Make, Deliver, Return, and Enable.	Concept Questions: 10.1–10.6 Problems: 11.5–11.11 Virtual Office Hours for Solved Problem: 11.1
ADDITIONAL MYLAB OPERATIONS MANAGEMENT RESOURCES	✔ Multiple Choice Case Questions (Premier Bicycle's COVID Problem) ✔ Recent Graduate Video: Nicholas Kostner, Supply Chain Manager, Kryton Engineered Metals ✔ Supply Chain Management Simulation	

Self **Test**

LO 11.1 The objective of supply chain management is to _____.

LO 11.2 The term *vertical integration* means to:
- **a)** develop the ability to produce products that complement or supplement the original product.
- **b)** produce goods or services previously purchased.
- **c)** develop the ability to produce the specified good more efficiently.
- **d)** all of the above.

LO 11.3 The bullwhip effect can be aggravated by:
- **a)** local optimization.
- **b)** sales incentives.
- **c)** quantity discounts.
- **d)** promotions.
- **e)** all of the above.

LO 11.4 Supplier selection requires:
- **a)** supplier evaluation and effective third-party logistics.
- **b)** supplier development and logistics.
- **c)** negotiations, supplier evaluation, supplier development, and contracts.
- **d)** an integrated supply chain.
- **e)** inventory and supply chain management.

LO 11.5 A major issue in logistics is:
- **a)** cost of purchases.
- **b)** supplier evaluation.
- **c)** product customization.
- **d)** cost of shipping alternatives.
- **e)** excellent e-procurement.

LO 11.6 Inventory turnover =
- **a)** Cost of goods sold ÷ Weeks of supply
- **b)** Weeks of supply ÷ Annual cost of goods sold
- **c)** Annual cost of goods sold ÷ 52 weeks
- **d)** Average inventory investment ÷ Cost of goods sold
- **e)** Cost of goods sold ÷ Average inventory investment

Answers: LO 11.1. build a chain of suppliers that focuses on maximizing value to the ultimate customer; LO 11.2. b; LO 11.3. e; LO 11.4. c; LO 11.5. d; LO 11.6. e.

Supply Chain Management Analytics

SUPPLEMENT 11

SUPPLEMENT OUTLINE

LEARNING OBJECTIVES

LO S11.1 ***Use*** a decision tree to determine the best number of suppliers to manage disaster risk 472

LO S11.2 ***Explain*** and measure the bullwhip effect 474

LO S11.3 ***Describe*** the factor-weighting approach to supplier evaluation 476

LO S11.4 ***Evaluate*** cost-of-shipping alternatives 477

LO S11.5 ***Allocate*** items to storage locations in a warehouse 479

The Tōhoku earthquake and tsunami devastated eastern sections of Japan. The economic impact was felt around the globe, as manufacturers had been relying heavily—in some cases exclusively—on suppliers located in the affected zones. In the month immediately following the earthquake, the Japanese-built vehicle outputs for both Toyota and Honda were down 63%. Plants in other countries ceased or reduced operations due to part shortages. Manufacturers in several industries worldwide took 6 months or longer before they saw their supply chains working normally again. Although disasters such as this one occur relatively infrequently, supply chain managers should consider their probabilities and repercussions when determining the makeup of the supply base.

mTaira/Shutterstock

Techniques for Evaluating Supply Chains

Many supply chain metrics exist that can be used to evaluate performance within a company as well as for its supply chain partners. This supplement introduces five techniques that are aimed at ways to build and evaluate performance of the supply chain.

Evaluating Disaster Risk in the Supply Chain

Disasters that disrupt supply chains can take many forms, including tornadoes, fires, hurricanes, typhoons, tsunamis, earthquakes, terrorism, and even pandemics. When you are deciding whether to purchase collision insurance for your car, the amount of insurance must be weighed against the probability of a minor accident occurring and the potential financial worst-case scenario if an accident happens (e.g., "totaling" of the car). Similarly, firms often use multiple suppliers for important components to mitigate the risks of total supply disruption.

LO S11.1 *Use a* decision tree to determine the best number of suppliers to manage disaster risk

As shown in Example S1, a decision tree can be used to help operations managers make this important decision regarding the number of suppliers. We will use the following notation for a given supply cycle:

S = the probability of a "super-event" that would disrupt *all* suppliers simultaneously
U = the probability of a "unique-event" that would disrupt only one supplier
L = the financial loss incurred in a supply cycle if *all* suppliers were disrupted
C = the marginal cost of managing a supplier

All suppliers will be disrupted simultaneously if either the super-event occurs or the super-event does not occur but a unique-event occurs for all of the suppliers. Assuming that the probabilities are all independent of each other, the probability of all n suppliers being disrupted simultaneously equals:

$$P(n) = S + (1 - S)U^n \tag{S11-1}$$

Example S1

HOW MANY SUPPLIERS ARE BEST FOR MANAGING RISK?

Xiaotian Geng, president of Shanghai Manufacturing Corp., wants to create a portfolio of suppliers for the motors used in her company's products that will represent a reasonable balance between costs and risks. While she knows that the single-supplier approach has many potential benefits with respect to quality management and just-in-time production, she also worries about the risk of fires, natural disasters, or other catastrophes at supplier plants disrupting her firm's performance. Based on historical data and climate and geological forecasts, Xiaotian estimates the probability of a "super-event" that would negatively impact all suppliers simultaneously to be 0.5% (i.e., probability = 0.005) during the supply cycle. She further estimates the "unique-event" risk for any of the potential suppliers to be 4% (probability = 0.04). Assuming that the marginal cost of managing an additional supplier is \$10,000, and the financial loss incurred if a disaster caused all suppliers to be down simultaneously is \$10,000,000, how many suppliers should Xiaotian use? Assume that up to three nearly identical suppliers are available.

APPROACH ▶ Use of a decision tree seems appropriate, as Shanghai Manufacturing Corp. has the basic data: a choice of decisions, probabilities, and payoffs (costs).

SOLUTION ▶ We draw a decision tree (Figure S11.1) with a branch for each of the three decisions (one, two, or three suppliers), assign the respective probabilities [using Equation (S11-1)] and payoffs for each branch, and then compute the respective expected monetary values (EMVs). The EMVs have been identified at each step of the decision tree.

Using Equation (S11-1), the probability of a total disruption equals:

One supplier: $0.005 + (1 - 0.005)0.04 = 0.005 + 0.0398 = 0.044800$, or 4.4800%
Two suppliers: $0.005 + (1 - 0.005)0.04^2 = 0.005 + 0.001592 = 0.006592$, or 0.6592%
Three suppliers: $0.005 + (1 - 0.005)0.04^3 = 0.005 + 0.000064 = 0.005064$, or 0.5064%

INSIGHT ▶ Even with significant supplier management costs and unlikely probabilities of disaster, a large enough financial loss incurred during a total supplier shutdown will suggest that multiple suppliers may be needed.

Figure **S11.1**

Decision Tree for Selection of Suppliers under Risk

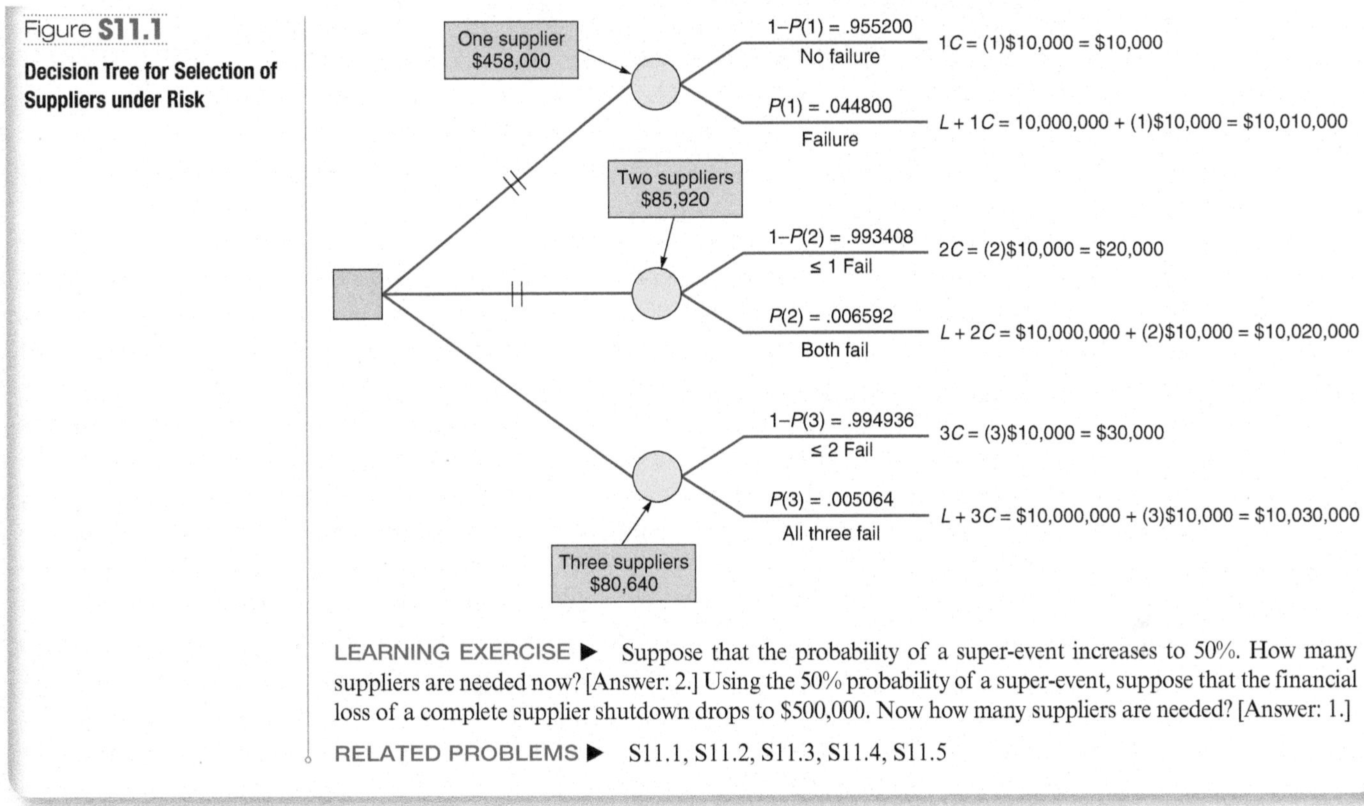

LEARNING EXERCISE ▶ Suppose that the probability of a super-event increases to 50%. How many suppliers are needed now? [Answer: 2.] Using the 50% probability of a super-event, suppose that the financial loss of a complete supplier shutdown drops to $500,000. Now how many suppliers are needed? [Answer: 1.]

RELATED PROBLEMS ▶ S11.1, S11.2, S11.3, S11.4, S11.5

An interesting implication of Equation (S11-1) is that as the probability of a super-event (S) increases, the advantage of having multiple suppliers diminishes (all would be knocked out anyway). On the other hand, large values of the unique event (U) increase the likelihood of needing more suppliers. These two phenomena taken together suggest that when multiple suppliers are used, managers may consider using ones that are geographically dispersed to lessen the probability of all failing simultaneously.

Managing the Bullwhip Effect

Figure S11.2 provides an example of the *bullwhip effect*, which describes the tendency for larger order size fluctuations as orders are relayed to the supply chain from retailers. "Bullwhip" fluctuations create unstable production schedules. This can result in expensive capacity change adjustments such as overtime, subcontracting, extra inventory, backorders, hiring and laying off

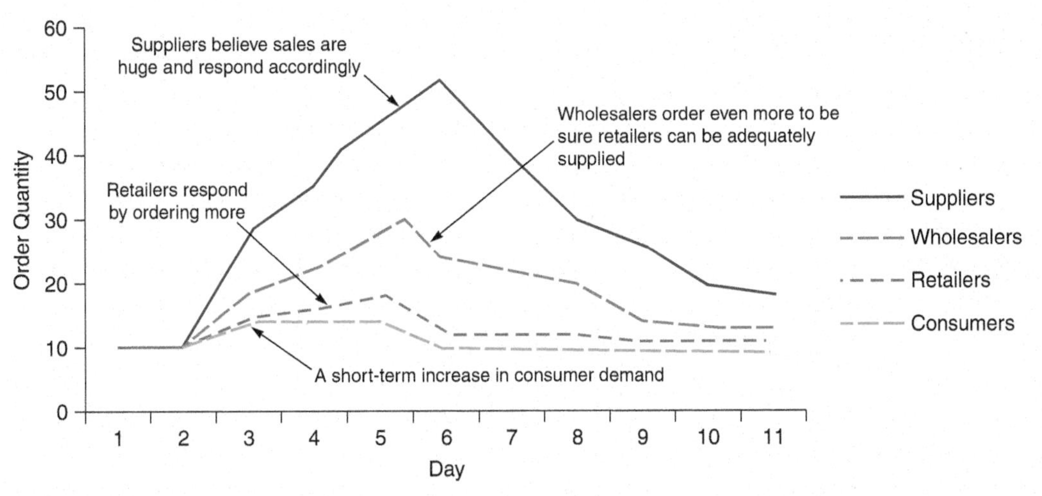

Figure **S11.2**

The Bullwhip Effect

The bullwhip effect causes members of the supply chain to overreact to changes in demand at the retail level. Minor demand changes at the consumer level may result in large ones at the supplier level.

TABLE S11.1 The Bullwhip Effect

CAUSE	REMEDY
Demand forecast errors (cumulative uncertainty in the supply chain)	Share demand information throughout the supply chain.
Order batching (large, infrequent orders leading suppliers to order even larger amounts)	Channel coordination: Determine lot sizes as though the full supply chain was one company.
Price fluctuations (buying in advance of demand to take advantage of low prices, discounts, or sales)	Price stabilization (everyday low prices).
Shortage gaming (hoarding supplies for fear of a supply shortage)	Allocate orders based on past demand.

of workers, and equipment modifications. Other impacts may include underutilization, longer lead times, and obsolescence of overproduced items.

Procter & Gamble found that although the use of Pampers diapers was steady and the retail-store orders had little fluctuation, as orders moved through the supply chain, fluctuations increased. By the time orders were initiated for raw material, the variability was substantial. Similar behavior has been observed and documented at many companies, including Campbell Soup, Hewlett-Packard, Barilla SpA, and Applied Materials. (See the *OM in Action* box, "The Bullwhip Effect Strikes Again.")

The bullwhip effect can occur when orders decrease as well as when they increase. Table S11.1 identifies some of the major causes and remedies of the bullwhip effect. Often the human tendency to overreact to stimuli causes managers to make decisions that exacerbate the phenomenon. The overarching solution to the bullwhip effect is simply for supply chain members to share information and work together, as seen in Figure S11.3.

Supplier coordination can help with demand shifts. During a recent worldwide recession, but prior to experiencing the economic recovery and increasing sales, Caterpillar started ordering more supplies. It also worked proactively with its suppliers to prepare them for a sharp increase in output. Caterpillar visited key suppliers individually. In some cases it helped suppliers obtain bank financing at favorable rates. As part of Caterpillar's risk-assessment activities, suppliers had to submit written plans describing their ability to ramp production back up once the economy improved. Careful, coordinated planning can help alleviate shortages and delays that might otherwise occur as the bullwhip snaps back upward.

LO S11.2 *Explain* and measure the bullwhip effect

A Bullwhip Effect Measure

A straightforward way to analyze the extent of the bullwhip effect at any link in the supply chain is to calculate the *bullwhip measure*:

$$\text{Bullwhip} = \frac{\text{Variance of orders}}{\text{Variance of demand}} = \frac{\sigma^2_{\text{orders}}}{\sigma^2_{\text{demand}}} \qquad \text{(S11-2)}$$

OM in Action — The Bullwhip Effect Strikes Again[1]

Supply chains typically get beaten up during reoccurring business cycles and rapid demand changes stemming from unexpected events such as the COVID-19 pandemic. As sales decline, companies draw down inventories to conserve cash instead of purchasing more parts and materials. Entire pipelines of supplies get cleaned out.

When demand improves, even modestly, suppliers may respond with an outsized increase in production to restock empty warehouses and assembly plants. The bullwhip effect ripples all along supply chains, generating unusually large orders for suppliers that are far from end customers. During the pandemic, the bullwhip effect was even more pronounced because demand for consumer products was extraordinarily high. At the same time, companies were placing supersized orders to compensate for the extra time it took to procure supplies from factories and freight operators constrained by global efforts to contain the coronavirus. That exacerbated the strain on supply chains.

A decades-long devotion to making supply chains leaner and more efficient made companies more vulnerable to the distortions of the bullwhip effect. To cut costs and boost profits, U.S. companies outsourced operations and whittled inventories. Many of their suppliers did the same. When demand increased unexpectedly during the pandemic, the same companies all placed orders at once into increasingly diffuse networks of far-flung suppliers. The result was a bullwhip crack more dramatic than usual. Everybody was saying, "I need to order a lot more."

Nautique Boats, for example, was preparing for a pandemic downturn as it stopped production. To the company's surprise, within weeks its dealers started reporting many new customers. When Nautique reopened, it ramped up production, but some suppliers were slow to respond. This led to shortages of components for engines, windshields, and wiring harnesses. The pandemic once again illustrated that lean supply chains leave little room for rapid change.

Variance *amplification* (i.e., the bullwhip effect) is present if the bullwhip measure is greater than 1. This means the size of a company's orders fluctuate more than the size of its incoming demand. If the measure equals 1, then no amplification is present. A value less than 1 would imply a *smoothing* or *dampening* scenario as orders move up the supply chain toward suppliers. Example S2 illustrates how to use Equation (S11-2) to analyze the extent of the bullwhip effect at each stage in the supply chain.

Example S2

CALCULATING THE BULLWHIP EFFECT

Chieh Lee Metals, Inc. orders sheet metal and transforms it into 50 formed tabletops that are sold to furniture manufacturers. The following table shows the weekly variance of demand and orders for each major company in this supply chain for tables. Each firm has one supplier and one customer, so the order variance for one firm will equal the demand variance for its supplier. Analyze the relative contributions to the bullwhip effect in this supply chain.

FIRM	VARIANCE OF DEMAND	VARIANCE OF ORDERS	BULLWHIP MEASURE
Furniture Mart, Inc.	100	110	110/100 = 1.10
Furniture Distributors, Inc.	110	180	180/110 = 1.64
Furniture Makers of America	180	300	300/180 = 1.67
Chieh Lee Metals, Inc.	300	750	750/300 = 2.50
Metal Suppliers Ltd.	750	2000	2000/750 = 2.67

APPROACH ▶ Use Equation (S11-2) to calculate the bullwhip measure for each firm in the chain.

SOLUTION ▶ The last column of the table displays the bullwhip measure for each firm.

INSIGHT ▶ This supply chain exhibits a classic bullwhip effect. Despite what might be a very stable demand pattern at the retail level, order sizes to suppliers vary significantly. Chieh Lee should attempt to identify the causes for her own firm's order amplification, and she should attempt to work with her supply chain partners to try to reduce amplification at every level of the chain.

LEARNING EXERCISE ▶ Suppose that Chieh Lee is able to reduce her bullwhip measure from 2.50 to 1.20. If the measure for all other firms remained the same, what would be the new reduced variance of orders from Metal Suppliers? [Answer: 961.]

RELATED PROBLEMS ▶ S11.6, S11.7, S11.8, S11.9

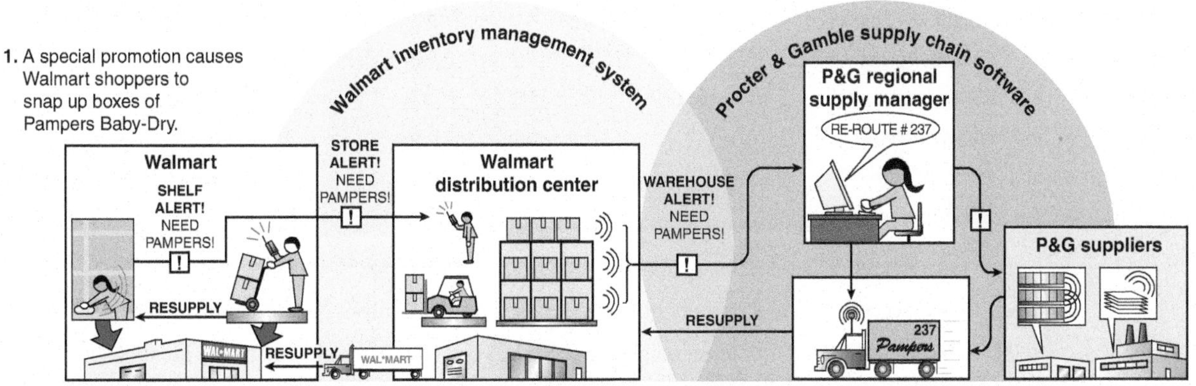
1. A special promotion causes Walmart shoppers to snap up boxes of Pampers Baby-Dry.
Walmart inventory management system
Procter & Gamble supply chain software
Walmart
SHELF ALERT! NEED PAMPERS!
RESUPPLY
STORE ALERT! NEED PAMPERS!
Walmart distribution center
RESUPPLY
WAREHOUSE ALERT! NEED PAMPERS!
RESUPPLY
P&G regional supply manager
RE-ROUTE # 237
P&G suppliers
237
Pampers
2. Each box of Pampers has an RFID tag. Shelf-mounted scanners alert the stockroom of urgent need for restock.
3. Walmart's inventory management system tracks and links its in-store stock and its warehouse stock, prompting quicker replenishment and providing accurate real-time data.
4. Walmart's systems are linked to the P&G supply chain management system. Demand spikes reported by RFID tags are immediately visible throughout the supply chain.
5. P&G's logistics software tracks its trucks with GPS locators, and tracks their contents with RFID tag readers. Regional managers can reroute trucks to fill urgent needs.
6. P&G suppliers also use RFID tags and readers on their raw materials, giving P&G visibility several tiers down the supply chain and giving suppliers the ability to accurately forecast demand and production.

Figure **S11.3**

RFID Helps Control the Bullwhip

Supply chains often break down when confronted by a sudden surge or rapid drop in demand. Radio frequency ID (RFID) tags can change that by providing real-time information about what's happening on store shelves. Here's how the system works for Procter & Gamble's (P&G's) Pampers.

STUDENT TIP
The factor-weighting model adds objectivity to decision making.

Supplier Selection Analysis

Selecting suppliers from among a multitude of candidates can be a daunting task. Choosing suppliers simply based on the lowest bid has become a somewhat rare approach. Various, sometimes competing, factors often play a role in the decision. Buyers may consider such supplier characteristics as product quality, delivery speed, delivery reliability, customer service, and financial performance.

VIDEO S11.1
Supply Chain Issues at Nautique Boat Company

LO S11.3 *Describe the factor-weighting approach to supplier evaluation*

The *factor-weighting* technique, presented here, simultaneously considers multiple supplier criteria. Each factor must be assigned an importance *weight*, and then each potential supplier is *scored* on each factor. The weights typically sum to 100%. Factors are scored using the same scale (e.g., 1 – 10). Sometimes a key is provided for supplier raters that converts qualitative ratings into numerical scores (e.g., "Very good" = 8). Example S3 illustrates the weighted criteria in comparing two competing suppliers.

Example S3

FACTOR-WEIGHTING APPROACH TO SUPPLIER EVALUATION

Erick Davis, president of Creative Toys in Palo Alto, California, is interested in evaluating suppliers who will work with him to make nontoxic, environmentally friendly paints and dyes for his line of children's toys. This is a critical strategic element of his supply chain, and he desires a firm that will contribute to his product.

APPROACH ▶ Erick has narrowed his choices to two suppliers: Faber Paint and Smith Dye. He will use the factor-weighting approach to supplier evaluation to compare the two.

SOLUTION ▶ Erick develops the following list of selection criteria. He then assigns the weights shown to help him perform an objective review of potential suppliers. His staff assigns the scores and computes the total weighted score.

		FABER PAINT		*SMITH DYE*	
CRITERION	WEIGHT	SCORE (1 – 5) (5 HIGHEST)	WEIGHT × SCORE	SCORE (1 – 5) (5 HIGHEST)	WEIGHT × SCORE
Engineering/innovation skills	.20	5	1.0	5	1.0
Production process capability	.15	4	0.6	5	0.75
Distribution capability	.05	4	0.2	3	0.15
Quality performance	.10	2	0.2	3	0.3
Facilities/location	.05	2	0.1	3	0.15
Financial strength	.15	4	0.6	5	0.75
Information systems	.10	2	0.2	5	0.5
Integrity	.20	5	1.0	3	0.6
Total	1.00		3.9		4.2

Smith Dye received the higher score of 4.2 and, based on this analysis, would be the preferred vendor.

INSIGHT ▶ The use of a factor-weighting approach can help firms systematically identify the features that are important to them and evaluate potential suppliers in an objective manner. A certain degree of subjectivity remains in the process, however, with regard to the criteria chosen, the weights applied to those criteria, and the supplier scores that are applied to each criterion.

LEARNING EXERCISE ▶ If Erick believes that integrity should be twice as important while production process capability and financial strength should both only be 1/3 as important, how does the analysis change? [Answer: Faber Paint's score becomes 4.1, while Smith Dye's score becomes 3.8, so Faber Paint is now the preferred vendor.]

RELATED PROBLEMS ▶ S11.10–S11.15

Transportation Mode Analysis

The longer a product is in transit, the longer the firm has its money invested. But faster shipping is usually more expensive than slow shipping. A simple way to obtain some insight into this trade-off is to evaluate holding cost against shipping options. We do this in Example S4.

Example S4

DETERMINING DAILY COST OF HOLDING

A shipment of new connectors for semiconductors needs to go from San Jose to Singapore for assembly. The value of the connectors is $1,750, and holding cost is 40% per year. One airfreight carrier can ship the connectors 1 day faster than its competitor, at an extra cost of $20.00. Which carrier should be selected?

APPROACH ▶ First we determine the daily holding cost and then compare the daily holding cost with the cost of faster shipment.

SOLUTION ▶ Daily cost of holding the product $= (\text{Annual holding cost} \times \text{Product value})/365$

$$= (.40 \times \$1,750)/365$$
$$= \$1.92$$

LO S11.4 *Evaluate* cost-of-shipping alternatives

Because the cost of saving one day is $20.00, which is much more than the daily holding cost of $1.92, we decide on the less costly of the carriers and take the extra day to make the shipment. This saves $18.08 ($20.00 − $1.92).

INSIGHT ▶ The solution becomes radically different if the 1-day delay in getting the connectors to Singapore delays delivery (making a customer angry) or delays payment of a $150,000 final product. (Even 1 day's interest on $150,000 or an angry customer makes a savings of $18.08 insignificant.)

LEARNING EXERCISE ▶ If the holding cost is 100% per year, what is the decision? [Answer: Even with a holding cost of $4.79 per day, the less costly carrier is selected.]

RELATED PROBLEMS ▶ S11.16 – S11.19

Example S4 looks only at holding cost versus shipping cost. For the operations or logistics manager there are many other considerations, including ensuring *on-time delivery*, coordinating shipments to maintain a schedule, getting a new product to market, and keeping a customer happy. Estimates of these other costs can be added to the estimate of the daily holding cost. Determining the impact and cost of these considerations makes the evaluation of shipping alternatives a challenging OM task.

Warehouse Storage

Storage represents a significant step for many items as they travel through their respective supply chains. The U.S. alone has more than 19,000 buildings dedicated to warehouse and storage. Some exceed the size of several connected football fields. In fact, more than 35% have over 100,000 square feet of floor space.

These storage facilities typically use substantial automation and very sophisticated software that tracks, locates, and optimally stores items in what may appear to be random locations (see the discussion in Chapter 9). Buried in that software will be code that will consider everything from damage potential, to size and weight, open locations, combining orders, picking sequences, frequency of use, travel time, and so forth. Placement of items makes a huge difference in both required storage space and order-filling time. In Example S5, we introduce one important trade-off of managing such a facility.

Example S5 DETERMINING STORAGE LOCATIONS IN A WAREHOUSE

Erika Marsillac manages a warehouse for a local chain of specialty hardware stores. As seen in Figure S11.4, the single-aisle rectangular warehouse has a dock for pickup and delivery, along with 16 equal-sized storage blocks for inventory items.

Figure **S11.4**

Storage Locations in the Warehouse

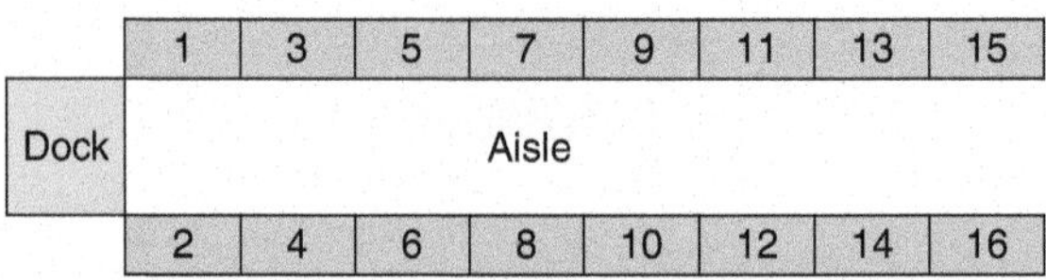

The following table shows: (1) the category of each item stored in the warehouse, (2) the estimated number of times per month (trips) that workers need to either store or retrieve those items, and (3) the area (number of specialized blocks) required to store the items. Erika wishes to assign items to the storage blocks to minimize average distance traveled.

ITEM	MONTHLY TRIPS TO STORAGE	BLOCKS OF STORAGE SPACE NEEDED
Lumber	600	5
Paint	260	2
Tools	150	3
Small hardware	400	2
Chemical bags	90	3
Lightbulbs	220	1

LO S11.5 *Allocate* items to storage locations in a warehouse

APPROACH ▶ For each item, calculate the ratio of the number of trips to blocks of storage area needed. Rank the items according to this ratio, and place the *highest*-ranked items closest to the dock.

SOLUTION ▶ The following table calculates the ratio for each item and ranks the items from highest to lowest. Based on the ranking, items are assigned to the remaining blocks that are as close to the dock as possible. (Where applicable, given a choice between two equidistant blocks, items should be placed next to items of the same type rather than across the aisle from them.)

ITEM	TRIPS/BLOCKS	RANKING	ASSIGNED BLOCKS
Lumber	600/5 = 120	4	6, 7, 8, 9, 10
Paint	260/2 = 130	3	3, 5
Tools	150/3 = 50	5	11, 12, 13
Small hardware	400/2 = 200	2	2, 4
Chemical bags	90/3 = 30	6	14, 15, 16
Lightbulbs	220/1 = 220	1	1

INSIGHT ▶ This procedure allocates items with the highest "bang-for-the-buck" first. The "bang" (value) here is the number of trips. Because we want to minimize travel, we would like to place items with high-frequency visits near the front. The storage space represents the "buck" (cost). We want items that take up a lot of space moved toward the back because if they were placed near the front, we would have to travel past their multiple blocks every time we needed to store or retrieve an item from a different category. This bang versus buck trade-off is neatly accommodated by using the trips/blocks ratio (column 2 of the solution table). In this example, even though lumber has the highest number of trips, the lumber takes up so much storage space that it is placed further back, toward the middle of the warehouse.

LEARNING EXERCISE ▶ Order frequency for paint is expected to increase to 410 trips per month. How will that change the storage plan? [Answer: Paint and small hardware will switch storage locations.]

RELATED PROBLEMS ▶ S11.20–S11.22

Summary

Myriad tools have been developed to help supply chain managers make well-informed decisions. We have provided a small sampling in this supplement. A decision tree can help determine the best number of suppliers to protect against supply disruption from potential disasters. The bullwhip measure can identify each supply chain member's contribution to exacerbating ordering fluctuations. The factor-weighting approach can be used to help select suppliers based on multiple criteria. Inventory holding costs can be computed for various shipping alternatives to better compare their overall cost impact. Finally, items can be ranked according to the ratio of (trips/blocks of storage) to determine their best placement in a warehouse.

Discussion Questions

1. What is the difference between "unique-event" risk and "super-event" risk?
2. If the probability of a "super-event" increases, does the "unique-event" risk increase or decrease in importance? Why?
3. If the probability of a "super-event" decreases, what happens to the likelihood of needing multiple suppliers?
4. Describe some ramifications of the bullwhip effect.
5. Describe causes of the bullwhip effect and their associated remedies.
6. Describe how the bullwhip measure can be used to analyze supply chains.
7. Describe some potentially useful categories to include in a factor-weighting analysis for supplier selection.
8. Describe some potential pitfalls in relying solely on the results of a factor-weighting analysis for supplier selection.
9. Describe some disadvantages of using a slow shipping method.
10. Besides warehouse layout decisions, what are some other applications where ranking items according to "bang/buck" might make sense?
11. Develop a vendor-rating form that represents your comparison of the education offered by universities in which you considered (or are considering) enrolling. Fill in the necessary data, and identify the "best" choice. Are you attending that "best" choice? If not, why not?

Solved Problems

SOLVED PROBLEM S11.1

Jon Jackson Manufacturing is searching for suppliers for its new line of equipment. Jon has narrowed his choices to two sets of suppliers. Believing in diversification of risk, Jon would select two suppliers under each choice. However, he is still concerned about the risk of both suppliers failing at the same time. The "San Francisco option" uses both suppliers in San Francisco. Both are stable, reliable, and profitable firms, so Jon calculates the "unique-event" risk for either of them to be 0.5%. However, because San Francisco is in an earthquake zone, he estimates the probability of an event that would knock out both suppliers to be 2%. The "North American option" uses one supplier in Canada and another in Mexico. These are upstart firms; John calculates the "unique-event" risk for either of them to be 10%. But he estimates the "super-event" probability that would knock out both of these suppliers to be only 0.1%. Purchasing costs would be $500,000 per year using the San Francisco option and $510,000 per year using the North American option. A total disruption would create an annualized loss of $800,000. Which option seems best?

SOLUTION

Using Equation (S11-1), the probability of a total disruption (i.e., the probability of incurring the $800,000 loss) equals:

$$\text{San Francisco option}: 0.02 + (1 - 0.02)0.005^2 = 0.02 + 0.0000245 = 0.0200245, \text{ or } 2.00245\%$$

$$\text{North American option}: 0.001 + (1 - 0.001)0.1^2 = 0.001 + 0.0099 = 0.01099, \text{ or } 1.099\%$$

$$\text{Total annual expected costs} = \text{Annual purchasing costs} + \text{Expected annualized disruption costs}$$

$$\text{San Francisco option}: \$500,000 + \$800,000(0.0200245) = \$500,000 + \$16,020 = \$516,020$$

$$\text{North American option}: \$510,000 + \$800,000(0.01099) = \$510,000 + \$8,792 = \$518,792$$

In this case, the San Francisco option appears to be slightly cheaper.

SOLVED PROBLEM S11.2

Over the past 10 weeks, demand for gears at Michael's Metals has been 140, 230, 100, 175, 165, 220, 200, and 178. Michael has placed weekly orders of 140, 250, 90, 190, 140, 240, 190, and 168 units.

The sample variance of a data set can be found by using the VAR.S function in Excel or by plugging each value (x) of the data set into the formula: $\text{Variance} = \dfrac{\sum(x - \bar{x})^2}{(n-1)}$ where $\bar{x}$ is the mean of the data set and n is the number of values in the set. Using Equation (S11-2), calculate the bullwhip measure for Michael's Metals over the 10-week period.

SOLUTION

Mean demand $= (140 + 230 + 100 + 175 + 165 + 220 + 200 + 178)/8 = 1{,}408/8 = 176$

Variance of demand

$$= \frac{(140-176)^2 + (230-176)^2 + (100-176)^2 + (175-176)^2 + (165-176)^2 + (220-176)^2 + (200-176)^2 + (178-176)^2}{(8-1)}$$

$$= \frac{36^2 + 54^2 + 76^2 + 1^2 + 11^2 + 44^2 + 24^2 + 2^2}{7} = \frac{1{,}296 + 2{,}916 + 5{,}776 + 1 + 121 + 1{,}936 + 576 + 4}{7}$$

$$= \frac{12{,}626}{7} = 1{,}804$$

Mean orders $= (140 + 250 + 90 + 190 + 140 + 240 + 190 + 168)/8 = 1{,}408/8 = 176$

Variance of orders

$$= \frac{(140-176)^2 + (250-176)^2 + (90-176)^2 + (190-176)^2 + (140-176)^2 + (240-176)^2 + (190-176)^2 + (168-176)^2}{(8-1)}$$

$$= \frac{36^2 + 74^2 + 86^2 + 14^2 + 36^2 + 64^2 + 14^2 + 8^2}{7} = \frac{1{,}296 + 5{,}476 + 7{,}396 + 196 + 1{,}296 + 4{,}096 + 196 + 64}{7}$$

$$= \frac{20{,}016}{7} = 2{,}859$$

From Equation (S11-2), the bullwhip measure = 2,859/1,804 = 1.58.
Since 1.58 > 1, Michael's Metals is contributing to the bullwhip effect in its supply chain.

SOLVED PROBLEM S11.3

Victor Pimentel, purchasing manager of Office Supply Center of Mexico, is searching for a new supplier for its paper. The most important supplier criteria for Victor include paper quality, delivery reliability, customer service, and financial condition, and he believes that paper quality is twice as important as each of the other three criteria. Victor has narrowed the choice to two suppliers, and his staff has rated each supplier on each criterion (using a scale of 1 to 100, with 100 being highest), as shown in the following table:

	PAPER QUALITY	DELIVERY RELIABILITY	CUSTOMER SERVICE	FINANCIAL CONDITION
Monterrey Paper	85	70	65	80
Papel Grande	80	90	95	75

Use the factor-weighting approach to determine the best supplier choice.

SOLUTION

To determine the appropriate weights for each category, create a simple algebraic relationship:

Let x = weight for criteria 2, 3, and 4.
Then $2x + x + x + x = 100\%$, i.e., $5x = 100\%$, or $x = 0.2 = 20\%$

Thus, paper quality has a weight of 2(20%) = 40%, and the other three criteria each have a weight of 20%.

The following table presents the factor-weighting analysis:

		MONTERREY PAPER		*PAPEL GRANDE*	
CRITERION	WEIGHT	SCORE (1 – 100) (100 HIGHEST)	WEIGHT × SCORE	SCORE (1 – 100) (100 HIGHEST)	WEIGHT × SCORE
Paper quality	.40	85	34	80	32
Delivery reliability	.20	70	14	90	18
Customer service	.20	65	13	95	19
Financial condition	.20	80	16	75	15
Total	1.00		77		84

Because 84 > 77, Papel Grande should be the chosen supplier according to the factor-weighting method.

SOLVED PROBLEM S11.4

A French car company ships 120,000 cars annually to the United Kingdom. The current method of shipment uses ferries to cross the English Channel and averages 10 days. The firm is considering shipping by rail through the Chunnel (the tunnel that goes under the English Channel) instead. That transport method would average approximately 2 days. Shipping through the Chunnel costs $80 more per vehicle. The firm has a holding cost of 25% per year. The average value of each car shipped is $20,000. Which transportation method should be selected?

SOLUTION

$$\text{Daily cost of holding the product} = (.25 \times \$20{,}000)/365 = \$13.70$$
$$\text{Total holding cost savings by using the Chunnel} = (10 - 2) \times \$13.70 = \$110 \text{ (rounded)}$$

Because the $110 savings exceeds the $80 higher shipping cost, the Chunnel option appears best.
This switch would save the firm (120,000)($110 − $80) = $3,600,000 per year.

Problems

Problems S11.1–S11.5 relate to Evaluating Disaster Risk in the Supply Chain

• **S11.1** How would you go about attempting to come up with the probability of a "super-event" or the probability of a "unique-event?" What factors would you consider?

•• **S11.2** Phillip Witt, president of Witt Input Devices, wishes to create a portfolio of local suppliers for his new line of keyboards. As the suppliers all reside in a location prone to hurricanes, tornadoes, flooding, and earthquakes, Phillip believes that the probability in any year of a "super-event" that might shut down all suppliers at the same time for at least 2 weeks is 3%. Such a total shutdown would cost the company approximately $400,000. He estimates the "unique-event" risk for any of the suppliers to be 5%. Assuming that the marginal cost of managing an additional supplier is $15,000 per year, how many suppliers should Witt Input Devices use? Assume that up to three nearly identical local suppliers are available.

•• **S11.3** Still concerned about the risk in Problem S11.2, suppose that Phillip is willing to use one local supplier and up to two more located in other territories within the country. This would reduce the probability of a "super-event" to 0.5%, but due to increased distance the annual costs for managing each of the distant suppliers would be $25,000 (still $15,000 for the local supplier). Assuming that the local supplier would be the first one chosen, how many suppliers should Witt Input Devices use now?

•• **S11.4** Risk manager Camila De León of Johnson Chemicals is considering two options for the firm's supplier portfolio. Option 1 uses two local suppliers. Each has a "unique-event" risk of 5%, and the probability of a "super-event" that would disable both at the same time is estimated to be 1.5%. Option 2 uses two suppliers located in different countries. Each has a "unique-event" risk of 13%, and the probability of a "super-event" that would disable both at the same time is estimated to be 0.2%.

a) What is the probability that both suppliers will be disrupted using option 1?
b) What is the probability that both suppliers will be disrupted using option 2?
c) Which option would provide the lowest risk of a total shutdown?

•• **S11.5** Bloom's Jeans is searching for new suppliers, and Debbie Bloom, the owner, has narrowed her choices to two sets. Debbie is very concerned about supply disruptions, so she has chosen to use three suppliers no matter what. For option 1, the suppliers are well established and located in the same country. Debbie calculates the "unique-event" risk for each of them to be 4%. She estimates the probability of a nationwide event that would knock out all three suppliers to be 2.5%. For option 2, the suppliers are newer but located in three different countries. Debbie calculates the "unique-event" risk for each of them to be 20%. She estimates the "super-event" probability that would knock out all three of these suppliers to be 0.4%. Purchasing and transportation costs would be $1,000,000 per year using option 1 and $1,010,000 per year using option 2. A total disruption would create an annualized loss of $500,000.

a) What is the probability that all three suppliers will be disrupted using option 1?
b) What is the probability that all three suppliers will be disrupted using option 2?
c) What is the total annual purchasing and transportation cost plus expected annualized disruption cost for option 1?
d) What is the total annual purchasing and transportation cost plus expected annualized disruption cost for option 2?
e) Which option seems best?

Problems S11.6–S11.9 relate to Managing the Bullwhip Effect

•• **S11.6** Consider the supply chain illustrated here:

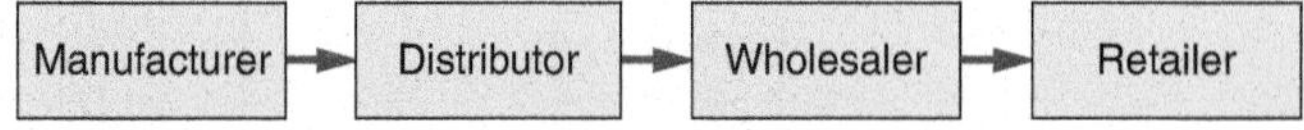

Last year, the retailer's weekly variance of demand was 200 units. The variance of orders was 500, 600, 750, and 1,350 units for the retailer, wholesaler, distributor, and manufacturer, respectively. (Note that the variance of orders equals the variance of demand for that firm's supplier.)

a) Calculate the bullwhip measure for the retailer.
b) Calculate the bullwhip measure for the wholesaler.
c) Calculate the bullwhip measure for the distributor.
d) Calculate the bullwhip measure for the manufacturer.
e) Which firm appears to be contributing the most to the bullwhip effect in this supply chain?

• **S11.7** Over the past 5 weeks, demand for wine at Winston's Winery has averaged 1,890 bottles, and the variance of demand has been 793,000 bottles. Owner Chiney Winston has ordered an average of 1,900 bottles per week over that time period, with a variance of orders of 1,805,000 bottles.

a) What is the bullwhip measure for glass bottles for Winston's Winery?
b) Is Winston's Winery providing an amplifying or smoothing effect?

•• **S11.8** Over the past 12 months, Super Toy Mart has experienced a demand variance of 10,000 units and has produced an order variance of 12,000 units.

a) What is the bullwhip measure for Super Toy Mart?
b) If Super Toy Mart had made a perfect forecast of demand over the past 12 months and had decided to order 1/12 of that annual demand each month, what would its bullwhip measure have been?

••• **S11.9** Consider a three-firm supply chain consisting of a retailer, manufacturer, and supplier. The retailer's demand over an 8-week period was 100 units each of the first 2 weeks, 200 units each of the second 2 weeks, 300 units each of the third 2 weeks, and 400 units each of the fourth 2 weeks. The provided table presents the orders placed by each firm in the supply chain. Notice, as is often the case in supply chains due to economies of scale, that total units are the same in each case, but firms further up the supply chain (away from the retailer) place larger, less frequent, orders.

a) What is the bullwhip measure for the retailer?
b) What is the bullwhip measure for the manufacturer?
c) What is the bullwhip measure for the supplier?
d) What conclusions can you draw regarding the impact that economies of scale may have on the bullwhip effect?

WEEK	RETAILER	MANUFACTURER	SUPPLIER
1	100	200	600
2	100		
3	200	400	
4	200		
5	300	600	1400
6	300		
7	400	800	
8	400		

Hint: Recall that the sample variance of a data set can be found by using the VAR.S function in Excel or by plugging each x *value of the data set into the formula: Variance* $= \frac{\sum (x - \bar{x})^2}{(n-1)}$ *where* $\bar{x}$ *is the mean of the data set and* n *is the number of values in the set.*

Problems S11.10–S11.15 relate to Supplier Selection Analysis

•• **S11.10** As purchasing agent for Eynan Enterprises in Richmond, Virginia, you ask your buyer Aldis Jakubovskis to provide you with a ranking of "excellent," "good," "fair," or "poor" for a variety of characteristics for two potential vendors. You suggest that the "Products" total be weighted 40% and the other three categories totals be weighted 20% each. Aldis has returned the rankings shown in Table S11.2.

Which of the two vendors would you select? **PX**

TABLE S11.2 Vendor Ratings for Problems S11.10 and S11.11

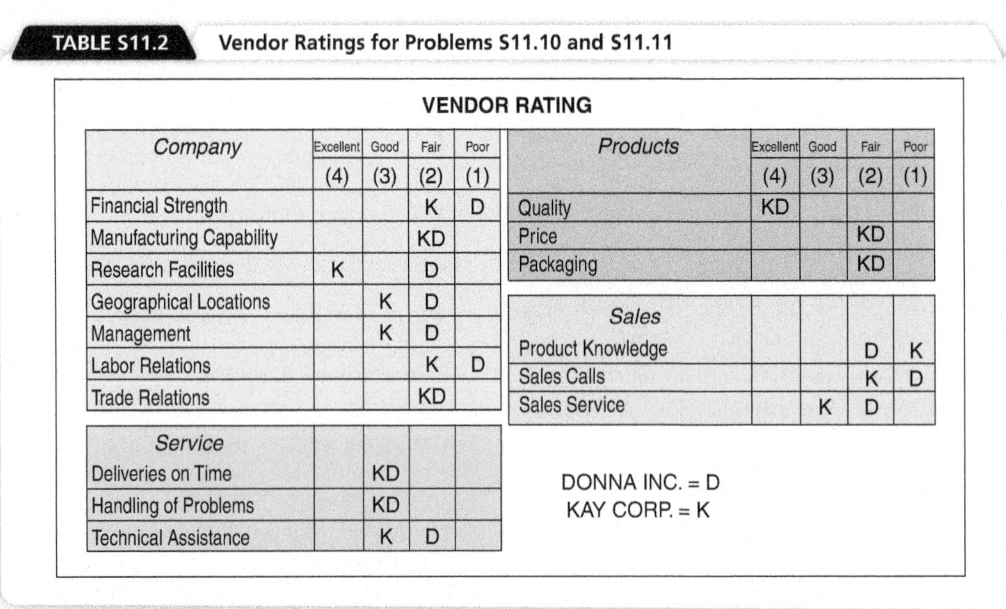

VENDOR RATING

Company	Excellent (4)	Good (3)	Fair (2)	Poor (1)
Financial Strength			K	D
Manufacturing Capability			KD	
Research Facilities	K		D	
Geographical Locations		K	D	
Management		K	D	
Labor Relations			K	D
Trade Relations			KD	

Service	Excellent (4)	Good (3)	Fair (2)	Poor (1)
Deliveries on Time		KD		
Handling of Problems		KD		
Technical Assistance		K	D	

Products	Excellent (4)	Good (3)	Fair (2)	Poor (1)
Quality	KD			
Price			KD	
Packaging			KD	

Sales	Excellent (4)	Good (3)	Fair (2)	Poor (1)
Product Knowledge			D	K
Sales Calls			K	D
Sales Service		K	D	

DONNA INC. = D
KAY CORP. = K

TABLE S11.3 Vendor Ratings for Problems S11.12, S11.13, and S11.14

		JOGLEKAR TECHNOLOGY, LLC	SIEMSEN SYSTEMS, INC.
VENDOR CRITERION	WEIGHT	SCORE (1–5) (5 HIGHEST)	SCORE (1–5) (5 HIGHEST)
Engineering Competence	20	5	4
Process Capability	20	4	4
Cost	10	1	5
Quality	20	4	4
Performance to Schedule	20	4	3
After-Sales Service	10	3	4
Total	100		

•• **S11.11** Using the data in Problem S11.10, assume that both Donna, Inc. and Kay Corp. are able to move all their "poor" ratings to "fair." How would you then rank the two firms? **PX**

• **S11.12** The supply chain manager for Keskin Industries, Sean Willems, has developed the data shown in Table S11.3 for two vendors. Based on the weights and rankings shown, which is the preferred vendor? **PX**

•• **S11.13** Referring to Problem S11.12, Sean Willems, in addition to the two possible vendors evaluated in Table S11.3, has found a third possible vendor. He now wants to evaluate all three vendors using the weights in Table S11.3. He scores the third vendor, Fricker V-Tech, Ltd., as: Engineering Competence 5, Process Capability 5, Cost 1, Quality 5, Performance to Schedule 4, and After-Sales Service 1. Which vendor has the highest score? **PX**

•• **S11.14** Sean Willems has evaluated the vendors in Table S11.3 as well as Fricker V-Tech, the vendor in Problem S11.13, but the Quality Control Manager says that new data suggest that Siemsen Systems, Inc.'s quality has deteriorated, and Siemsen should now score a 2 in the Quality category. Which of the three vendors now has the highest score? **PX**

•• **S11.15** The Director of Supply Chain Relations at Long Beach Retailers wants you to evaluate two vendors, Classic Ladies Ware (L) and International Fashions (I). She suggests the weights be: Products 0.35, Service 0.30, Sales Personnel 0.20, and Company Strength 0.15. Based on reports from buyers and work that she had done evaluating the suppliers, she has given you the following vendor ranking sheet, where Excellent ratings score 4 points, Good ratings score 3, Fair score 2, and Poor score 1. Which vendor would you choose? **PX**

L→CLASSIC LADIES WARE I→INTERNATIONAL FASHIONS	EXCNT 4	GOOD 3	FAIR 2	POOR 1
Company Strength			L, I	
Service	I	L		
Products	L		I	
Sales Personnel		I		L

Problems S11.16–S11.19 relate to Transportation Mode Analysis

•• **S11.16** Your options for shipping $100,000 of machine parts from Baltimore to Kuala Lumpur, Malaysia, are (1) use a ship that will take 30 days at a cost of $3,800 or (2) truck the parts to Los Angeles and then ship at a total cost of $4,800. The second option will take only 20 days. You are paid via a letter of credit the day the parts arrive. Your holding cost is estimated at 30% of the value per year.
a) Which option is more economical?
b) What customer issues are not included in the data presented?

•• **S11.17** If you have a third option for the data in Problem S11.16 and it costs only $4,000 and also takes 20 days, what is your most economical plan?

•• **S11.18** Monczka-Trent Shipping is the logistics vendor for Handfield Manufacturing Co. in Ohio. Handfield has daily shipments of a power-steering pump from its Ohio plant to an auto assembly line in Alabama. The value of the standard shipment is $250,000. Monczka-Trent has two options: (1) its standard 2-day shipment or (2) a subcontractor who will team drive overnight with an effective delivery of one day. The extra driver costs $175. Handfield's holding cost is 35% annually for this kind of inventory.
a) Which option is more economical?
b) What production issues are not included in the data presented?

••• **S11.19** Recently, Abercrombie & Fitch (A&F) began shifting a large portion of its Asian deliveries to the U.S. from air freight to slower but cheaper ocean freight. Shipping costs have been cut dramatically, but shipment times have gone from days to weeks. In addition to having less control over inventory and being less responsive to fashion changes, the holding costs have risen for the goods in transport. Meanwhile, Central America might offer an inexpensive manufacturing alternative that could reduce shipping time through the Panama Canal to, say, 6 days, compared to, say, 27 days from Asia. Suppose that A&F uses an annual holding rate of 30%. Suppose further that the product costs $20 to produce in Asia. Assume that the transportation cost via ocean liner would be approximately the same whether coming from Asia or Central America.

a) What is the product cost plus holding cost per unit from Asia?
b) If the product cost in Central America is x, what is the holding cost per unit when shipping from Central America (as a function of x)?
c) Given a) and b), what is the break-even point for x? In other words, what would the maximum production cost in Central America need to be in order for that to be a competitive source compared to the Asian producer?

Problems S11.20–S11.22 relate to Warehouse Storage

• **S11.20** Peter Bell of Bell Electronics maintains an extensive stock of computer components for the consumer market. He has hundreds of drawers/cabinets for the many small parts (mother boards, graphic cards, fans, etc.), and he wants to organize them so as to minimize his travel and maximize his time at the counter interacting with customers.

ITEM	WEEKLY TRIPS	AREA NEEDED (DRAWERS)
A	300	60
B	219	3
C	72	1
D	90	10
E	24	3

a) Which item should be stored at the very front (closest to the counter)?
b) Which item should be stored at the very back (furthest from the counter)?

•• **S11.21** Amy Zeng, owner of Zeng's Restaurant Distributions, supplies nonperishable goods to restaurants around the metro area. She stores all the goods in a storage area with a single dock and one aisle. The goods are divided into five categories according to the following table. The table indicates the number of trips per month to store or retrieve items in each category, as well as the number of storage blocks taken up by each.

ITEM CATEGORY	MONTHLY TRIPS	AREA NEEDED (BLOCKS)
Paper Products	50	2
Dishes, Glasses, and Silverware	16	4
Cleaning Agents	6	2
Cooking Oils and Seasonings	30	2
Pots and Pans	12	6

The following picture of the storage area provides an identification number for each of the 16 storage blocks. For each item category, indicate into which blocks it should be stored.

	1	3	5	7	9	11	13	15
Dock				Aisle				
	2	4	6	8	10	12	14	16

•• **S11.22** The food items listed in the following table are stored on a cruise ship for a 7-day cruise.

ITEM	TRIPS PER VOYAGE	AREA NEEDED (BLOCKS)
A	2	1
B	160	8
C	16	1
D	40	4
E	24	2
F	15	1
G	4	1

Using the following figure, indicate the best storage location for each item to minimize average distance traveled.

Kitchen				Aisle				

CASE STUDY

Supply Chain Issues at Nautique Boat Company

Video Case 

Like most manufacturers, Nautique Boat Company, the premier manufacturer of ski boats, wakeboard boats, and wake surfing boats, finds that it spends most of its revenue on purchased items. This is particularly reflected in its high-dollar purchases for engines ($43 million per year), fiberglass/resin ($13 million), and towers[†] ($11 million). Moreover, ongoing changes in colors, along with sundry innovations in materials for upholstery, audio, and instruments, require constant communication between Nautique and its suppliers. Nautique has noted that the more collaboration it can establish with its suppliers, the more both parties benefit. It finds that tight relationships with suppliers that reflect the desired innovation, quality, and timely delivery are a necessary part of its differentiation strategy. Nautique even sends its engineers to suppliers to help *them* improve design,

Correct Craft Holding Company, LLC

efficiency, and quality. President Greg Meloon believes these relationships are critical to Nautique's leading edge in design and performance.

Steve Carlton, Chief Designer and Director of Product Design and Development, initiates the desired innovation and collaboration with suppliers by having a purchasing agent imbedded in his department. This allows early and frequent communication with existing, new, and potential suppliers. With the strong innovative element inherent in Nautique boats, this early interface is not just desirable, but critical. Additionally, Materials and Supply Chain Manager Drew Pope realizes supplier selection is a critical task. Consequently, he formalizes that process by using a factor-weighting technique to help him clarify the vendor selection decision.

For more stable products Nautique has joined a dozen or so other boat manufacturers in a purchasing group known as the American Boatbuilders Association (ABA). This association aggregates the purchases of these boat builders, negotiating primarily for volume pricing.

Drew has also developed contractual relationships with local vendors that deliver many standard items such as hardware and fasteners (often classified as "C" items) to deliver items directly to storage areas adjacent to the assembly line. Nautique employees have direct access to these areas, which are often referred to in a manufacturing environment as "manufacturing supermarkets." Title typically transfers when the items are delivered to the property. This practice tends to drive down total inventory and the cost of that inventory, as well as avoiding large lot deliveries.

		SCORE (1–10) (10 HIGHEST)		
VENDOR CRITERION	WEIGHT	RADICAL	SUPERIOR TOWERS	BOLT ENGINEERING
Engineering/ Innovation Skills	.30	9	10	7
Production Process Capability	.15	10	7	5
Quality	.20	6	8	7
Delivery	.15	2	7	4
Financial Strength	.10	10	5	8
Information Systems/ Collaboration	.05	10	3	9
Integrity	.05	5	6	10

Discussion Questions*

1. What other techniques might Nautique use to improve supply chain management?
2. What are the difficulties one might expect as Nautique attempts to establish collaborative relationships with suppliers?
3. Why is the supply chain such an important element of Nautique's strategy and ultimate economic performance?
4. Referring to the table, which vendor for towers would Drew select?

*The Global Company Profile featuring Nautique Boat Company (which opens Chapter 5) provides further background on Nautique's operations as does the video that accompanies this case. You may wish to review both prior to answering these questions.

†A wakeboard tower is a frame that is mounted to the hull of an inboard boat. Its main purpose is to elevate the tow point, which is crucial to wakeboarding specifically (see photo).

Endnote

1. Sources: *The Wall Street Journal* (February 23, 2021); and *The Orlando Sentinel* (April 9, 2020).

Bibliography

Berger, P. D., A. Gerstenfeld, and A. Z. Zeng. "How Many Suppliers Are Best? A Decision-Analysis Approach." *Omega* 32, no. 1 (February 2004): 9–15.

Chase, C. W. "Neutralizing the Bullwhip Effect to Manage Extreme Demand Volatility." *Journal of Business Forecasting* 39, no. 4 (Winter 2020–2021): 16–19.

Chopra, S. *Supply Chain Management: Strategy, Planning, and Operation*, 7th ed. Boston: Pearson, 2019.

Disney, S. M., and M. R. Lambrecht. "On Replenishment Rules, Forecasting, and the Bullwhip Effect in Supply Chains." *Foundations and Trends in Technology, Information and Operations Management* 2, no. 1 (2007), 1–80.

Lee, C., and C. L. Munson. "A Predictive Global Sensitivity Analysis Approach to Monitoring and Modifying Operational Hedging Positions." *International Journal of Integrated Supply Management* 9, no. 3 (2015): 178–201.

Meena, P. L., and S. P. Sarmah. "Supplier Selection and Demand Allocation under Supply Disruption Risks." *The International Journal of Advanced Manufacturing Technology* 83, no. 1–4 (2016): 265–274.

Rahimi, I., A. H. Gandomi, S. J. Fong, and M. A. Ülkü (eds.). *Big Data Analytics in Supply Chain Management: Theory and Applications*. Boca Raton, FL: CRC Press, 2021.

Roodbergen, K. J., I. F. A. Vis, and G. D. Taylor Jr. "Simultaneous Determination of Warehouse Layout and Control Policies." *International Journal of Production Research* 53, no. 11 (2015): 3306–3326.

Supplement 11 ***Rapid*** Review

Main Heading	Review Material	MyLab Operations Management
TECHNIQUES FOR EVALUATING SUPPLY CHAINS	Many supply chain metrics exist that can be used to evaluate performance within a company and for its supply chain partners. The 2011 Tōhoku earthquake and tsunami devastated eastern sections of Japan. The economic impact was felt around the globe, as manufacturers had been relying heavily, in some cases exclusively, on suppliers located in the affected zones. Manufacturers in several industries worldwide took 6 months or longer before they saw their supply chains working normally again.	Concept Question: 1.1
EVALUATING DISASTER RISK IN THE SUPPLY CHAIN	Disasters that disrupt supply chains can take on many forms, including tornadoes, fires, hurricanes, typhoons, tsunamis, earthquakes, and terrorism. Firms often use multiple suppliers for important components to mitigate the risks of total supply disruption. *The probability of all n suppliers being disrupted simultaneously*: $P(n) = S + (1 - S)U^n$ (S11-1) where: S = probability of a "super-event" disrupting all suppliers simultaneously U = probability of a "unique-event" disrupting only one supplier L = financial loss incurred in a supply chain if all suppliers were disrupted C = marginal cost of managing a supplier All suppliers will be disrupted simultaneously if either the super-event occurs or the super-event does not occur but a unique-event occurs for all of the suppliers. As the probability of a super-event (S) increases, the advantage of having multiple suppliers diminishes (all would be knocked out anyway). On the other hand, large values of the unique event (U) increase the likelihood of needing more suppliers. These two phenomena taken together suggest that when multiple suppliers are used, managers may consider using ones that are geographically dispersed to lessen the probability of all failing simultaneously. A decision tree can be used to help operations managers make this important decision regarding number of suppliers. One supplier: 1–P(1) No failure → \$1C; P(1) Failure → \$$L$ + \$1C Two suppliers: 1–P(2) ≤ 1 Fail → \$2C; P(2) Both fail → \$$L$ + \$2C N suppliers: 1–P(N) ≤ (N–1) Fail → \$$NC$; P(N) All N fail → \$$L$ + \$$NC$	Concept Questions: 2.1–2.5 Problems: S11.1–S11.5 Virtual Office Hours for Solved Problem: S11.1
MANAGING THE BULLWHIP EFFECT	*Demand forecast updating, order batching, price fluctuations*, and *shortage gaming* can all produce inaccurate information, resulting in distortions and fluctuations in the supply chain and causing the *bullwhip effect*. ■ **Bullwhip effect**—The increasing fluctuation in orders that often occurs as orders move through the supply chain. "Bullwhip" fluctuations create unstable production schedules. This can result in expensive capacity change adjustments such as overtime, subcontracting, extra inventory, backorders, hiring and laying off of workers, and equipment modifications. Other impacts may include equipment underutilization, longer lead times, and obsolescence of overproduced items. The bullwhip effect can occur when orders decrease as well as when they increase. Often the human tendency to overreact to stimuli causes managers to make decisions that exacerbate the phenomenon. The overarching solution to the bullwhip effect is simply for supply chain members to share information and work together.	Concept Questions: 3.1–3.6 Problems: S11.6–S11.9 Virtual Office Hours for Solved Problem: S11.2

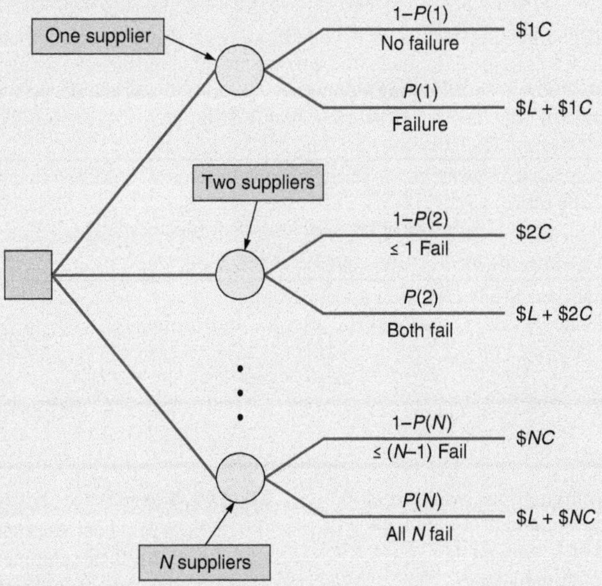

Main Heading	Review Material	MyLab Operations Management
	Specific remedies for the four primary causes include: Demand forecast errors → *Share demand information throughout the chain* Order batching → *Think of the supply chain as one firm when choosing order sizes* Price fluctuations → *Institute everyday low prices* Shortage gaming → *Allocate orders based on past demand* A straightforward way to measure the extent of the bullwhip effect at any link in the supply chain is to calculate the *bullwhip measure*: $$\text{Bullwhip} = \frac{\text{Variance of orders}}{\text{Variance of demand}} = \frac{\sigma^2_{\text{orders}}}{\sigma^2_{\text{demand}}} \quad \text{(S11-2)}$$ Variance *amplification* (i.e., the bullwhip effect) is present if the bullwhip measure is greater than 1. That means the size of a company's orders fluctuate more than the size of its incoming demand. If the measure equals 1, then no amplification is present. A value less than 1 would imply a *smoothing or dampening* scenario as orders move up the supply chain from the retailer toward suppliers.	
SUPPLIER SELECTION ANALYSIS	Choosing suppliers simply based on the lowest bid has become a somewhat rare approach. Various, sometimes competing, factors often play a role in the decision. Buyers may consider such supplier characteristics as product quality, delivery speed, delivery reliability, customer service, and financial performance. The *factor-weighting* technique simultaneously considers multiple supplier criteria. Each factor must be assigned an importance *weight*, and then each potential supplier is *scored* on each factor. The weights typically sum to 100%. Factors are scored using the same scale (e.g., 1–10). Sometimes a key is provided for supplier raters that converts qualitative ratings into numerical scores (e.g., "Very good" = 8).	Concept Questions: 4.1–4.2 Problems: S11.10–S11.15 **VIDEO S11.1** Supply Chain Issues at Nautique Boat Company Virtual Office Hours for Solved Problem: S11.3
TRANSPORTATION MODE ANALYSIS	The longer a product is in transit, the longer the firm has its money invested. But faster shipping is usually more expensive than slow shipping. A simple way to obtain some insight into this trade-off is to evaluate holding cost against shipping options. *Daily cost of holding the product:* $$\text{Daily holding cost} = (\text{Annual holding cost} \times \text{Product value})/365$$ There are many other considerations beyond holding vs. shipping costs when choosing the appropriate transportation mode and carrier, including ensuring *on-time delivery* (whether fast or slow), coordinating shipments to maintain a schedule, getting a new product to market, and keeping a customer happy. Estimates of these other costs can be added to the estimate of the daily holding cost.	Concept Questions: 5.1–5.2 Problems: S11.16–S11.19 Virtual Office Hours for Solved Problem: S11.4
WAREHOUSE STORAGE	When determining storage locations for items in a warehouse, rank the items according to the ratio: (Number of trips/Blocks of storage needed) Place the items with the *highest* ratios closest to the dock.	Concept Questions: 6.1–6.3 Problems: S11.20–S11.22
ADDITIONAL MYLAB OPERATIONS MANAGEMENT RESOURCES	✔ Additional Case Study (JIT after a Catastrophe) ✔ Multiple Choice Case Questions (JIT after a Catastrophe)	

Self Test

LO S11.1 Which of the following combinations would result in needing to utilize the largest number of suppliers?
- **a)** a high value of S and high value of U
- **b)** a high value of S and low value of U
- **c)** a low value of S and high value of U
- **d)** a low value of S and low value of U

LO S11.2 Typically, the bullwhip effect is most pronounced at which level of the supply chain?
- **a)** consumers
- **b)** suppliers
- **c)** wholesalers
- **d)** retailers

LO S11.3 Which of the following is not a characteristic of the factor-weighting approach to supplier evaluation?
- **a)** it applies quantitative scores to qualitative criteria
- **b)** the weights typically sum to 100%
- **c)** multiple criteria can be considered simultaneously
- **d)** subjective judgment is often involved
- **e)** it applies qualitative assessments to quantitative criteria

LO S11.4 A more expensive shipper tends to provide:
- **a)** faster shipments and lower holding costs
- **b)** faster shipments and higher holding costs
- **c)** slower shipments and lower holding costs
- **d)** slower shipments and higher holding costs

LO S11.5 Which of the following items is most likely to be stored at the back of a warehouse, furthest away from the shipping dock?
- **a)** low number of trips and low number of storage blocks
- **b)** low number of trips and high number of storage blocks
- **c)** high number of trips and low number of storage blocks
- **d)** high number of trips and high number of storage blocks

Answers: LO S11.1. c; LO S11.2. b; LO S11.3. e; LO S11.4. a; LO S11.5. b.

Inventory Management

CHAPTER 12

CHAPTER OUTLINE

GLOBAL COMPANY PROFILE: ***Amazon.com***

Kostikova Natalia/Shutterstock

LEGO makes 36 billion pieces a year in 11,000 shapes/forms. Whether you are buying a basic set of blocks or the 7,541-piece Millennium Falcon set, each LEGO piece is manufactured to an exacting degree of precision. Machine tolerances are 10 micrometers (half the width of a human hair). But to the customer, getting the right number of pieces is the most important issue. Sets with the wrong number of pieces or parts can result in some very unhappy customers, so inventory accuracy is paramount. During the packaging process, precision valves open and close automatically, dropping exactly the correct number of LEGOs into plastic bags. Then each bag is weighed to make sure the contents are correct. At the end of the process, packaging operators again check the sets.

10 OM STRATEGY DECISIONS

- Design of Goods and Services
- Managing Quality
- Process Strategies
- Location Strategies
- Layout Strategies
- Human Resources
- Supply Chain Management
- *Inventory Management*
  - ***Independent Demand (Ch. 12)***
  - Dependent Demand (Ch. 14)
  - Lean Operations (Ch. 16)
- Scheduling
- Maintenance

GLOBAL COMPANY PROFILE
Amazon.com

Inventory Management Provides Competitive Advantage at Amazon.com

When Jeff Bezos opened his revolutionary business in 1995, Amazon.com was intended to be a "virtual" retailer—no inventory, no warehouses, no overhead—just a bunch of computers taking orders for books and authorizing others to fill them. Things clearly didn't work out that way. About 350 million items are now available, amid hundreds of thousands of bins on shelves in hundreds of warehouses around the world. Additionally,

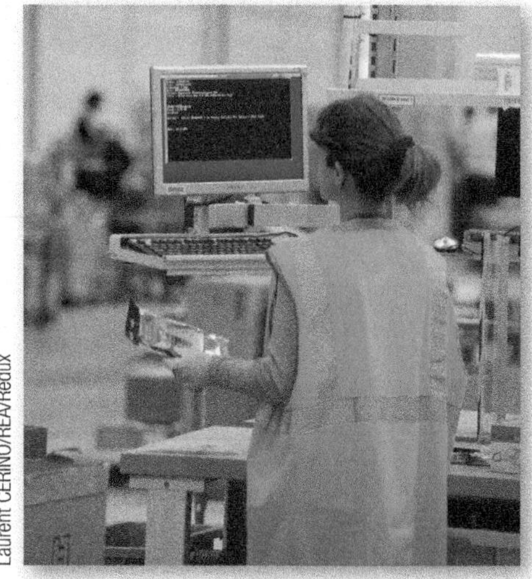

Laurent CERINO/REA/Redux

1. **You order three items, and a computer in Seattle takes charge.** A computer assigns your order—a book, a game, and a picture frame—to one of Amazon's massive U.S. distribution centers.

Brandon Bailey/AP Images

2. **A KIVA robot, moving under a stack of merchandise pods that can range up to 750 pounds, brings merchandise to an area where items can be merged for your order.** Amazon uses robots and technology to move racks of merchandise to workers (rather than workers going up and down aisles) who "pick" the items for the massive number of orders. The robots have raised the average picker's productivity from 100 items per hour to 300–400.

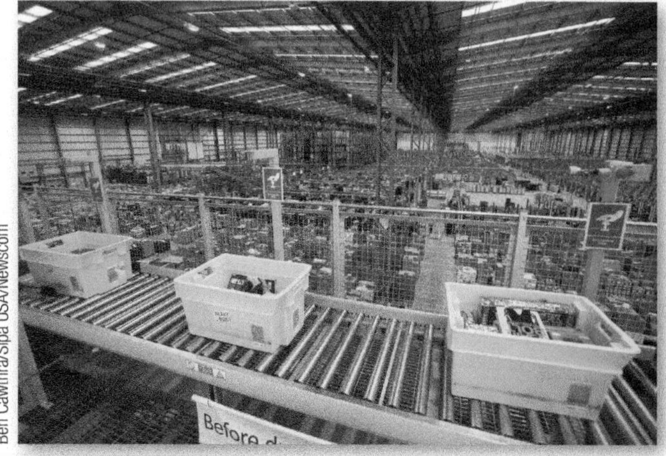

Ben Cawthra/Sipa USA/Newscom

3. **Your items are put into large yellow crates on moving conveyors.** When full, the crates ride a series of conveyors that wind more than 10 miles through the plant at a constant speed of 2.9 feet per second. The bar code on each item is scanned 15 times by machines and by many of the workers. The goal is to reduce errors to zero—returns are very expensive.

Tobias Schwarz/Reuters

4. **The box is packed, taped, weighed, and labeled before leaving the warehouse in a truck.** A typical plant is designed to ship as many as 200,000 pieces a day. The company delivers about two-thirds of the orders with its own trucks.

Amazon's software is so good that it sells its order taking, processing, and billing expertise to others.

The company wants the customer experience to be one that yields the lowest price, the fastest delivery, and an error-free order fulfillment process so no other contact with Amazon is necessary.

Managing this massive inventory precisely is the key for Amazon to be the world-class leader in warehouse automation and management. The time from the moment the customer clicks the send button to the time Amazon can ship is 15 minutes. And the labor required to receive, process, and position the stock in storage and to then accurately "pull" and package an order is less than 2 minutes. This underlines the high benchmark that Amazon has achieved. This is world-class performance in a company that obtains competitive advantage through inventory management.

Thomas Frey/dpa/Alamy Stock Photo

Jeramey Lende/Alamy Stock Photo

5. **Your order arrives at your doorstep.** In 1 or 2 days (or less) your order is delivered.

LEARNING OBJECTIVES

- **LO 12.1** ***Conduct*** an ABC analysis 494
- **LO 12.2** ***Explain*** and use cycle counting 495
- **LO 12.3** ***Explain*** and use the EOQ model for independent inventory demand 498
- **LO 12.4** ***Compute*** a reorder point and explain safety stock 503
- **LO 12.5** ***Apply*** the production order quantity model 504
- **LO 12.6** ***Explain*** and use the quantity discount model 506
- **LO 12.7** ***Understand*** service levels and probabilistic inventory models 512

The Importance of Inventory

As **Amazon.com** well knows, inventory is one of the most expensive assets of many companies, representing as much as 50% of total invested capital. Operations managers around the globe have long recognized that good inventory management is crucial. On the one hand, a firm can reduce costs by reducing inventory. On the other hand, production may stop and customers become dissatisfied when an item is out of stock. *The objective of inventory management is to strike a balance between inventory investment and customer service.* You can never achieve a low-cost strategy without good inventory management.

All organizations have some type of inventory planning and control system. A bank has methods to control its inventory of cash. A hospital has methods to control blood supplies and pharmaceuticals. Government agencies, schools, and, of course, virtually every manufacturing and production organization are concerned with inventory planning and control.

In cases involving physical products, the organization must determine whether to produce goods or to purchase them. Once this decision has been made, the next step is to forecast demand, as discussed in Chapter 4. Then operations managers determine the inventory necessary to service that demand. In this chapter, we discuss the functions, types, and management of inventory. We then address two basic inventory issues: how much to order and when to order.

Functions of Inventory

VIDEO 12.1
Managing Inventory at Frito-Lay

Inventory can serve several functions that add flexibility to a firm's operations. The four functions of inventory are:

1. *To provide a selection of goods for anticipated customer demand and to separate the firm from fluctuations in that demand.* Such inventories are typical in retail establishments.
2. To *decouple various parts of the production process*. For example, if a firm's supplies fluctuate, extra inventory may be necessary to decouple the production process from suppliers.
3. To *take advantage of quantity discounts*, because purchases in larger quantities may reduce the cost of goods or their delivery.
4. To *hedge against inflation* and upward price changes.

Types of Inventory

Raw material inventory
Materials that are usually purchased but have yet to enter the manufacturing process.

Work-in-process (WIP) inventory
Products or components that are no longer raw materials but have yet to become finished products.

Maintenance/repair/operating (MRO) inventory
Maintenance, repair, and operating materials.

To accommodate the functions of inventory, firms maintain four types of inventories: (1) raw material inventory, (2) work-in-process inventory, (3) maintenance/repair/operating supply (MRO) inventory, and (4) finished-goods inventory.

Raw material inventory has been purchased but not processed. This inventory can be used to decouple (i.e., separate) suppliers from the production process. However, the preferred approach is to eliminate supplier variability in quality, quantity, or delivery time so that separation is not needed. Work-in-process (WIP) inventory is components or raw material that have undergone some change but are not completed. WIP exists because of the time it takes for a product to be made (called *flow time*). Reducing flow time reduces inventory. Often this task is not difficult: during most of the time a product is "being made," it is in fact sitting idle. As Figure 12.1 shows, actual work time, or "run" time, is a small portion of the material flow time, perhaps as low as 5%.

MROs are inventories devoted to maintenance/repair/operating supplies necessary to keep machinery and processes productive. They exist because the need and timing for maintenance and

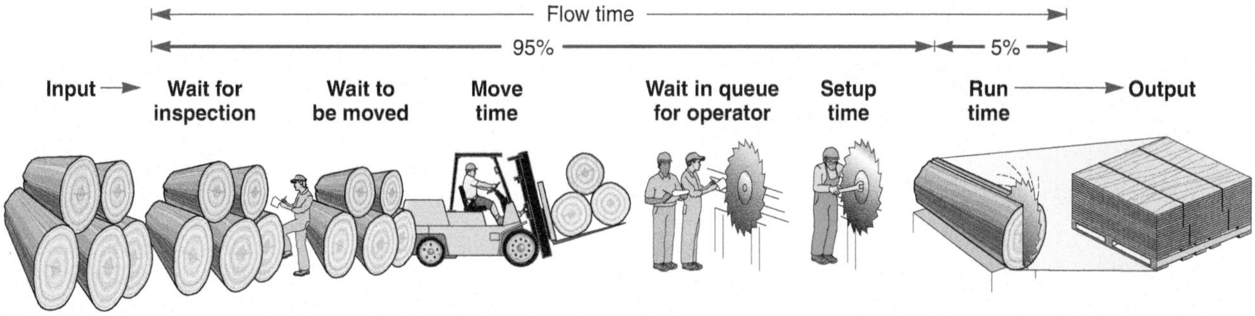

Flow time
95% 5%
Input → Wait for inspection | Wait to be moved | Move time | Wait in queue for operator | Setup time | Run time → Output

Figure **12.1**

The Material Flow Cycle

Most of the time that work is in-process (95% of the flow time) is not productive time.

repair of some equipment are unknown. Although the demand for MRO inventory is often a function of maintenance schedules, other unscheduled MRO demands must be anticipated. **Finished-goods inventory** is completed product awaiting shipment. Finished goods may be inventoried because future customer demands are unknown.

Finished-goods inventory
An end item ready to be sold, but still an asset on the company's books.

Managing Inventory

Operations managers establish systems for managing inventory. In this section, we briefly examine two ingredients of such systems: (1) how inventory items can be classified (called *ABC analysis*) and (2) how accurate inventory records can be maintained. We will then look at inventory control in the service sector.

ABC Analysis

ABC analysis divides on-hand inventory into three classifications on the basis of annual dollar volume. ABC analysis is an inventory application of what is known as the *Pareto principle* (named after Vilfredo Pareto, a 19th-century Italian economist). The Pareto principle states that there are a "critical few and trivial many." The idea is to establish inventory policies that focus resources on the *few critical* inventory parts and not the many trivial ones. It is not realistic to monitor inexpensive items with the same intensity as very expensive items.

ABC analysis
A method for dividing on-hand inventory into three classifications based on annual dollar volume.

To determine annual dollar volume for ABC analysis, we measure the *annual demand* of each inventory item times the *cost per unit*. *Class A* items are those on which the annual dollar volume is high. Although such items may represent only about 15% of the total inventory items, they represent 70% to 80% of the total dollar usage. *Class B* items are those inventory items of medium annual dollar volume. These items may represent about 30% of inventory items and 15% to 25% of the total value. Those with low annual dollar volume are *Class C*, which may represent only 5% of the annual dollar volume but about 55% of the total inventory items.

STUDENT TIP
A, B, and C categories need not be exact. The idea is to recognize that levels of control should match the risk.

Graphically, the inventory of many organizations would appear as presented in Figure 12.2.

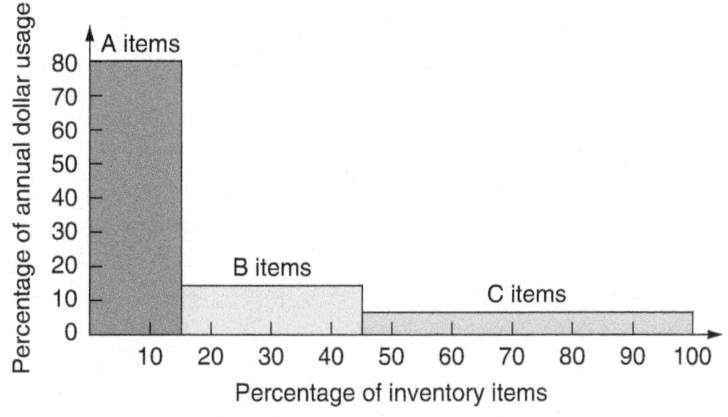

Figure **12.2**

Graphic Representation of ABC Analysis

An example of the use of ABC analysis is shown in Example 1.

Example 1

LO 12.1 *Conduct* an ABC analysis

ABC ANALYSIS FOR A CHIP MANUFACTURER

Silicon Chips, Inc., maker of superfast DRAM chips, wants to categorize its 10 major inventory items using ABC analysis.

APPROACH ▶ ABC analysis organizes the items on an annual dollar-volume basis. Shown below (in columns 1–4) are the 10 items (identified by stock numbers), their annual demands, and unit costs.

SOLUTION ▶ Annual dollar volume is computed in column 5, along with the percentage of the total represented by each item in column 6. Column 7 groups the 10 items into A, B, and C categories.

ABC Calculation

(1)	(2)	(3)		(4)		(5)	(6)		(7)
ITEM STOCK NUMBER	PERCENTAGE OF NUMBER OF ITEMS STOCKED	ANNUAL VOLUME (UNITS)	×	UNIT COST	=	ANNUAL DOLLAR VOLUME	PERCENTAGE OF ANNUAL DOLLAR VOLUME		CLASS
#10286	20%	1,000		$ 90.00		$ 90,000	38.8%	72%	A
#11526		500		154.00		77,000	33.2%		A
#12760	30%	1,550		17.00		26,350	11.3%	23%	B
#10867		350		42.86		15,001	6.4%		B
#10500		1,000		12.50		12,500	5.4%		B
#12572	50%	600		14.17		8,502	3.7%	5%	C
#14075		2,000		.60		1,200	.5%		C
#01036		100		8.50		850	.4%		C
#01307		1,200		.42		504	.2%		C
#10572		250		.60		150	.1%		C
		8,550				$232,057	100.0%		

INSIGHT ▶ The breakdown into A, B, and C categories is not hard and fast. The objective is to try to separate the "important" from the "unimportant."

LEARNING EXERCISE ▶ The unit cost for Item #10286 has increased from $90.00 to $120.00. How does this impact the ABC analysis? [Answer: The total annual dollar volume increases by $30,000, to $262,057, and the two A items now comprise 75% of that amount.]

RELATED PROBLEMS ▶ 12.1–12.3, 12.5–12.6

EXCEL OM Data File **Ch12Ex1.xls** can be found online.

Criteria other than annual dollar volume can determine item classification. For instance, high shortage or holding cost, anticipated engineering changes, delivery problems, or quality problems may dictate upgrading items to a higher classification. The advantage of dividing inventory items into classes allows policies and controls to be established for each class.

Policies that may be based on ABC analysis include the following:

1. Purchasing resources expended on supplier development should be much higher for individual A items than for C items.
2. A items, as opposed to B and C items, should have tighter physical inventory control; perhaps they belong in a more secure area, and perhaps the accuracy of inventory records for A items should be verified more frequently.
3. Forecasting A items may warrant more care than forecasting other items.

Better forecasting, physical control, supplier reliability, and an ultimate reduction in inventory can all result from classification systems such as ABC analysis.

Record Accuracy

Record accuracy is a prerequisite to inventory management, production scheduling, and, ultimately, sales. Accuracy can be maintained by either periodic or perpetual systems. *Periodic systems* require regular (periodic) checks of inventory to determine quantity on hand. Some small retailers and facilities with vendor-managed inventory (the vendor checks quantity on hand and resupplies as necessary) use these systems. However, the downside is lack of control between reviews and the necessity of carrying extra inventory to protect against shortages.

A variation of the periodic system is a *two-bin system*. In practice, a store manager sets up two containers (each with adequate inventory to cover demand during the time required to receive another order) and places an order when the first container is empty.

Alternatively, *perpetual inventory* tracks both receipts and subtractions from inventory on a continuing basis. Receipts are usually noted in the receiving department in some semiautomated way, such as via a bar-code reader, and disbursements are noted as items leave the stockroom or, in retailing establishments, at the point-of-sale (POS) cash register.

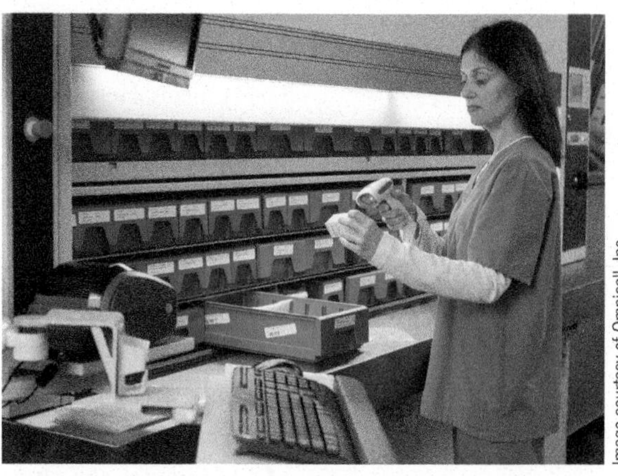
Image courtesy of Omnicell, Inc.

In this hospital, these vertically rotating storage carousels provide rapid access to hundreds of critical items and at the same time save floor space. This Omnicell inventory management carousel is also secure and has the added advantage of printing bar code labels.

Regardless of the inventory system, record accuracy requires good incoming and outgoing record keeping as well as good security. Stockrooms will have limited access, good housekeeping, and storage areas that hold fixed amounts of inventory. In both manufacturing and retail facilities, bins, shelf space, and individual items must be stored and labeled accurately. Meaningful decisions about ordering, scheduling, and shipping, are made only when the firm knows what it has on hand.

Cycle Counting

Even though an organization may have made substantial efforts to record inventory accurately, these records must be verified through a continuing audit. Such audits are known as **cycle counting**. Historically, many firms performed annual physical inventories. This practice often meant shutting down the facility and having inexperienced people count parts and material. Inventory records should instead be verified via cycle counting. Cycle counting uses inventory classifications developed through ABC analysis. With cycle counting procedures, items are counted, records are verified, and inaccuracies are periodically documented. The cause of inaccuracies is then traced and appropriate remedial action taken to ensure integrity of the inventory system. **A** items will be counted frequently, perhaps once a month; **B** items will be counted less frequently, perhaps once a quarter; and **C** items will be counted perhaps once every 6 months. Example 2 illustrates how to compute the number of items of each classification to be counted each day.

Cycle counting
A continuing reconciliation of inventory with inventory records.

LO 12.2 *Explain* and use cycle counting

Example 2

CYCLE COUNTING AT COLE'S TRUCKS, INC.

Cole's Trucks, Inc., a builder of high-quality refuse trucks, has about 5,000 items in its inventory. It wants to determine how many items to cycle count each day.

APPROACH ▶ After hiring Angie Clark, a bright young OM student, for the summer, the firm determined that it has 500 A items, 1,750 B items, and 2,750 C items. Company policy is to count all A items every month (every 20 working days), all B items every quarter (every 60 working days), and all C items every 6 months (every 120 working days). The firm then allocates some items to be counted each day.

SOLUTION ▶

ITEM CLASS	QUANTITY	CYCLE-COUNTING POLICY	NUMBER OF ITEMS COUNTED PER DAY
A	500	Each month (20 working days)	500/20 = 25/day
B	1,750	Each quarter (60 working days)	1,750/60 = 29/day
C	2,750	Every 6 months (120 working days)	2,750/120 = 23/day
			77/day

Each day, 77 items are counted.

INSIGHT ▶ This daily audit of 77 items is much more efficient and accurate than conducting a massive inventory count once a year.

LEARNING EXERCISE ▶ Cole's reclassifies some B and C items so there are now 1,500 B items and 3,000 C items. How does this change the cycle count? [Answer: B and C both change to 25 items each per day, for a total of 75 items per day.]

RELATED PROBLEM ▶ 12.4

In Example 2, the particular items to be cycle counted can be sequentially or randomly selected each day. Another option is to cycle count items when they are reordered.

Cycle counting also has the following advantages:

Shrinkage
Retail inventory that is unaccounted for between receipt and sale.

Pilferage
A small amount of theft.

1. Eliminates the shutdown and interruption of production necessary for annual physical inventories.
2. Eliminates annual inventory adjustments.
3. Trained personnel audit the accuracy of inventory.
4. Allows the cause of the errors to be identified and remedial action to be taken.
5. Maintains accurate inventory records.

Control of Service Inventories

Courtesy of Pro Glove

Inventory managers use bar code readers, such as this *ProGlove*, in order picking applications to save up to four seconds per pick. The device, with wireless connection and USB port, alerts workers of errors and tracks workflow. With rapid and accurate data, items are easily verified, improving inventory and shipment accuracy.

Although we may think of the service sector of our economy as not having inventory, that is seldom the case. Extensive inventory is held in wholesale and retail businesses, making inventory management crucial. In the food-service business, control of inventory is often the difference between success and failure. Moreover, inventory that is in transit or idle in a warehouse is lost value. Similarly, inventory damaged or stolen prior to sale is a loss. In retailing, inventory that is unaccounted for between receipt and time of sale is known as **shrinkage**. Shrinkage occurs from damage and theft as well as from sloppy paperwork. Inventory theft is also known as **pilferage**. Retail inventory loss of 1% of sales is considered good, with losses in many stores exceeding 3%. Because the impact on profitability is substantial, inventory accuracy and control are critical. Applicable techniques include the following:

1. *Good personnel selection, training, and discipline:* These are never easy but very necessary in food-service, wholesale, and retail operations, where employees have access to directly consumable merchandise.
2. *Tight control of incoming shipments:* This task is being addressed by many firms through the use of Universal Product Code (or bar code) and radio frequency ID (RFID) systems that read every incoming shipment and automatically check tallies against purchase orders. When properly designed, these systems—where each stock keeping unit (SKU; pronounced "skew") has its own identifier—can be very hard to defeat.
3. *Effective control of all goods leaving the facility:* This task is accomplished with bar codes, RFID tags, and codes in magnetic strips on merchandise and via direct observation. Direct observation can be personnel stationed at exits (as at Costco and Sam's Club wholesale stores) and in potentially high-loss areas, or it can take the form of one-way mirrors, video surveillance, or facial recognition.

A handheld reader can scan RFID tags, aiding control of both incoming and outgoing shipments.

Successful retail operations require very good store-level control with accurate inventory in its proper location. Major retailers lose 10% to 25% of overall profits due to poor or inaccurate inventory records.[2] Retailers with fewer SKUs tend to reduce both unsold inventory and stockouts. (See the *OM in Action* box, "150 Shades of Red at Mattel.")

VIDEO 12.2
Inventory Management at Celebrity Cruises

OM in Action 150 Shades of Red at Mattel[1]

Mattel has gotten to the heart of one of its problems: too many reds aren't a good thing. The company's designers until recently could choose from about 150 types of red when making Barbie dolls, Hot Wheels cars, or other toys in its stable. Each variation added storage costs and downtime at factories for cleaning equipment to swap out shades.

"Complexity is really a killer," said Mattel's chief supply chain officer. Mattel has chopped the choices of reds by more than a third and is doing the same for other colors, part of a broad edict to simplify the company's inventory SKUs. The goal is to improve, modernize, and ultimately tame a sprawling supply chain that operates 13 factories, employs 35,000 people, and delivers toys to 375,000 retail locations worldwide.

As part of its drive to reduce inventory and costs, Mattel says it plans to keep factories that are "strategically important" while consolidating plants that are underused. The company already changed how it sells and fulfills orders to retailers. In Europe, Mattel implemented an automated online ordering system for wholesale orders, eliminating the need for its sales team to manually process orders. It also increased the minimum order size so that it wasn't shipping orders valued at just a few hundred dollars into a fragmented retail market.

Alessandro Vecchi/Alamy Stock Photo

More broadly, Mattel is using new algorithms to tie its manufacturing output more closely to demand, helping the toy maker gauge the right inventory of toys for the holidays. Mattel also will be making fewer products. The company is planning to cut the number of items it sells by 30%, targeting the 45% of items it sells that only make up 6% of its revenue.

Inventory Models

We now examine a variety of inventory models and the costs associated with them.

Independent vs. Dependent Demand

Inventory control models assume that demand for an item is either independent of or dependent on the demand for other items. For example, the demand for refrigerators is *independent* of the demand for toaster ovens. However, the demand for toaster oven components is *dependent* on the requirements of toaster ovens.

VIDEO 12.3
Inventory at Nautique Boat Company

This chapter focuses on managing inventory where demand is *independent*. Chapter 14 presents *dependent* demand management.

Holding, Ordering, and Setup Costs

Holding costs are the costs associated with holding or "carrying" inventory over time. Therefore, holding costs also include obsolescence and costs related to storage, such as insurance, extra staffing, and interest payments. Table 12.1 shows the kinds of costs that need to be evaluated to determine holding costs. Many firms fail to include all the inventory holding costs. Consequently, inventory holding costs are often understated.

Holding cost
The cost to keep or carry inventory in stock.

Ordering cost includes costs of supplies, forms, order processing, purchasing, clerical support, and so forth. When orders are being manufactured, ordering costs also exist, but they are a part of what is called setup costs. **Setup cost** is the cost to prepare a machine or process for manufacturing an order. This includes time and labor to clean and change tools or holders. Operations managers can lower ordering costs by reducing setup costs and by using such efficient procedures as electronic ordering and payment.

Ordering cost
The cost of the ordering process.

Setup cost
The cost to prepare a machine or process for production.

In manufacturing environments, setup cost is highly correlated with **setup time**. Setups usually require a substantial amount of work even before a setup is actually performed at the work center. With proper planning, much of the preparation required by a setup can be done prior to shutting down the machine or process. Setup times can thus be reduced substantially. Machines and processes that traditionally have taken hours to set up are now being set up in less than a minute by the more imaginative world-class manufacturers. Reducing setup times is an excellent way to reduce inventory investment and to improve productivity.

Setup time
The time required to prepare a machine or process for production.

TABLE 12.1 Determining Inventory Holding Costs

CATEGORY	COST (AND RANGE) AS A PERCENTAGE OF INVENTORY VALUE
Housing costs (building rent or depreciation, operating cost, taxes, insurance)	6% (3–10%)
Material-handling costs (equipment lease or depreciation, power, operating cost)	3% (1–3.5%)
Labor cost (receiving, warehousing, security)	3% (3–5%)
Investment costs (borrowing costs, taxes, and insurance on inventory)	11% (6–24%)
Pilferage, scrap, and obsolescence (much higher in industries undergoing rapid change like tablets and smart phones)	3% (2–5%)
Overall carrying cost	26%

Note: All numbers are approximate, as they vary substantially depending on the nature of the business, location, and current interest rates.

STUDENT TIP An overall inventory carrying cost of less than 15% is very unlikely, but this cost can exceed 40%, especially in high-tech and fashion industries.

Inventory Models for Independent Demand

In this section, we introduce three inventory models that address two important questions: *when to order* and *how much to order*. These *independent* demand models are:

1. Basic economic order quantity (EOQ) model
2. Production order quantity model
3. Quantity discount model

The Basic Economic Order Quantity (EOQ) Model

Economic order quantity (EOQ) model An inventory-control technique that minimizes the total of ordering and holding costs.

The **economic order quantity (EOQ) model** is one of the most commonly used inventory-control techniques. This technique is relatively easy to use but is based on several assumptions:

1. Demand for an item is known, reasonably constant, and independent of decisions for other items.
2. Lead time—that is, the time between placement and receipt of the order—is known and consistent.
3. Receipt of inventory is instantaneous and complete. In other words, the inventory from an order arrives in one batch at one time.
4. Quantity discounts are not possible.
5. The only variable costs are the cost of setting up or placing an order (setup or ordering cost) and the cost of holding or storing inventory over time (holding or carrying cost). These costs were discussed in the previous section.
6. Stockouts (shortages) can be completely avoided if orders are placed at the right time.

LO 12.3 *Explain* and use the EOQ model for independent inventory demand

With these assumptions, the graph of inventory usage over time has a sawtooth shape, as in Figure 12.3. In Figure 12.3, Q represents the amount that is ordered. If this amount is 500 dresses,

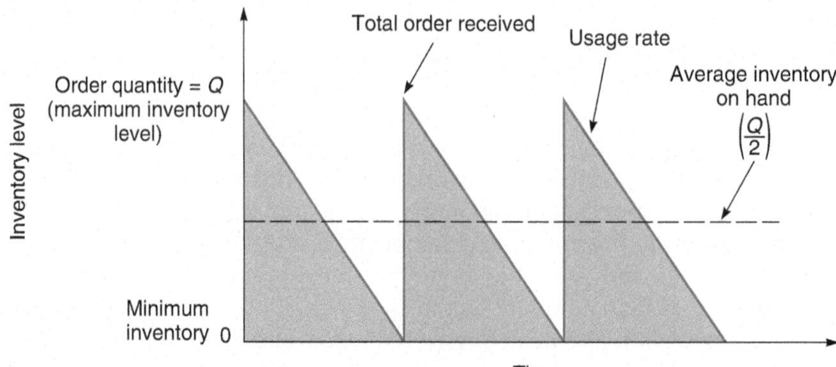

Figure 12.3 **Inventory Usage over Time**

STUDENT TIP If the maximum we can ever have is Q (say, 500 units) and the minimum is 0, then if inventory is used (or sold) at a fairly steady rate, the average $= (Q + 0)/2 = Q/2$.

all 500 dresses arrive at one time (when an order is received). Thus, the inventory level jumps from 0 to 500 dresses. In general, an inventory level increases from 0 to Q units when an order arrives.

Because demand is constant over time, inventory drops at a uniform rate over time. (Refer to the sloped lines in Figure 12.3.) Each time the inventory is received, the inventory level again jumps to Q units (represented by the vertical lines). This process continues indefinitely over time.

Minimizing Costs

The objective of most inventory models is to minimize total costs. With the assumptions just given, significant costs are setup (or ordering) cost and holding (or carrying) cost. All other costs, such as the cost of the inventory itself, are constant. Thus, if we minimize the sum of setup and holding costs, we will also be minimizing total costs. To help you visualize this, in Figure 12.4 we graph total costs as a function of the order quantity, Q. The optimal order size, Q^*, will be the quantity that minimizes the total costs. As the quantity ordered increases, the total number of orders placed per year will decrease. Thus, as the quantity ordered increases, the annual setup or ordering cost will decrease [Figure 12.4(a)]. But as the order quantity increases, the holding cost will increase due to the larger average inventories that are maintained [Figure 12.4(b)].

STUDENT TIP
Figure 12.4 is the heart of EOQ inventory modeling. We want to find the smallest total cost (top curve), which is the sum of the two curves below it.

As we can see in Figure 12.4(c), a reduction in either holding or setup cost will reduce the total cost curve. A reduction in the setup cost curve also reduces the optimal order quantity (lot size). In addition, smaller lot sizes have a positive impact on quality and production flexibility. At Toshiba, the $40 billion Japanese conglomerate, workers can make as few as 10 laptop computers before changing models. This lot-size flexibility has allowed Toshiba to move toward a "build-to-order" mass customization system, an important ability in an industry that has product life cycles measured in months, not years.

VIDEO 12.4
Inventory Control at Wheeled Coach

You should note that in Figure 12.4(c), the optimal order quantity occurs at the point where the ordering-cost curve and the carrying-cost curve intersect. This was not by chance. With the EOQ model, the optimal order quantity will occur at a point where the total setup cost is equal to the total holding cost.[3] We use this fact to develop equations that solve directly for Q^*. The necessary steps are:

1. Develop an expression for setup or ordering cost.
2. Develop an expression for holding cost.
3. Set setup (order) cost equal to holding cost.
4. Solve the equation for the optimal order quantity.

Using the following variables, we can determine setup and holding costs and solve for Q^*:

Q = Number of units per order
Q^* = Optimum number of units per order (EOQ)
D = Annual demand in units for the inventory item
S = Setup or ordering cost for each order
H = Holding or carrying cost per unit per year

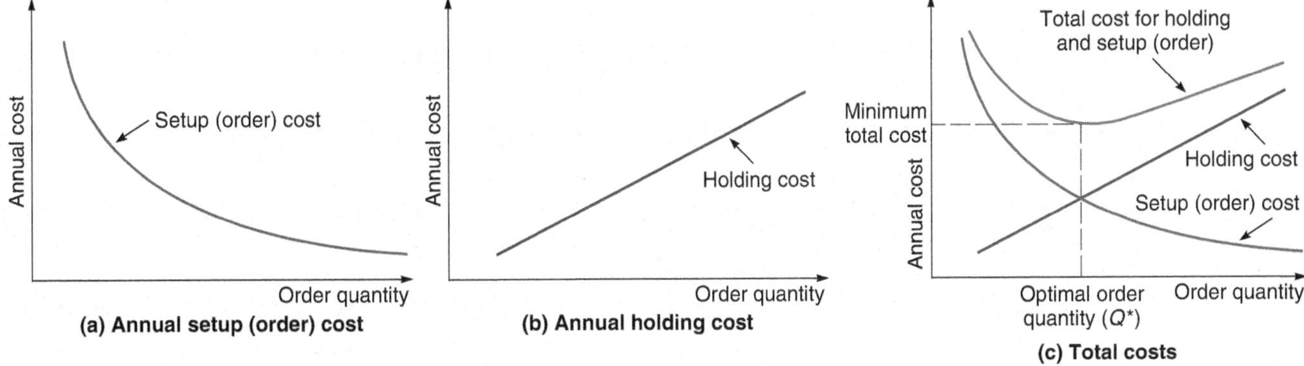

Figure 12.4
Costs as a Function of Order Quantity

1. Annual setup cost = (Number of orders placed per year) × (Setup or order cost per order)

$$= \left(\frac{\text{Annual demand}}{\text{Number of units in each order}}\right)(\text{Setup or order cost per order})$$

$$= \left(\frac{D}{Q}\right)(S) = \frac{D}{Q}S$$

2. Annual holding cost = (Average inventory level) × (Holding cost per unit per year)

$$= \left(\frac{\text{Order quantity}}{2}\right)(\text{Holding cost per unit per year})$$

$$= \left(\frac{Q}{2}\right)(H) = \frac{Q}{2}H$$

3. Optimal order quantity is found when annual setup (order) cost equals annual holding cost, namely:

$$\frac{D}{Q}S = \frac{Q}{2}H$$

4. To solve for Q^*, simply cross-multiply terms and isolate Q on the left of the equal sign:

$$2DS = Q^2H$$

$$Q^2 = \frac{2DS}{H}$$

$$Q^* = \sqrt{\frac{2DS}{H}} \qquad (12\text{-}1)$$

Now that we have derived the equation for the optimal order quantity, Q^*, it is possible to solve inventory problems directly, as in Example 3.

Example 3

FINDING THE OPTIMAL ORDER SIZE AT SHARP, INC.

Sharp, Inc., a company that markets painless hypodermic needles to hospitals, would like to reduce its inventory cost by determining the optimal number of hypodermic needles to obtain per order.

APPROACH ▶ The annual demand is 1,000 units; the setup or ordering cost is $10 per order; and the holding cost per unit per year is $.50.

SOLUTION ▶ Using these figures, we can calculate the optimal number of units per order:

$$Q^* = \sqrt{\frac{2DS}{H}}$$

$$Q^* = \sqrt{\frac{2(1{,}000)(10)}{0.50}} = \sqrt{40{,}000} = 200 \text{ units}$$

INSIGHT ▶ Sharp, Inc. now knows how many needles to order per order. The firm also has a basis for determining ordering and holding costs for this item, as well as the number of orders to be processed by the receiving and inventory departments.

LEARNING EXERCISE ▶ If D increases to 1,200 units, what is the new Q^*? [Answer: Q^* = 219 units.]

RELATED PROBLEMS ▶ 12.7–12.11, 12.14, 12.15, 12.17, 12.29, 12.31, 12.32, 12.33a, 12.35a

EXCEL OM Data File **Ch12Ex3.xls** can be found online.

ACTIVE MODEL 12.1 This example is further illustrated in Active Model 12.1 found online.

We can also determine the expected number of orders placed during the year (N) and the expected time between orders (T), as follows:

$$\text{Expected number of orders} = N = \frac{\text{Demand}}{\text{Order quantity}} = \frac{D}{Q^*} \tag{12-2}$$

$$\text{Expected time between orders} = T = \frac{\text{Number of working days per year}}{N} \tag{12-3}$$

Example 4 illustrates this concept.

Example 4

COMPUTING NUMBER OF ORDERS AND TIME BETWEEN ORDERS AT SHARP, INC.

Sharp, Inc. (in Example 3) has a 250-day working year and wants to find the number of orders (N) and the expected time between orders (T).

APPROACH ▶ Using Equations (12-2) and (12-3), Sharp enters the data given in Example 3.

SOLUTION ▶

$$N = \frac{\text{Demand}}{\text{Order quantity}}$$

$$= \frac{1{,}000}{200} = 5 \text{ orders per year}$$

$$T = \frac{\text{Number of working days per year}}{\text{Expected number of orders}}$$

$$= \frac{250 \text{ working days per year}}{5 \text{ orders}} = 50 \text{ days between orders}$$

INSIGHT ▶ The company now knows not only how many needles to order per order but that the time between orders is 50 days and that there are five orders per year.

LEARNING EXERCISE ▶ If $D = 1{,}200$ units instead of 1,000, find N and T. [Answer: $N \cong 5.48$, $T = 45.62$.]

RELATED PROBLEMS ▶ 12.14, 12.15, 12.17, 12.35c,d

As mentioned previously in this section, the total annual variable inventory cost is the sum of setup and holding costs:

$$\text{Total annual cost} = \text{Setup (order) cost} + \text{Holding cost} \tag{12-4}$$

In terms of the variables in the model, we can express the total cost TC as:

$$TC = \frac{D}{Q}S + \frac{Q}{2}H \tag{12-5}$$

Example 5 shows how to use this formula.

Example 5

COMPUTING COMBINED COST OF ORDERING AND HOLDING

Sharp, Inc. (from Examples 3 and 4) wants to determine the combined annual ordering and holding costs.

APPROACH ▶ Apply Equation (12-5), using the data in Example 3.

SOLUTION ▶

$$TC = \frac{D}{Q}S + \frac{Q}{2}H$$

$$= \frac{1{,}000}{200}(\$10) + \frac{200}{2}(\$.50)$$

$$= (5)(\$10) + (100)(\$.50)$$

$$= \$50 + \$50 = \$100$$

INSIGHT ▶ These are the annual setup and holding costs. The $100 total does not include the actual cost of goods. Notice that in the EOQ model, holding costs always equal setup (order) costs.

LEARNING EXERCISE ▶ Find the total annual cost if $D = 1{,}200$ units in Example 3. [Answer: $109.54.]

RELATED PROBLEMS ▶ 12.11, 12.14, 12.15, 12.16, 12.33b,c, 12.35e, 12.36a,b

Inventory costs may also be expressed to include the actual cost of the material purchased. If we assume that the annual demand and the price per hypodermic needle are known values (e.g., 1,000 hypodermics per year at $P = \$10$) and total annual cost should include purchase cost, then Equation (12-5) becomes:

$$TC = \frac{D}{Q}S + \frac{Q}{2}H + PD$$

Because material cost does not depend on the particular order policy, we still incur an annual material cost of $D \times P = (1{,}000)(\$10) = \$10{,}000$. (Later in this chapter we will discuss the case in which this may not be true—namely, when a quantity discount is available.)[4]

Robust
Giving satisfactory answers even with substantial variation in the parameters.

Robust Model A benefit of the EOQ model is that it is robust. By **robust** we mean that it gives satisfactory answers even with substantial variation in its parameters. As we have observed, determining accurate ordering costs and holding costs for inventory is often difficult. Consequently, a robust model is advantageous. The total cost of the EOQ changes little in the neighborhood of the minimum. The curve is very shallow. This means that variations in setup costs, holding costs, demand, or even EOQ make relatively modest differences in total cost. Example 6 shows the robustness of EOQ.

Example 6

EOQ IS A ROBUST MODEL

Management in the Sharp, Inc. examples underestimates total annual demand by 50% (say demand is actually 1,500 needles rather than 1,000 needles) while using the same Q. How will the annual inventory cost be impacted?

APPROACH ▶ We will solve for annual costs twice. First, we will apply the wrong EOQ; then we will recompute costs with the correct EOQ.

SOLUTION ▶ If demand in Example 5 is actually 1,500 needles rather than 1,000, but management uses an order quantity of $Q = 200$ (when it should be $Q = 244.9$ based on $D = 1{,}500$), the sum of holding and ordering cost increases to $125:

$$\begin{aligned}\text{Annual cost} &= \frac{D}{Q}S + \frac{Q}{2}H \\ &= \frac{1{,}500}{200}(\$10) + \frac{200}{2}(\$.50) \\ &= \$75 + \$50 = \$125\end{aligned}$$

However, had we known that the demand was for 1,500 with an EOQ of 244.9 units, we would have spent $122.47, as shown:

$$\begin{aligned}\text{Annual cost} &= \frac{1{,}500}{244.9}(\$10) + \frac{244.9}{2}(\$.50) \\ &= 6.125(\$10) + 122.45(\$.50) \\ &= \$61.25 + \$61.22 = \$122.47\end{aligned}$$

INSIGHT ▶ Note that the expenditure of $125.00, made with an estimate of demand that was substantially wrong, is only 2% ($2.52/$122.47) higher than we would have paid had we known the actual demand and ordered accordingly. Note also that were it not due to rounding, the annual holding costs and ordering costs would be exactly equal.

LEARNING EXERCISE ▶ Demand at Sharp remains at 1,000, H is still \$.50, and we order 200 needles at a time (as in Example 5). But if the true order cost = S = \$15 (rather than \$10), what is the annual cost? [Answer: Annual order cost increases to \$75, and annual holding cost stays at \$50. So the total cost = \$125.]

RELATED PROBLEMS ▶ 12.10b, 12.16, 12.36a,b

We may conclude that the EOQ is indeed robust and that significant errors do not cost us very much. This attribute of the EOQ model is most convenient because our ability to accurately determine demand, holding cost, and ordering cost is limited.

Lead time
In purchasing systems, the time between placing an order and receiving it; in production systems, the wait, move, queue, setup, and run times for each component produced.

Reorder point (ROP)
The inventory level (point) at which action is taken to replenish the stocked item.

Safety stock (*ss*)
Extra stock to allow for uneven demand; a buffer.

Reorder Points

Now that we have decided *how much* to order, we will look at the second inventory question, *when* to order. Simple inventory models assume that receipt of an order is instantaneous. In other words, they assume (1) that a firm will place an order when the inventory level for that particular item reaches zero and (2) that it will receive the ordered items immediately. However, the time between placement and receipt of an order, called **lead time**, or delivery time, can be as short as a few hours or as long as months. Thus, the when-to-order decision is usually expressed in terms of a **reorder point (ROP)**—the inventory level at which an order should be placed (see Figure 12.5).

The reorder point (ROP) is given as:

$$\text{ROP} = \text{Demand per day} \times \text{Lead time for a new order in days}$$

$$\text{ROP} = d \times L \qquad (12\text{-}6)$$

This equation for ROP *assumes that demand during lead time and lead time itself are constant*. When this is not the case, extra stock, often called **safety stock (*ss*)**, should be added. The reorder point with safety stock then becomes:

$$\text{ROP} = \text{Expected demand during lead time} + \text{Safety stock}$$

The demand per day, d, is found by dividing the annual demand, D, by the number of working days in a year:

$$d = \frac{D}{\text{Number of working days in a year}}$$

Computing the reorder point is demonstrated in Example 7.

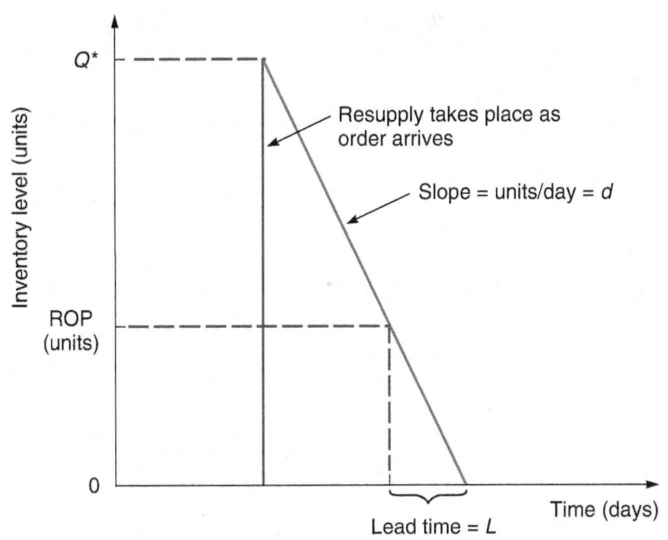

Figure **12.5**

The Reorder Point (ROP)

Q^* is the optimum order quantity, and lead time represents the time between placing and receiving an order.

Example 7

COMPUTING REORDER POINTS (ROP) FOR IPHONES WITH AND WITHOUT SAFETY STOCK

LO 12.4 *Compute* a reorder point and explain safety stock

An Apple store has a demand (D) for 8,000 iPhones per year. The firm operates a 250-day working year. On average, delivery of an order takes 3 working days, but has been known to take as long as 4 days. The store wants to calculate the reorder point without a safety stock and then with a one-day safety stock.

APPROACH ▶ First compute the daily demand and then apply Equation (12-6) for the ROP. Then compute the ROP with safety stock.

SOLUTION ▶

$$d = \frac{D}{\text{Number of working days in a year}} = \frac{8{,}000}{250} = 32 \text{ units}$$

$$\text{ROP} = \text{Reorder point} = d \times L = 32 \text{ units per day} \times 3 \text{ days} = 96 \text{ units}$$

ROP with safety stock adds 1 day's demand (32 units) to the ROP (for 128 units).

INSIGHT ▶ When iPhone inventory stock drops to 96 units, an order should be placed. If the safety stock for a possible 1-day delay in delivery is added, the ROP is 128 (= 96 + 32).

LEARNING EXERCISE ▶ If there are only 200 working days per year, what is the correct ROP, without safety stock and with safety stock? [Answer: 120 iPhones without safety stock and 160 with safety stock.]

RELATED PROBLEMS ▶ 12.11d, 12.12, 12.13, 12.15f, 12.33d, 12.34, 12.35f, 12.36c

When demand is not constant or variability exists in the supply chain, safety stock can be critical. We discuss safety stock in more detail later in this chapter.

The Production Order Quantity Model

In the previous inventory model, we assumed that the entire inventory order was received at one time. There are times, however, when the firm may receive its inventory over a period of time. Such cases require a different model, one that does not require the instantaneous-receipt assumption. This model is applicable under two situations: (1) when inventory continuously flows or builds up over a period of time after an order has been placed or (2) when units are produced and sold simultaneously. Under these circumstances, we take into account daily production (or inventory-flow) rate and daily demand rate. Figure 12.6 shows inventory levels as a function of time (and inventory dropping to zero between orders).

Production order quantity model
An economic order quantity technique applied to production orders.

Because this model is especially suitable for the production environment, it is commonly called the **production order quantity model**. It is useful when inventory continuously builds up over time, and traditional economic order quantity assumptions are valid. We derive this model by setting ordering or setup costs equal to holding costs and solving for optimal order size, Q^*. Using the following symbols, we can determine the expression for annual inventory holding cost for the production order quantity model:

LO 12.5 *Apply* the production order quantity model

Q = Number of units per order
H = Holding cost per unit per year
p = Daily production rate
d = Daily demand rate, or usage rate
t = Length of the production run in days

1. $$\begin{pmatrix}\text{Annual inventory}\\\text{holding cost}\end{pmatrix} = (\text{Average inventory level}) \times \begin{pmatrix}\text{Holding cost}\\\text{per unit per year}\end{pmatrix}$$

2. $$(\text{Average inventory level}) = (\text{Maximum inventory level})/2$$

STUDENT TIP Note in Figure 12.6 that inventory buildup is not instantaneous but gradual. So the formula reduces the average inventory and thus the holding cost by the ratio of that buildup.

3. $$\begin{pmatrix}\text{Maximum}\\\text{inventory level}\end{pmatrix} = \begin{pmatrix}\text{Total produced during}\\\text{the production run}\end{pmatrix} - \begin{pmatrix}\text{Total used during}\\\text{the production run}\end{pmatrix}$$
$$= pt - dt$$

However, Q = total produced = pt, and thus $t = Q/p$. Therefore:

$$\text{Maximum inventory level} = p\left(\frac{Q}{p}\right) - d\left(\frac{Q}{p}\right) = Q - \frac{d}{p}Q$$
$$= Q\left(1 - \frac{d}{p}\right)$$

Figure **12.6**

Change in Inventory Levels over Time for the Production Model

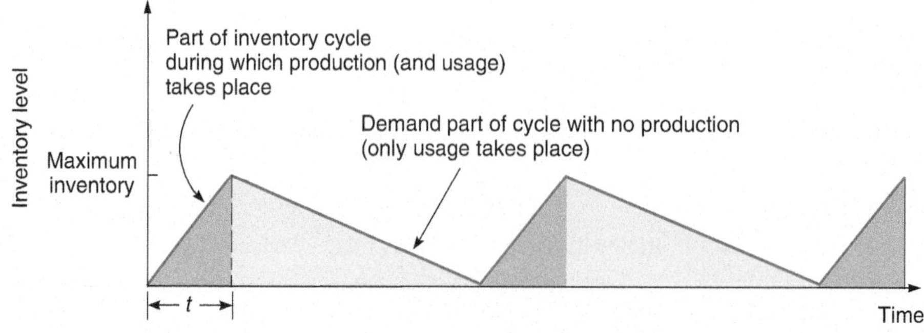

4. Annual inventory holding cost (or simply holding cost) =

$$\frac{\text{Maximum inventory level}}{2}(H) = \frac{Q}{2}\left[1 - \left(\frac{d}{p}\right)\right]H$$

Using this expression for holding cost and the expression for setup cost developed in the basic EOQ model, we solve for the optimal number of pieces per order by equating setup cost and holding cost:

Setup cost $= (D/Q)S$ Holding cost $= \frac{1}{2}HQ[1 - (d/p)]$

Set ordering cost equal to holding cost to obtain Q_p^*:

$$\frac{D}{Q}S = \frac{1}{2}HQ\left[1 - (d/p)\right]$$

$$Q^2 = \frac{2DS}{H[1 - (d/p)]}$$

$$Q_p^* = \sqrt{\frac{2DS}{H[1 - (d/p)]}} \quad (12\text{-}7)$$

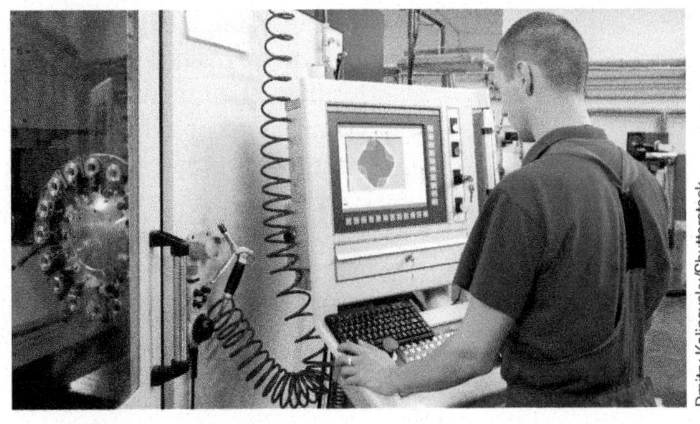

Each order may require a change in the way a machine or process is set up. Reducing setup time usually means a reduction in setup cost, and reductions in setup costs make smaller batches (lots) economical to produce. Increasingly, setup (and operation) is performed by computer-controlled machines, such as this one, operating from previously written programs.

Dmitry Kalinovsky/Shutterstock

In Example 8, we use Equation (12-7), Q_p^*, to solve for the optimum order or production quantity when inventory is consumed as it is produced.

Example 8

A PRODUCTION ORDER QUANTITY MODEL

Nathan Manufacturing, Inc. makes and sells specialty wheels for the retail automobile aftermarket. Nathan's forecast for its wire wheels is 1,000 units next year, with an average daily demand of 4 units. However, the production process is most efficient at 8 units per day. So the company produces 8 per day but uses only 4 per day. The company wants to solve for the optimum number of units per order. (*Note:* This plant schedules production of this wheel only as needed, during the 250 days per year the shop operates.)

APPROACH ▶ Gather the cost data and apply Equation (12-7):

$$\begin{aligned} \text{Annual demand} &= D = 1{,}000 \text{ units} \\ \text{Setup costs} &= S = \$10 \\ \text{Holding cost} &= H = \$0.50 \text{ per unit per year} \\ \text{Daily production rate} &= p = 8 \text{ units daily} \\ \text{Daily demand rate} &= d = 4 \text{ units daily} \end{aligned}$$

SOLUTION ▶

$$Q_p^* = \sqrt{\frac{2DS}{H[1 - (d/p)]}}$$

$$Q_p^* = \sqrt{\frac{2(1{,}000)(10)}{0.50[1 - (4/8)]}}$$

$$= \sqrt{\frac{20{,}000}{0.50(1/2)}} = \sqrt{80{,}000} = 282.8 \text{ wheels, or } 283 \text{ wheels}$$

INSIGHT ▶ The difference between the production order quantity model and the basic EOQ model is that the effective annual holding cost per unit is reduced in the production order quantity model because the entire order does not arrive at once.

LEARNING EXERCISE ▶ If Nathan can increase its daily production rate from 8 to 10, how does Q_p^* change? [Answer: $Q_p^* = 258$.]

RELATED PROBLEMS ▶ 12.18, 12.19, 12.20, 12.30, 12.37

EXCEL OM Data File **Ch12Ex8.xls** can be found online.

ACTIVE MODEL 12.2 This example is further illustrated in Active Model 12.2 found online.

You may want to compare this solution with the answer in Example 3, which had identical D, S, and H values. Eliminating the instantaneous-receipt assumption, where $p = 8$ and $d = 4$, resulted in an increase in Q^* from 200 in Example 3 to 283 in Example 8. This increase in Q^* occurred because holding cost dropped from \$.50 to [\$.50 × (1 – d/p)], making a larger order quantity optimal. Also note that:

$$d = 4 = \frac{D}{\text{Number of days the plant is in operation}} = \frac{1{,}000}{250}$$

We can also calculate Q_p^* when *annual* data are available. When annual data are used, we can express Q_p^* as:

$$Q_p^* = \sqrt{\frac{2DS}{H\left(1 - \dfrac{\text{Annual demand rate}}{\text{Annual production rate}}\right)}} \qquad (12\text{-}8)$$

The Quantity Discount Model

Quantity discount
A reduced price for items purchased in large quantities.

Quantity discounts appear everywhere—you cannot go into a grocery store without seeing them on nearly every shelf. In fact, researchers have found that *most* companies either offer or receive quantity discounts for at least some of the products that they sell or purchase. A **quantity discount** is simply a reduced price (P) for an item when it is purchased in larger quantities. A typical quantity discount schedule appears in Table 12.2. As can be seen in the table, the normal price of the item is \$100. When 120 to 1,499 units are ordered at one time, the price per unit drops to \$98; when the quantity ordered at one time is 1,500 units or more, the price is \$96 per unit. The 120 quantity and the 1,500 quantity are called *price-break quantities* because they represent the first order amount that would lead to a new lower price. As always, management must decide when and how much to order. However, given these quantity discounts, how does the operations manager make these decisions?

As with other inventory models, the objective is to minimize total cost. Because the unit cost for the second discount in Table 12.2 is the lowest, you may be tempted to order 1,500 units. Placing an order for that quantity, however, even with the greatest discount price, may not minimize total inventory cost. This is because holding cost increases. Thus, the major trade-off when considering quantity discounts is between *reduced product cost* and *increased holding cost*. When we include the cost of the product, the equation for the total annual inventory cost can be calculated as follows:

$$\text{Total annual cost} = \text{Annual setup (ordering) cost} + \text{Annual holding cost} + \text{Annual product cost,}$$

or

LO 12.6 *Explain* and use the quantity discount model

$$TC = \frac{D}{Q}S + \frac{Q}{2}IP + PD \qquad (12\text{-}9)$$

where Q = Quantity ordered
D = Annual demand in units
S = Setup or ordering cost per order
P = Purchase price per unit
I = Holding cost per unit per year expressed as a percentage of price P

Note that holding cost is IP instead of H as seen in the regular EOQ model. Because the price of the item is a factor in annual holding cost, we do not assume that the holding cost is a constant when the price per unit changes for each quantity discount. Thus, it is common to

TABLE 12.2 A Quantity Discount Schedule

PRICE RANGE	QUANTITY ORDERED	PRICE PER UNIT *P*
Initial price	1–119	\$100
Discount price 1	120–1,499	\$ 98
Discount price 2	1,500 and greater	\$ 96

express the holding cost as a percentage (I) of unit price (P) when evaluating costs of quantity discount schedules.

The EOQ formula (12-1) is modified for the quantity discount problem as follows:

$$Q^* = \sqrt{\frac{2DS}{IP}} \tag{12-10}$$

The solution procedure uses the concept of a *feasible EOQ*. An EOQ is feasible if it lies in the quantity range that leads to the same price P used to compute it in Equation (12-10). For example, suppose that $D = 5{,}200$, $S = \$200$, and $I = 28\%$. Using Table 12.2 and Equation (12-10), the EOQ for the \$96 price equals $\sqrt{2(5{,}200)(200)/[(.28)(96)]} = 278$ units. Because $278 < 1{,}500$ (the price-break quantity needed to receive the \$96 price), the EOQ for the \$96 price is *not feasible*. On the other hand, the EOQ for the \$98 price equals 275 units. This amount is *feasible* because if 275 units were actually ordered, the firm would indeed receive the \$98 purchase price.

Now we have to determine the quantity that will minimize the total annual inventory cost. Because there are a few discounts, this process involves two steps. In Step 1, we identify all possible order quantities that could be the best solution. In Step 2, we calculate the total cost of all possible best order quantities, and the least expensive order quantity is selected.

Solution Procedure

STEP 1: Starting with the *lowest* possible purchase price in a quantity discount schedule and working toward the highest price, keep calculating Q^* from Equation (12-10) until the first feasible EOQ is found. The first feasible EOQ is a possible best order quantity, along with all price-break quantities for all *lower* prices.

STEP 2: Calculate the total annual cost TC using Equation (12-9) for each of the possible best order quantities determined in Step 1. Select the quantity that has the lowest total cost.

Note that no quantities need to be considered for any prices greater than the first feasible EOQ found in Step 1. This occurs because if an EOQ for a given price is feasible, then the EOQ for any *higher* price *cannot* lead to a lower cost (TC is guaranteed to be higher).

Figure 12.7 provides a graphical illustration of Step 1 using the three price ranges from Table 12.2. In that example, the EOQ for the lowest price is infeasible, but the EOQ for the

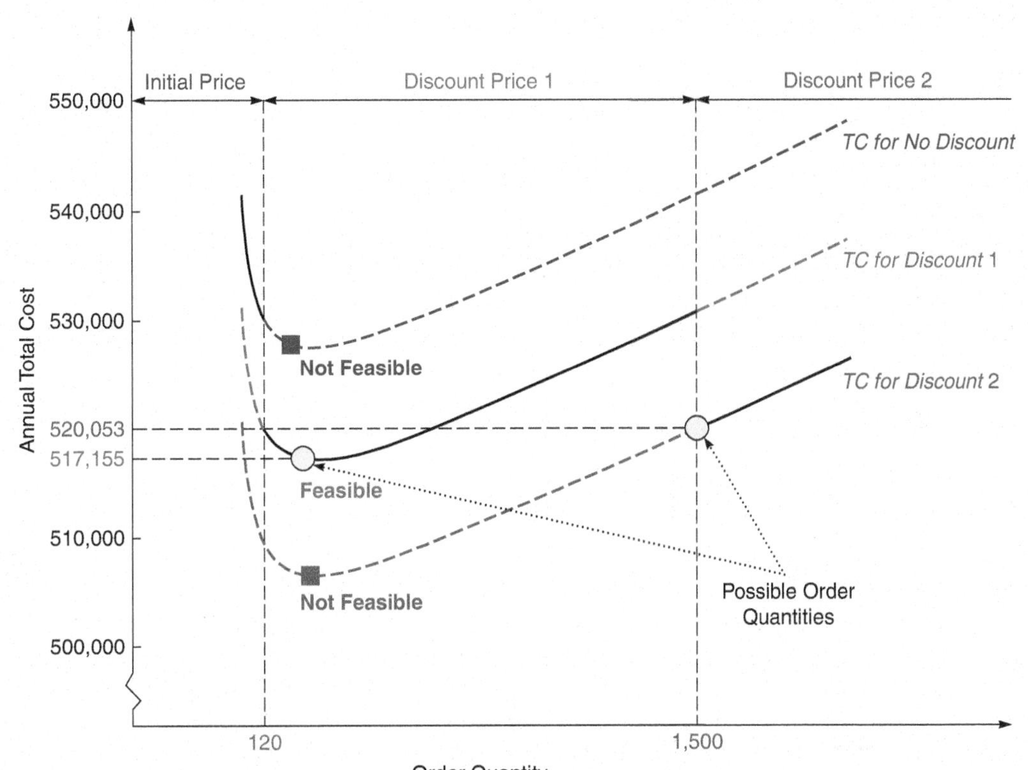

Figure 12.7

EOQs and Possible Best Order Quantities for the Quantity Discount Problem with Three Prices in Table 12.2

The solid black curves represent the realized total annual setup plus holding plus purchasing cost at the applicable order quantities. The black curve drops to the total cost curve for the next discount level when each price-break quantity is reached.

second-lowest price is feasible. So the EOQ for the second-lowest price, along with the price-break quantity for the lowest price, are the possible best order quantities. Finally, the highest price (no discount) can be ignored because a feasible EOQ has already been found for a lower price.

Example 9 illustrates how the full solution procedure can be applied.

Example 9

QUANTITY DISCOUNT MODEL

Chris Beehner Electronics stocks toy remote-control drones. Recently, the store has been offered a quantity discount schedule for these drones. This quantity schedule was shown in Table 12.2. Furthermore, setup cost is \$200 per order, annual demand is 5,200 units, and annual inventory carrying charge as a percentage of cost, I, is 28%. What order quantity will minimize the total inventory cost?

APPROACH ▶ We will follow the two steps just outlined for the quantity discount model.

SOLUTION ▶ First we calculate the Q^* for the lowest possible price of \$96, as we did earlier:

$$Q^*_{\$96} = \sqrt{\frac{2(5{,}200)(\$200)}{(.28)(\$96)}} = 278 \text{ drones per order}$$

Because $278 < 1{,}500$, this EOQ is *infeasible* for the \$96 price. So now we calculate Q^* for the next-higher price of \$98:

$$Q^*_{\$98} = \sqrt{\frac{2(5{,}200)(\$200)}{(.28)(\$98)}} = 275 \text{ drones per order}$$

Because 275 is between 120 and 1,499 units, this EOQ is *feasible* for the \$98 price. Thus, the possible best order quantities are 275 (the first feasible EOQ) and 1,500 (the price-break quantity for the lower price of \$96). We need not bother to compute Q^* for the initial price of \$100 because we found a feasible EOQ for a lower price.

Step 2 uses Equation (12-9) to compute the total cost for each of the possible best order quantities. This step is taken with the aid of Table 12.3.

TABLE 12.3 Total Cost Computations for Chris Beehner Electronics

ORDER QUANTITY	UNIT PRICE	ANNUAL ORDERING COST	ANNUAL HOLDING COST	ANNUAL PRODUCT COST	TOTAL ANNUAL COST
275	\$98	\$3,782	\$ 3,773	\$509,600	\$517,155
1,500	\$96	\$ 693	\$20,160	\$499,200	\$520,053

Because the total annual cost for 275 units is lower, 275 units should be ordered. The costs for this example are shown in Figure 12.7.

INSIGHT ▶ Even though Beehner Electronics could save more than \$10,000 in annual product costs, ordering 1,500 units (28.8% of annual demand) at a time would generate even more than that in increased holding costs. So in this example it is not in the store's best interest to order enough to attain the lowest possible purchase price per unit. On the other hand, if the price-break quantity for the \$96 had been 1,000 units rather than 1,500 units, then total annual costs would have been \$513,680, which would have been cheaper than ordering 275 units at \$98.

LEARNING EXERCISE ▶ Resolve the problem with $D = 2{,}000$, $S = \$5$, $I = 50\%$, discount price $1 = \$99$, and discount price $2 = \$98$. [Answer: only 20 units should be ordered each time, which is the EOQ at the \$100 price.]

RELATED PROBLEMS ▶ 12.21–12.28, 12.38–12.40

EXCEL OM Data file **Ch12Ex9.xls** can be found online.

In this section we have studied the most popular form of single-purchase quantity discount called the *all-units discount*. In practice, quantity discounts appear in a variety of forms. For example, *incremental quantity discounts* apply only to those units purchased beyond the price-break quantities rather than to all units. *Fixed fees*, such as a fixed shipping and processing cost for a catalog order or a \$5,000 tooling setup cost for any order placed with a manufacturer, encourage buyers to purchase more units at a time. Some discounts are *aggregated* over items or time. *Item aggregation* bases price breaks on total units or dollars purchased. *Time aggregation* applies to total items or dollars spent over a specific time period such as one year. *Truckload discounts, buy-one-get-one-free offers*, and *one-time-only sales* also represent types of quantity discounts in that they provide price incentives for buyers to purchase more units at one time. Most purchasing managers deal with some form of quantity discounts on a regular basis.

Probabilistic Models and Safety Stock

All the inventory models we have discussed so far make the assumption that demand for a product is constant and certain. We now relax this assumption. The following inventory models apply when product demand is not known but can be specified by means of a probability distribution. These types of models are called **probabilistic models**. Probabilistic models are a real-world adjustment because demand and lead time won't always be known and constant.

Probabilistic model
A statistical model applicable when product demand or any other variable is not known but can be specified by means of a probability distribution.

An important concern of management is maintaining an adequate service level in the face of uncertain demand. The **service level** is the *complement* of the probability of a stockout. For instance, if the probability of a stockout is 0.05, then the service level is .95. Uncertain demand raises the possibility of a stockout. One method of reducing stockouts is to hold extra units in inventory. As we noted previously, such inventory is referred to as safety stock. Safety stock involves adding a number of units as a buffer to the reorder point. As you recall:

Service level
The probability that demand will not be greater than supply during lead time. It is the complement of the probability of a stockout.

$$\text{Reorder point} = \text{ROP} = d \times L$$

where d = Daily demand
L = Order lead time, or number of working days it takes to deliver an order

The inclusion of safety stock (ss) changed the expression to:

$$\text{ROP} = d \times L + ss \qquad (12\text{-}11)$$

The amount of safety stock maintained depends on the cost of incurring a stockout and the cost of holding the extra inventory. Annual stockout cost is computed as follows:

$$\begin{aligned}\text{Annual stockout costs} = {} & \text{The sum of the units short for each demand level} \\ & \times \text{The probability of that demand level} \times \text{The stockout cost/unit} \\ & \times \text{The number of orders per year} \qquad (12\text{-}12)\end{aligned}$$

Example 10 illustrates this concept.

Example 10

DETERMINING SAFETY STOCK WITH PROBABILISTIC DEMAND AND CONSTANT LEAD TIME

David Rivera Optical has determined that its reorder point for eyeglass frames is 50 ($d \times L$) units. Its carrying cost per frame per year is \$5, and stockout (or lost sale) cost is \$40 per frame. The store has experienced the following probability distribution for inventory demand during the lead time (reorder period). The optimum number of orders per year is 6.

NUMBER OF UNITS	PROBABILITY
30	.2
40	.2
ROP → 50	.3
60	.2
70	.1
	1.0

How much safety stock should David Rivera keep on hand?

APPROACH ▶ The objective is to find the amount of safety stock that minimizes the sum of the additional inventory holding costs and stockout costs. The annual holding cost is simply the holding cost per unit multiplied by the units added to the ROP. For example, a safety stock of 20 frames, which implies that the new ROP, with safety stock, is 70 (= 50 + 20), raises the annual carrying cost by \$5(20) = \$100.

However, computing annual stockout cost is more interesting. For any level of safety stock, stockout cost is the expected cost of stocking out. We can compute it, as in Equation (12-12), by multiplying the number of frames short (Demand − ROP) by the probability of demand at that level, by the stockout cost, by the number of times per year the stockout can occur (which in our case is the number of orders per year). Then we add stockout costs for each possible stockout level for a given ROP.[5]

SOLUTION ▶ We begin by looking at zero safety stock. For this safety stock, a shortage of 10 frames will occur if demand is 60, and a shortage of 20 frames will occur if the demand is 70. Thus, the stockout costs for zero safety stock are:

$$(10 \text{ frames short})(.2)(\$40 \text{ per stockout})(6 \text{ possible stockouts per year}) + (20 \text{ frames short})(.1)(\$40)(6) = \$960$$

The following table summarizes the total costs for each of the three alternatives:

SAFETY STOCK	ADDITIONAL HOLDING COST	STOCKOUT COST	TOTAL COST
20	(20) ($5) = $100	$ 0	$100
10	(10) ($5) = $ 50	(10) (.1) ($40) (6) = $240	$290
0	$ 0	(10) (.2) ($40) (6) + (20) (.1) ($40) (6) = $960	$960

The safety stock with the lowest total cost is 20 frames. Therefore, this safety stock changes the reorder point to 50 + 20 = 70 frames.

INSIGHT ▶ The optical company now knows that a safety stock of 20 frames will be the most economical decision.

LEARNING EXERCISE ▶ David Rivera's holding cost per frame is now estimated to be $20, while the stockout cost is $30 per frame. Does the reorder point change? [Answer: Safety stock = 10 now, with a total cost of $380, which is the lowest of the three. ROP = 60 frames.]

RELATED PROBLEMS ▶ 12.43, 12.44, 12.45

When it is difficult or impossible to determine the cost of being out of stock, a manager may decide to follow a policy of keeping enough safety stock on hand to meet a prescribed customer service level. For instance, Figure 12.8 shows the use of safety stock when demand (for hospital resuscitation kits) is probabilistic. We see that the safety stock in Figure 12.8 is 16.5 units, and the reorder point is also increased by 16.5.

The manager may want to define the service level as meeting 95% of the demand (or, conversely, having stockouts only 5% of the time). Assuming that demand during lead time (the

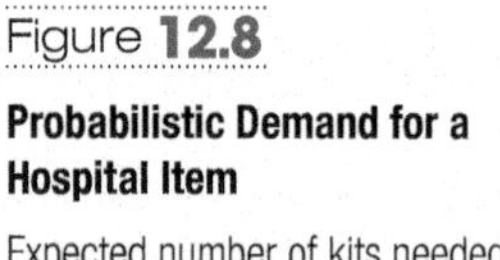

Figure 12.8

Probabilistic Demand for a Hospital Item

Expected number of kits needed during lead time is 350, but for a 95% service level, the reorder point should be raised to 366.5.

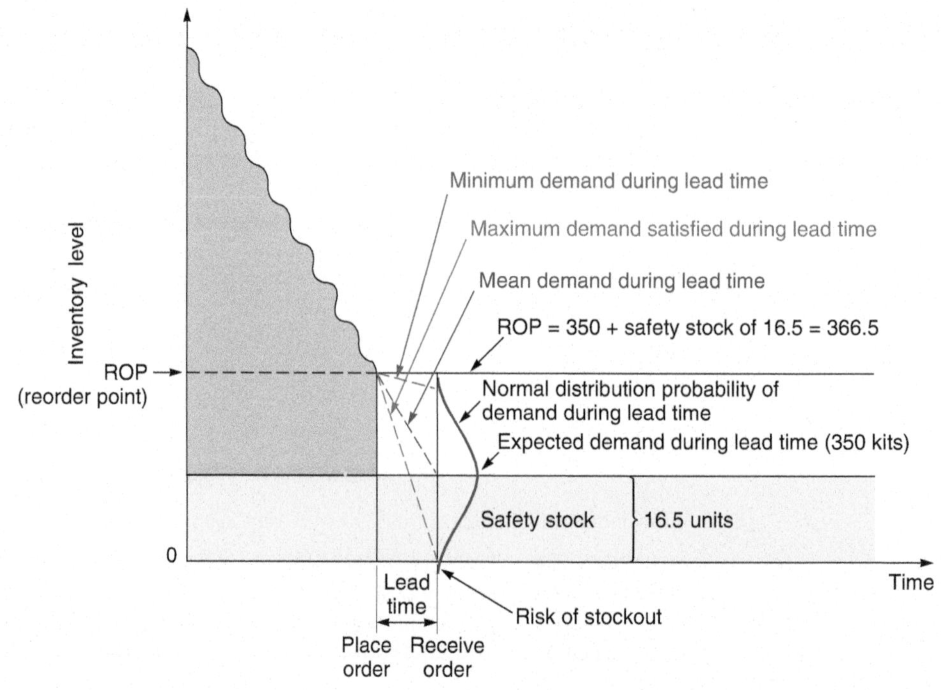

reorder period) follows a normal curve, only the mean and standard deviation are needed to define the inventory requirements for any given service level. Sales data are usually adequate for computing the mean and standard deviation. Example 11 uses a normal curve with a known mean (μ) and standard deviation (σ) to determine the reorder point and safety stock necessary for a 95% service level. We use the following formula:

$$\text{ROP} = \text{Expected demand during lead time} + Z\sigma_{dLT} \quad (12\text{-}13)$$

where Z = Number of standard deviations
σ_{dLT} = Standard deviation of demand during lead time

Example 11

SAFETY STOCK WITH PROBABILISTIC DEMAND

Memphis Regional Hospital stocks a "code blue" resuscitation kit that has a normally distributed demand during the reorder period. The mean (average) demand during the reorder period is 350 kits, and the standard deviation is 10 kits. The hospital administrator wants to follow a policy that results in stockouts only 5% of the time.

(a) What is the appropriate value of Z? (b) How much safety stock should the hospital maintain? (c) What reorder point should be used?

APPROACH ▶ The hospital determines how much inventory is needed to meet the demand 95% of the time. The figure in this example may help you visualize the approach. The data are as follows:

μ = Mean demand = 350 kits
σ_{dLT} = Standard deviation of demand during lead time = 10 kits
Z = Number of standard normal deviations

STUDENT TIP Recall that the service level is 1 minus the risk of a stockout.

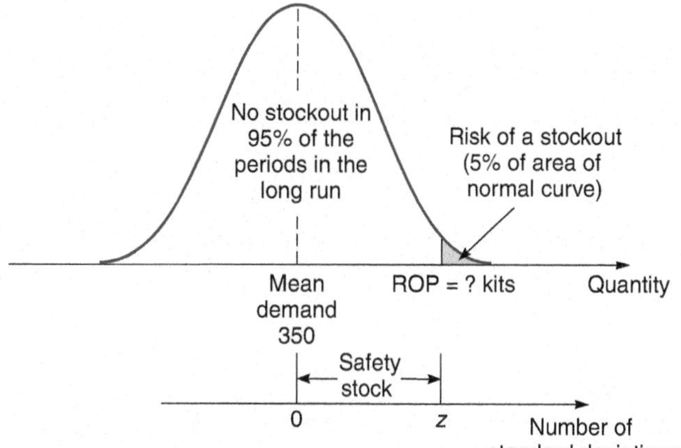

SOLUTION ▶

a) We use the properties of a standardized normal curve to get a Z-value for an area under the normal curve of .95 (or 1 − .05). Using a normal table (see Appendix I) or the Excel formula =NORMSINV(.95), we find a Z-value of 1.645 standard deviations from the mean.

b) Because: Safety stock $= x - \mu$

and: $$Z = \frac{x - \mu}{\sigma_{dLT}}$$

then: $$\text{Safety stock} = Z\sigma_{dLT} \quad (12\text{-}14)$$

c) Solving for safety stock, as in Equation (12-14), gives:

$$\text{Safety stock} = 1.645(10) = 16.5 \text{ kits}$$

This is the situation illustrated in Figure 12.8.

d) The reorder point is:

$$\begin{aligned}\text{ROP} &= \text{Expected demand during lead time} + \text{Safety stock}\\ &= 350 \text{ kits} + 16.5 \text{ kits of safety stock} = 366.5, \text{ or } 367 \text{ kits}\end{aligned}$$

INSIGHT ▶ The cost of the inventory policy increases dramatically (exponentially) with an increase in service levels.

LEARNING EXERCISE ▶ What policy results in stockouts 10% of the time? [Answer: $Z = 1.28$; safety stock = 12.8; ROP = 363 kits.]

RELATED PROBLEMS ▶ 12.41, 12.42, 12.49, 12.50

Other Probabilistic Models

Equations (12-13) and (12-14) assume that both an estimate of expected demand during lead times and its standard deviation are available. When data on lead time demand are *not* available, the preceding formulas cannot be applied. However, three other models are available. We need to determine which model to use for three situations:

LO 12.7 *Understand* service levels and probabilistic inventory models

1. Demand is variable and lead time is constant
2. Lead time is variable and demand is constant
3. Both demand and lead time are variable

All three models assume that demand and lead time are independent variables. Note that our examples use days, but weeks can also be used. Let us examine these three situations separately, because a different formula for the ROP is needed for each.

Demand Is Variable and Lead Time Is Constant (See Example 12.) When *only the demand is variable*, then:[6]

$$\text{ROP} = (\textit{Average}\text{ daily demand} \times \text{Lead time in days}) + Z\sigma_{dLT} \qquad (12\text{-}15)$$

where σ_{dLT} = Standard deviation of demand during lead time = $\sigma_d\sqrt{\text{Lead time}}$
and σ_d = Standard deviation of demand per day

Example 12

ROP FOR VARIABLE DEMAND AND CONSTANT LEAD TIME

The *average* daily demand for Lenovo laptop computers at a Circuit Town store is 15, with a standard deviation of 5 units. The lead time is constant at 2 days. Find the reorder point if management wants a 90% service level (i.e., risk stockouts only 10% of the time). How much of this is safety stock?

APPROACH ▶ Apply Equation (12-15) to the following data:

Average daily demand (normally distributed) = 15

Lead time in days (constant) = 2

Standard deviation of daily demand = σ_d = 5

Service level = 90%

SOLUTION ▶ From the normal table (Appendix I) or the Excel formula =NORMSINV(.90), we derive a Z-value for 90% of 1.28. Then:

$$\begin{aligned}\text{ROP} &= (15 \text{ units} \times 2 \text{ days}) + Z\sigma_d\sqrt{\text{Lead time}}\\ &= 30 + 1.28(5)(\sqrt{2})\\ &= 30 + 1.28(5)(1.41) = 30 + 9.02 = 39.02 \cong 39\end{aligned}$$

Thus, safety stock is about 9 Lenovo computers.

INSIGHT ▶ The value of Z depends on the manager's stockout risk level. The smaller the risk, the higher the Z.

LEARNING EXERCISE ▶ If the Circuit Town manager wants a 95% service level, what is the new ROP? [Answer: ROP = 41.63, or 42.]

RELATED PROBLEM ▶ 12.46

Lead Time Is Variable and Demand Is Constant When the demand is constant and *only the lead time is variable*, then:

$$\text{ROP} = (\text{Daily demand} \times \textit{Average}\text{ lead time in days}) + Z \times \text{Daily demand} \times \sigma_{LT} \qquad (12\text{-}16)$$

where σ_{LT} = Standard deviation of lead time in days

Example 13

ROP FOR CONSTANT DEMAND AND VARIABLE LEAD TIME

The Circuit Town store in Example 12 sells about 10 tablets a day (almost a constant quantity). Lead time for tablet delivery is normally distributed with a mean time of 6 days and a standard deviation of 1 day. A 98% service level is set. Find the ROP.

APPROACH ▶ Apply Equation (12-16) to the following data:

Daily demand = 10
Average lead time = 6 days
Standard deviation of lead time = σ_{LT} = 1 day

Service level = 98%, so Z (from Appendix I or the Excel formula = NORMSINV(.98)) = 2.054

SOLUTION ▶ From the equation we get:

$$\text{ROP} = (10 \text{ units} \times 6 \text{ days}) + 2.054(10 \text{ units})(1)$$
$$= 60 + 20.54 = 80.54$$

The reorder point is about 81 tablets.

INSIGHT ▶ Note how the very high service level of 98% drives the ROP up.

LEARNING EXERCISE ▶ If a 90% service level is applied, what does the ROP drop to? [Answer: ROP = 60 + (1.28) (10) (1) = 60 + 12.8 = 72.8 because the Z-value is only 1.28.]

RELATED PROBLEM ▶ 12.47

Both Demand and Lead Time Are Variable When both the demand and lead time are variable, the formula for reorder point becomes more complex:[7]

$$\text{ROP} = (\text{Average daily demand} \times \text{Average lead time in days}) + Z\sigma_{dLT} \qquad (12\text{-}17)$$

where

$$\sigma_d = \text{Standard deviation of demand per day}$$
$$\sigma_{LT} = \text{Standard deviation of lead time in days}$$
$$\text{and } \sigma_{dLT} = \sqrt{(\text{Average lead time} \times \sigma_d^2) + (\text{Average daily demand})^2 \sigma_{LT}^2}$$

Example 14

ROP FOR VARIABLE DEMAND AND VARIABLE LEAD TIME

The Circuit Town store's most popular item is six-packs of 9-volt batteries. About 150 packs are sold per day, following a normal distribution with a standard deviation of 16 packs. Batteries are ordered from an out-of-state distributor; lead time is normally distributed with an average of 5 days and a standard deviation of 1 day. To maintain a 95% service level, what ROP is appropriate?

APPROACH ▶ Determine a quantity at which to reorder by applying Equation (12-17) to the following data:

Average daily demand = 150 packs
Standard deviation of demand = σ_d = 16 packs
Average lead time = 5 days
Standard deviation of lead time = σ_{LT} = 1 day

Service level = 95%, so Z =1.645 (from Appendix I or the Excel formula = NORMSINV(.95))

SOLUTION ▶ From the equation we compute:

$$\text{ROP} = (150 \text{ packs} \times 5 \text{ days}) + 1.645\,\sigma_{dLT}$$

where

$$\sigma_{dLT} = \sqrt{(5 \text{ days} \times 16^2) + (150^2 \times 1^2)}$$
$$= \sqrt{(5 \times 256) + (22{,}500 \times 1)}$$
$$= \sqrt{1{,}280 + 22{,}500} = \sqrt{23{,}780} \cong 154$$

So ROP = (150 × 5) + 1.645(154) ≅ 750 + 253 = 1,003 packs

INSIGHT ▶ When both demand and lead time are variable, the formula looks quite complex. But it is just the result of squaring the standard deviations in Equations (12-15) and (12-16) to get their variances, then summing them, and finally taking the square root.

LEARNING EXERCISE ▶ For an 80% service level, what is the ROP? [Answer: $Z = .84$ and ROP = 879 packs.]

RELATED PROBLEM ▶ 12.48

Single-Period Model

Single-period inventory model
A system for ordering items that have little or no value at the end of a sales period (perishables).

A **single-period inventory model** describes a situation in which *one* order is placed for a product. At the end of the sales period, any remaining product has little or no value. This is a typical problem for seasonal goods such as Christmas trees or high-fashion apparel (see the *OM in Action Box*, "Wrestling with Unsold Clothing Inventory"), bakery goods, newspapers, and magazines. (Indeed, this inventory issue is often called the "newsstand problem.") In other words, even though items at a newsstand are ordered weekly or daily, they cannot be held over and used as inventory in the next sales period. So our decision is how much to order at the beginning of the period.

Because the exact demand for such seasonal products is never known, we consider a probability distribution related to demand. If the normal distribution is assumed, and we stocked and sold an average (mean) of 100 Christmas trees each season, then there is a 50% chance we would stock out and a 50% chance we would have trees left over. To determine the optimal stocking policy for trees before the season begins, we also need to know the standard deviation and consider these two marginal costs:

C_s = Cost of shortage (we underestimated) = Sales price per unit − Cost per unit

C_o = Cost of overage (we overestimated) = Cost per unit − Salvage value per unit (if there is any)

The service level, that is, the probability of *not* stocking out, is set at:

$$\text{Service level} = \frac{C_s}{C_s + C_o} \qquad (12\text{-}18)$$

OM in Action: Wrestling with Unsold Clothing Inventory[8]

Apparel companies, from elite fashion houses to mass-market chains, occasionally become saddled with an inventory glut following recessions and economic downturns (e.g., the pandemic). When that happens, and it does happen every few years, they try to get rid of the excess without angering waste-conscious consumers—or harming their brands. In the U.S., when brands and retailers have huge excess inventories, they tend to flood charities with unsold products in addition to sending goods to discount stores and liquidators.

Good360, a nonprofit that collects excess merchandise and distributes it to charities, recently received more than $660 million in donations. "Brands don't want their unsold products winding up at flea markets or on Craigslist," said the CEO of Good360.

LVMH—which owns Louis Vuitton, Dior, and other luxury brands—recently booked a $200 million write-down on its inventories because many products destined for the spring/summer fashion season were ordered just before much of the luxury market went into a downturn.

Big retailers sometimes destroy returned products rather than deal with the cost of trying to resell or even give them away. Brands that destroy unsold goods have sparked outrage from consumers, politicians, and environmental groups. In spite of public concern, French businesses destroyed $700 million in unsold goods in a recent year—six times more than they donated. High-end fashion companies fear angering clientele who would spend thousands on a designer dress or bag, only to see the same item a year later at a discount store selling for a fraction of the price.

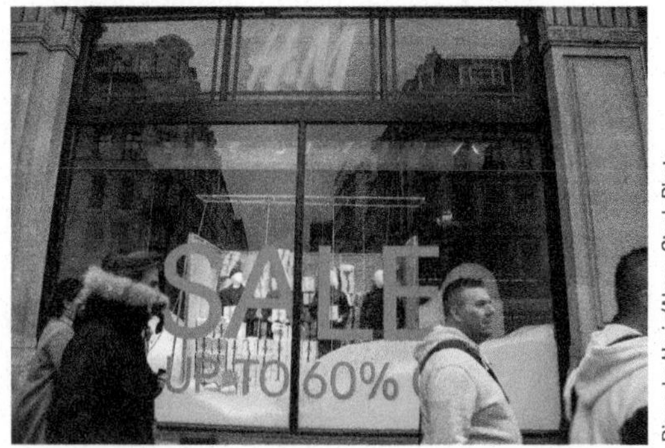

Dinendra Haria/Alamy Stock Photo

Therefore, we should consider increasing our order quantity until the service level is equal to or more than the ratio of $[C_s/(C_s + C_o)]$.

This model, illustrated in Example 15, is used in many service industries, from hotels to airlines to bakeries to clothing retailers.

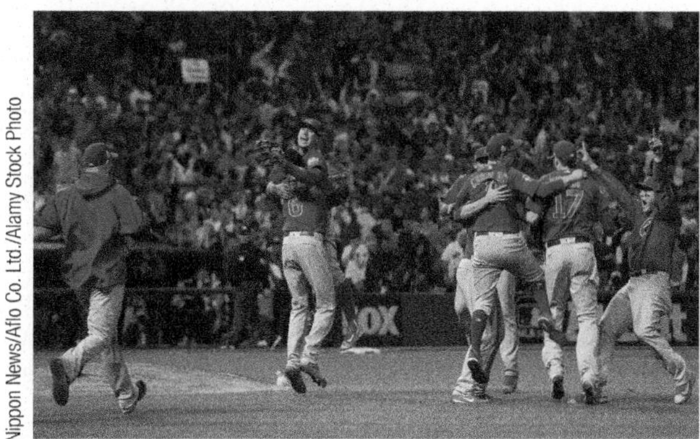

Nippon News/Aflo Co. Ltd./Alamy Stock Photo

The postseason run for sports teams can bring big business for merchandise companies—if they understand the *single-period inventory model*. Before the Chicago Cubs recently won the World Series, product forecasting was difficult because there were no licensed goods in 1908—the last time they achieved that position. "No one alive has seen Cubs World Series merchandise in their lifetime, so I believe that every Cubs fan will want something," said a merchandiser. "We can't *not* have enough to supply the fans. But what makes me nervous is if there's a number where you overbuy?" Companies need to compute the cost of underestimating (shortages) versus the cost of overestimating.

Example 15

SINGLE-PERIOD INVENTORY DECISION

Chris Ellis's newsstand, just outside the Smithsonian subway station in Washington, DC, usually sells 120 copies of the *Washington Post* each day. Chris believes the sale of the *Post* is normally distributed, with a standard deviation of 15 papers. He pays 70 cents for each paper, which sells for $1.25. The *Post* gives him a 30-cent credit for each unsold paper. He wants to determine how many papers he should order each day and the stockout risk for that quantity.

APPROACH ▶ Chris's data are as follows:

$$C_s = \text{cost of shortage} = \$1.25 - \$.70 = \$.55$$
$$C_o = \text{cost of overage} = \$.70 - \$.30\ (\text{salvage value}) = \$.40$$

Chris will apply Equation (12-18) and the normal table, using $\mu = 120$ and $\sigma = 15$.

SOLUTION ▶

a) $\text{Service level} = \dfrac{C_s}{C_s + C_o} = \dfrac{.55}{.55 + .40} = \dfrac{.55}{.95} = .579$

b) Chris needs to find the Z score for his normal distribution that yields a probability of .579.

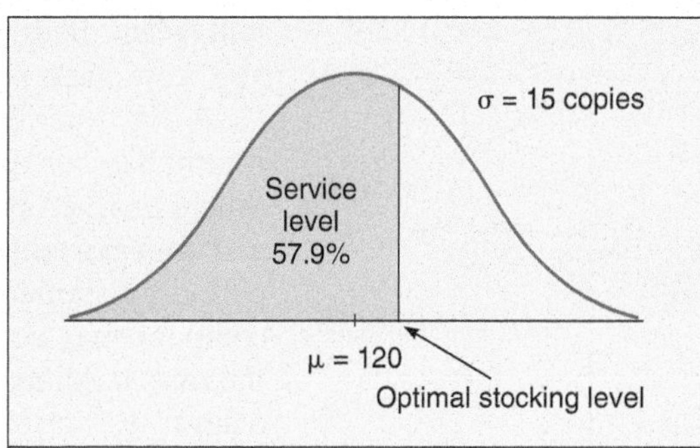

So 57.9% of the area under the normal curve must be to the left of the optimal stocking level.

c) Using Appendix I or the Excel formula = NORMSINV(.579), for an area of .579, the Z value ≅ .199.

$$\text{Then, the optimal stocking level} = 120\ \text{copies} + (.199)(\sigma)$$
$$= 120 + (.199)(15) = 120 + 3 = 123\ \text{papers}$$

The stockout risk if Chris orders 123 copies of the *Post* each day is
1 − Service level = 1 − 579 = .421 = 42.1%.

INSIGHT ▶ If the service level is ever under .50, Chris should order fewer than 120 copies per day.

LEARNING EXERCISE ▶ How does Chris's decision change if the *Post* changes its policy and offers *no credit* for unsold papers, a policy many publishers are adopting?

[Answer: Service level = .44, Z = −.15. Therefore, stock 120 + (−.15)(15) = 117.75, or 118 papers.]

RELATED PROBLEMS ▶ 12.51, 12.52, 12.53

Fixed-Period (*P*) Systems

Fixed-quantity (*Q*) system
An ordering system with the same order amount each time.

Perpetual inventory system
A system that keeps track of each withdrawal or addition to inventory continuously, so records are always current.

Fixed-period (*P*) system
A system in which inventory orders are made at regular time intervals.

The inventory models that we have considered so far are **fixed-quantity**, or ***Q*, systems**. That is, the same fixed amount is added to inventory every time an order for an item is placed. We saw that orders are event triggered. When inventory decreases to the reorder point (ROP), a new order for Q units is placed.

To use the fixed-quantity model, inventory must be continuously monitored.[9] This requires a **perpetual inventory system**. Every time an item is added to or withdrawn from inventory, records must be updated to determine whether the ROP has been reached. In a **fixed-period system** (also called a *periodic review*, or ***P* system**), on the other hand, inventory is ordered at the end of a given period. Then, and only then, is on-hand inventory counted. Only the amount necessary to bring total inventory up to a prespecified target level (T) is ordered. Figure 12.9 illustrates this concept.

Fixed-period systems have several of the same assumptions as the basic EOQ fixed-quantity system:

- The only relevant costs are the ordering and holding costs.
- Lead times are known and constant.
- Items are independent of one another.

STUDENT TIP
A fixed-period model potentially orders a different quantity each time.

The downward-sloped lines in Figure 12.9 again represent on-hand inventory levels. But now, when the time between orders (P) passes, we place an order to raise inventory up to the target quantity (T). The amount ordered during the first period may be Q_1, the second period Q_2, and so on. The Q_i value is the difference between current on-hand inventory and the target inventory level.

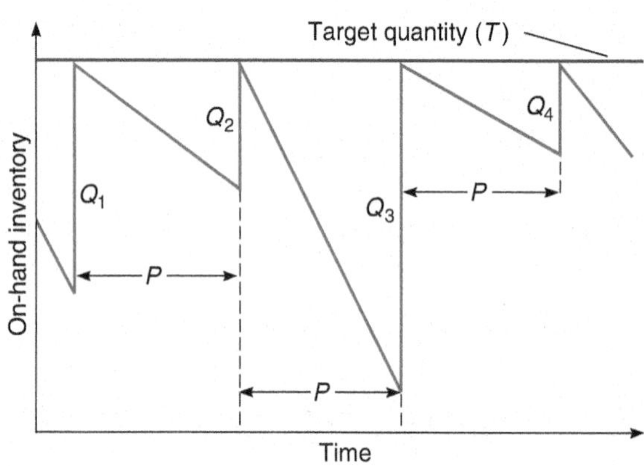

Figure 12.9

Inventory Level in a Fixed-Period (*P*) System
Various amounts (Q_1, Q_2, Q_3, etc.) are ordered at regular time intervals (P) based on the quantity necessary to bring inventory up to the target quantity (T).

The advantage of the fixed-period system is that there is no physical count of inventory items after an item is withdrawn—this occurs only when the time for the next review comes up. This procedure is also convenient administratively.

A fixed-period (P) system is appropriate when vendors make routine (i.e., at fixed-time interval) visits to customers to take fresh orders or when purchasers want to combine orders to save ordering and transportation costs (therefore, they will have the same review period for similar inventory items). For example, a vending machine company may come to refill its machines every Tuesday. This is also the case at Anheuser-Busch, whose sales reps may visit a store every 5 days.

The disadvantage of the P system is that because there is no tally of inventory during the review period, there is the possibility of a stockout during this time. This scenario is possible if a large order draws the inventory level down to zero right after an order is placed. Therefore, a higher level of safety stock (as compared to a fixed-quantity system) needs to be maintained to provide protection against stockout during both the time between reviews and the lead time.

Summary

Inventory represents a major investment for many firms. This investment is often larger than it should be because firms find it easier to have "just-in-case" inventory rather than "just-in-time" inventory. Inventories are of four types:

1. Raw material and purchased components
2. Work-in-process
3. Maintenance, repair, and operating (MRO)
4. Finished goods

In this chapter, we discussed independent inventory, ABC analysis, record accuracy, cycle counting, and inventory models used to control independent demands. The EOQ model, production order quantity model, and quantity discount model can all be solved using Excel, Excel OM, or POM for Windows software.

Key Terms

Raw material inventory (p. 492)
Work-in-process (WIP) inventory (p. 492)
Maintenance/repair/operating (MRO) inventory (p. 492)
Finished-goods inventory (p. 493)
ABC analysis (p. 493)
Cycle counting (p. 495)
Shrinkage (p. 496)
Pilferage (p. 496)
Holding cost (p. 497)
Ordering cost (p. 497)
Setup cost (p. 497)
Setup time (p. 497)
Economic order quantity (EOQ) model (p. 498)
Robust (p. 502)
Lead time (p. 503)
Reorder point (ROP) (p. 503)
Safety stock (*ss*) (p. 503)
Production order quantity model (p. 504)
Quantity discount (p. 506)
Probabilistic model (p. 509)
Service level (p. 509)
Single-period inventory model (p. 514)
Fixed-quantity (*Q*) system (p. 516)
Perpetual inventory system (p. 516)
Fixed-period (*P*) system (p. 516)

Ethical Dilemma

Wayne Hills Hospital in tiny Wayne, Nebraska, faces a problem common to large, urban hospitals as well as to small, remote ones like itself. That problem is deciding how much of each type of whole blood to keep in stock. Because blood is expensive and has a limited shelf life (up to 5 weeks under 1–6°C refrigeration), Wayne Hills naturally wants to keep its stock as low as possible. Unfortunately, past disasters such as a major tornado and a train wreck demonstrated that lives would be lost when not enough blood was available to handle massive needs. The hospital administrator wants to set an 85% service level based on demand over the past decade. Discuss the implications of this decision. What is the hospital's responsibility with regard to stocking lifesaving medicines with short shelf lives? How would you set the inventory level for a commodity such as blood?

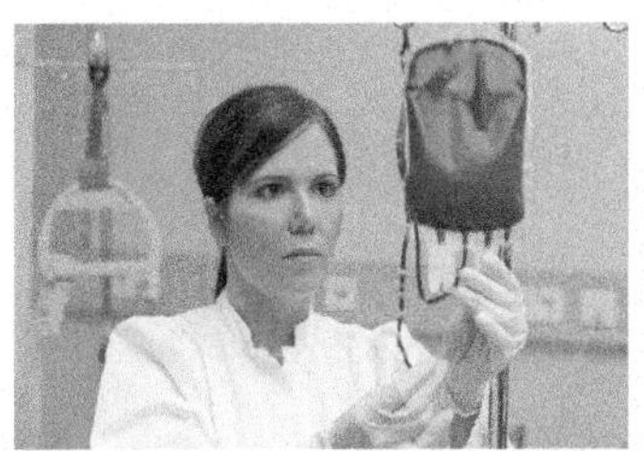
Ginasanders/123RF

Discussion Questions

1. Describe the four types of inventory.
2. With the advent of low-cost computing, do you see alternatives to the popular ABC classifications?
3. What is the purpose of the ABC classification system?
4. Identify and explain the types of costs that are involved in an inventory system.
5. Explain the major assumptions of the basic EOQ model.
6. What is the relationship of the economic order quantity to demand? To the holding cost? To the setup cost?
7. Explain why it is not necessary to include product cost (price or price times quantity) in the EOQ model, but the quantity discount model requires this information.
8. What are the advantages of cycle counting?
9. What impact does a decrease in setup time have on EOQ?
10. When quantity discounts are offered, why is it not necessary to check discount points that are below the EOQ or points above the EOQ that are not discount points?
11. What is meant by "service level"?
12. Explain the following: All things being equal, the production order quantity will be larger than the economic order quantity.
13. Describe the difference between a fixed-quantity (*Q*) and a fixed-period (*P*) inventory system.
14. Explain what is meant by the expression "robust model." Specifically, what would you tell a manager who exclaimed, "Uh-oh, we're in trouble! The calculated EOQ is wrong; actual demand is 10% greater than estimated."
15. What is "safety stock"? What does safety stock provide safety against?
16. When demand is not constant, the reorder point is a function of what four parameters?
17. How are inventory levels monitored in retail stores?
18. State a major advantage, and a major disadvantage, of a fixed-period (*P*) system.

Using Software to Solve Inventory Problems

This section presents three ways to solve inventory problems with computer software. First, you can create your own Excel spreadsheets. Second, you can use the Excel OM software accompanying this text and found online. Third, POM for Windows, also available online, can solve all problems marked with a **P**.

CREATING YOUR OWN EXCEL SPREADSHEETS

Program 12.1 illustrates how you can make an Excel model to solve Example 8, which is a production order quantity model.

Program 12.1

Using Excel for a Production Model, with Data from Example 8

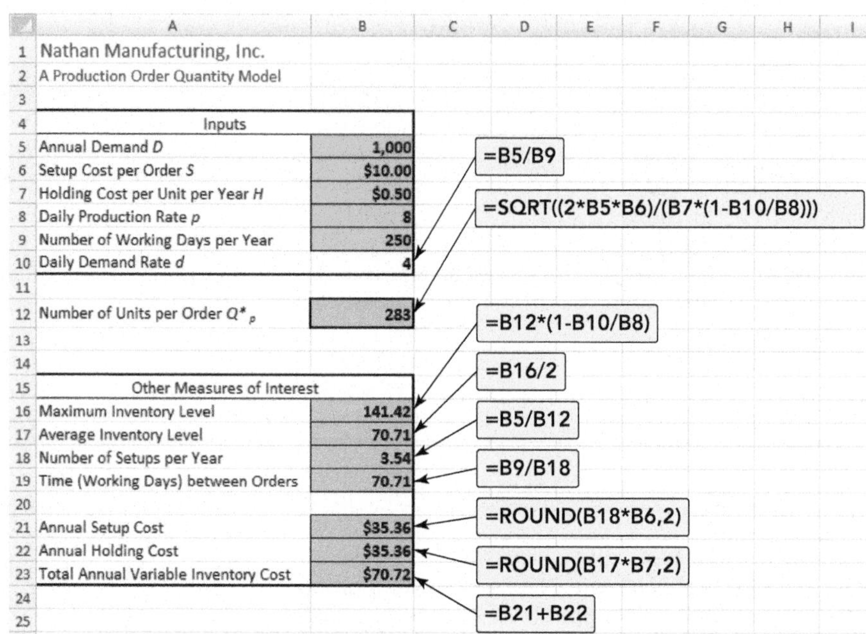

Program 12.2 illustrates how you can make an Excel model to solve Example 15, which is a single-period inventory model.

Program 12.2

Using Excel for a Single-Period Inventory Model, with Data from Example 15

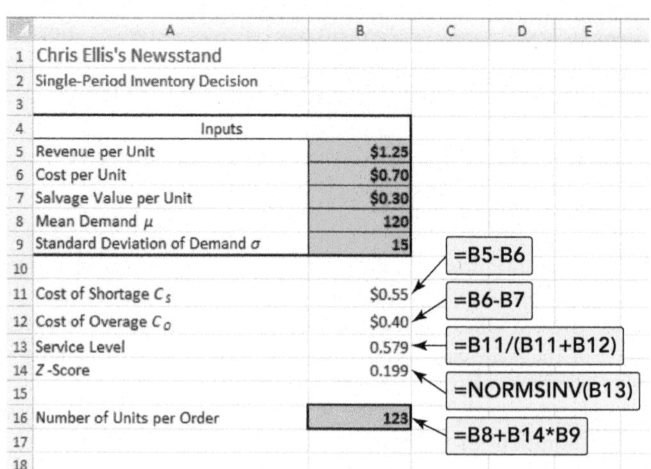

USING EXCEL OM

Excel OM allows us to easily model inventory problems ranging from ABC analysis, to the basic EOQ model, to the production model, to quantity discount situations.

Program 12.3 shows the input data, selected formulas, and results for an ABC analysis, using data from Example 1. After the data are entered, we use the *Data* and *Sort* Excel commands to rank the items from largest to smallest dollar volumes.

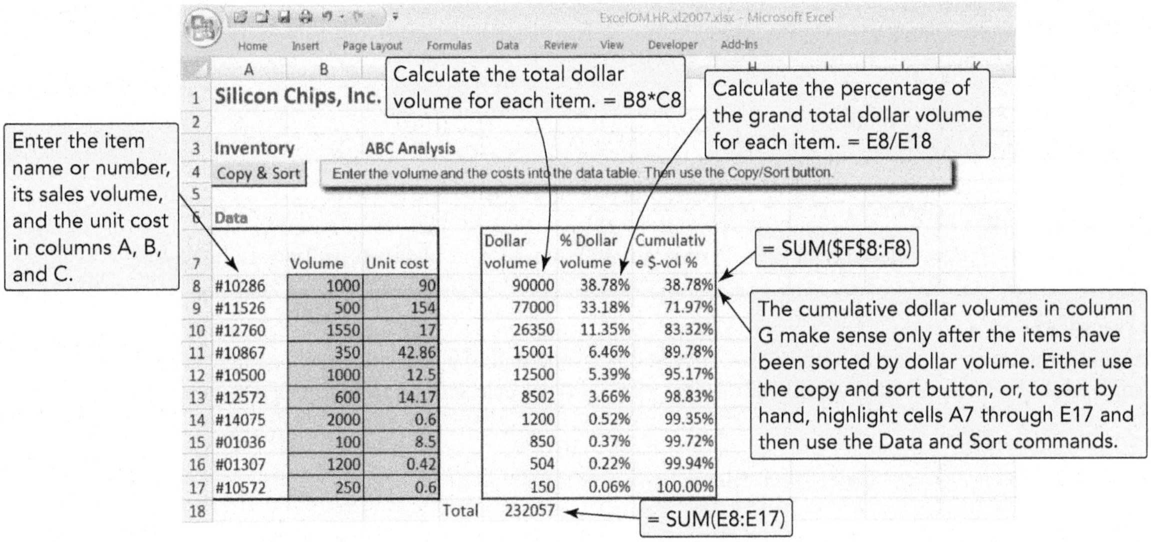

ExcelOM.HR.xl2007.xlsx - Microsoft Excel

Home Insert Page Layout Formulas Data Review View Developer Add-Ins

Enter the item name or number, its sales volume, and the unit cost in columns A, B, and C.

Calculate the total dollar volume for each item. = B8*C8

Calculate the percentage of the grand total dollar volume for each item. = E8/E18

1 Silicon Chips, Inc.

3 Inventory — ABC Analysis

4 Copy & Sort — Enter the volume and the costs into the data table. Then use the Copy/Sort button.

6 Data

		Volume	Unit cost	Dollar volume	% Dollar volume	Cumulative $-vol %
8	#10286	1000	90	90000	38.78%	38.78%
9	#11526	500	154	77000	33.18%	71.97%
10	#12760	1550	17	26350	11.35%	83.32%
11	#10867	350	42.86	15001	6.46%	89.78%
12	#10500	1000	12.5	12500	5.39%	95.17%
13	#12572	600	14.17	8502	3.66%	98.83%
14	#14075	2000	0.6	1200	0.52%	99.35%
15	#01036	100	8.5	850	0.37%	99.72%
16	#01307	1200	0.42	504	0.22%	99.94%
17	#10572	250	0.6	150	0.06%	100.00%
18			Total	232057		

= SUM($F$8:F8)

The cumulative dollar volumes in column G make sense only after the items have been sorted by dollar volume. Either use the copy and sort button, or, to sort by hand, highlight cells A7 through E17 and then use the Data and Sort commands.

= SUM(E8:E17)

Program **12.3**

Using Excel OM for an ABC Analysis, with Data from Example 1

P USING POM FOR WINDOWS

The POM for Windows Inventory module can also solve the entire EOQ family of problems. Please refer to Appendix II for further details.

Solved Problems

SOLVED PROBLEM 12.1

David Alexander has compiled the following table of six items in inventory at Angelo Products, along with the unit cost and the annual demand in units:

IDENTIFICATION CODE	UNIT COST ($)	ANNUAL DEMAND (UNITS)
XX1	5.84	1,200
B66	5.40	1,110
3CPO	1.12	896
33CP	74.54	1,104
R2D2	2.00	1,110
RMS	2.08	961

Use ABC analysis to determine which item(s) should be carefully controlled using a quantitative inventory technique and which item(s) should not be closely controlled.

SOLUTION

The item that needs strict control is 33CP, so it is an A item. Items that do not need to be strictly controlled are 3CPO, R2D2, and RMS; these are C items. The B items will be XX1 and B66.

CODE	ANNUAL DOLLAR VOLUME = UNIT COST × DEMAND
XX1	$ 7,008.00
B66	$ 5,994.00
3CPO	$ 1,003.52
33CP	$82,292.16
R2D2	$ 2,220.00
RMS	$ 1,998.88

Total cost $= \$100,516.56$

70% of total cost $= \$70,347.92$

SOLVED PROBLEM 12.2

The Warren W. Fisher Computer Corporation purchases 8,000 transistors each year as components in minicomputers. The unit cost of each transistor is $10, and the cost of carrying one transistor in inventory for a year is $3. Ordering cost is $30 per order.

What are (a) the optimal order quantity, (b) the expected number of orders placed each year, and (c) the expected time between orders? Assume that Fisher operates on a 200-day working year.

SOLUTION

a) $Q^* = \sqrt{\dfrac{2DS}{H}} = \sqrt{\dfrac{2(8{,}000)(30)}{3}} = 400 \text{ units}$

b) $N = \dfrac{D}{Q^*} = \dfrac{8{,}000}{400} = 20 \text{ orders}$

c) $\text{Time between orders} = T = \dfrac{\text{Number of working days}}{N} = \dfrac{200}{20} = 10 \text{ working days}$

With 20 orders placed each year, an order for 400 transistors is placed every 10 working days.

SOLVED PROBLEM 12.3

Annual demand for notebook binders at Meyer's Stationery Shop is 10,000 units. Brad Meyer operates his business 300 days per year and finds that deliveries from his supplier generally take 5 working days. Calculate the reorder point for the notebook binders.

SOLUTION

$$L = 5 \text{ days}$$
$$d = \frac{10{,}000}{300} = 33.3 \text{ units per day}$$
$$\text{ROP} = d \times L = (33.3 \text{ units per day})(5 \text{ days}) = 166.7 \text{ units}$$

Thus, Brad should reorder when his stock reaches 167 units.

SOLVED PROBLEM 12.4

Leonard Presby, Inc., has an annual demand rate of 1,000 units but can produce at an average production rate of 2,000 units. Setup cost is \$10; carrying cost is \$1. What is the optimal number of units to be produced each time?

SOLUTION

$$Q_p^* = \sqrt{\frac{2DS}{H\left(1 - \dfrac{\text{Annual demand rate}}{\text{Annual production rate}}\right)}} = \sqrt{\frac{2(1{,}000)(10)}{1\left[1 - (1{,}000/2{,}000)\right]}}$$
$$= \sqrt{\frac{20{,}000}{1/2}} = \sqrt{40{,}000} = 200 \text{ units}$$

SOLVED PROBLEM 12.5

Whole Nature Foods sells a gluten-free product for which the annual demand is 5,000 boxes. At the moment, it is paying \$6.40 for each box; carrying cost is 25% of the unit cost; ordering costs are \$25. A new supplier has offered to sell the same item for \$6.00 if Whole Nature Foods buys at least 3,000 boxes per order. Should the firm stick with the old supplier, or take advantage of the new quantity discount?

SOLUTION

Step 1, under the lowest possible price of \$6.00 per box:

Economic order quantity, using Equation (12-10):

$$Q^*_{\$6.00} = \sqrt{\frac{2(5{,}000)(25)}{(0.25)(6.00)}}$$
$$= 408.25, \text{ or } 408 \text{ boxes}$$

Because $408 < 3{,}000$, this EOQ is *infeasible* for the \$6.00 price. So now we calculate Q^* for the next-higher price of \$6.40, which equals 395 boxes (and is feasible). Thus, the best possible order quantities are 395 (the first feasible EOQ) and 3,000 (the price-break quantity for the lower price of \$6.00).

Step 2 uses Equation (12-9) to compute the total cost for both of the possible best order quantities:

$$TC_{395} = \frac{5{,}000}{395}(\$25) + \frac{395}{2}(0.25)(\$6.40) + \$6.40(5{,}000)$$
$$= \$316 + \$316 + \$32{,}000$$
$$= \$32{,}632$$

And under the quantity discount price of \$6.00 per box:

$$TC_{3{,}000} = \frac{5{,}000}{3{,}000}(\$25) + \frac{3{,}000}{2}(0.25)(\$6.00) + \$6.00(5{,}000)$$
$$= \$42 + \$2{,}250 + \$30{,}000$$
$$= \$32{,}292$$

Therefore, the new supplier with which Whole Nature Foods would incur a total cost of \$32,292 is preferable, but not by a large amount. If buying 3,000 boxes at a time raises problems of storage or freshness, the company may very well wish to stay with the current supplier.

SOLVED PROBLEM 12.6

Children's art sets are ordered once each year by Ashok Kumar, Inc., and the reorder point, without safety stock (dL), is 100 art sets. Inventory carrying cost is $10 per set per year, and the cost of a stockout is $50 per set per year. Given the following demand probabilities during the lead time, how much safety stock should be carried?

DEMAND DURING LEAD TIME	PROBABILITY
0	.1
50	.2
ROP → 100	.4
150	.2
200	.1
	1.0

SOLUTION

	INCREMENTAL COSTS		
SAFETY STOCK	CARRYING COST	STOCKOUT COST	TOTAL COST
0	0	50 × (50 × 0.2 + 100 × 0.1) = 1,000	$1,000
50	50 × 10 = 500	50 × (0.1× 50) = 250	750
100	100 × 10 = 1,000	0	1,000

The safety stock that minimizes total incremental cost is 50 sets. The reorder point then becomes 100 sets + 50 sets, or 150 sets.

SOLVED PROBLEM 12.7

What safety stock should Ron Satterfield Corporation maintain if mean sales are 80 during the reorder period, the standard deviation is 7, and Ron can tolerate stockouts 10% of the time?

SOLUTION

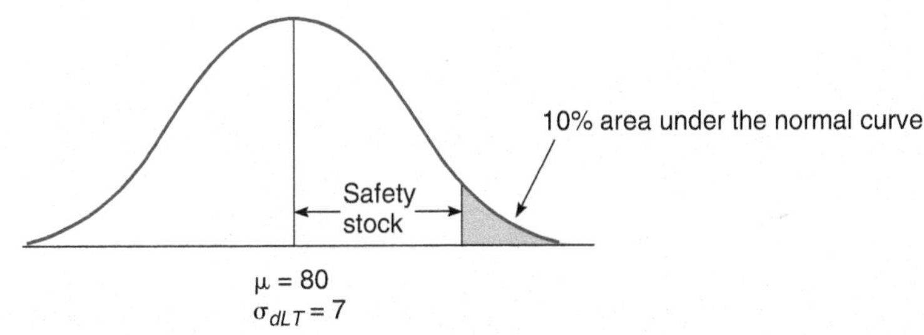

From Appendix I, Z at an area of .9 (or 1 – .10) = 1.28, and Equation (12-14):

$$\text{Safety stock} = Z\sigma_{dLT}$$
$$= 1.28(7) = 8.96 \text{ units, or } 9 \text{ units}$$

SOLVED PROBLEM 12.8

The daily demand for 52" flat-screen TVs at Sarah's Discount Emporium is normally distributed, with an average of 5 and a standard deviation of 2 units. The lead time for receiving a shipment of new TVs is 10 days and is fairly constant. Determine the reorder point and safety stock for a 95% service level.

SOLUTION

The ROP for this variable demand and constant lead time model uses Equation (12-15):

$$\text{ROP} = (\text{Average daily demand} \times \text{Lead time in days}) + Z\sigma_{dLT}$$

where $\sigma_{dLT} = \sigma_d\sqrt{\text{Lead time}}$

So, with $Z = 1.645$,

$$\text{ROP} = (5 \times 10) + 1.645(2)\sqrt{10}$$
$$= 50 + 10.4 = 60.4 \cong 60 \text{ TVs, or rounded up to 61 TVs}$$

The safety stock is 10.4, which can be rounded up to 11 TVs.

SOLVED PROBLEM 12.9

The demand at Arnold Palmer Hospital for a specialized surgery pack is 60 per week, virtually every week. The lead time from McKesson, its main supplier, is normally distributed, with a mean of 6 weeks for this product and a standard deviation of 2 weeks. A 90% weekly service level is desired. Find the ROP.

SOLUTION

Here the demand is constant and lead time is variable, with data given in weeks, not days. We apply Equation (12-16):

$$ROP = (\text{Weekly demand} \times \text{Average lead time in weeks}) + Z(\text{Weekly demand})\sigma_{LT}$$

where σ_{LT} = standard deviation of lead time in weeks = 2

So, with $Z = 1.28$, for a 90% service level:

$$ROP = (60 \times 6) + 1.28(60)(2)$$
$$= 360 + 153.6 = 513.6 \cong 514 \text{ surgery packs}$$

Problems

Note: **PX** means the problem may be solved with POM for Windows and/or Excel OM.

Problems 12.1–12.6 relate to Managing Inventory

•• **12.1** L. Houts Plastics is a large manufacturer of injection-molded plastics in North Carolina. An investigation of the company's manufacturing facility in Charlotte yields the information presented in the table below. How would the plant classify these items according to an ABC classification system? **PX**

L. Houts Plastics' Charlotte Inventory Levels

ITEM CODE #	AVERAGE INVENTORY (UNITS)	VALUE ($/UNIT)
1289	400	3.75
2347	300	4.00
2349	120	2.50
2363	75	1.50
2394	60	1.75
2395	30	2.00
6782	20	1.15
7844	12	2.05
8210	8	1.80
8310	7	2.00
9111	6	3.00

•• **12.2** Boreki Enterprises has the following 10 items in inventory. Theodore Boreki asks you, a recent OM graduate, to divide these items into ABC classifications.

ITEM	ANNUAL DEMAND	COST/UNIT
A2	3,000	$ 50
B8	4,000	12
C7	1,500	45
D1	6,000	10
E9	1,000	20
F3	500	500
G2	300	1,500
H2	600	20
I5	1,750	10
J8	2,500	5

a) Develop an ABC classification system for the 10 items.
b) How can Boreki use this information?
c) Boreki reviews the classification and then places item A2 into the A category. Why might he do so? **PX**

•• **12.3** Jean-Marie Bourjolly's restaurant has the following inventory items that it orders on a weekly basis:

INVENTORY ITEM	$ VALUE/CASE	# ORDERED/WEEK
Rib eye steak	135	3
Lobster tail	245	3
Pasta	23	12
Salt	3	2
Napkins	12	2
Tomato sauce	23	11
French fries	43	32
Pepper	3	3
Garlic powder	11	3
Trash can liners	12	3
Table cloths	32	5
Fish filets	143	10
Prime rib roasts	166	6
Oil	28	2
Lettuce (case)	35	24
Chickens	75	14
Order pads	12	2
Eggs (case)	22	7
Bacon	56	5
Sugar	4	2

a) Which is the most expensive item, using annual dollar volume?
b) Which are C items?
c) What is the annual dollar volume for all 20 items? **PX**

• **12.4** Lindsay Electronics, a small manufacturer of electronic research equipment, has approximately 7,000 items in its inventory and has hired Joan Blasco-Paul to manage its inventory. Joan has determined that 10% of the items in inventory are A items, 35% are B items, and 55% are C items. She would like to set up a system in which all A items are counted monthly (every 20 working days), all B items are counted quarterly (every 60 working days), and all C items are counted semiannually (every 120 working days). How many items need to be counted each day?

• **12.5** Barbara Flynn's company has compiled the following data on a small set of products:

ITEM	ANNUAL DEMAND	UNIT COST
A	100	$300
B	75	100
C	50	50
D	200	100
E	150	65

Perform an ABC analysis on her data. **PX**

•• **12.6** Lynn Fish opened a new beauty-products retail store. There are numerous items in inventory, and Lynn knows that there are costs associated with inventory. However, because her time is limited, she cannot carefully evaluate the inventory policy for all products. Lynn wants to classify the items according to dollars invested in them. The following table provides information about the 10 items that she carries:

ITEM NUMBER	UNIT COST	DEMAND (UNITS)
E102	$4	800
D23	$16	2,400
D27	$8	700
R02	$2	1,000
R19	$8	200
S107	$12	500
S123	$1	1,200
U11	$7	800
U23	$1	1,500
V75	$14	2,500

Use ABC analysis to classify these items into categories A, B, and C. **PX**

Problems 12.7–12.40 relate to Inventory Models for Independent Demand

• **12.7** William Beville's computer training school, in Richmond, stocks workbooks with the following characteristics:

$$\text{Demand } D = 19{,}500 \text{ units/year}$$
$$\text{Ordering cost } S = \$25/\text{order}$$
$$\text{Holding cost } H = \$4/\text{unit/year}$$

a) Calculate the EOQ for the workbooks.
b) What are the annual holding costs for the workbooks?
c) What are the annual ordering costs? **PX**

• **12.8** If $D = 8{,}000$ per month, $S = \$45$ per order, and $H = \$2$ per unit per month,
a) What is the economic order quantity?
b) How does your answer change if the holding cost doubles?
c) What if the holding cost drops in half? **PX**

•• **12.9** Leilani Lavender's law office has traditionally ordered ink refills 60 units at a time. The firm estimates that carrying cost is 40% of the $10 unit cost and that annual demand is about 240 units per year. The assumptions of the basic EOQ model are thought to apply.
a) For what value of ordering cost would its action be optimal?
b) If the true ordering cost turns out to be much greater than your answer to (a), what is the impact on the firm's ordering policy?

• **12.10** Matthew Liotine's Dream Store sells beds and assorted supplies. His best-selling bed has an annual demand of 400 units. Ordering cost is $40; holding cost is $5 per unit per year.
a) To minimize the total cost, how many units should be ordered each time an order is placed?
b) If the holding cost per unit was $6 instead of $5, what would be the optimal order quantity? **PX**

• **12.11** Southeastern Bell stocks a certain switch connector at its central warehouse for supplying field service offices. The yearly demand for these connectors is 15,000 units. Southeastern estimates its annual holding cost for this item to be $25 per unit. The cost to place and process an order from the supplier is $75. The company operates 300 days per year, and the lead time to receive an order from the supplier is 2 working days.
a) Find the economic order quantity.
b) Find the annual holding costs.
c) Find the annual ordering costs.
d) What is the reorder point? **PX**

• **12.12** Lead time for one of your fastest-moving products is 21 days. Demand during this period averages 100 units per day.
a) What would be an appropriate reorder point?
b) How does your answer change if demand during lead time doubles?
c) How does your answer change if demand during lead time drops in half?

• **12.13** Annual demand for the notebook binders at Duncan's Stationery Shop is 10,000 units. Dana Duncan operates her business 300 days per year and finds that deliveries from her supplier generally take 5 working days.
a) Calculate the reorder point for the notebook binders that she stocks.
b) Why is this number important to Duncan?

•• **12.14** Thomas Kratzer is the purchasing manager for the headquarters of a large insurance company chain with a central inventory operation. Thomas's fastest-moving inventory item has a demand of 6,000 units per year. The cost of each unit is $100, and the inventory carrying cost is $10 per unit per year. The average ordering cost is $30 per order. It takes about 5 days for an order to arrive, and the demand for 1 week is 120 units. (This is a corporate operation, and there are 250 working days per year.)
a) What is the EOQ?
b) What is the average inventory if the EOQ is used?
c) What is the optimal number of orders per year?
d) What is the optimal number of days in between any two orders?
e) What is the annual cost of ordering and holding inventory?
f) What is the total annual inventory cost, including the cost of the 6,000 units? **PX**

•• **12.15** Jordin Henry's machine shop uses 2,500 brackets during the course of a year. These brackets are purchased from a supplier 90 miles away. The following information is known about the brackets:

Annual demand:	2,500
Holding cost per bracket per year:	$1.50
Order cost per order:	$18.75
Lead time:	2 days
Working days per year:	250

a) Given the information, what would be the economic order quantity (EOQ)?
b) Given the EOQ, what would be the average inventory? What would be the annual inventory holding cost?

c) Given the EOQ, how many orders would be made each year? What would be the annual order cost?
d) Given the EOQ, what is the total annual cost of managing the inventory?
e) What is the time between orders?
f) What is the reorder point (ROP)? **PX**

•• **12.16** Abey Kuruvilla, of Parkside Plumbing, uses 1,200 of a certain spare part that costs $25 for each order, with an annual holding cost of $24.
a) Calculate the total cost for order sizes of 25, 40, 50, 60, and 100.
b) Identify the economic order quantity and consider the implications for making an error in calculating economic order quantity. **PX**

••• **12.17** M. Cotteleer Electronics supplies microcomputer circuitry to a company that incorporates microprocessors into refrigerators and other home appliances. One of the components has an annual demand of 250 units, and this is constant throughout the year. Carrying cost is estimated to be $1 per unit per year, and the ordering (setup) cost is $20 per order.
a) To minimize cost, how many units should be ordered each time an order is placed?
b) How many orders per year are needed with the optimal policy?
c) What is the average inventory if costs are minimized?
d) Suppose that the ordering (setup) cost is not $20, and Cotteleer has been ordering 150 units each time an order is placed. For this order policy (of $Q = 150$) to be optimal, determine what the ordering (setup) cost would have to be. **PX**

•• **12.18** Race One Motors is an Indonesian car manufacturer. At its largest manufacturing facility, in Jakarta, the company produces subcomponents at a rate of 300 per day, and it uses these subcomponents at a rate of 12,500 per year (of 250 working days). Holding costs are $2 per item per year, and ordering (setup) costs are $30 per order.
a) What is the economic production quantity?
b) How many production runs per year will be made?
c) What will be the maximum inventory level?
d) What percentage of time will the facility be producing components?
e) What is the annual cost of ordering and holding inventory? **PX**

•• **12.19** Radovilsky Manufacturing Company, in Hayward, California, makes flashing lights for toys. The company operates its production facility 300 days per year. It has orders for about 12,000 flashing lights per year and has the capability of producing 100 per day. Setting up the light production costs $50. The cost of each light is $1. The holding cost is $0.10 per light per year.
a) What is the optimal size of the production run?
b) What is the average holding cost per year?
c) What is the average setup cost per year?
d) What is the total cost per year, including the cost of the lights? **PX**

•• **12.20** Arthur Meiners is the production manager of Wheel-Rite, a small producer of metal parts. Wheel-Rite supplies Cal-Tex, a larger assembly company, with 10,000 wheel bearings each year. This order has been stable for some time. Setup cost for Wheel-Rite is $40, and holding cost is $.60 per wheel bearing per year. Wheel-Rite can produce 500 wheel bearings per day. Cal-Tex is a just-in-time manufacturer and requires that 50 bearings be shipped to it each business day.
a) What is the optimum production quantity?
b) What is the maximum number of wheel bearings that will be in inventory at Wheel-Rite?
c) How many production runs of wheel bearings will Wheel-Rite have in a year?
d) What is the total setup + holding cost for Wheel-Rite? **PX**

•• **12.21** Cesar Rego Computers, a Mississippi chain of computer hardware and software retail outlets, supplies both educational and commercial customers with memory and storage devices. It currently faces the following ordering decision relating to purchases of very high-density disks:

$$D = 36{,}000 \text{ disks}$$
$$S = \$25$$
$$I = 20\%$$
$$\text{Purchase price} = \$0.85$$
$$\text{Discount price} = \$0.82$$

Quantity needed to qualify for the discount = 6,000 disks

Should the discount be taken? **PX**

•• **12.22** Bell Computers purchases integrated chips at $350 per chip. The holding cost is $35 per unit per year, the ordering cost is $120 per order, and sales are steady, at 400 per month. The company's supplier, Rich Blue Chip Manufacturing, Inc., decides to offer price concessions in order to attract larger orders. The price structure is shown in the table.

Rich Blue Chip's Price Structure

QUANTITY PURCHASED	PRICE/UNIT
1–99 units	$350
100–199 units	$325
200 or more units	$300

a) What is the optimal order quantity and the minimum annual cost for Bell Computers to order, purchase, and hold these integrated chips?
b) Bell Computers wishes to use a 10% holding cost rather than the fixed $35 holding cost in (a). What is the optimal order quantity, and what is the optimal annual cost? **PX**

•• **12.23** Meena Distributors has an annual demand for an airport metal detector of 1,400 units. The cost of a typical detector to Meena is $400. Carrying cost is estimated to be 20% of the unit cost, and the ordering cost is $25 per order. If Purushottama Meena, the owner, orders in quantities of 300 or more, he can get a 5% discount on the cost of the detectors. Should Meena take the quantity discount? **PX**

•• **12.24** The catering manager of La Vista Hotel, Lisa Ferguson, is disturbed by the amount of silverware she is losing every week. Last Friday night, when her crew tried to set up for a banquet for 500 people, they did not have enough knives. She decides she needs to order some more silverware, but wants to take advantage of any quantity discounts her vendor will offer.

For a small order (2,000 or fewer pieces), her vendor quotes a price of $1.80/piece.

If she orders 2,001–5,000 pieces, the price drops to $1.60/piece. 5,001–10,000 pieces brings the price to $1.40/piece, and 10,001 and above reduces the price to $1.25.

Lisa's order costs are $200 per order, her annual holding costs are 5%, and the annual demand is 15,000 pieces. For the best option:
a) What is the optimal order quantity?
b) What is the annual holding cost?
c) What is the annual ordering (setup) cost?

d) What are the annual costs of the silverware itself with an optimal order quantity?
e) What is the total annual cost, including ordering, holding, and purchasing the silverware? **PX**

•• **12.25** Rocky Mountain Tire Center sells 20,000 go-kart tires per year. The ordering cost for each order is $40, and the holding cost is 20% of the purchase price of the tires per year. The purchase price is $20 per tire if fewer than 500 tires are ordered, $18 per tire if 500 or more—but fewer than 1,000—tires are ordered, and $17 per tire if 1,000 or more tires are ordered.
a) How many tires should Rocky Mountain order each time it places an order?
b) What is the total cost of this policy? **PX**

•• **12.26** M. P. VanOyen Manufacturing has gone out on bid for a regulator component. Expected demand is 700 units per month. The item can be purchased from either Allen Manufacturing or Baker Manufacturing. Their price lists are shown in the table. Ordering cost is $50, and annual holding cost per unit is $5.

ALLEN MFG.		BAKER MFG.	
QUANTITY	UNIT PRICE	QUANTITY	UNIT PRICE
1–499	$16.00	1–399	$16.10
500–999	15.50	400–799	15.60
1,000+	15.00	800+	15.10

a) What is the economic order quantity?
b) Which supplier should be used? Why?
c) What is the optimal order quantity and total annual cost of ordering, purchasing, and holding the component? **PX**

•• **12.27** Chris Sandvig Irrigation, Inc., has summarized the price list from four potential suppliers of an underground control valve. See the accompanying table. Annual usage is 2,400 valves; order cost is $10 per order; and annual inventory holding costs are $3.33 per unit.

Which vendor should be selected and what order quantity is best if Sandvig Irrigation wants to minimize total cost? **PX**

VENDOR A		VENDOR B	
QUANTITY	PRICE	QUANTITY	PRICE
1–49	$35.00	1–74	$34.75
50–74	34.75	75–149	34.00
75–149	33.55	150–299	32.80
150–299	32.35	300–499	31.60
300–499	31.15	500+	30.50
500+	30.75		

VENDOR C		VENDOR D	
QUANTITY	PRICE	QUANTITY	PRICE
1–99	$34.50	1–199	$34.25
100–199	33.75	200–399	33.00
200–399	32.50	400+	31.00
400+	31.10		

••• **12.28** Emery Pharmaceutical uses an unstable chemical compound that must be kept in an environment where both temperature and humidity can be controlled. Emery uses 800 pounds per month of the chemical, estimates the holding cost to be 50% of the purchase price (because of spoilage), and estimates order costs to be $50 per order. The cost schedules of two suppliers are as follows:

VENDOR 1		VENDOR 2	
QUANTITY	PRICE/LB	QUANTITY	PRICE/LB
1–499	$17.00	1–399	$17.10
500–999	16.75	400–799	16.85
1,000+	16.50	800–1,199	16.60
		1,200+	16.25

a) What is the economic order quantity for each supplier?
b) What quantity should be ordered, and which supplier should be used?
c) What is the total cost for the most economic order size?
d) What factor(s) should be considered besides total cost? **PX**

••• **12.29** Cherylene Brown has asked you to help her determine the best ordering policy for a new product. The demand for the new product has been forecasted to be about 1,000 units annually. To help you get a handle on the carrying and ordering costs, Cherylene has given you the list of last year's costs. She thought that these costs might be appropriate for the new product.

COST FACTOR	COST ($)	COST FACTOR	COST ($)
Taxes for the warehouse	2,000	Warehouse supplies	280
Receiving and incoming inspection	1,500	Research and development	2,750
New product development	2,500	Purchasing salaries & wages	30,000
Acct. Dept. costs to pay invoices	500	Warehouse salaries & wages	12,800
Inventory insurance	600	Pilferage of inventory	800
Product advertising	800	Purchase order supplies	500
Spoilage	750	Inventory obsolescence	300
Sending purchasing orders	800	Purchasing Dept. overhead	1,000

She also told you that these data were compiled for 10,000 inventory items that were carried or held during the year. You have also determined that 200 orders were placed last year. Your job as a new operations management graduate is to help Cherylene determine the economic order quantity for the new product.

•••• **12.30** Emarpy Appliance is a company that produces all kinds of major appliances. Bud Banis, the president of Emarpy, is concerned about the production policy for the company's best-selling refrigerator. The annual demand has been about 8,000 units each year, and this demand has been constant throughout the year. The production capacity is 200 units per day. Each time production starts, it costs the company $120 to move materials into place, reset the assembly line, and clean the equipment. The holding cost of a refrigerator is $50 per year. The current production plan calls for 400 refrigerators to be produced in each production run. Assume there are 250 working days per year.
a) What is the daily demand of this product?
b) If the company were to continue to produce 400 units each time production starts, how many days would production continue?

c) Under the current policy, how many production runs per year would be required? What would the annual setup cost be?
d) If the current policy continues, how many refrigerators would be in inventory when production stops? What would the average inventory level be?
e) If the company produces 400 refrigerators at a time, what would the total annual setup cost and holding cost be?
f) If Bud Banis wants to minimize the total annual inventory cost, how many refrigerators should be produced in each production run? How much would this save the company in inventory costs compared to the current policy of producing 400 in each production run? **PX**

• **12.31** If $D = 1{,}000$ per year, $S = \$62.50$ per order, and $H = \$.50$ per unit per year, what is the economic order quantity for purchasing manager Soomin Park? **PX**

•• **12.32** If the economic order quantity = 300, annual demand = 8,000 units, and order costs = $45 per order, what is the holding cost for P. S. Ravi Metalworks?

•• **12.33** Victor Pimentel uses 1,500 units per year of a certain subassembly that has an annual holding cost of $45 per unit. Each order placed costs Victor $150. He operates 300 days per year and has found that an order must be placed with his supplier 6 working days before he can expect to receive that order. For this subassembly, find:
a) Economic order quantity.
b) Annual holding cost.
c) Annual ordering cost.
d) Reorder point. **PX**

• **12.34** It takes approximately 2 weeks (14 days) for an order of steel bolts to arrive once the order has been placed. The demand for bolts is fairly constant; on the average, the manager, Michelle Wu, has observed that the hardware store sells 500 of these bolts each day. Because the demand is fairly constant, Michelle believes that she can avoid stockouts completely if she orders the bolts at the correct time. What is the reorder point? **PX**

•• **12.35** Rick Jerz is attempting to perform an inventory analysis on one of his most popular products. Annual demand for this product is 5,000 units; carrying cost is $50 per unit per year; order costs for his company typically run nearly $30 per order; and lead time averages 10 days. (Assume 250 working days per year.)
a) What is the economic order quantity?
b) What is the average inventory?
c) What is the optimal number of orders per year?
d) What is the optimal number of working days between orders?
e) What is the total annual inventory cost (carrying cost + ordering cost)?
f) What is the reorder point? **PX**

••• **12.36** Phillip Flamm's Computer Store in Texas sells a printer for $200. Demand is constant during the year, and annual demand is forecasted to be 600 units. Holding cost is $20 per unit per year, whereas the cost of ordering is $60 per order. Currently, the company is ordering 12 times per year (50 units each time). There are 250 working days per year, and the lead time is 10 days.
a) Given the current policy of ordering 50 units at a time, what is the total of the annual ordering cost and the annual holding cost?
b) If the company used the absolute best inventory policy, what would be the total of ordering and holding costs?
c) What is the reorder point? **PX**

• **12.37** Dan Bumblauskas is the owner of a small Iowa company that produces electric knives used to cut fabric. The annual demand is for 8,000 knives, and Dan produces the knives in batches. On average, Dan can produce 150 knives per day; during the production process, demand has been about 40 knives per day. The cost to set up the production process is $100, and it costs Dan $.80 to carry a knife for 1 year. How many knives should Dan produce in each batch? **PX**

•• **12.38** Given the following data on a hardware item stocked by Andreas Wieland's paint store in Copenhagen, should the quantity discount be taken?

- D = 2,000 units; S = $10; H = $1; P = $1
- Discount price = $.75
- Quantity needed to qualify for discount = 2,000 units **PX**

•• **12.39** Happy Pet, Inc., is a large pet store located in Long Beach Mall. Although the store specializes in dogs, it also sells fish, turtle, and bird supplies. The Everlast Leader, a leather lead for dogs, costs Happy Pet $7 each. There is an annual demand for 6,000 Everlast Leaders. The manager, Stephan Wagner, has determined that the ordering cost is $20 per order and the carrying cost, as a percentage of unit cost, is 15%. Happy Pet is now considering a new supplier of Everlast Leaders. Each lead would cost only $6.65, but, in order to get this discount, Happy Pet would have to buy shipments of 3,000 at a time. Should Happy Pet use the new supplier and take this discount for quantity buying? **PX**

•• **12.40** Huehn-Brown Products in St. Petersburg offers the following discount schedule for its 4-by-8-foot sheets of quality plywood.

ORDER	UNIT COST
9 sheets or less	$18.00
10 to 50 sheets	$17.50
More than 50 sheets	$17.25

Home Sweet Home Company orders plywood from Huehn-Brown. Home Sweet Home has an ordering cost of $45. Carrying cost is 20%, and annual demand is 100 sheets. What do you recommend? **PX**

Problems 12.41–12.50 relate to Probabilistic Models and Safety Stock

•• **12.41** Barbara Flynn is in charge of maintaining hospital supplies at General Hospital. During the past year, the mean lead time demand for bandage BX-5 was 60 (and was normally distributed). Furthermore, the standard deviation for BX-5 was 7. Ms. Flynn would like to maintain a 90% service level.
a) What safety stock level do you recommend for BX-5?
b) What is the appropriate reorder point? **PX**

•• **12.42** Based on available information, lead time demand for PC thumb drives averages 50 units (normally distributed), with a standard deviation of 5 drives. Management wants a 97% service level.
a) What value of Z should be applied?
b) How many drives should be carried as safety stock?
c) What is the appropriate reorder point? **PX**

••• **12.43** Authentic Thai rattan chairs (shown in the photo) are delivered to Gary Schwartz's chain of retail stores, called The Kathmandu Shop, once a year. The reorder point, without safety stock, is 200 chairs. Carrying cost is $30 per unit per year, and the cost of a stockout is $70 per chair per year. Given the following demand probabilities during the lead time, how much safety stock should be carried?

Courtesy of Barry Render

DEMAND DURING LEAD TIME	PROBABILITY
0	0.2
100	0.2
200	0.2
300	0.2
400	0.2

•• **12.44** Tobacco is shipped from North Carolina to a cigarette manufacturer in Cambodia once a year. The reorder point, without safety stock, is 200 kilos. The carrying cost is $15 per kilo per year, and the cost of a stockout is $70 per kilo per year. Given the following demand probabilities during the lead time, how much safety stock should be carried? **PX**

DEMAND DURING LEAD TIME (KILOS)	PROBABILITY
0	0.1
100	0.1
200	0.2
300	0.4
400	0.2

••• **12.45** Mr. Beautiful, an organization that sells weight training sets, has an ordering cost of $40 for the BB-1 set. (BB-1 stands for Body Beautiful Number 1.) The carrying cost for BB-1 is $5 per set per year. To meet demand, Mr. Beautiful orders large quantities of BB-1 7 times a year. The stockout cost for BB-1 is estimated to be $50 per set. Over the past several years, Mr. Beautiful has observed the following demand during the lead time for BB-1: **PX**

DEMAND DURING LEAD TIME	PROBABILITY
40	.1
50	.2
60	.2
70	.2
80	.2
90	.1
	1.0

The reorder point for BB-1 is 60 sets. What level of safety stock should be maintained for BB-1? **PX**

•• **12.46** Chicago's Hard Rock Hotel distributes a mean of 1,000 bath towels per day to guests at the pool and in their rooms. This demand is normally distributed with a standard deviation of 100 towels per day, based on occupancy. The laundry firm that has the linens contract requires a 2-day lead time. The hotel expects a 98% service level to satisfy high guest expectations.

a) What is the safety stock?
b) What is the ROP? **PX**

•• **12.47** First Printing has contracts with legal firms in San Francisco to copy their court documents. Daily demand is almost constant at 12,500 pages of documents. The lead time for paper delivery is normally distributed with a mean of 4 days and a standard deviation of 1 day. A 97% service level is expected. Compute First's ROP. **PX**

••• **12.48** Gainesville Cigar stocks Cuban cigars that have variable lead times because of the difficulty in importing the product: lead time is normally distributed with an average of 6 weeks and a standard deviation of 2 weeks. Demand is also a variable and normally distributed with a mean of 200 cigars per week and a standard deviation of 25 cigars.

a) For a 90% service level, what is the ROP?
b) What is the ROP for a 95% service level?
c) Explain what these two service levels mean. Which is preferable? **PX**

•••• **12.49** A gourmet coffee shop in downtown San Francisco is open 200 days a year and sells an average of 75 pounds of Kona coffee beans a day. (Demand can be assumed to be distributed normally, with a standard deviation of 15 pounds per day.) After ordering (fixed cost = $16 per order), beans are always shipped from Hawaii within exactly 4 days. Per-pound annual holding costs for the beans are $3.

a) What is the economic order quantity (EOQ) for Kona coffee beans?
b) What are the total annual holding costs of stock for Kona coffee beans?
c) What are the total annual ordering costs for Kona coffee beans?
d) Assume that management has specified that no more than a 1% risk during stockout is acceptable. What should the reorder point (ROP) be?
e) What is the safety stock needed to attain a 1% risk of stockout during lead time?
f) What is the annual holding cost of maintaining the level of safety stock needed to support a 1% risk?
g) If management specified that a 2% risk of stockout during lead time would be acceptable, would the safety stock holding costs decrease or increase?

• **12.50** Demand during lead time for one brand of TV is normally distributed with a mean of 36 TVs and a standard deviation of 15 TVs. What safety stock should be carried for Dennis Yu's Appliance Outlet to provide a 90% service level? What is the appropriate reorder point? **PX**

Problems 12.51–12.53 relate to Single-Period Model

•• **12.51** Cynthia Knott's oyster bar buys fresh Louisiana oysters for $5 per pound and sells them for $9 per pound. Any oysters not sold that day are sold to her cousin, who has a nearby grocery store, for $2 per pound. Cynthia believes that demand follows the normal distribution, with a mean of 100 pounds and a standard deviation of 15 pounds. How many pounds should she order each day?

•• **12.52** Jantel Mitchell's bakery prepares all its cakes between 4 A.M. and 6 A.M. so they will be fresh when customers arrive. Day-old cakes are virtually always sold, but at a 50% discount off the regular $10 price. The cost of baking a cake is $6, and demand is estimated to be normally distributed, with a mean of 25 and a standard deviation of 4. What is the optimal stocking level?

••• **12.53** University of Florida football programs are printed 1 week prior to each home game. Attendance averages 90,000 screaming and loyal Gators fans, of whom two-thirds usually buy the program, following a normal distribution, for $4 each. Unsold programs are sent to a recycling center that pays only 10 cents per program. The standard deviation is 5,000 programs, and the cost to print each program is $1.

a) What is the cost of underestimating demand for each program?
b) What is the overage cost per program?
c) How many programs should be ordered per game?
d) What is the stockout risk for this order size?

CASE STUDIES

Zhou Bicycle Company

Zhou Bicycle Company (ZBC), located in Seattle, is a wholesale distributor of bicycles and bicycle parts. Formed in 1981 by University of Washington Professor Yong-Pin Zhou, the firm's primary retail outlets are located within a 400-mile radius of the distribution center. These retail outlets receive the order from ZBC within 2 days after notifying the distribution center, provided that the stock is available. However, if an order is not fulfilled by the company, no backorder is placed; the retailers arrange to get their shipment from other distributors, and ZBC loses that amount of business.

The company distributes a wide variety of bicycles. The most popular model, and the major source of revenue to the company, is the AirWing. ZBC receives all the models from a single manufacturer in China, and shipment takes as long as 4 weeks from the time an order is placed. With the cost of communication, paperwork, and customs clearance included, ZBC estimates that each time an order is placed, it incurs a cost of $65. The purchase price paid by ZBC, per bicycle, is roughly 60% of the suggested retail price for all the styles available, and the inventory carrying cost is 1% per month (12% per year) of the purchase price paid by ZBC. The retail price (paid by the customers) for the AirWing is $170 per bicycle.

ZBC is interested in making an inventory plan for 2022. The firm wants to maintain a 95% service level with its customers to minimize the losses on the lost orders. The data collected for the past 2 years are summarized in the following table. A forecast for AirWing model sales in 2022 has been developed and will be used to make an inventory plan for ZBC.

Demands for AirWing Model

MONTH	2020	2021	FORECAST FOR 2022
January	6	7	8
February	12	14	15
March	24	27	31
April	46	53	59
May	75	86	97
June	47	54	60
July	30	34	39
August	18	21	24
September	13	15	16
October	12	13	15
November	22	25	28
December	38	42	47
Total	343	391	439

Source: Professor Kala Chand Seal, Loyola Marymount University.

Discussion Questions

1. Develop an inventory plan to help ZBC.
2. Discuss ROPs and total costs.
3. How can you address demand that is not level throughout the planning horizon?

Managing Inventory at Frito-Lay

Video Case 

Frito-Lay has flourished since its origin—the 1931 purchase of a small San Antonio firm for $100 that included a recipe, 19 retail accounts, and a hand-operated potato ricer. The multi-billion-dollar company, headquartered in Dallas, now has 41 products—21 with sales of over $100 million per year and 7 at over $1 billion in sales. Production takes place in 36 product-focused plants in the U.S. and Canada, with 48,000 employees.

Inventory is a major investment and an expensive asset in most firms. Holding costs often exceed 25% of product value, but in Frito-Lay's prepared food industry, holding cost can be much higher because the raw materials are perishable. In the food industry, inventory spoils. So poor inventory management is not only expensive but can also yield an unsatisfactory product that in the extreme can also ruin market acceptance.

Major ingredients at Frito-Lay are corn meal, corn, potatoes, oil, and seasoning. Using potato chips to illustrate rapid inventory flow: potatoes are moved via truck from farm, to regional plants for processing, to warehouse, to the retail store. This happens in a matter of hours—not days or weeks. This keeps freshness high and holding costs low.

Frequent deliveries of the main ingredients at the Florida plant, for example, take several forms:

- Potatoes are delivered in 10 truckloads per day, with 150,000 lbs consumed in one shift: the entire potato storage area will only hold 7½ hours' worth of potatoes.
- Oil inventory arrives by rail car, which lasts only 4½ days.
- Corn meal arrives from various farms in the Midwest, and inventory typically averages 4 days' production.

- Seasoning inventory averages 7 days.
- Packaging inventory averages 8 to 10 days.

Frito-Lay's product-focused facility represents a major capital investment. That investment must achieve high utilization to be efficient. The capital cost must be spread over a substantial volume to drive down total cost of the snack foods produced. This demand for high utilization requires reliable equipment and tight schedules. Reliable machinery requires an inventory of critical components: this is known as MRO, or maintenance, repair, and operating supplies. MRO inventory of motors, switches, gears, bearings, and other critical specialized components can be costly but is necessary.

Frito-Lay's non-MRO inventory moves rapidly. Raw material quickly becomes work-in-process, moving through the system and out the door as a bag of chips in about $1\frac{1}{2}$ shifts. Packaged finished products move from production to the distribution chain in less than 1.4 days.

Discussion Questions*

1. How does the mix of Frito-Lay's inventory differ from those at a machine or cabinet shop (a process-focused facility)?
2. What are the major inventory items at Frito-Lay, and how rapidly do they move through the process?
3. What are the four types of inventory? Give an example of each at Frito-Lay.
4. How would you rank the dollar investment in each of the four types (from the most investment to the least investment)?
5. Why does inventory flow so quickly through a Frito-Lay plant?
6. Why does the company keep so many plants open?
7. Why doesn't Frito-Lay make all its 41 products at each of its plants?

*You may wish to view the video that accompanies this case before addressing these questions.

Inventory Management at Celebrity Cruises

Video Case

Sandwiched between mass-market cruise line players such as Carnival and Disney, and luxury cruise lines such as Crystal Cruises, Celebrity is making a name for itself in the "premium" market, offering both an upscale experience and premium meals at "an intelligent price."

Among the many features and services of a premium cruise line is food quality. It is critical to the success of the cruise experience and the number-one reason customers return. Dining facilities must be managed to achieve the highest possible standards. Celebrity operates a variety of specialty restaurants in addition to the main buffet and dining rooms. The restaurants serve 3 meals a day, handling as many as 2,500 guests at dinner alone.

Celebrity uses restaurant and menu data collected over the past 10 years for every item on the menu to forecast food supplies for each sailing, be it 3, 4, 7, 10, or 12 days, or longer.

The quantity of food served on a typical 7-day Celebrity cruise is massive, as indicated in the following table. (The average cruise guest consumes 4.5 meals per day.) This is where Celebrity's outstanding supply chain and inventory management come into play. At the beginning of most cruises, the ship's food storage areas are stacked to the forecasted limit and, by the end, the goal is to be "down to two bananas," as one executive humorously states.

Typical 7-Day Cruise Food Inventory

ITEM CONSUMED	AMOUNT
Beef	9,250 pounds
Chicken	3,000 pounds
Fresh vegetables	26,000 pounds
Rice	2,250 pounds
Cereal	500 pounds
Jelly	300 pounds
Cookies	1,450 pounds
Ice cream	600 gallons
Coffee	1,000 pounds
Bottles of champagne	150
Bottles of vodka	250
Cans/bottles of beer	9,900
Bottles of wine	2,600

Source: Celebrity Cruises

A Celebrity ship is "like a small upscale city that moves all the time," says Associate VP for Supply Chain Management and Sourcing, Paul Litvinov. Tens of thousands of different items (SKUs) are stocked for each sailing of ships all over the world, but crews have only 8 hours to provision them during the period when disembarking passengers leave (usually 7–9 a.m.) and embarking passengers board (3–5 p.m.).

Strategic inventory planning and supply chain decisions are made at Celebrity's Miami headquarters by Litvinov and his staff. Celebrity keeps long-term contracts with some suppliers (such as Coca-Cola and Starbucks) but typically contracts for the short term for items such as meats, fish, and produce, whose prices are more volatile.

With more than a dozen ships scattered around scores of ports worldwide, Litvinov often buys frozen food supplies stateside and sends 6 to 10 cargo containers by commercial shippers to stock cruises departing Europe or Asia. Despite the delays stemming from 30- to 60-day shipping routes, Celebrity's cost analysis determines this to be the optimum strategy for many products. However, dairy and produce products are always sourced locally.

Tactical inventory decisions are made by the ship's inventory manager and the 6- to 9-member staff. This group is responsible for inspecting all foods being loaded at the start of each cruise, as well as stowing and managing the supplies on board. Inspection includes sampling all produce to make sure it is perfectly ripe and not bruised or wilted. Celebrity operates a computerized stock management system that allows goods to be controlled accurately. Inventory in the bars and storerooms can be easily checked to ensure there are no shortages or discrepancies.

Beverly Amer/Aspenleaf Productions

With more than $100 million a year in food and beverage purchases, Celebrity has strong buying power. It carefully vets vendors in the U.S. and at each port where supplies are loaded.

Discussion Questions*

1. What inventory adjustment might Celebrity make to prepare for an extreme event such as a hurricane or mechanical failure while at sea?
2. What inventory management techniques seen in this chapter could be used at Celebrity?
3. How should a cost analysis be performed to determine whether to buy at a foreign port or to ship supplies from Celebrity's U.S. base in Florida?
4. In many respects, the food inventory decision for each Celebrity cruise matches the single-period inventory model described in this chapter. Consider the decision about chicken. Suppose that Celebrity purchases chicken in bulk for $0.80 per pound, and any chicken left over after a cruise is donated to local food banks. Management estimates that a shortage penalty of $1.00 per pound is incurred due to lost customer goodwill from unsatisfied demand. If chicken demand for a 7-day cruise is normally distributed with a mean of 3,000 pounds and a standard deviation of 200 pounds, how many pounds should be brought onto the ship?

*You may wish to watch the video that accompanies this case before answering these questions.

Inventory at Nautique Boat Company

Video Case

Nautique Boat Company, headquartered on a 200-acre campus with two lakes, is the leading manufacturer of waterskiing, wakeboarding, and wake-surfing boats in the U.S. Since the 1962 introduction of its now iconic $80,000 Ski Nautique line, the company has expanded with new products every year, most recently with its top-of-the line $300,000 Paragon P23.

Known for setting industry standards in quality and innovation (see the Global Company Profile on Nautique's new product development that opens Chapter 5), the firm takes inventory control very seriously. It holds $10–$11 million in inventory at any given time, plus another $2 million in spare parts to support customer needs for boats no longer in production. There are a total of 2,600 SKUs in inventory for ongoing production (about 1,000–1,200 per boat) and past customer support.

Managing the massive warehouse, full from floor to 16-foot-high storage racks, is the responsibility of Materials and Supply Chain Manager Drew Pope. Drew, a young engineer who bought his first Nautique as a teen and spent four years rebuilding it, uses the ABC method to categorize his wide range of parts, their lead times, and their storage locations.

Drew's top SKUs in terms of annual dollar volume are: engines ($43 million), fiberglass/resins/composites ($13 million), towers ($12 million), audio systems ($8 million), Biminis and mooring covers ($6 million), navigation equipment ($5 million), plastics ($5 million), electronics/wiring ($5 million), nonskid decking ($4 million), and seat foam ($3 million).

Nautique's continuing stream of innovative and special ordered boats adds complexity to the inventory challenge, but with tight controls and cycle counts each Friday, Drew manages to turn inventory 18–19 times per year. As a result, Nautique has been able to obtain an inventory accuracy level of 99.7% of dollar value.

Table 12.4 provides a sampling of the 2,600 SKUS used in active production. Bulk goods, such as fiberglass, resins, and brackets, are all vendor-managed inventory (VMI), replenished at least twice a week.

TABLE 12.4 Sample of Nautique's Inventory Items

ITEM STOCK NUMBER	PRODUCT NAME	ANNUAL DOLLAR VOLUME
004096	CATALYST, AKZO CADOX L50A CLEAR	$28,679
130227	GELCOAT, BLACK INTERIOR	$17,752
130337	STEERING CABLE, 24 FOOT	$33,026
130414	COOSA BOARD - 1" BLUEWATER 20 (4'x8' SHEET)	$28,218
150101	SHIFT OVERRIDE HARNESS/SWITCH	$168,480
160140	BALLAST BAG G21 PORT WITH FITTINGS	$36,168
160222	DRIVER'S SEAT SLIDE SWIVEL	$295,161
170166	GS TOWER RAW	$1,368,531
170440	FUEL TANK, 54 GALLON, GS22	$174,212
180050	MOORING COVER, 230 W/TAPS ANTI-PULLING SYSTEM	$34,930
190047	UNDERWATER LIGHT, 12 LED WHITE 5500 LUMENS	$295,098
190067	AMP CABLE KIT, 210/230	$16,808
190088	BIMINI, BI-FOLD FRAME AND HARDWARE	$1,462,041
200204	FUEL TANK 69 GALLON	$174,822
200218	DECK PLENUM G-SERIES PORT	$29,677
200229	HARNESS, AMPLIFIER COCKPIT SPEAKERS	$17,038
200302	ENGINE, ZR4 V-DRIVE 1.48 W/REMOTE DEGAS BOTTLE, 210	$1,010,412
200338	SENSOR, BALLAST PRESSURE	$54,358
200425	DECAL, WINDSHIELD END CAP CHROMAX 2PC SET P-SERIES	$11,671
200633	NON-SKID, UNDERDECK KIT SAHARA/ANTHRACITE	$21,834

Discussion Questions*

1. Conduct an ABC analysis on the 20 SKUs listed in Table 12.4.
2. Why would Nautique use vendor-managed inventory for some of its SKUs?
3. Should Nautique consider vertically integrating suppliers of the top annual dollar volume inventory items? What are the advantages and disadvantages of this decision? (For example, Nautique's parent company has purchased Performance Custom Marine (PCM) engines.)
4. How could Nautique reduce the number of SKUs it manages?

*The Global Company Profile featuring Nautique Boat Company (which opens Chapter 5) provides further background on Nautique's operations as does the video that accompanies this case. You may wish to review both prior to answering these questions.

Endnotes

1. Sources: *The Wall Street Journal* (January 2, 2020); and *Supply Chain Digest* (January 8, 2020).
2. See E. Malykhina, "Retailers Take Stock," *Information Week* (February 7, 2005): 20–22, and A. Raman, N. DeHoratius, and Z. Ton, "Execution: The Missing Link in Retail Operations," *California Management Review* 43, no. 3 (Spring 2001): 136–141.
3. This is the case when holding costs are linear and begin at the origin—that is, when inventory costs do not decline (or they increase) as inventory volume increases and all holding costs are in small increments. In addition, there is probably some learning each time a setup (or order) is executed—a fact that lowers subsequent setup costs. Consequently, the EOQ model is probably a special case. However, we abide by the conventional wisdom that this model is a reasonable approximation.
4. The formula for the economic order quantity (Q^*) can also be determined by finding where the total cost curve is at a minimum (i.e., where the slope of the total cost curve is zero). Using calculus, we set the derivative of the total cost with respect to Q^* equal to 0.
   The calculations for finding the minimum of

   $$TC = \frac{D}{Q}S + \frac{Q}{2}H + PD$$

   are $\dfrac{d(TC)}{dQ} = \left(\dfrac{-DS}{Q^2}\right) + \dfrac{H}{2} + 0 = 0$

   Thus, $Q^* = \sqrt{\dfrac{2DS}{H}}$
5. The number of units short, Demand–ROP, is true only when Demand–ROP is non-negative.
6. Equations (12-15), (12-16), and (12-17) are expressed in days; however, they could equivalently be expressed in weeks, months, or even years. Just be consistent, and use the same time units for all terms in the equations.
7. Note that Equation (12-17) can also be expressed as:

   $$\text{ROP} = \text{Average daily demand} \times \text{Average lead time} + Z\sqrt{(\text{Average lead time} \times \sigma_d^2) + \bar{d}^{\,2}\sigma_{LT}^2}$$
8. Sources: *The Wall Street Journal* (August 14, 2020) and (June 24, 2020); and *Vogue Business* (January 5, 2020).
9. OM managers also call these *continuous review systems*.

Bibliography

Abernathy, F. H., et al. "Control Your Inventory in a World of Lean Retailing." *Harvard Business Review* 78, no. 6 (November–December 2000): 169–176.

Burt, D. N., S. Petcavage, and R. Pinkerton. *Supply Management*, 8th ed. Burr Ridge, IL: Irwin/McGraw, 2010.

Chapman, S., T. K. Arnold, A. K. Gatewood, and L. M. Clive. *Introduction to Materials Management*, 8th ed. Boston: Pearson, 2018.

Harris, C. and T. Parker. "Portion Control." *APICS* 24, no. 5 (September/October 2014): 26–28.

Jacobs, F. R., W. L. Berry, D. C. Whybark, and T. E. Vollmann. *Manufacturing Planning and Control for Supply Chain Management*, 6th ed. New York: McGraw-Hill, 2011.

Keren, B. "The Single Period Inventory Model." *Omega* 37, no. 4 (August 2009): 801.

Lee, H., and O. Ozer. "Unlocking the Value of RFID." *Production and Operations Management* 16, no. 1 (2007): 40–64.

McDonald, S. C. *Materials Management*. New York: Wiley, 2009.

Muller, M. *Essentials of Inventory Management*, 2nd ed. Seattle: Amazon Press, 2011.

Munson, C. L., and J. Jackson. "Quantity Discounts: An Overview and Practical Guide for Buyers and Sellers." *Foundations and Trends in Technology, Information and Operations Management* 8, nos. 1–2 (2014): 1–130.

Render, B., R. M. Stair, M. Hanna, and T. Hale. *Quantitative Analysis for Management*, 13th ed. Boston: Pearson, 2018.

Stein, A. C. "Warehouse Measurement and Control." *APICS* 24, no. 3 (May/June 2014): 42–45.

Teunter, R. H., M. Z. Babai, and A. A. Syntetos. "ABC Classification: Service Levels and Inventory Cost." *Production and Operations Management* 19, no. 3 (May–June 2010): 343–352.

Chapter 12 *Rapid* Review

Main Heading	Review Material	MyLab Operations Management
THE IMPORTANCE OF INVENTORY	Inventory is one of the most expensive assets of many companies. *The objective of inventory management is to strike a balance between inventory investment and customer service.* The two basic inventory issues are how much to order and when to order. ▪ **Raw material inventory**—Materials that are usually purchased but have yet to enter the manufacturing process. ▪ **Work-in-process (WIP) inventory**—Products or components that are no longer raw materials but have yet to become finished products. ▪ **MRO inventory**—Maintenance, repair, and operating materials. ▪ **Finished-goods inventory**—An end item ready to be sold but still an asset on the company's books.	Concept Questions: 1.1–1.6 **VIDEO 12.1** Managing Inventory at Frito-Lay
MANAGING INVENTORY	▪ **ABC analysis**—A method for dividing on-hand inventory into three classifications based on annual dollar volume. ▪ **Cycle counting**—A continuing reconciliation of inventory with inventory records. ▪ **Shrinkage**—Retail inventory that is unaccounted for between receipt and sale. ▪ **Pilferage**—A small amount of theft.	Concept Questions: 2.1–2.6 Problems: 12.1–12.6 Virtual Office Hours for Solved Problem: 12.1 **VIDEO 12.2** Inventory Management at Celebrity Cruises
INVENTORY MODELS	▪ **Holding cost**—The cost to keep or carry inventory in stock. ▪ **Ordering cost**—The cost of the ordering process. ▪ **Setup cost**—The cost to prepare a machine or process for production. ▪ **Setup time**—The time required to prepare a machine or process for production.	Concept Questions: 3.1–3.6 **VIDEO 12.3** Inventory at Nautique Boat Company
INVENTORY MODELS FOR INDEPENDENT DEMAND	▪ **Economic order quantity (EOQ) model**—An inventory-control technique that minimizes the total of ordering and holding costs: $Q^* = \sqrt{\frac{2DS}{H}}$ (12-1) $\text{Expected number of orders} = N = \frac{\text{Demand}}{\text{Order quantity}} = \frac{D}{Q^*}$ (12-2) $\text{Expected time between orders} = T = \frac{\text{Number of working days per year}}{N}$ (12-3) $\text{Total annual cost} = \text{Setup (order) cost} + \text{Holding cost}$ (12-4) $TC = \frac{D}{Q}S + \frac{Q}{2}H$ (12-5) ▪ **Robust**—Giving satisfactory answers even with substantial variation in the parameters. ▪ **Lead time**—In purchasing systems, the time between placing an order and receiving it; in production systems, the wait, move, queue, setup, and run times for each component produced. ▪ **Reorder point (ROP)**—The inventory level (point) at which action is taken to replenish the stocked item. *ROP for known demand:* $\text{ROP} = \text{Demand per day} \times \text{Lead time for a new order in days} = d \times L$ (12-6) ▪ **Safety stock (*ss*)**—Extra stock to allow for uneven demand; a buffer. ▪ **Production order quantity model**—An economic order quantity technique applied to production orders: $Q_p^* = \sqrt{\frac{2DS}{H[1-(d/p)]}}$ (12-7) $Q_p^* = \sqrt{\frac{2DS}{H\left(1 - \frac{\text{Annual demand rate}}{\text{Annual production rate}}\right)}}$ (12-8) ▪ **Quantity discount**—A reduced price for items purchased in large quantities: $TC = \frac{D}{Q}S + \frac{Q}{2}H + PD$ (12-9) $Q^* = \sqrt{\frac{2DS}{IP}}$ (12-10)	Concept Questions: 4.1–4.6 Problems: 12.7–12.40 **VIDEO 12.4** Inventory Control at Wheeled Coach Virtual Office Hours for Solved Problems: 12.2–12.5 ACTIVE MODELS 12.1, 12.2
PROBABILISTIC MODELS AND SAFETY STOCK	▪ **Probabilistic model**—A statistical model applicable when product demand or any other variable is not known but can be specified by means of a probability distribution. ▪ **Service level**—The complement of the probability of a stockout.	Concept Questions: 5.1–5.5

Main Heading	Review Material	MyLab Operations Management
	ROP for unknown demand: $$\text{ROP} = d \times L + ss \quad (12\text{-}11)$$ Annual stockout costs = The *sum* of the units short for each demand level × The probability of that demand level × The stockout cost/unit × The number of orders per year (12-12) *ROP for unknown demand and given service level:* $$\text{ROP} = \text{Expected demand during lead time} + Z\sigma_{dLT} \quad (12\text{-}13)$$ $$\text{Safety stock} = Z\sigma_{dLT} \quad (12\text{-}14)$$ *ROP for variable demand and constant lead time:* $$\text{ROP} = (\text{Average daily demand} \times \text{Lead time in days}) + Z\sigma_{dLT} \quad (12\text{-}15)$$ *ROP for constant demand and variable lead time:* $$\text{ROP} = (\text{Daily demand} \times \text{Avg. daily lead time}) + Z \times \text{Daily demand} \times \sigma_{LT} \quad (12\text{-}16)$$ *ROP for variable demand and variable lead time:* $$\text{ROP} = (\text{Average daily demand} \times \text{Average lead time in days}) + Z\sigma_{dLT} \quad (12\text{-}17)$$ In each case, $\sigma_{dLT} = \sqrt{(\text{Average lead time} \times \sigma_d^2) + \bar{d}^2\sigma_{LT}^2}$ but under constant demand: $\sigma_d^2 = 0$, and under constant lead time: $\sigma_{LT}^2 = 0$.	Problems: 12.41–12.50 Virtual Office Hours for Solved Problems: 12.6–12.9
SINGLE-PERIOD MODEL	■ **Single-period inventory model**—A system for ordering items that have little or no value at the end of the sales period: $$\text{Service level} = \frac{C_s}{C_s + C_o} \quad (12\text{-}18)$$	Concept Questions: 6.1–6.5 Problems: 12.51–12.53
FIXED-PERIOD (P) SYSTEMS	■ **Fixed-quantity (*Q*) system**—An ordering system with the same order amount each time. ■ **Perpetual inventory system**—A system that keeps track of each withdrawal or addition to inventory continuously, so records are always current. ■ **Fixed-period (*P*) system**—A system in which inventory orders are made at regular time intervals.	Concept Questions: 7.1–7.4
ADDITIONAL MYLAB OPERATIONS MANAGEMENT RESOURCES	✔ Videos for Creating Your Own Excel Spreadsheets (Examples 8 and 15) ✔ Additional Case Studies (Southwestern University (F), LaPlace Power and Light, Parker Hi-Fi Systems, and Inventory Control at Wheeled Coach) ✔ Southwestern University Case Studies are integrated in Chapters 3, 4, 6, 8, 12, and 13 and in Supplement 7 ✔ Multiple Choice Case Questions (Zhou Bicycle Company) ✔ Recent Graduate Video: Nicholas Delmonaco, Supply Chain Analyst, UPMC ✔ Inventory Management Simulation	

Self Test

LO 12.1 ABC analysis divides on-hand inventory into three classes, based on:
- **a)** unit price.
- **b)** the number of units on hand.
- **c)** annual demand.
- **d)** annual dollar values.

LO 12.2 Cycle counting:
- **a)** provides a measure of inventory turnover.
- **b)** assumes that all inventory records must be verified with the same frequency.
- **c)** is a process by which inventory records are periodically verified.
- **d)** all of the above.

LO 12.3 The two most important inventory-based questions answered by the typical inventory model are:
- **a)** when to place an order and the cost of the order.
- **b)** when to place an order and how much of an item to order.
- **c)** how much of an item to order and the cost of the order.
- **d)** how much of an item to order and with whom the order should be placed.

LO 12.4 Extra units in inventory to help reduce stockouts are called:
- **a)** reorder point.
- **b)** safety stock.
- **c)** just-in-time inventory.
- **d)** all of the above.

LO 12.5 The difference(s) between the basic EOQ model and the production order quantity model is(are) that:
- **a)** the production order quantity model does not require the assumption of known, constant demand.
- **b)** the EOQ model does not require the assumption of negligible lead time.
- **c)** the production order quantity model does not require the assumption of instantaneous delivery.
- **d)** all of the above.

LO 12.6 The EOQ model with quantity discounts attempts to determine:
- **a)** the lowest amount of inventory necessary to satisfy a certain service level.
- **b)** the lowest purchase price.
- **c)** whether to use a fixed-quantity or fixed-period order policy.
- **d)** how many units should be ordered.
- **e)** the shortest lead time.

LO 12.7 The appropriate level of safety stock is typically determined by:
- **a)** minimizing an expected stockout cost.
- **b)** choosing the level of safety stock that assures a given service level.
- **c)** carrying sufficient safety stock so as to eliminate all stockouts.
- **d)** annual demand.

Answers: LO 12.1. d; LO 12.2. c; LO 12.3. b; LO 12.4. b; LO 12.5. c; LO 12.6. d; LO 12.7. b.

Aggregate Planning and S&OP

CHAPTER 13

CHAPTER OUTLINE

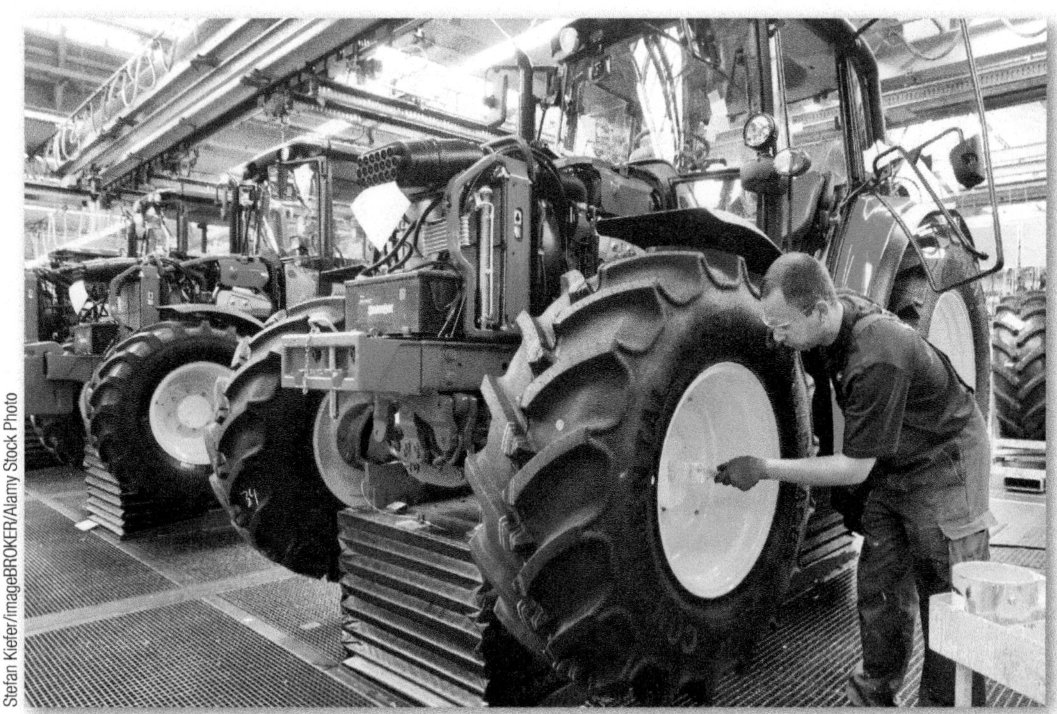

Stefan Kiefer/imageBROKER/Alamy Stock Photo

John Deere and Company, the "granddaddy" of farm equipment manufacturers, uses sales incentives to smooth demand. During the fall and winter off-seasons, sales are boosted with price cuts and other incentives. About 70% of Deere's big machines are ordered in advance of seasonal use—about double the industry rate. Incentives hurt margins, but Deere keeps its market share and controls costs by producing more steadily all year long. Similarly, service businesses such as L.L. Bean offer customers free shipping on orders placed before the Christmas rush. These are two of the aggregate planning strategies introduced in this chapter.

10 OM STRATEGY DECISIONS

- Design of Goods and Services
- Managing Quality
- Process Strategies
- Location Strategies
- Layout Strategies
- Human Resources
- Supply Chain Management
- Inventory Management
- *Scheduling*
  - *Aggregate/S&OP (Ch. 13)*
  - Short-Term (Ch. 15)
- Maintenance

GLOBAL COMPANY PROFILE
Frito-Lay

Aggregate Planning Provides a Competitive Advantage at Frito-Lay

Like other organizations throughout the world, Frito-Lay relies on effective aggregate planning to match fluctuating multi-billion-dollar demand to capacity in its 36 North American plants. Planning for the intermediate term (3 to 18 months) is the heart of aggregate planning. Effective aggregate planning combined with tight scheduling, effective maintenance, and efficient employee and facility scheduling are the keys to high plant utilization. High utilization is a critical factor in facilities such as Frito-Lay, where capital investment is substantial.

Frito-Lay has more than three dozen brands of snacks and chips, 21 of which sell more than $100 million annually and 7 of which sell over $1 billion. Its brands include such well-known names as Fritos, Lay's, Doritos, Sun Chips, Cheetos, Tostitos, Flat Earth, and Ruffles. Unique processes using specially designed equipment are required to produce each of these products. Because these specialized processes generate high fixed cost, they must operate at very high volume. But such product-focused facilities benefit by having low variable costs. High utilization and performance above the break-even point require a good match between demand and capacity. Idle equipment is disastrous.

At Frito-Lay's headquarters near Dallas, planners create a total demand profile. They use historical product sales, forecasts of new products, product innovations, product promotions,

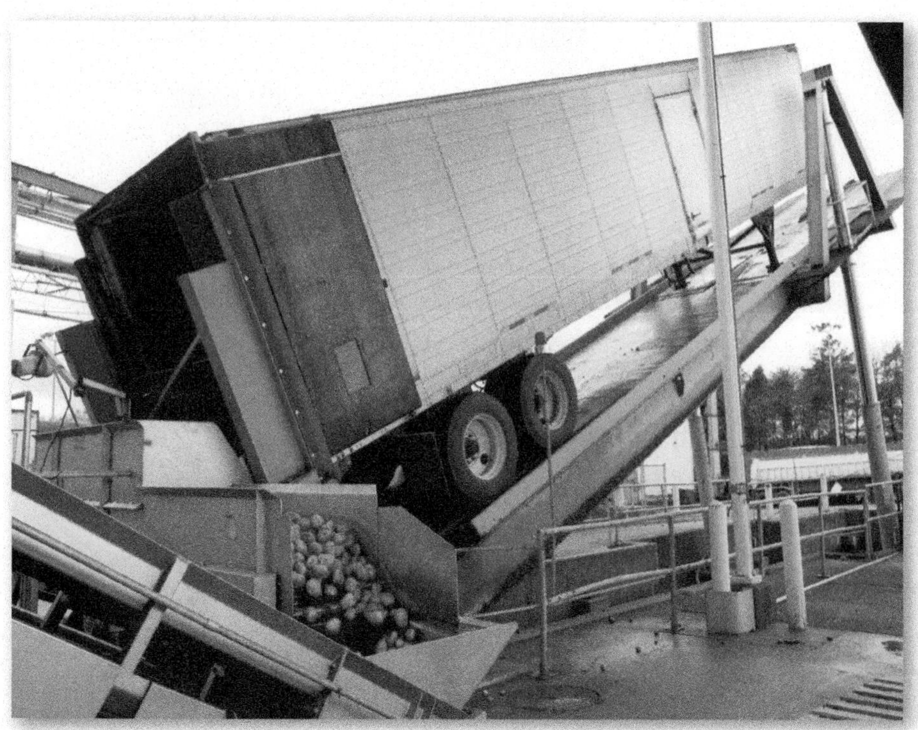

The aggregate plan adjusts for farm location, yield, and quantities for timely delivery of Frito-Lay's unique varieties of potatoes. During harvest times, potatoes go directly to the plant. During non-harvest months, potatoes are stored in climate-controlled environments to maintain quality, texture, and taste.

Frito-Lay North America

Frito-Lay North America

As potatoes arrive at the plant, they are promptly washed and peeled to ensure freshness and taste.

After peeling, potatoes are cut into thin slices, rinsed of excess starch, and cooked in sunflower and/or corn oil.

Frito-Lay North America

and dynamic local demand data from account managers to forecast demand. Planners then match the total demand profile to existing capacity, capacity expansion plans, and cost. This becomes the aggregate plan. The aggregate plan is communicated to each of the firm's 17 regions and to the 36 plants. Every quarter, headquarters and each plant modify the respective plans to incorporate changing market conditions and plant performance.

Each plant uses its quarterly plan to develop a 4-week plan, which in turn assigns specific products to specific product lines for production runs. Finally, each week raw materials and labor are assigned to each process. Effective aggregate planning is a major factor in high utilization and low cost. As the company's 60% market share indicates, excellent aggregate planning yields a competitive advantage at Frito-Lay.

Frito-Lay North America

After cooking is complete, inspection, bagging, weighing, and packing operations prepare Lay's potato chips for shipment to customers—all in a matter of hours.

LEARNING OBJECTIVES

LO 13.1 ***Define*** sales and operations planning 539

LO 13.2 ***Define*** aggregate planning 540

LO 13.3 ***Identify*** optional strategies for developing an aggregate plan 541

LO 13.4 ***Prepare*** a graphical aggregate plan 544

LO 13.5 ***Solve*** an aggregate plan via the transportation method 549

LO 13.6 ***Understand*** and solve a revenue management problem 554

The Planning Process

In Chapter 4, we saw that demand forecasting can address long-, medium-, and short-range decisions. Figure 13.1 illustrates how managers translate these forecasts into long-, intermediate-, and short-range plans. Long-range forecasts, the responsibility of top management, provide data for a firm's multi-year plans. These long-range plans require policies and strategies related to issues such as capacity and capital investment (Supplement 7), facility location (Chapter 8), new products (Chapter 5) and processes (Chapter 7), and supply chain development (Chapter 11).

Intermediate plans are designed to be consistent with top management's long-range plans and strategy, and they work within the resource constraints determined by earlier strategic decisions. The challenge is to have these plans match production to the ever-changing demands of the market. Intermediate plans are the job of the operations manager, working with other functional areas of the firm. In this chapter we deal with intermediate plans, typically measured in months.

Short-range plans are usually for less than 3 months. These plans are also the responsibility of operations personnel. Operations managers work with supervisors to translate the

Figure **13.1**

Planning Tasks and Responsibilities

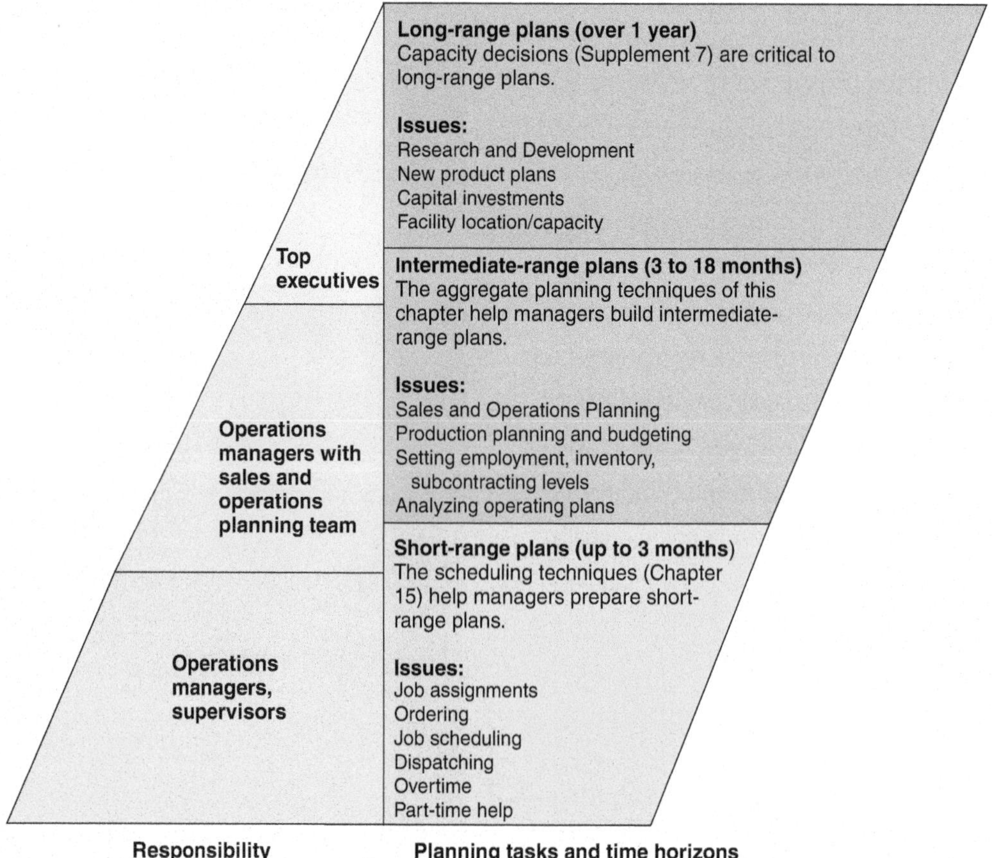

intermediate plan into short-term plans consisting of weekly, daily, and hourly schedules. Short-term planning techniques are discussed in Chapter 15.

Intermediate planning is initiated by a process known as *sales and operations planning (S&OP)*.

Sales and Operations Planning

Good intermediate planning requires the coordination of demand forecasts with functional areas of a firm and its supply chain. And because each functional part of a firm and the supply chain has its own limitations and constraints, the coordination can be difficult. This coordinated planning effort has evolved into a process known as **sales and operations planning (S&OP)**. As Figure 13.2 shows, S&OP receives input from a variety of sources both internal and external to the firm. Because of the diverse inputs, S&OP is typically done by cross-functional teams that align the competing constraints.

Sales and operations planning (S&OP)
A process of balancing resources and forecasted demand, aligning an organization's competing demands from supply chain to final customer, while linking strategic planning with operations over all planning horizons.

One of the tasks of S&OP is to determine which plans are feasible in the coming months and which are not. Any limitations, both within the firm and in the supply chain, must be reflected in an intermediate plan that brings day-to-day sales and operational realities together. When the resources appear to be substantially at odds with market expectations, S&OP provides advanced warning to top management. If the plan cannot be implemented in the short run, the planning exercise is useless. And if the plan cannot be supported in the long run, strategic changes need to be made. To keep aggregate plans current and to support its intermediate planning role, S&OP uses rolling forecasts that are frequently updated—often weekly or monthly.

LO 13.1 *Define* sales and operations planning

The output of S&OP is called an *aggregate plan*. The **aggregate plan** is concerned with determining the quantity and timing of production for the intermediate future, often from 3 to 18

Aggregate plan
A plan that includes forecast levels for families of products of finished goods, inventory, shortages, and changes in the workforce.

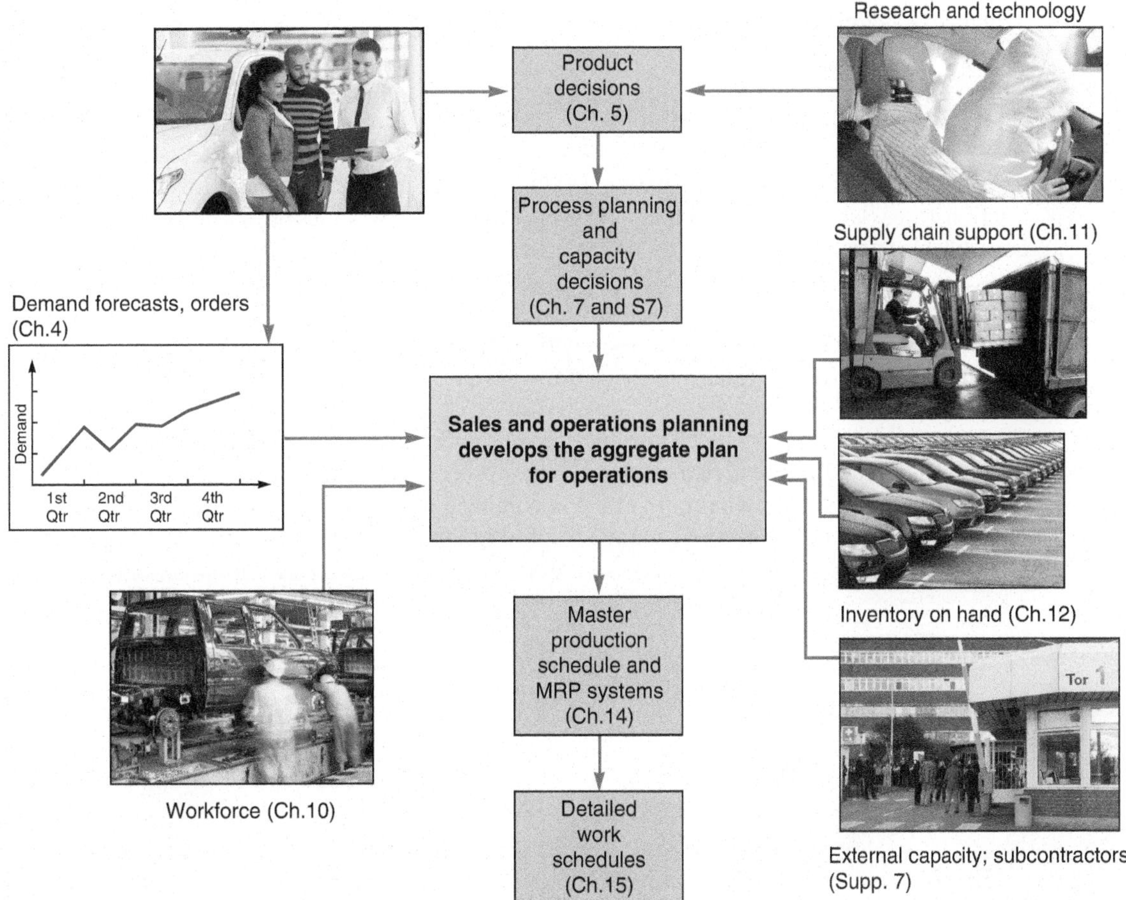

Figure **13.2**

Relationships of S&OP and the Aggregate Plan

Image credits (clockwise from top-left): Nestor Rizhniak/Shutterstock; Dmitry Vereshchagin/Fotolia; Dmitry Kalinovsky/Shutterstock; Industrieblick/Fotolia; Juergen Moers/vario images GmbH & Co.KG/Alamy Stock Photo; Gui Yong Nian/Fotolia

months ahead. Aggregate plans use information regarding product families or product lines rather than individual products. These plans are concerned with the total, or *aggregate*, of the individual product lines.

Rubbermaid, Office Max, and Rackspace have developed formal systems for S&OP, each with its own planning focus. Rubbermaid may use S&OP with a focus on production decisions; Office Max may focus S&OP on supply chain and inventory decisions; while Rackspace, a data storage firm, tends to have its S&OP focus on its critical and expensive investments in capacity. In *all* cases, though, the decisions must be tied to strategic planning and integrated with *all* areas of the firm over *all* planning horizons. Specifically, S&OP is aimed at (1) the coordination and integration of the internal and external resources necessary for a successful aggregate plan and (2) communication of the plan to those charged with its execution. The added advantage of S&OP and the aggregate plan is that they can be effective tools to engage members of the supply chain in achieving the firm's goals.

LO 13.2 *Define* aggregate planning

Besides being representative, timely, and comprehensive, an effective S&OP process needs these four additional features to generate a useful aggregate plan:

- A logical unit for measuring sales and output, such as pounds of Doritos at Frito-Lay, air-conditioning units at GE, or terabytes of storage at Rackspace
- A forecast of demand for a reasonable intermediate planning period in aggregate terms
- A method to determine the relevant costs
- A model that combines forecasts and costs so scheduling decisions can be made for the planning period

In this chapter we describe several techniques that managers use when developing an aggregate plan for both manufacturing and service-sector firms. For manufacturers, an aggregate schedule ties a firm's strategic goals to production plans. For service organizations, an aggregate schedule ties strategic goals to workforce schedules.

The Nature of Aggregate Planning

The S&OP team builds an aggregate plan that satisfies forecasted demand by adjusting production rates, labor levels, inventory levels, overtime work, subcontracting rates, and other controllable variables. The plan can be for Frito-Lay, Whirlpool, hospitals, colleges, or Pearson Education, the company that publishes this textbook. Regardless of the firm, *the objective of aggregate planning is usually to meet forecast demand while minimizing cost over the planning period*. However, other strategic issues may be more important than low cost. These strategies may be to smooth employment, to drive down inventory levels, or to meet a high level of service, regardless of cost.

Let's look at Snapper, which produces many different models of lawn mowers. Snapper makes walk-behind mowers, rear-engine riding mowers, garden tractors, and many more, for a total of 145 models. For each month in the upcoming 3 quarters, the aggregate plan for Snapper might have the following output (in units of production) for Snapper's "family" of mowers:

QUARTER 1			QUARTER 2			QUARTER 3		
Jan.	Feb.	March	April	May	June	July	Aug.	Sept.
150,000	120,000	110,000	100,000	130,000	150,000	180,000	150,000	140,000

Briggs & Stratton Power Products Marketing

S&OP builds an aggregate plan using the total expected demand for all of the family products, such as 145 models at Snapper (a few of which are shown above). Only when the forecasts are assembled in the aggregate plan does the company decide how to meet the total requirement with the available resources. These resource constraints include facility capacity, workforce size, supply chain limitations, inventory issues, and financial resources.

OM in Action — Building the Plan at Snapper[1]

Every bright-red Snapper lawn mower sold anywhere in the world comes from a factory in McDonough, Georgia. Ten years ago, the Snapper line had about 40 models of mowers, leaf blowers, and snow blowers. Today, reflecting the demands of mass customization, the product line is much more complex. Snapper designs, manufactures, and sells 145 models. This means that aggregate planning and the related short-term scheduling have become more complex, too.

In the past, Snapper met demand by carrying a huge inventory for 52 regional distributors and thousands of independent dealerships. It manufactured and shipped tens of thousands of lawn mowers, worth tens of millions of dollars, without quite knowing when they would be sold—a very expensive approach to meeting demand. Some changes were necessary. The new plan's goal is for each distribution center to receive only the minimum inventory necessary to meet demand. Today, operations managers at Snapper evaluate production capacity and use frequent data from the field as inputs to sophisticated software to forecast sales. The new system tracks customer demand and aggregates forecasts for every model in every region of the country. It even adjusts for holidays and weather. And the number of distribution centers has been cut from 52 to 4.

Once evaluation of the aggregate plan against capacity determines the plan to be feasible, Snapper's planners break down the plan into production needs for each model. Production by model is accomplished by building rolling monthly and weekly plans. These plans track the pace at which various units are selling. Then, the final step requires juggling work assignments to various work centers for each shift, such as 265 lawn mowers in an 8-hour shift. That's a new Snapper every 109 seconds.

Note that the plan looks at production *in the aggregate* (the family of mowers), not as a product-by-product breakdown. Likewise, an aggregate plan for BMW tells the auto manufacturer how many cars to make but not how many should be two-door vs. four-door or red vs. green. It tells Nucor Steel how many tons of steel to produce but does not differentiate grades of steel. (We extend the discussion of planning at Snapper in the *OM in Action* box, "Building the Plan at Snapper.")

In a manufacturing environment, the process of breaking the aggregate plan down into greater detail is called **disaggregation**. Disaggregation results in a **master production schedule**, which provides input to material requirements planning (MRP) systems. The master production schedule addresses the purchasing or production of major parts or components (see Chapter 14). It is *not* a sales forecast. Detailed work schedules for people and priority scheduling for products constitute the final step of the production planning system (and are discussed in Chapter 15).

Disaggregation
The process of breaking an aggregate plan into greater detail.

Master production schedule
A timetable that specifies what is to be made and when.

Aggregate Planning Strategies

When generating an aggregate plan, the operations manager must answer several questions:

STUDENT TIP
Managers can meet aggregate plans by adjusting either capacity or demand.

1. Should inventories be used to absorb changes in demand during the planning period?
2. Should changes be accommodated by varying the size of the workforce?
3. Should part-timers be used, or should overtime and idle time absorb fluctuations?
4. Should subcontractors be used on fluctuating orders so a stable workforce can be maintained?
5. Should prices or other factors be changed to influence demand?

All of these are legitimate planning strategies. They involve the manipulation of inventory, production rates, labor levels, capacity, and other controllable variables. We will now examine eight options in more detail. The first five are called *capacity options* because they do not try to change demand but attempt to absorb demand fluctuations. The last three are *demand options* through which firms try to smooth out changes in the demand pattern over the planning period.

LO 13.3 *Identify* optional strategies for developing an aggregate plan

Capacity Options

A firm can choose from the following basic capacity (production) options:

1. *Changing inventory levels:* Managers can increase inventory during periods of low demand to meet high demand in future periods. If this strategy is selected, costs associated with

storage, insurance, handling, obsolescence, pilferage, and capital invested will increase. On the other hand, with low inventory on hand and increasing demand, shortages can occur, resulting in longer lead times and poor customer service.

2. *Varying workforce size by hiring or layoffs:* One way to meet demand is to hire or lay off production workers to match production rates. However, new employees need to be trained, and productivity drops temporarily as they are absorbed into the workforce. Layoffs or terminations, of course, lower the morale of all workers and also lead to lower productivity.
3. *Varying production rates through overtime or idle time:* Keeping a constant workforce while varying working hours may be possible. Yet when demand is on a large upswing, there is a limit on how much overtime is realistic. Overtime pay increases costs, and too much overtime can result in worker fatigue and a drop in productivity. Overtime also implies added overhead costs to keep a facility open. On the other hand, when there is a period of decreased demand, the company must somehow absorb workers' idle time—often a difficult and expensive process.
4. *Subcontracting:* A firm can acquire temporary capacity by subcontracting work during peak demand periods. Subcontracting, however, has several pitfalls. First, it may be costly; second, it risks opening the door to a competitor. Third, developing the perfect subcontract supplier can be a challenge.
5. *Using part-time workers:* Especially in the service sector, part-time workers can fill labor needs. This practice is common in restaurants, retail stores, and supermarkets.

Demand Options

The basic demand options are:

1. *Influencing demand:* When demand is low, a company can try to increase demand through advertising, promotion, personal selling, and price cuts. Airlines and hotels have long offered weekend discounts and off-season rates; theaters cut prices for matinees; some colleges give discounts to senior citizens; and air conditioners are least expensive in winter. However, even special advertising, promotions, selling, and pricing are not always able to balance demand with production capacity.
2. *Backordering during high-demand periods:* Backorders are orders for goods or services that a firm accepts but is unable (either on purpose or by chance) to fill at the moment. If customers are willing to wait without loss of their goodwill or order, backordering is a possible strategy. Many firms backorder, but the approach often results in lost sales.
3. *Counterseasonal product and service mixing:* A widely used active smoothing technique among manufacturers is to develop a product mix of counterseasonal items. Examples include companies that make both furnaces and air conditioners or lawn mowers and snowblowers. However, companies that follow this approach may find themselves involved in products or services beyond their area of expertise or beyond their target market.

These eight options, along with their advantages and disadvantages, are summarized in Table 13.1.

Mixing Options to Develop a Plan

Although each of the five capacity options and three demand options previously discussed may produce an effective aggregate schedule, some combination of capacity options and demand options may be better.

Many manufacturers assume that the use of the demand options has been fully explored by the marketing department and those reasonable options incorporated into the demand forecast. The operations manager then builds the aggregate plan based on that forecast. However, using the five available capacity options, the operations manager can still draw from a multitude of possible plans. These plans can embody, at one extreme, a *chase strategy* and, at the other, a *level-scheduling strategy*. They may, of course, fall somewhere in between.

TABLE 13.1 Aggregate Planning Options: Advantages and Disadvantages

OPTION	ADVANTAGES	DISADVANTAGES	COMMENTS
Changing inventory levels	Changes in human resources are gradual or none; no abrupt production changes.	Inventory holding costs may increase. Shortages may result in lost sales.	Applies mainly to production, not service, operations.
Varying workforce size by hiring or layoffs	Avoids the costs of other alternatives.	Hiring, layoff, and training costs may be significant.	Used where size of labor pool is large.
Varying production rates through overtime or idle time	Matches seasonal fluctuations without hiring/training costs.	Overtime premiums; tired workers; may not meet demand.	Allows flexibility within the aggregate plan.
Subcontracting	Permits flexibility and smoothing of the firm's output.	Loss of quality control; reduced profits; potential loss of future business.	Applies mainly in production settings.
Using part-time workers	Is less costly and more flexible than full-time workers.	High turnover/training costs; quality suffers; scheduling difficult.	Good for unskilled jobs in areas with large temporary labor pools.
Influencing demand	Tries to use excess capacity. Discounts draw new customers.	Uncertainty in demand. Hard to match demand to supply exactly.	Creates marketing ideas. Overbooking used in some businesses.
Backordering during high-demand periods	May avoid overtime. Keeps capacity constant.	Customer must be willing to wait, but goodwill is lost.	Many companies backorder.
Counterseasonal product and service mixing	Fully utilizes resources; allows stable workforce.	May require skills or equipment outside firm's areas of expertise.	Risky finding products or services with opposite demand patterns.

Chase Strategy A chase strategy typically attempts to achieve output rates for each period that match the demand forecast for that period. This strategy can be accomplished in a variety of ways. For example, the operations manager can vary workforce levels by hiring or laying off or can vary output by means of overtime, idle time, part-time employees, or subcontracting. Many service organizations favor the chase strategy because the changing inventory levels option is difficult or impossible to adopt. Industries that have moved toward a chase strategy include education, hospitality, and construction.

Chase strategy
A planning strategy that sets production equal to forecast demand.

Level Strategy A level strategy (or level scheduling) is an aggregate plan in which production is uniform from period to period. Firms like Toyota and Nissan attempt to keep production at uniform levels and may (1) let the finished-goods inventory vary to buffer the difference between demand and production or (2) find alternative work for employees. Their philosophy is that a stable workforce leads to a better-quality product, less turnover and absenteeism, and more employee commitment to corporate goals. Other hidden savings include more experienced employees, easier scheduling and supervision, and fewer dramatic startups and shutdowns. Level scheduling works well when demand is reasonably stable.

Level scheduling
Maintaining a constant output rate, production rate, or workforce level over the planning horizon.

For most firms, neither a chase strategy nor a level strategy is likely to prove ideal, so a combination of the eight options (called a mixed strategy) must be investigated to achieve minimum cost. However, because there are a huge number of possible mixed strategies, managers find that aggregate planning can be a challenging task. Finding the one "optimal" plan is not always possible, but as we will see in the next section, a number of techniques have been developed to aid the aggregate planning process.

Mixed strategy
A planning strategy that uses two or more controllable variables to set a feasible production plan.

Methods for Aggregate Planning

In this section, we introduce techniques that operations managers use to develop aggregate plans. They range from the widely used graphical method to the transportation method of linear programming.

Graphical Methods

Graphical techniques
Aggregate planning techniques that work with a few variables at a time to allow planners to compare projected demand with existing capacity.

Graphical techniques are popular because they are easy to understand and use. These plans work with a few variables at a time to allow planners to compare projected demand with existing capacity. They are trial-and-error approaches that do not guarantee an optimal production plan, but they require only limited computations and can be performed by clerical staff. Following are the five steps in the graphical method:

LO 13.4 *Prepare a graphical aggregate plan*

1. Determine the demand in each period.
2. Determine capacity for regular time, overtime, and subcontracting each period.
3. Find labor costs, hiring and layoff costs, and inventory holding costs.
4. Consider company policy that may apply to the workers or to stock levels.
5. Develop alternative plans and examine their total costs.

These steps are illustrated in Examples 1 through 4.

Example 1

GRAPHICAL APPROACH TO AGGREGATE PLANNING FOR A ROOFING SUPPLIER

A Juarez, Mexico, manufacturer of roofing supplies has developed monthly forecasts for a family of products. Data for the 6-month period January to June are presented in Table 13.2. The firm would like to begin development of an aggregate plan.

TABLE 13.2 Monthly Forecasts

MONTH	EXPECTED DEMAND	PRODUCTION DAYS	DEMAND PER DAY (COMPUTED)
Jan.	900	22	41
Feb.	700	18	39
Mar.	800	21	38
Apr.	1,200	21	57
May	1,500	22	68
June	1,100	20	55
	6,200	124	

APPROACH ▶ Plot daily and average demand to illustrate the nature of the aggregate planning problem.

SOLUTION ▶ First, compute demand per day by dividing the expected monthly demand by the number of production days (working days) each month and drawing a graph of those forecast demands (Figure 13.3). Second, draw a dotted line across the chart that represents the production rate required to meet average demand over the 6-month period. The chart is computed as follows:

$$\text{Average requirement} = \frac{\text{Total expected demand}}{\text{Number of production days}} = \frac{6,200}{124} = 50 \text{ units per day}$$

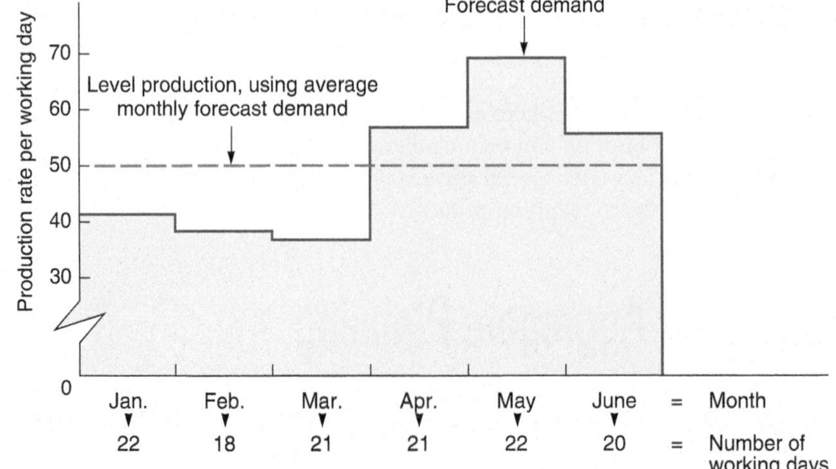

Figure **13.3**
Graph of Forecast and Average Forecast Demand

INSIGHT ▶ Changes in the production rate become obvious when the data are graphed. Note that in the first 3 months, expected demand is lower than average, while expected demand in April, May, and June is above average.

LEARNING EXERCISE ▶ If demand for June increases to 1,200 (from 1,100), what is the impact on Figure 13.3? [Answer: The daily rate for June will go up to 60, and average production will increase to 50.8 (6,300/124).]

RELATED PROBLEM ▶ 13.1

The graph in Figure 13.3 illustrates how the forecast differs from the average demand. Some strategies for meeting the forecast were listed earlier. The firm, for example, might staff in order to yield a production rate that meets *average* demand (as indicated by the dashed line). Or it might produce a steady rate of, say, 30 units and then subcontract excess demand to other roofing suppliers. Other plans might combine overtime work with subcontracting to absorb demand or vary the workforce by hiring and laying off. Examples 2 to 4 illustrate three possible strategies.

Example 2

PLAN 1 FOR THE ROOFING SUPPLIER—A CONSTANT WORKFORCE

One possible strategy (call it plan 1) for the manufacturer described in Example 1 is to maintain a constant workforce throughout the 6-month period. A second (plan 2) is to maintain a constant workforce at a level necessary to meet the lowest demand month (March) and to meet all demand above this level by subcontracting. Both plan 1 and plan 2 have level production and are, therefore, called *level strategies*. Plan 3 is to hire and lay off workers as needed to produce exact monthly requirements—*a chase strategy*. Table 13.3 provides cost information necessary for analyzing these three alternatives.

TABLE 13.3 Cost Information

Inventory carrying cost	\$ 5 per unit per month
Subcontracting cost per unit	\$ 20 per unit
Average pay rate	\$ 10 per hour (\$80 per day)
Overtime pay rate	\$ 17 per hour (above 8 hours per day)
Labor-hours to produce a unit	1.6 hours per unit
Cost of increasing daily production rate (hiring and training)	\$300 per unit
Cost of decreasing daily production rate (layoffs)	\$600 per unit

ANALYSIS OF PLAN 1 APPROACH ▶ Here we assume that 50 units are produced per day and that we have a constant workforce, no overtime or idle time, no safety stock, and no subcontractors. The firm accumulates inventory during the slack period of demand, January through March, and depletes it during the higher-demand warm season, April through June. We assume beginning inventory = 0 and planned ending inventory = 0.

SOLUTION ▶ We construct the table below and accumulate the costs:

MONTH	PRODUCTION DAYS	PRODUCTION AT 50 UNITS PER DAY	DEMAND FORECAST	MONTHLY INVENTORY CHANGE	ENDING INVENTORY
Jan.	22	1,100	900	+200	200
Feb.	18	900	700	+200	400
Mar.	21	1,050	800	+250	650
Apr.	21	1,050	1,200	−150	500
May	22	1,100	1,500	−400	100
June	20	1,000	1,100	−100	0
					1,850

Total units of inventory carried over from one month to the next month = 1,850 units
Workforce required to produce 50 units per day = 10 workers

Because each unit requires 1.6 labor-hours to produce, each worker can make 5 units in an 8-hour day. Therefore, to produce 50 units, 10 workers are needed.

Finally, the costs of plan 1 are computed as follows:

COST		CALCULATIONS
Inventory carrying	$ 9,250	(= 1,850 units carried × $5 per unit)
Regular-time labor	99,200	(= 10 workers × $80 per day × 124 days)
Other costs (overtime, hiring, layoffs, subcontracting)	0	
Total cost	$108,450	

INSIGHT ▶ Note the significant cost of carrying the inventory.

LEARNING EXERCISE ▶ If demand for June decreases to 1,000 (from 1,100), what is the change in cost? [Answer: Total inventory carried will increase to 1,950 at $5, for an inventory cost of $9,750 and total cost of $108,950.]

RELATED PROBLEMS ▶ 13.2–13.12, 13.19

EXCEL OM Data File **Ch13Ex2.xls** can be found online.

ACTIVE MODEL 13.1 This example is further illustrated in Active Model 13.1 found online.

The graph for Example 2 was shown in Figure 13.3. Some planners prefer a *cumulative* graph to display visually how the forecast deviates from the average requirements. Such a graph is provided in Figure 13.4. Note that both the level production line and the forecast line produce the same total production.

Figure 13.4

Cumulative Graph for Plan 1

STUDENT TIP We saw another way to graph this data in Figure 13.3.

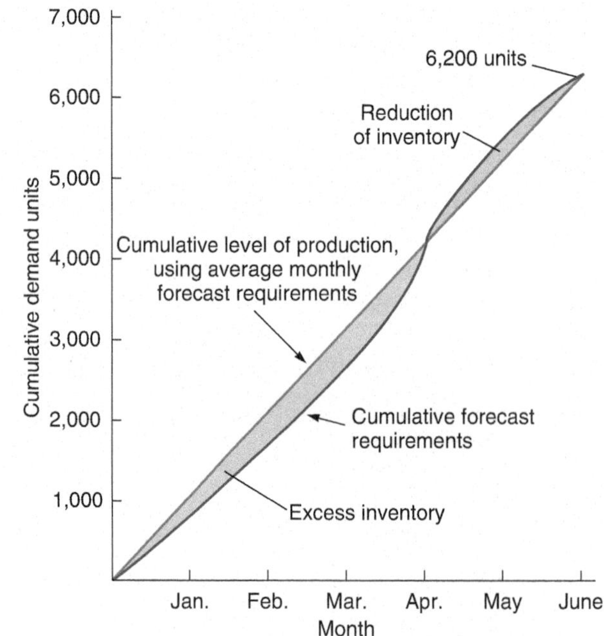

Example 3

PLAN 2 FOR THE ROOFING SUPPLIER—USE OF SUBCONTRACTORS WITHIN A CONSTANT WORKFORCE

ANALYSIS OF PLAN 2 APPROACH ▶ Although a constant workforce is also maintained in plan 2, it is set low enough to meet demand only in March, the lowest demand-per-day month. To produce 38 units per day $(800/21)$ in-house, 7.6 workers are needed. (You can think of this as 7 full-time workers and 1 part-timer.) *All* other demand is met by subcontracting. Subcontracting is thus required in every other month. No inventory holding costs are incurred in plan 2.

SOLUTION ▶ Because 6,200 units are required during the aggregate plan period, we must compute how many can be made by the firm and how many must be subcontracted:

$$\text{In-house production} = 38 \text{ units per day} \times 124 \text{ production days}$$
$$= 4{,}712 \text{ units}$$
$$\text{Subcontract units} = 6{,}200 - 4{,}712 = 1{,}488 \text{ units}$$

The costs of plan 2 are computed as follows:

COST		CALCULATIONS
Regular-time labor	$ 75,392	(= 7.6 workers × $80 per day × 124 days)
Subcontracting	29,760	(= 1,488 units × $20 per unit)
Total cost	$105,152	

INSIGHT ▶ Note the lower cost of regular labor but the added subcontracting cost.

LEARNING EXERCISE ▶ If demand for June increases to 1,200 (from 1,100), what is the change in cost? [Answer: Subcontracting requirements increase to 1,588 at $20 per unit, for a subcontracting cost of $31,760 and a total cost of $107,152.]

RELATED PROBLEMS ▶ 13.2–13.12, 13.19

Example 4

PLAN 3 FOR THE ROOFING SUPPLIER—HIRING AND LAYOFFS

ANALYSIS OF PLAN 3 APPROACH ▶ The final strategy, plan 3, involves varying the workforce size by hiring and layoffs as necessary. The production rate will equal the demand, and there is no change in production from the previous month, December.

SOLUTION ▶ Table 13.4 shows the calculations and the total cost of plan 3. Recall that it costs $600 per unit produced to reduce production from the previous month's daily level and $300 per unit change to increase the daily rate of production through hirings.

TABLE 13.4 Cost Computations for Plan 3

MONTH	FORECAST (UNITS)	DAILY PRODUCTION RATE	BASIC PRODUCTION COST (DEMAND × 1.6 HR PER UNIT × $10 PER HR)	EXTRA COST OF INCREASING PRODUCTION (HIRING COST)	EXTRA COST OF DECREASING PRODUCTION (LAYOFF COST)	TOTAL COST
Jan.	900	41	$14,400	—	—	$ 14,400
Feb.	700	39	11,200	—	$1,200 (= 2 × $600)	12,400
Mar.	800	38	12,800	—	$ 600 (= 2 × $600)	13,400
Apr.	1,200	57	19,200	$5,700 (= 19 × $300)	—	24,900
May	1,500	68	24,000	$3,300 (= 11 × $300)	—	27,300
June	1,100	55	17,600	—	$7,800 (= 13 × $600)	25,400
			$99,200	$9,000	$9,600	$117,800

Thus, the total cost, including production, hiring, and layoff, for plan 3 is $117,800.

INSIGHT ▶ Note the substantial cost associated with changing (both increasing and decreasing) the production levels.

LEARNING EXERCISE ▶ If demand for June increases to 1,200 (from 1,100), what is the change in cost? [Answer: Daily production for June is 60 units, which is a decrease of 8 units in the daily production rate from May's 68 units, so the new June layoff cost is $4,800(= 8 × $600), but an additional production cost for 100 units is $1,600(100 × 1.6 × $10) with a total plan 3 cost of $116,400.]

RELATED PROBLEMS ▶ 13.2–13.12, 13.19

The final step in the graphical method is to compare the costs of each proposed plan and to select the approach with the least total cost. A summary analysis is provided in Table 13.5. We see that because plan 2 has the lowest cost, it is the best of the three options.

TABLE 13.5 Comparison of the Three Plans

COST	PLAN 1 (CONSTANT WORKFORCE OF 10 WORKERS)	PLAN 2 (WORKFORCE OF 7.6 WORKERS PLUS SUBCONTRACTORS)	PLAN 3 (HIRING AND LAYOFFS TO MEET DEMAND)
Inventory carrying	$ 9,250	$ 0	$ 0
Regular labor	99,200	75,392	99,200
Overtime labor	0	0	0
Hiring	0	0	9,000
Layoffs	0	0	9,600
Subcontracting	0	29,760	0
Total cost	$108,450	$105,152	$117,800

Of course, many other feasible strategies can be considered in a problem like this, including combinations that use some overtime. Although graphing is a popular management tool, its help is in evaluating strategies, not generating them. To generate strategies, a systematic approach that considers all costs and produces an effective solution is needed.

Mathematical Approaches

This section briefly describes mathematical approaches to aggregate planning.

Transportation method of linear programming
A way of solving for the optimal solution to an aggregate planning problem.

The Transportation Method of Linear Programming When an aggregate planning problem is viewed as one of allocating operating capacity to meet forecast demand, it can be formulated in a linear programming format. The **transportation method of linear programming** is not a trial-and-error approach like graphing but rather produces an optimal plan for minimizing costs. It is also flexible in that it can specify regular and overtime production in each time period, the number of units to be subcontracted, extra shifts, and the inventory carryover from period to period.

In Example 5, the supply consists of on-hand inventory and units produced by regular time, overtime, and subcontracting. Costs per unit, in the upper-right corner of each cell of the matrix in Table 13.7, relate to units produced in a given period or units carried in inventory from an earlier period.

Example 5

AGGREGATE PLANNING WITH THE TRANSPORTATION METHOD

Farnsworth Tire Company would like to develop an aggregate plan via the transportation method. Data that relate to production, demand, capacity, and cost at its West Virginia plant are shown in Table 13.6.

TABLE 13.6 Farnsworth's Production, Demand, Capacity, and Cost Data

	SALES PERIOD		
	MAR.	APR.	MAY
Demand	800	1,000	750
Capacity:			
Regular	700	700	700
Overtime	50	50	50
Subcontracting	150	150	130
Beginning inventory	100 tires		

COSTS	
Regular time	$40 per tire
Overtime	$50 per tire
Subcontract	$70 per tire
Carrying cost	$ 2 per tire per month

APPROACH ▶ Solve the aggregate planning problem by minimizing the costs of matching production in various periods to future demands.

SOLUTION ▶ Table 13.7 illustrates the structure of the transportation table and an initial feasible solution.

LO 13.5 *Solve* an aggregate plan via the transportation method

TABLE 13.7 FARNSWORTH'S Transportation Table[a]

SUPPLY FROM		DEMAND FOR: Period 1 (Mar.)	Period 2 (Apr.)	Period 3 (May)	Unused Capacity (dummy)	TOTAL CAPACITY AVAILABLE (supply)
Beginning inventory		0 100	2	4	0	100
Period 1	Regular time	40 700	42	44	0	700
	Overtime	50	52 50	54	0	50
	Subcontract	70	72 150	74	0	150
Period 2	Regular time	×	40 700	42	0	700
	Overtime	×	50 50	52	0	50
	Subcontract	×	70 50	72	0 100	150
Period 3	Regular time	×	×	40 700	0	700
	Overtime	×	×	50 50	0	50
	Subcontract	×	×	70	0 130	130
TOTAL DEMAND		800	1,000	750	230	2,780

[a]Cells with an *x* indicate that backorders are not used at Farnsworth. When using Excel OM or POM for Windows to solve, you must insert a *very* high cost (e.g., 9999) in each cell that is not used for production.

When setting up and analyzing this table, you should note the following:

1. Carrying costs are $2/tire per month. Tires produced in 1 period and held for 1 month will have a $2 higher cost. Because holding cost is linear, 2 months' holdover costs $4. So when you move across a row from left to right, regular time, overtime, and subcontracting costs are lowest when output is used in the same period it is produced. If goods are made in one period and carried over to the next, holding costs are incurred. Beginning inventory, however, is generally given a unit cost of 0 if it is used to satisfy demand in period 1.
2. Transportation problems require that supply equals demand, so a dummy column called "unused capacity" has been added. Costs of not using capacity are zero.
3. Because backordering is not a viable alternative for this particular company, no production is possible in those cells that represent production in a period to satisfy demand in a past period (i.e., those periods with an "×"). If backordering is allowed, costs of expediting, loss of goodwill, and loss of sales revenues are summed to estimate backorder cost.
4. Quantities in red in each column of Table 13.7 designate the levels of inventory needed to meet demand requirements (shown in the bottom row of the table). Demand of 800 tires in March is met by using 100 tires from beginning inventory and 700 tires from regular time.
5. In general, to complete the table, allocate as much production as you can to a cell with the smallest cost without exceeding the unused capacity in that row or demand in that column. If there is still some demand left in that row, allocate as much as you can to the next-lowest-cost cell. You then repeat this process for periods 2 and 3 (and beyond, if necessary). When you are finished, the sum of all your entries in a row must equal the total row capacity, and the sum of all entries in a column must equal the demand for that period. (This step can be accomplished by the transportation method or by using POM for Windows or Excel OM software.)

Try to confirm that the cost of this initial solution is $105,900. The initial solution is not optimal, however. See if you can find the production schedule that yields the least cost (which turns out to be $105,700) using software or by hand.

INSIGHT ▶ The transportation method is flexible when costs are linear but does not work when costs are nonlinear.

LEARNING EXAMPLE ▶ What is the impact on this problem if there is no beginning inventory? [Answer: Total capacity (units) available is reduced by 100 units and the need to subcontract increases by 100 units.]

RELATED PROBLEMS ▶ 13.13–13.18, 13.20–13.22

EXCEL OM Data File **Ch13Ex5.xls** can be found online.

The transportation method of linear programming described in the preceding example works well when analyzing the effects of holding inventories, using overtime, and subcontracting. However, it does not work when nonlinear or negative factors are introduced. Thus, when other factors such as hiring and layoffs are introduced, the more general method of linear programming must be used. Similarly, computer simulation models look for a minimum-cost combination of values.

A number of commercial S&OP software packages that incorporate the techniques of this chapter are available to ease the mechanics of aggregate planning. These include Arkieva's *S&OP Workbench* for process industries, Demand Solutions' *S&OP Software*, and Steelwedge's *S&OP Suite*.

Aggregate Planning in Services

Some service organizations conduct aggregate planning in exactly the same way as we did in Examples 1 through 5 in this chapter, but with demand management taking a more active role. Because most services pursue *combinations* of the eight capacity and demand options discussed earlier, they usually formulate mixed aggregate planning strategies. In industries such as banking, trucking, and fast foods, aggregate planning may be easier than in manufacturing.

Controlling the cost of labor in service firms is critical. Successful techniques include:

1. Accurate scheduling of labor-hours to ensure quick response to customer demand
2. An on-call labor resource that can be added or deleted to meet unexpected demand
3. Flexibility of individual worker skills that permits reallocation of available labor
4. Flexibility in rate of output or hours of work to meet changing demand

These options may seem demanding, but they are not unusual in service industries, in which labor is the primary aggregate planning vehicle. For instance:

STUDENT TIP
The major variable in capacity management for services is labor.

- Excess capacity is used to provide study and planning time by real estate and auto salespersons.
- Police and fire departments have provisions for calling in off-duty personnel for major emergencies. Where the emergency is extended, police or fire personnel may work longer hours and extra shifts.
- When business is unexpectedly light, restaurants and retail stores send personnel home early.
- Supermarket stock clerks work cash registers when checkout lines become too lengthy.
- Experienced servers increase their pace and efficiency of service as crowds of customers arrive.

Approaches to aggregate planning differ by the type of service provided. Here we discuss five service scenarios.

Restaurants

In a business with a highly variable demand, such as a restaurant, aggregate scheduling is directed toward (1) smoothing the production rate and (2) finding the optimal size of the workforce. The general approach usually requires building very modest levels of inventory during slack periods and depleting inventory during peak periods, but using labor to accommodate most of the changes in demand. Because this situation is very similar to those found in manufacturing, traditional aggregate planning methods may be applied to restaurants as well. One difference is that even modest amounts of inventory may be perishable. In addition, the relevant units of time may be much smaller than in manufacturing. For example, in fast-food restaurants, peak and slack periods may be measured in fractions of an hour, and the "product" may be inventoried for as little as 10 minutes.

Hospitals

Hospitals face aggregate planning problems in allocating money, staff, and supplies to meet the demands of patients. Michigan's Henry Ford Hospital, for example, plans for bed capacity and personnel needs in light of a patient-load forecast developed by moving averages. The necessary labor focus of its aggregate plan has led to the creation of a new floating staff pool serving each nursing pod.

National Chains of Small Service Firms

With the advent of national chains of small service businesses such as funeral homes, oil change outlets, and photocopy/printing centers, the question of aggregate planning versus independent planning at each business establishment becomes an issue. Both purchases and production capacity may be centrally planned when demand can be influenced through special promotions. This approach to aggregate scheduling is often advantageous because it reduces costs and helps manage cash flow at independent sites.

Miscellaneous Services

Most "miscellaneous" services—financial, transportation, and many communication and recreation services—provide intangible output. Aggregate planning for these services deals mainly with planning for human resource requirements and managing demand. The twofold goal is to level demand peaks and to design methods for fully utilizing labor resources during low-demand periods. Example 6 illustrates such a plan for a legal firm.

Example 6

AGGREGATE PLANNING IN A LAW FIRM

Klasson and Avalon, a medium-sized Tampa law firm of 32 legal professionals, wants to develop an aggregate plan for the next quarter. The firm has developed 3 forecasts of billable hours for the next quarter for each of 5 categories of legal business it performs (column 1, Table 13.8). The 3 forecasts (best, likely, and worst) are shown in columns 2, 3, and 4 of Table 13.8.

TABLE 13.8 Labor Allocation at Klasson and Avalon, Forecasts for Coming Quarter (1 lawyer = 500 hours of labor)

FORECASTED LABOR-HOURS REQUIRED				CAPACITY CONSTRAINTS	
(1) CATEGORY OF LEGAL BUSINESS	(2) BEST (HOURS)	(3) LIKELY (HOURS)	(4) WORST (HOURS)	(5) MAXIMUM DEMAND FOR PERSONNEL	(6) NUMBER OF QUALIFIED PERSONNEL
Trial work	1,800	1,500	1,200	3.6	4
Legal research	4,500	4,000	3,500	9.0	32
Corporate law	8,000	7,000	6,500	16.0	15
Real estate law	1,700	1,500	1,300	3.4	6
Criminal law	3,500	3,000	2,500	7.0	12
Total hours	19,500	17,000	15,000		
Lawyers needed	39	34	30		

APPROACH ▶ If we make some assumptions about the workweek and skills, we can provide an aggregate plan for the firm. Assuming a 40-hour workweek and that 100% of each lawyer's hours are billed, about 500 billable hours are available from each lawyer this fiscal quarter.

SOLUTION ▶ We divide hours of billable time (which is the demand) by 500 to provide a count of lawyers needed (lawyers represent the capacity) to cover the estimated demand. Capacity then is shown to be 39, 34, and 30 for the three forecasts, best, likely, and worst, respectively. For example, the best-case scenario of 19,500 total hours, divided by 500 hours per lawyer, equals 39 lawyers needed. Because all 32 lawyers at Klasson and Avalon are qualified to perform basic legal research, this skill has maximum scheduling flexibility (column 6). The most highly skilled (and capacity-constrained) categories are trial work and corporate law. The firm's best-case forecast just barely covers trial work, with 3.6 lawyers needed (see column 5) and 4 qualified (column 6). And corporate law is short 1 full person.

Overtime may be used to cover the excess this quarter, but as business expands, it may be necessary to hire or develop talent in both of these areas. Available staff adequately covers real estate and criminal practice, as long as other needs do not use their excess capacity. With its current legal staff of 32, Klasson and Avalon's best-case forecast will increase the workload by [(39 − 32)/32 =] 21.8% (assuming no new hires). This represents 1 extra day of work per lawyer per week. The worst-case scenario will result in about a 6% underutilization of talent. For both of these scenarios, the firm has determined that available staff will provide adequate service.

INSIGHT ▶ While our definitions of demand and capacity are different than for a manufacturing firm, aggregate planning is as appropriate, useful, and necessary in a service environment as in manufacturing.

LEARNING EXERCISE ▶ If the criminal law best-case forecast increases to 4,500 hours, what happens to the number of lawyers needed? [Answer: The demand for lawyers increases to 41.]

RELATED PROBLEMS ▶ 13.23, 13.24

Source: Based on Glenn Bassett, *Operations Management for Service Industries* (Westport, CT: Quorum Books): 110.

Airline Industry

Airlines and auto-rental firms also have unique aggregate scheduling problems. Consider an airline that has its headquarters in New York, two hub sites in cities such as Atlanta and Dallas, and 150 offices in airports throughout the country. This planning is considerably more complex than aggregate planning for a single site or even for a number of independent sites.

Aggregate planning consists of schedules for (1) number of flights into and out of each hub; (2) number of flights on all routes; (3) number of passengers to be serviced on all flights; (4) number of air personnel and ground personnel required at each hub and airport; and (5) determining the seats to be allocated to various fare classes. Techniques for determining seat allocation are called revenue (or yield) management, our next topic.

Revenue Management

STUDENT TIP
Revenue management changes the focus of aggregate planning from capacity management to demand management.

Most operations models, like most business models, assume that firms charge all customers the same price for a product. In fact, many firms work hard at charging different prices. The idea is to match capacity and demand by charging different prices based on the customer's willingness to pay. The management challenge is to identify those differences and price accordingly. The technique for multiple price points is called revenue management.

Revenue (or yield) management
Capacity decisions that determine the allocation of resources to maximize revenue or yield.

Revenue (or yield) management is the aggregate planning process of allocating the company's scarce resources to customers at prices that will maximize revenue. Popular use of the technique dates to the 1980s, when American Airlines' reservation system (called SABRE) allowed the airline to alter ticket prices, in real time and on any route, based on demand information. If it looked like demand for expensive seats was low, more discounted seats were offered. If demand for full-fare seats was high, the number of discounted seats was reduced.

American Airlines' success in revenue management spurred many other companies and industries to adopt the concept. Revenue management in the hotel industry began in the late 1980s at Marriott International, which now claims an additional $400 million a year in profit from its management of revenue. The competing Omni hotel chain uses software that performs more than 100,000 calculations every night at each facility. The Dallas Omni, for example, charges its highest rates on weekdays but heavily discounts on weekends. Its sister hotel in San Antonio, which is in a more tourist-oriented destination, reverses this rating scheme, with better deals for its consumers on weekdays. Similarly, Walt Disney World has multiple prices: an annual admission "premium" pass for an adult was recently quoted at $1,273, but for a Florida resident, $766, with different discounts for AAA members and active-duty military. The *OM in Action* box, "Revenue Management Makes Disney the

OM in Action Revenue Management Makes Disney the "King" of the Broadway Jungle[2]

Disney accomplished the unthinkable for long-running Broadway musicals: *The Lion King* transformed from a declining moneymaker into the top-grossing Broadway show. How? Hint: It's not because the show added performances after 25 years.

The show's producers are using a previously undisclosed computer algorithm to recommend the highest ticket prices that audiences would be likely to pay for each of the 1,700 seats. Other shows also employ this dynamic pricing model to raise seat prices during tourist-heavy holiday weeks, but only Disney has reached the level of sophistication achieved in the airline and hotel industries. By continually using its algorithm to calibrate prices based on ticket demand and purchasing patterns, Disney has been able to achieve sales records.

By charging $10 more here, $20 more there, *The Lion King* has stunned Broadway as the No. 1 earner (over $1.7 billion), bumping off the prior champ, *Wicked*. And Disney even managed to do it by charging half as much for top tickets as some rivals. "Credit the management science experts at Disney's corporate offices—a data army that no Broadway producer could ever match—for helping develop the winning formula," writes *The New York Times*. Disney's algorithm, a software tool that draws on *Lion King* data for

JTB Photo/JTB Media Creation Inc/Alamy Stock Photo

11.5 million past customers, recommends prices for multiple categories of performances—peak dates such as Christmas, off-peak dates such as a weeknight in February, and various periods in between. "The Lion King" is widely believed to have sold far more seats for $227 than most Broadway shows sell at their top rates, a situation that bolsters its grosses.

VIDEO 13.1
Using Revenue Management to Set Orlando Magic Ticket Prices

'King' of the Broadway Jungle," describes this practice in the live theatre industry. The *Video Case Study* at the end of this chapter addresses revenue management for the Orlando Magic.

Organizations that have *perishable inventory*, such as airlines, hotels, car rental agencies, cruise lines, and even electrical utilities, have the following shared characteristics that make yield management of interest:

1. Service or product can be sold in advance of consumption
2. Fluctuating demand
3. Relatively fixed resource (capacity)
4. Segmentable demand
5. Low variable costs and high fixed costs

LO 13.6 *Understand* and solve a revenue management problem

Example 7 illustrates how revenue management works in a hotel.

Example 7

REVENUE MANAGEMENT

The Cleveland Downtown Inn is a 100-room hotel that has historically charged one set price for its rooms, $150 per night. The variable cost of a room being occupied is low. Management believes the cleaning, air-conditioning, and incidental costs of soap, shampoo, and so forth, are $15 per room per night. Sales average 50 rooms per night. Figure 13.5 illustrates the current pricing scheme. Net sales are $6,750 per night with a single price point.

APPROACH ▶ Analyze pricing from the perspective of revenue management. We note in Figure 13.5 that some guests would have been willing to spend more than $150 per room—"money left on the table." Others would be willing to pay more than the variable cost of $15 but less than $150—"passed-up contribution."

Figure **13.5**

Hotel Sets Only One Price Level

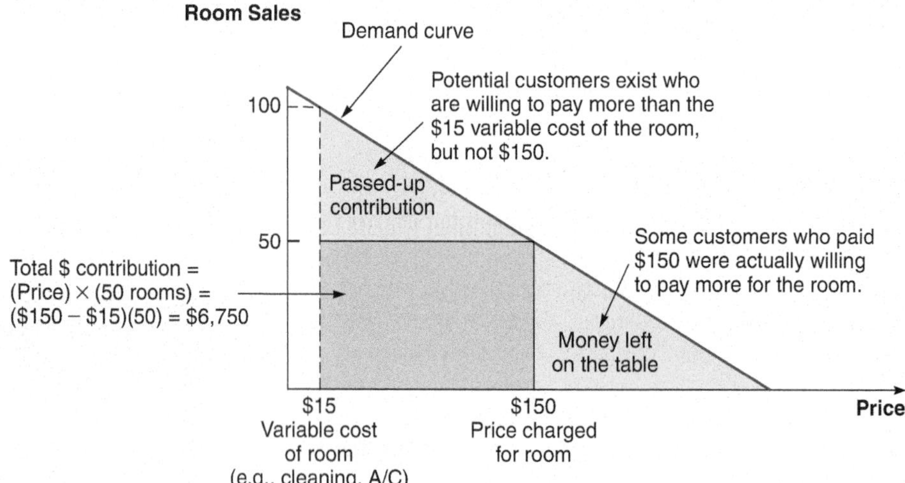

SOLUTION ▶ In Figure 13.6, the inn decides to set *two* price levels. It estimates that 30 rooms per night can be sold at $100 and another 30 rooms at $200, using revenue management software that is widely available.

INSIGHT ▶ Revenue management has increased total contribution to $8,100 ($2,550 from $100 rooms and $5,550 from $200 rooms). It may be that even more price levels are called for at Cleveland Downtown Inn.

LEARNING EXERCISE ▶ If the hotel develops a third price of $150 and can sell half of the $100 rooms at the increased rate, what is the contribution? [Answer: $8,850 = (15 × $85) + (15 × $135) + (30 × $185).]

RELATED PROBLEMS ▶ 13.25, 13.26

Figure 13.6
Hotel with Two Price Levels

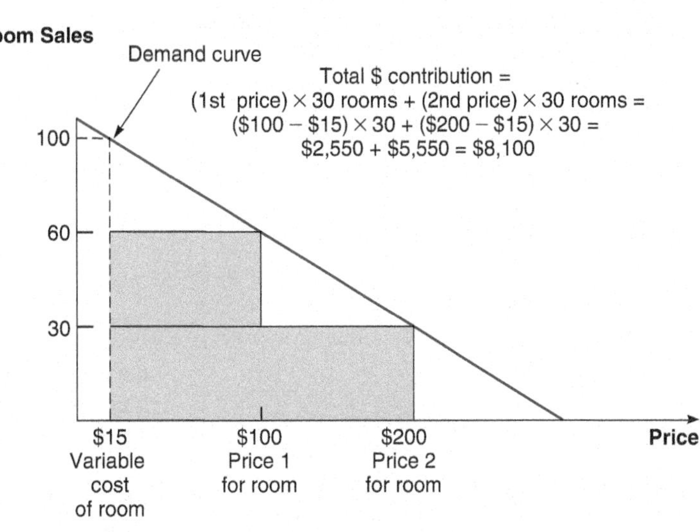

Industries traditionally associated with revenue management include hotels, airlines, and rental cars. They are able to apply variable pricing for their product and control product use or availability (number of airline seats or hotel rooms sold at economy rates). Others, such as movie theaters, arenas, or performing arts centers have less pricing flexibility but still use time (evening or matinee) and location (orchestra, side, or balcony) to manage revenue. In both cases, management has control over the amount of the resource used—both the quantity and the duration of the resource.

The manager's job is more difficult in facilities such as restaurants and on golf courses because the duration and the use of the resource are less controllable. However, with imagination, managers are using excess capacity even for these industries. For instance, the golf course may sell less desirable tee times at a reduced rate, and the restaurant may have an "early bird" special to generate business before the usual dinner hour.

To make revenue management work, the company needs to manage three issues:

1. *Multiple pricing structures:* These structures must be feasible and appear logical (and preferably fair) to the customer. Such justification may take various forms, for example, first-class seats on an airline or the preferred starting time at a golf course. (See the Ethical Dilemma at the end of this chapter.)
2. *Forecasts of the use and duration of the use:* How many economy seats should be available? How much will customers pay for a room with an ocean view?
3. *Changes in demand:* This means managing the increased use as more capacity is sold. It also means dealing with issues that occur because the pricing structure may not seem logical and fair to all customers. Finally, it means managing new issues, such as overbooking because the forecast was not perfect.

Precise pricing through revenue management has substantial potential, and several firms sell software available to address the issue. These include NCR's *Teradata*, SPS, *DemandTec*, and Oracle with *Profit Logic*.

Summary

Sales and operations planning (S&OP) can be a strong vehicle for coordinating the functional areas of a firm as well as for communication with supply chain partners. The output of S&OP is an *aggregate plan*. An aggregate plan provides both manufacturing and service firms the ability to respond to changing customer demands and produce with a winning strategy.

Aggregate schedules set levels of inventory, production, subcontracting, and employment over an intermediate time range, usually 3 to 18 months. This chapter describes two aggregate planning techniques: the popular graphical approach and the transportation method of linear programming.

The aggregate plan is an important responsibility of an operations manager and a key to efficient use of existing resources. It leads to the more detailed master production schedule, which becomes the basis for disaggregation, detail scheduling, and MRP systems.

Restaurants, airlines, and hotels are all service systems that employ aggregate plans. They also have an opportunity to implement revenue management.

Regardless of the industry or planning method, the S&OP process builds an aggregate plan that a firm can implement and suppliers endorse.

Key Terms

Sales and operations planning (S&OP) (p. 539)
Aggregate plan (p. 539)
Disaggregation (p. 541)
Master production schedule (p. 541)
Chase strategy (p. 543)
Level scheduling (p. 543)
Mixed strategy (p. 543)
Graphical techniques (p. 544)
Transportation method of linear programming (p. 548)
Revenue (or yield) management (p. 553)

Ethical Dilemma

Airline passengers today stand in numerous lines, are crowded into small seats on mostly full airplanes, and often spend time on taxiways because of air-traffic problems or lack of open gates. But what gripes travelers almost as much as these annoyances is finding out that the person sitting next to them paid a much lower fare than they did for their seat. This concept of revenue management results in ticket pricing that can range from free to thousands of dollars on the same plane. Figure 13.7 illustrates what passengers recently paid for various seats on an 11:35 A.M. flight from Minneapolis to Anaheim, California, on an Airbus A320.

Make the case for, and then against, this pricing system. Does the general public seem to accept revenue management? What would happen if you overheard the person in front of you in line getting a better room rate at a Hilton Hotel? How do customers manipulate the airline systems to get better fares?

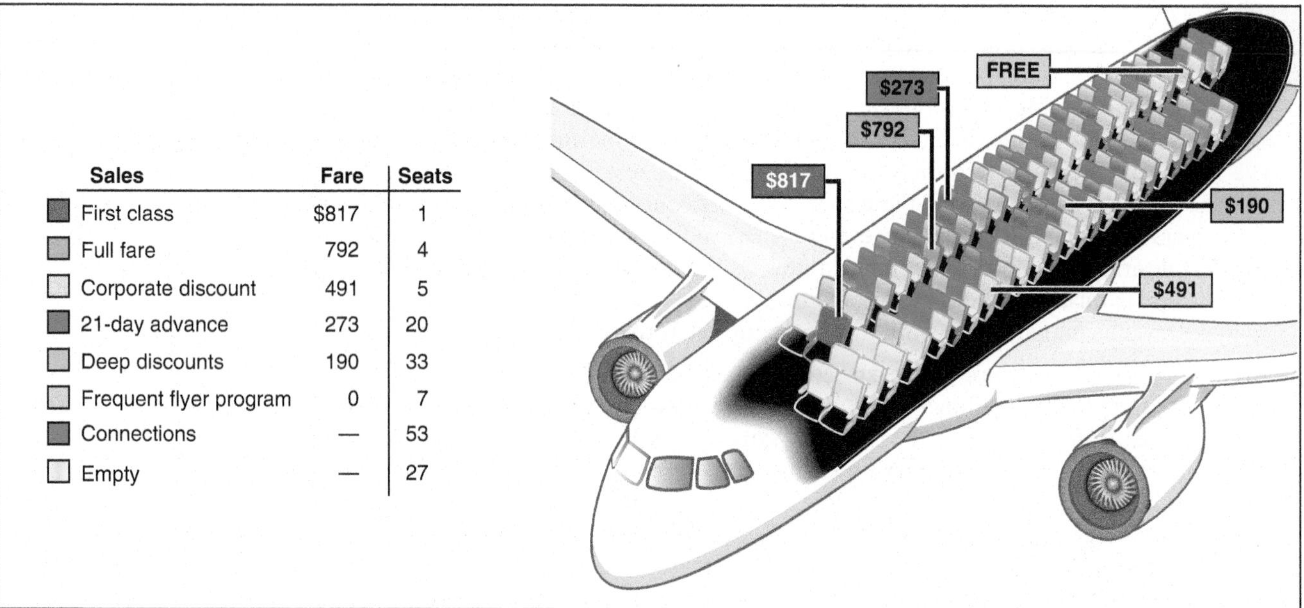

Sales	Fare	Seats
First class	$817	1
Full fare	792	4
Corporate discount	491	5
21-day advance	273	20
Deep discounts	190	33
Frequent flyer program	0	7
Connections	—	53
Empty	—	27

Figure **13.7**

Revenue Management Seat Costs on a Typical Flight

Discussion Questions

1. Define *sales* and *operations planning*.
2. Why are S&OP teams typically cross-functional?
3. Define *aggregate planning*.
4. Explain what the term *aggregate* in "aggregate planning" means.
5. List the strategic objectives of aggregate planning. Which one of these is most often addressed by the quantitative techniques of aggregate planning? Which one of these is generally the most important?
6. Define *chase strategy*.
7. What is level scheduling? What is the basic philosophy underlying it?
8. Define *mixed strategy*. Why would a firm use a mixed strategy instead of a simple pure strategy?
9. What are the advantages and disadvantages of varying the size of the workforce to meet demand requirements each period?
10. How does aggregate planning in service differ from aggregate planning in manufacturing?
11. What is the relationship between the aggregate plan and the master production schedule?
12. Why are graphical aggregate planning methods useful?
13. What are major limitations of using the transportation method for aggregate planning?
14. How does revenue management impact an aggregate plan?

Using Software for Aggregate Planning

This section illustrates the use of Excel, Excel OM, and POM for Windows in aggregate planning.

CREATING YOUR OWN EXCEL SPREADSHEETS

Program 13.1 illustrates how you can make an Excel model to solve Example 5, which uses the transportation method for aggregate planning.

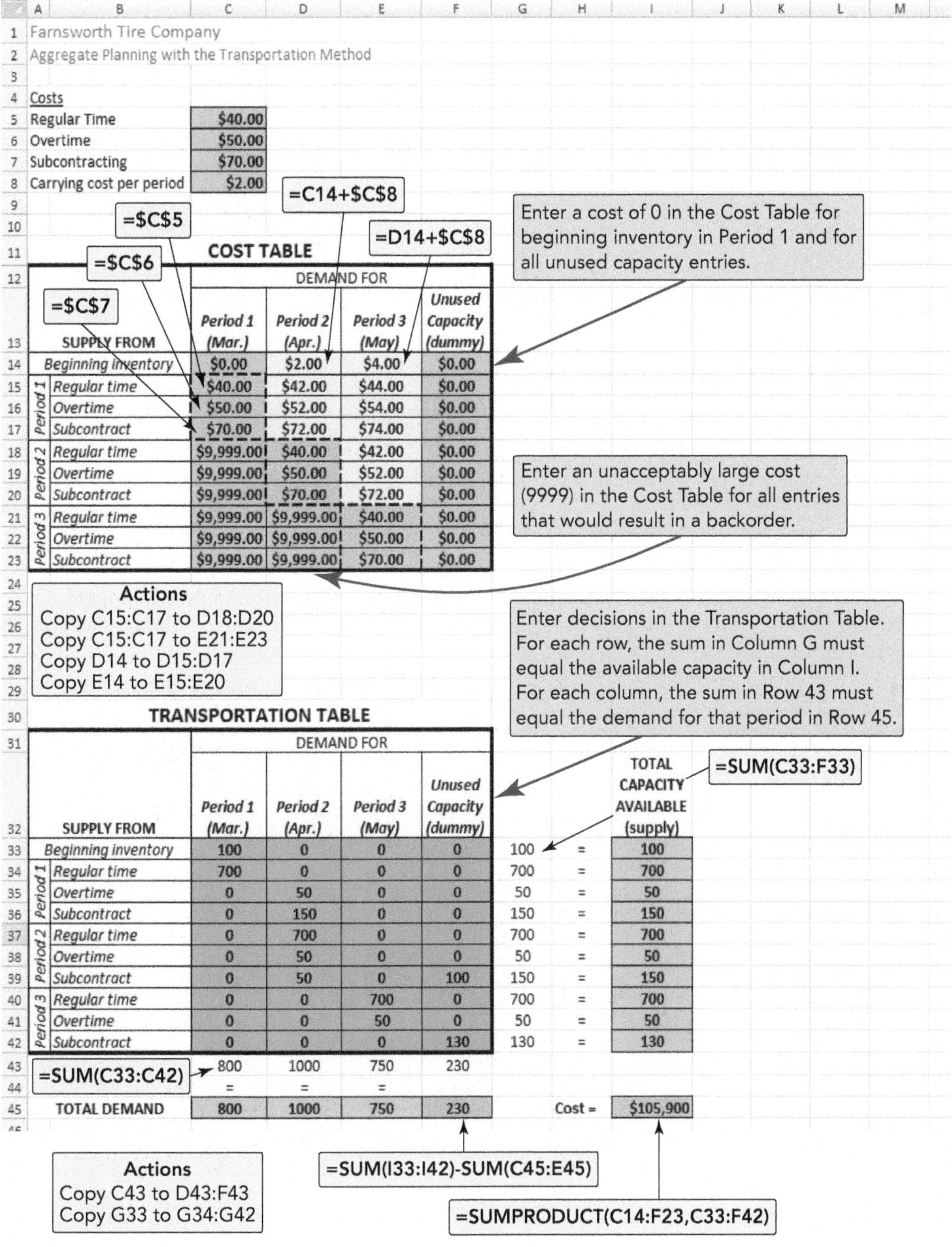

Farnsworth Tire Company
Aggregate Planning with the Transportation Method

Costs	
Regular Time	$40.00
Overtime	$50.00
Subcontracting	$70.00
Carrying cost per period	$2.00

COST TABLE

SUPPLY FROM		Period 1 (Mar.)	Period 2 (Apr.)	Period 3 (May)	Unused Capacity (dummy)
Beginning inventory		$0.00	$2.00	$4.00	$0.00
Period 1	Regular time	$40.00	$42.00	$44.00	$0.00
	Overtime	$50.00	$52.00	$54.00	$0.00
	Subcontract	$70.00	$72.00	$74.00	$0.00
Period 2	Regular time	$9,999.00	$40.00	$42.00	$0.00
	Overtime	$9,999.00	$50.00	$52.00	$0.00
	Subcontract	$9,999.00	$70.00	$72.00	$0.00
Period 3	Regular time	$9,999.00	$9,999.00	$40.00	$0.00
	Overtime	$9,999.00	$9,999.00	$50.00	$0.00
	Subcontract	$9,999.00	$9,999.00	$70.00	$0.00

TRANSPORTATION TABLE

SUPPLY FROM		Period 1 (Mar.)	Period 2 (Apr.)	Period 3 (May)	Unused Capacity (dummy)			TOTAL CAPACITY AVAILABLE (supply)
Beginning inventory		100	0	0	0	100	=	100
Period 1	Regular time	700	0	0	0	700	=	700
	Overtime	0	50	0	0	50	=	50
	Subcontract	0	150	0	0	150	=	150
Period 2	Regular time	0	700	0	0	700	=	700
	Overtime	0	50	0	0	50	=	50
	Subcontract	0	50	0	100	150	=	150
Period 3	Regular time	0	0	700	0	700	=	700
	Overtime	0	0	50	0	50	=	50
	Subcontract	0	0	0	130	130	=	130
		800	1000	750	230			
		=	=	=				
TOTAL DEMAND		800	1000	750	230		Cost =	$105,900

Program **13.1**

Using Excel for Aggregate Planning via the Transportation Method, with Data from Example 5

Excel comes with an Add-in called Solver that offers the ability to analyze linear programs such as the transportation problem. Solver should appear on the far right of the Data tab. If Solver is not on the Data tab, then if using Windows click on **File**, then **Options**, then **Add-ins**. Next to **Manage:** at the bottom, make sure that **Excel Add-ins** is selected, and click on the **Go…** button. Check **Solver Add-in**, and click **OK**. (Or if using Excel for Mac, click on **Tools** on the top (Excel) menu, then **Add-ins**. Check Solver **Add-in** and click **OK**.) The Solver dialog box will appear by clicking on **Data**, then **Solver**. The following screen shot shows how to use Solver to find the optimal (very best) solution to Example 4. Click on **Solve**, and the solution will automatically appear in the Transportation Table, yielding a cost of $105,700.

X USING EXCEL OM

Excel OM's Aggregate Planning module is demonstrated in Program 13.2. Using data from Example 2, Program 13.2 provides input and some of the formulas used to compute the costs of regular time, overtime, subcontracting, holding, shortage, and increase or decrease in production. The user must provide the production plan for Excel OM to analyze.

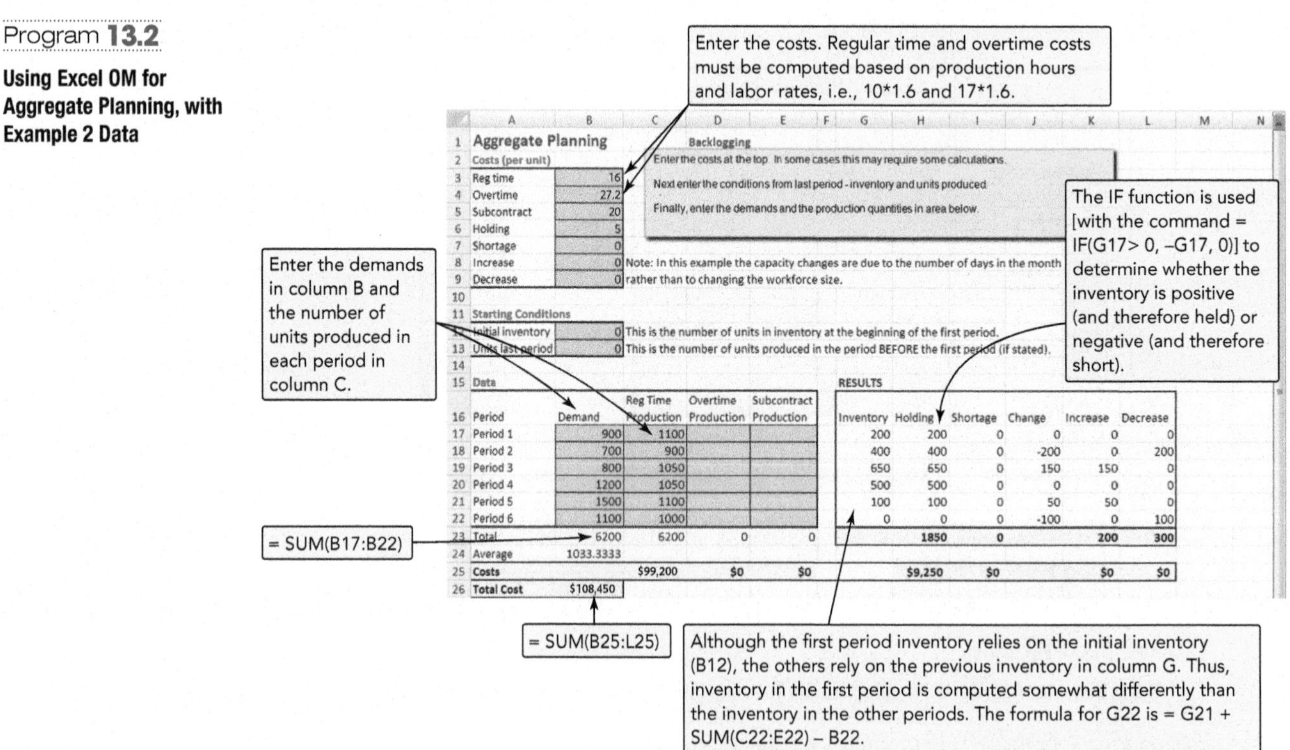

Period	Demand	Reg Time Production	Overtime Production	Subcontract Production	Inventory	Holding	Shortage	Change	Increase	Decrease
Period 1	900	1100			200	200	0	0	0	0
Period 2	700	900			400	400	0	-200	0	200
Period 3	800	1050			650	650	0	150	150	0
Period 4	1200	1050			500	500	0	0	0	0
Period 5	1500	1100			100	100	0	50	50	0
Period 6	1100	1000			0	0	0	-100	0	100
Total	6200	6200	0	0		1850	0		200	300
Average	1033.3333									
Costs		$99,200	$0	$0		$9,250	$0		$0	$0
Total Cost	$108,450									

Program 13.2
Using Excel OM for Aggregate Planning, with Example 2 Data

P USING POM FOR WINDOWS

The POM for Windows Aggregate Planning module performs aggregate or production planning for up to 90 time periods. Given a set of demands for future periods, you can try various plans to determine the lowest-cost plan based on holding, shortage, production, and changeover costs. Four methods are available for planning. More help is available on each after you choose the method. See Appendix II for further details.

Solved Problems

SOLVED PROBLEM 13.1

The roofing supplies manufacturer described in Examples 1 to 4 of this chapter wishes to consider yet a fourth planning strategy (plan 4). This one maintains a constant workforce of eight people and uses overtime whenever necessary to meet demand. Use the information found in Table 13.3. Again, assume beginning and ending inventories are equal to zero.

SOLUTION

Employ eight workers and use overtime when necessary. Note that carrying costs will be encountered in this plan.

MONTH	PRODUCTION DAYS	PRODUCTION AT 40 UNITS PER DAY	BEGINNING-OF-MONTH INVENTORY	FORECAST DEMAND THIS MONTH	OVERTIME PRODUCTION NEEDED	ENDING INVENTORY
Jan.	22	880	—	900	20 units	0 units
Feb.	18	720	0	700	0 units	20 units
Mar.	21	840	20	800	0 units	60 units
Apr.	21	840	60	1,200	300 units	0 units
May	22	880	0	1,500	620 units	0 units
June	20	800	0	1,100	300 units	0 units
					1,240 units	80 units

$$\text{Carrying cost totals} = 80 \text{ units} \times \$5/\text{unit}/\text{month} = \$400$$

Regular pay:

$$8 \text{ workers} \times \$80/\text{day} \times 124 \text{ days} = \$79,360$$

Overtime pay:

To produce 1,240 units at overtime rate requires $1,240 \times 1.6 \text{ hours/unit} = 1,984 \text{ hours}$.

$$\text{Overtime cost} = \$17/\text{hour} \times 1,984 \text{ hours} = \$33,728$$

Plan 4

COSTS (WORKFORCE OF 8 PLUS OVERTIME)		
Carrying cost	$ 400	(80 units carried × $5/unit)
Regular labor	79,360	(8 workers × $80/day × 124 days)
Overtime	33,728	(1,984 hours × $17/hour)
Hiring or firing	0	
Subcontracting	0	
Total costs	$113,488	

Plan 2, at $105,152, is still preferable.

SOLVED PROBLEM 13.2

A Dover, Delaware, plant has developed the accompanying supply, demand, cost, and inventory data. The firm has a constant workforce and meets all its demand. Allocate production capacity to satisfy demand at a minimum cost. What is the cost of this plan?

Supply Capacity Available (units)

PERIOD	REGULAR TIME	OVERTIME	SUBCONTRACT
1	300	50	200
2	400	50	200
3	450	50	200

Demand Forecast

PERIOD	DEMAND (UNITS)
1	450
2	550
3	750

Other data

Initial inventory	50 units
Regular-time cost per unit	$50
Overtime cost per unit	$65
Subcontract cost per unit	$80
Carrying cost per unit per period	$ 1
Backorder cost per unit per period	$ 4

SOLUTION

SUPPLY FROM		DEMAND FOR *Period 1*	*Period 2*	*Period 3*	*Unused Capacity (dummy)*	TOTAL CAPACITY AVAILABLE (supply)
Beginning inventory		[0] 50	[1]	[2]	[0]	50
Period 1	*Regular time*	[50] 300	[51]	[52]	[0]	300
	Overtime	[65] 50	[66]	[67]	[0]	50
	Subcontract	[80] 50	[81]	[82]	[0] 150	200
Period 2	*Regular time*	[54]	[50] 400	[51]	[0]	400
	Overtime	[69]	[65] 50	[66]	[0]	50
	Subcontract	[84]	[80] 100	[81] 50	[0] 50	200
Period 3	*Regular time*	[58]	[54]	[50] 450	[0]	450
	Overtime	[73]	[69]	[65] 50	[0]	50
	Subcontract	[88]	[84]	[80] 200	[0]	200
TOTAL DEMAND		450	550	750	200	1,950

Cost of plan:

$$\begin{aligned}
&\text{Period 1: } 50(\$0)+300(\$50)+50(\$65)+50(\$80) = \$22{,}250 \\
&\text{Period 2: } 400(\$50)+50(\$65)+100(\$80) = \$31{,}250 \\
&\text{Period 3: } 50(\$81)+450(\$50)+50(\$65)+200(\$80) = \underline{\$45{,}800^*} \\
&\text{Total cost } \$99{,}300
\end{aligned}$$

*Includes 50 units of subcontract and carrying cost.

Problems

Note: **PX** means the problem may be solved with POM for Windows and/or Excel OM.

Problems 13.1–13.22 relate to Methods for Aggregate Planning

• **13.1** Prepare a graph of the monthly forecasts and average forecast demand for Chicago Paint Corp., a manufacturer of specialized paint for artists.

MONTH	PRODUCTION DAYS	DEMAND FORECAST
January	22	1,000
February	18	1,100
March	22	1,200
April	21	1,300
May	22	1,350
June	21	1,350
July	21	1,300
August	22	1,200
September	21	1,100
October	22	1,100
November	20	1,050
December	20	900

•• **13.2** Develop another plan for the Mexican roofing manufacturer described in Examples 1 to 4 and Solved Problem 13.1.

a) For this plan, plan 5, the firm wants to maintain a constant workforce of six, using subcontracting to meet remaining demand. Is this plan preferable?
b) The same roofing manufacturer in Examples 1 to 4 and Solved Problem 13.1 has yet a sixth plan. A constant workforce of seven is selected, with the remainder of demand filled by subcontracting.
c) Is this better than plans 1–5? **PX**

••• **13.3** The president of Hill Enterprises, Terri Hill, projects the firm's aggregate demand requirements over the next 8 months as follows:

Jan.	1,400	May	2,200
Feb.	1,600	June	2,200
Mar.	1,800	July	1,800
Apr.	1,800	Aug.	1,800

Her operations manager is considering a new plan, which begins in January with 200 units on hand. Stockout cost of lost sales is $100 per unit. Inventory holding cost is $20 per unit per month. Ignore any idle-time costs. The plan is called plan A.

Plan A: Vary the workforce level to execute a strategy that produces the quantity demanded in the *prior* month. The December demand and rate of production are both 1,600 units per month. The cost of hiring additional workers is $5,000 per 100 units. The cost of laying off workers is $7,500 per 100 units. Evaluate this plan. **PX**

Note: Both hiring and layoff costs are incurred in the month of the change. For example, going from 1,600 in January to 1,400 in February incurs a cost of layoff for 200 units in February.

•• **13.4** Using the information in Problem 13.3, develop plan B. Produce at a constant rate of 1,400 units per month, which will meet minimum demands. Then use subcontracting, with additional units at a premium price of $75 per unit. Evaluate this plan by computing the costs for January through August. **PX**

•• **13.5** Hill is now considering plan C: Keep a stable workforce by maintaining a constant production rate equal to the average requirements and allow varying inventory levels. Beginning inventory, stockout costs, and holding costs are provided in Problem 13.3.

Plot the demand with a graph that also shows average requirements. Conduct your analysis for January through August. **PX**

••• **13.6** Hill's operations manager (see Problems 13.3 through 13.5) is also considering two mixed strategies for January–August: Produce in overtime or subcontracting only when there is no inventory.

- *Plan D:* Keep the current workforce stable at producing 1,600 units per month. Permit a maximum of 20% overtime at an additional cost of $50 per unit. A warehouse now constrains the maximum allowable inventory on hand to 400 units or less.
- *Plan E:* Keep the current workforce, which is producing 1,600 units per month, and subcontract to meet the rest of the demand. **PX**

Evaluate plans D and E and make a recommendation.

Note: Do not produce in overtime if production or inventory are adequate to cover demand.

••• **13.7** Consuelo Chua, Inc., is a Columbian disk drive manufacturer in need of an aggregate plan for July through December. The company has gathered the following data:

COSTS	
Holding cost	$8/disk/month
Subcontracting	$80/disk
Regular-time labor	$12/hour
Overtime labor	$18/hour for hours above 8 hours/worker/day
Hiring cost	$40/worker
Layoff cost	$80/worker

DEMAND*					
July	Aug.	Sept.	Oct.	Nov.	Dec.
400	500	550	700	800	700

*No costs are incurred for unmet demand, but unmet demand (backorders) must be handled in the following period. If half or more of a worker is needed, round up.

OTHER DATA	
Current workforce (June)	8 people
Labor-hours/disk	4 hours
Workdays/month	20 days
Beginning inventory	150 disks**
No requirement for ending inventory	0 disks

**Note that there is no holding cost for June.

What will each of the two following strategies cost?

a) Vary the workforce so that production meets demand. Chua had eight workers on board in June.
b) Vary overtime only and use a constant workforce of eight.
c) What action should management take to reduce costs? **PX**

•• **13.8** You manage a consulting firm down the street from Consuelo Chua, Inc., and to get your foot in the door, you have told Ms. Chua (see Problem 13.7) that you can do a better job at aggregate planning than her current staff. She said, "Fine. You do that, and you have a one-year contract." You now have to make good on your boast using the data in Problem 13.7. You decide to hire 5 workers in August and 5 more in October. Your results?

••• **13.9** The S&OP team at Kansas Furniture, led by David Angelow, has received the following estimates of demand requirements:

July	Aug.	Sept.	Oct.	Nov.	Dec.
1,000	1,200	1,400	1,800	1,800	1,800

a) Assuming one-time stockout costs for lost sales of $100 per unit, inventory carrying costs of $25 per unit per month, and zero beginning and ending inventory, evaluate these two plans on an *incremental* cost basis:
   - *Plan A:* Produce at a steady rate (equal to minimum requirements) of 1,000 units per month and subcontract additional units at a $60 per unit premium cost.
   - *Plan B:* Vary the workforce, to produce the prior month's demand. The firm produced 1,300 units in June. The cost of hiring additional workers is $3,000 per 100 units produced. The cost of layoffs is $6,000 per 100 units cut back.

   Note: Both hiring and layoff costs are incurred in the month of the change, (i.e., going from production of 1,300 in July to 1,000 in August requires a layoff [and related costs] of 300 units in August, just as going from production of 1,000 in August to 1,200 in September requires hiring [and related costs] of 200 units in September). **PX**
b) Which plan is best and why?

••• **13.10** The S&OP team (see Problem 13.9) is considering two more mixed strategies. Using the data in Problem 13.9, compare plans C and D with plans A and B and make a recommendation.

- *Plan C:* Keep the current workforce steady at a level producing 1,300 units per month. Subcontract the remainder to meet demand. Assume that 300 units remaining from June are available in July.
- *Plan D:* Keep the current workforce at a level capable of producing 1,300 units per month. Permit a maximum of 20% overtime at a premium of $40 per unit. Assume that warehouse limitations permit no more than a 180-unit carryover from month to month. This plan means that any time inventories reach 180, the plant is kept idle. Idle time per unit is $60. Any additional needs are subcontracted at a cost of $60 per incremental unit.

••• **13.11** Deb Bishop Health and Beauty Products has developed a new shampoo, and you need to develop its aggregate schedule. The cost accounting department has supplied you the costs relevant to the aggregate plan, and the marketing department has provided a four-quarter forecast. All are shown as follows:

QUARTER	FORECAST
1	1,400
2	1,200
3	1,500
4	1,300

COSTS	
Previous quarter's output	1,500 units
Beginning inventory	0 units
Stockout cost for backorders	$50 per unit
Inventory holding cost	$10 per unit for every unit held at the end of the quarter
Hiring workers	$40 per unit
Layoff workers	$80 per unit
Unit cost	$30 per unit
Overtime	$15 extra per unit
Subcontracting	Not available

Your job is to develop an aggregate plan for the next four quarters.

a) First, try hiring and layoffs (to meet the forecast) as necessary.
b) Then try a plan that holds employment steady.
c) Which is the more economical plan for Deb Bishop Health and Beauty Products? **PX**

••• **13.12** Southeast Soda Pop, Inc., has a new fruit drink for which it has high hopes. John Mittenthal, the production planner, has assembled the following cost data and demand forecast:

QUARTER	FORECAST
1	1,800
2	1,100
3	1,600
4	900

COSTS/OTHER DATA
Previous quarter's output = 1,300 cases
Beginning inventory = 0 cases
Stockout cost = $150 per case
Inventory holding cost = $40 per case at end of quarter
Hiring employees = $40 per case
Terminating employees = $80 per case
Subcontracting cost = $60 per case
Unit cost on regular time = $30 per case
Overtime cost = $15 extra per case
Capacity on regular time = 1,800 cases per quarter

John's job is to develop an aggregate plan. The three initial options he wants to evaluate are:

- *Plan A:* a strategy that hires and fires personnel as necessary to meet the forecast.
- *Plan B:* a level strategy.
- *Plan C:* a level strategy that produces 1,200 cases per quarter and meets the forecast demand with inventory and subcontracting.

a) Which strategy is the lowest-cost plan?
b) If you are John's boss, the VP for operations, which plan do you implement and why? **PX**

•• **13.13** Ram Roy's firm has developed the following supply, demand, cost, and inventory data. Allocate production capacity to meet demand at a minimum cost using the transportation method. What is the cost? Assume that the initial inventory has no holding cost in the first period and backorders are not permitted.

Supply Available

PERIOD	REGULAR TIME	OVERTIME	SUBCONTRACT	DEMAND FORECAST
1	30	10	5	40
2	35	12	5	50
3	30	10	5	40

Initial inventory	20 units
Regular-time cost per unit	$100
Overtime cost per unit	$150
Subcontract cost per unit	$200
Carrying cost per unit per month	$ 4

PX

•• **13.14** Jerusalem Medical Ltd., an Israeli producer of portable kidney dialysis units and other medical products, develops a 4-month aggregate plan. Demand and capacity (in units) are forecast as follows:

CAPACITY SOURCE	MONTH 1	MONTH 2	MONTH 3	MONTH 4
Labor				
Regular time	235	255	290	300
Overtime	20	24	26	24
Subcontract	12	15	15	17
Demand	255	294	321	301

The cost of producing each dialysis unit is $985 on regular time, $1,310 on overtime, and $1,500 on a subcontract. Inventory carrying cost is $100 per unit per month. There is to be no beginning or ending inventory in stock and backorders are not permitted. Set up a production plan that minimizes cost using the transportation method. **PX**

•• **13.15** The production planning period for flat-screen monitors at Louisiana's Rao Electronics, Inc., is 4 months. Cost data are as follows:

Regular-time cost per monitor	$ 70
Overtime cost per monitor	$110
Subcontract cost per monitor	$120
Carrying cost per monitor per month	$ 4

For each of the next 4 months, capacity and demand for flat-screen monitors are as follows:

	PERIOD			
	MONTH 1	MONTH 2	MONTH 3[a]	MONTH 4
Demand	2,000	2,500	1,500	2,100
Capacity				
Regular time	1,500	1,600	750	1,600
Overtime	400	400	200	400
Subcontract	600	600	600	600

[a]Factory closes for 2 weeks of vacation.

CEO Mohan Rao expects to enter the planning period with 500 monitors in stock. Backordering is not permitted (meaning, for example, that monitors produced in the second month cannot be used to cover first month's demand). Develop a production plan that minimizes costs using the transportation method. **PX**

•• **13.16** A large St. Louis feed mill, Robert Orwig Processing, prepares its 6-month aggregate plan by forecasting demand for 50-pound bags of cattle feed as follows: January, 1,000 bags; February, 1,200; March, 1,250; April, 1,450; May, 1,400; and June, 1,400. The feed mill plans to begin the new year with no inventory left over from the previous year, and backorders are not permitted. It projects that capacity (during regular hours) for producing bags of feed will remain constant at 800 until the end of April, and then increase to 1,100 bags per month when a planned expansion is completed on May 1. Overtime capacity is set at 300 bags per month until the expansion, at which time it will increase to 400 bags per month. A friendly competitor in Sioux City, Iowa, is also available as a backup source to meet demand—but can provide only 500 bags total during the 6-month period. Develop a 6-month production plan for the feed mill using the transportation method.

Cost data are as follows:

Regular-time cost per bag (until April 30)	$12.00
Regular-time cost per bag (after May 1)	$11.00
Overtime cost per bag (during entire period)	$16.00
Cost of outside purchase per bag	$18.50
Carrying cost per bag per month	$ 1.00

PX

•• **13.17** Yu Amy Xia has developed a specialized airtight vacuum bag to extend the freshness of seafood shipped to restaurants. She has put together the following demand cost data:

QUARTER	FORECAST (UNITS)	REGULAR TIME	OVERTIME	SUB-CONTRACT
1	500	400	80	100
2	750	400	80	100
3	900	800	160	100
4	450	400	80	100

Initial inventory = 250 units	Subcontracting cost = $2.00/unit
Regular time cost = $1.00/unit	Carrying cost = $0.20/unit/quarter
Overtime cost = $1.50/unit	Backorder cost = $0.50/unit/quarter

Yu decides that the initial inventory of 250 units will incur the $0.20/unit cost from each prior quarter (unlike the situation in most other companies, where a 0 unit cost is assigned).

a) Find the optimal plan using the transportation method.
b) What is the cost of the plan?
c) Does any regular time capacity go unused? If so, how much in which periods?
d) What is the extent of backordering in units and dollars? PX

••• **13.18** José Martinez of El Paso has developed a polished stainless steel tortilla machine that makes it a "showpiece" for display in Mexican restaurants. He needs to develop a 5-month aggregate plan. His forecast of capacity and demand follows:

	MONTH				
	1	2	3	4	5
Demand	150	160	130	200	210
Capacity					
Regular	150	150	150	150	150
Overtime	20	20	10	10	10

Subcontracting: 100 units available over the 5-month period
Beginning inventory: 0 units
Ending inventory required: 20 units

COSTS	
Regular-time cost per unit	$100
Overtime cost per unit	$125
Subcontract cost per unit	$135
Inventory holding cost per unit per month	$ 3

Assume that backorders are not permitted. Using the transportation method, what is the total cost of the optimal plan? PX

•••• **13.19** Dwayne Cole, owner of a Florida firm that manufactures display cabinets, develops an 8-month aggregate plan. Demand and capacity (in units) are forecast as follows:

CAPACITY SOURCE (UNITS)	JAN.	FEB.	MAR.	APR.	MAY	JUNE	JULY	AUG.
Regular time	235	255	290	300	300	290	300	290
Overtime	20	24	26	24	30	28	30	30
Subcontract	12	16	15	17	17	19	19	20
Demand	255	294	321	301	330	320	345	340

The cost of producing each unit is $1,000 on regular time, $1,300 on overtime, and $1,800 on a subcontract. Inventory carrying cost is $200 per unit per month. There is no beginning or ending inventory in stock, and no backorders are permitted from period to period.

Let the production (workforce) vary by using regular time first, then overtime, and then subcontracting.

a) Set up a production plan that minimizes cost by producing exactly what the demand is each month. This plan allows no backorders or inventory. What is this plan's cost?
b) Through better planning, regular-time production can be set at exactly the same amount, 275 units, per month. If demand cannot be met there is no cost assigned to shortages and they will not be filled. Does this alter the solution?
c) If overtime costs per unit rise from $1,300 to $1,400, will your answer to (a) change? What if overtime costs then fall to $1,200?

•• **13.20** Consider the following aggregate planning problem for one quarter:

	REGULAR TIME	OVERTIME	SUB-CONTRACTING
Production capacity/month	1,000	200	150
Production cost/unit	$5	$7	$8

Assume that there is no initial inventory and a forecasted demand of 1,250 units in each of the 3 months. Carrying cost is $1 per unit per month. Solve this aggregate planning problem for Production Manager Victor Shi using the transportation method. PX

•• **13.21** Maple Leaf Foundry, owned by Ahmet Satir, produces cast-iron ingots according to a 3-month capacity plan. The cost of labor averages $100 per regular shift hour and $140 per overtime (O.T.) hour. Inventory carrying cost is thought to be $4 per labor-hour of inventory carried. There are 50 direct labor-hours of inventory left over from March. For the next 3 months, demand and capacity (in labor-hours) are as follows:

	CAPACITY		
MONTH	REGULAR LABOR (HOURS)	O.T. LABOR (HOURS)	DEMAND
April	2,880	355	3,000
May	2,780	315	2,750
June	2,760	305	2,950

Develop an aggregate plan for the 3-month period using the transportation method. PX

••• **13.22** The Tamara Walker Chemical Supply Company manufactures and packages expensive vials of mercury. Given the following demand, supply, cost, and inventory data, allocate production capacity to meet demand at minimum cost using the transportation method. A constant workforce is expected. Backorders are permitted. PX

	SUPPLY CAPACITY (IN UNITS)			
PERIOD	REGULAR TIME	OVERTIME	SUBCONTRACT	DEMAND (IN UNITS)
1	25	5	6	32
2	28	4	6	32
3	30	8	6	40
4	29	6	7	40

OTHER DATA	
Initial inventory	4 units
Ending inventory desired	3 units
Regular-time cost per unit	$2,000
Overtime cost per unit	$2,475
Subcontracting cost per unit	$3,200
Carrying cost per unit per period	$200
Backorder cost per unit per period	$600

Problems 13.23–13.24 relate to Aggregate Planning in Services

••• **13.23** Forrester and Cohen is a small accounting firm, managed by Joseph Cohen since the retirement in December of his partner Brad Forrester. Cohen and his three CPAs can together bill 640 hours per month. When Cohen or another accountant bills more than 160 hours per month, he or she gets an additional "overtime" pay of $62.50 for each of the extra hours: this is above and beyond the $5,000 salary each draws during the month. (Cohen draws the same base pay as his employees.) Cohen strongly discourages any CPA from working (billing) more than 240 hours in any given month. The demand for billable hours for the firm over the next 6 months is estimated below:

MONTH	ESTIMATE OF BILLABLE HOURS
Jan.	600
Feb.	500
Mar.	1,000

MONTH	ESTIMATE OF BILLABLE HOURS
Apr.	1,200
May	650
June	590

Cohen has an agreement with Forrester, his former partner, to help out during the busy tax season, if needed, for an hourly fee of $125. Cohen will not even consider laying off one of his colleagues in the case of a slow economy. He could, however, hire another CPA at the same salary, as business dictates.

a) Develop an aggregate plan for the 6-month period.
b) Compute the cost of Cohen's plan of using overtime and Forrester.
c) Should the firm remain as is, with a total of four CPAs?

•• **13.24** Refer to the CPA firm in Problem 13.23. In planning for next year, Cohen estimates that billable hours will increase by 10% in each of the 6 months. He therefore proceeds to hire a fifth CPA. The same regular time, overtime, and outside consultant (i.e., Forrester) costs still apply.

a) Develop the new aggregate plan and compute its costs.
b) Comment on the staffing level with five accountants. Was it a good decision to hire the additional accountant?

Problems 13.25–13.26 relate to Revenue Management

•• **13.25** Southeastern Airlines' daily flight from Atlanta to Charlotte uses a Boeing 737, with all-coach seating for 120 people. In the past, the airline has priced every seat at $140 for the one-way flight. An average of 80 passengers are on each flight. The variable cost of a filled seat is $25. Aysajan Eziz, the new operations manager, has decided to try a yield revenue approach, with seats priced at $80 for early bookings and at $190 for bookings within 1 week of the flight. He estimates that the airline will sell 65 seats at the lower price and 35 at the higher price. Variable cost will not change. Which approach is preferable to Mr. Eziz?

•• **13.26** The exclusive Swink Golf Driving Range has had a standard price of $15 per hour. The facility has 36 golfing stations, with average usage of 50%, 10 hours a day, 7 days a week. Morgan Swink, the owner, would like to enhance revenue. He proposes new pricing at $10 per hour on weekdays and $20 per hour on weekends. He estimates that weekday usage will increase to 60% and weekend usage will remain at 50%, even with the price increase. Variable cost is a consistent $3 per hour. Which strategy is better?

CASE STUDIES

Andrew-Carter, Inc.

Andrew-Carter, Inc. (A-C), is a major Canadian producer and distributor of outdoor lighting fixtures. Its products are distributed throughout South and North America and have been in high demand for several years. The company operates three plants to manufacture fixtures and distribute them to five distribution centers (warehouses).

During the present global slowdown, A-C has seen a major drop in demand for its products, largely because the housing market has declined. Based on the forecast of interest rates, the head of operations feels that demand for housing and thus for A-C's products will remain depressed for the foreseeable future. A-C is considering closing one of its plants, as it is now operating with a forecasted excess capacity of 34,000 units per week. The forecasted demands (left table) and capacities (right table), in units per week, for the coming year are as follows:

Warehouse 1	9,000 units
Warehouse 2	13,000
Warehouse 3	11,000
Warehouse 4	15,000
Warehouse 5	8,000

Plant 1, regular time	27,000 units
Plant 1, on overtime	7,000
Plant 2, regular time	20,000
Plant 2, on overtime	5,000
Plant 3, regular time	25,000
Plant 3, on overtime	6,000

If A-C shuts down any plants, its weekly costs will change because fixed costs will be lower for a nonoperating plant. Table 13.9 shows production costs at each plant, both variable at regular time and overtime, and fixed when operating and shut down. Table 13.10 shows distribution costs from each plant to each distribution center.

TABLE 13.9 Andrew-Carter, Inc., Variable Costs and Fixed Production Costs per Week

		FIXED COST PER WEEK	
PLANT	VARIABLE COST (PER UNIT)	OPERATING	NOT OPERATING
1, regular time	$2.80	$14,000	$6,000
1, overtime	3.52	—	—
2, regular time	2.78	12,000	5,000
2, overtime	3.48	—	—
3, regular time	2.72	15,000	7,500
3, overtime	3.42	—	—

TABLE 13.10 Andrew-Carter, Inc., Distribution Costs per Unit

	TO DISTRIBUTION CENTERS				
FROM PLANTS	W1	W2	W3	W4	W5
1	$.50	$.44	$.49	$.46	$.56
2	.40	.52	.50	.56	.57
3	.56	.53	.51	.54	.35

Source: Reprinted by permission of Professor Michael Ballot, University of the Pacific, Stockton, CA. Copyright © by Michael Ballot.

Discussion Questions

1. Evaluate the various configurations of operating and closed plants that will meet weekly demand. Determine which configuration minimizes total costs.
2. Discuss the implications of closing a plant.

Using Revenue Management to Set Orlando Magic Ticket Prices

Video Case

Revenue management was once the exclusive domain of the airline industry. But it has since spread its wings into the hotel business, auto rentals, and now even professional sports, with the San Francisco Giants, Boston Celtics, and Orlando Magic as leaders in introducing dynamic pricing into their ticketing systems. Dynamic pricing means looking at unsold tickets for every single game, every day, to see if the current ticket price for a particular seat needs to be lowered (because of slow demand) or raised (because of higher-than-expected demand).

Pricing can be impacted by something as simple as bad weather or by whether the team coming to play in the arena is on a winning streak or has just traded for a new superstar player. For example, a few years ago, a basketball star was traded in midseason to the Denver Nuggets; this resulted in an immediate runup in unsold ticket prices for the teams the Nuggets were facing on the road. Had the Nuggets been visiting the Orlando Magic 2 weeks after the trade and the Magic not raised prices, they would have been "leaving money on the table" (as shown in Figure 13.5).

As the Magic became more proficient in revenue management, they evolved from (1) setting the price for each seat at the start of the season and never changing it; to (2) setting the prices for each seat at season onset, based on the popularity of the opponent, the day of the week, and the time of season (see the *Video Case Study* in Chapter 4)—but keeping the prices frozen once the season began (see Table 13.11); to (3) pricing tickets based on projected demand, but adjusting them frequently to match market demand as the season progressed.

To track market demand, the Magic use listed prices on Stub Hub and other online ticket exchange services. The key is to sell out all 18,500 seats every home game, keeping the pressure on Anthony Perez, the director of business strategy, and Chris Dorso, the Magic's vice president of sales.

Perez and Dorso use every tool available to collect information on demand, including counting unique page views at the Ticketmaster Web site. If, for example, there are 5,000 page views for the Miami Heat game near Thanksgiving, it indicates enough demand that prices of unsold seats can be notched up. If there are only 150 Ticketmaster views for the Utah Jazz game 3 days later, there may not be sufficient information to make any changes yet.

TABLE 13.11 An Example of Variable Pricing for a $68 Terrace V Seat in Zone 103

OPPONENT POPULARITY RATING	NUMBER OF GAMES IN THIS CATEGORY	PRICE
Tier I	3	$187
Tier II	3	$170
Tier III	4	$ 85
Tier IV	6	$ 75
Tier V	14	$ 60
Tier VI	9	$ 44
Tier VII	6	$ 40
Average		$ 68

With a database of 650,000, the Magic can use e-mail blasts to react quickly right up to game day. The team may discount seat prices, offer other perks, or just point out that prime seats are still available for a game against an exciting opponent.

Discussion Questions*

1. After researching revenue (yield) management in airlines, describe how the Magic system differs from that of American or other airline carriers.
2. The Magic used its original pricing systems of several years ago and set the price for a Terrace V, Zone 103 seat at $68 per game. There were 230 such seats *not* purchased as part of season ticket packages and thus available to the public. If the team switched to the 7-price dynamic system (illustrated in Table 13.11), how would the profit-contribution for the 45-game season change? (Note that the 45-game season includes 4 preseason games.)
3. What are some concerns the team needs to consider when using dynamic pricing with frequent changes in price?

*You may wish to view the video that accompanies this case before addressing these questions.

Endnotes

1. Sources: *Fair Disclosure Wire* (January 17, 2008); *The Wall Street Journal* (July 14, 2006); *Fast Company* (January/February 2006); and **snapper.com**.
2. Sources: **TheWrap.com** (October 15, 2020); and **CBR.com** (October 6, 2020).

Bibliography

APICS. *2011 Sales and Operations Planning Practices and Challenges.* Chicago: APICS, 2011.

Bilginer, Ö., and F. Erhun. "Production and Sales Planning in Capacitated New Product Introductions." *Production and Operations Management* 24, no. 1 (January 2015): 42–53.

Buxey, G. "Aggregate Planning for Seasonal Demand." *International Journal of Operations & Production Management* 25, no. 11 (2005): 1083–1100.

Chen, F. "Salesforce Initiative, Market Information, and Production/Inventory Planning." *Management Science* 51, no. 1 (January 2005): 60–75.

Hotze, T. "Executing a Value-Chain Flow Plan with S&OP to Manage Unprecedented Supply and Demand Shocks." *Journal of Business Forecasting* 39, no. 4 (Winter 2020-2021): 30–34.

Jacobs, F. R., W. L. Berry, D. C. Whybark, and T. E. Vollmann. *Manufacturing Planning and Control for Supply Chain Management*, 6th ed. Burr Ridge, IL: Irwin, 2011.

Kimes, S. E., and G. M. Thompson. "Restaurant Revenue Management at Chevy's." *Decision Sciences* 35, no. 3 (Summer 2004): 371–393.

Metters, R., et al. "The 'Killer Application' of Revenue Management: Harrah's Cherokee Casino and Hotel." *Interfaces* 38, no. 3 (May–June 2008): 161–178.

Mukhopadhyay, S., S. Samaddar, and G. Colville. "Improving Revenue Management Decision Making for Airlines." *Decision Sciences* 38, no. 2 (May 2007): 309–327.

Palmatier, G. E., and C. Crum. *Enterprise Sales and Operations Planning: Synchronizing Demand, Supply and Resources for Peak Performance*. Boca Raton, FL: J. Ross Publishing, 2002.

Rasmi, S. A. B., and M. Türkay. *Aggregate Planning: Strategies, Models, and Analysis*. Heidelberg: Springer, 2021.

Strik, J. *Market Development and Aggregate Production Planning.* Saarbrucken, Germany: VDM Verlag, 2011.

Chapter 13 *Rapid* Review

Main Heading	Review Material	MyLab Operations Management
THE PLANNING PROCESS	■ *Long-range plans* develop policies and strategies related to location, capacity, products and process, supply chain, research, and capital investment. ■ *Intermediate planning* develops plans that match production to demand. ■ *Short-run planning* translates intermediate plans into weekly, daily, and hourly schedules.	Concept Questions: 1.1–1.6
SALES AND OPERATIONS PLANNING	■ **Sales and operation planning (S&OP)**—Balances resources and forecasted demand, aligning the organization's competing demands, from supply chain to final customer, while linking strategic planning with operations over all planning horizons. ■ **Aggregate planning**—An approach to determine the quantity and timing of production for the intermediate future (usually 3 to 18 months ahead). Four things are needed for aggregate planning: 1. A logical unit for measuring sales and output 2. A forecast of demand for a reasonable intermediate planning period in these aggregate terms 3. A method for determining the relevant costs 4. A model that combines forecasts and costs so that scheduling decisions can be made for the planning period	Concept Questions: 2.1–2.4
THE NATURE OF AGGREGATE PLANNING	Usually, *the objective of aggregate planning is to meet forecasted demand while minimizing cost over the planning period.* An aggregate plan looks at production *in the aggregate* (a family of products), not as a product-by-product breakdown. ■ **Disaggregation**—The process of breaking an aggregate plan into greater detail. ■ **Master production schedule**—A timetable that specifies what is to be made and when.	Concept Questions: 3.1–3.4
AGGREGATE PLANNING STRATEGIES	The basic aggregate planning capacity (production) options are: ■ *Changing inventory levels* ■ *Varying workforce size by hiring or layoffs* ■ *Varying production rates through overtime or idle time* ■ *Subcontracting* ■ *Using part-time workers* The basic aggregate planning demand options are: ■ *Influencing demand* ■ *Backordering during high-demand periods* ■ *Counterseasonal product and service mixing* ■ **Chase strategy**—A planning strategy that sets production equal to forecast demand. Many service organizations favor the chase strategy because the inventory option is difficult or impossible to adopt. ■ **Level scheduling**—Maintaining a constant output rate, production rate, or workforce level over the planning horizon. Level scheduling works well when demand is reasonably stable. ■ **Mixed strategy**—A planning strategy that uses two or more controllable variables to set a feasible production plan.	Concept Questions: 4.1–4.6
METHODS FOR AGGREGATE PLANNING	■ **Graphical techniques**—Aggregate planning techniques that work with a few variables at a time to allow planners to compare projected demand with existing capacity. Graphical techniques are trial-and-error approaches that do not guarantee an optimal production plan, but they require only limited computations. A *cumulative* graph displays visually how the forecast deviates from the average requirements. ■ **Transportation method of linear programming**—A way of solving for the optimal solution to an aggregate planning problem. The transportation method of linear programming is flexible in that it can specify regular and overtime production in each time period, the number of units to be subcontracted, extra shifts, and the inventory carryover from period to period. Transportation problems require that supply equals demand, so when it does not, a dummy column called "unused capacity" may be added. Costs of not using capacity are zero.	Concept Questions: 5.1–5.6 Problems: 13.1–13.22 Virtual Office Hours for Solved Problems: 13.1, 13.2 **ACTIVE MODEL 13.1**

Main Heading	Review Material	MyLab Operations Management
	Demand requirements are shown in the bottom row of a transportation table. Total capacity available (supply) is shown in the far-right column. In general, to complete a transportation table, allocate as much production as you can to a cell with the smallest cost, without exceeding the unused capacity in that row or demand in that column. If there is still some demand left in that row, allocate as much as you can to the next-lowest-cost cell. You then repeat this process for periods 2 and 3 (and beyond, if necessary). When you are finished, the sum of all your entries in a row must equal total row capacity, and the sum of all entries in a column must equal the demand for that period. The transportation method does not work when nonlinear or negative factors are introduced.	
AGGREGATE PLANNING IN SERVICES	Successful techniques for controlling the cost of labor in service firms include: 1. Accurate scheduling of labor-hours to ensure quick response to customer demand. 2. An on-call labor resource that can be added or deleted to meet unexpected demand. 3. Flexibility of individual worker skills that permits reallocation of available labor. 4. Flexibility in rate of output or hours of work to meet changing demand.	Concept Questions: 6.1–6.6 Problems: 13.23–13.24
REVENUE MANAGEMENT	■ **Revenue** (or **yield) management**—Capacity decisions that determine the allocation of resources to maximize revenue. Organizations that have *perishable inventory*, such as airlines, hotels, car rental agencies, and cruise lines, have the following shared characteristics that make revenue management of interest: 1. Service or product can be sold in advance of consumption 2. Fluctuating demand 3. Relatively fixed resource (capacity) 4. Segmentable demand 5. Low variable costs and high fixed costs To make revenue management work, the company needs to manage three issues: 1. *Multiple pricing structures.* 2. *Forecasts of the use and duration of the use.* 3. *Changes in demand.*	Concept Questions: 7.1–7.5 Problems: 13.25–13.26 **VIDEO 13.1** Using Revenue Management to Set Orlando Magic Ticket Prices
ADDITIONAL MYLAB OPERATIONS MANAGEMENT RESOURCES	✔ Video for Creating Your Own Excel Spreadsheets (Example 5) ✔ Additional Case Studies (Cornwell Glass & Southwestern University (G)) ✔ Southwestern University Case Studies are integrated in Chapters 3, 4, 6, 8, 12, and 13 and in Supplement 7 ✔ Multiple Choice Case Questions (Andrew-Carter, Inc.)	

Self Test

LO 13.1 The outputs from an S&OP process are:
- **a)** long-run plans.
- **b)** detail schedules.
- **c)** aggregate plans.
- **d)** revenue management plans.
- **e)** short-run plans.

LO 13.2 Aggregate planning is concerned with determining the quantity and timing of production in the:
- **a)** short term.
- **b)** intermediate term.
- **c)** long term.
- **d)** all of the above.

LO 13.3 Aggregate planning deals with a number of constraints. These typically are:
- **a)** job assignments, job ordering, dispatching, and overtime help.
- **b)** part-time help, weekly scheduling, and SKU production scheduling.
- **c)** subcontracting, employment levels, inventory levels, and capacity.
- **d)** capital investment, expansion or contracting capacity, and R&D.
- **e)** facility location, production budgeting, overtime, and R&D.

LO 13.4 Which of the following is not one of the graphical method steps?
- **a)** Determine the demand in each period.
- **b)** Determine capacity for regular time, overtime, and subcontracting each period.
- **c)** Find labor costs, hiring and layoff costs, and inventory holding costs.
- **d)** Construct the transportation table.
- **e)** Consider company policy that may apply to the workers or stock levels.
- **f)** Develop alternative plans and examine their total costs.

LO 13.5 When might a dummy column be added to a transportation table?
- **a)** When supply does not equal demand
- **b)** When overtime is greater than regular time
- **c)** When subcontracting is greater than regular time
- **d)** When subcontracting is greater than regular time plus overtime
- **e)** When production needs to spill over into a new period

LO 13.6 Revenue management requires management to deal with:
- **a)** multiple pricing structures.
- **b)** changes in demand.
- **c)** forecasts of use.
- **d)** forecasts of duration of use.
- **e)** all of the above.

Answers: LO 13.1. c; LO 13.2. b; LO 13.3. c; LO 13.4. d; LO 13.5. a; LO 13.6. e.

Material Requirements Planning (MRP) and ERP

CHAPTER 14

CHAPTER OUTLINE

Nautique Boat Company in Orlando, Florida, primarily delivers custom-ordered boats, each with over 1,000 unique SKUs. Like most repetitive manufacturers, Nautique uses an MRP system to maintain tight schedules and on-time delivery. MRP is designed to answer four questions: (1) What are we going to build? (2) What material do we need to build the items? (3) What material do we have? and (4) What is on order, and when will we receive it? Here a long-lead time item, an engine with the distinct specifications required for this particular boat, is being installed.

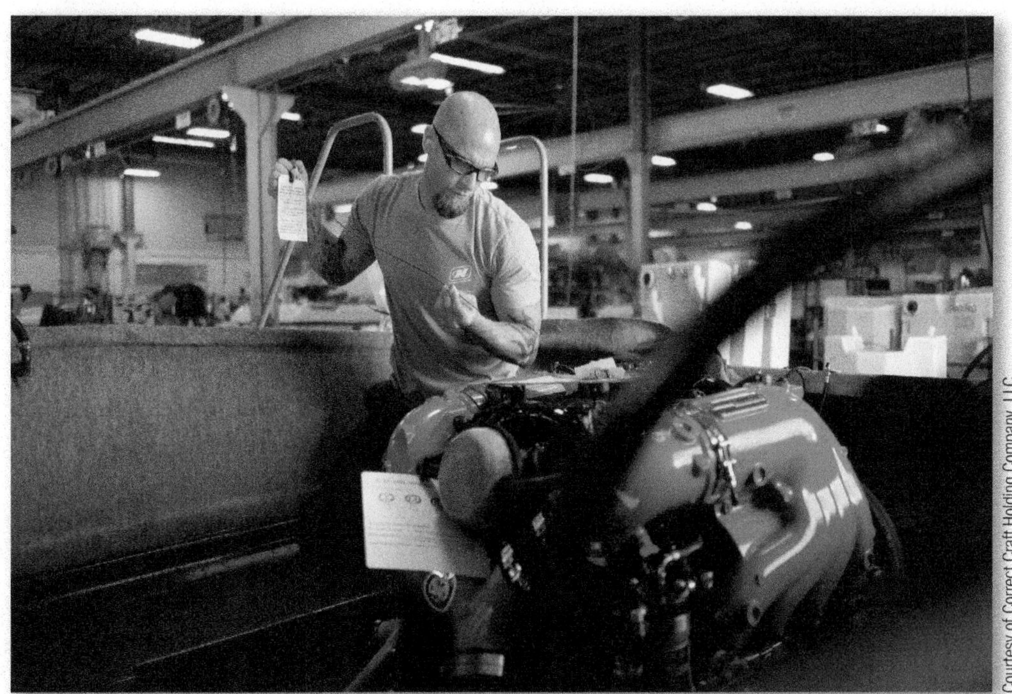

Courtesy of Correct Craft Holding Company, LLC

10 OM STRATEGY DECISIONS

- Design of Goods and Services
- Managing Quality
- Process Strategies
- Location Strategies
- Layout Strategies
- Human Resources
- Supply Chain Management
- *Inventory Management*
  - Independent Demand (Ch. 12)
  - ***Dependent Demand (Ch. 14)***
  - Lean Operations (Ch. 16)
- Scheduling
- Maintenance

GLOBAL COMPANY PROFILE
Wheeled Coach

MRP Provides a Competitive Advantage for Wheeled Coach

Wheeled Coach, headquartered in Winter Park, Florida, is the largest manufacturer of ambulances in the world. The $200 million firm is an international competitor that sells more than 25% of its vehicles to markets outside the U.S. Twelve major ambulance designs are produced on assembly lines (i.e., a repetitive process) at the Florida plant, using 18,000 different inventory items,

This cutaway of one ambulance interior indicates the complexity of the product, which for some rural locations may be the equivalent of a hospital emergency room in miniature. To complicate production, virtually every ambulance is custom ordered. This customization necessitates precise orders, excellent bills of materials, exceptional inventory control from supplier to assembly, and an MRP system that works.

Wheeled Coach Industries Incorporated

Wheeled Coach uses work cells to feed the assembly line. It maintains a complete carpentry shop (to provide interior cabinetry), a paint shop (to prepare, paint, and detail each vehicle), an electrical shop (to provide for the complex electronics in a modern ambulance), an upholstery shop (to make interior seats and benches), and as shown here, a metal fabrication shop (to construct the shell of the ambulance).

Wheeled Coach Industries Incorporated

of which 6,000 are manufactured and 12,000 purchased. Most of the product line is custom designed and assembled to meet the specific and often unique requirements demanded by the ambulance's application and customer preferences.

This variety of products and the nature of the process demand good material requirements planning (MRP). Effective use of an MRP system requires accurate bills of material and inventory records. The Wheeled Coach system provides daily updates and has reduced inventory by more than 30% in just two years.

Wheeled Coach insists that four key tasks be performed properly. First, the material plan must meet both the requirements of the master schedule and the capabilities of the production facility. Second, the plan must be executed as designed. Third, inventory investment must be minimized through effective "time-phased" material deliveries, consignment inventories, and a constant review of purchase methods. Finally, excellent record integrity must be maintained. Record accuracy is recognized as a fundamental ingredient of Wheeled Coach's successful MRP program. Its cycle counters are charged with material audits that not only correct errors but also investigate and correct problems.

Wheeled Coach Industries uses MRP as the catalyst for low inventory, high quality, tight schedules, and accurate records. Wheeled Coach has found competitive advantage via MRP.

On five parallel lines, ambulances move forward each day to the next workstation. The MRP system makes certain that just the materials needed at each station arrive overnight for assembly the next day.

Wheeled Coach Industries Incorporated

Here an employee is installing the wiring for an ambulance. There are an average of 15 miles of wire in a Wheeled Coach vehicle. This compares to 17 miles of wire in a sophisticated F-16 fighter jet.

Wheeled Coach Industries Incorporated

LEARNING OBJECTIVES

Dependent Demand

STUDENT TIP "Dependent demand" means the demand for one item is related to the demand for another item.

Wheeled Coach, the subject of the *Global Company Profile*, and many other firms have found important benefits in material requirements planning (MRP). These benefits include (1) better response to customer orders as the result of improved adherence to schedules, (2) faster response to market changes, (3) improved utilization of facilities and labor, and (4) reduced inventory levels. Better response to customer orders and to the market wins orders and market share. Better utilization of facilities and labor yields higher productivity and return on investment. Less inventory frees up capital and floor space for other uses. These benefits are the result of a strategic decision to use a *dependent* inventory scheduling system. Demand for every component of an ambulance is dependent.

Demand for items is dependent when the relationship between the items can be determined. Therefore, once management receives an order or makes a forecast for the final product, quantities for all components can be computed. All components are dependent items. The Boeing Aircraft operations manager who schedules production of one plane per week, for example, knows the requirements down to the last rivet. For any product, all components of that product are dependent demand items. *More generally, for any well-defined product for which a schedule can be established, dependent techniques should be used.*

When the requirements of MRP are met, dependent models are preferable to the models for independent demand (EOQ) described in Chapter 12.[1] Dependent models are better not only for manufacturers and distributors but also for a wide variety of firms from restaurants to hospitals. The dependent technique used in a production environment is called material requirements planning (MRP).

Material requirements planning (MRP)
A dependent demand technique that uses a bill-of-material, inventory, expected receipts, and a master production schedule to determine material requirements.

Because MRP provides such a clean structure for dependent demand, it has evolved as the basis for enterprise resource planning (ERP). ERP is an information system for identifying and planning the enterprise-wide resources needed to take, make, ship, and account for customer orders. We will discuss ERP in the latter part of this chapter.

Dependent Inventory Model Requirements

Effective use of dependent inventory models requires that the operations manager know the following:

1. Master production schedule (what is to be made and when)
2. Specifications or bill of material (materials and parts required to make the product)
3. Inventory availability (what is in stock)
4. Purchase orders outstanding (what is on order, also called expected receipts)
5. Lead times (how long it takes to get various components)

We now discuss each of these requirements in the context of material requirements planning.

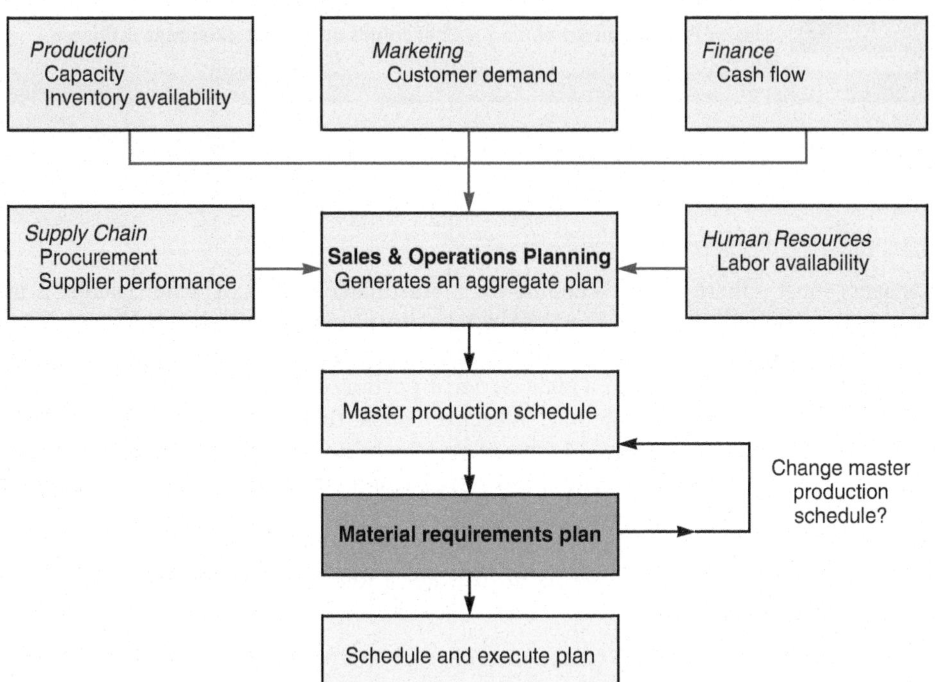

Figure **14.1**

The Planning Process

Master Production Schedule

A **master production schedule (MPS)** specifies what is to be made (e.g., the number of finished products or items) and when. The schedule must be in accordance with an aggregate plan. The aggregate plan sets the overall level of output in broad terms (e.g., product families, standard hours, or dollar volume). The plan, usually developed by the sales and operations planning team, includes a variety of inputs, including financial data, customer demand, labor availability, inventory fluctuations, supplier performance, and other considerations. Each of these inputs contributes in its own way to the aggregate plan, as shown in Figure 14.1.

Master production schedule (MPS)
A timetable that specifies what is to be made (usually finished goods) and when.

As the planning process moves from the aggregate plan to execution, each of the lower-level plans must be feasible. When one is not, feedback to the next higher level is required to make the necessary adjustment. One of the major strengths of MRP is its ability to determine precisely the feasibility of a schedule within aggregate capacity constraints. This planning process can yield excellent results. The aggregate plan sets the upper and lower bounds on the master production schedule.

The master production schedule tells us how to satisfy demand by specifying what items to make and when. It *disaggregates* the aggregate plan. While the *aggregate plan* (as discussed in Chapter 13) is established in gross terms such as families of products or tons of steel, the *master production schedule* is established in terms of specific products. Figure 14.2 shows the master production schedules for three stereo models that flow from the aggregate plan (sales & operations plan) for a family of stereo amplifiers.

Aggregate Plan (S&OP)
(Shows the total quantity of amplifiers)

Master Production Schedule
(Shows the specific type and quantity of amplifier to be produced)

Months	January				February			
Total amplifiers	1,500				1,200			
Weeks	1	2	3	4	5	6	7	8
240-watt amplifier	100		100		100		100	
150-watt amplifier		500		500		450		450
75-watt amplifier			300				100	

Figure **14.2**

The Aggregate Plan Is the Basis for Development of the Master Production Schedule

TABLE 14.1 Master Production Schedule for Chef John's Buffalo Chicken Mac & Cheese

GROSS REQUIREMENTS FOR CHEF JOHN'S BUFFALO CHICKEN MAC & CHEESE									
Day	6	7	8	9	10	11	12	13	14	and so on
Quantity	450		200	350	525		235	375		

Managers must adhere to the schedule for a reasonable length of time (usually a major portion of the production cycle—the time it takes to produce a product). Many organizations establish a master production schedule and establish a policy of not changing ("fixing") the near-term portion of the plan. This near-term portion of the plan is then referred to as the "fixed," "firm," or "frozen" schedule. Wheeled Coach, the subject of the *Global Company Profile* for this chapter, fixes the last 14 days of its schedule. Only changes farther out, beyond the fixed schedule, are permitted. The master production schedule is a "rolling" production schedule. For example, a fixed 7-week plan has an additional week added to it as each week is completed, so a 7-week fixed schedule is maintained. Note that the master production schedule is a statement of *what is to be produced;* it is *not* a forecast. The master schedule can be expressed in the following terms:

- As *customer orders* in a job shop (make-to-order) process (examples: print shops, machine shops, fine-dining restaurants)
- *As modules* in a repetitive (assemble-to-order or forecast) process (examples: Harley-Davidson motorcycles, TVs, fast-food restaurants)
- As *end items* in a continuous (stock-to-forecast) process (examples: steel, beer, bread, light bulbs, paper)

VIDEO 14.1
When 18,500 Orlando Magic Fans Come to Dinner

A master production schedule for Chef John's "Buffalo Chicken Mac & Cheese" at the Orlando Magic's Amway Center is shown in Table 14.1.

Bills of Material

Bill of material (BOM)
A listing of the components, their description, and the quantity of each required to make one unit of a product.

Defining what goes into a product may seem simple, but it can be difficult in practice. As we noted in Chapter 5, to aid this process, manufactured items are defined via a bill of material. A **bill of material (BOM)** is a list of quantities of components, ingredients, and materials required to make a product. Individual drawings describe not only physical dimensions but also any special processing as well as the raw material from which each part is made. Chef John's recipe for Buffalo Chicken Mac & Cheese specifies ingredients and quantities, just as Wheeled Coach has a full set of drawings for an ambulance. Both are bills of material (although we call one a recipe, and they do vary somewhat in scope).

VIDEO 14.2
MRP at Wheeled Coach Ambulances

One way a bill of material defines a product is by providing a product structure. Example 1 shows how to develop the product structure and "explode" it to reveal the requirements for each component. A bill of material for item A in Example 1 consists of items B and C. Items above any level are called *parents*; items below any level are called *components* or *children.* By convention, the top level in a BOM is the 0 level.

Example 1

DEVELOPING A PRODUCT STRUCTURE AND GROSS REQUIREMENTS

Speaker Kits, Inc., packages high-fidelity components for mail order. Components for the top-of-the-line speaker kit, "Awesome" (A), include 2 Bs and 3 Cs.

Each B consists of 2 Ds and 2 Es. Each of the Cs has 2 Fs and 2 Es. Each F includes 2 Ds and 1 G. It is an *awesome* sound system. (Most purchasers require hearing aids within 3 years, and at least one court case is pending because of structural damage to a men's dormitory.) As we can see, the demand for B, C, D, E, F, and G is completely dependent on the master production schedule for A—the Awesome speaker kits.

LO 14.1 *Develop* a product structure

APPROACH ▶ Given the preceding information, we construct a product structure and "explode" the requirements.

SOLUTION ▶ This structure has four levels: 0, 1, 2, and 3. There are four parents: A, B, C, and F. Each parent item has at least one level below it. Items B, C, D, E, F, and G are components because each item has at least one level above it. In this structure, B, C, and F are both parents and components. The number in parentheses indicates how many units of that particular item are needed to make the item immediately above it. Thus, $B_{(2)}$ means that it takes two units of B for every unit of A, and $F_{(2)}$ means that it takes two units of F for every unit of C.

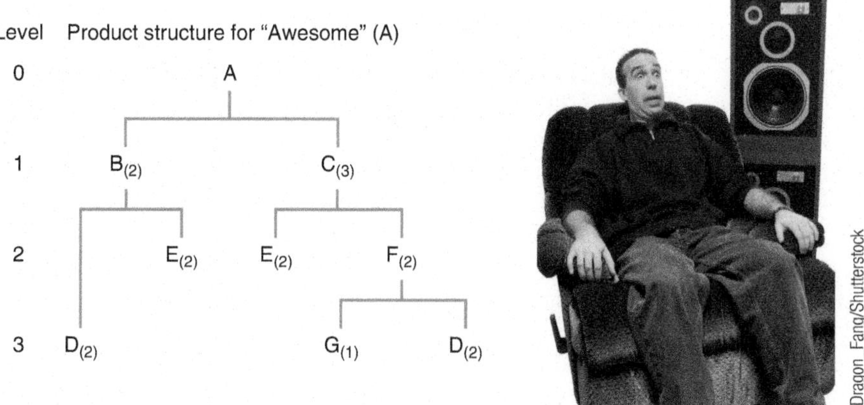

Once we have developed the product structure, we can determine the number of units of each item required to satisfy demand for a new order of 50 Awesome speaker kits. We "explode" the requirements as shown:

Part B:	$2 \times$ number of As =	$(2)(50) =$	100
Part C:	$3 \times$ number of As =	$(3)(50) =$	150
Part D:	$2 \times$ number of Bs $+ 2 \times$ number of Fs =	$(2)(100) + (2)(300) =$	800
Part E:	$2 \times$ number of Bs $+ 2 \times$ number of Cs =	$(2)(100) + (2)(150) =$	500
Part F:	$2 \times$ number of Cs =	$(2)(150) =$	300
Part G:	$1 \times$ number of Fs =	$(1)(300) =$	300

INSIGHT ▶ We now have a visual picture of the Awesome speaker kit requirements and knowledge of the quantities required. Thus, for 50 units of A, we will need 100 units of B, 150 units of C, 800 units of D, 500 units of E, 300 units of F, and 300 units of G.

LEARNING EXERCISE ▶ If there are 100 Fs in stock, how many Ds do you need? [Answer: 600.]

RELATED PROBLEMS ▶ 14.1–14.4, 14.5a,b, 14.13a,b, 14.14a, 14.18a,b, 14.20a,b

Bills of material not only specify requirements but also are useful for costing, and they can serve as a list of items to be issued to production or assembly personnel. When bills of material are used in this way, they are usually called *pick lists*.

Modular Bills Bills of material may be organized around product modules (see Chapter 5). *Modules* are not final products to be sold, but are components that can be produced and assembled into units. They are often major components of the final product or product options. Bills of material for modules are called **modular bills**. Modular bills are convenient because production scheduling and production are often facilitated by organizing around relatively few modules rather than a multitude of final assemblies. For instance,

Modular bills
Bills of material organized by major subassemblies or by product options.

a firm may make 138,000 different final products but may have only 40 modules that are mixed and matched to produce those 138,000 final products. The firm builds an aggregate production plan and prepares its master production schedule for the 40 modules, not the 138,000 configurations of the final product. This approach allows the MPS to be prepared for a reasonable number of items. The 40 modules can then be configured for specific orders at final assembly.

Planning bills (or kits)
Material groupings created in order to assign an artificial parent to a bill of material; also called "pseudo" bills.

Phantom bills of material
Bills of material for components, usually assemblies, that exist only temporarily; they are never inventoried.

Planning Bills and Phantom Bills Two other special kinds of bills of material are planning bills and phantom bills. Planning bills (sometimes called "pseudo" bills, or super bills) are created in order to assign an artificial parent to the bill of material. Such bills are used (1) when we want to group subassemblies so the number of items to be scheduled is reduced and (2) when we want to issue "kits" to the production department. For instance, it may not be efficient to issue inexpensive items such as washers and cotter pins with each of numerous subassemblies, so we call this a *kit* and generate a planning bill. The planning bill specifies the *kit* to be issued. Consequently, a planning bill may also be known as kitted material, or kit. Phantom bills of material are bills of material for components, usually subassemblies, that exist only temporarily. These components go directly into another assembly and are never inventoried. Therefore, components of phantom bills of material are coded to receive special treatment; lead times are zero, and they are handled as an integral part of their parent item. An example is a transmission shaft with gears and bearings assembly that is placed directly into a transmission.

Low-level coding
A number that identifies items at the lowest level at which they occur.

Low-Level Coding Low-level coding of an item in a BOM is necessary when identical items exist at various levels in the BOM. Low-level coding means that the item is coded at the lowest level at which it occurs. For example, item D in Example 1 is coded at the lowest level at which it is used. Item D could be coded as part of B and occur at level 2. However, because D is also part of F, and F is level 2, item D becomes a level-3 item. Low-level coding is a convention to allow easy computing of the requirements of an item.

Accurate Inventory Records

As we saw in Chapter 12, knowledge of what is in stock is the result of good inventory management. Good inventory management is an absolute necessity for an MRP system to work. If the firm does not exceed 99% record accuracy, then material requirements planning will not work.[2]

Lead time
In purchasing systems, the time between recognition of the need for an order and receiving it; in production systems, it is the order, wait, move, queue, setup, and run times for each component.

Purchase Orders Outstanding

Knowledge of outstanding orders exists as a by-product of well-managed purchasing and inventory-control departments. When purchase orders are executed, records of those orders and their scheduled delivery dates must be available to production personnel. Only with good purchasing data can managers prepare meaningful production plans and effectively execute an MRP system.

TABLE 14.2
Lead Times for Awesome Speaker Kits (As)

COMPONENT	LEAD TIME
A	1 week
B	2 weeks
C	1 week
D	1 week
E	2 weeks
F	3 weeks
G	2 weeks

Lead Times for Components

Once managers determine when products are needed, they determine when to acquire them. The time required to acquire (that is, purchase, produce, or assemble) an item is known as lead time. Lead time for a manufactured item consists of *move, setup*, and *assembly* or *run times* for each component. For a purchased item, the lead time includes the time between recognition of need for an order and when it is available for production.

When the bill of material for Awesome speaker kits (As), in Example 1, is turned on its side and modified by adding lead times for each component (see Table 14.2), we then have a *time-phased product structure*. Time in this structure is shown on the horizontal axis of Figure 14.3 with item A due for completion in week 8. Each component is then offset to accommodate lead times.

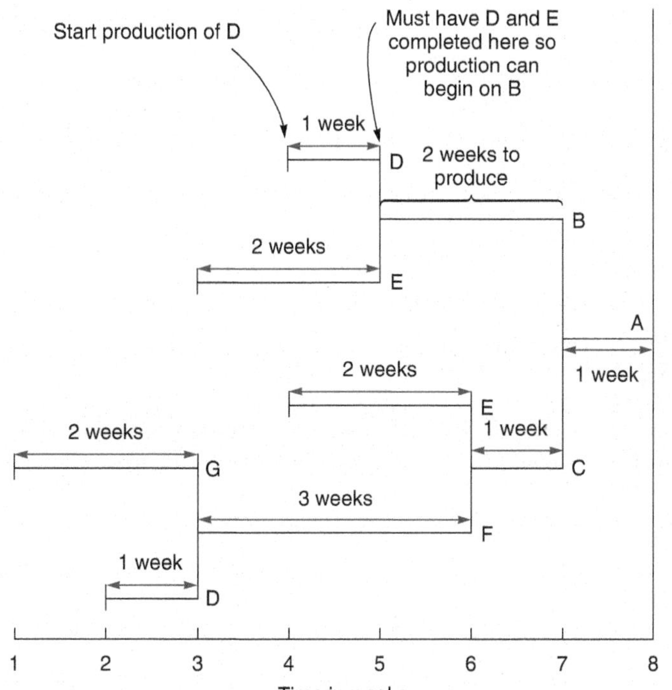

Figure 14.3

Time-Phased Product Structure

STUDENT TIP

This is a product structure on its side, with lead times.

MRP Structure

Although most MRP systems are computerized, the MRP procedure is straightforward, and we can illustrate a small one by hand. A master production schedule, a bill of material, inventory and purchase records, and lead times for each item are the ingredients of a material requirements planning system (see Figure 14.4).

Once these ingredients are available and accurate, the next step is to construct a gross material requirements plan. The **gross material requirements plan** is a schedule, as shown in Example 2. It combines a master production schedule (that requires one unit of A in week 8) and the time-phased schedule (Figure 14.3). It shows when an item must be ordered from suppliers if there is no inventory on hand or when the production of an item must be started to satisfy demand for the finished product by a particular date.

Gross material requirements plan

A schedule that shows the total demand for an item (prior to subtraction of on-hand inventory and scheduled receipts) and (1) when it must be ordered from suppliers, or (2) when production must be started to meet its demand by a particular date.

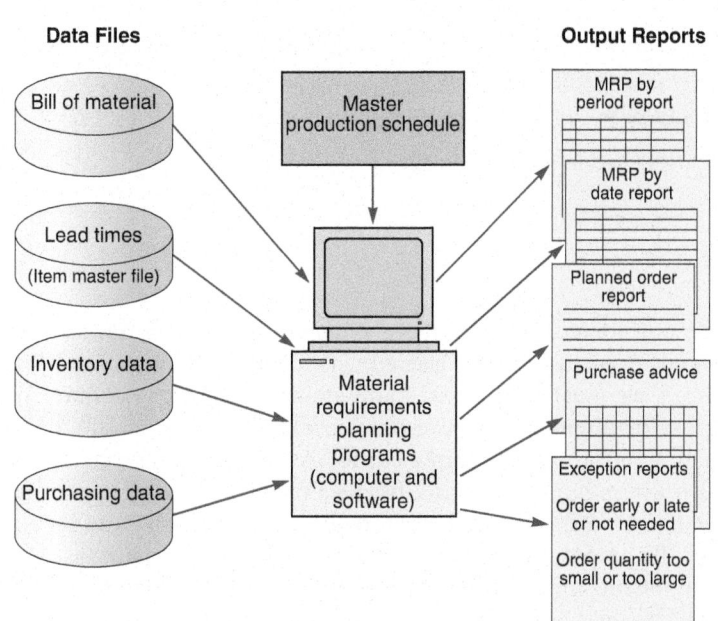

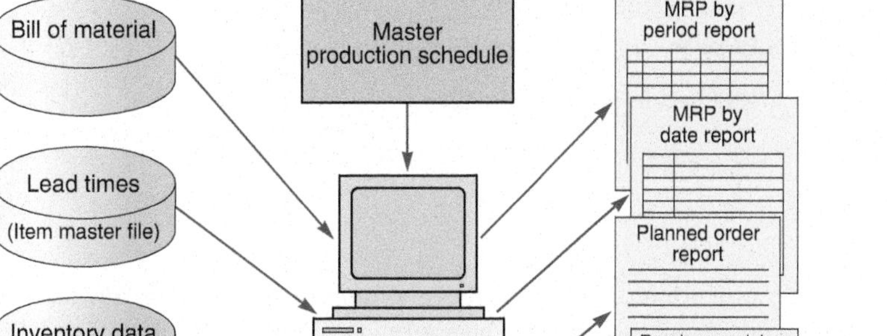

Figure 14.4

Structure of the MRP System

STUDENT TIP

MRP software programs are popular because manual approaches are slow and error prone.

Example 2

BUILDING A GROSS REQUIREMENTS PLAN

Each Awesome speaker kit (item A of Example 1) requires all the items in the product structure for A. Lead times are shown in Table 14.2.

APPROACH ▶ Using the information in Example 1 and Table 14.2, we construct the gross material requirements plan with a production schedule that will satisfy the demand of 50 units of A by week 8.

SOLUTION ▶ We prepare a schedule as shown in Table 14.3.

TABLE 14.3 Gross Material Requirements Plan for 50 Awesome Speaker Kits (As) with Order Release Dates Also Shown

	WEEK								
	1	2	3	4	5	6	7	8	LEAD TIME
A. Required date								50	
Order release date							50		1 week
B. Required date							100		
Order release date					100				2 weeks
C. Required date							150		
Order release date						150			1 week
E. Required date					200	300			
Order release date			200	300					2 weeks
F. Required date						300			
Order release date			300						3 weeks
D. Required date			600		200				
Order release date		600		200					1 week
G. Required date			300						
Order release date	300								2 weeks

LO 14.2 *Build* a gross requirements plan

You can interpret the gross material requirements shown in Table 14.3 as follows: If you want 50 units of A at week 8, you must start assembling A in week 7. Thus, in week 7, you will need 100 units of B and 150 units of C. These two items take 2 weeks and 1 week, respectively, to produce. Production of B, therefore, should start in week 5, and production of C should start in week 6 (lead time subtracted from the required date for these items). Working backward, we can perform the same computations for all of the other items. Because D and E are used in two different places in Awesome speaker kits, there are two entries in each data record.

INSIGHT ▶ The gross material requirements plan shows when production of each item should begin and end in order to have 50 units of A at week 8. Management now has an initial plan.

LEARNING EXERCISE ▶ If the lead time for G decreases from 2 weeks to 1 week, what is the new order release date for G? [Answer: 300 in week 2.]

RELATED PROBLEMS ▶ 14.6, 14.8, 14.10a, 14.11a, 14.20c

EXCEL OM Data File **Ch14Ex2.xls** can be found online.

Net requirements plan
The result of adjusting gross requirements for inventory on hand and scheduled receipts.

So far, we have considered *gross material requirements*, which assumes that there is no inventory on hand. A **net requirements plan** adjusts for on-hand inventory. When considering on-hand inventory, we must realize that many items in inventory contain subassemblies or parts. If the gross requirement for Awesome speaker kits (As) is 100 and there are 20 of those speakers on hand, the net requirement for As is 80 (that is, 100 – 20). However, each Awesome speaker kit on hand contains 2 Bs. As a result, the requirement for Bs drops by 40 Bs (20 A kits on hand × 2 Bs per A). Therefore, if inventory is on hand for a parent item, the requirements for the parent item and all its components decrease because each Awesome kit contains the components for lower-level items. Example 3 shows how to create a net requirements plan.

Example 3

DETERMINING NET REQUIREMENTS

LO 14.3 *Build* a net requirements plan

Speaker Kits, Inc., developed a product structure from a bill of material in Example 1. Example 2 developed a gross requirements plan. Given the following on-hand inventory, Speaker Kits, Inc., now wants to construct a net requirements plan. The gross requirement remains 50 units in week 8, and component requirements are as shown in the product structure in Example 1.

ITEM	ON HAND	ITEM	ON HAND
A	10	E	10
B	15	F	5
C	20	G	0
D	10		

Net Material Requirements Plan for 50 Units of Product A in Week 8. ***(The superscript is the source of the demand)***

Lot Size	Lead Time (weeks)	On Hand	Safety Stock	Allo-cated	Low-Level Code	Item Identi-fication			Week 1	2	3	4	5	6	7	8
Lot-for-Lot	1	10	—	—	0	A	Gross Requirements									50
							Scheduled Receipts									
							Projected On Hand	10	10	10	10	10	10	10	10	10
							Net Requirements									40
							Planned Order Receipts									40
							Planned Order Releases								40	
Lot-for-Lot	2	15	—	—	1	B	Gross Requirements								80[A]	
						2 × number of As = 80	Scheduled Receipts									
							Projected On Hand	15	15	15	15	15	15	15	15	
							Net Requirements								65	
							Planned Order Receipts								65	
							Planned Order Releases						65			
Lot-for-Lot	1	20	—	—	1	C	Gross Requirements								120[A]	
						3 × number of As = 120	Scheduled Receipts									
							Projected On Hand	20	20	20	20	20	20	20	20	
							Net Requirements								100	
							Planned Order Receipts								100	
							Planned Order Releases							100		
Lot-for-Lot	2	10	—	—	2	E	Gross Requirements						130[B]	200[C]		
						2 × number of Bs = 130 2 × number of Cs = 200	Scheduled Receipts									
							Projected On Hand	10	10	10	10	10	10			
							Net Requirements						120	200		
							Planned Order Receipts						120	200		
							Planned Order Releases				120	200				
Lot-for-Lot	3	5	—	—	2	F	Gross Requirements							200[C]		
						2 × number of Cs = 200	Scheduled Receipts									
							Projected On Hand	5	5	5	5	5	5	5		
							Net Requirements							195		
							Planned Order Receipts							195		
							Planned Order Releases				195					
Lot-for-Lot	1	10	—	—	3	D	Gross Requirements				390[F]		130[B]			
						2 × number of Bs = 130 2 × number of Fs = 390	Scheduled Receipts									
							Projected On Hand	10	10	10	10					
							Net Requirements				380		130			
							Planned Order Receipts				380		130			
							Planned Order Releases			380		130				
Lot-for-Lot	2	0	—	10	3	G	Gross Requirements				205[F]					
						1 × number of Fs + 10 allocated = 205	Scheduled Receipts									
							Projected On Hand				0					
							Net Requirements				205					
							Planned Order Receipts				205					
							Planned Order Releases		205							

APPROACH ▶ A net material requirements plan includes gross requirements, on-hand inventory, net requirements, planned order receipt, and planned order release for each item. We begin with A and work backward through the components.

SOLUTION ▶ Shown in the MRP format is the net material requirements plan for product A.

Constructing a net requirements plan is similar to constructing a gross requirements plan. Starting with item A, we work backward to determine net requirements for all items. To do these computations, we refer to the product structure, on-hand inventory, and lead times. The gross requirement for A is 50 units in week 8. Ten items are on hand; therefore, the net requirements and the scheduled **planned order receipt** are both 40 items in week 8. Because of the 1-week lead time, the **planned order release** is 40 items in week 7 (see the arrow connecting the order receipt and order release). Referring to week 7 and the product structure in Example 1, we can see that 80 (2 × 40) items of B and 120 (3 × 40) items of C are required in week 7 to have a total for 50 items of A in week 8. The letter superscripted A to the right of the gross figure for items B and C was generated as a result of the demand for the parent, A. Performing the same type of analysis for B and C yields the net requirements for D, E, F, and G. Note the on-hand inventory in row E in week 6 is zero. It is zero because the on-hand inventory (10 units) was used to make B in week 5. By the same token, the inventory for D was used to make F in week 3.

Planned order receipt
The quantity planned to be received at a future date.

Planned order release
The scheduled date for an order to be released.

INSIGHT ▶ Once a net requirement plan is completed, management knows the quantities needed, an ordering schedule, and a production schedule for each component.

LEARNING EXERCISE ▶ If the on-hand inventory quantity of component F is 95 rather than 5, how many units of G will need to be ordered in week 1? [Answer: 105 units.]

RELATED PROBLEMS ▶ 14.9, 14.10b, 14.11b, 14.12, 14.13c, 14.14b, 14.15a,b, 14.16a,b, 14.17, 14.19, 14.21

ACTIVE MODEL 14.1 This example is further illustrated in Active Model 14.1 found online.

EXCEL OM Data File **Ch14Ex3.xls** can be found online.

STUDENT TIP MRP gross requirements can combine multiple products, spare parts, and items sold directly.

Examples 2 and 3 considered only product A, the Awesome speaker kit, and its completion only in week 8. Fifty units of A were required in week 8. Normally, however, there is a demand for many products over time. For each product, management must prepare a master production schedule (as we saw earlier, in Table 14.1). Scheduled production of each product is added to the master schedule and ultimately to the net material requirements plan. Figure 14.5 shows how several product schedules, including requirements for components sold directly, can contribute to one gross material requirements plan.

Most inventory systems also note the number of units in inventory that have been assigned to specific future production but not yet used or issued from the stockroom. Such items are often referred to as *allocated* items. Allocated items increase requirements as shown for part G in Example 3, where gross requirements have been increased from 195 to 205 to reflect the 10 allocated items.

Figure 14.5
Several Schedules Contributing to a Gross Requirements Schedule for B
One B is in each A, and one B is in each S; in addition, 10 Bs sold directly are scheduled in week 1, and 10 more that are sold directly are scheduled in week 2.

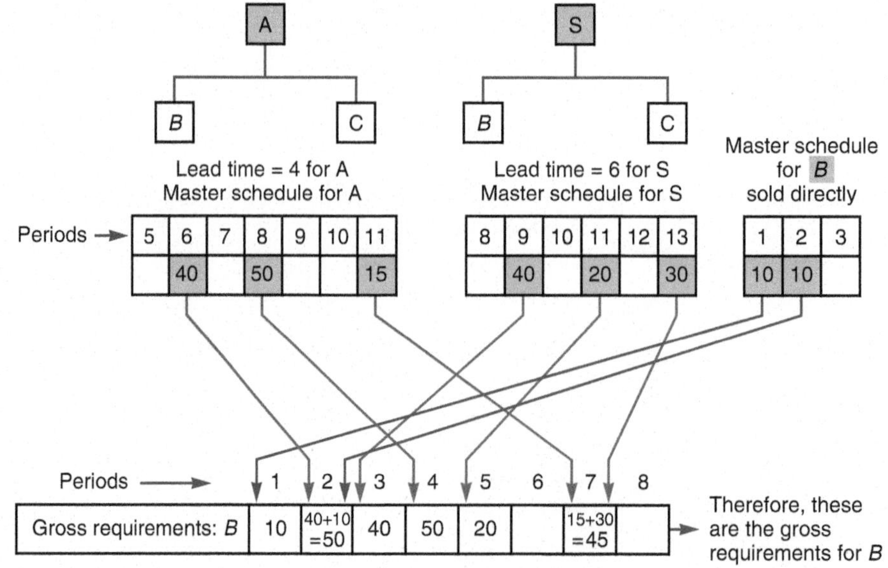

Lot-Sizing Techniques

An MRP system is an excellent way to perform production planning and determine net requirements. But net requirements still demand a decision about *how much and when* to order dependent items. This decision is called a **lot-sizing decision**. There are a variety of ways to determine lot sizes in an MRP system; commercial MRP software usually includes the choice of several lot-sizing techniques. We now review a few of them.

Lot-sizing decision
The process of, or techniques used in, determining lot size.

Lot-for-Lot In Example 3, we used a lot-sizing technique known as **lot-for-lot**, which produced exactly what was required. This decision is consistent with the objective of an MRP system, which is to meet the requirements of *dependent* demand. Thus, an MRP system should produce units only as needed, with no safety stock and no anticipation of further orders. When frequent orders are economical (i.e., when setup costs are low) and just-in-time inventory techniques implemented, lot-for-lot can be very efficient. However, when setup costs are significant, lot-for-lot can be expensive. Example 4 uses the lot-for-lot criteria and determines cost for 10 weeks of demand.

Lot-for-lot
A lot-sizing technique that generates exactly what is required to meet the plan.

Example 4

LOT SIZING WITH LOT-FOR-LOT

Speaker Kits, Inc., wants to compute its ordering and carrying cost of inventory on lot-for-lot criteria.

APPROACH ▶ With lot-for-lot, we order material only as it is needed. Once we have the cost of ordering (setting up), the cost of holding each unit for a given time period, and the production schedule, we can assign orders to our net requirements plan.

SOLUTION ▶ Speaker Kits has determined that, for **component B**, setup cost is $100 and holding cost is $1 per period. The production schedule, as reflected in net requirements for assemblies, is as follows:

MRP Lot Sizing: Lot-for-Lot Technique*

WEEK		1	2	3	4	5	6	7	8	9	10
Gross requirements		35	30	40	0	10	40	30	0	30	55
Scheduled receipts											
Projected on hand	35	35	0	0	0	0	0	0	0	0	0
Net requirements		0	30	40	0	10	40	30	0	30	55
Planned order receipts			30	40		10	40	30		30	55
Planned order releases		30	40		10	40	30		30	55	

*Holding costs = $1/unit/week; setup cost = $100; gross requirements average per week = 27; lead time = 1 week.

The lot-sizing solution using the lot-for-lot technique is shown in the table. The holding cost is zero as there is never any end-of-period inventory. (Inventory in the first period is used immediately and therefore has no holding cost.) But seven separate setups (one associated with each order) yield a total cost of $700. (Holding cost $= 0 \times 1 = 0$; ordering cost $= 7 \times 100 = 700$.)

INSIGHT ▶ When supply is reliable and frequent orders are inexpensive, but holding cost or obsolescence is high, lot-for-lot ordering can be very efficient.

LEARNING EXERCISE ▶ What is the impact on total cost if holding cost is $2 per period rather than $1? [Answer: Total holding cost remains zero, as no units are held from one period to the next with lot-for-lot.]

RELATED PROBLEMS ▶ 14.22, 14.25, 14.26a, 14.27a, 14.28b

This Nissan line in Smyrna, Tennessee, has little inventory because Nissan schedules to a razor's edge. At Nissan, MRP helps reduce inventory to world-class standards. World-class automobile assembly requires that purchased parts have a turnover of slightly more than once a day and that overall turnover approaches 150 times per year.

John Russell/AP Images

LO 14.4 *Determine* lot sizes for lot-for-lot, EOQ, and POQ

Economic Order Quantity (EOQ) We now extend our discussion of EOQ in Chapter 12 to use it as a lot-sizing technique for MRP systems. As we indicated there, EOQ is useful when we have relatively constant demand. However, demand may change every period in MRP systems. Therefore, EOQ lot sizing often does not perform well in MRP. Operations managers should take advantage of demand information when it is known, rather than assuming a constant demand. EOQ is used to do lot sizing in Example 5 for comparison purposes.

Example 5

LOT SIZING WITH EOQ

With a setup cost of \$100 and a holding cost per week of \$1, Speaker Kits, Inc., wants to examine its cost for **component B**, with lot sizes based on an EOQ criteria.

APPROACH ▶ Using the same cost and production schedule as in Example 4, we determine net requirements and EOQ lot sizes.

SOLUTION ▶ Ten-week usage equals a gross requirement of 270 units; therefore, weekly usage equals 27, and 52 weeks (annual usage) equals 1,404 units. From Chapter 12, the EOQ model is:

$$Q^* = \sqrt{\frac{2DS}{H}}$$

where D = annual usage = 1,404
S = setup cost = \$100
H = holding (carrying) cost, on an annual basis per unit
= \$1 × 52 weeks = \$52

$$Q^* = 73 \text{ units}$$

Therefore, place an order of 73 units, as necessary, to avoid a stockout.

MRP Lot Sizing: EOQ Technique*

WEEK		1	2	3	4	5	6	7	8	9	10
Gross requirements		35	30	40	0	10	40	30	0	30	55
Scheduled receipts											
Projected on hand	35	35	0	43	3	3	66	26	69	69	39
Net requirements		0	30	0	0	7	0	4	0	0	16
Planned order receipts			73			73		73			73
Planned order releases		73			73		73			73	

*Holding costs = \$1/unit/week; setup cost = \$100; gross requirements average per week = 27; lead time = 1 week.

For the 10-week planning period:

$$\text{Holding cost} = 375 \text{ units} \times \$1 = \$375 \text{ (includes 57 remaining at the end of week 10)}$$

$$\text{Ordering cost} = 4 \times \$100 = \$400$$

$$\text{Total} = \$375 + \$400 = \$775$$

INSIGHT ▶ EOQ can be a reasonable lot-sizing technique when demand is relatively constant. However, notice that actual holding cost will vary substantially depending on the rate of actual usage. If any stockouts had occurred, these costs too would need to be added to our actual EOQ cost of $775.

LEARNING EXERCISE ▶ What is the impact on total cost if holding cost is $2 per period rather than $1? [Answer: The EOQ quantity becomes 52, the theoretical annual total cost becomes $5,404, and the 10-week cost is $1,039 ($5,404 × (10/52).]

RELATED PROBLEMS ▶ 14.23, 14.25, 14.26b, 14.27b, 14.28a

Periodic Order Quantity Periodic order quantity (POQ) is a lot-sizing technique that orders the quantity needed during a predetermined time between orders, such as every 3 weeks. We define the *POQ interval* as the EOQ divided by the average demand per period (e.g., one week).[3] The POQ is the order quantity that covers the specific demand for that interval. *Each order quantity is recalculated at the time of the order release*, never leaving extra inventory. An application of POQ is shown in Example 6.

Periodic order quantity (POQ)
An inventory-ordering technique that issues orders on a predetermined time interval, with the order quantity covering the total of the interval's requirements.

Example 6

LOT SIZING WITH POQ

With a setup cost of $100 and a holding cost per week of $1, Speaker Kits, Inc., wants to examine its cost for **component B**, with lot sizes based on POQ.

APPROACH ▶ Using the same cost and production schedule as in Example 5, we determine net requirements and POQ lot sizes.

SOLUTION ▶ Ten-week usage equals a gross requirement of 270 units; therefore, average weekly usage equals 27, and from Example 5, we know the EOQ is 73 units.

We set the *POQ interval* equal to the EOQ divided by the average weekly usage.
Therefore:

$$\text{POQ interval} = \text{EOQ/Average weekly usage} = 73/27 = 2.7, \text{or } 3 \text{ weeks.}$$

The *POQ order size* will vary by the quantities required in the respective weeks, as shown in the following table, with first planned order release in week 1.

Note: Orders are postponed if no demand exists, which is why week 7's order release is postponed until week 8.

MRP Lot Sizing: POQ Technique*

WEEK		1	2	3	4	5	6	7	8	9	10
Gross requirements		35	30	40	0	10	40	30	0	30	55
Scheduled receipts											
Projected on hand	35	35	0	40	0	0	70	30	0	0	55
Net requirements		0	30	0	0	10	0	0	0	55	0
Planned order receipts			70			80				85	
Planned order releases		70			80				85		

*Holding costs = $1/unit/week; setup cost = $100; gross requirements average per week = 27; lead time = 1 week.

$$\text{Setup cost} = 3 \times \$100 = \$300$$

$$\text{Holding cost} = (40 + 70 + 30 + 55)\text{ units} \times \$1\text{ each} = \$195$$

$$\text{The POQ solution yields a computed 10-week cost of } \$300 + \$195 = \$495$$

INSIGHT ▶ Because POQ tends to produce a balance between holding and ordering costs with no excess inventory, POQ typically performs much better than EOQ. Notice that even with frequent recalculations, actual holding cost can vary substantially, depending on the demand fluctuations. We are assuming no stockouts. In this and similar examples, we are also assuming no safety stock; such costs would need to be added to our actual cost.

LEARNING EXERCISE ▶ What is the impact on total cost if holding cost is \$2 per period rather than \$1? [Answer: EOQ = 52; POQ interval = 52/27 = 1.93 ≈ 2 weeks; holding cost = \$270; setups = \$400. The POQ total cost becomes \$670.]

RELATED PROBLEMS ▶ 14.24, 14.25, 14.26c, 14.27c

Other lot-sizing techniques, known as *dynamic lot-sizing*, are similar to periodic order quantity as they attempt to balance the lot size against the setup cost. These are *part period balancing* (also called *least total cost*), *least unit cost*, and *least period cost* (also called *Silver-Meal*). Another technique, *Wagner-Whitin*, takes a different approach by using dynamic programming to optimize ordering over a finite time horizon.[4]

Lot-Sizing Summary In the three speaker kits lot-sizing examples, we found the following costs:

	COSTS		
	SETUP	HOLDING	TOTAL
Lot-for-lot	\$700	\$ 0	\$700
Economic order quantity (EOQ)	\$400	\$375	\$775
Periodic order quantity (POQ)	\$300	\$195	\$495

These examples should not, however, lead operations personnel to hasty conclusions about the preferred lot-sizing technique. In theory, new lot sizes should be computed whenever there is a schedule or lot-size change anywhere in the MRP hierarchy. In practice, such changes cause the instability and system nervousness referred to earlier in this chapter. Consequently, such frequent changes are not made. This means that all lot sizes are wrong because the production system cannot and should not respond to frequent changes. Note that there are no "shortage" (out of stock) charges in any of these lot-sizing techniques. This limitation places added demands on accurate forecasts and "time fences."

In general, the lot-for-lot approach should be used whenever low-cost setup can be achieved. Lot-for-lot is the goal. Lots can be modified as necessary for scrap allowances, process constraints (for example, a heat-treating process may require a lot of a given size), or raw material purchase lots (for example, a truckload of chemicals may be available in only one lot size). However, caution should be exercised prior to any modification of lot size because the modification can cause substantial distortion of actual requirements at lower levels in the MRP hierarchy. When setup costs are significant and demand is reasonably smooth, POQ or even EOQ should provide satisfactory results. Too much concern with lot sizing yields false accuracy because of MRP dynamics. A correct lot size can be determined only after the fact, based on what actually happened in terms of requirements.

MRP Management

In this section we examine demand-driven MRP, capacity planning, system nervousness, and MRP II.

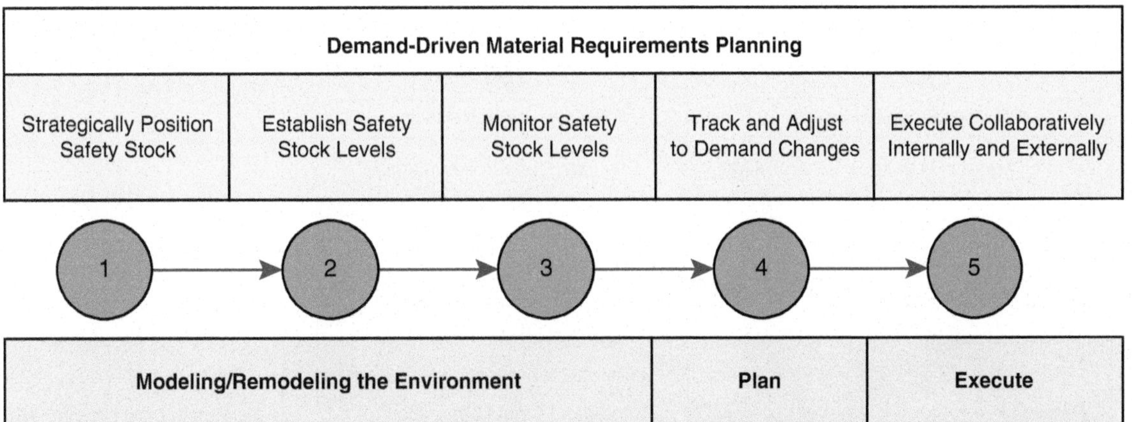

Figure 14.6

The Five Components of Demand-Driven MRP

Demand-Driven MRP

For successful MRP performance, accurate inputs are critical. The bills of material must be timely and accurate; the vendor quality, quantity, and lead times precise; inventory counts accurate; and production processes reliable. Seldom are all precisely aligned. Consequently, consideration of safety stock (buffer inventory) is prudent. But because of the significant domino effect of any change in the MRP system, any change that alters lead time or safety stock should be carefully considered. When safety stock is deemed absolutely necessary, the usual policy is to build (increase) safety stock into the inventory requirements within the MRP logic. Distortion of actual requirements tends to be minimized when safety stock is held primarily at the finished-goods, module, and purchased-item levels.

However, stretching lead times or adding inventory can be very expensive and may undermine an efficient MRP system. Consequently, strategically altering lead times and precisely placing and adjusting safety-stock (buffer-inventory) levels have led to what is known as **demand-driven MRP (DDMRP)**.[5] DDMRP implementation relies on the five primary components noted below and shown in Figure 14.6.

Demand-driven MRP (DDMRP)
Strategically alters lead times and precisely places safety stock within the BOM structure to improve MRP performance.

1. Determine where within the BOM structure to position safety stock, based on considerations such as demand and supply variability and which points provide the most leverage and flexibility if provided more inventory.
2. Determine, based on demand and supply variability, initial safety-stock levels.
3. Monitor conditions to adjust safety-stock levels as needed.
4. Identify, track, and prioritize forecasted demand and adjust safety-stock levels accordingly.
5. Use the information designed into the DDMRP system for increased communication and collaboration.

When properly implemented, DDMRP has been found to reduce stockouts and improve stability in challenging MRP environments.

Capacity Planning

MRP is a planning technique that identifies short-range capacity requirements. It provides information for **capacity planning**, including the master production schedule, the production plan, and production activity control (as shown on the left side of Figure 14.7). MRP is an excellent tool for product-focused and repetitive facilities, but it has limitations in process focused (make-to-order) environments. MRP is a planning technique with *fixed* lead times that typically loads work into *infinite*-sized "buckets." The **buckets** are time units that can be hours, days, weeks, or even months. MRP systems provide feedback about the workload at each work center, putting work into buckets without regard to capacity. Consequently, MRP is considered an *infinite* scheduling technique.

Capacity planning
Determining short-range capacity requirements.

Buckets
Time units in a material requirements planning system.

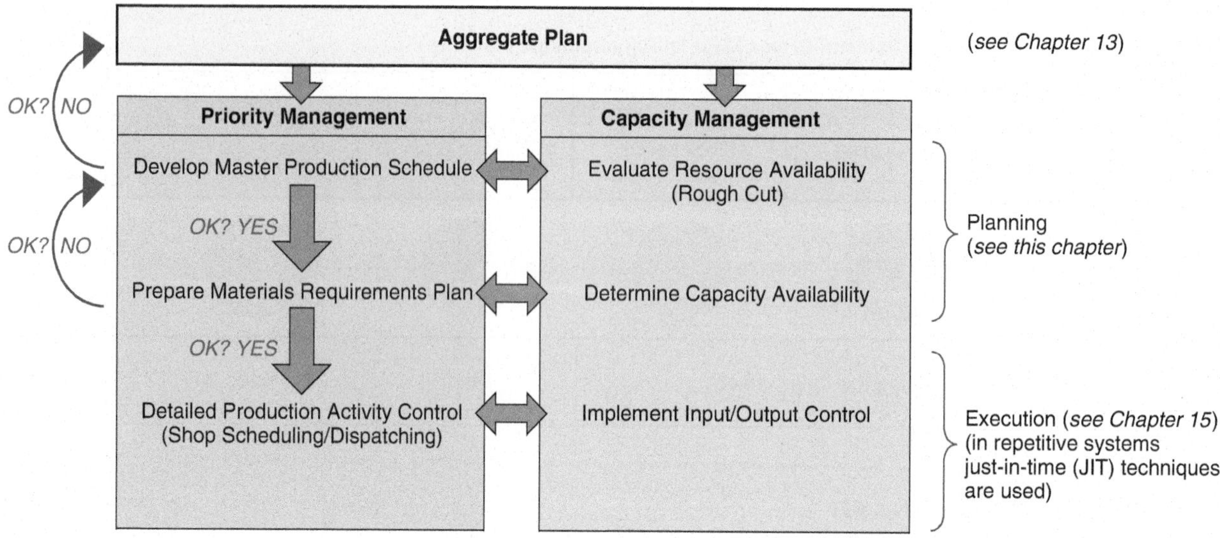

Figure **14.7**

Capacity Planning with Material Requirements Planning

MRP systems are being developed that constrain (limit) the size of buckets, moving the excess demand to alternative buckets. However, once the size of the bucket is limited, the issue becomes where to assign excess demand. To be a useful alternative, the new bucket must fill all of the requirements for production. Specifically, to be useful, alternative buckets must have capacity, material availability, account for material lead time, or identify legitimate alternative work centers, etc. Based on these constraints, MRP systems provide tentative finite solutions via a *load or dispatch report*. (Other approaches to finite scheduling techniques are discussed in Chapter 15.)

Load or dispatch report
A report showing the resource requirements in a work center for all work currently assigned as well as all planned and expected orders.

Load or dispatch reports show the resource requirements in a work center for all work currently assigned, planned, or expected. Beyond the normal production requirements of material and capacity, production planners must deal with the many changing real-world issues that impact the production process. This includes issues such as material quality, employee absences, machine down time, and scheduling or logistics issues. Therefore, some human intervention is usually necessary even for MRP systems with finite planning. Production planners, recognizing these variables, move the work by creatively seeking an ideal balance between capacity, cost, and customer expectations. (This is the "Detailed Production Activity Control" part of Figure 14.7.)

LO 14.5 *Describe* the purpose of the load or dispatch report

Figure 14.8(a) shows that the initial load in the milling center exceeds capacity on days 2, 3, and 5. The production planner uses techniques such as the following to smooth demand as shown in Figure 14.8(b).

1. *Overlapping*, which reduces the lead time, sends pieces to the second operation before the entire lot is completed on the first operation.

Figure **14.8**

(a) Initial Resource Requirements Profile for a Work Center

(b) Smoothed Resource Requirements Profile for a Work Center

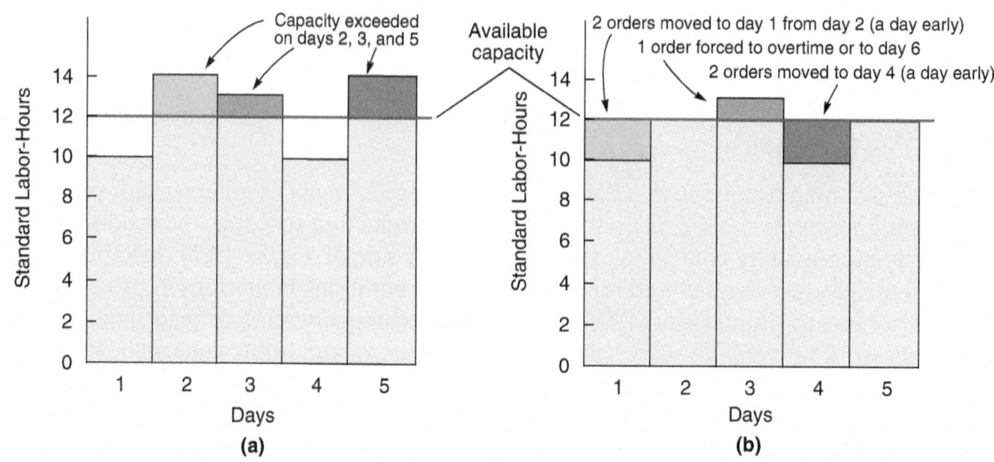

2. *Operations splitting* sends the lot to two different machines for the same operation. This involves an additional setup but results in shorter throughput times because only part of the lot is processed on each machine.
3. *Order splitting*, or *lot splitting*, involves breaking up the order and running part of it earlier (or later) in the schedule.

The production planner then has the MRP system reschedule all items in the net requirements plan.

Example 7 shows a brief detailed capacity scheduling example using order splitting to improve utilization.

Example 7

ORDER SPLITTING

Kevin Watson, the production planner at Wiz Products, needs to develop a capacity plan for a work center. The current load report shows production orders for the next 5 days. There are 12 hours available in the work cell each day. Raw material is on hand. Overtime, while expensive, is possible, and extra time and talent are available for extra machine setups. The parts being produced require 1 hour each.

Day	1	2	3	4	5
Orders	10	14	13	10	14

APPROACH ▶ Compute the time available in the work center and the time necessary to complete the production requirements.

SOLUTION ▶

DAY	UNITS ORDERED	CAPACITY REQUIRED (HOURS)	CAPACITY AVAILABLE (HOURS)	UTILIZATION: OVER/(UNDER) (HOURS)	PRODUCTION PLANNER'S ACTION	NEW PRODUCTION SCHEDULE
1	10	10	12	(2)		12
2	14	14	12	2	Split order: move 2 units to day 1	12
3	13	13	12	1	Split order: move 1 unit to day 6 or request overtime	13
4	10	10	12	(2)		12
5	14	14	12	2	Split order: move 2 units to day 4	12
	61					

INSIGHT ▶ By moving orders, the production planner is able to utilize capacity more effectively and still meet the order requirements, with only 1 order produced on overtime in day 3.

LEARNING EXERCISE ▶ If the units ordered for day 5 increase to 16, what are the production planner's options? [Answer: In addition to moving 2 units to day 4, move 2 units of production to day 6, or request overtime.]

RELATED PROBLEMS ▶ 14.29, 14.30

When the workload consistently exceeds work-center capacity, the tactics just discussed are not adequate. This may mean adding capacity via personnel, machinery, overtime, or subcontracting.

System Nervousness

Even under the best of circumstances, the inputs to MRP (the master schedule, BOM, lead times, and inventory) frequently change. Conveniently, a central strength of MRP systems is timely and accurate replanning. However, many firms find they do not want to respond to minor scheduling or quantity changes even if they are aware of them. These frequent changes generate what is called **system nervousness** and can create havoc in purchasing and production departments if implemented. Consequently, OM personnel reduce such nervousness by evaluating the need and impact of changes prior to disseminating requests to other departments. Two tools are particularly helpful when trying to reduce MRP system nervousness.

System nervousness
Frequent changes in an MRP system.

Time fences
A means for allowing a segment of the master schedule to be designated as "not to be rescheduled."

Pegging
In material requirements planning systems, tracing upward the bill of material from the component to the parent item.

The first is time fences. **Time fences** allow a segment of the master schedule to be designated as "not to be rescheduled." This segment of the master schedule is therefore not changed during the periodic regeneration of schedules. The second tool is pegging. **Pegging** means tracing upward in the BOM from the component to the parent item. By pegging upward, the production planner can determine the cause for the requirement and make a judgment about the necessity for a change in the schedule.

With MRP, the operations manager *can* react to the dynamics of the real world. If the nervousness is caused by legitimate changes, then the proper response may be to investigate the production environment—not adjust via MRP.

Material Requirements Planning II (MRP II)

Material requirements planning II (MRP II)
A system that allows, with MRP in place, inventory data to be augmented by other resource variables; in this case, MRP becomes *material resource planning.*

LO 14.6 *Describe* MRP II

Material requirements planning II is an extremely powerful technique. Once a firm has MRP in place, requirements data can be enriched by resources other than just components. When MRP is used this way, *resource* is usually substituted for *requirements*, and MRP becomes **MRP II**. It then stands for material resource planning.

So far in our discussion of MRP, we have scheduled products and their components. However, products require many resources, such as energy and money, beyond the product's tangible components. In addition to these resource inputs, *outputs* can be generated as well. Outputs can be as diverse as scrap, packaging waste, effluent, and carbon emissions. As OM becomes increasingly sensitive to environmental and sustainability issues, identifying and managing by-products takes on more significance. MRP II provides a vehicle for doing so. Table 14.4 shows an example of labor-hours, machine-hours, grams of greenhouse gas emissions, pounds of scrap, and cash, in the format of a gross requirements plan. With MRP II, management can identify both the inputs and outputs as well as the relevant schedule. MRP II provides another tool in OM's battle for sustainable operations.

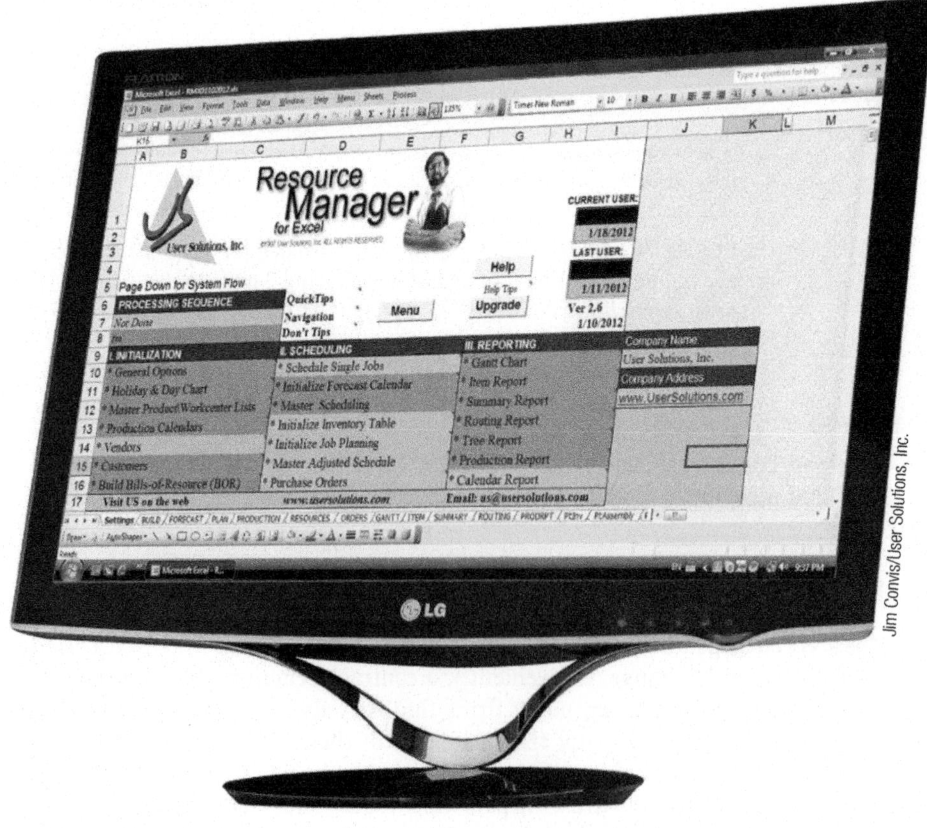

Jim Convis/User Solutions, Inc.

Many MRP programs, such as *Resource Manager for Excel*, are commercially available. *Resource Manager's* initial menu screen is shown here.

A demo program is available for student use at **www.usersolutions.com/students/**.

TABLE 14.4 Material Resource Planning (MRP II)

	LEAD TIME	WEEKS 5	6	7	8
Computer *Labor-hours:* .2 each *Machine-hours:* .2 each *GHG Emissions:* .25 each *Scrap:* 1 ounce fiberglass each *Payables:* $0	1				100 20 20 25 grams 6.25 lb $0
PC board (1 each) *Labor-hours:* .15 each *Machine-hours:* .1 each *GHG Emissions:* 2.5 each *Scrap:* .5 ounces copper each *Payables:* raw material at $5 each	2			100 15 10 250 grams 3.125 lb $500	
Processors (5 each) *Labor-hours:* .2 each *Machine-hours:* .2 each *GHG Emissions:* .50 each *Scrap:* .01 ounces of acid waste each *Payables:* processor components at $10 each	4	500 100 100 25,000 grams 0.3125 lb $5,000			

By utilizing the logic of MRP, resources such as labor, machine-hours, greenhouse gas emissions, scrap, and cost can be accurately determined and scheduled. Weekly demand for labor, machine-hours, greenhouse gas emissions, scrap, and payables for 100 computers are shown.

MRP II systems are seldom stand-alone programs. Most are tied into other computer software. Purchasing, production scheduling, capacity planning, inventory, and warehouse management systems are a few examples of this data integration.

MRP in Services

The demand for many services or service items is classified as dependent demand when it is directly related to or derived from the demand for other services. Such services often require product-structure trees, bills of material and labor, and scheduling. Variations of MRP systems can make a major contribution to operational performance in such services. Examples from restaurants, hospitals, and hotels follow.

Restaurants In restaurants, ingredients and side dishes (bread, vegetables, and condiments) are typically meal components. These components are dependent on the demand for meals. The meal is an end item in the master schedule. Figure 14.9 shows (a) a product-structure tree and (b) bill of material (here called a *product specification*) for six portions of *Buffalo Chicken Mac & Cheese*, a popular dish prepared by Chef John for Orlando Magic fans at the Amway Center.

Hospitals MRP is also applied in hospitals, especially when dealing with surgeries that require known equipment, materials, and supplies. Houston's Park Plaza Hospital and many hospital suppliers, for example, use the technique to improve the scheduling and management of expensive surgical inventory.

Hotels Marriott develops a bill of material and a bill of labor when it renovates each of its hotel rooms. Marriott managers explode the BOM to compute requirements for materials, furniture, and decorations. MRP then provides net requirements and a schedule for use by purchasing and contractors.

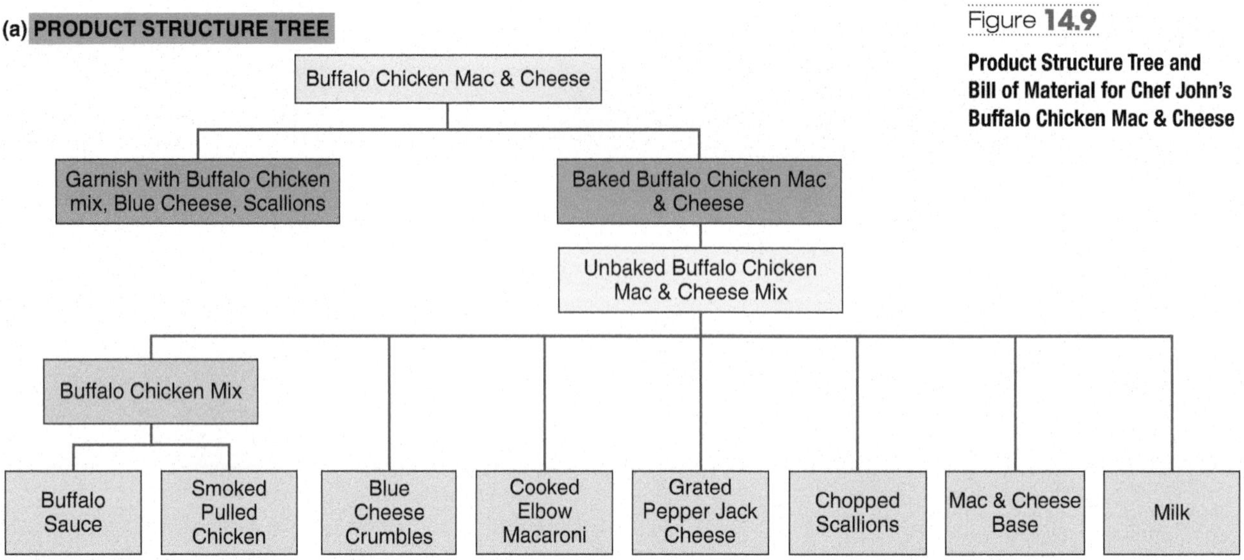

Figure **14.9**

Product Structure Tree and Bill of Material for Chef John's Buffalo Chicken Mac & Cheese

(b) **BILL OF MATERIALS**

Production Specification	**Buffalo Chicken Mac & Cheese (6 portions)**				
Ingredients	*Quantity*	*Measure*	*Unit Cost*	*Total Cost*	*Labor Hrs.*
Elbow Macaroni (large, uncooked)	20.00	oz.	$ 0.09	$ 1.80	
Cheese—Pepper Jack (grated)	10.00	oz.	0.17	1.70	
Mac and Cheese Base (from refrigerator)	32.00	oz.	0.80	25.60	
Milk	4.00	oz.	0.03	0.12	
Smoked Pulled Chicken	2.00	lb.	2.90	5.80	
Buffalo Sauce	8.00	oz.	0.09	0.72	
Blue Cheese Crumbles	4.00	oz.	0.19	0.76	
Scallions	2.00	oz.	0.18	0.36	
Total Labor Hours					0.2 hrs

Distribution Resource Planning (DRP)

Distribution resource planning (DRP)
A time-phased stock-replenishment plan for all levels of a distribution network.

When dependent techniques are used in the supply chain, they are called distribution resource planning (DRP). **Distribution resource planning (DRP)** is a time-phased stock-replenishment plan for all levels of the supply chain.

DRP procedures and logic are analogous to MRP. With DRP, expected demand becomes gross requirements. Net requirements are determined by allocating available inventory to gross requirements. The DRP procedure starts with the forecast at the retail level (or the most distant point of the distribution network being supplied). All other levels are computed. As is the case with MRP, inventory is then reviewed with an aim to satisfying demand. So that stock will arrive when it is needed, net requirements are offset by the necessary lead time. A planned order release quantity becomes the gross requirement at the next level down the distribution chain.

DRP *pulls* inventory through the system. Pulls are initiated when the retail level orders more stock. Allocations are made to the retail level from available inventory and production after being adjusted to obtain shipping economies. Effective use of DRP requires an integrated information system to rapidly convey planned order releases from one level to the next. The goal of the DRP system is small and frequent replenishment within the bounds of economical ordering and shipping.

Enterprise resource planning (ERP)
An information system for identifying and planning the enterprise-wide resources needed to take, make, ship, and account for customer orders.

Enterprise Resource Planning (ERP)

Advances in MRP II systems that tie customers and suppliers to MRP II have led to the development of enterprise resource planning (ERP) systems. **Enterprise resource planning (ERP)** is software that allows companies to (1) automate and integrate many of their business processes,

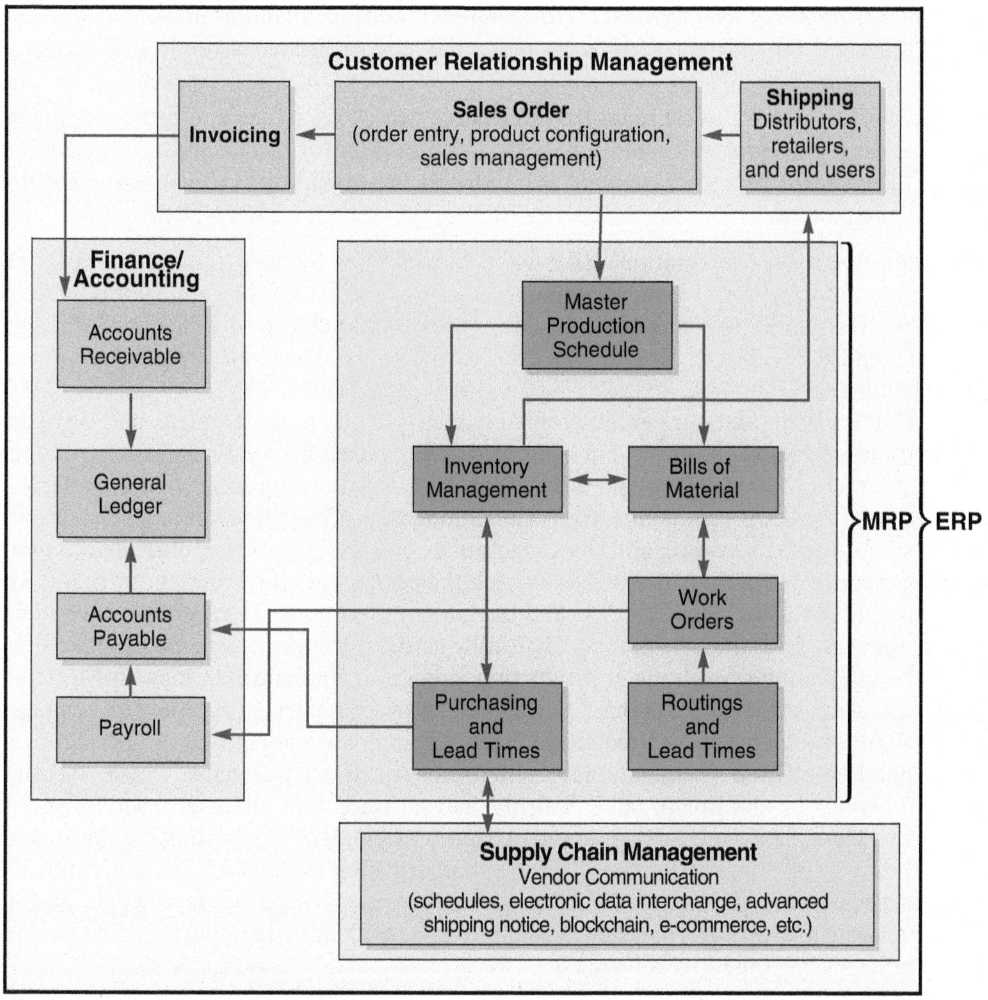

Figure 14.10

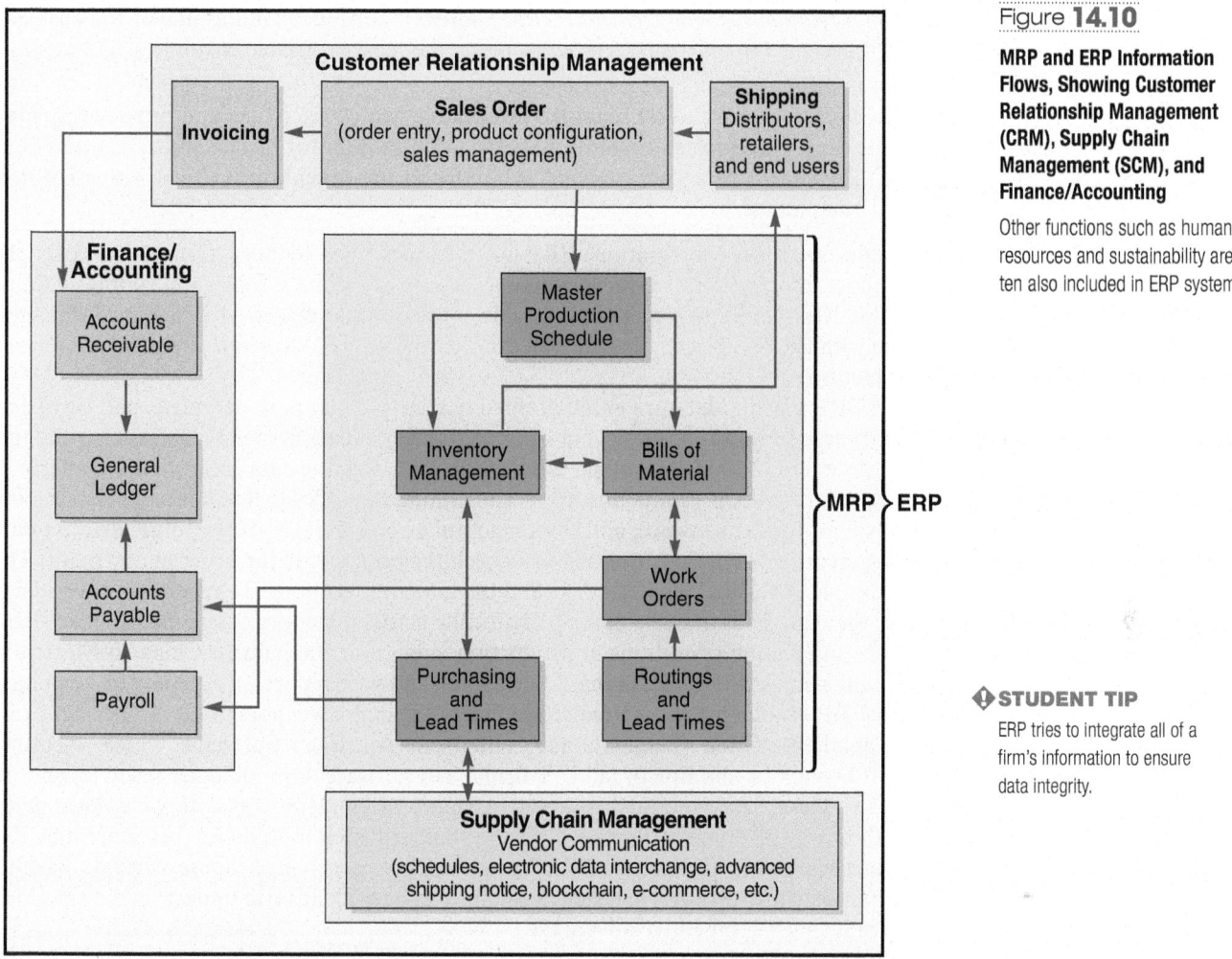

MRP and ERP Information Flows, Showing Customer Relationship Management (CRM), Supply Chain Management (SCM), and Finance/Accounting

Other functions such as human resources and sustainability are often also included in ERP systems.

STUDENT TIP

ERP tries to integrate all of a firm's information to ensure data integrity.

(2) share a common database and business practices throughout the enterprise, and (3) produce information in real time. A schematic showing some of these relationships for a manufacturing firm appears in Figure 14.10.

The objective of an ERP system is to coordinate a firm's entire business, from supplier evaluation to customer invoicing. This objective is seldom achieved, but ERP systems are umbrella systems that tie together a variety of specialized systems. This is accomplished by using a centralized database to assist the flow of information among business functions. Exactly what is tied together, and how, varies on a case-by-case basis. In addition to the traditional components of MRP, ERP systems usually provide financial and human resource (HR) management information. ERP systems may also include:

LO 14.7 *Describe* ERP

- *Supply chain management (SCM)* software to support sophisticated vendor communication, e-commerce, and those activities necessary for efficient warehousing and logistics. The idea is to tie operations (MRP) to procurement, to materials management, and to suppliers, providing the tools necessary for effective management of all four areas.
- *Blockchain* software replaces paper-based transaction records with digital-based records to reduce the complexity of tracking documentation. Blockchains use a digital ledger (block) to continuously track transaction information, making the transaction links more verifiable and secure. Firms such as Walmart are using blockchains to improve visibility of the sequence of transactions for food shipments. Similarly, global shipping giant Maersk uses the technology to track shipping containers and move them through customs faster. Eastman Kodak uses the technology to help photographers track use of their images. And enterprise software firms, such as Oracle, include blockchain in their offerings.

- *Customer relationship management (CRM)* software for the incoming side of the business. CRM is designed to aid analysis of sales, target the most profitable customers, and manage the sales force.
- *Sustainability* software to tie together sustainable workforce issues and provide transparency for supply chain sustainability issues, as well as monitor health and safety activities, energy use and efficiency, emissions (carbon footprint, greenhouse gases), and environmental compliance.

In addition to data integration, ERP software promises reduced transaction costs and fast, accurate information. A strategic emphasis on just-in-time systems and supply chain integration drives the desire for enterprise-wide software. The *OM in Action* box, "Managing Benetton with ERP Software," provides an example of how ERP software helps integrate company operations.

In an ERP system, data are entered only once into a common, complete, and consistent database shared by all applications. For example, when a Nike salesperson enters an order into his ERP system for 20,000 pairs of sneakers for Foot Locker, the data are instantly available on the manufacturing floor. Production crews start filling the order if it is not in stock, accounting prints Foot Locker's invoice, and shipping notifies Foot Locker of the future delivery date. The salesperson, or even the customer, can check the progress of the order at any point. This is all accomplished using the same data and common applications. To reach this consistency, however, the data fields must be defined identically across the entire enterprise. In Nike's case, this means integrating operations at production sites from Vietnam to China to Mexico, at business units across the globe, in many currencies, and with reports in a variety of languages.

Each ERP vendor produces unique products. The major vendors, SAP (a German firm), BEA (Canadian), Odoo (Belgian), Sage (English), American Software, Oracle NetSuite, Microsoft Dynamics, and Epicor (all U.S. firms), sell software or modules designed for specific industries (a set of SAP's modules is shown in Figure 14.11). However, companies must determine if their way of doing business will fit the standard ERP module. If they determine that the product will not fit the standard ERP product, they can change the way they do business to accommodate the software. But such a change can have an adverse impact on their business process, reducing a competitive advantage.

OM in Action Managing Benetton with ERP Software[6]

Thanks to ERP, the Italian sportswear company Benetton can probably claim to have the world's fastest factory and the most efficient distribution in the garment industry. Located in Ponzano, Italy, Benetton makes and ships 150 million pieces of clothing each year. That is 90,000 boxes every day—boxes that must be filled with exactly the items ordered going to the correct store of the 5,000 Benetton outlets in 120 countries. This highly automated distribution center uses only 19 people. Without ERP, hundreds of people would be needed.

Here is how ERP software works:

IgorGolovniov/Shutterstock

1. *Ordering:* A salesperson in the south Boston store finds that she is running out of a best-selling blue sweater. Using a laptop PC, her local Benetton sales agent taps into the ERP sales module.
2. *Availability:* ERP's inventory software simultaneously forwards the order to the mainframe in Italy and finds that half the order can be filled immediately from the Italian warehouse. The rest will be manufactured and shipped in 4 weeks.
3. *Production:* Because the blue sweater was originally created by computer-aided design (CAD), ERP manufacturing software passes the specifications to a knitting machine. The knitting machine makes the sweaters.
4. *Warehousing:* The blue sweaters are boxed with a radio frequency identification (RFID) tag addressed to the Boston store and placed in one of the 300,000 slots in the Italian warehouse. A robot flies by, reading RFID tags, picks out any and all boxes ready for the Boston store, and loads them for shipment.
5. *Order tracking:* The Boston salesperson logs onto the ERP system through the Internet and sees that the sweater (and other items) are completed and being shipped.
6. *Planning:* Based on data from ERP's forecasting and financial modules, Benetton's chief buyer decides that blue sweaters are in high demand and quite profitable. She decides to add three new hues.

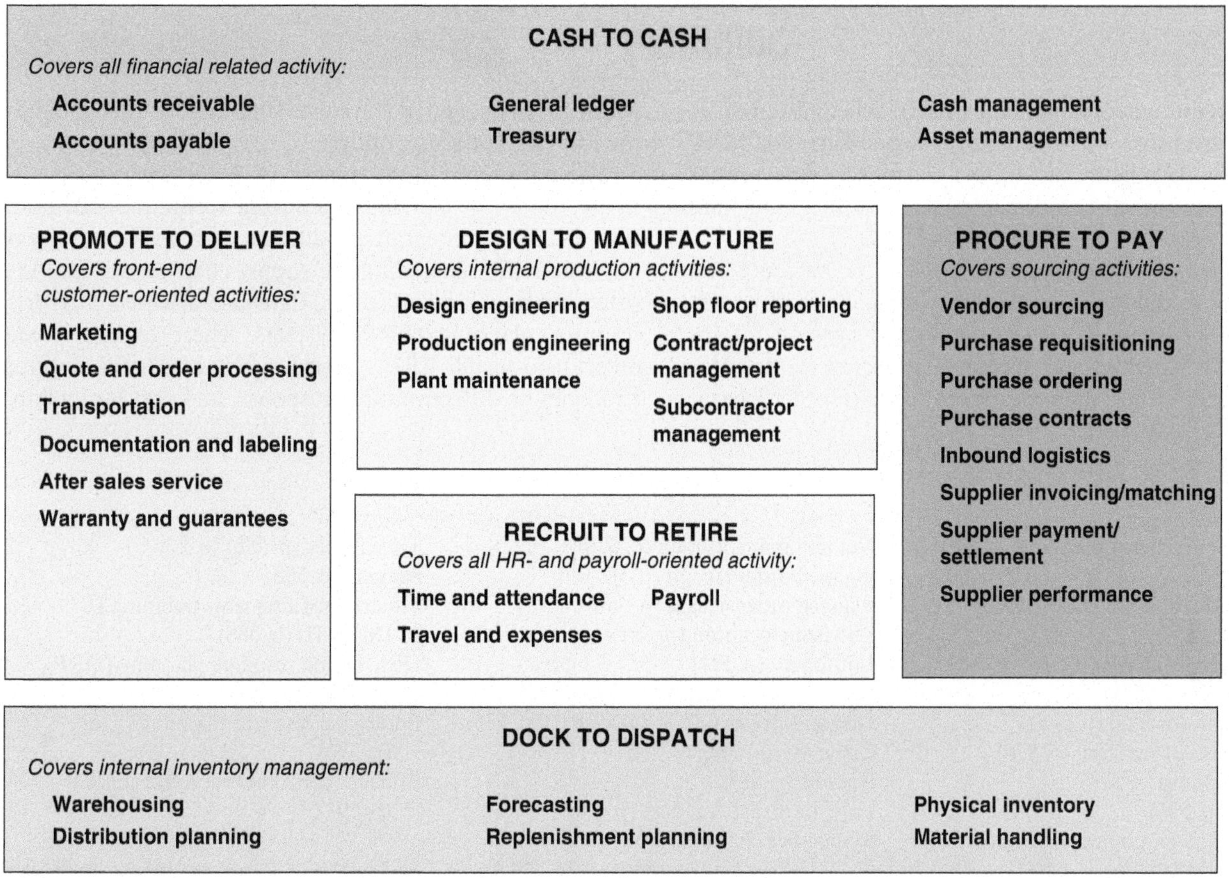

CASH TO CASH
Covers all financial related activity:
Accounts receivable
Accounts payable
General ledger
Treasury
Cash management
Asset management

PROMOTE TO DELIVER
Covers front-end customer-oriented activities:
Marketing
Quote and order processing
Transportation
Documentation and labeling
After sales service
Warranty and guarantees

DESIGN TO MANUFACTURE
Covers internal production activities:
Design engineering
Production engineering
Plant maintenance
Shop floor reporting
Contract/project management
Subcontractor management

RECRUIT TO RETIRE
Covers all HR- and payroll-oriented activity:
Time and attendance
Travel and expenses
Payroll

PROCURE TO PAY
Covers sourcing activities:
Vendor sourcing
Purchase requisitioning
Purchase ordering
Purchase contracts
Inbound logistics
Supplier invoicing/matching
Supplier payment/settlement
Supplier performance

DOCK TO DISPATCH
Covers internal inventory management:
Warehousing
Distribution planning
Forecasting
Replenishment planning
Physical inventory
Material handling

Figure **14.11**

SAP's Modules for ERP

Source: Based on SAP, **www.sap.com**.

Alternatively, ERP software can be customized to meet their specific process requirements. Although the vendors build the software to keep the customization process simple, many companies spend up to five times the cost of the software to customize it. In addition to the expense, the major downside of customization is that when ERP vendors provide an upgrade or enhancement to the software, the customized part of the code must be rewritten to fit into the new version. ERP programs cost from a minimum of $300,000 for a small company to hundreds of millions of dollars for global giants like Ford and Coca-Cola. It is easy to see, then, that ERP systems are expensive, full of hidden issues, and time-consuming to install.

ERP in the Service Sector

ERP vendors have developed a series of service modules for such markets as health care, government, retail stores, and financial services. Springer-Miller Systems, for example, has created an ERP package for the hotel market with software that handles all front- and back-office functions. This system integrates tasks such as maintaining guest histories, booking room and dinner reservations, scheduling golf tee times, and managing multiple properties in a chain. PeopleSoft/Oracle combines ERP with supply chain management to coordinate airline meal preparation. In the grocery industry, these supply chain systems are known as *efficient consumer response* (ECR) systems. **Efficient consumer response** systems tie sales to buying, to inventory, to logistics, and to production.

Efficient consumer response (ECR)
Supply chain management systems in the grocery industry that tie sales to buying, to inventory, to logistics, and to production.

Summary

Material requirements planning (MRP) schedules production and inventory when demand is dependent. For MRP to work, management must have a master schedule, precise requirements for all components, accurate inventory and purchasing records, and accurate lead times.

When properly implemented, MRP can contribute in a major way to reduction in inventory while improving customer-service levels. MRP techniques allow the operations manager to schedule and replenish stock on a "need-to-order" basis rather than simply a "time-to-order" basis. Many firms using MRP systems find that lot-for-lot can be the low-cost lot-sizing option.

The continuing development of MRP systems has led to its use with lean manufacturing techniques. In addition, MRP can integrate production data with a variety of other activities, including the supply chain and sales. As a result, we now have integrated database-oriented enterprise resource planning (ERP) systems. These expensive and difficult-to-install ERP systems, when successful, support strategies of differentiation, response, and cost leadership.

Key Terms

Material requirements planning (MRP) (p. 572)
Master production schedule (MPS) (p. 573)
Bill of material (BOM) (p. 574)
Modular bills (p. 575)
Planning bills (or kits) (p. 576)
Phantom bills of material (p. 576)
Low-level coding (p. 576)
Lead time (p. 576)
Gross material requirements plan (p. 577)
Net requirements plan (p. 578)
Planned order receipt (p. 580)
Planned order release (p. 580)
Lot-sizing decision (p. 581)
Lot-for-lot (p. 581)
Periodic order quantity (POQ) (p. 583)
Demand-driven MRP (DDMRP) (p. 585)
Capacity planning (p. 585)
Buckets (p. 585)
Load or dispatch report (p. 586)
System nervousness (p. 587)
Time fences (p. 588)
Pegging (p. 588)
Material requirements planning II (MRP II) (p. 588)
Distribution resource planning (DRP) (p. 590)
Enterprise resource planning (ERP) (p. 590)
Efficient consumer response (ECR) (p. 593)

Ethical Dilemma

For many months your prospective ERP customer has been analyzing the hundreds of assumptions built into the $900,000 ERP software you are selling. So far, you have knocked yourself out to try to make this sale. If the sale goes through, you will reach your yearly quota and get a nice bonus. On the other hand, loss of this sale may mean you start looking for other employment.

The accounting, human resource, supply chain, and marketing teams put together by the client have reviewed the specifications and finally recommended purchase of the software. However, as you looked over their shoulders and helped them through the evaluation process, you began to realize that their purchasing procedures—with much of the purchasing being done at hundreds of regional stores—were not a good fit for the software. At the very least, the customizing will add $250,000 to the implementation and training cost. The team is not aware of the issue, and you know that the necessary $250,000 is not in the budget.

What do you do?

Discussion Questions

1. What is the difference between a *gross* requirements plan and a *net* requirements plan?
2. Once a material requirements plan (MRP) has been established, what other managerial applications might be found for the technique?
3. What are the similarities between MRP and DRP?
4. How does MRP II differ from MRP?
5. Which is the best lot-sizing policy for manufacturing organizations?
6. What impact does ignoring carrying cost in the allocation of stock in a DRP system have on lot sizes?
7. MRP is more than an inventory system; what additional capabilities does MRP possess?
8. What are the options for the production planner who has:
   a) scheduled more than capacity in a work center next week?
   b) a consistent lack of capacity in that work center?
9. Master schedules are expressed in three different ways depending on whether the process is continuous, a job shop, or repetitive. What are these three ways?
10. What functions of the firm affect an MRP system? How?
11. What is the rationale for (a) a phantom bill of material, (b) a planning bill of material, and (c) a pseudo bill of material?
12. Identify five specific requirements of an effective MRP system.
13. What are the typical benefits of ERP?
14. What are the distinctions between MRP, DRP, and ERP?
15. As an approach to inventory management, how does MRP differ from the approach taken in Chapter 12, dealing with economic order quantities (EOQs)?
16. What are the disadvantages of ERP?
17. Use the Web or other sources to:
   a) Find stories that highlight the advantages of an ERP system.
   b) Find stories that highlight the difficulties of purchasing, installing, or failure of an ERP system.

18. Use the Web or other sources to identify what an ERP vendor (SAP, PeopleSoft/Oracle, American Software, etc.) includes in these software features:
   a) Customer relationship management.
   b) Supply chain management.
   c) Product life cycle management.
   d) Blockchain.

19. The structure of MRP systems suggests "buckets" and infinite loading. What is meant by these two terms?

Using Software to Solve MRP Problems

There are many commercial MRP software packages, for companies of all sizes. MRP software for small and medium-sized companies includes User Solutions, Inc., a demo of which is available at **www.usersolutions.com**, and MAX, from Exact Software North America, Inc. Software for larger systems is available from SAP, CMS, BEA, Oracle, i2 Technologies, and many others. The Excel OM software that accompanies this text can be found online and includes an MRP module, as does POM for Windows. The use of both is explained in the following sections.

X USING EXCEL OM

Using Excel OM's MRP module requires the careful entry of several pieces of data. The initial MRP screen is where we enter (1) the total number of occurrences of items in the BOM (including the top item), (2) what we want the BOM items to be called (e.g., Item no., Part), (3) total number of periods to be scheduled, and (4) what we want the periods called (e.g., days, weeks).

Excel OM's second MRP screen provides the data entry for an indented bill of material. Here we enter (1) the name of each item in the BOM, (2) the quantity of that item in the assembly, and (3) the correct indent (e.g., parent/child relationship) for each item. The indentations are critical, as they provide the logic for the BOM explosion. The indentations should follow the logic of the product structure tree with indents for each assembly item in that assembly.

Excel OM's third MRP screen repeats the indented BOM and provides the standard MRP tableau for entries. This is shown in Program 14.1 using the data from Examples 1, 2, and 3.

P USING POM FOR WINDOWS

The POM for Windows MRP module can also solve Examples 1 to 3. Up to 18 periods can be analyzed. Here are the inputs required:

1. *Item names:* The item names are entered in the left column. The same item name will appear in more than one row if the item is used by two parent items. Each item must follow its parents.
2. *Item level:* The level in the indented BOM must be given here. The item *cannot* be placed at a level more than one below the item immediately above.
3. *Lead time:* The lead time for an item is entered here. The default is 1 week.
4. *Number per parent:* The number of units of this subassembly needed for its parent is entered here. The default is 1.
5. *On hand:* List current inventory on hand once, even if the subassembly is listed twice.
6. *Lot size:* The lot size can be specified here. A 0 or 1 will perform lot-for-lot ordering. If another number is placed here, then all orders for that item will be in integer multiples of that number.

Program **14.1**

Using Excel OM's MRP Module to Solve Examples 1, 2, and 3

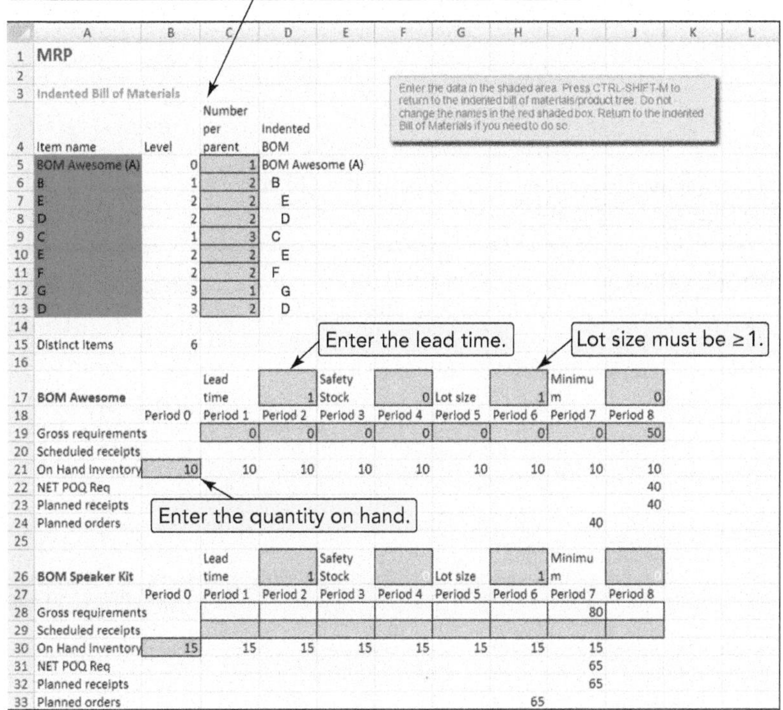

7. *Demands:* The demands are entered in the end item row in the period in which the items are demanded.
8. *Scheduled receipts:* If units are scheduled to be received in the future, they should be listed in the appropriate time period (column) and item (row). (An entry here in level 1 is a demand; all other levels are receipts.)

Further details regarding POM for Windows are seen in Appendix II.

Solved Problems

SOLVED PROBLEM 14.1

Determine the low-level coding and the quantity of each component necessary to produce 10 units of an assembly we will call Alpha. The product structure and quantities of each component needed for each assembly are noted in parentheses.

SOLUTION

Redraw the product structure with low-level coding. Then multiply down the structure until the requirements of each branch are determined. Then add across the structure until the total for each is determined.

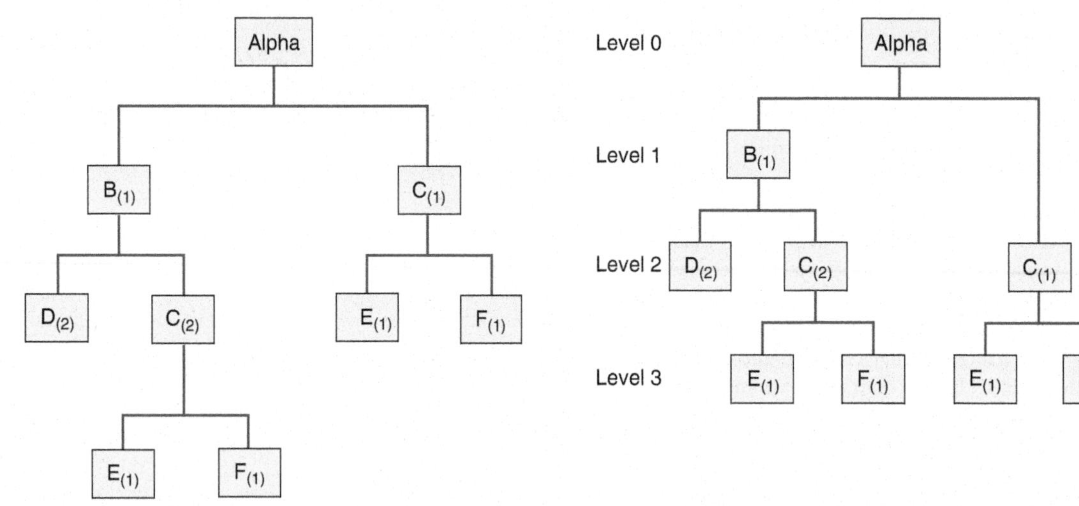

SOLUTION (CONT.)

Es required for left branch:

$$(1_{\text{alpha}} \times 1_B \times 2_C \times 1_E) = 2 \text{ Es}$$

and Es required for right branch:

$$(1_{\text{alpha}} \times 1_C \times 1_E) = \underline{1 \text{ E}}$$
$$3 \text{ Es required in total}$$

Then "explode" the requirement by multiplying each by 10, as shown in the table to the right:

LEVEL	ITEM	QUANTITY PER UNIT	TOTAL REQUIREMENTS FOR 10 ALPHA
0	Alpha	1	10
1	B	1	10
2	C	3	30
2	D	2	20
3	E	3	30
3	F	3	30

SOLVED PROBLEM 14.2

Using the product structure for Alpha in Solved Problem 14.1, and the following lead times, quantity on hand, and master production schedule, prepare a net MRP table for Alphas.

ITEM	LEAD TIME	QUANTITY ON HAND
Alpha	1	10
B	2	20
C	3	0
D	1	100
E	1	10
F	1	50

Master Production Schedule for Alpha

PERIOD	6	7	8	9	10	11	12	13
Gross requirements			50			50		100

SOLUTION

See the chart called "Net Material Requirements Planning Sheet for Alpha for Solved Problem 14.2."

Lot Size	Lead Time (# of Periods)	On Hand	Safety Stock	Allo-cated	Low-Level Code	Item ID		Period (week, day) 1	2	3	4	5	6	7	8	9	10	11	12	13
Lot-for-Lot	1	10	—	—	0	Alpha (A)	Gross Requirements								50			50		100
							Scheduled Receipts													
							Projected On Hand 10								10					
							Net Requirements								40			50		100
							Planned Order Receipts								40			50		
							Planned Order Releases							40			50		100	
Lot-for-Lot	2	20	—	—	1	B	Gross Requirements							40(A)			50(A)		100(A)	
							Scheduled Receipts													
							Projected On Hand 20							20						
							Net Requirements							20			50		100	
							Planned Order Receipts							20			50		100	
							Planned Order Releases					20			50		100			
Lot-for-Lot	3	0	—	—	2	C	Gross Requirements					40(B)		40(A)	100(B)		200(B) + 50(A)		100(A)	
							Scheduled Receipts													
							Projected On Hand 0													
							Net Requirements					40		40	100		250		100	
							Planned Order Receipts					40		40	100		250		100	
							Planned Order Releases		40		40	100		250		100				
Lot-for-Lot	1	100	—	—	2	D	Gross Requirements					40(B)			100(B)		200(B)			
							Scheduled Receipts													
							Projected On Hand 100					100			60					
							Net Requirements					0			40		200			
							Planned Order Receipts					0			40		200			
							Planned Order Releases				0			40		200				
Lot-for-Lot	1	10	—	—	3	E	Gross Requirements		40(C)		40(C)	100(C)		250(C)		100(C)				
							Scheduled Receipts													
							Projected On Hand 10		10											
							Net Requirements		30		40	100		250		100				
							Planned Order Receipts		30		40	100		250		100				
							Planned Order Releases	30		40	100		250		100					
Lot-for-Lot	1	50	—	—	3	F	Gross Requirements		40(C)		40(C)	100(C)		250(C)		100(C)				
							Scheduled Receipts													
							Projected On Hand 50		50		10					—				
							Net Requirements		0		30	100		250		100				
							Planned Order Receipts				30	100		250		100				
							Planned Order Releases			30	100		250		100					

Net Material Requirements Planning Sheet for Alpha for Solved Problem 14.2

The letter in parentheses (A) is the source of the demand.

SOLVED PROBLEM 14.3

Hip Replacements, Inc., has a master production schedule for its newest model, as shown below, a setup cost of \$50, a holding cost per week of \$2, beginning inventory of 0, and lead time of 1 week. What are the costs of using lot-for-lot for this 10-week period?

SOLUTION

Holding cost = \$0 (as there is never any end-of-period inventory)

Ordering costs = 4 orders × \$50 = \$200

Total cost for lot-for-lot = \$0 + \$200 = \$200

WEEK		1	2	3	4	5	6	7	8	9	10
Gross requirements		0	0	50	0	0	35	15	0	100	0
Scheduled receipts											
Projected on hand	0	0	0	0	0	0	0	0	0	0	0
Net requirements		0	0	50	0	0	35	15	0	100	
Planned order receipts				50			35	15		100	
Planned order releases			50			35	15		100		

SOLVED PROBLEM 14.4

Hip Replacements, Inc., has a master production schedule for its newest model, as shown in Solved Problem 14.3, a setup cost of \$50, a holding cost per week of \$2, beginning inventory of 0, and lead time of 1 week. What are the costs of using (a) EOQ and (b) POQ for this 10-week period?

SOLUTION

a) For the **EOQ** lot size, first determine the EOQ.

Annual usage = 200 units for 10 weeks; weekly usage = 200/10 weeks = 20 per week. Therefore, 20 units × 52 weeks (annual demand) = 1,040 units. From Chapter 12, the EOQ model is:

$$Q^* = \sqrt{\frac{2DS}{H}}$$

where D = annual demand = 1,040
S = Setup cost = \$50
H = holding (carrying) cost, on an annual basis per unit = \$2 × 52 = \$104
Q^* = 31.62 ≈ 32 units (order the EOQ or in multiples of the EOQ)

WEEK		1	2	3	4	5	6	7	8	9	10	
Gross requirements		0	0	50	0	0	35	15	0	100	0	
Scheduled receipts												
Projected on hand	0	0	0	0	14	14	14	11	28	28	24	24
Net requirements		0	0	50	0	0	21	0	0	72	0	
Planned order receipts				64			32	32		96		
Planned order releases			64			32	32		96			

Holding cost = 157 units × \$2 = \$314 (note the 24 units available in period 11, for which there is an inventory charge as they are in on-hand inventory at the end of period 10)

Ordering costs = 4 orders × \$50 = \$200

Total cost for EOQ lot sizing = \$314 + \$200 = \$514

b) For the **POQ** lot size we use the EOQ computed above to find the time period between orders:

Period interval = EOQ / average weekly usage = 32/20 = 1.6 ≈ 2 periods

POQ order size = Demand required in the 2 periods, postponing orders in periods with no demand.

WEEK		1	2	3	4	5	6	7	8	9	10
Gross requirements		0	0	50	0	0	35	15	0	100	0
Scheduled receipts											
Projected on hand	0	0	0	0	0	0	0	15	0	0	
Net requirements		0	0	50	0	0	50	0	0	100	0
Planned order receipts				50			50			100	
Planned order releases			50			50			100		

Holding cost = 15 units × \$2 = \$30

Ordering costs = 3 orders × \$50 = \$150

Total cost for POQ lot sizing = \$30 + \$150 = \$180

Problems

Note: **PX** means the problem may be solved with POM for Windows and/or Excel OM. Additionally, further understanding of commercial MRP software is available by examining *Resource Manager for Excel*, a system made available by User Solutions, Inc. Access to a trial version of the software and a set of notes for the user are available at www.usersolutions.com; click on the Free Software Trials button then the Download Now link, then complete the request form and press submit.

Problems 14.1–14.4 relate to Dependent Inventory Model Requirements

• **14.1** You have developed the following simple product structure of items needed for your gift bag for a rush party for prospective pledges in your organization. You forecast 200 attendees. Assume that there is no inventory on hand of any of the items. Explode the bill of material. (Subscripts indicate the number of units required.)

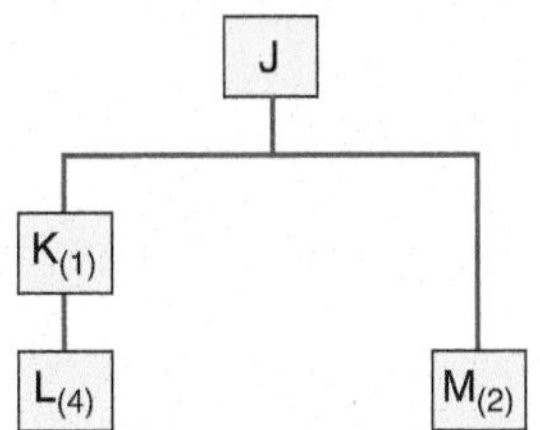

•• **14.2** You are expected to have the gift bags in Problem 14.1 ready at 5 P.M. However, you need to personalize the items (monogrammed pens, note pads, literature from the printer, etc.). The lead time is 1 hour to assemble 200 Js once the other items are prepared. The other items will take a while as well. Given the volunteers you have, the other time estimates are item K (2 hours), item L (1 hour), and item M (4 hours). Develop a time-phased assembly plan to prepare the gift bags.

•• **14.3** As the production planner for Xiangling Hu Products, Inc., you have been given a bill of material for a bracket that is made up of a base, two springs, and four clamps. The base is assembled from one clamp and two housings. Each clamp has one handle and one casting. Each housing has two bearings and one shaft. There is no inventory on hand.

a) Design a product structure noting the quantities for each item and show the low-level coding.
b) Determine the gross quantities needed of each item if you are to assemble 50 brackets.
c) Compute the net quantities needed if there are 25 of the base and 100 of the clamp in stock. **PX**

•• **14.4** Your boss at Xiangling Hu Products, Inc., has just provided you with the schedule and lead times for the bracket in Problem 14.3. The unit is to be prepared in week 10. The lead times for the components are bracket (1 week), base (1 week), spring (1 week), clamp (1 week), housing (2 weeks), handle (1 week), casting (3 weeks), bearing (1 week), and shaft (1 week).

a) Prepare the time-phased product structure for the bracket.
b) In what week do you need to start the castings? **PX**

Problems 14.5–14.21 relate to MRP Structure

•• **14.5** The demand for subassembly S is 100 units in week 7. Each unit of S requires 1 unit of T and 2 units of U. Each unit of T requires 1 unit of V, 2 units of W, and 1 unit of X. Finally, each unit of U requires 2 units of Y and 3 units of Z. One firm manufactures all items. It takes 2 weeks to make S, 1 week to make T, 2 weeks to make U, 2 weeks to make V, 3 weeks to make W, 1 week to make X, 2 weeks to make Y, and 1 week to make Z.

a) Construct a product structure. Identify all levels, parents, and components.
b) Prepare a time-phased product structure.

•• **14.6** Using the information in Problem 14.5, construct a gross material requirements plan. **PX**

•• **14.7** Using the information in Problem 14.5, construct a net material requirements plan using the following on-hand inventory. **PX**

ITEM	ON-HAND INVENTORY	ITEM	ON-HAND INVENTORY
S	20	W	30
T	20	X	25
U	40	Y	240
V	30	Z	40

•• **14.8** Refer again to Problems 14.5 and 14.6. In addition to 100 units of S, there is also a demand for 20 units of U, which is a component of S. The 20 units of U are needed for maintenance purposes. These units are needed in week 6. Modify the *gross material requirements plan* to reflect this change. **PX**

•• **14.9** Refer again to Problems 14.5 and 14.7. In addition to 100 units of S, there is also a demand for 20 units of U, which is

Master Production Schedule for X1

PERIOD	7	8	9	10	11	12
Gross requirements		50		20		100

ITEM	LEAD TIME	ON HAND	ITEM	LEAD TIME	ON HAND
X1	1	50	C	1	0
B1	2	20	D	1	0
B2	2	20	E	3	10
A1	1	5			

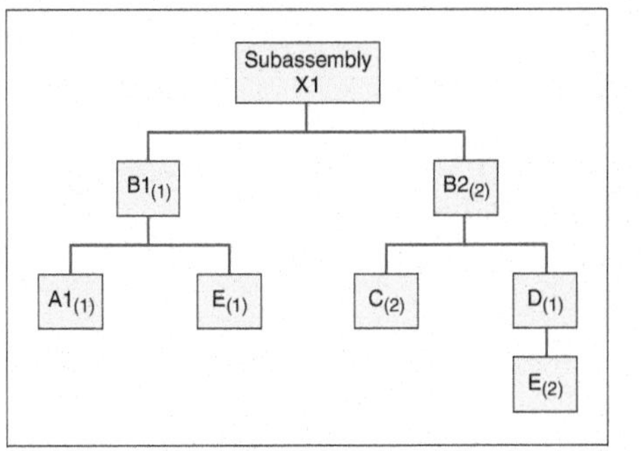

Figure **14.12**

Information for Problem 14.10

Figure 14.13

Information for Problems 14.11 and 14.12

PERIOD	8	9	10	11	12
Gross requirements: A	100		50		150

ITEM	ON HAND	LEAD TIME	ITEM	ON HAND	LEAD TIME
A	0	1	E	75	2
B	100	2	F	75	2
C	50	2	G	75	1
D	50	1			

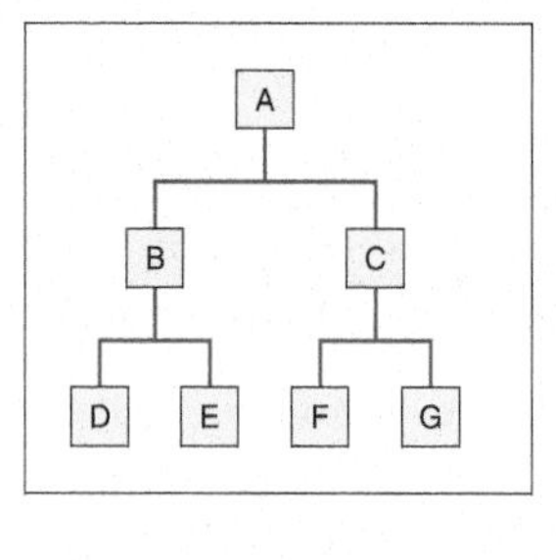

a component of S. The 20 units of U are needed for maintenance purposes. These units are needed in week 6. Additionally, a new order has been received for 80 units of S due in week 8. Modify the *net requirements plan* to reflect these changes. **PX**

•• **14.10**

a) Given the product structure and master production schedule shown in Figure 14.12, develop a gross requirements plan for all items.

b) Given the product structure, master production schedule, and inventory status shown in Figure 14.12, develop a net requirements plan (planned order release) for all items. **PX**

•• **14.11** Given the product structure, master production schedule, and inventory status in Figure 14.13, and assuming the requirement for A is 1, using lot-for-lot:

a) develop a gross requirements plan for Item C;

b) develop a net requirements plan for Item C. **PX**

••• **14.12** Based on the data in Figure 14.13, complete a net material requirements schedule using lot-for-lot for:

a) All items (7 schedules in all), assuming the requirement for A is 1.

b) All 7 items, assuming the requirement for all items is 1, except B, C, and F, which require *2 each*. **PX**

••• **14.13** Electro Fans has just received an order for one thousand 20-inch fans due week 7. Each fan consists of a housing assembly, two grills, a fan assembly, and an electrical unit. The housing assembly consists of a frame, two supports, and a handle. The fan assembly consists of a hub and five blades. The electrical unit consists of a motor, a switch, and a knob. The following table gives lead times, on-hand inventory, and scheduled receipts.

a) Construct a product structure.

b) Construct a time-phased product structure.

c) Prepare a net material requirements plan. **PX**

Data Table for Problem 14.13

COMPONENT	LEAD TIME	ON-HAND INVENTORY	LOT SIZE*	SCHEDULED RECEIPT
20" Fan	1	100	—	
Housing	1	100	—	
Frame	2	—	—	
Supports (2)	1	50	100	
Handle	1	400	500	
Grills (2)	2	200	500	
Fan Assembly	3	150	—	
Hub	1	—	—	
Blades (5)	2	—	100	
Electrical Unit	1	—	—	
Motor	1	—	—	
Switch	1	20	12	
Knob	1	—	25	200 knobs in week 2

*Lot-for-lot unless otherwise noted.

••• **14.14** A part structure, lead time (weeks), and on-hand quantities for product A are shown in Figure 14.14. From the information shown, generate:

a) An indented bill of material for product A (see Figure 5.9 in Chapter 5 as an example of a BOM).

b) Net requirements for each part to produce 10 As in week 8 using lot-for-lot. **PX**

••• **14.15** You are product planner for product A (in Problem 14.14 and Figure 14.14). The field service manager, Mark Moon, has just called and told you that the requirements for B and F should each be increased by 10 units for his repair requirements in the field.

a) Prepare a list showing the quantity of each part required to produce the requirements for the service manager *and* the production request of 10 Bs and Fs.

b) Prepare a net requirement plan by date for the new requirements (for both production and field service), assuming that

Figure 14.14

Information for Problems 14.14, 14.15, 14.16, and 14.17

PART	INVENTORY ON HAND
A	0
B	2
C	10
D	5
E	4
F	5
G	1
H	10

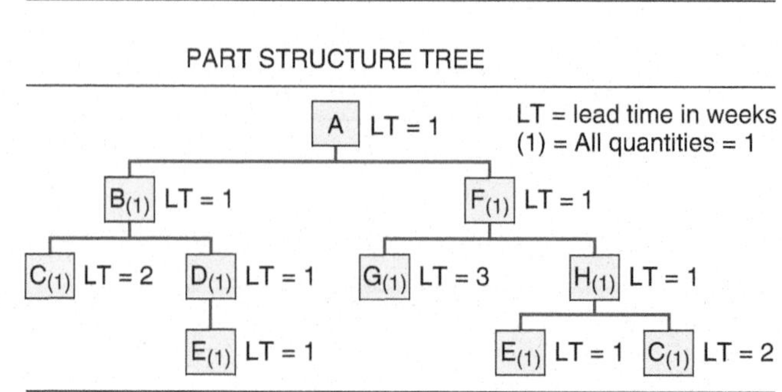

the field service manager wants his 10 units of B and F in week 6 and the 10 production units of A in week 8. **PX**

••• **14.16** You have just been notified that the lead time for component G of product A (Problem 14.15 and Figure 14.14) has been increased to 5 weeks.

a) Which items have changed, and why?
b) What are the implications for the production plan?
c) As production planner, what can you do? **PX**

••• **14.17** Your stockroom manager, Mehmet Altag, arrived at your desk just after you had completed the net requirements plan for product A (use data in Figure 14.14 and Problem 14.15), exclaiming that the cycle counter should be fired. It seems that the cycle counter was wrong; there are 3 As available now, not 0, as the original data showed; moreover, 5 Es are also available. About then, your boss, Joe Orlicky, who overheard the discussion, says, "You might as well extend the net requirements plan out to 16 weeks because we just received an order for 10 more As in week 12 and 5 more in week 15. Additionally, count on the field service department wanting 3 more Bs in week 16, as well as those 10 units of B and F in week 6. And item G now has a lead time of 4 weeks." Your assignment is to prepare a new net requirements plan, based on the actual inventory (as reported) and the new schedule. **PX**

••• **14.18** Heather Adams, production manager for a Colorado exercise equipment manufacturer, needs to schedule an order for 50 UltimaSteppers, which are to be shipped in week 8. Subscripts indicate quantity required for each parent. Assume lot-for-lot ordering. Below is information about the steppers:

ITEM	LEAD TIME	ON-HAND INVENTORY	COMPONENTS
Stepper	2	20	$A_{(1)}$, $B_{(3)}$, $C_{(2)}$
A	1	10	$D_{(1)}$, $F_{(2)}$
B	2	30	$E_{(1)}$, $F_{(3)}$
C	3	10	$D_{(2)}$, $E_{(3)}$
D	1	15	
E	2	5	
F	2	20	

a) Develop a product structure for Heather.
b) Develop a time-phased structure. **PX**

•• **14.19** Heather Adams, using the data in Problem 14.18, now wants a net material requirements plan for item F. Develop one for her. **PX**

••• **14.20** You need to schedule 10 units of product Alpha for delivery in 6 weeks. Three units of D and 2 units of F are required for each Alpha. The lead time for Alpha is 1 week. Lead time for D is 1 week, and lead time for F is 2 weeks.

a) Construct a product structure.
b) Prepare a time-phased product structure.
c) Prepare a gross requirements plan for Alpha. **PX**

••• **14.21** Using the information in Problem 14.20, construct a net material requirements plan. There are 2 Alphas on hand and 4 Ds. **PX**

Problems 14.22–14.28 relate to Lot-Sizing Techniques

Data Table for Problems 14.22 through 14.25*

PERIOD	1	2	3	4	5	6	7	8	9	10	11	12
Gross requirements	30		40		30	70	20		10	80		50

*Holding cost = $2.50/unit/week; setup cost = $150; lead time = 1 week; beginning inventory = 40; stockout cost = $10.

••• **14.22** Develop a lot-for-lot solution and calculate total relevant costs for the data in the preceding table. **PX**

••• **14.23** Develop an EOQ solution and calculate total relevant costs for the data in the preceding table. **PX**

••• **14.24** Develop a POQ solution and calculate total relevant costs for the data in the preceding table. **PX**

••• **14.25** Using your answers for the lot sizes computed in Problems 14.22, 14.23, and 14.24, which is the best technique and why?

•• **14.26** M. de Koster, of Rene Enterprises, has the master production plan shown here:

Period (weeks)	1	2	3	4	5	6	7	8	9
Gross requirements		15		20		10			25

Lead time = 1 period; setup cost = $200; holding cost = $10 per week; stockout cost = $10 per week. Your job is to develop an ordering plan and costs for:

a) Lot-for-lot. b) EOQ.
c) POQ. d) Which plan has the lowest cost? **PX**

••• **14.27** Grace Greenberg, production planner for Science and Technology Labs, in New Jersey, has the master production plan shown here:

Period (weeks)	1	2	3	4	5	6	7	8	9	10	11	12
Gross requirements		35		40		10			25	10		45

Lead time = 1 period; setup costs = $200; holding cost = $10 per week; stockout cost = $10 per week. Develop an ordering plan and costs for Grace, using these techniques:

a) Lot-for-lot. b) EOQ.
c) POQ. d) Which plan has the lowest cost? **PX**

••• **14.28** Keebock, a maker of outstanding running shoes, keeps the soles of its size 13 running shoes in inventory for one period at a cost of $.25 per unit. The setup costs are $50. Beginning inventory is zero, and lead time is 1 week. Shown here are the net requirements per period.

Period	1	2	3	4	5	6	7	8	9	10	11
Net requirements		35	30	45	0	10	40	30	0	30	55

Determine Keebock's cost based on

a) EOQ. b) Lot-for-lot. **PX**

Problems 14.29–14.32 relate to MRP Management

••• **14.29** Karl Knapps, Inc., has received the following orders:

Period	1	2	3	4	5	6	7	8	9	10
Order size	0	40	30	40	10	70	40	10	30	60

The entire fabrication for these units is scheduled on one machine. There are 2,250 usable minutes in a week, and each unit will take 65 minutes to complete. Develop a capacity plan, using lot splitting, for the 10-week time period.

••• **14.30** Coleman Rich, Ltd., has received the following orders:

Period	1	2	3	4	5	6	7	8	9	10
Order size	60	30	10	40	70	10	40	30	40	0

The entire fabrication for these units is scheduled on one machine. There are 2,250 usable minutes in a week, and each unit will take 65 minutes to complete. Develop a capacity plan, using lot splitting, for the 10-week time period.

••• **14.31** Courtney Kamauf schedules production of a popular Rustic Coffee Table at Kamauf Enterprises, Inc. The table requires a top, four legs, $\frac{1}{8}$ gallon of stain, $\frac{1}{16}$ gallon of glue, 2 short braces between the legs and 2 long braces between the legs, and a brass cap

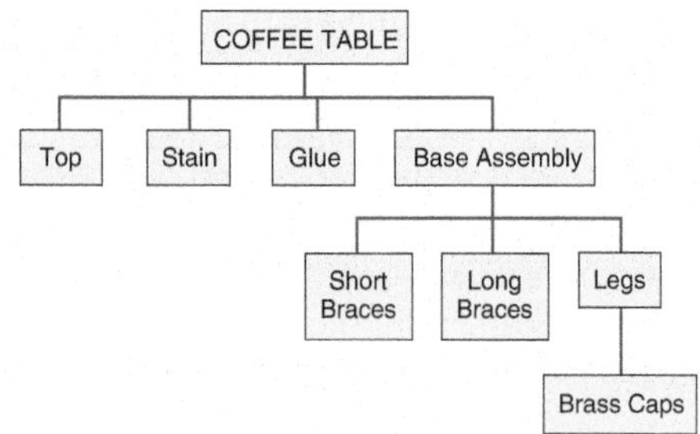

that goes on the bottom of each leg. She has 100 gallons of glue in inventory, but none of the other components. All items except the brass caps, stain, and glue are ordered on a lot-for-lot basis. The caps are purchased in quantities of 1,000, stain and glue by the gallon. Lead time is 1 day for each item. Schedule the order releases necessary to produce 640 coffee tables on days 5 and 6, and 128 on days 7 and 8. (Note: This problem integrates lot sizing with net requirements planning.) **PX**

•••• **14.32** Using the data for the coffee table in Problem 14.31, build a labor schedule when the labor standard for each top is 2 labor-hours; each leg including brass cap installation requires $\frac{1}{4}$ hour, as does each pair of braces. Base assembly requires 1 labor-hour, and final assembly requires 2 labor-hours. What is the total number of labor-hours required each day, and how many employees are needed each day at 8 hours per day? (Hint: Use the MRP II format shown in Table 14.4.)

CASE STUDIES

Hill's Automotive, Inc.

Hill's Automotive, Inc., is an aftermarket producer and distributor of automotive replacement parts. Art Hill has slowly expanded the business, which began as a supplier of hard-to-get auto air-conditioning units for classic cars and hot rods. The firm has limited manufacturing capability, but a state-of-the-art MRP system and extensive inventory and assembly facilities. Components are purchased, assembled, and repackaged. Among its products are private-label air-conditioning units, carburetors, and ignition kits. The downturn in the economy, particularly the company's discretionary segment, has put downward pressure on volume and margins. Profits have fallen considerably. In addition, customer service levels have declined, with late deliveries now exceeding 25% of orders. And to make matters worse, customer returns have been rising at a rate of 3% per month.

Wally Hopp, vice president of sales, claims that most of the problem lies with the assembly department. He says that although the firm has accurate bills of material, indicating what goes into each product, it is not producing the proper mix of the product. He also believes the firm has poor quality control and low productivity, and as a result its costs are too high.

Melanie Thompson, treasurer, believes that problems are due to investing in the wrong inventories. She thinks that marketing has too many options and products. Melanie also thinks that purchasing department buyers have been hedging their inventories and requirements with excess purchasing commitments.

The assembly manager, Kalinga Jagoda, says, "The symptom is that we have a lot of parts in inventory but no place to assemble them in the production schedule. When we have the right part, it is not very good, but we use it anyway to meet the schedule."

Marshall Fisher, manager of purchasing, has taken the stance that purchasing has not let Hill's Automotive down. He has stuck by his old suppliers, used historical data to determine requirements, maintained what he views as excellent prices from suppliers, and evaluated new sources of supply with a view toward lowering cost. Where possible, Marshall reacted to the increased pressure for profitability by emphasizing low cost and early delivery.

Discussion Questions

1. Prepare a plan for Art Hill that gets the firm back on a course toward improved profitability. Be sure to identify the symptoms, the problems, and the specific changes you would implement.
2. Explain how MRP plays a role in this plan.

When 18,500 Orlando Magic Fans Come to Dinner

With vast experience at venues such as the American Airlines Arena (in Miami), the Kentucky Derby, and Super Bowls, Chef John Nicely now also plans huge culinary events at Orlando's Amway Center, home of the Orlando Magic basketball team. With his unique talent and exceptional operations skills, Nicely serves tens of thousands of cheering fans at some of the world's largest events. And when more than 18,500 basketball fans show up for a game, expecting great food and great basketball, he puts his creative as well as operations talent to work.

Chef John must be prepared. This means determining not only a total demand for all 18,500 fans, but also translating that demand into specific menu items and beverages. He prepares a forecast from current ticket sales, history of similar events at other venues, and his own records, which reflect the demand with this particular opponent, night of week, time of year, and even time of day. He then breaks the demand for specific menu items and quantities into items to be available at each of the 22 concession stands, 7 restaurants, and 68 suites. He must also be prepared to accommodate individual requests from players on both teams.

Chef John frequently changes the menu to keep it interesting for the fans who attend many of the 41 regular season home games each season. Even the culinary preference of the opponent's fans who may be attending influences the menu. Additionally, when entertainment other than the Magic is using the Amway Center, the demographic mix is likely to be different, requiring additional tweaking of the menu. The size of the wait staff and the kitchen staff change to reflect the size of the crowd; Chef John may be supervising as many as 90 people working in the kitchen. Similarly, the concessions stands, 40% of which have their own grills and fryers, present another challenge, as they are managed by volunteers from nonprofit organizations. The use of these volunteers adds the need for special training and extra enforcement of strict quality standards.

Once deciding on the overall demand and the menu, Chef John must prepare the production specifications (a bill of material) for each item. For the evening game with the Celtics, Chef John is preparing his unique *Cheeto Crusted Mac & Cheese* dish. The ingredients, quantity, costs, and labor requirements are shown here:

Production Specifications

CHEETO CRUSTED MAC & CHEESE (6 PORTIONS)					
INGREDIENTS	QUANTITY	MEASURE	UNIT COST	TOTAL COST	LABOR-HOURS
Elbow macaroni (large, uncooked)	20.00	oz.	$0.09	$ 1.80	
Cheese—cheddar shredded	10.00	oz.	0.16	1.60	
Mac and cheese base (see recipe)	44.00	oz.	0.80	35.20	
Milk	4.00	oz.	0.03	0.12	
Cheetos, crushed	6.00	oz.	0.27	1.62	
Sliced green onion—garnish	0.50	oz.	0.18	0.09	
Whole Cheetos—garnish	2.00	oz.	0.27	0.54	
Total labor hours					0.2 hour

The yield on this dish is six portions, and labor cost is $15 per hour, with fringes. The entire quantity required for the evening is prepared prior to the game and kept in warming ovens until needed. Demand for each basketball game is divided into five periods: prior to the game, first quarter, second quarter, half-time, and second half. At the Magic vs. Celtics game next week, the demand (number of portions) in each period is 60, 36, 48, 60, and 12 for the Cheeto Crusted Mac & Cheese dish, respectively.

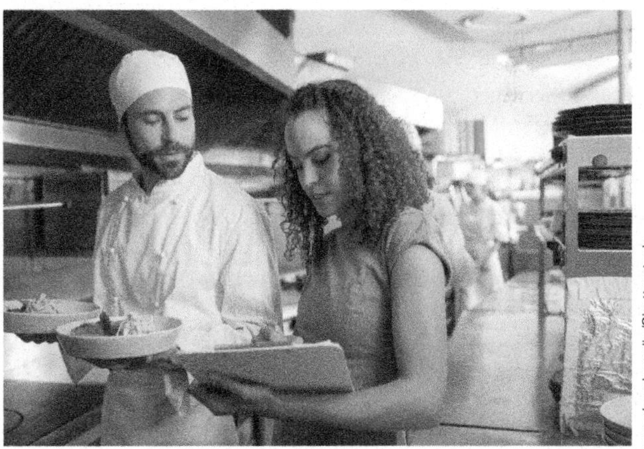
wavebreakmedia/Shutterstock

Discussion Questions*

1. Prepare a bill of material explosion and total cost for the 216 portions of Cheeto Crusted Mac & Cheese.
2. What is the cost per portion? How much less expensive is the Cheeto Crusted Mac & Cheese than Chef John's alternative creation, the Buffalo Chicken Mac & Cheese, shown in Figure 14.9 of this chapter?
3. Assuming that there is no beginning inventory of the Cheeto Crusted Mac & Cheese and cooking time for the entire 216 portions is 0.6 hours, when must preparation begin?

*You may wish to view the video that accompanies this case before answering the questions.

MRP at Wheeled Coach

Video Case

Wheeled Coach, the world's largest manufacturer of ambulances, builds thousands of different and constantly changing configurations of its products. The custom nature of its business means lots of options and special designs—and a potential scheduling and inventory nightmare. Wheeled Coach addressed such problems, and succeeded in solving a lot of them, with an MRP system (described in the *Global Company Profile* that opens this chapter). As with most MRP installations, however, solving one set of problems uncovers a new set.

One of the new issues that had to be addressed by plant manager Lynn Whalen was newly discovered excess inventory. Managers discovered a substantial amount of inventory that was not called for in any finished products. Excess inventory was evident because of the increased level of inventory accuracy required by the MRP system. The other reason was a new series of inventory reports generated by the MRP system installed by Wheeled Coach. One of those reports indicates where items are used and is known as the "Where Used" report. Interestingly, many inventory items were not called out on bills of material for any current products. In some cases, the reason some parts were in the stockroom remained a mystery.

The discovery of this excess inventory led to renewed efforts to ensure that the bills of material, as well as the receiving reports and inventory, were accurate. With substantial work, the accuracy of all three was increased. Additionally, the increased accuracy of the bills of material had the added advantage of reducing the number of engineering change notices (ECNs). Similarly, purchase-order accuracy, with regard to both part numbers and quantities ordered, was improved, reducing costs and ultimately improving quality and performance to schedule.

Eventually, Lynn Whalen concluded that the residual amounts of excess inventory were the result, at least in part, of rapid changes in ambulance design and technology. Another source was customer changes made after specifications had been determined and materials ordered. This latter excess occurs because, even though Wheeled Coach's own throughput time is only 17 days, many of the items that it purchases require much longer lead times.

Discussion Questions*

1. Why is accurate inventory such an important issue at Wheeled Coach?
2. Why does Wheeled Coach have excess inventory, and what kind of a plan would you suggest for dealing with it?
3. Be specific in your suggestions for reducing inventory and how to implement them.

*You may wish to view the video that accompanies this case before answering the questions.

Endnotes

1. The inventory models (EOQ) discussed in Chapter 12 assumed that the demand for one item was independent of the demand for another item. For example, EOQ assumes the demand for refrigerator parts is *independent* of the demand for refrigerators and that demand for parts is constant. MRP makes neither of these assumptions.
2. Record accuracy of 99% may sound good, but note that even when each component has an availability of 99% and a product has only seven components, the likelihood of a product being completed is only .932 (because $.99^7 = .932$).
3. Using EOQ is a convenient approach for determining the time between orders, but other rules can be used.
4. *Part period balancing, Silver-Meal,* and *Wagner-Whitin* are included in the software POM for Windows and ExcelOM, found online.
5. See Ptak, Carol A., and Chad J. Smith, *Orlicky's Material Requirements Planning*, 3rd ed., New York: McGraw-Hill, 2011.
6. Sources: *WritePass Journal* (November 19, 2016); *Forbes* (December 2, 2011); and **madeinitaly.com**.

Bibliography

Barba-Gutierrez, Y., B. Adenso-Diaz, and S. M. Gupta. "Lot Sizing in Reverse MRP for Scheduling Disassembly." *International Journal of Production Economics* 111, no. 2 (February 2008): 741.

Bell, S. "Time Fence Secrets." *APICS* 16, no. 4 (April 2006): 44–48.

Bolander, S., and S. G. Taylor. "Scheduling Techniques: A Comparison of Logic." *Production and Inventory Management Journal* 41, no. 1 (1st Quarter 2000): 1–5.

Crandall, R. E. "The Epic Life of ERP." *APICS* 16, no. 2 (February 2006): 17–19.

DeMatteis J. J. "An Economic Lot-Sizing Technique, I: The Part-Period Algorithm." *IBM Systems Journal* 7 (1968): 30–38.

Diehl G. W., and A. J. Armstrong "Making MRP Work." *Industrial Engineer* 43, no. 11 (November 2011): 35–40.

Kanet, J., and V. Sridharan. "The Value of Using Scheduling Information in Planning Material Requirements." *Decision Sciences* 29, no. 2 (Spring 1998): 479–498.

Kirchmier, B., G. Plenert, and G. Quinn. *Finite Capacity Scheduling: Optimizing a Constrained Supply Chain.* New York: Wiley, 2013.

Louly, M.-A., and A. Dolgui. "Optimal Time Phasing and Periodicity for MRP with POQ Policy." *International Journal of Production Economics* 131, no. 1 (May 2011): 76.

Norris, G. *E-Business & ERP.* New York: Wiley, 2005.

O'Sullivan, J., and G. Caiola. *Enterprise Resource Planning*, 2nd ed. New York: McGraw-Hill, 2008.

Ptak, C. A., and C. Smith. *Demand Driven Material Requirements Planning*. South Norwalk CT: Industrial Press, 2016.

Ptak, C. A, and C. J. Smith, *Orlicky's Material Requirements Planning,* 3rd ed. New York: McGraw-Hill, 2011.

Rossi, T., R. Pozzi, M. Pero, and R. Cigolini. "Improving Production Planning through Finite-Capacity MRP." *International Journal of Production Research* 55, no. 2 (2017): 377–391.

Steger-Jensen, K., H. Hvolby, P. Nielsen, and I. Nielsen. "Advanced Planning and Scheduling Technology." *Production Planning and Control* 22, no. 8 (December 2011): 800–808.

Tian, F., and S. X. Xu. "How do Enterprise Resource Planning Systems Affect Firm Risk? Post-Implementation Impact." *MIS Quarterly* 39, no. 1 (March 2015): 39–60.

Turbide, D. "New-Fashioned MRP." *APICS* 24, no. 3 (July/August 2015): 40–44.

Wagner, H. M., and T. M. Whitin. "Dynamic Version of the Economic Lot Size Model." *Management Science* 5, no.1 (1958): 89–96.

Wu, J.-H., et al. "Using Multiple Variables Decision-Making Analysis for ERP Selection." *International Journal of Manufacturing Technology and Management* 18, no. 2 (2009): 228.

Chapter 14 *Rapid* Review

Main Heading	Review Material	MyLab Operations Management
DEPENDENT DEMAND	Demand for items is *dependent* when the relationship between the items can be determined. For any product, all components of that product are dependent demand items. ■ **Material requirements planning (MRP)**—A dependent demand technique that uses a bill-of-material, inventory, expected receipts, and a master production schedule to determine material requirements.	Concept Questions: 1.1–1.5
DEPENDENT INVENTORY MODEL REQUIREMENTS	Dependent inventory models require that the operations manager know the: (1) Master production schedule; (2) Specifications or bill of material; (3) Inventory availability; (4) Purchase orders outstanding; and (5) Lead times. ■ **Master production schedule (MPS)**—A timetable that specifies what is to be made and when. The MPS is a statement of *what is to be produced*, not a forecast of demand. ■ **Bill of material (BOM)**—A listing of the components, their description, and the quantity of each required to make one unit of a product. Items above any level in a BOM are called *parents*; items below any level are called *components*, or *children*. The top level in a BOM is the 0 level. ■ **Modular bills**—Bills of material organized by major subassemblies or by product options. ■ **Planning bills (or kits)**—Material groupings created in order to assign an artificial parent to a bill of material; also called "pseudo" bills. ■ **Phantom bills of material**—Bills of material for components, usually subassemblies, that exist only temporarily; they are never inventoried. ■ **Low-level coding**—A number that identifies items at the lowest level at which they occur. ■ **Lead time**—In purchasing systems, the time between recognition of the need for an order and receiving it; in production systems, it is the order, wait, move, queue, setup, and run times for each component. When a bill of material is turned on its side and modified by adding lead times for each component, it is called a *time-phased product structure*.	Concept Questions: 2.1–2.6 Problems: 14.1–14.4 Virtual Office Hours for Solved Problem: 14.1 **VIDEO 14.1** When 18,500 Orlando Magic Fans Come to Dinner **VIDEO 14.2** MRP at Wheeled Coach Ambulances
MRP STRUCTURE	■ **Gross material requirements plan**—A schedule that shows the total demand for an item (prior to subtraction of on-hand inventory and scheduled receipts) and (1) when it must be ordered from suppliers, or (2) when production must be started to meet its demand by a particular date. ■ **Net material requirements**—The result of adjusting gross requirements for inventory on hand and scheduled receipts. ■ **Planned order receipt**—The quantity planned to be received at a future date. ■ **Planned order release**—The scheduled date for an order to be released. Net requirements = Gross requirements + Allocations − (On hand + Scheduled receipts)	Concept Questions: 3.1–3.5 Problems: 14.5–14.21 Virtual Office Hours for Solved Problem: 14.2 **ACTIVE MODEL 14.1**
LOT-SIZING TECHNIQUES	■ **Lot-sizing decision**—The process of, or techniques used in, determining lot size. ■ **Lot-for-lot**—A lot-sizing technique that generates exactly what is required to meet the plan. ■ **Periodic order quantity (POQ)**—A lot-sizing technique that issues orders on a predetermined time interval with an order quantity equal to all of the interval's requirements. In general, the lot-for-lot approach should be used whenever low-cost deliveries setup can be achieved.	Concept Questions: 4.1–4.5 Problems: 14.22–14.28
MRP MANAGEMENT	■ **Demand-driven MRP (DDMRP)**—Strategically alters lead times and precisely places inventory within the BOM structure to improve MRP performance. ■ **Capacity planning**—Determining short-range capacity requirements. ■ **Buckets**—Time units in a material requirements planning system. ■ **Load or dispatch report**—A report for showing the resource requirements in a work center for all work currently assigned as well as all planned and expected orders.	Concept Questions: 5.1–5.11 Problems: 14.29–14.32

Chapter 14 **Rapid Review** *continued*

Main Heading	Review Material	MyLab Operations Management
	■ **System nervousness**—Frequent changes in an MRP system. ■ **Time fences**—A means for allowing a segment of the master schedule to be designated as "not to be rescheduled." ■ **Pegging**—In material requirements planning systems, tracing upward the bill of material from the component to the parent item. ■ **Material requirements planning II (MRP II)**—A system that allows, with MRP in place, inventory data to be augmented by other resource variables; in this case, MRP becomes *material resource planning*. Tactics for smoothing the load and minimizing the impact of changed lead time include: *overlapping, operations splitting*, and *order splitting* or *lot splitting*.	
MRP IN SERVICES	■ **Distribution resource planning (DRP)**—A time-phased stock-replenishment plan for all levels of a distribution network.	Concept Questions: 6.1–6.4
ENTERPRISE RESOURCE PLANNING (ERP)	■ **Enterprise resource planning (ERP)**—An information system for identifying and planning the enterprise-wide resources needed to take, make, ship, and account for customer orders. In an ERP system, data are entered only once into a common, complete, and consistent database shared by all applications. ■ **Efficient consumer response (ECR)**—Supply chain management systems in the grocery industry that tie sales to buying, to inventory, to logistics, and to production.	Concept Questions: 7.1–7.5
ADDITIONAL MYLAB OPERATIONS MANAGEMENT RESOURCES	✔ Additional Case Study (OSI's Attempt at ERP) ✔ Multiple Choice Case Questions (Hill's Automotive, Inc.)	

Self **Test**

LO 14.1 In a product structure diagram:
- **a)** parents are found only at the top level of the diagram.
- **b)** parents are found at every level in the diagram.
- **c)** children are found at every level of the diagram except the top level.
- **d)** all items in the diagrams are both parents and children.
- **e)** all of the above.

LO 14.2 The difference between a gross material requirements plan (gross MRP) and a net material requirements plan (net MRP) is:
- **a)** the gross MRP may not be computerized, but the net MRP must be computerized.
- **b)** the gross MRP includes consideration of the inventory on hand, whereas the net MRP doesn't include the inventory consideration.
- **c)** the net MRP includes consideration of the inventory on hand, whereas the gross MRP doesn't include the inventory consideration.
- **d)** the gross MRP doesn't take taxes into account, whereas the net MRP includes the tax considerations.
- **e)** the net MRP is only an estimate, whereas the gross MRP is used for actual production scheduling.

LO 14.3 Net requirements =
- **a)** Gross requirements + Allocations − On-hand inventory + Scheduled receipts.
- **b)** Gross requirements − Allocations − On-hand inventory − Scheduled receipts.
- **c)** Gross requirements − Allocations − On-hand inventory + Scheduled receipts.
- **d)** Gross requirements + Allocations − On-hand inventory − Scheduled receipts.

LO 14.4 A lot-sizing procedure that orders on a predetermined time interval with the order quantity equal to the total of the interval's requirement is:
- **a)** periodic order quantity.
- **b)** part period balancing.
- **c)** economic order quantity.
- **d)** all of the above.

LO 14.5 A _____ report provides work center information to the production plan.
- **a)** dynamic
- **b)** closed-loop
- **c)** pegging
- **d)** retrospective
- **e)** load

LO 14.6 MRP II stands for:
- **a)** material resource planning.
- **b)** management requirements planning.
- **c)** management resource planning.
- **d)** material revenue planning.
- **e)** material risk planning.

LO 14.7 Which system extends MRP II to tie in customers and suppliers?
- **a)** MRP III
- **b)** JIT
- **c)** IRP
- **d)** ERP
- **e)** Enhanced MRP II

Answers: LO 14.1. c; LO 14.2. c; LO 14.3. d; LO 14.4. a; LO 14.5. e; LO 14.6. a; LO 14.7. d.

Short-Term Scheduling

CHAPTER 15

CHAPTER OUTLINE

Scheduling servers, chefs, bartenders, dishwashers, and cleaning staff for nearly round-the-clock activities is a critical issue on luxury ships such as those of Celebrity Cruises. Here, on Celebrity's newest ship, the Apex, over 1,300 crew members have to be scheduled to serve 2,900 passengers in the 29 restaurants, bars, and lounges onboard.

Beverly Amer/Aspenleaf Productions

Beverly Amer/Aspenleaf Productions

Beverly Amer/Aspenleaf Productions

10 OM STRATEGY DECISIONS

- Design of Goods and Services
- Managing Quality
- Process Strategies
- Location Strategies
- Layout Strategies
- Human Resources
- Supply Chain Management
- Inventory Management
- *Scheduling*
  - Aggregate/S&OP (Ch. 13)
  - ***Short-Term (Ch. 15)***
- Maintenance

GLOBAL COMPANY PROFILE
Alaska Airlines

Scheduling Flights When Weather Is the Enemy

Seattle–Tacoma International Airport (Sea-Tac) is the 9th busiest in the U.S. in passenger traffic. Served by 28 airlines that fly non-stop to 81 domestic and 24 international destinations, it is a weather forecaster's nightmare, raining 5 inches a month in the winter season. But it is also the top-ranked U.S. airport in on-time departures, at 85.8%. Much of the credit goes to Alaska Airlines, which dominates traffic at Sea-Tac with over 50% of all domestic flights. Alaska's scheduling is critical to efficiency and passenger service.

4 A.M.	10 A.M.	1:30 P.M.	5 P.M.	10 P.M.
FORECAST: Rain with a chance of light snow for Seattle.	FORECAST: Freezing rain after 5 P.M.	FORECAST: Rain changing to snow.	FORECAST: Less snow than expected.	FORECAST: Snow tapering off.
ACTION: Discuss status of planes and possible need for cancellations.	ACTION: Ready deicing trucks; develop plans to cancel 50% to 80% of flights after 6 P.M.	ACTION: Cancel half the flights from 6 P.M. to 10 A.M.; notify passengers and reroute planes.	ACTION: Continue calling passengers and arrange alternate flights.	ACTION: Find hotels for 600 passengers stranded by the storm.

This is typical of what Alaska Air officials had to do one December day when a storm bore down on Seattle.

Managers at airlines, such as Alaska, learn to expect the unexpected. Events that require rapid rescheduling are a regular part of life. Throughout the ordeals of hurricanes, tornadoes, ice storms, snowstorms, and more, airlines around the globe struggle to cope with delays, cancellations, and furious passengers. The inevitable schedule changes often create a ripple effect that impacts passengers at dozens of airports.

Courtesy of Alaska Airlines

To improve flight rescheduling efforts, Alaska Air employees monitor numerous screens that display flights in progress, meteorological charts, and weather patterns at its Flight Operations Department in Seattle. Note the many *andon* signal lights used to indicate "status OK" (green), "needs attention" (yellow), or "major issue—emergency" (red).

Weather-related disruptions can create major scheduling and expensive snow removal issues for airlines (right), just as they create major inconveniences for passengers (left).

To maintain schedules, Alaska Airlines uses elaborate equipment and motivated personnel for snow and ice removal.

Alaska Air's quest to provide passenger and freight service to the state of Alaska complicates its scheduling even more than that of other airlines. Here are just three examples: (1) Juneau's airport is surrounded by mountains, so the approach is often buffeted by treacherous wind shears; (2) Sitka's one small runway is on a narrow strip of land surrounded by water; and (3) in Kodiak, the landing strip ends abruptly at a mountainside. The airport approach is so tricky that first officers are not allowed to land there—only captains are trusted to do so.

Alaska Air takes the sting out of the scheduling nightmares that come from weather-related problems by using the latest technology on its planes and in its Flight Operations Department, located near Sea-Tac airport. From computers to telecommunications systems to deicers, the department reroutes flights, gets its jets in the air, and quickly notifies customers of schedule changes. The department's job is to keep flights flowing despite the disruptions. Alaska estimates that it saves $18 million a year by using its technology to reduce cancellations and delays.

With mathematical scheduling models such as the ones described in this text, Alaska quickly develops alternate schedules and route changes. This may mean coordinating incoming and outgoing aircraft, ensuring crews are on hand, and making sure information gets to passengers as soon as possible. Weather may be the enemy, but Alaska Airlines has learned how to manage it.

LEARNING OBJECTIVES

LO 15.1 ***Explain*** the relationship among short-term scheduling, capacity planning, aggregate planning, and a master schedule 611

LO 15.2 ***Draw*** Gantt loading and scheduling charts 615

LO 15.3 ***Apply*** the assignment method for loading jobs 616

LO 15.4 ***Name*** and describe each of the priority sequencing rules 621

LO 15.5 ***Use*** Johnson's rule 624

LO 15.6 ***Define*** finite capacity scheduling 625

LO 15.7 ***Use*** the cyclical scheduling technique 627

The Importance of Short-Term Scheduling

Alaska Airlines doesn't just schedule its 156 aircraft every day; it also schedules over 6,000 pilots and flight attendants to accommodate passengers seeking timely arrival at their destinations. This schedule, developed with huge computer programs, plays a major role in meeting customer expectations. Alaska finds competitive advantage with its ability to make last-minute adjustments to demand fluctuations and weather disruptions.

Scheduling decisions for five organizations—an airline, a hospital, a college, a sports arena, and a manufacturer—are shown in Table 15.1. These decisions all deal with the timing of operations.

When manufacturing firms make schedules that match resources to customer demands, scheduling competence focuses on making parts on a just-in-time basis, with low setup times, little work-in-process, and high facility utilization. Efficient scheduling is how manufacturing companies drive down costs and meet promised due dates.

The strategic importance of scheduling is clear:

- Internally effective scheduling means faster movement of goods and services through a facility and greater use of assets. The result is greater capacity per dollar invested, which translates into lower costs.
- Externally good scheduling provides faster throughput, added flexibility, and more dependable delivery, improving customer service.

STUDENT TIP Scheduling decisions range from years, for capacity planning, to minutes/hours/days, called short-term scheduling. This chapter focuses on the latter.

Scheduling Issues

Figure 15.1 shows that a series of decisions affects scheduling. Schedule decisions begin with planning capacity, which defines the facility and equipment resources available (discussed in Supplement 7). *Capacity plans* are usually made over a period of years as new equipment

TABLE 15.1 Scheduling Decisions

ORGANIZATION	MANAGERS SCHEDULE THE FOLLOWING
Alaska Airlines	Maintenance of aircraft Departure timetables Flight crews, catering, gate, and ticketing personnel
Arnold Palmer Hospital	Operating room use Patient admissions Nursing, security, maintenance staffs Outpatient treatments
University of Alabama	Classrooms and audiovisual equipment Student and instructor schedules Graduate and undergraduate courses
Amway Center	Ushers, ticket takers, food servers, security personnel Delivery of fresh foods and meal preparation Orlando Magic games, concerts, arena football
Lockheed Martin factory	Production of goods Purchases of materials Workers

VIDEO 15.1 From the Eagles to the Magic: Converting the Amway Center

Capacity Planning
(Long term; years)
Changes in facilities
Changes in equipment
See Chapter 7 and Supplement 7

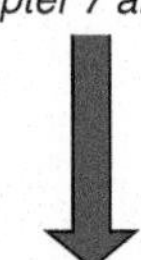

Capacity Plan for New Facilities
Adjust capacity to the demand suggested by strategic plan

Aggregate Planning
(Intermediate term; quarterly or monthly)
Facility utilization
Personnel changes
Subcontracting
See Chapter 13

Aggregate Production Plan for All Bikes
(Determine personnel or subcontracting necessary to match aggregate demand to existing facilities/capacity)

Month	1	2
Bike Production	800	850

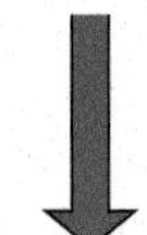

Master Schedule
(Intermediate term; weekly)
Material requirements planning
Disaggregate the aggregate plan
See Chapters 13 and 14

Master Production Schedule for Bike Models
(Determine weekly capacity schedule)

	Month 1				Month 2			
Week	1	2	3	4	5	6	7	8
Model 22		200		200		200		200
Model 24	100		100		150		100	
Model 26	100		100		100		100	

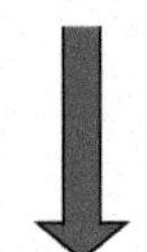

Short-Term Scheduling
(Short term; days, hours, minutes)
Work center loading
Job sequencing/dispatching
See this chapter

Work Assigned to Specific Personnel and Work Centers
Make finite capacity schedule by matching specific tasks to specific people and machines

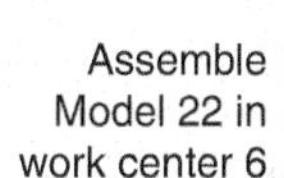

Figure **15.1**

The Relationship among Capacity Planning, Aggregate Planning, Master Schedule, and Short-Term Scheduling for a Bike Company

and facilities are designed, built, purchased, or shut down. *Aggregate plans* (Chapter 13) are the result of a Sales and Operating Planning team that makes decisions regarding the use of facilities, inventory, people, and outside contractors. Aggregate plans are typically for 3 to 18 months, and resources are allocated in terms of an aggregate measure such as total units, tons, or shop hours. The *master schedule* breaks down the aggregate plan and develops weekly schedules for specific products or product lines. *Short-term schedules* then translate capacity decisions, aggregate (intermediate) plans, and master schedules into job sequences and specific assignments of personnel, materials, and machinery. In this chapter, we focus on this last step, scheduling goods and services in the *short run* (that is, matching daily or hourly demands to specific personnel and equipment capacity). See the *OM in Action* box, "Prepping for the Orlando Magic Basketball Game."

LO 15.1 *Explain* the relationship among short-term scheduling, capacity planning, aggregate planning, and a master schedule

The objective of scheduling is to allocate and prioritize demand (generated by either forecasts or customer orders) to available facilities. Three factors are pervasive in scheduling: (1) generating the schedule forward or backward, (2) finite and infinite loading, and (3) the criteria (priorities) for sequencing jobs. We discuss these topics next.

Forward and Backward Scheduling

Scheduling can be initiated forward or backward. Forward scheduling starts the schedule *as soon as the job requirements are known*. Forward scheduling is used in organizations such as hospitals, clinics, restaurants, and machine tool manufacturers. In these facilities, jobs are performed to customer order, and delivery is typically scheduled at the earliest possible date.

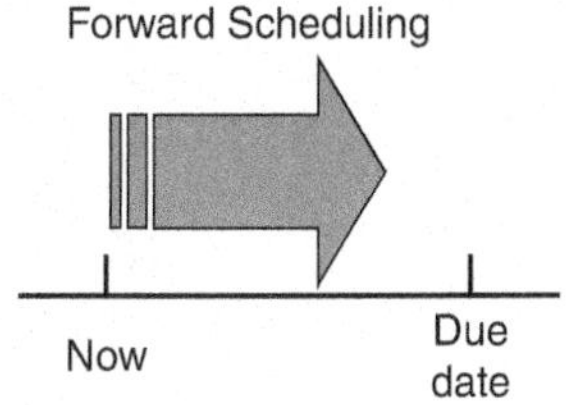

OM in Action Prepping for the Orlando Magic Basketball Game[1]

Tuesday. It's time for John Nicely to make a grocery list. He is serving dinner on Sunday, so he will need a few things... 200 pounds of chicken and steak, ingredients for 800 servings of mac 'n' cheese, 500 spring rolls, and 75 pounds of shrimp. Plus a couple hundred pizzas and a couple thousand hot dogs—just enough to feed the Orlando Magic basketball players and the 18,500 guests expected. You see, Nicely is the executive chef of the Amway Center in Orlando, and on Sunday the Magic are hosting the Boston Celtics.

How do you feed huge crowds good food in a short time? It takes good scheduling, combined with creativity and improvisation. With 42 facilities serving food and beverages, "the Amway Center," Nicely says, "is its own beast."

Wednesday. Shopping Day.

Thursday–Saturday. The staff prepares whatever it can. Chopping vegetables, marinating meats, mixing salad dressings—everything but cooking the food. Nicely also begins his shopping lists for next Tuesday's game against the Miami Heat and for a Lady Gaga concert 3 days later.

Sunday. 4 P.M. Crunch time. Suddenly the kitchen is a joke-free zone. In 20 minutes, Nicely's first clients, 120 elite ticket holders who belong to the Ritz Carlton Club, expect their meals—from a unique menu created for each game.

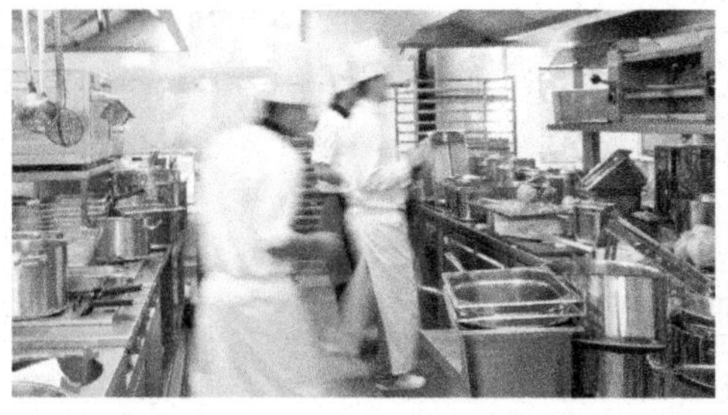
STOCK4B GmbH/Alamy Stock Photo

5 P.M. As the Magic and Celtics start warming up, the chefs move their operation in a brisk procession of hot boxes and cold-food racks to the satellite kitchens.

6:12 P.M. Nicely faces surprises at three concession stands: a shortage of cashiers and a broken cash register.

Halftime. There is a run on rice pilaf in the upscale Jernigan's restaurant. But Nicely has thought ahead and has anticipated. The backup dishes arrive before customers even notice.

For Nicely, successful scheduling means happy guests as a result of a thousand details having been identified, planned, and executed. Just another night of delivering restaurant-quality meals and top-grade fast food to a sold-out arena crowd in a span of a few hours.

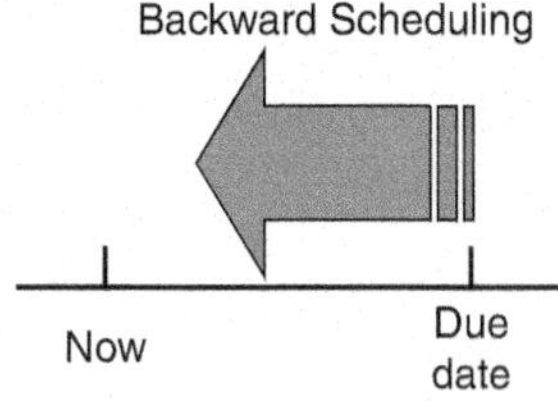

Backward scheduling *begins with the due date,* scheduling the *final* operation first. Steps within the job are then scheduled, one at a time, in reverse order. By subtracting the time needed for each item, the start time is obtained. Backward scheduling is used in manufacturing environments, as well as service environments such as catering a banquet or scheduling surgery. In practice, a combination of forward and backward scheduling is often used to find a reasonable trade-off between capacity constraints and customer expectations.

Finite and Infinite Loading

Loading
The assigning of jobs to work or processing centers.

Loading is the process of assigning jobs to work stations or processes. Scheduling techniques that load (or assign) work only up to the capacity of the process are called *finite loading*. The advantage of finite loading is that, in theory, all of the work assigned can be accomplished. However, because only work that can be accomplished is loaded into workstations—when in fact there may be more work than capacity—the due dates may be pushed out to an unacceptable future time.

Techniques that load work without regard for the capacity of the process are *infinite loading*. All the work that needs to be accomplished in a given time period is assigned. The capacity of the process is not considered. Most material requirements planning (MRP) systems (discussed in Chapter 14) are infinite loading systems. The advantage of infinite loading is an initial schedule that meets due dates. Of course, when the workload exceeds capacity, either the capacity or the schedule must be adjusted.

Scheduling Criteria

The correct scheduling technique depends on the volume of orders, the nature of operations, and the overall complexity of jobs, as well as the importance placed on each of four criteria:

1. *Minimize completion time:* Evaluated by determining the average completion time.
2. *Maximize utilization:* Evaluated by determining the percent of the time the facility is utilized.
3. *Minimize work-in-process (WIP) inventory:* Evaluated by determining the average number of jobs in the system. The relationship between the number of jobs in the system

TABLE 15.2 Different Processes Suggest Different Approaches to Scheduling

Process-focused facilities (job shops) ◆ Scheduling to customer orders where changes in both volume and variety of jobs/clients/patients are frequent. ◆ Schedules are often due-date focused, with loading refined by finite loading techniques. ◆ *Examples:* foundries, machine shops, cabinet shops, print shops, many restaurants, and the fashion industry.
Repetitive facilities (assembly lines) ◆ Schedule module production and product assembly based on frequent forecasts. ◆ Finite loading with a focus on generating a forward-looking schedule. ◆ JIT techniques are used to schedule components that feed the assembly line. ◆ *Examples:* assembly lines for washing machines at Whirlpool and automobiles at Ford.
Product-focused facilities (continuous) ◆ Schedule high-volume finished products of limited variety to meet a reasonably stable demand within existing fixed capacity. ◆ Finite loading with a focus on generating a forward-looking schedule that can meet known setup and run times for the limited range of products. ◆ *Examples:* huge paper machines at International Paper, beer in a brewery at Anheuser-Busch, and potato chips at Frito-Lay.

and WIP inventory will be high. Therefore, the fewer the number of jobs that are in the system, the lower the inventory.

4. *Minimize customer waiting time:* Evaluated by determining the average number of late periods (e.g., days or hours).

These four criteria are used in this chapter, as they are in industry, to evaluate scheduling performance. In addition, good scheduling techniques should be simple, clear, easily understood, easy to carry out, flexible, and realistic.

Scheduling is further complicated by machine breakdowns, absenteeism, quality problems, shortages, and other factors. Consequently, assignment of a date does not ensure that the work will be performed according to the schedule. Many specialized techniques have been developed to aid in preparing reliable schedules. Table 15.2 provides an overview of approaches to scheduling for three different processes.

In this chapter, we first examine the scheduling of process-focused facilities and then the challenge of scheduling employees in the service sector.

Scheduling Process-Focused Facilities

Process-focused facilities (also known as *intermittent* or *job-shop facilities*) are common in high-variety, low-volume manufacturing and service organizations. These facilities produce make-to-order products or services and include everything from auto repair garages and hospitals to beauty salons. The production items themselves differ considerably, as do the talents, material, and equipment required to make them. Scheduling requires that the sequence of work (its routing), time required for each item, and the capacity and availability of each work center be known. The variety of products and unique requirements means that scheduling is often complex. In this section we look at some of the tools available to managers for loading and sequencing work for these facilities.

Loading Jobs

Operations managers assign jobs to work centers so that costs, idle time, or completion times are kept to a minimum. "Loading" work centers takes two forms. One is oriented to capacity; the second is related to assigning specific jobs to work centers.

First, we examine loading from the perspective of capacity via a technique known as *input–output* control. Then, we present two approaches used for loading: *Gantt charts* and the *assignment method* of linear programming.

Input–Output Control

Many firms have difficulty scheduling (that is, achieving effective throughput) because they overload the production processes. This often occurs because they do not know actual performance in the work centers. Effective scheduling depends on matching the schedule to performance. Lack of knowledge about capacity and performance causes reduced throughput.

Input–output control
A system that allows operations personnel to manage facility work flows.

Input–output control is a technique that allows operations personnel to manage facility work flows. If the work is arriving faster than it is being processed, the facility is overloaded, and a backlog develops. Overloading causes crowding in the facility, leading to inefficiencies and quality problems. If the work is arriving at a slower rate than jobs are being performed, the facility is underloaded, and the work center may run out of work. Underloading the facility results in idle capacity and wasted resources. Example 1 shows the use of input–output controls.

Example 1

INPUT–OUTPUT CONTROL

Bronson Machining, Inc., manufactures driveway security fences and gates. It wants to develop an input–output control report for its welding work center for 5 weeks (weeks 6/6 through 7/4). The planned input is 280 standard hours per week. The actual input is close to this figure, varying between 250 and 285. Output is scheduled at 320 standard hours, which is the assumed capacity. A backlog exists in the work center.

APPROACH ▶ Bronson uses schedule information to create Figure 15.2, which monitors the workload–capacity relationship at the work center.

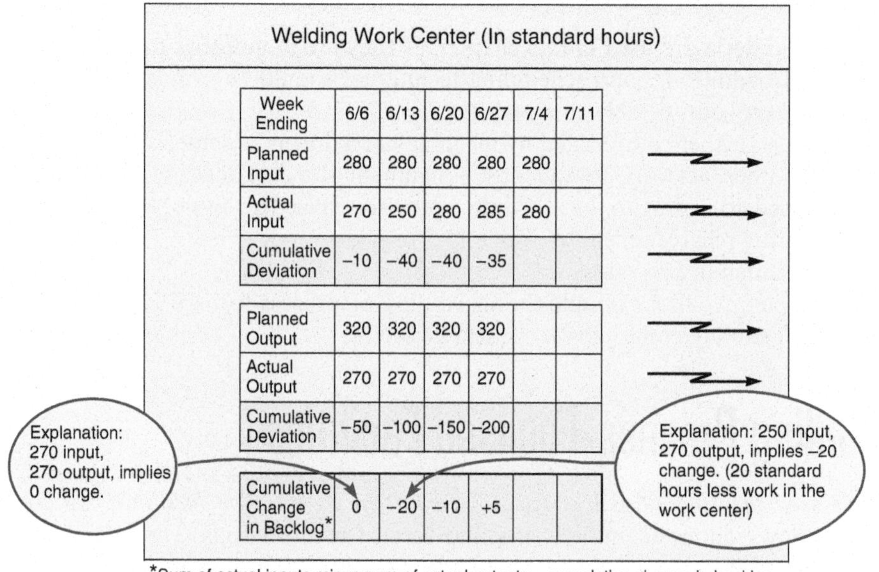

Welding Work Center (In standard hours)

Week Ending	6/6	6/13	6/20	6/27	7/4	7/11
Planned Input	280	280	280	280	280	
Actual Input	270	250	280	285	280	
Cumulative Deviation	−10	−40	−40	−35		
Planned Output	320	320	320	320		
Actual Output	270	270	270	270		
Cumulative Deviation	−50	−100	−150	−200		
Cumulative Change in Backlog*	0	−20	−10	+5		

Explanation: 270 input, 270 output, implies 0 change.
Explanation: 250 input, 270 output, implies −20 change. (20 standard hours less work in the work center)

*Sum of actual inputs minus sum of actual outputs = cumulative change in backlog

Figure 15.2
Input–Output Control

SOLUTION ▶ The deviations between scheduled input and actual output are shown in Figure 15.2. Actual output (270 hours) is substantially less than planned. Therefore, neither the input plan nor the output plan is being achieved.

INSIGHT ▶ The backlog of work in this work center has actually increased by 5 hours by week 6/27. This increases work-in-process inventory, complicating the scheduling task and indicating the need for manager action.

LEARNING EXERCISE ▶ If actual output for the week of 6/27 was 275 (instead of 270), what changes? [Answer: Output cumulative deviation now is −195, and cumulative change in backlog is 0.]

RELATED PROBLEM ▶ 15.10

ConWIP cards
Cards that control the amount of work in a work center, aiding input–output control.

Input–output control can be maintained by a system of **ConWIP cards**, which control the amount of work in a work center. ConWIP is an acronym for *constant work-in-process*. The ConWIP card travels with a job (or batch) through the work center. When the job is finished, the card is released and returned to the initial workstation, authorizing the entry of a new batch into the work center.

The ConWIP card effectively limits the amount of work in the work center, controls lead time, and monitors the backlog.

Gantt Charts

Gantt charts are visual aids that are useful in loading and scheduling. The name is derived from Henry Gantt, who developed them in the late 1800s. The charts show the use of resources, such as work centers and labor.

Gantt charts
Planning charts used to schedule resources and allocate time.

When used in *loading*, Gantt charts show the loading and idle times of several departments, machines, or facilities. They display the relative workloads in the system so that the manager knows what adjustments are appropriate. For example, when one work center becomes overloaded, employees from a low-load center can be transferred temporarily to increase the workforce. Or if waiting jobs can be processed at different work centers, some jobs at high-load centers can be transferred to low-load centers. Versatile equipment may also be transferred among centers. Example 2 illustrates a simple Gantt load chart.

Example 2

GANTT LOAD CHART

A New Orleans washing machine manufacturer accepts special orders for machines to be used in such unique facilities as submarines, hospitals, and large industrial laundries. The production of each machine requires varying tasks and durations. The company wants to build a load chart for the week of March 8.

APPROACH ▶ The Gantt chart is selected as the appropriate graphical tool.

SOLUTION ▶ Figure 15.3 shows the completed Gantt chart.

Figure **15.3**
Gantt Load Chart for the Week of March 8

Work Center \ Day	Monday	Tuesday	Wednesday	Thursday	Friday
Metalworks	Job 349		← Job 350 →		
Mechanical		← Job 349 →		Job 408	
Electronics	Job 408			Job 349	
Painting	← Job 295 →		Job 408		Job 349

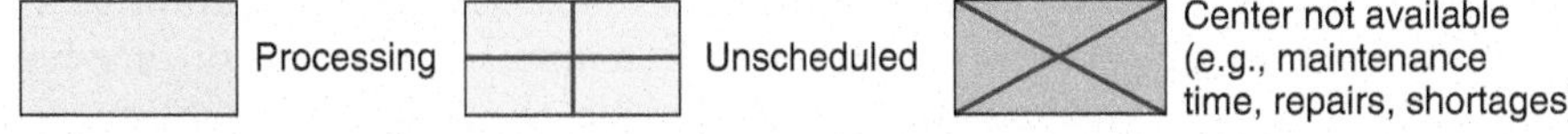

INSIGHT ▶ The four work centers process several jobs during the week. This particular chart indicates that the metalworks and painting centers are completely loaded for the entire week. The mechanical and electronic centers have some idle time scattered during the week. We also note that the metalworks center is unavailable on Tuesday, and the painting center is unavailable on Thursday, perhaps for preventive maintenance.

LEARNING EXERCISE ▶ What impact results from the electronics work center closing on Tuesday for preventive maintenance? [Answer: None.]

RELATED PROBLEM ▶ 15.1b

The Gantt *load chart* has a major limitation: it does not account for production variability such as unexpected breakdowns or human errors that require reworking a job. Consequently, the chart must also be updated regularly to account for new jobs and revised time estimates.

A Gantt *schedule chart* is used to monitor jobs in progress (and is also used for project scheduling). It indicates which jobs are on schedule and which are ahead of or behind schedule. In practice, many versions of the chart are found. The schedule chart in Example 3 places jobs in progress on the vertical axis and time on the horizontal axis.

LO 15.2 *Draw* Gantt loading and scheduling charts

Example 3

GANTT SCHEDULING CHART

First Printing in Winter Park, Florida, wants to use a Gantt chart to show the scheduling of three orders, jobs A, B, and C.

APPROACH ▶ In Figure 15.4, each pair of brackets on the time axis denotes the estimated starting and finishing of a job enclosed within it. The solid bars reflect the actual status or progress of the job. We are just finishing day 5.

SOLUTION ▶

Figure **15.4**

Gantt Scheduling Chart for Jobs A, B, and C at First Printing

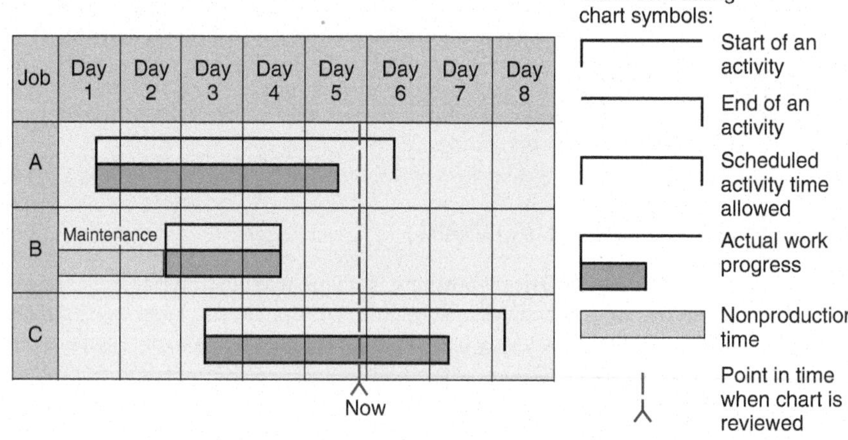

INSIGHT ▶ Figure 15.4 illustrates that job A is about a half-day behind schedule at the end of day 5. Job B was completed after equipment maintenance. We also see that job C is ahead of schedule.

LEARNING EXERCISE ▶ Redraw the Gantt chart to show that job A is a half-day *ahead* of schedule. [Answer: The orangish bar now extends all the way to the end of the activity.]

RELATED PROBLEMS ▶ 15.1a, 15.2

Assignment Method

Assignment method
A special class of linear programming models that involves assigning tasks or jobs to resources.

The **assignment method** involves assigning tasks or jobs to resources. Examples include assigning jobs to machines, contracts to bidders, people to projects, and salespeople to territories. The objective is most often to minimize total costs or time required to perform the tasks at hand. One important characteristic of assignment problems is that only one job (or worker) is assigned to one machine (or project).

Each assignment problem uses a table. The numbers in the table will be the costs or times associated with each particular assignment. For example, if First Printing has three available typesetters (A, B, and C) and three new jobs to be completed, its table might appear as follows. The dollar entries represent the firm's estimate of what it will cost for each job to be completed by each typesetter.

LO 15.3 *Apply* the assignment method for loading jobs

	TYPESETTER		
JOB	A	B	C
R-34	$11	$14	$ 6
S-66	$ 8	$10	$11
T-50	$ 9	$12	$ 7

The assignment method involves adding and subtracting appropriate numbers in the table to find the lowest *opportunity cost*[2] for each assignment. There are four steps to follow:

1. Subtract the smallest number in each row from every number in that row and then, from the resulting matrix, subtract the smallest number in each column from every number in that column. This step has the effect of reducing the numbers in the table until a series

of zeros, meaning *zero opportunity costs*, appear. Even though the numbers change, this reduced problem is equivalent to the original one, and the same solution will be optimal.

2. Draw the minimum number of vertical and horizontal straight lines necessary to cover all zeros in the table. If the number of lines equals either the number of rows or the number of columns in the table, then we can make an optimal assignment (see Step 4). If the number of lines is less than the number of rows or columns, we proceed to Step 3.
3. Subtract the smallest number not covered by a line from every other uncovered number. Add the same number to any number(s) lying at the intersection of any two lines. Do not change the value of the numbers that are covered by only one line. Return to Step 2 and continue until an optimal assignment is possible.
4. Optimal assignments will always be at zero locations in the table. One systematic way of making a valid assignment is first to select a row or column that contains only one zero square. We can make an assignment to that square and then draw lines through its row and column. From the uncovered rows and columns, we choose another row or column in which there is only one zero square. We make that assignment and continue the procedure until we have assigned each person or machine to one task.

Example 4 shows how to use the assignment method.

Example 4

ASSIGNMENT METHOD

First Printing wants to find the minimum total cost assignment of 3 jobs to 3 typesetters.

APPROACH ▶ The cost table shown earlier in this section is repeated here, and steps 1 through 4 are applied.

STUDENT TIP

You can also tackle assignment problems with our Excel OM or POM software or with Excel's Solver add-in.

JOB \ TYPESETTER	A	B	C
R-34	$11	$14	$ 6
S-66	$ 8	$10	$11
T-50	$ 9	$12	$ 7

SOLUTION ▶

Step 1A: Using the previous table, subtract the smallest number in each row from every number in the row. The result is shown in the table on the left.

JOB \ TYPESETTER	A	B	C
R-34	5	8	0
S-66	0	2	3
T-50	2	5	0

JOB \ TYPESETTER	A	B	C
R-34	5	6	0
S-66	0	0	3
T-50	2	3	0

Step 1B: Using the above left table, subtract the smallest number in each column from every number in the column. The result is shown in the table on the right.

Step 2: Draw the minimum number of vertical and horizontal straight lines needed to cover all zeros. Because two lines suffice, the solution is not optimal.

JOB \ TYPESETTER	A	B	C
R-34	5	6	0
S-66	0	0	3
T-50	(2)	3	0

Smallest uncovered number

Step 3: Subtract the smallest uncovered number (2 in this table) from every other uncovered number and add it to numbers at the intersection of two lines.

JOB \ TYPESETTER	A	B	C
R-34	3	4	0
S-66	0	0	5
T-50	0	1	0

Return to step 2. Cover the zeros with straight lines again.

JOB \ TYPESETTER	A	B	C
R-34	3	4	0
S-66	0	0	5
T-50	0	1	0

Because three lines are necessary, an optimal assignment can be made (see Step 4). Assign R-34 to person C, S-66 to person B, and T-50 to person A. Referring to the original cost table, we see that:

$$\text{Minimum cost} = \$6 + \$10 + \$9 = \$25$$

INSIGHT ▶ If we had assigned S-66 to typesetter A, we could not assign T-50 to a zero location.

LEARNING EXERCISE ▶ If it costs $10 for Typesetter C to complete Job R-34 (instead of $6), how does the solution change? [Answer: R-34 to A, S-66 to B, T-50 to C; cost = $28.]

RELATED PROBLEMS ▶ 15.3–15.9, 15.11–15.14

EXCEL OM Data File **Ch15Ex4.Xls** can be found online.

Some assignment problems entail *maximizing* profit, effectiveness, or payoff of an assignment of people to tasks or of jobs to machines. An equivalent minimization problem can be obtained by converting every number in the table to an *opportunity loss*. To convert a maximizing problem to an equivalent minimization problem, we create a minimizing table by subtracting every number in the original payoff table from the largest single number in that table. We then proceed to step 1 of the four-step assignment method. Minimizing the opportunity loss produces the same assignment solution as the original maximization problem.

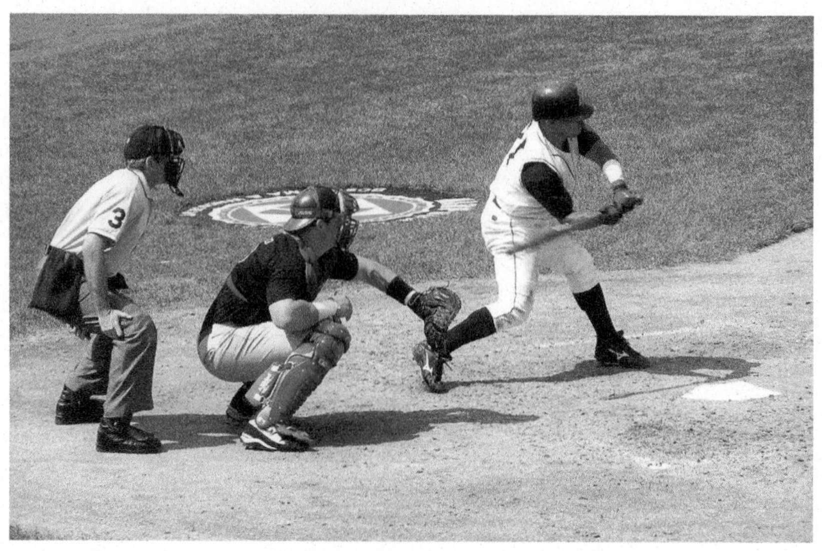

The problem of scheduling major league baseball umpiring crews from one series of games to the next is complicated by many restrictions on travel. The league strives to achieve two conflicting objectives: (1) balance crew assignments relatively evenly among all teams over the course of a season and (2) minimize travel costs. Using the assignment method, the time it takes the league to generate a schedule has been significantly decreased, and the quality of the schedule has improved.

Nicholas D. Cacchione/Shutterstock

Sequencing Jobs

Once jobs are *loaded* in a work center, as we just discussed, managers decide the *sequence* in which they are to be completed. Sequencing (often called *dispatching*) is accomplished by specifying the priority rules to use to release (dispatch) jobs to each work center.

Sequencing
Determining the order in which jobs should be done at each work center.

Priority Rules for Sequencing Jobs

Priority rules are especially applicable for process-focused facilities such as clinics, print shops, and manufacturing job shops. The most popular priority rules are:

Priority rules
Rules used to determine the sequence of jobs in process-oriented facilities.

- **FCFS:** first come, first served. Jobs are completed in the order they arrived.
- **SPT:** shortest processing time. Jobs with the shortest processing times are assigned first.
- **EDD:** earliest due date. Earliest due date jobs are assigned first.
- **LPT:** longest processing time. Jobs with the longest processing time are assigned first.

Performance Criteria The choice of which priority rule to choose depends in part on how each rule performs on four criteria: the priority rules try to minimize *completion time,* maximize *facility utilization,* minimize *number of jobs in the system,* and minimize *job lateness.* These performance criteria incorporate the concept of flow time, which measures the time each job spends waiting plus time being processed. For example, if Job B waits 6 days for Job A to be processed and then takes 2 more days of operation time itself, its flow time would be 6 + 2 = 8 days. The performance criteria are measured as:

Flow time
The time between the release of a job to a work center until the job is finished.

$$\text{Average completion time} = \frac{\text{Total flow time}}{\text{Number of jobs}} \quad (15\text{-}1)$$

$$\text{Utilization metric} = \frac{\text{Total job work (processing) time}}{\text{Total flow time}} \quad (15\text{-}2)$$

$$\text{Average number of jobs in the system} = \frac{\text{Total flow time}}{\text{Total job work (processing) time}} \quad (15\text{-}3)$$

$$\text{Average job lateness} = \frac{\text{Total late days}}{\text{Number of jobs}} \quad (15\text{-}4)$$

Computing the lateness of a particular job involves assumptions about the start time during the day and the timing of delivering a completed job. Equation (15-5) assumes that today is a work day, work has not yet begun today, and a job finished by the end of a day can be delivered to the customer that same day.

$$\text{Job lateness} = \text{Max}\{0, \text{yesterday} + \text{flow time} - \text{due date}\} \quad (15\text{-}5)$$

For example, suppose that today is day 20 (thus yesterday was day 19). Job A is due tomorrow (day 21) and has a flow time of 1 day. That job would be considered to be completed on time, i.e., not late:

$$\text{Max}\{0, 19 + 1 - 21\} = \text{Max}\{0, -1\} = 0 \text{ days late.}$$

Meanwhile, Job B is due on day 32 and has a flow time of 15 days. The lateness of Job B would be:

$$\text{Max}\{0, 19 + 15 - 32\} = \text{Max}\{0, 2\} = 2 \text{ days late.}$$

We will examine four of the most popular priority rules in Example 5.

Example 5

PRIORITY RULES FOR DISPATCHING

Five architectural rendering jobs are waiting to be assigned at Avanti Sethi Architects. Their work (processing) times and due dates are given in the following table. The firm wants to determine the sequence of processing according to (1) FCFS, (2) SPT, (3) EDD, and (4) LPT rules. Jobs were assigned a letter in the order they arrived. Today is day 1, and work begins today.

Job	Job Work (Processing) Time (Days)	Job Due Date (Days)
A	6	8
B	2	6
C	8	18
D	3	15
E	9	23

APPROACH ▶ Each of the four priority rules is examined in turn. Four measures of effectiveness can be computed for each rule and then compared to see which rule is best for the company.

SOLUTION ▶

1. The *FCFS* sequence shown in the next table is simply A–B–C–D–E.

Job Sequence	Job Work (Processing) Time	Flow Time	Job Due Date	Job Lateness
A	6	6	8	0
B	2	8	6	2
C	8	16	18	0
D	3	19	15	(4)
E	9	28	23	5
	28	77		11

=Max{0, 0+19–15} from Eq. (15-5)

The FCFS rule results in the following measures of effectiveness:

a. Average completion time $= \dfrac{\text{Total flow time}}{\text{Number of jobs}}$

$= \dfrac{77\text{ days}}{5} = 15.4\text{ days}$

b. Utilization metric $= \dfrac{\text{Total job work (processing) time}}{\text{Total flow time}}$

$= \dfrac{28}{77} = 36.4\%$

c. Average number of jobs in the system $= \dfrac{\text{Total flow time}}{\text{Total job work (processing) time}}$

$= \dfrac{77\text{ days}}{28\text{ days}} = 2.75\text{ jobs}$

d. Average job lateness $= \dfrac{\text{Total late days}}{\text{Number of jobs}} = \dfrac{11}{5} = 2.2\text{ days}$

2. The *SPT* rule shown in the next table results in the sequence B–D–A–C–E. Orders are sequenced according to processing time, with the highest priority given to the shortest job.

Job Sequence	Job Work (Processing) Time	Flow Time	Job Due Date	Job Lateness
B	2	2	6	0
D	3	5	15	0
A	6	11	8	3
C	8	19	18	1
E	9	28	23	5
	28	65		9

Measurements of effectiveness for SPT are:

a. Average completion time $= \dfrac{65}{5} = 13$ days

b. Utilization metric $= \dfrac{28}{65} = 43.1\%$

c. Average number of jobs in the system $= \dfrac{65}{28} = 2.32$ jobs

d. Average job lateness $= \dfrac{9}{5} = 1.8$ days

LO 15.4 *Name* and describe each of the priority sequencing rules

3. The *EDD* rule shown in the next table gives the sequence B–A–D–C–E. Note that jobs are ordered by earliest due date first.

Job Sequence	Job Work (Processing) Time	Flow Time	Job Due Date	Job Lateness
B	2	2	6	0
A	6	8	8	0
D	3	11	15	0
C	8	19	18	1
E	9	28	23	5
	28	68		6

Measurements of effectiveness for EDD are:

a. Average completion time $= \frac{68}{5} = 13.6$ days

b. Utilization metric $= \frac{28}{68} = 41.2\%$

c. Average number of jobs in the system $= \frac{68}{28} = 2.43$ jobs

d. Average job lateness $= \frac{6}{5} = 1.2$ days

4. The *LPT* rule shown in the next table results in the order E–C–A–D–B.

Job Sequence	Job Work (Processing) Time	Flow Time	Job Due Date	Job Lateness
E	9	9	23	0
C	8	17	18	0
A	6	23	8	15
D	3	26	15	11
B	2	28	6	22
	28	103		48

Measures of effectiveness for LPT are:

a. Average completion time $= \frac{103}{5} = 20.6$ days

b. Utilization metric $= \frac{28}{103} = 27.2\%$

c. Average number of jobs in the system $= \frac{103}{28} = 3.68$ jobs

d. Average job lateness $= \frac{48}{5} = 9.6$ days

The results of these four rules are summarized in the following table:

Rule	Average Completion Time (Days)	Utilization Metric (%)	Average Number of Jobs in System	Average Lateness (Days)
FCFS	15.4	36.4	2.75	2.2
SPT	**13.0**	**43.1**	**2.32**	**1.8**
EDD	13.6	41.2	2.43	1.2
LPT	20.6	27.2	3.68	9.6

INSIGHT ▶ LPT is the least effective measurement for sequencing for the Avanti Sethi firm. SPT is superior in three measures, and EDD is superior in the fourth (average lateness).

LEARNING EXERCISE ▶ If job A takes 7 days (instead of 6), how do the four measures of effectiveness change under the FCFS rule? [Answer: 16.4 days, 35.4%, 2.83 jobs, 2.8 days late.]

RELATED PROBLEMS ▶ 15.15, 15.17a–d, 15.18, 15.19, 15.24–15.27

EXCEL OM Data File **Ch15Ex5.Xls** can be found online.

ACTIVE MODEL 15.1 This example is further illustrated in Active Model 15.1 found online.

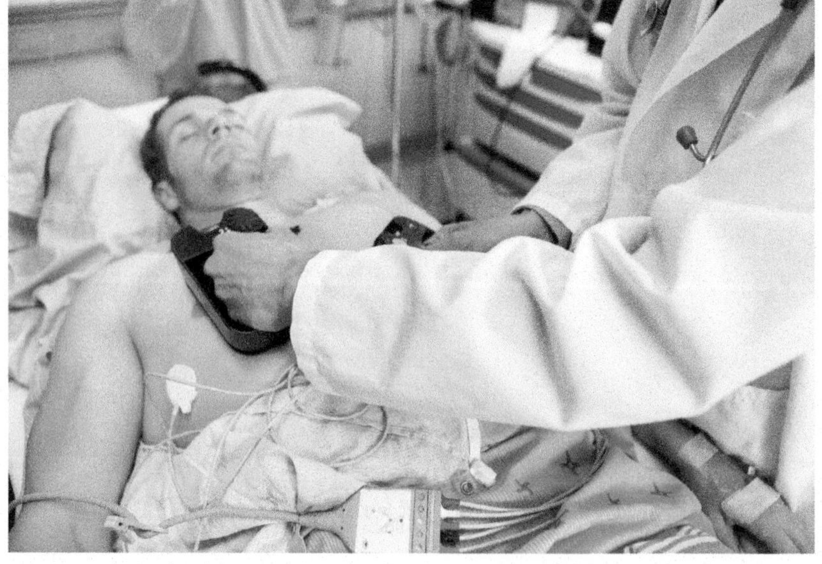

Your doctor may use a first-come, first-served priority rule satisfactorily. However, such a rule may be less than optimal for this emergency room. What priority rule might be best, and why? What priority rule is often used on TV hospital dramas?

Tyler Olson/Fotolia

The results in Example 5 are typically true in the real world also. No one sequencing rule always excels on all criteria. Experience indicates the following:

1. **Shortest processing time** is generally the best technique for minimizing job flow and minimizing the average number of jobs in the system. Its chief disadvantage is that long-duration jobs may be continuously pushed back in priority in favor of short-duration jobs. Customers may view this dimly, and a periodic adjustment for longer jobs must be made.
2. **First come, first served** does not score well on most criteria (but neither does it score particularly poorly). It has the advantage, however, of appearing fair to customers, which is important in service systems.
3. **Earliest due date** minimizes maximum tardiness, which may be necessary for jobs that have a very heavy penalty after a certain date. In general, EDD works well when lateness is an issue.

Critical Ratio

Critical ratio (CR)
A sequencing rule that is an index number computed by dividing the time remaining until due date by the work time remaining.

For organizations that have due dates (such as manufacturers and many firms like your local printer and furniture re-upholsterer), the critical ratio for sequencing jobs is beneficial. The **critical ratio (CR)** is an index number computed by dividing the time remaining until due date by the work time remaining. As opposed to the priority rules, critical ratio is dynamic and easily updated. It tends to perform better than FCFS, SPT, EDD, or LPT on the average job-lateness criterion.

The critical ratio gives priority to jobs that must be done to keep shipping on schedule. A job with a low critical ratio (less than 1.0) is one that is falling behind schedule. If CR is exactly 1.0, the job is on schedule. A CR greater than 1.0 means the job is ahead of schedule and has some slack.

The formula for critical ratio is:

$$\text{CR} = \frac{\text{Time remaining}}{\text{Workdays remaining}} = \frac{\text{Due date} - \text{Today's date}}{\text{Work (lead) time remaining}} \quad (15\text{-}6)$$

Example 6 shows how to use the critical ratio.

Example 6

CRITICAL RATIO

Today is day 25 on Zyco Medical Testing Laboratories' production schedule. Three jobs are on order, as indicated here:

JOB	DUE DATE	WORKDAYS REMAINING
A	30	4
B	28	5
C	27	2

APPROACH ▶ Zyco wants to compute the critical ratios, using the formula for CR.

SOLUTION ▶

JOB	CRITICAL RATIO	PRIORITY ORDER
A	$(30 - 25)/4 = 1.25$	3
B	$(28 - 25)/5 = .60$	1
C	$(27 - 25)/2 = 1.00$	2

INSIGHT ▶ Job B has a critical ratio of less than 1, meaning it will be late unless expedited. Thus, it has the highest priority. Job C is on time, and job A has some slack. Once job B has been completed, we would recompute the critical ratios for jobs A and C to determine whether their priorities have changed.

LEARNING EXERCISE ▶ Today is day 24 (a day earlier) on Zyco's schedule. Recompute the CRs and determine the priorities. [Answer: 1.5, 0.8, 1.5; B is still number 1, but now jobs A and C are tied for second.]

RELATED PROBLEMS ▶ 15.16, 15.17e, 15.21

In most production scheduling systems, the critical-ratio rule can help do the following:

1. Determine the status of a specific job.
2. Establish relative priority among jobs on a common basis.
3. Adjust priorities (and revise schedules) automatically for changes in both demand and job progress.
4. Dynamically track job progress.

Sequencing *N* Jobs on Two Machines: Johnson's Rule

The next step in complexity is the case in which *N* jobs (where *N* is 2 or more) must go through two different machines or work centers in the same order. (Each work center only works on one job at a time.) This is called the *N*/2 problem.

Johnson's rule can be used to minimize the time for sequencing a group of jobs through two work centers. It also minimizes total idle time on the machines. *Johnson's rule* involves four steps:

Johnson's rule
An approach that minimizes the total time for sequencing a group of jobs through two work centers while minimizing total idle time in the work centers.

1. All jobs are to be listed, and the time that each requires on a machine is to be shown.
2. Select the job with the shortest activity time. If the shortest time lies with the first machine, the job is scheduled first. If the shortest time lies with the second machine, schedule the job last. Ties in activity times can be broken arbitrarily.
3. Once a job is scheduled, eliminate it.
4. Apply steps 2 and 3 to the remaining jobs, working toward the center of the sequence.

Example 7 shows how to apply Johnson's rule.

Example 7

JOHNSON'S RULE

Five specialty jobs at a La Crosse, Wisconsin, tool and die shop must be processed through two work centers (drill press and lathe). The time for processing each job follows:

Work (processing) Time for Jobs (hours)

JOB	WORK CENTER 1 (DRILL PRESS)	WORK CENTER 2 (LATHE)
A	5	2
B	3	6
C	8	4
D	10	7
E	7	12

The owner, Niranjan Pati, wants to set the sequence to minimize his total time for the five jobs.

LO 15.5 *Use* Johnson's rule

APPROACH ▶ Pati applies the four steps of Johnson's rule.

SOLUTION ▶

1. The job with the shortest processing time is A, in work center 2 (with a time of 2 hours). Because it is at the second center, schedule A last. Eliminate it from consideration.

				A

2. Job B has the next shortest time (3 hours). Because that time is at the first work center, we schedule it first and eliminate it from consideration.

B				A

3. The next shortest time is job C (4 hours) on the second machine. Therefore, it is placed as late as possible.

B			C	A

4. There is a tie (at 7 hours) for the shortest remaining job. We can place E, which was on the first work center, first. Then D is placed in the last sequencing position:

B	E	D	C	A

The sequential times are:

Work center 1	3	7	10	8	5
Work center 2	6	12	7	4	2

The time-phased flow of this job sequence is best illustrated graphically:

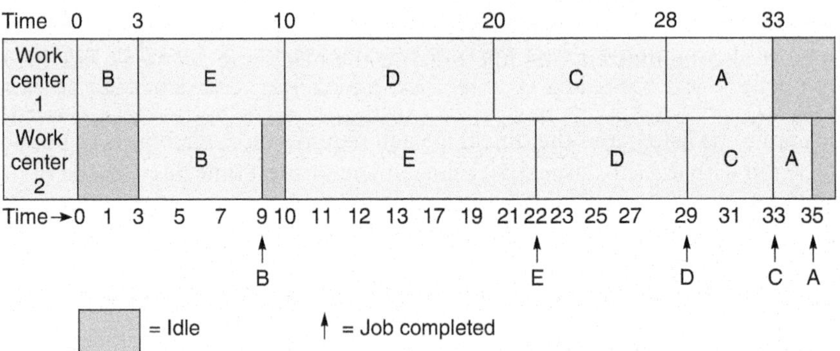

Thus, the five jobs are completed in 35 hours.

INSIGHT ▶ The second work center will wait 3 hours for its first job, and it will also wait 1 hour after completing job B.

LEARNING EXERCISE ▶ If job C takes 8 hours in work center 2 (instead of 4 hours), what sequence is best? [Answer: B–E–C–D–A.]

RELATED PROBLEMS ▶ 15.20, 15.22, 15.23, 15.28

EXCEL OM Data File **Ch15Ex7.Xls** can be found online.

Limitations of Rule-Based Sequencing Systems

The scheduling techniques just discussed are rule-based techniques, but rule-based systems have a number of limitations. Among these are the following:

1. Scheduling is dynamic; therefore, rules need to be revised to adjust to changes in orders, process, equipment, product mix, and so forth.

2. Rules do not look upstream or downstream; idle resources and bottleneck resources in other departments may not be recognized.
3. Rules do not look beyond due dates. For instance, two orders may have the same due date. One order involves restocking a distributor and the other is a custom order that will shut down the customer's factory if not completed. Both may have the same due date, but clearly the custom order is more important.

Despite these limitations, schedulers often use sequencing rules such as SPT, EDD, or critical ratio. They apply these methods at each work center and then modify the sequence to deal with a multitude of real-world variables. They may do this manually or with finite capacity scheduling software.

Finite Capacity Scheduling (FCS)

Short-term scheduling systems are also called finite capacity scheduling.[3] **Finite capacity scheduling (FCS)** overcomes the disadvantages of systems based exclusively on rules by providing the scheduler with interactive computing and graphic output. In dynamic scheduling environments such as job shops (with high variety, low volume, and shared resources) we expect changes. But changes disrupt schedules. Operations managers are moving toward FCS systems that allow virtually instantaneous change by the operator. Improvements in communication on the shop floor are also enhancing the accuracy and speed of information necessary for effective control in job shops. Computer-controlled machines can monitor events and collect information in near real time. This means the scheduler can make schedule changes based on up-to-the-minute information. These schedules are often displayed in Gantt chart form. In addition to including priority rule options, many of the current FCS systems also combine an "expert system" or simulation techniques and allow the scheduler to assign costs to various options. The scheduler has the flexibility to handle any situation, including order, labor, or machine changes.

Finite capacity scheduling (FCS)
Computerized short-term scheduling that overcomes the disadvantage of rule-based systems by providing the user with graphical interactive computing.

LO 15.6 *Define* finite capacity scheduling

The combining of planning and FCS data, priority rules, models to assist analysis, and Gantt chart output is shown in Figure 15.5.

Finite capacity scheduling allows delivery requirements to be based on today's conditions and today's orders, not according to some predefined rule. The scheduler determines what constitutes a "good" schedule. FCS software packages such as Lekin (shown in Figure 15.6), ProPlanner, Preactor, Asprova, Schedlyzer, and Jobplan are currently used at over 60% of U.S. plants.

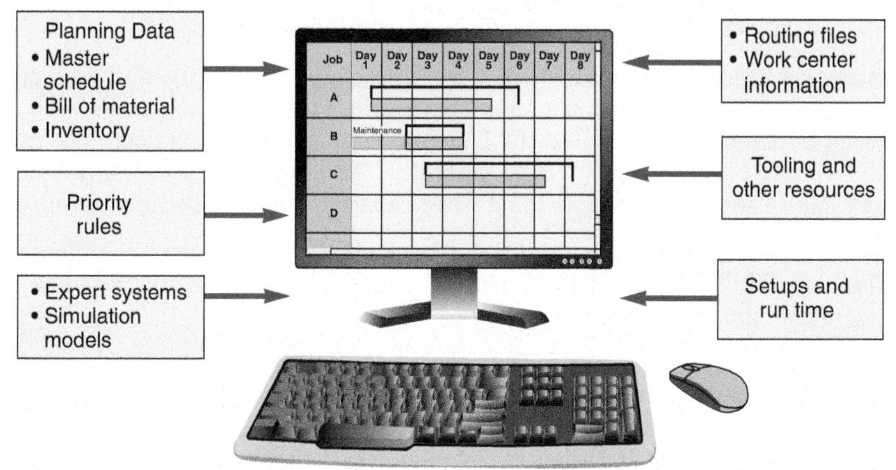

Figure 15.5
Finite Capacity Scheduling Systems Use Production Data to Generate Gantt Load Charts, and Work-in-Process Data That Can Be Manipulated by the User to Evaluate Schedule Alternatives

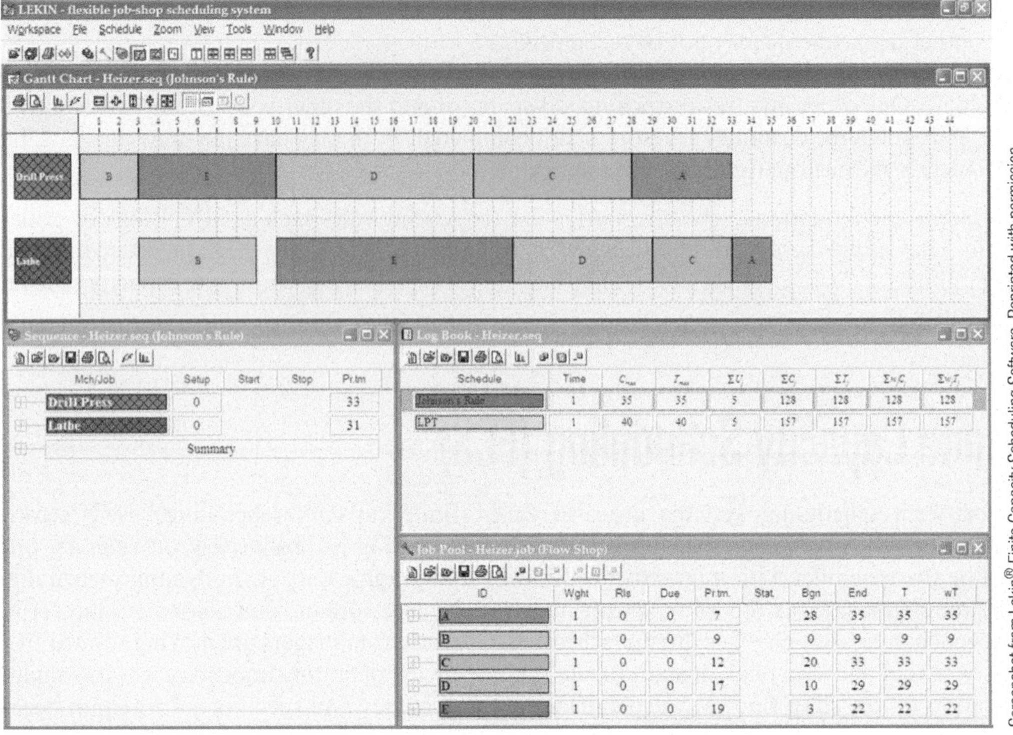

Screenshot from Lekin® Finite Capacity Scheduling Software. Reprinted with permission.

Figure 15.6

Finite Capacity Scheduling (FCS) System

This Lekin® finite capacity scheduling software presents a schedule of the five jobs and the two work centers shown in Example 7 in Gantt chart form. The software is capable of using a variety of priority rules and many jobs. The Lekin software is available for free at **http://community.stern.nyu.edu/om/software/lekin/** and can solve many of the problems at the end of this chapter.

Scheduling Services

STUDENT TIP Scheduling people to perform services can be even more complex than scheduling machines.

Scheduling service systems differs from scheduling manufacturing systems in several ways:

- In manufacturing, the scheduling emphasis is on machines and materials; in services, it is on staffing levels.
- Inventories can help smooth demand for manufacturers, but many service systems do not maintain inventories.
- Services are labor intensive, and the demand for this labor can be highly variable.
- Legal considerations, such as wage and hour laws and union contracts that limit hours worked per shift, week, or month, constrain scheduling decisions.
- Because services usually schedule people (rather than material), social, fatigue, seniority, and status issues complicate scheduling. This is especially so in the National Basketball Association (see the *OM in Action* box), where basketball players suffer from highly structured and inflexible scheduling of games during the season.

VIDEO 15.2 Scheduling at Hard Rock Cafe

The following examples note the complexity of scheduling services.

Hospitals A hospital is an example of a service facility that may use a scheduling system every bit as complex as one found in a job shop. Hospitals seldom use a machine shop priority system such as first come, first served (FCFS) for treating emergency patients, but they often use FCFS *within* a priority class, a "triage" approach. And they often schedule products (such as surgeries) just like a factory, maintaining excess capacity to meet wide variations in demand.

Banks Cross-training of the workforce in a bank allows loan officers and other managers to provide short-term help for tellers if there is a surge in demand. Banks also employ part-time personnel to provide a variable capacity.

Retail Stores Scheduling optimization systems, such as Workbrain, Cybershift, and Kronos, are used at retailers including Walmart, Payless Shoes, and Target. These systems track individual store sales, transactions, units sold, and customer traffic in 15-minute increments to create work schedules. Walmart's 2.3 million and Target's 360,000 employees used to take thousands of managers' hours to schedule; now staffing is drawn up nationwide in a few hours, and customer checkout experience has improved dramatically.

OM in Action: The NBA's Scheduling Secret[4]

It's the afternoon of February 26, during a 3-games-in-4-nights stretch, and Miami Heat center Hassan Whiteside feels fatigued. Tomorrow night, his Heat will host the Golden State Warriors and then fly to Houston to face the Rockets on February 28. But now he's rattling off what time the Warriors game will end (10 P.M.), when they'll board their flight (after 11:30), when they'll land in Houston (2 A.M.), and when they'll arrive at the hotel (3 A.M.) before playing the Rockets later that day.

Sleep matters, Whiteside says—it matters a lot. It "could be the difference between you having a career game or playing terrible." Is it possible within the current NBA schedule to obtain consistent, quality sleep? "Nah," Whiteside says. "It's impossible. It's impossible."

Fatigue has long been a reality of life in the NBA, a league with teams that play 82 games in under 6 months and fly up to 50,000 miles per season—enough to circle the globe twice. Over a typical season, the average NBA team plays every 2 days, has 13 back-to-back sets, and flies the equivalent of 250 miles a day for 25 straight weeks.

Despite the league's best efforts—lengthening its schedule in recent years, reducing back-to-backs, eliminating 4-in-5 stretches, reducing the nationally televised games that tip off at 10:30 P.M., and creating more rest days—sleep deprivation remains "our biggest issue without a solution. It's the dirty little secret that everybody knows about," says an NBA exec.

"I think in a couple years," Tobias Harris says, "sleep deprivation will be an issue that's talked about, like the NFL with concussions." During the season, it is estimated that players get 5 hours sleep per night. Chronic sleep loss has been associated with higher risk for cancer, diabetes, obesity, heart disease, heart attacks, Alzheimer's disease, dementia, depression, stroke, psychosis, and suicide.

Cory Thoman/Shutterstock

Airlines Two of the constraints airlines face when scheduling flight crews are: (1) a complex set of FAA work-time limitations and (2) union contracts that guarantee crew pay for some number of hours each day or each trip. Planners must also make efficient use of their other expensive resource: aircraft. These schedules are typically built using linear programming models (see Module B).

24/7 Operations Emergency hotlines, police/fire departments, telephone operations, and mail-order businesses (such as L.L. Bean) schedule employees 24 hours a day, 7 days a week. To allow management flexibility in staffing, sometimes part-time workers can be employed. This provides both benefits (in using odd shift lengths or matching anticipated workloads) and difficulties (from the large number of possible alternatives in terms of days off, lunch hour times, rest periods, starting times). Most companies use computerized scheduling systems to cope with these complexities.

Scheduling Service Employees with Cyclical Scheduling

LO 15.7 *Use the* cyclical scheduling technique

A number of techniques and algorithms exist for scheduling service-sector employees when staffing needs vary. This is typically the case for police officers, nurses, restaurant staff, tellers, and retail sales clerks. Managers, trying to set a timely and efficient schedule that keeps personnel happy, can spend substantial time each month developing employee schedules. Such schedules often consider a fairly long planning period (say, 6 weeks). One approach that is workable yet simple is *cyclical scheduling*.[5]

Patricia McDonnell/AP Images

Good scheduling in the healthcare industry can help keep nurses happy and costs contained. Here, nurses in Boston protest nurse-staffing levels in Massachusetts hospitals. Shortages of qualified nurses is a chronic problem.

Cyclical Scheduling Cyclical scheduling focuses on developing varying (inconsistent) schedules with the minimum number of workers. In these cases, each employee is assigned to a shift and has prescribed time off. Let's look at Example 8.

Example 8

CYCLICAL SCHEDULING

Hospital administrator Doris Laughlin wants to staff the oncology ward using a standard 5-day work-week with 2 consecutive days off, but also wants to minimize the staff. However, as in most hospitals, she faces an inconsistent demand. Weekends have low usage. Doctors tend to work early in the week, and patients peak on Wednesday then taper off.

APPROACH ▶ Doris must first establish staffing requirements. Then the following five-step process is applied.

SOLUTION ▶

1. Doris has determined that the necessary daily staffing requirements are:

DAY	MONDAY	TUESDAY	WEDNESDAY	THURSDAY	FRIDAY	SATURDAY	SUNDAY
Staff required	5	5	6	5	4	3	3

2. Identify the 2 consecutive days that have the *lowest total requirement* and circle these. Assign these 2 days off to the first employee. In this case, the first employee has Saturday and Sunday off because 3 plus 3 is the *lowest sum* of any 2 days. In the case of a tie, choose the days with the lowest adjacent requirement, or by first assigning Saturday and Sunday as an "off" day. If there are more than one, make an arbitrary decision.
3. We now have an employee working each of the uncircled days; therefore, make a new row for the next employee by subtracting 1 from the first row (because one day has been worked)—except for the circled days (which represent the days not worked) and any day that has a zero. That is, do not subtract from a circled day or a day that has a value of zero.
4. In the new row, identify the two consecutive days that have the lowest total requirement and circle them. Assign the next employee to the remaining days.
5. Repeat the process (Steps 3 and 4) until all staffing requirements are met.

STUDENT TIP Days off are circled.

	MONDAY	TUESDAY	WEDNESDAY	THURSDAY	FRIDAY	SATURDAY	SUNDAY
Employee 1	5	5	6	5	4	(3)	(3)
Employee 2	4	4	5	4	3	(3)	(3)
Employee 3	3	3	4	3	(2)	(3)	3
Employee 4	2	2	3	(2)	(2)	3	2
Employee 5	(1)	(1)	2	2	2	2	1
Employee 6	1	1	1	1	1	(1)	(0)
Employee 7						1	
Capacity (measured in number of employees)	5	5	6	5	4	3	3
Excess capacity	0	0	0	0	0	1	0

Doris needs six full-time employees to meet the staffing needs and one employee to work Saturday.

Notice that capacity (number of employees) equals requirements, provided an employee works overtime on Saturday, or a part-time employee is hired for Saturday.

INSIGHT ▶ Doris has implemented an efficient scheduling system that accommodates 2 consecutive days off for every employee.

LEARNING EXERCISE ▶ If Doris meets the staffing requirement for Saturday with a full-time employee, how does she schedule that employee? [Answer: That employee can have any 2 days off, except Saturday, and capacity will exceed requirements by 1 person each day the employee works (except Saturday).]

RELATED PROBLEMS ▶ 15.29–15.32

Using the approach in Example 8, Colorado General Hospital saved an average of 10 to 15 hours a month and found these added advantages: (1) no computer was needed, (2) the nurses were happy with the schedule, (3) the cycles could be changed seasonally to accommodate avid skiers, and (4) recruiting was easier because of predictability and flexibility. This approach yields an optimum, although there may be multiple optimal solutions.

Other cyclical scheduling techniques have been developed to aid service scheduling. Some approaches use linear programming: This is how Hard Rock Cafe schedules its services (see the *Video Case Study* at the end of this chapter). There is a natural bias in scheduling to use tools that are understood and yield solutions that are accepted.

Summary

Scheduling involves the timing of operations to achieve the efficient movement of units through a system. This chapter addressed the issues of short-term scheduling in process-focused and service environments. We saw that process-focused facilities are production systems in which products are made to order and that scheduling tasks in them can become complex. Several aspects and approaches to scheduling, loading, and sequencing of jobs were introduced. These ranged from Gantt charts and the assignment method of scheduling to a series of priority rules, the critical-ratio rule, Johnson's rule for sequencing, and finite capacity scheduling.

Service systems generally differ from manufacturing systems. This leads to the use of first-come, first-served rules and appointment and reservation systems, as well as linear programming for matching capacity to demand in service environments.

Key Terms

Loading (p. 612)
Input–output control (p. 614)
ConWIP cards (p. 614)
Gantt charts (p. 615)
Assignment method (p. 616)
Sequencing (p. 619)
Priority rules (p. 619)
Flow time (p. 619)
Critical ratio (CR) (p. 622)
Johnson's rule (p. 623)
Finite capacity scheduling (FCS) (p. 625)

Ethical Dilemma

Scheduling people to work second and third shifts (evening and "graveyard") is a problem in almost every 24-hour company. Medical and ergonomic data indicate the body does not respond well to significant shifts in its natural circadian rhythm of sleep. There are also significant long-run health issues with frequent changes in work and sleep cycles.

Consider yourself the manager of a nonunion steel mill that must operate 24-hour days, and where the physical demands are such that 8-hour days are preferable to 10- or 12-hour days. Your empowered employees have decided that they want to work weekly rotating shifts. That is, they want a repeating work cycle of 1 week, 7 A.M. to 3 P.M., followed by a second week from 3 P.M. to 11 P.M., and the third week from 11 P.M. to 7 P.M. You are sure this is not a good idea in terms of both productivity and the long-term health of the employees. If you do not accept their decision, you undermine the work empowerment program, generate a morale issue, and perhaps, more significantly, generate several more votes for a union. What is the ethical position and what do you do?

NDAB Creativity/Shutterstock

Discussion Questions

1. What is the overall objective of scheduling?
2. List the four criteria for determining the effectiveness of a *scheduling* decision. How do these criteria relate to the four criteria for *sequencing* decisions?
3. Describe what is meant by "loading" work centers. What are the two ways work centers can be loaded? What are two techniques used in loading?
4. Name five priority sequencing rules. Explain how each works to assign jobs.

5. What are the advantages and disadvantages of the shortest processing time (SPT) rule?
6. What is a due date?
7. Explain the terms *flow time* and *lateness*.
8. Which shop-floor scheduling rule would you prefer to apply if you were the leader of the only team of experts charged with defusing several time bombs scattered throughout your building? You can see the bombs; they are of different types. You can tell how long each one will take to defuse. Discuss.
9. When is Johnson's rule best applied in job-shop scheduling?
10. State the four effectiveness measures for dispatching rules.
11. What are the steps of the assignment method of linear programming?
12. What are the advantages to finite capacity scheduling?
13. What is input–output control?

Using Software for Short-Term Scheduling

In addition to the commercial software we noted in this chapter, short-term scheduling problems can be solved with the Excel OM software found online. POM for Windows also includes a scheduling module. The use of each of these programs is explained next.

USING EXCEL OM

Excel OM has two modules that help solve short-term scheduling problems: Assignment and Job Shop Scheduling. The Assignment module is illustrated in Programs 15.1 and 15.2. The input screen, using the Example 4 data, appears first, as Program 15.1. Once the data are all entered, we choose the **Data tab** command, followed by the **Solver** command. Excel's Solver uses linear programming to optimize assignment problems. (So select Simplex LP.) The constraints are also shown in Program 15.1. We then select the **Solve** command; the solution appears in Program 15.2.

Excel OM's Job Shop Scheduling module is illustrated in Program 15.3. Program 15.3 uses Example 5's data. Because jobs are listed in the sequence in which they arrived (see column A), the results are for the FCFS rule. Program 15.3 also shows some of the formulas (columns F, G, H, I, J) used in the calculations.

To solve with the SPT rule, we need four intermediate steps: (1) Select (that is, highlight) the data in columns A, B, C for all jobs; (2) invoke the **Data** command; (3) invoke the **Sort** command; and (4) sort by **Time** (column C) in *ascending* order. To solve for EDD, Step 4 changes to sort by **Due Date** (column D) in *ascending* order. Finally, for an LPT solution, Step 4 becomes sort by **Due Date** (column D) in *descending* order.

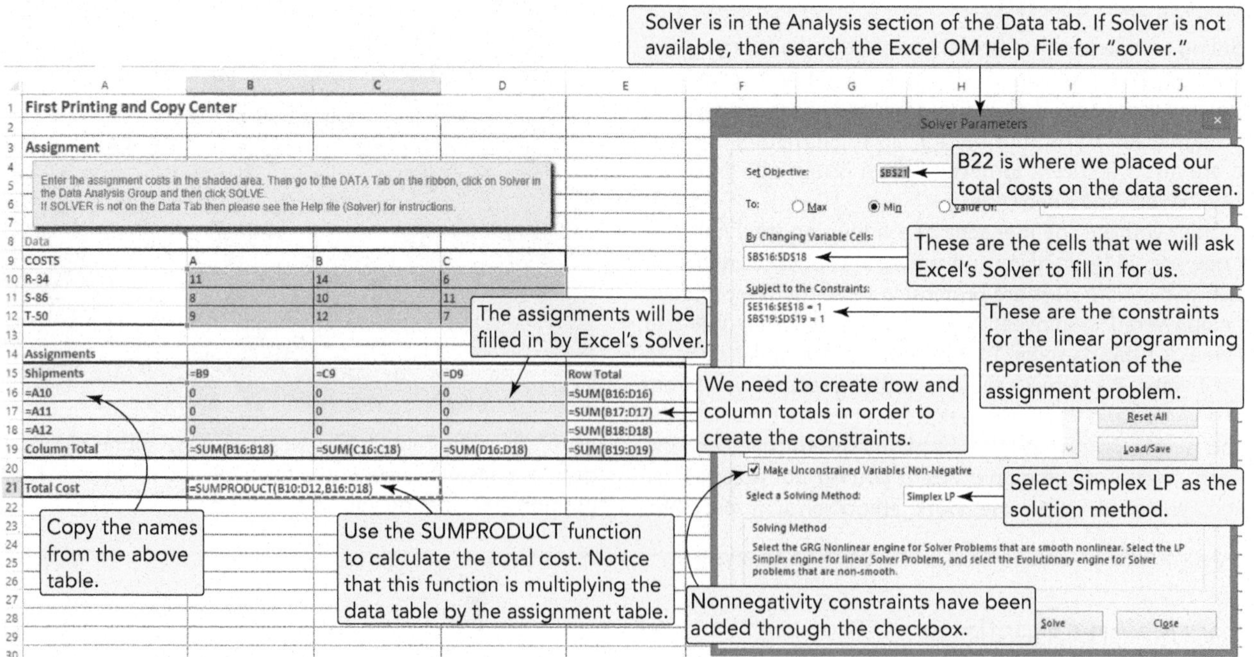

Program **15.1**

Excel OM's Assignment Module Using Example 4's Data

After entering the problem data in the yellow area, select Data, then Solver.

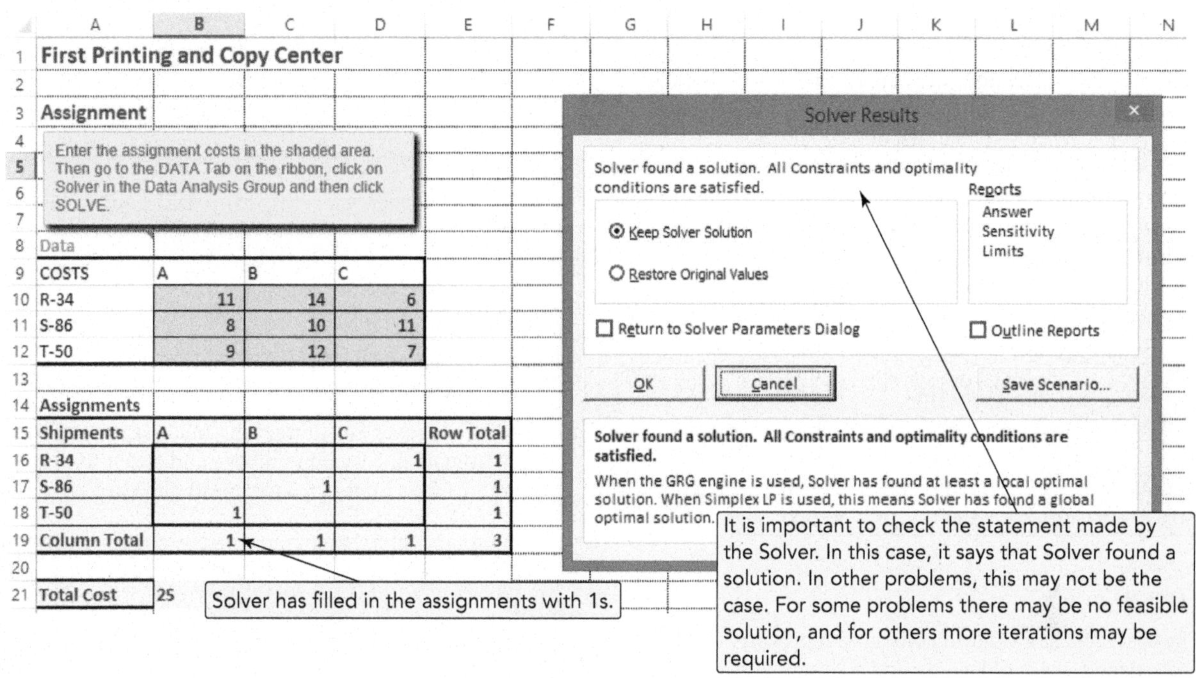

Program 15.2

Excel OM Output Screen for Assignment Problem Described in Program 15.1

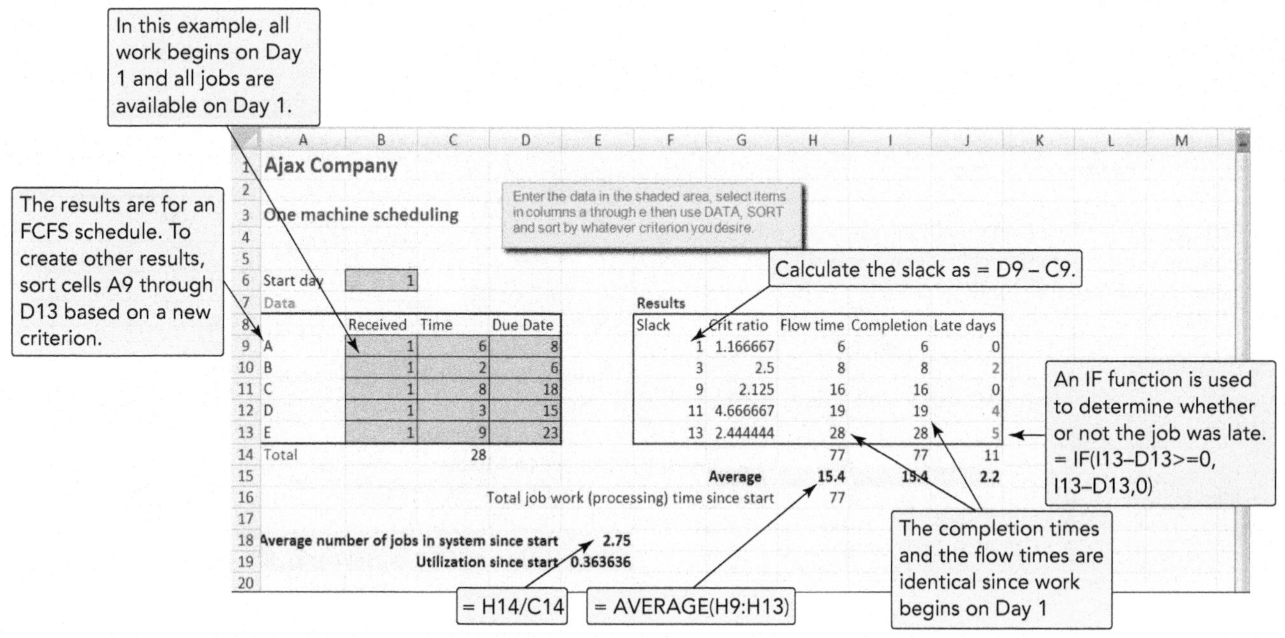

Program 15.3

Excel OM's Job Shop Scheduling Module Applied to Example 5's Data

P USING POM FOR WINDOWS

POM for Windows can handle both categories of scheduling problems we see in this chapter. Its Assignment module is used to solve the traditional one-to-one assignment problem of people to tasks, machines to jobs, and so on. Its Job Shop Scheduling module can solve a one- or two-machine job-shop problem. Available priority rules include SPT, FCFS, EDD, and LPT. Each can be examined in turn once the data are all entered. Refer to Appendix II for specifics regarding POM for Windows.

Solved Problems

SOLVED PROBLEM 15.1

King Finance Corporation, headquartered in New York, wants to assign three recently hired college graduates, Julie Jones, Al Smith, and Pat Wilson, to regional offices. However, the firm also has an opening in New York and would send one of the three there if it were more economical than a move to Omaha, Dallas, or Miami. It will cost $1,000 to relocate Jones to New York, $800 to relocate Smith there, and $1,500 to move Wilson. What is the optimal assignment of personnel to offices?

OFFICE / HIREE	OMAHA	MIAMI	DALLAS
Jones	$800	$1,100	$1,200
Smith	$500	$1,600	$1,300
Wilson	$500	$1,000	$2,300

SOLUTION

a) The cost table has a fourth column to represent New York. To "balance" the problem, we add a "dummy" row (person) with a zero relocation cost to each city.

OFFICE / HIREE	OMAHA	MIAMI	DALLAS	NEW YORK
Jones	$800	$1,100	$1,200	$1,000
Smith	$500	$1,600	$1,300	$ 800
Wilson	$500	$1,000	$2,300	$1,500
Dummy	0	0	0	0

b) Subtract the smallest number in each row and cover all zeros (column subtraction of each column's zero will give the same numbers and therefore is not necessary):

OFFICE / HIREE	OMAHA	MIAMI	DALLAS	NEW YORK
Jones	0	300	400	200
Smith	0	1,100	800	300
Wilson	0	500	1,800	1,000
Dummy	0	0	0	0

c) Only 2 lines cover, so subtract the smallest uncovered number (200) from all uncovered numbers, and add it to each square where two lines intersect. Then cover all zeros:

OFFICE / HIREE	OMAHA	MIAMI	DALLAS	NEW YORK
Jones	0	100	200	0
Smith	0	900	600	100
Wilson	0	300	1,600	800
Dummy	200	0	0	0

d) Only 3 lines cover, so subtract the smallest uncovered number (100) from all uncovered numbers, and add it to each square where two lines intersect. Then cover all zeros:

OFFICE / HIREE	OMAHA	MIAMI	DALLAS	NEW YORK
Jones	0	0	100	0
Smith	0	800	500	100
Wilson	0	200	1,500	800
Dummy	300	0	0	100

e) Still only 3 lines cover, so subtract the smallest uncovered number (100) from all uncovered numbers, add it to squares where two lines intersect, and cover all zeros:

OFFICE / HIREE	OMAHA	MIAMI	DALLAS	NEW YORK
Jones	100	0	100	0
Smith	0	700	400	0
Wilson	0	100	1,400	700
Dummy	400	0	0	100

f) Because it takes four lines to cover all zeros, an optimal assignment can be made at zero squares. We assign:
Wilson to Omaha
Jones to Miami
Dummy (no one) to Dallas
Smith to New York

$$\text{Cost} = \$500 + \$1{,}100 + \$0 + \$800 = \$2{,}400$$

SOLVED PROBLEM 15.2

A defense contractor in Dallas has six jobs awaiting processing. Processing time and due dates are given in the table. Assume that jobs arrive in the order shown. Set the processing sequence according to FCFS and evaluate. Start date is day 1.

JOB	JOB PROCESSING TIME (DAYS)	JOB DUE DATE (DAYS)
A	6	22
B	12	14
C	14	30
D	2	18
E	10	25
F	4	34

SOLUTION

FCFS has the sequence A–B–C–D–E–F.

Job Sequence	Job Processing Time	Flow Time	Due Date	Job Lateness
A	6	6	22	0
B	12	18	14	4
C	14	32	30	2
D	2	34	18	16
E	10	44	25	19
F	4	48	34	14
	48	182		55

1. Average completion time = 182/6 = 30.33 days
2. Average number of jobs in system = 182/48 = 3.79 jobs
3. Average job lateness = 55/6 = 9.16 days
4. Utilization = 48/182 = 26.4%

SOLVED PROBLEM 15.3

The Dallas firm in Solved Problem 15.2 also wants to consider job sequencing by the SPT priority rule. Apply SPT to the same data, and provide a recommendation.

SOLUTION

SPT has the sequence D–F–A–E–B–C.

Job Sequence	Job Processing Time	Flow Time	Due Date	Job Lateness
D	2	2	18	0
F	4	6	34	0
A	6	12	22	0
E	10	22	25	0
B	12	34	14	20
C	14	48	30	18
	48	124		38

1. Average completion time = 124/6 = 20.67 days
2. Average number of jobs in system = 124/48 = 2.58 jobs
3. Average job lateness = 38/6 = 6.33 days
4. Utilization = 48/124 = 38.7%

SPT is superior to FCFS in this case on all four measures. If we were to also analyze EDD, we would, however, find its average job lateness to be lowest at 5.5 days. SPT is a good recommendation. SPT's major disadvantage is that it makes long jobs wait, sometimes for a long time.

SOLVED PROBLEM 15.4

Use Johnson's rule to find the optimum sequence for processing the jobs shown through two work centers. Times at each center are in hours.

Job	Work Center 1	Work Center 2
A	6	12
B	3	7
C	18	9
D	15	14
E	16	8
F	10	15

SOLUTION

B	A	F	D	C	E

The sequential times are:

Work center 1	3	6	10	15	18	16
Work center 2	7	12	15	14	9	8

SOLVED PROBLEM 15.5

Illustrate the throughput time and idle time at the two work centers in Solved Problem 15.4 by constructing a time-phased chart.

SOLUTION

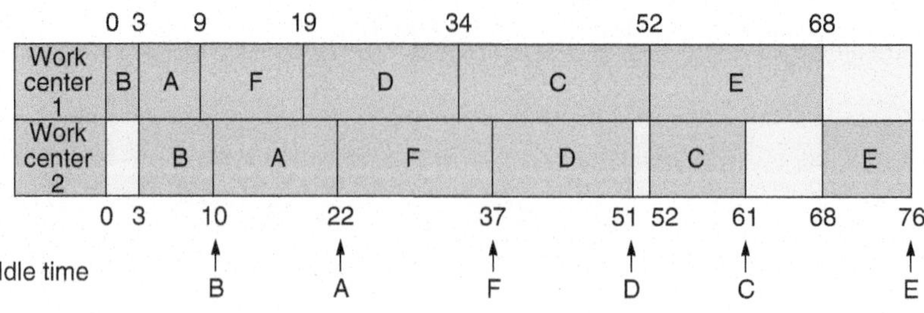

Problems

Note: **PX** means the problem may be solved with POM for Windows and/or Excel OM.

Problems 15.1–15.14 relate to Loading Jobs

•• **15.1** Ron Satterfield's excavation company uses both Gantt scheduling charts and Gantt load charts.

a) Today, which is the end of day 7, Ron is reviewing the Gantt chart depicting these schedules:
   - Job #151 was scheduled to begin on day 3 and to take 6 days. As of now, it is 1 day ahead of schedule.
   - Job #177 was scheduled to begin on day 1 and take 4 days. It is currently on time.
   - Job #179 was scheduled to start on day 7 and take 2 days. It actually got started on day 6 and is progressing according to plan.
   - Job #211 was scheduled to begin on day 5, but missing equipment delayed it until day 6. It is progressing as expected and should take 3 days.
   - Job #215 was scheduled to begin on day 4 and take 5 days. It got started on time but has since fallen behind 2 days.

Draw the Gantt scheduling chart for the activities described.

b) Ron now wants to use a Gantt load chart to see how much work is scheduled in each of his three work teams: Able, Baker, and Charlie. Five jobs constitute the current workload for these three work teams: Job #250, requiring 48 hours and #275 requiring 32 hours for Work Team Able; Jobs #210 and #280, requiring 16 and 24 hours, respectively, for Team Baker; and Job #225, requiring 40 hours, for Team Charlie.

Prepare the Gantt load chart for these activities.

•• **15.2** First Printing and Copy Center has four more jobs to be scheduled, in addition to those shown in Example 3 in the chapter. Production scheduling personnel are reviewing the Gantt chart at the end of day 4.

- Job D was scheduled to begin early on day 2 and to end on the middle of day 9. As of now (the review point after day 4), it is 2 days ahead of schedule.
- Job E should begin on day 1 and end on day 3. It is on time.
- Job F was to begin on day 3, but maintenance forced a delay of 1½ days. The job should now take 5 full days. It is now on schedule.
- Job G is a day behind schedule. It started at the beginning of day 2 and should require 6 days to complete.

Develop a Gantt schedule chart for First Printing and Copy Center.

• **15.3** The Green Cab Company has a taxi waiting at each of four cabstands in Evanston, Illinois. Four customers have called and requested service. The distances, in miles, from the waiting taxis to the customers are given in the following table. Find the optimal assignment of taxis to customers so as to minimize total driving distances to the customers.

	CUSTOMER			
CAB SITE	A	B	C	D
Stand 1	7	3	4	8
Stand 2	5	4	6	5
Stand 3	6	7	9	6
Stand 4	8	6	7	4

PX

• **15.4** J.C. Howard's medical testing company in Kansas wishes to assign a set of jobs to a set of machines. The following table provides the production data of each machine when performing the specific job:

	MACHINE			
JOB	A	B	C	D
1	7	9	8	10
2	10	9	7	6
3	11	5	9	6
4	9	11	5	8

a) Determine the assignment of jobs to machines that will *maximize* total production.

b) What is the total production of your assignments? **PX**

• **15.5** The Johnny Ho Manufacturing Company in Columbus, Ohio, is putting out four new electronic components. Each of Ho's four plants has the capacity to add one more product to its current line of electronic parts. The unit-manufacturing costs for producing the different parts at the four plants are shown in the accompanying table. How should Ho assign the new products to the plants to minimize manufacturing costs?

	PLANT			
ELECTRONIC COMPONENT	1	2	3	4
C53	$0.10	$0.12	$0.13	$0.11
C81	0.05	0.06	0.04	0.08
D5	0.32	0.40	0.31	0.30
D44	0.17	0.14	0.19	0.15

PX

• **15.6** Jamison Day Consultants has been entrusted with the task of evaluating a business plan that has been divided into four sections—marketing, finance, operations, and human resources. Chris, Steve, Juana, and Rebecca form the evaluation team. Each of them has expertise in a certain field and tends to finish that section faster. The estimated times taken by each team member for each section have been outlined in the following table. Further information states that each of these individuals is paid $60/hour.

a) Assign each member to a different section such that Jamison Consultants' overall cost is minimized.
b) What is the total cost of these assignments?

Times Taken by Team Members for Different Sections (minutes)

	MARKETING	FINANCE	OPERATIONS	HR
Chris	80	120	125	140
Steve	20	115	145	160
Juana	40	100	85	45
Rebecca	65	35	25	75

PX

•• **15.7** The Baton Rouge Police Department has five detective squads available for assignment to five open crime cases. The chief of detectives, Jose Noguera, wishes to assign the squads so that the total time to conclude the cases is minimized. The average number of days, based on past performance, for each squad to complete each case is as follows:

	CASE				
SQUAD	A	B	C	D	E
1	14	7	3	7	27
2	20	7	12	6	30
3	10	3	4	5	21
4	8	12	7	12	21
5	13	25	24	26	8

Each squad is composed of different types of specialists, and whereas one squad may be very effective in certain types of cases, it may be almost useless in others.

a) Solve the problem by using the assignment method.
b) Assign the squads to the above cases, but with the constraint that squad 5 cannot work on case E because of a conflict. PX

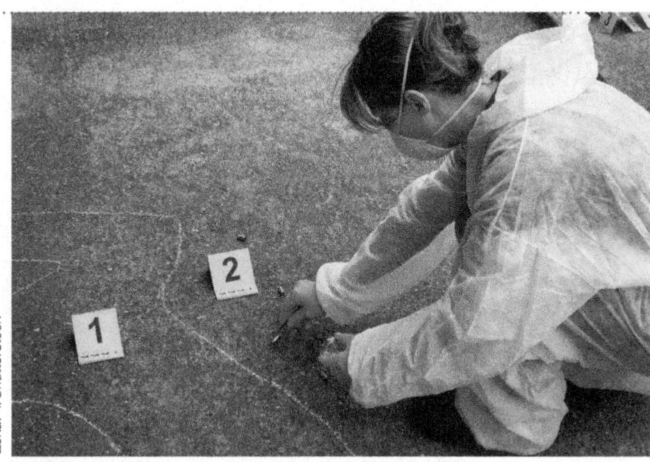

Zoka74/Shutterstock

• **15.8** Tigers Sports Club has to select four separate coed doubles teams to participate in an interclub table tennis tournament. The preselection results in the selection of a group of four men—Raul, Jack, Gray, and Ajay—and four women—Barbara, Dona, Stella, and Jackie. Now, the task ahead lies in pairing these men and women in the best fashion. The following table shows a matrix that has been designed for this purpose, indicating how each of the men complements the game of each of the women. A higher score indicates a higher degree of compatibility in the games of the two individuals concerned. Find the best pairs.

Game Compatibility Matrix

	BARBARA	DONA	STELLA	JACKIE
Raul	30	20	10	40
Jack	70	10	60	70
Gray	40	20	50	40
Ajay	60	70	30	90

PX

••• **15.9** Daniel Glaser, chairman of the College of San Antonio's business department, needs to assign professors to courses next semester. As a criterion for judging who should teach each course, Professor Glaser reviews the past 2 years' teaching evaluations (which were filled out by students). Since each of the four professors taught each of the four courses at one time or another during the 2-year period, Glaser is able to record a course rating for each instructor. These ratings are shown in the following table.

a) Find the assignment of professors to courses to maximize the overall teaching rating.
b) Assign the professors to the courses with the exception that Professor Fisher cannot teach statistics. PX

	COURSE			
PROFESSOR	STATISTICS	MANAGEMENT	FINANCE	ECONOMICS
W. W. Fisher	90	65	95	40
D. Golhar	70	60	80	75
Z. Hug	85	40	80	60
N. K. Rustagi	55	80	65	55

•• **15.10** Lifang Wu owns an automated machine shop that makes precision auto parts. He has just compiled an input–output report for the grinding work center. Complete this report and analyze the results.

Input–Output Report

PERIOD	1	2	3	4	TOTAL
Planned input	80	80	100	100	
Actual input	85	85	85	85	
Deviation					
Planned output	90	90	90	90	
Actual output	85	85	80	80	
Deviation					
Initial backlog: 30					

• **15.11** Carlita Garcia's company wishes to assign a set of jobs to a set of machines. The following table provides data as to the cost of each job when performed on a specific machine.

a) Determine the assignment of jobs to machines that will *minimize* Carlita's total cost.
b) What is the total cost of your assignments? PX

	MACHINE			
JOB	A	B	C	D
1	7	9	8	10
2	10	9	7	6
3	11	5	9	6
4	9	11	5	8

• **15.12** Ravi Behara, the managing partner at a large law firm in Virginia, must assign three clients to three attorneys. Cost data are presented in the following table:

	ATTORNEY		
CLIENT	1	2	3
Divorce case	$800	$1,100	$1,200
Felony case	$500	$1,600	$1,300
Discrimination case	$500	$1,000	$2,300

Use the assignment algorithm to solve this problem. PX

• **15.13** Jerry Wei, the hospital administrator at St. Charles General, must appoint head nurses to four newly established departments: urology, cardiology, orthopedics, and obstetrics. In anticipation of this staffing problem, he has hired four nurses, considered their backgrounds, personalities, and talents, and developed a cost scale ranging from 0 to 100 to be used in the assignment. A 0 for a nurse being assigned to the cardiology unit implies that the nurse would be perfectly suited to that task. A value close to 100, on the other hand, would imply that the nurse is not at all suited to head that unit. The accompanying table gives the complete set of cost figures that the hospital administrator feels represent all possible assignments. Which nurse should be assigned to which unit? **PX**

	DEPARTMENT			
NURSE	UROLOGY	CARDIOLOGY	ORTHOPEDICS	OBSTETRICS
Kitty Forman	28	18	15	75
Carol Hathaway	32	48	23	38
Jack McFarland	51	36	24	36
Margaret Houlihan	25	38	55	12

•• **15.14** The Doan Modianos Manufacturing Company is putting out seven new electronic components. Each of Modianos' eight plants has the capacity to add one more product to its current line of electronic parts. The unit manufacturing costs for producing the different parts at the eight plants are shown in the accompanying table. How should Modianos assign the new products to the plants in order to minimize manufacturing costs? **PX**

	PLANTS							
ELECTRONIC COMPONENTS	1	2	3	4	5	6	7	8
C53	$0.10	$0.12	$0.13	$0.11	$0.10	$0.06	$0.16	$0.12
C81	0.05	0.06	0.04	0.08	0.04	0.09	0.06	0.06
D5	0.32	0.40	0.31	0.30	0.42	0.35	0.36	0.49
D44	0.17	0.14	0.19	0.15	0.10	0.16	0.19	0.12
E2	0.06	0.07	0.10	0.05	0.08	0.10	0.11	0.05
E35	0.08	0.10	0.12	0.08	0.09	0.10	0.09	0.06
G99	0.55	0.62	0.61	0.70	0.62	0.63	0.65	0.59

Problems 15.15–15.25 relate to Sequencing Jobs

••• **15.15** The following jobs are waiting to be processed at the same machine center. Jobs are logged as they arrive:

JOB	DUE DATE	DURATION (DAYS)
A	313	8
B	312	16
C	325	40
D	314	5
E	314	3

In what sequence would the jobs be ranked according to the following decision rules: (a) FCFS, (b) EDD, (c) SPT, and (d) LPT? All dates are specified as manufacturing planning calendar days. Assume that all jobs arrive on day 275. Which decision is best and why? **PX**

• **15.16** The following five overhaul jobs are waiting to be processed at Myisha Adam's Engine Repair Inc. These jobs were logged as they arrived. All dates are specified as planning calendar days. Assume that all jobs arrived on day 180; today's date is 200.

JOB	DUE DATE	REMAINING TIME (DAYS)
103	214	10
205	223	7
309	217	11
412	219	5
517	217	15

Using the critical ratio scheduling rule, in what sequence would the jobs be processed? **PX**

••• **15.17** An Alabama lumberyard has four jobs on order, as shown in the following table. Today is day 205 on the yard's schedule.

JOB	DUE DATE	REMAINING TIME (DAYS)
A	212	6
B	209	3
C	208	3
D	210	8

PX

In what sequence would the jobs be ranked according to the following decision rules:

a) FCFS b) SPT c) LPT
d) EDD e) Critical ratio

Which is best and why? Which has the minimum lateness?

••• **15.18** The following jobs are waiting to be processed at Rick Solano's machine center. Solano's machine center has a relatively long backlog and sets a fresh schedule every 2 weeks, which does not disturb earlier schedules. Below are the jobs received during the previous 2 weeks. They are ready to be scheduled today, which is day 241 (day 241 is a work day). Job names refer to names of clients and contract numbers.

JOB	DATE JOB RECEIVED	PRODUCTION DAYS NEEDED	DATE JOB DUE
BR-02	228	15	300
CX-01	225	25	270
DE-06	230	35	320
RG-05	235	40	360
SY-11	231	30	310

a) Complete the following table. (Show your supporting calculations.)
b) Which dispatching rule has the best score for flow time?
c) Which dispatching rule has the best score for utilization metric?
d) Which dispatching rule has the best score for lateness?
e) Which dispatching rule would you select? Support your decision.

DISPATCHING RULE	JOB SEQUENCE	FLOW TIME	UTILIZATION METRIC	AVERAGE NUMBER OF JOBS	AVERAGE LATENESS
EDD					
SPT					
LPT					
FCFS					

PX

••• **15.19** The following jobs are waiting to be processed at Julie Morel's machine center:

JOB	DATE ORDER RECEIVED	PRODUCTION DAYS NEEDED	DATE ORDER DUE
A	110	20	180
B	120	30	200
C	122	10	175
D	125	16	230
E	130	18	210

In what sequence would the jobs be ranked according to the following rules: (a) FCFS, (b) EDD, (c) SPT, and (d) LPT? All dates are according to shop calendar days. Today on the planning calendar is day 130, and none of the jobs have been started or scheduled. Which rule is best? PX

•• **15.20** Sunny Park Tailors has been asked to make three different types of wedding suits for separate customers. The following table highlights the time taken in hours for (1) cutting and sewing and (2) delivery of each of the suits. Which schedule finishes sooner: first come, first served (123) or a schedule using Johnson's rule?

Times Taken for Different Activities (hours)

SUIT	CUT AND SEW	DELIVER
1	4	2
2	7	7
3	6	5

PX

•• **15.21** The following jobs are waiting to be processed at Jeremy LaMontagne's machine center. Today is day 250.

JOB	DATE JOB RECEIVED	PRODUCTION DAYS NEEDED	DATE JOB DUE
1	215	30	260
2	220	20	290
3	225	40	300
4	240	50	320
5	250	20	340

Using the critical ratio scheduling rule, in what sequence would the jobs be processed? PX

•••• **15.22** The following set of seven jobs is to be processed through two work centers at George Heinrich's printing company. The sequence is first printing, then binding. Processing time at each of the work centers is shown in the following table:

JOB	PRINTING (HOURS)	BINDING (HOURS)
T	15	3
U	7	9
V	4	10
W	7	6
X	10	9
Y	4	5
Z	7	8

a) What is the optimal sequence for these jobs to be scheduled?
b) Chart these jobs through the two work centers.
c) What is the total length of time of this optimal solution?
d) What is the idle time in the binding shop, given the optimal solution?
e) How much would the binding machine's idle time be cut by splitting Job Z in half? PX

••• **15.23** Six jobs are to be processed through a two-step operation. The first operation involves sanding, and the second involves painting. Processing times are as follows:

JOB	OPERATION 1 (HOURS)	OPERATION 2 (HOURS)
A	10	5
B	7	4
C	5	7
D	3	8
E	2	6
F	4	3

Determine a sequence that will minimize the total completion time for these jobs. PX

•• **15.24** The following jobs are waiting to be processed at a small machine center:

JOB	DUE DATE	DURATION (DAYS)
10	260	30
20	258	16
30	260	8
40	270	20
50	275	10

All dates are specified as manufacturing planning calendar days. Assume that all jobs arrived on day 210 (yesterday) in the order shown but are not scheduled to begin until day 211 (today).

a) Sequence the jobs according to LPT.
b) What is the average completion (flow) time?
c) What is the average job lateness?
d) What is the average number of jobs in the system? PX

••• **15.25** Kamauf Custom Glass, Inc., manufactures custom windows primarily for homes and churches. Small repair jobs may take a few hours, but larger jobs may take hundreds of hours. How to schedule the jobs is a continuing issue. The scheduler has decided to see what insight can be gained by examining some common sequencing rules, namely FCFS, EDD, SPT, and LPT, on the criteria of average completion (flow) time, average lateness, and average number of jobs in the system. Her sample is the six jobs on the scheduler's desk Monday morning. Assume that all jobs on the scheduler's desk arrived today (day 100, which is a workday) in the order shown below. The estimated times to complete the jobs and their due dates are noted.

JOB	DUE DATE (DAYS)	DURATION (DAYS REQUIRED)
Alter piece (A)	130	5
Baptist Church (B)	350	100
Church of Christ (C)	140	17
Deacon's window (D)	150	15
Entry way (special order) (E)	195	20
Frank's Construction (kitchen window) (F)	175	35

a) Sequence the jobs according to the following decision rules: FCFS, EDD, SPT, and LPT.
b) For the FCFS sequence, compute the average completion (flow) time, average lateness, and average number of jobs in the system.
c) For the EDD sequence, compute the average completion (flow) time, average lateness, and average number of jobs in the system.
d) For the SPT sequence, compute the average completion (flow) time, average lateness, and average number of jobs in the system.
e) For the LPT sequence, compute the average completion (flow) time, average lateness, and average number of jobs in the system.
f) Which is the best decision rule based on average completion (flow) time?
g) Which is the best decision rule based on average lateness? PX

•• **15.26** Courtney is a programmer receiving requests each week to analyze a large database. Five jobs were on her desk Monday morning, and she must decide in what order to write the code. Assume that all jobs arrived today (day 1 and hour 1) in the order shown below. Courtney has assigned the number of hours required to do the coding as noted. Note that Courtney works an 8-hour day and today is a workday.

JOB	DUE DATE (HOURS HENCE)	DURATION (HOURS REQUIRED)
A	8	5
B	16	2
C	24	10
D	32	16
E	12	6

a) Sequence the jobs according to EDD.
b) What is the average completion (flow) time?
c) What is the average job lateness?
d) What is the average number of jobs in the system? **PX**

••• **15.27** At the campus copy shop, six jobs have arrived in the order shown (A, B, C, D, E, and F) at the beginning of the day (today, which is a workday). All of the jobs are due at various times today. Your task is to evaluate four common sequencing rules.

JOB	DUE (IN HOURS)	DURATION (HOURS REQUIRED)
A	4 (this job is due at noon, in 4 hours)	1.90
B	6.5 (this job is due in 6.5 hours)	1.20
C	4.5 (this job is due after lunch in 4.5 hours)	0.55
D	6 (this job is due in 6 hours)	1.25
E	7 (this job is due in 7 hours)	1.60
F	8 (this job is due in 8 hours)	1.80

a) Sequence the jobs according to the following decision rules: FCFS, EDD, SPT, and LPT.
b) For the FCFS sequence, compute the average completion (flow) time, average lateness, and average number of jobs in the system.
c) For the EDD sequence, compute the average completion (flow) time, average lateness, and average number of jobs in the system.
d) For the SPT sequence, compute the average completion (flow) time, average lateness, and average number of jobs in the system.
e) For the LPT sequence, compute the average completion (flow) time, average lateness, and average number of jobs in the system.
f) Which is the best decision rule based on average completion (flow) time?
g) Which job is late with the SPT rule? **PX**

•• **15.28** Drew Rosen Automation Company estimates the data entry and verifying times for four jobs as follows:

JOB	DATA ENTRY (HOURS)	VERIFY (HOURS)
A	2.5	1.7
B	3.8	2.6
C	1.9	1.0
D	1.8	3.0

In what order should the jobs be done if the company has one operator for each job? Illustrate the time-phased flow of this job sequence graphically. **PX**

Problems 15.29–15.32 relate to Scheduling Services

•• **15.29** Daniel's Barber Shop at Newark Airport is open 7 days a week but has fluctuating demand. Daniel Ball is interested in treating his barbers as well as he can with steady work and preferably 5 days of work with 2 consecutive days off. His analysis of his staffing needs resulted in the following plan. Schedule Daniel's staff with the minimum number of barbers.

	DAY						
	MON.	TUE.	WED.	THU.	FRI.	SAT.	SUN.
Barbers needed	6	5	5	5	6	4	3

•• **15.30** Given the following demand for waiters and waitresses at S. Ghosh Grill, determine the minimum wait staff needed with a policy of 2 consecutive days off.

	DAY						
	MON.	TUE.	WED.	THU.	FRI.	SAT.	SUN.
Wait staff needed	3	4	4	5	6	7	4

•• **15.31** The Winter Green, Nevada, Fire Department has a full-time staff of six highly trained firefighters and EMTs: Judith Greene, Kyan Browne, Mekell White, Carlos Black, Jorge Blue, and Mimi Gray. Chief Val Karan must schedule each employee 5 days on and 2 consecutive days off.

	DAY						
	MON.	TUE.	WED.	THU.	FRI.	SAT.	SUN.
Firefighters needed	3	3	4	3	5	6	5

••• **15.32** Juanita Flores schedules volunteers (Denzel, Ronnell, Rosanne, Napoleon, Barry, and Rosemary) at the Pueblo, Colorado, Free Food Pantry, which is open from 5 to 9 P.M., 7 days a week. She wants a schedule so that each volunteer has 4 days on and 3 consecutive days off. Note that the rules in Example 8 now apply to 3 consecutive days off.

	DAY						
	MON.	TUE.	WED.	THU.	FRI.	SAT.	SUN.
Volunteers needed	3	4	3	3	5	2	3

CASE STUDIES

Old Oregon Wood Store

In 2021, George Wright started the Old Oregon Wood Store to manufacture Old Oregon tables. Each table is carefully constructed by hand using the highest-quality oak. Old Oregon tables can support more than 500 pounds, and since the start of the Old Oregon Wood Store, not one table has been returned because of faulty workmanship or structural problems. In addition to being rugged, each table is beautifully finished using a urethane varnish that George developed over years of working with wood-finishing materials.

The manufacturing process consists of four steps: preparation, assembly, finishing, and packaging. Each step is performed by one person. In addition to overseeing the entire operation, George does all of the finishing. Tom Surowski performs the preparation step, which involves cutting and forming the basic components of the tables. Leon Davis is in charge of the assembly, and Cathy Stark performs the packaging.

Although each person is responsible for only one step in the manufacturing process, everyone can perform any one of the steps. It is George's policy that occasionally everyone should complete several tables on his or her own without any help or assistance. A small competition is used to see who can complete an entire table in the least amount of time. George maintains average total and intermediate completion times. The data are shown in Figure 15.7.

It takes Cathy longer than the other employees to construct an Old Oregon table. In addition to being slower than the other employees, Cathy is also unhappy about her current responsibility of packaging, which leaves her idle most of the day. Her first preference is finishing, and her second preference is preparation.

In addition to quality, George is concerned with costs and efficiency. When one of the employees misses a day, it causes major scheduling problems. In some cases, George assigns another employee overtime to complete the necessary work. At other times, George simply waits until the employee returns to work to complete his or her step in the manufacturing process. Both solutions cause problems. Overtime is expensive, and waiting causes delays and sometimes stops the entire manufacturing process.

To overcome some of these problems, Randy Lane was hired. Randy's major duties are to perform miscellaneous jobs and to help out if one of the employees is absent. George has given Randy training in all phases of the manufacturing process, and he is pleased with the speed at which Randy has been able to learn how to completely assemble Old Oregon tables. Randy's average total and intermediate completion times are given in Figure 15.8.

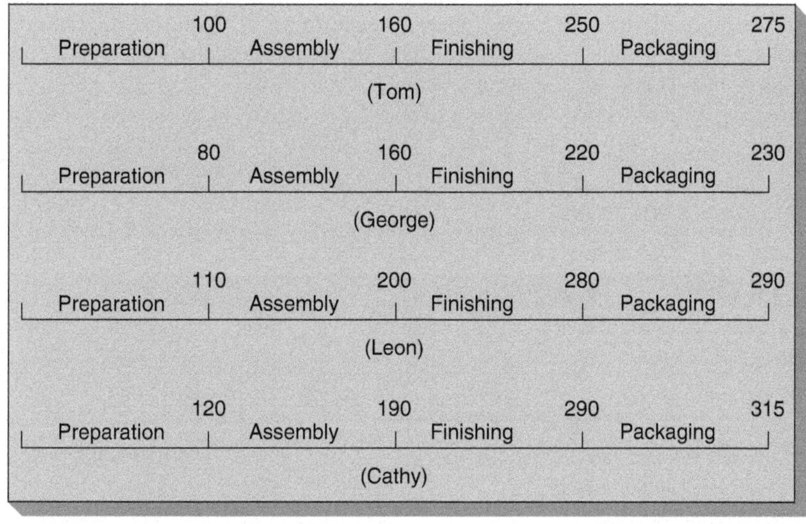

Figure **15.7**

Manufacturing Time in Minutes

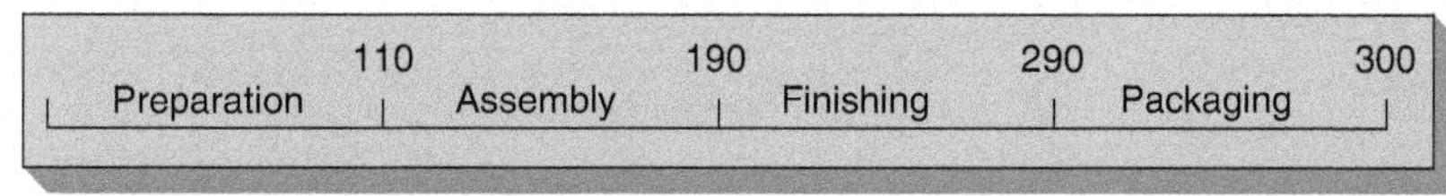

Figure **15.8**

Randy's Completion Times in Minutes

Discussion Questions

1. What is the fastest way to manufacture Old Oregon tables using the original crew? How many could be made per day?
2. Would production rates and quantities change significantly if George would allow Randy to perform one of the four functions and make one of the original crew the backup person?
3. What is the fastest time to manufacture a table with the original crew if Cathy is moved to either preparation or finishing?
4. Whoever performs the packaging function is severely underutilized. Can you find a better way of utilizing the four- or five-person crew than either giving each a single job or allowing each to manufacture an entire table? How many tables could be manufactured per day with this scheme?

From the Eagles to the Magic: Converting the Amway Center

Video Case 

The massive 875,000-square-foot Amway Center in Orlando, Florida, is a state-of-the-art sports entertainment center. While it is the home of the Orlando Magic basketball team, it is a flexible venue designed to accommodate a vast array of entertainment. The facility is used for everything from a concert by the Eagles or Britney Spears, to ice hockey, to arena football, to conventions, as well as 41 regular season home games played by its major tenant, the National Basketball Association's Orlando Magic.

The building is a LEED-certified (Leadership in Energy and Environmental Design), sustainable, environmentally friendly design, with unmatched technology. Dispersed throughout the building are over 1,000 digital monitors, the latest in broadcasting technology, and the tallest high-definition video board in an NBA venue. To fully utilize this nearly $500 million complex, conversions from one event to the next must be done rapidly—often in a matter of hours. Letting the facility sit idle because of delays in conversion is not an option.

Well-executed conversions help maximize facility revenue and at the same time minimize expenses. Fast and efficient conversions are critical. Like any other process, a conversion can be analyzed and separated into its component activities, each requiring its own human and capital resources. The operations manager must determine when to do the conversion, how to train and schedule the crew, which tools and capital equipment are necessary, and the specific steps necessary to break down the current event and set up for the next. In addition to trying to maintain a stable crew (typically provided by local staffing companies) and to maintain control during the frenzied pace of a conversion, managers divide the workforce into cross-trained crews, with each crew operating in its own uniquely colored shirt.

At the Amway Center, Charlie Leone makes it happen. Charlie is the operations manager, and as such, he knows that any conversion is loaded with complications and risks. Concerts add a

Mark Brown/Stringer/Getty Images

special risk because each concert has its own idiosyncrasies—and the breakdown for the Eagles concert will be unique. Charlie and his crews must anticipate and eliminate any potential problems. Charlie's immediate issue is making a schedule for converting the Eagles' concert venue to an NBA basketball venue. The activities and the time for various tasks have been determined and are shown in Table 15.3.

TABLE 15.3 CONCERT-TO-BASKETBALL CONVERSION TASKS

Available crew size = 16, including two fork truck drivers

TIME ALLOWED	TASKS	CREW AND TIME REQUIRED
3 to 4 hr	**11:20 P.M.** Performance crew begins teardown of concert stage & equipment	Concert's Responsibility
45 min	**11:20 P.M. Clear Floor Crew**	
	Get chair carts from storage	10 for 15 min
	Clear all chairs on floor, loading carts starting at south end, working north	16 for 30 min
	Move chair carts to north storage and stack as they become full	(includes 1 fork truck operator)
15 min	**11:50 P.M.** *(Or as soon as area under rigging is cleared)*	6 for 15 min
	Set up retractable basketball seating on north end	
	Take down railing above concert stage	
	Place railings on cart and move to storage	
2.5 hr	**12:05 A.M. Basketball Floor Crew**	8
	Position 15 basketball floor carts on floor	
	Mark out arena floor for proper placement of basketball floor	
	Position basketball floor by section	
	Assemble/join flooring/lay carpets over concrete	
	Position basketball nets in place	
	Set up scorer tables	
	Install risers for all courtside seating	
	Install 8-ft tables on east side of court	
2.5 hr	**Seating Unit Crew** *Starts same time as Basketball Floor Crew*	8
	Set up retractable basketball seating on north end	(includes 2 fork truck operators)
	Set up retractable basketball seating on south end	
	(Can only be done after concert stage and equipment is out of way)	
	Install stairs to Superstar Seating	
2 hr	**Board Crew** *Starts after Seating Unit Crew finishes*	4
	Install dasher board on south end	
	Move stairs to storage	

(Continued)

TABLE 15.3 *Continued*

TIME ALLOWED	TASKS	CREW AND TIME REQUIRED
2 hr	**Chair Crew** *Starts after Seating Unit Crew finishes*	12
	Get chair carts from storage	
	Position chair carts on floor	
	Position chairs behind goals, courtside, and scorer tables	
	Clean, sweep, and place carts in order	
45 min	**End-of-Shift Activities** *Starts after Chair Crew finishes*	12
	Perform checklist items	
	Ensure that steps and stairways and railings are in place and tight	
	Check all seats are in upright position and locked in place	
	Report any damaged seats or armrests in need of repair	
	Verify exact number of chairs behind goals, courtside, and scorer tables	
15 min	**Check Out** *Starts after End-of-Shift Activities*	16
	Check for next conversion date and time and inform crew	
	Report any injuries Punch out all employees before leaving **8:00 A.M.** Floor ready for Magic practice	

Discussion Questions*

1. Make a Gantt chart to help Charlie organize his crew to perform the concert-to-basketball conversion. *Note*: Do not include the teardown of the concert stage and equipment, as that is the responsibility of the concert crew.
2. What time will the floor be ready?
3. Does Charlie have any extra personnel or a shortage of personnel? If so how many?

*You may wish to view the video that accompanies this case before answering the questions.

Scheduling at Hard Rock Café

Video Case 

Whether it's scheduling nurses at Mayo Clinic, pilots at Southwest Airlines, classrooms at UCLA, or servers at a Hard Rock Cafe, it's clear that good scheduling is important. Proper schedules use an organization's assets (1) more effectively, by serving customers promptly, and (2) more efficiently, by lowering costs.

Hard Rock Cafe at Universal Studios, Orlando, is the world's largest restaurant, with 1,100 seats on two main levels. With typical turnover of employees in the restaurant industry at 80% to 100% per year, Hard Rock General Manager Ken Hoffman takes scheduling very seriously. Hoffman wants his 160 servers to be effective, but he also wants to treat them fairly. He has done so with scheduling software and flexibility that has increased productivity while contributing to turnover that is half the industry average. His goal is to find the fine balance that gives employees financially productive daily work shifts while setting the schedule tight enough so as to not overstaff between lunch and dinner.

The weekly schedule begins with a sales forecast. "First, we examine last year's sales at the cafe for the same day of the week," says Hoffman. "Then we adjust our forecast for this year based on a variety of closely watched factors. For example, we call the Orlando Convention Bureau every week to see what major groups will be in town. Then we send two researchers out to check on the occupancy of nearby hotels. We watch closely to see what concerts are scheduled at Hard Rock Live—the 3,000-seat concert stage next door. From the forecast, we calculate how many people we need to have on duty each day for the kitchen, the bar, as hosts, and for table service."

Once Hard Rock determines the number of staff needed, servers submit request forms, which are fed into the software's linear programming mathematical model. Individuals are given priority rankings from 1 to 9, based on their seniority and how important they are to fill each day's schedule. Schedules are then posted by day and by workstation. Trades are handled between employees, who understand the value of each specific shift and station.

Hard Rock employees like the system, as does the general manager, since sales per labor-hour are rising and turnover is dropping.

Discussion Questions*

1. Name and justify several factors that Hoffman could use in forecasting weekly sales.
2. What can be done to lower turnover in large restaurants?
3. Why is seniority important in scheduling servers?
4. How does the schedule impact productivity?

*You may wish to view the video that accompanies this case before answering the questions.

Endnotes

1. Source: Based on interview with Chef John Nicely and Orlando Magic executives.
2. Opportunity costs are those profits forgone or not obtained.
3. Finite capacity scheduling (FCS) systems go by a number of names, including finite scheduling and advanced planning and scheduling (APS). The name manufacturing execution systems (MES) may also be used, but MES tends to suggest an emphasis on the reporting system from shop operations back to the scheduling activity.
4. Sources: **ESPN.com** (October 14, 2019); **NBCsports.com/NBA** (October 14, 2019); and **medium.com/basketball-university** (February 13, 2021).
5. This problem was first solved by Rajen Tibrewala, D. Philippe, and J. J. Browne, "Optimal Scheduling of Two Consecutive Idle Periods," *Management Science*, 19(1): 71–75.

Bibliography

Baker, K. A., and D. Trietsch. *Principles of Sequencing and Scheduling.* New York: Wiley, 2009.

Chapman, S. *Fundamentals of Production Planning and Control.* Upper Saddle River, NJ: Prentice Hall, 2006.

Dietrich, B., G. A. Paleologo, and L. Wynter. "Revenue Management in Business Services." *Production and Operations Management* 17, no. 4 (July–August 2008): 475–480.

Farmer, A., J. S. Smith, and L. T. Miller. "Scheduling Umpire Crews for Professional Tennis Tournaments." *Interfaces* 37, no. 2 (March– April 2007): 187–196.

Kellogg, D. L., and S. Walczak. "Nurse Scheduling." *Interfaces* 37, no. 4 (July–August 2007): 355–369.

Lopez, P., and F. Roubellat. *Production Scheduling*. New York: Wiley, 2008.

Morton, T. E., and D. W. Pentico. *Heuristic Scheduling Systems.* New York: Wiley, 1993.

Pinedo, M. *Scheduling: Theory, Algorithms, and Systems*, 5th ed. New York: Springer, 2016.

Plenert, G., and B. Kirchmier. *Finite Capacity Scheduling*. New York: Wiley, 2000.

Render, B., R. M. Stair, M. Hanna and T. Hale. *Quantitative Analysis for Management*, 13th ed. Boston: Pearson, 2018.

Robidoux, L., and P. Donnelly. "Automated Employee Scheduling." *Nursing Management* 42, no. 12 (December 2011): 41–43.

Chapter 15 *Rapid* Review

Main Heading	Review Material	MyLab Operations Management
THE IMPORTANCE OF SHORT-TERM SCHEDULING	The strategic importance of scheduling is clear: ■ Effective scheduling means *faster movement* of goods and services through a facility. This means greater use of assets and hence greater capacity per dollar invested, which, in turn, *lowers cost*. ■ Added capacity, faster throughput, and the related flexibility mean better customer service through *faster delivery*. ■ Good scheduling contributes to realistic commitments, hence *dependable delivery*.	Concept Questions: 1.1–1.2
SCHEDULING ISSUES	*The objective of scheduling is to allocate and prioritize demand (generated by either forecasts or customer orders) to available facilities.* ■ Forward scheduling—Begins the schedule as soon as the requirements are known. ■ Backward scheduling—Begins with the due date by scheduling the final operation first and the other job steps in reverse order. ■ **Loading**—The assigning of jobs to work or processing centers. The four scheduling criteria are (1) *minimize completion time*, (2) *maximize utilization*, (3) *minimize work-in-process (WIP) inventory*, and (4) *minimize customer waiting time*.	Concept Questions: 2.1–2.6 **VIDEO 15.1** From the Eagles to the Magic: Converting the Amway Center
SCHEDULING PROCESS-FOCUSED FACILITIES	A process-focused facility is a high-variety, low-volume system commonly found in manufacturing and services. It is also called an intermittent, or job shop, facility.	Concept Questions: 3.1–3.4
LOADING JOBS	■ **Input–output control**—A system that allows operations personnel to manage facility work flows by tracking work added to a work center and its work completed. ■ **ConWIP cards**—Cards that control the amount of work in a work center, aiding input–output control. ConWIP is an acronym for *constant work-in-process*. A ConWIP card travels with a job (or batch) through the work center. When the job is finished, the card is released and returned to the initial workstation, authorizing the entry of a new batch into the work center. ■ **Gantt charts**—Planning charts used to schedule resources and allocate time. The Gantt *load chart* shows the loading and idle times of several departments, machines, or facilities. It displays the relative workloads in the system so that the manager knows what adjustments are appropriate. The Gantt *schedule chart* is used to monitor jobs in progress (and is also used for project scheduling). It indicates which jobs are on schedule and which are ahead of or behind schedule. ■ **Assignment method**—A special class of linear programming models that involves assigning tasks or jobs to resources. In assignment problems, only one job (or worker) is assigned to one machine (or project). The assignment method involves adding and subtracting appropriate numbers in the table to find the lowest *opportunity cost* for each assignment.	Concept Questions: 4.1–4.6 Problems: 15.1–15.14 Virtual Office Hours for Solved Problem: 15.1
SEQUENCING JOBS	■ **Sequencing**—Determining the order in which jobs should be done at each work center. ■ **Priority rules**—Rules used to determine the sequence of jobs in process-oriented facilities. ■ First come, first served (FCFS)—Jobs are completed in the order in which they arrived. ■ Shortest processing time (SPT)—Jobs with the shortest processing times are assigned first. ■ Earliest due date—Earliest due date jobs are performed first. ■ Longest processing time (LPT)—Jobs with the longest processing time are completed first. $\text{Average completion time} = \frac{\text{Total flow time}}{\text{Number of jobs}}$ (15-1) $\text{Utilization metric} = \frac{\text{Total job work (processing) time}}{\text{Total flow time}}$ (15-2)	Concept Questions: 5.1–5.6 Problems: 15.15–15.28 Virtual Office Hours for Solved Problems: 15.2–15.5 **ACTIVE MODEL 15.1**

Chapter 15 **Rapid Review** *continued*

Main Heading	Review Material	MyLab Operations Management
	$$\text{Average number of jobs in the system} = \frac{\text{Total flow time}}{\text{Total job work (processing) time}} \quad (15\text{-}3)$$ $$\text{Average job lateness} = \frac{\text{Total late days}}{\text{Number of jobs}} \quad (15\text{-}4)$$ $$\text{Job lateness} = \text{Max}\{0, \text{yesterday} + \text{flow time} - \text{due date}\} \quad (15\text{-}5)$$ SPT is the best technique for minimizing job flow and average number of jobs in the system. FCFS performs about average on most criteria, and it appears fair to customers. EDD minimizes maximum tardiness. ■ **Flow time**—The time each job spends waiting plus the time being processed. ■ **Critical ratio (CR)**—A sequencing rule that is an index number computed by dividing the time remaining until due date by the work time remaining: $$\text{CR} = \frac{\text{Time remaining}}{\text{Workdays remaining}} = \frac{\text{Due date} - \text{Today's date}}{\text{Work (lead) time remaining}} \quad (15\text{-}6)$$ As opposed to the priority rules, the critical ratio is dynamic and easily updated. It tends to perform better than FCFS, SPT, EDD, or LPT on the average job-lateness criterion. ■ **Johnson's rule**—An approach that minimizes processing time for sequencing a group of jobs through two work centers while minimizing total idle time in the work centers. Rule-based scheduling systems have the following limitations: (1) Scheduling is dynamic, (2) rules do not look upstream or downstream, and (3) rules do not look beyond due dates.	
FINITE CAPACITY SCHEDULING (FCS)	■ **Finite capacity scheduling (FCS)**—Computerized short-term scheduling that overcomes the disadvantage of rule-based systems by providing the user with graphical interactive computing.	Concept Questions: 6.1–6.2
SCHEDULING SERVICES	Cyclical scheduling with inconsistent staffing needs is often the case in services. The objective focuses on developing a schedule with the minimum number of workers. In these cases, each employee is assigned to a shift and has time off.	Concept Questions: 7.1–7.6 **VIDEO 15.2** Scheduling at Hard Rock Cafe Problems: 15.29–15.32
ADDITIONAL MYLAB OPERATIONS MANAGEMENT RESOURCES	✔ Additional Case Study (Payroll Planning, Inc.) ✔ Multiple Choice Case Questions (Old Oregon Wood Store)	

Self Test

LO 15.1 Which of the following decisions covers the longest time period?
a) Short-term scheduling
b) Capacity planning
c) Aggregate planning
d) A master schedule

LO 15.2 A visual aid used in loading and scheduling jobs is a:
a) Gantt chart.
b) planning file.
c) bottleneck.
d) load-schedule matrix.
e) level material chart.

LO 15.3 The assignment method involves adding and subtracting appropriate numbers in the table to find the lowest _____ for each assignment.
a) profit
b) number of steps
c) number of allocations
d) range per row
e) opportunity cost

LO 15.4 The most popular priority rules include:
a) FCFS
b) EDD.
c) SPT.
d) all of the above.

LO 15.5 The job that should be scheduled last when using Johnson's rule is the job with the:
a) largest total processing time on both machines.
b) smallest total processing time on both machines.
c) longest activity time if it lies with the first machine.
d) longest activity time if it lies with the second machine.
e) shortest activity time if it lies with the second machine.

LO 15.6 What is computerized short-term scheduling that overcomes the disadvantage of rule-based systems by providing the user with graphical interactive computing?
a) LPT
b) FCS
c) CSS
d) FCFS
e) GIC

LO 15.7 Cyclical scheduling is used to schedule:
a) jobs.
b) machines.
c) shipments.
d) employees.

Answers: LO 15.1. b; LO 15.2. a; LO 15.3. e; LO 15.4. d; LO 15.5. e; LO 15.6. b; LO 15.7. d.

Lean Operations

CHAPTER 16

CHAPTER OUTLINE

Many services have adopted Lean techniques as a normal part of their business. Restaurants such as Olive Garden expect and receive JIT deliveries. Both buyer and supplier expect fresh, high-quality produce delivered without fail just when it is needed. The system doesn't work any other way.

Wally Skalij/Los Angeles Times/Getty Images

10 OM STRATEGY DECISIONS

- Design of Goods and Services
- Managing Quality
- Process Strategies
- Location Strategies
- Layout Strategies
- Human Resources
- Supply Chain Management
- *Inventory Management*
  - Independent Demand (Ch. 12)
  - Dependent Demand (Ch. 14)
  - ***Lean Operations (Ch. 16)***
- Scheduling
- Maintenance

GLOBAL COMPANY PROFILE
Toyota Motor Corporation

Achieving Competitive Advantage with Lean Operations at Toyota Motor Corporation

Toyota Motor Corporation, with $273 billion from annual sales of nearly 10 million cars and trucks, is one of the largest vehicle manufacturers in the world. Two Lean techniques, just-in-time (JIT) and the Toyota Production System (TPS), have been instrumental in its growth. Toyota, with a wide range of vehicles, competes head-to-head with successful, long-established companies in Europe and the U.S. Taiichi Ohno, a former vice president of Toyota, created the basic framework for two of the world's most discussed systems for improving productivity, JIT and TPS. These two concepts provide much of the foundation for Lean operations:

- Central to JIT is a philosophy of continuous problem solving. In practice, JIT means making only what is needed, when it is needed. JIT provides an excellent vehicle for finding and eliminating problems because problems are easy to find in a system that eliminates the slack that inventory generates. When excess inventory is eliminated, shortcomings related to quality, layout, scheduling, and supplier performance become immediately evident—as does excess production.
- Central to TPS is employee learning and a continuing effort to create and produce products under ideal conditions. Ideal conditions exist only when management brings facilities,

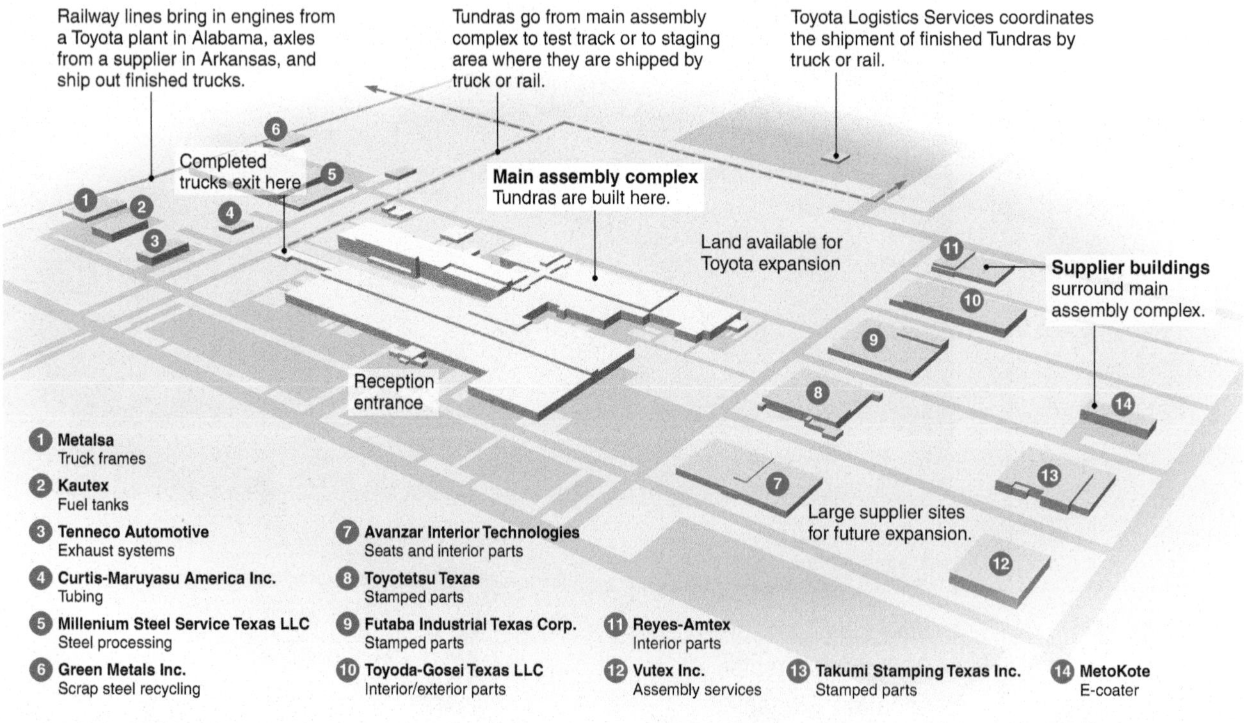
Railway lines bring in engines from a Toyota plant in Alabama, axles from a supplier in Arkansas, and ship out finished trucks.
Tundras go from main assembly complex to test track or to staging area where they are shipped by truck or rail.
Toyota Logistics Services coordinates the shipment of finished Tundras by truck or rail.
Completed trucks exit here
Main assembly complex
Tundras are built here.
Land available for Toyota expansion
Supplier buildings surround main assembly complex.
Reception entrance
Large supplier sites for future expansion.
1 Metalsa
Truck frames
2 Kautex
Fuel tanks
3 Tenneco Automotive
Exhaust systems
4 Curtis-Maruyasu America Inc.
Tubing
5 Millenium Steel Service Texas LLC
Steel processing
6 Green Metals Inc.
Scrap steel recycling
7 Avanzar Interior Technologies
Seats and interior parts
8 Toyotetsu Texas
Stamped parts
9 Futaba Industrial Texas Corp.
Stamped parts
10 Toyoda-Gosei Texas LLC
Interior/exterior parts
11 Reyes-Amtex
Interior parts
12 Vutex Inc.
Assembly services
13 Takumi Stamping Texas Inc.
Stamped parts
14 MetoKote
E-coater
14 suppliers outside the main plant

Outside: Toyota has a 2,000-acre site with 14 of the 21 on-site suppliers, adjacent rail lines, and a nearby interstate highway. The site provides expansion space for both Toyota and for its suppliers — and provides an environment for just-in-time.

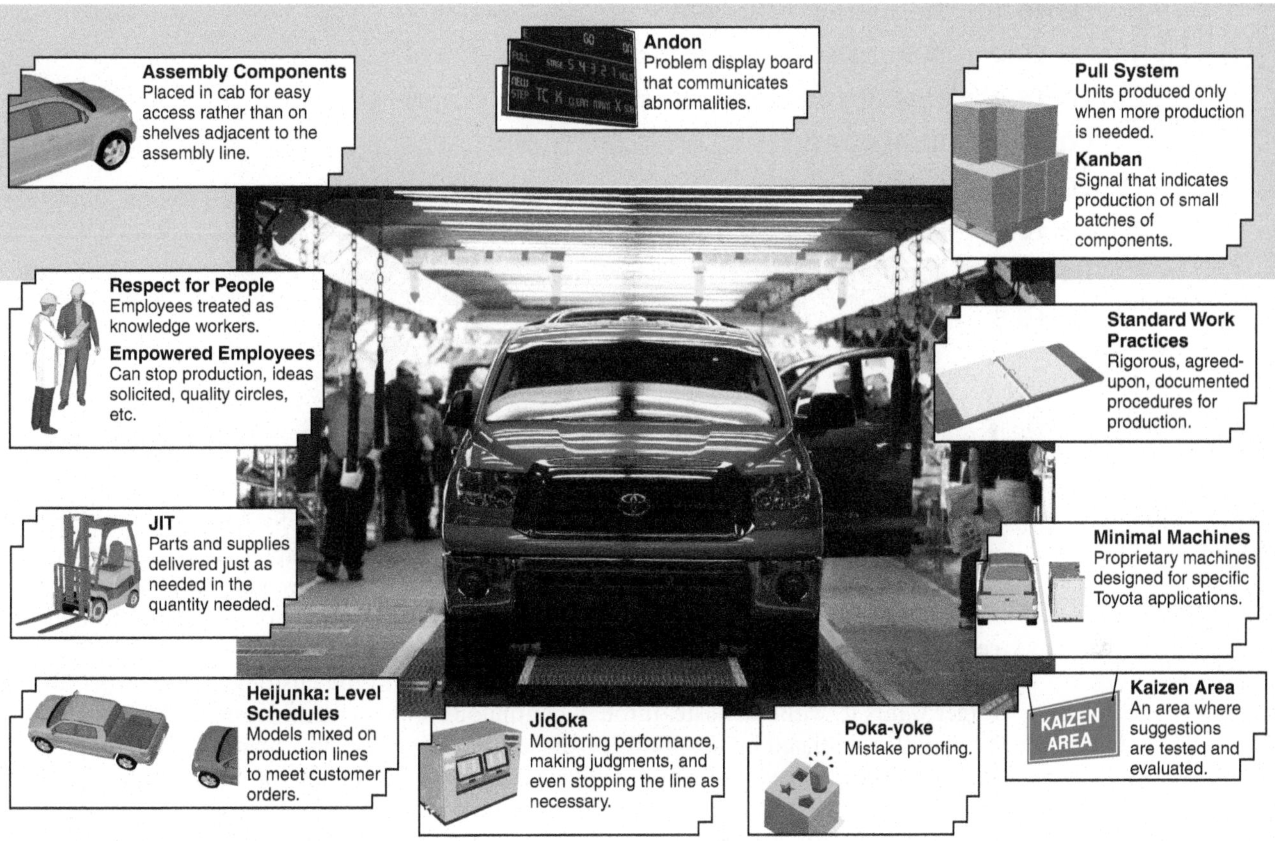

Andon
Problem display board that communicates abnormalities.

Assembly Components
Placed in cab for easy access rather than on shelves adjacent to the assembly line.

Pull System
Units produced only when more production is needed.

Kanban
Signal that indicates production of small batches of components.

Respect for People
Employees treated as knowledge workers.

Empowered Employees
Can stop production, ideas solicited, quality circles, etc.

Standard Work Practices
Rigorous, agreed-upon, documented procedures for production.

JIT
Parts and supplies delivered just as needed in the quantity needed.

Minimal Machines
Proprietary machines designed for specific Toyota applications.

Heijunka: Level Schedules
Models mixed on production lines to meet customer orders.

Jidoka
Monitoring performance, making judgments, and even stopping the line as necessary.

Poka-yoke
Mistake proofing.

Kaizen Area
An area where suggestions are tested and evaluated.

KAIZEN AREA

1 **AGC Automotive Americas** Glass assemblies

2 **ARK Inc.** Industrial waste management, recycling

3 **HERO Assemblers LLP** Assembly of tire onto wheel

4 **HERO Logistics LLP** Logistics

5 **PPG Industries Inc.** Glass assemblies

6 **Reyes Automotive Group** Interior/exterior parts

7 **Tokai Rika** Functional parts

Seven suppliers inside the main plant

Toyota's San Antonio plant has about 2 million interior sq. ft., providing facilities within the final assembly building for 7 of the 21 on-site suppliers, and capacity to build 200,000 pick-up trucks annually. But most importantly, Toyota practices the world-class Toyota Production System and expects its suppliers to do the same thing, wherever they are.

Toyota image: ZUMA Press Inc/Alamy Stock Photo; Poka-yoke image: Nemanja Cosovic/Shutterstock

machines, and people together to add value without waste. Waste undermines productivity by diverting resources to excess inventory, unnecessary processing, and poor quality. Respect for people, extensive training, cross-training, and standard work practices of empowered employees focusing on driving out waste are fundamental to TPS.

Toyota's implementation of TPS and JIT is present at its 2,000-acre San Antonio, Texas, facility, the largest Toyota land site for an automobile assembly plant in the U.S. Interestingly, despite its large site and annual production capability of 200,000, a throughput time of 20½ hours, and the output of a truck every 63 seconds, the building itself is one of the smallest in the industry. Modern automobiles have 30,000 parts, but at Toyota, independent suppliers combine many of these parts into subassemblies. Twenty-one of these suppliers are on site at the San Antonio facility and transfer components to the assembly line on a JIT basis.

Operations such as these taking place in the San Antonio plant are why Toyota continues to perform near the top in quality and maintain the lowest labor-hour assembly time in the industry. Lean operations *do* work—and they provide a competitive advantage for Toyota Motor Corporation.

LEARNING OBJECTIVES

Lean Operations

LO 16.1 *Define* Lean operations

As shown in the *Global Company Profile*, the Toyota Production System (TPS) contributes to a world-class operation at Toyota Motor Corporation. In this chapter, we discuss Lean operations, including JIT and TPS, as approaches to continuous improvement. All three approaches, in their own way, strive to increase the knowledge and capability of employees to build processes that improve productivity and lead to world-class operations.

Lean operations
Eliminates waste through continuous improvement and focus on exactly what the customer wants.

Just-in-time (JIT)
Continuous and forced problem solving via a focus on throughput and reduced inventory.

Toyota Production System (TPS)
Focus on continuous improvement, respect for people, and standard work practices.

Lean operations supply the customer with exactly what the customer wants when the customer wants it, without waste, through continuous improvement. Lean operations are driven by workflow initiated by the "pull" of the customer's order. **Just-in-time (JIT)** is an approach of continuous and forced problem solving via a focus on throughput and reduced inventory. The **Toyota Production System (TPS)**, with its emphasis on continuous improvement, respect for people, and standard work practices, is particularly suited for assembly lines. (See the *OM in Action* box, "Toyota's New Challenge.")

In this chapter we use the term *Lean operations* to encompass all the related approaches and techniques of both JIT and TPS. When implemented as a comprehensive operations strategy, Lean sustains competitive advantage and results in increased overall returns to stakeholders.

Regardless of the approach and label, operations managers address three issues that are fundamental to operations improvement: *eliminate waste, remove variability,* and *improve throughput.* We now introduce these three issues and then discuss the major attributes of Lean operations. Finally, we look at Lean applied to services.

Eliminate Waste

Lean producers set their sights on perfection: *no* bad parts, *no* inventory, *only* value-added activities, and *no* waste. Any activity that does not add value in the eyes of the customer is a waste. The customer defines product value. If the customer does not want to pay for it, it is a waste. Taiichi Ohno, noted for his work on the Toyota Production System, identified seven categories of waste. These categories have become popular in Lean organizations and cover many of the ways organizations waste or lose money. Ohno's **seven wastes** are:

Seven wastes
Overproduction
Queues
Transportation
Inventory
Motion
Overprocessing
Defective product

- *Overproduction:* Producing more than the customer orders or producing early (before it is demanded) is waste.
- *Queues:* Idle time, storage, and waiting are wastes (they add no value).
- *Transportation:* Moving material between plants or between work centers and handling it more than once is waste.
- *Inventory:* Unnecessary raw material, work-in-process (WIP), finished goods, and excess operating supplies add no value and are wastes.
- *Motion:* Movement of equipment or people that adds no value is waste.
- *Overprocessing:* Work performed on the product that adds no value is waste.
- *Defective product:* Returns, warranty claims, rework, and scrap are wastes.

LO 10.2 *Define* the seven wastes and the 5Ss

A broader perspective—one that goes beyond immediate production—suggests that other resources, such as energy, water, and air, are often wasted but should not be. Efficient, sustainable production minimizes inputs and maximizes outputs, wasting nothing.

OM in Action Toyota's New Challenge[1]

With the generally high value of the yen, making a profit on cars built in Japan but sold in foreign markets is a challenge. As a result, Honda and Nissan are moving plants overseas, closer to customers. But Toyota, despite marginal profit on cars produced for export, is maintaining its current Japanese capacity. Toyota, which led the way with JIT and the TPS, is doubling down on its manufacturing prowess and continuous improvement. For an organization that traditionally does things slowly and step-by-step, the changes are radical. With its new plant in Japan, Toyota believes it can once again set new production benchmarks. It is drastically reforming its production processes in a number of ways:

- The assembly line has cars sitting side-by-side, rather than bumper-to-bumper, shrinking the length of the line by 35% and requiring fewer steps by workers.
- Instead of having car chassis dangling from overhead conveyors, they are perched on raised platforms, reducing heating and cooling costs by 40%.

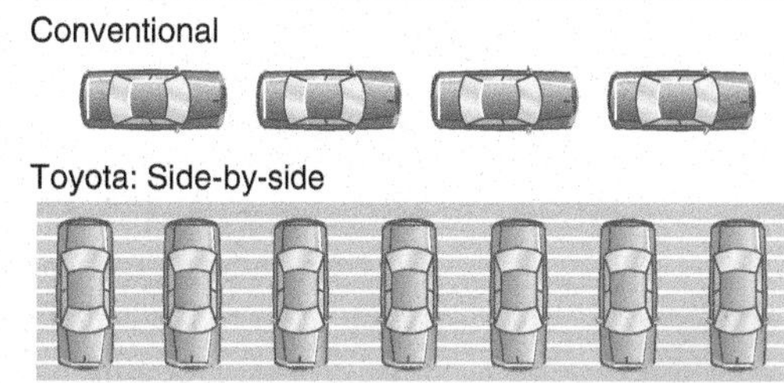

- Retooling permits faster changeovers, allowing for shorter product runs of components, supporting level scheduling.
- The assembly line uses quiet friction rollers with fewer moving parts, requiring less maintenance than conventional lines and reducing worker fatigue.

These TPS innovations, efficient production with small lot sizes, rapid changeover, level scheduling, half the workers, and half the square footage, are being duplicated in Toyota's new plant in Blue Springs, Mississippi.

For over a century, managers have pursued "housekeeping" for a neat, orderly, and efficient workplace and as a means of reducing waste by applying the simple concept of "a place for everything and everything in its place." Operations managers have embellished "housekeeping" to include a checklist—now known as the 5Ss.[2] The Japanese developed the initial 5Ss. Not only are the 5Ss a good checklist for Lean operations, but they also provide an easy vehicle with which to assist the culture change that is often necessary to bring about Lean operations. The 5Ss follow:

5Ss
A Lean production checklist:
Sort
Simplify
Shine
Standardize
Sustain

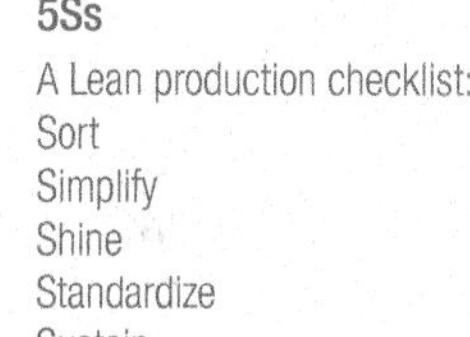

- *Sort/segregate:* Keep what is needed and remove everything else from the work area; when in doubt, throw it out. Identify nonvalue items and remove them. Getting rid of unneeded items makes space available and usually improves workflow.
- *Simplify/straighten:* Arrange and use methods analysis tools (see Chapter 7 and Chapter 10) to improve workflow and reduce wasted motion. Consider long-run and short-run ergonomic issues. Label and display for easy use only what is needed in the immediate work area. (For examples of visual displays, see Chapter 10, Figure 10.8 and the adjacent photo of equipment located within prescribed lines on the tarmac at Seattle's airport.)
- *Shine/sweep:* Clean daily; eliminate all forms of dirt, contamination, and clutter from the work area.
- *Standardize:* Remove variations from the process by developing standard operating procedures and checklists; good standards make the abnormal obvious. Standardize equipment and tooling so that cross-training time and cost are reduced. Train and retrain the work team so that when deviations occur, they are readily apparent to all.
- *Sustain/self-discipline:* Review periodically to recognize efforts and to motivate to sustain progress. Use visuals wherever possible to communicate and sustain progress.

In keeping with 5S, airports, like many other facilities, specify with painted guidelines exactly where tools and equipment such as this fuel pump are to be positioned.

Courtesy of Alaska Airlines

U.S. managers often add two additional Ss that contribute to establishing and maintaining a Lean workplace:

- *Safety:* Build good safety practices into the preceding five activities.
- *Support/maintenance:* Reduce variability, unplanned downtime, and costs. Integrate daily *shine* tasks with preventive maintenance.

The Ss support continuous improvement and provide a vehicle with which employees can identify. Operations managers need think only of the examples set by a well-run hospital emergency room or the spit-and-polish of a fire department for a benchmark. Offices and retail stores, as well as manufacturers, have successfully used the 5Ss in their respective efforts to eliminate waste and move to Lean operations. A place for everything and everything in its place does make a difference in a well-run office. And retail stores successfully use the Ss to reduce misplaced merchandise and improve customer service. An orderly workplace reduces waste, releasing assets for other, more productive, purposes.

Remove Variability

Variability
Any deviation from the optimum process that delivers a perfect product on time, every time.

Managers seek to remove variability caused by both internal and external factors. **Variability** is any deviation from the optimum process that delivers a perfect product on time, every time. Variability is a polite word for problems. The less variability in a system, the less waste in the system. Management's job is to crush variability. Among the many sources of variability are:

- Poor processes that allow employees and suppliers to produce improper quantities or nonconforming units
- Inadequate maintenance of facilities and processes
- Unknown and changing customer demands
- Incomplete or inaccurate drawings, specifications, and bills of material

Inventory reduction via JIT is an effective tool for identifying causes of variability. The precise timing of JIT makes variability evident, just as reducing inventory exposes variability. Defeating variability allows managers to move good materials on schedule, add value at each step of the process, drive down costs, and ultimately win orders.

Improve Throughput

Throughput
The rate at which units move through a process.

Throughput is the rate at which units move through a process. Each minute that products remain on the books, costs accumulate, and competitive advantage is lost. Time is money. For example, using a tight responsive supplier network, giant fashion retailer Inditex SA can now design, prototype, approve, and manufacture thousands of coats in a matter of weeks, not months. Its responsiveness to changes in the volatile fashion market is in direct proportion to the improvement in throughput.

Pull system
A concept that results in material being produced only when requested and moved to where it is needed just as it is needed.

A technique for increasing throughput is a pull system. A **pull system** *pulls* a unit to where it is needed just as it is needed. Pull systems are a standard tool of Lean. Pull systems use signals to request production and delivery from supplying stations to stations that have production capacity available. The pull concept is used both within the immediate production process and with suppliers. By *pulling* material through the system in very small lots—just as it is needed—waste and inventory are removed. As inventory is removed, clutter is reduced, problems become evident, and continuous improvement is emphasized. Removing the cushion of inventory also reduces both investment in inventory and flow time. A *push* system dumps orders on the next downstream workstation, regardless of timeliness and resource availability. Push systems are the antithesis of Lean. Pulling material through a production process as it is needed rather than in a "push" mode typically lowers cost and improves schedule performance, enhancing customer satisfaction.

Lean and Just-in-Time

Just-in-time (JIT), with its focus on rapid throughput and reduced inventory, is a powerful component of Lean. With the inclusion of JIT in Lean, materials arrive *where* they are needed only *when* they are needed. When good units do not arrive just as needed, a "problem" has been identified. This is the reason this aspect of Lean is so powerful—it focuses attention on *problems*. By driving out waste and delay, JIT reduces inventory, cuts variability and waste, and improves throughput. Every moment material is held, an activity that adds value should be occurring. Consequently, as Figure 16.1 suggests, JIT often yields a competitive advantage.

A well-executed Lean program requires a meaningful buyer–supplier partnership.

STUDENT TIP
JIT places added demands on performance, but that is why it pays off.

Supplier Partnerships

Supplier partnerships exist when a supplier and a purchaser work together with open communication and a goal of removing waste and driving down costs. Trust and close collaboration are critical to the success of Lean. Figure 16.2 shows the characteristics of supplier partnerships. Some specific goals are:

Supplier partnerships
Partnerships of suppliers and purchasers that remove waste and drive down costs for mutual benefits.

- *Removal of unnecessary activities*, such as receiving, incoming inspection, and paperwork related to bidding, invoicing, and payment.
- *Removal of in-plant inventory* by delivery in small lots directly to the using department as needed.
- *Removal of in-transit inventory* by encouraging suppliers to locate nearby and provide frequent small shipments. The shorter the flow of material in the resource pipeline, the less inventory. Inventory can also be reduced through a technique known as *consignment*. **Consignment inventory**, a variation of vendor-managed inventory (Chapter 11), means the supplier maintains the title to the inventory until it is used.
- *Obtain improved quality and reliability* through long-term commitments, communication, and cooperation.

Consignment inventory
An arrangement in which the supplier maintains title to the inventory until it is used.

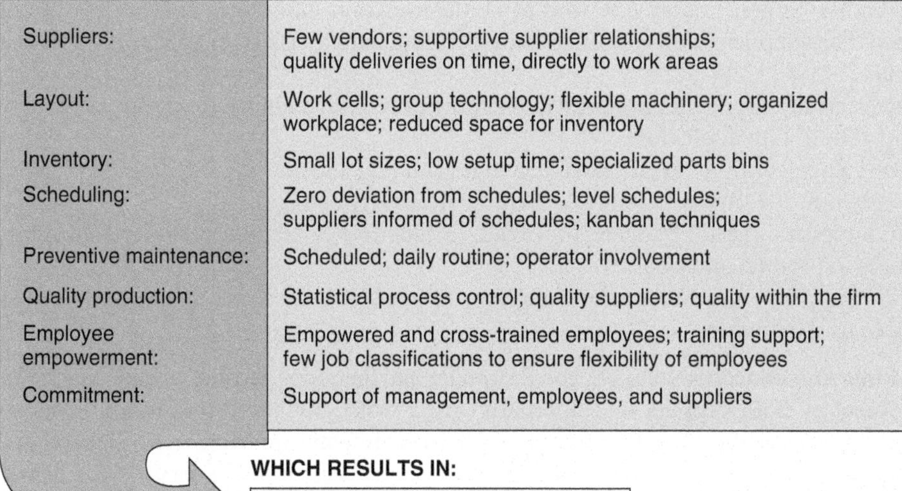

JIT TECHNIQUES:
Suppliers: Few vendors; supportive supplier relationships; quality deliveries on time, directly to work areas
Layout: Work cells; group technology; flexible machinery; organized workplace; reduced space for inventory
Inventory: Small lot sizes; low setup time; specialized parts bins
Scheduling: Zero deviation from schedules; level schedules; suppliers informed of schedules; kanban techniques
Preventive maintenance: Scheduled; daily routine; operator involvement
Quality production: Statistical process control; quality suppliers; quality within the firm
Employee empowerment: Empowered and cross-trained employees; training support; few job classifications to ensure flexibility of employees
Commitment: Support of management, employees, and suppliers
WHICH RESULTS IN:
Rapid throughput frees assets
Quality improvement reduces waste
Cost reduction adds pricing flexibility
Variability reduction
Rework reduction
WHICH WINS ORDERS BY:
Faster response to the customer at lower cost and higher quality—
A Competitive Advantage

Figure 16.1
Lean Contributes to Competitive Advantage

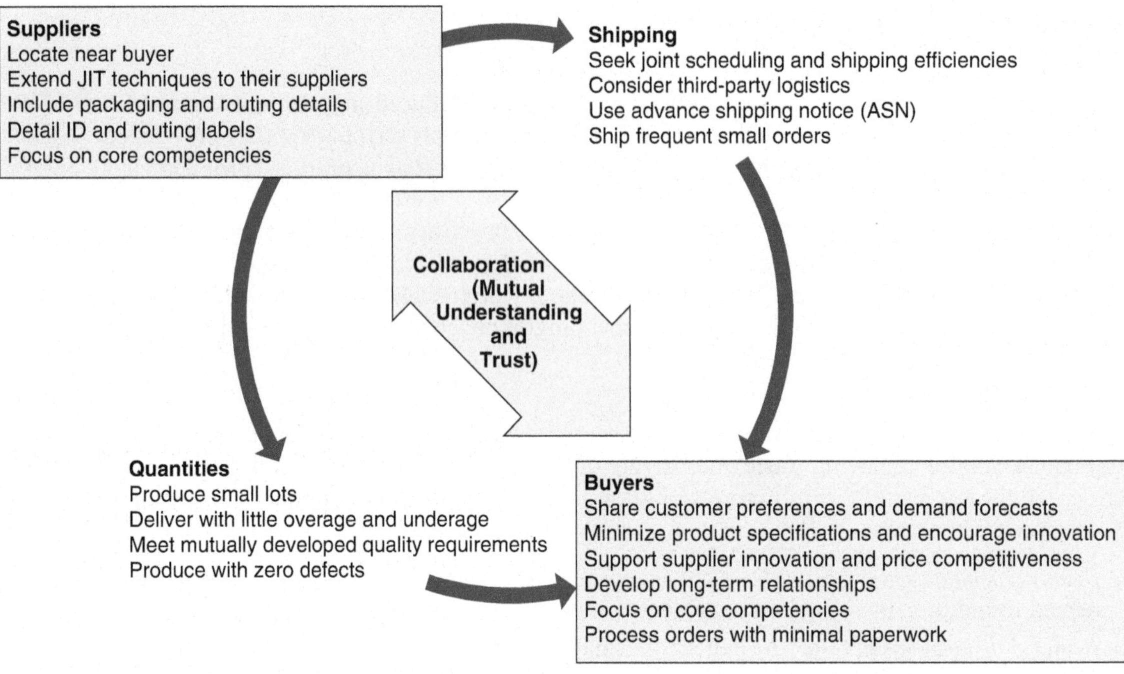

Figure **16.2**

Characteristics of Supplier Partnerships

Leading organizations view suppliers as extensions of their own organizations and expect suppliers to be fully committed to constant improvement. However, supplier concerns can be significant and must be addressed. These concerns include:

LO 16.3 *Identify* the concerns of suppliers when moving to supplier partnerships

1. *Diversification:* Suppliers may not want to tie themselves to long-term contracts with one customer. The suppliers' perception is that they reduce their risk if they have a variety of customers.
2. *Scheduling:* Many suppliers have little faith in the purchaser's ability to produce orders to a smooth, coordinated schedule.
3. *Lead time:* Engineering or specification changes can play havoc with JIT because of inadequate lead time for suppliers to implement the necessary changes.
4. *Quality:* Suppliers' capital budgets, processes, or technology may limit ability to respond to changes in product and quality.
5. *Lot sizes:* Suppliers may see frequent delivery in small lots as a way to transfer buyers' holding costs to suppliers.

As the foregoing concerns suggest, good supplier partnerships require a high degree of trust and respect by both supplier and purchaser—in a word, collaboration. Many firms establish this trust and collaborate very successfully. Two such firms are McKesson-General and Baxter International, who provide surgical supplies for hospitals on a JIT basis. They deliver prepackaged surgical supplies based on hospital operating schedules. Moreover, the surgical packages themselves are prepared so supplies are available in the sequence in which they will be used during surgery.

Lean Layout

Lean layouts reduce another kind of waste—movement. The movement of material on a factory floor (or paper in an office) does not add value. Consequently, managers want flexible layouts that reduce the movement of both people and material. Lean layouts place material directly in the location where needed. For instance, an assembly line should be designed with delivery points next to the line so material need not be delivered first to a receiving department and then moved again. Toyota has gone one step further and places components in the chassis of each vehicle moving down the assembly line. This is not only convenient, but it also allows Toyota to save space and opens areas adjacent to the assembly line previously occupied by shelves. When a layout reduces distance, firms often save labor and space and may have the added bonus of eliminating potential areas for accumulation of unwanted inventory. Table 16.1 provides a list of Lean layout tactics.

TABLE 16.1

LEAN LAYOUT TACTICS
Build work cells for families of products
Include a large number of operations in a small area
Minimize distance
Design little space for inventory
Improve employee communication
Use *poka-yoke* devices
Build flexible or movable equipment
Cross-train workers to add flexibility

Distance Reduction Reducing distance is a major contribution of work cells, work centers, and focused factories (see Chapter 9). The days of long production lines and huge economic lots, with goods passing through monumental, single-operation machines, are gone. Now firms use work cells, often arranged in a U shape, containing several machines performing different operations. These work cells are often based on group technology codes (as discussed in Chapter 5). Group technology codes help identify components with similar characteristics so they can be grouped into families. Once families are identified, work cells are built for them. The result can be thought of as a small product-oriented facility where the "product" is actually a group of similar products—a family of products. The cells produce one good unit at a time, and ideally, they produce the units *only* after a customer orders them.

Increased Flexibility Modern work areas are designed so they can be easily rearranged to adapt to changes in volume and product changes. Almost nothing is bolted down. This concept of layout flexibility applies to both factory and office environments. Not only is furniture and equipment movable, but so are walls, computer connections, and telecommunications. Equipment is modular. Layout flexibility aids the changes that result from product *and* process improvements that are inevitable at a firm with a philosophy of continuous improvement.

Impact on Employees When layouts provide for sequential operations, feedback, including quality issues, can be immediate, allowing employees working together to tell each other about problems and opportunities for improvement. When workers produce units one at a time, they test each product or component at each subsequent production stage. Work processes with self-testing *poka-yoke* functions detect defects automatically. Before Lean, defective products were replaced from inventory. Because surplus inventory is not kept in Lean facilities, there are no such buffers. Employees learn that getting it right the first time is critical. Indeed, Lean layouts allow cross-trained employees to bring flexibility and efficiency to the work area, reducing defects. Defects are waste.

Reduced Space and Inventory Because Lean layouts reduce travel distance, they also reduce inventory. When there is little space, inventory travels less and must be moved in very small lots or even single units. Units are always moving because there is no storage. For instance, each month a Bank of America focused facility sorts millions of checks and statements. With a Lean layout, mail-processing time has been reduced by 33%, annual salary costs by tens of thousands of dollars, floor space by 50%, and in-process waiting lines by 75% to 90%. Storage, including shelves and drawers, has been removed.

TABLE 16.2

LEAN INVENTORY TACTICS
Use a pull system to move inventory
Reduce lot size
Develop just-in-time delivery systems with suppliers
Deliver directly to the point of use
Perform to schedule
Reduce setup time
Use group technology

Lean Inventory

Inventories in production and distribution systems often exist "just in case" something goes wrong. That is, they are used just in case some variation from the production plan occurs. The "extra" inventory is then used to cover variations or problems. Lean inventory tactics require "just in time," not "just in case." **Lean inventory** is the minimum inventory necessary to keep a system running perfectly. With Lean inventory, the exact amount of goods arrives at the moment it is needed, not a minute before or a minute after. Some useful Lean inventory tactics are shown in Table 16.2 and discussed in more detail in the following sections.

Lean inventory
The minimum inventory necessary to keep a system running perfectly.

Reduce Inventory and Variability Operations managers move toward Lean by first reducing inventory. The idea is to eliminate variability in the production system hidden by inventory. Reducing inventory uncovers the "rocks" in Figure 16.3(a) that represent the variability and problems currently being tolerated. With reduced inventory, management chips away at the exposed problems. After the lake is lowered, managers make additional cuts in inventory

STUDENT TIP
Accountants book inventory as an asset, but operations managers know it is a cost.

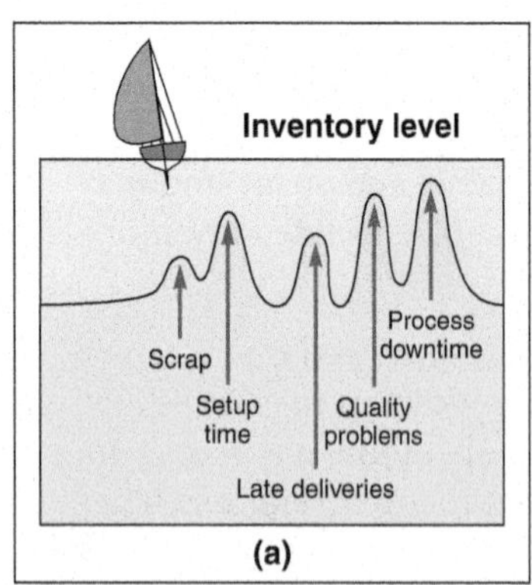

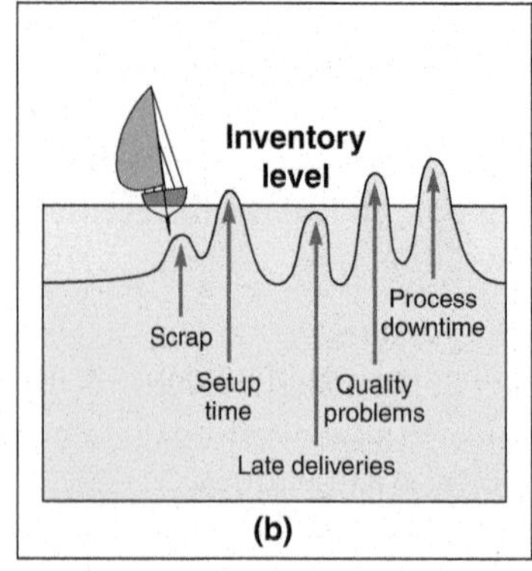

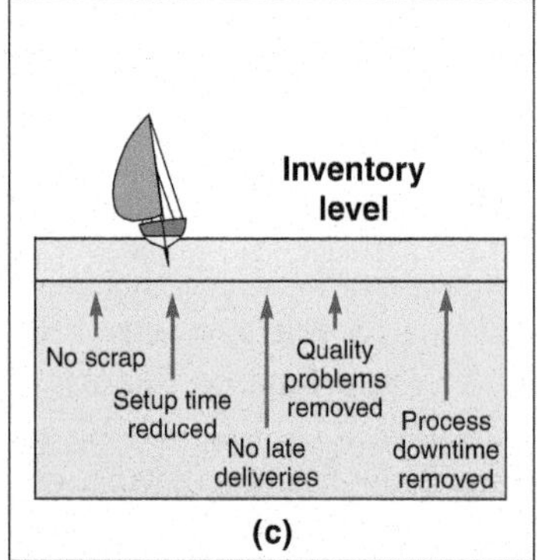

Figure **16.3**

High levels of inventory hide problems (a), but as we reduce inventory, problems are exposed (b), and finally after reducing inventory and removing problems, we have lower inventory, lower costs, and smooth sailing (c).

"Inventory is evil."
S. Shingo

and continue to chip away at the next level of exposed problems [see Figure 16.3(b, c)]. Ultimately, there will be little inventory and few problems (variability).

Firms with technology-sensitive products estimate that the rapid product innovations can cost as much as $\frac{1}{2}$% to 2% of the values of inventory *each week*. Shigeo Shingo, co-developer of the Toyota JIT system, says, "Inventory is evil." He is not far from the truth. If inventory itself is not evil, it hides evil at great cost.

Reduce Lot Sizes Lean also reduces waste by cutting the investment in inventory. A key to slashing inventory is to produce good product in small lot sizes. Reducing the size of batches can be a major help in reducing inventory and inventory costs. As we saw in Chapter 12, when inventory usage is constant, the average inventory level is the sum of the maximum inventory plus the minimum inventory divided by 2. Figure 16.4 shows that lowering the order size increases the number of orders, but drops inventory levels.

Determining Optimal Setup Time Ideally, in a Lean environment, order size is one and single units are being pulled from one adjacent process to another. More realistically, analysis of the process, transportation time, and physical attributes such as size of containers used for transport are considered when determining lot size. Such analysis typically results in a small lot size, but a lot size larger than one. Once a lot size has been determined, the EOQ production order quantity model can be modified to determine the desired setup time. We saw in Chapter 12 that the production order quantity model takes the form:

$$Q_p^* = \sqrt{\frac{2DS}{H[1 - (d/p)]}} \quad (16\text{-}1)$$

where D = Annual demand, S = Setup cost, H = Holding cost, d = Daily demand, p = Daily production

Figure **16.4**

Frequent Orders Reduce Average Inventory

A lower order size increases the number of orders and total ordering cost but reduces average inventory and total holding cost.

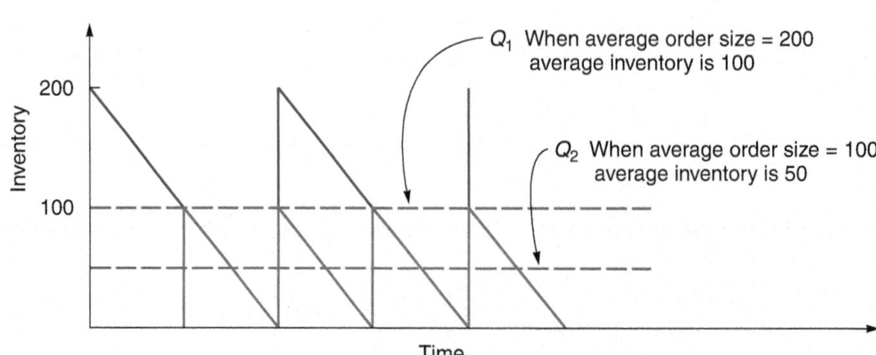

Example 1 shows how to determine the desired setup time.

Example 1

DETERMINING OPTIMAL SETUP TIME

Crate Furniture, Inc., a firm that produces rustic furniture, desires to move toward a reduced lot size. Crate Furniture's production analyst, Aleda Roth, determined that a 2-hour production cycle would be acceptable between two departments. Further, she concluded that a setup time that would accommodate the 2-hour cycle time should be achieved.

APPROACH ▶ Roth developed the following data and procedure to determine optimum setup time analytically:

D = Annual demand = 400,000 units
d = Daily demand = 400,000 per 250 days = 1,600 units per day
p = Daily production rate = 4,000 units per day
Q_p = EOQ desired = 400 (which is the 2-hour demand; that is, 1,600 per day per four 2-hour periods)
H = Holding cost = \$20 per unit per year
S = Setup cost (to be determined)

Hourly labor rate = \$30.00

LO 16.4 *Determine* optimal setup time

SOLUTION ▶ Roth determines that the cost and related time per setup should be:

$$Q_p = \sqrt{\frac{2DS}{H(1-d/p)}}$$

$$Q_p^2 = \frac{2DS}{H(1-d/p)}$$

$$S = \frac{Q_p^2(H)(1-d/p)}{2D} \tag{16-2}$$

$$= \frac{(400)^2(20)(1-1{,}600/4{,}000)}{2(400{,}000)} = \frac{(3{,}200{,}000)(0.6)}{800{,}000} = \$2.40$$

$$\begin{aligned}\text{Setup time} &= \$2.40/(\text{hourly labor rate})\\ &= \$2.40/(\$30 \text{ per hour})\\ &= 0.08 \text{ hour, or } 4.8 \text{ minutes}\end{aligned}$$

INSIGHT ▶ Now, rather than produce components in large lots, Crate Furniture can produce in a 2-hour cycle with the advantage of an inventory turnover of four *per day*.

LEARNING EXERCISE ▶ If labor cost goes to \$40 per hour, what should be the setup time? [Answer: 0.06 hours, or 3.6 minutes.]

RELATED PROBLEMS ▶ 16.1, 16.2, 16.3

Only two changes need to be made for small-lot material flow to work. First, material handling and work flow need to be improved. With short production cycles, there can be very little wait time. Improving material handling is usually easy and straightforward. The second change is more challenging, and that is a radical reduction in setup times. We discuss setup reduction next.

Reduce Setup Costs Both the quantity of inventory and the cost of holding it go down as the inventory-reorder quantity and the maximum inventory level drop. However, because inventory requires incurring an ordering or setup cost that is applied to the units produced, managers tend to purchase (or produce) large orders; the larger the order the less the cost to be absorbed by each unit. Consequently, the way to drive down lot sizes *and* reduce inventory cost is to reduce setup cost, which in turn lowers the optimum order size.

STUDENT TIP
Reduced lot sizes must be accompanied by reduced setup times.

The effect of reduced setup costs on total cost and lot size is shown in Figure 16.5. Moreover, smaller lot sizes hide fewer problems. In many environments, setup cost is highly correlated with setup time. In a manufacturing facility, setups usually require a substantial amount of preparation. Much of the preparation required by a setup can be done prior to

Figure 16.5

Lower Setup Costs Will Lower Total Cost

More frequent orders require reducing setup costs; otherwise, inventory costs will rise. As the setup costs are lowered (from S_1 to S_2), total inventory costs also fall (from T_1 to T_2).

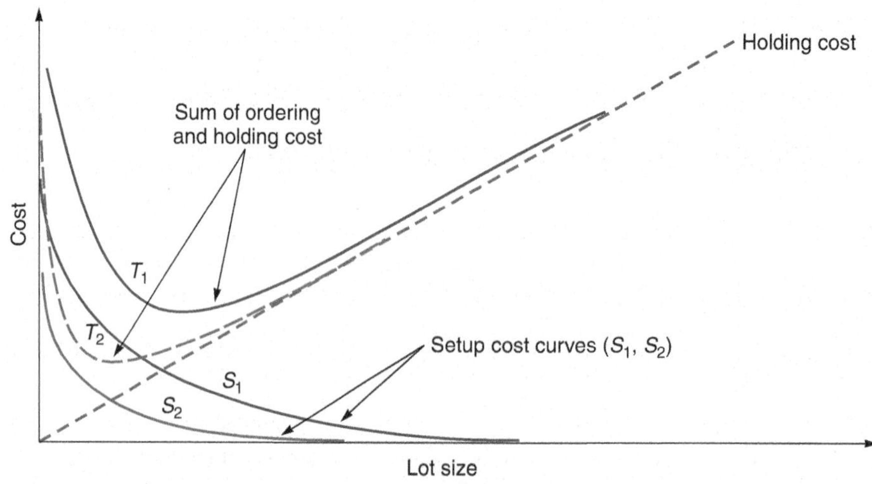

shutting down the machine or process. Setup times can be reduced substantially, as shown in Figure 16.6. For example, a photoelectric products plant in Mexico reduced the setup time to change a bearing from 12 hours to 6 minutes! This is the kind of progress that is typical of world-class manufacturers.

Just as setup costs can be reduced at a machine in a factory, setup time can also be reduced during the process of getting the order ready in the office. Driving down factory setup time from hours to minutes does little good if orders are going to take weeks to process or "set up" in the office. This is exactly what happens in organizations that forget that Lean concepts have applications in offices as well as in the factory. Reducing setup time (and cost) is an excellent way to reduce inventory investment, improve productivity, and speed throughput.

STUDENT TIP Effective scheduling is required for effective use of capital and personnel.

Lean Scheduling

Effective schedules, communicated to those within the organization as well as to outside suppliers, support Lean. Better scheduling also improves the ability to meet customer orders, drives down inventory by allowing smaller lot sizes, and reduces work-in-process. For instance, many companies, such as Ford, now tie suppliers to their final assembly schedule. Ford communicates its schedules to bumper manufacturer Polycon Industries from the Ford production control system. The scheduling system describes the style and color of the bumper needed for each vehicle moving down the final assembly line. The scheduling system transmits the information to portable terminals carried by Polycon warehouse personnel, who load the bumpers onto conveyors leading to the loading dock. The bumpers are then trucked 50 miles to the Ford plant. Total time is 4 hours. However, as we saw in our opening *Global Company Profile*, Toyota has moved its bumper supplier *inside* the new Tundra plant; techniques such as this drive down delivery time even further.

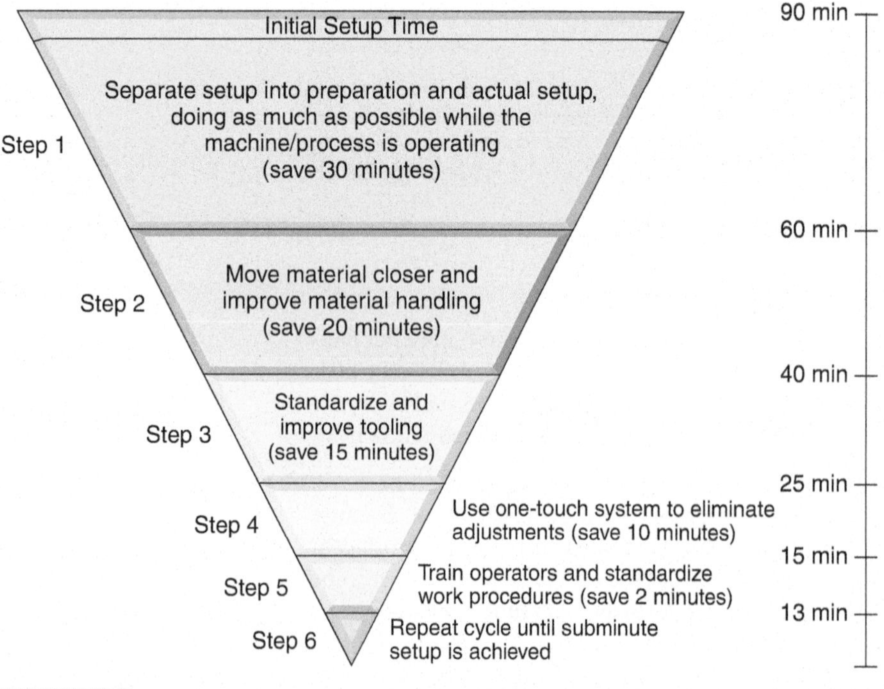

Figure 16.6

Steps for Reducing Setup Times

Reduced setup times are a major component of Lean.

JIT Level Material-Use Approach

AA BBB C AA BBB C AA BBB C AA BBB C AA BBB C AA BBB C AA BBB C AA BBB C

Large-Lot Approach

AAAAAA BBBBBBBBB CCC AAAAAA BBBBBBBBB CCC AAAAAA BBBBBBBBB CCC

Time

Figure **16.7**

Scheduling Small Lots of Parts A, B, and C Increases Flexibility to Meet Customer Demand and Reduces Inventory

The Lean approach to scheduling, described as hei-junka by the Japanese, produces just as many of each model per time period as the large-lot approach, provided setup times are lowered.

Table 16.3 suggests several items that can contribute to achieving these goals, but two techniques (in addition to communicating schedules) are paramount. They are *level schedules* and *kanban*.

Level schedules
Scheduling products so that each day's production meets the demand for that day.

Level Schedules Level schedules process frequent small batches rather than a few large batches. Figure 16.7 contrasts a traditional large-lot approach using large batches with a level schedule using many small batches. The operations manager's task is to make and move small lots so the level schedule is economical. This requires success with the issues discussed in this chapter that allow small lots. As lots get smaller, the constraints may change and become increasingly challenging. At some point, processing a unit or two may not be feasible. The constraint may be the way units are sold and shipped (four to a carton), or an expensive paint changeover (on an automobile assembly line), or the proper number of units in a sterilizer (for a food-canning line).

The scheduler may find that *freezing*, that is holding a portion of the schedule near due dates constant, allows the production system to function and the schedule to be met. Operations managers expect the schedule to be achieved with no deviations.

TABLE 16.3

LEAN SCHEDULING TACTICS
Make level schedules
Use kanbans
Communicate schedules to suppliers
Freeze part of the schedule
Perform to schedule
Seek one-piece-make and one-piece-move
Eliminate waste
Produce in small lots
Each operation produces a perfect part

Kanban One way to achieve small lot sizes is to move inventory through the shop only as needed rather than *pushing* it on to the next workstation whether or not the personnel there are ready for it. As noted earlier, when inventory is moved only as needed, it is referred to as a *pull* system, and the ideal lot size is one. The Japanese call this system *kanban*. Kanbans allow arrivals at a work center to match (or nearly match) the processing time.

Kanban is a Japanese word for *card*. In their effort to reduce inventory, the Japanese use systems that "pull" inventory through work centers. They often use a "card" to signal the need for another container of material—hence the name *kanban.* The card is the authorization for the next container of material to be produced. A kanban is simply a signal to make a specific quantity of a specific item, with a specific due date. Typically, a kanban signal exists for each container of items to be obtained. An order for the container is then initiated by each kanban and "pulled" from the producing department or supplier. A sequence of kanbans "pulls" the material through the plant.

Kanban
The Japanese word for *card*, which has come to mean "signal"; a kanban system moves parts through production via a "pull" from a signal.

LO 16.5 *Define* kanban

The system has been modified in many facilities so that even though it is called a *kanban*, the card itself does not exist. In some cases, an empty position on the floor is sufficient indication that the next container is needed. In other cases, some sort of signal, such as a flag or rag (Figure 16.8), alerts that it is time for the next container. Because there is an optimum lot size, the producing department may make several containers at a time.

Several additional points regarding kanbans may be helpful:

- When the producer and user are not in visual contact, a card can be used; otherwise, a light, flag, or empty spot on the floor may be adequate.
- Usually each card controls a specific quantity of parts, although multiple card systems are used if the work cell produces several components or if the lot size is different from the move size.

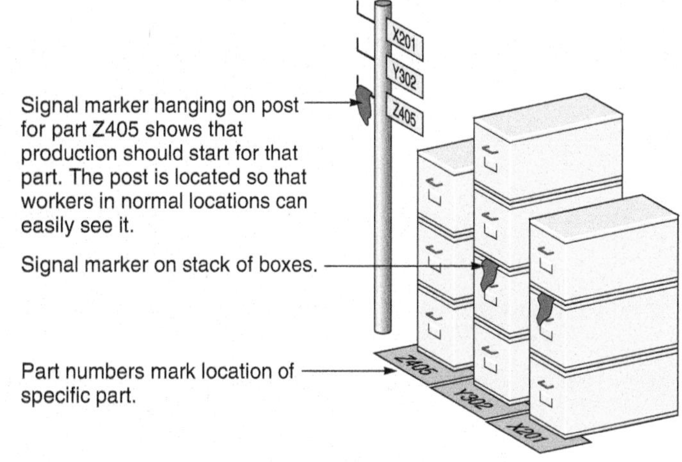
Signal marker hanging on post for part Z405 shows that production should start for that part. The post is located so that workers in normal locations can easily see it.
Signal marker on stack of boxes.
Part numbers mark location of specific part.
X201
Y302
Z405

Figure **16.8**

Diagram of Storage Area with Warning-Signal Marker

A kanban need not be as formal as signal lights or empty carts. The cook in a fast-food restaurant knows that when six cars are in line, eight meat patties and six orders of french fries should be cooking.

- The kanban cards provide a direct control (limit) on the amount of work-in-process between cells.

Determining the Number of Kanban Cards or Containers The number of kanban cards, or containers, sets the amount of authorized inventory. To determine the number of containers moving back and forth between the using area and the producing areas, management first sets the size of each container. This is done by computing the lot size, using a model such as the production order quantity model [discussed in Chapter 12 and shown again in Equation (16-1)]. Setting the number of containers involves knowing: (1) lead time needed to produce a container of parts and (2) the amount of safety stock needed to account for variability or uncertainty in the system. The number of kanban cards is computed as follows:

LO 16.6 *Compute* the required number of kanbans

$$\text{Number of kanbans (containers)} = \frac{\text{Demand during lead time} + \text{Safety stock}}{\text{Size of container}} \qquad (16\text{-}3)$$

Example 2 illustrates how to calculate the number of kanbans needed.

Example 2

DETERMINING THE NUMBER OF KANBAN CONTAINERS

Hobbs Bakery produces short runs of cakes that are shipped to grocery stores. The owner, Ken Hobbs, wants to try to reduce inventory by changing to a kanban system. He has developed the following data and asked you to finish the project.

$$\text{Production lead time} = \text{Wait time} + \text{Material handling time} + \text{Processing time} = 2\text{ days}$$
$$\text{Daily demand} = 500\text{ cakes}$$
$$\text{Safety stock} = \frac{1}{2}\text{ day}$$
$$\text{Container size (determined on a production order size EOQ basis)} = 250\text{ cakes}$$

APPROACH ▶ Having determined that the EOQ size is 250, we then determine the number of kanbans (containers) needed.

SOLUTION ▶ Demand during lead time =

$$\text{Lead time} \times \text{Daily demand} = 2\text{ days} \times 500\text{ cakes} = 1{,}000$$
$$\text{Safety stock} = \frac{1}{2} \times \text{Daily demand} = 250$$

Number of kanbans (containers) needed =

$$\frac{\text{Demand during lead time} + \text{Safety stock}}{\text{Container size}} = \frac{1{,}000 + 250}{250} = 5$$

INSIGHT ▶ Once the reorder point is hit, five containers should be released.

LEARNING EXERCISE ▶ If lead time drops to 1 day, how many containers are needed? [Answer: 3.]

RELATED PROBLEMS ▶ 16.4–16.12

Containers are typically very small, usually a matter of a few hours' worth of production. Such a system requires tight schedules, with small quantities being produced several times a day. The process must run smoothly with little variability in quality or lead time because any shortage has an almost immediate impact on the entire system. Kanban places added emphasis on meeting schedules, reducing the time and cost required by setups, and economical material handling.

Lean Quality

STUDENT TIP
Good quality costs less.

There is no Lean without quality. And Lean's "pull" production, smaller batch sizes, and low inventory all enhance quality by exposing bad quality. Savings occur because scrap, rework, inventory investment, and poor product are no longer buried in inventory. This means fewer bad units are produced. In short, whereas inventory *hides* bad quality, Lean *exposes* it.

As Lean shrinks queues and lead time, it keeps evidence of errors fresh and limits the number of potential sources of error. In effect, Lean creates an early warning system for quality problems so that fewer bad units are produced and feedback is immediate. This advantage accrues both within the firm and with goods received from outside vendors.

In addition, better quality means fewer buffers are needed, and therefore, a better, easier-to-maintain inventory system can exist. Often the purpose of keeping inventory is to protect against unreliable quality. But, when consistent quality exists, Lean firms can reduce all costs associated with inventory. Table 16.4 suggests some tactics for quality in a Lean environment.

TABLE 16.4

LEAN QUALITY TACTICS
Use statistical process control
Empower employees
Build fail-safe methods (poka-yoke, checklists, etc.)
Expose poor quality with small lots
Provide immediate feedback

Lean and the Toyota Production System

Toyota Motor's Eiji Toyoda and Taiichi Ohno are given credit for the Toyota Production System (TPS; see the *Global Company Profile* that opens this chapter). Three components of TPS are *continuous improvement, respect for people*, and *standard work practice*, which are now considered an integral part of Lean.

Continuous Improvement

Continuous improvement under TPS means building an organizational culture and instilling in its people a value system stressing that processes can be improved—indeed, that improvement is an integral part of every employee's job. This process is formalized in TPS by **kaizen**, the Japanese word for change for the good, or what is more generally known as *continuous improvement.* Kaizen is often implemented by a kaizen event. A **kaizen event** occurs when members of a work cell group or team meet to develop innovative ways to immediately implement improvements in the work area or process. In application, kaizen means making a multitude of small or incremental changes as one seeks elusive perfection. (See the *OM in Action* box, "Dr Pepper's Move to Kaizen.")

Kaizen
A focus on continuous improvement.

Kaizen event
Members of a work cell or team meet to develop improvements in the process.

Respect for People

STUDENT TIP
Respect for people brings the entire person to work.

Instilling the mantra of continuous improvement begins at personnel recruiting and continues through extensive and continuing training. One of the reasons continuous improvement works at Toyota is because of another core value at Toyota: Toyota's respect for people.

OM in Action — Dr Pepper's Move to Kaizen[3]

At its Plano, Texas, headquarters and in manufacturing plants in the U.S. and Mexico, Dr Pepper's mantra is **RCI**, or *rapid continuous improvement* (although some executives there use the Japanese word for improvement, *kaizen*). At "kaizen events," teams of Dr Pepper employees spend several days dissecting every step of their work flow in search of waste.

Here are some details, based on an interview with Dr Pepper's CFO:

We've done 575 kaizen events. RCI is about taking the existing baseline and improving it by finding the waste. It starts with walking the entire process. We call it "Going to Gemba." The goal is always to shorten cycle times. You would be surprised. You put a bunch of people in a room to describe how a process works, and they don't all agree with each other—and they all work on the same process!

We have 32 people in the RCI group. They aren't there to make improvements themselves but to facilitate teams. We've issued 6,500 certificates for participating in kaizen events. Through a number of projects, we improved inventory turnover by 35%, or 1.5 million square feet. We've also learned how to create flexibility, including setup reduction in our fountain-syrup line. Sanitizing lines to [receive] the next flavor used to take an average of 32 minutes. We figured out we could do it in 13. Some of the changes are as simple as: "He walked from the machine to get a tool. Why is the tool not at the machine?"

We walk by waste every day. A team watched the process of fountain-syrup bags being assembled and packed into the cardboard boxes used to ship the bags. Somebody asked, "Why does that box have the maroon Dr Pepper logo on it when the box isn't a consumer package?" You call on the box supplier and ask, "If we took that off, how much could we save a year?" They said $60,000, and we said, "Great!"

Like other Lean organizations, Toyota recruits, trains, and treats people as knowledge workers. Aided by aggressive cross-training and few job classifications, Lean firms engage the mental as well as physical capacities of employees in the challenging task of improving operations. Employees are empowered. They are empowered not only to make improvements, but also to stop machines and processes when quality problems exist. Indeed, empowered employees are an integral part of Lean. This means that those tasks that have traditionally been assigned to staff are moved to employees. Toyota recognizes that employees know more about their jobs than anyone else. Lean firms respect employees by giving them the opportunity to enrich both their jobs and their lives.

Krisztian Bocsi/Bloomberg via Getty Images

This Porsche assembly line, like most other Lean facilities, empowers employees so they can stop the entire production line, what the Japanese call *jidoka*, if any quality problems are spotted.

Processes and Standard Work Practice

Building effective and efficient processes requires establishing what Toyota calls standard work practices. The underlying principles are:

- Work is completely specified as to content, sequence, timing, and outcome; this is fundamental to a good process.
- Supplier connections for both internal and external customers are direct, specifying personnel, methods, timing, and quantity.
- Material and service flows are simple and directed to a specific person or machine.
- Process improvements are made only after rigorous analysis at the lowest possible level in the organization.

Lean requires that activities, connections, and flows include built-in tests (or poka-yokes) to signal problems. When a problem or defect occurs, production is stopped. Japanese call the practice of stopping production because of a defect, *jidoka*. The dual focus on (1) education and training of employees and (2) the responsiveness of the system to problems make the seemingly rigid system flexible and adaptable. The result is continuous improvement.

Lean Organizations

STUDENT TIP Lean drives out nonvalue-added activities.

Lean organizations understand processes and costs as well as the customer and the customer's expectations. The functional areas communicate and collaborate to verify that customer expectations are not only understood, but met efficiently. This means identifying costs and delivering the customer's value expectation by implementing the tools of Lean throughout the organization.

Activity-Based Costing

Another feature of successful Lean systems is the ability to identify actual costs. System improvements are difficult if the specific activities that determine costs are not known and applied precisely. Traditional accounting systems, because they are designed to meet many needs, often fail to do this. Costs of service departments, such as Human Resources, Accounting, Marketing, Purchasing, and Research and Development, are often categorized as *overhead*. Then this overhead cost is often allocated to departments or products based on the labor component (*direct labor*) of the product. However, with a historical trend of increasing automation, direct labor cost is on a long-term downward trend and is now as low as 7% to 10% of total cost in many industries. Consequently, overhead allocation based on direct labor can be very wrong. This complicates our understanding of actual costs.

For instance, a product requiring substantial research and development and state-of the art automation may incur a huge engineering and capital investment cost but use little direct labor. Its assigned overhead may therefore be low. But a department with a stable product and relatively low investment may have a high direct labor cost and be burdened with substantial overhead charges. Such an allocation can generate substantial distortion from actual costs.

The preferred approach is to identify and assign cost of resources based on actual use or *activity* of the applicable resources. **Activity-based costing** identifies actual costs and assigns only that portion of the resource actually consumed by that product or service. With activity-based costing, the cost of each traceable activity in the supporting service (e.g., each capital investment, purchase order, material delivery, customer interaction, etc.) is identified, determined, and assigned to the product.

Activity-based costing
Precisely identifies and assigns costs of resources to products and services based on the use of those resources.

Activity-based costing has implications beyond product cost. Managers may want to know the real cost of many functions in the firm such as qualifying a new vendor or identifying profitable customers. Knowing actual costs is necessary as we build Lean organizations.

Building a Lean Organization

Building Lean organizations is difficult, requiring exceptional leadership. Such leaders imbue the organization not just with the tools of Lean, but with a *culture* of continuous improvement. Building such a culture requires open communication and destroying isolated functional disciplines that act as independent "silos." There is no substitute for open two-way communication that fosters effective and efficient processes. Such an organizational culture will have a demonstrated respect for people and a management willing to fully understand how and where the work is performed. Lean firms sometimes use the Japanese term **Gemba or Gemba walk** to refer to going to where the work is actually performed.

Gemba or Gemba walk
Going to where the work is actually performed.

Building organizational cultures that foster ongoing improvement and that accept the constant change and improvement that makes improvement habitual is a challenge. However, such organizations exist. They understand the customer and drive out activities that do not add value in the eyes of the customer. They include industry leaders such as United Parcel Service, Alaska Airlines, and, of course, Toyota. Even traditionally idiosyncratic organizations such as hospitals (see the *OM in Action* box, "Lean Delivers the Medicine") find improved

VIDEO 16.1
Lean Operations at Alaska Airlines

OM in Action — Lean Delivers the Medicine[4]

Using kaizen techniques straight out of Lean, a team of employees at San Francisco General Hospital target and then analyze a particular area within the hospital for improvement. Hospitals today are focusing on throughput and quality in the belief that excelling on these measures will drive down costs and push up patient satisfaction. Doctors and nurses now work together in teams that immerse themselves in a weeklong kaizen event. These events generate plans that make specific improvements in flow, quality, costs, or the patients' experience.

One recent kaizen event focused on the number of minutes it takes from the moment a patient is wheeled into the operating room to when the first incision is made. A team spent a week coming up with ways to whittle 10 minutes off this "prep" time. Every minute saved reduces labor cost and opens up critical facilities. Another kaizen event targeted the Urgent Care Center, dropping the average wait from 5 hours down to 2.5, primarily by adding an on-site X-ray machine instead of requiring patients to walk 15 minutes to the main radiology department. Similarly, wait times in the Surgical Clinic dropped from 2.5 hours to 70 minutes. The operating room now uses a *5S* protocol and has implemented *Standard Work* for the preoperation process.

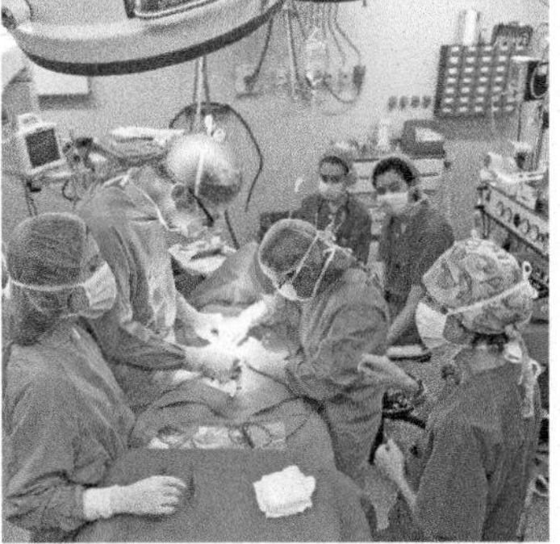
Franck Boston/Fotolia

As hospitals focus on improving medical quality and patient satisfaction, they are exposed to some Japanese terms associated with Lean, many of which do not have a direct English translation: **Gemba,** the place where work is actually performed; **Hansei,** a period of critical self-reflection; **Heijunka,** a level production schedule that provides balance and smooths day-to-day variation; **Jidoka,** using both human intelligence and technology to stop a process at the first sign of a potential problem; **Kaizen,** continuous improvement; and **Muda,** anything that consumes resources but provides no value.

Lean systems are increasingly being adopted by hospitals as they try to reduce costs while improving quality and increasing patient satisfaction—and as San Francisco General has demonstrated, Lean techniques are working.

productivity with Lean operations. Lean operations adopt a philosophy of minimizing waste by striving for perfection through continuous learning, creativity, and teamwork. They tend to share the following attributes:

LO 16.7 *Identify* six attributes of Lean organizations

- *Respect and develop employees* by improving job design, providing constant training, instilling commitment, and building teamwork.
- *Empower employees* with jobs that are made challenging by pushing responsibility to the lowest level possible.
- *Develop worker flexibility* through cross-training and reducing job classifications.
- *Build processes* that destroy variability by helping employees produce a perfect product every time.
- *Develop collaborative partnerships with suppliers*, helping them not only to understand the needs of the ultimate customer, but also to accept responsibility for satisfying those needs.
- *Eliminate waste by performing only value-added activities.* Material handling, inspection, inventory, travel time, wasted space, and rework are targets, as they do not add value.

Success requires leadership as well as the full commitment and involvement of managers, employees, and suppliers. The rewards that Lean producers reap are spectacular. Lean producers often become benchmark performers.

Lean Sustainability

Lean and sustainability are two sides of the same coin. Both seek to maximize resource utilization and economic efficiency. However, if Lean focuses on only the immediate process and system, then managers may miss the sustainability issues beyond the firm. As we discussed in Supplement 5, sustainability requires examining the systems in which the firm and its stakeholders operate. When this is done, both Lean and sustainability achieve higher levels of performance.

Lean drives out waste because waste adds nothing for the customer. Sustainability drives out waste because waste is both expensive and has an adverse effect on the environment. Driving out waste is the common ground of Lean sustainability.

STUDENT TIP Lean began in factories but is now also used in services throughout the world.

Lean in Services

The features of Lean apply to services—from hospitals to amusement parks and airlines—directly influencing the customers' received value. The Lean attributes of respect for people, efficient processes with rigorous standard practices that drive out waste, and a focus on continuous improvement are pervasive vehicles for consistently generating value for all stakeholders. If there is any change in focus of Lean between manufacturing and services, it may be that the high level of customer interaction places added emphasis on enabling people through training, motivation, and empowerment to contribute to their fullest. However, in addition to the customer interaction aspect of services, here are some specific applications of Lean applied to suppliers, layout, inventory, and scheduling in the service sector.

LO 16.8 *Explain* how Lean applies to services

Suppliers

Virtually every restaurant deals with its suppliers on a JIT basis. Those that do not are usually unsuccessful. The waste is too evident—food spoils, and customers complain, get sick, and may die. Similarly, JIT is basic to the financial sector that processes your deposits, withdrawals, and brokerage activities on a JIT basis. That is the industry standard.

Layouts

Lean layouts are required in restaurant kitchens, where cold food must be served cold and hot food hot. McDonald's, for example, has reconfigured its kitchen layout, at great expense, to drive seconds out of the production process, thereby speeding delivery to customers. With the new process, McDonald's can produce made-to-order hamburgers in 45 seconds. Layouts also make a difference at Alaska Airline's baggage claim, where customers expect their bags in 20 minutes or less.

Inventory

Stockbrokers drive inventory down to nearly zero every day. Most sell and buy orders occur on an immediate basis because an unexecuted sell or buy order is not acceptable to the client. A broker may be in serious trouble if left holding an unexecuted trade. Similarly, McDonald's reduces inventory waste by maintaining a time-stamped finished-goods inventory of only a few minutes; after that, it is thrown away. Hospitals, such as Arnold Palmer (described in this chapter's *Video Case Study*), manage JIT inventory and low safety stocks for many items. For instance, critical supplies such as pharmaceuticals may be held to low levels by developing community networks as backup. In this manner, if one pharmacy runs out of a needed drug, another member of the network can supply it until the next day's shipment arrives.

VIDEO 16.2
JIT at Arnold Palmer Hospital

Scheduling

Airlines must adjust to fluctuations in customer demand. But rather than adjusting by changes in inventory, demand is satisfied by personnel availability. Through elaborate scheduling, personnel show up just in time to cover peaks in customer demand. In other words, rather than "things" being inventoried, personnel are scheduled. At a salon, the focus is only slightly different: prompt service is assured by scheduling both the *customer* and the staff. At McDonald's and Walmart, scheduling of personnel is down to 15-minute increments, based on precise forecasting of demand. Notice that in these organizations scheduling is a key ingredient of Lean. Excellent forecasts drive those schedules. Such forecasts may be very elaborate, with seasonal, daily, and even hourly components in the case of the airline ticket counter (holiday sales, flight time, etc.), seasonal and weekly components at the salon (holidays and Fridays create special problems), and down to a few minutes (to respond to the daily meal cycle) at McDonald's.

To deliver goods and services to customers under continuously changing demand, suppliers need to be reliable, inventories low, cycle times short, and schedules nimble. Lean engages and empowers employees to create and deliver the customer's perception of value, eliminating whatever does not contribute to this goal. Lean techniques are widely used in both goods-producing and service-producing firms; they just look different.

Summary

Lean operations, including JIT and TPS, focuses on continuous improvement to eliminate waste. Because waste is found in anything that does not add value, organizations that implement these techniques are adding value more efficiently than other firms. The expectation of Lean firms is that empowered employees work with committed management to build systems that respond to customers with ever-increasing efficiency and higher quality.

Key Terms

Lean operations (p. 648)
Just-in-time (JIT) (p. 648)
Toyota Production System (TPS) (p. 648)
Seven wastes (p. 648)
5Ss (p. 649)
Variability (p. 650)
Throughput (p. 650)
Pull system (p. 650)
Supplier partnerships (p. 651)
Consignment inventory (p. 651)
Lean inventory (p. 653)
Level schedules (p. 657)
Kanban (p. 657)
Kaizen (p. 659)
Kaizen event (p. 659)
Activity-based costing (p. 661)
Gemba or Gemba walk (p. 661)

Ethical Dilemma

In this Lean operations world, in an effort to lower handling costs, speed delivery, and reduce inventory, retailers are forcing their suppliers to do more and more in the way of preparing their merchandise for their cross-docking warehouses, shipment to specific stores, and shelf presentation. Your company, a small manufacturer of aquarium decorations, is in a tough position. First, Mega-Mart wanted you to develop bar-code technology, then special packaging, then small individual shipments bar coded for each store. (This way when the merchandise hits the warehouse, it is cross-docked immediately to the truck destined for that store, and upon arrival the merchandise is ready for shelf placement.) And now Mega-Mart wants you to develop radio frequency identification (RFID)—immediately. Mega-Mart has made it clear that suppliers that cannot keep up with the technology will be dropped.

Earlier, when you didn't have the expertise for bar codes, you had to borrow money and hire an outside firm to do the development, purchase the technology, and train your shipping clerk. Then, meeting the special packaging requirement drove you into a loss for several months, resulting in a loss for last year. Now it appears that the RFID request is impossible. Your business, under the best of conditions, is marginally profitable, and the bank may not be willing to bail you out again. Over the years, Mega-Mart has slowly become your major customer and without it, you are probably out of business. What are the ethical issues, and what do you do?

Discussion Questions

1. What is a Lean producer?
2. What is JIT?
3. What is TPS?
4. What is level scheduling?
5. JIT attempts to remove delays, which do not add value. How, then, does JIT cope with weather and its impact on crop harvest and transportation times?
6. What are three ways in which Lean and quality are related?
7. What is kaizen, and what is a kaizen event?
8. What are the characteristics of supplier partnerships with respect to suppliers?
9. Discuss how the Japanese word for *card* has application in the study of JIT.
10. Standardized, reusable containers have obvious benefits for shipping. What is the purpose of these devices within the plant?
11. Does Lean production work in the service sector? Provide an example.
12. Which Lean techniques work in both the manufacturing *and* service sectors?
13. Why is activity-based costing important to Lean management?

Solved Problem

SOLVED PROBLEM 16.1

Krupp Refrigeration, Inc., is trying to reduce inventory and wants you to install a kanban system for compressors on one of its assembly lines. Determine the size of the kanban and the number of kanbans (containers) needed.

Setup cost = \$10
Daily production = 200 compressors
Lead time = 3 days
Annual holding cost per compressor = \$100
Safety stock = $\frac{1}{2}$ day's production of compressors
Annual usage = 25,000 (50 weeks × 5 days each × daily usage of 100 compressors)

SOLUTION

First, we must determine kanban container size. To do this, we determine the production order quantity [see discussion in Chapter 12 or Equation (16-1)], which determines the kanban size:

$$Q_p^* = \sqrt{\frac{2DS}{H\left(1-\frac{d}{p}\right)}} = \sqrt{\frac{2(25{,}000)(10)}{H\left(1-\frac{d}{p}\right)}} = \sqrt{\frac{500{,}000}{100\left(1-\frac{100}{200}\right)}} = \sqrt{\frac{500{,}000}{50}}$$

$= \sqrt{10{,}000} = 100$ compressors. So the production order size and the size of the kanban container = 100.

Then we determine the number of kanbans:

$$\text{Demand during lead time} = 300\ (= 3 \text{ days} \times \text{daily usage of } 100)$$

$$\text{Safety stock} = 100\ (= \tfrac{1}{2} \times \text{daily production of } 200)$$

$$\text{Number of kanbans} = \frac{\text{Demand during lead time} + \text{Safety stock}}{\text{Size of container}}$$

$$= \frac{300 + 100}{100} = \frac{400}{100} = 4 \text{ containers}$$

Problems

Note: **PX** means the problem may be solved with POM for Windows and/or Excel OM.

Problems 16.1–16.12 relate to Lean and Just-in-Time

••• **16.1** Carol Cagle has a repetitive manufacturing plant producing trailer hitches in Arlington, Texas. The plant has an average inventory turnover of only 12 times per year. She has therefore determined that she will reduce her component lot sizes. She has developed the following data for one component, the safety chain clip:

$$\begin{aligned}\text{Annual demand} &= 31{,}200 \text{ units}\\ \text{Daily demand} &= 120 \text{ units}\\ \text{Daily production (in 8 hours)} &= 960 \text{ units}\\ \text{Desired lot size (1 hour of production)} &= 120 \text{ units}\\ \text{Holding cost per unit per year} &= \$12\\ \text{Setup labor cost per hour} &= \$20\end{aligned}$$

How many minutes of setup time should she have her plant manager aim for regarding this component?

••• **16.2** Given the following information about a product at Michael Gibson's firm, what is the appropriate setup time?

$$\begin{aligned}\text{Annual demand} &= 39{,}000 \text{ units}\\ \text{Daily demand} &= 150 \text{ units}\\ \text{Daily production} &= 1{,}000 \text{ units}\\ \text{Desired lot size} &= 150 \text{ units}\\ \text{Holding cost per unit per year} &= \$10\\ \text{Setup labor cost per hour} &= \$40\end{aligned}$$

••• **16.3** Rick Wing has a repetitive manufacturing plant producing automobile steering wheels. Use the following data to prepare for a reduced lot size. The firm uses a work year of 305 days.

Annual demand for steering wheels	30,500
Daily demand	100
Daily production (8 hours)	800
Desired lot size (2 hours of production)	200
Holding cost per unit per year	$10

a) What is the setup cost, based on the desired lot size?
b) What is the setup time, based on $40 per hour setup labor?

• **16.4** Hartley Electronics, Inc., in Nashville, produces short runs of custom airwave scanners for the defense industry. The owner, Janet Hartley, has asked you to reduce inventory by introducing a kanban system. After several hours of analysis, you develop the following data for scanner connectors used in one work cell. How many kanbans do you need for this connector?

Daily demand	1,000 connectors
Lead time	2 days
Safety stock	$\frac{1}{2}$ day
Kanban size	500 connectors

• **16.5** Tej Dhakar's company wants to establish kanbans to feed a newly established work cell. The following data have been provided. How many kanbans are needed?

Daily demand	250 units
Lead time	$\frac{1}{2}$ day
Safety stock	$\frac{1}{4}$ day
Kanban size	50 units

•• **16.6** Pauline Found Manufacturing, Inc., is moving to kanbans to support its telephone switching-board assembly lines. Determine the size of the kanban for subassemblies and the number of kanbans needed.

$$\begin{aligned}\text{Setup cost} &= \$30\\ \text{Annual holding cost} &= \$120 \text{ per subassembly}\\ \text{Daily production} &= 20 \text{ subassemblies}\\ \text{Annual usage} &= 2{,}500 \text{ (50 weeks} \times 5 \text{ days each}\\ &\quad \times \text{daily usage of 10 subassemblies)}\\ \text{Lead time} &= 16 \text{ days}\\ \text{Safety stock} &= 4 \text{ days' production of subassemblies}\end{aligned}$$ **PX**

•• **16.7** Maggie Moylan Motorcycle Corp. uses kanbans to support its transmission assembly line. Determine the size of the kanban for the mainshaft assembly and the number of kanbans needed.

$$\begin{aligned}\text{Setup cost} &= \$20\\ \text{Annual holding cost of mainshaft assembly} &= \$250 \text{ per unit}\\ \text{Daily production} &= 300 \text{ mainshafts}\\ \text{Annual usage} &= 20{,}000 \ (= 50 \text{ weeks} \times 5 \text{ days each}\\ &\quad \times \text{daily usage of 80 mainshafts)}\\ \text{Lead time} &= 3 \text{ days}\\ \text{Safety stock} &= 12\\ \text{Safety stock} &= \tfrac{1}{2} \text{ day's production of mainshafts}\end{aligned}$$ **PX**

• **16.8** Discount-Mart, a major East Coast retailer, wants to determine the economic order quantity (see Chapter 12 for EOQ formulas) for its halogen lamps. It currently buys all halogen lamps from Specialty Lighting Manufacturers in Atlanta. Annual demand is 2,000 lamps, ordering cost per order is $30, and annual carrying cost per lamp is $12.

a) What is the EOQ?
b) What are the total annual costs of holding and ordering (managing) this inventory?
c) How many orders should Discount-Mart place with Specialty Lighting per year? **PX**

••• **16.9** Discount-Mart (see Problem 16.8), as part of its new Lean program, has signed a long-term contract with Specialty Lighting and will place orders electronically for its halogen lamps. Ordering costs will drop to $.50 per order, but Discount-Mart also reassessed its carrying costs and raised them to $20 per lamp.

a) What is the new economic order quantity?
b) How many orders will now be placed?
c) What is the total annual cost of managing the inventory with this policy? **PX**

•• **16.10** How do your answers to Problems 16.8 and 16.9 provide insight into a collaborative purchasing strategy?

•• **16.11** Ferdows Electronics, Inc. (FEI), produces short runs of custom microwave radios for railroads and other industrial clients. You have been asked to reduce inventory by introducing a kanban system. After several hours of analysis, you develop the following data for connectors used in one work cell. How many kanbans do you need for this connector? **PX**

Daily demand = 1,500 radios
Production lead time = 1 day
Safety stock = 1/2 day
Kanban size = 250 radios

•• **16.12** Flores Manufacturing, Inc., is moving to kanbans to support its electronic-board assembly lines. Determine the size of the kanban for subassemblies and the number of kanbans needed. **PX**

Setupcost = $25
Annual holding cost = $200 per subassembly
Daily production = 400 subassembliesa
Annual usage = 50,000 (= 50 weeks × 5 days each × daily usage of 200 subassemblies)
Lead time = 6 days
Safety stock = 1 day's production of subassemblies

CASE STUDIES

Lean Operations at Alaska Airlines

Video Case

Alaska Airlines operates in a land of rugged beauty, crystal clear lakes, spectacular glaciers, majestic mountains, and bright blue skies. But equally awesome is its operating performance. Alaska Airlines consistently provides the industry's number-one overall ranking and best on-time performance. A key ingredient of this excellent performance is Alaska Airlines' Lean initiative.

With an aggressive implementation of Lean, Ben Minicucci, Executive VP for Operations, is finding ever-increasing levels of performance. He pushes this initiative throughout the company with: (1) a focus on continuous improvement, (2) metrics that measure performance against targets, and (3) making performance relevant to Alaska Airlines' empowered employees.

With leadership training that includes a strong focus on participative management, Minicucci has created a seven-person Lean Department. The department provides extensive training in Lean via one-week courses, participative workshops, and two-week classes that train employees to become a Six Sigma Green Belt. Some employees even pursue the next step, Black Belt certification.

A huge part of any airline's operations is fuel cost, but capital utilization and much of the remaining cost is dependent upon ground equipment and crews that handle aircraft turnaround and maintenance, in-flight services, and customer service.

As John Ladner, Director of Seattle Airport Operations, has observed, "Lean eliminates waste, exposes non-standard work, and is forcing a focus on variations in documented best practices and work time."

Lean is now part of the Alaska Airlines corporate culture, with some 60 ongoing projects. Kaizen events (called "Accelerated Improvement Workshops" at Alaska Airlines), Gemba Walks (called "waste walks" by Alaska Airlines), and 5S are now a part of everyday conversation at Alaska Airlines. Lean projects have included:

- Applying 5S to identify aircraft ground equipment and its location on the tarmac.
- Improving preparation for and synchronization of the arrival and departure sequences; time to open the front door after arrival has been reduced from 4.5 to 1 min.
- Redefining the disconnect procedure for tow bars used to "push back" aircraft at departure time; planes now depart 2–3 minutes faster.
- Revising the deicing process, meaning less time for the plane to be on the tarmac.
- Improving pilot staffing, making Alaska's pilot productivity the highest in the industry. Every 1% improvement in productivity leads to a $5 million savings on a recurring basis. Alaska Airlines has achieved a 7% productivity improvement over the last five years.

Another current Lean project is passenger unloading and loading. Lean instructor Allison Fletcher calls this "the most unique project I have worked on." One exciting aspect of deplaning is Alaska's solar-powered "switchback" staircase for unloading passengers through the rear door (see photo). Alaska is saving two minutes, or nearly 17%, off previous unloading time with this new process. Alaska Airlines' Lean culture has made it a leader in the industry.

Discussion Questions*

1. What are the key ingredients of Lean, as identified at Alaska Airlines?
2. As an initial phase of a kaizen event, discuss the many ways passengers can be loaded and unloaded from airplanes.
3. Document the research that is being done on the aircraft passenger-loading problem.

* You may wish to view the video that accompanies this case before addressing these questions.

Courtesy of Alaska Airlines

JIT at Arnold Palmer Hospital

Video Case

Orlando's Arnold Palmer Hospital, founded in 1989, specializes in treatment of women and children and is renowned for its high-quality rankings (top 10% of benchmarked hospitals), its labor and delivery volume (about 14,000 births per year), and its neonatal intensive care unit (one of the highest survival rates in the nation). But quality medical practices and high patient satisfaction require costly inventory—some $30 million per year and thousands of SKUs.* With pressure on medical care to manage and reduce costs, Arnold Palmer Hospital has turned toward controlling its inventory with just-in-time (JIT) techniques.

Within the hospital, for example, drugs are now distributed at the nursing stations via dispensing machines (almost like vending machines) that electronically track patient usage and post the related charge to each patient. Each night, based on patient demand and prescriptions written by doctors, the dispensing stations are refilled.

To address JIT issues externally, Arnold Palmer Hospital turned to a major distribution partner, McKesson General Medical, which as a first-tier supplier provides the hospital with about one-quarter of all its medical/surgical inventory. McKesson supplies sponges, basins, towels, Mayo stand covers, syringes, and hundreds of other medical/surgical items. To ensure coordinated daily delivery of inventory purchased from McKesson, an account executive has been assigned to the hospital on a full-time basis, as well as two other individuals who address customer service and product issues. The result has been a drop in Central Supply average daily inventory from $400,000 to $114,000 since JIT.

JIT success has also been achieved in the area of *custom surgical packs*. Custom surgical packs are the sterile coverings, disposable plastic trays, gauze, and the like, specialized to each type of surgical procedure. Arnold Palmer Hospital uses 10 different custom packs for various surgical procedures. "Over 50,000 packs are used each year, for a total cost of about $1.5 million," says George DeLong, head of Supply Chain Management.

The packs are not only delivered in a JIT manner, but they are packed that way as well. That is, they are packed in the reverse order they are used so each item comes out of the pack in the sequence it is needed. The packs are bulky, are expensive, and must remain sterile. Reducing the inventory and handling while maintaining an ensured sterile supply for scheduled surgeries presents a challenge to hospitals.

Here is how the supply chain works: Custom packs are *assembled* by a packing company with *components supplied* primarily from manufacturers selected by the hospital, and *delivered* by McKesson from its local warehouse. Arnold Palmer Hospital works with its own surgical staff (through the Medical Economics Outcome Committee) to identify and standardize the custom packs to reduce the number of custom pack SKUs. With this integrated system, pack safety stock inventory has been cut to one day.

The procedure to drive the custom surgical pack JIT system begins with a "pull" from the doctors' daily surgical schedule. Then, Arnold Palmer Hospital initiates an electronic order to McKesson between 1:00 and 2:00 P.M. daily. At 4:00 A.M. the next day, McKesson delivers the packs. Hospital personnel arrive at 7:00 A.M. and stock the shelves for scheduled surgeries. McKesson then reorders from the packing company, which in turn "pulls" necessary inventory for the quantity of packs needed from the manufacturers.

Arnold Palmer Hospital's JIT system reduces inventory investment, expensive traditional ordering, and bulky storage and supports quality with a sterile delivery.

Discussion Questions**

1. What do you recommend be done when an error is found in a pack as it is opened for an operation?
2. How might the procedure for custom surgical packs described here be improved?
3. When discussing JIT in services, the text notes that suppliers, layout, inventory, and scheduling are all used. Provide an example of each of these at Arnold Palmer Hospital.
4. When a doctor proposes a new surgical procedure, how do you recommend the SKU for a new custom pack be entered into the hospital's supply-chain system?

*SKU = stock keeping unit
**You may wish to view the video that accompanies this case before answering these questions.

Endnotes

1. Sources: **TheDrive.com** (March 14, 2019); and *The Wall Street Journal* (November 29, 2011).
2. 5S comes from the Japanese words seiri (*sort* and clear out), seiton (*straighten* and configure), seiso (*scrub* and clean up), seiketsu (maintain *sanitation* and cleanliness of self and workplace), and shitsuke (*self-discipline and standardization* of practices).
3. Sources: *The Wall Street Journal* (February 22, 2016) and *Manufacturing Business Technology* (May 26, 2011). Reprinted with permission of *The Wall Street Journal*, Copyright © 2016 Dow Jones & Company, Inc. All Rights Reserved Worldwide.
4. Sources: *International Journal for Quality in Health Care* (April 2016); *NEJM Catalyst* (April 27, 2018); and *San Francisco General Hospital & Trauma Center Annual Report*.

Bibliography

Black, I. T., and D. T. Phillips. "The Lean to Green Evolution." *Industrial Engineer* 42, no. 6 (June 2010): 46–51.

Graban, M. *Lean Hospitals*. 2nd ed. New York: CRC Press, 2012.

Hall, R. W. "Lean and the Toyota Production System." *Target* 20, no. 3 (2004): 22–27.

Liker, J. K. *The Toyota Way: 14 Management Principles from the World's Greatest Manufacturer*, 2nd ed. New York: McGraw-Hill, 2021.

Mann, D. *Creating a Lean Culture*. 2nd ed. New York: Productivity Press, 2014.

Myerson, P. *Lean and Technology*. Boston: Pearson FT Press, 2017.

Radnor, Z. J., M. Holweg, and J. Waring, "Lean in Healthcare: The Unfilled Promise?" *Social Science & Medicine* 74, no. 3 (February 2012): 364–371.

Sayre, K. "A Lean Introduction to Sustainability." *Industrial Engineer* 40, no. 9 (September 2008): 34.

Schonberger, R. J. "Lean Extended." *Industrial Engineer* (December 2005): 26–31.

Wilson, L.. *How to Implement Lean Manufacturing*, 2nd ed. New York: McGraw-Hill, 2015.

Womack, J. P., and D. T. Jones. "Lean Consumption." *Harvard Business Review* 83 (March 2005): 58–68.

Womack, J. P., and D. T. Jones. *Lean Solutions: How Companies and Customers Can Create Value and Wealth Together*. New York: The Free Press, 2005.

Chapter 16 *Rapid* Review

Main Heading	Review Material	MyLab Operations Management
LEAN OPERATIONS	■ **Lean operations**—Eliminates waste through continuous improvement and focus on exactly what the customer wants. ■ **Just-in-time (JIT)**—Continuous and forced problem solving via a focus on throughput and reduced inventory. ■ **Toyota Production System (TPS)**—Focus on continuous improvement, respect for people, and standard work practices. *When implemented as a comprehensive manufacturing strategy, Lean, JIT, and TPS systems sustain competitive advantage and result in increased overall returns.* ■ **Seven wastes**—Overproduction, queues, transportation, inventory, motion, overprocessing, and defective product. ■ **5Ss**—A Lean production checklist: sort, simplify, shine, standardize, and sustain. U.S. managers often add two additional *S*s to the 5 original ones: *safety* and *support/maintenance*. ■ **Variability**—Any deviation from the optimum process that delivers perfect product on time, every time. Both JIT and inventory reduction are effective tools for identifying causes of variability. ■ **Throughput**—The rate at which units move through a process. ■ **Pull system**—A concept that results in material being produced only when requested and moved to where it is needed just as it is needed. Pull systems use signals to request production and delivery from supplying stations to stations that have production capacity available.	Concept Questions: 1.1–1.6
LEAN AND JUST-IN-TIME	■ **Supplier partnerships**—Suppliers and purchasers work together to remove waste and drive down costs for mutual benefit. Some specific goals of supplier partnerships are *removal of unnecessary activities, removal of in-plant inventory, removal of in-transit inventory,* and *obtain improved quality and reliability*. ■ **Consignment inventory**—An arrangement in which the supplier maintains title to the inventory until it is used. Concerns of suppliers in suppler partnerships include (1) *diversification*, (2) *scheduling*, (3) *lead time*, (4) *quality*, and (5) *lot sizes*. *Lean layout tactics* include building work cells for families of products, including a large number of operations in a small area, minimizing distance, designing little space for inventory, improving employee communication, using poka-yoke devices, building flexible or movable equipment, and cross-training workers to add flexibility. ■ **Lean inventory**—The minimum inventory necessary to keep a system running perfectly. The idea behind JIT is to eliminate inventory that hides variability in the production system. *Lean inventory tactics* include using a pull system to move inventory, reducing lot size, developing just-in-time delivery systems with suppliers, delivering directly to the point of use, performing to schedule, reducing setup time, and using group technology.	
	$$Q_p^* = \sqrt{\frac{2DS}{H[1-(d/p)]}} \quad (16\text{-}1)$$	Problems: 16.1–16.3
	Using Equation (16-1), for a given desired lot size, Q, we can solve for the optimal setup cost, S: $$S = \frac{(Q^2)(H)(1-d/p)}{2D} \quad (16\text{-}2)$$ *Lean scheduling tactics* include communicate schedules to suppliers, make level schedules, freeze part of the schedule, perform to schedule, seek one-piece-make and one-piece-move, eliminate waste, produce in small lots, use kanbans, and make each operation produce a perfect part. ■ **Level schedules**—Scheduling products so that each day's production meets the demand for that day. ■ **Kanban**—The Japanese word for *card*, which has come to mean "signal"; a kanban system moves parts through production via a "pull" from a signal.	
	$$\text{Number of kanbans (containers)} = \frac{\text{Demand during lead time + Safety stock}}{\text{Size of container}} \quad (16\text{-}3)$$ *Lean quality*—whereas inventory *hides* bad quality, Lean immediately *exposes* it. Lean quality tactics include using statistical process control, empowering employees, building fail-safe methods (poka-yoke, checklists, etc.), exposing poor quality with small lots, and providing immediate feedback.	Problems: 16.4–16.9, 16.11, 16.12 Virtual Office Hours for Solved Problem: 16.1

Main Heading	Review Material	MyLab Operations Management
LEAN AND THE TOYOTA PRODUCTION SYSTEM	▪ **Kaizen**—A focus on continuous improvement. ▪ **Kaizen event**—Members of a work cell or team meet to develop improvements in the process. Toyota recruits, trains, and treats people as knowledge workers. They are empowered. TPS employs aggressive cross-training and few job classifications.	Concept Questions: 3.1–3.6
LEAN ORGANIZATIONS	Lean operations tend to share the following attributes: *respect and develop employees* by improving job design, providing constant training, instilling commitment, and building teamwork; *empower employees* by pushing responsibility to the lowest level possible; *develop worker flexibility* through cross-training and reducing job classifications; *build processes* that destroy variability; *develop collaborative partnerships with suppliers* to help them accept responsibility for satisfying end customer needs; and *eliminate waste by performing only value-added activities.* ▪ **Activity-based costing**—Precisely identifies and assigns costs of resources to products and services based on the use of those resources. ▪ **Gemba** or **Gemba walk**—Going to where the work is actually performed.	Concept Questions: 4.1–4.6 **VIDEO 16.1** Lean Operations at Alaska Airlines
LEAN IN SERVICES	The features of Lean operations apply to services just as they do in other sectors. Forecasts in services may be very elaborate, with seasonal, daily, hourly, or even shorter components.	Concept Questions: 5.1–5.6 **VIDEO 16.2** JIT at Arnold Palmer Hospital
ADDITIONAL MYLAB OPERATIONS MANAGEMENT RESOURCES	✔ Multiple Choice Case Questions (Lean Operations at Alaska Airlines)	

Self Test

LO 16.1 Match Lean Operations, JIT, and TPS with the concepts shown below:
- **a)** Continuous improvement and a focus on exactly what the customer wants, and when.
- **b)** Supply the customer with exactly what the customer wants when the customer wants it, without waste, through continuous improvement.
- **c)** Emphasis on continuous improvement, respect for people, and standard work practices.

LO 16.2 Define the seven wastes and the 5Ss. The seven wastes are _____, _____, _____, _____, _____. _____, and _____, and the 5Ss are _____, _____, _____, _____, and _____.

LO 16.3 Concerns of suppliers when moving to supplier partnerships include:
- **a)** small lots sometimes seeming economically prohibitive.
- **b)** realistic quality demands.
- **c)** changes without adequate lead time.
- **d)** erratic schedules.
- **e)** all of the above.

LO 16.4 What is the formula for optimal setup time?
- **a)** $\sqrt{2DQ/[H(1-d/p)]}$
- **b)** $\sqrt{Q^2H(1-d/p)/(2D)}$
- **c)** $QH(1-d/p)/(2D)$
- **d)** $Q^2\,H(1-d/p)/(2D)$
- **e)** $H(1-d/p)$

LO 16.5 Kanban is the Japanese word for:
- **a)** car.
- **b)** pull.
- **c)** card.
- **d)** continuous improvement.
- **e)** level schedule.

LO 16.6 The required number of kanbans equals:
- **a)** 1.
- **b)** Demand during lead time/Q
- **c)** Size of container.
- **d)** Demand during lead time.
- **e)** Demand during lead time + Safety stock/Size of container

LO 16.7 The six attributes of Lean organizations are: _____, _____, _____, _____, _____, and _____.

LO 16.8 Lean applies to services:
- **a)** only in rare instances.
- **b)** except in terms of the supply chain.
- **c)** except in terms of employee issues.
- **d)** except in terms of both supply chain issues and employee issues.
- **e)** just as it applies to manufacturing.

Answers: LO 16.1. Lean = a, JIT = b, TPS = c; LO 16.2. overproduction, queues, transportation, inventory, motion, overprocessing, defective product; sort, simplify, shine, standardize, sustain; LO 16.3. e; LO 16.4. d; LO 16.5. c; LO 16.6. e; LO 16.7. respect and develop people, empower employees, develop worker flexibility, build excellent processes, develop collaborative partnerships with suppliers, eliminate waste; LO 16.8. e.

Maintenance and Reliability

CHAPTER 17

CHAPTER OUTLINE

GLOBAL COMPANY PROFILE: *Orlando Utilities Commission*

From replacing a bulb at the top of a cell phone tower, to ensuring integrity of a jet engine, to checking the cables for window washers, to testing electrical systems, maintenance not only keeps systems working by removing variability but also saves lives.

aapsky/Shutterstock; UrbanImages/Alamy Stock Photo; Solid photos/Shutterstock; Kampan/Shutterstock

10 OM STRATEGY DECISIONS

- Design of Goods and Services
- Managing Quality
- Process Strategies
- Location Strategies
- Layout Strategies
- Human Resources
- Supply Chain Management
- Inventory Management
- Scheduling
- *Maintenance*

GLOBAL COMPANY PROFILE
Orlando Utilities Commission

Maintenance Provides a Competitive Advantage for the Orlando Utilities Commission

The Orlando Utilities Commission (OUC) owns and operates power plants that supply power to two central Florida counties. Every year, OUC takes each one of its power-generating units off-line for 1 to 3 weeks to perform maintenance work.

In addition, each unit is also taken off-line every 3 years for a complete overhaul and turbine generator inspection. Overhauls are scheduled for spring and fall, when the weather is mildest and demand for power is low. These overhauls last from 6 to 8 weeks.

Units at OUC's Stanton Energy Center require that maintenance personnel perform approximately 12,000 repair and preventive maintenance tasks a year. To accomplish these tasks efficiently, many of these jobs are scheduled daily via a computerized maintenance management program. The computer generates preventive maintenance work orders and lists of required materials.

Every day that a plant is down for maintenance costs OUC about $110,000 extra for the replacement cost of power that must be generated elsewhere. However, these costs pale beside the costs associated with a forced outage. An unexpected outage could cost OUC an additional $350,000 to $600,000 each day!

Chris O'Meara/AP Images

The Stanton Energy Center in Orlando.

Scheduled overhauls are not easy; each one has 1,800 distinct tasks and requires 72,000 labor-hours. But the value of preventive maintenance was illustrated by the first overhaul of a new turbine generator. Workers discovered a cracked rotor blade, which could have destroyed a $27 million piece of equipment. To find such cracks, which are invisible to the naked eye, metals are examined using dye tests, X-rays, and ultrasound.

At OUC, preventive maintenance is worth its weight in gold. As a result, OUC's electric distribution system has been ranked number one in the Southeast U.S. by PA Consulting Group—a leading consulting firm. Effective maintenance provides a competitive advantage for the Orlando Utilities Commission.

Courtesy Orlando Utilities Commission

This inspector is examining a low-pressure section of turbine. The tips of these turbine blades will travel at supersonic speeds of 1,300 miles per hour when the plant is in operation. A crack in one of the blades can cause catastrophic failure.

Monty Rakusen Cultura/Newscom

Maintenance of capital-intensive facilities requires good planning to minimize downtime. Here, turbine overhaul is under way. Organizing the thousands of parts and pieces necessary for a shutdown is a major effort.

LEARNING OBJECTIVES

The Strategic Importance of Maintenance and Reliability

Managers at Orlando Utilities Commission (OUC), the subject of the chapter-opening *Global Company Profile*, fight for reliability to avoid the undesirable results of equipment failure. At OUC, a generator failure is very expensive for both the company and its customers. Power outages are instantaneous, with potentially devastating consequences. Similarly, managers at Frito-Lay, Walt Disney Company, and United Parcel Service (UPS) are intolerant of failures or breakdowns. Maintenance is critical at Frito-Lay to achieve high plant utilization and excellent sanitation. At Disney, sparkling-clean facilities and safe rides are necessary to retain its standing as one of the most popular vacation destinations in the world. Likewise, UPS's famed maintenance strategy keeps its delivery vehicles operating and looking as good as new for 20 years or more.

VIDEO 17.1
Maintenance Drives Profits at Frito-Lay

STUDENT TIP
If a system is not reliable, the other OM decisions are more difficult.

These companies, like most others, know that poor maintenance can be disruptive, inconvenient, wasteful, and expensive in dollars and even in lives. As Figure 17.1 illustrates, the interdependency of operator, machine, and mechanic is a hallmark of successful maintenance and reliability. Good maintenance and reliability management enhances a firm's performance and protects its investment.

Maintenance
The activities involved in maintaining capability of the system.

Reliability
The probability that a machine part or product will function properly for a specified time under stated conditions.

The objective of maintenance and reliability is to maintain the capability of the system. Good maintenance removes variability. Systems must be designed and maintained to reach expected performance and quality standards. **Maintenance** includes all activities involved in maintaining capability of the system. **Reliability** is the probability that a machine part or product will function properly for a specified time under stated conditions.

In this chapter, we examine four important tactics for improving the reliability and maintenance not only of products and equipment but also of the systems that produce them. The four tactics are organized around reliability and maintenance.

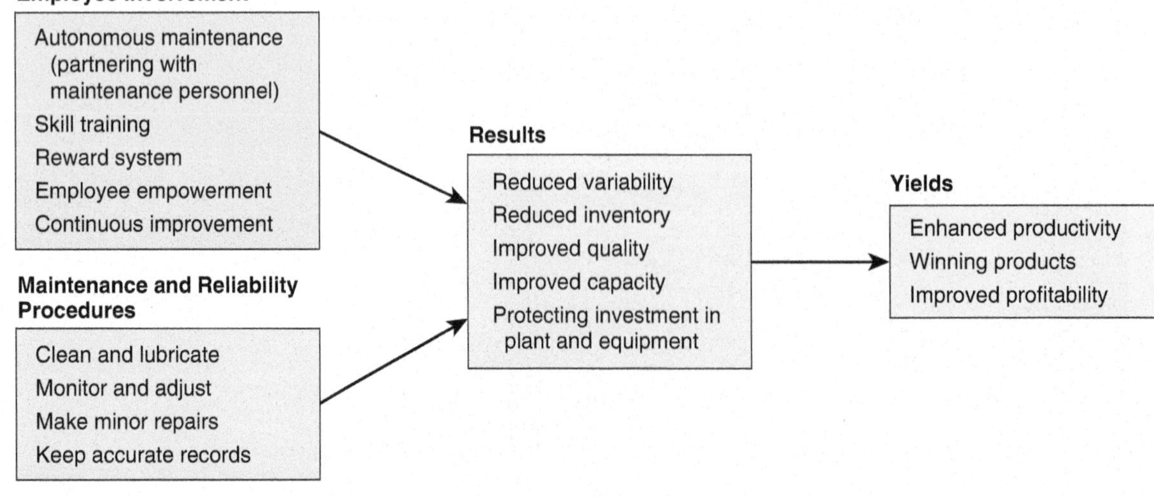

Figure **17.1**

Good Maintenance and Reliability Management Requires Employee Involvement and Good Procedures

The reliability tactics are:

1. Improving individual components
2. Providing redundancy

The maintenance tactics are:

1. Implementing or improving preventive maintenance
2. Increasing repair capabilities or speed

We will now discuss these tactics.

Reliability

Systems are composed of a series of individual interrelated components, each performing a specific job. If any *one* component fails to perform, for whatever reason, the overall system (for example, an airplane or machine) can fail. First, we discuss system reliability and then improvement via redundancy.

System Reliability

STUDENT TIP
Designing for reliability is an excellent place to start reducing variability.

Because failures do occur in the real world, understanding their occurrence is an important reliability concept. We now examine the impact of failure in a series. Figure 17.2 shows that as the number of components in a *series* increases, the reliability of the whole system declines very quickly. A system of $n = 50$ interacting parts, each of which has a 99.5% reliability, has an overall reliability of 78%. If the system or machine has 100 interacting parts, each with an individual reliability of 99.5%, the overall reliability will be only about 60%!

To measure reliability in a system in which each component may have its own unique reliability, we cannot use the reliability curve in Figure 17.2. However, the method of computing system reliability (R_s) is simple. It consists of finding the product of individual reliabilities as follows:

$$R_s = R_1 \times R_2 \times R_3 \times \ldots \times R_n \tag{17-1}$$

LO 17.1 *Describe* how to improve system reliability

where R_1 = reliability of component 1
R_2 = reliability of component 2

and so on.

Equation (17-1) assumes that the reliability of an individual component does not depend on the reliability of other components (that is, each component is *independent*). In addition, in this equation, as in most reliability discussions, reliabilities are presented as *probabilities*. Thus, a .90 reliability means that the unit will perform as intended 90% of the time. It also means that

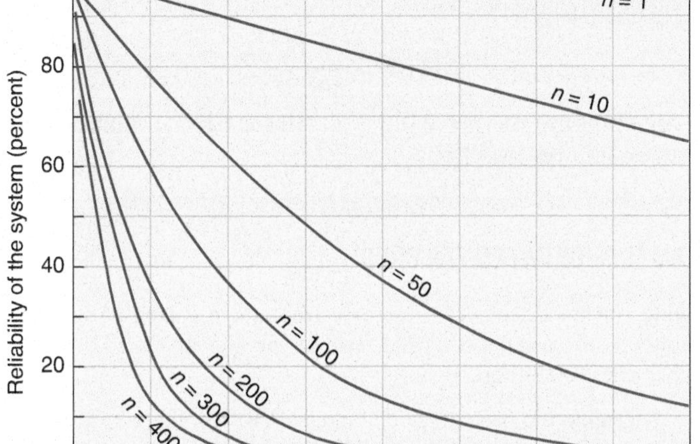

Figure 17.2
Overall System Reliability as a Function of Number of n Components (Each with the Same Reliability) and Component Reliability with Components in a Series

it will fail 1 − .90 = .10 = 10% of the time. We can use this method to evaluate the reliability of a service or a product, such as the one we examine in Example 1.

Example 1

RELIABILITY IN A SERIES

The National Bank of Greeley, Colorado, processes loan applications through three clerks (each checking different sections of the application in series), with reliabilities of .90, .80, and .99. It wants to find the system reliability.

APPROACH ▶ Apply Equation (17-1) to solve for R_s.

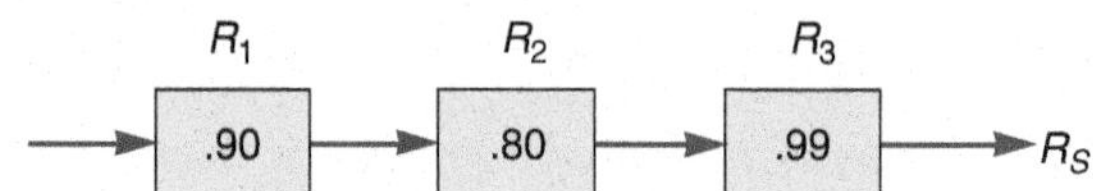

SOLUTION ▶ The reliability of the loan process is:

$$R_s = R_1 \times R_2 \times R_3 = (.90)(.80)(.99) = .713, \text{ or } 71.3\%$$

LO 17.2 *Determine* system reliability

INSIGHT ▶ Because each clerk in the series is less than perfect, the error probabilities are cumulative and the resulting reliability for this series is .713, which is less than any one clerk.

LEARNING EXERCISE ▶ If the lowest-performing clerk (.80) is replaced by a clerk performing at .95 reliability, what is the new expected reliability? [Answer: .846.]

RELATED PROBLEMS ▶ 17.1, 17.2, 17.3, 17.9, 17.15, 17.16, 17.17

ACTIVE **MODEL** 17.1 This example is further illustrated in Active Model 17.1 found online.

EXCEL **OM** Data File **Ch17Ex1.xls** can be found online.

The basic unit of measure for reliability is the *product failure rate* (FR). Firms producing high-technology equipment often provide failure-rate data on their products. As shown in Equations (17-2) and (17-3), the failure rate measures the percentage of failures among the total number of products tested, FR(%), or a number of failures during a period of operating time, FR(N):

$$\text{FR}(\%) = \frac{\text{Number of failures}}{\text{Number of units tested}} \times 100\% \qquad (17\text{-}2)$$

$$\text{FR}(N) = \frac{\text{Number of failures}}{\text{Number of unit-hours of operating time}} \qquad (17\text{-}3)$$

where Number of unit-hours of operating time = Total time − Nonoperating time

Mean time between failures (MTBF)
The expected time between a repair and the next failure of a component, machine, process, or product.

Perhaps the most common term in reliability analysis is the **mean time between failures (MTBF)**, which is the reciprocal of FR(N):

$$\text{MTBF} = \frac{1}{\text{FR}(N)} \qquad (17\text{-}4)$$

In Example 2, we compute the percentage of failure FR(%), number of failures FR(N), and mean time between failures (MTBF).

Example 2

DETERMINING MEAN TIME BETWEEN FAILURES

Twenty air-conditioning systems designed for use by astronauts in Russia's Soyuz spacecraft were each operated for 1,000 hours at a Russian test facility. Two of the systems failed during the test—one after 200 hours and the other after 600 hours.

APPROACH ▶ To determine the percentage of failures [FR(%)], the number of failures per unit of time [FR(N)], and the mean time between failures (MTBF), we use Equations (17-2), (17-3), and (17-4), respectively.

SOLUTION ▶ Percentage of failures:

$$FR(\%) = \frac{\text{Number of failures}}{\text{Number of units tested}} = \frac{2}{20}(100\%) = 10\%$$

Number of failures per unit-hour of operating time:

$$FR(N) = \frac{\text{Number of failures}}{\text{Number of unit-hours of operating time}}$$

where

$$\begin{aligned}\text{Total time} &= (1{,}000\text{ hrs})(20\text{ units})\\ &= 20{,}000\text{ unit-hour}\\ \text{Nonoperating time} &= 800\text{ hrs for 1st failure} + 400\text{ hrs for 2nd failure}\\ &= 1{,}200\text{ unit-hour}\\ \text{Number of unit-hours of operating time} &= \text{Total time} - \text{Nonoperating time}\\ &= .000106\text{ failure/unit-hour}\end{aligned}$$

Because $\text{MTBF} = \frac{1}{FR(N)}$:

LO 17.3 *Determine* mean time between failures (MTBF)

$$\text{MTBF} = \frac{1}{.000106} = 9{,}434\text{ hrs}$$

If the typical Soyuz shuttle trip to the International Space Station lasts 6 days, Russia may note that the failure rate per trip is:

$$\begin{aligned}\text{Failure rate} &= (\text{Failures/unit-hr})(24\text{ hrs/day})(6\text{ days/trip})\\ &= (.000106)(24)(6)\\ &= .0153\text{ failure/trip}\end{aligned}$$

INSIGHT ▶ Mean time between failures (MTBF) is the standard means of stating reliability.

LEARNING EXERCISE ▶ If nonoperating time drops to 800, what is the new MTBF? [Answer: 9,606 hr.]

RELATED PROBLEMS ▶ 17.4, 17.5

If the failure rate recorded in Example 2 is too high, Russia will have to increase systems reliability by either increasing the reliability of individual components or by redundancy.

Providing Redundancy

To increase the reliability of systems, **redundancy** is added in the form of *backup* components or *parallel paths*. Redundancy is provided to ensure that if one component or path fails, the system has recourse to another.

Redundancy
The use of backup components or parallel paths to raise reliability.

Backup Redundancy Assume that reliability of a component is .80 and we back it up with another component with reliability of .75. The resulting reliability is the probability of the first component working plus the probability of the backup component working multiplied by the probability of needing the backup component (1 – .8 = .2). Therefore:

$$R_s = \begin{pmatrix}\text{Probability}\\ \text{of first}\\ \text{component}\\ \text{working}\end{pmatrix} + \left[\begin{pmatrix}\text{Probability}\\ \text{of second}\\ \text{component}\\ \text{working}\end{pmatrix} \times \begin{pmatrix}\text{Probability}\\ \text{of needing}\\ \text{second}\\ \text{component}\end{pmatrix}\right] = \qquad (17\text{-}5)$$

$$(.8) \quad + \quad [(.75) \quad \times \quad (1 - .8)] \quad = .8 + .15 = .95$$

Example 3 shows how redundancy, in the form of backup components, can improve the reliability of the loan process presented in Example 1.

Example 3

RELIABILITY WITH BACKUP

The National Bank is disturbed that its loan-application process has a reliability of only .713 (see Example 1) and would like to improve this situation.

APPROACH ▶ The bank decides to provide redundancy for the two least reliable clerks, with clerks of equal competence.

SOLUTION ▶ This procedure results in the following system:

$$\begin{array}{ccc} R_1 & R_2 & R_3 \\ 0.90 & 0.80 & \\ \downarrow & \downarrow & \\ 0.90 \rightarrow & 0.80 \rightarrow & 0.99 \end{array}$$

$$R_s = [.9 + .9(1 - .9)] \times [.8 + .8(1 - .8)] \times .99$$
$$= [.9 + (.9)(.1)] \times [.8 + (.8)(.2)] \times .99$$
$$= .99 \times .96 \times .99 = .94$$

INSIGHT ▶ By providing redundancy for two clerks, National Bank has increased reliability of the loan process from .713 to .94.

LEARNING EXERCISE ▶ What happens when the bank replaces both R_2 clerks with one new clerk who has a reliability of .90? [Answer: $R_s = .88$.]

RELATED PROBLEMS ▶ 17.7, 17.10, 17.12, 17.13, 17.14

ACTIVE MODEL 17.2 This example is further illustrated in Active Model 17.2 found online.

EXCEL OM Data File **Ch17Ex3.xls** can be found online.

Parallel Redundancy Another way to enhance reliability is to provide parallel paths. In a parallel system, the paths are assumed to be independent; therefore, success on any one path allows the system to perform. In Example 4, we determine the reliability of a process with three parallel paths.

Example 4

RELIABILITY WITH PARALLEL REDUNDANCY

A new iPad design that is more reliable because of its parallel circuits is shown below. What is its reliability?

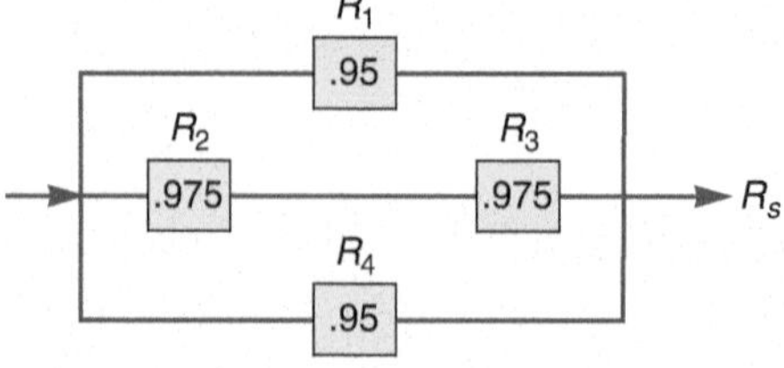

APPROACH ▶ Identify the reliability of each path, then compute the likelihood of needing additional paths (likelihood of failure), and finally subtract the product of those failures from 1.

SOLUTION ▶

Reliability for the middle path $= R_2 \times R_3 = .975 \times .975 = .9506$

Then determine the probability of *failure* for all 3 paths $= (1 - 0.95) \times (1 - .9506) \times (1 - 0.95)$
$= (.05) \times (.0494) \times (.05) = .00012$

Therefore the reliability of the new design is 1 minus the probability of failures, or

$$= 1 - .00012 = .99988$$

INSIGHT ▶ Even in a system where no component has reliability over .975, the parallel design increases reliability to over .999. Parallel paths can add substantially to reliability.

LEARNING EXERCISE ▶ If reliability of *all* components is only .90, what is the new reliability? [Answer: .9981.]

RELATED PROBLEMS ▶ 17.6, 17.8, 17.11

ACTIVE MODEL 17.3 This example is further illustrated in Active Model 17.3 found online.

Managers often use a combination of backup components or parallel paths to improve reliability.

Maintenance

Traditionally, there have been two types of maintenance: preventive maintenance and breakdown maintenance. **Preventive maintenance** involves monitoring equipment and facilities along with performing routine inspections and service to keep equipment and facilities reliable. It involves designing technical and human systems that will keep the productive process working within tolerance; and thus allows the system to perform as designed. These activities are intended to build a system that extends the life of capital investment and removes variability in system performance. **Breakdown maintenance** occurs when preventive maintenance fails and equipment/facilities must be repaired on an emergency or priority basis.

More recently, the digitalization of information and the current generation of sophisticated sensors allow managers to design maintenance procedures that cannot only detect, but also predict, equipment failure. Even the slightest unusual vibration, minute changes in temperature or pressure, and small variations in oil viscosity or chemical components can predict potential problems. These sensors extend to infrared thermography, sophisticated frequency analysis, and oil and water analysis. Application of these advanced technologies leads to a maintenance procedure known as **predictive maintenance**.

Preventive maintenance
A plan that involves monitoring equipment and facilities, and performing routine inspections and service to keep equipment and facilities reliable.

Breakdown maintenance
Remedial maintenance that occurs when preventive maintenance fails and equipment/facilities must be repaired on an emergency or priority basis.

Predictive maintenance
Using advanced technologies to monitor and predict equipment failure.

Implementing Preventive Maintenance

Preventive maintenance implies that we can determine when a system will need service or repair. Therefore, to perform preventive maintenance, we must have some indication of service life or probability of failure. Failures occur at different rates during the life of a product.

A high initial failure rate, known as **infant mortality**, may exist for many products. This is why many electronic firms "burn in" their products prior to shipment: that is to say, they execute a variety of tests (such as a full wash cycle at Whirlpool) to detect "startup" problems prior to shipment. Firms may also provide 90-day warranties. We should note that many infant mortality failures are not product failures per se, but rather failure due to improper use. This fact points up the importance in many industries of operations management's building an after-sales service system that includes installing and training.

Infant mortality
The failure rate early in the life of a product or process.

Once the product, machine, or process "settles in," a study can be made of the MTBF (mean time between failures) distribution. Such distributions often follow a normal curve. When these distributions exhibit small standard deviations, then we know we have a candidate for preventive maintenance, even if the maintenance is expensive.

LO 17.4 *Distinguish between preventive and breakdown maintenance*

Once our firm has a candidate for preventive maintenance, we want to determine *when* preventive maintenance is economical. Typically, the more expensive the maintenance, the narrower must be the MTBF distribution (that is, have a small standard deviation). In addition, if the process is no more expensive to repair when it breaks down than the cost of preventive maintenance, perhaps we should let the process break down and then do the repair. However, the consequence of the breakdown must be fully considered. Even some relatively minor breakdowns have catastrophic consequences. At the other extreme, preventive maintenance costs may be so incidental that preventive maintenance is appropriate even if the MTBF distribution is rather flat (that is, it has a large standard deviation).

With good reporting techniques, firms can maintain records of individual processes, machines, or equipment. Such records can provide a profile of both the kinds of maintenance

Figure 17.3

A Computerized Maintenance System

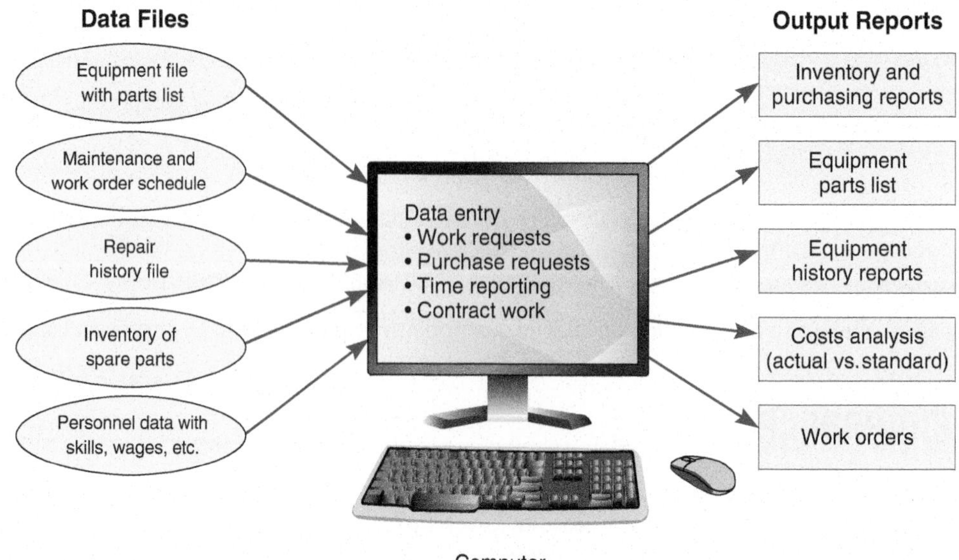

required and the timing of maintenance needed. Maintaining equipment history is an important part of a preventive maintenance system, as is a record of the time and cost to make the repair. Such records can also provide information about the family of equipment and suppliers.

LO 17.5 *Describe* how to improve maintenance

Reliability and maintenance are of such importance that most maintenance management systems are now computerized. Figure 17.3 shows the major components of such a system with files to be maintained on the left and reports generated on the right.

Companies from Boeing to Ford are improving product reliability via their maintenance information systems. Boeing monitors the health of planes in flight by relaying relevant information in real-time to the ground. This provides a head start on reliability and maintenance issues. Similarly, with wireless satellite service, millions of car owners are alerted to thousands of diagnostic issues, from faulty airbag sensors to the need for an oil change. These real-time systems provide immediate data that are used to head off quality issues before customers even notice a problem. The technology enhances reliability and customer satisfaction. And catching problems early saves millions of dollars in warranty costs.

Figure 17.4(a) shows a traditional view of the relationship between preventive maintenance and breakdown maintenance. In this view, operations managers consider a *balance* between the two costs. Allocating more resources to preventive maintenance will reduce the number of breakdowns. At some point, however, the decrease in breakdown maintenance costs may be less than the increase in preventive maintenance costs. At this point, the total cost curve begins to rise. Beyond this optimal point, the firm will be better off waiting for breakdowns to occur and repairing them when they do.

Figure 17.4

Maintenance Costs

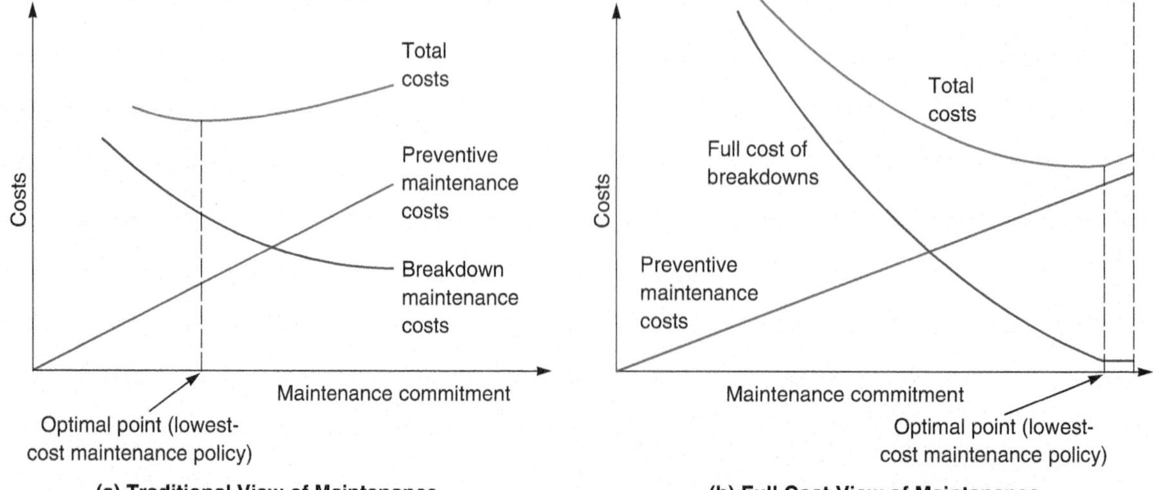

Unfortunately, cost curves such as in Figure 17.4(a) seldom consider the *full costs of a breakdown*. Many costs are ignored because they are not *directly* related to the immediate breakdown. For instance, the cost of inventory maintained to compensate for downtime is not typically considered. Moreover, downtime can have a devastating effect on safety and morale. Employees may also begin to believe that "performance to standard" and maintaining equipment are not important. Finally, downtime adversely affects delivery schedules, destroying customer relations and future sales. When the full impact of breakdowns is considered, Figure 17.4(b) may be a better representation of maintenance costs. In Figure 17.4(b), total costs are at a minimum when the system only breaks down due to unanticipated extraordinary events.

STUDENT TIP
When all breakdown costs are considered, much more maintenance may be advantageous.

Assuming that all potential costs associated with downtime have been identified, the operations staff can compute the optimal level of maintenance activity on a theoretical basis. Such analysis, of course, also requires accurate historical data on maintenance costs, breakdown probabilities, and repair times. Example 5 shows how to compare preventive and breakdown maintenance costs to select the least expensive maintenance policy.

Example 5

COMPARING PREVENTIVE AND BREAKDOWN MAINTENANCE COSTS

Farlen & Halikman is a CPA firm specializing in payroll preparation. The firm has been successful in automating much of its work, using high-speed printers for check processing and report preparation. The computerized approach, however, has problems. Over the past 20 months, the printers have broken down at the rate indicated in the following table:

NUMBER OF BREAKDOWNS	NUMBER OF MONTHS THAT BREAKDOWNS OCCURRED
0	2
1	8
2	6
3	4
	Total: 20

Each time the printers break down, Farlen & Halikman estimates that it loses an average of $300 in production time and service expenses. One alternative is to purchase a service contract for preventive maintenance. Even if Farlen & Halikman contracts for preventive maintenance, there will still be breakdowns, *averaging* one breakdown per month. The price for this service is $150 per month.

LO 17.6 *Compare* preventive and breakdown maintenance costs

APPROACH ▶ To determine if the CPA firm should follow a "run until breakdown" policy or contract for preventive maintenance, we follow a 4-step process:

Step 1 Compute the *expected number* of breakdowns (based on past history) if the firm continues as is, without the service contract.

Step 2 Compute the expected breakdown cost per month with no preventive maintenance contract.

Step 3 Compute the cost of preventive maintenance.

Step 4 Compare the two options and select the one that will cost less.

SOLUTION ▶

Step 1

NUMBER OF BREAKDOWNS	FREQUENCY	NUMBER OF BREAKDOWNS	FREQUENCY
0	2/20 = .1	2	6/20 = 0.3
1	8/20 = .4	3	4/20 = 0.2

$$\begin{pmatrix}\text{Expected number}\\ \text{of breakdowns}\end{pmatrix} = \sum\left[\begin{pmatrix}\text{Number of}\\ \text{breakdowns}\end{pmatrix} \times \begin{pmatrix}\text{Corresponding}\\ \text{frequency}\end{pmatrix}\right]$$

$$= (0)(.1) + (1)(.4) + (2)(.3) + (3)(.2)$$
$$= 0 + .4 + .6 + .6$$
$$= 1.6 \text{ breakdowns/month}$$

Step 2

$$\text{Expected breakdown cost} = \begin{pmatrix}\text{Expected number}\\ \text{of breakdowns}\end{pmatrix} \times \begin{pmatrix}\text{Cost per}\\ \text{breakdown}\end{pmatrix}$$
$$= (1.6)(\$300)$$
$$= \$480/\text{month}$$

Step 3

$$\begin{pmatrix}\text{Preventive}\\ \text{maintenance cost}\end{pmatrix} = \begin{pmatrix}\text{Cost of expected}\\ \text{breakdowns if service}\\ \text{contract signed}\end{pmatrix} + \begin{pmatrix}\text{Cost of}\\ \text{service contract}\end{pmatrix}$$
$$= (1\,\text{breakdown/month})(\$300) + \$150/\text{month}$$
$$= \$450/\text{month}$$

Step 4 Because it is less expensive overall to hire a maintenance service firm ($450) than to not do so ($480), Farlen & Halikman should hire the service firm.

INSIGHT ▶ Determining the expected number of breakdowns for each option is crucial to making a good decision. This typically requires good maintenance records.

LEARNING EXERCISE ▶ What is the best decision if the preventive maintenance contract cost increases to $195 per month? [Answer: At $495 (= $300 + $195) per month, "run until breakdown" becomes less expensive (assuming that all costs are included in the $300 per breakdown cost).]

RELATED PROBLEMS ▶ 17.18–17.23

Using variations of the technique shown in Example 5, operations managers can examine maintenance policies.

Increasing Repair Capabilities

Because reliability and preventive maintenance are seldom perfect, most firms opt for some level of repair capability. Enlarging repair facilities or improving maintenance competence may be an excellent way to get the system back in operation faster. For instance, AT&T is developing an augmented reality capability to increase the effectiveness of its 17,000 technicians.

However, not all repairs can be done in the firm's facility. Managers must, therefore, decide where repairs are to be performed. Figure 17.5 provides a continuum of options and how they rate in terms of speed, cost, and competence. Moving to the right in Figure 17.5 may improve the competence of the repair work, but at the same time it increases costs and replacement time.

LO 17.7 *Define* autonomous maintenance

Autonomous Maintenance

Autonomous maintenance
Operators partner with maintenance personnel to observe, check, adjust, clean, and notify.

Preventive maintenance policies and techniques must include an emphasis on employees accepting responsibility for the "observe, check, adjust, clean, and notify" type of equipment maintenance. Such policies are consistent with the advantages of employee empowerment. This approach is known as autonomous maintenance. Employees can predict failures, prevent breakdowns, and prolong equipment life. With autonomous maintenance, the manager is making a step toward both employee empowerment and maintaining system performance.

Figure **17.5**

The Operations Manager Determines How Maintenance Will Be Performed

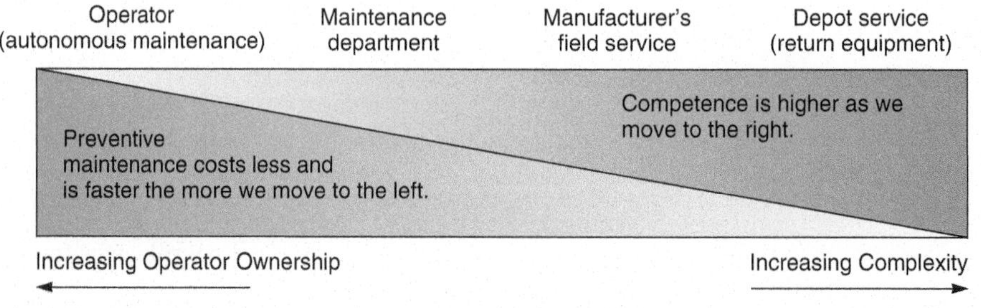

Total Productive Maintenance

Many firms have moved to bring total quality management concepts to the practice of preventive maintenance with an approach known as **total productive maintenance (TPM)**. It involves the concept of reducing variability through autonomous maintenance and excellent maintenance practices. Total productive maintenance includes:

- Designing machines that are reliable, easy to operate, and easy to maintain
- Emphasizing total cost of ownership when purchasing machines, so that service and maintenance are included in the cost
- Developing preventive maintenance plans that utilize the best practices of operators, maintenance departments, and depot service
- Training for autonomous maintenance so operators maintain their own machines and partner with maintenance personnel

Total productive maintenance (TPM)
Combines total quality management with a strategic view of maintenance from process and equipment design to preventive maintenance.

High utilization of facilities, tight scheduling, low inventory, and consistent quality demand reliability. Total productive maintenance, which continues to improve with recent advances in the use of simulation, expert systems, and sensors, is the key to reducing variability and improving reliability.

STUDENT TIP
Maintenance improves productivity.

Summary

Operations managers focus on design improvements, backup components, and parallel paths to improve reliability. Reliability improvements also can be obtained through the use of preventive maintenance and excellent repair facilities.

Firms give employees "ownership" of their equipment. When workers repair or do preventive maintenance on their own machines, breakdowns are less common. Well-trained and empowered employees ensure reliable systems through preventive maintenance. In turn, reliable, well-maintained equipment not only provides higher utilization but also improves quality and performance to schedule. Top firms build and maintain systems that drive out variability so that customers can rely on products and services to be produced to specifications and on time.

Key Terms

Ethical Dilemma

The space shuttle *Columbia* disintegrated over Texas on its 2003 return to Earth. The *Challenger* exploded shortly after launch in 1986. And the *Apollo 1* spacecraft imploded in fire on the launch pad in 1967. In each case, the lives of all crew members were lost. The hugely complex shuttle may have looked a bit like an airplane but was very different. In reality, its overall statistical reliability is such that about 1 out of every 50 flights had a major malfunction. As one aerospace manager stated, "Of course, you can be perfectly safe and never get off the ground."

Given the huge reliability and maintenance issues NASA faced (seals cracking in cold weather, heat shielding tiles falling off, tools left in the capsule), should astronauts have been allowed to fly? (In earlier *Atlas* rockets, men were inserted not out of necessity but because test pilots and politicians thought they should be there.) What are the pros and cons of staffed space exploration from an ethical perspective? Should the U.S. spend billions of dollars to return an astronaut to the moon or send one to Mars?

Discussion Questions

1. What is the objective of maintenance and reliability?
2. How does one identify a candidate for preventive maintenance?
3. Explain the notion of "infant mortality" in the context of product reliability.
4. How could simulation be a useful technique for maintenance problems?
5. What is the trade-off between operator-performed maintenance versus supplier-performed maintenance?
6. How can a manager evaluate the effectiveness of the maintenance function?
7. How does machine design contribute to either increasing or alleviating the maintenance problem?
8. What roles can computerized maintenance management systems play in the maintenance function?
9. During an argument as to the merits of preventive maintenance at Windsor Printers, the company owner asked, "Why fix it before it breaks?" How would you, as the director of maintenance, respond?
10. Will preventive maintenance eliminate *all* breakdowns?

Using Software to Solve Reliability Problems

PX Excel OM and POM for Windows may be used to solve reliability problems. The reliability module allows us to enter (1) number of systems (components) in the series (1 through 10); (2) number of backup, or parallel, components (1 through 12); and (3) component reliability for both series and parallel data.

Solved Problems

SOLVED PROBLEM 17.1

The semiconductor used in the Sullivan Wrist Calculator has five circuits, each of which has its own reliability rate. Component 1 has a reliability of .90; component 2, .95; component 3, .98; component 4, .90; and component 5, .99. What is the reliability of one semiconductor?

SOLUTION

$$\begin{aligned}\text{Semiconductor reliability, } R_s &= R_1 \times R_2 \times R_3 \times R_4 \times R_5 \\ &= (.90)(.95)(.98)(.90)(.99) \\ &= .7466\end{aligned}$$

SOLVED PROBLEM 17.2

A recent engineering change at Sullivan Wrist Calculator places a backup component in each of the two least reliable transistor circuits. The new circuits will look like the following:

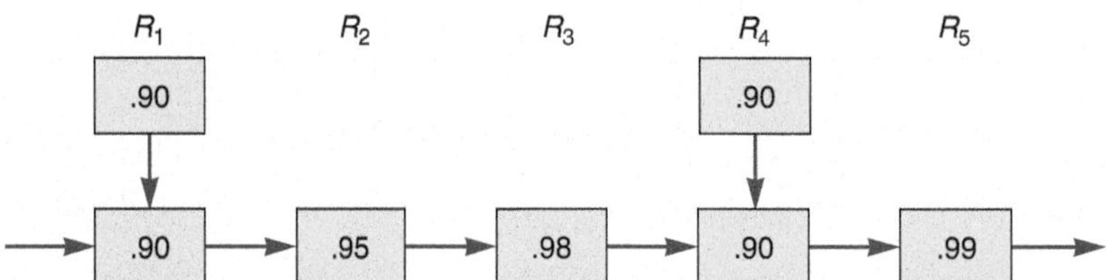

What is the reliability of the new system?

SOLUTION

$$\begin{aligned}\text{Reliability} &= [.9 + (1 - .9) \times .9] \times .95 \times .98 \times [.9 + (1 - .9) \times .9] \times .99 \\ &= [.9 + .09] \times .95 \times .98 \times [.9 + .09] \times .99 \\ &= .99 \times .95 \times .98 \times .99 \times .99 \\ &= .903\end{aligned}$$

Problems

Note: **PX** means the problem may be solved with POM for Windows and/or Excel OM.

Problems 17.1–17.17 relate to Reliability

• **17.1** The Beta II computer's electronic processing unit contains 50 components in series. The average reliability of each component is 99.0%. Using Figure 17.2, determine the overall reliability of the processing unit.

• **17.2** A testing process at Boeing Aircraft has 400 components in series. The average reliability of each component is 99.5%. Use Figure 17.2 to find the overall reliability of the whole testing process.

•• **17.3** A new aircraft control system is being designed that must be 98% reliable. This system consists of three components in series. If all three of the components are to have the same level of reliability, what level of reliability is required? **PX**

•• **17.4** Robert Klassen Manufacturing, a medical equipment manufacturer, subjected 100 heart pacemakers to 5,000 hours of testing. Halfway through the testing, 5 pacemakers failed. What was the failure rate in terms of the following:

a) Percentage of failures?
b) Number of failures per unit-hour of operating time?
c) Number of failures per unit-year?
d) If 1,100 people receive pacemaker implants, how many units can we expect to fail during the following year?

•• **17.5** A manufacturer of disk drives for notebook computers wants an MTBF of at least 50,000 hours. Recent test results for 10 units were one failure at 10,000 hours, another at 25,000 hours, and two more at 45,000 hours. The remaining units were still running at 60,000 hours. Determine the following:
a) Percentage of failures
b) Number of failures per unit-hour of operating time
c) MTBF at this point in the testing

•• **17.6** What is the reliability of the following parallel production process? $R_1 = 0.95$, $R_2 = 0.90$, $R_3 = 0.98$.

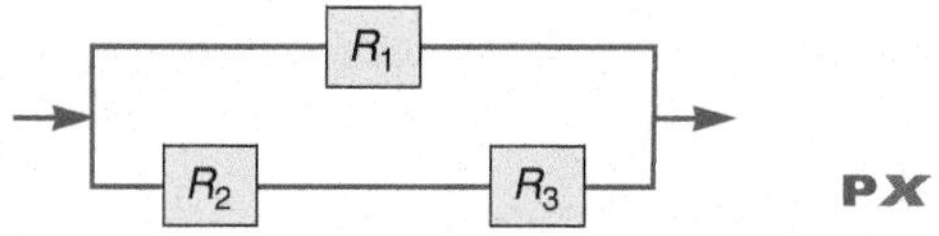

PX

•• **17.7** What is the overall reliability that bank loans will be processed accurately if each of the five clerks shown in the chart has the reliability shown?

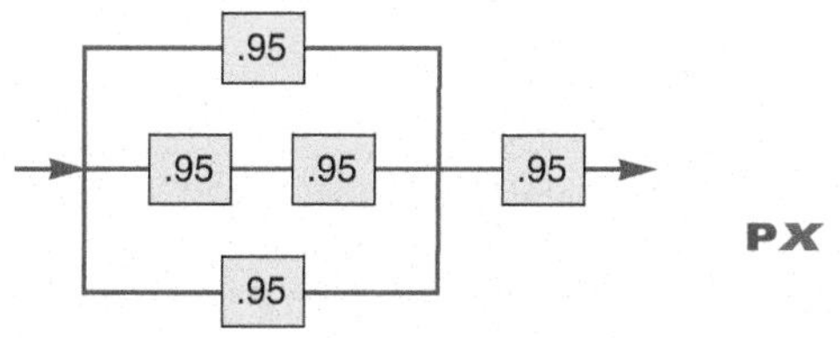

PX

Hint: The three paths are done in parallel, followed by an additional independent step.

•• **17.8** Merrill Kim Sharp has a system composed of three components in parallel. The components have the following reliabilities:

$$R_1 = 0.90,\ R_2 = 0.95,\ R_3 = 0.85$$

What is the reliability of the system? (*Hint:* See Example 4.) **PX**

• **17.9** A medical control system has three components in series with individual reliabilities (R_1, R_2, R_3) as shown:

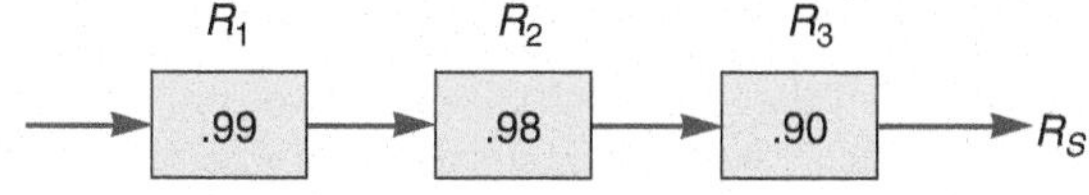

What is the reliability of the system? **PX**

•• **17.10** What is the reliability of the system shown?

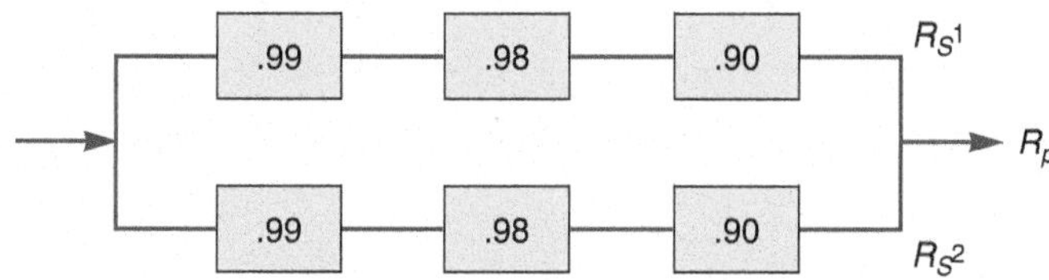

• **17.11** How much would reliability improve if the medical control system shown in Problem 17.9 changed to the redundant parallel system shown in Problem 17.10?

•• **17.12** Elizabeth Irwin's design team has proposed the following system with component reliabilities as indicated:

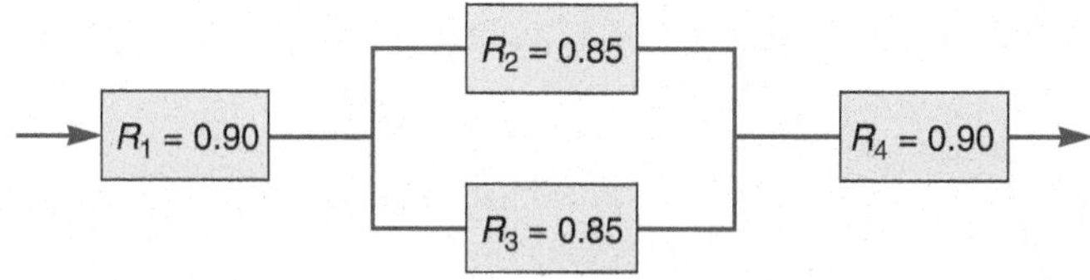

What is the reliability of the system? **PX**
Hint: The system functions if either R_2 or R_3 work.

•• **17.13** Rick Wing, salesperson for Wave Soldering Systems, Inc. (WSSI), has provided you with a proposal for improving the temperature control on your present machine. The machine uses a hot-air knife to cleanly remove excess solder from printed circuit boards; this is a great concept, but the hot-air temperature control lacks reliability. According to Wing, engineers at WSSI have improved the reliability of the critical temperature controls. The new system still has the four sensitive integrated circuits controlling the temperature, but the new machine has a backup for each. The four integrated circuits have reliabilities of .90, .92, .94, and .96. The four backup circuits all have a reliability of .90.
a) What is the reliability of the new temperature controller?
b) If you pay a premium, Wing says he can improve all four of the backup units to .93. What is the reliability of this option? **PX**

••• **17.14** As VP for operations at Méndez-Piñero Engineering, you must decide which product design, A or B, has the higher reliability. B is designed with backup units for components R_3 and R_4. What is the reliability of each design? **PX**

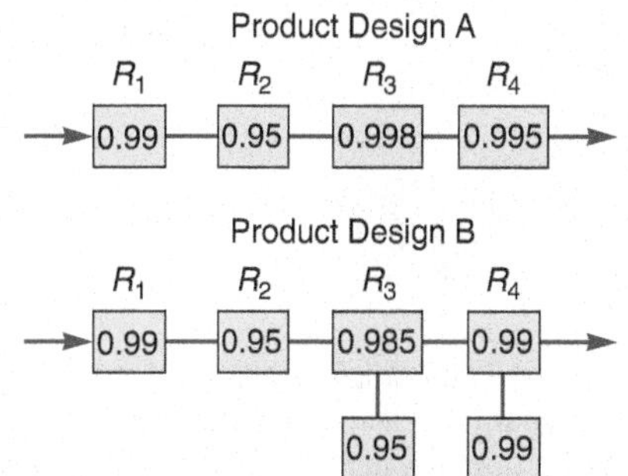

•• **17.15** A typical retail transaction consists of several smaller steps, which can be considered components subject to failure. A list of such components might include:

COMPONENT	DESCRIPTION	DEFINITION OF FAILURE
1	Find product in proper size, color, etc.	Can't find product
2	Enter cashier line	No lines open; lines too long; line experiencing difficulty
3	Scan product UPC for name, price, etc.	Won't scan; item not on file; scans incorrect name or price
4	Calculate purchase total	Wrong weight; wrong extension; wrong data entry; wrong tax
5	Make payment	Customer lacks cash; check not acceptable; credit card refused
6	Make change	Makes change incorrectly
7	Bag merchandise	Damages merchandise while bagging; bag splits
8	Conclude transaction and exit	No receipt; unfriendly, rude, or aloof clerk

Let the eight probabilities of success be .92, .94, .99, .99, .98, .97, .95, and .96. What is the reliability of the system; that is, the probability that there will be a satisfied customer? If you were the store manager, what do you think should be an acceptable value for this probability? Which components would be good candidates for backup, which for redesign?

• **17.16** The credit card issuing process at Atlanta Bank's VISA program consists of 10 steps performed in series by different bank employees. The average reliability of each employee is 98%. Use Figure 17.2 in the text to help manager Yusen Xia find the overall reliability of the credit card issuing process. **PX**

• **17.17** You have a system composed of a serial connection of four components with the following reliabilities:

COMPONENT	RELIABILITY
1	0.90
2	0.95
3	0.80
4	0.85

What is the reliability of the system? **PX**

Problems 17.18–17.23 relate to Maintenance

• **17.18** What are the *expected* number of yearly breakdowns for the power generator at Orlando Utilities that has exhibited the following data over the past 20 years? **PX**

NUMBER OF BREAKDOWNS	0	1	2	3	4	5	6
NUMBER OF YEARS IN WHICH BREAKDOWN OCCURRED	2	2	5	4	5	2	0

• **17.19** Each breakdown of a graphic plotter table at Airbus Industries costs $50. Find the expected daily breakdown cost, given the following data: **PX**

NUMBER OF BREAKDOWNS	0	1	2	3	4
DAILY BREAKDOWN PROBABILITY	.1	.2	.4	.2	.1

•• **17.20** Maureen Hall, chief of the maintenance department at Mechanical Dynamics, has presented you with the following failure curve. What does it suggest?

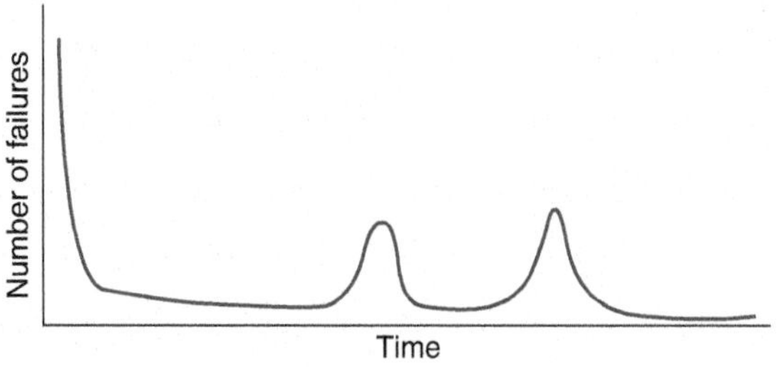

••• **17.21** The fire department has a number of failures with its oxygen masks and is evaluating the possibility of outsourcing preventive maintenance to the manufacturer. Because of the risk associated with a failure, the cost of each failure is estimated at $2,000. The current maintenance policy (with station employees performing maintenance) has yielded the following history:

NUMBER OF BREAKDOWNS	0	1	2	3	4	5
NUMBER OF YEARS IN WHICH BREAKDOWNS OCCURRED	4	3	1	5	5	0

This manufacturer will guarantee repairs on any and all failures as part of a service contract. The cost of this service is $5,000 per year.

a) What is the expected number of breakdowns per year with station employees performing maintenance?
b) What is the cost of the current maintenance policy?
c) What is the more economical policy?

• **17.22** Given the probabilities that follow for Bonnie Richardson's print shop, find the expected breakdown cost.

NUMBER OF BREAKDOWNS	0	1	2	3
DAILY FREQUENCY	0.3	0.2	0.2	0.3

The cost per breakdown is $10. **PX**

••• **17.23** Wharton Manufacturing Company operates its 23 large and expensive grinding and lathe machines from 7 A.M. to 11 P.M., 7 days a week. For the past year, the firm has been under contract with Simkin and Sons for daily preventive maintenance (lubrication, cleaning, inspection, and so on). Simkin's crew works between 11 P.M. and 2 A.M. so as not to interfere with the daily manufacturing crew. Simkin charges $645 per week for this service. Since signing the maintenance contract, Wharton Manufacturing has noted an average of only three breakdowns per week. When a grinding or lathe machine does break down during a working shift, it costs Wharton about $250 in lost production and repair costs.

After reviewing past breakdown records (for the period before signing a preventive maintenance contract with Simkin and Sons), Wharton's production manager was able to summarize the following patterns:

NUMBER OF BREAKDOWNS PER WEEK	0	1	2	3	4	5	6	7	8
NUMBER OF WEEKS IN WHICH BREAKDOWNS OCCURRED	1	1	3	5	9	11	7	8	5
	Total weeks of historical data: 50								

The production manager is not certain that the contract for preventive maintenance with Simkin is in Wharton's best financial interest. He recognizes that much of his breakdown data are old but is fairly certain that they are representative of the present picture.

a) What is the weekly cost of the breakdowns without the maintenance contract?
b) What is the cost per week of the breakdowns with the maintenance contract?
c) Is the contract for preventive maintenance with Simkin in Wharton's best financial interest? **PX**

CASE STUDY

Maintenance Drives Profits at Frito-Lay

Video Case

Frito-Lay, the multi-billion-dollar subsidiary of food and beverage giant PepsiCo, maintains 36 plants in the U.S. and Canada. These facilities produce dozen of snacks, including the well-known Lay's, Fritos, Cheetos, Doritos, Ruffles, and Tostitos brands, each of which sells over $1 billion per year.

Frito-Lay plants produce in the high-volume, low-variety process model common to commercial baked goods, steel, glass, and beer industries. In this environment, preventive maintenance of equipment takes a major role by avoiding costly downtime. Tom Rao, vice president for Florida operations, estimates that each 1% of downtime has a negative annual profit impact of $200,000. He is proud of the $1\frac{1}{2}$% unscheduled downtime his plant is able to reach—well below the 2% that is considered the "world-class" benchmark. This excellent performance is possible because the maintenance department takes an active role in setting the parameters for preventive maintenance. This is done with weekly input to the production schedule.

Maintenance policy impacts energy use as well. The Florida plant's technical manager, Jim Wentzel, states, "By reducing production interruptions, we create an opportunity to bring energy and utility use under control. Equipment maintenance and a solid production schedule are keys to utility efficiency. With every production interruption, there is substantial waste."

As a part of its total productive maintenance (TPM) program,* Frito-Lay empowers employees with what it calls the "Run Right" system. Run Right teaches employees to "identify and do." This means each shift is responsible for identifying problems and making the necessary corrections, when possible. This is accomplished through (1) a "power walk" at the beginning of the shift to ensure that equipment and process settings are performing to standard, (2) mid-shift and post-shift reviews of standards and performance, and (3) posting of any issues on a large whiteboard in the shift office. Items remain on the whiteboard until corrected, which is seldom more than a shift or two.

With good workforce scheduling and tight labor control to hold down variable costs, making time for training is challenging. But supervisors, including the plant manager, are available to fill in on the production line when that is necessary to free an employee for training.

The 30 maintenance personnel hired to cover 24/7 operations at the Florida plant all come with multi-craft skills (e.g., welding, electrical, plumbing). "Multi-craft maintenance personnel are harder to find and cost more," says Wentzel, "but they more than pay for themselves."

*At Frito-Lay, preventive maintenance, autonomous maintenance, and total productive maintenance are part of a Frito-Lay program known as total productive manufacturing.

Discussion Questions**

1. What might be done to help take Frito-Lay to the next level of outstanding maintenance? Consider factors such as sophisticated software.
2. What are the advantages and disadvantages of giving more responsibility for machine maintenance to the operator?
3. Discuss the pros and cons of hiring multi-craft maintenance personnel.

**You may wish to view the video that accompanies this case before answering these questions.

Bibliography

Ait-Kadi, D., J. B. Menge, and H. Kane. "Resource Assignment Model in Maintenance Activities Scheduling." *International Journal of Production Research* 49, no. 22 (November 2011): 6677–6689.

Bauer, E., X. Zhang, and D. A. Kimber. *Practical System Reliability.* New York: Wiley, 2009.

Blank, R. *The Basics of Reliability.* University Park, IL: Productivity Press, 2004.

Finigen, T., and J. Humphries. "Maintenance Gets Lean." *IE Industrial Systems* 38, no. 10 (October 2006): 26–31.

Phillippi, B., "3 Keys to Predictive Maintenance Success." *Flow Control* 21, no. 5 (May 2015): 7–8.

Rahman, A., "Maintenance Contract Model for Complex Asset/ Equipment." *International Journal of Reliability, Quality, & Safety Engineering* 21, no. 1 (February 2014): 1–11.

Stephens, M. P. *Productivity and Reliability-Based Maintenance Management.* Upper Saddle River, NJ: Prentice Hall, 2004.

Weil, M. "Beyond Preventive Maintenance." *APICS* 16, no. 4 (April 2006): 40–43.

Chapter 17 *Rapid* Review

Main Heading	Review Material	MyLab Operations Management
THE STRATEGIC IMPORTANCE OF MAINTENANCE AND RELIABILITY	Poor maintenance can be disruptive, inconvenient, wasteful, and expensive in dollars and even in lives. The interdependency of operator, machine, and mechanic is a hallmark of successful maintenance and reliability. Good maintenance and reliability management requires employee involvement and good procedures; it enhances a firm's performance and protects its investment. *The objective of maintenance and reliability is to maintain the capability of the system.* ■ **Maintenance**—All activities involved in maintaining capability of the system. ■ **Reliability**—The probability that a machine part or product will function properly for a specified time under stated conditions. The two main tactics for improving reliability are: 1. Improving individual components 2. Providing redundancy The two main tactics for improving maintenance are: 1. Implementing or improving preventive maintenance 2. Increasing repair capabilities or speed	Concept Questions: 1.1–1.5 **VIDEO 17.1** Maintenance Drives Profits at Frito-Lay
RELIABILITY	A system is composed of a series of individual interrelated components, each performing a specific job. If any *one* component fails to perform, the overall system can fail. As the number of components in a *series* increases, the reliability of the whole system declines very quickly: $R_s = R_1 \times R_2 \times R_3 \times \ldots \times R_n$ (17-1) where R_1 = reliability of component 1, R_2 = reliability of component 2, and so on. Equation (17-1) assumes that the reliability of an individual component does not depend on the reliability of other components. A .90 reliability means that the unit will perform as intended 90% of the time, and it will fail 10% of the time. The basic unit of measure for reliability is the *product failure rate* (FR). FR(%) is the percentage of failures among the total number of products tested, and FR(N) is the number of failures during a period of time: $\text{FR}(\%) = \frac{\text{Number of failures}}{\text{Number of units tested}} \times 100\%$ (17-2) $\text{FR}(N) = \frac{\text{Number of failures}}{\text{Number of unit-hours of operating time}}$ (17-3) ■ **Mean time between failures (MTBF)**—The expected time between a repair and the next failure of a component, machine, process, or product. $\text{MTBF} = \frac{1}{\text{FR}(N)}$ (17-4) ■ **Redundancy**—The use of components in parallel to raise reliability. The reliability of a component along with its backup equals: (Probability that 1st component works) + [(Prob. that backup works) × (Prob. that 1st fails)] (17-5)	Concept Questions: 2.1–2.6 Problems: 17.1–17.17 Virtual Office Hours for Solved Problems: 17.1, 17.2 **ACTIVE MODELS 17.1, 17.2, 17.3**
MAINTENANCE	■ **Preventive maintenance**—Involves monitoring equipment and facilities, and performing routine inspections and service to keep equipment and facilities reliable. ■ **Breakdown maintenance**—Remedial maintenance that occurs when preventive maintenance fails and equipment/facilities must be repaired on an emergency or priority basis. ■ **Predictive maintenance**—Using advanced technologies to monitor and predict equipment failure. ■ **Infant mortality**—The failure rate early in the life of a product or process. Consistent with job enrichment practices, machine operators must be held responsible for preventive maintenance of their own equipment and tools. Reliability and maintenance are of such importance that most maintenance systems are now computerized. Costs of a breakdown that may get ignored include:	Concept Questions: 3.1–3.6 Problems: 17.18–17.19, 17.21–17.23

Chapter 17 Rapid Review *continued*

Main Heading	Review Material	MyLab Operations Management
	1. The cost of inventory maintained to compensate for downtime 2. Downtime, which can have a devastating effect on safety and morale and which adversely affects delivery schedules, destroying customer relations and future sales ▪ **Autonomous maintenance**—Operators partner with maintenance personnel to observe, check, adjust, clean, and notify. Employees can predict failures, prevent breakdowns, and prolong equipment life. With autonomous maintenance, the manager is making a step toward both employee empowerment and maintaining system performance.	
TOTAL PRODUCTIVE MAINTENANCE	▪ **Total productive maintenance (TPM)**—Combines total quality management with a strategic view of maintenance from process and equipment design to preventive maintenance. Total productive maintenance includes: 1. Designing machines that are reliable, easy to operate, and easy to maintain 2. Emphasizing total cost of ownership when purchasing machines, so that service and maintenance are included in the cost 3. Developing preventive maintenance plans that utilize the best practices of operators, maintenance departments, and depot service 4. Training for autonomous maintenance so operators maintain their own machines and partner with maintenance personnel Three techniques that have proven beneficial to effective maintenance are simulation, expert systems, and sensors.	Concept Questions: 4.1–4.4
ADDITIONAL MYLAB OPERATIONS MANAGEMENT RESOURCES	✔ Additional Case Study (Cartak's Department Store) ✔ Multiple Choice Case Questions (Cartak's Department Store)	

Self Test

LO 17.1 The two main tactics for improving reliability are _____ and _____.

LO 17.2 The reliability of a system with n independent components equals:
- **a)** the sum of the individual reliabilities.
- **b)** the minimum reliability among all components.
- **c)** the maximum reliability among all components.
- **d)** the product of the individual reliabilities.
- **e)** the average of the individual reliabilities.

LO 17.3 What is the formula for the mean time between failures?
- **a)** Number of failures ÷ Number of unit-hours of operating time
- **b)** Number of unit-hours of operating time ÷ Number of failures
- **c)** (Number of failures ÷ Number of units tested) × 100%
- **d)** (Number of units tested ÷ Number of failures) × 100%
- **e)** 1 ÷ FR(%)

LO 17.4 The process that involves monitoring equipment and facilities, and performing routine inspections and service to keep equipment and facilities reliable is known as:
- **a)** breakdown maintenance.
- **b)** failure maintenance.
- **c)** preventive maintenance.
- **d)** all of the above.

LO 17.5 The two main tactics for improving maintenance are _____ and _____.

LO 17.6 The appropriate maintenance policy is developed by balancing preventive maintenance costs with breakdown maintenance costs. The problem is that:
- **a)** preventive maintenance costs are very difficult to identify.
- **b)** full breakdown costs are seldom considered.
- **c)** preventive maintenance should be performed, regardless of the cost.
- **d)** breakdown maintenance must be performed, regardless of the cost.

LO 17.7 _____ maintenance partners operators with maintenance personnel to observe, check, adjust, clean, and notify.
- **a)** Partnering
- **b)** Operator
- **c)** Breakdown
- **d)** Six Sigma
- **e)** Autonomous

Answers: LO 17.1. improving individual components, providing redundancy; LO 17.2. d; LO 17.3. b; LO 17.4. c; LO 17.5. implementing or improving preventive maintenance, increasing repair capabilities or speed; LO 17.6. b; LO 17.7. e.

Decision-Making Tools

MODULE A

MODULE OUTLINE

LEARNING OBJECTIVES

LO A.1 ***Create*** a simple decision tree 693

LO A.2 ***Build*** a decision table 693

LO A.3 ***Explain*** when to use each of the three types of decision-making environments 694

LO A.4 ***Calculate*** an expected monetary value (EMV) 695

LO A.5 ***Compute*** the expected value of perfect information (EVPI) 696

LO A.6 ***Evaluate*** the nodes in a decision tree 698

LO A.7 ***Create*** a decision tree with sequential decisions 700

Stockbyte/Getty Images

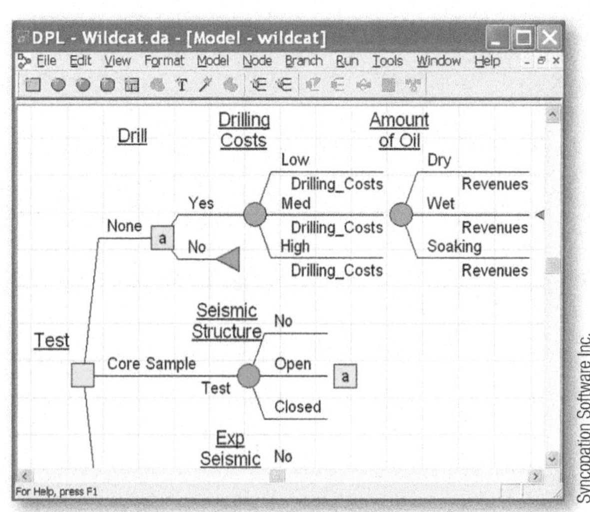

Syncopation Software Inc.

When Tomco Oil had to decide which of its new Kentucky lease areas to drill for oil, it turned to decision-tree analysis. The 74 different factors, including geological, engineering, economic, and political factors, became much clearer. Decision tree software such as DPL (shown here), Tree Plan, and Supertree allow decision problems to be analyzed with less effort and greater depth than ever before.

The Decision Process in Operations

Operations managers are decision makers. To achieve the goals of their organizations, managers must understand how decisions are made and know which decision-making tools to use. To a great extent, the success or failure of both people and companies depends on the quality of their decisions. Overcoming uncertainty is a manager's challenge.

What makes the difference between a good decision and a bad decision? A "good" decision—one that uses analytic decision making—is based on logic and considers all available data and possible alternatives. It also follows these six steps:

1. Clearly define the problem and the factors that influence it.
2. Develop specific and measurable objectives.
3. Develop a model—that is, a relationship between objectives and variables (which are measurable quantities).
4. Evaluate each alternative solution based on its merits and drawbacks.
5. Select the best alternative.
6. Implement and evaluate the decision and then set a timetable for completion.

So analytic decision making requires models, objectives, and quantifiable variables, often in the form of probabilities and payoffs. This module provides an introduction to the challenges facing managers by introducing two of the tools of decision making—decision tables and decision trees. These two tools are used in numerous OM situations, ranging from new-product analysis, to capacity planning, to location planning, to supply chain disaster planning, to scheduling, and to maintenance planning.

Fundamentals of Decision Making

Regardless of the complexity of a decision or the sophistication of the technique used to analyze it, all decision makers are faced with alternatives and "states of nature." The following notation will be used in this module:

1. Terms:
   a. *Alternative*—A course of action or strategy that may be chosen by a decision maker (e.g., not carrying an umbrella tomorrow).
   b. *State of nature*—An occurrence or a situation over which the decision maker has little or no control (e.g., tomorrow's weather).
2. Symbols used in a decision tree:
   a. ☐—Decision node from which one of several alternatives may be selected.
   b. ○—A state-of-nature node out of which one state of nature will occur.

To present a manager's decision alternatives, we can develop *decision trees* using the above symbols. When constructing a decision tree, we must be sure that all alternatives and states of nature are in their correct and logical places and that we include *all* possible alternatives and states of nature.

Example A1

A SIMPLE DECISION TREE

Getz Products Company is investigating the possibility of producing and marketing backyard storage sheds. Undertaking this project would require the construction of either a large or a small manufacturing plant. The market for the product produced—storage sheds—could be either favorable or unfavorable. Getz, of course, has the option of not developing the new product line at all.

APPROACH ▶ Getz decides to build a decision tree.

SOLUTION ▶ Figure A.1 illustrates Getz's decision tree.

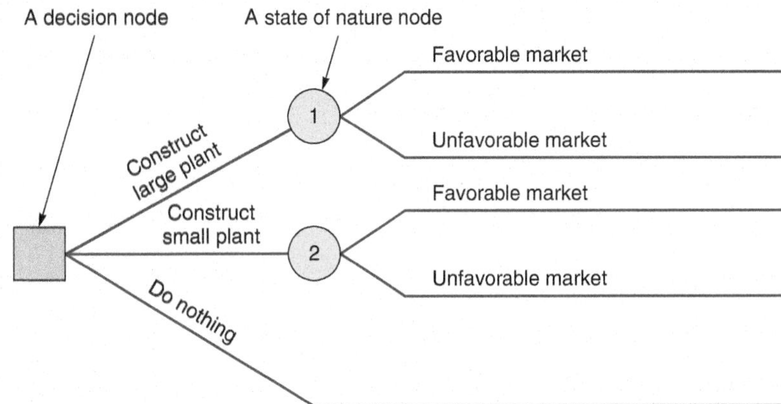

Figure A.1
Getz Products' Decision Tree

INSIGHT ▶ We never want to overlook the option of "doing nothing," as that is usually a possible decision.

LO A.1 *Create* a simple decision tree

LEARNING EXERCISE ▶ Getz now considers constructing a medium-sized plant as a fourth option. Redraw the tree in Figure A.1 to accommodate this. [Answer: Your tree will have a new node and branches between "Construct large plant" and "Construct small plant."]

RELATED PROBLEMS ▶ A.2e, A.8b, A.22–A.25

Decision Tables

We may also develop a decision or payoff table to help Getz Products define its alternatives. For any alternative and a particular state of nature, there is a *consequence* or *outcome*, which is usually expressed as a monetary value. This is called a *conditional value*. Note that all of the alternatives in Example A2 are listed down the left side of the table, that states of nature (outcomes) are listed across the top, and that conditional values (payoffs) are in the body of the **decision table**.

Decision table
A tabular means of analyzing decision alternatives and states of nature.

Example A2

A DECISION TABLE

Getz Products now wishes to organize the following information into a table. With a favorable market, a large facility will give Getz Products a net profit of $200,000. If the market is unfavorable, a $180,000 net loss will occur. A small plant will result in a net profit of $100,000 in a favorable market, but a net loss of $20,000 will be encountered if the market is unfavorable.

LO A.2 *Build* a decision table

APPROACH ▶ These numbers become conditional values in the decision table. We list alternatives in the left column and states of nature across the top of the table.

SOLUTION ▶ The completed table is shown in Table A.1.

TABLE A.1 Decision Table with Conditional Values for Getz Products

	STATES OF NATURE	
ALTERNATIVES	FAVORABLE MARKET	UNFAVORABLE MARKET
Construct large plant	$200,000	–$180,000
Construct small plant	$100,000	–$ 20,000
Do nothing	$ 0	$ 0

STUDENT TIP
Decision tables force logic into decision making.

INSIGHT ▶ The toughest part of decision tables is obtaining the data to analyze.

LEARNING EXERCISE ▶ In Examples A3 and A4, we see how to use decision tables to make decisions.

Types of Decision-Making Environments

LO A.3 *Explain* when to use each of the three types of decision-making environments

The types of decisions people make depend on how much knowledge or information they have about the situation. There are three decision-making environments:

- Decision making under uncertainty
- Decision making under risk
- Decision making under certainty

Decision Making Under Uncertainty

When there is complete *uncertainty* as to which state of nature in a decision environment may occur (i.e., when we cannot even assess probabilities for each possible outcome), we rely on three decision methods:

Maximax
A criterion that finds an alternative that maximizes the maximum outcome.

Maximin
A criterion that finds an alternative that maximizes the minimum outcome.

Equally likely
A criterion that assigns equal probability to each state of nature.

1. **Maximax**: This method finds an alternative that *max*imizes the *max*imum outcome for every alternative. First, we find the maximum outcome within every alternative, and then we pick the alternative with the maximum number. Because this decision criterion locates the alternative with the *highest* possible *gain*, it has been called an "optimistic" decision criterion.
2. **Maximin**: This method finds the alternative that *max*imizes the *min*imum outcome for every alternative. First, we find the minimum outcome within every alternative, and then we pick the alternative with the maximum number. Because this decision criterion locates the alternative that has the *least* possible *loss*, it has been called a "pessimistic" decision criterion.
3. **Equally likely**: This method finds the alternative with the highest average outcome. First, we calculate the average outcome for every alternative, which is the sum of all outcomes divided by the number of outcomes. We then pick the alternative with the maximum number. The equally likely approach assumes that each state of nature is equally likely to occur.

Example A3

A DECISION TABLE ANALYSIS UNDER UNCERTAINTY

Getz Products Company would like to apply each of these three approaches now.

APPROACH ▶ Given Getz's decision table from Example A2, he determines the maximax, maximin, and equally likely decision criteria.

SOLUTION ▶ Table A.2 provides the solution.

TABLE A.2 Decision Table for Decision Making Under Uncertainty

	STATES OF NATURE				
ALTERNATIVES	FAVORABLE MARKET	UNFAVORABLE MARKET	MAXIMUM IN ROW	MINIMUM IN ROW	ROW AVERAGE
Construct large plant	$200,000	–$180,000	$200,000 ◀ Maximax	–$180,000	$10,000
Construct small plant	$100,000	–$ 20,000	$100,000	–$ 20,000	$40,000 ◀ Equally likely
Do nothing	$ 0	$ 0	$ 0	$ 0 ◀ Maximin	$ 0
			Maximax	Maximin	Equally likely

1. The maximax choice is to construct a large plant. This is the *max*imum of the *max*imum number within each row, or alternative.
2. The maximin choice is to do nothing. This is the *max*imum of the *min*imum number within each row, or alternative.
3. The equally likely choice is to construct a small plant. This is the maximum of the average outcome of each alternative. This approach assumes that all outcomes for any alternative are *equally likely*.

INSIGHT ▶ There are optimistic decision makers ("maximax") and pessimistic ones ("maximin"). Maximax and maximin present best case–worst case planning scenarios.

LEARNING EXERCISE ▶ Getz reestimates the outcome for constructing a large plant when the market is favorable and raises it to \$250,000. What numbers change in Table A.2? Do the decisions change? [Answer: The maximax is now \$250,000, and the row average is \$35,000 for large plant. No decision changes.]

RELATED PROBLEMS ▶ A.1, A.2b–d, A.4, A.6, A.15

Decision Making Under Risk

Decision making under risk, a more common occurrence, relies on probabilities. Several possible states of nature may occur, each with an assumed probability. The states of nature must be mutually exclusive and collectively exhaustive and their probabilities must sum to 1.[1] Given a decision table with conditional values and probability assessments for all states of nature, we can determine the **expected monetary value (EMV)** for each alternative. This figure represents the expected value or *mean* return for each alternative *if we could repeat this decision (or similar types of decisions) a large number of times.*

Expected monetary value (EMV)
The expected payout or value of a variable that has different possible states of nature, each with an associated probability.

The EMV for an alternative is the sum of all possible payoffs from the alternative, each weighted by the probability of that payoff occurring:

$$\begin{aligned} \text{EMV (Alternative } i) = &\ (\text{Payoff of 1st state of nature}) \\ &\times (\text{Probability of 1st state of nature}) \\ &+ (\text{Payoff of 2nd state of nature}) \\ &\times (\text{Probability of 2nd state of nature}) \\ &+ \ldots + (\text{Payoff of last state of nature}) \\ &\times (\text{Probability of last state of nature}) \end{aligned}$$

LO A.4 *Calculate* an expected monetary value (EMV)

Example A4 illustrates how to compute the maximum EMV.

Example A4

EXPECTED MONETARY VALUE

Getz would like to find the EMV for each alternative.

APPROACH ▶ Getz Products' operations manager believes that the probability of a favorable market is 0.6, and that of an unfavorable market is 0.4. He can now determine the EMV for each alternative (see Table A.3).

SOLUTION ▶

1. $\text{EMV}(A_1) = (0.6)(\$200{,}000) + (0.4)(-\$180{,}000) = \$48{,}000$
2. $\text{EMV}(A_2) = (0.6)(\$100{,}000) + (0.4)(-\$20{,}000) = \$52{,}000$
3. $\text{EMV}(A_3) = (0.6)(\$0) + (0.4)(\$0) = \$0$

TABLE A.3 Decision Table for Getz Products

	STATES OF NATURE	
ALTERNATIVES	FAVORABLE MARKET	UNFAVORABLE MARKET
Construct large plant (A_1)	$200,000	–$180,000
Construct small plant (A_2)	$100,000	–$ 20,000
Do nothing (A_3)	$ 0	$ 0
Probabilities	0.6	0.4

INSIGHT ▶ The maximum EMV is seen in alternative A_2. Thus, according to the EMV decision criterion, Getz would build the small facility.

LEARNING EXERCISE ▶ What happens to the three EMVs if Getz increases the conditional value on the "large plant/favorable market" result to $250,000? [Answer: EMV $(A_1) = \$78,000$. A_1 is now the preferable decision.]

RELATED PROBLEMS ▶ A.2e, A.3a, A.5a, A.7a, A.8a, A.9a, A.10, A.11, A.12, A.13a, A.14, A.16a, A.17, A.18–A.20, A.22b,c, A.31c,d

EXCEL OM Data File **ModaexA4.xls** can be found online.

Decision Making Under Certainty

Now suppose that the Getz operations manager has been approached by a marketing research firm that proposes to help him make the decision about whether to build the plant to produce storage sheds. The marketing researchers claim that their technical analysis will tell Getz with certainty whether the market is favorable for the proposed product. In other words, it will change Getz's environment from one of decision making *under risk* to one of decision making *under certainty*. This information could prevent Getz from making a very expensive mistake. The marketing research firm would charge Getz $65,000 for the information. What would you recommend? Should the operations manager hire the firm to make the study? Even if the information from the study is perfectly accurate, is it worth $65,000? What might it be worth? Although some of these questions are difficult to answer, determining the value of such *perfect information* can be very useful. It places an upper bound on what you would be willing to spend on information, such as that being sold by a marketing consultant. This is the concept of the expected value of perfect information (EVPI), which we now introduce.

LO A.5 *Compute* the expected value of perfect information (EVPI)

Expected value of perfect information (EVPI)
The difference between the payoff under perfect information and the payoff under risk.

Expected Value of Perfect Information (EVPI)

If managers were able to determine which state of nature would occur, then they would know which decision to make. Once managers know which decision to make, the payoff increases because the payoff is now a certainty, not a probability. Because the payoff will increase with knowledge of which state of nature will occur, this knowledge has value. Therefore, we now look at how to determine the value of this information. We call this difference between the payoff under perfect information and the payoff under risk the **expected value of perfect information (EVPI)**.

$$\text{EVPI} = \text{Expected value with perfect information} - \text{Maximum EMV}$$

To find the EVPI, we must first compute the **expected value *with* perfect information (EVwPI)**, which is the expected (average) return if we have perfect information before a decision has to be made. To calculate this value, we choose the best alternative for each state of nature and multiply its payoff times the probability of occurrence of that state of nature:

Expected value *with* perfect information (EVwPI)
The expected (average) return if perfect information is available.

$$\begin{aligned}
&\text{Expected value } \textit{with} \\
&\text{perfect information (EVwPI)} = (\text{Best outcome or consequence for 1st state of nature}) \\
&\qquad \times (\text{Probability of 1st state of nature}) \\
&\qquad + (\text{Best outcome for 2nd state of nature}) \\
&\qquad \times (\text{Probability of 2nd state of nature}) \\
&\qquad + \ldots + (\text{Best outcome for last state of nature}) \\
&\qquad \times (\text{Probability of last state of nature})
\end{aligned}$$

In Example A5 we use the data and decision table from Example A4 to examine the expected value of perfect information.

Example A5

EXPECTED VALUE OF PERFECT INFORMATION

The Getz operations manager would like to calculate the maximum that he would pay for information—that is, the expected value of perfect information, or EVPI.

APPROACH ▶ Referring to Table A.3 in Example 4, he follows a two-stage process. First, the expected value *with* perfect information (EVwPI) is computed. Then, using this information, the EVPI is calculated.

SOLUTION ▶

1. The best outcome for the state of nature "favorable market" is "build a large facility" with a payoff of \$200,000. The best outcome for the state of nature "unfavorable market" is "do nothing" with a payoff of \$0. Expected value *with* perfect information $= (\$200,000)(0.6) + (\$0)(0.4) = \$120,000$. Thus, if we had perfect information, we would expect (on the average) \$120,000 if the decision could be repeated many times.
2. The maximum EMV is \$52,000 for A_2, which is the expected outcome without perfect information. Thus:

$$\begin{aligned}
\text{EVPI} &= \text{EVwPI} - \text{Maximum EMV} \\
&= \$120,000 - \$52,000 = \$68,000
\end{aligned}$$

INSIGHT ▶ The *most* Getz should be willing to pay for perfect information is \$68,000. This conclusion, of course, is again based on the assumption that the probability of the first state of nature is 0.6 and the second is 0.4.

LEARNING EXERCISE ▶ How does the EVPI change if the "large plant/favorable market" conditional value is \$250,000? [Answer: EVPI $= \$72,000$.]

RELATED PROBLEMS ▶ A.3b, A.5b, A.7, A.9, A.13, A.16b, A.22c

STUDENT TIP
EVPI places an upper limit on what you should pay for information.

Decision Trees

STUDENT TIP Decision trees can become complex, so we illustrate two of them in this section.

Decisions that lend themselves to display in a decision table also lend themselves to display in a decision tree. We will therefore analyze some decisions using decision trees. Although the use of a decision table is convenient in problems having one set of decisions and one set of states of nature, many problems include *sequential* decisions and states of nature.

Decision tree
A graphical means of analyzing decision alternatives and states of nature.

When there are two or more sequential decisions, and later decisions are based on the outcome of prior ones, the decision tree approach becomes appropriate. A **decision tree** is a graphic display of the decision process that indicates decision alternatives, states of nature and their respective probabilities, and payoffs for each combination of decision alternative and state of nature.

Expected monetary value (EMV) is the most commonly used criterion for decision tree analysis. One of the first steps in such analysis is to graph the decision tree and to specify the monetary consequences of all outcomes for a particular problem.

Analyzing problems with *decision trees* involves five steps:

1. Define the problem.
2. Structure or draw the decision tree.
3. Assign probabilities to the states of nature.
4. Estimate payoffs for each possible combination of decision alternatives and states of nature.
5. Solve the problem by computing the expected monetary values (EMV) for each state-of-nature node. This is done by working *backward*—that is, by starting at the right of the tree and working back to decision nodes on the left.

Example A6

SOLVING A TREE FOR EMV

Getz wants to develop a completed and solved decision tree.

LO A.6 *Evaluate* the nodes in a decision tree

APPROACH ▶ The payoffs are placed at the right-hand side of each of the tree's branches (see Figure A.2). The probabilities (first used by Getz in Example A4) are placed in parentheses next to each state of nature. The expected monetary values for each state-of-nature node are then calculated and placed by their respective nodes. The EMV of the first node is $48,000. This represents the branch from the decision node to "construct a large plant." The EMV for node 2, to "construct a small plant," is $52,000. The option of "doing nothing" has, of course, a payoff of $0.

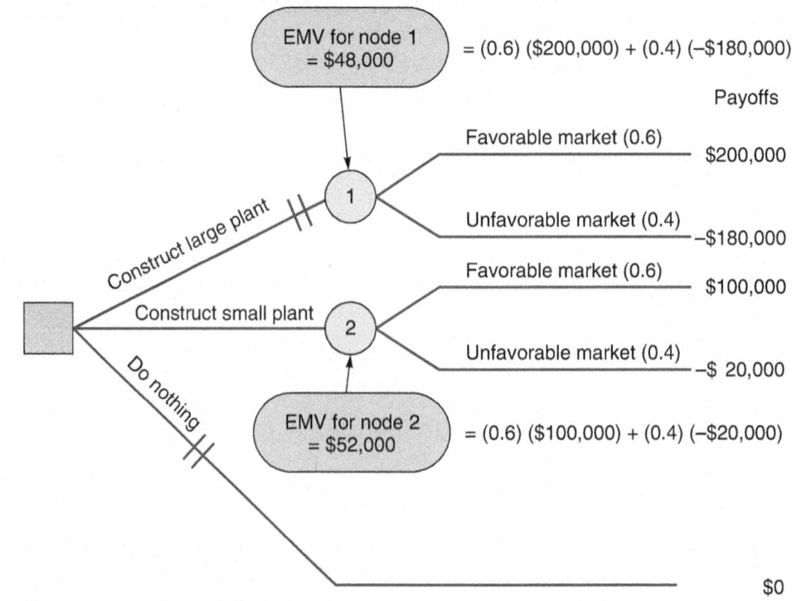

Figure **A.2**
Completed and Solved Decision Tree for Getz Products

SOLUTION ▶ The branch leaving the decision node leading to the state-of-nature node with the highest EMV will be chosen. In Getz's case, a small plant should be built.

INSIGHT ▶ This graphical approach is an excellent way for managers to understand all the options in making a major decision. Visual models are often preferred over tables.

LEARNING EXERCISE ▶ Correct Figure A.2 to reflect a $250,000 payoff for "construct large plant/favorable market." [Answer: Change one payoff and recompute the EMV for node 1.]

RELATED PROBLEMS ▶ A.2e, A.8b, A.22a,b, A.23, A.24, A.30, A.31

EXCEL OM Data File **MODAEXA6.XLS** can be found online.

A More Complex Decision Tree

When a *sequence* of decisions must be made, decision trees are much more powerful tools than are decision tables. Let's say that Getz Products has two decisions to make, with the second decision dependent on the outcome of the first. Before deciding about building a new plant, Getz has the option of conducting its own marketing research survey, at a cost of $10,000. The information from this survey could help it decide whether to build a large plant, to build a small plant, or not to build at all. Getz recognizes that although such a survey will not provide it with *perfect* information, it may be extremely helpful.

Getz's new decision tree is represented in Figure A.3 of Example A7. Take a careful look at this more complex tree. Note that *all possible outcomes and alternatives* are included in their logical sequence. This procedure is one of the strengths of using decision trees. Managers are forced to examine all possible outcomes, including unfavorable ones. They are also forced to make decisions in a logical, sequential manner.

Example A7

A DECISION TREE WITH SEQUENTIAL DECISIONS

Getz Products wishes to develop the new tree for this sequential decision.

APPROACH ▶ Examining the tree in Figure A.3, we see that Getz's first decision point is whether to conduct the $10,000 market survey. If it chooses not to do the study (the lower part of the tree), it can either build a large plant, a small plant, or no plant. This is Getz's second decision point. If the decision is to build, the market will be either favorable (0.6 probability) or unfavorable (0.4 probability). The payoffs for each of the possible consequences are listed along the right-hand side. As a matter of fact, this lower portion of Getz's tree is *identical* to the simpler decision tree shown in Figure A.2.

SOLUTION ▶ The upper part of Figure A.3 reflects the decision to conduct the market survey. State-of-nature node number 1 has 2 branches coming out of it. Let us say there is a 45% chance that the survey results will indicate a favorable market for the storage sheds. We also note that the probability is 0.55 that the survey results will be negative.

The rest of the probabilities shown in parentheses in Figure A.3 are all *conditional* probabilities. For example, 0.78 is the probability of a favorable market for the sheds given a favorable result from the market survey. Of course, you would expect to find a high probability of a favorable market given that the research indicated that the market was good. Don't forget, though: There is a chance that Getz's $10,000 market survey did not result in perfect or even reliable information. Any market research study is subject to error. In this case, there remains a 22% chance that the market for sheds will be unfavorable given positive survey results.

Likewise, we note that there is a 27% chance that the market for sheds will be favorable given negative survey results. The probability is much higher, 0.73, that the market will actually be unfavorable given a negative survey.

Finally, when we look to the payoff column in Figure A.3, we see that $10,000—the cost of the marketing study—has been subtracted from each of the top 10 tree branches. Thus, a large plant constructed in a favorable market would normally net a $200,000 profit. Yet because the market study was conducted, this figure is reduced by $10,000. In the unfavorable case, the loss of $180,000 would increase to $190,000. Similarly, conducting the survey and building *no plant* now results in a –$10,000 payoff.

Figure A.3
Getz Products Decision Tree with Probabilities and EMVs Shown

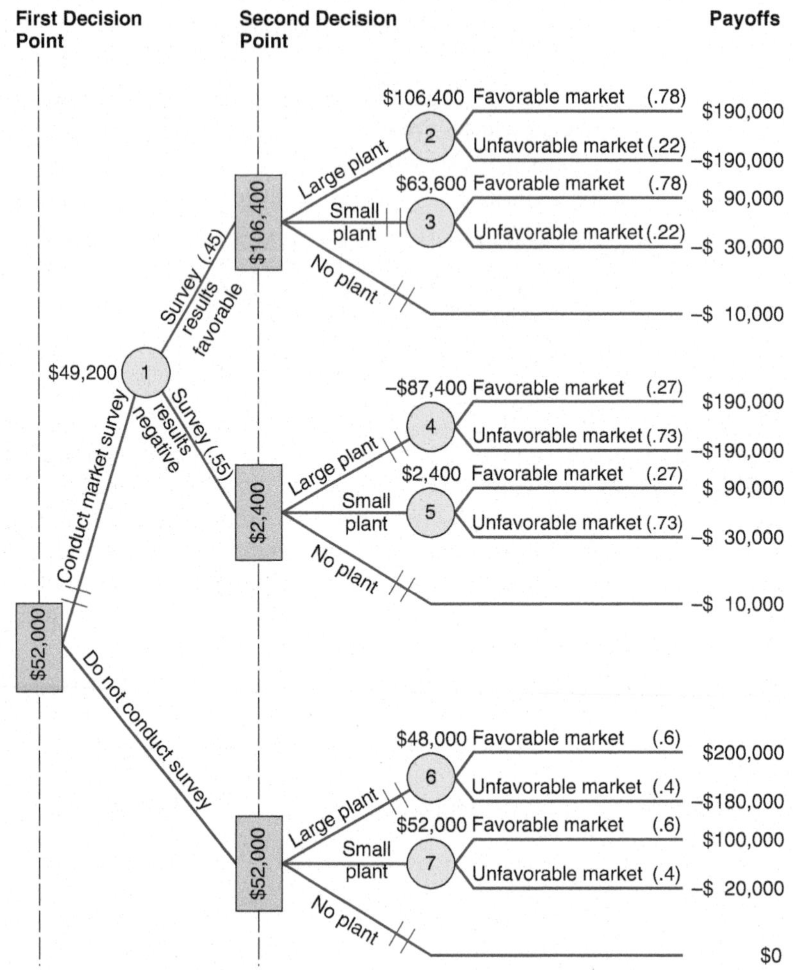

STUDENT TIP
The short parallel lines mean "prune" that branch, as it is less favorable than another available option and may be dropped.

LO A.7 *Create* a decision tree with sequential decisions

With all probabilities and payoffs specified, we can start calculating the expected monetary value of each branch. We begin at the end or right-hand side of the decision tree and work back toward the origin. When we finish, the best decision will be known.

1. Given favorable survey results:

$$\text{EMV(node 2)} = (.78)(\$190{,}000) + (.22)(-\$190{,}000) = \$106{,}400$$
$$\text{EMV(node 3)} = (.78)(\$90{,}000) + (.22)(-\$30{,}000) = \$63{,}600$$

   The EMV of no plant in this case is –$10,000. Thus, if the survey results are favorable, a large plant should be built.

2. Given negative survey results:

$$\text{EMV(node 4)} = (.27)(\$190{,}000) + (.73)(-\$190{,}000) = -\$87{,}400$$
$$\text{EMV(node 5)} = (.27)(\$90{,}000) + (.73)(-\$30{,}000) = \$2{,}400$$

   The EMV of no plant is again –$10,000 for this branch. Thus, given a negative survey result, Getz should build a small plant with an expected value of $2,400.

3. Continuing on the upper part of the tree and moving backward, we compute the expected value of conducting the market survey:

$$\text{EMV(node 1)} = (.45)(\$106{,}400) + (.55)(\$2{,}400) = \$49{,}200$$

4. If the market survey is *not* conducted:

$$\text{EMV(node 6)} = (.6)(\$200{,}000) + (.4)(-\$180{,}000) = \$48{,}000$$
$$\text{EMV(node 7)} = (.6)(\$100{,}000) + (.4)(-\$20{,}000) = \$52{,}000$$

   The EMV of no plant is $0. Thus, building a small plant is the best choice, given the marketing research is not performed.

5. Because the expected monetary value of not conducting the survey is \$52,000—vs. an EMV of \$49,200 for conducting the study—the best choice is to *not seek marketing information*. Getz should build the small plant.

INSIGHT ▶ You can reduce complexity in a large decision tree by viewing and solving a number of smaller trees—start at the end branches of a large one. Take one decision at a time.

LEARNING EXERCISE ▶ Getz estimates that if he conducts a market survey, there is really only a 35% chance the results will indicate a favorable market for the sheds. How does the tree change? [Answer: The EMV of conducting the survey = \$38,800, so Getz should still not do it.]

RELATED PROBLEMS ▶ A.21, A.25–A.29

Summary

This module examines two of the most widely used decision techniques—decision tables and decision trees. These techniques are especially useful for making decisions under risk. Many decisions in research and development, plant and equipment, and even new buildings and structures can be analyzed with these decision models. Problems in inventory control, aggregate planning, maintenance, scheduling, and production control also lend themselves to decision table and decision tree applications.

Key Terms

Decision table (p. 693)
Maximax (p. 694)
Maximin (p. 694)
Equally likely (p. 694)
Expected monetary value (EMV) (p. 695)
Expected value of perfect information (EVPI) (p. 696)
Expected value *with* perfect information (EVwPI) (p. 697)
Decision tree (p. 698)

Discussion Questions

1. Identify the six steps in the decision process.
2. Give an example of a good decision you made that resulted in a bad outcome. Also give an example of a bad decision you made that had a good outcome. Why was each decision good or bad?
3. What is the *equally likely* decision model?
4. Discuss the differences between decision making under certainty, under risk, and under uncertainty.
5. What is a decision tree?
6. Explain how decision trees might be used in several of the 10 OM decisions.
7. What is the expected value of perfect information (EVPI)?
8. What is the expected value *with* perfect information (EVwPI)?
9. Identify the five steps in analyzing a problem using a decision tree.
10. Why are the maximax and maximin strategies considered to be optimistic and pessimistic, respectively?
11. The expected value criterion is considered to be the rational criterion on which to base a decision. Is this true? Is it rational to consider risk?
12. When are decision trees most useful?

Using Software for Decision Models

Analyzing decision tables is straightforward with Excel, Excel OM, and POM for Windows. When decision trees are involved, Excel OM or commercial packages such as DPL, Tree Plan, Precision Tree, and Supertree provide flexibility, power, and ease. POM for Windows will also analyze trees but does not have graphic capabilities.

CREATING AN EXCEL SPREADSHEET TO EVALUATE A DECISION TABLE

In Program A.1, we illustrate how you can build your own Excel spreadsheet to analyze decision making under uncertainty and under risk. The data from Getz Products in Examples A3 and A4 are used. Maximax, maximin, equally likely, and EMV are computed, along with EVPI.

P USING POM FOR WINDOWS

POM for Windows can be used to calculate all of the information described in the decision tables and decision trees in this module. For details on how to use this software, please refer to Appendix II.

Program **A.1**

An Excel Spreadsheet for Analyzing Data in Examples A3 and A4 for Getz Products

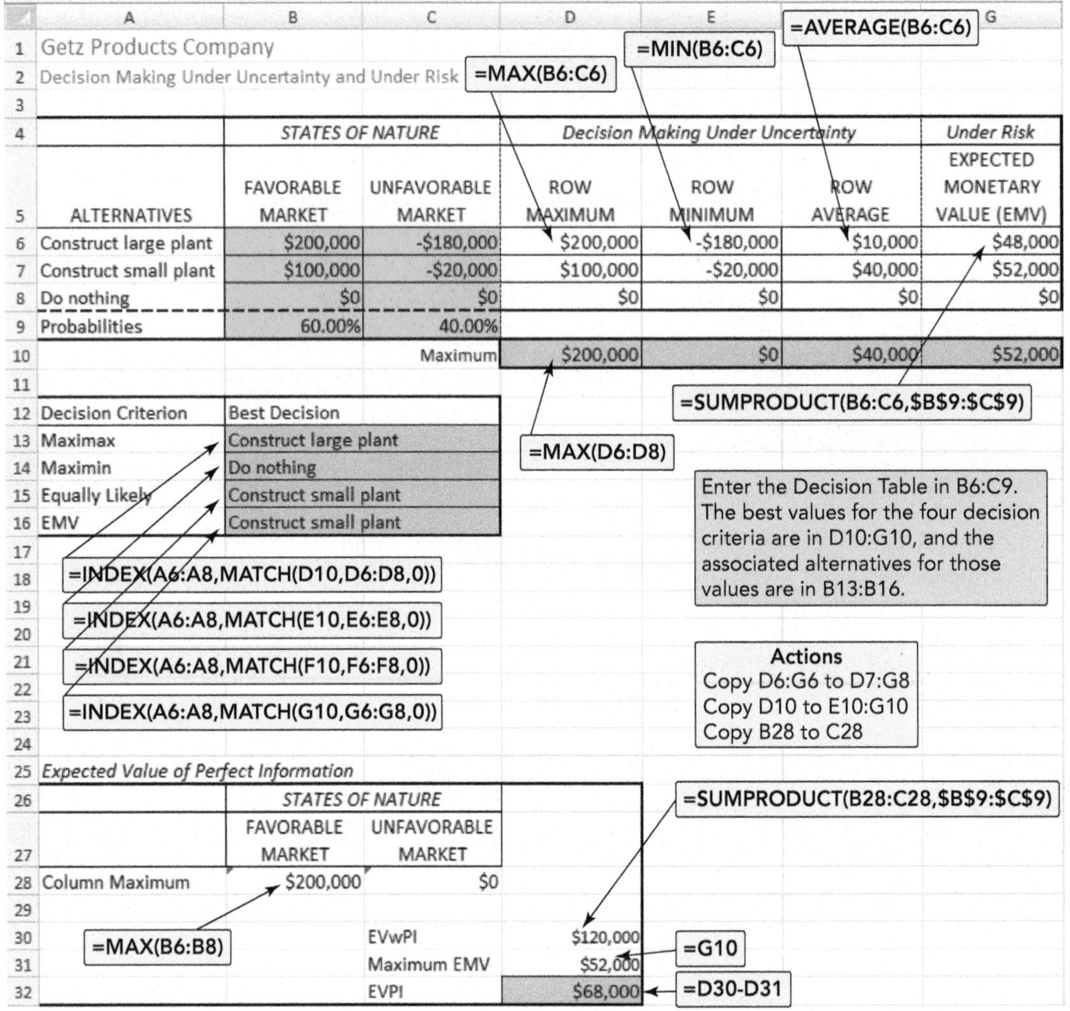

Getz Products Company
Decision Making Under Uncertainty and Under Risk
=MAX(B6:C6)
=MIN(B6:C6)
=AVERAGE(B6:C6)
STATES OF NATURE
Decision Making Under Uncertainty
Under Risk
ALTERNATIVES
FAVORABLE MARKET
UNFAVORABLE MARKET
ROW MAXIMUM
ROW MINIMUM
ROW AVERAGE
EXPECTED MONETARY VALUE (EMV)
Construct large plant $200,000 -$180,000 $200,000 -$180,000 $10,000 $48,000
Construct small plant $100,000 -$20,000 $100,000 -$20,000 $40,000 $52,000
Do nothing $0 $0 $0 $0 $0 $0
Probabilities 60.00% 40.00%
Maximum $200,000 $0 $40,000 $52,000
=SUMPRODUCT(B6:C6,$B$9:$C$9)
Decision Criterion
Best Decision
Maximax Construct large plant
Maximin Do nothing
Equally Likely Construct small plant
EMV Construct small plant
=MAX(D6:D8)
Enter the Decision Table in B6:C9. The best values for the four decision criteria are in D10:G10, and the associated alternatives for those values are in B13:B16.
=INDEX(A6:A8,MATCH(D10,D6:D8,0))
=INDEX(A6:A8,MATCH(E10,E6:E8,0))
=INDEX(A6:A8,MATCH(F10,F6:F8,0))
=INDEX(A6:A8,MATCH(G10,G6:G8,0))
Actions
Copy D6:G6 to D7:G8
Copy D10 to E10:G10
Copy B28 to C28
Expected Value of Perfect Information
STATES OF NATURE
FAVORABLE MARKET
UNFAVORABLE MARKET
=SUMPRODUCT(B28:C28,$B$9:$C$9)
Column Maximum $200,000 $0
=MAX(B6:B8)
EVwPI $120,000
Maximum EMV $52,000
EVPI $68,000
=G10
=D30-D31

USING EXCEL OM

Excel OM allows decision makers to evaluate decisions quickly and to perform sensitivity analysis on the results. Program A.2 uses Excel OM to create the decision tree for Getz Products shown previously in Example A6. The tool to create the tree is seen in the window on the right.

Program **A.2**

Getz Products' Decision Tree Using Excel OM

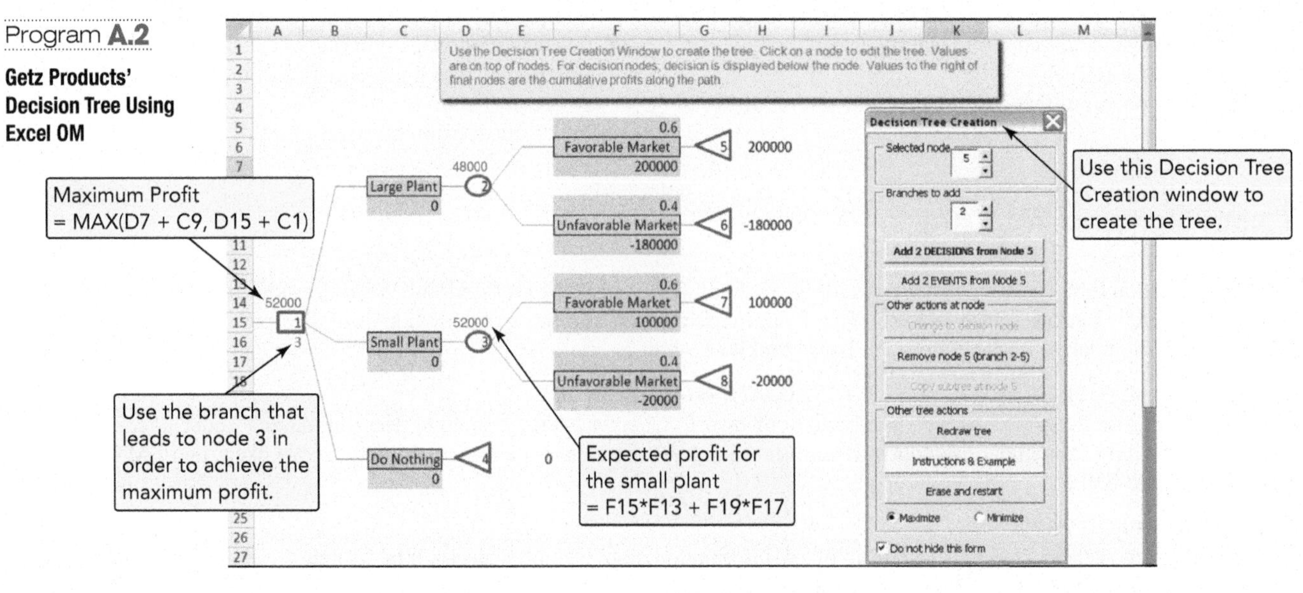

Maximum Profit = MAX(D7 + C9, D15 + C1)

Use the branch that leads to node 3 in order to achieve the maximum profit.
Expected profit for the small plant = F15*F13 + F19*F17

Instructions & Example

Solved Problems

SOLVED PROBLEM A.1

Stella Yan Hua is considering the possibility of opening a small dress shop on Fairbanks Avenue, a few blocks from the university. She has located a good mall that attracts students. Her options are to open a small shop, a medium-sized shop, or no shop at all. The market for a dress shop can be good, average, or bad. The probabilities for these three possibilities are .2 for a good market, .5 for an average market, and .3 for a bad market. The net profit or loss for the medium-sized or small shops for the various market conditions are given in the adjacent table. Building no shop at all yields no loss and no gain. What do you recommend?

	STATES OF NATURE		
ALTERNATIVES	GOOD MARKET	AVERAGE MARKET	BAD MARKET
Small shop	$ 75,000	$25,000	−$40,000
Medium-sized shop	$100,000	$35,000	−$60,000
No shop	$ 0	$ 0	$ 0
Probabilities	.20	.50	.30

SOLUTION

The problem can be solved by computing the expected monetary value (EMV) for each alternative:

$$\text{EMV (Small shop)} = (.2)(\$75{,}000) + (.5)(\$25{,}000) + (.3)(-\$40{,}000) = \$15{,}500$$
$$\text{EMV (Medium-sized shop)} = (.2)(\$100{,}000) + (.5)(\$35{,}000) + (.3)(-\$60{,}000) = \$19{,}500$$
$$\text{EMV (No shop)} = (.2)(\$0) + (.5)(\$0) + (.3)(\$0) = \$0$$

As you can see, the best decision is to build the medium-sized shop. The EMV for this alternative is $19,500.

SOLVED PROBLEM A.2

T.S. Amer's Ski Shop in Nevada has a 100-day season. T.S. has established the probability of various store traffic, based on historical records of skiing conditions, as indicated in the table to the right. T.S. has four merchandising plans, each focusing on a popular name brand. Each plan yields a daily net profit as noted in the table. He also has a meteorologist friend who, for a small fee, will accurately tell tomorrow's weather so T.S. can implement one of his four merchandising plans.

a) What is the expected monetary value (EMV) under risk?
b) What is the expected value *with* perfect information (EVwPI)?
c) What is the expected value of perfect information (EVPI)?

	TRAFFIC IN STORE BECAUSE OF SKI CONDITIONS (STATES OF NATURE)			
DECISION ALTERNATIVES (MERCHANDISING PLAN FOCUSING ON:)	1	2	3	4
Patagonia	$40	$92	$20	$48
North Face	50	84	10	52
Cloud Veil	35	80	40	64
Columbia	45	72	10	60
Probabilities	.20	.25	.30	.25

SOLUTION

a) The highest expected monetary value under risk is:

$$\text{EMV (Patagonia)} = .20(40) + .25(92) + .30(20) + .25(48) = \$49$$
$$\text{EMV (North Face)} = .20(50) + .25(84) + .30(10) + .25(52) = \$47$$
$$\text{EMV (Cloud Veil)} = .20(35) + .25(80) + .30(40) + .25(64) = \$55$$
$$\text{EMV (Columbia)} = .20(45) + .25(72) + .30(10) + .25(60) = \$45$$

So the maximum EMV = $55

b) The expected value *with* perfect information is:

$$\text{EVwPI} = .20(50) + .25(92) + .30(40) + .25(64)$$
$$= 10 + 23 + 12 + 16 = \$61$$

c) The expected value of perfect information is:

$$\text{EVPI} = \text{EVwPI} - \text{Maximum EMV} = 61 - 55 = \$6$$

SOLVED PROBLEM A.3

Daily demand for cases of Tidy Bowl cleaner at Ravinder Nath's Supermarket has always been 5, 6, or 7 cases. Develop a decision tree that illustrates her decision alternatives as to whether to stock 5, 6, or 7 cases.

SOLUTION

The decision tree is shown in Figure A.4.

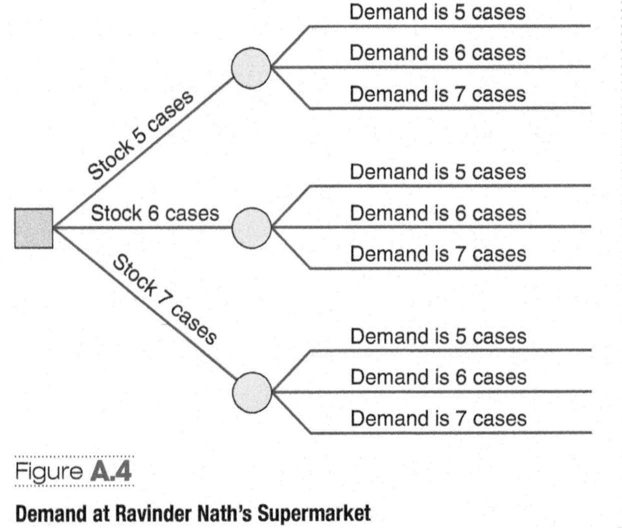

Figure A.4

Demand at Ravinder Nath's Supermarket

Problems

Note: **PX** means the problem may be solved with POM for Windows and/or Excel OM.

Problems A.1–A.20 relate to Types of Decision-Making Environments

• **A.1** Given the following conditional value table, determine the appropriate decision under uncertainty using:

a) Maximax
b) Maximin
c) Equally likely **PX**

	STATES OF NATURE		
ALTERNATIVES	VERY FAVORABLE MARKET	AVERAGE MARKET	UNFAVORABLE MARKET
Build new plant	$350,000	$240,000	–$300,000
Subcontract	$180,000	$ 90,000	–$ 20,000
Overtime	$110,000	$ 60,000	–$ 10,000
Do nothing	$ 0	$ 0	$ 0

••• **A.2** Even though independent gasoline stations have been having a difficult time, Ian Langella has been thinking about starting his own independent gasoline station. Ian's problem is to decide how large his station should be. The annual returns will depend on both the size of his station and a number of marketing factors related to the oil industry and demand for gasoline. After a careful analysis, Ian developed the following table:

SIZE OF FIRST STATION	GOOD MARKET	FAIR MARKET	POOR MARKET
Small	$ 50,000	$20,000	–$ 10,000
Medium	$ 80,000	$30,000	–$ 20,000
Large	$100,000	$30,000	–$ 40,000
Very large	$300,000	$25,000	–$160,000

For example, if Ian constructs a small station and the market is good, he will realize a profit of $50,000.

a) Develop a decision table for this decision, like the one illustrated in Table A.2.
b) What is the maximax decision?
c) What is the maximin decision?
d) What is the equally likely decision?
e) Develop a decision tree. Assume each outcome is equally likely, then find the highest EMV. **PX**

• **A.3** Andrea Dawson, a sandwich vendor at her town's annual Hard Rock Festival, created a table of conditional values for the various alternatives (stocking decision) and states of nature (size of crowd):

	STATES OF NATURE (DEMAND)		
ALTERNATIVES	BIG	AVERAGE	SMALL
Large stock	$22,000	$12,000	–$2,000
Average stock	$14,000	$10,000	$6,000
Small stock	$ 9,000	$ 8,000	$4,000

The probabilities associated with the states of nature are 0.3 for a big demand, 0.5 for an average demand, and 0.2 for a small demand.

a) Determine the alternative that provides Andrea the greatest expected monetary value (EMV).
b) Compute the expected value of perfect information (EVPI).

•• **A.4** Maria Sanchez owns a health and fitness center called Bulk-Up in Harrisburg. She is considering adding more floor space to meet increasing demand. She will add either no floor space (N), a moderate area of floor space (M), a large area of floor space (L), or an area of floor space that doubles the size of the facility (D). Demand will either stay fixed, increase slightly, or increase greatly. The following are the changes in Bulk-Up's annual profits under each combination of expansion level and demand change level:

	EXPANSION LEVEL			
DEMAND CHANGE	N	M	L	D
Fixed	$ 0	–$4,000	–$10,000	–$50,000
Slight increase	$2,000	$8,000	$ 6,000	$ 4,000
Major increase	$3,000	$9,000	$20,000	$40,000

Maria is risk averse and wishes to use the maximin criterion.

a) What are her decision alternatives and what are the states of nature?
b) What should she do? **PX**

• **A.5** Howard Weiss, Inc., is considering building a sensitive new radiation scanning device. His managers believe that there is a probability of 0.4 that the ATR Co. will come out with a competitive product. If Weiss adds an assembly line for the product and ATR Co. does not follow with a competitive product, Weiss's expected profit is $40,000; if Weiss adds an assembly line and ATR follows suit, Weiss still expects $10,000 profit. If Weiss adds a new plant addition and ATR does not produce a competitive product, Weiss expects a profit of $600,000; if ATR does compete for this market, Weiss expects a loss of $100,000.

a) Determine the EMV of each decision.
b) Compute the expected value of perfect information. **PX**

••• **A.6** Jerry Bildery's factory is considering three approaches for meeting an expected increase in demand. These three approaches are increasing capacity, using overtime, and buying more equipment. Demand will increase either slightly (S), moderately (M), or greatly (G). The profits for each approach under each possible scenario are as follows:

	DEMAND SCENARIO		
APPROACH	S	M	G
Increasing capacity	$700,000	$700,000	$ 700,000
Using overtime	$500,000	$600,000	$1,000,000
Buying equipment	$600,000	$800,000	$ 800,000

Because the goal is to maximize, and Jerry is risk-neutral, he decides to use the *equally likely* decision criterion to make the decision as to which approach to use. According to this criterion, which approach should be used?

• **A.7** The following payoff table provides profits based on various possible decision alternatives and various levels of demand at Robert Klassan's print shop:

	DEMAND	
	LOW	HIGH
Alternative 1	$10,000	$30,000
Alternative 2	$ 5,000	$40,000
Alternative 3	–$ 2,000	$50,000

The probability of low demand is 0.4, whereas the probability of high demand is 0.6.

a) What is the highest possible expected monetary value?
b) What is the expected value *with* perfect information (EVwPI)?
c) Calculate the expected value of perfect information for this situation. **PX**

• **A.8** Leah Johnson, director of Urgent Care of Brookline, wants to increase capacity to provide low-cost flu shots but must decide whether to do so by hiring another full-time nurse or by using part-time nurses. The following table shows the expected *costs* of the two options for three possible demand levels:

	STATES OF NATURE		
ALTERNATIVES	LOW DEMAND	MEDIUM DEMAND	HIGH DEMAND
Hire full-time	$300	$500	$ 700
Hire part-time	$ 0	$350	$1,000
Probability	.2	.5	.3

a) Using expected value, what should Ms. Johnson do?
b) Draw an appropriate decision tree showing payoffs and probabilities. **PX**

•• **A.9** Zhu Manufacturing is considering the introduction of a family of new products. Long-term demand for the product group is somewhat predictable, so the manufacturer must be concerned with the risk of choosing a process that is inappropriate. Faye Zhu is VP of operations. She can choose among batch manufacturing or custom manufacturing, or she can invest in group technology. Faye won't be able to forecast demand accurately until after she makes the process choice. Demand will be classified into four compartments: poor, fair, good, and excellent. The following table indicates the payoffs (profits) associated with each process/demand combination, as well as the probabilities of each long-term demand level:

	POOR	FAIR	GOOD	EXCELLENT
Probability	.1	.4	.3	.2
Batch	–$ 200,000	$1,000,000	$1,200,000	$1,300,000
Custom	$ 100,000	$ 300,000	$ 700,000	$ 800,000
Group technology	–$1,000,000	–$ 500,000	$ 500,000	$2,000,000

a) Based on expected value, what choice offers the greatest gain?
b) What would Faye Zhu be willing to pay for a forecast that would accurately determine the level of demand in the future? **PX**

•• **A.10** Consider the following decision table, which Joe Blackburn has developed for Vanderbilt Enterprises:

	STATES OF NATURE		
DECISION ALTERNATIVES	LOW	MEDIUM	HIGH
A	$40	$100	$60
B	$85	$ 60	$70
C	$60	$ 70	$70
D	$65	$ 75	$70
E	$70	$ 65	$80
Probability	.40	.20	.40

Which decision alternative maximizes the expected value of the payoff? **PX**

•• **A.11** The University of Miami bookstore stocks textbooks in preparation for sales each semester. It normally relies on departmental forecasts and preregistration records to determine how many copies of a text are needed. Preregistration shows 90 operations management students enrolled, but bookstore manager Vaidy Jayaraman has second thoughts, based on his intuition and some historical evidence. Vaidy believes that the distribution of sales may range from 70 to 90 units, according to the following probability model:

Demand	70	75	80	85	90
Probability	.15	.30	.30	.20	.05

This textbook costs the bookstore $82 and sells for $112. Any unsold copies can be returned to the publisher, less a restocking fee and shipping, for a net refund of $36.

a) Construct the table of conditional profits.
b) How many copies should the bookstore stock to achieve highest expected value? **PX**

•• **A.12** Palmer Jam Company is a small manufacturer of several different jam products. One product is an organic jam that has no preservatives, sold to retail outlets. Susan Palmer must decide how many cases of jam to manufacture each month. The probability that demand will be 6 cases is .1, for 7 cases it is .3, for 8 cases it is .5, and for 9 cases it is .1. The cost of every case is $45, and the price Susan gets for each case is $95. Unfortunately, any cases not sold by the end of the month are of no value as a result of spoilage. How many cases should Susan manufacture each month? **PX**

•• **A.13** Deborah Hollwager, a concessionaire for the Amway Center in Orlando, has developed a table of conditional values for the various alternatives (stocking decisions) and states of nature (size of crowd):

	STATES OF NATURE (SIZE OF CROWD)		
ALTERNATIVES	LARGE	AVERAGE	SMALL
Large inventory	$20,000	$10,000	–$2,000
Average inventory	$15,000	$12,000	$6,000
Small inventory	$ 9,000	$ 6,000	$5,000

If the probabilities associated with the states of nature are 0.3 for a large crowd, 0.5 for an average crowd, and 0.2 for a small crowd, determine:

a) The alternative that provides the greatest expected monetary value (EMV).
b) The expected value of perfect information (EVPI). **PX**

•••• **A.14** The city of Belgrade, Serbia, is contemplating building a second airport to relieve congestion at the main airport and is considering two potential sites, X and Y. Hard Rock Hotels would like to purchase land to build a hotel at the new airport. The value of land has been rising in anticipation and is expected to skyrocket once the city decides between sites X and Y. Consequently, Hard Rock would like to purchase land now. Hard Rock will sell the land if the city chooses not to locate the airport nearby. Hard Rock has four choices: (1) buy land at X, (2) buy land at Y, (3) buy land at both X and Y, or (4) do nothing. Hard Rock has collected the following data (which are in millions of euros):

	SITE X	SITE Y
Current purchase price	27	15
Profits if airport and hotel built at this site*	45	30
Sale price if airport not built at this site	9	6

*The second row of the table represents net operating profits from the hotel, not including the upfront cost of land.

Hard Rock determines there is a 45% chance the airport will be built at X (hence, a 55% chance it will be built at Y).

a) Set up the decision table.
b) What should Hard Rock decide to do to maximize total net profit? **PX**

• **A.15** Given the following conditional value table, determine the appropriate decision under uncertainty using

a) maximax
b) maximin
c) equally likely **PX**

	STATES OF NATURE		
ALTERNATIVES	VERY FAVORABLE MARKET	AVERAGE MARKET	UNFAVORABLE MARKET
Large plant	$275,000	$100,000	–$150,000
Small plant	$200,000	$ 60,000	–$ 10,000
Overtime	$100,000	$ 40,000	–$ 1,000
Do nothing	$ 0	$ 0	$ 0

•• **A.16** The following payoff table provides profits based on various possible decision alternatives and various levels of demand:

	DEMAND		
	LOW	MEDIUM	HIGH
Alternative 1	80	120	140
Alternative 2	90	90	90
Alternative 3	50	70	150

The probability of low demand is 0.4, whereas the probability of medium and high demand is each 0.3.

a) What is the highest possible expected monetary value?
b) Calculate the expected value of perfect information for this situation. **PX**

•• **A.17** Given the following conditional value table, determine the appropriate decision assuming that each state of nature has an equal likelihood of occurring: **PX**

	STATES OF NATURE		
ALTERNATIVES	VERY FAVORABLE MARKET	AVERAGE MARKET	UNFAVORABLE MARKET
Large plant	$275,000	$100,000	–$150,000
Small plant	$200,000	$ 60,000	–$ 10,000
Overtime	$100,000	$ 40,000	–$ 1,000
Do nothing	$ 0	$ 0	$ 0

Additional problems **A.18–A.20** *are available in* **MyLab Operations Management.** *Problem A.18 involves an inventory stocking problem for a product with a short shelf life. Problem A.19 is an EMV problem with two alternatives and three states of nature. Problem A.20 uses EMV to select the best of three technologies.*

Problems A.21–A.32 relate to Decision Trees

••• **A.21** Ronald Lau, chief engineer at South Dakota Electronics, has to decide whether to build a new state-of-the-art processing facility. If the new facility works, the company could realize a profit of $200,000. If it fails, South Dakota Electronics could lose $180,000. At this time, Lau estimates a 60% chance that the new process will fail.

The other option is to build a pilot plant and then decide whether to build a complete facility. The pilot plant would cost $10,000 to build. Lau estimates a 50–50 chance that the pilot plant will work. If the pilot plant works, there is a 90% probability that the complete plant, if it is built, will also work. If the pilot plant does not work, there is only a 20% chance that the complete project (if it is constructed) will work. Lau faces a dilemma. Should he build the plant? Should he build the pilot project and then make a decision? Help Lau by analyzing this problem. **PX**

•• **A.22** Dwayne Whitten, president of Whitten Industries, is considering whether to build a manufacturing plant in north Texas. His decision is summarized in the following table:

ALTERNATIVES	FAVORABLE MARKET	UNFAVORABLE MARKET
Build large plant	$400,000	–$300,000
Build small plant	$ 80,000	–$ 10,000
Don't build	$ 0	$ 0
Market probability	0.4	0.6

a) Construct a decision tree.
b) Determine the best strategy using expected monetary value (EMV).
c) What is the expected value of perfect information (EVPI)? **PX**

•• **A.23** Deborah Kellogg buys Breathalyzer test sets for the Winter Park Police Department. The quality of the test sets from her two suppliers is indicated in the following table:

PERCENT DEFECTIVE	PROBABILITY FOR WINTER PARK TECHNOLOGY	PROBABILITY FOR DAYTON ENTERPRISES
1	.70	.30
3	.20	.30
5	.10	.40

For example, the probability of getting a batch of tests that are 1% defective from Winter Park Technology is .70. Because Kellogg orders 10,000 tests per order, this would mean that there is a .70 probability of getting 100 defective tests out of the 10,000 tests if Winter Park Technology is used to fill the order. A defective Breathalyzer test set can be repaired for $0.50. Although the quality of the test sets of the second supplier, Dayton Enterprises, is lower, it will sell an order of 10,000 test sets for $37 less than Winter Park.

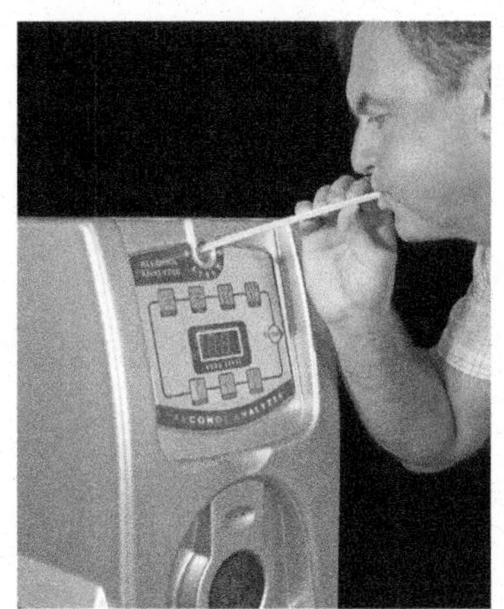
Jabiru/Shutterstock

a) Develop a decision tree.
b) Which supplier should Kellogg use? **PX**

• **A.24** Joseph Biggs owns his own ice cream truck and lives 30 miles from a Florida beach resort. The sale of his products is highly dependent on his location and on the weather. At the resort, his profit will be $120 per day in fair weather, $10 per day in bad weather. At home, his profit will be $70 in fair weather and $55 in bad weather. Assume that on any particular day, the weather service suggests a 40% chance of foul weather.
a) Construct Joseph's decision tree.
b) What decision is recommended by the expected value criterion? **PX**

•• **A.25** Jonatan Jelen is considering opening a bicycle shop in New York City. Jonatan enjoys biking, but this is to be a business endeavor from which he expects to make a living. He can open a small shop, a large shop, or no shop at all. Because there will be a 5-year lease on the building that Jonatan is thinking about using, he wants to make sure he makes the correct decision. Jonatan is also thinking about hiring his old marketing professor to conduct a marketing research study to see if there is a market for his services. The results of such a study could be either favorable or unfavorable. Develop a decision tree for Jonatan. **PX**

•• **A.26** Cheryl Drummond Retailers, Inc., must decide whether to build a small or a large facility at a new location in Fairfax. Demand at the location will either be low or high, with probabilities 0.4 and 0.6, respectively. If Cheryl builds a small facility and demand proves to be high, she then has the option of expanding the facility. If a small facility is built and demand proves to be high, and then the retailer expands the facility, the payoff is $270,000. If a small facility is built and demand proves to be high, but Cheryl then decides not to expand the facility, the payoff is $223,000.

If a small facility is built and demand proves to be low, then there is no option to expand and the payoff is $200,000. If a large facility is built and demand proves to be low, Cheryl then has the option of stimulating demand through local advertising. If she does not exercise this option, then the payoff is $40,000. If she does exercise the advertising option, then the response to advertising will either be modest or sizable, with probabilities of 0.3 and 0.7, respectively. If the response is modest, the payoff is $20,000. If it is sizable, the payoff is $220,000. Finally, if a large facility is built and demand proves to be high, then no advertising is needed and the payoff is $800,000.
a) What should Cheryl do to maximize her expected payoff?
b) What is the value of this expected payoff?

••• **A.27** Philip Musa can build either a large electronics section or a small one in his Birmingham drugstore. He can also gather additional information or simply do nothing. If he gathers additional information, the results could suggest either a favorable or an unfavorable market, but it would cost him $3,000 to gather the information. Musa believes that there is a 50–50 chance that the information will be favorable. If the market is favorable, Musa will earn $15,000 with a large section or $5,000 with a small one. With an unfavorable electronics market, however, Musa could lose $20,000 with a large section or $10,000 with a small section. Without gathering additional information, Musa estimates that the probability of a favorable market is .7. A favorable report from the study would increase the probability of a favorable market to .9. Furthermore, an unfavorable report from the additional information would decrease the probability of a favorable market to .4. Of course, Musa could ignore these numbers and do nothing. What is your advice to Musa?

Additional problem **A.28,** *a decision tree with five sequential decisions, is available in* MyLab Operations Management.

•••• **A.29** Louisiana is busy designing new lottery scratch-off games. In the latest game, Bayou Boondoggle, the player is instructed to scratch off one spot: A, B, or C. A can reveal "Loser," "Win $1," or "Win $50." B can reveal "Loser" or "Take a Second Chance." C can reveal "Loser" or "Win $500." On the second chance, the player is instructed to scratch off D or E. D can reveal "Loser" or "Win $1." E can reveal "Loser" or "Win $10." The probabilities at A are .9, .09, and .01. The probabilities at B are .8 and .2. The probabilities at C are .999 and .001. The probabilities at D are .5 and .5. Finally, the probabilities at E are .95 and .05. Draw the decision tree that represents this scenario. Use proper symbols and label all branches clearly. Calculate the expected value of this game.

•• **A.30** Bakery Products is considering the introduction of a new line of pastries. In order to produce the new line, the bakery is considering either a major or a minor renovation of its current plant. Bill Wicker, head of operations, has developed the following conditional values table:

ALTERNATIVES	FAVORABLE MARKET	UNFAVORABLE MARKET
Major renovation	$100,000	−$90,000
Minor renovation	$ 40,000	−$20,000
Do nothing	$ 0	$ 0

Under the assumption that the probability of a favorable market is equal to the probability of an unfavorable market, determine:

a) The appropriate decision tree showing payoffs and probabilities.
b) The best alternative using expected monetary value (EMV). **PX**

•• **A.31** Chris Suit is an administrator for Lowell Hospital. She is trying to determine whether to build a large wing on the existing hospital, a small wing, or no wing at all. If the population of Lowell continues to grow, a large wing could return $150,000 to the hospital each year. If a small wing were built, it would return $60,000 to the hospital each year if the population continues to grow. If the population of Lowell remains the same, the hospital would encounter a loss of $85,000 with a large wing and a loss of $45,000 with a small wing. Unfortunately, Suit does not have any information about the future population of Lowell.

a) Construct a decision tree.
b) Construct a decision table.
c) Assuming that each state of nature has the same likelihood, determine the best alternative.
d) If the likelihood of growth is .6 and that of remaining the same is .4 and the decision criterion is expected monetary value, which decision should Suit make? **PX**

CASE STUDY

Tom Thompson's Liver Transplant

Tom Thompson, a robust 50-year-old executive living in the northern suburbs of St. Paul, has been diagnosed by a University of Minnesota internist as having a decaying liver. Although he is otherwise healthy, Thompson's liver problem could prove fatal if left untreated.

Firm research data are not yet available to predict the likelihood of survival for a man of Thompson's age and condition without surgery. However, based on her own experience and recent medical journal articles, the internist tells him that if he elects to avoid surgical treatment of the liver problem, chances of survival will be approximately as follows: only a 60% chance of living 1 year, a 20% chance of surviving for 2 years, a 10% chance for 5 years, and a 10% chance of living to age 58. She places his probability of survival beyond age 58 without a liver transplant to be extremely low.

The transplant operation, however, is a serious surgical procedure. Five percent of patients die during the operation or its recovery stage, with an additional 45% dying during the first year. Twenty percent survive for 5 years, 13% survive for 10 years, and 8%, 5%, and 4% survive, respectively, for 15, 20, and 25 years.

Discussion Questions

1. Do you think that Thompson should select the transplant operation?
2. What other factors might be considered?

Endnote

1. To review these other statistical terms, refer to Tutorial 1, "Statistical Review for Managers," found online.

Bibliography

Balakrishnan, R., B. Render, R. M. Stair, and C. Munson. *Managerial Decision Modeling: Business Analytics with Spreadsheets*, 4th ed. Boston: DeGruyter, 2017.

Buchannan, L., and A. O'Connell. "A Brief History of Decision Making." *Harvard Business Review* 84, no. 1 (January, 2006): 32–41.

Hammond, J. S., R. L. Kenney, and H. Raiffa. "The Hidden Traps in Decision Making." *Harvard Business Review* 84, no. 1 (January 2006): 118–126.

Miller, C. C., and R. D. Ireland. "Intuition in Strategic Decision Making." *Academy of Management Executive* 19, no. 1 (February 2005): 19.

Parmigiani, G., and L. Inoue. *Decision Theory: Principles and Approaches*. New York: Wiley, 2010.

Peterson, M. *An Introduction to Decision Theory*, 2nd ed. Cambridge: Cambridge University Press, 2017.

Render, B., R. M. Stair Jr., M. Hanna, and T. Hale. *Quantitative Analysis for Management*, 13th ed. Boston: Pearson, 2018.

Module A *Rapid* Review

Main Heading	Review Material	MyLab Operations Management
THE DECISION PROCESS IN OPERATIONS	To achieve the goals of their organizations, managers must understand how decisions are made and know which decision-making tools to use. Overcoming uncertainty is a manager's mission. Decision tables and decision trees are used in a wide number of OM situations.	Concept Questions: 1.1–1.4
FUNDAMENTALS OF DECISION MAKING	*Alternative*—A course of action or strategy that may be chosen by a decision maker. *State of nature*—An occurrence or a situation over which a decision maker has little or no control. Symbols used in a decision tree: **1.** □—A decision node from which one of several alternatives may be selected. **2.** ○—A state-of-nature node out of which one state of nature will occur. When constructing a decision tree, we must be sure that all alternatives and states of nature are in their correct and logical places and that we include *all* possible alternatives and states of nature, usually including the "do nothing" option.	Concept Questions: 2.1–2.4
DECISION TABLES	■ **Decision table**—A tabular means of analyzing decision alternatives and states of nature. A decision table is sometimes called a *payoff table*. For any alternative and a particular state of nature, there is a *consequence*, or an *outcome*, which is usually expressed as a monetary value; this is called the *conditional value*.	Concept Questions: 3.1–3.5
TYPES OF DECISION-MAKING ENVIRONMENTS	There are three decision-making environments: (1) decision making under uncertainty, (2) decision making under risk, and (3) decision making under certainty. When there is complete *uncertainty* about which state of nature in a decision environment may occur (i.e., when we cannot even assess probabilities for each possible outcome), we rely on three decision methods: (1) maximax, (2) maximin, and (3) equally likely. ■ **Maximax**—A criterion that finds an alternative that maximizes the maximum outcome. ■ **Maximin**—A criterion that finds an alternative that maximizes the minimum outcome. ■ **Equally likely**—A criterion that assigns equal probability to each state of nature. Maximax is also called an "optimistic" decision criterion, while maximin is also called a "pessimistic" decision criterion. Maximax and maximin present best case/worst case planning scenarios. Decision making under risk relies on probabilities. The states of nature must be mutually exclusive and collectively exhaustive, and their probabilities must sum to 1. ■ **Expected monetary value (EMV)**—The expected payout or value of a variable that has different possible states of nature, each with an associated probability. The EMV represents the expected value or *mean* return for each alternative *if we could repeat this decision (or similar types of decisions) a large number of times.* The EMV for an alternative is the sum of all possible payoffs from the alternative, each weighted by the probability of that payoff occurring: $$\text{EMV (Alternative } i) = (\text{Payoff of 1st state of nature}) \times (\text{Probability of 1st of state of nature}) + (\text{Payoff of 2nd state of nature}) \times (\text{Probability of 2nd state of nature}) + \ldots + (\text{Payoff of last state of nature}) \times (\text{Probability of last state of nature})$$ ■ **Expected value of perfect information (EVPI)**—The difference between the payoff under perfect information and the payoff under risk. ■ **Expected value *with* perfect information (EVwPI)**—The expected (average) return if perfect information is available. EVPI represents an upper bound on what you would be willing to spend on state-of-nature information: $$\text{EVPI} = \text{EVwPI} - \text{Maximum EMV}$$ $$\text{EVwPI} = (\text{Best outcome for 1st state of nature}) \times (\text{Probability of 1st state of nature}) + (\text{Best outcome for 2nd state of nature}) \times (\text{Probability of 2nd state of nature}) + \ldots + (\text{Best outcome for last state of nature}) \times (\text{Probability of last state of nature})$$	Concept Questions: 4.1–4.6 Problems: A.1–A.20 Virtual Office Hours for Solved Problems: A.1, A.2

Module A *Rapid* Review *continued*

Main Heading	Review Material	MyLab Operations Management
DECISION TREES	When there are two or more sequential decisions, and later decisions are based on the outcome of prior ones, the decision tree (as opposed to decision table) approach becomes appropriate. ■ **Decision tree**—A graphical means of analyzing decision alternatives and states of nature. Analyzing problems with *decision trees* involves five steps: 1. Define the problem. 2. Structure or draw the decision tree. 3. Assign probabilities to the states of nature. 4. Estimate payoffs for each possible combination of decision alternatives and states of nature. 5. Solve the problem by computing the expected monetary values (EMV) for each state-of-nature node. This is done by working *backward*—that is, by starting at the right of the tree and working back to decision nodes on the left: $5,200; $5,200 (.4) $10,000; (.6) $2,000; $4,800 (.4) $6,000; (.6) $4,000; $4,400 (.4) $20,000; (.6) –$6,000 Decision trees force managers to examine all possible outcomes, including unfavorable ones. A manager is also forced to make decisions in a logical, sequential manner. Short parallel lines on a decision tree mean "prune" that branch, as it is less favorable than another available option and may be dropped.	Concept Questions: 5.1–5.5 Problems: A.21–A.31 Virtual Office Hours for Solved Problem: A.3
ADDITIONAL MYLAB OPERATIONS MANAGEMENT RESOURCES	✔ Video for Creating Your Own Excel Spreadsheets (Examples A3 and A4) ✔ Additional Case Studies (Arctic, Inc.; Ski Right Corp.; & Warehouse Tenting at the Port of Miami) ✔ Multiple Choice Case Questions (Tom Thompson's Liver Transplant) ✔ Additional Homework Problems (A.18–A.20, A.28)	

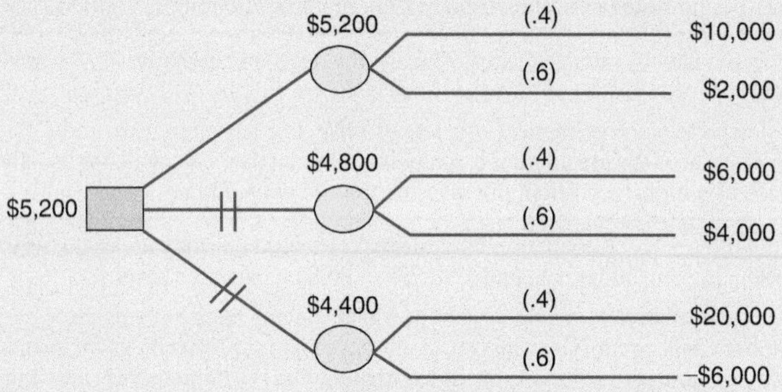

Self Test

LO A.1 On a decision tree, at each state-of-nature node:
a) the alternative with the greatest EMV is selected.
b) an EMV is calculated.
c) all probabilities are added together.
d) the branch with the highest probability is selected.

LO A.2 In decision table terminology, a course of action or a strategy that may be chosen by a decision maker is called a(n):
a) payoff. **b)** alternative.
c) state of nature. **d)** all of the above.

LO A.3 If probabilities are available to the decision maker, then the decision-making environment is called:
a) certainty. **b)** uncertainty.
c) risk. **d)** none of the above.

LO A.4 What is the EMV for Alternative 1 in the following decision table?

	STATE OF NATURE	
Alternative	S1	S2
A1	$15,000	$20,000
A2	$10,000	$30,000
Probability	0.30	0.70

a) $15,000 **b)** $17,000
c) $17,500 **d)** $18,500
e) $20,000

LO A.5 The most that a person should pay for perfect information is:
a) the EVPI.
b) the maximum EMV minus the minimum EMV.
c) the minimum EMV.
d) the maximum EMV.

LO A.6 On a decision tree, once the tree has been drawn and the payoffs and probabilities have been placed on the tree, the analysis (computing EMVs and selecting the best alternative):
a) is done by working backward (starting on the right and moving to the left).
b) is done by working forward (starting on the left and moving to the right).
c) is done by starting at the top of the tree and moving down.
d) is done by starting at the bottom of the tree and moving up.

LO A.7 A decision tree is preferable to a decision table when:
a) a number of sequential decisions are to be made.
b) probabilities are available.
c) the maximax criterion is used.
d) the objective is to maximize regret.

Answers: LO A.1. b; LO A.2. b; LO A.3. c; LO A.4. d; LO A.5. a; LO A.6. a; LO A.7. a.

Linear Programming

MODULE B

MODULE OUTLINE

LEARNING OBJECTIVES

LO B.1 ***Formulate*** linear programming models, including an objective function and constraints 713

LO B.2 ***Graphically*** solve an LP problem with the iso-profit line method 716

LO B.3 ***Graphically*** solve an LP problem with the corner-point method 717

LO B.4 ***Interpret*** sensitivity analysis and shadow prices 718

LO B.5 ***Construct*** and solve a minimization problem 721

LO B.6 ***Formulate*** production-mix, diet, and labor scheduling problems 721

LO B.7 ***Incorporate*** logical conditions into LPs using binary variables 726

The storm front closed in quickly on Boston's Logan Airport, shutting it down without warning. The heavy snowstorms and poor visibility sent airline passengers and ground crew scurrying. Because airlines use linear programming (LP) to schedule flights, hotels, crews, and refueling, LP has a direct impact on profitability. If an airline gets a major weather disruption at one of its hubs, a lot of flights may get canceled, which means a lot of crews and airplanes in the wrong places. LP is the tool that helps airlines unsnarl and cope with this weather mess.

Paul Italiano/Alamy Stock Photo

Why Use Linear Programming?

Many operations management decisions involve trying to make the most effective use of an organization's resources. Resources typically include machinery (such as planes, in the case of an airline), labor (such as pilots), money, time, and raw materials (such as jet fuel). These resources may be used to produce products (such as machines, furniture, food, or clothing) or services (such as airline schedules, advertising policies, or investment decisions). **Linear programming (LP)** is a widely used mathematical technique designed to help operations managers plan and make the decisions necessary to allocate resources.

Linear programming (LP)
A mathematical technique designed to help operations managers plan and make decisions necessary to allocate resources.

A few examples of problems in which LP has been successfully applied in operations management are:

1. Scheduling school buses to *minimize* the total distance traveled when carrying students
2. Allocating police patrol units to high crime areas to *minimize* response time to 911 calls
3. Scheduling tellers at banks so that needs are met during each hour of the day while *minimizing* the total cost of labor
4. Selecting the product mix in a factory to make the best use of machine- and labor-hours available while *maximizing* the firm's profit
5. Picking blends of raw materials in feed mills to produce finished feed combinations at *minimum* cost
6. Determining the distribution system that will *minimize* total shipping cost from several warehouses to various market locations
7. Developing a production schedule that will satisfy future demands for a firm's product and at the same time *minimize* total production and inventory costs
8. Allocating space for a tenant mix in a new shopping mall to *maximize* revenues to the leasing company

VIDEO B.1
Scheduling Challenges at Alaska Airlines

Requirements of a Linear Programming Problem

All LP problems have four requirements: an objective, constraints, alternatives, and linearity:

1. LP problems seek to *maximize* or *minimize* some quantity (usually profit or cost). We refer to this property as the **objective function** of an LP problem. The major objective of a typical firm is to maximize dollar profits in the long run. In the case of a trucking or airline distribution system, the objective might be to minimize shipping costs.
2. The presence of restrictions, or **constraints**, limits the degree to which we can pursue our objective. For example, deciding how many units of each product in a firm's product line to manufacture is restricted by available labor and machinery. We want, therefore, to maximize or minimize a quantity (the objective function) subject to limited resources (the constraints).
3. There must be *alternative courses of action* to choose from. For example, if a company produces three different products, management may use LP to decide how to allocate among them its limited production resources (of labor, machinery, and so on). If there were no alternatives to select from, we would not need LP.
4. The objective and constraints in linear programming problems must be expressed in terms of *linear equations* or inequalities. Linearity implies proportionality and additivity. If x_1 and x_2 are decision variables, there can be no products (e.g., x_1x_2) or powers (e.g., x_1^3) in the objective or constraints. For example, the expression $5x_1 + 8x_2 \leq 250$ is linear; however, the expression $5x_1 + 8x_2 - 2x_1x_2 \leq 300$ is not linear.

Objective function
A mathematical expression in linear programming that maximizes or minimizes some quantity (often profit or cost, but any goal may be used).

Constraints
Restrictions that limit the degree to which a manager can pursue an objective.

Formulating Linear Programming Problems

STUDENT TIP Here we set up an LP example that we will follow for most of this module.

One of the most common linear programming applications is the *product-mix problem*. Two or more products are usually produced using limited resources. The company would like to determine how many units of each product it should produce to maximize overall profit given its limited resources. Let's look at an example.

Glickman Electronics Example

The Glickman Electronics Company in Washington, DC, produces two products: (1) the Glickman x-pod and (2) the Glickman BlueBerry. The production process for each product is similar in that both require a certain number of hours of electronic work and a certain number of labor-hours in the assembly department. Each x-pod takes 4 hours of electronic work and 2 hours in the assembly shop. Each BlueBerry requires 3 hours in electronics and 1 hour in assembly. During the current production period, 240 hours of electronic time are available, and 100 hours of assembly department time are available. Each x-pod sold yields a profit of \$7; each BlueBerry produced may be sold for a \$5 profit.

Glickman's problem is to determine the best possible combination of x-pods and BlueBerrys to manufacture to reach the maximum profit. This product-mix situation can be formulated as a linear programming problem.

Active Model B.1
This example is further illustrated in Active Model B.1 found online.

We begin by summarizing the information needed to formulate and solve this problem (see Table B.1). Further, let's introduce some simple notation for use in the objective function and constraints. Let:

$$X_1 = \text{number of x-pods to be produced}$$

$$X_2 = \text{number of BlueBerrys to be produced}$$

Now we can create the LP *objective function* in terms of X_1 and X_2:

$$\text{Maximize profit} = \$7X_1 + \$5X_2$$

Our next step is to develop mathematical relationships to describe the two constraints in this problem. One general relationship is that the amount of a resource used is to be less than or equal to (≤) the amount of resource *available*.

First constraint: Electronic time used is ≤ Electronic time available.

$$4X_1 + 3X_2 \leq 240 \text{ (hours of electronic time)}$$

Second constraint: Assembly time used is ≤ Assembly time available.

LO B.1 *Formulate* linear programming models, including an objective function and constraints

$$2X_1 + 1X_2 \leq 100 \text{ (hours of assembly time)}$$

Both these constraints represent production capacity restrictions and, of course, affect the total profit. For example, Glickman Electronics cannot produce 70 x-pods during the production period because if $X_1 = 70$, both constraints will be violated. It also cannot make $X_1 = 50$ x-pods and $X_2 = 10$ BlueBerrys. This constraint brings out another important aspect of linear programming; that is, certain interactions will exist between variables. The more units of one product that a firm produces, the fewer it can make of other products.

TABLE B.1 Glickman Electronics Company Problem Data

	HOURS REQUIRED TO PRODUCE ONE UNIT		
DEPARTMENT	X-PODS (X_1)	BLUEBERRYS (X_2)	AVAILABLE HOURS THIS WEEK
Electronic	4	3	240
Assembly	2	1	100
Profit per unit	\$7	\$5	

Graphical Solution to a Linear Programming Problem

Graphical solution approach
A means of plotting a solution to a two-variable problem on a graph.

Decision variables
Choices available to a decision maker.

The easiest way to solve a small LP problem such as that of the Glickman Electronics Company is the **graphical solution approach**. The graphical procedure can be used only when there are two **decision variables** (such as number of x-pods to produce, X_1, and number of BlueBerrys to produce, X_2). When there are more than two variables, it is *not* possible to plot the solution on a two-dimensional graph; we then must turn to more complex approaches described later in this module.

Graphical Representation of Constraints

To find the optimal solution to a linear programming problem, we must first identify a set, or region, of feasible solutions. The first step in doing so is to plot the problem's constraints on a graph.

The variable X_1 (x-pods, in our example) is usually plotted as the horizontal axis of the graph, and the variable X_2 (BlueBerrys) is plotted as the vertical axis. The complete problem may be restated as:

$$\text{Maximize profit} = \$7X_1 + \$5X_2$$

Subject to the constraints:

STUDENT TIP
We named the decision variables X_1 and X_2 here, but any notations (e.g., *x-p* and *B* or *X* and *Y*) would do as well.

$$4X_1 + 3X_2 \leq 240 \text{ (electronics constraint)}$$
$$2X_1 + 1X_2 \leq 100 \text{ (assembly constraint)}$$
$$X_1 \geq 0 \text{ (number of x-pods produced is greater than or equal to 0)}$$
$$X_2 \geq 0 \text{ (number of BlueBerrys produced is greater than or equal to 0)}$$

(These last two constraints are also called *nonnegativity constraints.*)

The first step in graphing the constraints of the problem is to convert the constraint *inequalities* into *equalities* (or equations):

Related Homework Problems:
B.2–B.21

$$\text{Constraint A:} \quad 4X_1 + 3X_2 = 240$$
$$\text{Constraint B:} \quad 2X_1 + 1X_2 = 100$$

The equation for constraint A is plotted in Figure B.1 and for constraint B in Figure B.2.

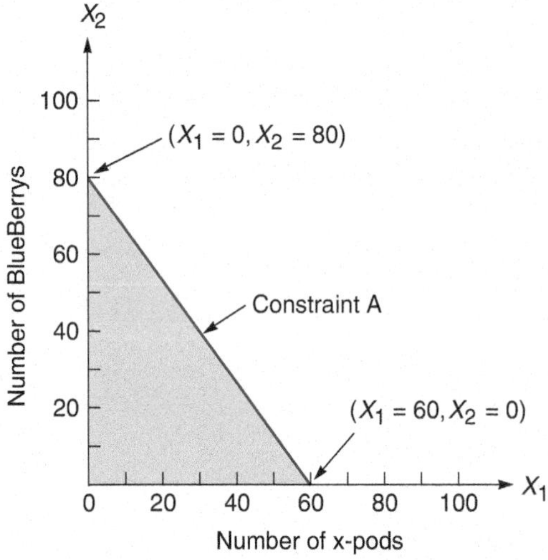

Figure **B.1**
Constraint A

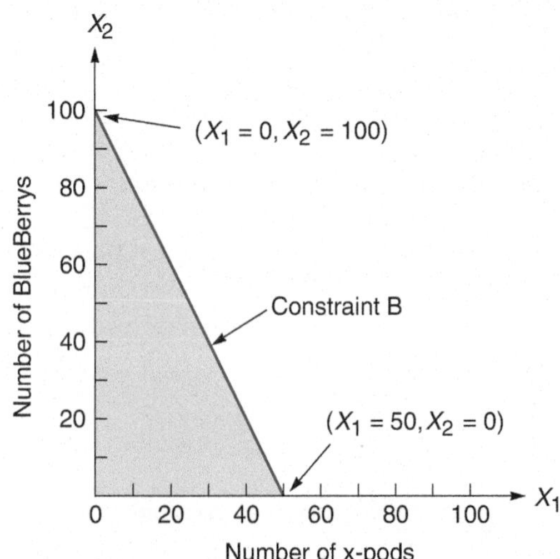

Figure **B.2**
Constraint B

To plot the line in Figure B.1, all we need to do is to find the points at which the line $4X_1 + 3X_2 = 240$ intersects the X_1 and X_2 axes. When $X_1 = 0$ (the location where the line touches the X_2 axis), it implies that $3X_2 = 240$ and that $X_2 = 80$. Likewise, when $X_2 = 0$, we see that $4X_1 = 240$ and that $X_1 = 60$. Thus, constraint A is bounded by the line running from ($X_1 = 0, X_2 = 80$) to ($X_1 = 60, X_2 = 0$). The shaded area represents all points that satisfy the original *inequality*.

Constraint B is illustrated similarly in Figure B.2. When $X_1 = 0$, then $X_2 = 100$; and when $X_2 = 0$, then $X_1 = 50$. Constraint B, then, is bounded by the line between ($X_1 = 0, X_2 = 100$) and ($X_1 = 50, X_2 = 0$). The shaded area represents the original inequality.

Figure B.3 shows both constraints together (along with the nonnegativity constraints). The shaded region is the part that satisfies all restrictions. The shaded region in Figure B.3 is called the *area of feasible solutions*, or simply the **feasible region**. This region must satisfy *all* conditions specified by the program's constraints and is thus the region where all constraints overlap. Any point in the region would be a *feasible solution* to the Glickman Electronics Company problem. Any point outside the shaded area would represent an *infeasible solution*. Hence, it would be feasible to manufacture 30 x-pods and 20 BlueBerrys ($X_1 = 30, X_2 = 20$), but it would violate the constraints to produce 70 x-pods and 40 BlueBerrys. This can be seen by plotting these points on the graph of Figure B.3.

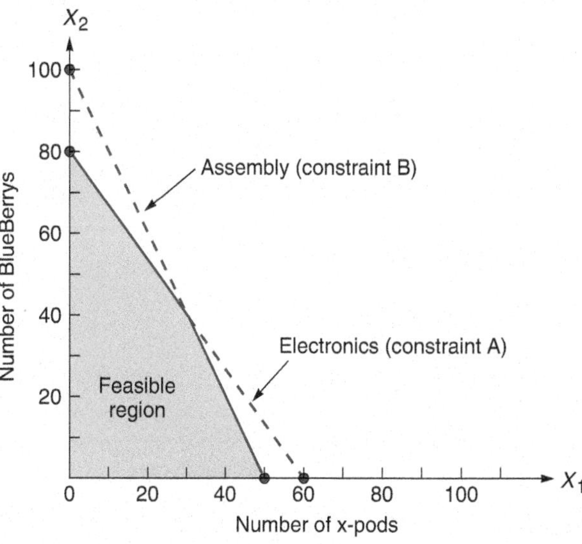

Figure **B.3**

Feasible Solution Region for the Glickman Electronics Company Problem

Feasible region
The set of all feasible combinations of decision variables.

Iso-Profit Line Solution Method

Now that the feasible region has been graphed, we can proceed to find the *optimal* solution to the problem. The optimal solution is the point lying in the feasible region that produces the highest profit.

Once the feasible region has been established, several approaches can be taken in solving for the optimal solution. The speediest one to apply is called the **iso-profit line method**.[1]

Iso-profit line method
An approach to solving a linear programming maximization problem graphically.

We start by letting profits equal some arbitrary but small dollar amount. For the Glickman Electronics problem, we may choose a profit of \$210. This is a profit level that can easily be obtained without violating either of the two constraints. The objective function can be written as $\$210 = 7X_1 + 5X_2$.

This expression is just the equation of a line; we call it an *iso-profit line*. It represents all combinations (of X_1, X_2) that will yield a total profit of \$210. To plot the profit line, we proceed exactly as we did to plot a constraint line. First, let $X_1 = 0$ and solve for the point at which the line crosses the X_2 axis:

$$\$210 = \$7(0) + \$5X_2$$
$$X_2 = 42 \text{ BlueBerrys}$$

Then let $X_2 = 0$ and solve for X_1:

$$\$210 = \$7X_1 + \$5(0)$$
$$X_1 = 30 \text{ x-pods}$$

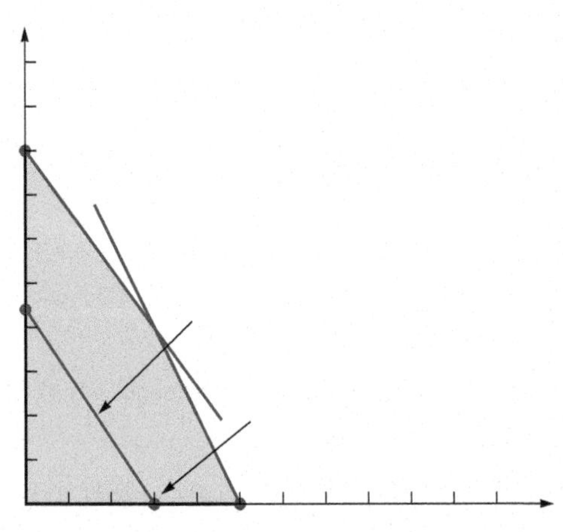

Figure B.4

A Profit Line of $210 Plotted for the Glickman Electronics Company

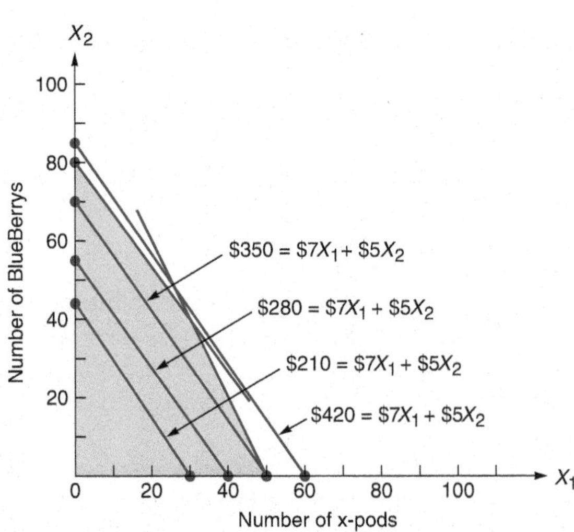

Figure B.5

Four Iso-Profit Lines Plotted for the Glickman Electronics Company

We can now connect these two points with a straight line. This profit line is illustrated in Figure B.4. All points on the line represent feasible solutions that produce a profit of $210.

We see, however, that the iso-profit line for $210 does not produce the highest possible profit to the firm. In Figure B.5, we try graphing three more lines, each yielding a higher profit. The middle equation, $\$280 = \$7X_1 + \$5X_2$, was plotted in the same fashion as the lower line. When $X_1 = 0$:

$$\$280 = \$7(0) + \$5X_2$$
$$X_2 = 56 \text{ BlueBerrys}$$

When $X_2 = 0$:

$$\$280 = \$7X_1 + \$5(0)$$
$$X_1 = 40 \text{ x-pods}$$

LO B.2 *Graphically solve an LP problem with the iso-profit line method*

Again, any combination of x-pods (X_1) and BlueBerrys (X_2) on this iso-profit line will produce a total profit of $280.

Note that the third line generates a profit of $350, even more of an improvement. The farther we move from the 0 origin, the higher our profit will be. Another important point to note is that these iso-profit lines are parallel. We now have two clues how to find the optimal solution to the original problem. We can draw a series of parallel profit lines (by carefully moving our ruler in a plane parallel to the first profit line). The highest profit line that still touches some point of the feasible region will pinpoint the optimal solution. Notice that the fourth line ($420) is too high to count because it does not touch the feasible region.

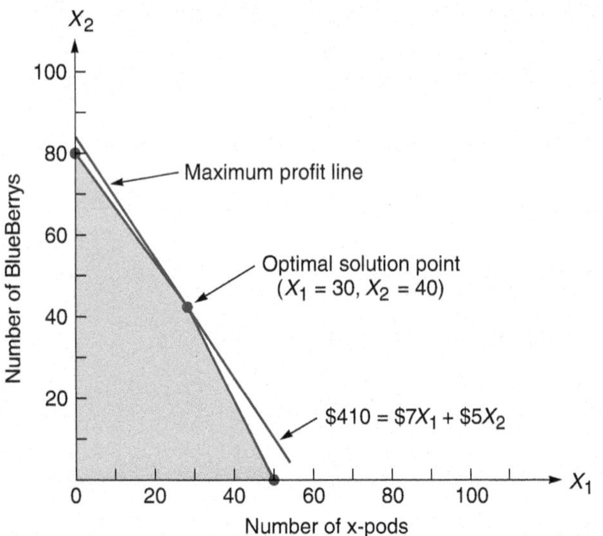

Figure B.6

Optimal Solution for the Glickman Electronics Problem

The highest possible iso-profit line is illustrated in Figure B.6. It touches the tip of the feasible region at the point where the two resource constraints intersect. To find its coordinates *accurately*, we will have to solve for the intersection of the two constraint lines. As you may recall from algebra, we can apply the method of *simultaneous equations* to the two constraint equations:

$$4X_1 + 3X_2 = 240 \quad \textit{(electronics time)}$$
$$2X_1 + 1X_2 = 100 \quad \textit{(assembly time)}$$

To solve these equations simultaneously, we multiply the second equation by –2:

$$-2(2X_1 + 1X_2 = 100) = -4X_1 - 2X_2 = -200$$

and then add it to the first equation:

$$\begin{array}{r} +4X_1 + 3X_2 = 240 \\ -4X_1 - 2X_2 = -200 \\ \hline +1X_2 = 40 \end{array}$$

or:

$$X_2 = 40$$

Doing this has enabled us to eliminate one variable, X_1, and to solve for X_2. We can now substitute 40 for X_2 in either of the original constraint equations and solve for X_1. Let us use the first equation. When $X_2 = 40$, then:

$$\begin{aligned} 4X_1 + 3(40) &= 240 \\ 4X_1 + 120 &= 240 \\ 4X_1 &= 120 \\ X_1 &= 30 \end{aligned}$$

Thus, the optimal solution has the coordinates ($X_1 = 30$, $X_2 = 40$). The profit at this point is $\$7(30) + \$5(40) = \$410$.

Corner-Point Solution Method

A second approach to solving linear programming problems employs the corner-point method. This technique is simpler in concept than the iso-profit line approach, but it involves looking at the profit at every corner point of the feasible region.

Corner-point method
A method for solving graphical linear programming problems.

The mathematical theory behind linear programming states that an optimal solution to any problem (that is, the values of X_1, X_2 that yield the maximum profit) will lie at a *corner point*, or *extreme point*, of the feasible region. Hence, it is necessary to find only the values of the variables at each corner; the maximum profit or optimal solution will lie at one (or more) of them.

Once again we can see (in Figure B.7) that the feasible region for the Glickman Electronics Company problem is a four-sided polygon with four corner, or extreme, points. These points are labeled ①, ②, ③, and ④ on the graph. To find the (X_1, X_2) values producing the maximum profit, we find out what the coordinates of each corner point are, then determine and compare their profit levels. (We showed how to find the coordinates for point ③ in the previous section describing the iso-profit line solution method.)

Figure B.7

The Four Corner Points of the Feasible Region

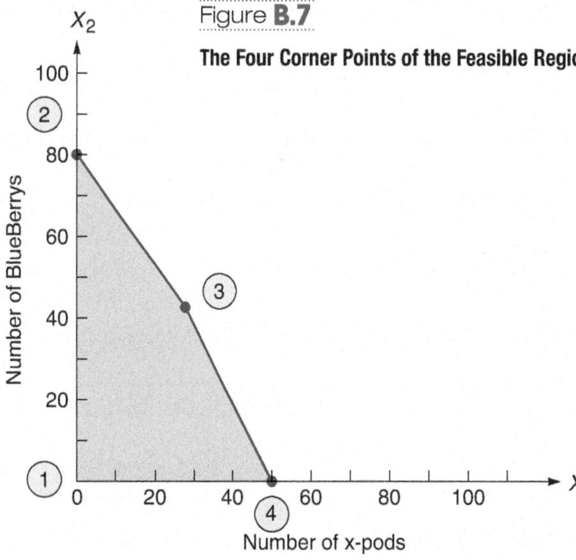

Point ①: ($X_1 = 0$, $X_2 = 0$) Profit $\$7(0) + \$5(0) = \$0$

Point ②: ($X_1 = 0$, $X_2 = 80$) Profit $\$7(0) + \$5(80) = \$400$

Point ③: ($X_1 = 30$, $X_2 = 40$) Profit $\$7(30) + \$5(40) = \$410$

Point ④: ($X_1 = 50$, $X_2 = 0$) Profit $\$7(50) + \$5(0) = \$350$

Because point ③ produces the highest profit of any corner point, the product mix of $X_1 = 30$ x-pods and $X_2 = 40$ BlueBerrys is the optimal solution to the Glickman Electronics problem. This solution will yield a profit of \$410 per production period; it is the same solution we obtained using the iso-profit line method.

LO B.3 *Graphically* solve an LP problem with the corner-point method

Sensitivity Analysis

Operations managers are usually interested in more than the optimal solution to an LP problem. In addition to knowing the value of each decision variable (the X_is) and the value of the objective function, they want to know how sensitive these answers are to input parameter changes. For example, what happens if the coefficients of the objective function are not exact, or if they change by 10% or 15%? What happens if the right-hand-side values of the constraints

Parameter
Numerical value that is given in a model.

Sensitivity analysis
An analysis that projects how much a solution may change if there are changes in the variables or input data.

change? Because solutions are based on the assumption that input parameters are constant, the subject of sensitivity analysis comes into play. **Sensitivity analysis**, or postoptimality analysis, is the study of how sensitive solutions are to parameter changes.

There are two approaches to determining just how sensitive an optimal solution is to changes. The first is simply a trial-and-error approach. This approach usually involves resolving the entire problem, preferably by computer, each time one input data item or parameter is changed. It can take a long time to test a series of possible changes in this way.

The approach we prefer is the analytic postoptimality method. After an LP problem has been solved, we determine a range of changes in problem parameters that will not affect the optimal solution or change the variables in the solution. This is done without resolving the whole problem. LP software, such as Excel's Solver or POM for Windows, has this capability. Let us examine several scenarios relating to the Glickman Electronics example.

STUDENT TIP
Here we look at the sensitivity of the final answers to changing inputs.

Program B.1 is part of the Excel Solver computer-generated output available to help a decision maker know whether a solution is relatively insensitive to reasonable changes in one or more of the parameters of the problem. (The complete computer run for these data, including input and full output, is illustrated in Programs B.3 and B.4 later in this module.)

Sensitivity Report

LO B.4 *Interpret* sensitivity analysis and shadow prices

The Excel *Sensitivity Report* for the Glickman Electronics example in Program B.1 has two distinct components: (1) a table titled Variable Cells and (2) a table titled Constraints. These tables permit us to answer several what-if questions regarding the problem solution.

It is important to note that while using the information in the sensitivity report to answer what-if questions, we assume that we are considering a change to only a *single* input data value at a time. That is, the sensitivity information does not always apply to simultaneous changes in several input data values.

Related Homework Problems:
B.22–B.24, B.40

The *Variable Cells* table presents information regarding the impact of changes to the objective function coefficients (i.e., the unit profits of $7 and $5) on the optimal solution. The *Constraints* table presents information related to the impact of changes in constraint right-hand-side (RHS) values (i.e., the 240 hours and 100 hours) on the optimal solution. Although different LP software packages may format and present these tables differently, the programs all provide essentially the same information.

Changes in the Resources or Right-Hand-Side Values

The right-hand-side values of the constraints often represent resources available to the firm. The resources could be labor-hours or machine time or perhaps money or production materials available. In the Glickman Electronics example, the two resources are hours available of

Program **B.1**
Sensitivity Analysis for Glickman Electronics, Using Excel's Solver

Microsoft Excel Sensitivity Report

The solution values for the variables appear. We should make 30 x-pods and 40 BlueBerrys.

Variable Cells

Cell	Name	Final Value	Reduced Cost	Objective Coefficient	Allowable Increase	Allowable Decrease
$B$5	Variable Values x-pods	30	0	7	3	0.333333333
$C$5	Variable Values BlueBerrys	40	0	5	0.25	1.5

Constraints

Cell	Name	Final Value	Shadow Price	Constraint R.H. Side	Allowable Increase	Allowable Decrease
$D$8	Electronic Time Available	240	1.5	240	60	40
$D$9	Assembly Time Available	100	0.5	100	20	20

We will use 240 hours and 100 hours of Electronics and Assembly time, respectively.

If we use 1 more Electronics hour, our profit will increase by $1.50. This is true for up to 60 more hours. The profit will fall by $1.50 for each Electronics hour less than 240 hours, down to as low as 200 hours.

electronics time and hours of assembly time. If additional hours were available, a higher total profit could be realized. How much should the company be willing to pay for additional hours? Is it profitable to have some additional electronics hours? Should we be willing to pay for more assembly time? Sensitivity analysis about these resources will help us answer these questions.

If the right-hand side of a constraint is changed, the feasible region will change (unless the constraint is redundant), and often the optimal solution will change. In the Glickman example, there were 100 hours of assembly time available each week and the maximum possible profit was $410. If the available assembly hours are *increased* to 110 hours, the new optimal solution seen in Figure B.8(a) is (45,20) and the profit is $415. Thus, the extra 10 hours of time resulted in an increase in profit of $5 or $0.50 per hour. If the hours are *decreased* to 90 hours as shown in Figure B.8(b), the new optimal solution is (15,60) and the profit is $405. Thus, reducing the hours by 10 results in a decrease in profit of $5 or $0.50 per hour. This $0.50 per hour change in profit that resulted from a change in the hours available is called the shadow price, or dual value. The shadow price for a constraint is the improvement in the objective function value that results from a one-unit increase in the right-hand side of the constraint.

Shadow price (or dual value)
The value of one additional unit of a scarce resource in LP.

Validity Range for the Shadow Price Given that Glickman Electronics' profit increases by $0.50 for each additional hour of assembly time, does it mean that Glickman can do this indefinitely, essentially earning infinite profit? Clearly, this is illogical. How far can Glickman increase its assembly time availability and still earn an extra $0.50 profit per hour? That is, for what level of increase in the RHS value of the assembly time constraint is the shadow price of $0.50 valid?

The shadow price of $0.50 is valid as long as the available assembly time stays in a range within which all current corner points continue to exist. The information to compute the upper and lower limits of this range is given by the entries labeled Allowable Increase and Allowable Decrease in the *Sensitivity Report* in Program B.1. In Glickman's case, these values show that the shadow price of $0.50 for assembly time availability is valid for an increase of up to 20 hours from the current value and a decrease of up to 20 hours. That is, the available assembly time can range from a low of 80 (= 100 – 20) to a high of 120 (= 100 + 20) for the shadow price of $0.50 to be valid. Note that the allowable decrease implies that for each hour of assembly time that Glickman loses (up to 20 hours), its profit decreases by $0.50.

Changes in the Objective Function Coefficient

Let us now focus on the information provided in Program B.1 titled Variable Cells. Each row in the Variable Cells table contains information regarding a decision variable (i.e., x-pods or BlueBerrys) in the LP model.

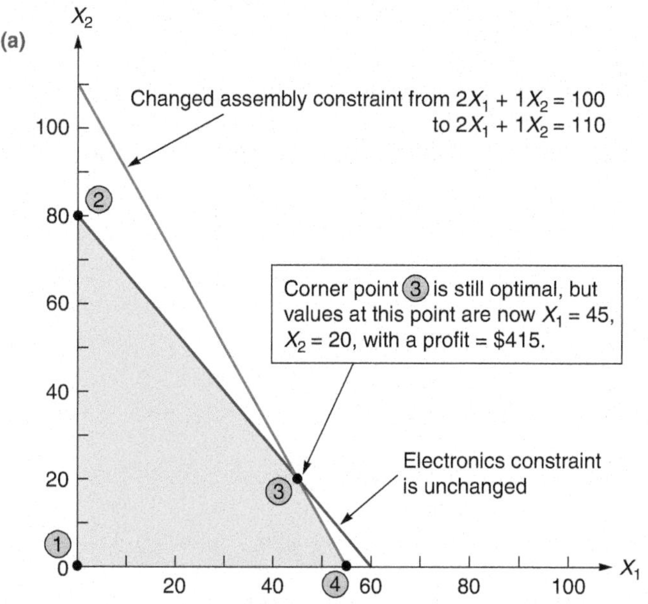

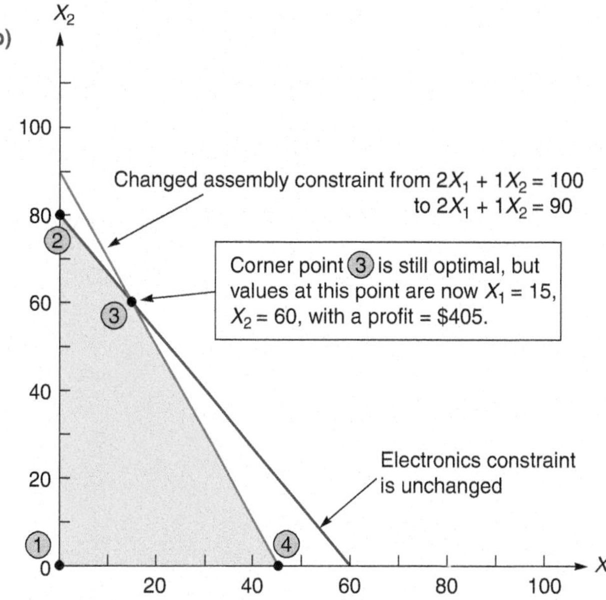

Figure **B.8**

Glickman Electronics Sensitivity Analysis on the Right-Hand Side (RHS) of the Assembly Resource Constraint

Allowable Ranges for Objective Function Coefficients As the unit profit contribution of either product changes, the slope of the iso-profit lines we saw earlier in Figure B.5 changes. The size of the feasible region, however, remains the same. That is, the locations of the corner points do not change.

The limits to which the profit coefficient of x-pods or BlueBerrys can be changed without affecting the optimality of the current solution is revealed by the values in the Allowable Increase and Allowable Decrease columns of the *Sensitivity Report* in Program B.1. The allowable increase in the objective function coefficient for BlueBerrys is only $0.25. In contrast, the allowable decrease is $1.50. Hence, if the unit profit of BlueBerrys drops to $4 (i.e., a decrease of $1 from the current value of $5), it is still optimal to produce 30 x-pods and 40 BlueBerrys. The total profit will drop to $370 (from $410) because each BlueBerry now yields less profit (of $1 per unit). However, if the unit profit drops below $3.50 per BlueBerry (i.e., a decrease of more than $1.50 from the current $5 profit), the current solution is no longer optimal. The LP problem will then have to be resolved using Solver, or other software, to find the new optimal corner point.

STUDENT TIP LP problems can be structured to minimize costs as well as maximize profits.

Solving Minimization Problems

Many linear programming problems involve *minimizing* an objective such as cost instead of maximizing a profit function. A restaurant, for example, may wish to develop a work schedule to meet staffing needs while minimizing the total number of employees. Also, a manufacturer may seek to distribute its products from several factories to its many regional warehouses in a way that minimizes total shipping costs.

Iso-cost An approach to solving a linear programming minimization problem graphically.

Minimization problems can be solved graphically by first setting up the feasible solution region and then using either the corner-point method or an iso-cost line approach (which is analogous to the iso-profit approach in maximization problems) to find the values of X_1 and X_2 that yield the minimum cost.

Example B1 shows how to solve a minimization problem.

Example B1

A MINIMIZATION PROBLEM WITH TWO VARIABLES

Cohen Chemicals, Inc., produces two types of photo-developing fluids. The first, a black-and-white picture chemical, costs Cohen $2,500 per ton to produce. The second, a color photo chemical, costs $3,000 per ton.

Based on an analysis of current inventory levels and outstanding orders, Cohen's production manager has specified that at least 30 tons of the black-and-white chemical and at least 20 tons of the color chemical must be produced during the next month. In addition, the manager notes that an existing inventory of a highly perishable raw material needed in both chemicals must be used within 30 days. To avoid wasting the expensive raw material, Cohen must produce a total of at least 60 tons of the photo chemicals in the next month.

APPROACH ▶ Formulate this information as a minimization LP problem.

Let:

$$X_1 = \text{number of tons of black-and-white photo chemical produced}$$

$$X_2 = \text{number of tons of color photo chemical produced}$$

$$\text{Objective: Minimize cost} = \$2,500X_1 + \$3,000X_2$$

Subject to:

$$\begin{aligned} X_1 &\geq 30 \text{ tons of black-and-white chemical} \\ X_2 &\geq 20 \text{ tons of color chemical} \\ X_1 + X_2 &\geq 60 \text{ tons total} \\ X_1, X_2 &\geq 0 \text{ nonnegativity requirements} \end{aligned}$$

SOLUTION ▶ To solve the Cohen Chemicals problem graphically, we construct the problem's feasible region, shown in Figure B.9.

Figure B.9

Cohen Chemicals' Feasible Region

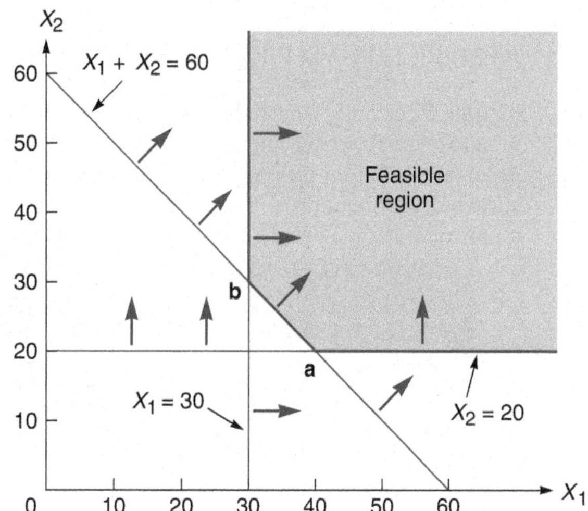

LO B.5 *Construct* and solve a minimization problem

Minimization problems are often unbounded outward (that is, on the right side and on the top), but this characteristic causes no problem in solving them. As long as they are bounded inward (on the left side and the bottom), we can establish corner points. The optimal solution will lie at one of the corners.

In this case, there are only two corner points, **a** and **b**, in Figure B.9. It is easy to determine that at point **a**, $X_1 = 40$ and $X_2 = 20$, and that at point **b**, $X_1 = 30$ and $X_2 = 30$. The optimal solution is found at the point yielding the lowest total cost.

Thus:

$$\begin{aligned}\text{Total cost at } \mathbf{a} &= 2{,}500X_1 + 3{,}000X_2 \\ &= 2{,}500(40) + 3{,}000(20) \\ &= \$160{,}000 \\ \text{Total cost at } \mathbf{b} &= 2{,}500X_1 + 3{,}000X_2 \\ &= 2{,}500(30) + 3{,}000(30) \\ &= \$165{,}000\end{aligned}$$

The lowest cost to Cohen Chemicals is at point **a**. Hence the operations manager should produce 40 tons of the black-and-white chemical and 20 tons of the color chemical.

INSIGHT ▶ The area is either not bounded to the right or above in a minimization problem (as it is in a maximization problem).

LEARNING EXERCISE ▶ Cohen's second constraint is recomputed and should be $X_2 \geq 15$. Does anything change in the answer? [Answer: Now $X_1 = 45, X_2 = 15$, and total cost $= \$157{,}500$.]

RELATED PROBLEMS ▶ B.25–B.33

EXCEL OM Data File **ModBExB1.xls** can be found online.

Linear Programming Applications

The foregoing examples each contained just two variables (X_1 and X_2). Most real-world problems contain many more variables, however. These can be solved via computer (see *Using Software to Solve LP Problems* later in this module). Let's use the principles already developed to formulate a few more-complex problems. The practice you will get by "paraphrasing" the following LP situations should help develop your skills for applying linear programming to other common operations situations.

STUDENT TIP
Now we look at three larger problems—ones that have more than two decision variables each and therefore are not graphed.

Production-Mix Example

LO B.6 *Formulate* production-mix, diet, and labor scheduling problems

Example B2 involves another *production-mix* decision. Limited resources must be allocated among various products that a firm produces. The firm's overall objective is to manufacture the selected products in such quantities as to maximize total profits.

Example B2

A PRODUCTION-MIX PROBLEM

Failsafe Electronics Corporation primarily manufactures four highly technical products, which it supplies to aerospace firms that hold NASA contracts. Each of the products must pass through the following departments before they are shipped: wiring, drilling, assembly, and inspection. The time requirements in each department (in hours) for each unit produced and its corresponding profit value are summarized in this table:

	DEPARTMENT				
PRODUCT	WIRING	DRILLING	ASSEMBLY	INSPECTION	UNIT PROFIT
XJ201	.5	3	2	.5	$ 9
XM897	1.5	1	4	1.0	$12
TR29	1.5	2	1	.5	$15
BR788	1.0	3	2	.5	$11

The production time available in each department each month and the minimum monthly production requirement to fulfill contracts are as follows:

DEPARTMENT	CAPACITY (HOURS)	PRODUCT	MINIMUM PRODUCTION LEVEL
Wiring	1,500	XJ201	150
Drilling	2,350	XM897	100
Assembly	2,600	TR29	200
Inspection	1,200	BR788	400

APPROACH ▶ Formulate this production-mix situation as an LP problem. The production manager first specifies production levels for each product for the coming month. He lets:

$$
\begin{aligned}
X_1 &= \text{number of units of XJ201 produced} \\
X_2 &= \text{number of units of XM897 produced} \\
X_3 &= \text{number of units of TR29 produced} \\
X_4 &= \text{number of units of BR788 produced}
\end{aligned}
$$

SOLUTION ▶ The LP formulation is:

$$\text{Objective: Maximize profit} = 9X_1 + 12X_2 + 15X_3 + 11X_4$$

$$
\begin{aligned}
\text{subject to:}\quad & .5X_1 + 1.5X_2 + 1.5X_3 + 1X_4 \le 1{,}500 \text{ hours of wiring available} \\
& 3X_1 + 1X_2 + 2X_3 + 3X_4 \le 2{,}350 \text{ hours of drilling available} \\
& 2X_1 + 4X_2 + 1X_3 + 2X_4 \le 2{,}600 \text{ hours of assembly available} \\
& .5X_1 + 1X_2 + 5X_3 + 5X_4 \le 1{,}200 \text{ hours of inspection} \\
& X_1 \ge 150 \text{ units of XJ201} \\
& X_2 \ge 100 \text{ units of XM897} \\
& X_3 \ge 200 \text{ units of TR29} \\
& X_4 \ge 400 \text{ units of BR788} \\
& X_1, X_2, X_3, X_4 \ge 0
\end{aligned}
$$

INSIGHT ▶ There can be numerous constraints in an LP problem. The constraint right-hand sides may be in different units, but the objective function uses one common unit—dollars of profit, in this case. Because there are more than two decision variables, this problem is not solved graphically.

LEARNING EXERCISE ▶ Solve this LP problem as formulated. What is the solution? [Answer: $X_1 = 150$, $X_2 = 300$, $X_3 = 200$, $X_4 = 400$, profit = \$12,350.]

RELATED PROBLEMS ▶ B.5–B.8, B.10–B.13, B.15, B.17, B.19, B.21, B.24

Diet Problem Example

Example B3 illustrates the *diet problem*, which was originally used by hospitals to determine the most economical diet for patients. Known in agricultural applications as the *feed-mix problem*, the diet problem involves specifying a food or feed ingredient combination that will satisfy stated nutritional requirements at a minimum cost level.

Example B3

A DIET PROBLEM

The Feed 'N Ship feedlot fattens cattle for local farmers and ships them to meat markets in Kansas City and Omaha. The owners of the feedlot seek to determine the amounts of cattle feed to buy to satisfy minimum nutritional standards and, at the same time, minimize total feed costs.

Each grain stock contains different amounts of four nutritional ingredients: A, B, C, and D. Here are the ingredient contents of each grain, in *ounces per pound of grain*:

	FEED		
INGREDIENT	STOCK X	STOCK Y	STOCK Z
A	3 oz	2 oz	4 oz
B	2 oz	3 oz	1 oz
C	1 oz	0 oz	2 oz
D	6 oz	8 oz	4 oz

The cost per pound of grains X, Y, and Z is $0.02, $0.04, and $0.025, respectively. The minimum requirement per cow per month is 64 ounces of ingredient A, 80 ounces of ingredient B, 16 ounces of ingredient C, and 128 ounces of ingredient D.

The feedlot faces one additional restriction—it can obtain only 500 pounds of stock Z per month from the feed supplier, regardless of its need. Because there are usually 100 cows at the Feed 'N Ship feedlot at any given time, this constraint limits the amount of stock Z for use in the feed of each cow to no more than 5 pounds, or 80 ounces, per month.

APPROACH ▶ Formulate this as a minimization LP problem.

Let: X_1 = number of pounds of stock X purchased per cow each month
X_2 = number of pounds of stock Y purchased per cow each month
X_3 = number of pounds of stock Z purchased per cow each month

SOLUTION ▶ Objective: Minimize cost $= .02X_1 + .04X_2 + .025X_3$

subject to: Ingredient A requirement: $3X_1 + 2X_2 + 4X_3 \geq 64$
Ingredient B requirement: $2X_1 + 3X_2 + 1X_3 \geq 80$
Ingredient C requirement: $1X_1 + 0X_2 + 2X_3 \geq 16$
Ingredient D requirement: $6X_1 + 8X_2 + 4X_3 \geq 128$
Stock Z limitation: $X_3 \leq 5$
$X_1, X_2, X_3 \geq 0$

The cheapest solution is to purchase 40 pounds of grain X, at a cost of $0.80 per cow.

INSIGHT ▶ Because the cost per pound of stock X is so low, the optimal solution excludes grains Y and Z.

LEARNING EXERCISE ▶ The cost of a pound of stock X just increased by 50%. Does this affect the solution? [Answer: Yes, when the cost per pound of grain *X* is $0.03, $X_1 = 16$ pounds, $X_2 = 16$ pounds, $X_3 = 0$, and cost = $1.12 per cow.]

RELATED PROBLEMS ▶ B.27, B.28, B.40

Labor Scheduling Example

Labor scheduling problems address staffing needs over a specific time period. They are especially useful when managers have some flexibility in assigning workers to jobs that require overlapping or interchangeable talents. Large banks and hospitals frequently use LP to tackle their labor scheduling. Example B4 describes how one bank uses LP to schedule tellers.

Example B4

SCHEDULING BANK TELLERS

Mexico City Bank of Commerce and Industry is a busy bank that has requirements for between 10 and 18 tellers depending on the time of day. Lunchtime, from noon to 2 P.M., is usually heaviest. The following table indicates the workers needed at various hours that the bank is open.

TIME PERIOD	NUMBER OF TELLERS REQUIRED	TIME PERIOD	NUMBER OF TELLERS REQUIRED
9 A.M.–10 A.M.	10	1 P.M.–2 P.M.	18
10 A.M.–11 A.M.	12	2 P.M.–3 P.M.	17
11 A.M.–Noon	14	3 P.M.–4 P.M.	15
Noon–1 P.M.	16	4 P.M.–5 P.M.	10

The bank now employs 12 full-time tellers, but many people are on its roster of available part-time employees. A part-time employee must put in exactly 4 hours per day but can start anytime between 9 A.M. and 1 P.M. Part-timers are a fairly inexpensive labor pool because no retirement or lunch benefits are provided to them. Full-timers, on the other hand, work from 9 A.M. to 5 P.M. but are allowed 1 hour for lunch. (Half the full-timers eat at 11 A.M., the other half at noon.) Full-timers thus provide 35 hours per week of productive labor time.

By corporate policy, the bank limits part-time hours to a maximum of 50% of the day's total requirement.

Part-timers earn the U.S. equivalent of $6 per hour (or $24 per day) on average, whereas full-timers earn the U.S. equivalent of $75 per day in salary and benefits on average.

APPROACH ▶ The bank would like to set a schedule, using LP, that would minimize its total manpower costs. It will release 1 or more of its full-time tellers if it is profitable to do so.

We can let:

F = full-time tellers
P_1 = part-timers starting at 9 A.M. (leaving at 1 P.M.)
P_2 = part-timers starting at 10 A.M. (leaving at 2 P.M.)
P_3 = part-timers starting at 11 A.M. (leaving at 3 P.M.)
P_4 = part-timers starting at noon (leaving at 4 P.M.)
P_5 = part-timers starting at 1 P.M. (leaving at 5 P.M.)

SOLUTION ▶ Objective function:

$$\text{Minimize total daily manpower cost} = \$75F + \$24(P_1 + P_2 + P_3 + P_4 + P_5)$$

Constraints: For each hour, the available labor-hours must be at least equal to the required labor-hours:

$$F + P_1 \geq 10 \quad \textit{(9 A.M. to 10 A.M. needs)}$$
$$F + P_1 + P_2 \geq 12 \quad \textit{(10 A.M. to 11 A.M. needs)}$$
$$\tfrac{1}{2}F + P_1 + P_2 + P_3 \geq 14 \quad \textit{(11 A.M. to noon needs)}$$
$$\tfrac{1}{2}F + P_1 + P_2 + P_3 + P_4 \geq 16 \quad \textit{(noon to 1 P.M. needs)}$$
$$F + P_2 + P_3 + P_4 + P_5 \geq 18 \quad \textit{(1 P.M. to 2 P.M. needs)}$$
$$F + P_3 + P_4 + P_5 \geq 17 \quad \textit{(2 P.M. to 3 P.M. needs)}$$
$$F + P_4 + P_5 \geq 15 \quad \textit{(3 P.M. to 4 P.M. needs)}$$
$$F + P_5 \geq 10 \quad \textit{(4 P.M. to 5 P.M. needs)}$$

Only 12 full-time tellers are available, so:

$$F \leq 12$$

Part-time worker-hours cannot exceed 50% of total hours required each day, which is the sum of the tellers needed each hour:

$$4(P_1 + P_2 + P_3 + P_4 + P_5) \leq .50(10 + 12 + 14 + 16 + 18 + 17 + 15 + 10)$$

or:

$$4P_1 + 4P_2 + 4P_3 + 4P_4 + 4P_5 \leq 0.50(112)$$
$$F, P_1, P_2, P_3, P_4, P_5 \geq 0$$

There are several alternative optimal schedules that Mexico City Bank can follow. One of them is to employ only 10 full-time tellers ($F = 10$) and to start 7 part-timers at 10 A.M. ($P_2 = 7$), 2 part-timers at 11 A.M. and noon ($P_3 = 2$ and $P_4 = 2$), and 3 part-timers at 1 P.M. ($P_5 = 3$). No part-timers would begin at 9 A.M. The cost of this policy is $1,086 per day.

INSIGHT ▶ It is not unusual for multiple optimal solutions to exist in large LP problems. In this case, it gives management the option of selecting, at the same cost, between schedules. To find an alternate optimal solution, you may have to enter the constraints in a different sequence.

LEARNING EXERCISE ▶ The bank decides to give part-time employees a raise to $7 per hour. Does the solution change? [Answer: There are still alternate optima. The new cost = $1,142.]

RELATED PROBLEM ▶ B.36

The *OM in Action* box, "Art and Science of Scheduling the NFL," illustrates another view of how LP can be used in scheduling.

OM in Action Art and Science of Scheduling the NFL[2]

"We're geniuses one day and absolute morons the next," says Howard Katz, director of scheduling for the National Football League (NFL). That's because Katz must consider a confounding array of factors, from the NFL's expanded Thursday night package, which gives each team a game in a short week, to potential baseball playoff situations that could affect the availability of stadiums and parking lots in October.

For the networks that pay billions of dollars to carry NFL games, Katz's staff has been mostly geniuses. NFL games were watched by an average of 17.5 million viewers recently. NFL games accounted for 23 of the 25 most-watched television shows among all programming and the 16 most-watched shows on cable last fall.

Designing a schedule that generates those ratings while also guaranteeing competitive fairness is more complicated than ever, even though software spits out 400,000 complete or partial schedules (once done entirely by hand) from a possible 824 trillion game combinations. Katz starts with thousands of tentative schedules in which a handful of critical games with attractive story lines are placed in select spots. Then the computers generate possibilities around those games.

The NFL also feeds the computer with penalties for situations it prefers to avoid—three-game trips, for example, or teams starting with two road games. There are requests not to play at home on certain holidays—the Jets and the Giants typically ask not to play home games during the Jewish High Holy Days. This year, the software generated 14,000 playable schedules, which were reduced to 150 with an eyeball test. Katz reviewed those 150 by hand, scoring them for each team and each network.

Linear programming may be at the heart of scheduling, but the process is definitely part art and part science.

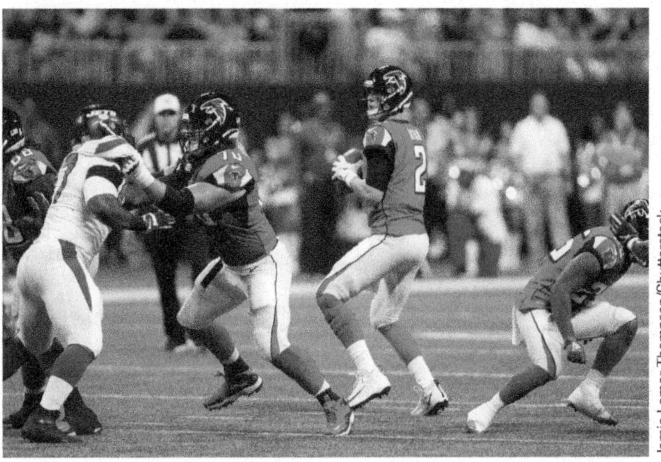

Jamie Lamor Thompson/Shutterstock

The Simplex Method of LP

Most real-world linear programming problems have more than two variables and thus are too complex for graphical solution. A procedure called the **simplex method** may be used to find the optimal solution to such problems. The simplex method is actually an algorithm (or a set of instructions) with which we examine corner points in a methodical fashion until we arrive at the best solution—highest profit or lowest cost. Computer programs (such as Excel OM and POM for Windows) and Excel spreadsheets are available to solve linear programming problems via the simplex method.

Simplex method
An algorithm for solving linear programming problems of all sizes.

For details regarding the algebraic steps of the simplex algorithm, see Tutorial 3 found online, or refer to a management science textbook.[3]

Integer and Binary Variables

Binary variables
Decision variables that can only take on the value of 0 or 1.

LO B.7 *Incorporate* logical conditions into LPs using binary variables

All the examples we have seen in this module so far have produced integer solutions. But it is very common to see LP solutions where the decision variables are not whole numbers. Computer software provides a simple way to guarantee only integer solutions. In addition, computers allow us to create special decision variables called **binary variables** that can only take on the values of 0 or 1. Binary variables allow us to introduce "yes-or-no" decisions into our linear programs and to introduce special logical conditions.

Creating Integer and Binary Variables

If we wish to ensure that decision variable values are integers rather than fractions, it is generally *not* good practice to simply round the solutions to the nearest integer values. The rounded solutions may not be optimal and, in fact, may not even be feasible. Fortunately, all LP software programs have simple ways to add constraints that enforce some or all of the decision variables to be either integer or binary. The main disadvantage of introducing such constraints is that larger programs may take longer to solve. The same LP that may take 3 seconds to solve on a computer could take several hours or more to solve if many of its variables are forced to be integer or binary. For relatively small programs, though, the difference may be unnoticeable.

Using Excel's Solver (see *Using Software to Solve LP Problems* later in this module), integer and binary constraints can be added by clicking **Add** from the main Solver dialog box. Using the Add Constraint dialog box (see Program B.2), highlight the decision variables themselves under Cell Reference:. Then select **int** or **bin** to ensure that those variables are integer or binary, respectively, in the optimal solution.

Program **B.2**

Excel's Solver Dialog Box to Add Integer or Binary Constraints on Variables

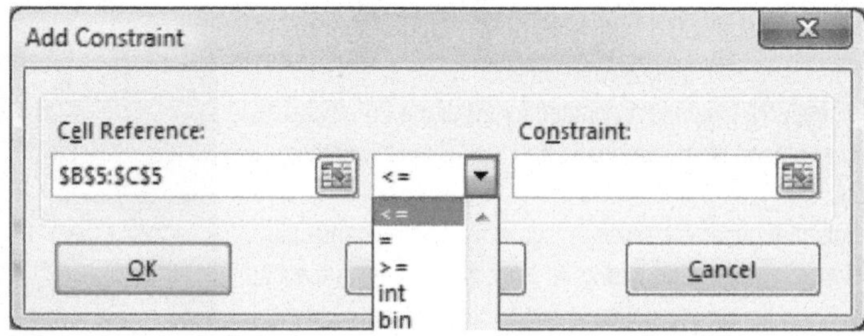

Linear Programming Applications with Binary Variables

In the written formulation of a linear program, binary variables are usually defined using the following form:

$$Y = \begin{cases} 1 & \text{if some condition holds} \\ 0 & \text{otherwise} \end{cases}$$

Sometimes we designate decision variables as binary if we are making a yes-or-no decision; for example, "Should we undertake this particular project?" "Should we buy that machine?" or "Should we locate a facility in Arkansas?" Other times, we create binary variables to introduce additional logic into our programs.

Limiting the Number of Alternatives Selected One common use of 0-1 variables involves limiting the number of projects or items that are selected from a group. Suppose a firm is required to select no more than two of three potential projects. This could be modeled with the following constraint:

$$Y_1 + Y_2 + Y_3 \le 2$$

If we wished to force the selection of *exactly* two of the three projects for funding, the following constraint should be used:

$$Y_1 + Y_2 + Y_3 = 2$$

This forces exactly two of the variables to have values of 1, whereas the other variable must have a value of 0.

Dependent Selections At times the selection of one project depends in some way on the selection of another project. This situation can be modeled with the use of 0-1 variables. Suppose G.E.'s new catalytic converter could be purchased ($Y_1 = 1$) only if the software was also purchased ($Y_2 = 1$). The following constraint would force this to occur:

$$Y_1 \leq Y_2$$

or, equivalently,

$$Y_1 - Y_2 \leq 0$$

Thus, if the software is not purchased, the value of Y_2 is 0, and the value of Y_1 must also be 0 because of this constraint. However, if the software is purchased ($Y_2 = 1$), then it is possible that the catalytic converter could also be purchased ($Y_1 = 1$), although this is not required.

If we wished for the catalytic converter and the software projects to either both be selected or both not be selected, we should use the following constraint:

$$Y_1 = Y_2$$

or, equivalently,

$$Y_1 - Y_2 = 0$$

Thus, if either of these variables is equal to 0, the other must also be 0. If either of these is equal to 1, the other must also be 1.

A Fixed-Charge Integer Programming Problem

Often businesses are faced with decisions involving a fixed charge that will affect the cost of future operations. Building a new factory or entering into a long-term lease on an existing facility would involve a fixed cost that might vary depending on the size of the facility and the location. Once a factory is built, the variable production costs will be affected by the labor cost in the particular city where it is located. Example B5 provides an illustration.

Example B5

A FIXED-CHARGE PROBLEM USING BINARY VARIABLES

Sitka Manufacturing is planning to build at least one new plant, and three cities are being considered: Baytown, Texas; Lake Charles, Louisiana; and Mobile, Alabama. Once the plant or plants have been constructed, the company wishes to have sufficient capacity to produce at least 38,000 units each year. The costs associated with the possible locations are given in the following table.

SITE	ANNUAL FIXED COST	VARIABLE COST PER UNIT	ANNUAL CAPACITY
Baytown, TX	$340,000	$32	21,000
Lake Charles, LA	$270,000	$33	20,000
Mobile, AL	$290,000	$30	19,000

APPROACH ▶ In modeling this as an integer program, the objective function is to minimize the total of the fixed costs and the variable costs. The constraints are: (1) total production capacity is at least 38,000; (2) the number of units produced at the Baytown plant is 0 if the plant is not built, and it is no more than 21,000 if the plant is built; (3) the number of units produced at the Lake Charles plant is 0 if the plant is not built, and it is no more than 20,000 if the plant is built; and (4) the number of units produced at the Mobile plant is 0 if the plant is not built, and it is no more than 19,000 if the plant is built.

Then we define the decision variables as

$$Y_1 = \begin{cases} 1 & \text{if factory is built in Baytown} \\ 0 & \text{otherwise} \end{cases}$$

$$Y_2 = \begin{cases} 1 & \text{if factory is built in Lake Charles} \\ 0 & \text{otherwise} \end{cases}$$

$$Y_3 = \begin{cases} 1 & \text{if factory is built in Mobile} \\ 0 & \text{otherwise} \end{cases}$$

X_1 = number of units produced at the Baytown plant
X_2 = number of units produced at the Lake Charles plant
X_3 = number of units produced at the Mobile plant

SOLUTION ▶ The integer programming problem formulation becomes

Objective: Minimize cost $= 340{,}000\ Y_1 + 270{,}000\ Y_2 + 290{,}000\ Y_3 + 32\ X_1 + 33\ X_2 + 30\ X_3$

subject to:

$$X_1 + X_2 + X_3 \geq 38{,}000$$
$$X_1 \leq 21{,}000Y_1 \text{ i.e., } X_1 - 21{,}000Y_1 \leq 0$$
$$X_2 \leq 20{,}000Y_2 \text{ i.e., } X_2 - 20{,}000Y_2 \leq 0$$
$$X_3 \leq 19{,}000Y_3 \text{ i.e., } X_3 - 19{,}000Y_3 \leq 0$$
$$X_1, X_2, X_3 \geq 0 \text{ and integer}$$
$$Y_1, Y_2, Y_3 = 0 \text{ or } 1$$

INSIGHT ▶ Examining the second constraint, the objective function will try to set the binary variable Y_1 equal to 0 because it wants to minimize cost. However, if $Y_1 = 0$, then the constraint will force X_1 to equal 0, in which case no units will be produced, and the plant will not be opened. Alternatively, if the rest of the program deems it worthwhile or necessary to produce some units of X_1, then Y_1 will have to equal 1 for the constraint to hold. And when $Y_1 = 1$, the firm will be charged the fixed cost of $340,000, and production will be limited to the capacity of 21,000 units. The same logic applies for constraints 3 and 4.

LEARNING EXERCISE ▶ Solve this integer program as formulated. What is the solution? [Answer: $Y_1 = 0, Y_2 = 1, Y_3 = 1, X_1 = 0, X_2 = 19{,}000, X_3 = 19{,}000$; Total Cost = $1,757,000.]

RELATED PROBLEMS ▶ B.41, B.42

Summary

This module introduces a special kind of model, linear programming. LP has proven to be especially useful when trying to make the most effective use of an organization's resources.

The first step in dealing with LP models is problem formulation, which involves identifying and creating an objective function and constraints. The second step is to solve the problem. If there are only two decision variables, the problem can be solved graphically, using the corner-point method or the iso-profit/iso-cost line method. With either approach, we first identify the feasible region, then find the corner point yielding the greatest profit or least cost. LP is used in a wide variety of business applications, as the examples and homework problems in this module reveal.

Key Terms

Linear programming (LP) (p. 712)
Objective function (p. 712)
Constraints (p. 712)
Graphical solution approach (p. 714)
Decision variables (p. 714)
Feasible region (p. 715)
Iso-profit line method (p. 715)
Corner-point method (p. 717)
Parameter (p. 717)
Sensitivity analysis (p. 718)
Shadow price (or dual value) (p. 719)
Iso-cost (p. 720)
Simplex method (p. 725)
Binary variables (p. 726)

Discussion Questions

1. List at least four applications of LP problems.
2. What is a "corner point"? Explain why solutions to LP problems focus on corner points.
3. Define the feasible region of a graphical LP problem. What is a feasible solution?
4. Each LP problem that has a feasible region has an infinite number of solutions. Explain.
5. Under what circumstances is the objective function more important than the constraints in an LP model?
6. Under what circumstances are the constraints more important than the objective function in an LP model?
7. Why is the diet problem, in practice, applicable for animals but not particularly for people?
8. How many feasible solutions are there in a linear program? Which ones do we need to examine to find the optimal solution?
9. Define shadow price (or dual value).
10. Explain how to use the iso-cost line in a graphical minimization problem.
11. Compare how the corner-point and iso-profit line methods work for solving graphical problems.
12. Where a constraint crosses the vertical or horizontal axis, the quantity is fairly obvious. How does one go about finding the quantity coordinates where two constraints cross, not at an axis?
13. Suppose an LP (maximization) problem has been solved and that the optimal value of the objective function is $300. Suppose an additional constraint is added to this problem. Explain how this might affect each of the following:
    a) The feasible region.
    b) The optimal value of the objective function.

Using Software to Solve LP Problems

All LP problems can be solved with the simplex method, using software such as Excel, Excel OM, or POM for Windows.

CREATING YOUR OWN EXCEL SPREADSHEETS

Excel offers the ability to analyze linear programming problems using built-in problem-solving tools. Excel's tool is named *Solver*.

We use Excel to set up the Glickman Electronics problem in Program B.3. The objective and constraints are repeated here:

$$\text{Objective function: Maximize profit} = \$7(\text{No. of x-pods}) + \$5(\text{No. of BlueBerrys})$$

$$\text{Subject to: } 4(\text{x-pods}) + 3(\text{BlueBerrys}) \le 240$$

$$2(\text{x-pods}) + 1(\text{BlueBerry}) \le 100$$

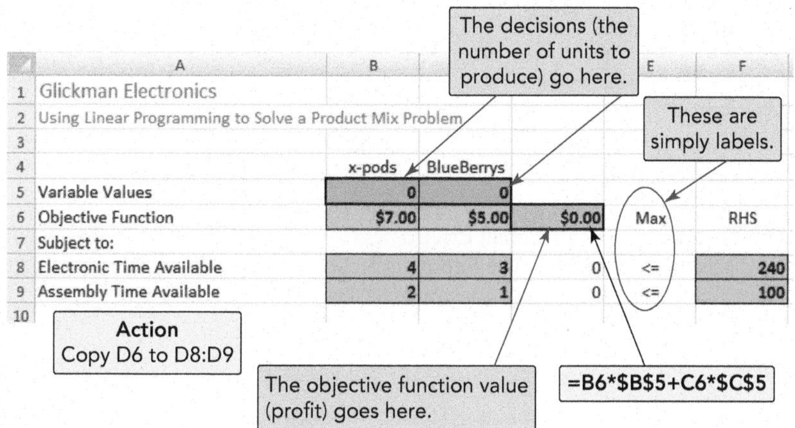

Program **B.3**

Using Excel to Formulate the Glickman Electronics Problem

Solver should appear on the far right of the Data tab. If Solver is not on the Data tab, then if using Windows, click on File, then Options, then Add-Ins. Next to Manage: at the bottom, make sure that Excel Add-Ins is selected, and click on the Go... button. Check Solver Add-In, and click OK. (Or if using Excel for Mac, click on Tools on the top (Excel) menu, then Add-ins. Check Solver Add-in and click OK.) The Solver dialog box will appear by clicking on Data, then Solver. Program B.4 shows how to use Solver to find the optimal (very best) solution to the Glickman Electronics problem. Click on Solve, and the solution will automatically appear in the spreadsheet in the green and blue cells.

Program **B.4**

Solver Dialog Boxes for the Glickman Electronics Problem

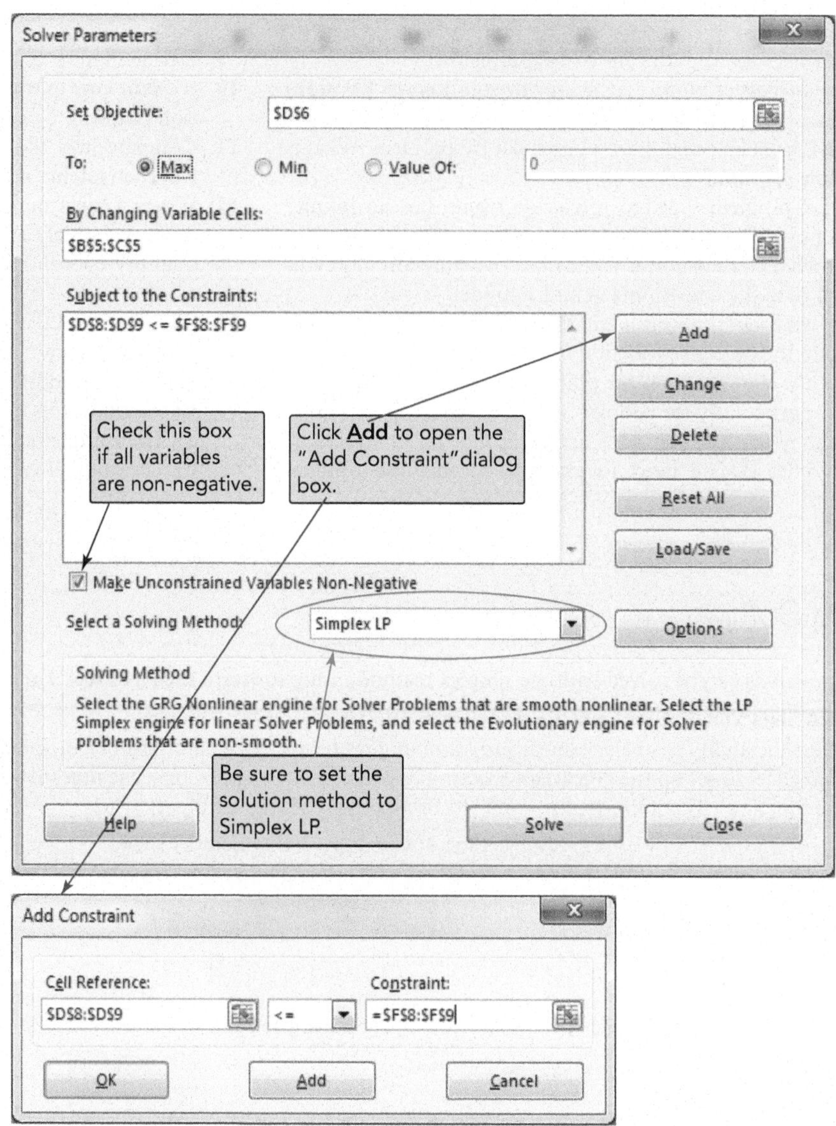

The Excel screen in Program B.5 shows Solver's solution to the Glickman Electronics Company problem. Note that the optimal solution is now shown in cells B5 and C5, which serve as the variables. The Reports selections perform more extensive analysis of the solution and its environment. Excel's sensitivity analysis capability was illustrated earlier in Program B.1.

Program **B.5**

Excel Solution to Glickman Electronics LP Problem

	A	B	C	D	E	F
1	Glickman Electronics					
2	Using Linear Programming to Solve a Product Mix Problem					
3						
4		x-pods	BlueBerrys			
5	Variable Values	30	40			
6	Objective Function	$7.00	$5.00	$410.00	Max	RHS
7	Subject to:					
8	Electronic Time Available	4	3	240	<=	240
9	Assembly Time Available	2	1	100	<=	100

PX USING EXCEL OM AND POM FOR WINDOWS

Excel OM and POM for Windows can handle relatively large LP problems. As output, the software provides optimal values for the variables, optimal profit or cost, and sensitivity analysis. In addition, POM for Windows provides graphical output for problems with only two variables. Appendix II provides further details.

Solved Problems

SOLVED PROBLEM B.1

Smith's, a Niagara, New York, clothing manufacturer that produces men's shirts and pajamas, has two primary resources available: sewing-machine time (in the sewing department) and cutting-machine time (in the cutting department). Over the next month, owner Barbara Smith can schedule up to 280 hours of work on sewing machines and up to 450 hours of work on cutting machines. Each shirt produced requires 1.00 hour of sewing time and 1.50 hours of cutting time. Producing each pair of pajamas requires .75 hours of sewing time and 2 hours of cutting time.

To express the LP constraints for this problem mathematically, we let:

$$X_1 = \text{number of shirts produced}$$
$$X_2 = \text{number of pajamas produced}$$

SOLUTION

First constraint: $1X_1 + .75X_2 \le 280$ hours of sewing-machine time available—our first scarce resource

Second constraint: $1.5X_1 + 2X_2 \le 450$ hours of cutting-machine time available—our second scarce resource

Note: This means that each pair of pajamas takes 2 hours of the cutting resource. Smith's accounting department analyzes cost and sales figures and states that each shirt produced will yield a \$4 contribution to profit and that each pair of pajamas will yield a \$3 contribution to profit.

This information can be used to create the LP *objective function* for this problem:

$$\text{Objective function: Maximize total contribution to profit} = \$4X_1 + \$3X_2$$

SOLVED PROBLEM B.2

We want to solve the following LP problem for Kevin Caskey Wholesale Inc. using the corner-point method:

$$\text{Objective: Maximize profit} = \$9X_1 + \$7X_2$$

$$\text{Constraints:} \quad 2X_1 + 1X_2 \le 40$$
$$X_1 + 3X_2 \le 30$$
$$X_1, X_2 \ge 0$$

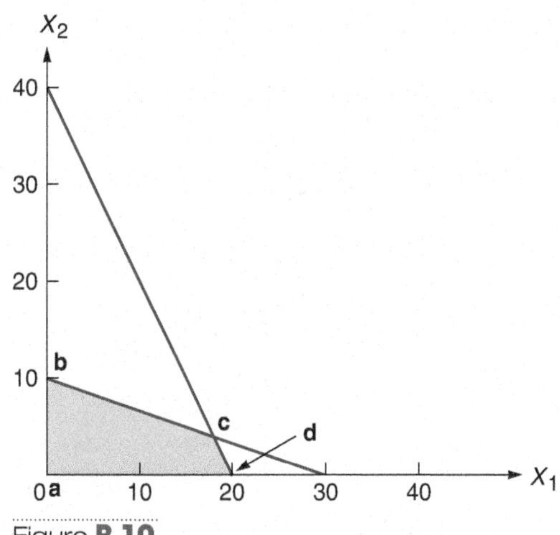

Figure **B.10**

K. Caskey Wholesale Inc.'s Feasible Region

SOLUTION

Figure B.10 illustrates these constraints:

Corner-point **a**: $(X_1 = 0, X_2 = 0)$ Profit $= 0$

Corner-point **b**: $(X_1 = 0, X_2 = 10)$ Profit $= 9(0) + 7(10) = \$70$

Corner-point **d**: $(X_1 = 20, X_2 = 0)$ Profit $= 9(20) + 7(0) = \$180$

Corner-point **c** is obtained by solving equations $2X_1 + 1X_2 = 40$ and $X_1 + 3X_2 = 30$ simultaneously. Multiply the second equtation by −2 and add it to the first.

$$\begin{aligned} 2X_1 + 1X_2 &= 40 \\ -2X_1 - 6X_2 &= -60 \\ \hline -5X_2 &= -20 \\ \text{Thus } X_2 &= 4 \end{aligned}$$

And $X_1 + 3(4) = 30$ or $X_1 + 12 = 30$ or $X_1 = 18$

Corner-point **c**: $(X_1 = 18, X_2 = 4)$ Profit $= 9(18) + 7(4) = \$190$

Hence the optimal solution is:

$$(X_1 = 18,\ X_2 = 4) \quad \text{Profit} = \$190$$

SOLVED PROBLEM B.3

Holiday Meal Turkey Ranch is considering buying two different types of turkey feed. Each feed contains, in varying proportions, some or all of the three nutritional ingredients essential for fattening turkeys. Brand Y feed costs the ranch \$.02 per pound. Brand Z costs \$.03 per pound. The rancher would like to determine the lowest-cost diet that meets the minimum monthly intake requirement for each nutritional ingredient.

The following table contains relevant information about the composition of brand Y and brand Z feeds, as well as the minimum monthly requirement for each nutritional ingredient per turkey.

COMPOSITION OF EACH POUND OF FEED			
INGREDIENT	BRAND Y FEED	BRAND Z FEED	MINIMUM MONTHLY REQUIREMENT
A	5 oz	10 oz	90 oz
B	4 oz	3 oz	48 oz
C	.5 oz	0 oz	1.5 oz
Cost/lb	\$.02	\$.03	

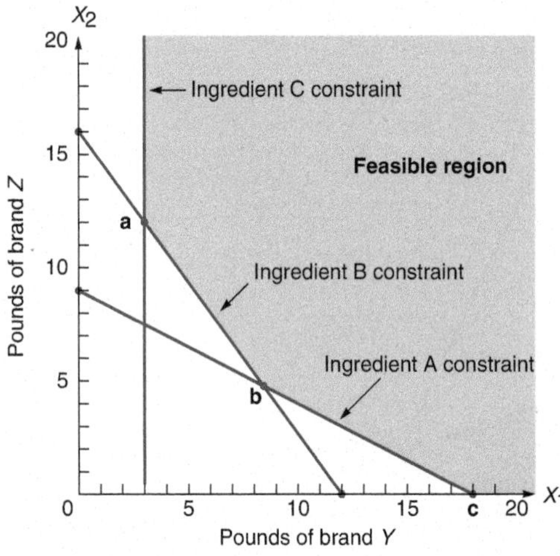

Figure **B.11**

Feasible Region for the Holiday Meal Turkey Ranch Problem

STUDENT TIP

Note that the last line parallel to the 54¢ iso-cost line that touches the feasible region indicates the optimal corner point.

SOLUTION

If we let:

$$X_1 = \text{number of pounds of brand Y feed purchased}$$
$$X_2 = \text{number of pounds of brand Z feed purchased}$$

then we may proceed to formulate this linear programming problem as follows:

$$\text{Objective: Minimize cost (in cents)} = 2X_1 + 3X_2$$

subject to these constraints:

$$5X_1 + 10X_2 \geq 90 \text{ oz} \quad (\textit{ingredient A constraint})$$
$$4X_1 + 3X_2 \geq 48 \text{ oz} \quad (\textit{ingredient B constraint})$$
$$\tfrac{1}{2}X_1 \geq 1\tfrac{1}{2} \text{ oz} \quad (\textit{ingredient C constraint})$$

Figure B.11 illustrates these constraints.

The iso-cost line approach may be used to solve LP minimization problems such as that of the Holiday Meal Turkey Ranch. As with iso-profit lines, we need not compute the cost at each corner point, but instead draw a series of parallel cost lines. The last cost point to touch the feasible region provides us with the optimal solution corner.

For example, we start in Figure B.12 by drawing a 54¢ cost line, namely, $54 = 2X_1 + 3X_2$. Obviously, there are many points in the feasible region that would yield a lower total cost. We proceed to move our iso-cost line toward the lower left, in a plane parallel to the 54¢ solution line. The last point we touch while still in contact with the feasible region is the same as corner point **b** of Figure B.11. It has the coordinates ($X_1 = 8.4$, $X_2 = 4.8$) and an associated cost of 31.2 cents.

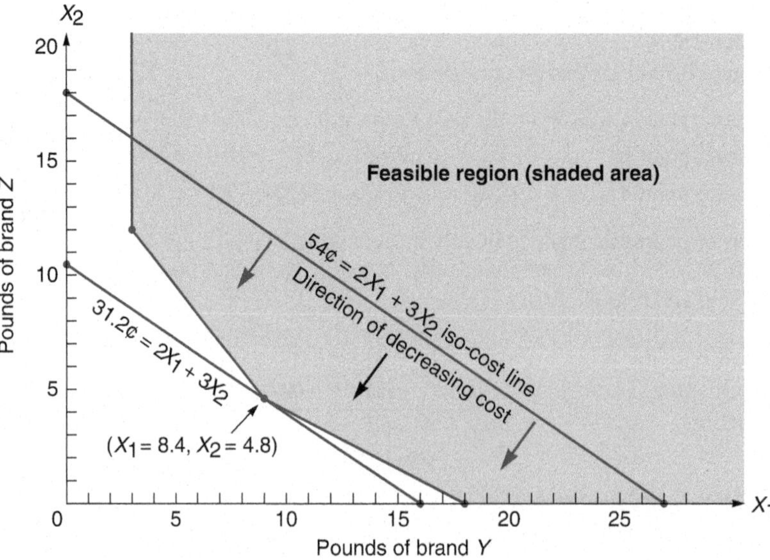

Figure **B.12**

Graphical Solution to the Holiday Meal Turkey Ranch Problem Using the Iso-Cost Line

Problems

Note: **PX** means the problem may be solved with POM for Windows and/or Excel OM.

Problem B.1 relates to Requirements of a Linear Programming Problem

• **B.1** The LP relationships that follow were formulated by Richard Martin at the Long Beach Chemical Company. Which ones are invalid for use in an LP problem, and why?

$$\text{Maximize} = 6X_1 + \tfrac{1}{2}X_1X_2 + 5X_3$$

$$\text{Subject to:} \quad 4X_1X_2 + 2X_3 \le 70$$

$$7.9X_1 - 4X_2 \ge 15.6$$

$$3X_1 + 3X_2 + 3X_3 \ge 21$$

$$19X_2 - \tfrac{1}{3}X_3 = 17$$

$$-X_1 - X_2 + 4X_3 = 5$$

$$4X_1 + 2X_2 + 3\sqrt{X_3} \le 80$$

Problems B.2–B.21 relate to Graphical Solution to a Linear Programming Problem

• **B.2** Solve the following LP problem graphically:

$$\text{Maximize profit} = 4X + 6Y$$

$$\text{Subject to:} \quad X + 2Y \le 8$$

$$5X + 4Y \le 20$$

$$X,\ Y \ge 0$$ **PX**

• **B.3** Solve the following LP problem graphically:

$$\text{Maximize profit} = X + 10Y$$

$$\text{Subject to:} \quad 4X + 3Y \le 36$$

$$2X + 4Y \le 40$$

$$Y \ge 3$$

$$X,\ Y \ge 0$$ **PX**

•• **B.4** Consider the following LP problem:

$$\text{Maximize profit} = 30X_1 + 10X_2$$

$$\text{Subject to:} \quad 3X_1 + X_2 \le 300$$

$$X_1 + X_2 \le 200$$

$$X_1 \le 100$$

$$X_2 \ge 50$$

$$X_1,\ X_2 \ge 0$$

a) Solve the problem graphically.
b) Is there more than one optimal solution? Explain. **PX**

• **B.5** The Attaran Corporation manufactures two electrical products: portable air conditioners and portable heaters. The assembly process for each is similar in that both require a certain amount of wiring and drilling. Each air conditioner takes 3 hours of wiring and 2 hours of drilling. Each heater must go through 2 hours of wiring and 1 hour of drilling. During the next production period, 240 hours of wiring time are available and up to 140 hours of drilling time may be used. Each air conditioner sold yields a profit of $25. Each heater assembled may be sold for a $15 profit.

Formulate and solve this LP production-mix situation, and find the best combination of air conditioners and heaters that yields the highest profit. **PX**

• **B.6** The Christina Alvarez Company manufactures two lines of designer yard gates, called model A and model B. Every gate requires blending a certain amount of steel and zinc; the company has available a total of 25,000 lb of steel and 6,000 lb of zinc. Each model A gate requires a mixture of 125 lb of steel and 20 lb of zinc, and each yields a profit of $90. Each model B gate requires 100 lb of steel and 30 lb of zinc and can be sold for a profit of $70.

Find by graphical LP the best production mix of yard gates. **PX**

•• **B.7** Green Vehicle Inc. manufactures electric cars and small delivery trucks. It has just opened a new factory where the C1 car and the T1 truck can both be manufactured. To make either vehicle, processing in the assembly shop and in the paint shop are required. It takes 1/40 of a day and 1/60 of a day to paint a truck of type T1 and a car of type C1 in the paint shop, respectively. It takes 1/50 of a day to assemble either type of vehicle in the assembly shop.

A T1 truck and a C1 car yield profits of $300 and $220, respectively, per vehicle sold.

a) Define the objective function and constraint equations.
b) Graph the feasible region.
c) What is a maximum-profit daily production plan at the new factory?
d) How much profit will such a plan yield, assuming whatever is produced is sold? **PX**

• **B.8** The Lifang Wu Corporation manufactures two models of industrial robots, the Alpha 1 and the Beta 2. The firm employs 5 technicians, working 160 hours each per month, on its assembly line. Management insists that full employment (that is, *all* 160 hours of time) be maintained for each worker during next month's operations. It requires 20 labor-hours to assemble each Alpha 1 robot and 25 labor-hours to assemble each Beta 2 model. Wu wants to see at least 10 Alpha 1s and at least 15 Beta 2s produced during the production period. Alpha 1s generate a $1,200 profit per unit, and Beta 2s yield $1,800 each.

Determine the most profitable number of each model of robot to produce during the coming month. **PX**

•• **B.9** Consider the following LP problem developed at Zafar Malik's Carbondale, Illinois, optical scanning firm:

$$\text{Maximize profit} = \$1X_1 + \$1X_2$$

$$\text{Subject to:} \quad 2X_1 + 1X_2 \le 100$$

$$1X_1 + 2X_2 \le 100$$

a) What is the optimal solution to this problem? Solve it graphically.
b) If a technical breakthrough occurred that raised the profit per unit of X_1 to $3, would this affect the optimal solution?
c) Instead of an increase in the profit coefficient X_1 to $3, suppose that profit was overestimated and should only have been $1.25. Does this change the optimal solution? **PX**

• **B.10** A craftsman named William Barnes builds two kinds of birdhouses, one for wrens and a second for bluebirds. Each wren birdhouse takes 4 hours of labor and 4 units of lumber. Each bluebird house requires 2 hours of labor and 12 units of lumber. The craftsman has available 60 hours of labor and 120 units of lumber. Wren houses yield a profit of $6 each, and bluebird houses yield a profit of $15 each.

a) Write out the objective and constraints.
b) Solve graphically. **PX**

•• **B.11** Each coffee table produced by Kevin Watson Designers nets the firm a profit of $9. Each bookcase yields a $12 profit. Watson's firm is small and its resources limited. During any given production period (of 1 week), 10 gallons of varnish and 12 lengths of high-quality redwood are available. Each coffee table requires approximately 1 gallon of varnish and 1 length of redwood. Each bookcase takes 1 gallon of varnish and 2 lengths of wood.

Formulate Watson's production-mix decision as a linear programming problem, and solve. How many tables and bookcases should be produced each week? What will the maximum profit be? **PX**

• **B.12** Par, Inc., produces a standard golf bag and a deluxe golf bag on a weekly basis. Each golf bag requires time for cutting and dyeing and time for sewing and finishing, as shown in the following table:

	HOURS REQUIRED PER BAG	
PRODUCT	CUTTING AND DYEING	SEWING AND FINISHING
Standard bag	1/2	1
Deluxe bag	1	2/3

The profits per bag and weekly hours available for cutting and dyeing and for sewing and finishing are as follows:

PRODUCT	PROFIT PER UNIT ($)
Standard bag	10
Deluxe bag	8

ACTIVITY	WEEKLY HOURS AVAILABLE
Cutting and dyeing	300
Sewing and finishing	360

Par, Inc., will sell whatever quantities it produces of these two products.

a) Find the mix of standard and deluxe golf bags to produce per week that maximizes weekly profit from these activities.
b) What is the value of the profit? **PX**

•• **B.13** The Denver advertising agency promoting the new Breem dishwashing detergent wants to get the best exposure possible for the product within the $100,000 advertising budget ceiling placed on it. To do so, the agency needs to decide how much of the budget to spend on each of its two most effective media: (1) television spots during the afternoon hours and (2) large ads in the city's Sunday newspaper. Each television spot costs $3,000; each Sunday newspaper ad costs $1,250. The expected exposure, based on industry ratings, is 35,000 viewers for each TV commercial and 20,000 readers for each newspaper advertisement. The agency director, Deborah Kellogg, knows from experience that it is important to use both media in order to reach the broadest spectrum of potential Breem customers. She decides that at least 5 but no more than 25 television spots should be ordered, and that at least 10 newspaper ads should be contracted. How many times should each of the two media be used to obtain maximum exposure while staying within the budget? Use the graphical method to solve. **PX**

Additional problem **B.14** *is available in* **MyLab Operations Management.** *It involves a hospital trying to maximize revenues.*

• **B.15** Walter Wallace is trying to determine how many units each of two commercial multiline telephones to produce each day. One of these is the standard model; the other one is the deluxe model. The profit per unit on the standard model is $40, on the deluxe model $60. Each unit requires 30 minutes assembly time. The standard model requires 10 minutes of inspection time, the deluxe model 15 minutes. The company must fill an order for six standard phones. There are 450 minutes of assembly time and 180 minutes of inspection time available each day. How many units of each product should be manufactured to maximize profits? **PX**

• **B.16** Solve the following linear programming problem graphically:

$$\begin{aligned} \text{Maximize } Z &= 3X + 5Y \\ \text{Subject to: } 4X + 4Y &\le 48 \\ 1X + 2Y &\le 20 \\ Y &\ge 2 \\ X, Y &\ge 0 \end{aligned}$$ **PX**

• **B.17** The Outdoor Furniture Corporation manufactures two products, benches and picnic tables, for use in yards and parks. The firm has two main resources: its carpenters (labor force) and a supply of redwood for use in furniture. During the next production cycle, 1,200 hours of labor are available under a union agreement. The firm also has a stock of 3,500 board feet of quality redwood. Each bench that Outdoor Furniture produces requires 4 labor-hours and 10 board feet of redwood; each picnic table takes 6 labor-hours and 35 board feet of redwood. Completed benches will yield a profit of $9 each, and tables will result in a profit of $20 each. How many benches and tables should operations manager Shilei Yang produce in order to obtain the largest possible profit? Use the graphical LP approach. **PX**

• **B.18** Solve the following LP problem graphically:

$$\begin{aligned} \text{Maximize } Z &= 4X_1 + 4X_2 \\ \text{Subject to: } 3X_1 + 5X_2 &\le 150 \\ X_1 - 2X_2 &\le 10 \\ 5X_1 + 3X_2 &\le 150 \\ X_1, X_2 &\ge 0 \end{aligned}$$ **PX**

•• **B.19** Gupta Furniture manufactures two different types of china cabinets, a French provincial model and a Danish modern model. Each cabinet produced must go through three departments: carpentry, painting, and finishing. The accompanying table contains all relevant information concerning production times per cabinet produced and production capacities for each operation per day, along with net revenue per unit produced. The firm has a contract with a Miami distributor to produce a minimum of 300 of each cabinet per week (or 60 cabinets per day). Owner Sushil Gupta would like to determine a product mix to maximize his daily revenue.

CABINET STYLE	CARPENTRY (HOURS PER CABINET)	PAINTING (HOURS PER CABINET)	FINISHING (HOURS PER CABINET)	NET REVENUE PER CABINET
French provincial	3	1.5	0.75	$ 28
Danish modern	2	1	0.75	$ 25
Department capacity (hours)	360	200	125	

Formulate this as an LP problem and solve. **PX**

• **B.20** Solve the following LP problem graphically. Indicate the corner points on your graph.

$$\begin{aligned} \text{Maximize profit } &= \$3X_1 + \$5X_2 \\ \text{Subject to: } X_2 &\le 6 \\ 3X_1 + 2X_2 &\le 18 \\ X_1, X_2 &\ge 0 \end{aligned}$$ **PX**

••• **B.21** Aretha's Bicycle Company (ABC) has the hottest new product on the upscale toy market—boys' and girls' bikes in bright fashion colors, with oversize hubs and axles; shell design safety tires;

strong padded frames; chrome-plated chains, brackets, and valves; and non-slip handlebars. Due to the seller's market for high-quality toys for the newest baby boomers, ABC can sell all the bicycles it manufactures at the following prices: boys' bikes, \$220; girls' bikes, \$175. This is the price payable to ABC at its Orlando plant.

The firm's accountant, V. R. Dondeti, has determined that direct labor costs will be 45% of the price that ABC receives for the boys' model and 40% of the price received for the girls' model. Production costs, other than labor but excluding painting and packaging, are \$44 per boys' bicycle and \$30 per girls' bicycle. Painting and packaging are \$20 per bike, regardless of model.

The Orlando plant's overall production capacity is 390 bicycles per day. Each boys' bike requires 2.5 labor-hours, each girls' model 2.4 labor-hours. ABC currently employs 120 workers, each of whom puts in an 8-hour day. The firm has no desire to hire or fire to affect labor availability, for it believes its stable workforce is one of its biggest assets.

Using a graphic approach, determine the best product mix for ABC. **PX**

Problems B.22–B.24 relate to Sensitivity Analysis

•• **B.22** Kalyan Singhal Corp. makes three products, and it has three machines available as resources as given in the following LP problem:

$$\text{Maximize contribution} = 4X_1 + 4X_2 + 7X_3$$

$$\begin{aligned} \text{Subject to: } 1X_1 + 7X_2 + 4X_3 &\le 100 \text{ (hours on machine 1)} \\ 2X_1 + 1X_2 + 7X_3 &\le 110 \text{ (hours on machine 2)} \\ 8X_1 + 4X_2 + 1X_3 &\le 100 \text{ (hours on machine 3)} \end{aligned}$$

a) Determine the optimal solution using LP software.
b) Is there unused time available on any of the machines with the optimal solution?
c) What would it be worth to the firm to make an additional hour of time available on the third machine?
d) How much would the firm's profit increase if an extra 10 hours of time were made available on the second machine at no extra cost? **PX**

Additional problems **B.23–B.24** *are available in* MyLab Operations Management. *B.23 involves analyzing an Excel Solver printout. B.24 examines sensitivity of 15 manufactured products.*

Problems B.25–B.33 relate to Solving Minimization Problems

• **B.25** Solve the following linear program graphically:

$$\begin{aligned} \text{Minimize cost} &= X_1 + X_2 \\ \text{Subject to: } 8X_1 + 16X_2 &\ge 64 \\ X_1 &\ge 0 \\ X_2 &\ge -2 \end{aligned}$$ **PX**

(*Note:* X_2 values can be negative in this problem.)

• **B.26** Solve the following LP problem graphically:

$$\begin{aligned} \text{Minimize cost} &= 24X + 15Y \\ \text{Subject to: } 7X + 11Y &\ge 77 \\ 16X + 4Y &\ge 80 \\ X, Y &\ge 0 \end{aligned}$$ **PX**

•• **B.27** Kaleena Turner Food Processors wishes to introduce a new brand of dog biscuits composed of chicken- and liver-flavored biscuits that meet certain nutritional requirements. The liver-flavored biscuits contain 1 unit of nutrient A and 2 units of nutrient B; the chicken-flavored biscuits contain 1 unit of nutrient A and 4 units of nutrient B. According to federal requirements, there must be at least 40 units of nutrient A and 60 units of nutrient B in a package of the new mix. In addition, the company has decided that there can be no more than 15 liver-flavored biscuits in a package. If it costs 1¢ to make 1 liver-flavored biscuit and 2¢ to make 1 chicken-flavored, what is the optimal product mix for a package of the biscuits to minimize the firm's cost?

a) Formulate this as a linear programming problem.
b) Solve this problem graphically, giving the optimal values of all variables.
c) What is the total cost of a package of dog biscuits using the optimal mix? **PX**

• **B.28** The Sweet Smell Fertilizer Company markets bags of manure labeled "not less than 60 lb dry weight." The packaged manure is a combination of compost and sewage wastes. To provide good-quality fertilizer, each bag should contain at least 30 lb of compost but no more than 40 lb of sewage. Each pound of compost costs Sweet Smell 5¢ and each pound of sewage costs 4¢. Use a graphical LP method to determine the least-cost blend of compost and sewage in each bag. **PX**

• **B.29** Consider Paulina Jordan's following LP formulation:

$$\begin{aligned} \text{Minimize cost} &= \$1X_1 + \$2X_2 \\ \text{Subject to: } X_1 + 3X_2 &\ge 90 \\ 8X_1 + 2X_2 &\ge 160 \\ 3X_1 + 2X_2 &\ge 120 \\ X_2 &\le 70 \end{aligned}$$

a) Graphically illustrate the feasible region to indicate to Paulina which corner point produces the optimal solution.
b) What is the cost of this solution? **PX**

• **B.30** Solve the following LP problem graphically:

$$\begin{aligned} \text{Minimize cost} &= 4X_1 + 5X_2 \\ \text{Subject to: } X_1 + 2X_2 &\ge 80 \\ 3X_1 + X_2 &\ge 75 \\ X_1, X_2 &\ge 0 \end{aligned}$$ **PX**

•• **B.31** How many corner points are there in the feasible region of the following problem?

$$\begin{aligned} \text{Minimize cost} &= X - Y \\ \text{Subject to: } X &\le 4 \\ -X &\le 2 \\ X + 2Y &\le 6 \\ -X + 2Y &\le 8 \\ Y &\ge 0 \end{aligned}$$

(*Note: X* values can be negative in this problem.)

•• **B.32** This is the slack time of year at JES, Inc. The firm would actually like to shut down the plant, but if it laid off its core employees, they would probably go to work for a competitor. JES could keep its core (full-time, year-round) employees busy by making 10,000 round tables per month, or by making 20,000 square tables per month (or some ratio thereof). JES does, however, have a contract with a supplier to buy a minimum of 5,000 square tables per month. Handling and storage costs per round table will be \$10; these costs would be \$8 per square table.

Draw a graph, algebraically describe the constraint inequalities and the objective function, identify the points bounding the feasible solution area, and find the cost at each point and the optimum solution. Let X_1 equal the thousands of round tables per month and X_2 equal the thousands of square tables per month. **PX**

•• **B.33** Lizao Zhang, the advertising director for Diversey Paint and Supply, a chain of four retail stores on Chicago's North Side, is considering two media possibilities. One plan is for a series of half-page ads in the Sunday *Chicago Tribune* newspaper, and the other is for advertising time on Chicago TV. The stores are expanding their line of do-it-yourself tools, and the advertising director is interested in an exposure level of at least 40% within the city's neighborhood and 60% in northwest suburban areas.

The TV viewing time under consideration has an exposure rating per spot of 5% in city homes and 3% in the northwest suburbs. The Sunday newspaper has corresponding exposure rates of 4% and 3% per ad. The cost of a half-page *Tribune* advertisement is $925; a television spot costs $2,000.

Diversey Paint would like to select the least costly advertising strategy that would meet the desired exposure levels. Formulate and solve this LP problem. **PX**

Problems B.34–B.40 relate to Linear Programming Applications

••• **B.34** The Hills County, Michigan, superintendent of education is responsible for assigning students to the three high schools in his county. He recognizes the need to bus a certain number of students, because several sectors, A–E, of the county are beyond walking distance to a school. The superintendent partitions the county into five geographic sectors as he attempts to establish a plan that will minimize the total number of student miles traveled by bus. He also recognizes that if students happen to live in a certain sector and are assigned to the high school in that sector, there is no need to bus them because they can walk to school. The three schools are located in sectors B, C, and E.

Hans Magelssen/Shutterstock

The accompanying table reflects the number of high-school-age students living in each sector and the distance in miles from each sector to each school:

	DISTANCE TO SCHOOL			
SECTOR	SCHOOL IN SECTOR B	SCHOOL IN SECTOR C	SCHOOL IN SECTOR E	NUMBER OF STUDENTS
A	5	8	6	700
B	0	4	12	500
C	4	0	7	100
D	7	2	5	800
E	12	7	0	400
				2,500

Each high school has a capacity of 900 students.

a) Set up the objective function and constraints of this problem using LP so that the total number of student miles traveled by bus is minimized.
b) Solve the problem. **PX**

•• **B.35** The Rio Credit Union has $250,000 available to invest in a 12-month commitment. The money can be placed in Brazilian treasury notes yielding an 8% return or in riskier high-yield bonds at an average rate of return of 9%. Credit union regulations require diversification to the extent that at least 50% of the investment be placed in Treasury notes. It is also decided that no more than 40% of the investment be placed in bonds. How much should the Rio Credit Union invest in each security so as to maximize its return on investment? **PX**

•• **B.36** Wichita's famous Sethi Restaurant is open 24 hours a day. Servers report for duty at 3 A.M., 7 A.M., 11 A.M., 3 P.M., 7 P.M., or 11 P.M., and each works an 8-hour shift. The following table shows the minimum number of workers needed during the 6 periods into which the day is divided:

PERIOD	TIME	NUMBER OF SERVERS REQUIRED
1	3 A.M.–7 A.M.	3
2	7 A.M.–11 A.M.	12
3	11 A.M.–3 P.M.	16
4	3 P.M.–7 P.M.	9
5	7 P.M.–11 P.M.	11
6	11 P.M.–3 A.M.	4

Owner Avanti Sethi's scheduling problem is to determine how many servers should report for work at the start of each time period in order to minimize the total staff required for one day's operation. (*Hint:* Let X_i equal the number of servers beginning work in time period i, where $i = 1, 2, 3, 4, 5, 6$.) **PX**

•• **B.37** Leach Distributors packages and distributes industrial supplies. A standard shipment can be packaged in a class A container, a class K container, or a class T container. A single class A container yields a profit of $9; a class K container, a profit of $7; and a class T container, a profit of $15. Each shipment prepared requires a certain amount of packing material and a certain amount of time.

RESOURCES NEEDED PER STANDARD SHIPMENT		
CLASS OF CONTAINER	PACKING MATERIAL (POUNDS)	PACKING TIME (HOURS)
A	2	2
K	1	6
T	3	4
Total resource available each week:	130 pounds	240 hours

Hugh Leach, head of the firm, must decide the optimal number of each class of container to pack each week. He is bound by the previously mentioned resource restrictions but also decides that he must keep his 6 full-time packers employed all 240 hours (6 workers × 40 hours) each week.

Formulate and solve this problem using LP software. **PX**

••• **B.38** Tri-State Manufacturing has three factories (1, 2, and 3) and three warehouses (A, B, and C). The following table shows the shipping costs between each factory and warehouse, the factory manufacturing capabilities (in thousands), and the warehouse capacities (in thousands). Management would like to keep the warehouses filled to capacity in order to generate demand.

TO / FROM	WAREHOUSE A	WAREHOUSE B	WAREHOUSE C	PRODUCTION CAPABILITY
Factory 1	$ 6	$ 5	$ 3	6
Factory 2	$ 8	$10	$ 8	8
Factory 3	$11	$14	$18	10
Capacity	7	12	5	

a) Write the objective function and the constraint equations. Let X_{1A} = 1,000s of units shipped from factory 1 to warehouse A, and so on.
b) Solve by computer. **PX**

Additional problem **B.39** *is available in* MyLab Operations Management. *It uses LP to solve a project management crashing problem for a construction company.*

•••• **B.40** You have just been hired as a planner for the municipal school system, and your first assignment is to redesign the subsidized lunch program. In particular, you are to formulate the least expensive lunch menu that will still meet all state and federal nutritional guidelines.

The guidelines are as follows: A meal must be between 500 and 800 calories. It must contain at least 200 calories of protein, at least 200 calories of carbohydrates, and no more than 400 calories of fat. It also needs to have at least 200 calories of a food classified as a fruit or vegetable.

Table B.2 provides a list of the foods you can consider as possible menu items, with contract-determined prices and nutritional information. Note that all percentages sum to 100% per food—as all calories are protein, carbohydrate, or fat calories. For example, a serving of applesauce has 100 calories, all of which are carbohydrates, and it counts as a fruit/veg food. You are allowed to use fractional servings, such as 2.25 servings of turkey breast and a 0.33 portion of salad. Costs and nutritional attributes scale likewise: e.g., a 0.33 portion of salad costs $.30 and has 33 calories.

TABLE B.2 Data for Problem B.40

FOOD	COST/SERVING	CALORIES/SERVING	% PROTEIN	% CARBS	% FAT	FRUIT/VEG
Applesauce	$0.30	100	0%	100%	0%	Y
Canned corn	$0.40	150	20%	80%	0%	Y
Fried chicken	$0.90	250	55%	5%	40%	N
French fries	$0.20	400	5%	35%	60%	N
Mac and cheese	$0.50	430	20%	30%	50%	N
Turkey breast	$1.50	300	67%	0%	33%	N
Garden salad	$0.90	100	15%	40%	45%	Y

Formulate and solve as a linear problem. Print out your formulation in Excel showing the objective function coefficients and constraint matrix in standard form.

- Display, on a separate page, the full *Answer Report* as generated by Excel Solver.
- Highlight *and label as Z* the objective value for the optimal solution on the Answer Report.
- Highlight the nonzero decision variables for the optimal solution on the Answer Report.
- Display, on a separate page, the full *Sensitivity Report* as generated by Excel Solver. **PX**

Problems B.41–B.42 relate to Integer and Binary Variables

•• **B.41** Rollins Publishing needs to decide what textbooks from the following table to publish.

TEXTBOOK	DEMAND	FIXED COST	VARIABLE COST	SELLING PRICE
Book 1	9,000	$12,000	$19	$40
Book 2	8,000	$21,000	$28	$60
Book 3	5,000	$15,000	$30	$52
Book 4	6,000	$10,000	$20	$34
Book 5	7,000	$18,000	$20	$45

For each book, the maximum demand, fixed cost of publishing, variable cost, and selling price are provided. Rollins has the capacity to publish a total of 20,000 books.
a) Formulate this problem to determine which books should be selected and how many of each should be published to maximize profit.
b) Solve using computer software. **PX**

•• **B.42** Porter Investments needs to develop an investment portfolio for Mrs. Singh from the following list of possible investments:

INVESTMENT	COST	EXPECTED RETURN
A	$10,000	$ 700
B	$12,000	$1,000
C	$ 3,500	$ 390
D	$ 5,000	$ 500
E	$ 8,500	$ 750
F	$ 8,000	$ 640
G	$ 4,000	$ 300

Mrs. Singh has a total of $60,000 to invest. The following conditions must be met: (1) If investment F is chosen, then investment G must also be part of the portfolio, (2) at least four investments should be chosen, and (3) of investments A and B, exactly one must be included. Formulate and solve this problem using LP software to determine which stocks should be included in Mrs. Singh's portfolio. **PX**

CASE STUDIES

Quain Lawn and Garden, Inc.

Bill and Jeanne Quain spent a career as a husband-and-wife real estate investment partnership in Atlantic City, New Jersey. When they finally retired to a 25-acre farm in nearby Cape May County, they became ardent amateur gardeners. Bill planted shrubs and fruit trees, and Jeanne spent her hours potting all sizes of plants. When the volume of shrubs and plants reached the point that the Quains began to think of their hobby in a serious vein, they built a greenhouse adjacent to their home and installed heating and watering systems.

By 2018, the Quains realized their retirement from real estate had really only led to a second career—in the plant and shrub business—and they filed for a New Jersey business license. Within a matter of months, they asked their attorney to file incorporation documents and formed the firm Quain Lawn and Garden, Inc.

Early in the new business's existence, the Quains recognized the need for a high-quality commercial fertilizer that they could blend themselves, both for sale and for their own nursery. Their goal was to keep their costs to a minimum while producing a top-notch product that was especially suited to the New Jersey climate.

Working with chemists at Rutgers University, the Quains blended "Quain-Grow." It consists of four chemical compounds, C-30, C-92, D-21, and E-11. The cost per pound for each compound is indicated in the following table:

CHEMICAL COMPOUND	COST PER POUND
C-30	\$.12
C-92	.09
D-21	.11
E-11	.04

The specifications for Quain-Grow are established as:

a) Chemical E-11 must constitute at least 15% of the blend.
b) C-92 and C-30 must together constitute at least 45% of the blend.
c) D-21 and C-92 can together constitute no more than 30% of the blend.
d) Quain-Grow is packaged and sold in 50-lb bags.

Discussion Questions

1. Formulate an LP problem to determine what blend of the four chemicals will allow the Quains to minimize the cost of a 50-lb bag of the fertilizer.
2. Solve to find the best solution.

Scheduling Challenges at Alaska Airlines

Video Case 

Good airline scheduling is essential to delivering outstanding customer service with high plane utilization rates. Airlines must schedule pilots, flight attendants, aircraft, baggage handlers, customer service agents, and ramp crews. At Alaska Airlines, it all begins with seasonal flight schedules that are developed 330 days in advance.

Revenue and marketing goals drive the potential routing decisions, but thousands of constraints impact these schedules. Using SABRE scheduling optimizer software, Alaska considers the number of planes available, seat capacity, ranges, crew availability, union contracts that dictate hours that crews can fly, and maintenance regulations that regularly take planes out of service, just to name a few. Alaska's scheduling department sends preliminary schedules to the human resources, maintenance, operations, customer service, marketing, and other departments for feedback before finalizing flight schedules.

Alaska Airlines' historic mission is to serve its extremely loyal customer base in the remote and unreachable small towns in Alaska. Serving many airports in Alaska is especially complex because the airline requires its pilots to have special skills to deal with extremely adverse weather, tight mountain passes, and short runways. Some airports lack full-time TSA agents or strong ground support and may not even be open 24 hours per day. In some cases, runways are not plowed because the village plow is busy clearing the roads for school buses. Navigational technology developed by Alaska Airlines has significantly reduced weather-related cancelled flights as Alaska can now land where many other carriers cannot.

After the SABRE optimizer schedules thousands of flights, scheduling activity turns to the next step: crew optimizing. The crew optimizer (Alaska uses Jeppersen software developed by Boeing and based on linear programming) attempts to eliminate unnecessary layovers and crew idle time while adhering to FAA and union restrictions. Alaska leads the industry in pilot "hard time" (i.e., the amount of time a pilot is being paid when passengers are actually being moved). After the crew requirements for every flight are determined, the 3,000 flight attendants and 1,500 pilots rank their preferred routings on a monthly basis. Personnel are assigned to each flight using seniority and feasibility.

Interestingly, not every pilot or flight attendant always bids on the Hawaii routes (about 20% of all flights), the long-haul East Coast routes, or the Mexico flights. Some prefer the flying challenge of the "milk run" flights to Ketchikan, Sitka, Juneau, Fairbanks, Anchorage, and back to Seattle, which are in keeping with the culture and contact with local residents.

As an airline that accentuates risk taking and empowers employees to think "out of the box," Alaska recently decided to experiment with a schedule change on its Seattle-to-Chicago route. Given crew restrictions on flying hours per day, the flight had previously included a crew layover in Chicago. When a company analyst documented the feasibility of running the same crew on the two 4-hour legs of the round trip (which implied an *extremely* tight turnaround schedule in Chicago), his data indicated that on 98.7% of the round trip flights, the crew would not "time out." His boss gave the go-ahead.

Discussion Questions*

1. Why is scheduling for Alaska more complex than for other airlines?
2. What operational considerations may prohibit Alaska from adding flights and more cities to its network?
3. What were the risks of keeping the same crew on the Seattle—Chicago—Seattle route?
4. Estimate the direct costs to the airline should the crew "time out" and not be able to fly its Boeing 737 back to Seattle from Chicago on the same day. These direct variable costs should include moving and parking the plane overnight along with hotel and meal costs for the crew and passengers. Do you think this is more advantageous than keeping a spare crew in Chicago?

*You may wish to view the video that accompanies this case before addressing these questions.

Endnotes

1. *Iso* means "equal" or "similar." Thus, an iso-profit line represents a line with all profits the same, in this case $210.
2. Sources: *The New York Times* (April 20, 2012); and *Sports Illustrated* (April 21, 2017).
3. See, for example, B. Render, R. M. Stair, M. Hanna, and T. Hale, *Quantitative Analysis for Management*, 13th ed. (Boston: Pearson, 2018): Chapters 7–9; or R. Balakrishnan, B. Render, R. M. Stair, and C. Munson, *Managerial Decision Modeling: Business Analytics with Spreadsheets,* 4th ed. (Boston: DeGruyter, 2017): Chapters 2–4.

Bibliography

Aboelmagd, Y. M. "Linear Programming Applications in Construction Sites." *Alexandria Engineering Journal* 57, no. 4 (December 2018), 4177–4187.

Anderson, R., et al. "Kidney Exchange and the Alliance for Paired Donation: Operations Research Changes the Way Kidneys Are Transplanted." *Interfaces* 45, no. 1 (January–February 2015), 26–42.

Balakrishman, R., B. Render, R. M. Stair, and C. Munson. *Managerial Decision Modeling: Business Analytics with Spreadsheets*, 4th ed. Boston: DeGruyter, 2017.

Harrod, S. "A Spreadsheet-Based, Matrix Formulation Linear Programming Lesson." *Decision Sciences Journal of Innovative Education* 7, no. 1 (January 2009): 249.

Render, B., R. M. Stair, M. Hanna, and T. Hale. *Quantitative Analysis for Management*, 13th ed. Boston: Pearson, 2018.

Sodhi, M. S., and S. Norri. "A Fast and Optimal Modeling Approach Applied to Crew Rostering at London Underground." *Annals of OR* 127 (March 2004): 259.

Taha, H. A. *Operations Research: An Introduction*, 9th ed. Boston: Pearson, 2011.

Thie, P., and G. E. Keough. *An Introduction to Linear Programming and Game Theory*, 3rd ed. New York: Wiley, 2011.

Tian, Z., P. Kouvelis, and C. L. Munson. "Understanding and Managing Product Line Complexity: Applying Sensitivity Analysis to a Large-Scale MILP Model to Price and Schedule New Customer Orders." *IIE Transactions* 47, no. 4 (2015), 307–328.

van Dooren, C. "A Review of the Use of Linear Programming to Optimize Diets, Nutritiously, Economically and Environmentally." *Frontiers in Nutrition* 5 (June 21, 2018), 48.

Module B *Rapid* Review

Main Heading	Review Material	MyLab Operations Management
WHY USE LINEAR PROGRAMMING?	■ **Linear programming (LP)**—A mathematical technique designed to help operations managers plan and make decisions relative to allocation of resources.	Concept Questions: 1.1–1.5 **VIDEO B.1** Scheduling Challenges at Alaska Airlines
REQUIREMENTS OF A LINEAR PROGRAMMING PROBLEM	✔ **Objective function**—A mathematical expression in linear programming that maximizes or minimizes some quantity (often profit or cost, but any goal may be used). ■ **Constraints**—Restrictions that limit the degree to which a manager can pursue an objective. All LP problems have four properties in common: 1. LP problems seek to *maximize* or *minimize* some quantity. We refer to this property as the *objective function* of an LP problem. 2. The presence of restrictions, or *constraints*, limits the degree to which we can pursue our objective. We want, therefore, to maximize or minimize a quantity (the objective function) subject to limited resources (the constraints). 3. There must be *alternative courses of action* to choose from. 4. The objective and constraints in linear programming problems must be expressed in terms of *linear equations* or inequalities.	Concept Questions: 2.1–2.6 Problem: B.1
FORMULATING LINEAR PROGRAMMING PROBLEMS	One of the most common linear programming applications is the *product-mix problem*. Two or more products are usually produced using limited resources. For example, a company might like to determine how many units of each product it should produce to maximize overall profit, given its limited resources. An important aspect of linear programming is that certain interactions will exist between variables. The more units of one product that a firm produces, the fewer it can make of other products.	Concept Questions: 3.1–3.4 Virtual Office Hours for Solved Problem: B.1 ACTIVE MODEL B.1
GRAPHICAL SOLUTION TO A LINEAR PROGRAMMING PROBLEM	✔ **Graphical solution approach**—A means of plotting a solution to a two-variable problem on a graph. ■ **Decision variables**—Choices available to a decision maker. Constraints of the form $X \geq 0$ are called *nonnegativity constraints*. ✔ **Feasible region**—The set of all feasible combinations of decision variables. Any point inside the feasible region represents a *feasible solution*, while any point outside the feasible region represents an *infeasible solution*. ■ **Iso-profit line method**—An approach to identifying the optimum point in a graphic linear programming problem. The line that touches a particular point of the feasible region will pinpoint the optimal solution. ■ **Corner-point method**—Another method for solving graphical linear programming problems. The mathematical theory behind linear programming states that an optimal solution to any problem will lie at a *corner point*, or an *extreme point*, of the feasible region. Hence, it is necessary to find only the values of the variables at each corner; the optimal solution will lie at one (or more) of them. This is the corner-point method. X_2 (2) (3) (1) (4) X_1	Concept Questions: 4.1–4.6 Problems: B.2–B.21 Virtual Office Hours for Solved Problem: B.2
SENSITIVITY ANALYSIS	✔ **Parameter**—A numerical value that is given in a model. ■ **Sensitivity analysis**—An analysis that projects how much a solution may change if there are changes in the variables or input data. Sensitivity analysis is also called *postoptimality analysis*. There are two approaches to determining just how sensitive an optimal solution is to changes: (1) a trial-and-error approach and (2) the analytic postoptimality method. To use the analytic postoptimality method, after an LP problem has been solved, we determine a range of changes in problem parameters that will not affect the optimal solution or change the variables in the solution. LP software has this capability.	Concept Questions: 5.1–5.5 Problems: B.22–B.24

Module B **Rapid Review** *continued*

Main Heading	Review Material	MyLab Operations Management
	While using the information in a sensitivity report to answer what-if questions, we assume that we are considering a change to only a *single* input data value at a time. That is, the sensitivity information does not generally apply to simultaneous changes in several input data values. ▪ **Shadow price** (or **dual value)**—The value of one additional unit of a scarce resource in LP. The shadow price is valid as long as the right-hand side of the constraint stays in a range within which all current corner points continue to exist. The information to compute the upper and lower limits of this range is given by the entries labeled Allowable Increase and Allowable Decrease in the sensitivity report.	
SOLVING MINIMIZATION PROBLEMS	▪ **Iso-cost**—An approach to solving a linear programming minimization problem graphically. The iso-cost line approach to solving minimization problems is analogous to the iso-profit approach for maximization problems, but successive iso-cost lines are drawn *inward* instead of outward.	Concept Questions: 6.1–6.3 Problems: B.25–B.33 Virtual Office Hours for Solved Problem: B.3
LINEAR PROGRAMMING APPLICATIONS	The *diet problem*, known in agricultural applications as the *feed-mix problem*, involves specifying a food or feed ingredient combination that will satisfy stated nutritional requirements at a minimum cost level. *Labor scheduling problems* address staffing needs over a specific time period. They are especially useful when managers have some flexibility in assigning workers to jobs that require overlapping or interchangeable talents.	Concept Questions: 7.1–7.3 Problems: B.34–B.39
THE SIMPLEX METHOD OF LP	▪ **Simplex method**—An algorithm for solving linear programming problems of all sizes. The simplex method is actually a set of instructions with which we examine corner points in a methodical fashion until we arrive at the best solution—highest profit or lowest cost. Computer programs (such as Excel OM and POM for Windows) and Excel's Solver add-in are available to solve linear programming problems via the simplex method.	Concept Questions: 8.1–8.2 Virtual Office Hours for Solved Problem: C.1 (note that this Module C video is an LP application of the transportation problem)
INTEGER AND BINARY VARIABLES	▪ **Binary variables**—Decision variables that can only take on the value of 0 or 1. Using computer software, decision variables for linear programs can be forced to be integer or even binary. Binary variables extend the flexibility of linear programs to include such options as mutually exclusive alternatives, either-or constraints, contingent decisions, fixed-charge problems, and threshold levels.	Concept Questions: 9.1–9.6
ADDITIONAL MYLAB OPERATIONS MANAGEMENT RESOURCES	✔ Video for Creating Your Own Excel Spreadsheets (Glickman Electronics Problem) ✔ Additional Case Study (Coastal States Chemical) ✔ Multiple Choice Case Questions (Coastal States Chemical) ✔ Additional Homework Problems (B.14, B.23–B.24, B.39)	

Self **Test**

LO B.1 Which of the following is *not* a valid LP constraint formulation?
- **a)** $3X + 4Y \le 12$
- **b)** $2X \times 2Y \le 12$
- **c)** $3Y + 2Z = 18$
- **d)** $100 \ge X + Y$
- **e)** $2.5X + 1.5Z = 30.6$

LO B.2 Using a *graphical solution procedure* to solve a maximization problem requires that we:
- **a)** move the iso-profit line up until it no longer intersects with any constraint equation.
- **b)** move the iso-profit line down until it no longer intersects with any constraint equation.
- **c)** apply the method of simultaneous equations to solve for the intersections of constraints.
- **d)** find the value of the objective function at the origin.

LO B.3 Consider the following linear programming problem:

$$\begin{aligned} &\text{Maximize } 4X + 10Y \\ &\text{Subject to: } 3X + 4Y \le 480 \\ &\qquad 4X + 2Y \le 360 \\ &\qquad X,\ Y \ge 0 \end{aligned}$$

The feasible corner points are (48,84), (0,120), (0,0), and (90,0). What is the maximum possible value for the objective function?
- **a)** 1,032
- **b)** 1,200
- **c)** 360
- **d)** 1,600
- **e)** 840

LO B.4 A zero shadow price for a resource ordinarily means that:
- **a)** the resource is scarce.
- **b)** the resource constraint was redundant.
- **c)** the resource has not been used up.
- **d)** something is wrong with the problem formulation.
- **e)** none of the above.

LO B.5 For these two constraints, which point is in the feasible region of this minimization problem?

$$14x + 6y \ge 42 \quad \text{and} \quad x + y \ge 3$$

- **a)** $x = -1,\ y = 1$
- **b)** $x = 0,\ y = 4$
- **c)** $x = 2,\ y = 1$
- **d)** $x = 5,\ y = 1$
- **e)** $x = 2,\ y = 0$

LO B.6 When applying LP to diet problems, the objective function is usually designed to:
- **a)** maximize profits from blends of nutrients.
- **b)** maximize ingredient blends.
- **c)** minimize production losses.
- **d)** maximize the number of products to be produced.
- **e)** minimize the costs of nutrient blends.

LO B.7 Using binary variables A and B, what LP constraint specifies that A can only occur if B occurs?

Answers: LO B.1. b; LO B.2. a; LO B.3. b; LO B.4. c; LO B.5. d; LO B.6. e; LO B.7 $A \le B$.

Transportation Models

MODULE C

MODULE OUTLINE

LEARNING OBJECTIVES

LO C.1 ***Develop*** an initial solution to a transportation model with the northwest-corner and intuitive lowest-cost methods 745

LO C.2 ***Solve*** a problem with the stepping-stone method 748

LO C.3 ***Balance*** a transportation problem 751

LO C.4 ***Deal*** with a problem that has degeneracy 751

The problem facing rental companies like Avis, Hertz, and National is cross-country travel. Lots of it. Cars rented in New York end up in Chicago, cars from L.A. come to Philadelphia, and cars from Boston come to Miami. The scene is repeated in over 100 cities around the U.S. As a result, there are too many cars in some cities and too few in others. Operations managers have to decide how many of these rentals should be trucked (by costly auto carriers) from each city with excess capacity to each city that needs more rentals. The process requires quick action for the most economical routing, so rental car companies turn to transportation modeling.

Vibrant Image Studio/Shutterstock

Transportation Modeling

Because location of a new factory, warehouse, or distribution center is a strategic issue with substantial cost implications, most companies consider and evaluate several locations. With a wide variety of objective and subjective factors to be considered, rational decisions are aided by a number of techniques. One of those techniques is transportation modeling.

The transportation models described in this module prove useful when considering alternative facility locations *within the framework of an existing distribution system.* Each new potential plant, warehouse, or distribution center will require a different allocation of shipments, depending on its own production and shipping costs and the costs of each existing facility. The choice of a new location depends on which will yield the minimum cost *for the entire system.*

Transportation modeling

An iterative procedure for solving problems that involves minimizing the cost of shipping products from a series of sources to a series of destinations.

Transportation modeling finds the least-cost means of shipping supplies from several origins to several destinations. *Origin points* (or *sources*) can be factories, warehouses, car rental agencies like Avis, or any other points from which goods are shipped. *Destinations* are any points that receive goods. To use the transportation model, we need to know the following:

1. The origin points and the capacity or supply per period at each.
2. The destination points and the demand per period at each.
3. The cost of shipping one unit from each origin to each destination.

The transportation model is one form of the linear programming models discussed in Business Analytics Module B. Software is available to solve both transportation problems and the more general class of linear programming problems. To fully use such programs, though, you need to understand the assumptions that underlie the model. To illustrate the transportation problem, we now look at a company called Arizona Plumbing, which makes, among other products, a full line of bathtubs. In our example, the firm must decide which of its factories should supply which of its warehouses. Relevant data for Arizona Plumbing are presented in Table C.1 and Figure C.1. Table C.1 shows, for example, that it costs Arizona Plumbing $5 to ship one bathtub from its Des Moines factory to its Albuquerque warehouse, $4 to Boston, and $3 to Cleveland.

Likewise, we see in Figure C.1 that the 300 units required by Arizona Plumbing's Albuquerque warehouse may be shipped in various combinations from its Des Moines, Evansville, and Fort Lauderdale factories.

The first step in the modeling process is to set up a *transportation matrix*. Its purpose is to summarize all relevant data and to keep track of algorithm computations. Using the information displayed in Figure C.1 and Table C.1, we can construct a transportation matrix as shown in Figure C.2.

TABLE C.1 Transportation Costs per Bathtub for Arizona Plumbing

TO / FROM	ALBUQUERQUE	BOSTON	CLEVELAND
Des Moines	$5	$4	$3
Evansville	$8	$4	$3
Fort Lauderdale	$9	$7	$5

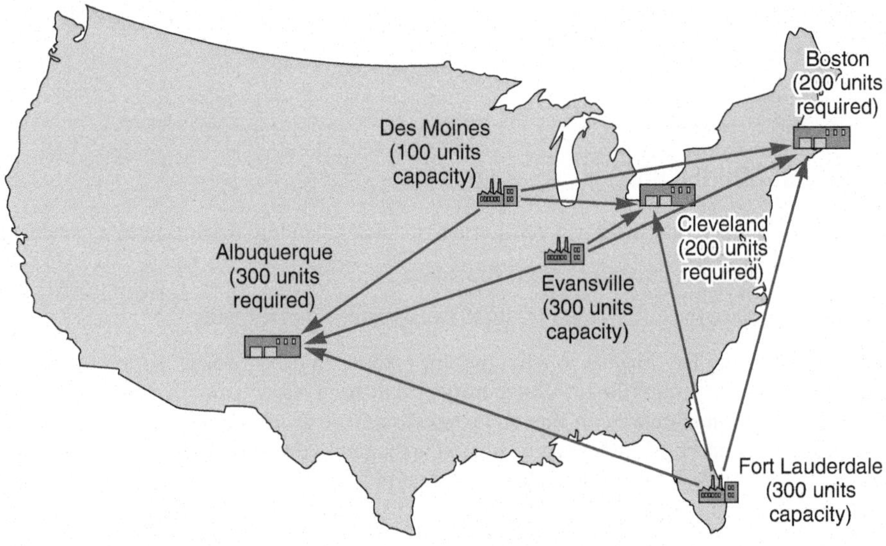

Figure **C.1**

Transportation Problem

From \ To	Albuquerque	Boston	Cleveland	Factory capacity
Des Moines	$5	$4	$3	100
Evansville	$8	$4	$3	300
Fort Lauderdale	$9	$7	$5	300
Warehouse requirement	300	200	200	700

Des Moines capacity constraint (→ 100)

Cell representing a possible source-to-destination shipping assignment (Evansville to Cleveland)

Cost of shipping 1 unit from Fort Lauderdale factory to Boston warehouse (→ $7)

Cleveland warehouse demand (→ 200)

Total demand and total supply (→ 700)

Figure **C.2**

Transportation Matrix for Arizona Plumbing

Developing an Initial Solution

LO C.1 *Develop* an initial solution to a transportation model with the northwest-corner and intuitive lowest-cost methods

Once the data are arranged in tabular form, we must establish an initial feasible solution to the problem. A number of different methods have been developed for this step. We now discuss two of them, the northwest-corner rule and the intuitive lowest-cost method.

The Northwest-Corner Rule

The **northwest-corner rule** requires that we start in the upper-left-hand cell (or northwest corner) of the table and allocate units to shipping routes as follows:

Northwest-corner rule
A procedure in the transportation model where one starts at the upper-left-hand cell of a table (the northwest corner) and systematically allocates units to shipping routes.

1. Exhaust the supply (factory capacity) of each row (e.g., Des Moines: 100) before moving down to the next row.
2. Exhaust the (warehouse) requirements of each column (e.g., Albuquerque: 300) before moving to the next column on the right.
3. Check to ensure that all supplies and demands are met.

Example C1 applies the northwest-corner rule to our Arizona Plumbing problem.

Example C1

THE NORTHWEST-CORNER RULE

Arizona Plumbing wants to use the northwest-corner rule to find an initial solution to its problem.

APPROACH ▶ Follow the three steps listed above. See Figure C.3.

SOLUTION ▶ To make the initial solution, these five assignments are made:

1. Assign 100 tubs from Des Moines to Albuquerque (exhausting Des Moines' supply).
2. Assign 200 tubs from Evansville to Albuquerque (exhausting Albuquerque's demand).
3. Assign 100 tubs from Evansville to Boston (exhausting Evansville's supply).
4. Assign 100 tubs from Fort Lauderdale to Boston (exhausting Boston's demand).
5. Assign 200 tubs from Fort Lauderdale to Cleveland (exhausting Cleveland's demand and Fort Lauderdale's supply).

Figure **C.3**

Northwest-Corner Solution to Arizona Plumbing Problem

From \ To	(A) Albuquerque	(B) Boston	(C) Cleveland	Factory capacity
(D) Des Moines	$5 / 100	$4	$3	100
(E) Evansville	$8 / 200	$4 / 100	$3	300
(F) Fort Lauderdale	$9	$7 / (100)	$5 / 200	300
Warehouse requirement	300	200	200	700

Means that the firm is shipping 100 bathtubs from Fort Lauderdale to Boston

The total cost of this shipping assignment is $4,200 (see Table C.2).

TABLE C.2 Computed Shipping Cost

ROUTE				
FROM	TO	TUBS SHIPPED	COST PER UNIT	TOTAL COST
D	A	100	$5	$ 500
E	A	200	8	1,600
E	B	100	4	400
F	B	100	7	700
F	C	200	5	1,000
				Total: $4,200

INSIGHTS ▶ The solution given is feasible because it satisfies all demand and supply constraints. The northwest-corner rule is easy to use, but it totally ignores costs, and therefore should only be considered as a starting position.

LEARNING EXERCISE ▶ Does the shipping assignment change if the cost from Des Moines to Albuquerque increases from $5 per unit to $10 per unit? Does the total cost change? [Answer: The initial assignment is the same, but cost = $4,700.]

RELATED PROBLEMS ▶ C.1a, C.3a, C.15

The Intuitive Lowest-Cost Method

The **intuitive method** makes initial allocations based on lowest cost. This straightforward approach uses the following steps:

Intuitive method
A cost-based approach to finding an initial solution to a transportation problem.

1. Identify the cell with the lowest cost. Break any ties for the lowest cost arbitrarily.
2. Allocate as many units as possible to that cell without exceeding the supply or demand. Then cross out that row or column (or both) that is exhausted by this assignment.
3. Find the cell with the lowest cost from the remaining (not crossed out) cells.
4. Repeat Steps 2 and 3 until all units have been allocated.

Example C2 THE INTUITIVE LOWEST-COST APPROACH

Arizona Plumbing now wants to apply the intuitive lowest-cost approach.

APPROACH ▶ Apply the 4 steps listed above to the data in Figure C.2.

SOLUTION ▶ When the firm uses the intuitive approach on the data (rather than the northwest-corner rule) for its starting position, it obtains the solution seen in Figure C.4.

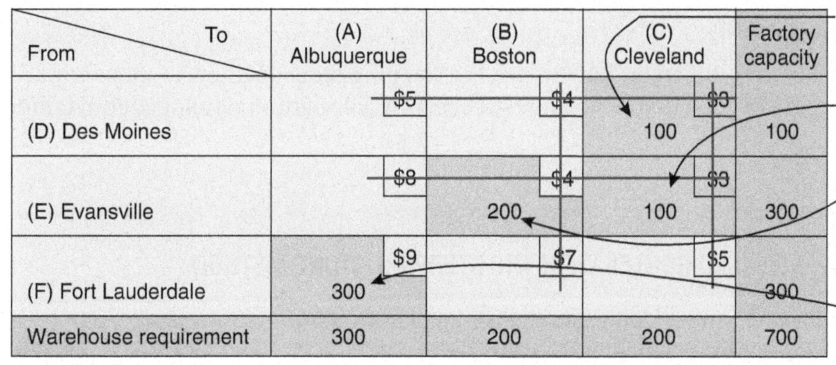

From \ To	(A) Albuquerque	(B) Boston	(C) Cleveland	Factory capacity
(D) Des Moines	$5	$4	$3 100	100
(E) Evansville	$8	$4 200	$3 100	300
(F) Fort Lauderdale	$9 300	$7	$5	300
Warehouse requirement	300	200	200	700

Figure **C.4**

Intuitive Lowest-Cost Solution to Arizona Plumbing Problem

The total cost of this approach = \$3(100) + \$3(100) + \$4(200) + \$9(300) = \$4,100
(D to C) (E to C) (E to B) (F to A)

INSIGHT ▶ This method's name is appropriate as most people find it intuitively correct to include costs when making an initial assignment.

LEARNING EXERCISE ▶ If the cost per unit from Des Moines to Cleveland is not \$3, but rather \$6, does this initial solution change? [Answer: Yes, now D – B = 100, D – C = 0, E – B = 100, E – C = 200, F – A = 300. Others unchanged at zero. Total cost stays the same.]

RELATED PROBLEMS ▶ C.1b, C.2, C.3b

Although the likelihood of a minimum-cost solution *does* improve with the intuitive method, we would have been fortunate if the intuitive solution yielded the minimum cost. In this case, as in the northwest-corner solution, it did not. Because the northwest-corner and the intuitive lowest-cost approaches are meant only to provide us with a starting point, we often will have to employ an additional procedure to reach an *optimal* solution.

The Stepping-Stone Method

Stepping-stone method
An iterative technique for moving from an initial feasible solution to an optimal solution in the transportation method.

LO C.2 *Solve* a problem with the stepping-stone method

The **stepping-stone method** will help us move from an initial feasible solution to an optimal solution. It is used to evaluate the cost effectiveness of shipping goods via transportation routes not currently in the solution. When applying it, we test each unused cell, or square, in the transportation table by asking: What would happen to total shipping costs if one unit of the product (for example, one bathtub) was tentatively shipped on an unused route? We conduct the test as follows:

1. Select any unused square to evaluate.
2. Beginning at this square, trace a closed path back to the original square via squares that are currently being used (only horizontal and vertical moves are permissible). You may, however, step over either an empty or an occupied square.
3. Beginning with a plus (+) sign at the unused square, place alternating minus signs and plus signs on each corner square of the closed path just traced.
4. Calculate an improvement index by first adding the unit-cost figures found in each square containing a plus sign and then by subtracting the unit costs in each square containing a minus sign.
5. Repeat Steps 1 through 4 until you have calculated an improvement index for all unused squares. If all indices computed are *greater than or equal to zero*, you have reached an optimal solution. If not, the current solution can be improved further to decrease total shipping costs.

Example C3 illustrates how to use the stepping-stone method to move toward an optimal solution. We begin with the northwest-corner initial solution developed in Example C1.

Example C3

CHECKING UNUSED ROUTES WITH THE STEPPING-STONE METHOD

Arizona Plumbing wants to evaluate unused shipping routes.

APPROACH ▶ Start with Example C1's Figure C.3 and follow the 5 steps listed above. As you can see, the four currently unassigned routes are Des Moines to Boston, Des Moines to Cleveland, Evansville to Cleveland, and Fort Lauderdale to Albuquerque.

SOLUTION ▶ **Steps 1 and 2.** Beginning with the Des Moines–Boston route, trace a closed path *using only currently occupied squares* (see Figure C.5). Place alternating plus and minus signs in the corners of this path. In the upper-left square, for example, we place a minus sign because we have *subtracted* 1 unit from the original 100. Note that we can use only squares currently used for shipping to turn the corners of the route we are tracing. Hence, the path Des Moines–Boston to Des Moines–Albuquerque to Fort Lauderdale–Albuquerque to Fort Lauderdale–Boston to Des Moines–Boston would not be acceptable because the Fort Lauderdale–Albuquerque square is empty. It turns out that *only one closed route exists for each empty square*. Once this one closed path is identified, we can begin assigning plus and minus signs to these squares in the path.

Step 3. How do we decide which squares get plus signs and which squares get minus signs? The answer is simple. Because we are testing the cost-effectiveness of the Des Moines–Boston shipping route, we try shipping 1 bathtub from Des Moines to Boston. This is 1 *more* unit than we *were* sending between the two cities, so place a plus sign in the box. However, if we ship 1 more unit than before from Des Moines to Boston, we end up sending 101 bathtubs out of the Des Moines factory. Because the Des Moines factory's capacity is only 100 units, we must ship 1 bathtub less from Des Moines to Albuquerque. This change prevents us from violating the capacity constraint.

To indicate that we have reduced the Des Moines–Albuquerque shipment, place a minus sign in its box. As you continue along the closed path, notice that we are no longer meeting our Albuquerque warehouse requirement for 300 units. In fact, if we reduce the Des Moines–Albuquerque shipment to 99 units, we must increase the Evansville–Albuquerque load by 1 unit, to 201 bathtubs. Therefore, place a plus sign in that box to indicate the increase. You may also observe that those squares in which we turn a corner (and only those squares) will have plus or minus signs.

Figure C.5

Stepping-Stone Evaluation of Alternative Routes for Arizona Plumbing

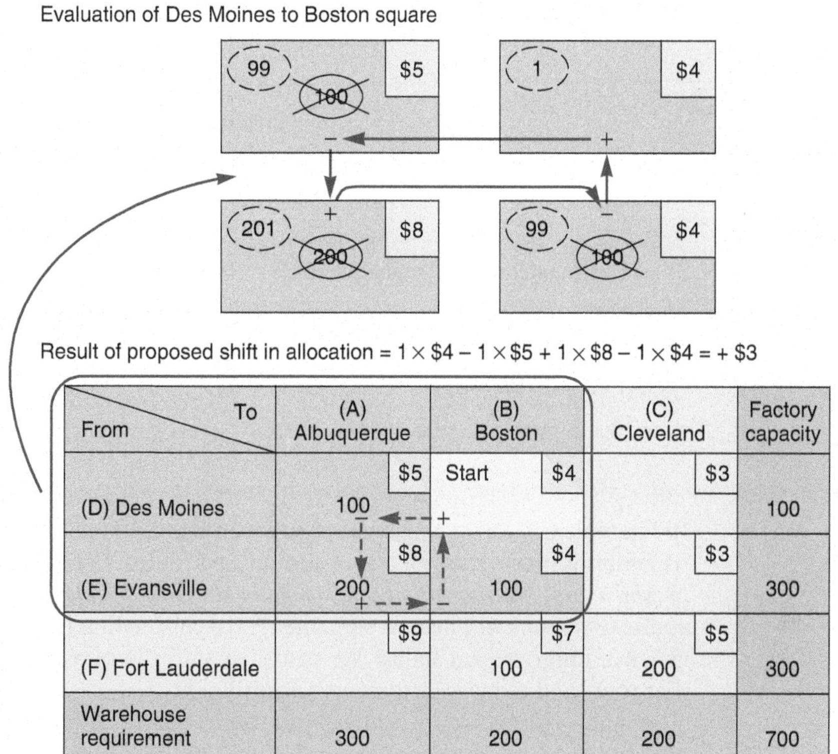

From \ To	(A) Albuquerque	(B) Boston	(C) Cleveland	Factory capacity
(D) Des Moines	$5 100 –	Start $4 +	$3	100
(E) Evansville	$8 200 +	$4 100 –	$3	300
(F) Fort Lauderdale	$9	$7 100	$5 200	300
Warehouse requirement	300	200	200	700

Finally, note that if we assign 201 bathtubs to the Evansville–Albuquerque route, then we must reduce the Evansville–Boston route by 1 unit, to 99 bathtubs, to maintain the Evansville factory's capacity constraint of 300 units. To account for this reduction, we thus insert a minus sign in the Evansville–Boston box. By so doing, we have balanced supply limitations among all four routes on the closed path.

Step 4. Compute an improvement index for the Des Moines–Boston route by adding unit costs in squares with plus signs and subtracting costs in squares with minus signs.

$$\text{Des Moines} - \text{Boston index} = \$4 - \$5 + \$8 - \$4 = +\$3$$

This means that for every bathtub shipped via the Des Moines–Boston route, total transportation costs will increase by $3 over their current level.

Let us now examine the unused Des Moines–Cleveland route, which is slightly more difficult to trace with a closed path (see Figure C.6). Again, notice that we turn each corner along the path only at squares on the existing route. Our path, for example, can go through the Evansville–Cleveland box but cannot turn a corner; thus we cannot place a plus or minus sign there. We may use occupied squares only as stepping-stones:

$$\text{Des Moines} - \text{Cleveland index} = \$3 - \$5 + \$8 - \$4 + \$7 - \$5 = +\$4$$

Figure C.6

Testing Des Moines to Cleveland

From \ To	(A) Albuquerque	(B) Boston	(C) Cleveland	Factory capacity
(D) Des Moines	$5 100 –	$4	Start $3 +	100
(E) Evansville	$8 200 +	$4 – 100	$3	300
(F) Fort Lauderdale	$9	$7 + 100	$5 – 200	300
Warehouse requirement	300	200	200	700

Again, opening this route fails to lower our total shipping costs.

Two other routes can be evaluated in a similar fashion:

$$\text{Evansville – Cleveland index} = \$3 - \$4 + \$7 - \$5 = +\$1$$
$$(\text{Closed path} = EC - EB + FB - FC)$$
$$\text{Fort Lauderdale – Albuquerque index} = \$9 - \$7 + \$4 - \$8 = -\$2$$
$$(\text{Closed path} = FA - FB + EB - EA)$$

INSIGHT ▶ Because this last index is negative, we can realize cost savings by using the (currently unused) Fort Lauderdale–Albuquerque route.

LEARNING EXERCISE ▶ What would happen to total cost if Arizona used the shipping route from Des Moines to Cleveland? [Answer: Total cost of the current solution would increase by $400.]

RELATED PROBLEMS ▶ C.1c, C.3c, C.4–C.13

EXCEL OM Data File **ModCExC3.Xls** can be found online.

In Example C3, we see that a better solution is indeed possible because we can calculate a negative improvement index on one of our unused routes. *Each negative index represents the amount by which total transportation costs could be decreased if one unit was shipped by the source–destination combination.* The next step, then, is to choose that route (unused square) with the *largest* negative improvement index. We can then ship the maximum allowable number of units on that route and reduce the total cost accordingly.

What is the maximum quantity that can be shipped on our new money-saving route? That quantity is found by referring to the closed path of plus signs and minus signs drawn for the route and then selecting the *smallest number found in the squares containing minus signs*. To obtain a new solution, we add this number to all squares on the closed path with plus signs and subtract it from all squares on the path to which we have assigned minus signs.

One iteration of the stepping-stone method is now complete. Again, of course, we must test to see if the solution is optimal or whether we can make any further improvements. We do this by evaluating each unused square, as previously described. Example C4 continues our effort to help Arizona Plumbing arrive at a final solution.

Example C4

IMPROVEMENT INDICES

Arizona Plumbing wants to continue solving the problem.

APPROACH ▶ Use the improvement indices calculated in Example C3. We found in Example C3 that the largest (and only) negative index is on the Fort Lauderdale–Albuquerque route (which is the route depicted in Figure C.7).

SOLUTION ▶ The maximum quantity that may be shipped on the newly opened route, Fort Lauderdale–Albuquerque (FA), is the smallest number found in squares containing minus signs—in this case, 100 units. Why 100 units? Because the total cost decreases by $2 per unit shipped, we know we would like to ship the maximum possible number of units. Previous stepping-stone calculations indicate that each unit shipped over the FA route results in an increase of 1 unit shipped from Evansville (E) to Boston (B) and a decrease of 1 unit in amounts shipped both from F to B (now 100 units) and from E to A (now 200 units). Hence, the maximum we can ship over the FA route is 100 units. This solution results in zero units being shipped from F to B. Now we take the following four steps:

1. Add 100 units (to the zero currently being shipped) on route FA.
2. Subtract 100 from route FB, leaving zero in that square (though still balancing the row total for F).
3. Add 100 to route EB, yielding 200.
4. Finally, subtract 100 from route EA, leaving 100 units shipped.

Note that the new numbers still produce the correct row and column totals as required. The new solution is shown in Figure C.8.

Total shipping cost has been reduced by $(100 \text{ units}) \times (\$2 \text{ saved per unit}) = \$200$ and is now $4,000. This cost figure, of course, can also be derived by multiplying the cost of shipping each unit by the number of units transported on its respective route, namely: $100(\$5) + 100(\$8) + 200(\$4) + 100(\$9) + 200(\$5) = \$4{,}000$.

Figure C.7

Transportation Table: Route FA

From \ To	(A) Albuquerque	(B) Boston	(C) Cleveland	Factory capacity
(D) Des Moines	\$5 100	\$4	\$3	100
(E) Evansville	\$8 200	\$4 100	\$3	300
(F) Fort Lauderdale	\$9	\$7 100	\$5 200	300
Warehouse demand	300	200	200	700

STUDENT TIP

FA has a negative index:
FA (+9) to FB (−7) to EB (+4) to EA (−8) = −\$2

Figure C.8

Solution at Next Iteration (Still Not Optimal)

From \ To	(A) Albuquerque	(B) Boston	(C) Cleveland	Factory capacity
(D) Des Moines	\$5 100	\$4	\$3	100
(E) Evansville	\$8 100	\$4 200	\$3	300
(F) Fort Lauderdale	\$9 100	\$7	\$5 200	300
Warehouse demand	300	200	200	700

INSIGHT ▶ Looking carefully at Figure C.8, however, you can see that it, too, is not yet optimal. Route EC (Evansville–Cleveland) has a negative cost improvement index of –\$1. Closed path = EC − EA + FA − FC.

LEARNING EXERCISE ▶ Find the final solution for this route on your own. [Answer: Programs C.1 and C.2, at the end of this module, provide an Excel OM solution.]

RELATED PROBLEMS ▶ C.4–C.13

Special Issues in Modeling

Demand Not Equal to Supply

LO C.3 *Balance* a transportation problem

A common situation in real-world problems is the case in which total demand is not equal to total supply. We can easily handle these so-called unbalanced problems with the solution procedures that we have just discussed by introducing *dummy sources* or *dummy destinations.* If total supply is greater than total demand, we make demand exactly equal the surplus by creating a dummy destination. Conversely, if total demand is greater than total supply, we introduce a dummy source (factory) with a supply equal to the excess of demand. Because these units will not in fact be shipped, we assign cost coefficients of zero to each square on the dummy location. In each case, then, the cost is zero.

Related Homework Problems: C.14–C.18

Degeneracy

LO C.4 *Deal* with a problem that has degeneracy

To apply the stepping-stone method to a transportation problem, we must observe a rule about the number of shipping routes being used: *The number of occupied squares in any solution (initial or later) must be equal to the number of rows in the table plus the number of columns minus 1.* Solutions that do not satisfy this rule are called *degenerate.*

Degeneracy
An occurrence in transportation models in which too few squares or shipping routes are being used, so that tracing a closed path for each unused square becomes impossible.

Degeneracy occurs when too few squares or shipping routes are being used. As a result, it becomes impossible to trace a closed path for one or more unused squares. The Arizona Plumbing problem we just examined was *not* degenerate, as it had 5 assigned routes (3 rows or factories + 3 columns or warehouses − 1).

To handle degenerate problems, we must artificially create an occupied cell: That is, we place a zero or a *very* small amount (representing a fake shipment) in one of the unused squares and *then treat that square as if it were occupied.* The chosen square must be in such a position as to allow all stepping-stone paths to be closed.

Summary

The transportation model, a form of linear programming, is used to help find the least-cost solutions to systemwide shipping problems. The northwest-corner method (which begins in the upper-left corner of the transportation table) or the intuitive lowest-cost method may be used for finding an initial feasible solution. The stepping-stone algorithm is then used for finding optimal solutions. Unbalanced problems are those in which the total demand and total supply are not equal. Degeneracy refers to the case in which the number of rows + the number of columns −1 is not equal to the number of occupied squares. The transportation model approach is one of the four location models described earlier in Chapter 8 and is one of the two aggregate planning models discussed in Chapter 13. Additional solution techniques are presented in Tutorial 4 found online.

Key Terms

Discussion Questions

1. What are the three information needs of the transportation model?
2. What are the steps in the intuitive lowest-cost method?
3. Identify the three "steps" in the northwest-corner rule.
4. How do you know when an optimal solution has been reached?
5. Which starting technique generally gives a better initial solution, and why?
6. The more sources and destinations there are for a transportation problem, the smaller the percentage of all cells that will be used in the optimal solution. Explain.
7. All of the transportation examples appear to apply to long distances. Is it possible for the transportation model to apply on a much smaller scale, for example, within the departments of a store or the offices of a building? Discuss.
8. Develop a *northeast*-corner rule and explain how it would work. Set up an initial solution for the Arizona Plumbing problem analyzed in Example C1.
9. What is meant by an unbalanced transportation problem, and how would you balance it?
10. How many occupied cells must all solutions use?
11. Explain the significance of a negative improvement index in a transportation-minimizing problem.
12. How can the transportation method address production costs in addition to transportation costs?
13. Explain what is meant by the term *degeneracy* within the context of transportation modeling.

Using Software to Solve Transportation Problems

Excel, Excel OM, and POM for Windows may all be used to solve transportation problems. Excel uses Solver, which requires that you enter your own constraints. Excel OM also uses Solver but is prestructured so that you need enter only the actual data. POM for Windows similarly requires that only demand data, supply data, and shipping costs be entered.

USING EXCEL OM

Excel OM's Transportation module uses Excel's built-in Solver routine to find optimal solutions to transportation problems. Program C.1 illustrates the input data (from Arizona Plumbing) and total-cost formulas. Be certain that the solving method is "Simplex LP." The output appears in Program C.2.

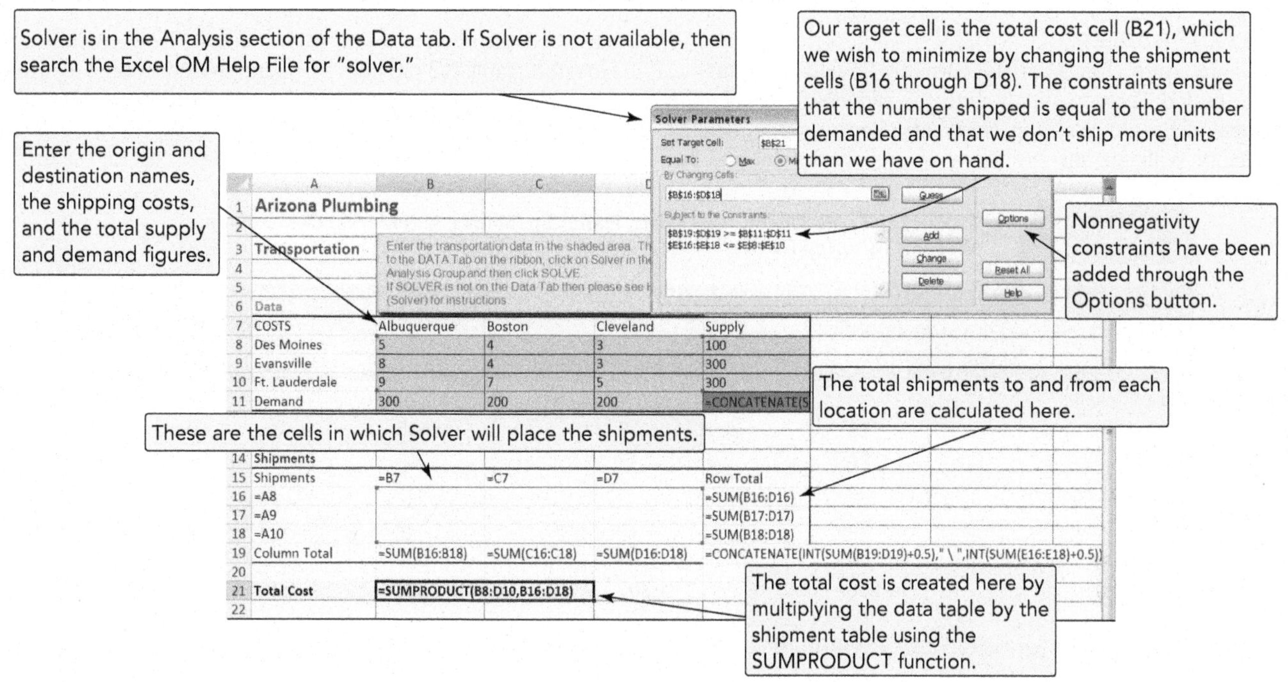

Program **C.1**

Excel OM Input Screen and Formulas, Using Arizona Plumbing Data

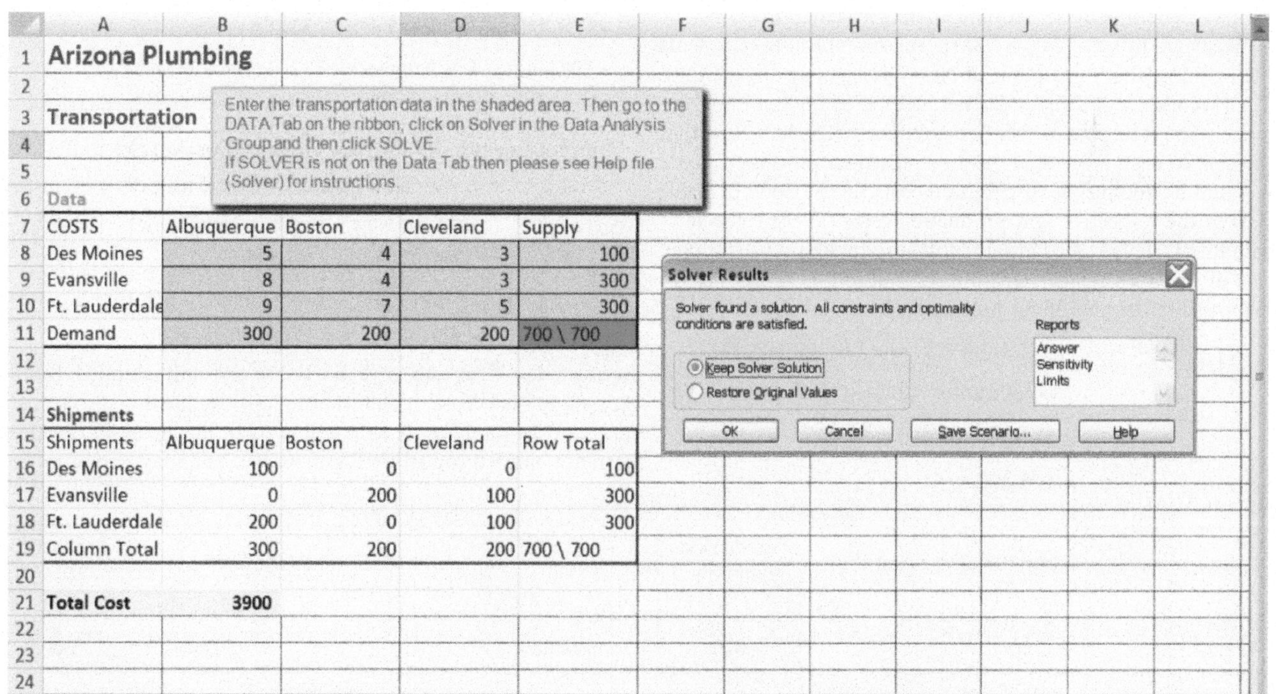

COSTS	Albuquerque	Boston	Cleveland	Supply
Des Moines	5	4	3	100
Evansville	8	4	3	300
Ft. Lauderdale	9	7	5	300
Demand	300	200	200	700 \ 700

Shipments	Albuquerque	Boston	Cleveland	Row Total
Des Moines	100	0	0	100
Evansville	0	200	100	300
Ft. Lauderdale	200	0	100	300
Column Total	300	200	200	700 \ 700

Total Cost	3900

Program **C.2**

Output from Excel OM with Optimal Solution to Arizona Plumbing Problem

P USING POM FOR WINDOWS

The POM for Windows Transportation module can solve both maximization and minimization problems by a variety of methods. Input data are the demand data, supply data, and unit shipping costs. See Appendix II for further details.

Solved Problems

SOLVED PROBLEM C.1

Williams Auto Top Carriers currently maintains plants in Atlanta and Tulsa to supply auto top carriers to distribution centers in Los Angeles and New York. Because of expanding demand, Williams has decided to open a third plant and has narrowed the choice to one of two cities—New Orleans and Houston. Table C.3 provides pertinent production and distribution costs as well as plant capacities and distribution demands.

Which of the new locations, in combination with the existing plants and distribution centers, yields a lower cost for the firm?

TABLE C.3 Production Costs, Distribution Costs, Plant Capabilities, and Market Demands for Williams Auto Top Carriers

	TO DISTRIBUTION CENTERS			
FROM PLANTS	LOS ANGELES	NEW YORK	NORMAL PRODUCTION	UNIT PRODUCTION COST
Existing plants				
Atlanta	\$8	\$5	600	\$6
Tulsa	\$4	\$7	900	\$5
Proposed locations				
New Orleans	\$5	\$6	500	\$4 (anticipated)
Houston	\$4	\$6[a]	500	\$3 (anticipated)
Forecast demand	800	1,200	2,000	

[a]Indicates distribution cost (shipping, handling, storage) will be \$6 per carrier between Houston and New York.

SOLUTION

To answer this question, we must solve two transportation problems, one for each combination. We will recommend the location that yields a lower total cost of distribution and production in combination with the existing system.

We begin by setting up a transportation table that represents the opening of a third plant in New Orleans (see Figure C.9). Then we use the northwest-corner method to find an initial solution. The total cost of this first solution is \$23,600. Note that the cost of each individual "plant-to-distribution-center" route is found by adding the distribution costs (in the body of Table C.3) to the respective unit production costs (in the right-hand column of Table C.3). Thus, the total production-plus-shipping cost of one auto top carrier from Atlanta to Los Angeles is \$14 (\$8 for shipping plus \$6 for production).

$$\begin{aligned}\text{Total cost} &= (600 \text{ units} \times \$14) + (200 \text{ units} \times \$9) \\ &\quad + (700 \text{ units} \times \$12) + (500 \text{ units} \times \$10) \\ &= \$8,400 + \$1,800 + \$8,400 + \$5,000 \\ &= \$23,600\end{aligned}$$

Is this initial solution (in Figure C.9) optimal? We can use the stepping-stone method to test it and compute improvement indices for unused routes:

Improvement index for Atlanta–New York route:

$$\begin{aligned}&= +\$11(\text{Atlanta} - \text{New York}) - \$14(\text{Atlanta} - \text{Los Angeles}) \\ &\quad + \$9(\text{Tulsa} - \text{Los Angeles}) - \$12(\text{Tulsa} - \text{New York}) \\ &= -\$6\end{aligned}$$

Improvement index for New Orleans–Los Angeles route:

$$\begin{aligned}&= +\$9(\text{New Orleans} - \text{Los Angeles}) \\ &\quad -\$10(\text{New Orleans} - \text{New York}) \\ &\quad +\$12(\text{Tulsa} - \text{New York}) \\ &\quad -\$9(\text{Tulsa} - \text{Los Angeles}) \\ &= \$2\end{aligned}$$

Because the firm can save \$6 for every unit shipped from Atlanta to New York, it will want to improve the initial solution and send as many units as possible (600, in this case) on this currently unused route (see Figure C.10). You may also want to

From \ To	Los Angeles	New York	Production capacity
Atlanta	\$14 — 600	\$11	600
Tulsa	\$9 — 200	\$12 — 700	900
New Orleans	\$9	\$10 — 500	500
Demand	800	1,200	2,000

Figure C.9
Initial Williams Transportation Table for New Orleans

From \ To	Los Angeles	New York	Production capacity
Atlanta	\$14	\$11 — 600	600
Tulsa	\$9 — 800	\$12 — 100	900
New Orleans	\$9	\$10 — 500	500
Demand	800	1,200	2,000

Figure C.10
Improved Transportation Table for Williams

confirm that the total cost is now \$20,000, a savings of \$3,600 over the initial solution.

Next, we must test the two unused routes to see if their improvement indices are also negative numbers:

Index for Atlanta–Los Angeles:

$$= \$14 - \$11 + \$12 - \$9 = \$6$$

Index for New Orleans–Los Angeles:

$$= \$9 - \$10 + \$12 - \$9 = \$2$$

Because both indices are greater than zero, we have already reached our optimal solution for the New Orleans location. If Williams elects to open the New Orleans plant, the firm's total production and distribution cost will be \$20,000.

This analysis, however, provides only half the answer to Williams's problem. The same procedure must still be followed to determine the minimum cost if the new plant is built in Houston. Determining this cost is left as a homework problem. You can help provide complete information and recommend a solution by solving Problem C.7.

SOLVED PROBLEM C.2

In Solved Problem C.1, we examined the Williams Auto Top Carriers problem by using a transportation table. An alternative approach is to structure the same decision analysis using linear programming (LP), which we explained in detail in Business Analytics Module B.

SOLUTION

Using the data in Figure C.9, we write the objective function and constraints as follows:

$$\text{Minimize total cost} = \$14X_{\text{Atl,LA}} + \$11X_{\text{Atl,NY}} + \$9X_{\text{Tul,LA}} + \$12X_{\text{Tul,NY}} + \$9X_{\text{NO,LA}} + \$10X_{\text{NO,NY}}$$

Subject to:

$$X_{\text{Atl,LA}} + X_{\text{Atl,NY}} \le 600 \quad \text{(production capacity at Atlanta)}$$
$$X_{\text{Tul,LA}} + X_{\text{Tul,NY}} \le 900 \quad \text{(production capacity at Tulsa)}$$
$$X_{\text{NO,LA}} + X_{\text{NO,NY}} \le 500 \quad \text{(production capacity at New Orleans)}$$
$$X_{\text{Atl,LA}} + X_{\text{Tul,LA}} + X_{\text{NO,LA}} \ge 800 \quad \text{(Los Angeles demand constraint)}$$
$$X_{\text{Atl,NY}} + X_{\text{Tul,NY}} + X_{\text{NO,NY}} \ge 1200 \quad \text{(New York demand constraint)}$$

Problems

Note: **PX** means the problem may be solved with POM for Windows and/or Excel OM.

Problems C.1–C.3 relate to Developing an Initial Solution

• **C.1** Find an initial solution to the following transportation problem for Supply Chain Manager Gabriela Valdez.

	TO			
FROM	LOS ANGELES	CALGARY	PANAMA CITY	SUPPLY
Mexico City	\$ 6	\$18	\$ 8	100
Detroit	\$17	\$13	\$19	60
Ottawa	\$20	\$10	\$24	40
Demand	50	80	70	

a) Use the northwest-corner method. What is its total cost?
b) Use the intuitive lowest-cost approach. What is its total cost?
c) Using the stepping-stone method, find the optimal solution. Compute the total cost. **PX**

• **C.2** Consider the transportation table that follows. Unit costs for each shipping route are in dollars. What is the total cost of the basic feasible solution that the intuitive lowest-cost method would find for this problem? **PX**

	Destination					
Source	A	B	C	D	E	Supply
1	12	8	5	10	4	18
2	6	11	3	7	9	14
Demand	6	8	12	4	2	

• **C.3** Refer to the table that follows.

a) Use the northwest-corner method to find an initial feasible solution. What must you do before beginning the solution steps?
b) Use the intuitive lowest-cost approach to find an initial feasible solution. Is this approach better than the northwest-corner method?
c) Find the optimal solution using the stepping-stone method.

FROM	TO A	B	C	SUPPLY
X	$10	$18	$12	100
Y	$17	$13	$ 9	50
Z	$20	$18	$14	75
Demand	50	80	70	

Problems C.4–C.13 relate to The Stepping-Stone Method

• **C.4** Consider the transportation table below. The solution displayed was obtained by performing some iterations of the transportation method on this problem. What is the total cost of the shipping plan that would be obtained by performing *one more iteration* of the stepping-stone method on this problem?

Destination

Source	Denver	Yuma	Miami	Supply
Houston	10 [$2]	[$8]	[$1]	10
St. Louis	10 [$4]	10 [$5]	[$6]	20
Chicago	[$6]	10 [$3]	20 [$2]	30
Demand	20	20	20	

•• **C.5** The following table is the result of one or more iterations.

From \ To	1	2	3	Capacity
A	40 [30]	[30]	10 [5]	50
B	[10]	30 [10]	[10]	30
C	[20]	30 [10]	45 [25]	75
Demand	40	60	55	155

a) Complete the next iteration using the stepping-stone method.
b) Calculate the "total cost" incurred if your results were to be accepted as the final solution. **PX**

•• **C.6** The three blood banks in Seminole County, Florida, are coordinated through a central office, managed by Jantel Saballly, that facilitates blood delivery to four hospitals in the region. The cost to ship a standard container of blood from each bank to each hospital is shown in the following table. Also given are the biweekly number of containers available at each bank and the biweekly number of containers of blood needed at each hospital. How many shipments should be made biweekly from each blood bank to each hospital so that total shipment costs are minimized? **PX**

FROM	TO HOSP. 1	HOSP. 2	HOSP. 3	HOSP. 4	SUPPLY
Bank 1	$ 8	$ 9	$11	$16	50
Bank 2	$12	$ 7	$ 5	$ 8	80
Bank 3	$14	$10	$ 6	$ 7	120
Demand	90	70	40	50	250

•• **C.7** In Solved Problem C.1, Williams Auto Top Carriers proposed opening a new plant in either New Orleans or Houston. Management found that the total system cost (of production plus distribution) would be $20,000 for the New Orleans site. What would be the total cost if Williams opened a plant in Houston? At which of the two proposed locations (New Orleans or Houston) should Williams open the new facility? **PX**

•• **C.8** The Donna Mosier Clothing Group owns factories in three towns (W, Y, and Z), which distribute to three retail dress shops in three other cities (A, B, and C). The following table summarizes factory availabilities, projected store demands, and unit shipping costs:

From \ To	Dress Shop A	Dress Shop B	Dress Shop C	Factory availability
Factory W	[$4]	[$3]	[$3]	35
Factory Y	[$6]	[$7]	[$6]	50
Factory Z	[$8]	[$2]	[$5]	50
Store demand	30	65	40	135

a) Complete the analysis, determining the optimal solution for shipping at the Mosier Clothing Group.
b) How do you know whether it is optimal or not? **PX**

••• **C.9** Captain Borders Corp. manufacturers fishing equipment. Currently, the company has a plant in Los Angeles and a plant in New Orleans. William Borders, the firm's owner, is deciding where to build a new plant—Philadelphia or Seattle. Use the following table to find the total shipping costs for each potential site. Which should Borders select? **PX**

PLANT	WAREHOUSE PITTSBURGH	ST. LOUIS	DENVER	CAPACITY
Los Angeles	$100	$75	$50	150
New Orleans	$ 80	$60	$90	225
Philadelphia	$ 40	$50	$90	350
Seattle	$110	$70	$30	350
Demand	200	100	400	

•• **C.10** Dana Johnson Corp. is considering adding a fourth plant to its three existing facilities in Decatur, Minneapolis, and Carbondale. Both St. Louis and East St. Louis are being considered. Evaluating only the transportation costs per unit as shown in the table, decide which site is best. **PX**

	FROM EXISTING PLANTS			
TO	DECATUR	MINNEAPOLIS	CARBONDALE	DEMAND
Blue Earth	$20	$17	$21	250
Ciro	$25	$27	$20	200
Des Moines	$22	$25	$22	350
Capacity	300	200	150	

	FROM PROPOSED PLANTS	
TO	EAST ST. LOUIS	ST. LOUIS
Blue Earth	$29	$27
Ciro	$30	$28
Des Moines	$30	$31
Capacity	150	150

•• **C.11** Using the data from Problem C.10 and the unit production costs in the following table, show which locations yield the lowest cost. **PX**

LOCATION	PRODUCTION COSTS ($)
Decatur	$50
Minneapolis	$60
Carbondale	$70
East St. Louis	$40
St. Louis	$50

Additional problems **C.12–C.13** *are available in* **MyLab Operations Management**. *Problem C.12 takes Example C4's data to the next iteration. Problem C.13 begins after the first iteration and seeks an optimal solution.*

Problems C.14–C.18 relate to Special Issues in Modeling

•• **C.14** Allen Air Conditioning manufactures room air conditioners at plants in Houston, Phoenix, and Memphis. These are sent to regional distributors in Dallas, Atlanta, and Denver. The shipping costs vary, and the company would like to find the least-cost way to meet the demands at each of the distribution centers. Dallas needs to receive 800 air conditioners per month, Atlanta needs 600, and Denver needs 200. Houston has 850 air conditioners available each month, Phoenix has 650, and Memphis has 300. The shipping cost per unit from Houston to Dallas is $8, to Atlanta $12, and to Denver $10. The cost per unit from Phoenix to Dallas is $10, to Atlanta $14, and to Denver $9. The cost per unit from Memphis to Dallas is $11, to Atlanta $8, and to Denver $12. How many units should owner Stephen Allen ship from each plant to each regional distribution center? What is the total transportation cost? (Note that a "dummy" destination is needed to balance the problem.) **PX**

•• **C.15** For the following Gregory Bier Corp. data, find the starting solution and initial cost using the northwest-corner method. What must you do to balance this problem? **PX**

	TO				
FROM	W	X	Y	Z	SUPPLY
A	$132	$116	$250	$110	220
B	$220	$230	$180	$178	300
C	$152	$173	$196	$164	435
Demand	160	120	200	230	

•• **C.16** The following table presents cost, capacity, and demand data for a transportation problem in Jinfeng Yue's furniture company. Set up the appropriate transportation table and find the initial solution using the northwest-corner rule. Note that a "dummy" source is needed to balance the problem. **PX**

FROM\TO	1	2	3	CAPACITY
A	$30	$10	$ 5	20
B	$10	$10	$10	30
C	$20	$10	$25	75
Demand	40	60	55	

Additional problems **C.17–C.18** *are available in* **MyLab Operations Management**. *Problem C.17 requires a dummy plant. Problem C.18 involves adding a new plant to an existing three plants.*

CASE STUDY

Custom Vans, Inc.

Custom Vans, Inc., specializes in converting standard vans into campers. Depending on the amount of work and customizing to be done, the customizing can cost from less than $1,000 to more than $5,000. In less than 4 years, Tony Rizzo was able to expand his small operation in Gary, Indiana, to other major outlets in Chicago, Milwaukee, Minneapolis, and Detroit.

Innovation was the major factor in Tony's success in converting a small van shop into one of the largest and most profitable custom van operations in the Midwest. Tony seemed to have a special ability to design and develop unique features and devices that were always in high demand by van owners. An example was Shower-Rific, which was developed by Tony only 6 months after Custom Vans, Inc., was started. These small showers were completely self-contained, and they could be placed in almost any type of van and in a number of different locations within a van. Shower-Rific was made of fiberglass, and contained towel racks, built-in soap and shampoo holders, and a unique plastic door. Each Shower-Rific took 2 gallons of fiberglass and 3 hours of labor to manufacture.

Most of the Shower-Rifics were manufactured in Gary in the same warehouse where Custom Vans, Inc., was founded. The manufacturing plant in Gary could produce 300 Shower-Rifics in a month, but this capacity never seemed to be enough. Custom Van shops in all locations were complaining about not getting enough Shower-Rifics, and because Minneapolis was farther away from Gary than the other locations, Tony was always inclined to ship Shower-Rifics to the other locations before Minneapolis. This infuriated the manager of Custom Vans at Minneapolis, and after many heated discussions, Tony decided to start another manufacturing plant for Shower-Rifics at Fort Wayne, Indiana. The manufacturing plant at Fort Wayne could produce 150 Shower-Rifics per month.

The manufacturing plant at Fort Wayne was still not able to meet current demand for Shower-Rifics, and Tony knew that the demand for his unique camper shower would grow rapidly in the next year. After consulting with his lawyer and banker, Tony concluded that he should open two new manufacturing plants as soon as possible. Each plant would have the same capacity as the Fort

Wayne manufacturing plant. An initial investigation into possible manufacturing locations was made, and Tony decided that the two new plants should be located in Detroit, Michigan; Rockford, Illinois; or Madison, Wisconsin. Tony knew that selecting the best location for the two new manufacturing plants would be difficult. Transportation costs and demands for the various locations would be important considerations.

The Chicago shop was managed by Bill Burch. This shop was one of the first established by Tony, and it continued to outperform the other locations. The manufacturing plant at Gary was supplying 200 Shower-Rifics each month, although Bill knew that the demand for the showers in Chicago was 300 units. The transportation cost per unit from Gary was $10, and although the transportation cost from Fort Wayne was double that amount, Bill was always pleading with Tony to get an additional 50 units from the Fort Wayne manufacturer. The two additional manufacturing plants would certainly be able to supply Bill with the additional 100 showers he needed. The transportation costs would, of course, vary, depending on which two locations Tony picked. The transportation cost per shower would be $30 from Detroit, $5 from Rockford, and $10 from Madison.

Wilma Jackson, manager of the Custom Van shop in Milwaukee, was the most upset about not getting an adequate supply of showers. She had a demand for 100 units, and at the present time, she was only getting half of this demand from the Fort Wayne manufacturing plant. She could not understand why Tony didn't ship her all 100 units from Gary. The transportation cost per unit from Gary was only $20, while the transportation cost from Fort Wayne was $30. Wilma was hoping that Tony would select Madison for one of the manufacturing locations. She would be able to get all the showers needed, and the transportation cost per unit would only be $5. If not in Madison, a new plant in Rockford would be able to supply her total needs, but the transportation cost per unit would be twice as much as it would be from Madison. Because the transportation cost per unit from Detroit would be $40, Wilma speculated that even if Detroit became one of the new plants, she would not be getting any units from Detroit.

Custom Vans, Inc., of Minneapolis was managed by Tom Poanski. He was getting 100 showers from the Gary plant. Demand was 150 units. Tom faced the highest transportation costs of all locations. The transportation cost from Gary was $40 per unit. It would cost $10 more if showers were sent from the Fort Wayne location. Tom was hoping that Detroit would not be one of the new plants, as the transportation cost would be $60 per unit. Rockford and Madison would have a cost of $30 and $25, respectively, to ship one shower to Minneapolis.

The Detroit shop's position was similar to Milwaukee's—only getting half of the demand each month. The 100 units that Detroit did receive came directly from the Fort Wayne plant. The transportation cost was only $15 per unit from Fort Wayne, while it was $25 from Gary. Dick Lopez, manager of Custom Vans, Inc., of Detroit, placed the probability of having one of the new plants in Detroit fairly high. The factory would be located across town, and the transportation cost would be only $5 per unit. He could get 150 showers from the new plant in Detroit and the other 50 showers from Fort Wayne. Even if Detroit was not selected, the other two locations were not intolerable. Rockford had a transportation cost per unit of $35, and Madison had a transportation cost of $40.

Tony pondered the dilemma of locating the two new plants for several weeks before deciding to call a meeting of all the managers of the van shops. The decision was complicated, but the objective was clear—to minimize total costs. The meeting was held in Gary, and everyone was present except Wilma.

Tony:	Thank you for coming. As you know, I have decided to open two new plants at Rockford, Madison, or Detroit. The two locations, of course, will change our shipping practices, and I sincerely hope that they will supply you with the Shower-Rifics that you have been wanting. I know you could have sold more units, and I want you to know that I am sorry for this situation.
Dick:	Tony, I have given this situation a lot of consideration, and I feel strongly that at least one of the new plants should be located in Detroit. As you know, I am now only getting half of the showers that I need. My brother, Leon, is very interested in running the plant, and I know he would do a good job.
Tom:	Dick, I am sure that Leon could do a good job, and I know how difficult it has been since the recent layoffs by the auto industry. Nevertheless, we should be considering total costs and not personalities. I believe that the new plants should be located in Madison and Rockford. I am farther away from the other plants than any other shop, and these locations would significantly reduce transportation costs.
Dick:	That may be true, but there are other factors. Detroit has one of the largest suppliers of fiberglass, and I have checked prices. A new plant in Detroit would be able to purchase fiberglass for $2 per gallon less than any of the other existing or proposed plants.
Tom:	At Madison, we have an excellent labor force. This is due primarily to the large number of students attending the University of Wisconsin. These students are hard workers, and they will work for $1 less per hour than the other locations that we are considering.
Bill:	Calm down, you two. It is obvious that we will not be able to satisfy everyone in locating the new plants. Therefore, I would like to suggest that we vote on the two best locations.
Tony:	I don't think that voting would be a good idea. Wilma was not able to attend, and we should be looking at all of these factors together in some type of logical fashion.

Discussion Question

Where would you locate the two new plants? Why?

Source: From *Quantitative Analysis for Management*, B. Render, R. M. Stair, M. Hanna, and T. Hale. 13th ed.

Bibliography

Drezner, Z., and H. W. Hamacher. *Facility Location: Applications and Theory*. Berlin: Springer-Verlag, 2004.

Render, B., R. M. Stair, M. Hanna, and T. Hale. *Quantitative Analysis for Management*, 13th ed. Boston: Pearson, 2018.

Module C *Rapid* Review

Main Heading	Review Material	MyLab Operations Management
TRANSPORTATION MODELING	The transportation models described in this module prove useful when considering alternative facility locations *within the framework of an existing distribution system.* The choice of a new location depends on which will yield the minimum cost for the entire system. ■ **Transportation modeling**—An iterative procedure for solving problems that involves minimizing the cost of shipping products from a series of sources to a series of destinations. *Origin points* (or *sources*) can be factories, warehouses, car rental agencies, or any other points from which goods are shipped. *Destinations* are any points that receive goods. To use the transportation model, we need to know the following: 1. The origin points and the capacity or supply per period at each. 2. The destination points and the demand per period at each. 3. The cost of shipping one unit from each origin to each destination. The transportation model is a type of linear programming model. A *transportation matrix* summarizes all relevant data and keeps track of algorithm computations. Shipping costs from each origin to each destination are contained in the appropriate cross-referenced box.	Concept Questions: 1.1–1.6

FROM \ TO	DESTINATION 1	DESTINATION 2	DESTINATION 3	CAPACITY
Source A				
Source B				
Source C				
Demand				

Main Heading	Review Material	MyLab Operations Management
DEVELOPING AN INITIAL SOLUTION	Two methods for establishing an initial feasible solution to the problem are the northwest-corner rule and the intuitive lowest-cost method. ■ **Northwest-corner rule**—A procedure in the transportation model where one starts at the upper-left-hand cell of a table (the northwest corner) and systematically allocates units to shipping routes. The northwest-corner rule requires that we: 1. Exhaust the supply (origin capacity) of each row before moving down to the next row. 2. Exhaust the demand requirements of each column before moving to the next column to the right. 3. Check to ensure that all supplies and demands are met. The northwest-corner rule is easy to use and generates a feasible solution, but it totally ignores costs and therefore should be considered only as a starting position. ■ **Intuitive method**—A cost-based approach to finding an initial solution to a transportation problem. The intuitive method uses the following steps: 1. Identify the cell with the lowest cost. Break any ties for the lowest cost arbitrarily. 2. Allocate as many units as possible to that cell, without exceeding the supply or demand. Then cross out that row or column (or both) that is exhausted by this assignment. 3. Find the cell with the lowest cost from the remaining (not crossed out) cells. 4. Repeat Steps 2 and 3 until all units have been allocated.	Concept Questions: 2.1–2.6 Problems: C.1–C.3, C.15
THE STEPPING-STONE METHOD	■ **Stepping-stone method**—An iterative technique for moving from an initial feasible solution to an optimal solution in the transportation method. The stepping-stone method is used to evaluate the cost-effectiveness of shipping goods via transportation routes not currently in the solution. When applying it, we test each unused cell, or square, in the transportation table by asking: What would happen to total shipping costs if one unit of the product were tentatively shipped on an unused route? We conduct the test as follows: 1. Select any unused square to evaluate. 2. Beginning at this square, trace a closed path back to the original square via squares that are currently being used (only horizontal and vertical moves are permissible). You may, however, step over either an empty or an occupied square.	Concept Questions: 3.1–3.5 Problems: C.4–C.13 Virtual Office Hours for Solved Problem: C.1

Main Heading	Review Material	MyLab Operations Management
	3. Beginning with a plus (+) sign at the unused square, place alternative minus signs and plus signs on each corner square of the closed path just traced. 4. Calculate an improvement index by first adding the unit-cost figures found in each square containing a plus sign and then subtracting the unit costs in each square containing a minus sign. 5. Repeat Steps 1 through 4 until you have calculated an improvement index for all unused squares. If all indices computed are *greater than or equal to zero*, you have reached an optimal solution. If not, the current solution can be improved further to decrease total shipping costs. *Each negative index represents the amount by which total transportation costs could be decreased if one unit was shipped by the source–destination combination.* The next step, then, is to choose that route (unused square) with the *largest* negative improvement index. We can then ship the maximum allowable number of units on that route and reduce the total cost accordingly. That maximum quantity is found by referring to the closed path of plus signs and minus signs drawn for the route and then selecting the *smallest number found in the squares containing minus signs*. To obtain a new solution, we add this number to all squares on the closed path with plus signs and subtract it from all squares on the path to which we have assigned minus signs. From this new solution, a new test of unused squares needs to be conducted to see if the new solution is optimal or whether we can make further improvements.	
SPECIAL ISSUES IN MODELING	*Dummy sources*—Artificial shipping source points created when total demand is greater than total supply to effect a supply equal to the excess of demand over supply. *Dummy destinations*—Artificial destination points created when the total supply is greater than the total demand; they serve to equalize the total demand and supply. Because units from dummy sources or to dummy destinations will not in fact be shipped, we assign cost coefficients of zero to each square on the dummy location. If you are solving a transportation problem by hand, be careful to decide first whether a dummy source (row) or a dummy destination (column) is needed. When applying the stepping-stone method, *the number of occupied squares in any solution (initial or later) must be equal to the number of rows in the table plus the number of columns minus 1.* Solutions that do not satisfy this rule are called *degenerate*. ▪ **Degeneracy**—An occurrence in transportation models in which too few squares or shipping routes are being used, so that tracing a closed path for each unused square becomes impossible. To handle degenerate problems, we must artificially create an occupied cell: That is, we place a zero (representing a fake shipment) into one of the unused squares and *then treat that square as if it were occupied.* Remember that the chosen square must be in such a position as to allow all stepping-stone paths to be closed.	Concept Questions: 4.1–4.4 Problems: C.14–C.18 Virtual Office Hours for Solved Problem: C.2
ADDITIONAL MYLAB OPERATIONS MANAGEMENT RESOURCES	✔ Additional Case Study (Consolidated Bottling [B]) ✔ Multiple Choice Case Questions (Custom Vans, Inc.) ✔ Additional Homework Problems (C.12–C.13, C.17–C.18)	

Self Test

LO C.1 With the transportation technique, the initial solution can be generated in any fashion one chooses. The only restriction(s) is that:
- **a)** the solution be optimal.
- **b)** one use the northwest-corner method.
- **c)** the edge constraints for supply and demand be satisfied.
- **d)** the solution not be degenerate.
- **e)** all of the above.

LO C.2 The purpose of the stepping-stone method is to:
- **a)** develop the initial solution to a transportation problem.
- **b)** identify the relevant costs in a transportation problem.
- **c)** determine whether a given solution is feasible.
- **d)** assist one in moving from an initial feasible solution to the optimal solution.
- **e)** overcome the problem of degeneracy.

LO C.3 The purpose of a *dummy source* or a *dummy destination* in a transportation problem is to:
- **a)** provide a means of representing a dummy problem.
- **b)** obtain a balance between total supply and total demand.
- **c)** prevent the solution from becoming degenerate.
- **d)** make certain that the total cost does not exceed some specified figure.
- **e)** change a problem from maximization to minimization.

LO C.4 If a solution to a transportation problem is degenerate, then:
- **a)** it will be impossible to evaluate all empty cells without removing the degeneracy.
- **b)** a dummy row or column must be added.
- **c)** there will be more than one optimal solution.
- **d)** the problem has no feasible solution.
- **e)** increase the cost of each cell by 1.

Answers: LO C.1. c; LO C.2. d; LO C.3. b; LO C.4. a.

Waiting-Line Models

MODULE OUTLINE

LEARNING OBJECTIVES

LO D.1 ***Describe*** the characteristics of arrivals, waiting lines, and service systems 762

LO D.2 ***Apply*** the single-server queuing model equations 768

LO D.3 ***Conduct*** a cost analysis for a waiting line 770

LO D.4 ***Apply*** the multiple-server queuing model formulas 770

LO D.5 ***Apply*** the constant-service-time model equations 775

LO D.6 ***Perform*** a finite-population model analysis 776

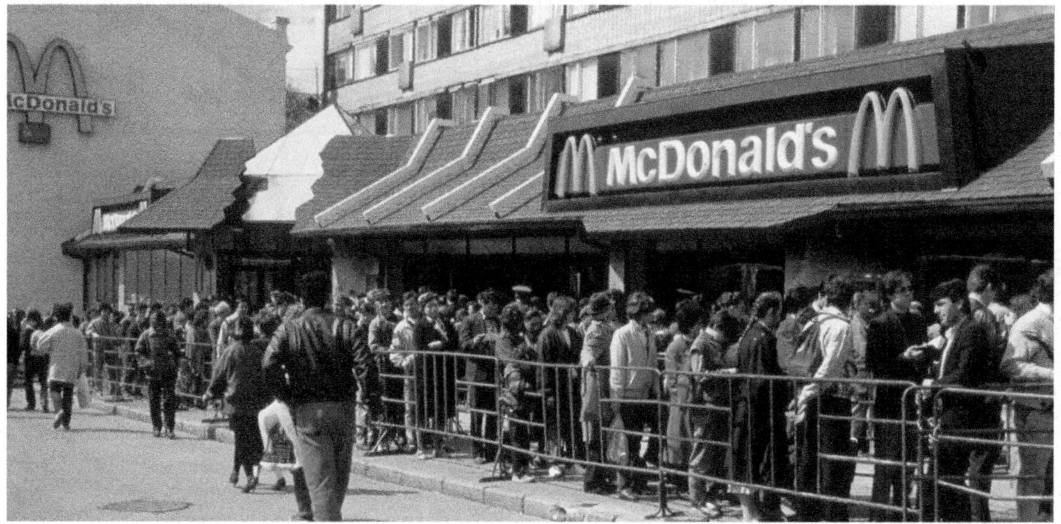

The giant Moscow McDonald's located in Pushkin Square may be the most popular McDonald's in the entire world. It boasts 900 seats and 800 employees. Americans would balk at the average waiting time of 45 minutes, but Russians are used to such long lines. McDonald's represents good service in Moscow. Some people have even had their wedding receptions there.

Roy/Explorer/Science Source

Queuing Theory

Queuing theory
A body of knowledge about waiting lines.

Waiting line (queue)
Items or people in a line awaiting service.

The body of knowledge about waiting lines, often called **queuing theory**, is an important part of operations and a valuable tool for the operations manager. **Waiting lines** are a common situation—they may, for example, take the form of cars waiting for repair at a Midas Muffler Shop, copying jobs waiting to be completed at a FedEx office, or vacationers waiting to enter a ride at Disney. Table D.1 lists just a few OM uses of waiting-line models.

Waiting-line models are useful in both manufacturing and service areas. Analysis of queues in terms of waiting-line length, average waiting time, and other factors helps us to understand service systems (such as bank teller stations), maintenance activities (that might repair broken machinery), and shop-floor control activities. Indeed, patients waiting in a doctor's office and broken drill presses waiting in a repair facility have a lot in common from an OM perspective. Both use human and equipment resources to restore valuable production assets (people and machines) to good condition.

TABLE D.1 Common Queuing Situations

SITUATION	ARRIVALS IN QUEUE	SERVICE PROCESS
Supermarket	Grocery shoppers	Checkout clerks at cash register
Highway toll booth	Automobiles	Collection of tolls at booth
Doctor's office	Patients	Treatment by doctors and nurses
Computer system	Programs to be run	Computer processes jobs
Telephone company	Callers	Switching equipment forwards calls
Bank	Customers	Transactions handled by teller
Machine maintenance	Broken machines	Repair people fix machines
Harbor	Ships and barges	Dock workers load and unload

Characteristics of a Waiting-Line System

In this section, we take a look at the three parts of a waiting-line, or queuing, system (as shown in Figure D.1):

LO D.1 *Describe* the characteristics of arrivals, waiting lines, and service systems

1. Arrivals or inputs to the system: These have characteristics such as population size, behavior, and a statistical distribution.
2. Queue discipline, or the waiting line itself: Characteristics of the queue include whether it is limited or unlimited in length and the discipline of people or items in it.
3. The service facility: Its characteristics include its design and the statistical distribution of service times.

We now examine each of these three parts.

Arrival Characteristics

The input source that generates arrivals or customers for a service system has three major characteristics:

1. Size of the arrival population
2. Behavior of arrivals
3. Pattern of arrivals (statistical distribution)

Unlimited, or infinite, population
A queue in which a virtually unlimited number of people (arrivals) request the services, or in which the number of customers or arrivals on hand at any given moment is a very small portion of potential arrivals.

Limited, or finite, population
A queue in which there are only a limited number of potential users of the service.

Size of the Arrival (Source) Population Population sizes are considered either unlimited (essentially infinite) or limited (finite). When the number of customers or arrivals on hand at any given moment is just a small portion of all potential arrivals, the arrival population is considered **unlimited**, or **infinite**. Examples of unlimited populations include cars arriving at a big-city car wash, shoppers arriving at a supermarket, and students arriving to register for classes at a large university. Most queuing models assume such an infinite arrival population. An example of a **limited**, or **finite**, population is found in a copying shop that has, say, eight

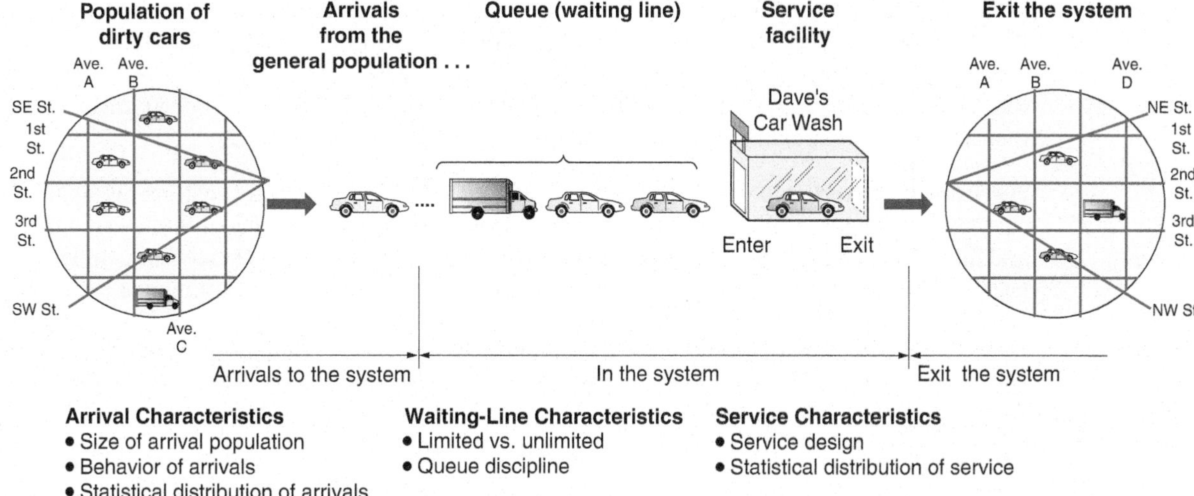

Figure **D.1**

Three Parts of a Waiting Line, or Queuing System, at Dave's Car Wash

copying machines. Each of the copiers is a potential "customer" that may break down and require service.

Pattern of Arrivals at the System Customers arrive at a service facility either according to some known schedule (for example, one patient every 15 minutes or one student every half hour) or else they arrive *randomly*. Arrivals are considered random when they are independent of one another and their occurrence cannot be predicted exactly. Frequently in queuing problems, the number of arrivals per unit of time can be estimated by a probability distribution known as the **Poisson distribution**.[1] For any given arrival time (such as 2 customers per hour or 4 trucks per minute), a discrete Poisson distribution can be established by using the formula:

Poisson distribution
A discrete probability distribution that often describes the arrival rate in queuing theory.

$$P(x) = \frac{e^{-\lambda}\lambda^x}{x!} \text{ for } x = 0, 1, 2, 3, 4, \ldots \qquad \text{(D-1)}$$

where $P(x)$ = probability of x arrivals
x = number of arrivals per unit of time
λ = average arrival rate
e = 2.7183 (which is the base of the natural logarithm)

Figure D.2 illustrates the Poisson distribution for $\lambda = 2$ and $\lambda = 4$. This means that if the average arrival rate is $\lambda = 2$ customers per hour, the probability of 0 customers arriving in any

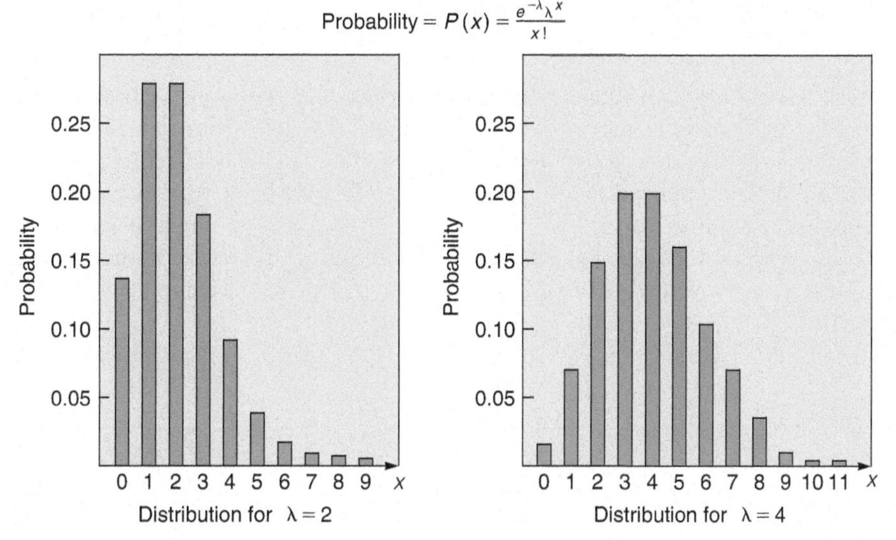

Figure **D.2**

Two Examples of the Poisson Distribution for Arrival Times

STUDENT TIP
Notice that even though the mean arrival rate might be $\lambda = 2$ per hour, there is still a small chance that as many as 9 customers arrive in an hour.

random hour is about 0.13 (13%), probability of 1 customer is about 27%, 2 customers about 27%, 3 customers about 18%, 4 customers about 9%, and so on. The chances that 9 or more will arrive are virtually nil. Arrivals, of course, are not always Poisson distributed (they may follow some other distribution). Patterns, therefore, should be examined to make certain that they are well approximated by Poisson before that distribution is applied.

Behavior of Arrivals Most queuing models assume that an arriving customer is a patient customer. Patient customers are people or machines that wait in the queue until they are served and do not switch between lines. Unfortunately, life is complicated by the fact that people have been known to balk or to renege. Customers who *balk* refuse to join the waiting line because it is too long to suit their needs or interests. *Reneging* customers are those who enter the queue but then become impatient and leave without completing their transaction. Actually, both of these situations just serve to highlight the need for queuing theory and waiting-line analysis.

Waiting-Line Characteristics

The waiting line itself is the second component of a queuing system. The length of a line can be either limited or unlimited. A queue is *limited* when it cannot, either by law or because of physical restrictions, increase to an infinite length. A small barbershop, for example, will have only a limited number of waiting chairs. Queuing models are treated in this module under an assumption of *unlimited* queue length. A queue is *unlimited* when its size is unrestricted, as in the case of the toll booth serving arriving automobiles.

A second waiting-line characteristic deals with *queue discipline*. This refers to the rule by which customers in the line are to receive service. Most systems use a queue discipline known as the first-in, first-out (FIFO) rule. In a hospital emergency room or an express checkout line at a supermarket, however, various assigned priorities may preempt FIFO. Patients who are critically injured will move ahead in treatment priority over patients with broken fingers or noses. Shoppers with fewer than 10 items may be allowed to enter the express checkout queue (but are *then* treated as first-come, first-served). Computer-programming runs also operate under priority scheduling. In most large companies, when computer-produced paychecks are due on a specific date, the payroll program gets highest priority.[2]

First-in, first-out (FIFO) rule
A queue discipline in which the first customers in line receive the first service.

Service Characteristics

The third part of any queuing system is the service characteristics. Two basic properties are important: (1) design of the service system and (2) the distribution of service times.

Basic Queuing System Designs Service systems are usually classified in terms of their number of *servers* (number of channels) and number of *phases* (number of service stops that must be made). See Figure D.3. A single-server (or single-channel) queuing system, with one server, is typified by the drive-in bank with only one open teller. If, on the other hand, the bank has several tellers on duty, with each customer waiting in one common line for the first available teller, then we would have a multiple-server (or multiple channel) queuing system. Most banks today are multiple-server systems, as are most large barbershops, airline ticket counters, and post offices.

In a single-phase system, the customer receives service from only one station and then exits the system. A fast-food restaurant in which the person who takes your order also brings your food and takes your money is a single-phase system. So is a driver's license agency in which the person taking your application also grades your test and collects your license fee. However, say the restaurant requires you to place your order at one station, pay at a second, and pick up your food at a third. In this case, it is a multiphase system. Likewise, if the driver's license agency is large or busy, you will probably have to wait in one line to complete your application

Single-server queuing system
A service system with one line and one server.

Multiple-server queuing system
A service system with one waiting line but with several servers.

Single-phase system
A system in which the customer receives service from only one station and then exits the system.

Multiphase system
A system in which the customer receives services from several stations before exiting the system.

(the first service stop), queue again to have your test graded, and finally go to a third counter to pay your fee. To help you relate the concepts of servers and phases, Figure D.3 presents these four possible configurations.

Service Time Distribution Service patterns are like arrival patterns in that they may be either constant or random. If service time is constant, it takes the same amount of time to take care of each customer. This is the case in a machine-performed service operation such as an automatic car wash. More often, service times are randomly distributed. In many cases, we can assume that random service times are described by the **negative exponential probability distribution**.

Negative exponential probability distribution
A continuous probability distribution often used to describe the service time in a queuing system.

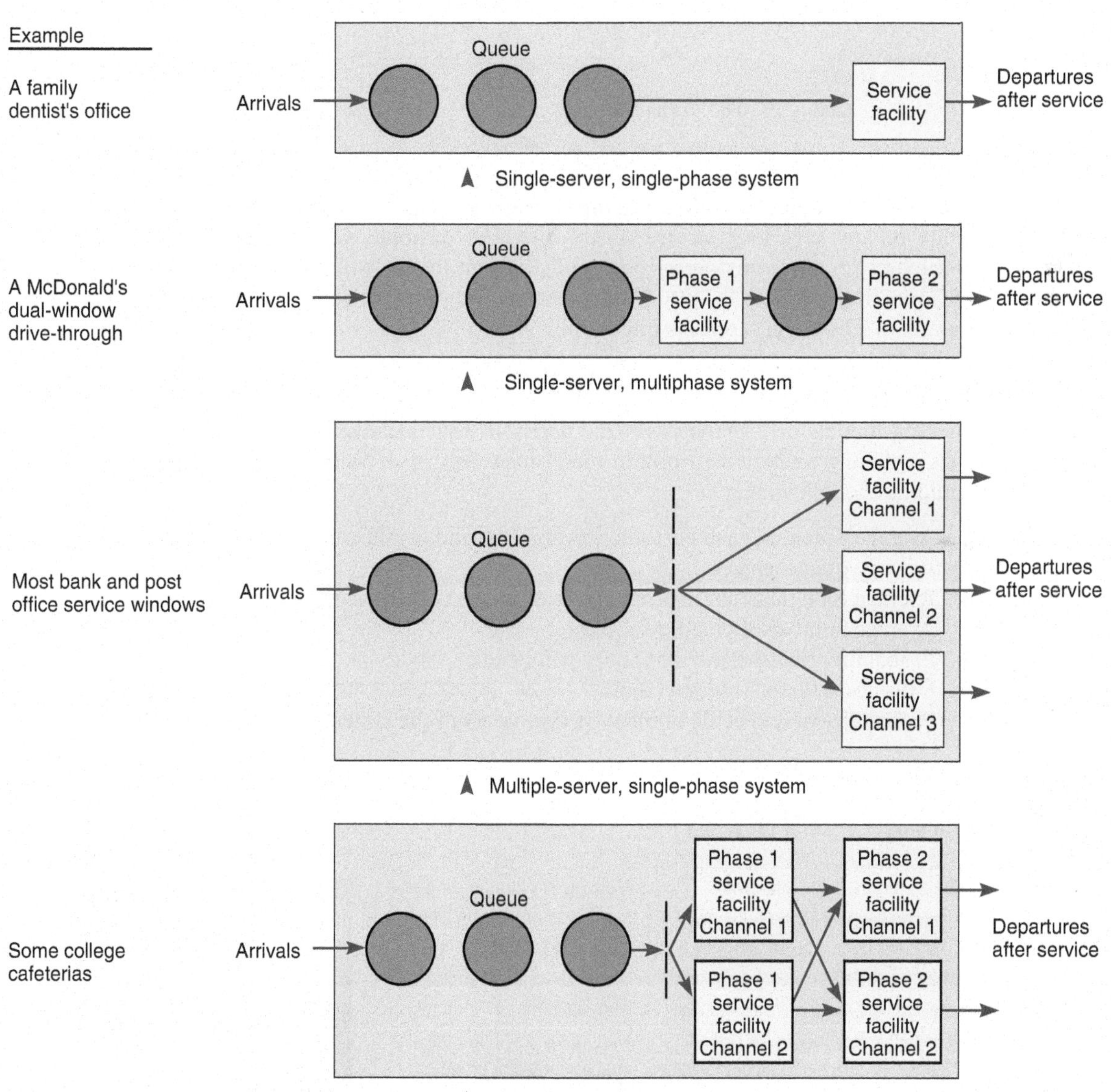

Figure **D.3**

Basic Queuing System Designs

Figure D.4

Two Examples of the Negative Exponential Distribution for Service Times

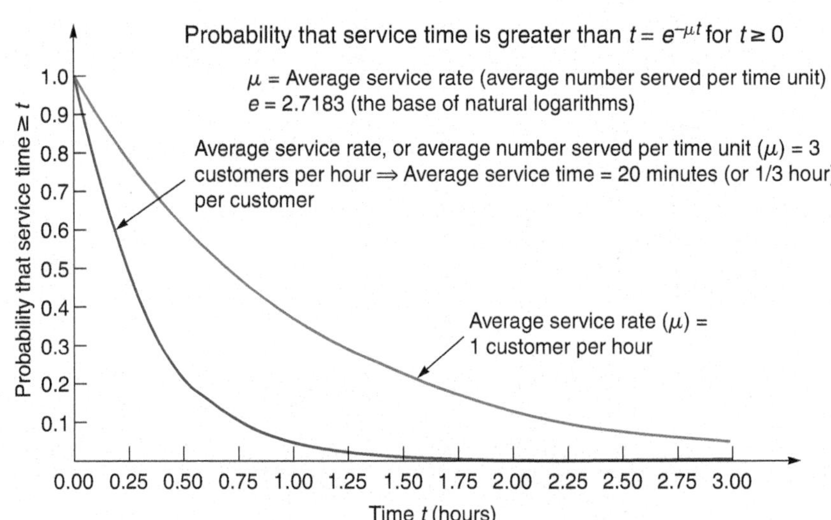

STUDENT TIP

Although Poisson and exponential distributions are commonly used to describe arrival rates and service times, other probability distributions are valid in some cases.

Figure D.4 shows that if *service times* follow a negative exponential distribution, the probability of any very long service time is low. For example, when an average service time is 20 minutes (or three customers per hour), seldom if ever will a customer require more than 1.5 hours in the service facility. If the mean service time is 1 hour, the probability of spending more than 3 hours in service is quite low.

Measuring a Queue's Performance

Queuing models help managers make decisions that balance service costs with waiting-line costs. Queuing analysis can obtain many measures of a waiting-line system's performance, including the following:

1. Average time that each customer or object spends in the queue
2. Average queue length
3. Average time that each customer spends in the system (waiting time plus service time)
4. Average number of customers in the system
5. Probability that the service facility will be idle
6. Utilization factor for the system
7. Probability of a specific number of customers in the system

OM in Action The High Cost of Long ER Waits[3]

Crowded emergency rooms have long been a problem in the U.S. In our discussions of queuing theory in this module, we typically focus on the many attributes of a waiting line—length, time, cost. However, a recent study by a South Carolina professor shows that when a new emergency room (ER) opens, crowding at nearby facilities instantly falls an average of 10%. When comparing mortality rates at the older ERs before and after the change, the research found that a 10% drop in patient volume leads to a 24% reduction in mortality rates in the first 30 days and a 17% reduction over 6 months.

In ERs across the U.S., many patients wait for hours to be seen, and about one in 50 leaves before receiving treatment. ER patients awaiting admission to the hospital often have to wait in hallways on gurneys, while ambulances may be turned away from busy facilities. Researchers have long sought to quantify these costs of crowding.

The drop in mortality rates could be attributed to fewer people leaving against medical advice. Ten percent fewer patients in the ER reduced the number of patients walking out by about 51%. That is important because about 46% of people who leave the ER without being seen still need immediate medical attention. In fact, 11% are hospitalized in the following week. Since patients often come back for care soon after they leave, that could help explain why the drop in mortality rate was most significant in the first 30 days.

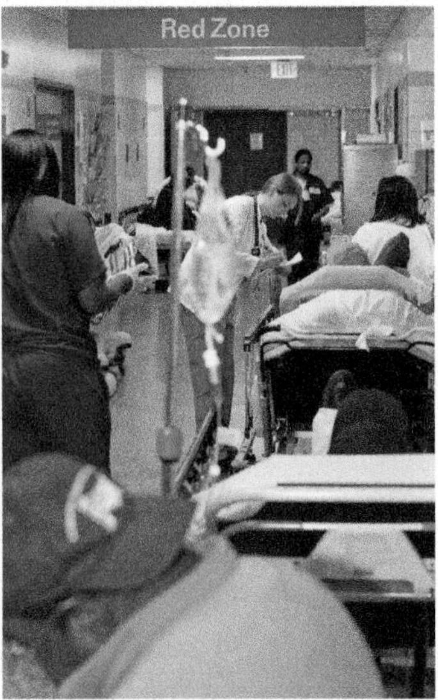

Ric Feld/AP Images

Queuing Costs

STUDENT TIP
The two costs we consider here are cost of servers and cost of lost time waiting.

As described in the *OM in Action* box, "The High Cost of Long ER Waits," operations managers must recognize the trade-off that takes place between two costs: the cost of providing good service and the cost of customer or machine waiting time. Managers want queues that are short enough so that customers do not become unhappy and either leave without buying, or buy, but never return. However, managers may be willing to allow some waiting if it is balanced by a significant savings in service costs.

One means of evaluating a service facility is to look at total expected cost. Total cost is the sum of expected service costs plus expected waiting costs.

As you can see in Figure D.5, service costs increase as a firm attempts to raise its level of service. Managers in *some* service centers can vary capacity by having standby personnel and machines that they can assign to specific service stations to prevent or shorten excessively long lines. In grocery stores, for example, managers and stock clerks can open extra checkout counters. In banks and airport check-in points, part-time workers may be called in to help.

As the level of service improves (that is, speeds up), however, the cost of time spent waiting in lines decreases. (Refer to Figure D.5.) Waiting cost may reflect lost productivity of workers while tools or machines await repairs, or it may simply be an estimate of the cost of customers lost because of poor service and long queues. In some service systems (for example, an emergency ambulance service), the cost of long waiting lines may be intolerably high.

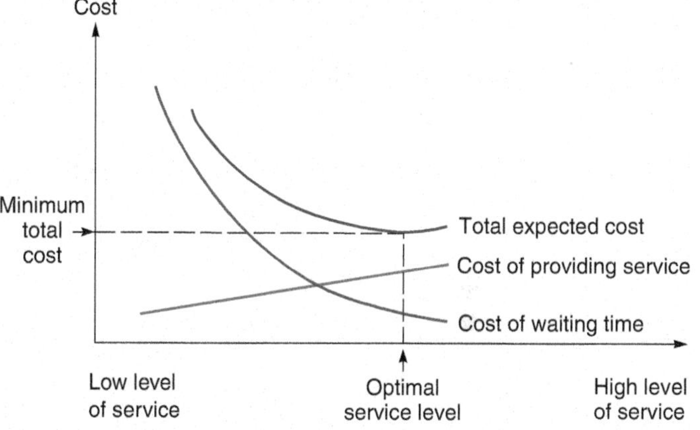

Figure **D.5**

The Trade-off between Waiting Costs and Service Costs

The Variety of Queuing Models

A wide variety of queuing models may be applied in operations management. We will introduce you to four of the most widely used models. These are outlined in Table D.2, and examples of each follow in the next few sections. More complex models are described in queuing theory textbooks or can be developed through the use of simulation (the topic of Module F). Note that all four queuing models listed in Table D.2 have three characteristics in common. They all assume:

1. Poisson distribution arrivals
2. FIFO discipline
3. A single-service phase

STUDENT TIP
This is the main section of Module D. We illustrate four important queuing models.

TABLE D.2 Queuing Models Described in This Module

MODEL	NAME (TECHNICAL NAME IN PARENTHESES)	EXAMPLE	NUMBER OF SERVERS (CHANNELS)	NUMBER OF PHASES	ARRIVAL RATE PATTERN	SERVICE TIME PATTERN	POPULATION SIZE	QUEUE DISCIPLINE
A	Single-server system (M/M/1)	Checkout at 7-Eleven	Single	Single	Poisson	Negative exponential	Unlimited	FIFO
B	Multiple-server (M/M/S)	Airline ticket counter	Multi-server	Single	Poisson	Negative exponential	Unlimited	FIFO
C	Constant service (M/D/1)	Automated car wash	Single	Single	Poisson	Constant	Unlimited	FIFO
D	Finite population (M/M/1 with finite source)	Shop with only a dozen machines that might break	Single	Single	Poisson	Negative exponential	Limited	FIFO

In addition, they all describe service systems that operate under steady, ongoing conditions. This means that arrival and service rates remain stable during the analysis.

Model A (M/M/1): Single-Server Queuing Model with Poisson Arrivals and Exponential Service Times

LO D.2 *Apply* the single-server queuing model equations

The most common case of queuing problems involves the *single-server* (also called *single-channel*) waiting line. In this situation, arrivals form a single line to be serviced by a single station (see Figure D.3). We assume that the following conditions exist in this type of system:

1. Arrivals are served on a first-in, first-out (FIFO) basis, and every arrival waits to be served, regardless of the length of the line or queue.
2. Arrivals are independent of preceding arrivals, but the average number of arrivals (*arrival rate*) does not change over time.
3. Arrivals are described by a Poisson probability distribution and come from an infinite (or very, very large) population.
4. Service times vary from one customer to the next and are independent of one another, but their average rate is known and follows the negative exponential distribution.
5. The service rate is faster than the arrival rate.

When these conditions are met, the series of equations shown in Table D.3 can be developed. Examples D1 and D2 illustrate how Model A (which in technical journals is known as the *M/M/1 model*) may be used.[4]

TABLE D.3 Queuing Formulas for Model A: Single-Server System, Also Called M/M/1

λ = average number of arrivals per time
μ = average number of people or items served per time period (average service rate)
L_s = average number of units (customers) in the system (waiting and being served) $= \dfrac{\lambda}{\mu - \lambda}$
W_s = average time a unit spends in the system (waiting time plus service time) $= \dfrac{1}{\mu - \lambda}$
L_q = average number of units waiting in the queue $= \dfrac{\lambda^2}{\mu(\mu - \lambda)}$
W_q = average time a unit spends waiting in the queue $= \dfrac{\lambda}{\mu(\mu - \lambda)} = \dfrac{L_q}{\lambda}$
ρ = utilization factor for the system $= \dfrac{\lambda}{\mu}$
P_0 = probability of 0 units in the system (that is, the service unit is idle) $= 1 - \dfrac{\lambda}{\mu}$
$P_{n>k}$ = probability of more than k units in the system, where n is the number of units in the system $= \left(\dfrac{\lambda}{\mu}\right)^{k+1}$

Example D1

A SINGLE-SERVER QUEUE

Tom Jones, the mechanic at Golden Muffler Shop, is able to install new mufflers at an average rate of 3 per hour (or about 1 every 20 minutes), according to a negative exponential distribution. Customers seeking this service arrive at the shop on the average of 2 per hour, following a Poisson distribution. They are served on a first-in, first-out basis and come from a very large (almost infinite) population of possible buyers.

We would like to obtain the operating characteristics of Golden Muffler's queuing system.

APPROACH ▶ This is a single-server (M/M/1) system, and we apply the formulas in Table D.3.

SOLUTION ▶

$$\lambda = 2 \text{ cars arriving per hour}$$

$$\mu = 3 \text{ cars serviced per hour}$$

$$L_s = \frac{\lambda}{\mu - \lambda} = \frac{2}{3-2} = \frac{2}{1}$$

$$= 2 \text{ cars in the system, on average}$$

$$W_s = \frac{1}{\mu - \lambda} = \frac{1}{3-2} = 1$$

$$= 1\text{-hour average time in the system}$$

$$L_q = \frac{\lambda^2}{\mu(\mu - \lambda)} = \frac{2^2}{3(3-2)} = \frac{4}{3(1)} = \frac{4}{3}$$

$$= 1.33 \text{ cars waiting in line, on average}$$

$$W_q = \frac{\lambda}{\mu(\mu - \lambda)} = \frac{2}{3(3-2)} = \frac{2}{3} \text{ hour}$$

$$= 40\text{-minute average waiting time per car}$$

$$\rho = \frac{\lambda}{\mu} = \frac{2}{3}$$

$$= 66.7\% \text{ of the time the mechanic is busy}$$

$$P_0 = 1 - \frac{\lambda}{\mu} = 1 - \frac{2}{3} = \frac{1}{3}$$

$$= .333 \text{ probability there are 0 cars in the system}$$

Probability of More Than *K* Cars in the System

K	$P_{n>k} = (2/3)^{k+1}$
0	.667 ← Note that this is equal to $1 - P_0 = 1 - .333 = .667$.
1	.444
2	.296
3	.198 ← Implies that there is a 19.8% chance that more than 3 cars are in the system.
4	.132
5	.088
6	.058
7	.039

INSIGHT ▶ Recognize that arrival and service times are converted to the same rate. For example, a service time of 20 minutes is stated as an average *rate* of 3 mufflers *per hour*. It's also important to differentiate between time in the *queue* and time in the *system*.

LEARNING EXERCISE ▶ If $\mu = 4$ cars/hour instead of the current 3 arrivals, what are the new values of L_s, W_s, L_q, W_q, and P_0? [Answer: 1 car, 30 min., .5 cars, 15 min., 50%, .50.]

RELATED PROBLEMS ▶ D.1–D.2, D.4, D.6–D.8, D.9a–e, D.10, D.11a–c, D.12a–d, D.31–D.33, D.34a–e, D.35a–e, D.38–D.39

EXCEL OM Data File **ModDExD1.xls** can be found online.

ACTIVE MODEL D.1 This example is further illustrated in Active Model D.1 found online.

Once we have computed the operating characteristics of a queuing system, it is often important to do an economic analysis of their impact. Although the waiting-line model just described is valuable in predicting potential waiting times, queue lengths, idle times, and so on, it does not identify optimal decisions or consider cost factors. As we saw earlier, the solution to a queuing problem may require management to make a trade-off between the increased cost of providing better service and the decreased waiting costs derived from providing that service.

Example D2 examines the costs involved in Example D1.

Example D2

ECONOMIC ANALYSIS OF EXAMPLE D1

Golden Muffler Shop's owner is interested in cost factors as well as the queuing parameters computed in Example D1. She estimates that the cost of customer waiting time, in terms of customer dissatisfaction and lost goodwill, is $15 per hour spent *waiting* in line. Jones, the mechanic, is paid $22 per hour.

APPROACH ▶ First compute the average daily customer waiting time, then the daily salary for Jones, and finally the total expected cost.

LO D.3 *Conduct* a cost analysis for a waiting line

SOLUTION ▶ Because the average car has a $\frac{2}{3}$-hour wait (W_q) and because there are approximately 16 cars serviced per day (2 arrivals per hour times 8 working hours per day), the total number of hours that customers spend waiting each day for mufflers to be installed is:

$$\frac{2}{3}(16) = \frac{32}{3} = 10\frac{2}{3} \text{ hours}$$

Hence, in this case:

$$\text{Customer waiting-time cost} = \$15\left(10\frac{2}{3}\right) = \$160 \text{ per day}$$

The only other major cost that Golden's owner can identify in the queuing situation is the salary of Jones, the mechanic, who earns $22 per hour, or $176 per day. Thus:

$$\begin{aligned}\text{Total expected cost} &= \$160 + \$176 \\ &= \$336 \text{ per day}\end{aligned}$$

This approach will be useful in Solved Problem D.

INSIGHT ▶ L_q and W_q are the two most important queuing parameters when it comes to cost analysis. Calculating customer wait times, we note, is based on average time waiting in the queue (W_q) times the number of arrivals per hour (λ) times the number of hours per day. This is because this example is set on a daily basis. This is the same as using L_q because $L_q = W_q\lambda$.

LEARNING EXERCISE ▶ If the customer waiting time is actually $20 per hour and Jones gets a wage increase to $30 per hour, what is the total daily expected cost? [Answer: $453.33.]

RELATED PROBLEMS ▶ D.12e–f, D.13, D.22, D.23, D.24, D.37

Model B (M/M/S): Multiple-Server Queuing Model

LO D.4 *Apply* the multiple-server queuing model formulas

Now let's turn to a multiple-server (multiple-channel) queuing system in which two or more servers are available to handle arriving customers. We still assume that customers awaiting service form one single line and then proceed to the first available server. Multiple-server, single-phase waiting lines are found in many banks today: a common line is formed, and the customer at the head of the line proceeds to the first free teller. (Refer to Figure D.3 for a typical multiple-server configuration.)

Jeff Zelevansky JAZ/HB/Reuters

To shorten lines (or wait times), each Costco register is staffed with two employees. This approach has improved efficiency by 20–30%.

The multiple-server system presented in Example D3 again assumes that arrivals follow a Poisson probability distribution and that service times are negative exponentially distributed. Service is first-come, first-served, and all servers are assumed to perform at the same rate. Other assumptions listed earlier for the single-server model also apply.

The queuing equations for Model B (which also has the technical name *M*/*M*/*S*) are shown in Table D.4. These equations are obviously more complex than those used in the single-server model, yet they are used in exactly the same fashion and provide the same type of information as the simpler model. (*Note:* The POM for Windows and Excel OM software described later in this chapter can prove very useful in solving multiple-server and other queuing problems.)

TABLE D.4 Queuing Formulas for Model B: Multiple-Server System, Also Called M/M/S

M = number of servers (channels) open
λ = average number of arrivals per time period (average arrival rate)
μ = average service rate at each server (channel)

The probability that there are zero people or units in the system is:

$$P_0 = \frac{1}{\left[\sum_{n=0}^{M-1} \frac{1}{n!}\left(\frac{\lambda}{\mu}\right)^n\right] + \frac{1}{M!}\left(\frac{\lambda}{\mu}\right)^M \frac{M\mu}{M\mu - \lambda}} \quad \text{for } M\mu > \lambda$$

The average number of people or units in the system is:

$$L_s = \frac{\lambda\mu(\lambda/\mu)^M}{(M-1)!(M\mu - \lambda)^2} P_0 + \frac{\lambda}{\mu}$$

The average time a unit spends in the waiting line and being serviced (namely, in the system) is:

$$W_s = \frac{\mu(\lambda/\mu)^M}{(M-1)!(M\mu - \lambda)^2} P_0 + \frac{1}{\mu} = \frac{L_s}{\lambda}$$

The average number of people or units in line waiting for service is:

$$L_q = L_s - \frac{\lambda}{\mu}$$

The average time a person or unit spends in the queue waiting for service is:

$$W_q = W_s - \frac{1}{\mu} = \frac{L_q}{\lambda}$$

Example D3

A MULTIPLE-SERVER QUEUE

The Golden Muffler Shop has decided to open a second garage bay and hire a second mechanic to handle installations. Customers, who arrive at the rate of about $\lambda = 2$ per hour, will wait in a single line until 1 of the 2 mechanics is free. Each mechanic installs mufflers at the rate of about $\mu = 3$ per hour.

The company wants to find out how this system compares with the old single-server waiting-line system.

APPROACH ▶ Compute several operating characteristics for the $M = 2$ server system, using the equations in Table D.4, and compare the results with those found in Example D1.

SOLUTION ▶

$$P_0 = \frac{1}{\left[\sum_{n=0}^{1} \frac{1}{n!}\left(\frac{2}{3}\right)^n\right] + \frac{1}{2!}\left(\frac{2}{3}\right)^2 \frac{2(3)}{2(3)-2}}$$

$$= \frac{1}{1 + \frac{2}{3} + \frac{1}{2}\left(\frac{4}{9}\right)\left(\frac{6}{6-2}\right)} = \frac{1}{1 + \frac{2}{3} + \frac{1}{3}} = \frac{1}{2}$$

$= .5$ probability of zero cars in the system

Then:

$$L_s = \frac{(2)(3)(2/3)^2}{1![2(3)-2]^2}\left(\frac{1}{2}\right) + \frac{2}{3} = \frac{8/3}{16}\left(\frac{1}{2}\right) + \frac{2}{3} = \frac{3}{4}$$

$= .75$ average number of cars in the system

$$W_s = \frac{L_s}{\lambda} = \frac{3/4}{2} = \frac{3}{8} \text{ hour}$$

$= 22.5$ minutes average time a car spends in the system

$$L_q = L_s - \frac{\lambda}{\mu} = \frac{3}{4} - \frac{2}{3} = \frac{9}{12} - \frac{8}{12} = \frac{1}{12}$$

$= .083$ average number of cars in the queue (waiting)

$$W_q = \frac{L_q}{\lambda} = \frac{.083}{2} = .0415 \text{ hour}$$

$= 2.5$ minutes average time a car spends in the queue (waiting)

INSIGHT ▶ It is very interesting to see the big differences in service performance when an additional server is added.

LEARNING EXERCISE ▶ If $\mu = 4$ per hour, instead of $\mu = 3$, what are the new values for P_0, L_s, W_s, L_q, and W_q? [Answers: 0.6, .53 cars, 16 min, .033 cars, 1 min.]

RELATED PROBLEMS ▶ D.7h, D.9f, D.11d, D.15, D.20, D.35f, D.36

EXCEL OM Data File **ModDEx.xls** can be found online.

ACTIVE MODEL D.2 This example is further illustrated in Active Model D.2 found online.

We can summarize the characteristics of the two-server model in Example D3 and compare them to those of the single-server model in Example D1 as follows:

MEASURE		SINGLE SERVER	TWO SERVERS (CHANNELS)
Probability of 0 units in the system	P_0	.33	.5
Number of units in the system	L_s	2 cars	.75 car
Average time in the system	W_s	60 minutes	22.5 minutes
Average number in the queue	L_q	1.33 cars	.083 car
Average time in the queue	W_q	40 minutes	2.5 minutes

The increased service has a dramatic effect on almost all characteristics. For instance, note that the time spent waiting in line drops from 40 minutes to only 2.5 minutes.

Use of Waiting-Line Tables Imagine the work a manager would face in dealing with M = 3-, 4-, or 5- server waiting-line models if a computer was not readily available. The arithmetic becomes increasingly troublesome. Fortunately, much of the burden of manually examining multiple-server queues can be avoided by using Table D.5. This table, the result of hundreds of computations, represents the relationship between three things: (1) a ratio, λ/μ, (2) number of servers open, and (3) the average number of customers in the queue, L_q (which is what we'd like to find). For any combination of the ratio λ/μ and $M = 1, 2, 3, 4$, or 5 servers, you can quickly look in the body of the table to read off the appropriate value for L_q.

TABLE D.5 Values of L_q for M = 1–5 Servers (channels) and Selected Values of λ/μ

POISSON ARRIVALS, EXPONENTIAL SERVICE TIMES					
	NUMBER OF SERVERS (CHANNELS), M				
λ/μ	1	2	3	4	5
.10	.0111				
.15	.0264	.0008			
.20	.0500	.0020			
.25	.0833	.0039			
.30	.1285	.0069			
.35	.1884	.0110			
40	.2666	.0166			
.45	.3681	.0239	.0019		
.50	.5000	.0333	.0030		
.55	.6722	.0449	.0043		
.60	.9000	.0593	.0061		
.65	1.2071	.0767	.0084		
.70	1.6333	.0976	.0112		
.75	2.2500	.1227	.0147		
.80	3.2000	.1523	.0189		
.85	4.8166	.1873	.0239	.0031	
.90	8.1000	.2285	.0300	.0041	
.95	18.0500	.2767	.0371	.0053	
1.0		.3333	.0454	.0067	
1.2		.6748	.0904	.0158	
1.4		1.3449	.1778	.0324	.0059
1.6		2.8444	.3128	.0604	.0121
1.8		7.6734	.5320	.1051	.0227
2.0			.8888	.1739	.0398
2.2			1.4907	.2770	.0659
2.4			2.1261	.4305	.1047
2.6			4.9322	.6581	.1609
2.8			12.2724	1.0000	.2411
3.0				1.5282	.3541
3.2				2.3856	.5128
3.4				3.9060	.7365
3.6				7.0893	1.0550
3.8				16.9366	1.5184
4.0					2.2164
4.2					3.3269
4.4					5.2675
4.6					9.2885
4.8					21.6384

Example D4 illustrates the use of Table D.5.

Example D4

USE OF WAITING-LINE TABLES

Alaska National Bank is trying to decide how many drive-in teller windows to open on a busy Saturday. CEO Ted Eschenbach estimates that customers arrive at a rate of about $\lambda = 18$ per hour, and that each teller can service about $\mu = 20$ customers per hour.

APPROACH ▶ Ted decides to use Table D.5 to compute L_q and W_q.

SOLUTION ▶ The ratio is $\lambda/\mu = \frac{18}{20} = .90$. Turning to the table, under $\lambda/\mu = .90$, Ted sees that if only $M = 1$ service window is open, the average number of customers in line will be 8.1. If two windows are open, L_q drops to .2285 customer, to .03 for $M = 3$ tellers, and to .0041 for $M = 4$ tellers. Adding more open windows at this point will result in an average queue length of 0.

It is also a simple matter to compute the average waiting time in the queue, W_q, since $W_q = L_q/\lambda$. When one service window is open, $W_q = 8.1$ customers/(18 customers per hour) = .45 hour = 27 minutes waiting time; when two tellers are open, $W_q = .2285\,\text{customer}/(18\,\text{customers per hour}) = .0127$ hour $\cong \frac{3}{4}$ minute; and so on.

INSIGHT ▶ If a computer is not readily available, Table D.5 makes it easy to find L_q and to then compute W_q. Table D.5 is especially handy to compare L_q for different numbers of servers (M).

LEARNING EXERCISE ▶ The number of customers arriving on a Thursday afternoon at Alaska National is 15/hour. The service rate is still 20 customers/hour. How many people are in the queue if there are 1, 2, or 3 servers? [Answer: 2.25, .1227, .0147.]

RELATED PROBLEMS ▶ D.3, D.5

You might also wish to check the calculations in Example D3 against tabled values just to practice the use of Table D.5. You may need to interpolate if your exact value is not found in the first column. Other common operating characteristics besides L_q are published in tabular form in queuing theory textbooks.

Elenathewise/Fotolia

Stephen Brashear/AP Images

Long check-in lines (left photo) such as at Los Angeles International (LAX) are a common airport sight. This is an M/M/S model—passengers wait in a single queue for one of several agents. But at Anchorage International Airport (right photo), Alaska Airlines, like several other airlines, has jettisoned the traditional wall of ticket counters. Instead, millions of passengers per year use self-service check-in machines and staffed "bag drop" stations. Looking nothing like a typical airport, the system doubled the airline's check-in capacity and cut staff needs in half, all while speeding travelers through in less than 15 minutes, even during peak hours.

Model C (M/D/1): Constant-Service-Time Model

LO D.5 *Apply* the constant-service-time model equations

Some service systems have constant, instead of exponentially distributed, service times. When customers or equipment are processed according to a fixed cycle, as in the case of an automatic car wash or an amusement park ride, constant service times are appropriate. Because constant rates are certain, the values for L_q, W_q, L_s, and W_s are always less than they would be in Model A, which has variable service rates. As a matter of fact, both the average queue length and the average waiting time in the queue are halved with Model C. Constant-service-model formulas are given in Table D.6. Model C also has the technical name *M/D/1* in the literature of queuing theory.

TABLE D.6 Queuing Formulas for Model C: Constant Service, Also Called M/D/1

Average length of queue: $L_q = \dfrac{\lambda^2}{2\mu(\mu - \lambda)}$
Average waiting time in queue: $W_q = \dfrac{\lambda}{2\mu(\mu - \lambda)}$
Average number of customers in system: $L_s = L_q + \dfrac{\lambda}{\mu}$
Average time in system: $W_s = W_q + \dfrac{1}{\mu}$

Example D5 gives a constant-service-time analysis.

Example D5

A CONSTANT-SERVICE-TIME MODEL

Inman Recycling, Inc., collects and compacts aluminum cans and glass bottles in Reston, Louisiana. Its truck drivers currently wait an average of 15 minutes before emptying their loads for recycling. The cost of driver and truck time while they are in queues is valued at \$60 per hour. A new automated compactor can be purchased to process truckloads at a *constant* rate of 12 trucks per hour (that is, 5 minutes per truck). Trucks arrive according to a Poisson distribution at an average rate of 8 per hour. If the new compactor is put in use, the cost will be amortized at a rate of \$3 per truck unloaded.

APPROACH ▶ CEO Tony Inman hires a summer college intern to conduct an analysis to evaluate the costs versus benefits of the purchase. The intern uses the equation for W_q in Table D.6.

SOLUTION ▶

$$\text{Current Waiting cost/trip} = (1/4\,\text{hr waiting now})(\$60/\text{hr cost}) = \$15/\text{trip}$$

$$\text{New system: } \lambda = 8 \text{ trucks/hr arriving} \qquad \mu = 12 \text{ trucks/hr served}$$

$$\text{Average waiting time in queue} = W_q = \frac{\lambda}{2\mu(\mu - \lambda)} = \frac{8}{2(12)(12 - 8)} = \frac{1}{12}\text{ hr}$$

$$\text{Waiting cost/trip with new compactor} = (1/12\,\text{hr wait})(\$60/\text{hr cost}) = \$5/\text{trip}$$

$$\text{Savings with new equipment} = \$15(\text{current system}) - \$5(\text{new system}) = \$10/\text{trip}$$

$$\text{Cost of new equipment amortized:} \qquad = \$\ 3/\text{trip}$$

$$\text{Net savings} \qquad = \$\ 7/\text{trip}$$

INSIGHT ▶ Constant service times, usually attained through automation, help control the variability inherent in service systems. This can lower average queue length and average waiting time. Note the 2 in the denominator of the equations for L_q and W_q in Table D.6.

LEARNING EXERCISE ▶ With the new constant-service-time system, what are the average waiting time in the queue, average number of trucks in the system, and average waiting time in the system? [Answer: 0.0833 hour, 1.33 trucks, 0.1667 hour.]

RELATED PROBLEMS ▶ D.14, D.16, D.21, D.34f

EXCEL OM Data File **ModDExD5.xls** can be found online.

ACTIVE MODEL D.3 This example is further illustrated in Active Model D.3 found online.

Little's Law

A practical and useful relationship in queuing for any system in a *steady state* is called Little's Law. A steady state exists when a queuing system is in its normal operating condition (e.g., after customers waiting at the door when a business opens in the morning are taken care of). Little's Law can be written as either:

$$L_s = \lambda W_s \text{ (which is the same as } W_s = L_s/\lambda) \quad \text{(D-2)}$$

or:

$$L_q = \lambda W_q \text{ (which is the same as } W_q = L_q/\lambda) \quad \text{(D-3)}$$

The advantage of these formulas is that once two of the parameters are known, the other one can easily be found. This is important because in certain waiting-line situations, one of these might be easier to determine than the other.

Example D6

LITTLE'S LAW

Customers walk into the local U.S. Post Office at an average rate of 20 per hour. On average, there are 5 people waiting in line to be served. The probability distributions that describe arrival and service times are unknown. The manager, Vicky Luo, wishes to determine how long customers are waiting in line.

APPROACH ▶ Even though the probability distributions and even the number of servers are unknown, Vicky can use Little's Law to quickly determine the average waiting time.

SOLUTION ▶ $\lambda = 20$ customers per hour

$L_q = 5$ customers

Using Equation (D-3), $W_q = L_q/\lambda$

$= 5/20 = 0.25$ hour

And $(0.25 \text{ hours})(60 \text{ minutes/hour}) = 15$ minutes

INSIGHT ▶ It can be relatively easy to count the number of arriving customers per hour, and the average number of customers in line can be estimated by counting the line length throughout the day and taking the average of those lengths. However, it would take more effort to keep track of the time that the customers enter the facility and the time that they begin being served. Little's Law eliminates the need to track actual waiting times.

LEARNING EXERCISE ▶ Vicky believes that 15 minutes is an unreasonable waiting time. She adds a server to help during busy times, and the average number of customers in line reduces to 1.2 customers. Now how long do customers wait? [Answer: 3.6 minutes.]

RELATED PROBLEMS ▶ D.25–D.30

Little's Law is also important because it makes no assumptions about the probability distributions for arrivals and service times, the number of servers, or service priority rules. The law applies to all the queuing systems discussed in this module, except the finite-population model, which we discuss next.

Model D (M/M/1 with Finite Source): Finite-Population Model

LO D.6 *Perform* a finite-population model analysis

When there is a limited (or finite) population of potential customers for a service facility, we must consider a different queuing model. This model would be used, for example, if we were considering equipment repairs in a factory that has 5 machines, if we were in charge of maintenance for a fleet of 10 commuter airplanes, or if we ran a hospital ward that has 20 beds. The finite-population model allows any number of repair people (servers) to be considered.

This model differs from the three earlier queuing models because there is now a *dependent* relationship between the length of the queue and the arrival rate. Let's illustrate the extreme situation: If your factory had five machines and all were broken and awaiting repair, the arrival rate would drop to zero. In general, then, as the *waiting line* becomes longer in the finite population model, the *arrival rate* of customers or machines drops.

The finite calling population model has the following assumptions:

1. There is only one server.
2. The population of units seeking service is finite.[5]
3. Arrivals follow a Poisson distribution, and service times are negative exponentially distributed.
4. Customers are served on a first-come, first-served basis.

Table D.7 displays the queuing formulas for the finite-population model.

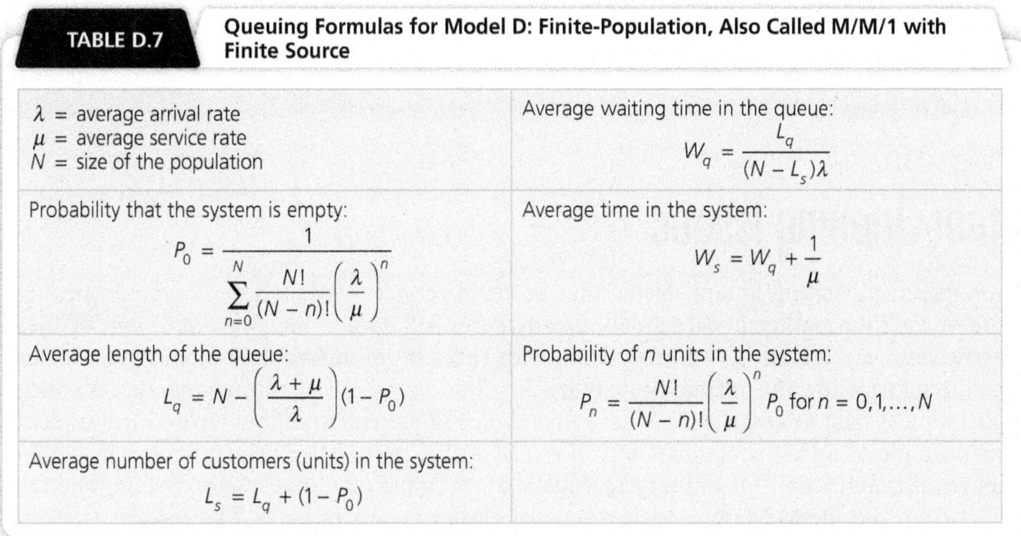

TABLE D.7 Queuing Formulas for Model D: Finite-Population, Also Called M/M/1 with Finite Source

λ = average arrival rate μ = average service rate N = size of the population	Average waiting time in the queue: $W_q = \dfrac{L_q}{(N - L_s)\lambda}$
Probability that the system is empty: $P_0 = \dfrac{1}{\sum\limits_{n=0}^{N} \dfrac{N!}{(N-n)!}\left(\dfrac{\lambda}{\mu}\right)^n}$	Average time in the system: $W_s = W_q + \dfrac{1}{\mu}$
Average length of the queue: $L_q = N - \left(\dfrac{\lambda + \mu}{\lambda}\right)(1 - P_0)$	Probability of n units in the system: $P_n = \dfrac{N!}{(N-n)!}\left(\dfrac{\lambda}{\mu}\right)^n P_0$ for $n = 0,1,\ldots,N$
Average number of customers (units) in the system: $L_s = L_q + (1 - P_0)$	

Example D7 illustrates Model D.

Example D7

A FINITE-POPULATION MODEL

Past records indicate that each of the 5 massive laser computer printers at the U.S. Department of Energy (DOE), in Washington, DC, needs repair after about 20 hours of use. Breakdowns have been determined to be Poisson distributed. The one technician on duty can service a printer in an average of 2 hours, following an exponential distribution. Printer downtime costs $120 per hour. The technician is paid $25 per hour.

APPROACH ▶ To compute the system's operation characteristics we note that the mean arrival rate is $\lambda = 1/20 = 0.05$ printers/hour. The mean service rate is $\mu = 1/2 = 0.50$ printer/hour.

SOLUTION ▶

1. $P_0 = \dfrac{1}{\sum\limits_{n=0}^{5} \dfrac{5!}{(5-n)!}\left(\dfrac{0.05}{0.5}\right)^n} = 0.564$ (we leave these calculations for you to confirm)

2. $L_q = 5 - \left(\dfrac{0.05 + 0.5}{0.05}\right)(1 - P_0) = 5 - (11)(1 - 0.564) = 5 - 4.8 = 0.2$ printer

3. $L_s = 0.2 + (1 - 0.564) = 0.64$ printer

4. $W_q = \dfrac{0.2}{(5 - 0.64)(0.05)} = \dfrac{0.2}{0.22} = 0.91$ hour

5. $W_s = 0.91 + \dfrac{1}{0.50} = 2.91$ hours

We can also compute the total cost per hour:

$$\text{Total hourly cost} = (\text{Average number of printers down})(\text{Cost per downtime hour}) + \text{Cost per technician hour}$$
$$= (0.64)(\$120) + \$25 = \$76.80 + \$25.00 = \$101.80$$

INSIGHT ▶ Management can now determine whether these costs and wait times are acceptable. Perhaps it is time to add a second technician.

LEARNING EXERCISE ▶ DOE has just replaced its printers with a new model that seems to break down after about 18 hours of use. Recompute the total hourly cost. [Answer: $L_s = .25$, so cost = (.25)($120) + $25 = $30 + $25 = $55.]

RELATED PROBLEMS ▶ D.17, D.18, D.19

EXCEL OM Data File **ModDExD7.xls** can be found online.

Other Queuing Issues

STUDENT TIP When the assumptions of the four models we just introduced do not hold true, there are other approaches still available to us.

Many practical waiting-line problems that occur in service systems have characteristics like those of the four mathematical models already described. Often, however, *variations* of these specific cases are present in an analysis. Service times in an automobile repair shop, for example, tend to follow the normal probability distribution instead of the exponential. A college registration system in which seniors have first choice of courses and hours over other students is an example of a first-come, first-served model with a preemptive priority queue discipline. A physical examination for military recruits is an example of a multiphase system, one that differs from the single-phase models discussed earlier in this module. The recruits first line up to have blood drawn at one station, then wait for an eye exam at the next station, talk to a psychiatrist at the third, and are examined by a doctor for medical problems at the fourth. At each phase, the recruits must enter another queue and wait their turn. Many models, some very complex, have been developed to deal with situations such as these.

The Psychology of Waiting

Waiting represents one of the most stress-inducing experiences for humans. No matter what the context, frustration quickly mounts, resulting in angry customers, shouting matches, switching to a competitor, or even road rage. Social psychologists have identified the tools below to help organizations manage the waiting process and reduce customer anxiety as much as possible.

- *Make waiting more comfortable* (chairs; air conditioning; refreshments)
- *Establish virtual queues* (pagers; Disney's Genie+)
- *Distract customers' attention* (mirrors near elevators; videos)
- *Start service early* (take drink orders; let customers look at the menu before being seated)
- *Explain reasons for the wait* (reduces uncertainty; creates understanding and empathy)
- *Provide pessimistic estimates of waiting time* (customers are pleasantly surprised with a shorter wait)
- *Compensate for extraordinary waiting* (free drinks; coupons)
- *Don't make unrealistic promises* (avoids even greater anger later)
- *Be fair!* (customers more willing to "share the pain" of waiting if others have similar wait times)

Summary

Queues are an important part of the world of operations management. In this module, we describe several common queuing systems and present mathematical models for analyzing them.

The most widely used queuing models include Model A (M/M/1), the basic single-server, single-phase system with Poisson arrivals and negative exponential service times; Model B (M/M/S), the multiple-server equivalent of Model A; Model C (M/D/1), a constant-service-rate model; and Model D, a finite-population system (M/M/1 with finite source). All four models allow for Poisson arrivals; first-in, first-out service; and a single-service phase. Typical operating characteristics we examine include average time spent waiting in the queue and system, average number of customers in the queue and system, idle time, and utilization rate.

A variety of queuing models exists for which all the assumptions of the traditional models need not be met. In these cases, we use more complex mathematical models or turn to a technique called *simulation*. The application of simulation to problems of queuing systems is addressed in Module F.

Key Terms

Queuing theory (p. 762)
Waiting line (queue) (p. 762)
Unlimited, or infinite, population (p. 762)
Limited, or finite, population (p. 762)
Poisson distribution (p. 763)
First-in, first-out (FIFO) rule (p. 764)
Single-server queuing system (p. 764)
Multiple-server queuing system (p. 764)
Single-phase system (p. 764)
Multiphase system (p. 764)
Negative exponential probability distribution (p. 765)

Discussion Questions

1. Name the three parts of a typical queuing system.
2. When designing a waiting line system, what "qualitative" concerns need to be considered?
3. Name the three factors that govern the structure of "arrivals" in a queuing system.
4. State the seven common measures of queuing system performance.
5. State the assumptions of the "basic" single-server queuing model (Model A, or M/M/1).
6. Is it good or bad to operate a supermarket bakery system on a strict first-come, first-served basis? Why?
7. Describe what is meant by the waiting-line terms *balk* and *renege*. Provide an example of each.
8. Which is larger, W_s or W_q? Explain.
9. Briefly describe three situations in which the first-in, first-out (FIFO) discipline rule is not applicable in queuing analysis.
10. Describe the behavior of a waiting line where $\lambda > \mu$. Use both analysis and intuition.
11. Discuss the likely outcome of a waiting line system where $\mu > \lambda$ but only by a tiny amount (e.g., $\mu = 4.1$, $\lambda = 4$).
12. Provide examples of four situations in which there is a limited, or finite, waiting line.
13. What are the components of the following queuing systems? Draw and explain the configuration of each.
    a) Barbershop
    b) Car wash
    c) Laundromat
    d) Small grocery store
14. Do doctors' offices generally have random arrival rates for patients? Are service times random? Under what circumstances might service times be constant?
15. What happens if two single-server systems have the same mean arrival and service rates, but the service time is constant in one and exponential in the other?
16. What dollar value do you place on yourself per hour that you spend waiting in lines? What value do your classmates place on themselves? Why do the values differ?
17. Why is Little's Law a useful queuing concept?

Using Software to Solve Queuing Problems

Both Excel OM and POM for Windows may be used to analyze most of the homework problems in this module.

X USING EXCEL OM

Excel OM's Waiting-Line program handles all four of the models developed in this module. Program D.1 illustrates our first model, the M/M/1 system, using the data from Example D1.

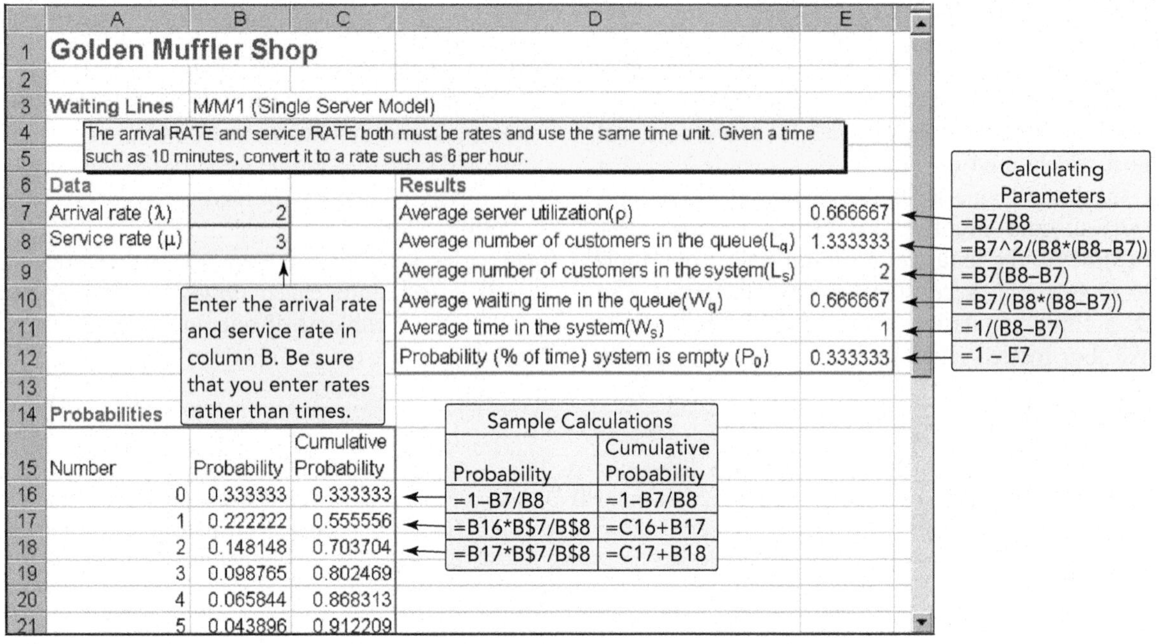

Program **D.1**

Using Excel OM for Queuing

Example D1's (Golden Muffler Shop) data are illustrated in the M/M/1 model.

P USING POM FOR WINDOWS

There are several POM for Windows queuing models from which to select in that program's Waiting-Line module. The program can include an economic analysis of cost data, and, as an option, you may display probabilities of various numbers of people/items in the system. See Appendix II for further details.

Solved Problems

SOLVED PROBLEM D.1

Sid Das Brick Distributors in Jamaica currently employs 1 worker whose job is to load bricks on outgoing company trucks. An average of 24 trucks per day, or 3 per hour, arrive at the loading platform, according to a Poisson distribution. The worker loads them at a rate of 4 trucks per hour, following approximately the exponential distribution in his service times.

Das believes that adding an additional brick loader will substantially improve the firm's productivity. He estimates that a 2-person crew loading each truck will double the loading rate (μ) from 4 trucks per hour to 8 trucks per hour. Analyze the effect on the queue of such a change, and compare the results to those achieved with one worker. What is the probability that there will be more than 3 trucks either being loaded or waiting?

SOLUTION

	NUMBER OF BRICK LOADERS	
	1	2
Truck arrival rate (λ)	3/hr	3/hr
Loading rate (μ)	4/hr	8/hr
Average number in system (L_s)	3 trucks	.6 truck
Average time in system (W_s)	1 hr	.2 hr
Average number in queue (L_q)	2.25 trucks	.225 truck
Average time in queue (W_q)	.75 hr	.075 hr
Utilization rate ρ	.75	.375
Probability system empty (P_0)	.25	.625

Probability of More than *K* Trucks in System

	PROBABILITY *N* > *K*	
K	1 LOADER	2 LOADERS
0	.75	.375
1	.56	.141
2	.42	.053
3	.32	.020

These results indicate that when only one loader is employed, the average truck must wait three-quarters of an hour before it is loaded. Furthermore, there is an average of 2.25 trucks waiting in line to be loaded. This situation may be unacceptable to management. Note also the decline in queue size after the addition of a second loader.

SOLVED PROBLEM D.2

Truck drivers working for Sid Das in Jamaica (see Solved Problem D.1) earn an average of \$10 per hour. Brick loaders receive about \$6 per hour. Truck drivers waiting *in the queue or at the loading platform* are drawing a salary but are productively idle and unable to generate revenue during that time. What would be the *hourly* cost savings to the firm if it employed 2 loaders instead of 1?

Referring to the data in Solved Problem D.1, we note that the average number of trucks *in the system* is 3 when there is only 1 loader and .6 when there are 2 loaders.

SOLUTION

	NUMBER OF LOADERS	
	1	2
Truck driver idle time costs [(Average number of trucks) × (Hourly rate)] = (3)(\$10) =	\$30	\$ 6 = (.6)(\$10)
Loading costs	6	12 = (2)(\$6)
Total expected cost per hour	\$36	\$18

The firm will save \$18 per hour by adding another loader.

SOLVED PROBLEM D.3

Sid Das is considering building a second platform or gate to speed the process of loading trucks. This system, he thinks, will be even more efficient than simply hiring another loader to help out on the first platform (as in Solved Problem D.1).

Assume that the worker at each platform will be able to load 4 trucks per hour each and that trucks will continue to arrive at the rate of 3 per hour. Then apply the appropriate equations to find the waiting line's new operating conditions. Is this new approach indeed speedier than hiring a second loader, as Das has considered in the preceding Solved Problems?

SOLUTION

$$P_0 = \frac{1}{\left[\sum_{n=0}^{1} \frac{1}{n!}\left(\frac{3}{4}\right)^n\right] + \frac{1}{2!}\left(\frac{3}{4}\right)^2 \frac{2(4)}{2(4) - 3}}$$

$$= \frac{1}{1 + \frac{3}{4} + \frac{1}{2}\left(\frac{3}{4}\right)^2\left(\frac{8}{8-3}\right)} = .4545$$

$$L_s = \frac{3(4)(3/4)^2}{(1)!(8-3)^2}(.4545) + \frac{3}{4} = .873$$

$$W_s = \frac{.873}{3} = .291\,\text{hr}$$

$$L_q = .873 - 3/4 = .123$$

$$W_q = \frac{.123}{3} = .041\,\text{hr}$$

Looking back at Solved Problem D.1, we see that although length of the *queue* and average time in the queue are lowest when a second platform is open, the average number of trucks in the *system* and average time spent waiting in the system are smallest when two workers are employed at a *single* platform. Thus, we would probably recommend not building a second platform.

SOLVED PROBLEM D.4

Mount Sinai Hospital's orthopedic care unit has 5 beds, which are virtually always occupied by patients who have just undergone orthopedic surgery. One registered nurse is on duty in the unit in each of the three 8-hour shifts. About every 2 hours (following a Poisson distribution), one of the patients requires a nurse's attention. The nurse will then spend an average of 30 minutes (negative exponentially distributed) assisting the patient and updating medical records regarding the problem and care provided.

Because immediate service is critical to the 5 patients, two important questions are: What is the average number of patients either waiting for or being attended by the nurse? What is the average time that a patient spends waiting for the nurse to arrive?

SOLUTION

λ = .5 arrival/hour

μ = 2 served/hour

N = 5 patients

$$P_0 = \frac{1}{\sum_{n=0}^{5} \frac{5!}{(5-n)!}\left(\frac{.5}{2}\right)^n} = 0.20$$

$$L_q = 5 - \left(\frac{.5+2}{.5}\right)(1 - 0.20) = 1 \text{ patient}$$

$$L_s = 1 + (1 - 0.20) = 1.8 \text{ patients}$$

$$W_q = \frac{1}{(5-1.8)(.5)} = .62 \text{ hour} = 37.28 \text{ min.}$$

$$W_s = .62 + \frac{1}{2} = 1.12 \text{ hours} = 67.28 \text{ min}$$

So the average number of patients in the system = 1.8

Average wait time in the queue = .62 hour = 37.28 minutes

Problems

Note: **PX** means the problem may be solved with POM for Windows and/or Excel OM.

Problems D.1–D.39 relate to The Variety of Queuing Models

• **D.1** Customers arrive at Rich Dunn's Styling Shop at a rate of 3 per hour, distributed in a Poisson fashion. Rich's service times follow a negative exponential distribution, and Rich can complete an average of 5 haircuts per hour.

a) Find the average number of customers waiting for haircuts.
b) Find the average number of customers in the shop.
c) Find the average time customers wait until it is their turn.
d) Find the average time customers spend in the shop.
e) Find the percentage of time that Rich is busy. **PX**

• **D.2** There is only one copying machine in the student lounge of the business school. Students arrive at the rate of $\lambda = 40$ per hour (according to a Poisson distribution). Copying takes an average of 40 seconds, or $\mu = 90$ per hour (according to a negative exponential distribution). Compute the following:

a) The percentage of time that the machine is used.
b) The average length of the queue.
c) The average number of students in the system.
d) The average time spent waiting in the queue.
e) The average time in the system. **PX**

•• **D.3** Twenty-four customers arrive every hour (Poisson distributed) at Andy Johnson's food truck when it parks outside the Orlando Courthouse from 11–2 P.M. on weekdays. It takes Andy a mean time of 2 minutes to fill each order (following a negative exponential distribution). Using Waiting Line Table D.5 and $W_q = L_q/\lambda$:

a) What is the average number of customers in the queue?
b) How long will a person wait in line on average?
c) If Andy's wife joins him in serving meals (at the same speed as Andy), how will your answers to parts (a) and (b) change?

• **D.4** Dr. Tarun Gupta, a Michigan vet, is running a rabies vaccination clinic for dogs at the local grade school. Tarun can "shoot" a dog every 3 minutes. It is estimated that the dogs will arrive independently and randomly throughout the day at a rate of one dog every 6 minutes according to a Poisson distribution. Also assume that Tarun's shooting times are negative exponentially distributed. Compute the following:

a) The probability that Tarun is idle.
b) The proportion of the time that Tarun is busy.
c) The average number of dogs being vaccinated and waiting to be vaccinated.
d) The average number of dogs waiting to be vaccinated.
e) The average time a dog waits before getting vaccinated.
f) The average amount of time a dog spends waiting in line and being vaccinated. **PX**

•• **D.5** The pharmacist at Arnold Palmer Hospital, Wende Huehn-Brown, receives 12 requests for prescriptions each hour, Poisson distributed. It takes her a mean time of 4 minutes to fill each, following a negative exponential distribution. Use the waiting-line table, Table D.5, and $W_q = L_q/\lambda$, to answer these questions.

a) What is the average number of prescriptions in the queue?
b) How long will the average prescription spend in the queue?
c) Wende decides to hire a second pharmacist, Ajay Aggerwal, with whom she went to school and who operates at the same speed in filling prescriptions. How will the answers to parts (a) and (b) change?

• **D.6** Calls arrive at Lynn Ann Fish's hotel switchboard at a rate of 2 per minute. The average time to handle each is 20 seconds. There is only one switchboard operator at the current time. The Poisson and negative exponential distributions appear to be relevant in this situation.

a) What is the probability that the operator is busy?
b) What is the average time that a customer must wait before reaching the operator?
c) What is the average number of calls waiting to be answered? **PX**

•• **D.7** Automobiles arrive at the drive-through window at the downtown Baton Rouge, Louisiana, post office at the rate of 4 every 10 minutes. The average service time is 2 minutes. The Poisson distribution is appropriate for the arrival rate and service times are negative exponentially distributed.

a) What is the average time a car is in the system?
b) What is the average number of cars in the system?
c) What is the average number of cars waiting to receive service?
d) What is the average time a car is in the queue?
e) What is the probability that there are no cars at the window?
f) What percentage of the time is the postal clerk busy?
g) What is the probability that there are exactly 2 cars in the system?
h) By how much would your answer to part (a) be reduced if a second drive-through window, with its own server, were added? **PX**

• **D.8** Virginia's Ron McPherson Electronics Corporation retains a service crew to repair machine breakdowns that occur on average $\lambda = 3$ per 8-hour workday (approximately Poisson in nature). The crew can service an average of $\mu = 8$ machines per workday, with a repair time distribution that resembles the negative exponential distribution.

a) What is the utilization rate of this service system?
b) What is the average downtime for a broken machine?
c) How many machines are waiting to be serviced at any given time?
d) What is the probability that more than 1 machine is in the system? The probability that more than 2 are broken and waiting to be repaired or being serviced? More than 3? More than 4? **PX**

•• **D.9** Neve Commercial Bank is the only bank in the town of York, Pennsylvania. On a typical Friday, an average of 10 customers per hour arrive at the bank to transact business. There is currently one teller at the bank, and the average time required to transact business is 4 minutes. It is assumed that service times may be described by the negative exponential distribution. If a single teller is used, find:

a) The average time in the line.
b) The average number in the line.
c) The average time in the system.
d) The average number in the system.
e) The probability that the bank is empty.
f) CEO Benjamin Neve is considering adding a second teller (who would work at the same rate as the first) to reduce the waiting time for customers. A single line would be used, and the customer at the front of the line would go to the first available bank teller. He assumes that this will cut the waiting time in half. If a second teller is added, find the new answers to parts (a) to (e). **PX**

Additional problem **D.10**, *which involves arrivals at a theatre complex, is available in* **MyLab Operations Management.**

•• **D.11** Bill Youngdahl has been collecting data at the TU student grill. He has found that, between 5:00 P.M. and 7:00 P.M., students arrive at the grill at a rate of 25 per hour (Poisson distributed) and service time takes an average of 2 minutes (negative exponential distribution). There is only 1 server, who can work on only 1 order at a time.

a) What is the average number of students in line?
b) What is the average time a student is in the grill area?
c) Suppose that a second server can be added to team up with the first (and, in effect, act as 1 faster server). This would reduce the average service time to 90 seconds. How would this affect the average time a student is in the grill area?
d) Suppose a second server is added and the 2 servers act independently, with *each* taking an average of 2 minutes. What would be the average time a student is in the system?

••• **D.12** The wheat harvesting season in the American Midwest is short, and farmers deliver their truckloads of wheat to a giant central storage bin within a 2-week span. Because of this, wheat-filled trucks waiting to unload and return to the fields have been known to back up for a block at the receiving bin. The central bin is owned cooperatively, and it is to every farmer's benefit to make the unloading/storage process as efficient as possible. The cost of grain deterioration caused by unloading delays and the cost of truck rental and idle driver time are significant concerns to the cooperative members. Although farmers have difficulty quantifying crop damage, it is easy to assign a waiting and unloading cost for truck and driver of $18 per hour. During the 2-week harvest season, the storage bin is open and operated 16 hours per day, 7 days per week, and can unload 35 trucks per hour according to a negative exponential distribution. Full trucks arrive all day long (during the hours the bin is open) at a rate of about 30 per hour, following a Poisson pattern.

To help the cooperative get a handle on the problem of lost time while trucks are waiting in line or unloading at the bin, find the following:

a) The average number of trucks in the unloading system
b) The average time per truck in the system
c) The utilization rate for the bin area
d) The probability that there are more than three trucks in the system at any given time
e) The total daily cost to the farmers of having their trucks tied up in the unloading process
f) As mentioned, the cooperative uses the storage bin heavily only 2 weeks per year. Farmers estimate that enlarging the bin would cut unloading costs by 50% next year. It will cost $9,000 to do so during the off-season. Would it be worth the expense to enlarge the storage area? **PX**

••• **D.13** Janson's Department Store in Stark, Ohio, maintains a successful catalog sales department in which a clerk takes orders by telephone. If the clerk is occupied on one line, incoming phone calls to the catalog department are answered automatically by a recording machine and asked to wait. As soon as the clerk is free, the party who has waited the longest is transferred and serviced first. Calls come in at a rate of about 12 per hour. The clerk can take an order in an average of 4 minutes. Calls tend to follow a Poisson distribution, and service times tend to be negative exponential.

The cost of the clerk is $10 per hour, but because of lost goodwill and sales, Janson's loses about $25 per hour of customer time spent waiting for the clerk to take an order.

a) What is the average time that catalog customers must wait before their calls are transferred to the order clerk?
b) What is the average number of customers waiting to place an order?
c) Pamela Janson is considering adding a second clerk to take calls. The store's cost would be the same $10 per hour. Should she hire another clerk? Explain your decision. **PX**

• **D.14** Altug's Coffee Shop decides to install an automatic coffee vending machine outside one of its stores to reduce the number of people standing in line inside. Mehmet Altug charges $3.50 per cup. However, it takes too long for people to make change. The service time is a constant 3 minutes, and the arrival rate is 15 per hour (Poisson distributed).

a) What is the average wait in line?
b) What is the average number of people in line?
c) Mehmet raises the price to $5 per cup and takes 60 seconds off the service time. However, because the coffee is now so expensive, the arrival rate drops to 10 per hour. Now what are the average wait time and the average number of people in the queue (waiting)? **PX**

••• **D.15** The typical subway station in Washington, DC, has six turnstiles, each of which can be controlled by the station manager to be used for either entrance or exit control—but never for both. The manager must decide at different times of the day how many turnstiles to use for entering passengers and how many to use for exiting passengers.

At the George Washington University (GWU) Station, passengers enter the station at a rate of about 84 per minute between the hours of 7 A.M. and 9 A.M. Passengers exiting trains at the stop reach the exit turnstile area at a rate of about 48 per minute during the same morning rush hours. Each turnstile can allow an average of 30 passengers per minute to enter or exit. Arrival and service times have been thought to follow Poisson and negative exponential distributions, respectively. Assume riders form a common queue at both entry and exit turnstile areas and proceed to the first empty turnstile.

The GWU station manager, Gerald Aase, does not want the average passenger at his station to have to wait in a turnstile line for more than 6 seconds, nor does he want more than 8 people in any queue at any average time.

Kuosumo/Fotolia

a) How many turnstiles should be opened in each direction every morning?
b) Discuss the assumptions underlying the solution of this problem using queuing theory. **PX**

•• **D.16** Renuka Jain's Car Wash takes a constant time of 4.5 minutes in its automated car wash cycle. Autos arrive following a Poisson distribution at the rate of 10 per hour. Renuka wants to know:

a) The average waiting time in line.
b) The average length of the line.

••• **D.17** Debra Bishop's cabinet-making shop, in Des Moines, has five tools that automate the drilling of holes for the installation of hinges. Each machine needs an average of 3 "resettings" every 8-hour day, following the Poisson distribution. There is a single technician for setting these machines. Her service times are negative exponential, averaging 2 hours each.

a) What is the probability this system is empty?
b) What is the average number of machines in the system (i.e., being reset or waiting to be reset)?
c) What is the average waiting time in the queue? **PX**

••• **D.18** A technician monitors a group of 5 computers that run an automated manufacturing facility. It takes an average of 15 minutes (negative exponentially distributed) to adjust a computer that develops a problem. On average, one of the computers requires adjustment every 85 minutes. Determine the following:

a) The average number of computers waiting for adjustment (i.e., in the queue)
b) The average number in the system
c) The probability no computer needs adjustment **PX**

••• **D.19** One mechanic services 5 drilling machines for a steel plate manufacturer. Machines break down on an average of once every 6 working days, and breakdowns tend to follow a Poisson

distribution. The mechanic can handle an average of one repair job per day. Repairs follow a negative exponential distribution.
a) On the average, how many machines are waiting for service?
b) On the average, what is the *waiting* time to be serviced? **PX**

••• **D.20** Ted Glickman, the administrator at D.C. General Hospital emergency room, faces the problem of providing treatment for patients who arrive at different rates during the day. There are four doctors available to treat patients when needed. If not needed, they can be assigned other responsibilities (such as doing lab tests, reports, X-ray diagnoses) or else rescheduled to work at other hours.

It is important to provide quick and responsive treatment, and Ted thinks that, on the average, patients should not have to sit in the waiting area for more than 5 minutes before being seen by a doctor. Patients are treated on a first-come, first-served basis and see the first available doctor after waiting in the queue. The arrival pattern for a typical day is as follows:

TIME	ARRIVAL RATE
9 A.M.–3 P.M.	6 patients/hour
3 P.M.–8 P.M.	4 patients/hour
8 P.M.–midnight	12 patients/hour

Arrivals follow a Poisson distribution, and treatment times, 12 minutes on the average, follow the negative exponential pattern.
a) How many doctors should be on duty during each period to maintain the level of patient care expected?
b) What condition would exist if only one doctor were on duty between 9 A.M. and 3 P.M.? **PX**

••• **D.21** The Pontchartrain Bridge is a 16-mile toll bridge that crosses Lake Pontchartrain in New Orleans. Currently, there are 7 toll booths, each staffed by an employee. Since Hurricane Katrina, the Port Authority has been considering replacing the employees with machines. Many factors must be considered because the employees are unionized. However, one of the Port Authority's concerns is the effect that replacing the employees with machines will have on the times that drivers spend in the system. Customers arrive to any one toll booth at a rate of 10 per minute. In the exact change lanes with employees, the service time is essentially constant at 5 seconds for each driver. With machines, the average service time would still be 5 seconds, but it would be negative exponential rather than constant because it takes time for the coins to rattle around in the machine. Contrast the two systems for a single lane. **PX**

••• **D.22** The registration area has just opened at a large convention of dentists in Tallahassee, Florida. There are 200 people arriving per hour (Poisson distributed), and the cost of their waiting time in the queue is valued at \$100 per person per hour. The Tallahassee Convention Center provides servers to register guests at a fee of \$15 per person per hour. It takes about one minute to register an attendee (negative exponentially distributed). A single waiting line, with multiple servers, is set up.
a) What is the minimum number of servers for this system?
b) What is the optimal number of servers for this system?
c) What is the cost for the system, per hour, at the optimum number of servers?
d) What is the server utilization rate with the minimum number of servers? **PX**

•• **D.23** Refer to Problem D.22. A new registration manager, Dwayne Cole, is hired who initiates a program to entertain the people in line with a juggler whom he pays \$15/hour. This reduces the waiting costs to \$50 per hour.
a) What is the optimal number of servers?
b) What is the cost for the system, per hour, at the optimal service level?

•••• **D.24** The Chattanooga Furniture store gets an average of 50 customers per shift. Marilyn Helms, the manager, wants to calculate whether she should hire 1, 2, 3, or 4 salespeople. She has determined that average waiting times will be 7 minutes with one salesperson, 4 minutes with two salespeople, 3 minutes with three salespeople, and 2 minutes with four salespeople. She has estimated the cost per minute that customers wait at \$1. The cost per salesperson per shift (including fringe benefits) is \$70.

How many salespeople should be hired?

•• **D.25** During the afternoon peak hours the First Bank of Dubuque has an average of 40 customers arriving every hour. There is also an average of 8 customers at First Bank at any time. The probability of the arrival distribution is unknown. How long does the average customer spend in the bank?

•• **D.26** An average of 9 cars can be seen in the system (both the drive-through line and the drive-through window) at Burger Universe. Approximately every 20 seconds, a car attempts to enter the drive-through line; however, 40% of cars simply leave the restaurant because they're discouraged by the length of the line. On average, how long does a car spend going through the drive-through at Burger Universe?

•• **D.27** Lobster World stores approximately 1,000 pounds of fish on average. In a typical day, the busy restaurant cooks and sells 360 (raw) pounds of fish. How long do the fish stay in storage on average?

•• **D.28** Gamma Bank processes a typical loan application in 2.4 weeks. Customers fill out 30 loan applications per week. On average how many loan applications are being processed somewhere in the system at Gamma Bank?

•• **D.29** Fisher's Furniture Store sells \$800,000 worth of furniture to customers on credit each month. The Accounts Receivable balance in the accounting books averages \$2 million. On average, how long are customers taking to pay their bills?

•• **D.30** Vacation Inns, a chain of hotels operating in the southeastern region of the U.S., uses a toll-free telephone number to take reservations for all of its hotels. An average of 12 calls are received per hour. The probability distribution that describes the arrivals is unknown. Over a period of time, it is determined that the average caller spends 6 minutes on hold waiting for service. Find the average number of callers in the queue by using Little's Law.

• **D.31** Customers arrive at a local 7–11 store at the rate of $\lambda = 40$ per hour (and follow a Poisson process). The only employee in the store can check them out at a rate of $\mu = 60$ per hour (following an exponential distribution). Compute the following:
a) The percentage of time that the employee is busy with checkouts.
b) The average length of the queue.
c) The average number of customers in the system.
d) The average time spent waiting in the queue.
e) The average time in the system. **PX**

• **D.32** Due to a recent increase in business, a paralegal must now process an average of 20 legal briefs per day (assume a Poisson distribution). It takes the paralegal approximately 20 minutes to prepare each brief (assume an exponential distribution). Assuming the paralegal works 8 hours per day:
a) What is the paralegal's utilization rate?
b) What is the average waiting time before the paralegal processes a brief?

c) What is the average number of briefs waiting to be done?
d) What is the probability that the paralegal has more than 5 briefs to do? PX

• **D.33** At the start of football season, the ticket office gets very busy the day before the first game. Customers arrive at the rate of four every 10 minutes, and the average time to transact business is 2 minutes.
a) What is the average number of people in line?
b) What is the average time that a person will spend at the ticket office?
c) What proportion of time is the server busy? PX

••• **D.34** Cynthia Knott's Car Wash is open 6 days a week, but its busiest day is always Saturday. From historical data, Cynthia estimates that dirty cars arrive at the rate of 20 per hour all day Saturday. With a full crew working the hand-wash line, she figures that cars can be cleaned at the rate of one every 2 minutes. One car at a time is cleaned in this example of a single-channel waiting line.

Assuming Poisson arrivals and exponential service times, find the following:
a) The average number of cars in line.
b) The average time that a car waits before it is washed.
c) The average time that a car spends in the service system.
d) The utilization rate of the car wash.
e) The probability that no cars are in the system.
f) Cynthia is thinking of switching to an all-automated car wash that uses no crew. The equipment under consideration washes one car every minute at a constant rate. How will your answers to parts (a) through (e) change with the new system? PX

Additional problems **D.35–D.39** *are available in* **MyLab Operations Management**. *D.35 deals with a university cafeteria, with 1 or 2 cashiers. D.36 involves waiting lines at a styling salon. D.37 describes a department store wishing to evaluate the cost-benefit of various staffing levels. D.38 evaluates a single-channel car wash. D.39 computes queuing parameters at a hospital pharmacy.*

CASE STUDIES

New England Foundry

For more than 75 years, New England Foundry, Inc. (NEFI), has manufactured wood stoves for home use. In recent years, with increasing energy prices, president George Mathison has seen sales triple. This dramatic increase has made it difficult for George to maintain quality in all his wood stoves and related products.

Unlike other companies manufacturing wood stoves, NEFI is in the business of making *only* stoves and stove-related products. Its major products are the Warmglo I, the Warmglo II, the Warmglo III, and the Warmglo IV. The Warmglo I is the smallest wood stove, with a heat output of 30,000 BTUs, and the Warmglo IV is the largest, with a heat output of 60,000 BTUs.

The Warmglo III outsold all other models by a wide margin. Its heat output and available accessories were ideal for the typical home. The Warmglo III also had a number of other outstanding features that made it one of the most attractive and heat-efficient stoves on the market. These features, along with the accessories, resulted in expanding sales and prompted George to build a new factory to manufacture the Warmglo III model. An overview diagram of the factory is shown in Figure D.6.

The new foundry used the latest equipment, including a new Disamatic that helped in manufacturing stove parts. Regardless of new equipment or procedures, casting operations have remained basically unchanged for hundreds of years. To begin with, a wooden pattern is made for every cast-iron piece in the stove. The wooden pattern is an exact duplicate of the cast-iron piece that is to be manufactured. All NEFI patterns are made by Precision Patterns, Inc. and are stored in the pattern shop and maintenance room. Next, a specially formulated sand is molded around the wooden pattern. There can be two or more sand molds for each pattern. The sand is mixed and the molds are made in the molding room. When the wooden pattern is removed, the resulting sand molds form a negative image of the desired casting. Next, molds are transported to the casting room, where molten iron is poured into them and allowed to cool. When the iron has solidified, molds are moved into the cleaning, grinding, and preparation room, where they are dumped into large vibrators that shake most of the sand from the casting. The rough castings are then subjected to both sandblasting to remove the rest of the sand and grinding to finish some of their surfaces. Castings are then painted with a special heat-resistant paint, assembled into workable stoves, and inspected for manufacturing defects that may have gone undetected. Finally, finished stoves are moved to storage and shipping, where they are packaged and transported to the appropriate locations.

At present, the pattern shop and the maintenance department are located in the same room. One large counter is used by both maintenance personnel, who store tools and parts (which are mainly used by the casting department), and sand molders, who need various patterns for the molding operation. Pete Nawler and Roberta De Santo, who work behind the counter, can service a total of 10 people per hour (about 5 per hour each). On average, 4 people from casting and 3 from molding arrive at the counter each hour. People from molding and casting departments arrive randomly, and to be served, they form a single line.

Pete and Roberta have always had a policy of first come, first served. Because of the location of the pattern shop and maintenance department, it takes an average of 3 minutes for an individual from the casting department to walk to the pattern and maintenance room, and it takes about 1 minute for an individual to walk from the molding department to the pattern and maintenance room.

After observing the operation of the pattern shop and maintenance room for several weeks, George decided to make some changes to the factory layout. An overview of these changes appears in Figure D.7.

Separating the maintenance shop from the pattern shop would have a number of advantages. It would take people from the casting

Figure **D.6**

Overview of Factory

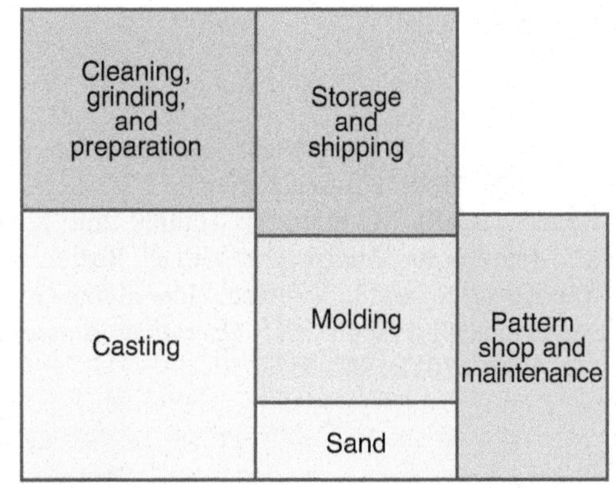

Figure D.7

Overview of Factory after Changes

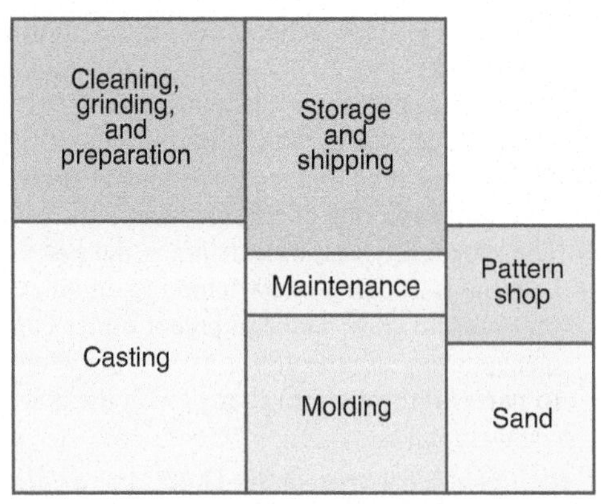

department only 1 minute instead of 3 to get to the new maintenance room. The time from molding to the pattern shop would be unchanged. Using motion and time studies, George was also able to determine that improving the layout of the maintenance room would allow Roberta to serve 6 people from the casting department per hour; improving the layout of the pattern department would allow Pete to serve 7 people from the molding shop per hour.

Discussion Questions

1. How much time would the new layout save?
2. If casting personnel were paid \$19.00 per hour and molding personnel were paid \$23.50 per hour, how much could be saved per hour with the new factory layout?
3. Should George have made the change in layout?

The Winter Park Hotel

Lori Cook, manager of the Winter Park Hotel, is considering how to restructure the front desk to reach an optimum level of staff efficiency and guest service. At present, the hotel has five clerks on duty, each with a separate waiting line, during peak check-in time of 3:00 P.M. to 5:00 P.M. Observation of arrivals during this period shows that an average of 90 guests arrive each hour (although there is no upward limit on the number that could arrive at any given time). It takes an average of 3 minutes for the front-desk clerk to register each guest.

Ms. Cook is considering three plans for improving guest service by reducing the length of time that guests spend waiting in line. The first proposal would designate one employee as a quick-service clerk for guests registering under corporate accounts, a market segment that fills about 30% of all occupied rooms. Because corporate guests are preregistered, their registration takes just 2 minutes. With these guests separated from the rest of the clientele, the average time for registering a typical guest would climb to 3.4 minutes. Under this plan, noncorporate guests would choose any of the remaining four lines.

The second plan is to implement a single-line system. All guests could form a single waiting line to be served by whichever of five clerks became available. This option would require sufficient lobby space for what could be a substantial queue.

The use of an automated kiosk for check-ins is the basis of the third proposal. This kiosk would provide about the same service rate as would a clerk. Cook estimates that 20% of customers, primarily frequent guests, would be willing to use the machines. (This might be a conservative estimate if guests perceive direct benefits from using the kiosk.) Ms. Cook would set up a single queue for customers who prefer human check-in clerks. This line would be served by the five clerks, although Cook is hopeful that the kiosk will allow a reduction to four.

Discussion Questions

1. Determine the average amount of time that a guest spends checking in. How would this change under each of the stated options?
2. Which option do you recommend?

Endnotes

1. When the arrival rates follow a Poisson process with mean arrival rate, λ, the time between arrivals follows a negative exponential distribution with mean time between arrivals of $1/\lambda$. The negative exponential distribution, then, is also representative of a Poisson process but describes the time between arrivals and specifies that these time intervals are completely random.
2. The term *FIFS* (first-in, first-served) is often used in place of FIFO. Another discipline, LIFS (last-in, first-served), also called last-in, first-out (LIFO), is common when material is stacked or piled so that the items on top are used first.
3. Sources: *The Wall Street Journal* (June 9, 2020); *The Washington Post* (May 30, 2020); and *Harvard Business Review* (February 6, 2019).
4. In queuing notation, the first letter refers to the arrivals (where M stands for Poisson distribution); the second letter refers to service (where M is again a Poisson distribution, which is the same as an exponential rate for service—and D is a constant service rate); the third symbol refers to the number of servers. So an M/D/1 system (our Model C) has Poisson arrivals, constant service, and one server.
5. Although there is no definite number that we can use to divide finite from infinite populations, the general rule of thumb is this: If the number in the queue is a significant proportion of the calling population, use a finite queuing model.

Bibliography

Balakrishnan, R., B. Render, R. M. Stair, and C. Munson. *Managerial Decision Modeling: Business Analytics with Spreadsheets*, 4th ed. Boston: DeGruyter, 2017.

Cochran, J. K., and K. Roche. "A Queueing-Based Decision Support Methodology to Estimate Hospital Inpatient Bed Demand." *Journal of the Operational Research Society* 59, no. 11 (November 2008): 1471–1483.

Gross, D., J. F. Shortle, J. M. Thompson, and C. M. Harris. *Fundamentals of Queuing Theory*, 4th ed. New York: Wiley, 2008.

Prabhu, N. U. *Foundations of Queuing Theory.* Dordecht, Netherlands: Kluwer Academic Publishers, 1997.

Render, B., R. M. Stair, M. Hanna, and T. Hale. *Quantitative Analysis for Management*, 13th ed. Boston: Pearson, 2018.

Yan Kovic, N., and L. V. Green. "Identifying Good Nursing Levels: A Queuing Approach." *Operations Research* 59, no. 4 (July/August 2011): 942–955.

Module D *Rapid* Review

<table>
<tr><th>Main Heading</th><th>Review Material</th><th>MyLab Operations Management</th></tr>
<tr><td>QUEUING THEORY</td><td>

- **Queuing theory**—A body of knowledge about waiting lines.
- **Waiting line (queue)**—Items or people in a line awaiting service.

</td><td>Concept Questions: 1.1–1.2</td></tr>
<tr><td>CHARACTERISTICS OF A WAITING-LINE SYSTEM</td><td>

The three parts of a waiting-line, or queuing, system are: *Arrivals or inputs to the system; queue discipline, or the waiting line itself;* and *the service facility.*

- **Unlimited, or infinite, population**—A queue in which a virtually unlimited number of people (arrivals) request the services, or in which the number of customers or arrivals on hand at any given moment is a very small portion of potential arrivals.
- **Limited, or finite, population**—A queue in which there are only a limited number of potential users of the service.
- **Poisson distribution**—A discrete probability distribution that often describes the arrival rate in queuing theory:

$$P(x) = \frac{e^{-\lambda}\lambda^x}{x!} \text{ for } x = 0,1,2,3,4,\ldots \quad \text{(D-1)}$$

A queue is *limited* when it cannot, either by law or because of physical restrictions, increase to an infinite length. A queue is *unlimited* when its size is unrestricted.

Queue discipline refers to the rule by which customers in the line are to receive service:

- **First-in, first-out (FIFO) rule**—A queue discipline in which the first customers in line receive the first service.
- **Single-server (single-channel) queuing system**—A service system with one line and one server.
- **Multiple-server (multiple-channel) queuing system**—A service system with one waiting line but with more than one server (channel).
- **Single-phase system**—A system in which the customer receives service from only one station and then exits the system.
- **Multiphase system**—A system in which the customer receives services from several stations before exiting the system.
- Negative exponential probability distribution—A continuous probability distribution often used to describe the service time in a queuing system.

</td><td>Concept Questions: 2.1–2.6</td></tr>
<tr><td>QUEUING COSTS</td><td>Operations managers must recognize the trade-off that takes place between two costs: the cost of providing good service and the cost of customer or machine waiting time.</td><td>Concept Questions: 3.1–3.5</td></tr>
<tr><td>THE VARIETY OF QUEUING MODELS</td><td>

***Model A:** Single-Server System (M/M/1):*

Queuing Formulas:

λ = mean number of arrivals per time period

μ = mean number of people or items served per time period

L_s = average number of units in the system = $\lambda/(\mu - \lambda)$

W_s = average time a unit spends in the system = $1/(\mu - \lambda)$

L_q = average number of units waiting in the queue = $\lambda^2/[\mu(\mu - \lambda)$

W_q = average time a unit spends waiting in the queue = $\lambda/[\mu(\mu - \lambda)] = L_q/\lambda$

ρ = utilization factor for the system = λ/μ

P_0 = probability of 0 units in the system (i.e., the service unit is idle) = $1 - (\lambda/\mu)$

$P_{n>k}$ = probability of $>k$ units in the system = $(\lambda/\mu)^{k+1}$

***Model B**: Multiple-Server System (M/M/S):*

$$P_0 = \frac{1}{\left[\sum_{n=0}^{M-1} \frac{1}{n!}\left(\frac{\lambda}{\mu}\right)^n\right] + \frac{1}{M!}\left(\frac{\lambda}{\mu}\right)^M \frac{M\mu}{M\mu - \lambda}} \text{ for } M\mu > \lambda$$

$$L_s = \frac{\lambda\mu(\lambda/\mu)^M}{(M-1)!(M\mu - \lambda)^2} P_0 + \frac{\lambda}{\mu}$$

$$W_s = L_s/\lambda \quad L_q = L_s - (\lambda/\mu) \quad W_q = L_q/\lambda$$

***Model C**: Constant Service (M/D/1):*

$$L_q = \lambda^2/[2\mu(\mu - \lambda)] \qquad W_q = \lambda/[2\mu(\mu - \lambda)]$$

$$L_s = L_q + (\lambda/\mu) \qquad W_s = W_q + (1/\mu)$$

</td><td>

Concept Questions: 4.1–4.6

Problems: D.1–D.14, D.16–D.21, D.24–D.39

Virtual Office Hours for Solved Problems: D.1–D.4

ACTIVE MODELS D.1, D.2, D.3

</td></tr>
</table>

Main Heading	Review Material	MyLab Operations Management
	Little's Law A useful relationship in queuing for any system in a steady state is called Little's Law: $L_s = \lambda W_s$ (which is the same as $W_s = L_s/\lambda$) (D-2) $L_q = \lambda W_q$ (which is the same as $W_q = L_q/\lambda$) (D-3) ***Model D:*** *Finite Population (M/M/1 with finite source)* With a limited, or finite, population, there is a *dependent* relationship between the length of the queue and the arrival rate. As the *waiting* line becomes longer, the *arrival rate* drops. N = size of the population $P_0 = \dfrac{1}{\sum_{n=0}^{N} \dfrac{N!}{(N-n)!}\left(\dfrac{\lambda}{\mu}\right)^n}$ $L_q = N - \left(\dfrac{\lambda+\mu}{\lambda}\right)(1-P_0)$ $L_s = L_q + (1-P_0)$ $W_q = \dfrac{L_q}{(N-L_s)\lambda}$ $W_s = W_q + \dfrac{1}{\mu}$ $P_n = \dfrac{N!}{(N-n)!}\left(\dfrac{\lambda}{\mu}\right)^n P_0$ for $n = 0, 1, \ldots, N$	
OTHER QUEUING ISSUES	Often, *variations* of the four basic queuing models are present in an analysis. Many models, some very complex, have been developed to deal with such variations.	Concept Questions: 5.1–5.4
ADDITIONAL MYLAB OPERATIONS MANAGEMENT RESOURCES	✔ Additional Case Study (Pantry Shopper) ✔ Multiple Choice Case Questions (New England Foundry) ✔ Additional Homework Problems (D.10, D.35–D.39)	

Self Test

LO D.1 Which of the following is *not* a key operating characteristic for a queuing system?
- **a)** Utilization rate
- **b)** Percentage of idle time
- **c)** Average time spent waiting in the system and in the queue
- **d)** Average number of customers in the system and in the queue
- **e)** Average number of customers who renege

LO D.2 Customers enter the waiting line at a cafeteria's only cash register on a first-come, first-served basis. The arrival rate follows a Poisson distribution, while service times follow an exponential distribution. If the average number of arrivals is 6 per minute and the average service rate of a single server is 10 per minute, what is the average number of customers in the system?
- **a)** 0.6
- **b)** 0.9
- **c)** 1.5
- **d)** 0.25
- **e)** 1.0

LO D.3 In performing a cost analysis of a queuing system, the waiting time cost is sometimes based on the time in the queue and sometimes based on the time in the system. The waiting cost should be based on time in the system for which of the following situations?
- **a)** Waiting in line to ride an amusement park ride
- **b)** Waiting to discuss a medical problem with a doctor
- **c)** Waiting for a picture and an autograph from a rock star
- **d)** Waiting for a computer to be fixed so it can be placed back in service

LO D.4 Which of the following is *not* an assumption in a multiple-server queuing model?
- **a)** Arrivals come from an infinite, or very large, population.
- **b)** Arrivals are Poisson distributed.
- **c)** Arrivals are treated on a first-in, first-out basis and do not balk or renege.
- **d)** Service times follow the exponential distribution.
- **e)** Servers each perform at their own individual speeds.

LO D.5 If everything else remains the same, including the mean arrival rate and service rate, except that the service time becomes constant instead of exponential:
- **a)** the average queue length will be halved.
- **b)** the average waiting time will be doubled.
- **c)** the average queue length will increase.
- **d)** we cannot tell from the information provided.

LO D.6 A company has one computer technician who is responsible for repairs on the company's 20 computers. As a computer breaks, the technician is called to make the repair. If the repairperson is busy, the machine must wait to be repaired. This is an example of:
- **a)** a multiple-server system.
- **b)** a finite population system.
- **c)** a constant service rate system.
- **d)** a multiphase system.
- **e)** all of the above.

Answers: LO D.1. e; LO D.2. c; LO D.3. d; LO D.4. e; LO D.5. a; LO D.6. b.

Learning Curves

MODULE OUTLINE

LEARNING OBJECTIVES

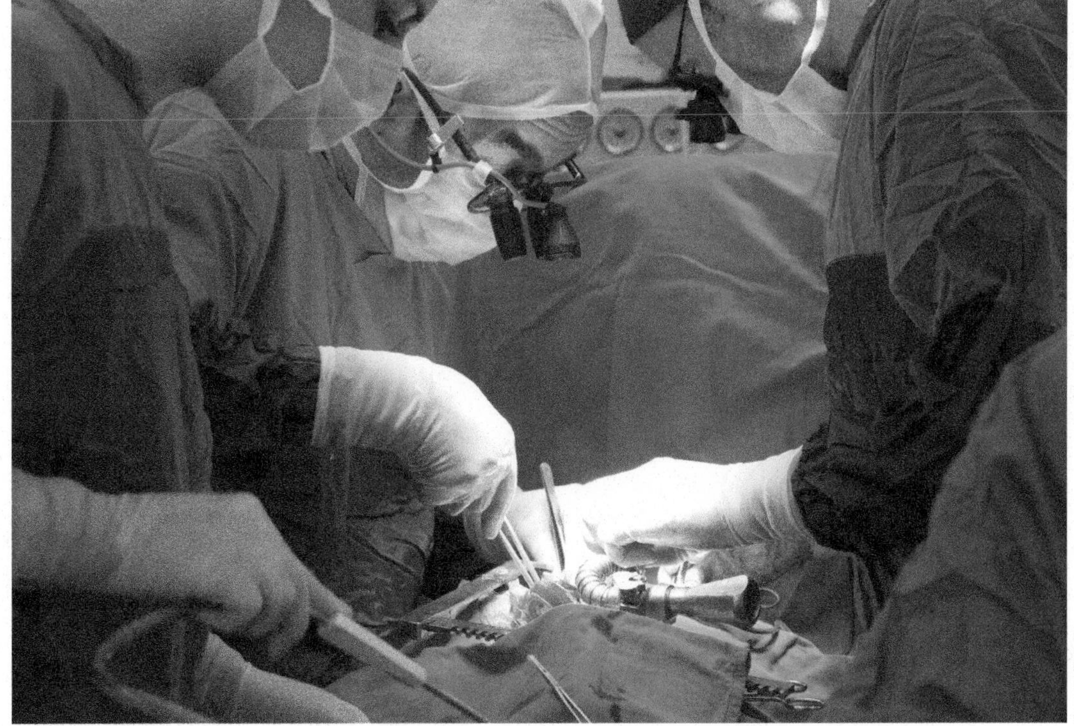

Medical procedures such as heart surgery follow a learning curve. Research indicates that the death rate from heart transplants drops at a 79% learning curve, a learning rate not unlike that in many industrial settings. It appears that as doctors and medical teams improve with experience, so do your odds as a patient. If the death rate is halved every three operations, practice may indeed make perfect.

Derege/Fotolia

What Is a Learning Curve?

Most organizations learn and improve over time. As firms and employees perform a task over and over, they learn how to perform more efficiently. This means that task times and costs decrease.

Learning curves
The premise that people and organizations get better at their tasks as the tasks are repeated; sometimes called experience curves.

Learning curves are based on the premise that people and organizations become better at their tasks as the tasks are repeated. A learning curve graph (illustrated in Figure E.1) displays cost (or time) per unit versus the cumulative number of units produced. From it we see that the time needed to produce a unit decreases, usually following a negative exponential curve (part a), as the person or company produces more units. In other words, *it takes less time to complete each additional unit a firm produces*. However, we also see in Figure E.1 that the time *savings* in completing each subsequent unit *decreases*. These are the major attributes of the learning curve.

Learning curves were first applied to industry in a report by T. P. Wright of Curtis-Wright Corp. in 1936. Wright described how direct labor costs of making a particular airplane decreased with learning, a theory since confirmed by other aircraft manufacturers. Regardless of the time needed to produce the first plane, learning curves are found to apply to various categories of air frames (e.g., jet fighters versus passenger planes versus bombers). Learning curves have since been applied not only to labor but also to a wide variety of other costs, including material and purchased components. The power of the learning curve is so significant that it plays a major role in many strategic decisions related to employment levels, costs, capacity, and pricing.

LO E.1 *Define* learning curve

The learning curve is based on a *doubling* of production: That is, when production doubles, the decrease in time per unit affects the rate of the learning curve. So, if the learning curve is an 80% rate, the second unit takes 80% of the time of the first unit, the fourth unit takes 80% of the time of the second unit, the eighth unit takes 80% of the time of the fourth unit, and so forth. This principle is shown as:

$$T \times L_n = \text{Time required for the } n\text{th unit} \quad \text{(E-1)}$$

where T = unit cost or unit time of the first unit
L = learning curve rate
n = number of times T is doubled

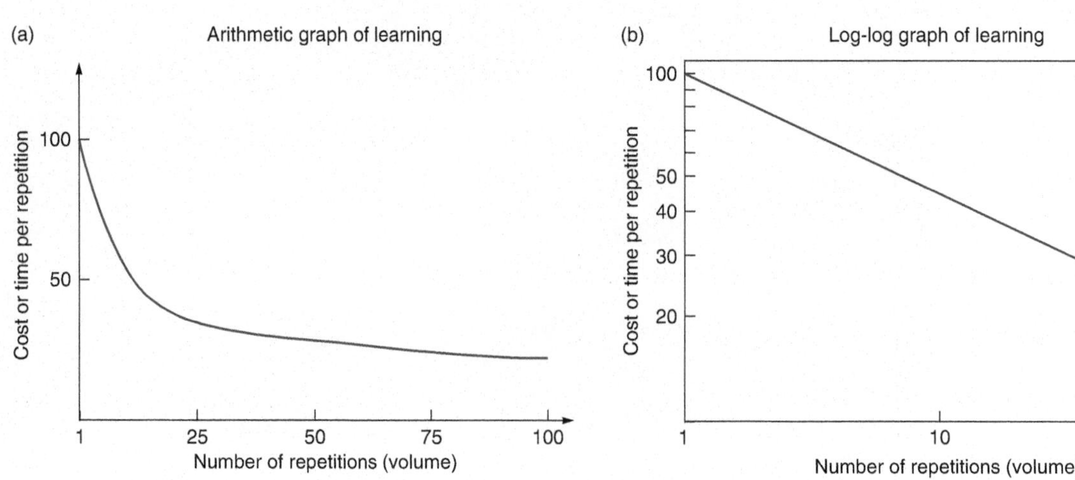

Figure **E.1**

The Learning-Curve Effect States That Time per Repetition Decreases as the Number of Repetitions Increases

Both curves show that the labor-hours to build an airplane decline by 20% each time the production volume doubles. The left graph (a) shows the exponential decline using an arithmetic scale. The log-log graph (b) yields a straight line that is easier to extrapolate.

If the first unit of a particular product took 10 labor-hours, and if a 70% learning curve is present, the hours the fourth unit will take require doubling twice—from 1 to 2 to 4. Therefore, the formula is:

$$\text{Hours required for unit } 4 = 10 \times (.7)^2 = 4.9 \text{ hours}$$

Learning Curves in Services and Manufacturing

STUDENT TIP Learning is a universal concept, but rates of learning differ widely.

Different organizations—indeed, different products—have different learning curves. The rate of learning varies depending on the quality of management and the potential of the process and product. *Any change in process, product, or personnel disrupts the learning curve.* Therefore, caution should be exercised in assuming that a learning curve is continuing and permanent.

As you can see in Table E.1, industry learning curves vary widely. The lower the number (say, 70% compared to 90%), the steeper the slope and the faster the drop in costs. By tradition, learning curves are defined in terms of the *complements* of their improvement rates. For example, a 70% learning curve implies a 30% decrease in time each time the number of repetitions is doubled. A 90% curve means there is a corresponding 10% rate of improvement.

Stable, standardized products and processes tend to have costs that decline more steeply than others. Between 1920 and 1955, for instance, the steel industry was able to reduce labor-hours per unit to 79% each time cumulative production doubled.

Learning curves have application in services as well as industry. As was noted in the caption for the opening photograph, 1-year death rates of heart transplant patients at Temple University Hospital follow a 79% learning curve. The results of that hospital's 3-year study of 62 patients receiving transplants found that every three operations resulted in a halving of the 1-year death rate. As more hospitals face pressure from both insurance companies and the government to enter fixed-price negotiations for their services, their ability to learn from experience becomes increasingly critical. In addition to having applications in both services and industry, learning curves are useful for a variety of purposes. These include:

1. **Internal:** Labor forecasting, scheduling, establishing costs and budgets (see the *OM in Action* box, "The Navy's Learning Curve Challenge").
2. **External:** Supply chain negotiations (see the SMT case study at the end of this module).
3. **Strategic:** Evaluation of company and industry performance, including costs and pricing.

The consequences of learning curves can be far-reaching. For instance, for Boeing's 787 (one of the world's most popular two-aisle commercial jets) to reach break-even at 1,100 planes, an aggressive learning curve rate of 80% must be reached. (Boeing lost an average of $69 million per plane on its first four hundred and fifteen 787s.)

TABLE E.1 Examples of Learning-Curve Effects

EXAMPLE	IMPROVING PARAMETER	CUMULATIVE PARAMETER	LEARNING-CURVE SLOPE (%)
1. Model-T Ford production	Price	Units produced	86
2. Aircraft assembly	Direct labor-hours per unit	Units produced	80
3. Equipment maintenance at GE	Average time to replace a group of parts	Number of replacements	76
4. Steel production	Production worker labor-hours per unit produced	Units produced	79
5. Integrated circuits	Average price per unit	Units produced	72
6. Handheld calculator	Average factory selling price	Units produced	74
7. Disk memory drives	Average price per bit	Number of bits	76
8. Heart transplants	1-year death rates	Transplants completed	79
9. Cesarean section baby deliveries	Average operation time	Number of surgeries	93

OM in Action The Navy's Learning Curve Challenge[1]

Alliance Images/Alamy Stock Photo

Huntington Ingalls Industries, the sole U.S. builder of aircraft carriers, continues to fall short of the navy's demand to cut labor expenses to stay within an \$11.39 billion cost cap mandated by Congress on the second in a new class of warships. With about 47% of construction complete on the USS *John F. Kennedy*, the navy figures show the contractor isn't yet meeting the goal it negotiated with the service: reducing labor hours by 18% from the first carrier, the USS *Gerald Ford*, which at \$13 billion has become the costliest warship ever. They're the first two of a planned four-carrier, \$55 billion program.

It took about 49 million hours of labor to build the first carrier, the *Ford*. The Navy's goal for the *Kennedy*, the second of the four carriers, is to reduce labor hours to about 40 million. Huntington Ingalls's performance "remains stable at approximately a 16% improvement," said a navy spokesman. "Key production milestones and the ship's preliminary acceptance date remain on track," and there are "ample opportunities for improvement with nearly 4 years until contract delivery and over 70% of assembly work remaining on the vessel." Navy officials have cited what they describe as progress on the *Kennedy* as one justification for buying the third and fourth carriers under a single contract.

The navy assesses that, although difficult, the shipbuilder can still attain the 18% reduction goal. The navy secretary, who's been closely monitoring the carrier program, said that Huntington Ingalls has been on "an impressive learning curve" in reducing labor hours. But a director with the General Accounting Office, who monitors navy shipbuilding, said, "With so much of the program underway, it is unlikely that the navy will improve efficiency. In later phases of a shipbuilding contract, performance typically degrades, not improves."

There may also be scheduling issues if the learning improvement is not considered, with labor and facilities underutilized a portion of the time. Firms may also refuse more work because they ignore their own efficiency improvements.

Applying the Learning Curve

STUDENT TIP Here are the three ways of solving learning curve problems.

A mathematical relationship enables us to express the time required to produce a certain unit. This relationship is a function of how many units have been produced before the unit in question and how long it took to produce them. To gain a mastery of this relationship, we will work through learning curve scenarios using three different methods: the doubling approach, formula approach, and learning curve table approach.

LO E.2 *Use* the doubling concept to estimate times

Doubling Approach

The doubling approach is the simplest approach to learning-curve problems. As noted earlier, each time production doubles, labor per unit declines by a constant factor, known as the learning curve rate. So, if we know that the learning curve rate is 80% and that the first unit produced took 100 hours, the hours required to produce the 2nd, 4th, 8th, and 16th units are as follows:

*N*TH UNIT PRODUCED	HOURS FOR *N*TH UNIT
1	100.0
2	$80.0 = (.8 \times 100)$
4	$64.0 = (.8 \times 80)$
8	$51.2 = (.8 \times 64)$
16	$41.0 = (.8 \times 51.2)$

As long as we wish to find the hours required to produce N units and N is one of the doubled values, then this approach works. The doubling approach does not tell us how many hours will be needed to produce other units. For this flexibility, we turn to the formula approach.

Formula Approach

The formula approach allows us to determine labor for *any* unit, T_N, by the formula:

$$T_N = T_1(N^b) \qquad \text{(E-2)}$$

where T_N = time for the Nth unit
T_1 = time to produce the first unit
b = (log of the learning rate)/(log 2) = slope of the learning curve

Some of the values for b are presented in Table E.2. Example E1 shows how this formula works.

TABLE E.2
Learning-Curve Values of *b*

Learning Rate (%)	*b*
70	–.515
75	–.415
80	–.322
85	–.234
90	–.152

Example E1

USING LOGS TO COMPUTE LEARNING CURVES

The learning-curve rate for a typical CPA to conduct a dental practice audit is 80%. Greg Lattier, a new graduate of Lee College, completed his first audit in 100 hours. If the dental offices he audits are about the same, how long should he take to finish his third job?

APPROACH ▶ We will use the formula approach in Equation (E-2).

SOLUTION ▶

$$T_N = T_1(N^b)$$

$$T_3 = (100\text{ hours})(3^b)$$

$$= (100)(3^{\log .8/\log 2})$$

$$= (100)(3^{-.322}) = 70.2 \text{ labor-hours}$$

INSIGHT ▶ Greg improved quickly from his first to his third audit. An 80% learning-curve rate means that from just the first to second jobs, his time decreased by 20%.

LEARNING EXERCISE ▶ If Greg's learning-curve rate were only 90%, how long would the third audit take? [Answer: 84.621 hours.]

RELATED PROBLEMS ▶ E.1, E.2, E.9, E.10, E.11, E.16

EXCEL OM Data File **ModEExE1.xls** can be found online.

The formula approach allows us to determine the hours required for *any* unit produced, but there *is* a simpler method.

Learning-Curve Table Approach

LO E.3 *Compute* learning-curve effects with the formula and learning-curve table approaches

The learning-curve table technique uses Table E.3 (to provide the coefficient C) and the following equation:

$$T_N = T_1 C \qquad \text{(E-3)}$$

where T_N = number of labor-hours required to produce the Nth unit
T_1 = number of labor-hours required to produce the first unit
C = learning-curve coefficient found in Table E.3

The learning-curve coefficient, C, depends on both the learning curve rate (70%, 75%, 80%, and so on) and the unit number of interest.

Example E2 uses the preceding equation and Table E.3 to calculate learning-curve effects.

Example E2

USING LEARNING-CURVE COEFFICIENTS

It took a Korean shipyard 125,000 labor-hours to produce the first of several tugboats that you expect to purchase for your shipping company, Great Lakes, Inc. Boats 2 and 3 have been produced by the Koreans with a learning factor of 85%. At $40 per hour, what should you, as purchasing agent, expect to pay for the fourth unit?

APPROACH ▶ First, search Table E.3 for the fourth unit and a learning-curve rate of 85%. The learning-curve coefficient, C, is .7225.

SOLUTION ▶ To produce the fourth unit, then, takes:

$$T_N = T_1 C$$
$$T_4 = (125{,}000 \text{ hours})(.7225)$$
$$= 90{,}312.5 \text{ hours}$$

To find the cost, multiply by $40:

$$90{,}312.5 \text{ hours} \times \$40 \text{ per hour} = \$3{,}612{,}500$$

INSIGHT ▶ The learning-curve table approach is very easy to apply. If we had not factored learning into our cost estimates, the price would have been 125,000 hours × $40 per hour (same as the first boat) = $6,000,000.

LEARNING EXERCISE ▶ If the learning factor improved to 80%, how would the cost change? [Answer: It would drop to $3,200,000.]

RELATED PROBLEMS ▶ E.1, E.2, E.3a, E.5a,c, E.6a,b, E.9, E.10, E.11, E.14, E.16, E.22, E.26, E.28, E.30, E.31

EXCEL OM Data File **ModEExE2.xls** can be found online.

ACTIVE MODEL E.1 This example is further illustrated in **Active MODEL E.1** found online.

Table E.3 also shows *cumulative values*. These allow us to compute the total number of hours needed to complete a specified number of units. Again, the computation is straightforward. Just multiply the table coefficient value by the time required for the first unit. Example E3 illustrates this concept.

Example E3

USING CUMULATIVE COEFFICIENTS

Example E2 computed the time to complete the fourth tugboat that Great Lakes plans to buy. How long will *all four* boats require?

APPROACH ▶ We look at the "Total Time Coefficient" column in Table E.3 and find that the cumulative coefficient for 4 boats with an 85% learning-curve factor is 3.3454.

SOLUTION ▶ The time required is:

$$T_N = T_1 C$$
$$T_4 = (125{,}000)(3.3454) = 418{,}175 \text{ hours in total for all 4 boats}$$

INSIGHT ▶ For an illustration of how Excel OM can be used to solve Examples E2 and E3, see Program E.1 at the end of this module.

LEARNING EXERCISE ▶ What is the value of T_4 if the learning-curve factor is 80% instead of 85%? [Answer: 392,762.5 hours.]

RELATED PROBLEMS ▶ E.3b, E.4, E.5b,c, E.6c, E.7, E.15, E.19, E.20a

Using Table E.3 requires that we know how long it takes to complete the first unit. Yet, what happens if our most recent or most reliable information available pertains to some other unit? The answer is that we must use these data to find a revised estimate for the first unit and then apply the table coefficient to that number. Example E4 illustrates this concept.

Example E4

REVISING LEARNING-CURVE ESTIMATES

Great Lakes, Inc., believes that unusual circumstances in producing the first boat (see Example E2) imply that the time estimate of 125,000 hours is not as valid a base as the time required to produce the third boat. Boat number 3 was completed in 100,000 hours. It wants to solve for the revised estimate for boat number 1.

APPROACH ▶ We return to Table E.3, with a unit value of $N = 3$ and a learning-curve coefficient of $C = .7729$ in the 85% column.

SOLUTION ▶ To find the revised estimate, divide the actual time for boat number 3, 100,000 hours, by $C = .7729$:

$$\frac{100,000}{.7729} = 129,383 \text{ hours}$$

So 129,383 hours is the new (revised) estimate for boat 1.

INSIGHT ▶ Any change in product, process, or personnel will change the learning curve. The new estimate for boat 1 suggests that related cost and volume estimates need to be revised.

LEARNING EXERCISE ▶ Boat 4 was just completed in 90,000 hours. Great Lakes thinks the 85% learning-curve rate is valid but isn't sure about the 125,000 hours for the first boat. Find a revised estimate for boat 1. [Answer: 124,567.5, suggesting that boat 1's time was fairly accurate after all.]

RELATED PROBLEMS ▶ E.8, E.12, E.13, E.17, E.18, E.20b, E.21

EXCEL OM Data File **ModEExE4.xls** can be found online.

Examples E1 through E4 all assume that the learning curve rate is known. For a new product, this can be a major assumption. If a firm has observed the cost or time of any two products already produced, it's easy to work backward from Equation (E-3) and Table E.3 to impute the actual learning curve *rate*. Example E5 illustrates this concept.

Example E5

COMPUTING THE LEARNING-CURVE RATE FROM OBSERVED PRODUCTION

Boeing completed production on its forty-fifth 787 airliner at a cost of $184 million. The first plane off the assembly line cost $448 million. What is Boeing's learning-curve rate for this model?

APPROACH ▶ We use Equation (E-3), with costs for T_1 and T_{45} known, and then find the learning-curve coefficient (C) in Table E.3.

SOLUTION ▶ Equation (E-3) is $T_N = T_1C$. We solve for $C = \frac{T_N}{T_1}$.

$$C = \frac{184}{448} = .41$$

In Table E.3, we follow the "Unit Number" row for $N = 45$, and we see that .41 falls under the 85% learning-curve rate for unit times (or costs, in this case).

INSIGHT ▶ Boeing's goal is to reach an 80% learning-curve rate, so OM must begin to lower costs dramatically. Progress should be checked with each plane from this point on.

LEARNING EXERCISE ▶ Let's say Boeing's fifth 787 cost $350 million. What was the learning-curve rate at that time relative to plane number 1? [Answer: C = $350 million / $448 million = .78. This suggests a 90% learning-curve rate, so Boeing's performance has improved since the fifth unit.]

RELATED PROBLEMS ▶ E.20, E.23, E.27, E.29

TABLE E.3 Learning-Curve Coefficients, Where Coefficient C = $N^{(\text{LOG OF LEARNING RATE/LOG 2})}$

	70%		75%		80%		85%		90%	
UNIT NUMBER (*N*)	UNIT TIME CO-EFFICIENT	TOTAL TIME CO-EFFICIENT	UNIT TIME CO-EFFICIENT	TOTAL TIME CO-EFFICIENT	UNIT TIME CO-EFFICIENT	TOTAL TIME CO-EFFICIENT	UNIT TIME CO-EFFICIENT	TOTAL TIME CO-EFFICIENT	UNIT TIME CO-EFFICIENT	TOTAL TIME CO-EFFICIENT
1	1.0000	1.0000	1.0000	1.0000	1.0000	1.0000	1.0000	1.0000	1.0000	1.0000
2	.7000	1.7000	.7500	1.7500	.8000	1.8000	.8500	1.8500	.9000	1.9000
3	.5682	2.2682	.6338	2.3838	.7021	2.5021	.7729	2.6229	.8462	2.7462
4	.4900	2.7582	.5625	2.9463	.6400	3.1421	.7225	3.3454	.8100	3.5562
5	.4368	3.1950	.5127	3.4591	.5956	3.7377	.6857	4.0311	.7830	4.3392
6	.3977	3.5928	.4754	3.9345	.5617	4.2994	.6570	4.6881	.7616	5.1008
7	.3674	3.9601	.4459	4.3804	.5345	4.8339	.6337	5.3217	.7439	5.8447
8	.3430	4.3031	.4219	4.8022	.5120	5.3459	.6141	5.9358	.7290	6.5737
9	.3228	4.6260	.4017	5.2040	.4929	5.8389	.5974	6.5332	.7161	7.2898
10	.3058	4.9318	.3846	5.5886	.4765	6.3154	.5828	7.1161	.7047	7.9945
11	.2912	5.2229	.3696	5.9582	.4621	6.7775	.5699	7.6860	.6946	8.6890
12	.2784	5.5013	.3565	6.3147	.4493	7.2268	.5584	8.2444	.6854	9.3745
13	.2672	5.7685	.3449	6.6596	.4379	7.6647	.5480	8.7925	.6771	10.0516
14	.2572	6.0257	.3344	6.9941	.4276	8.0923	.5386	9.3311	.6696	10.7211
15	.2482	6.2739	.3250	7.3190	.4182	8.5105	.5300	9.8611	.6626	11.3837
16	.2401	6.5140	.3164	7.6355	.4096	8.9201	.5220	10.3831	.6561	12.0398
17	.2327	6.7467	.3085	7.9440	.4017	9.3218	.5146	10.8977	.6501	12.6899
18	.2260	6.9727	.3013	8.2453	.3944	9.7162	.5078	11.4055	.6445	13.3344
19	.2198	7.1925	.2946	8.5399	.3876	10.1037	.5014	11.9069	.6392	13.9735
20	.2141	7.4065	.2884	8.8284	.3812	10.4849	.4954	12.4023	.6342	14.6078
25	.1908	8.4040	.2629	10.1907	.3548	12.3086	.4701	14.8007	.6131	17.7132
30	.1737	9.3050	.2437	11.4458	.3346	14.0199	.4505	17.0907	.5963	20.7269
35	.1605	10.1328	.2286	12.6179	.3184	15.6428	.4345	19.2938	.5825	23.6660
40	.1498	10.9024	.2163	13.7232	.3050	17.1935	.4211	21.4252	.5708	26.5427
45	.1410	11.6245	.2060	14.7731	.2936	18.6835	.4096	23.4955	.5607	29.3658
50	.1336	12.3069	.1972	15.7761	.2838	20.1217	.3996	25.5131	.5518	32.1420

LO E.4 *Describe* the strategic implications of learning curves

Strategic Implications of Learning Curves

So far, we have shown how operations managers can forecast labor-hour requirements for a product. We have also shown how purchasing agents can determine a supplier's cost, knowledge that can help in price negotiations. Another important application of learning curves concerns strategic planning.

An example of a company cost line and industry price line are so labeled in Figure E.2. These learning curves are straight because both scales are log scales. When the *rate* of change is constant, a log-log graph yields a straight line. If an organization believes its cost line to be the "company cost" line, and the industry price is indicated by the dashed horizontal line, then the company must have costs at the points below the dashed line (for example, point *a* or *b*) or else operate at a loss (point *c*).

Lower costs are not automatic; they must be managed down. When a firm's strategy is to pursue a learning curve steeper than the industry average (the company cost line in Figure E.2), it does this by:

1. Following an aggressive pricing policy
2. Focusing on continuing cost reduction and productivity improvement

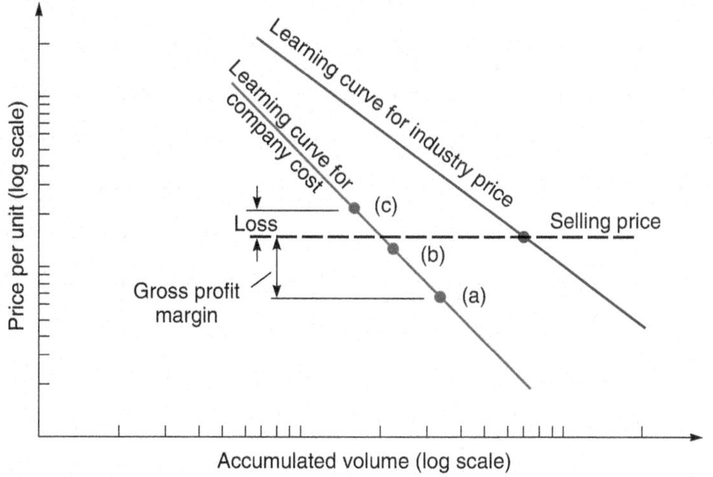

STUDENT TIP
Both the vertical and horizontal axes of this figure are log scales in this log-log graph.

Figure **E.2**

Industry Learning Curve for Price Compared with Company Learning Curve for Cost

3. Building on shared experience
4. Keeping capacity growing ahead of demand

Costs may drop as a firm pursues the learning curve, but volume must increase for the learning curve to exist. Moreover, managers must understand competitors before embarking on a learning-curve strategy. Weak competitors are undercapitalized, stuck with high costs, or do not understand the logic of learning curves. However, strong and dangerous competitors control their costs, have solid financial positions for the large investments needed, and have a track record of using an aggressive learning-curve strategy. Taking on such a competitor in a price war may help only the consumer.

Limitations of Learning Curves

Before using learning curves, some cautions are in order:

- Because learning curves differ from company to company, as well as industry to industry, estimates for each organization should be developed rather than applying someone else's.
- Learning curves are often based on the time necessary to complete the early units; therefore, those times must be accurate. As current information becomes available, reevaluation is appropriate.
- Any changes in personnel, design, or procedure can be expected to alter the learning curve, causing the curve to spike up for a short time, even if it is going to drop in the long run.
- While workers and processes may improve, the same learning curves do not always apply to indirect labor and material.
- The culture of the workplace, as well as resource availability and changes in the process, may alter the learning curve. For instance, as a project nears its end, worker interest and effort may drop, curtailing progress down the curve.

STUDENT TIP
Determining accurate rates of learning requires careful analysis.

Summary

The learning curve is a powerful tool for an operations manager. This tool can assist operations managers in determining future cost standards for items produced as well as purchased. In addition, the learning curve can provide understanding about company and industry performance. We saw three approaches to learning curves: the doubling approach, formula approach, and learning-curve table approach. Software can also help analyze learning curves.

Key Term

Learning curves (p. 790)

Discussion Questions

1. What are some of the limitations of learning curves?
2. Identify three applications of the learning curve.
3. What are the approaches to solving learning-curve problems?
4. Refer to Example E2. What are the implications for Great Lakes, Inc., if the engineering department wants to change the engine in the third and subsequent tugboats that the firm purchases?
5. Why isn't the learning-curve concept as applicable in a high-volume assembly line as it is in most other human activities?
6. What are the elements that can disrupt the learning curve?
7. Explain the concept of the doubling effect in learning curves.
8. What techniques can a firm use to move to a steeper learning curve?

Using Software for Learning Curves

Excel, Excel OM, and POM for Windows may all be used in analyzing learning curves. You can use the ideas in the following section on Excel OM to build your own Excel spreadsheet if you wish.

USING EXCEL OM

Program E.1 shows how Excel OM develops a spreadsheet for learning-curve calculations. The input data come from Examples E2 and E3. In cell B7, we enter the unit number for the base unit (which does not have to be 1), and in B8 we enter the time for this unit. Learning-curve rates can also be developed from observed times or costs, as illustrated in Example E5.

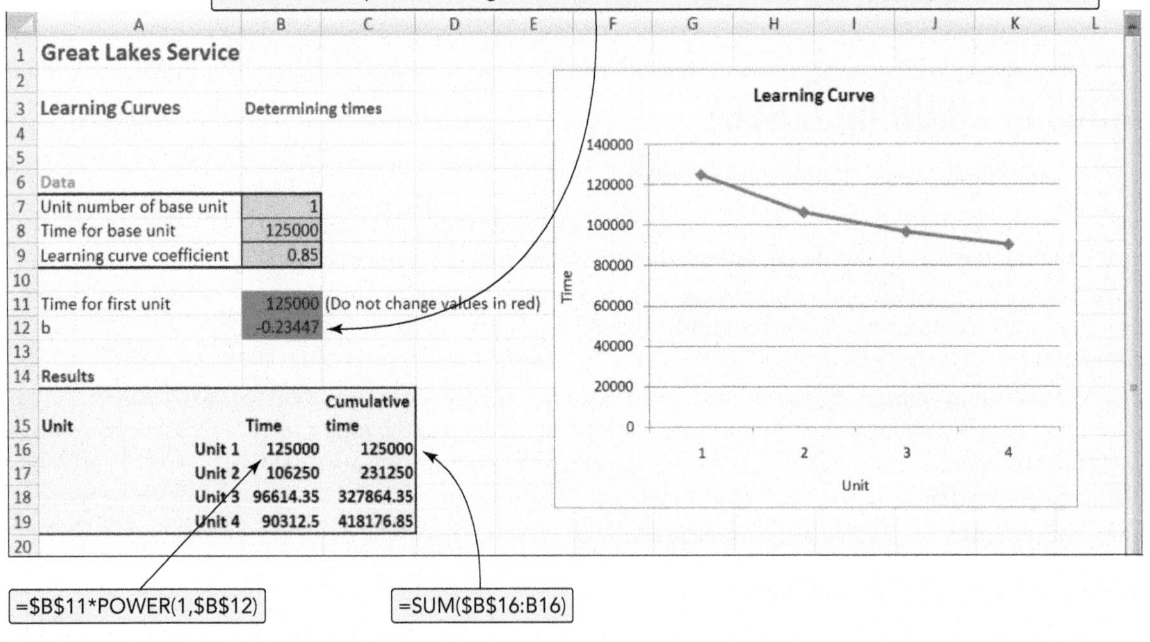

Program **E.1**

Excel OM's Learning Curve Module, Using Data from Examples E2 and E3

USING POM FOR WINDOWS

The POM for Windows Learning Curve module computes the length of time that future units will take, given the time required for the base unit and the learning rate (expressed as a number between 0 and 1). As an option, if the times required for the first and *N*th units are already known, the learning *rate* can be computed. See Appendix II for further details.

Solved Problems

SOLVED PROBLEM E.1

Digicomp produces a new telephone system with built-in TV screens. Its learning-curve rate is 80%.

a) If the first one took 56 hours, how long will it take Digicomp to make the eleventh system?

b) How long will the first 11 systems take in total?

c) As a purchasing agent, you expect to buy units 12 through 15 of the new phone system. What would be your expected cost for the units if Digicomp charges $30 for each labor-hour?

SOLUTION

a) $T_N = T_1 C$ (← from Table E.3, coefficient for 80% unit time)

$T_{11} = (56\,\text{hours})(.4621) = 25.9\,\text{hours}$

b) Total time for the first 11 units = (56 hours)(6.7775) = 379.5 hours

(from Table E.3, coefficient for 80% total time)

c) To find the time for units 12 through 15, we take the total cumulative time for units 1 to 15 and subtract the total time for units 1 to 11, which was computed in part (b). Total time for the first 15 units = (56 hours)(8.5105) = 476.6 hours. So the time for units 12 through 15 is 476.6 – 379.5 = 97.1 hours. (This figure could also be confirmed by computing the times for units 12, 13, 14, and 15 separately using the unit-time coefficient column and then adding them.) Expected cost for units 12 through 15 = (97.1 hours)($30 per hour) = $2,913.

SOLVED PROBLEM E.2

If the first time you performed a job took 60 minutes, how long will the eighth job take if you are on an 80% learning curve?

SOLUTION

Three doublings from 1 to 2 to 4 to 8 implies $.8^3$. Therefore, we have:

$$60 \times (.8)^3 = 60 \times .512 = 30.72\,\text{minutes}$$

or, using Table E.3, we have $C = .512$. Therefore:

$$60 \times .512 = 30.72\,\text{minutes}$$

Problems

Note: **PX** means the problem may be solved with POM for Windows and/or Excel OM.

Problems E.1–E.31 relate to Applying the Learning Curve

• **E.1** Susan Sherer, an IRS auditor, took 45 minutes to process her first tax return. The IRS uses an 85% learning curve. How long will the:
a) 2nd return take?
b) 4th return take?
c) 8th return take? **PX**

• **E.2** Temple Trucking Co. just hired Ed Rosenthal to verify daily invoices and accounts payable. He took 9 hours and 23 minutes to complete his task on the first day. Prior employees in this job have tended to follow a 90% learning curve. How long will the task take at the end of:
a) the 2nd day?
b) the 4th day?
c) the 8th day?
d) the 16th day? **PX**

• **E.3** If Professor Laurie Macdonald takes 15 minutes to grade the first exam and follows an 80% learning curve, how long will it take her:
a) to grade the 25th exam?
b) to grade the first 10 exams? **PX**

• **E.4** If it took 563 minutes to complete a hospital's first cornea transplant, and the hospital uses a 90% learning rate, what is the cumulative time to complete:
a) the first 3 transplants?
b) the first 6 transplants?
c) the first 8 transplants?
d) the first 16 transplants? **PX**

•• **E.5** Beth Zion Hospital has received initial certification from the state of California to become a center for liver transplants. The hospital, however, must complete its first 18 transplants under great scrutiny and at no cost to the patients. The very first transplant, just completed, required 30 hours. On the basis of research at the hospital, Beth Zion estimates that it will have an 80% learning curve. Estimate the time it will take to complete:
a) the 5th liver transplant.
b) all of the first 5 transplants.
c) the 18th transplant.
d) all 18 transplants. **PX**

•• **E.6** Refer to Problem E.5. Beth Zion Hospital has just been informed that only the first 10 transplants must be performed at the hospital's expense. The cost per hour of surgery is estimated to be $5,000. Again, the learning rate is 80% and the first surgery took 30 hours.

a) How long will the 10th surgery take?
b) How much will the 10th surgery cost?
c) How much will all 10 cost the hospital? PX

• **E.7** Manceville Air has just produced the first unit of a large industrial compressor that incorporated new technology in the control circuits and a new internal venting system. The first unit took 112 hours of labor to manufacture. The company knows from past experience that this labor content will decrease significantly as more units are produced. In reviewing past production data, it appears that the company has experienced a 90% learning curve when producing similar designs. The company is interested in estimating the total time to complete the next 7 units. Your job as the production cost estimator is to prepare the estimate. PX

• **E.8** Olivia Rhee, a student at SUNY, bought 6 bookcases for her dorm room. Each required unpacking of parts and assembly, which included some nailing and bolting. Olivia completed the first bookcase in 5 hours and the second in 4 hours.

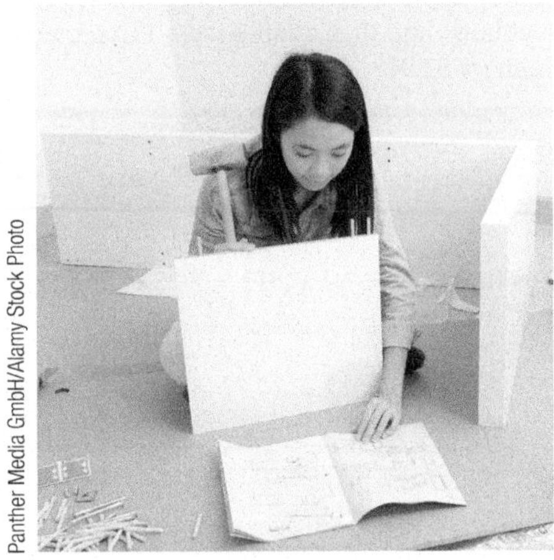
Panther Media GmbH/Alamy Stock Photo

a) What is her learning rate?
b) Assuming that the same rate continues, how long will the 3rd bookcase take?
c) The 4th, 5th, and 6th cases?
d) All 6 cases? PX

•• **E.9** Professor Mary Beth Marrs took 6 hours to prepare the first lecture in a new course. Traditionally, she has experienced a 90% learning curve. How much time should it take her to prepare the 15th lecture? PX

• **E.10** The first vending machine that Willona Kine, Inc., assembled took 80 labor-hours. Estimate how long the fourth machine will require for each of the following learning rates:
a) 95%
b) 87%
c) 72% PX

• **E.11** D. Shimshak Systems is installing networks for Advantage Insurance. The first installation took 46 labor-hours to complete. Estimate how long the 4th and the 8th installations will take for each of the following learning rates:
a) 92%
b) 84%
c) 77% PX

••• **E.12** Providence Assessment Center screens and trains employees for a computer assembly firm in Boston. The progress of all trainees is tracked, and those not showing the proper progress are moved to less demanding programs. By the tenth repetition trainees must be able to complete the assembly task in 1 hour or less. Susan Sweaney has just spent 5 hours on the fourth unit and 4 hours completing her eighth unit, while another trainee, Julie Burgmeier, took 4 hours on the third and 3 hours on the sixth unit. Should you encourage either or both of the trainees to continue? Why? PX

•• **E.13** The better students at Providence Assessment Center (see Problem E.12) have an 80% learning curve and can do a task in 20 minutes after just six times. You would like to identify the weak students sooner and decide to evaluate them after the third unit. How long should the third unit take? PX

•• **E.14** Suad Alwan, the purchasing agent for Dubai Airlines, is interested in determining what he can expect to pay for airplane number 4 if the third plane took 20,000 hours to produce. What would Alwan expect to pay for plane number 5? Number 6? Use an 85% learning curve and a $40-per-hour labor charge. PX

•• **E.15** Using the data from Problem E.14, how long will it take to complete the 12th plane? The 15th plane? How long will it take to complete planes 12 through 15 inclusive? At $40 per hour, what can Alwan, as purchasing agent, expect to pay for planes 12 through 15? PX

•• **E.16** Central Electronics Corp. produces semiconductors and has a learning curve of .7. The price per bit is 100 millicents when the volume is $.7 \times 10^{12}$ bits. What is the expected price at 1.4×10^{12} bits? What is the expected price at 89.6×10^{12} bits? PX

•• **E.17** Regional Power owns 25 small power generating plants. It has contracted with Genco Services to overhaul the power turbines of each of the plants. The number of hours that Genco billed Regional to complete the third turbine was 460. Regional pays Genco $60 per hour for its services. As the maintenance manager for Regional, you are trying to estimate the cost of overhauling the fourth turbine. How much would you expect to pay for the overhaul of number 5 and number 6? All the turbines are similar, and an 80% learning curve is appropriate. PX

•• **E.18** If it took Boeing 28,718 hours to produce the eighth 787 jet and the learning-curve factor is 80%, how long did it take to produce the tenth 787? PX

•• **E.19** Richard Dulski's firm is about to bid on a new radar system. Although the product uses new technology, Dulski believes that a learning rate of 75% is appropriate. The first unit is expected to take 700 hours, and the contract is for 40 units. PX
a) What is the total amount of hours to build the 40 units?
b) What is the average time to build each of the 40 units?
c) Assume that a worker works 2,080 hours per year. How many workers should be assigned to this contract to complete it in a year? PX

••• **E.20** As the estimator for Rajendra Tibrewala Enterprises, your job is to prepare an estimate for a potential customer service contract. The contract is for the service of diesel locomotive cylinder heads. The shop has done some of these in the past on a sporadic basis. The time required to service the first cylinder head in each job has been exactly 4 hours, and similar work has been accomplished at an 85% learning curve. The customer wants you to quote the total time for a contract for 12 and a contract for 20.

a) Prepare the quote.
b) After preparing the quote, you find a labor ticket for this customer for five locomotive cylinder heads. From the notations on the labor ticket, you conclude that the fifth unit took 2.5 hours. What do you conclude about the learning curve and your quote? **PX**

•• **E.21** Girish Shambu and Delores Reisel are teammates at a discount store; their new job is assembling bicycles for customers. Assembly of a bike has a learning rate of 90%. They forgot to time their effort on the first bike, but spent 4 hours on the second set. They have 6 more bikes to do. Determine approximately how much time will be (was) required for:

a) the 1st unit
b) the 8th unit
c) all 8 units **PX**

•• **E.22** Kelly-Lambing, Inc., a builder of government-contracted small ships, has a steady work force of 10 very skilled craftspeople. These workers can supply 2,500 labor-hours each per year. Kelly-Lambing is about to undertake a new contract, building a new style of boat. The first boat is expected to take 6,000 hours to complete. The firm thinks that 90% is the expected learning rate.

a) What is the firm's "capacity" to make these boats—that is, how many units can the firm make in 1 year?
b) If the operations manager can increase the learning rate to 85% instead of 90%, how many units can the firm make?

••• **E.23** The service times for a new data entry clerk have been measured and sequentially recorded as shown below:

REPORT	TIME (MINUTES)
1	66
2	56
3	53
4	48
5	47
6	45
7	44
8	41

a) What is the learning curve rate, based on this information?
b) Using an 85% learning curve rate and the above times, estimate the length of time the clerk will take to complete the 48th report. **PX**

•• **E.24** If the first unit of a production run takes 1 hour and the firm is on an 80% learning curve, how long will unit 100 take? (*Hint:* Apply the coefficient in Table E.3 twice.) **PX**

••• **E.25** Boeing spent $270 million to make the eleventh 787 in its production line. The first 787 cost $448 million. What was the learning curve rate at this point?

• **E.26** Dr. Che Mahmood took 563 minutes to complete a hospital's first cornea transplant, and the hospital uses a 90% learning rate. How long should

a) the third transplant take?
b) the sixth transplant take?
c) the eighth transplant take?
d) the sixteenth transplant take? **PX**

••• **E.27** If the fourth oil change and lube job at Trendo-Lube took 18 minutes and the second took 20 minutes, estimate how long

a) the first job took.
b) the third job took.
c) the eighth job will take.
d) the actual learning rate is. **PX**

• **E.28** Cleaning a toxic landfill took EPA contractor, Henry Garcia, 300 labor-days. If Garcia follows an 85% learning rate, how long will the firm take, in total, to clean the next five (that is, landfills two through six)? **PX**

•• **E.29** Professor Hayes took 20 minutes to grade the first exam. He estimates that it will take him two hours to grade all 10 exams. What learning factor is he assuming? **PX**

• **E.30** It takes 80,000 hours to produce the first jet engine at T.R.'s aerospace division and the learning factor is 90%. Help production manager Gopal Easwaran determine how long it takes to produce the eighth engine. **PX**

•• **E.31** If the first unit of a production run takes 1 hour and the firm is on a 90% learning curve, how long will unit 100 take? (*Hint:* Apply the coefficient in Table E.3 twice.) **PX**

Problem E.32 relates to Strategic Implications of Learning Curves

•••• **E.32** Using the accompanying log-log graph, answer the following questions:

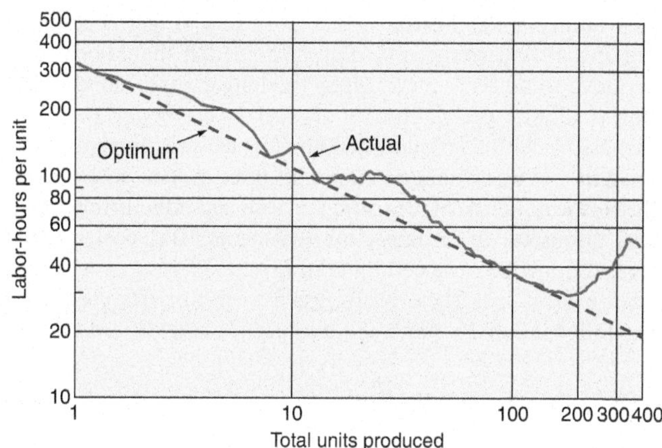

a) What are the implications for management if it has forecast its cost on the optimum line?
b) What could be causing the fluctuations above the optimum line?
c) If management forecasted the 10th unit on the optimum line, what was that forecast in hours?
d) If management built the 10th unit as indicated by the actual line, how many hours did it take?

CASE STUDY

SMT's Negotiation with IBM

IBM asked SMT and one other, much larger company to bid on 80 more units of a particular computer product. The request for quote (RFQ) asked that the overall bid be broken down to show the hourly rate, the parts and materials component in the price, and any charges for subcontracted services. SMT quoted $1.62 million and supplied the cost breakdown as requested. The second company submitted only one total figure, $5 million, with no cost breakdown. The decision was made to negotiate with SMT.

The IBM negotiating team included two purchasing managers and two cost engineers. One cost engineer had developed manufacturing cost estimates for every component, working from engineering drawings and cost-data books that he had built up from previous experience and that contained time factors, both setup and run times, for a large variety of operations. He estimated material costs by working both from data supplied by the IBM corporate purchasing staff and from purchasing journals. He visited SMT facilities to see the tooling available so that he would know what processes were being used. He assumed that there would be perfect conditions and trained operators, and he developed cost estimates for the 158th unit (previous orders were for 25, 15, and 38 units). He added 5% for scrap-and-flow loss; 2% for the use of temporary tools, jigs, and fixtures; 5% for quality control; and 9% for purchasing burden. Then, using an 85% learning curve, he backed up his costs to get an estimate for the first unit. He next checked the data on hours and materials for the 25, 15, and 38 units already made and found that his estimate for the first unit was within 4% of actual cost. His check, however, had indicated a 90% learning-curve effect on hours per unit.

In the negotiations, SMT was represented by one of the two owners of the business, two engineers, and one cost estimator. The sessions opened with a discussion of learning curves. The IBM cost estimator demonstrated that SMT had in fact been operating on a 90% learning curve. But, he argued, it should be possible to move to an 85% curve, given the longer runs, reduced setup time, and increased continuity of workers on the job that would be possible with an order for 80 units. The owner agreed with this analysis and was willing to reduce his price by 4%.

However, as each operation in the manufacturing process was discussed, it became clear that some IBM cost estimates were too low because certain crating and shipping expenses had been overlooked. These oversights were minor, however, and in the following discussions, the two parties arrived at a common understanding of specifications and reached agreements on the costs of each manufacturing operation.

At this point, SMT representatives expressed great concern about the possibility of inflation in material costs. The IBM negotiators volunteered to include a form of price escalation in the contract, as previously agreed among themselves. IBM representatives suggested that if overall material costs changed by more than 10%, the price could be adjusted accordingly. However, if one party took the initiative to have the price revised, the other could require an analysis of *all* parts and materials invoices in arriving at the new price.

Another concern of the SMT representatives was that a large amount of overtime and subcontracting would be required to meet IBM's specified delivery schedule. IBM negotiators thought that a relaxation in the delivery schedule might be possible if a price concession could be obtained. In response, the SMT team offered a 5% discount, and this was accepted. As a result of these negotiations, the SMT price was reduced almost 20% below its original bid price.

In a subsequent meeting called to negotiate the prices of certain pipes to be used in the system, it became apparent to an IBM cost estimator that SMT representatives had seriously underestimated their costs. He pointed out this apparent error because he could not understand why SMT had quoted such a low figure. He wanted to be sure that SMT was using the correct manufacturing process. In any case, if SMT estimators had made a mistake, it should be noted. It was IBM's policy to seek a fair price both for itself and for its suppliers. IBM procurement managers believed that if a vendor was losing money on a job, there would be a tendency to cut corners. In addition, the IBM negotiator felt that by pointing out the error, he generated some goodwill that would help in future sessions.

Discussion Questions

1. What are the advantages and disadvantages to IBM and SMT from this approach?
2. How does SMT's proposed learning rate compare with that of other industries?
3. What are the limitations of the learning curve in this case?

Source: Based on E. Raymond Corey, *Procurement Management: Strategy, Organization, and Decision Making* (New York: Van Nostrand Reinhold).

Endnote

1. Sources: *Industry Week* (August 17, 2018); *Defense News* (November 7, 2019); and *USNI News* (October 22, 2019).

Bibliography

Boh, W. F., S. A. Slaughter, and J. A. Espinosa. "Learning from Experience in Software Development." *Management Science* 53, no. 8 (August 2007): 1315–1332.

Jaber, M. Y. *Learning Curves: Theory, Models, and Applications*. London: CRC Press, 2017.

McDonald, A., and L. Schrattenholzer. "Learning Curves and Technology Assessment." *International Journal of Technology Management* 23 (2002): 718.

Morrison, J. B. "Putting the Learning Curve into Context." *Journal of Business Research* 61, no. 1 (November 2008): 1182.

Smunt, T. L., and C. A. Watts. "Improving Operations Planning with Learning Curves." *Journal of Operations Management* 21 (January 2003): 93.

Wright, T. P. "Factors Affecting the Cost of Airplanes." *Journal of the Aeronautical Sciences* 3, no. 4 (February 1936).

Module E Rapid Review

Main Heading	Review Material	MyLab Operations Management
WHAT IS A LEARNING CURVE?	■ **Learning curves**—The premise that people and organizations get better at their tasks as the tasks are repeated; sometimes called experience curves. Learning usually follows a negative exponential curve. *It takes less time to complete each additional unit a firm produces;* however, the time *savings* in completing each subsequent unit *decreases*. Learning curves were first applied to industry in a report by T. P. Wright of Curtis-Wright Corp. in 1936. Wright described how direct labor costs of making a particular airplane decreased with learning. Learning curves have been applied not only to labor but also to a wide variety of other costs, including material and purchased components. The power of the learning curve is so significant that it plays a major role in many strategic decisions related to employment levels, costs, capacity, and pricing. The learning curve is based on a *doubling* of production: That is, when production doubles, the decrease in time per unit affects the rate of the learning curve. $$T \times L^n = \text{Time required for the } n\text{th unit} \quad \text{(E-1)}$$ where T = unit cost or time of the first unit L = learning curve rate n = number of times T is doubled	Concept Questions: 1.1–1.6
LEARNING CURVES IN SERVICES AND MANUFACTURING	Different organizations—indeed, different products—have different learning curves. The rate of learning varies, depending on the quality of management and the potential of the process and product. *Any change in process, product, or personnel disrupts the learning curve.* Therefore, caution should be exercised in assuming that a learning curve is continuing and permanent. The steeper the slope of the learning curve, the faster the drop in costs. By tradition, learning curves are defined in terms of the *complements* of their improvement rates (i.e., a 75% learning rate is better than an 85% learning rate). Stable, standardized products and processes tend to have costs that decline more steeply than others. Learning curves are useful for a variety of purposes, including: 1. *Internal:* Labor forecasting, scheduling, establishing costs and budgets 2. *External:* Supply-chain negotiations 3. *Strategic:* Evaluation of company and industry performance, including costs and pricing	Concept Questions: 2.1–2.6
APPLYING THE LEARNING CURVE	If learning curve improvement is ignored, potential problems could arise, such as scheduling mismatches, leading to idle labor and productive facilities, refusal to accept new orders because capacity is assumed to be full, or missing an opportunity to negotiate with suppliers for lower purchase prices as a result of large orders. Three ways to approach the mathematics of learning curves are (1) doubling approach, (2) formula approach, and (3) learning-curve table approach. The doubling approach uses the production doubling Equation (E-1). The formula approach allows us to determine labor for *any* unit, T_N, by the formula: $$T_N = T_1\left(N^b\right) \quad \text{(E-2)}$$ where T_N = time for the Nth unit T_1 = time to produce the first unit $b = (\text{log of the learning rate})/(\log 2)$ = slope of the learning curve The learning-curve table approach makes use of Table E.3 and uses the formula: $$T_N = T_1 C \quad \text{(E-3)}$$ where T_N = number of labor-hours required to produce the Nth unit T_1 = number of labor-hours required to produce the first unit C = learning-curve coefficient found in Table E.3 The learning-curve coefficient, C, depends on both the learning rate and the unit number of interest.	Concept Questions: 3.1–3.5 Problems: E.1–E.31 Virtual Office Hours for Solved Problems: E.1, E.2 ACTIVE MODEL E.1

Main Heading	Review Material	MyLab Operations Management
	Formula (E-3) can also use the "Total Time Coefficient" columns of Table E.3 to provide the total cumulative number of hours needed to complete the specified number of units. If the most recent or most reliable information available pertains to some unit other than the first, these data should be used to find a revised estimate for the first unit, and then the applicable formulas should be applied to that revised number.	
STRATEGIC IMPLICATIONS OF LEARNING CURVES	When a firm's strategy is to pursue a learning cost curve steeper than the industry average, it can do this by: 1. Following an aggressive pricing policy 2. Focusing on continuing cost reduction and productivity improvement 3. Building on shared experience 4. Keeping capacity growing ahead of demand Managers must understand competitors before embarking on a learning-curve strategy. For example, taking on a strong competitor in a price war may help only the consumer.	Concept Questions: 4.1–4.3 Problem: E.32
LIMITATIONS OF LEARNING CURVES	Before using learning curves, some cautions are in order: ▪ Because learning curves differ from company to company, as well as industry to industry, estimates for each organization should be developed rather than applying someone else's. ▪ Learning curves are often based on the time necessary to complete the early units; therefore, those times must be accurate. As current information becomes available, reevaluation is appropriate. ▪ Any changes in personnel, design, or procedure can be expected to alter the learning curve, causing the curve to spike up for a short time, even if it is going to drop in the long run. ▪ While workers and process may improve, the same learning curves do not always apply to indirect labor and material. ▪ The culture of the workplace, as well as resource availability and changes in the process, may alter the learning curve. For instance, as a project nears its end, worker interest and effort may drop, curtailing progress down the curve.	Concept Questions: 5.1–5.4
ADDITIONAL MYLAB OPERATIONS MANAGEMENT RESOURCES	✔ Multiple Choice Case Questions (SMT's Negotiation with IBM)	

Self **Test**

LO E.1 A learning curve describes:

a) the rate at which an organization acquires new data.
b) the amount of production time per unit as the total number of units produced increases.
c) the increase in production time per unit as the total number of units produced increases.
d) the increase in number of units produced per unit time as the total number of units produced increases.

LO E.2 A surgical procedure with a 90% learning curve required 20 hours for the initial patient. The fourth patient should require approximately how many hours?

a) 18
b) 16.2
c) 28
d) 30
e) 54.2

LO E.3 The first transmission took 50 hours to rebuild at Bob's Auto Repair, and the learning rate is 80%. How long will it take to rebuild the third unit? (Use at least three decimals in the exponent if you use the formula approach.)

a) under 30 hours
b) about 32 hours
c) about 35 hours
d) about 60 hours
e) about 45 hours

LO E.4 Which one of the following courses of action would *not* be taken by a firm wanting to pursue a learning curve steeper than the industry average?

a) Following an aggressive pricing policy
b) Focusing on continuing cost reduction
c) Keeping capacity equal to demand to control costs
d) Focusing on productivity improvement
e) Building on shared experience

Answers: LO E.1. b; LO E.2. b; LO E.3. c; LO E.4. c.

Simulation

MODULE OUTLINE

LEARNING OBJECTIVES

LO F.1 ***List*** the advantages and disadvantages of modeling with simulation 807

LO F.2 ***Perform*** the five steps in a Monte Carlo simulation 808

LO F.3 ***Simulate*** an inventory problem 812

LO F.4 ***Use*** Excel spreadsheets to create a simulation 814

Kaiser/Agencja Fotograficzna Caro/Alamy Stock Photo

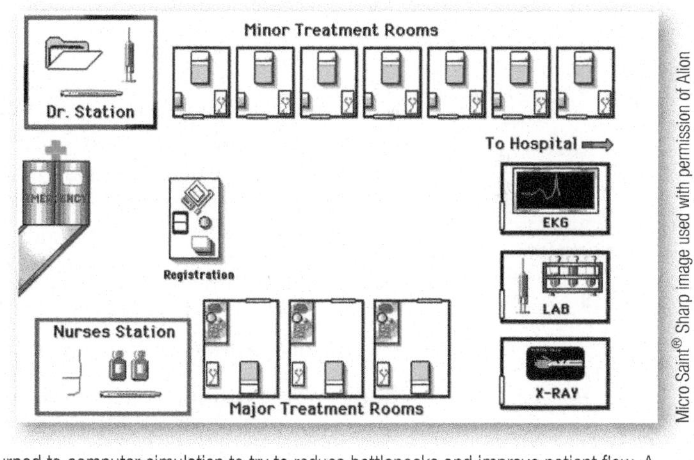

Micro Saint® Sharp image used with permission of Alion Science and Technology Corp.

When Bay Medical Center faced severe overcrowding at its outpatient clinic, it turned to computer simulation to try to reduce bottlenecks and improve patient flow. A simulation language called Micro Saint analyzed current data relating to patient service times between clinic rooms. By simulating different numbers of doctors and staff, simulating the use of another clinic for overflow, and simulating a redesign of the existing clinic, Bay Medical Center was able to make decisions based on an understanding of both costs and benefits. This resulted in better patient service at lower cost.

What Is Simulation?

Simulation models abound in our world. The city of Atlanta, for example, uses them to control traffic. Europe's Airbus Industries uses them to test the aerodynamics of proposed jets. The U.S. Army simulates war games on computers. Business students use management gaming to simulate realistic business competition. And thousands of organizations like Bay Medical Center develop simulation models to help make operations decisions.

Most of the large companies in the world use simulation models. Table F.1 lists just a few areas in which simulation is now being applied.

TABLE F.1 Some Applications of Simulation

Ambulance location and dispatching	Bus scheduling
Assembly-line balancing	Design of library operations
Parking lot and harbor design	Taxi, truck, and railroad dispatching
Distribution system design	Production facility scheduling
Scheduling aircraft	Plant layout
Labor-hiring decisions	Capital investments
Personnel scheduling	Production scheduling
Traffic-light timing	Sales forecasting
Voting pattern prediction	Inventory planning and control

Simulation
The attempt to duplicate the features, appearance, and characteristics of a real system, usually via a computerized model.

Simulation is the attempt to duplicate the features, appearance, and characteristics of a real system. In this module, we will show how to simulate part of an operations management system by building a mathematical model that comes as close as possible to representing the reality of the system. The model will then be used to estimate the effects of various actions. The idea behind simulation is threefold:

1. To imitate a real-world situation mathematically
2. Then to study its properties and operating characteristics
3. Finally, to draw conclusions and make action decisions based on the results of the simulation

In this way, a real-life system need not be touched until the advantages and disadvantages of a major policy decision are first measured on the model.

To use simulation, an OM manager should:

1. Define the problem.
2. Introduce the important variables associated with the problem.
3. Construct a numerical model.
4. Set up possible courses of action for testing by specifying values of variables.
5. Run the experiment.
6. Consider the results (possibly modifying the model or changing data inputs).
7. Decide what course of action to take.

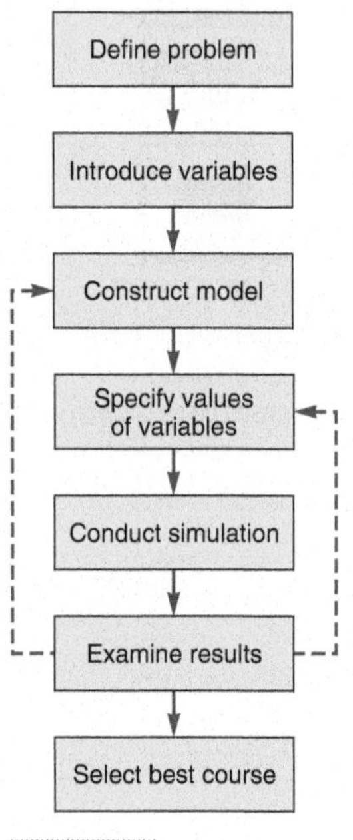

Figure F.1
The Process of Simulation

These steps are illustrated in Figure F.1.

The problems tackled by simulation may range from very simple to extremely complex, from bank-teller lines to an analysis of the U.S. economy. Although small simulations can be conducted by hand, effective use of the technique requires a computer. Large-scale models, simulating perhaps years of business decisions, are virtually all handled by computer.

In this module, we examine the basic principles of simulation and then tackle some problems in the areas of waiting-line analysis and inventory control. Why do we use simulation in these areas when mathematical models described in other chapters can solve similar problems? The answer is that simulation provides an alternative approach for problems that are very complex mathematically. It can handle, for example, inventory problems in which demand or lead time is not constant.

Advantages and Disadvantages of Simulation

Simulation is a tool that has become widely accepted by managers for several reasons. The main *advantages* of simulation are as follows:

STUDENT TIP
There are many reasons it's better to simulate a real-world system than to experiment with it.

LO F.1 *List* the advantages and disadvantages of modeling with simulation

1. It can be used to analyze large and complex real-world situations that cannot be solved by conventional operations management models.
2. Real-world complications can be included that most OM models cannot permit. For example, simulation can use *any* probability distribution the user defines; it does not require standard distributions.
3. "Time compression" is possible. The effects of OM policies over many months or years can be obtained by computer simulation in a short time.
4. Simulation allows "what-if?" types of questions. Managers like to know in advance what options will be most attractive. With a computerized model, a manager can try out several policy decisions within a matter of minutes.
5. Simulations do not interfere with real-world systems. It may be too disruptive, for example, to experiment physically with new policies or ideas in a hospital or manufacturing plant.

The main *disadvantages* of simulation are as follows:

1. Good simulation models can take a long time to develop.
2. It is a repetitive approach that may produce different solutions in repeated runs. It does not generate optimal solutions to problems (as does linear programming).
3. Managers must generate all of the conditions and constraints for solutions that they want to examine. The simulation model does not produce answers without adequate, realistic input.
4. Each simulation model is unique. Its solutions and inferences are not usually transferable to other problems.

Donna Shader

Computer simulation models have been developed to address a variety of productivity issues at fast-food restaurants such as Burger King. In one, the ideal distance between the drive-through order station and the pickup window was simulated. For example, because a longer distance reduced waiting time, 12 to 13 additional customers could be served per hour—a benefit of about $20,000 in extra sales per restaurant per year. In another simulation, a second drive-through window was considered. This model predicted a sales increase of 15%.

Monte Carlo Simulation

Monte Carlo method
A simulation technique that uses random elements when chance exists in their behavior.

When a system contains elements that exhibit *chance* in their behavior, the Monte Carlo method of simulation may be applied. The basis of Monte Carlo simulation is experimentation on chance (or *probabilistic*) elements by means of random sampling.

The technique breaks down into five simple steps:

LO F.2 *Perform* the five steps in a Monte Carlo simulation

1. Setting up a probability distribution for important variables.
2. Building a cumulative probability distribution for each variable.
3. Establishing an interval of random numbers for each variable.
4. Generating random numbers.
5. Actually simulating a series of trials.

Let's examine these steps in turn.

Step 1. Establishing Probability Distributions. The basic idea in the Monte Carlo simulation is to generate values for the variables making up the model under study. In real-world systems, a lot of variables are probabilistic in nature. To name just a few: inventory demand; lead time for orders to arrive; times between machine breakdowns; times between customer arrivals at a service facility; service times; times required to complete project activities; and number of employees absent from work each day.

One common way to establish a *probability distribution* for a given variable is to examine historical outcomes. We can find the probability, or relative frequency, for each possible outcome of a variable by dividing the frequency of observation by the total number of observations. Here's an example.

The daily demand for radial tires at Barry's Auto Tire over the past 200 days is shown in columns 1 and 2 of Table F.2. Assuming that past arrival rates will hold in the future, we can convert this demand to a probability distribution by dividing each demand frequency by the total demand, 200. The results are shown in column 3.

TABLE F.2 Demand for Barry's Auto Tire

(1) DEMAND FOR TIRES	(2) FREQUENCY	(3) PROBABILITY OF OCCURRENCE	(4) CUMULATIVE PROBABILITY
0	10	10/200 = .05	.05
1	20	20/200 = .10	.15
2	40	40/200 = .20	.35
3	60	60/200 = .30	.65
4	40	40/200 = .20	.85
5	30	30/200 = .15	1.00
	200 days	200/200 = 1.00	

STUDENT TIP
To establish a probability distribution for tires, we assume that historical demand is a good indicator of future demand.

Cumulative probability distribution
The accumulation of individual probabilities of a distribution.

Random-number intervals
A set of numbers to represent each possible value or outcome in a computer simulation.

Random number
A series of digits that have been selected by a totally random process.

Step 2. Building a Cumulative Probability Distribution for Each Variable. The conversion from a regular probability distribution, such as in column 3 of Table F.2, to a cumulative probability distribution is an easy job. In column 4, we see that the cumulative probability for each level of demand is the sum of the number in the probability column (column 3) added to the previous cumulative probability.

Step 3. Setting Random-Number Intervals. Once we have established a cumulative probability distribution for each variable in the simulation, we must assign a set of numbers to represent each possible value or outcome. These are referred to as random-number intervals. Basically, a random number is a series of digits (say, two digits from 01, 02,..., 98, 99, 00) that have been selected by a totally random process—a process in which each random number has an equal chance of being selected.

TABLE F.3 The Assignment of Random-Number Intervals for Barry's Auto Tire

DAILY DEMAND	PROBABILITY	CUMULATIVE PROBABILITY	INTERVAL OF RANDOM NUMBERS
0	.05	.05	01 through 05
1	.10	.15	06 through 15
2	.20	.35	16 through 35
3	.30	.65	36 through 65
4	.20	.85	66 through 85
5	.15	1.00	86 through 00

STUDENT TIP

You may start random-number intervals at either 01 or 00, but the text starts at 01 so that the top of each range is the cumulative probability.

If, for example, there is a 5% chance that demand for Barry's radial tires will be 0 units per day, then we will want 5% of the random numbers available to correspond to a demand of 0 units. If a total of 100 two-digit numbers is used in the simulation, we could assign a demand of 0 units to the first 5 random numbers: 01, 02, 03, 04, and 05.[1] Then a simulated demand for 0 units would be created every time one of the numbers 01 to 05 was drawn. If there is also a 10% chance that demand for the same product will be 1 unit per day, we could let the next 10 random numbers (06, 07, 08, 09, 10, 11, 12, 13, 14, and 15) represent that demand—and so on for other demand levels.

Similarly, we can see in Table F.3 that the length of each interval on the right corresponds to the probability of 1 of each of the possible daily demands. Thus, in assigning random numbers to the daily demand for 3 radial tires, the range of the random-number interval (36 through 65)

TABLE F.4 Table of 2-Digit Random Numbers

52	06	50	88	53	30	10	47	99	37	66	91	35	32	00	84	57	07
37	63	28	02	74	35	24	03	29	60	74	85	90	73	59	55	17	60
82	57	68	28	05	94	03	11	27	79	90	87	92	41	09	25	36	77
69	02	36	49	71	99	32	10	75	21	95	90	94	38	97	71	72	49
98	94	90	36	06	78	23	67	89	85	29	21	25	73	69	34	85	76
96	52	62	87	49	56	59	23	78	71	72	90	57	01	98	57	31	95
33	69	27	21	11	60	95	89	68	48	17	89	34	09	93	50	44	51
50	33	50	95	13	44	34	62	64	39	55	29	30	64	49	44	30	16
88	32	18	50	62	57	34	56	62	31	15	40	90	34	51	95	26	14
90	30	36	24	69	82	51	74	30	35	36	85	01	55	92	64	09	85
50	48	61	18	85	23	08	54	17	12	80	69	24	84	92	16	49	59
27	88	21	62	69	64	48	31	12	73	02	68	00	16	16	46	13	85
45	14	46	32	13	49	66	62	74	41	86	98	92	98	84	54	33	40
81	02	01	78	82	74	97	37	45	31	94	99	42	49	27	64	89	42
66	83	14	74	27	76	03	33	11	97	59	81	72	00	64	61	13	52
74	05	81	82	93	09	96	33	52	78	13	06	28	30	94	23	37	39
30	34	87	01	74	11	46	82	59	94	25	34	32	23	17	01	58	73
59	55	72	33	62	13	74	68	22	44	42	09	32	46	71	79	45	89
67	09	80	98	99	25	77	50	03	32	36	63	65	75	94	19	95	88
60	77	46	63	71	69	44	22	03	85	14	48	69	13	30	50	33	24
60	08	19	29	36	72	30	27	50	64	85	72	75	29	87	05	75	01
80	45	86	99	02	34	87	08	86	84	49	76	24	08	01	86	29	11
53	84	49	63	26	65	72	84	85	63	26	02	75	26	92	62	40	67
69	84	12	94	51	36	17	02	15	29	16	52	56	43	26	22	08	62
37	77	13	10	02	18	31	19	32	85	31	94	81	43	31	58	33	51

Source: A *Million Random Digits with 100,000 Normal Deviates.* New York: The Free Press, 1995. Used by permission.

corresponds *exactly* to the probability (or proportion) of that outcome. A daily demand for 3 radial tires occurs 30% of the time. All of the 30 random numbers greater than 35 up to and including 65 are assigned to that event.

Step 4. Generating Random Numbers. Random numbers may be generated for simulation problems in two ways. If the problem is large and the process under study involves many simulation trials, computer programs are available to generate the needed random numbers. If the simulation is being done by hand, the numbers may be selected from a table of random digits.

Step 5. Simulating the Experiment. We may simulate outcomes of an experiment by simply selecting random numbers from Table F.4. Beginning anywhere in the table, we note the interval in Table F.3 into which each number falls. For example, if the random number chosen is 81 and the interval 66 through 85 represents a daily demand for 4 tires, then we select a demand of 4 tires. Example F1 carries the simulation further.

Example F1

SIMULATING DEMAND

Barry's Auto Tire wants to simulate 10 days of demand for radial tires.

APPROACH ▶ Earlier, we went through Steps 1 and 2 in the Monte Carlo method (in Table F.2) and Step 3 (in Table F.3). Now we need to generate random numbers (Step 4) and simulate demand (Step 5).

SOLUTION ▶ We select the random numbers needed from Table F.4, starting in the upper-left-hand corner and continuing down the first column, and record the corresponding daily demand:

DAY NUMBER	RANDOM NUMBER	SIMULATED DAILY DEMAND
1	52	3
2	37	3
3	82	4
4	69	4
5	98	5
6	96	5
7	33	2
8	50	3
9	88	5
10	90	5
		39 Total 10-day demand
		39/10 = 3.9 = tires average daily demand

INSIGHT ▶ It is interesting to note that the average demand of 3.9 tires in this 10-day simulation differs substantially from the *expected* daily demand, which we may calculate from the data in Table F.3:

$$\begin{aligned}\text{Expected demand} &= \sum_{i=0}^{5}(\text{probability of } i \text{ units}) \times (\text{demand of } i \text{ units})\\ &= (.05)(0) + (.10)(1) + (.20)(2) + (.30)(3) + (.20)(4) + (.15)(5)\\ &= 0 + .1 + .4 + .9 + .8 + .75\\ &= 2.95 \text{ tires}\end{aligned}$$

However, if this simulation was repeated hundreds or thousands of times, the average *simulated* demand would be nearly the same as the *expected* demand.

LEARNING EXERCISE ▶ Resimulate the 10 days, this time with random numbers from column 2 of Table F.4. What is the average daily demand? [Answer: 2.5.]

RELATED PROBLEMS ▶ F.1–F.15

OM in Action Simulation Takes the Kinks out of Starbucks' Lines[2]

The animation on the computer screen is not encouraging. Starbucks is running a digital simulation of customers ordering new warm sandwiches and pastries at a "virtual" store.

At first, things seem to go well, as animated workers rush around, preparing orders. But then they can't keep up. Soon the customers are stacking up in line, and the goal of serving each person in less than 3 minutes is blown. The line quickly reaches the point at which customers decide the snack or drink isn't worth the wait—called the "balking point" in queuing theory.

Fortunately for Starbucks, the customers departing without their Frappuccinos and decaf slim lattes are digital. The simulation helps operations managers find out what caused the backup before the scene repeats itself in the real world.

Simulation software is also used to find the point where capital expenditures will pay off. In large chains such as Starbucks, adding even a minor piece of equipment can add up. A $200 blender in each of Starbucks' more than 33,000 stores in 76 countries can cost the firm millions.

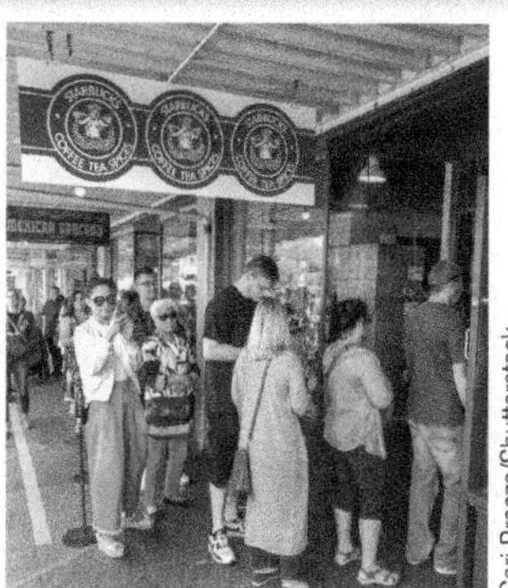

Ceri Breeze/Shutterstock

Naturally, it would be risky to draw any hard and fast conclusions about the operation of a firm from only a short simulation like Example F1. Seldom would anyone actually want to go to the effort of simulating such a simple model containing only one variable. Simulating by hand does, however, demonstrate the important principles involved and may be useful in small-scale studies.

Simulation with Two Decision Variables

Often, there is more than one variable to be simulated. In Example F1 it was demand for tires. But many OM decisions have multiple variables. In a queuing situation, it may be arrival times and service times. In an inventory problem, as we saw in Chapter 12, both demand and lead time might be variable (not constant).

STUDENT TIP
Most real-world inventory systems have probabilistic events and benefit from a simulation approach.

In this section, we present an inventory problem with two decision variables and two probabilistic components. The owner of the hardware store in Example F2 would like to establish *order quantity* and *reorder point* decisions for a particular product that has probabilistic (uncertain) daily demand and reorder lead time. He wants to make a series of simulation runs, trying out various order quantities and reorder points, to minimize his total inventory cost for the item. Inventory costs in this case will include ordering, holding, and stockout costs.

Example F2

AN INVENTORY SIMULATION WITH TWO VARIABLES

Simkin's Hardware Store, in Reno, sells the Ace model electric drill. Daily demand for this particular product is relatively low but subject to some variability. Lead times tend to be variable as well. Mark Simkin wants to develop a simulation to test an inventory policy of ordering 10 drills, with a reorder point of 5. In other words, every time the on-hand inventory level at the end of the day is 5 or less, Simkin will call his supplier that evening and place an order for 10 more drills. Simkin notes that if the lead time is 1 day, the order will not arrive the next morning but rather at the beginning of the following workday. Stockouts become lost sales, not backorders.

APPROACH ▶ Simkin wants to follow the 5 steps in the Monte Carlo simulation process.

SOLUTION ▶ Over the past 300 days, Simkin has observed the sales shown in column 2 of Table F.5. He converts this historical frequency into a probability distribution for the variable daily demand (column 3). A cumulative probability distribution is formed in column 4 of Table F.5. Finally, Simkin establishes an interval of random numbers to represent each possible daily demand (column 5).

When Simkin places an order to replenish his inventory of drills, there is a delivery lag of from 1 to 3 days. This means that lead time may also be considered a probabilistic variable. The number of days

TABLE F.5 Probabilities and Random-Number Intervals for Daily Ace Drill Demand

(1) DEMAND FOR ACE DRILL	(2) FREQUENCY	(3) PROBABILITY	(4) CUMULATIVE PROBABILITY	(5) INTERVAL OF RANDOM NUMBERS
0	15	.05	.05	01 through 05
1	30	.10	.15	06 through 15
2	60	.20	.35	16 through 35
3	120	.40	.75	36 through 75
4	45	.15	.90	76 through 90
5	30	.10	1.00	91 through 00
	300 days	1.00		

LO F.3 *Simulate* an inventory problem

that it took to receive the past 50 orders is presented in Table F.6. In a fashion similar to the creation of the demand variable, Simkin establishes a probability distribution for the lead time variable (column 3 of Table F.6), computes the cumulative distribution (column 4), and assigns random-number intervals for each possible time (column 5).

TABLE F.6 Probabilities and Random-Number Intervals for Reorder Lead Time

(1) LEAD TIME (DAYS)	(2) FREQUENCY	(3) PROBABILITY	(4) CUMULATIVE PROBABILITY	(5) RANDOM-NUMBER INTERVAL
1	10	.20	.20	01 through 20
2	25	.50	.70	21 through 70
3	15	.30	1.00	71 through 00
	50 orders	1.00		

The entire process is simulated in Table F.7 for a 10-day period. We assume that beginning inventory (column 3) is 10 units on day 1. We took the random numbers (column 4) from column 2 of Table F.4.

TABLE F.7 Simkin Hardware's Inventory Simulation. Order Quantity = 10 Units; Reorder Point = 5 Units

(1) DAY	(2) UNITS RECEIVED	(3) BEGINNING INVENTORY	(4) RANDOM NUMBER	(5) DEMAND	(6) ENDING INVENTORY	(7) LOST SALES	(8) ORDER?	(9) RANDOM NUMBER	(10) LEAD TIME
1		10	06	1	9	0	No		
2	0	9	63	3	6	0	No		
3	0	6	57	3	3[a]	0	Yes	02[b]	1
4	0	3	94[c]	5	0	2	No[d]		
5	10[e]	10	52	3	7	0	No		
6	0	7	69	3	4	0	Yes	33	2
7	0	4	32	2	2	0	No		
8	0	2	30	2	0	0	No		
9	10[f]	10	48	3	7	0	No		
10	0	7	88	4	3	0	Yes	14	1
					Totals: 41	2			

[a]This is the first time inventory dropped to the reorder point of five drills. Because no prior order was outstanding, an order is placed.
[b]The random number 02 is generated to represent the first lead time. It was drawn from column 2 of Table F.4 as the next number in the list being used. A separate column could have been used from which to draw lead-time random numbers if we had wanted to do so, but in this example, we did not do so.
[c]Again, notice that the random digits 02 were used for lead time (see footnote *b*). So the next number in the column is 94.
[d]No order is placed on day 4 because there is an order outstanding from the previous day that has not yet arrived.
[e]The lead time for the first order placed is 1 day, but as noted in the text, an order does not arrive the next morning but rather the beginning of the following day. Thus, the first order arrives at the start of day 5.
[f]This is the arrival of the order placed at the close of business on day 6. Fortunately for Simkin, no lost sales occurred during the 2-day lead time before the order arrived.

Table F.7 was filled in by proceeding 1 day (or line) at a time, working from left to right. It is a four-step process:

1. Begin each simulated day by checking to see whether any ordered inventory has just arrived. If it has, increase current inventory by the quantity ordered (10 units, in this case).
2. Generate a daily demand from the demand probability distribution for the selected random number.
3. Compute: Ending inventory = Beginning inventory minus Demand. If on-hand inventory is insufficient to meet the day's demand, satisfy as much demand as possible and note the number of lost sales.
4. Determine whether the day's ending inventory has reached the reorder point (5 units). If it has, and if there are no outstanding orders, place an order. Lead time for a new order is simulated for the selected random number corresponding to the distribution in Table F.6.

INSIGHTS ▶ Simkin's inventory simulation yields some interesting results. The average daily ending inventory is:

$$\text{Average ending inventory} = \frac{41 \text{ total units}}{10 \text{ days}} = 4.1 \text{ units/day}$$

We also note the average lost sales and number of orders placed per day:

$$\text{Average lost sales} = \frac{2 \text{ sales lost}}{10 \text{ days}} = .2 \text{ unit/day}$$

$$\text{Average number of orders placed} = \frac{3 \text{ orders}}{10 \text{ days}} = .3 \text{ order/day}$$

LEARNING EXERCISE ▶ How would these 3 averages change if the random numbers for day 10 were 04 and 93 instead of 88 and 14? [Answer: 4.4, .2 (no change), and .2.]

RELATED PROBLEMS ▶ F.16–F.25

Now that we have worked through Example F2 we want to emphasize something very important: This simulation should be extended many more days before we draw any conclusions as to the cost of the order policy being tested. If a hand simulation is being conducted, 100 days would provide a better representation. If a computer is doing the calculations, 1,000 days would be helpful in reaching accurate cost estimates. (Moreover, remember that even with a 1,000-day simulation, the generated distribution should be compared with the desired distribution to ensure valid results.)

Summary

Simulation involves building mathematical models that attempt to act like real operating systems. In this way, a real-world situation can be studied without imposing on the actual system. Although simulation models can be developed manually, simulation by computer is generally more desirable. The Monte Carlo approach uses random numbers to represent variables, such as inventory demand or people waiting in line, which are then simulated in a series of trials. Simulation is widely used as an operations tool because its advantages usually outweigh its disadvantages.

Key Terms

Simulation (p. 806)
Monte Carlo method (p. 808)
Cumulative probability distribution (p. 808)
Random-number intervals (p. 808)
Random number (p. 808)

Discussion Questions

1. State the seven steps, beginning with "Defining the Problem," that an operations manager should perform when using simulation to analyze a problem.
2. List the advantages of simulation.
3. List the disadvantages of simulation.
4. Explain the difference between *simulated* average demand and *expected* average demand.
5. What is the role of random numbers in a Monte Carlo simulation?
6. Why might the results of a simulation differ each time you make a run?
7. What is Monte Carlo simulation? What principles underlie its use, and what steps are followed in applying it?
8. List six ways that simulation can be used in business.
9. Why is simulation such a widely used technique?
10. What are the advantages of special-purpose simulation languages (see below)?
11. In the simulation of an order policy for drills at Simkin's hardware (Example F2), would the results (of Table F.7) change significantly if a longer period were simulated? Why is the 10-day simulation valid or invalid?

12. Why is a computer necessary in conducting a real-world simulation?
13. Why might a manager be forced to use simulation instead of an analytical model in dealing with a problem of:
    a) inventory order policy?
    b) ships docking in a port to unload?
    c) bank-teller service windows?
    d) the U.S. economy?

Using Software in Simulation

Computers are critical in simulating complex tasks. They can generate random numbers, simulate thousands of time periods in a matter of seconds or minutes, and provide management with reports that improve decision making. A computer approach is almost a necessity in order to draw valid conclusions from a simulation.

Computer programming languages can help the simulation process. *General-purpose languages*, such as BASIC or C ++, constitute one approach. *Special-purpose simulation languages*, such as GPSS and SIMSCRIPT, have a few advantages: (1) they require less programming time for large simulations, (2) they are usually more efficient and easier to check for errors, and (3) random-number generators are already built in as subroutines.

Commercial, easy-to-use prewritten simulation programs are also available. Some are generalized to handle a wide variety of situations ranging from queuing to inventory. These include programs such as Extend, Modsim, Witness, MAP/1, Enterprise Dynamics, Simfactory, ProModel, Micro Saint, and ARENA.

Spreadsheet software such as Excel can also be used to develop simulations quickly and easily. Such packages have built-in random-number generators and develop outputs through "data-fill" table commands.

USING EXCEL SPREADSHEETS

The ability to generate random numbers and then "look up" these numbers in a table to associate them with a specific event makes spreadsheets excellent tools for conducting simulations. Program F.1 illustrates an Excel simulation for Example F1.

Notice that the cumulative probabilities are calculated in column E of Program F.1. This procedure reduces the chance of error and is useful in larger simulations involving more levels of demand.

The **VLOOKUP** function in column I looks up the random number (generated in column H) in the leftmost column of the defined lookup table. The **VLOOKUP** function moves downward through this column until it finds a cell that is bigger than the random number. It then goes to the previous row and gets the value from column B of the table.

LO F.4 *Use* Excel spreadsheets to create a simulation

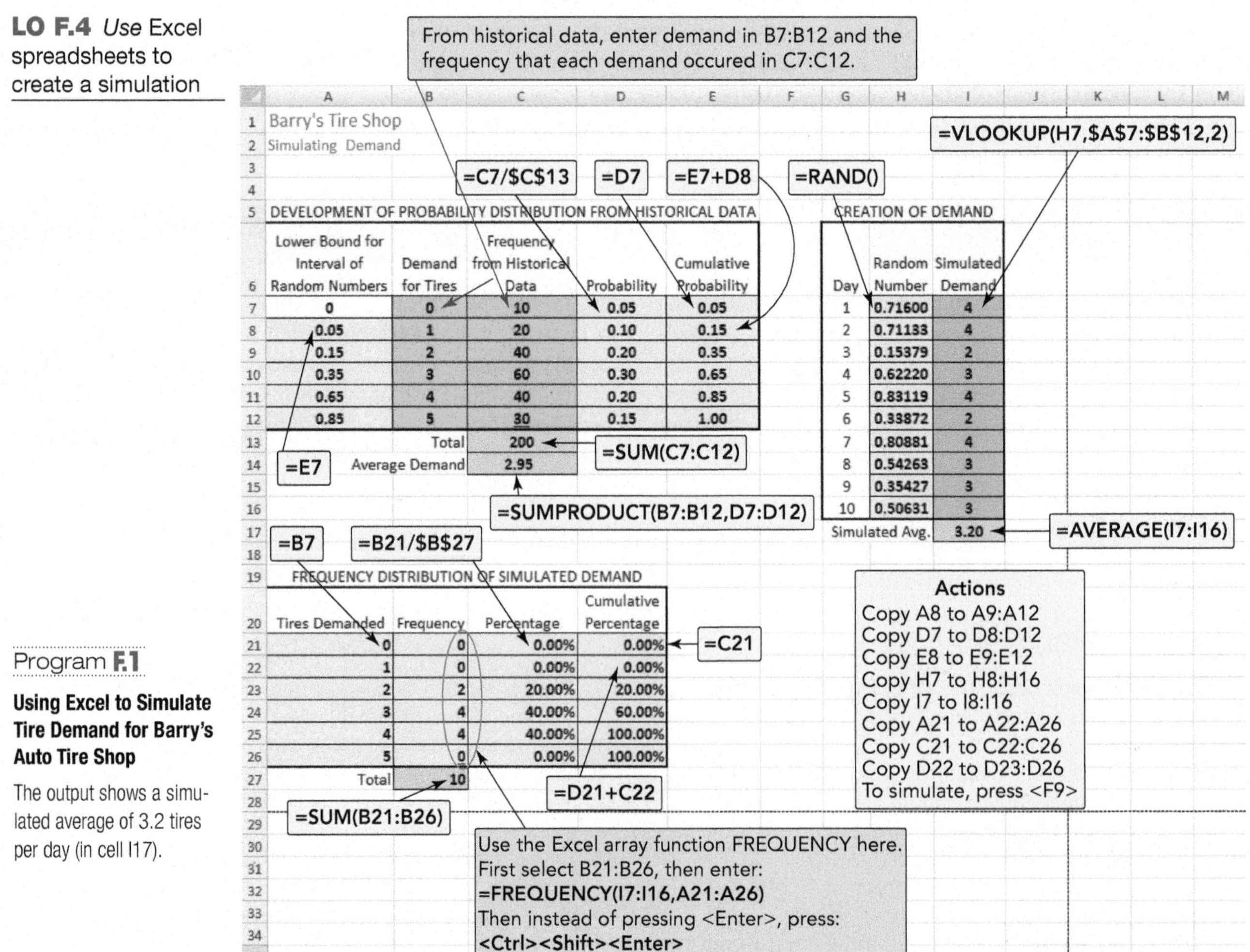

Program F.1

Using Excel to Simulate Tire Demand for Barry's Auto Tire Shop

The output shows a simulated average of 3.2 tires per day (in cell I17).

In column H, for example, the first random number shown is .716. Excel looked down the left-hand column of the lookup table ($A$7:$B$12) of Program F.1 until it found .85. From the previous row it retrieved the value in column B which is 4. Pressing the F9 function key recalculates the random numbers and the simulation.

PX USING POM FOR WINDOWS AND EXCEL OM

POM for Windows and Excel OM are capable of handling any simulation that contains only one random variable, such as Example F1. For further details, please refer to Appendix II.

Solved Problems

SOLVED PROBLEM F.1

Higgins Plumbing and Heating maintains a stock of 30-gallon water heaters that it sells to homeowners and installs for them. Owner Jim Higgins likes the idea of having a large supply on hand to meet any customer demand. However, he also recognizes that it is expensive to do so. He examines water heater sales over the past 50 weeks and notes the following:

WATER HEATER SALES PER WEEK	NUMBER OF WEEKS THIS NUMBER WAS SOLD
4	6
5	5
6	9
7	12
8	8
9	7
10	3
	50 weeks total data

a) If Higgins maintains a constant supply of 8 water heaters in any given week, how many times will he stockout during a 20-week simulation? We use random numbers from the 7th column of Table F.4, beginning with the random digit 10.

b) What is the average number of sales per week over the 20-week period?

c) Using an analytic nonsimulation technique, determine the expected number of sales per week. How does this compare with the answer in part (b)?

SOLUTION

HEATER SALES	PROBABILITY	CUMULATIVE PROBABILITY	RANDOM-NUMBER INTERVALS
4	.12	.12	01 through 12
5	.10	.22	13 through 22
6	.18	.40	23 through 40
7	.24	.64	41 through 64
8	.16	.80	65 through 80
9	.14	.94	81 through 94
10	.06	1.00	95 through 00
	1.00		

a)

WEEK	RANDOM NUMBER	SIMULATED SALES	WEEK	RANDOM NUMBER	SIMULATED SALES
1	10	4	11	08	4
2	24	6	12	48	7
3	03	4	13	66	8
4	32	6	14	97	10
5	23	6	15	03	4
6	59	7	16	96	10
7	95	10	17	46	7
8	34	6	18	74	8
9	34	6	19	77	8
10	51	7	20	44	7

With a supply of 8 heaters, Higgins will stock out three times during the 20-week period (in weeks 7, 14, and 16).

b) Average sales by simulation = total sales/20 weeks = 135/20 = 6.75 per week

c) Using expected values, we obtain:

E (sales) = .12(4 heaters) + .10(5) + .18(6) + .24(7) + .16(8) + .14(9) + .06(10) = 6.88 heaters

With a longer simulation, these two approaches will lead to even closer values.

SOLVED PROBLEM F.2

Random numbers may be used to simulate continuous distributions. As a simple example, assume that fixed cost equals \$300, profit contribution equals \$10 per item sold, and you expect an equally likely chance of 0 to 99 units to be sold. That is, profit equals $-\$300 + \$10X$, where X is the number sold. The mean amount you expect to sell is 49.5 units.

a) Calculate the expected value.
b) Simulate the sale of 5 items, using the following double-digit randomly-selected numbers of items sold:
   37 77 13 10 85
c) Calculate the expected value of (b) and compare with the results of (a).

SOLUTION

a) Expected value $= -300 + 10(49.5) = \$195$

b) $-300 + \$10(37) = \$70$
   $-300 + \$10(77) = \$470$
   $-300 + \$10(13) = -\$170$
   $-300 + \$10(10) = -\$200$
   $-300 + \$10(85) = \$550$

c) The mean of these simulated sales is \$144. If the sample size were larger, we would expect the two values to be closer.

Problems

Note: **PX** means the problem may be solved with POM for Windows and/or Excel OM or Excel.

The problems that follow involve simulations that can be done by hand. However, to obtain accurate and meaningful results, long periods must be simulated. This task is usually handled by a computer. If you are able to program some of the problems in Excel or a computer language with which you are familiar, we suggest you try to do so. If not, the hand simulations will still help you understand the simulation process.

Problems F.1–F.15 relate to Monte Carlo Simulation

• **F.1** The daily demand for tuna sandwiches at an Ohio University cafeteria vending machine is 8, 9, 10, or 11, with probabilities 0.4, 0.3, 0.2, or 0.1, respectively. Assume the following random numbers have been generated: 09, 55, 73, 67, 53, 59, 04, 23, 88, and 84. Using these numbers, generate daily sandwich sales for 10 days. **PX**

• **F.2** The number of machine breakdowns per day at Yuwen Chen's factory is 0, 1, or 2, with probabilities 0.5, 0.3, or 0.2, respectively. The following random numbers have been generated: 13, 14, 02, 18, 31, 19, 32, 85, 31, and 94. Use these numbers to generate the number of breakdowns for 10 consecutive days. What proportion of these days had at least one breakdown? **PX**

• **F.3** The following table shows the partial results of a Monte Carlo simulation. Assume that the simulation began at 8:00 A.M., and there is only one server.

CUSTOMER NUMBER	ARRIVAL TIME	SERVICE TIME
1	8:01	6
2	8:06	7
3	8:09	8
4	8:15	6
5	8:20	6

a) When does service begin for customer number 3?
b) When will customer number 5 leave?
c) What is the average waiting time in line?
d) What is the average time in the system?

• **F.4** Barbara Flynn sells *The Financial Daily* papers at a newspaper stand near Wall Street for \$3.50. The papers cost her \$2.50, giving her a \$1.00 profit on each one she sells. From past experience Barbara knows that:

a) 20% of the time she sells 100 papers.
b) 20% of the time she sells 150 papers.
c) 30% of the time she sells 200 papers.
d) 30% of the time she sells 250 papers.

Assuming that Barbara believes the cost of a lost sale to be \$.50 and any unsold papers cost her \$2.50, simulate her profit outlook over 5 days if she orders 200 papers for each of the 5 days. Use the following random numbers: 52, 06, 50, 88, and 53. **PX**

•• **F.5** Arnold Palmer Hospital is studying the number of emergency surgery kits that it uses on weekends. Over the past 40 weekends, the number of kits used was as follows:

NUMBER OF KITS	FREQUENCY
4	4
5	6
6	10
7	12
8	8

The following random numbers have been generated: 11, 52, 59, 22, 03, 03, 50, 86, 85, 15, 32, 47. Simulate 12 weekends of emergency kit usage. What is the average number of kits used during these 12 weekends? **PX**

•• **F.6** Susan Sherer's grocery store has noted the following figures with regard to the number of people who arrive at the store's three checkout stands and the time it takes to check them out:

ARRIVALS/MINUTE	FREQUENCY
0	.3
1	.5
2	.2

SERVICE TIME (MINUTES)	FREQUENCY
1	.1
2	.3
3	.4
4	.2

Simulate the utilization of the three checkout stands over 5 minutes. For arrivals, use the random numbers 07, 60, 49, 95, and 14. For service times, use the random numbers 77, 76, 51, 16, 05, 99, 95, 67, 34, and 20 (not all may be needed). Record the results at the end of 5 minutes. Start at time = 0. **PX**

• **F.7** A warehouse manager at Gihan Edirisinghe Corp. needs to simulate the demand placed on a product that does not fit standard models. The concept being measured is "demand during lead time," where both lead time and daily demand are variable. The historical record for this product, along with the cumulative distribution, appear in the table. Random numbers have been generated to simulate the next 5 order cycles; they are 91, 45, 37, 65, and 51. What are the five demand values? What is their average?

DEMAND DURING LEAD TIME	PROBABILITY	CUMULATIVE PROBABILITY
100	.01	.01
120	.15	.16
140	.30	.46
160	.15	.61
180	.04	.65
200	.10	.75
220	.25	1.00

PX

• **F.8** Phantom Controls monitors and repairs control circuit boxes on elevators installed in multistory buildings in downtown Chicago. The company has the contract for 108 buildings. When a box malfunctions, Phantom installs a new one and rebuilds the failed unit in its repair facility in Gary, Indiana. The data for failed boxes over the last 2 years is shown in the following table:

NUMBER OF FAILED BOXES PER MONTH	PROBABILITY
0	.10
1	.14
2	.26
3	.20
4	.18
5	.12

Simulate 2 years (24 months) of operation for Phantom and determine the average number of failed boxes per month from the simulation. Was it common to have fewer than 7 failures over 3 months of operation? (Start your simulation at the top of the 10th column of Table F.4, $RN = 37$, and go down in the table.) **PX**

• **F.9** The number of cars arriving at Patti Miles's Car Wash, in Orono, Maine, during the last 200 hours of operation is observed to be the following:

NUMBER OF CARS ARRIVING	FREQUENCY
3 or fewer	0
4	20
5	30
6	50
7	60
8	40
9 or more	0
	200

a) Set up a probability and cumulative-probability distribution for the variable of car arrivals.
b) Establish random-number intervals for the variable.
c) Simulate 15 hours of car arrivals and compute the average number of arrivals per hour. Select the random numbers needed from column 1, Table F.4, beginning with the digits 52. **PX**

•• **F.10** Leonard Presby's dog racing track uses naive forecasting to order tomorrow's programs. The number of programs ordered corresponds to the previous day's demand. Today's demand for programs was 22. Presby buys the programs for \$.20 and sells them for \$.50. Whenever there is unsatisfied demand, Presby estimates the lost goodwill cost at \$.10. Complete the accompanying table and answer the questions that follow.

DEMAND	PROBABILITY
21	.25
22	.15
23	.10
24	.20
25	.30

DAY	AMOUNT ORDERED	RANDOM NUMBER	DEMAND	REVENUE	COST	GOOD-WILL COST	NET PROFIT
1	22	37					
2		19					
3		52					
4		8					
5		22					
6		61					

a) What is the demand on day 3?
b) What is the total net profit at the end of the 6 days?
c) What is the lost goodwill on day 6?
d) What is the net profit on day 2?
e) How many programs has Presby ordered for day 5? **PX**

•• **F.11** Every home football game for the past 8 years at Southwestern University has been sold out. The revenues from ticket sales are significant, but the sale of food, beverages, and souvenirs has contributed greatly to the overall profitability of the football program. One particular souvenir is the football program for each game. The number of programs sold at each game is described by the probability distribution given in the following table.

NUMBERS OF PROGRAMS SOLD	PROBABILITY
2,300	0.15
2,400	0.22
2,500	0.24
2,600	0.21
2,700	0.18

Each program costs \$.80 to produce and sells for \$2.00. Any programs that are not sold are donated to a recycling center and do not produce any revenue.

a) Simulate the sales of programs at 10 football games. Use the last column in the random-number table (Table F.4) and begin at the top of the column.
b) If the university decided to print 2,500 programs for each game, what would the average profits be for the 10 games that were simulated?

c) If the university decided to print 2,600 programs for each game, what would the average profits be for the 10 games that were simulated? **PX**

• **F.12** Refer to the data in Solved Problem F.1, which deals with Higgins Plumbing and Heating. Higgins has now collected 100 weeks of data and finds the following distribution for sales:

WATER HEATER SALES PER WEEK	NUMBER OF WEEKS THIS NUMBER WAS SOLD	WATER HEATER SALES PER WEEK	NUMBER OF WEEKS THIS NUMBER WAS SOLD
3	2	8	12
4	9	9	12
5	10	10	10
6	15	11	5
7	25		100

a) Assuming that Higgins maintains a constant supply of 8 heaters, simulate the number of stockouts incurred over a 20-week period (using the seventh column of Table F.4).
b) Conduct this 20-week simulation two more times and compare your answers with those in (a). Did they change significantly? Why or why not?
c) What is the new expected number of sales per week? **PX**

• **F.13** Adventure Rafting runs rafts on the Colorado River. It has eight rafts in its inventory. The demand for rafts during the busy months of June and July has been either 4, 5, 6, 7 or 8, with probabilities of 0.1, 0.3, 0.3, 0.2, or 0.1, respectively. Use Table F.4 to simulate the number of rafts the company will need for 10 consecutive days. Start at the top of column number 4 (random number = 88) and move down in the table (second number = 02) to locate the remaining numbers. **PX**

• **F.14** The number of cars arriving at Margarita Cardozo's self-service gasoline station during the last 50 hours of operation are as follows:

NUMBER OF CARS ARRIVING	6	7	8	9
FREQUENCY	10	12	20	8

The following random numbers have been generated: 44, 30, 26, 09, 49, 13, 33, 89, 13, and 37. Simulate 10 hours of arrivals. What is the average number of arrivals during this period? **PX**

• **F.15** Woodworth Property Management is responsible for the maintenance, rental, and day-to-day operation of a large apartment complex in El Paso. Bruce Woodworth is especially concerned about the cost projections for replacing air conditioner compressors. He would like to simulate the number of compressor failures each year over the next 20 years. Using data from a similar apartment building that he also manages, Woodworth establishes the following table of relative frequency of failures during a year:

NUMBER OF COMPRESSOR FAILURES	0	1	2	3	4	5	6
PROBABILITY (RELATIVE FREQUENCY)	0.06	0.13	0.25	0.28	0.20	0.07	0.01

He decides to simulate the 20-year period by selecting 2-digit random numbers from column 3 of Table F.4 (starting with the random number 50). Conduct the simulation for Woodworth. Is it common to have 3 or more consecutive years of operation with 2 or fewer compressor failures per year? **PX**

Problems F.16–F.25 relate to Simulation with Two Decision Variables

•• **F.16** The time between arrivals at the drive-through window of Brittany Beard's fast-food restaurant follows the distribution given in the table. The service-time distribution is also given. Use the random numbers provided to simulate the activity of the first 4 arrivals. Assume that the window opens at 11:00 A.M. and that the first arrival occurs afterward, based on the first interarrival time generated.

TIME BETWEEN ARRIVALS	PROBABILITY	SERVICE TIME	PROBABILITY
1	.2	1	.3
2	.3	2	.5
3	.3	3	.2
4	.2		

Random numbers for arrivals: 14, 74, 27, 03
Random numbers for service times: 88, 32, 36, 24
At what time does the fourth customer leave the system? **PX**

••• **F.17** Central Hospital in York, Pennsylvania, has an emergency room that is divided into six departments: (1) an initial exam station to treat minor problems or to make a diagnosis; (2) an X-ray department; (3) an operating room; (4) a cast-fitting room; (5) an observation room (for recovery and general observation before final diagnosis or release); and (6) an outprocessing department (where clerks check out patients and arrange for payment or insurance forms).

The probabilities that a patient will go from one department to another are presented in the following table:

FROM	TO	PROBABILITY
Initial exam at emergency room entrance	X-ray department	.45
	Operating room	.15
	Observation room	.10
	Outprocessing clerk	.30
X-ray department	Operating room	.10
	Cast-fitting room	.25
	Observation room	.35
	Outprocessing clerk	.30
Operating room	Cast-fitting room	.25
	Observation room	.70
	Outprocessing clerk	.05
Cast-fitting room	Observation room	.55
	X-ray department	.05
	Outprocessing clerk	.40
Observation room	Operating room	.15
	X-ray department	.15
	Outprocessing clerk	.70

a) Simulate the trail followed by 10 emergency room patients. Proceed, one patient at a time, from each one's entry at the initial exam station until he or she leaves through outprocessing. You should be aware that a patient can enter the same department more than once.
b) Using your simulation data, determine the chances that a patient enters the X-ray department twice.

Additional problem **F.18**, *a tanning bed company evaluating expansion, is available in* **MyLab Operations Management.**

••• **F.19** Kathryn Marley owns and operates the largest Mercedes-Benz auto dealership in Pittsburgh. In the past 36 months, her sales have ranged from a low of 6 new cars to a high of 12 new cars, as reflected in the following table:

SALES OF NEW CARS/MONTH	FREQUENCY
6	3
7	4
8	6
9	12
10	9
11	1
12	1
	36 months

Marley believes that sales will continue during the next 24 months at about the same historical rates, and that delivery times will also continue to follow the following pace (stated in probability form):

DELIVERY TIME (MONTHS)*	PROBABILITY
1	.44
2	.33
3	.16
4	.07
	1.00

*With cars arriving at the end of the month (i.e., orders placed at the end of the 1st month, with a lead time of 2 months, will arrive at the end of the 3rd month, too late for sales in the 3rd month).

Marley's current policy is to order 14 cars at a time (two full truckloads, with 7 autos on each truck), and to place a new order whenever the stock on hand reaches 12 autos. Use random numbers starting from the top of the right-hand column of Table F.4.

a) What are the results of this policy when simulated over the next 2 years?
b) Marley establishes the following relevant costs: (1) carrying cost per Mercedes per month is $600; (2) cost of a lost sale averages $4,350; and (3) cost of placing an order is $570. What is the total inventory cost of this policy?

•• **F.20** Dumoor Appliance Center sells and services several brands of major appliances. Past sales for a particular model of refrigerator have resulted in the following probability distribution for demand:

DEMAND PER WEEK	0	1	2	3	4
PROBABILITY	0.20	0.40	0.20	0.15	0.05

The lead-time in weeks is described by the following distribution:

LEAD TIME (WEEKS)	1	2	3
PROBABILITY	0.15	0.35	0.50

Based on cost considerations as well as storage space, the company has decided to order 10 of these each time an order is placed. The holding cost is $1 per week for each unit that is left in inventory at the end of the week. The stockout cost has been set at $40 per stockout. The company has decided to place an order whenever there are only two refrigerators left at the end of the week. Simulate 10 weeks of operation for Dumoor Appliance, assuming that there are currently 5 units in inventory. Determine what the weekly stockout cost and weekly holding cost would be for the problem. Use the random numbers in the first column of Table F.4 for demand and the second column for lead time.

•• **F.21** Repeat the simulation in Problem F.20, assuming that the reorder point is 4 units rather than 2. Compare the costs for these two situations. Use the same random numbers as in Problem F.20.

Additional problems **F.22–F.25** *are available in* **MyLab Operations Management**. *F.22 is an inventory simulation with demands and lead times as variables. F.23 resimulates Example F.2 in this module. F.24 simulates a student's income and expenses. F.25 evaluates a commercial graphics plotter breakdown.*

CASE STUDY

Alabama Airlines' Call Center

Alabama Airlines opened its doors in 2020 as a commuter service with its headquarters and hub located in Birmingham. The airline was started and managed by two former pilots, Sheila Hawkins and Leilani Parker. It acquired a fleet of 12 used prop-jet planes and the airport gates vacated by Delta Air Lines in 2020.

TABLE F.8 Incoming Call Distribution

TIME BETWEEN CALLS (MINUTES)	PROBABILITY
1	.11
2	.21
3	.22
4	.20
5	.16
6	.10

With business growing quickly, Sheila turned her attention to Alabama Air's "800" reservations system. Between midnight and 6:00 A.M., only one telephone reservations agent had been on duty. The time between incoming calls during this period is distributed as shown in Table F.8. Carefully observing and timing the agent, Sheila estimated that the time required to process passenger inquiries is distributed as shown in Table F.9.

All customers calling Alabama Air go "on hold" and are served in the order of the calls received unless the reservations agent is

TABLE F.9 Service-Time Distribution

TIME TO PROCESS CUSTOMER INQUIRIES (MINUTES)	PROBABILITY
1	.20
2	.19
3	.18
4	.17
5	.13
6	.10
7	.03

available for immediate service. Sheila is deciding whether a second agent should be on duty to cope with customer demand. To maintain customer satisfaction, Alabama Air wants a customer to be "on hold" for no more than 3 to 4 minutes; it also wants to maintain a "high" operator utilization.

Furthermore, the airline is planning a new TV advertising campaign. As a result, it expects an increase in "800" line phone inquiries. Based on similar campaigns in the past, the incoming call distribution from midnight to 6:00 A.M. is expected to be as shown in Table F.10. (The same service-time distribution will apply.)

TABLE F.10 Incoming Call Distribution

TIME BETWEEN CALLS (MINUTES)	PROBABILITY
1	.22
2	.25
3	.19
4	.15
5	.12
6	.07

Discussion Questions

1. Given the original call distribution, what would you advise Alabama Air to do for the current reservation system? Create a simulation model to investigate the scenario. Describe the model carefully, and justify the duration of the simulation, assumptions, and measures of performance.
2. What are your recommendations regarding operator utilization and customer satisfaction if the airline proceeds with the advertising campaign?

Source: Professor Zbigniew H. Przasnyski, Loyola Marymount University. Reprinted by permission.

Endnotes

1. Alternatively, we could have assigned the random numbers 00, 01, 02, 03, and 04 to represent a demand of 0 units. The 2 digits 00 can be thought of as either 0 or 100. As long as 5 numbers out of 100 are assigned to the 0 demand, it does not make any difference which 5 they are.
2. Sources: **news.starbucks.com/news**; **cnn.com/business** (January 27, 2017); and **slideshare.net** (April 25, 2015).

Bibliography

Balakrishnan, R., B. Render, R. M. Stair, and C. Munson. *Managerial Decision Modeling: Business Analytics with Spreadsheets*, 4th ed. Boston: DeGruyter, 2017.

Banks, J., and R. R. Gibson. "The ABC's of Simulation Practice." *Analytics* (Spring 2009): 16–23.

Banks, J., J. S. Carson, B. L. Nelson, and D. M. Nicol. *Discrete-Event System Simulation*, 5th ed. Upper Saddle River, NJ: Prentice Hall, 2010.

Beaverstock, M., E. Lavory, and A. Greenwood. *Applied Simulation*, 4th ed. Orem, Utah: Flexsim Software, 2015.

Huang, H. C., et al. "Sim Man—A Simulation Model for Workforce Capacity Planning." *Computers & Operations Research* 196, no. 3 (August 1, 2009): 1147.

Kelton, W. D., R. P. Sadowski, and N. Zupick. *Simulation with Arena*, 6th ed. New York: McGraw-Hill, 2014.

Render, B., R. M. Stair, M. Hanna, and T. Hale. *Quantitative Analysis for Management*, 13th ed. Boston: Pearson, 2018.

Rossetti, M. D. *Simulation Modeling and ARENA*, 2nd ed. New York: Wiley, 2015.

Taylor, S. J. E., et al. "Simulation Modelling Is 50! Do We Need a Reality Check?" *Journal of the Operational Research Society* 60, no. S1 (May 2009): S69–S82.

Zhang, A., and X. Xie. "Simulation-Based Optimization for Surgery Appointment Scheduling of Multiple Operating Rooms." *IIE Transactions* 47, no. 9 (September 2015): 998–1012.

Module F *Rapid* Review

Main Heading	Review Material	MyLab Operations Management
WHAT IS SIMULATION?	Most of the large companies in the world use simulation models. ■ **Simulation**—The attempt to duplicate the features, appearance, and characteristics of a real system, usually via a computerized model. The idea behind simulation is threefold: 1. To imitate a real-world situation mathematically 2. Then to study its properties and operating characteristics 3. Finally, to draw conclusions and make action decisions based on the results of the simulation In this way, a real-life system need not be touched until the advantages and disadvantages of a major policy decision are first measured on the model. To use simulation, an OM manager should: 1. Define the problem. 2. Introduce the important variables associated with the problem. 3. Construct a numerical model. 4. Set up possible courses of action for testing by specifying values of variables. 5. Run the experiment. 6. Consider the results (possibly modifying the model or changing data inputs). 7. Decide what course of action to take.	Concept Questions: 1.1–1.5

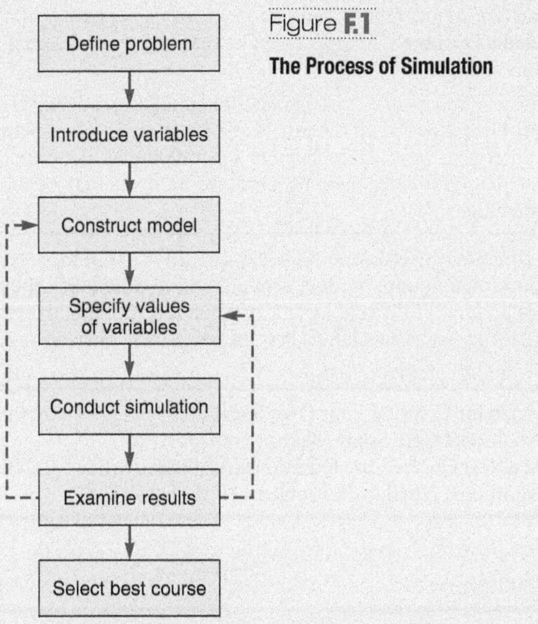

Figure F.1
The Process of Simulation

Main Heading	Review Material	MyLab Operations Management
ADVANTAGES AND DISADVANTAGES OF SIMULATION	The main *advantages* of simulation are: 1. It can be used to analyze large and complex real-world situations that cannot be solved using conventional operations management models. 2. Real-world complications can be included that most OM models cannot permit. For example, simulation can use any probability distribution the user defines; it does not require standard distributions. 3. "Time compression" is possible. The effects of OM policies over many months or years can be obtained by computer simulation in a short time. 4. Simulation allows "what-if?" types of questions. Managers like to know in advance what options will be most attractive. With a computerized model, a manager can try out several policy decisions within a matter of minutes. 5. Simulations do not interfere with real-world systems. It may be too disruptive, for example, to experiment physically with new policies or ideas. The main *disadvantages* of simulation are: 1. Good simulation models can be very expensive; they may take many months to develop. 2. It is a repetitive approach that may produce different solutions in repeated runs. It does not generate optimal solutions to problems.	Concept Questions: 2.1–2.5

Main Heading	Review Material	MyLab Operations Management
	3. Managers must generate all of the conditions and constraints for solutions that they want to examine. The simulation model does not produce answers without adequate, realistic input. 4. Each simulation model is unique. Its solutions and inferences are not usually transferable to other problems.	
MONTE CARLO SIMULATION	■ **Monte Carlo method**—A simulation technique that selects random numbers assigned to a distribution. The Monte Carlo method breaks down into five simple steps: 1. Setting up a probability distribution for important variables 2. Building a cumulative probability distribution for each variable 3. Establishing an interval of random numbers for each variable 4. Generating random numbers 5. Actually simulating a series of trials One common way to establish a *probability distribution* for a given variable is to examine historical outcomes. We can find the probability, or relative frequency, for each possible outcome of a variable by dividing the frequency of observation by the total number of observations. ■ **Cumulative probability distribution**—The accumulation (summary) of probabilities of a distribution. ■ **Random-number intervals**—A set of numbers to represent each possible value or outcome in a computer simulation. ■ **Random number**—A series of digits that have been selected using a totally random process. Random numbers may be generated for simulation problems in two ways: (1) If the problem is large and the process under study involves many simulation trials, computer programs are available to generate the needed random numbers; or (2) if the simulation is being done by hand, the numbers may be selected from a table of random digits.	Concept Questions: 3.1–3.6 Problems: F.1–F.15 Virtual Office Hours for Solved Problems: F.1, F.2
SIMULATION WITH TWO DECISION VARIABLES	The commonly used economic order quantity (EOQ) models are based on the assumption that both product demand and reorder lead time are known, constant values. In most real-world inventory situations, though, demand and lead time are variables, so accurate analysis becomes extremely difficult to handle by any means other than simulation.	Concept Questions: 4.1–4.4 Problems: F.16–F.25
ADDITIONAL MYLAB OPERATIONS MANAGEMENT RESOURCES	✔ Video for Creating Your Own Excel Spreadsheets (Example F1) ✔ Additional Case Study (Saigon Transport) ✔ Multiple Choice Case Questions (Alabama Airlines' Call Center) ✔ Additional Homework Problems (F.18, F.22–F.25)	

Self **Test**

LO F.1 Which of the following is *not* an advantage of simulation?
- **a)** Simulation is relatively straightforward and flexible.
- **b)** Good simulation models are usually inexpensive to develop.
- **c)** "Time compression" is possible.
- **d)** Simulation can study the interactive effects of individual variables.
- **e)** Simulations do not interfere with real-world systems.

LO F.2 The five steps required to implement the Monte Carlo simulation technique are _____, _____, _____, _____, and _____.

LO F.3 Two particularly good candidates to be probabilistic components in the simulation of an inventory problem are:
- **a)** order quantity and reorder point.
- **b)** setup cost and holding cost.
- **c)** daily demand and reorder lead time.
- **d)** order quantity and reorder lead time.
- **e)** reorder point and reorder lead time.

LO F.4 One important reason that spreadsheets are excellent tools for conducting simulations is that they can:
- **a)** generate random numbers.
- **b)** easily provide animation of the simulation.
- **c)** provide more security than manual simulations.
- **d)** prohibit "time compression" from corrupting the results.
- **e)** be easily programmed.

Answers: LO F.1. b; LO F.2. set up a probability distribution for each of the important variables, build a cumulative probability distribution for each of the important variables, establish an interval of random numbers for each variable, generate sets of random numbers, actually simulate a set of trials; LO F.3. c; LO F.4. a.

Applying Analytics to Big Data in Operations Management

MODULE G

MODULE OUTLINE

LEARNING OBJECTIVES

LO G.1 ***Describe*** the three categories of business analytics *824*

LO G.2 ***Apply*** Excel's Conditional Formatting tool to clean a dataset *827*

LO G.3 ***Create*** a heat map using Excel's Conditional Formatting tool *830*

LO G.4 ***Describe*** advanced business analytics tools *832*

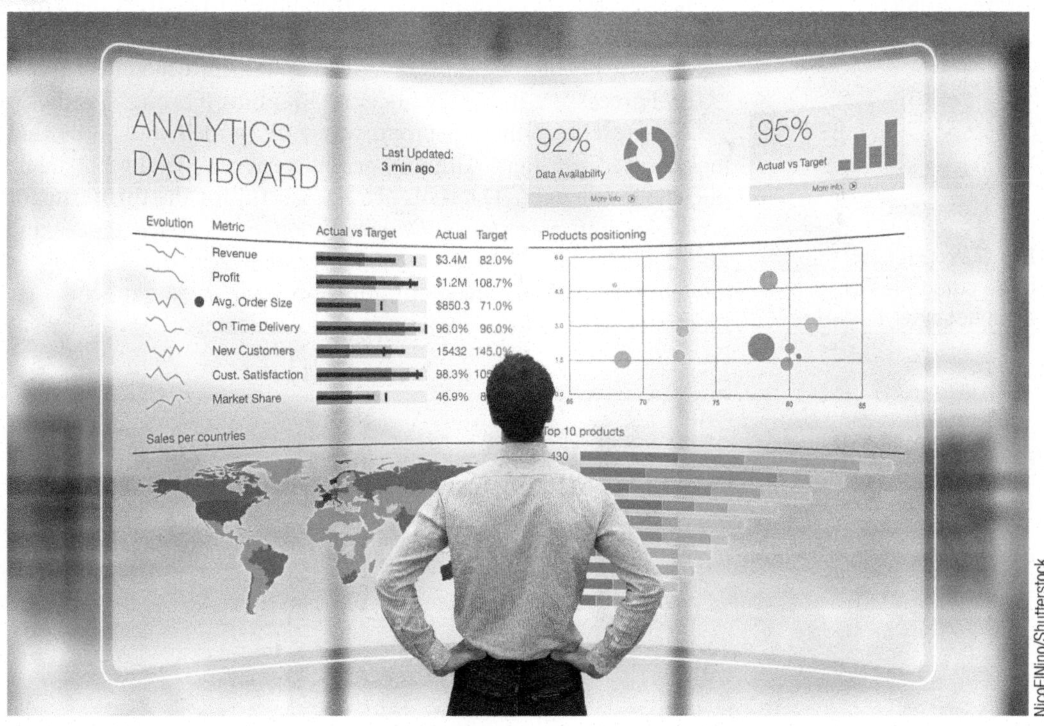

Organizations combine big data with visualization to generate analytics dashboards that present current levels of key success indicators.

NicoElNino/Shutterstock

Introduction to Big Data and Business Analytics

Today's digital world allows us to store huge amounts of data in digital form, generated from such locations as:

- web sites (number of hits, time on the site, number of keystrokes)
- credit cards (number of transactions, date, amount, vendor)
- point-of-sale records (credit card number, date, place of transaction)
- social media (location, source recipient, trending topics, content analysis).

Digital data are also collected in operations environments, where they may be organized by part or serial number, workstation, quality tests, maintenance, logistical moves, assembly line, or even time.

Data collection is so massive that it is now measured in terabytes (one terabyte equals 2^{40} or approximately one trillion bytes). Walmart, for example, handles more than 1 million customer transactions every hour, which are imported into databases that contain more than 2.5 petabytes (2560 terabytes) of data. (By comparison, the entire U.S. Library of Congress book collection contains only 235 terabytes of data.) Similarly, Amazon data are collected for over 300 million customers and 500 million products. **Big data** is the term used to describe this huge amount of data, which can be difficult to process efficiently by traditional data techniques. Although big data is a potential source of valuable information, it requires sophistication in how it is stored, processed, and analyzed. Managers, when considering how to evaluate and use big data, may find it helpful to think of data in terms of five Vs: volume (How much?), variety (How many kinds?), veracity (How accurate?), velocity (How fast is it arriving?), and value (What is its worth?).

Big data
The huge amount of production, consumer, and social media data collected in digital form.

Business analytics uses tools and techniques to convert data into summary information and business insights. The management task is all about gathering the *right* data in a usable form and then *applying* the best tools to make well-informed decisions. (See the *OM in Action* box, "UPS Forecasting Improves Logistics Planning Through Predictive Analysis.")

Business analytics
Uses tools and techniques to convert data into summary information and business insights for decision making.

This text introduces a broad range of models and tools that help operations managers make better decisions. Although, of necessity, we deal with smaller datasets, the tools are applicable to big data. The tools are also useful for the three categories of analytics: *descriptive, predictive*, and *prescriptive. Descriptive analytics* characterizes and summarizes data to facilitate understanding. *Predictive analytics* analyzes past data to predict the future. *Prescriptive analytics* invokes advanced optimization tools to recommend a strategy or action. Figure G.1 illustrates the relationships among the three categories.

LO G.1 *Describe* the three categories of business analytics

Figure **G.1**
Three Categories of Analytics

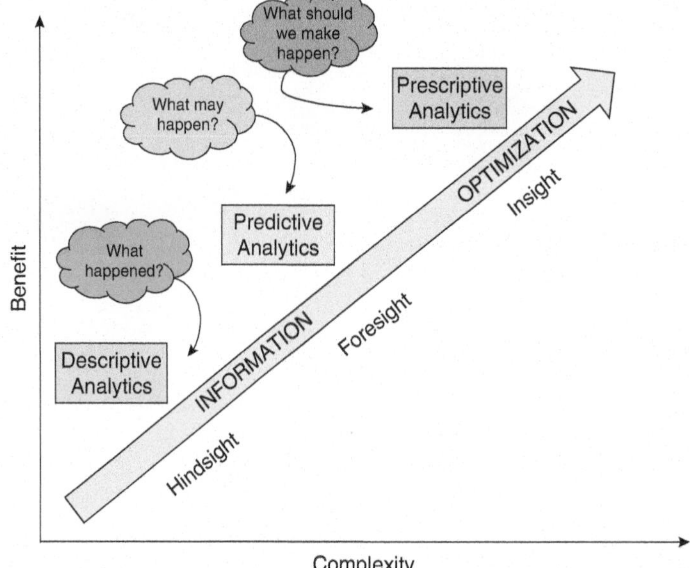

OM in Action UPS Forecasting Improves Logistics Planning Through Predictive Analysis[1]

United Parcel Service (UPS) operates in more than 220 countries and territories with 1,800 facilities and a delivery fleet of over 119,000 vehicles and 500 planes. So you can imagine the company's challenge in building an analytics system to gather logistics data to better predict package flow, volume, and delivery status. The predictive analytics software analyzes more than 1,200 events per second–1 billion data points per day! This includes data about package weight, shape, and size, as well as forecast, capacity, and customer data–and allows UPS to know exactly what's going where, and when it's going to arrive.

The project is an example of how UPS is upgrading analytics as it faces heavy competition from rivals amidst ever-growing e-commerce demands. The company is opening new automated facilities and working on technology upgrades as part of a $20 billion capital spending plan.

UPS now provides its staff with more accurate forecasts about volume and process needs at UPS facilities on any given day. That information gives employees the lead time necessary to match resources at package and sorting facilities to volume. Predictive analytics should also help eliminate supply chain bottlenecks because of unforeseen weather or emergency situations. Knowing how upcoming inclement weather will impact the supply chain days in advance will support better planning.

dpa picture alliance archive/Alamy Stock Photo

Developing the tool was an ambitious feat because of the massive number of data points needed to be consolidated into one single platform. Until now, forecast, capacity, customer, and package data were housed in different software applications. The tool is available to UPS employees via a smartphone, desktop, and tablet application.

In this module, we introduce several topics related to all three analytics categories. First, we discuss the importance of working with good data (i.e., *data management*). Then we introduce the power of *pivot tables* to drill down into a dataset, followed by the insight provided by effectively visualizing data. We illustrate these methods using Excel. (Extremely large datasets may necessitate the use of specialized software.) Finally, we identify several additional tools that are often considered part of business analytics but are not covered in this text.

Data Management

First, we must get the data right. Incorrect or inconsistent data lead to false conclusions and bad decisions. Organizations often want to analyze data from many diverse sources, internal and external to the firm. The collection of vast amounts of data can lead to less than perfect data. Consequently, managers need to *clean* data to avoid erroneous decisions.

The immediate challenge is to help the decision maker close the gap between raw data and good data. This requires effective **data management**. Data must be *organized*, that is structured, in such a way as to ensure data integrity. Data integrity requires that data be: *complete, consistent, and accurate*.

Data management
Overall management of data's integrity, including completeness, consistency, and accuracy.

1. *Completeness:* the degree to which all required data are present. Incomplete records suggest investigation. With huge databases, omission of a few records may be insignificant, but this may not be the case with small datasets and sample data. We want to first examine the data for missing values. With small datasets, visual inspection of each cell in each file in the record may be adequate, but large datasets suggest a software solution. Once incomplete records have been identified, judgments must be made about either investigation to seek correction for the missing data or deletion of the entire record. Once we conclude that our data are complete, we look for consistency.
2. *Consistency:* the degree to which data are equivalent across systems. Inconsistency occurs when two data items in the dataset contradict each other (e.g., a sale is recorded in two different systems). Data can be stored three ways: character, pattern, or numeric. Character data, such as the variable 'P,' may be used to designate a primary supplier and

	A	B	C	D	E	F	G	H	I	J
1	Thompson Industries Monthly Supplier Report - April									
2										
3			Supplier		Change from				No. of	New
4		Supplier	Category	Total	Previous	Quality	On-time	Percentage	Rejected	Supplier
5	Supplier Name	Number	(P,S)	Purchases	Month	Rating (1-5)	Percentage	Returns	Shipments	(Y, N)
6	Maverick Supplies	6071	P	$39,117	$5,000	5	100%	0%	0	N
7	Cowboy Pete's Lumber	6894	P	$25,554	$400	5	90%	0%	0	N
8	ABC Metals	5054	S	$13,051	$13,051	3	0%	75%	2	Y
9	Perkins Plastics	5528	P	$27,886	$32,564	4	100%	3%	0	N
10	Everything Brass	4413	P	$29,300	($4,658)	5	1000%	0%	0	N
11	Jones Aluminum & Copper	5858	P	$16,862	$16,862	4	50%	5%	1	Y
12	Rick's Wrought Iron	9733	S	$29,078	($3,121)	1	60%	80%	6	N
13	Penelope's Paper Products	4718		$25,716	$8,795	3	100%	12%	0	N
14	Anthony's Adhesives	9944	P	$22,750	$1,268	4	90%		0	N
15	Columbus Manufacturing	1343	S	$20,103	$469	4	100%	0%	0	N
16	Smart Electronics, Inc.	4785	P	$30,062	$30,062	5	33%	50%	0	Y
17	Watson Woodworking	2730		$32,589	$471	5	25%	1%	0	N
18	Fred's Foam Molding	2949	S	$47,886	$47,886	5	100%	0.50%	0	Y
19	Impenetrable Insulation, Inc.	1443	P	$24,971	($6,895)	4	15%	4%	0	N
20	Suzy's Shipping Supplies	8039	P	$23,851	$1,269	5	50%	0%	0	N

Figure **G.2**

Raw Data for Thompson Industries' Suppliers

(Data file can be found online.)

'S' for a secondary supplier. Meanwhile, a "maverick" supplier might be coded as 'MAV' or 'M' or even 'Maverick.' Alternatively, we find that unique patterns (formats) of data (e.g., dates, social security and phone numbers, or company-specific product codes) can also be checked through visual inspection or computerized search techniques to ensure consistency.

3. *Accuracy:* the degree of conformity of a measure to a standard value (e.g., the dimension, strength, or price of a part). Numeric data, such as unusual variations in a part's dimension, may be identified by inspecting for data values outside a predetermined range. We want to identify "outliers" prior to them being accepted into the dataset.

Figure G.2 represents a monthly report of raw data on Thompson Industries' 15 suppliers in an Excel spreadsheet. Like any dataset, the data should be checked and cleaned before analysis begins.

Graphical Techniques for Cleaning Data

Graphs of the raw data can sometimes visually reveal inaccuracies. Figure G.3(a) displays a histogram of the percentage of on-time performance from Figure G.2 (column G). We can quickly see on the right side of the chart that problems may exist, as we have a supplier in the "More" bucket (over 100%) indicating an error. Similarly, on the left side of the chart, the 0% on-time performance should be investigated. While it is possible that the supplier was always late, it is also possible that the supplier's performance was not entered correctly.

Figure G.3(b) displays a scatter plot relating percentage returns to the quality rating of each supplier from Figure G.2 (column F). In a graph such as this, we draw a rough curve through the middle of the points from left-to-right and then look for points that are far away from that curve. Here we add a red curve to indicate the general "average" that should be observed. We now see that two points in particular may need further investigation. The point (1.00%, 0) must be inaccurate because quality ratings can only be between 1 and 5. In addition, the point (50.00%, 5) seems inaccurate. How can the firm rate a supplier as having top quality when half of its shipments are returned? Visual displays of data such as these provide excellent tools to identify outliers and potential inaccuracies.

Excel Techniques for Cleaning Data

Microsoft Excel has several features that can aid in data cleanup. In fact, several different tools can accomplish the same task.

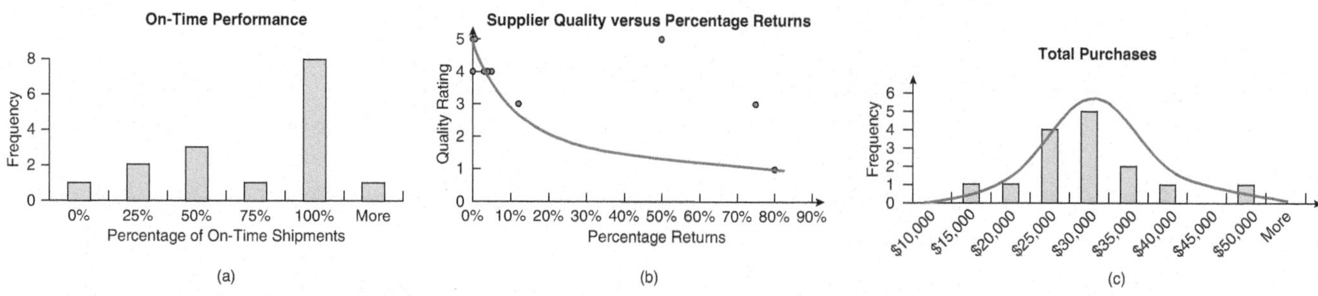

Figure **G.3**

Graphs of Data from Figure G.2 to Check for Inaccuracies

Finding Inconsistent Data Perhaps the simplest way to identify missing records in an Excel database such as that in Figure G.2 is to use a combination of the <End> key and the <Down Arrow> key. If the cell cursor is placed at the top of a column of numbers or characters, and then the <End> key is pressed, followed directly by pressing the <Down Arrow> key, the cell cursor will go all the way to the bottom of that column to the last cell that contains a nonblank entry.

A quick way to see if there are any empty cells in a database is to use the COUNTIF function. For example, for Figure G.2, put this formula into cell H21:

=COUNTIF(H6:H20,"")

The two quotation marks right next to each other indicate a cell is empty. The entire database could be checked by changing the formula to =COUNTIF(A6:J20,""). This would return a value of 3 indicating three empty cells (C13, C17, and H14).

The Excel functions MIN, SMALL, MAX, and LARGE can facilitate quick accuracy checks on the extreme values in a field. More formally, similar to statistical process control techniques (Supplement 6), the analyst can use AVERAGE and STDEV.P to determine the range containing data points that are 3 standard deviations above and below the mean.[2] For example, the limits for the Total Purchases field in Figure G.2 would be computed as:

$$\text{Lower limit} = \text{AVERAGE}(\text{D6:D20})\text{-}3 * \text{STDEV.P}(\text{D6:D20})$$

$$= \$27{,}252 - 3(\$8{,}221.4) = \$2{,}588$$

$$\text{Upper limit} = \text{AVERAGE}(\text{D6:D20}) + 3 * \text{STDEV.P}(\text{D6:D20})$$

$$= \$27{,}252 + 3(\$8{,}221.4) = \$51{,}916$$

Because all purchases in Figure G.2 fall within those limits, they pass this particular accuracy test. These data are plotted in Figure G.3(c).

Conditional Formatting Excel provides a powerful, yet easy-to-use, tool that helps to visually identify characteristics of data via formatting (especially colors or icons). **Conditional Formatting** is a handy technique for emphasizing unusual values in a database. Examples could include the following:

Conditional Formatting
An Excel tool to visually identify characteristics of data using formatting.

LO G.2 *Apply* Excel's Conditional Formatting tool to clean a dataset

- What are the exceptions to profit over time?
- What was the maximum profit? The minimum?
- Who sold more than $50,000 this month?
- Which products had demand decrease by more than 12% this month?
- Which products had the most rejected shipments?

Conditional Formatting is also an excellent tool for cleaning databases because it can be used to check for inconsistencies and inaccuracies. The logic for Conditional Formatting is straightforward. If the condition is met, the formatting is applied; if the condition is not met, the formatting is not applied. Example G1 illustrates a few ways to use Conditional Formatting to identify 'bad' data for cleaning.

Example G1

USE OF CONDITIONAL FORMATTING FOR CLEANING DATA

Vicky Luo, analytics manager for Thompson Industries, wishes to clean the data in Figure G.2 prior to analyzing it. She wants to check the following four items:

1. Any blank entries need to be identified and filled.
2. The percentages in columns G and H must be between 0% and 100%.
3. The value in column E for a supplier cannot exceed the value in column D for that supplier.
4. The value in column F for a supplier should not be 5 if the value in column H exceeds 10%.

	A	B	C	D	E	F	G	H	I	J
1	Thompson Industries Monthly Supplier Report - April									
2										
3			Supplier		Change from				No. of	New
4		Supplier	Category	Total	Previous	Quality	On-time	Percentage	Rejected	Supplier
5	Supplier Name	Number	(P,S)	Purchases	Month	Rating (1-5)	Percentage	Returns	Shipments	(Y, N)
6	Maverick Supplies	6071	P	$39,117	$5,000	5	100%	0%	0	N
7	Cowboy Pete's Lumber	6894	P	$25,554	$400	5	90%	0%	0	N
8	ABC Metals	5054	S	$13,051	$13,051	3	0%	75%	2	Y
9	Perkins Plastics	5528	P	$27,886	$32,564	4	100%	3%	0	N
10	Everything Brass	4413	P	$29,300	($4,658)	5	1000%	0%	0	N
11	Jones Aluminum & Copper	5858	P	$16,862	$16,862	4	50%	5%	1	Y
12	Rick's Wrought Iron	9733	S	$29,078	($3,121)	1	60%	80%	6	N
13	Penelope's Paper Products	4718		$25,716	$8,795	3	100%	12%	0	N
14	Anthony's Adhesives	9944	P	$22,750	$1,268	4	90%		0	N
15	Columbus Manufacturing	1343	S	$20,103	$469	4	100%	0%	0	N
16	Smart Electronics, Inc.	4785	P	$30,062	$30,062	5	33%	50%	0	Y
17	Watson Woodworking	2730		$32,589	$471	5	25%	1%	0	N
18	Fred's Foam Molding	2949	S	$47,886	$47,886	5	100%	0.50%	0	Y
19	Impenetrable Insulation, Inc.	1443	P	$24,971	($6,895)	4	15%	4%	0	N
20	Suzy's Shipping Supplies	8039	P	$23,851	$1,269	5	50%	0%	0	N

Figure **G.4**

Thompson Industries Spreadsheet after Applying Conditional Formatting

APPROACH ▶ Conditional Formatting can be applied for each item as described below.

1. Select A6:J20; Rule type: Format only cells that contain, select Blanks; Color purple.
2. Select G6:H20; Rule type: Format only cells that contain, select Cell value, then not between, then 0 and 1; Color light green.
3. Select E6:E20; Rule type: Formula, enter =E6>D6; Color maroon.
4. Select F6:F20; Rule type: Formula, enter =AND(F6=5,H6>0.1); Color dark green.

After applying these four Conditional Formatting rules, the spreadsheet would look like Figure G.4.

INSIGHT ▶ Creative use of Excel functions allows analysts to apply Conditional Formatting in a wide variety of ways to ensure data that are complete, consistent, and accurate. Perhaps the biggest challenge is trying to identify all the possible ways that the data might not be correct.

LEARNING EXERCISE ▶ Vicky also wants to flag any records whose number of rejected shipments exceeds 5. What Conditional Formatting could she use? [Answer: Select I6:I20; Rule type: Format only cells that contain, select Cell value, then greater than, then 5; Choose a format.]

RELATED PROBLEMS ▶ G.3–G.5, G.7–G.8

Once we are assured that we have clean data, we can begin to use the tools of business analytics.

Pivot table

A tool to facilitate in-depth analysis of numeric data by applying filters and providing summary computations for categories and subcategories of the dataset.

Using Excel's PivotTable Tool

Big data contains a substantial amount of information, and the trick for the analyst is "teasing it out." A **pivot table** can help. It is useful to filter out, expand, and collapse levels of data to drill down to categories and subcategories of interest. An experienced analyst can generate pivot tables to respond to unanticipated questions quickly.

	A	B	C	D	E	F	G	H	I	J
1	**Thompson Industries Monthly Supplier Report - April**									
2										
3	**SupplierName**	**SupNo**	**SupCat**	**Purchases**	**Change**	**Quality**	**On-Time**	**Returns**	**RejShip**	**NewSup**
4	Maverick Supplies	6071	P	$39,117	$5,000	5	100%	0%	0	N
5	Cowboy Pete's Lumber	6894	P	$25,554	$400	5	90%	0%	0	N
6	ABC Metals	5054	S	$13,051	$13,051	3	0%	75%	2	Y
7	Perkins Plastics	5528	P	$27,886	$3,256	4	100%	3%	0	N
8	Everything Brass	4413	P	$29,300	($4,658)	5	100%	0%	0	N
9	Jones Aluminum & Copper	5858	P	$16,862	$16,862	4	50%	5%	1	Y
10	Rick's Wrought Iron	9733	S	$29,078	($3,121)	1	60%	80%	6	N
11	Penelope's Paper Products	4718	P	$25,716	$8,795	3	100%	12%	0	N
12	Anthony's Adhesives	9944	P	$22,750	$1,268	4	90%	4%	0	N
13	Columbus Manufacturing	1343	S	$20,103	$469	4	100%	0%	0	N
14	Smart Electronics, Inc.	4785	P	$30,062	$30,062	2	33%	50%	0	Y
15	Watson Woodworking	2730	S	$32,589	$471	5	25%	1%	0	N
16	Fred's Foam Molding	2949	S	$47,886	$47,886	5	100%	0.50%	0	Y
17	Impenetrable Insulation, Inc.	1443	P	$24,971	($6,895)	4	15%	4%	0	N
18	Suzy's Shipping Supplies	8039	P	$23,851	$1,269	5	50%	0%	0	N

Figure **G.5**

Thompson Industries Spreadsheet after Cleaning the Data in Preparation for PivotTable Analysis

Excel's PivotTable is quite easy to use. In general, the user clicks and drags field headings into the appropriate buckets and then applies any desired filters. We will illustrate the tool using Figure G.5, which represents Thompson Industries' supplier file after Vicky Luo cleaned up the omissions, inconsistencies, and errors found in Example G1. For PivotTable to work properly, each field should have a heading in the cell above the first record. Figure G.7 provides converted headings from Figure G.2.

To create a PivotTable, place the cell cursor inside the table and click Insert, then PivotTable (under Tables). The dialog box should automatically select the entire range of the database, including the single-cell headers. Click Ok to place the PivotTable in a new sheet. On the far right of the new sheet, you will see all of the PivotTable fields listed, and below that you will see space to click-and-drag desired fields into one of the four categories (areas). The four areas are:

- *Filters:* filters the whole dataset according to user specifications (similar to Excel's Data Filtering tool).
- *Rows:* all entries with the same value for that field will be summarized in one row in the table.
- *Columns:* all entries with the same value for that field will be summarized in one column in the table.
- *Values:* contain the numeric fields that the user wishes to summarize.

Fields in both the *Rows* and *Columns* areas can be filtered. For example, if the *SupCat* field were placed in the *Rows*, the user could just show results for the 'P' suppliers and not the 'S' suppliers. Note that PivotTables are easiest to read if there is at most one field in the *Rows* area and one field in the *Columns* area. It is possible to include more than one field in each area, but the table becomes more difficult to read.

Example G2 PIVOTTABLE FOR THOMPSON INDUSTRIES

Analytics manager Vicky Luo has been asked to analyze total purchases grouped by quality level and subtotaled by supplier category ('P' or 'S').

APPROACH ▶ Excel's PivotTable tool is applied to the set of supplier records in Figure G.5.
The *Quality* field is placed in the *Rows* area, the *SupCat* field is placed in the *Columns* area, and the *Purchases* field is placed in the *Values* area (automatically summed by default). Figure G.6 shows the PivotTable.

Figure G.6

PivotTable for Thompson Industries Showing Total Purchases Grouped by Quality and Subtotaled by Supplier Category

	A	B	C	D
1				
2				
3	Sum of Purchases	Column Labels		
4	Row Labels	P	S	Grand Total
5	1		$29,078	$29,078
6	2	$30,062		$30,062
7	3	$25,716	$13,051	$38,767
8	4	$92,469	$20,103	$112,572
9	5	$117,822	$80,475	$198,297
10	Grand Total	$266,069	$142,707	$408,776

INSIGHT ▶ From Figure G.6, we can quickly see that Thompson Industries purchases approximately twice as much from primary ('P') suppliers as secondary ('S') suppliers.

LEARNING EXERCISE ▶ Vicky has also been asked to present the number of rejected shipments, grouped by quality level (sorted highest-to-lowest) and subtotaled by new supplier status. [Answer: The *Quality* field is placed in the *Rows* area, the *NewSup* field is placed in the *Columns* area, and the *RejShip* field is placed in the *Values* area. To reverse the default lowest-to-highest ordering for quality ratings, click on the **Row Labels** selection arrow and select Sort Largest to Smallest.]

▶ RELATED PROBLEMS G.6, G.9

Data Visualization

Most data software packages, including Excel, provide a myriad of options for graphing data. But a few standard graph types are used for most applications. Figure G.7 displays some of the most common: (a) *bar graphs*, (b) *line graphs*, (c) *scatter diagrams*, (d) *slope graphs*, (e) *pie charts*, (f) *histograms*, (g) *regression lines*, and (h) *network diagrams*.

Graphing Tips Keep the following tips in mind as you prepare graphs for presentation.

- Focus on the message and be clear about the takeaways.
- Graphs should be self-explanatory. Do the obvious helpful things, such as providing a meaningful descriptive title and labeling axes with units of measure evident.
- Ensure that multiple datasets on the same graph each have a key or legend.
- Be efficient and reduce clutter. Do not overwhelm the viewer with too much data in a figure.
- Use symbols where they work.
- Avoid 3-dimensional graphics. They tend to provide no added value and may distort size perceptions.
- Incorporate no more than five unique variables.

Dashboards

Graphically present data for important metrics in one location.

Many business analytics tools lend themselves to a visual presentation, and business analysts should be resourceful and use these visual tools when possible. Companies are regularly using **dashboards**, which present an overview of the most important metrics (i.e., *key success indicators*) all in one place, often presented in graphical form and regularly updated. Software applications such as Tableau, SAS, and Power BI all have dashboard capabilities.

Using Excel's Visualization Tools

LO G.3 *Create* a heat map using Excel's Conditional Formatting tool

Excel incorporates many visualization tools—too many to cover in detail in this module. Furthermore, each chart includes a myriad of options that you can explore to make the most effective visual presentations possible. In Example G3, we provide a short illustration of how to use Excel's Conditional Formatting to superimpose a heat map onto a column of data. A *heat map* uses colors to represent values, with darker colors applied to more extreme values and lighter colors applied to more moderate values.

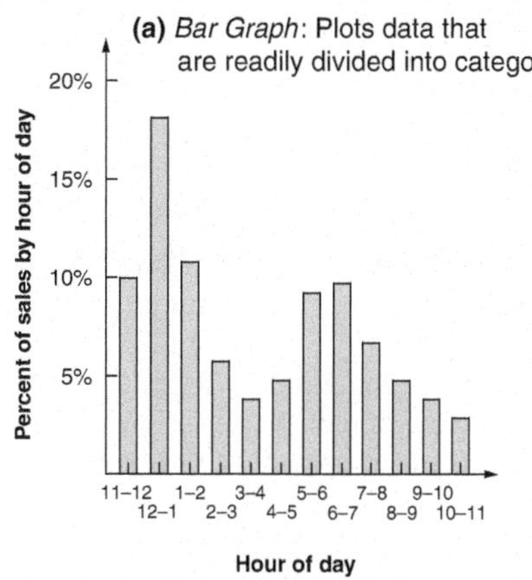

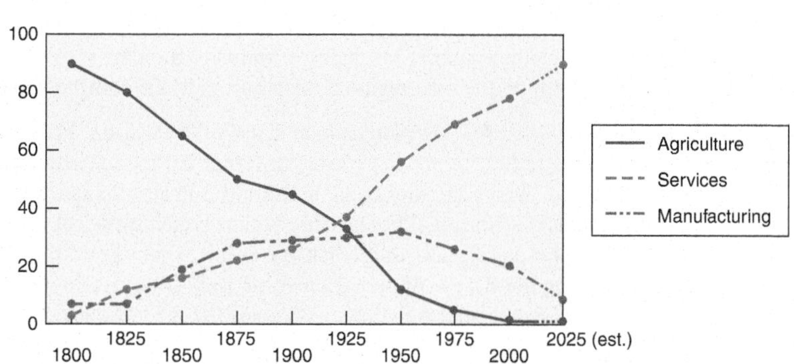

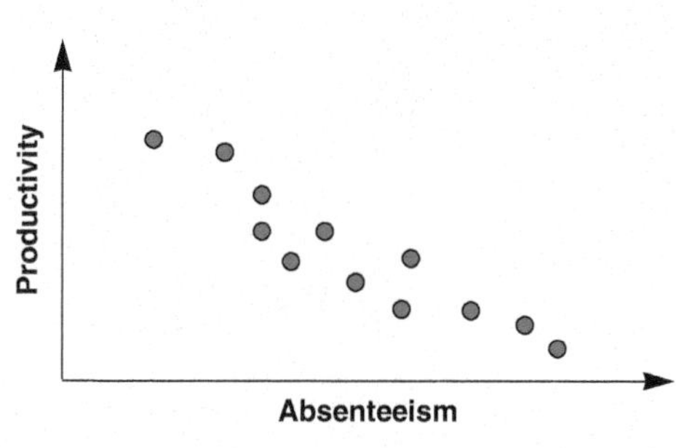

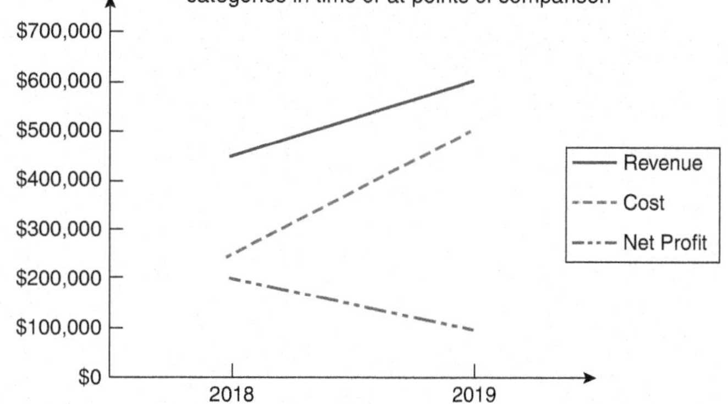

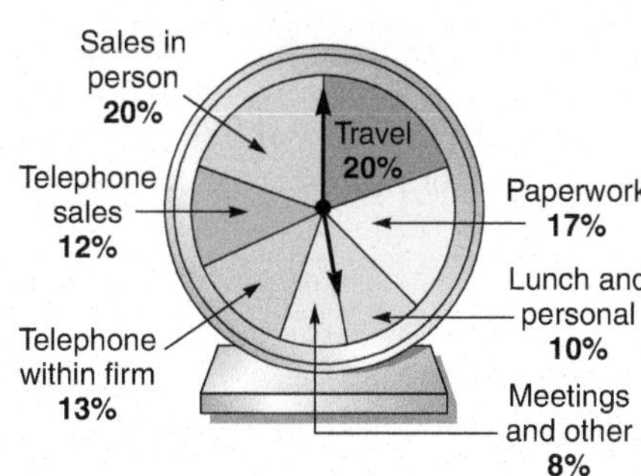

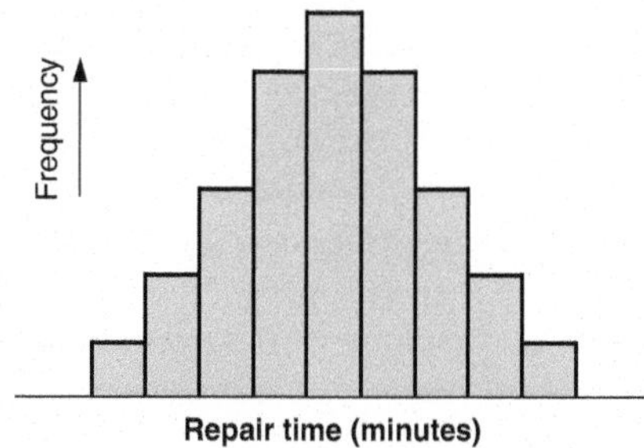

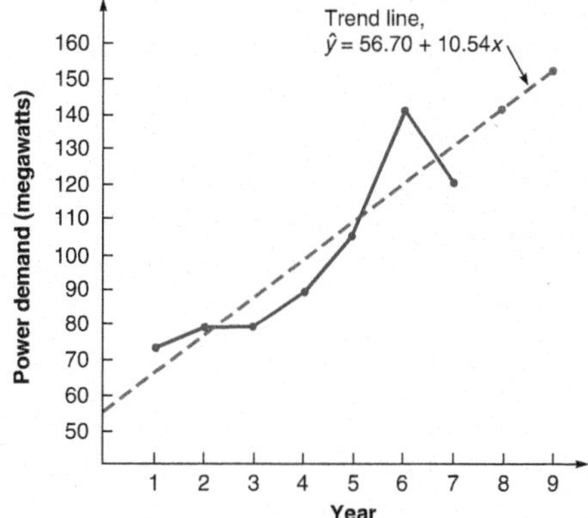

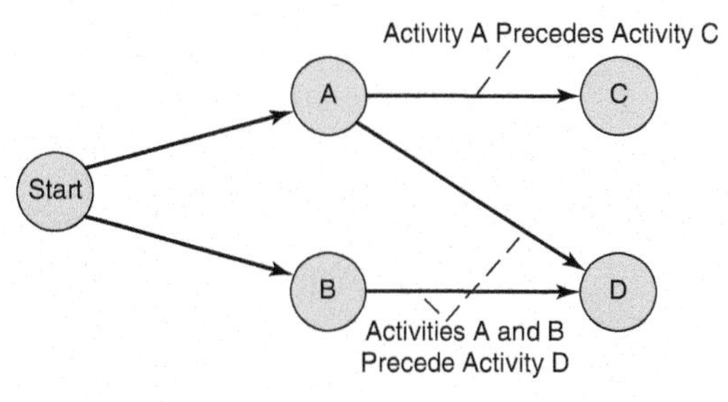

Figure **G.7**

Examples of Data Visualization Graphs

Example G3

HEAT MAP

Supply chain manager Jon Jackson wants to indicate visually the relative values of annual inventory turnover levels of the nine regional distributors in his company's supply chain (data shown in Figure G.8(b)).

APPROACH ▶ A *color scale* in Excel's Conditional Formatting changes the *Fill* color of every cell by using a gradation of two or three different colors. Darker means more extreme. This technique creates a *heat map*, which highlights the highest and lowest values.

To use Conditional Formatting within Excel, we select the cell or cells to conditionally format (B2:B10 in this example), and then click on Home, then Conditional Formatting (under the Styles group). By choosing the Blue – White – Red color scale (Figure G.8(a)), the heat map in Figure G.8(b) is produced.

Figure **G.8**

Conditional Formatting with the Blue – White – Red Color Scale

In Fig. (a), the 2nd row, 1st column under Color Scales is selected; (b) shows the resulting heat map.

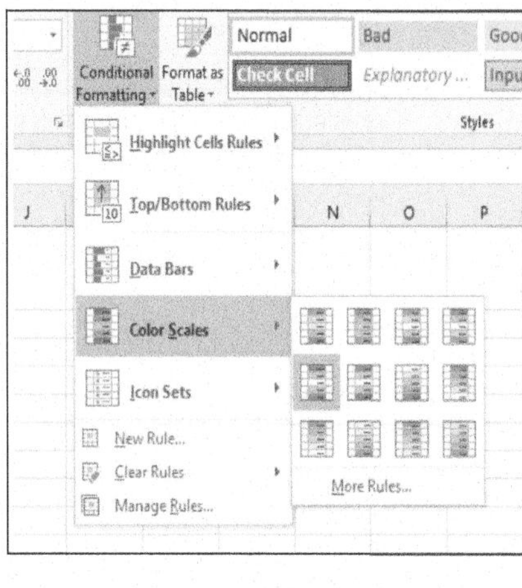

(a)

	A	B
1	Distributor	Annual Inventory Turnover
2	Mid-Atlantic	45
3	Southeast	60
4	Northwest	95
5	North Central	34
6	Northeast	25
7	California/Nevada	85
8	Mountain West	92
9	Southwest	58
10	South Central	12

(b)

INSIGHT ▶ A quick glance at the colors in Figure G.8(b) suggests that the western distributors are performing best, while the south central and northeast distributors are struggling by comparison.

LEARNING EXERCISE ▶ Jon wants to see how the table would look with icons instead of cell colors. [Answer: By selecting Icon Sets under Conditional Formatting, Jon has a variety of icon choices that automatically place an icon in each cell (e.g., arrows, circles, stop lights, or flags) that are colored green for high values, yellow for medium values, and red for low values.]

RELATED PROBLEM ▶ G.10

Predictive and Prescriptive Business Analytics Tools

Sophisticated predictive and prescriptive tools have existed for many decades. However, the power of today's computers coupled with the plethora of big data has illuminated the value of these tools in the eyes of corporate managers worldwide. We are truly entering an *Age of Analytics*.

A rich variety of analytical tools are included in this text. These include forecasting/time series analysis (Chapter 4), correlation and regression analysis (Chapter 4), decision tables and trees (Module A), linear programming (Modules B and C), queuing theory (Module D), learning curves (Module E), and simulation (Module F).

Other Business Analytics Tools Not Covered in This Text

LO G.4 *Describe* advanced business analytics tools

Other business analytics techniques, not included in this text, include data mining, mapping and tracking, cohort analysis, cluster analysis, and neural networks and machine learning. We briefly introduce them now.

Data Mining This process is designed to explore very large datasets (i.e., big data), looking for relevant patterns or relationships that may provide insights for improved decision making.

Data mining can go down three paths. One is *text analysis* to tag or annotate terms or ideas, such as how often a new product is mentioned in the press. A variation is *sentiment analysis* or *opinion mining*, which seeks to extract 'sentiment' from text or audio data, such as the success or lack of success in dealing with recurring customer complaints to a call center. A third approach is *pattern analytics*, which extracts information from images such as photographs, medical images (MRI, x-rays), or even facial recognition. Research now indicates that data mining can yield better results than individual doctors reviewing MRIs. The data cloud software produced by SAS is an excellent tool for text and sentiment analysis.

Mapping and Tracking Electronic sensors or video can track employee and customer movement. Tracing part or product movement in a factory or warehouse has traditionally been used to analyze and improve facility layout. Given the consistent pattern of part movement, this is a fairly straightforward task. But tracking becomes much more challenging as we seek to identify the dynamic patterns of individual customer movements. Identifying customer movements in a store allows us to understand more about who is visiting, who is buying, and their pattern of movement.

Cohort Analysis This technique focuses on the study of the behavior of groups over time. For instance, how does the financial standing of your suppliers change over time?

Cluster Analysis This technique identifies and organizes data into groups with similar attributes. For example, local distributors might constitute one cluster, regional distributors a second cluster, and national distributors a third.

Neural Networks and Machine Learning Neural networks are computer systems, which, guided by statistical techniques, take large volumes of data and potential variables to form groupings of variables to identify complex paths and associations. Machine learning often builds on these paths and associations to make science fiction a reality by creating speech recognition and self-driving cars. As applied to analytics, machine learning provides a vehicle to sift through vast amounts of big data to provide insight.

Students may find a variety of software tools available for advanced business analytics techniques including many features of SAS, as well as Tableau, Analytic Solver Platform, XLMiner, R, Excel, Excel VBA, and SQL.

Summary

The emergence of big data has brought the world's attention to the field of business analytics—a broad term incorporating statistics, computer science, and operations research techniques that attempt to turn data into decisions.

Prior to analyzing data, managers should 'clean' the data to ensure that they are complete, consistent, and accurate. Straightforward Excel tools can be applied to data cleaning, including judicious use of Excel functions or Conditional Formatting.

With big data, pivot tables represent a powerful way to dig deep into a dataset to extract and summarize information for categories of interest. Managers can use Excel's PivotTable tool to perform such analysis by placing the categories into proper buckets and applying appropriate filters.

Effective data visualization techniques have become vital to communicate important features of the vast amount of big data to constituents. Excel has numerous charting, histogram, and other tools to facilitate visualization.

The manager of tomorrow needs to be well-versed in the varied and growing business analytics techniques to stay ahead in this new data-rich *Age of Analytics*.

Key Terms

Big data (p. 824)
Business analytics (p. 824)
Data management (p. 825)
Conditional Formatting (p. 827)
Pivot table (p. 828)
Dashboards (p. 830)

Discussion Questions

1. Describe the three categories of business analytics.
2. Describe the three major issues for data management and clean data.
3. What are some ways to use Excel to clean data?
4. Describe the purpose of a heat map.
5. Consider Example G1. Think of three additional items for which the manager might want to check to ensure that the data are clean.
6. Describe the three paths of data mining.

Solved Problems

SOLVED PROBLEM G.1

The supply chain manager for the distributors from Figure G.8(b) wishes to see icons that indicate when the annual inventory turnover exceeds 90 and when it is below 30.

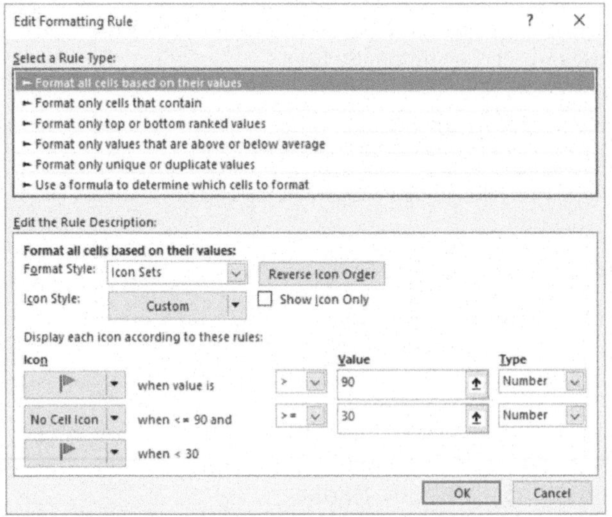

SOLUTION

When using icon sets, not every cell has to contain an icon. Here we place a green flag next to all inventory turnover amounts greater than 90 and a red flag for all amounts less than 30. Click on More rules... at the bottom of the icon set box to open the applicable dialog box.

	A	B	
1	Distributor		Annual Inventory Turnover
2	Mid-Atlantic		45
3	Southeast		60
4	Northwest	▶	95
5	North Central		34
6	Northeast	▶	25
7	California/Nevada		85
8	Mountain West	▶	92
9	Southwest		58
10	South Central	▶	12

SOLVED PROBLEM G.2

Vicky Luo has been asked to identify the number of primary ('P') suppliers for Thompson Industries (Figure G.5) that are not new and have a quality rating of 4 or 5.

SOLUTION

There are several ways to display this using PivotTable. For example, after creating a PivotTable from Figure G.5, complete these five steps:

1. Place *NewSup* in the *Filters* area. Then click on the selection arrow and select **N** then click Ok (to include only the suppliers that are not new).
2. Place *SupCat* in the *Rows* area. Then click on the **Row Labels** selection arrow and select **P** then click Ok (to include only primary suppliers).
3. Place *Quality* in the *Columns* area. Then click on the **Column Labels** selection arrow and select **4** and **5,** then click Ok (to include only suppliers with a quality rating of 4 or 5).
4. To count the number of records that satisfy the respective criteria, any other category can be placed in the *Values* area. Use the Count of *SupplierName.*
5. Type over the **Row Labels** heading in cell A4 and the **Column Labels** heading in cell B3 and replace those with their actual respective field headings.

	A	B	C	D
1	NewSup	N		
2				
3	**Count of SupplierName**	**Quality**		
4	**SupCat**	**4**	**5**	**Grand Total**
5	P	3	4	7
6	**Grand Total**	**3**	**4**	**7**

Problems

Problems G.1–G.9 relate to Data Management

•• **G.1** For the 'Change' column (E) in Figure G.5:

a) Determine the upper and lower limits for a 3-standard-deviation accuracy check.

b) Which, if any, values fall outside the limits computed in part (a)?

•• **G.2** For the 'RejShip' column (I) in Figure G.5:

a) Determine the upper and lower limits for a 3-standard-deviation accuracy check.

b) Which, if any, values fall outside the limits computed in part (a)?

••• **G.3** Using Figure G.2, enter \$6,000 into cell E8. In that spreadsheet, any new supplier (with a 'Y' in column J) should have its total purchases in column D equal to its change from the previous month in column E. Analytics manager Vicky Luo wishes to set up Conditional Formatting in Excel to identify any new suppliers whose purchases do not meet that condition. What formula should go into the box **Format values where this formula is true:**? *Hint: The Excel symbol for "not equals to" is <>.*

• **G.4** Using Figure G.2, analytics manager Vicky Luo wishes to set up Conditional Formatting in Excel to ensure that all the suppliers have different supplier numbers in column B. Which Conditional Formatting 'Rule Type' should she select?

• • • **G.5** Using the data in Figure G.8(b), supply chain manager Jon Jackson wishes to use Excel's Conditional Formatting to highlight any entries in column B that are at least as large as this year's threshold level for distributor bonuses contained in cell C2 (not shown in the figure). What formula should go into the box **Format values where this formula is true:**?

• • **G.6** Thompson Industries' supply chain manager, Xun Xu, wishes to understand the makeup of his supply base better. He suggests that you, using the data in Figure G.5, use Excel to create a PivotTable that displays the number of suppliers, grouped by purchases (in groups of $15,000, starting at $1), and subtotaled by quality level. *Hint: To group rows by a range of values, right-click on any value in column A of the PivotTable and select* Group.... How many suppliers have a purchase amount between $15,000 and $30,000 and a quality level less than 5?

• • **G.7** Thompson Industries' supply chain manager, Xun Xu, wishes to identify suppliers with the largest and smallest purchase levels, so he knows where to focus his purchasing resources. He suggests that you, using the data in Figure G.5, use Excel to apply Conditional Formatting to the 'Purchases' column. Format all cells with a value in at least the top 75% of the range with a green circle, and format all cells with a value in the bottom 30% of the range with a red circle. Which supplier or suppliers have a green circle?

• • **G.8** Thompson Industries' CEO, Mary Beth Marrs, wishes to identify the top-performing suppliers to present with a performance award. She suggests that you, using the data in Figure G.5, use Excel to apply Conditional Formatting to the 'On-Time' column. Format the top 10% of cells with the 'yellow fill with dark yellow text' option. Which supplier or suppliers receive the formatting (and the award)?

• • **G.9** Thompson Industries' supply chain manager, Xun Xu, wishes to better understand the quality and on-time performance of his supply base. He suggests that you, using the data in Figure G.5, use Excel to create a PivotTable that displays the number of suppliers, grouped by quality level (sorted largest-to-smallest), and subtotaled by percentage of on-time shipments (but only for suppliers with more than 50% on-time shipments). Show values as **% of Column Total**. What value appears in the cell for 100% on-time shipping percentage and quality level of 3?

Problem G.10 relates to Data Visualization

• **G.10** Operations manager Phil Witt wishes to apply the Blue – White – Red color scale using Excel's Conditional Formatting to the following data:

PLANT	PRODUCTIVITY (UNITS PER LABOR-HOUR)
St. Louis	30
Bentonville	42
Yakima	28
Phoenix	35
Auburn	18

a) Which plant will be colored the darkest shade of blue?
b) Which plant will be colored the darkest shade of red?

CASE STUDY

Labor Concerns at Zapco Industries

"What are you people doing?" exclaimed an exacerbated CEO Dawn Jacobs as she stormed into the production meeting. "Wage costs were nearly $12,000 over budget last month, and defect numbers are soaring. Output seems pretty good, but I see extra workers staying late all the time fixing defects. Surely those folks would rather be enjoying time with their families than sticking around here. Let's take a look at the numbers and get rid of the deadbeats."

Production Manager Juan Velasquez put last month's (22-day) worker data on the viewing monitor. Zapco Industries runs 3 shifts employing 8 line workers per shift. Workers range in experience from 0 to 30 years with the company. The union contract sets wages at $20.00 per hour plus $0.50 per hour per year of experience. Workers are given a production target each day. If they exceed that target, they receive a pay bonus based on the percentage completed above target. For example, a worker achieving a labor standard of 110% would receive 10% higher pay. Workers missing the target do not receive a salary deduction. In addition, as part of its total quality management program, each worker is responsible to remain at work after the shift ends and fix his or her own defects. The labor standard for error correction is 30 minutes.

"What is this telling us, Juan?" asked Dawn. "Should we start by firing everyone that cost us more than $500 extra last month? Or should we get rid of those averaging more than one error per day?"

"It may not be that simple," replied Juan. "Some of our most experienced workers have a higher wage rate. So naturally their bonuses should be higher as well."

"Remind me," directed Dawn. "How do we penalize people for producing defects?"

"They have to fix those after hours," responded Juan.

"We don't have to pay them for that, do we? After all, the mistakes are their fault," said Dawn.

Juan replied, "That may be true, but the union contract absolutely specifies time-and-a-half pay for overtime work."

Shift 1 Line Manager Mario Reyes added, "Look at Suzy Walsh. She made a lot of errors last month but was also extremely productive. And we paid her an extra $1,757 last month. I wonder if something fishy is going on."

"Several things may be happening with these workers," noted Dawn. "It's too hard to tell from looking at all the data at once. Mario, as our analytics expert, please create some pivot tables to determine the average performance and average monthly errors split in the following ways: (1) by shift, (2) by experience (in blocks of 5 years each), (3) by number of absences, and (4) by gender. Let's dig deeper into this."

	A	B	C	D	E	F	G	H	I	J
1	Zapco Industries									
2	Monthly Labor Report - April									
3										
4	Name	Gender	Shift	Experience	LaborStd	Errors	Absences	Overtime	Wage	ExtraPay
5	Alfeha, Huda	F	3	4	90%	1	1	0.5	$22.00	$16.50
6	Blackwell, Austin	M	3	22	100%	0	2	0	$31.00	$0.00
7	Gonzales, Manuel	M	2	30	120%	19	1	9.5	$35.00	$1,730.75
8	Hu, Lizao	F	1	29	105%	25	1	12.5	$34.50	$950.48
9	Jefferson, Michael	M	1	3	106%	1	0	0.5	$21.50	$243.17
10	Jerge, David	M	1	8	114%	3	5	1.5	$24.00	$645.36
11	Jerge, Robin	F	1	10	110%	0	6	0	$25.00	$440.00
12	Johnson, Shantel	F	1	0	104%	20	0	10	$20.00	$440.80
13	Jones, Avery	M	1	28	110%	22	0	11	$34.00	$1,159.40
14	Jordan, Joquin	F	3	6	98%	3	2	1.5	$23.00	$51.75
15	Kazemi, Amir	M	3	0	80%	30	3	15	$20.00	$450.00
16	Kuno, Angie	F	3	16	84%	2	0	1	$28.00	$42.00
17	Luo, Xinchang	M	3	15	101%	3	2	1.5	$27.50	$110.28
18	Miller, Charles	M	2	10	105%	1	5	0.5	$25.00	$238.75
19	Munson, Kim	F	3	23	90%	0	1	0	$31.50	$0.00
20	Nardai, Federico	M	3	7	87%	0	2	0	$23.50	$0.00
21	Rosa, Maria	F	2	17	98%	4	0	2	$28.50	$85.50
22	Rosenblatt, Amit	M	1	0	85%	16	1	8	$20.00	$240.00
23	Rosenblatt, Zehava	F	2	11	108%	2	6	1	$25.50	$397.29
24	Smith, Vincent	M	2	1	96%	28	0	14	$20.50	$430.50
25	Starkey, David	M	2	26	112%	18	2	9	$33.00	$1,142.46
26	Starkey, Pam	F	2	12	110%	2	1	1	$26.00	$496.60
27	Tang, Yixuan	F	1	1	108%	25	2	12.5	$20.50	$673.02
28	Walsh, Suzy	F	2	30	120%	20	1	10	$35.00	$1,757.00
29	Total							122.5		$11,741.59

Discussion Questions

1. What insights are obtained after creating the four pivot tables requested by Dawn?
2. How effective is the current incentive system at Zapco?
3. What recommendations should be made for the company?

Endnotes

1. Sources: *Predictive Analytics Times* (June 7, 2017); *SmallBiz Technology* (July 18, 2018); and *Tech Republic* (May 11, 2020).
2. As we are checking for internal consistency of a particular set of data, we assume that we are dealing with the entire population; hence, we use the Excel formula for *population* standard deviation, STDEV.P. The formula is: $\sqrt{\frac{\sum_{i=1}^{n}(x_i - \bar{x})}{n}}$, where x_i is the value of observation i, $\bar{x}$ is the average of the observations, and n is the number of observations. If only a subset of the dataset is being used to calculate the standard deviation, then use the STDEV.S Excel formula, which uses $n - 1$ in the denominator instead of n.

Bibliography

Albright, S. C., *VBA for Modelers: Developing Decision Support Systems with Microsoft® Office Excel®*, 5th ed. Boston, MA: Cengage Learning, 2016.

Camm, J. D., J. J. Cochran, M. J. Fry, J. W. Ohlmann, D. R. Anderson, D. J. Sweeney, and T. A. Williams. *Business Analytics*, 3rd ed. Boston: Cengage, 2019.

Cody, R. *Cody's Data Cleaning Techniques Using SAS®*, 3rd ed. Cary, North Carolina: SAS Institute Inc., 2017.

Davenport, T. H., and J. G. Harris. *Competing on Analytics: The New Science of Winning*. Boston: Harvard Business Review Press, 2017.

Erl, T., W. Khattak, and P. Buhler. *Big Data Fundamentals: Concepts, Drivers & Techniques*. Boston: Pearson, 2016.

Harvard Business Review. *HBR Guide to Data Analytics Basics for Managers*. Boston: Harvard Business Review Press, 2018.

Knaflic, C. N. *Storytelling with Data: A Data Visualization Guide for Business Professionals*. New York: Wiley, 2015.

Lustig, I., B. Dietrich, C. Johnson, and C. Dziekan. "The Analytics Journey." *Analytics Magazine* (November/December 2010): 11–18.

Rose, R. "Defining Analytics: A Conceptual Framework." *OR/MS Today* 43, no. 3 (June 2016).

Sanders, N. R. *Big Data Driven Supply Chain Management: A Framework for Implementing Analytics and Turning Information into Intelligence.* Upper Saddle River, NJ: Pearson, 2014.

Tufte, E. *The Visual Display of Quantitative Information*. Cheshire, CT: Graphics Press, 2001.

Module G *Rapid* Review

Main Heading	Review Material	MyLab Operations Management
INTRODUCTION TO BIG DATA AND BUSINESS ANALYTICS	■ **Big data**—The huge amount of production, consumer, and social media data collected in digital form. ■ **Business analytics**—Uses tools and techniques to convert data into summary information and business insights for decision making. The three categories of analytics include *descriptive analytics* (characterizing and summarizing data to facilitate understanding), *predictive analytics* (analyzing past data to predict the future), and *prescriptive analytics* (invoking advanced optimization tools to recommend a strategy or action).	Concept Questions: 1.1–1.6
DATA MANAGEMENT	■ **Data management**—Overall management of data's integrity, including completeness, consistency, and accuracy. The combination of the <End> and <Down Arrow> keys in Excel can be used to quickly locate missing records in an Excel database. A quick method to see if there are any empty cells in an Excel database is to use the COUNTIF(range,"") function. To check an Excel dataset for accuracy, outliers can be identified that do not fall within these ranges: Lower limit =AVERAGE(*range*)-3*STDEV.P(*range*) Upper limit =AVERAGE(*range*)+3*STDEV.P(*range*) ■ **Conditional Formatting**—An Excel tool to visually identify characteristics of data using formatting. Conditional Formatting is an excellent tool for finding inconsistencies and inaccuracies in a dataset. Numerous formatting choices exist, including color scales, icon sets, and coloring cells that meet the stated condition. ■ **Pivot table**—A tool to facilitate in-depth analysis of numeric data by applying filters and providing summary computations for categories and subcategories of the dataset. The four Excel PivotTable categories (areas) are: 1. *Filters*: filters the whole dataset according to user specifications (similar to Excel's Data Filtering tool) 2. *Rows*: all entries with the same value for that field will be summarized in one row in the table. 3. *Columns*: all entries with the same value for that field will be summarized in one column in the table. 4. *Values*: contain the numeric fields that the user wishes to summarize. After inserting an Excel PivotTable, the user clicks and drags field headings into the appropriate categories and then applies any desired filters.	Concept Questions: 2.1–2.6 Problems: G.1–G.9 Virtual Office Hours for Solved Problems G.1–G.2
DATA VISUALIZATION	Eight of the most common data visualization graphs include: 1. *Bar graphs*—plot data that are readily divided into categories 2. *Line graphs*—show data that vary continuously 3. *Scatter diagrams*—note the value of one variable vs. another variable 4. *Slope graphs*—illustrate relative increases or decreases of categories in time or at points of comparison 5. *Pie charts*—represent percentages or portions of a whole 6. *Histograms*—distributions that show the frequency of occurrences of a variable 7. *Regression lines*—provide a *best-fit* line through the central tendency of points on a graph 8. *Network diagrams*—use arcs and nodes to provide representations of flow through a system Graphing Tips: 1. Graphs should be self-explanatory. Do the obvious helpful things, such as providing a meaningful descriptive title and labeled axes with units of measure evident. 2. Ensure that multiple datasets on the same graph each have a key or legend. 3. Do not overwhelm the viewer with too much data in a figure. 4. Avoid 3-dimensional graphics. ■ **Dashboards**—Graphically present data for important metrics in one location.	Concept Questions: 3.1–3.4 Problem G.10

Module G **Rapid Review** *continued*

Main Heading	Review Material	MyLab Operations Management
	Excel's Conditional Formatting can be used to superimpose a *heat map* onto a column of data. A *heat map* uses colors to represent values, with darker colors applied to more extreme values and lighter colors applied to more moderate values.	

Figure G.8(b)
Heat Map

	A	B
1	Distributor	Annual Inventory Turnover
2	Mid-Atlantic	45
3	Southeast	60
4	Northwest	95
5	North Central	34
6	Northeast	25
7	California/Nevada	85
8	Mountain West	92
9	Southwest	58
10	South Central	12

(b)

Main Heading	Review Material	MyLab Operations Management
PREDICTIVE AND PRESCRIPTIVE BUSINESS ANALYTICS TOOLS	*Data mining* explores very large datasets (i.e., big data), looking for relevant patterns or relationships that may provide insights for improved decision making. The three data mining paths include *text analysis, sentiment analysis* or *opinion mining*, and *pattern analysis*. With *mapping* or *tracking*, electronic sensors or video can track employee and customer movement. *Cohort analysis* studies the behavior of groups over time. *Cluster analysis* identifies and organizes data into groups with similar attributes. *Neural networks* are computer systems, which, guided by statistical techniques, take large volumes of data and potential variables to form groupings of variables to identify complex paths and associations. *Machine learning* often builds on these paths and associations to make science fiction a reality by creating speech recognition and self-driving cars. As applied to analytics, machine learning provides a vehicle to sift through vast amounts of big data to provide insight.	Concept Questions: 4.1–4.5
ADDITIONAL MYLAB OPERATIONS MANAGEMENT RESOURCES	✔ Recent Graduate Video: Charles Render, Data Analyst, Shutterstock ✔ Multiple Choice Case Questions (Labor Concerns at Zapco Industries)	

Self **Test**

LO G.1 Which category of analytics recommends a strategy or action?
- **a)** descriptive analytics
- **b)** predictive analytics
- **c)** prescriptive analytics

LO G.2 Which Excel Conditional Formatting *Rule Type* should be used to format values in one column based on the condition of values in another column?
- **a)** Format all cells based on their values
- **b)** Format only cells that contain
- **c)** Format only top or bottom ranked values
- **d)** Format only unique or duplicate values
- **e)** Use a formula to determine which cells to format

LO G.3 Which formatting choice for Excel's Conditional Formatting tool can be used to create a heat map?
- **a)** icon sets
- **b)** changing the fill color of cells
- **c)** changing font type
- **d)** data bars
- **e)** color scales

LO G.4 Which business analytics tool studies the behavior of groups over time?
- **a)** cohort analysis
- **b)** tracking
- **c)** neural networks
- **d)** cluster analysis
- **e)** data mining

Answers: LO G.1. c; LO G.2. e; LO G.3. e; LO G.4. a.

Appendixes

APPENDIX I

Normal Curve Areas

APPENDIX II

Using Excel OM and POM for Windows

APPENDIX III

Solutions to Even-Numbered Problems

APPENDIX I
NORMAL CURVE AREAS

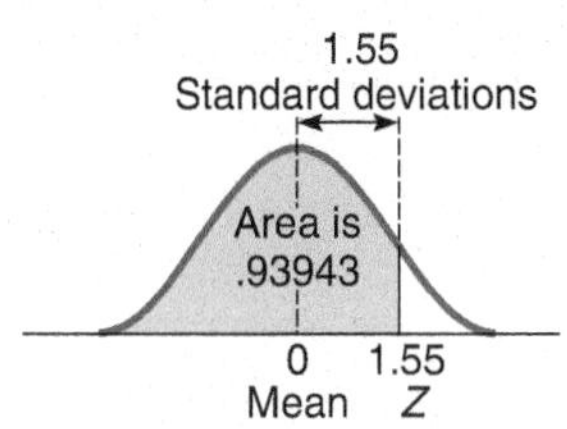

To find the area under the normal curve, you can apply either Table I.1 or Table I.2. In Table I.1, you must know how many standard deviations that point is to the right of the mean. Then, the area under the normal curve can be read directly from the normal table. For example, the total area under the normal curve for a point that is 1.55 standard deviations to the right of the mean is .93943.

TABLE I.1

Z	.00	.01	.02	.03	.04	.05	.06	.07	.08	.09
.0	.50000	.50399	.50798	.51197	.51595	.51994	.52392	.52790	.53188	.53586
.1	.53983	.54380	.54776	.55172	.55567	.55962	.56356	.56749	.57142	.57535
.2	.57926	.58317	.58706	.59095	.59483	.59871	.60257	.60642	.61026	.61409
.3	.61791	.62172	.62552	.62930	.63307	.63683	.64058	.64431	.64803	.65173
.4	.65542	.65910	.66276	.66640	.67003	.67364	.67724	.68082	.68439	.68793
.5	.69146	.69497	.69847	.70194	.70540	.70884	.71226	.71566	.71904	.72240
.6	.72575	.72907	.73237	.73565	.73891	.74215	.74537	.74857	.75175	.75490
.7	.75804	.76115	.76424	.76730	.77035	.77337	.77637	.77935	.78230	.78524
.8	.78814	.79103	.79389	.79673	.79955	.80234	.80511	.80785	.81057	.81327
.9	.81594	.81859	.82121	.82381	.82639	.82894	.83147	.83398	.83646	.83891
1.0	.84134	.84375	.84614	.84849	.85083	.85314	.85543	.85769	.85993	.86214
1.1	.86433	.86650	.86864	.87076	.87286	.87493	.87698	.87900	.88100	.88298
1.2	.88493	.88686	.88877	.89065	.89251	.89435	.89617	.89796	.89973	.90147
1.3	.90320	.90490	.90658	.90824	.90988	.91149	.91309	.91466	.91621	.91774
1.4	.91924	.92073	.92220	.92364	.92507	.92647	.92785	.92922	.93056	.93189
1.5	.93319	.93448	.93574	.93699	.93822	.93943	.94062	.94179	.94295	.94408
1.6	.94520	.94630	.94738	.94845	.94950	.95053	.95154	.95254	.95352	.95449
1.7	.95543	.95637	.95728	.95818	.95907	.95994	.96080	.96164	.96246	.96327
1.8	.96407	.96485	.96562	.96638	.96712	.96784	.96856	.96926	.96995	.97062
1.9	.97128	.97193	.97257	.97320	.97381	.97441	.97500	.97558	.97615	.97670
2.0	.97725	.97784	.97831	.97882	.97932	.97982	.98030	.98077	.98124	.98169
2.1	.98214	.98257	.98300	.98341	.98382	.98422	.98461	.98500	.98537	.98574
2.2	.98610	.98645	.98679	.98713	.98745	.98778	.98809	.98840	.98870	.98899
2.3	.98928	.98956	.98983	.99010	.99036	.99061	.99086	.99111	.99134	.99158
2.4	.99180	.99202	.99224	.99245	.99266	.99286	.99305	.99324	.99343	.99361
2.5	.99379	.99396	.99413	.99430	.99446	.99461	.99477	.99492	.99506	.99520
2.6	.99534	.99547	.99560	.99573	.99585	.99598	.99609	.99621	.99632	.99643
2.7	.99653	.99664	.99674	.99683	.99693	.99702	.99711	.99720	.99728	.99736
2.8	.99744	.99752	.99760	.99767	.99774	.99781	.99788	.99795	.99801	.99807
2.9	.99813	.99819	.99825	.99831	.99836	.99841	.99846	.99851	.99856	.99861
3.0	.99865	.99869	.99874	.99878	.99882	.99886	.99889	.99893	.99896	.99900
3.1	.99903	.99906	.99910	.99913	.99916	.99918	.99921	.99924	.99926	.99929
3.2	.99931	.99934	.99936	.99938	.99940	.99942	.99944	.99946	.99948	.99950
3.3	.99952	.99953	.99955	.99957	.99958	.99960	.99961	.99962	.99964	.99965
3.4	.99966	.99968	.99969	.99970	.99971	.99972	.99973	.99974	.99975	.99976
3.5	.99977	.99978	.99978	.99979	.99980	.99981	.99981	.99982	.99983	.99983
3.6	.99984	.99985	.99985	.99986	.99986	.99987	.99987	.99988	.99988	.99989
3.7	.99989	.99990	.99990	.99990	.99991	.99991	.99992	.99992	.99992	.99992
3.8	.99993	.99993	.99993	.99994	.99994	.99994	.99994	.99995	.99995	.99995
3.9	.99995	.99995	.99996	.99996	.99996	.99996	.99996	.99996	.99997	.99997

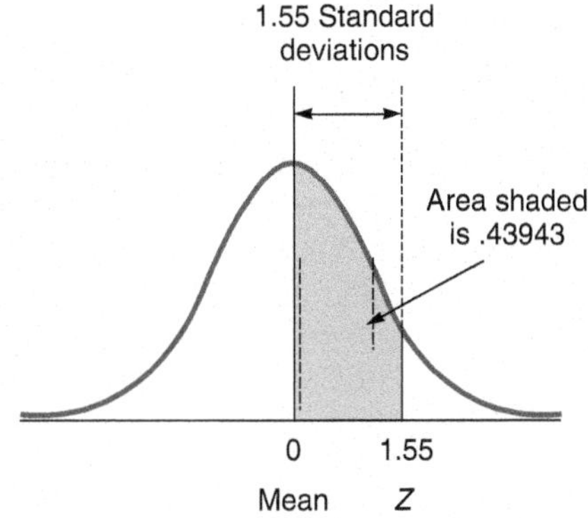

As an alternative to Table I.1, the numbers in Table I.2 represent the proportion of the total area away from the mean, μ, to one side. For example, the area between the mean and a point that is 1.55 standard deviations to its right is .43943.

TABLE I.2

Z	.00	.01	.02	.03	.04	.05	.06	.07	.08	.09
0.0	.00000	.00399	.00798	.01197	.01595	.01994	.02392	.02790	.03188	.03586
0.1	.03983	.04380	.04776	.05172	.05567	.05962	.06356	.06749	.07142	.07535
0.2	.07926	.08317	.08706	.09095	.09483	.09871	.10257	.10642	.11026	.11409
0.3	.11791	.12172	.12552	.12930	.13307	.13683	.14058	.14431	.14803	.15173
0.4	.15542	.15910	.16276	.16640	.17003	.17364	.17724	.18082	.18439	.18793
0.5	.19146	.19497	.19847	.20194	.20540	.20884	.21226	.21566	.21904	.22240
0.6	.22575	.22907	.23237	.23565	.23891	.24215	.24537	.24857	.25175	.25490
0.7	.25804	.26115	.26424	.26730	.27035	.27337	.27637	.27935	.28230	.28524
0.8	.28814	.29103	.29389	.29673	.29955	.30234	.30511	.30785	.31057	.31327
0.9	.31594	.31859	.32121	.32381	.32639	.32894	.33147	.33398	.33646	.33891
1.0	.34134	.34375	.34614	.34850	.35083	.35314	.35543	.35769	.35993	.36214
1.1	.36433	.36650	.36864	.37076	.37286	.37493	.37698	.37900	.38100	.38298
1.2	.38493	.38686	.38877	.39065	.39251	.39435	.39617	.39796	.39973	.40147
1.3	.40320	.40490	.40658	.40824	.40988	.41149	.41309	.41466	.41621	.41174
1.4	.41924	.42073	.42220	.42364	.42507	.42647	.42786	.42922	.43056	.43189
1.5	.43319	.43448	.43574	.43699	.43822	.43943	.44062	.44179	.44295	.44408
1.6	.44520	.44630	.44738	.44845	.44950	.45053	.45154	.45254	.45352	.45449
1.7	.45543	.45637	.45728	.45818	.45907	.45994	.46080	.46164	.46246	.46327
1.8	.46407	.46485	.46562	.46638	.46712	.46784	.46856	.46926	.46995	.47062
1.9	.47128	.47193	.47257	.47320	.47381	.47441	.47500	.47558	.47615	.47670
2.0	.47725	.47778	.47831	.47882	.47932	.47982	.48030	.48077	.48124	.48169
2.1	.48214	.48257	.48300	.48341	.48382	.48422	.48461	.48500	.48537	.48574
2.2	.48610	.48645	.48679	.48713	.48745	.48778	.48809	.48840	.48870	.48899
2.3	.48928	.48956	.48983	.49010	.49036	.49061	.49086	.49111	.49134	.49158
2.4	.49180	.49202	.49224	.49245	.49266	.49286	.49305	.49324	.49343	.49361
2.5	.49379	.49396	.49413	.49430	.49446	.49461	.49477	.49492	.49506	.49520
2.6	.49534	.49547	.49560	.49573	.49585	.49598	.49609	.49621	.49632	.49643
2.7	.49653	.49664	.49674	.49683	.49693	.49702	.49711	.49720	.49728	.49736
2.8	.49744	.49752	.49760	.49767	.49774	.49781	.49788	.49795	.49801	.49807
2.9	.49813	.49819	.49825	.49831	.49836	.49841	.49846	.49851	.49856	.49861
3.0	.49865	.49869	.49874	.49878	.49882	.49886	.49889	.49893	.49897	.49900
3.1	.49903	.49906	.49910	.49913	.49916	.49918	.49921	.49924	.49926	.49929

APPENDIX II

USING EXCEL OM AND POM FOR WINDOWS

Two approaches to computer-aided decision making are provided with this text: **Excel OM** and **POM** (Production and Operations Management) **for Windows**. These are the two most user-friendly software packages available to help you learn and understand operations management. Both programs can be used either to solve homework problems identified with an icon in the textbook or to check answers you have developed by hand. POM uses the standard Windows interface and runs on any computer operating Windows 7 or later. Excel OM is available for Windows operating systems running Excel 2010 or later and Mac running Excel 2016 or later. Both software packages are available in MyLab Operations Management and can be found online. If you have technical difficulties with either of these two software packages, please contact hweiss@comcast.net.

EXCEL OM

Excel OM has been designed to help you to better learn and understand both OM and Excel. Even though the software contains 26 modules and more than 75 submodules, the screens for every module are consistent and easy to use. The Chapter menu (Excel 2010 and later for Windows) and the OM menu (Excel 2016 for Mac) lists the modules in *chapter* order, as illustrated for Excel 2010 and later for PCs in Program II.1(a). The Alphabetical menu (Excel 2010 and later for Windows) or Excel OM menu (Excel 2016 and later for Mac) lists the modules in alphabetical order, as illustrated for Excel OM for Mac using Excel 2016 in Program II.1(b).

To install Excel OM, download the appropriate file for your PC or Mac from the Download Center. In the PC installation, default values have been assigned in the setup program, but you

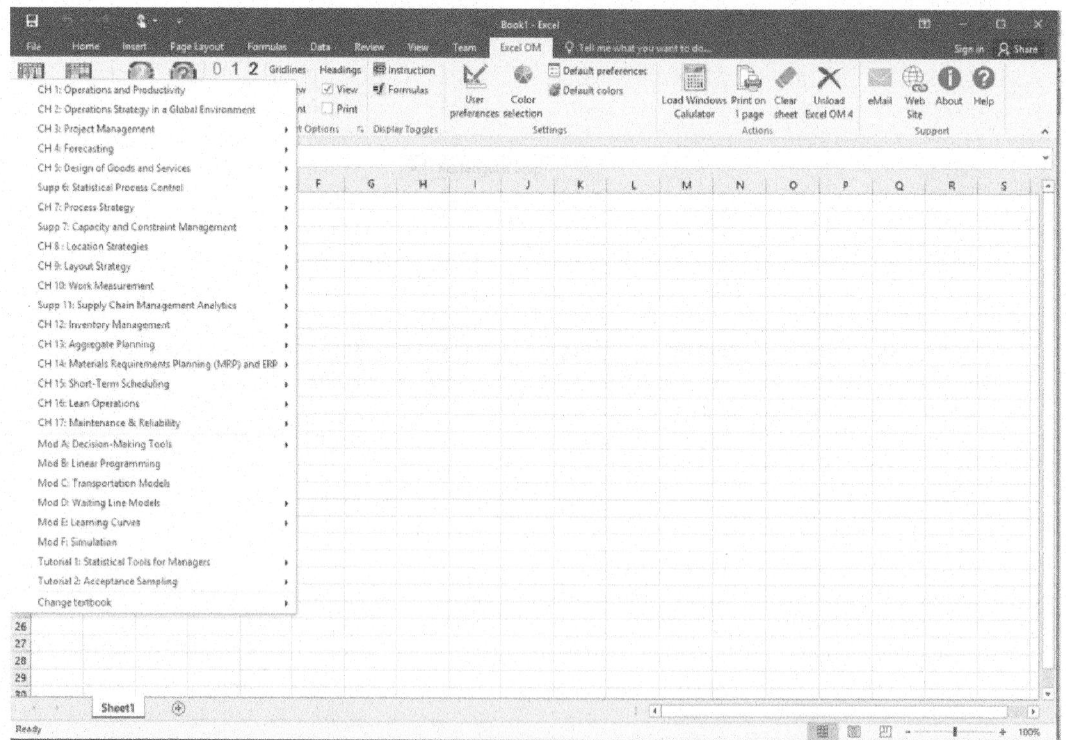

Program **II.1(a)**

Excel OM Modules Menu in Excel OM Tab in Excel 2010 and later for PCs

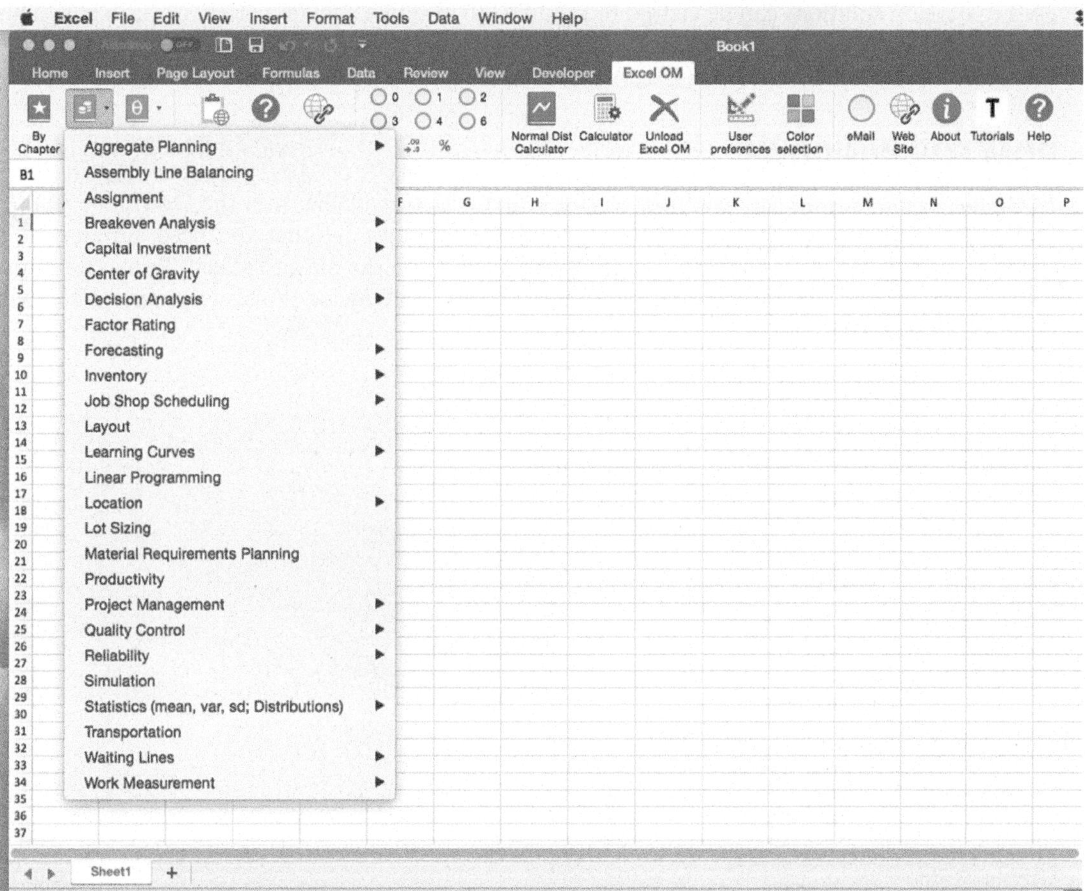

Program **II.1(b)**

Excel OM Modules Menu in OM for Mac 2016

may change them if you like. Generally speaking, it is simply necessary to click *Next* each time the installation asks a question. For Mac, simply unzip the zip file that is provided.

Starting the Program To start Excel OM using Windows, double-click on the Excel OM V5 shortcut placed on the desktop during installation. The Excel OM options will appear in an Excel OM tab that will be added to the Excel ribbon, as displayed in Program II.1(a). To start Excel OM on the Mac, open the file excelomqmv5.xlam. Be certain that you open the file from the folder that has the excelomqmv5.lic file. You may want to create an Alias for excelomqmv5.xlam and place it on your desktop to make startup easier.

Please note that Excel OM is not installed as a permanent Excel add-in but must be started in the fashion above each time you use it. If you have Excel 2010 or later and do not see an Excel OM tab on the Ribbon, then your Excel security settings need to be revised to enable Excel OM V5.

Excel OM serves two purposes in the learning process. First, it helps you solve problems. You enter the appropriate data, and the program provides numerical solutions. POM for Windows operates on the same principle. However, Excel OM allows for a second approach: the Excel *formulas* used to develop solutions can be viewed or modified to examine a wider variety of problems. This "open" approach enables you to observe, understand, and even change the formulas underlying the Excel calculations—conveying Excel's power as an OM analysis tool.

POM FOR WINDOWS

POM for Windows is decision support software that is also available from the Download Center in MyLab Operations Management. If you receive a message that you need administrator permission to install POM then right-click on the installation.exe file and select "Run as administrator". Once you follow the standard setup instructions, a POM for Windows program icon will be added to your desktop.

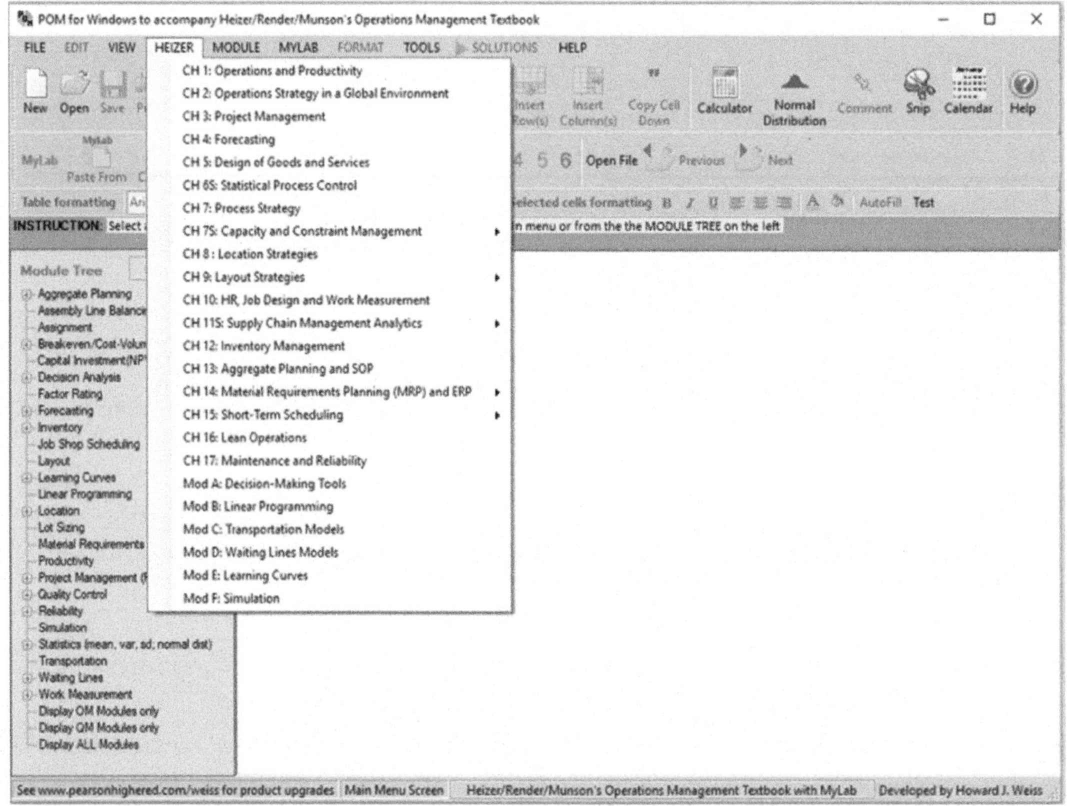

Program **II.2**

POM for Windows Module List

Starting the Program To start the program, double-click on the POM icon that was placed on the desktop. Program II.2 shows a list of the 25 OM modules that can be accessed from the menu tree on the left. The modules can also be accessed from the MODULE menu where they are in alphabetical order or the HEIZER menu as shown in Program II.2 where they are in chapter order. As in Excel OM, the screens for every module are consistent and easy to use. POM for Windows is not available on a Mac unless you are running Windows on your Mac.

USING EXCEL OM AND POM FOR WINDOWS WITH MYLAB OPERATIONS MANAGEMENT

Both Excel OM and POM for Windows have MyLab tools (hidden under the dropdown menus in the two figures above) that make it very easy to copy tables of data from MyLab Operations Management and paste the data into Excel OM or POM for Windows. In addition, there are tools that make it easy to set the specific number of decimals for which MyLab Operations Management asks for any problem. To copy data from MyLab Operations Management: after creating a model in Excel OM or POM, go to MyLab Operations Management and click on the Copy icon. Then, in POM or in Excel OM, click on the **Paste from MyLab** icon on the toolbar or in the context (right-click) menu from any cell in the empty table of data.

APPENDIX III

SOLUTIONS TO EVEN-NUMBERED PROBLEMS

Chapter 1

1.2 **(a)** 2 valves/hr.; **(b)** 2.25 valves/hr.; **(c)** 12.5%
1.4 **(a)** 20 pkg/hour; **(b)** 26.6 pkg/hour; **(c)** 33.0%
1.6 **(a)** .078 fewer resources (7.8% improvement)
1.8 **(a)** 2.5 tires/hour; **(b)** 0.025 tires/dollar; **(c)** 2.56% increase
1.10 .000375 autos per dollar of inputs
1.12 4 workers
1.14 1.6%
1.16 $57.00 per labor hour

Chapter 2

2.2 Venezuela, China, United States, Switzerland, Denmark
2.4 Differentiation is evident when comparing most restaurants or restaurant chains.
2.6 **(a)** Focus more on standardization, make fewer product changes, find optimum capacity, and stabilize manufacturing process are a few possibilities
(b) New human resource skills, added capital investment for new equipment/processes
(c) Same as (b)
2.8 **(a)** Canada, 1.7; **(b)** No change
2.10 **(a)** Worldwide, 81.5 weighted *average*, 815 weighted *total*
(b) No change
(c) Overnight Shipping now preferred, weighted total = 880
2.12 Company C, $1.0 \le w \le 25.0$

Chapter 3

3.2 Here are some detailed activities for the first two activities for Day's WBS:
1.1.1 Set initial goals for fundraising.
1.1.2 Set strategy, including identifying sources and solicitation.
1.1.3 Raise the funds.
1.2.1 Identify voters' concerns.
1.2.2 Analyze competitor's voting record.
1.2.3 Establish position on issues.
3.4 **(a)** AON network:

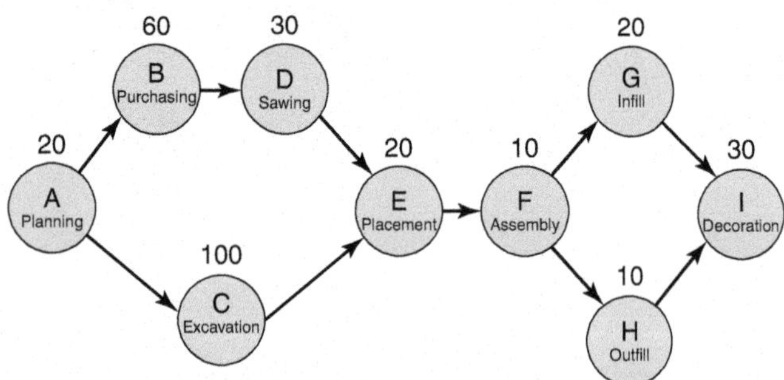

(b) AOA network:

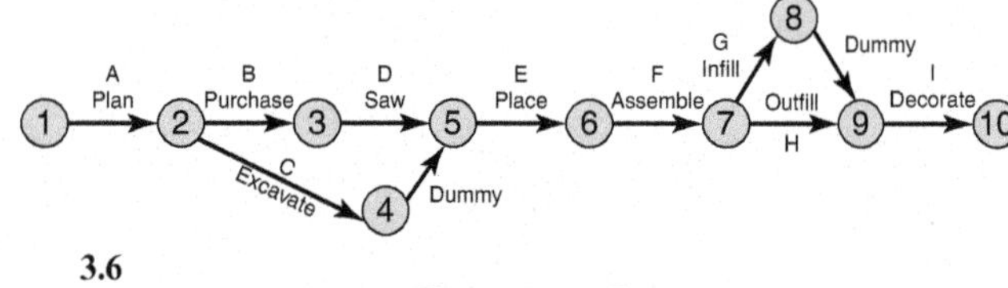

3.6

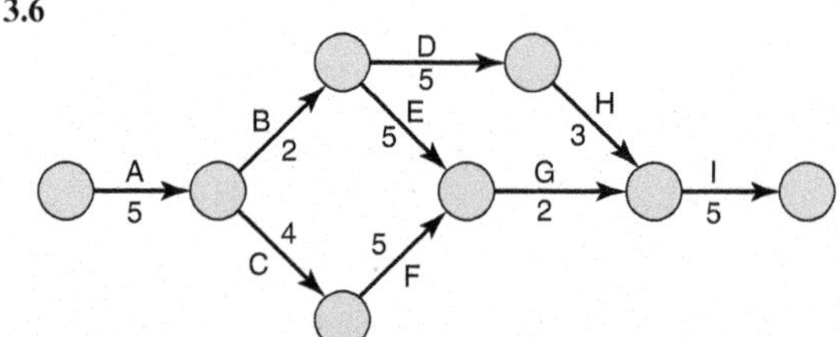

A–C–F–G–I is critical path; 21 days.
This is an AOA network.

3.8 **(a)**

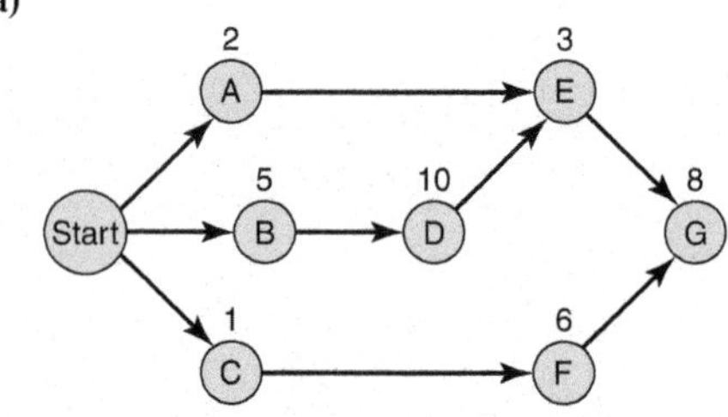

(b) B–D–E–G; **(c)** 26 days
(d)

Activity	Slack
A	13
B	0
C	11
D	0
E	0
F	11
G	0

3.10 **(a)**

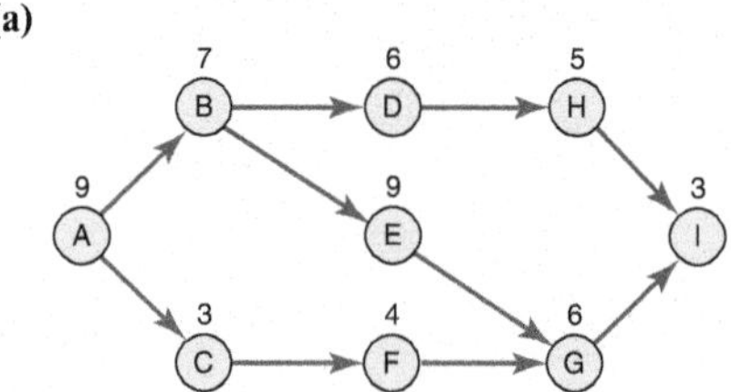

(b) A–B–E–G–I is critical path; **(c)** 34

3.12

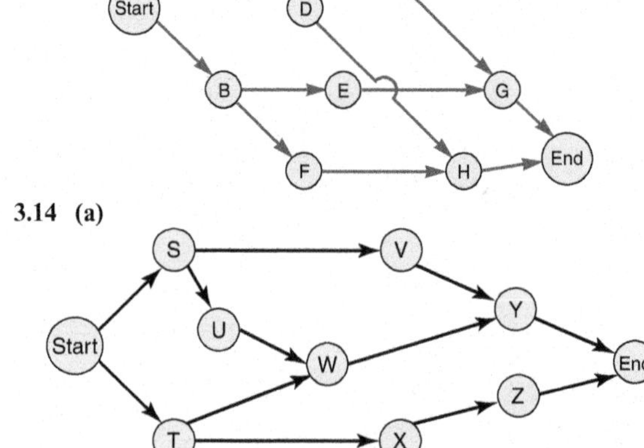

3.14 **(a)**

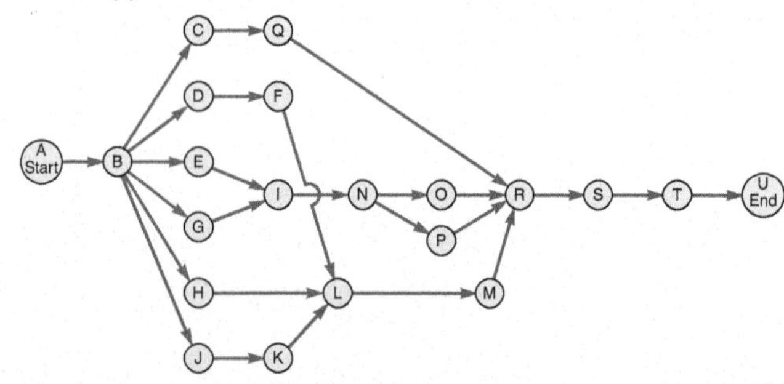

(b) Critical path is S–U–W–Y. Completion time is 17 weeks.

3.16 **(a)**

(b) Critical path is A–B–J–K–L–M–R–S–T–U for 18 days.
(c) i. No, transmissions and drivetrains are not on the critical path.
ii. No, halving engine-building time will reduce the critical path by only 1 day.
iii. No, it is not on the critical path.
(d) Reallocating workers not involved with critical-path activities to activities along the critical path will reduce the critical path length.

3.18 A: 15, 1.33; B: 32, 2.33; C: 18, 0
D: 13.17, 1.83; E: 18.17, 0.5; F: 19, 1.0

3.20 **(a)** A: 10, 0.11; B: 10, 4; C: 10, 0.11; D: 8, 1
(b) A–C: 20 weeks; B–D: 18 weeks
(c) .22, 5; **(d)** 1.0; **(e)** 0.963

3.22 **(a)** 32 weeks, C-H-M-O; **(b)** 50%; **(c)** No impact

3.24 **(a)** A = 5, B = 6, C = 7, D = 6, E = 3, Time = 15
(b) 1, 1, 1, 4, 0; Project variance = 2

3.26 **(a)** Critical path = A–C–D. Project length = 27 days.
(b) 89.97%
(c) 84.61%
(d) No. The non-critical path has an even lower probability than the critical path.

3.28 Crash C to 3 weeks at $200 total for one week. Now both paths are critical. Not worth it to crash further.

3.30 **(a)** Reduce A, cheapest path @ $600 per time period
(b) Reduce B, best choice now @ $900
(c) Total cost = $1,500

3.32 **(a)** Slacks are: 0, 2, 11, 0, 2, 11, 0
(b) First, crash D by 2 weeks. Then crash D and E by 2 weeks each.
(c) Minimum completion time = 7. Crash cost = $1,550.

Chapter 4

4.2 **(a)** None obvious.
(b) 7, 7.67, 9, 10, 11, 11, 11.33, 11, 9
(c) 6.4, 7.8, 11, 9.6, 10.9, 12.2, 10.5, 10.6, 8.4
(d) The 3-yr. moving average.

4.4 **(a)** 41.6; **(b)** 42.3; **(c)** Banking industry's seasonality.

4.6 **(b)** Naïve = 23; 3-mo. moving = 21.33; 6-mo. weighted = 20.6; exponential smoothing = 20.62; trend = 20.67
(c) Trend projection

4.8 **(a)** 91.3; **(b)** 89; **(c)** MAD = 2.7; **(d)** MSE = 13.35
(e) MAPE = 2.99%

4.10 **(a)** 4.67, 5.00, 6.33, 7.67, 8.33, 8.00, 9.33, 11.67, 13.7
(b) 4.50, 5.00, 7.25, 7.75, 8.00, 8.25, 10.00, 12.25, 14.0
(c) Forecasts are about the same.

4.12 72

4.14 Method 1: MAD = .125; MSE = .021
Method 2: MAD = .1275; MSE = .018

4.16 **(a)** $y = 421 + 33.6x$. When $x = 6$, $y = 622.8$.
(b) MAD = 5.6
(c) MSE = 32.88

4.18 alpha = .25; forecast = 49

4.20 $\alpha = .1$, $\beta = .8$, August forecast = $71,303; MSE = 12.7 for $\beta = .8$ vs. MSE = 18.87 for $\beta = .2$ in Problem 4.19.

4.22 Confirm that you match the numbers in Table 4.1.

4.24 $y = 5 + 20x$, $y = 105$

4.26 1,680 sailboats

4.28 96.344, 132.946, 169.806, 85.204

4.30 $y = 29.76 + 3.28x$
Year 11 = 65.8, Year 12 = 69.1 (patients)
$r^2 = 0.853$

4.32 Forecasts: 50.00, 50.00, 44.60, 36.64, 35.77, 37.21, and 35.09
Both the MAD and the week 7 forecast are better with the trend adjustment.

4.34 9.4

4.36 7.86

4.38 **(a)** 13.67, MAD = 2.20; **(b)** 13.17, MAD = 2.72

4.40 150,000; 126,000; 120,000; 198,000

4.42 **(a)** 7,000; **(b)** 9,000

4.44 **(a)** 337; **(b)** 380; **(c)** 423

4.46 **(a)** $y = 50 + 18x$; **(b)** $410

4.48 **(a)** 83,502; **(d)** 0.397

4.50 **(a)** $y = -.158 + .1308x$; **(b)** 2.719; **(c)** $r = .966$; $r^2 = .934$

4.52 **(b)** $y = 0.511 + 0.159x$
(c) 2,101,000 riders
(d) 511,000 people
(e) .404 (rounded to .407 in POM software)
(f) $r^2 = 0.840$

4.54 **(a)** Sales (y) = −9.349 + .1121 (contracts)
(b) $r = .8963$; $S_{xy} = 1.3408$

4.56 **(a)** $y = 1 + 1x$; **(b)** 0.45; **(c)** 3.65

4.58 Games lost = 6.41 + 0.533 × rainy days

4.60 MAD = 10.875; Tracking signal = 3.586

Chapter 5

5.2 *Possible strategies:*
Tablet (growth phase):
Increase capacity and improve balance of production system.
Attempt to make production facilities more efficient.
Smart watch (introductory phase):
Increase R&D to better define required product characteristics.
Modify and improve production process.
Develop supplier and distribution systems.
Hand calculator (decline phase):
Concentrate on production and distribution cost reduction.

5.4 The following diagram is a house of quality for a sports watch in the under $50 market.

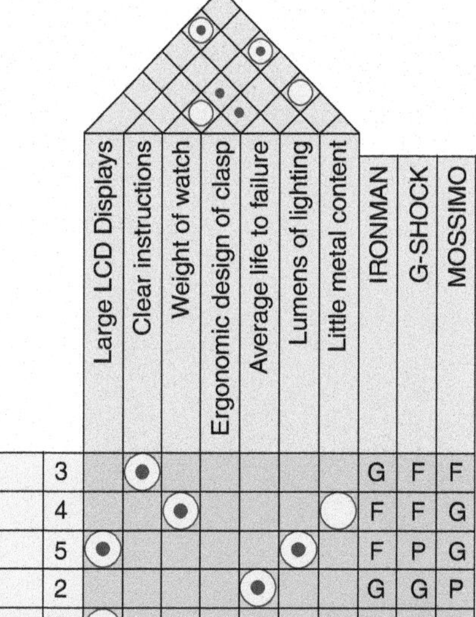

Large LCD Displays
Clear instructions
Weight of watch
Ergonomic design of clasp
Average life to failure
Lumens of lighting
Little metal content
IRONMAN
G-SHOCK
MOSSIMO
Easy to program 3 G F F
Lightweight 4 F F G
Easy to read 5 F P G
Reliable 2 G G P
Digital readouts 5 G G F
Easy to fasten 1 G F G
Our importance ratings 40 15 20 3 10 25 12
G = Good
F = Fair
P = Poor

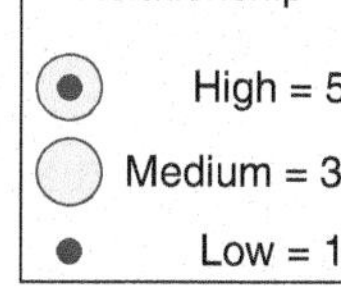

5.6 Build a house of quality similar to the one shown in Example 1 in the text.
Consider customer requirements such as:
Effective Luring, Reliability, Kills Quickly, Finger Safe, etc.
Consider Manufacturing issues such as:
Luring Radius, Dead Mouse Ratio, Time to Kill, Cost, etc.

5.8 House of Quality Sequence for Ice Cream

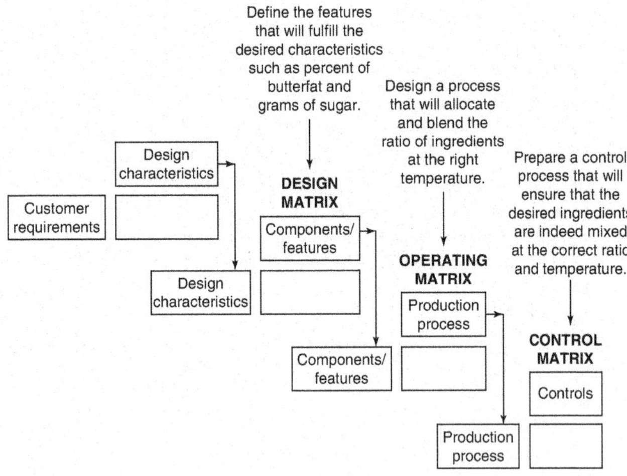

5.10 Assembly chart for the eyeglasses:

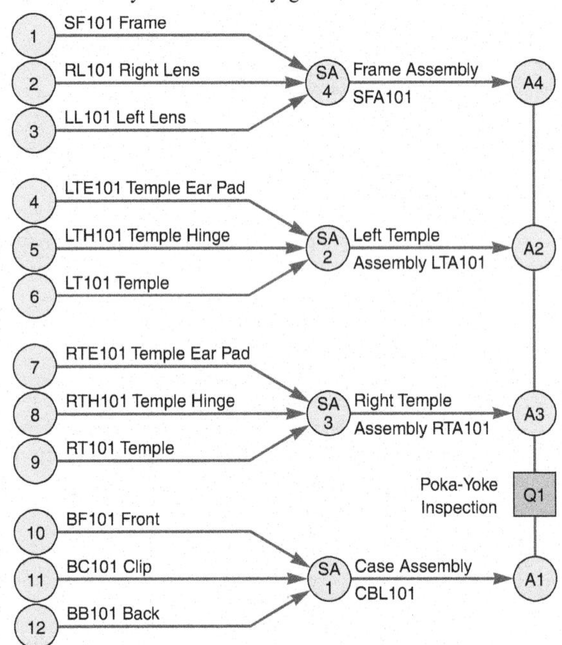

5.12 Assembly chart for a table lamp:

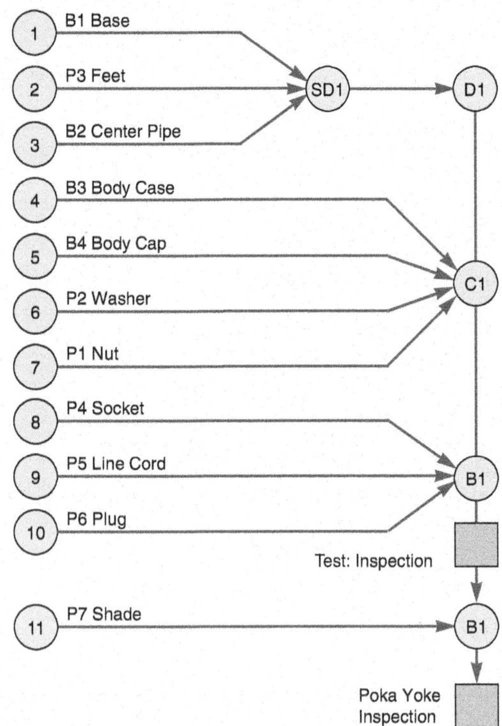

5.14 Bill of material for a wooden pencil with eraser:

Description	Quantity
Pencil	1
Wood half	2
Graphite rod	1
Band	1
Eraser	1
Yellow paint	1 gram
Glue	1 gram

5.16 The major assemblies in the bill of material for a computer mouse:

Bill of Material for a Computer Mouse

Part Number	Description	Quantity
M1001	Computer Mouse	1
SC004	Phillips Head No. 12 0.5 inch. Screw	1
TA101	Top Mouse Assembly	1
BA101	Base Assembly	1
ML101	Mouse Label	1

5.18 Use low-technology, cost of $145,000.
5.20 Produce the deluxe, EMV = $4,000.
5.22 Joint design provides the lowest cost, EMV = $103,800.
5.24 Use K1, EMV = $27,500.
5.26 Test land. If test is positive, drill. If test is negative, sell. Expected profit = $565,000.

Chapter 5 Supplement

S5.2 Brew Master revenue retrieval = $5.31 provides higher opportunity.
S5.4 $66,809
S5.6 3.57 years
S5.8 3.48 years
S5.10 66,667 miles
S5.12 **(a)** $4.53; **(b)** $3.23; **(c)** GF Deluxe
S5.14 **(a)** $4.53, GF Deluxe
(b) $5.06, Premium Mate
(c) Premium Mate model
S5.16 **(a)** 45,455 miles;
(b) gas vehicle
S5.18 42,105 miles

Chapter 6

6.2 Individual answer, in the style of Figure 6.6(b).
6.4 Individual answer, in the style of Figure 6.6(f).
6.6 Partial flowchart for planning a party:

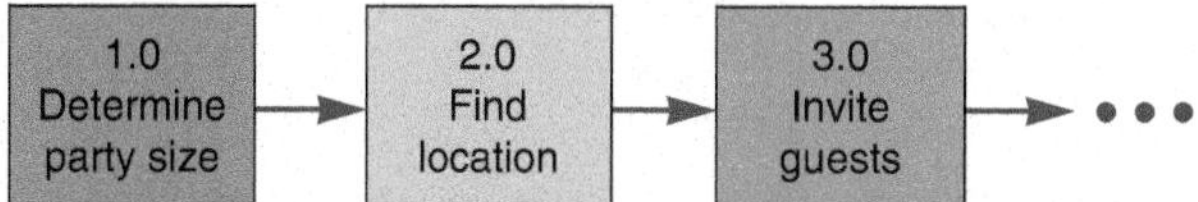

6.8 See figure below for a partial fish-bone. Individual answer in the style of Figure 6.7 in the chapter.
6.10 Individual answer, in the style of Figure 6.7 in the chapter.
6.12 Pareto chart, in the style of Example 1 with parking/drives most frequent, pool second, etc.
6.14 See figure below.
Issues: Materials: 4, 12, 14; Methods: 3, 7, 15, 16; Manpower: 1, 5, 6, 11; Machines: 2, 8, 9, 10, 13.
6.16 **(a)** A scatter diagram in the style of Figure 6.6(b) that shows a strong positive relationship between shipments and defects
(b) A scatter diagram in the style of Figure 6.6(b) that shows a mild relationship between shipments and turnover
(c) A Pareto chart in the style of Figure 6.6(d) that shows frequency of each type of defect
(d) A fishbone chart in the style of Figure 6.6(c) with the 4 *M*s showing possible causes of increasing defects in shipments

▼ Figure for Problem 6.8.

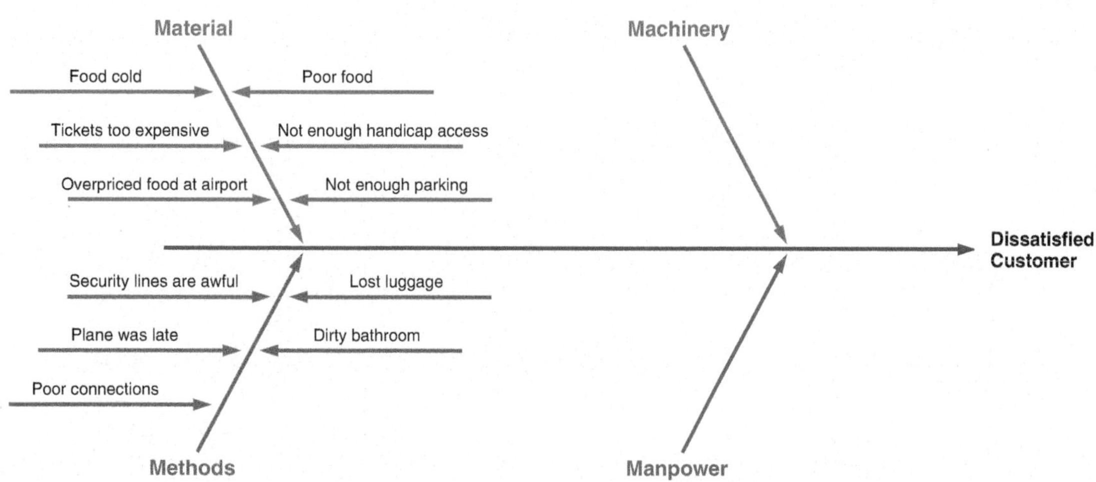

▼ Figure for Problem 6.14.

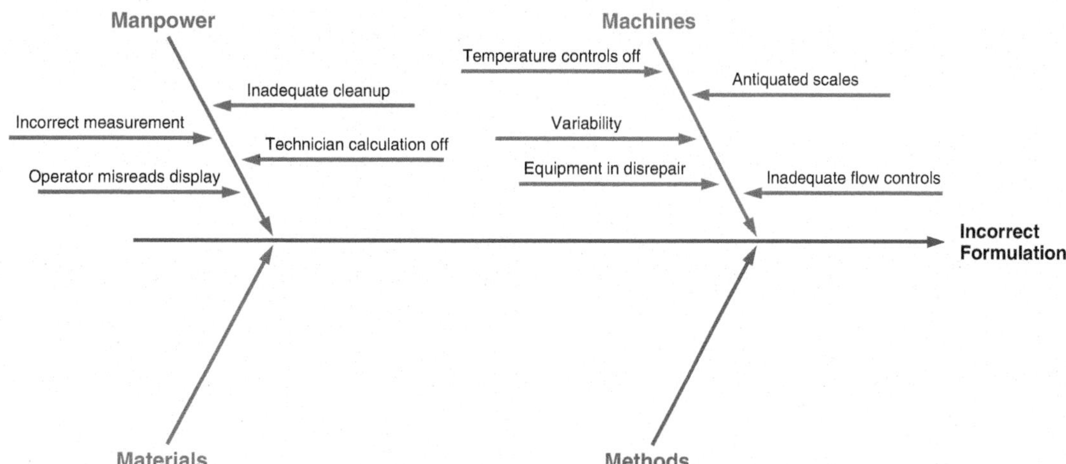

6.18 See figure below.

6.20 Stitching
Seams alignment
Buttons and buttonholes
Collar alignment
Hem

▼ Figure for Problem 6.18.

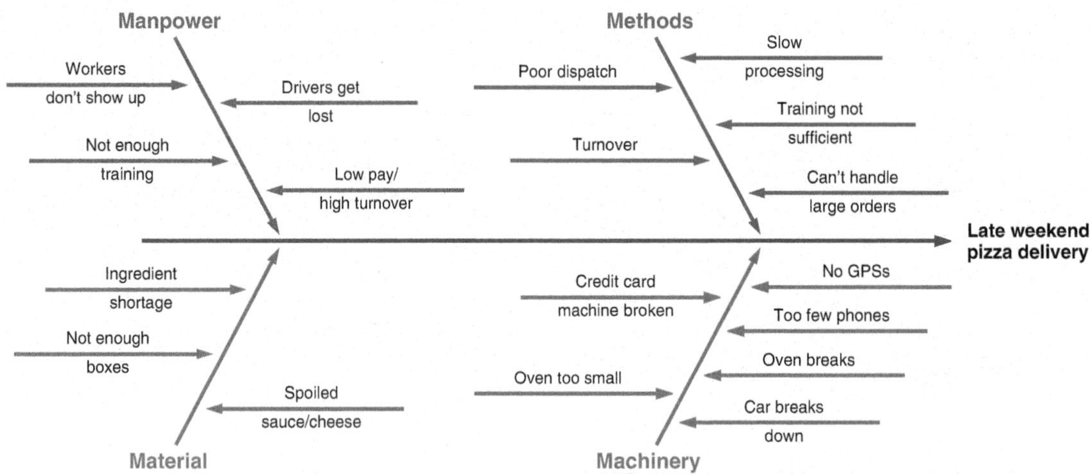

Chapter 6 Supplement

S6.2 **(a)** $UCL_{\bar{x}} = 52.31$
$LCL_{\bar{x}} = 47.69$
(b) $UCL_{\bar{x}} = 51.54$
$LCL_{\bar{x}} = 48.46$

S6.4 **(a)** $UCL_{\bar{x}} = 440$ calories
$LCL_{\bar{x}} = 400$ calories
(b) $UCL_{\bar{x}} = 435$ calories
$LCL_{\bar{x}} = 405$ calories

S6.6 $UCL_{\bar{x}} = 3.728$
$LCL_{\bar{x}} = 2.236$
$UCL_R = 2.336$
$LCL_R = 0.0$
The process is in control.

S6.8 **(a)** $UCL_{\bar{x}} = 16.08$
$LCL_{\bar{x}} = 15.92$
(b) $UCL_{\bar{x}} = 16.12$
$LCL_{\bar{x}} = 15.88$

S6.10 **(a)** $\sigma_{\bar{x}} = 0.61$
(b) Using $\sigma_{\bar{x}}$, $UCL_{\bar{x}} = 11.83$, and $LCL_{\bar{x}} = 8.17$.
Using A_2, $UCL_{\bar{x}} = 11.90$, and $LCL_{\bar{x}} = 8.10$.
(c) $UCL_R = 6.98$; $LCL_R = 0$
(d) Yes

S6.12 $UCL_R = 6.058$; $LCL_R = 0.442$
Averages are increasing.

S6.14 $UCL_c = 4$; $LCL_c = 0$

S6.16 $UCL_p = .0313$; $LCL_p = 0$

S6.18 $UCL_p = 0.077$; $LCL_p = 0.003$

S6.20 **(a)** $UCL_p = .0581$
$LCL_p = 0$
(b) in control
(c) $UCL_p = .1154$
$LCL_p = 0$

S6.22 **(a)** c-chart
(b) $UCL_c = 13.35$
$LCL_c = 0$
(b) in control
(c) not in control

S6.24 **(a)** $UCL_c = 26.063$
$LCL_c = 3.137$
(b) No point out of control.

S6.26 $UCL_{\bar{x}} = 46.966$, $LCL_{\bar{x}} = 45.034$, $UCL_R = 4.008$, $LCL_R = 0$

S6.28 $UCL_{\bar{x}} = 76.85$ $LCL_{\bar{x}} = 73.15$

S6.30 $UCL_{\bar{x}} = 60.924$ $LCL_{\bar{x}} = 59.076$
$UCL_R = 5.331$ $LCL_R = 0.669$

S6.32 **(a)** $UCL_{\bar{x}} = 20.15$ $LCL_{\bar{x}} = 19.65$
(b) $UCL_R = 0.78$ $LCL_R = 0$

S6.34 At least 29 holes that meet tolerance, but no more than 88 holes before being replaced.

S6.36 $UCL_p = 0.0209$ $LCL_p = 0.0011$

S6.38 $UCL_p = 0.0637$ $LCL_p = 0.0000$

S6.40 $C_p = 1.0$. The process is barely capable.

S6.42 $C_{pk} = 1.125$. Process *is* centered and will produce within tolerance.

S6.44 $C_{pk} = .1667$

S6.46 $C_{pk} = 0.33$

S6.48 Machine 1 has index of $C_{pk} = 0.83$ (not capable)
Machine 2 has index of $C_{pk} = 1.0$ (capable)

S6.50 $C_p = 1.667$ (very capable)

S6.52 AOQ = 0.02 or 2%

S6.54 AOQ = 0.0117 or 1.17%

Chapter 7

7.2 GPE is best below 100,000.
FMS is best between 100,000 and 300,000.
DM is best over 300,000.

7.4 Optimal process will change at 100,000 and 300,000.

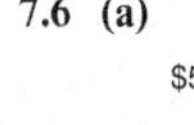

7.6 **(a)**

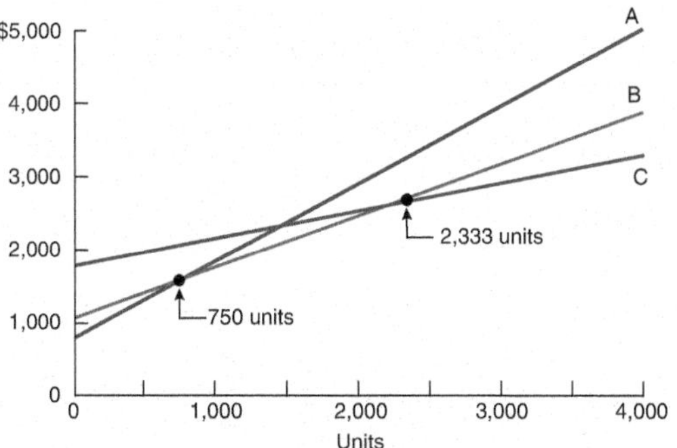

(b) Plan c; **(c)** Plan b

7.8 Rent HP software since projected volume of 80 is above the crossover point of 75.

7.10 **(a)** 82,000 units; **(b)** Loss of $10,000; **(c)** Profit of $1,000

7.12 **(a)** 7,750 units; **(b)** Proposal A

7.14

Present Method [X] PROCESS CHART Proposed Method [ ]

SUBJECT CHARTED Shoe Shine DATE

CHART BY J.C.

DEPARTMENT SHEET NO. 1 OF 1

DIST. IN FEET	TIME IN MINS.	CHART SYMBOLS	PROCESS DESCRIPTION
	0.67		Clean/Brush Shoes
1.	0.05		Obtain Polish
	0.5		Open and Apply Polish
	0.75		Buff
	0.05		Inspect
	0.25		Collect Payment
1.	2.27	4 1 1	Totals

7.16

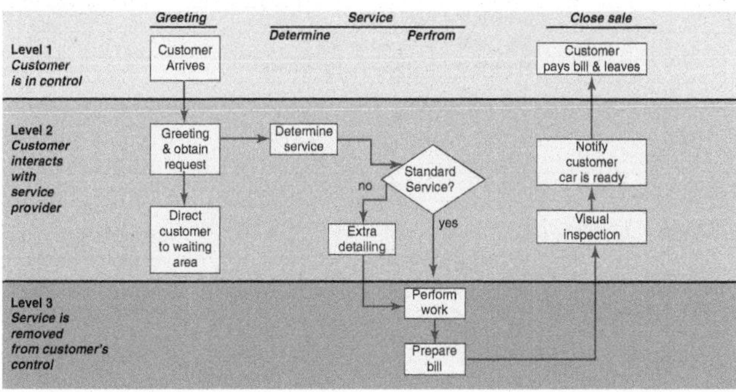

Chapter 7 Supplement

S7.2 69.2%

S7.4 88.9%

S7.6 Design = 88,920
Fabrication = 160,680
Finishing = 65,520

S7.8 5.17 (or 6) bays

S7.10 15 min/unit

S7.12 **(a)** Throughput time = 40 min
(b) Bottleneck time = 12 min.
(c) Station 2
(d) Weekly capacity = 240 units

S7.14 **(a)** Work station C at 20 min/unit; **(b)** 3 units/hr

S7.16 **(a)** 2,000 units; **(b)** $1,500

S7.18 **(a)** $150,000; **(b)** $160,000

S7.20 **(a)** $BEP_A = 1,667$;
$BEP_B = 2,353$
(b, c) Oven A slightly more profitable
(d) 13,333 pizzas

S7.22 **(a)** $18,750; **(b)** 375,000

S7.24 Yes, purchase new equipment and raise price. Profit = $2,500

S7.26 $BEP_\$ = \$7,584.83$ per mo
Daily meals = 13

S7.28 **(a)** 1,875; **(b)** 1,700

S7.30 **(a)** 1,000

S7.32 Option **B**; $74,000

S7.34 $4,590

S7.36 NPV = $1,764

S7.38 **(a)** Purchase two large ovens.
(b) Equal quality, equal capacity.
(c) Payments are made at end of each time period, and future interest rates are known.

S7.40 Investment A; payoff = $24,234

S7.42 (11 percent) should not purchase; NPV = –$7,678 (4 percent) should purchase; NPV = $5,379

S7.44 Machine B; NPV = $85,983

Chapter 8

8.2 China, $2.89

8.4 India is $.05 less than elsewhere.

8.6 **(a)** Mobile = 53; Jackson = 60; select Jackson.
(b) Jackson now = 66.

8.8 **(a)** Hyde Park, with 54.5 points.
(b) Present location = 51 points.

8.10 **(a)** Location C, with a total *weighted* score of 1,530.
(b) Location B = 1,360
(c) B can never be in first place.

8.12 **(a)** Great Britain, at 4.6.
(b) Great Britain is now 3.6.

8.14 **(a)** Italy is highest.
(b) Spain always lowest.

8.16 **(a)** Site 1 up to 125, site 2 from 125 to 233, site 3 above 233
(b) Site 2

8.18 **(a)** Above 10,000 batteries, site C is lowest cost.
(b) Site A optimal from 0 to 10,000 batteries.
(c) Site B is never optimal.

8.20 **(a)** (5.15, 7.31); **(b)** (5.13, 7.67)

8.22 **(a)** (6.23, 6.08); **(b)** safety, etc.

8.24 **(a)** Site C is best, with a score of 374
(b) For all positive values of w_7 such that $w_7 \leq 14$

8.26 Downtown rating = 2.24
Shopping mall rating = 3.24 (best)
Coral Gables rating = 2.42
When grade A = 4, B = 3, C = 2, D = 1

8.28 Site 1 = 78.125
Site 2 = 75.0
Site 3 = 86.56 (highest)
Site 4 = 80.94

8.30 **(a)** Atlanta TC = 125,000 + 6x
Burlington TC = 75,000 + 5x
Cleveland TC = 100,000 + 4x
Denver TC = 50,000 + 12x
(b) Denver best from 0 to 3,571 units
(c) At 5,000 units, Burlington best

8.32 (7.97, 6.69)

8.34 Spain is the best choice (1.3). Poland is second-best (0.9).

Chapter 9

9.2 **(a)** $23,400; **(b)** $20,600; **(c)** $22,000; **(d)** Plan B

9.4 Benders to area 1; Materials to 2; Welders to 3; Drills to 4; Grinder to 5; and Lathes to 6; Trips × Distance = 13,000 ft.

9.6 Layout #1, distance = 600 with areas fixed
Layout #2, distance = 602 with areas fixed

9.8 Layout #4, distance = 609

9.10 **(a)** Two approaches yield 47,900 ft. and 44,400 ft.
(b) Better layout is 43,880 ft. using A-D-F-C-B-E.

9.12 (a) 20 seconds
(b) 3
(c) Yes; Station 1 with A, C;
Station 2 with B, D;
Station 3 with E

9.14 (a) 4 stations
(b) Cannot be done with theoretical minimum; requires 5 stations
(c) 80% for 5 stations

9.16 (b) Station 1–A, G, B, with .5 min. left
Station 2–C, D, E with no time left
Station 3–F, H, I, J with .5 min left
(c) If stations 1 and 3 each had 0.5 min. work more to do, the line would be 100% efficient
(d) 3

9.18 (a) Station 1–A, C
Station 2–E
Station 3–B, D
Station 4–F, H
Station 5–G, I
(b) 97.6% with cycle time = 3.33 (theoretical efficiency)
87.6% is operating efficiency
(c) 4
(d) 2 min./boat

9.20 (b) 15 min.
(c) 144 units/day
(d) 5 stations
(e) 83.33%
(f) 10 min./cycle

9.22 (b) 3.75 patients/hour
(c) Medical exam station, 16 min.
(d) Paramedics are idle 2 minutes, doctors 10 minutes for each patient
(e) 5 patients/hour

9.24 (a) 6 minutes
(b) 4 workstations

(c)

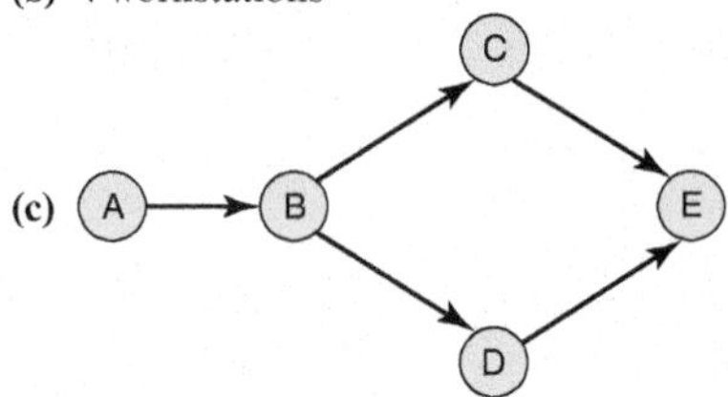

9.26 (a)

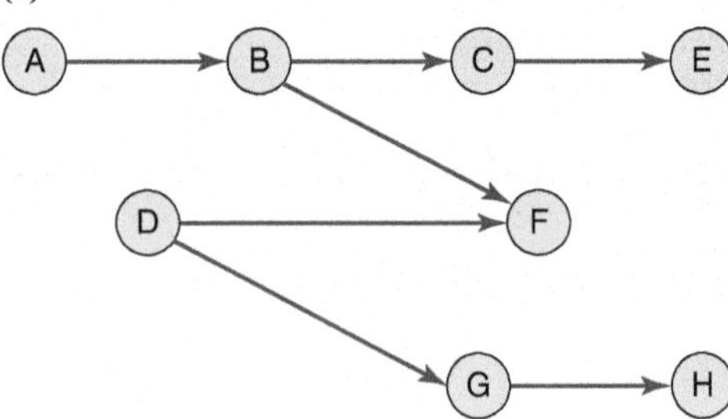

(b) Multiple alternatives are possible
(c) 92.5%

Chapter 10

10.2 The first 8 steps of the process chart are:

Process Chart

Charted by______

Date:____ Sheet______ of______

Problem:______

Summary	
Operation	26
Transport	4
Inspect	2
Delay	
Store	
Vert. Dist.	
Hor. Dist.	24
Time (min)	24.4

Distance (ft)	Time (mins)	Chart Symbols	Process Description
	1.0	○⇨□D▽	Park auto
	0.1	○⇨□D▽	Set parking brake
	0.1	○⇨□D▽	Set gear shift to park
	0.1	○⇨□D▽	Turn off engine
	0.2	○⇨□D▽	Exit vehicle
8	0.1	○⇨□D▽	Move to trunk of auto
	0.3	○⇨□D▽	Open trunk
	1.0	○⇨□D▽	Remove jack and spare tire

10.4 Individually prepared solution in the style of Figure 10.6

10.6 The first 8 steps of the process chart are:

Process Chart

Charted by______

Date:____ Sheet 1 of 1

Problem: adding a memory board to your computer

Summary	
Operation	14
Transport	4
Inspect	1
Delay	
Store	
Vert. Dist.	
Hor. Dist.	68
Time (min)	24.7

Distance (ft)	Time (mins)	Chart Symbols	Process Description
	0.2	○⇨□D▽	Turn computer off
	2.0	○⇨□D▽	Disconnect all cables
30	1.0	○⇨□D▽	Move computer to table top
	1.5	○⇨□D▽	Remove screws from cover
	1.0	○⇨□D▽	Remove cover
3	0.1	○⇨□D▽	Set cover on floor
	0.5	○⇨□D▽	Find board to be replaced
5	0.2	○⇨□D▽	Bring box with new board to table

10.8

Process Chart

Charted by *H. Molano*

Date ______ Sheet *1* of *1*

Problem *Pit crew jack man*

Summary	
Operation	*2*
Transport	*3*
Inspect	
Delay	*2*
Store	
Vert. Dist.	
Hor. Dist.	
Time (seconds)	*12.5*

Distance (feet)	Time (seconds)	Chart Symbols	Process Description
15	*2.0*	○⇨□D▽	*Move to right side of car*
	2.0	○⇨□D▽	*Raise car*
	1.0	○⇨□D▽	*Wait for tire exchange to finish*
10	*1.8*	○⇨□D▽	*Move to left side of car*
	2.0	○⇨□D▽	*Raise car*
	1.2	○⇨□D▽	*Wait for tire exchange to finish*
5	*2.5*	○⇨□D▽	*Move back over wall from left side*

10.10 The following shows the first portion of the activity chart.

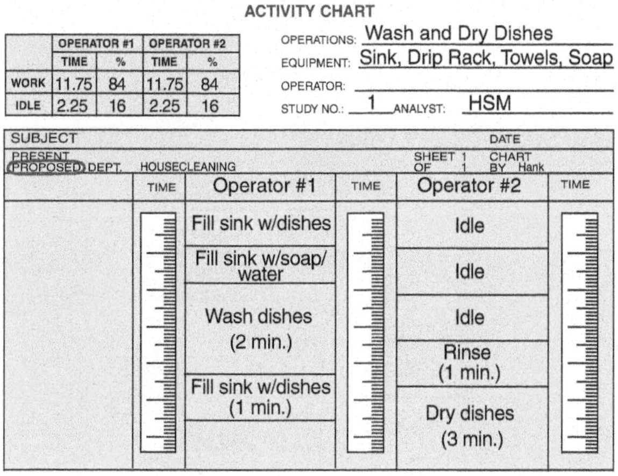

ACTIVITY CHART

	OPERATOR #1		OPERATOR #2	
	TIME	%	TIME	%
WORK	11.75	84	11.75	84
IDLE	2.25	16	2.25	16

OPERATIONS: Wash and Dry Dishes
EQUIPMENT: Sink, Drip Rack, Towels, Soap
OPERATOR:
STUDY NO.: 1 ANALYST: HSM

SUBJECT DATE
PRESENT (PROPOSED) DEPT. HOUSECLEANING SHEET 1 OF 1 CHART BY Hank

TIME	Operator #1	TIME	Operator #2	TIME
	Fill sink w/dishes		Idle	
	Fill sink w/soap/ water		Idle	
	Wash dishes (2 min.)		Idle	
			Rinse (1 min.)	
	Fill sink w/dishes (1 min.)		Dry dishes (3 min.)	

10.12 The following shows the first portion of the process chart.

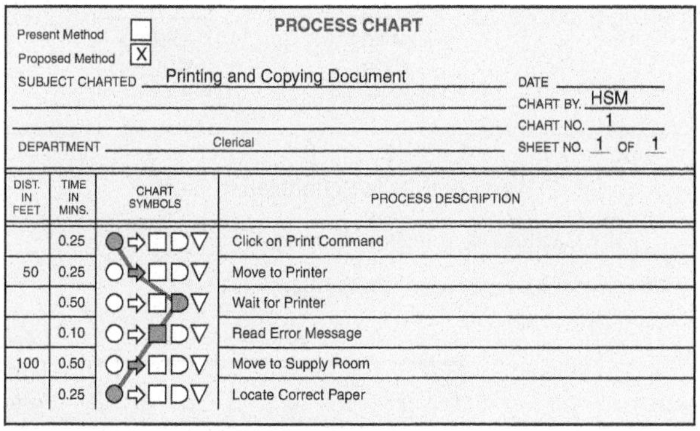

PROCESS CHART

Present Method ☐
Proposed Method ☒
SUBJECT CHARTED Printing and Copying Document DATE
CHART BY. HSM
CHART NO. 1
DEPARTMENT Clerical SHEET NO. 1 OF 1

DIST. IN FEET	TIME IN MINS.	CHART SYMBOLS	PROCESS DESCRIPTION
	0.25	●⇨□D▽	Click on Print Command
50	0.25	○➡□D▽	Move to Printer
	0.50	○⇨□D▽	Wait for Printer
	0.10	○⇨■D▽	Read Error Message
100	0.50	○➡□D▽	Move to Supply Room
	0.25	●⇨□D▽	Locate Correct Paper

10.14 NT = 7.65 sec; slower than normal
10.16 **(a)** 6.525 sec; **(b)** 6.2 sec; **(c)** 6.739 sec
10.18 **(a)** 12.6 min; **(b)** 15 min
10.20 **(a)** 12.0 sec; **(b)** 14.12 sec
10.22 10.12 min
10.24 **(a)** 3.24 min; **(b)** 4.208 min
10.26 n = 14.06, or 15 observations
10.28 **(a)** 45.36, 13.75, 3.6, 15.09; **(b)** 91.53 min; **(c)** 97 samples
10.30 **(a)** 47.6 min; **(b)** 75 samples
10.32 n = 348
10.34 73.8%
10.36 6.55 sec
10.38 **(a)** 240 min
(b) 150 hr
(c) Clean 8 rooms; refresh 16 rooms; 38 housekeepers
(d) 50 employees
10.40 **(a)** 11 min.; **(b)** 0.167, 13.2 min.
10.42 **(a)** 43.0 sec.; **(b)** 50.6 sec.
10.44 n = 225
10.46 n = 7 (rounded from 6.9)

Chapter 11

11.2 **(a)** 25% decrease in material costs; $45,000
(b) 75% increase in sales; $175,000
11.4 Problems include communication, product valuation, selecting virtual partners
11.6 **(a)** Weeks of supply = 3.85
(b) % of assets in inventory = 11.63%
(c) Turnover = 13.5
(d) No, but note they are in different industries
11.8 **(a)** Last year = 10.4
(b) This year = 9.67
(c) Yes
11.10 **(a)** 5.30%; **(b)** 6.13%; **(c)** 0.83%

Chapter 11 Supplement

S11.2 Two suppliers best, $42,970
S11.4 **(a)** $P(2)$ = 0.017463
(b) $P(2)$ = 0.018866
(c) Option 1 (2 local suppliers) has lower risk.
S11.6 **(a)** 2.5; **(b)** 1.2; **(c)** 1.25; **(d)** 1.8; **(e)** Retailer
S11.8 **(a)** 1.20; **(b)** Bullwhip = 0 if order sizes all the same.
S11.10 Donna Inc., 8.2; Kay Corp., 9.8
S11.12 Preferred vendor is Siemsen, score = 390.
S11.14 Preferred vendor is Fricker V-Tech, score = 400.
S11.16 **(a)** Alternative (1) (slower shipping)
(b) Customer satisfaction and interest earned on earlier payments
S11.18 **(a)** Alternative (2) (faster shipping)
(b) Potential delay in the production process
S11.20 **(a)** Item B; **(b)** Item A
S11.22

	B	B	B	B	C	E	D	D	G
Kitchen					Aisle				
	B	B	B	B	F	E	D	D	A

Chapter 12

12.2 **(a)** A items are G2 and F3; B items are A2, C7, and D1; all others are C.
12.4 108 items
12.6

(E102)	$3,200; C
(D23)	$38,400; A
(D27)	$5,600; B
(R02)	$2,000; C
(R19)	$1,600; C
(S107)	$6,000; B
(S123)	$1,200; C
(U11)	$5,600; B
(U23)	$1,500; C
(V75)	$35,000; A

12.8 **(a)** 600 units; **(b)** 424.26 units; **(c)** 848.53 units
12.10 **(a)** 80 units; **(b)** 73 units
12.12 **(a)** 2,100 units; **(b)** 4,200 units; **(b)** 1,050 units
12.14 **(a)** 189.74 units; **(b)** 94.87; **(c)** 31.62; **(d)** 7.91; **(e)** $1,897.30
(f) $601,897
12.16 **(a)** Order quantity variations have limited impact on total cost.
(b) EOQ = 50
12.18 **(a)** 671 units; **(b)** 18.63; **(c)** 559 = max. inventory
(d) 16.7%;
(e) $1,117.90
12.20 **(a)** 1,217 units; **(b)** 1,095 = max. inventory
(c) 8.22 production runs; **(d)** $657.30
12.22 **(a)** EOQ = 200, total cost = $1,446,380
(b) EOQ = 200, total cost = $1,445,880
12.24 **(a)** 10,001 units; **(b)** $312.53; **(c)** $299.97; **(d)** $18,750;
(e) $19,362.50.
12.26 **(a)** EOQ = 410
(b) Vendor Allen has slightly lower cost.
(c) Optimal order quantity = 1,000 @ total cost of $128,920
12.28 **(a)** EOQ (1) = 336; EOQ (2) = 335
(b) Order 1,200 from Vendor 2.
(c) At 1,200 lb., total cost = $161,275.
(d) Storage space and perishability.
12.30 **(a)** 32; **(b)** 2; **(c)** 20; $2,400; **(d)** 336; 168
(e) $10,800; **(f)** 214; $1,820
12.32 $8
12.34 7,000
12.36 **(a)** $1,220; **(b)** $1,200; **(c)** 24
12.38 (without discount) $2,200
(with discount) $2,510
No, do not take the discount.
12.40 Order more than 50 sheets; cost = $1,901.22

12.42 **(a)** $Z = 1.88$
(b) Safetystock $= Z\sigma = 1.88(5) = 9.4$ drives
(c) ROP = 59.4 drives
12.44 100 kilos of safety stock
12.46 **(a)** 291 towels; **(b)** 2,291 towels
12.48 **(a)** ROP = 1,718 cigars; **(b)** 1,868 cigars
(c) A higher service level means a lower probability of stocking out.
12.50 55
12.52 28 cakes

Chapter 13

13.2 **(a)** \$109,120 = total cost
(b) \$106,640 = total cost
(c) No, plan 2 is better at \$105,152.
13.4 Cost = \$244,000 for plan B
13.6 **(a)** Plan D, \$128,000; **(b)** Plan E is \$140,000
13.8 Extra total cost = \$2,960.
13.10 **(a)** Plan C, \$104,000; **(b)** plan D, \$93,800, assuming initial inventory = 0
13.12 **(a)** Plan A: Cost is \$314,000.
Plan B: Cost is \$329,000.
Plan C: Cost is \$222,000 (lowest cost).
(b) Plan C, with lowest cost and steady employment.
13.14 \$1,186,810
13.16 \$100,750
13.18 \$90,850
13.20 \$20,400
13.22 \$308,125
13.24 **(a)** Total cost = \$150,000 + \$43,750 + \$15,000 = \$208,750.
(b) Five accounts appear necessary (plus some overtime).
13.26 Current pricing model is better at \$15,120.

Chapter 14

14.2 The time-phased plan for the gift bags is:

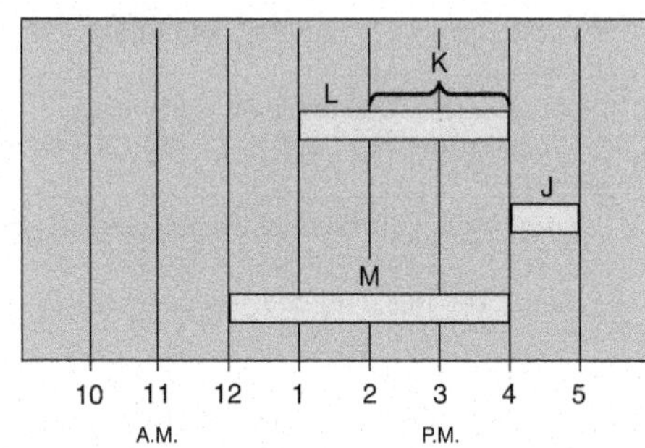

Someone should start on item M by noon.

14.4 **(a)** Time-phased product structure for bracket with start times:

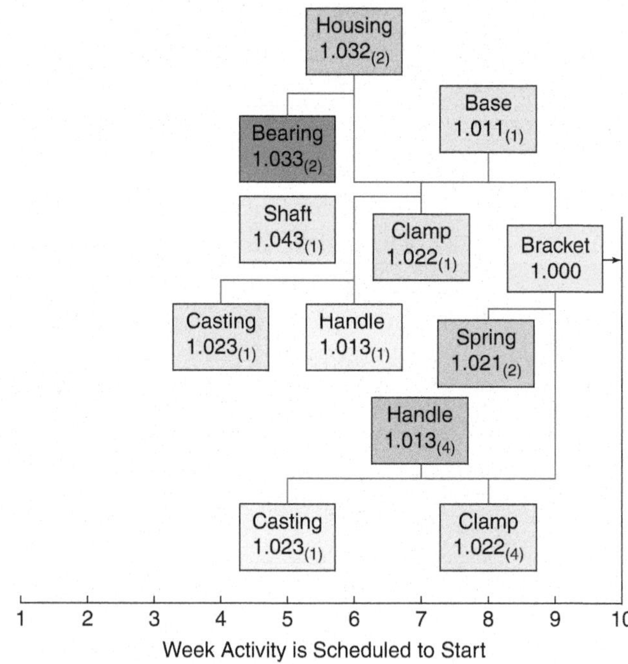

(b) Castings need to start in week 4.

14.6 Gross material requirements plan:

Item		Week 1	2	3	4	5	6	7	8	Lead Time (wk)
S	Gross req.							100		
	Order release					100				2
T	Gross req.					100				
	Order release				100					1
U	Gross req.					200				
	Order release			200						2
V	Gross req.				100					
	Order release		100							2
W	Gross req.				200					
	Order release	200								3
X	Gross req.				100					
	Order release			100						1
Y	Gross req.			400						
	Order release	400								2
Z	Gross req.			600						
	Order release		600							1

14.8 Gross material requirements plan, modified to include the 20 units of U required for maintenance purposes:

Item		Week 1	2	3	4	5	6	7	8	Lead Time (wk)
S	Gross req.							100		
	Order release					100				2
T	Gross req.					100				
	Order release				100					1
U	Gross req.					200	20			
	Order release			200	20					2
V	Gross req.				100					
	Order release		100							2
W	Gross req.				200					
	Order release	200								3
X	Gross req.				100					
	Order release			100						1
Y	Gross req.			400	40					
	Order release	400	40							2
Z	Gross req.			600	60					
	Order release		600	60						1

14.10 **(a)** Gross material requirements plan for the first three items:

Item		Week 1	2	3	4	5	6	7	8	9	10	11	12
X1	Gross req.								50		20		100
	Order release							50		20		100	
B1	Gross req.							50		20		100	
	Order release					50		20		100			
B2	Gross req.							100		40		200	
	Order release					100		40		200			

(b) The net materials requirement plan for the first two items:

Level: 0 | Parent: | Quantity:
Item: X1 | Lead Time: | Lot Size: L4L

Week No.	1	2	3	4	5	6	7	8	9	10	11	12
Gross Requirement								50		20		100
Scheduled Receipt												
On-hand Inventory								50		0		0
Net Requirement								0		20		100
Planned Order Receipt										20		100
Planned Order Release									20		100	

Level: 1 | Parent: X1 | Quantity: 1X
Item: B1 | Lead Time: 2 | Lot Size: L4L

Week No.	1	2	3	4	5	6	7	8	9	10	11	12
Gross Requirement									20		100	
Scheduled Receipt												
On-hand Inventory									20		0	
Net Requirement									0		100	
Planned Order Receipt											100	
Planned Order Release									100			

14.12 **(a)** Net material requirements schedule (only item A is shown):

	Week 1	2	3	4	5	6	7	8	9	10	11	12
A Gross Required								100		50		150
On Hand								0		0		0
Net Required								100		50		150
Order Receipt								100		50		150
Order Release							100		50		150	

(b) Net material requirements schedule (only items B and D are shown; schedule for item A remains the same as in part a).

		Week 4	5	6	7	8	9	10	11	12
B Gross Requirements					200		100		300	
Scheduled Receipts										
Projected on Hand	100				100		0		0	
Net Requirements					100		100		300	
Planned Order Receipts					100		100		300	
Planned Order Releases			100		100		300			
D Gross Requirements			100		100		300			
Scheduled Receipts										
Projected on Hand	50		50		100		300			
Net Requirements			50		100		300			
Planned Order Receipts			50		100		300			
Planned Order Releases		50		100		300				

14.14 **(a)**

Level	Description	Qty
0	A	1
1	B	1
2	C	1
2	D	1
3	E	1
1	F	1
2	G	1
2	H	1
3	E	1
3	C	1

Note: with low-level coding "C" would be a level-3 code

14.14 **(b)** Net material requirements schedule (only items A, B, and F are shown).

Lot Size	Lead Time	On Hand	Safety Stock	Allo-cated	Low-Level Code	Item ID		1	2	3	4	5	6	7	8
Lot	1	0	—	—	0	A	Gross Requirement								10
for							Scheduled Receipt								
Lot							Projected on Hand								0
							Net Requirement								10
							Planned Receipt								10
							Planned Release							10	
Lot	1	2	—	—	1	B	Gross Requirement							10	
for							Scheduled Receipt								
Lot							Projected on Hand	2	2	2	2	2	2	2	0
							Net Requirement							8	
							Planned Receipt							8	
							Planned Release						8		
Lot	1	5	—	—	1	F	Gross Requirement							10	
for							Scheduled Receipt								
Lot							Projected on Hand	5	5	5	5	5	5	5	0
							Net Requirement							5	
							Planned Receipt							5	
							Planned Release						5		

14.16 **(a)** Only item G changes.

(b) Component F and 4 units of A will be delayed one week.

(c) Options include: delaying 4 units of A for 1 week; asking supplier of G to expedite production.

(b)

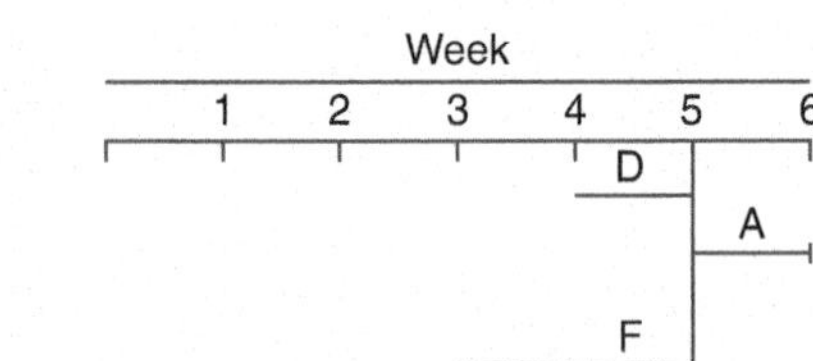

(c)

Week	2	3	4	5	6	
Required Date					10	A
Order Release				10		
Required Date				30		D
Order Release			30			
Required Date				20		F
Order Release		20				

14.18 **(a)**

(a)

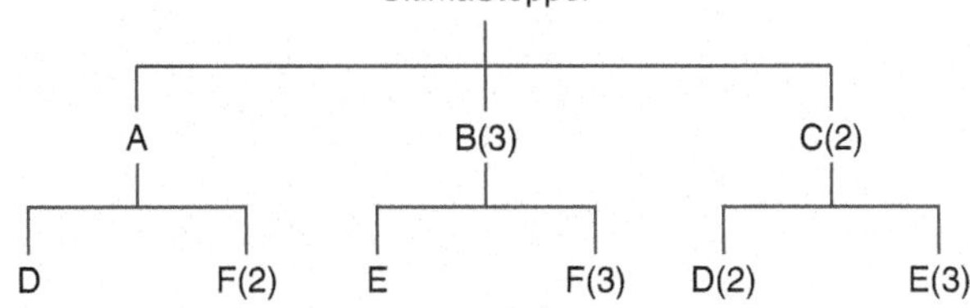

(b)

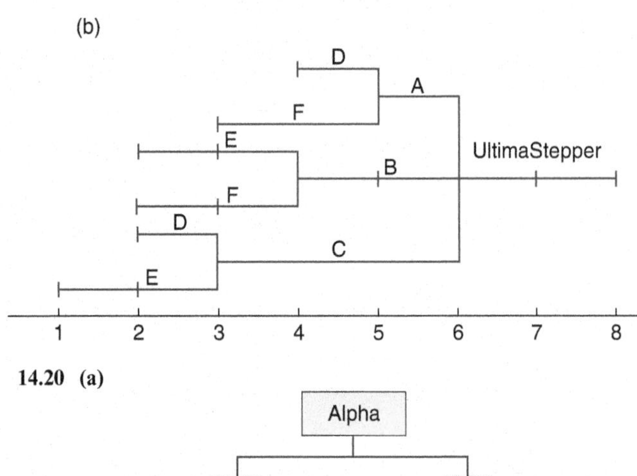

14.20 **(a)**

Alpha

D(3) F(2)

14.22 Lot-for-lot: Total cost = 7 orders × \$150/order + 20 units × \$2.50/unit/period = \$1,100.

14.24 POQ: Setup cost = 5 × \$150 = \$750; Holding cost = 2.50 × \$170 = \$425; Total \$1,175.

14.26 **(a)** Lot-for-lot: Setup cost = \$800; Holding cost = \$0.0; Total cost = \$800.

(b) EOQ: EOQ = 18, Setup cost = \$800; Holding cost = \$370; Total cost = \$1,170.

(c) POQ: EOQ = 18, POQ = 2, Setup cost = \$800; Holding cost = \$0.0; Total cost = \$800.

(d) Lot-for-lot and POQ have the same cost.

14.28 **(a)** EOQ = 105; orders released now and in weeks 3 and 7; Actual cost = \$307.50

(b) Lot-for-lot: orders released now and in weeks 1, 2, 4, 5, 6, 8, 9; Total cost = \$400.

14.30 Selection for first 5 weeks:

Week	Units	Capacity Required (time)	Capacity Available (time)	Over/ (Under)	Production Scheduler's Action
1	60	3,900	2,250	1,650	Lot split. Move 300 minutes (4.3 units) to week 2 and 1,350 minutes to week 3.
2	30	1,950	2,250	(300)	
3	10	650	2,250	(1,600)	
4	40	2,600	2,250	350	Lot split. Move 250 minutes to week 3. Operations split. Move 100 minutes to another machine, overtime, or subcontract.
5	70	4,550	2,250	2,300	Lot split. Move 1,600 minutes to week 6. Overlap operations to get product out door. Operations split. Move 700 minutes to another machine, overtime, or subcontract.

14.32

Coffee Table Master Schedule	Hrs Required	Lead Time	Day 1	Day 2	Day 3	Day 4	Day 5	Day 6	Day 7	Day 8
							640	640	128	128
Table Assembly	2	1				1,280	1,280	256	256	
Top Preparation	2	1			1,280	1,280	256	256		
Assemble Base	1	1			640	640	128	128		
Long Brace (2)	0.25	1		320	320	64	64			
Short Brace (2)	0.25	1		320	320	64	64			
Leg (4)	0.25	1		640	640	128	128			
Total Hours			0	1,280	3,200	3,456	1,920	640	256	
Employees needed @ 8 hrs. each			0	160	400	432	240	80	32	

Chapter 15

15.2

Job	Day 1	Day 2	Day 3	Day 4	Day 5	Day 6	Day 7	Day 8	Day 9
D									
E									
F									
G									

Now

15.4 **(a)** 1–D, 2–A, 3–C, 4–B
(b) 40

15.6 Chris–Finance, Steve–Marketing, Juana–H.R., Rebecca–Operations, $210

15.8 Ajay–Jackie, Jack–Barbara, Gray–Stella, Raul–Dona, 230

15.10

Period	1	2	3	4
Input deviation	+5	+5	−15	−15
Output deviation	−5	−5	−10	−10
Backlog	30	30	35	40

15.12 Divorce to Attorney 3
Felony to Attorney 1
Discrimination to Attorney 2
Total cost = $2,700

15.14 G99 to 1; E2 to 2; C81 to 3;
D5 to 4; D44 to 5; C53 to 6;
E35 to 8; no component to 7.
Total cost = $1.18

15.16 Sequence 517, 103, 309, 205, 412

15.18 **(b)** SPT for best flow time
(c) SPT for best utilization
(d) EDD for best lateness
(e) LPT scores poorly on all three criteria

15.20 Johnson's Rule finishes in 21 days
First-come, First-served finishes in 23 days

15.22 **(a)** V–Y–U–Z–X–W–T; **(c)** 57 hours
(d) 7 hours; **(e)** unchanged

15.24 **(a)** 10, 40, 20, 50, 30; **(b)** 61.20 days; **(c)** 12.60 days; **(d)** 3.64 jobs

15.26 **(a)** A, E, B, C, D; **(b)** 18.20 hours; **(c)** 1.40 hours; **(d)** 2.33 jobs

15.28 D, B, A, C

15.30 7 employees

15.32 Denzel R.: Tu–W–Th–F; Ronnell S.: F–Sa–Su–M; Rosanne B: Tu–W–Th–F; Napoleon A.: Sa–Su–M–Tu; Barry R: Tu–W–Th–F; Rosemary T.: F–Sa–Su–M

Chapter 16

16.2 Setup time = 3.675 minutes

16.4 Number of kanbans needed = 5

16.6 Production order quantity = 50; 5 Kanbans

16.8 **(a)** EOQ = 100 lamps; **(b)** TC = $1,200; **(c)** 20 orders/year

16.10 EOQ size decreases; orders increase; inventory costs can be expected to drop.

16.12 Kanban size = 158; number of kanbans needed = 10

Chapter 17

17.2 From Figure 17.2, about 13% overall reliability.

17.4 **(a)** 5.0%; **(b)** .00001026 failures/unit-hr.; **(c)** .08985; **(d)** 98.83

17.6 $R_s = .9941$

17.8 $R_p = .99925$

17.10 $R_p = .984$

17.12 $R = .7918$

17.14 System B is slightly higher, at .9397.

17.16 From Figure 17.2, about 82%

17.18 2.7 breakdowns

17.20 The figure suggests that there are likely to be at least three separate modes of failure; one or more causes of infant mortality, and two modes of failure which occur at later times.

17.22 1.5 breakdowns; $15

17.24 Current cost = $1,255/week; Contract cost = $1,395; therefore, eliminate maintenance contract.

Business Analytics Module A

A.2 (a)

Size of First Station	Good Market ($)	Fair Market ($)	Poor Market ($)	EV Under Equally Likely
Small	50,000	20,000	−10,000	20,000
Medium	80,000	30,000	−20,000	30,000
Large	100,000	30,000	− 40,000	30,000
Very large	300,000	25,000	−160,000	55,000

(b) Maximax: Build a very large station.
(c) Maximin: Build a small station.
(d) Equally likely: Build a very large station.
(e)

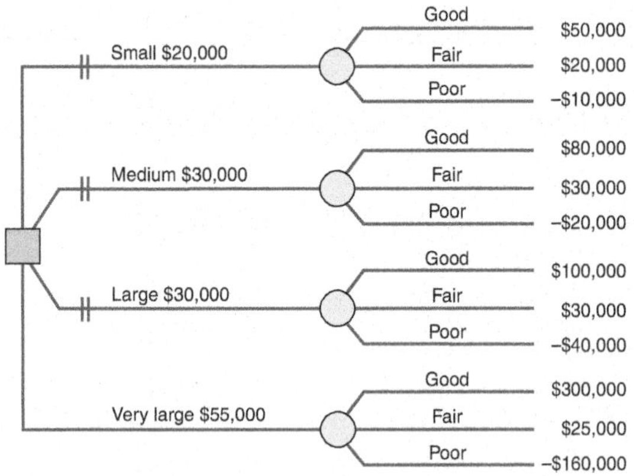

A.4 (a) Alternatives: N, M, L, D. States of nature: Fixed, Slight Increase, Major Increase
(b) Use maximin criterion. No floor space (N).

A.6 Buying equipment at $733,333

A.8 (a) E(cost full-time) = $520
E(cost part-timers) = $475

A.10 Alternative B; 74

A.12 8 cases; EMV = $352.50

A.14 (b) EMV(Y) = 4.2, which is best

A.16 (a) EMV(Alt. 1) = 110 = max. EMV
(b) EVPI = 7

A.18 (a) Stock 11 cases; EMV = $385.00
(d) Stock 13 cases; EMV = $411.25

A.20 Alternative A, with an EMV = $1,400,000

A.22 (a)

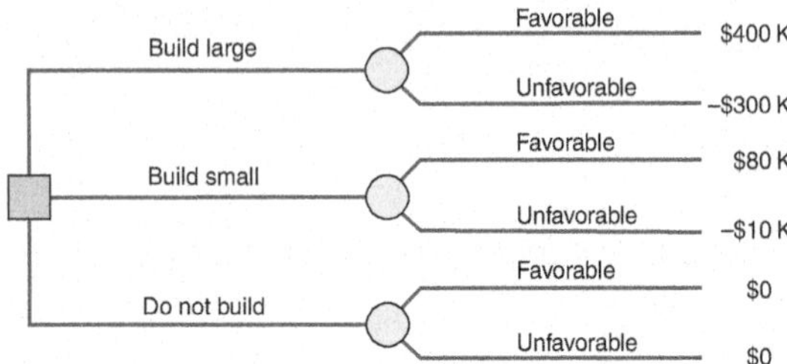

(b) Small plant with EMV = $26,000
(c) EVPI = $134,000

A.24 (a) Resort has higher EMV than home
(b) EMV (Resort) = $76

A.26 (a) Build large facility; (b) $544,000

A.28 Maximum expected profit = $33,000
Michael should wait 1 day. Then if an XP02 is available, he should buy it. Otherwise, he should stop pursuing an XP02 on the wholesale market.

A.30 (a) Minor renovation; EMV = $10,000

Business Analytics Module B

B.2 $X = 1.33$, $Y = 3.33$; profit = $25.33

B.4 (b) Yes; $P = \$3{,}000$ at (83.3333, 50) and (50, 150)

B.6 $x_1 = 200$, $x_2 = 0$, profit = $18,000

B.8 10 Alpha 1s, 24 Beta 2s, profit = $55,200

B.10 (a) Let X = wren houses, Y = bluebird houses
Maximize profit = $6X + 15Y$
Subject to:
$4X + 2Y \le 60$
$4X + 12Y \le 120$
$X,\ Y \ge 0$
(b) $X = 12$, $Y = 6$; profit = $162

B.12 (a) Standard = 240, Deluxe = 180; (b) $3,840

B.14 One approach results in 2,790 medical patients and 2,104 surgical patients, with a revenue of $9,551,659 per year (which can change slightly to $9,548,760 with rounding). This yields 61 integer medical beds and 29 integer surgical beds.

B.16 $X = 4$, $Y = 8$; profit = $52

B.18 $X_1 = 18.75$, $X_2 = 18.75$; profit = $150

B.20 $X_1 = 2$, $X_2 = 6$; profit = $36

B.22 (a) $x_1 = 7.95$, $x_2 = 5.95$, $x_3 = 12.59$, $P = \$143.76$
(b) No unused time
(c) 26¢
(d) $7.86

B.24 (a) A158
(b) Nonbinding; increasing; reducing
(c) 0
(d) should; $119.32
(e) Objective function; 17.91; 49.99; 24.00; 8.88; 77.01
(f) Constraints 7 through 11 become ≥0, and Constraint 12 is still ≥20.

B.26 $X = 3.86$, $Y = 4.54$; cost = $160.86

B.28 Let X = number of pounds of compost in each bag
Let Y = number of pounds of sewage waste in each bag
$X = 30$, $Y = 30$; cost = $2.70

B.30 $x_1 = 14$, $x_2 = 33$, cost = 221

B.32 Objective function: Minimize Z (in thousands) $10X_1 + 8X_2$
Constraints: $1X_2 \ge 5$
$2X_1 + 1X_2 \ge 20$
Solution: 7.5 thousand round tables, 5 thousand square tables; cost = $115 (in thousands)

B.34 (a) Let Xij = number of students bused from sector i to school j.
Objective: minimize total travel miles =
$5X_{AB} + 8X_{AC} + 6X_{AE}$
$+ 0X_{BB} + 4X_{BC} + 12X_{BE}$
$+ 4x_{CB} + 0X_{CC} + 7X_{CE}$
$+ 7X_{DB} + 2X_{DC} + 5x_{DE}$
$+ 12X_{EB} + 7X_{EC} + 0X_{EE}$
Subject to:
$X_{AB} + X_{AC} + X_{AE} = 700$ (number of students in sector A)
$X_{BB} + X_{BC} + X_{BE} = 500$ (number students in sector B)
$X_{CB} + X_{CC} + X_{CE} = 100$ (number students in sector C)
$X_{DB} + X_{DC} + X_{DE} = 800$ (number students in sector D)
$X_{EB} + X_{EC} + X_{EE} = 400$ (number of students in sector E)
$X_{AB} + X_{BB} + X_{CB} + X_{DB} + X_{EB} \le 900$ (school B capacity)
$X_{AC} + X_{BC} + X_{CC} + X_{DC} + X_{EC} \le 900$ (school C capacity)
$X_{AE} + X_{BE} + X_{CE} + X_{DE} + X_{EE} \le 900$ (school E capacity)

(b) Solution: $X_{AB} = 400$
$X_{AE} = 300$
$X_{BB} = 500$
$X_{CC} = 100$
$X_{DC} = 800$
$X_{EE} = 400$
Distance = 5,400 "student miles"

B.36 Hire 30 workers; three solutions are feasible; two of these are:
16 begin at 7 A.M.
9 begin at 3 P.M.
2 begin at 7 P.M.
3 begin at 11 P.M.
An alternate optimum is:
3 begin at 3 A.M.
9 begin at 7 A.M.
7 begin at 11 A.M.
2 begin at 3 P.M.
9 begin at 7 P.M.
0 begin at 11 P.M.

B.38 **(a)** Minimize = $6X_{1A} + 5X_{1B} + 3X_{1C} + 8X_{2A} + 10X_{2B} + 8X_{2C} + 11X_{3A} + 14X_{3B} + 18X_{3C}$
Subject to:
$X_{1A} + X_{2A} + X_{3A} = 7$
$X_{1B} + X_{2B} + X_{3B} = 12$
$X_{1C} + X_{2C} + X_{3C} = 5$
$X_{1A} + X_{1B} + X_{1C} \leq 6$
$X_{2A} + X_{2B} + X_{2C} \leq 8$
$X_{3A} + X_{3B} + X_{3C} \leq 10$
(b) Minimum cost = $219,000

B.40 Apple sauce = 0, Canned corn = 1.33, Fried chicken = 0.46, French fries = 0, Mac & Cheese = 1.13, Turkey = 0, Garden salad = 0, Cost = $1.51.

B.42 Include all but investment A.; total expected return = $3,580

Business Analytics Module C

C.2 $208
C.4 $170
C.6 $2,020
C.8 Total cost = $505
C.10 Optimal site is St. Louis, at $17,250
C.12 D–A, 100; E–B, 200; F–A, 200; E–C, 100; F–C, 100. $3,900 total cost.
C.14 $14,700, Houston-Dallas, 800; Houston-Atlanta, 50; Phoenix-Atlanta, 250; Phoenix-Denver, 200; Phoenix-Dummy, 200; Memphis-Atlanta, 300.
C.16 A–1, 20; B–1, 20; B–2, 10; C–2, 50; C–3, 25; Dummy–3, 30
C.18 Dublin, $1,535,000
Fountainbleau, $1,530,000 (lowest cost)

Business Analytics Module D

D.2 **(a)** 44%; **(b)** .36 people; **(c)** .8 people; **(d)** .53 min; **(e)** 1.2 min
D.4 **(a)** .5; **(b)** .5; **(c)** 1; **(d)** .5; **(e)** .05 hr; **(f)** .1 hr
D.6 **(a)** .667; **(b)** .667 min; **(c)** 1.33
D.8 **(a)** .375; **(b)** 1.6 hr (or .2 days)
(c) .225; **(d)** 0.141, 0.053, 0.020, 0.007
D.10 **(a)** 2.25; **(b)** .75; **(c)** .857 min. (.014 hr)
(d) .64 min. (.011 hr); **(e)** 42%, 32%, 24%
D.12 **(a)** 6 trucks; **(b)** 12 min; **(c)** .857; **(d)** .54; **(e)** $1,728/day
(f) Yes, save $3,096 in the first year.
D.14 **(a)** .075 hrs (4.5 min); **(b)** 1.125 people
(c) .0083 hrs (0.5 min), 0.083 people
D.16 **(a)** .113 hr. = 6.8 min; **(b)** 1.13 cars
D.18 **(a)** 0.577; **(b)** 1.24; **(c)** 33.6%
D.20 **(a)** 3, 2, 4 MDs, respectively
(b) Because $\lambda > \mu$, an indefinite queue buildup can occur.
D.22 **(a)** 4 servers; **(b)** 6 servers; **(c)** $109; **(d)** 83.33%
D.24 2 salespeople ($340)
D.26 5 min.
D.28 72 loans
D.30 1.2 callers
D.32 **(a)** 0.833; **(b)** 1.667 hours; **(c)** 4.167; **(d)** 0.335
D.34 **(a)** 1.33 cars; **(b)** 0.0667 hours; **(c)** 0.10 hours; **(d)** 0.667; **(e)** 0.33; **(f)** $L_q = 0.083$, $W_q = 0.0042$ hrs., $W_s = 0.0209$ hrs., $\rho = 0.333$, $P_0 = 0.667$
D.36 **(a)** 0.0377
(b) 4.53 persons
(c) 22.65 min.
(d) 7.62 min. (or 0.127 hrs.)
(e) 1.53 customers
(f) 0.9533
D.38 **(a)** 4.167 cars
(b) 0.4167 hrs.
(c) 0.5 hrs.
(d) 0.8333
(e) 0.1667

Business Analytics Module E

E.2 **(a)** 507 min; **(b)** 456 min; **(c)** 410 min; **(d)** 369 min
E.4 **(a)** 1,546 min; **(b)** 2,872 min; **(c)** 3,701 min; **(d)** 6,778 min
E.6 **(a)** 14.295 hr; **(b)** $71,475; **(c)** $947,310
E.8 **(a)** 80%; **(b)** 3.51; **(c)** 3.2, 2.98, 2.81; **(d)** 21.497
E.10 **(a)** 72.2 hr; **(b)** 60.55 hr; **(c)** 41.47 hr
E.12 Susan will take 3.723 hr and Julie 2.427 hr. Neither trainee will reach 1 hr by the 10th unit.
E.14 $747,840 for fourth, $709,760 for fifth, $680,040 for sixth
E.16 **(a)** 70 millicents/bit; **(b)** 8.2 millicents/bit
E.18 26,727 hours
E.20 **(a)** 32.978 hrs, 49.609 hrs; **(b)** Initial quote is high.
E.22 **(a)** Four boats can be completed.
(b) Five boats can be completed.
E.24 .227 hr
E.26 **(a)** 476.41; **(b)** 428.78; **(c)** 410.43; **(d)** 369.38
E.28 1,106.43 days
E.30 58,320 hours
E.32 **(a)** Actual cost is consistently above forecast; **(b)** Several possible causes; **(c)** About 110 hours; **(d)** About 135 hours.

Business Analytics Module F

F.2 0, 0, 0, 0, 0, 0, 0, 2, 0, 2
F.4 Profits = $200, –$150, $200, $175, $200; average equals $125.
F.6 At the end of 5 min, two checkouts are still busy and one is available.
F.8 Aver. no. of failures = 2.88 units/month; 7 units failed over each 3-month stretch
F.10 **(a)** 24; **(b)** $36.70; **(c)** $0.30; **(d)** $6.10; **(e)** 21
F.12 **(a)** 5 times; **(b)** 6.95 times; yes; **(c)** 7.16 heaters
F.14 Average = 7
F.16

Arrivals	*Arrival Time*	*Service Time*	*Departure Time*
1	11:01	3	11:04
2	11:04	2	11:06
3	11:06	2	11:08
4	11:07	1	11:09

F.18 During 4 hours, 7 customers balked.
Taboo missed 105 minutes = $21.00/day or $504/month. Additional bed not justified.
F.20 Average weekly stockout cost = $20; Weekly holding cost = $2.30

F.22 First order, total demand during lead time = 31.
No stockout.
Second order, total demand during lead time = 42.
One stockout.

F.24 Balance never drops below $400, she should be able to balance her account.

Business Analytics Module G

G.2 **(a)** upper limit = 5.22, lower limit = −4.02 (or 0);
(b) The observation of 6 rejected shipments for Rick's Wrought Iron is too high.

G.4 'Format only unique or duplicate values'

G.6 7 suppliers

G.8 Maverick Supplies, Perkins Plastics, Everything Brass, Penelope's Paper Products, Columbus Manufacturing, and Fred's Foam Molding

G.10 **(a)** Bentonville; **(b)** Auburn

Online Tutorial 1

T1.2 5.45; 4.06

T1.4 **(a)** .2743
(b) 0.5

T1.6 .1587; .2347; .1587

T1.8 **(a)** .0548
(b) .6554
(c) .6554
(d) .2119

Online Tutorial 2

T2.2 (selected values)

Fraction Defective	Mean of Poisson	P(x ≤ 1)
.01	.05	.999
.05	.25	.974
.10	.50	.910
.30	1.50	.558
.60	3.00	.199
1.00	5.00	.040

T2.4 The plan meets neither the producer's nor the consumer's requirement.

Online Tutorial 3

T3.2 **(a)** Max $3x_1 + 9x_2$
$x_1 + 4x_2 + s_1 = 24$
$x_1 + 2x_2 + s_2 = 16$
(b) See the steps in the tutorial.
(c) Second tableau:

c_j	Mix	x_1	x_2	s_1	s_2	Qty.
9	x_2	.25	1	.25	0	6
0	s_2	.50	0	−.50	1	4
	z_j	2.25	9	2.25	0	54
	$c_j - z_j$	.75	0	−2.25	0	

(d) $x_1 = 8$, $x_2 = 4$, Profit = $60

T3.4 Basis for 1st tableau:
$A_1 = 80$
$A_2 = 75$
Basis for 2nd tableau:
$A_1 = 55$
$X_1 = 25$
Basis for 3rd tableau:
$X_1 = 14$
$X_2 = 33$
Cost = $221 at optimal solution

T3.6 **(a)** x_1
(b) A_1

Online Tutorial 4

T4.2 Cost = $980; 1–A = 20; 1–B = 50; 2–C = 20; 2–Dummy = 30; 3–A = 20; 3–C = 40

T4.4 Total = 3,100 mi; Morgantown–Coaltown = 35; Youngstown–Coal Valley = 30; Youngstown–Coaltown = 5; Youngstown–Coal Junction = 25; Pittsburgh–Coaltown = 5; Pittsburgh–Coalsburg = 20

T4.6 **(a)** Using VAM, cost = $635; A–Y = 35; A–Z = 20; B–W = 10; B–X = 20; B–Y = 15; C–W = 30
(b) Using MODI, cost is also $635 (i.e., initial solution was optimal). An alternative optimal solution is A–X = 20; A–Y = 15; A–Z = 20; B–W = 10; B–Y = 35; C–W = 30

Online Tutorial 5

T5.2 **(a)** $I_{13} = 12$
(b) $I_{35} = 7$
(c) $I_{51} = 4$

T5.4 **(a)** Tour: 1–2–4–5–7–6–8–3–1; 37.9 mi
(b) Tour: 4–5–7–6–8–1–2–3–4; 30.2 mi

T5.6 **(a)** Vehicle 1: Tour 1–2–4–3–5–1 = $134
Vehicle 2: Tour 1–6–10–9–8–7–1 = $188

T5.8 The cost matrix is shown below:

	1	2	3	4	5	6	7	8
1	—	107.26	118.11	113.20	116.50	123.50	111.88	111.88
2		—	113.53	111.88	118.10	125.30	116.50	118.10
3			—	110.56	118.70	120.50	119.90	124.90
4				—	109.90	119.10	111.88	117.90
5					—	111.88	106.60	118.50
6						—	111.88	123.50
7							—	113.20
8								—

Name Index

General Index

Note: Page numbers beginning with a T refer to the Online Tutorial chapters that appear in MyLab Operations Management.

E

F

G

J

Q